1 MONTH OF
FREE
READING

at

www.ForgottenBooks.com

By purchasing this book you are
eligible for one month membership to
ForgottenBooks.com, giving you
unlimited access to our entire
collection of over 1,000,000 titles via
our web site and mobile apps.

To claim your free month visit:

www.forgottenbooks.com/free919014

ISBN 978-0-265-98316-4
PIBN 10919014

This book is a reproduction of an important historical work. Forgotten Books uses
state-of-the-art technology to digitally reconstruct the work, preserving the original format
whilst repairing imperfections present in the aged copy. In rare cases, an imperfection in
the original, such as a blemish or missing page, may be replicated in our edition. We do,
however, repair the vast majority of imperfections successfully; any imperfections that
remain are intentionally left to preserve the state of such historical works.

JOURNAL

OF

THE CHEMICAL SOCIETY.

TRANSACTIONS.

Committee of Publication:

H. E. ARMSTRONG, Ph.D., LL.D., F.R.S.
E. C. C. BALY.
A. W. CROSSLEY, D.Sc., Ph.D.
BERNARD DYER, D.Sc.
M. O. FORSTER, D.Sc., Ph.D., F.R.S.

R. MELDOLA, F.R.S.
E. J. MILLS, D.Sc., LL.D., F.R.S.
Sir W. RAMSAY, K.C.B., LL.D., F.R.S.
A. SCOTT, D.Sc., F.R.S.
W. A. TILDEN, D.Sc., F.R.S.
JOHN WADE, D.Sc.

Editors:

G. T. MORGAN, D.Sc.
J. C. CAIN, D.Sc.

Sub-Editor:

A. J. GREENAWAY.

Assistant Sub-Editor:

C. H. DESCH, D.Sc., Ph.D.

1906. Vol. LXXXIX.

LONDON:

GURNEY & JACKSON, 10, PATERNOSTER ROW.

1906.

RICHARD CLAY & SONS, LIMITED,
BREAD STREET HILL, E.C., AND
BUNGAY, SUFFOLK.

JOURNAL

OF

THE CHEMICAL SOCIETY.

TRANSACTIONS.

1906. Vol. LXXXIX. Part I.

LONDON

GURNEY & JACKSON, 10, PATERNOSTER ROW.

1906.

RICHARD CLAY & SONS, LIMITED.
BREAD STREET HILL, E.C., AND
BUNGAY, SUFFOLK

CONTENTS.

CONTENTS.

PAGE

1491

1496

1505

1512

1520

1527

1532

1552

56

JOURNAL

OF

THE CHEMICAL SOCIETY.

TRANSACTIONS.

I.—*The Preparation and Reactions of Benzoyl Nitrate.*

By Francis Ernest Francis.

HE investigation of the action of nitrates of silver and lead on ʼnzoyl chloride was carried out by B. Lachowicz (*Ber.*, 1884, **17**, ʼ81), who showed that nitrogen tetroxide and oxygen were evolved ʼd benzoic anhydride formed. The reaction appears to be general *ʼer.*, 1885, **18**, 2990), and mono- and di-basic acid chlorides of the ʼiphatic and aromatic series give good yields of the corresponding ʼhydrides.

But if the interaction of benzoyl chloride and silver nitrate takes ʼace at a much lower temperature, oxides of nitrogen are not evolved ʼd benzoyl nitrate is formed ; this change, which is nearly quantita-ʼe, may be expressed by the equation :

$$C_6H_5 \cdot CO \cdot Cl + AgO \cdot NO_2 = C_6H_5 \cdot CO \cdot O \cdot NO_2 + AgCl.$$

The greatest care has to be taken during this reaction to avoid the ʼsence of moisture, since this at once brings about the decomposition ʼhe nitrate into benzoic and nitric acids.

Thirty-five grams of benzoyl chloride are treated with excess of dry powdered silver nitrate in a flask provided with a phosphoric oxide drying tube, and kept at a temperature of $-15°$ with continual shaking for two and a half hours. At the end of that time, the change of benzoyl chloride into the nitrate is usually complete, and the contents of the flask are rapidly filtered through a layer of carefully dried asbestos, and, if then not free from solid silver salts, may be again filtered through dry filter paper, but should the latter be in the slightest degree moist explosive decomposition may set in. Thirty grams of a clear yellow oil are obtained, which it is only possible to preserve for any length of time by sealing up in glass tubes.

When benzoyl nitrate is carefully and gradually heated to about $100°$, oxides of nitrogen are rapidly evolved and benzoic anhydride formed in theoretical quantity, but if a small amount is heated with a free flame this decomposition sets in with explosive violence and the yield of anhydride is much diminished.

It is, therefore, clear that in the case of benzoyl chloride, and most probably with the other acid chlorides, benzoyl nitrate and corresponding derivatives are the intermediate products in the Lachowicz reaction for the preparation of anhydrides. The decomposition of the nitrate, a substance which may be regarded as the mixed anhydride of benzoic and nitric acids, is expressed by the following equation :

$$2C_6H_5 \cdot CO \cdot O \cdot NO_2 = (C_6H_5 \cdot CO)_2O + O + N_2O_4.$$

This reaction is similar to that undergone by the mixed anhydrides of both aliphatic and aromatic acids, which on distillation yield a mixture of the corresponding simple anhydrides.

The conditions under which benzoyl nitrate is converted into m-nitrobenzoic acid are at present being investigated, but it appears probable that with the pure nitrate this change only takes place with small velocity ; sealed up in glass tubes and kept at a temperature of $12°$, small quantities of solid only commenced to separate out after seven days. When impurities are present, however, the velocity of this change appears to be greatly increased. No other isomeric acid is formed, since, on fractionally recrystallising the barium salt, the various acids obtained showed no indication of the presence of the less fusible p-derivative or of the sweet o-nitrobenzoic acid.

An apparently similar change takes place in several inert solvents, but notably in nitrobenzene. A solution of the nitrate in this solvent left in a warm place deposited pure m-nitrobenzoic acid in the course of ten days. An investigation of the filtrate showed that no dinitrobenzene had been formed. Now, since benzoyl nitrate acts as a nitrating agent, as described later, it appears that the presence of a nitro-group in the benzene nucleus prevents the entrance of a second group, and that consequently *m*-nitrobenzoyl nitrate would not show a tendency to pass into 3 : 5-dinitrobenzoic acid. For this reason, a small amount of *m*-nitrobenzoyl nitrate was prepared in a manner similar to that previously described in the case of the benzoyl derivative, but since the acid chloride melts at 35° the temperature maintained during the reaction was between 30° and 40°. An oil was obtained which sets to a hard, white, crystalline mass on cooling. It fuses between 40° and 50°, and, owing to the extreme ease with which it absorbs moisture, was not obtained quite free from *m*-nitrobenzoic acid. It fumes in the air, decomposes slowly at the ordinary temperature, and with explosive violence if rapidly heated to 100°. But so far as the investigation of this substance has been carried out, no change into 3 : 5-dinitrobenzoic acid has, as yet, been observed.

Benzoyl nitrate is a most reactive substance, and the following decompositions with perfectly dry reagents take place at temperatures between 0° and −15°.

Ethyl alcohol gives ethyl nitrate and benzoic acid, the former substance being readily detected by its boiling point, odour, and insolubility in water. Benzene and, less readily, toluene are converted into their nitro-derivatives with the simultaneous production of benzoic acid. If the reagent is added to an excess of well cooled phenetole, benzoic acid separates, and, after extracting the substance with aqueous caustic potash, a quantitative yield of nitrophenetole, presumably the *o*-derivative, is obtained from the residue. The nitro-derivative boils between 264° and 269° and melts below 0°, and is apparently quite free from isomeric nitrophenetoles. The reaction with phenol takes place equally readily and *o*-nitrophenol is the chief product. Methylaniline dissolved in light petroleum and well cooled is converted into phenylmethylnitramine, $C_6H_5 \cdot N(CH_3) \cdot NO_2$, and benzoic acid, the former substance being recognised by its melting point and characteristic behaviour with sulphuric acid ; this reaction again takes place quantitatively. Dimethylaniline in the same solvent does not react so simply, but *p*-nitrodimethylaniline is among the products. These are some of the reactions which have been but partially investigated, and a more detailed examination of the whole question is in progress. But it appears very probable that in benzoyl nitrate we have a new nitrating agent, which will enable this operation to be carried out at low temperatures and in the absence of water.

The only reaction in which benzoyl nitrate shows its analogy to the chloride is in the case of aniline, which, under light petroleum, is quantitatively converted into aniline nitrate and benzanilide.

It is hoped that mixed anhydrides of organic and other inorganic acids may be isolated, and experiments on these lines are now being carried out.

My best thanks are due to Mr. T. H. Butler for the assistance he has given me in this investigation.

UNIVERSITY COLLEGE,
BRISTOL.

II.—The Diazo-derivatives of 1 : 5- and 1 : 8-Benzenesulphonylnaphthylenediamines.

By GILBERT THOMAS MORGAN and FRANCES MARY GORE MICKLETHWAIT.

THE interactions of the diamines of the benzene series with nitrous acid afford a striking illustration of the influence of orientation on the properties of aromatic compounds. The ortho-diamines and their mono-acyl derivatives yield cyclic diazoimides which are generally colourless * and very stable towards hydrolytic agents, and although until recently condensation had not been detected among the para-diamines, yet about a year ago the authors obtained from the aryl-sulphonyl derivatives of these bases a new series of diazoimides which differ from their ortho-isomerides in having a distinctly yellow colour, and in readily undergoing fission on treatment with acids, phenols, or aromatic amines, the acids regenerating the diazonium salt, whilst the phenols and aromatic amines give rise to azo-derivatives (Trans., 1905, **87**, 73, 921, 1302). The formation of these two series of diazoimides indicates very forcibly the profound mutual influence exercised by the substituents when they occupy the ortho- or para-positions of the aromatic nucleus, for in the meta-series this tendency for the diamines and their acyl derivatives to yield diazoimino-compounds by internal condensation seems to be absent.

The marked differences in the physical and chemical properties of the two series of diazoimino-derivatives point to fundamentally dissimilar configurations for the two groups of substances, and, as was stated in the first communication on this subject (*loc. cit.*, p. 73), the ortho-compounds which are at once formed, even in the presence of strong mineral acids, may be formulated either as diazoimides or as diazonium-imides, whereas the para-derivatives, which are produced only in faintly acid or neutral solutions, might be regarded as being

---

* 2 : 3-Diazoiminonaphthalene is, however, described as being a yellow compound (Friedländer and Zakrzewski, *Ber.*, 1894, **27**, 765).

either diazoimides' (I) or imido-*p*-quinonediazides (II), the conditions of formation in this case excluding the diazonium configuration.

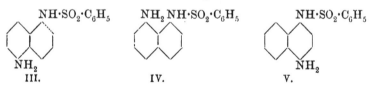

I.   II.

Hitherto no experimental evidence has been forthcoming to enable one to decide between these two formulæ for the para-diazoimides, for although the former sufficed to explain the formation and modes of fission of these substances, yet their colour and great reactivity suggested a quinonoid structure.

This question of the constitution of these coloured reactive para-diazoimides has now been put to the test of experiment by extending the investigation to the benzenesulphonyl derivatives of two of the heteronucleal naphthylenediamines.

*Benzenesulphonyl-*1 : 5-*naphthylenediamine* (III), a compound in which the substituents, although placed at the extremities of a chain of four carbon atoms, are nevertheless situated at the opposite ends of different benzene rings, shows no tendency to yield a diazoimide when treated with nitrous acid, this negative result indicating that the presence of one substituent in the 4- or δ-position with respect to the other does not in itself determine the formation of a diazo-anhydride by internal condensation, but that the orientation of the two groups in the aromatic nucleus is the more important factor.

$NH \cdot SO_2 \cdot C_6H_5$ NH_2 $NH \cdot SO_2 \cdot C_6H_5$ $NH \cdot SO_2 \cdot C_6H_5$

NH_2 NH_2

III.   IV.   V.

*Benzenesulphonyl-*1 : 8-*naphthylenediamine* (IV), which contains one substituent in the 3- or γ-position with respect to the other, these two groups being situated in adjacent positions in the two rings, yields a diazoanhydride quite as readily and completely as its homonucleal para-isomeride (V) (Trans., 1905, 85, 929). The new peri-diazoimide (VI) is yellow and also comparable in all other respects with the isomeric para-diazoimide (VII) already described (*loc. cit.*, p. 928), and as it contains its substituents in both rings of the naphthalene nucleus, it does not seem possible to construct its formula on the assumption that an intranucleal quinonoid rearrangement has occurred, and hence the cyclic structure (VI) adopted below is, in the present state of our knowledge, the simplest mode of representing the constitution of the compound. As, however, the properties of the para-diazoimides resemble so closely those of the peri-diazoimide, the adoption of a cyclic configuration (VI) for the latter compound leads by analogy to the acceptance of the cyclic structure (I) for the para-series in preference to the quinonoid formula (II).

Accordingly the isomeric benzenesulphonyl-*peri*- and -*para*-diazoimides of the naphthalene series should be represented respectively by the following formulæ :

$$N = N - N \cdot SO_2 \cdot C_6H_5$$

VI.

$$- N \cdot SO_2 \cdot C_6H_5$$

VII.

The quinonoid formula being excluded, the colour of these diazoimides must now be referred to the presence in their molecules of the chromophore $-N\!:\!N-$, a group which is generally assumed to be present in the coloured diazoamines, diazosulphonates, diazocyanides, and azo-compounds. Thus, the acceptance of the cyclic formula for the yellow diazoimides brings these compounds into line with other coloured azo- and diazo-derivatives. On the other hand, the ortho diazoimides, many of which are colourless, are also formulated as containing this group, and from this point of view it seems desirable that the constitution of these substances should be reconsidered.

Further evidence in support of the foregoing formula for the peri diazoimide has been obtained by applying the diazo-reaction to as-*benzenesulphonyl*-N-*methylnaphthylenediamine*,

$$NH_2 \quad N(CH_3) \cdot SO_2 \cdot C_6H_5$$

a substance which readily gives rise in succession to a diazonium salt and an azo-β-naphthol, but does not yield a diazoanhydride, thus proving that the production of the peri-diazoimide is due to the inter action of the diazo-group with the complex $NH \cdot SO_2 \cdot C_6H_5$, for when the labile hydrogen of the latter is replaced by methyl no condensation occurs.

1 : 8-Naphthylenediamine, when treated with nitrous acid, yield

1 : 8-diazoiminonaphthalene, $C_{10}H_6 \!\! < \!\! \substack{N \\ NH} \!\! > \!\! N$ (De Aguiar, *Ber.*, 187

7, 315), a red compound, which, unlike its foregoing benzenesulphonyl derivative, does not undergo fission when treated with cold concentrated hydrochloric acid, and accordingly resembles the stable orthodiazoimines. Hence, as regards its behaviour towards nitrous acid, *peri-* or 1 : 8-naphthylenediamine occupies a position intermediate between the ortho- and para-diamines. Like the former, it yields itself a stable diazoimino-compound not decomposed by acids, the peri-derivative, however, differing from the ortho-diazoimines in being intensely coloured. The relationship of the peri-base to the paradiamines is plainly indicated by the analogous behaviour of their benzenesulphonyl derivatives, as set forth in the foregoing discussion.

EXPERIMENTAL.

Preparation of 5- *and* 8-*Nitro-a-naphthylamines.*

The method of preparation employed, which was originally devised by Noelting, has already been described in this Journal (Meldola and Streatfeild, Trans., 1893, **63**, 1054). It consists in nitrating a well-cooled solution of 100 grams of a-naphthylamine in 1000 grams of concentrated sulphuric acid with 64 grams of nitric acid (sp. gr. 1·42) mixed with 128 grams of concentrated sulphuric acid. The reaction being liable to get out of control, the solution of the base in the concentrated sulphuric acid and the addition of the nitrating mixture must both be carried out very slowly at low temperatures ; the product, when left overnight and poured into iced water, yields a dark brown precipitate which contains 5-nitro-a-naphthylamine sulphate. The filtrate, when neutralised with sodium carbonate, yielded crude 8-nitro-a-naphthylamine, which was purified by repeating several times the operation of dissolving the base in sulphuric acid, filtering the solution from sparingly soluble sulphate, and reprecipitating with sodium carbonate. The crude sulphate of the 5-nitro-base was dissolved in boiling water and the base set free from the filtered solution by sodium carbonate.

Ultimately, the two isomerides were further purified by crystallisation from petroleum, when the peri-base was obtained from the solvent boiling at 80—100° in red leaflets and scales melting at 95—97°, whereas the 1 : 5-isomeride crystallised from the fraction (b. p. 100—120°) in dark red needles melting at 114—116°. The yield of the peri-base was about 5—6 per cent. of the weight of a-naphthylamine originally taken, and the nitration was repeated until about 30 grams of this material had accumulated.

Action of Benzenesulphonic Chloride on the Two Nitro-α-naphthylamines.

The nitro-base (1 gram) was dissolved in dry pyridine (10—20 c.c.), treated with 1·5 grams of benzenesulphonic chloride, and the solution heated to boiling for some time. With the 1 : 5-base, the reaction was practically completed in three hours, but with the 1 : 8-isomeride a large proportion of unaltered base was recovered, even after heating for 10—12 hours.* The product was poured on to crushed ice, acidified with dilute hydrochloric acid, and the insoluble residue dissolved in aqueous sodium carbonate. The crude benzenesulphonyl derivative, when reprecipitated from the alkaline solution with dilute hydrochloric acid, was crystallised from dilute alcohol (1 : 1).

Benzenesulphonyl-5-nitro-α-naphthylamine, $NO_2 \cdot C_{10}H_6 \cdot NH \cdot SO_2 \cdot C_6H_5$, crystallised in needles and melted at 183°.

0·2008 gave 15·6 c.c. nitrogen at 19° and 753 mm.   N = 8·85.

$C_{16}H_{12}O_4N_2S$ requires N = 8·53 per cent.

Benzenesulphonyl-8-nitro-α-naphthylamine, when crystallised first from dilute and then from strong alcohol, separated in almost colourless needles melting at 194°.

0·2128 gave 15·85 c.c. nitrogen at 19·5° and 744 mm.   N = 8·36.
0·1602 „   0·3434 CO_2 and 0·0539 H_2O.   C = 58·46 ; H = 3·74.
$C_{16}H_{12}O_4N_2S$ requires N = 8·53 ; C = 58·53 ; H = 3·66 per cent.

Reduction of the Benzenesulphonylnitro-α-naphthylamines.

Five grams of the nitro-derivative were suspended in 250 c.c. of warm water and treated with 10 grams of iron and 1·5 c.c. of glacial acetic acid; the mixture was boiled for half an hour, when the reduction, which took place with the same readiness for both isomerides, appeared to be complete. The mixture was rendered alkaline with sodium carbonate and filtered; in both cases the benzenesulphonyldi-amine remained in solution and was at once precipitated with dilute acetic acid.

Benzenesulphonyl-1 : 5-naphthylenediamine,

$$NH_2 \cdot C_{10}H_6 \cdot NH \cdot SO_2 \cdot C_6H_5,$$

crystallised well either from alcohol or from toluene containing a small proportion of petroleum (b. p. 80—100°); from the former medium, it separated in large aggregates of radiating, colourless, silky needles, some of the individual crystals being 1½ inches long. Its melting point was 161°.

* It was found that the colour of the recovered 1 : 8-base was much less intense than that of the original specimen, and that its melting point had risen to 98—100°.

0·1874 gave 15 c.c. nitrogen at 19° and 779 mm.   N = 9·44.

0·3700 „   0·2881 BaSO₄·   S = 10·69.

$C_{16}H_{14}O_2N_2S$ requires N = 9·39 ; S = 10·73 per cent.

Benzenesulphonyl-1 : 8-*naphthylenediamine* crystallised less readily than its isomeride from dilute alcohol or from toluene and petroleum, and separated in pale grey needles melting at 166°.

0·2304 gave 19·0 c.c. nitrogen at 19° and 744 mm.   N = 9·27 per cent.

Action of Nitrous Acid on the Benzenesulphonyl-1 : 5- *and* -1 : 8-*naphthylenediamines.*

One gram of the 1 : 5-isomeride, when suspended in 12 c.c. of glacial acetic acid and 8 c.c. of concentrated hydrochloric acid and treated at − 10° to − 5° with 2·8 c.c. of aqueous sodium nitrite (20 per cent.), yielded a soluble diazonium chloride together with a brown, amorphous product. The latter was removed by filtration, the filtrate diluted with water, and again filtered from a small proportion of tarry matter ; the final filtrate, when treated with a large excess of aqueous sodium acetate, remained clear and showed no indications of yielding an insoluble diazoimide. The diazotised product was shown to be still in the form of a diazonium salt by combining it with β-naphthol. The red sodium derivative of the azo-naphthol, which was deposited almost completely from the alkaline solution, when collected and treated with acetic acid, yielded *benzenesulphonyl-5-aminonaphthalene-1-azo-β-naphthol,*

$$NH·SO_2·C_6H_5$$

$$HO·C_{10}H_6·N_2$$

which separated from glacial acetic acid in red, felted needles and melted at 260°.

0·1600 gave 13·0 c.c. nitrogen at 19° and 755 mm.   N = 9·52.

$C_{26}H_{19}O_3N_3S$ requires N = 9·27 per cent.

The azo-compound developed an intense reddish-violet coloration with cold concentrated sulphuric acid.

Benzenesulphonyl-1 : 8-*naphthylenediazoimide,*

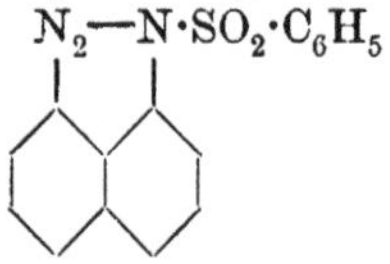

$$N_2\text{—}N\cdot SO_2\cdot C_6H_5$$

Benzenesulphonyl-1 : 8-naphthylenediamine (1·5 grams), suspende
in 12 c.c. of glacial acetic acid and 12·0 c.c. of concentrated hydr
chloric acid, cooled to −10°, and slowly treated with 3·6 c.c.
aqueous sodium nitrite (20 per cent.), gave rise to a soluble diazoniu
chloride together with a small proportion of a greenish-brown, amo
phous precipitate. The filtered solution yielded a further deposit
viscid impurity on dilution, and, after filtration, was cautiously treate
with dilute aqueous sodium acetate to remove last traces of resinoi
products. As soon as the acetate gave a slight permanent precipita
of cyclic diazoimide, the turbid solution was again filtered, and tl
clear, light yellow filtrate then treated with excess of concentrate
acetate solution, when the diazoimide separated as a flocculent, orang
yellow precipitate, which on stirring became crystalline and lighter i
colour.

The deposit, which was collected and thoroughly washed successive
with water, alcohol, and light petroleum, was dried in the desiccate
until its weight was constant. The alcohol removed a red impurit
but otherwise the diazoimide was quite insoluble either in this solve
or in water; when thoroughly dried, the compound, without furth
purification, gave the following numbers on analysis :

0·2353 gave 0·5368 CO_2 and 0·0754 H_2O. C = 62·22 ; H = 3·56.
0·2416 ,, 0·5481 CO_2 ,, 0·0804 H_2O. C = 61·88 ; H = 3·69.
0·2486 ,, 0·5687 CO_2. C = 62·39.
0·1524 ,, 17·2 c.c. nitrogen at 19° and 779 mm. N = 13·45.
0·2173 ,, 0·1673 $BaSO_4$. S = 10·57.
$C_{16}H_{11}O_2N_3S$ requires C = 62·14 ; H = 3·56 ; N = 13·59 ; S = 10·35
per cent.

When heated in the combustion tube, the substance decompos
energetically, evolving puffs of yellow smoke. It may be kept for
indefinite time in the dark, but is affected by light, rapidly becomi
brown.*

The sodium acetate filtrate from benzenesulphonyl-1 : 8-naphthylei
diazoimide was added to alkaline β-naphthol, but no azo-colour w

* The colour of the yellow diazoimides assumes a paler hue, but is not entir
destroyed when the compounds are cooled to the temperature of liquid oxygen.

produced, thus showing that the conversion of the diazonium salt into the 1 : 8-diazoanhydride was complete.

Fission of the peri-Diazoimide.

(1) *Fission with Acids.*—When suspended in 50 parts of cold glacial acetic acid, the diazoimide remained undissolved, and only passed into solution on warming on the water-bath. The deep red liquid, when poured into alkaline β-naphthol, yielded the alkali derivative of the azo-compound, which was treated with glacial acetic acid in order to set free the azo-β-naphthol.

The *peri*-diazoimide dissolved immediately in ice-cold concentrated hydrochloric acid to an almost colourless solution, which, when diluted with cold water and poured into alkaline β-naphthol, yielded the alkali azo-derivative, from which the free azo-naphthol was isolated by means of acetic acid.

Another portion of the solution in hydrochloric acid after dilution with water was treated with excess of platinic chloride, when a pale yellow, crystalline *diazonium platinichloride* was precipitated, which, when dried in the air, gave the following result on analysis :

0·2315 gave 0·0427 Pt. Pt = 18·44.

$(C_6H_5 \cdot SO_2 \cdot NH \cdot C_{10}H_6 \cdot N_2)_2PtCl_6$ requires Pt = 18·95 per cent.

(2) *Fission with Phenols.*—Equal parts of the *peri*-diazoimide and β-naphthol were dissolved in dry pyridine and heated for two and a half hours on the water-bath. The deep red solution thus obtained was allowed to evaporate slowly, the solid residue was treated with excess of aqueous caustic soda and collected, the free azo-naphthol being then set free by glacial acetic acid.

Benzenesulphonyl-8-aminonaphthalene-1-azo-β-naphthol,
$$HO \cdot C_{10}H_6 \cdot N \vdots N \quad NH \cdot SO_2 \cdot C_6H_5$$

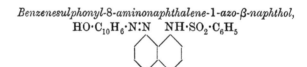

The azo-compound produced in the foregoing fission experiments was dissolved in benzene and precipitated from this solution by the addition of light petroleum.

0·1802 gave 14·2 c.c. nitrogen at 19·5° and 771 mm. N = 9·18.

$C_{26}H_{19}O_3N_3S$ requires N = 9·27 per cent.

This compound, which was not obtained crystalline, melted somewhat indefinitely at 170—180°, and developed an intense reddish-violet coloration with cold concentrated sulphuric acid.

Benzenesulphonyl-8-nitro-N-methyl-α-naphthylamine,
$$NO_2 \cdot C_{10}H_6 \cdot N(CH_3) \cdot SO_2 \cdot C_6H_5.$$

An alcoholic solution of 2 grams of recrystallised benzenesulphonyl
8-nitro-α-naphthylamine and 0·3 gram of caustic soda was boiled fo
six hours with excess of methyl iodide (2 grams) added in smal
portions. The crystalline residue obtained after evaporating off th
solvent was washed successively with aqueous caustic soda and watei
and when crystallised repeatedly from alcohol separated in pale yellov
needles melting at 179°.

0·2851 gave 20·8 c.c. nitrogen at 19° and 766 mm. N = 8·45.

$C_{17}H_{14}O_4N_2S$ requires N = 8·18 per cent.

The alkylation was practically quantitative ; the alkaline filtrate
from the *as*-benzenesulphonyl-8-nitro-*N*-methyl-α-naphthylamine gav
on acidifying no unmethylated nitro-compound.

as-Benzenesulphonyl-N-methyl-1 : 8-naphthylenediamine was readil:
obtained from the foregoing nitro-derivative by reducing 4 grams o
this substance with 16 grams of iron, 2 c.c. of glacial acetic acid, an
200 c.c. of warm water, the heating being continued for an hou
before the mixture was rendered alkaline with sodium carbonate
The alkaline filtrate from the iron oxide contained no organic base
the product being isolated from this insoluble residue by repeatec
extraction with alcohol. The new base did not crystallise well fron
this solvent, but was deposited in brownish-white, nodular crystal:
from its benzene solution on the addition of a small quantity of ligh
petroleum.

0·2898 gave 22·2 c.c. nitrogen at 19° and 772 mm. N = 8·95.

$C_{17}H_{16}O_2N_2S$ requires N = 8·97 per cent.

After repeated crystallisation, the substance melted at 161—162°.

as-Benzenesulphonyl-N-methyl-8-aminonaphthalene-1-azo-β-naphthol,

$$HO \cdot C_{10}H_6 \cdot N \vdots N \qquad N(CH_3) \cdot SO_2 \cdot C_6H_5$$

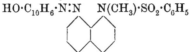

When diazotised with aqueous sodium nitrite in a mixture of con
centrated hydrochloric and acetic acids at − 10° to − 5°, the precedinɡ
methyl base yielded a soluble diazonium salt, from the filtered solutior
of which aqueous sodium acetate precipitated no insoluble diazo
anhydride. The clear solution of the diazo-salt, when poured into ar
alkaline solution of β-naphthol, yielded an insoluble bright scarlet azo·

compound which crystallised readily from glacial acetic acid in transparent, ruby-red nodules.

0·2359 gave 18 c.c. nitrogen at 19° and 770 mm. N = 8·89.

$C_{27}H_{21}O_3N_3S$ requires N = 8·99 per cent.

This azo-derivative, which melted at 215°, developed an intense reddish-violet coloration with cold concentrated sulphuric acid.

The authors' thanks are due to the Government Grant Committee of the Royal Society for a grant which has partly defrayed the expenses of this investigation.

ROYAL COLLEGE OF SCIENCE, LONDON,
SOUTH KENSINGTON, S.W.

III.—*Azo-derivatives of* 4 : 6-*Dimethylcoumarin.*

By JOHN THEODORE HEWITT and HERBERT VICTOR MITCHELL.

IN a recent communication by one of the authors, it has been shown that the benzeneazocoumarin described by Borsche (*Ber.*, 1904, **37**, 346, 4116), when dissolved in aqueous alkali with formation of the coumarinate, is precipitated from solution either by excess of hydrochloric acid or by carbon dioxide in the form of the azocoumarin, and not as an azocoumarinic acid (Mitchell, Trans., 1905, **87**, 1229). Such behaviour is more in accord with the hydroxyazo- than with the quinonehydrazone-structure of this and analogous compounds, and is not surprising, seeing that parahydroxyazo-compounds of the benzene series behave in practically all respects as if they possessed a hydroxyazo-structure, and are almost universally regarded as having this configuration.

In the case of orthohydroxyazo-compounds, the facts are by no means as clear, for whilst alkylation both in the ortho- and para-series leads to oxygen ethers, some difference of view has been expressed as to the constitution of the substances formed by acylating orthohydroxyazo-compounds. The formation of acetanilide by the reduction of benzene-azo-*p*-tolyl acetate and of Meldola and East's benzeneazo-β-naphthyl acetate observed by Goldschmidt and Brubacher (*Ber.*, 1891, **24**, 2300) certainly appears to favour a quinonehydrazone constitution for these substances, whilst the reduction of *p*-tolueneazo-β-naphthyl acetate studied by Meldola and Hawkins (Trans., 1893, **63**, 926) led to a result which also points to a hydrazone formula, aceto-*p*-toluidide being obtained as a chief product.

Other objections to a hydroxyazo-formulation of these compound may be seen in the insolubility of many o-hydroxyazo-compounds aqueous alkalis (compare Meldola, *Phil. Mag.*, 1888, [v], 26, 403 ; Meldo and Forster, Trans., 1891, 59, 710 ; and Meldola and Hawkins, Tran 1893, 63, 923, concerning the possibility of the oxygen atom becomir a member of a closed chain) and in the fact that ortho-hydroxyaz compounds do not appreciably associate in non-hydroxylic solven (Auwers and Orton, *Zeit. physikal. Chem.*, 1896, 21, 337).

Too much weight must not be placed on the above arguments ; th reduction of an acyl derivative of an azophenol should give rise fission products in which the acyl group is attached to nitrogen is n altogether unexpected, as not merely might intramolecular change tal place during the process of reduction, but also intermolecular chan between the products of complete fission, the reaction

$$R \cdot NH_2 + R' \cdot CO_2 R'' = R'' OH + R \cdot NH \cdot CO \cdot R'$$

being of very general occurrence.

Auwers and Orton's conclusions as to parahydroxyazo-compoun possessing a structure corresponding to their name, whilst the orth compounds are of quinonehydrazone type, are not justified by the own results. They find that ortho-substituted phenols generally sho less association in non-hydroxylic solvents than the correspondin meta- or para-compounds, so that the complete inhibition of associatio by the somewhat negative and large arylazo-group when in an orth position with respect to a phenolic hydroxyl is not unlikely.

Moreover, it must be remembered that benzeneazo-*p*-cresol, whe treated with bromine in acetic acid suspension, sodium acetate bein present, gives benzeneazobromo-*p*-cresol in nearly quantitative yield, fact strongly in favour of the hydroxyazo-formulation of the compoun (Hewitt and Phillips, Trans., 1901, 79, 160).

EXPERIMENTAL.

Some months ago an attempt was made to nitrate benzeneazo-*p* cresol with warm dilute nitric acid in the hope of confirming Hewi1 and Phillips' bromination results. So far the results have been di: appointing, and we have now turned our attention to the behaviour (azocoumarins containing the azo-group in the ortho-position wit respect to the oxygen atom of the lactonic ring. To obtain substanc of this type, it was necessary to use a coumarin in which the par: position to the oxygen atom was already substituted. We hem chose the 4 : 6-dimethylcoumarin described by von Pechmann an Cohen (*Ber.*, 1884, 17, 2188), as it is readily prepared by the coi densation of *p*-cresol with ethyl acetoacetate.

Benzeneazo-4 : 6-dimethylcoumarin,

Dimethylcoumarin (1·7 grams) was boiled with fairly concentrated potassium hydroxide until completely dissolved, when water and ice were added in considerable quantities. To this solution of potassium dimethylcoumarinate a phenyldiazonium chloride solution, prepared from 0·93 gram of aniline, 0·72 gram of sodium nitrite, 3 c.c. of fuming hydrochloric acid, and sufficient ice, was added. By acidification of the intensely red solution of the alkaline salt of benzeneazodimethyl-coumarinic acid, a yellow precipitate is deposited consisting not of the acid, but of the corresponding lactone. By twice crystallising from alcohol, the substance is obtained in lustrous, orange-red needles which melt at 199—200 .

0·1169 gave 0·3162 CO_2 and 0·0583 H_2O.   C = 73·7 ; H = 5·5.

$C_{17}H_{14}O_2N_2$ requires C = 73·4 ; H = 5·0 per cent.

Benzeneazodimethylcoumarin dissolves in alcohol, toluene, chloroform, and pyridine; it is insoluble in light petroleum. Like the nitro-derivatives about to be described, it resembles the acyl derivatives rather than the azophenols themselves.

Although insoluble in cold alkaline solutions, prolonged boiling brings about solution with formation of an azocoumarinate and development of an intense coloration.

o-Nitrobenzeneazo-4 : 6-dimethylcoumarin, $NO_2 \cdot C_6H_4 \cdot N:N \cdot C_{11}H_9O_2$.

This substance was prepared in the usual manner. The colour of the solution of its alkaline azocoumarinate is reddish-violet. The free azocoumarin separates from chloroform as scarlet needles which melt at 250° with decomposition. The substance is also soluble in pyridine, but dissolves sparingly in alcohol.

0·1082 gave 0·2507 CO_2 and 0·0404 H_2O.   C = 63·2 ; H = 4·1.
0·1465   ,,   16·4 c.c. nitrogen at 15° and 748 mm.   N = 13·0.

$C_{17}H_{13}O_4N_3$ requires C = 63·2 ; H = 4·0 ; N = 13·0 per cent.

m-Nitrobenzeneazo-4 : 6-dimethylcoumarin, prepared in the usual manner, separates from chloroform in large, transparent, reddish-

brown tablets. These soon become opaque owing to the volatilisation of one molecule of chloroform of crystallisation.

3·3354 lost 0·8919 at 120°.   Loss = 26·75 per cent.
$C_{17}H_{13}O_4N_3,CHCl_3$ requires $CHCl_3 = 27·05$ per cent.

The dried substance melted at 212° and gave the following figures on analysis :

0·1612 gave 18·2 c.c. nitrogen at 12° and 760 mm.   N = 13·4.
0·2482 „   28·1 c.c.   „   12° „ 733 mm.   N = 13·0.
$C_{17}H_{13}O_4N_3$ requires N = 13·0 per cent.

m-Nitrobenzeneazo-4 : 6-dimethylcoumarin is also soluble in acetic acid, somewhat sparingly so in alcohol. The solutions of the alkaline coumarinates are red in colour.

p-*Nitrobenzeneazo*-4 : 6-*dimethylcoumarin* was obtained by coupling potassium dimethylcoumarinate with p-nitrophenyldiazonium chloride ; its alkaline solution is intensely violet, the colour being far bluer in shade than is the case with the two isomerides. After the precipitated azocoumarin had been twice recrystallised from dilute acetic acid and once from chloroform, it was obtained as small, brown crystals melting at 229°.

0·1018 gave 11·4 c.c. nitrogen at 15° and 755 mm.   N = 13·0.
$C_{17}H_{13}O_4N_3$ requires N = 13·0 per cent.

p-Nitrobenzeneazodimethylcoumarin also dissolves in pyridine ; it is sparingly soluble in benzene and insoluble in light petroleum.

The immediate production of lactones on acidification indicates the presence of ready-formed hydroxyl groups, and attention may again be drawn to the fact that the passage of carbon dioxide into a solution of benzeneazocoumarin in alkali leads to the precipitation of benzeneazocoumarin (Mitchell, Trans., 1905, **87**, 1230). If we assume the equation

$$C_6H_5·N\!:\!N·C_6H_4ONa + H_2CO_3 = NaHCO_3 + C_6H_5·NH·N\!:\!C_6H_4\!:\!O$$

to be correct, we must express the first stage of the action of carbonic acid on the dipotassium salt of benzeneazocoumarinic acid by the equation

$$C_6H_5·N\!:\!N·C_6H_3(OK)·C_2H_2·CO_2K + H_2CO_3 = KHCO_3 +$$
$$C_6H_5·NH·N\!:\!C_6H_3(\!:\!O)·C_2H_2·CO_2K.$$

The last formula indicates the sodium salt of a fairly strong carboxylic acid from which the weak carbonic acid would only liberate very small quantities of the corresponding free acid. The formation of benzeneazocoumarin would then have to be represented by the following equations :

(a) $C_6H_5 \cdot NH \cdot N \colon C_6H_3(\colon O) \cdot C_2H_2 \cdot CO_2K + H_2CO_3 \rightleftharpoons$
$$KHCO_3 + C_6H_5 \cdot NH \cdot N \colon C_6H_3(\colon O) \cdot C_2H_2 \cdot CO_2H.$$

(b) $C_6H_5 \cdot NH \cdot N \colon C_6H_3(\colon O) \cdot C_2H_2 \cdot CO_2H \rightleftharpoons$
$$C_6H_5 \cdot N \colon N \cdot C_6H_3(OH) \cdot C_2H_2 \cdot CO_2H.$$

(c) $C_6H_5 \cdot N \colon N \cdot C_6\ _3(OH) \cdot C_2H_2 \cdot CO_2H \longrightarrow$
$$H_2O + C_6H_5 \cdot N \colon N \cdot C_6H_3 {<}^{C_2H_2}_{\underline{}O} {>} CO.$$

Since the product on the right-hand side of equation (a) can only be present in very small quantity, the equilibrium expressed in equation (b) must be established with enormous rapidity in order to explain anything more than the very slow formation of a lactone. Such an extremely rapid establishment of equilibrium, even if not definitely disproved, is at least improbable.

EAST LONDON COLLEGE.

IV.—*Azo-derivatives of 4-Methyl-a-naphthocoumarin.*

By JOHN THEODORE HEWITT and HERBERT VICTOR MITCHELL.

IN continuation of earlier work on azocoumarins containing the azo-group in the para-position with respect to the lactonic oxygen (Mitchell, Trans., 1905, 87, 1229), it has been considered advisable to examine representatives of the naphthalene series. Whilst benzene-azophenol cannot be directly prepared from quinone and phenyl-hydrazine, benzeneazo-a-naphthol was obtained by Zincke and Binde-wald (*Ber.*, 1884, 17, 3026) by the interaction of phenylhydrazine and a-naphthoquinone. Further support for the formulation of benzene-azo-a-naphthol as a-naphthoquinonehydrazone is afforded by the re-action between this substance and tetramethyldiaminobenzhydrol discovered by Möhlau and Kegel (*Ber.*, 1900, 33, 2858).

As a convenient substance for our experiments, we have chosen the 4-methyl-a-naphthocoumarin obtàined by Bartsch, who condensed a-naphthol with ethyl acetoacetate (*Ber.*, 1903, 36, 1966).

Benzeneazo-4-methyl-a-naphthocoumarin,

was obtained by the action of phenyldiazonium chloride on a solution of potassium 4-methylnaphthocoumarinate. The colour of the alkaline solution produced is intensely red, and, on addition of dilute sulphuric acid, the azocoumarin is deposited as a scarlet precipitate. By recrystallisation from pyridine, orange-brown, long, flaky needles are obtained, which melt at 207°. The substance is also soluble in chloroform and alcohol, but is insoluble in light petroleum.

0·0911 gave 0·2553 CO_2 and 0·0363 H_2O.   C = 76·4 ; H = 4·4.

$C_{20}H_{14}O_2N_2$ requires C = 76·4 ; H = 4·5 per cent.

o-*Nitrobenzeneazo-4-methyl-α-naphthocoumarin* was obtained from diazotised o-nitroaniline in the usual manner; the alkaline solution is reddish-violet. The azocoumarin crystallises from pyridine in small, brown flakes melting at 268°.

0·1274 gave 13·6 c.c. nitrogen at 20° and 763 mm.   N = 12·1.

$C_{20}H_{13}O_4N_3$ requires N = 11·8 per cent.

The substance dissolves in chloroform, less readily in alcohol, and is insoluble in light petroleum.

m-*Nitrobenzeneazo-4-methyl-α-naphthocoumarin* gives alkaline solutions possessing a colour similar to that of cobalt nitrate, although not blue in shade. On crystallisation from chloroform, light brown needles are obtained which melt at 239°. Whilst soluble in pyridine, the substance is nearly insoluble in alcohol.

0·1224 gave 0·3004 CO_2 and 0·0426 H_2O.   C = 66·9 ; H = 3·8.

$C_{20}H_{13}O_4N_3$ requires C = 66·9 ; H = 3·6 per cent.

p-*Nitrobenzeneazo-4-methyl-α-naphthocoumarin* is remarkable, in that its alkaline solutions are indigo-blue in colour. On acidifying the alkaline solution, the azocoumarin is precipitated as a reddish-brown powder which, when recrystallised from toluene, melts at 270—271°.

0·1170 gave 0·2886 CO_2 and 0·0393 H_2O.   C = 67·3 ; H = 3·7.

$C_{20}H_{13}O_4N_3$ requires C = 66·9 ; H = 3·6 per cent.

The substance dissolves fairly readily in toluene, and only sparingly in alcohol ; it is gradually dissolved by solutions of hot alkalis with production of the indigo coloration. So extremely marked is the coloration that it seems advisable to inquire into the cause, especially in view of the fact that the p-nitrobenzene-4-azo-α-naphthol discovered by Bamberger (*Ber.*, 1895, **28**, 848; compare Hantzsch, *ibid.*, **28**, 1124), although brownish-red in colour, gives violet alkaline solutions. Such difference in colour between an azophenol and its alkali salts only seems to occur when a nitro-group is present in the substituent arylazo-group in the para-position.

We therefore prepared a specimen of p-nitrobenzene-4-azo-α-naphthol-2-carboxylic acid, which, itself brown in colour, dissolved in alkalis with a bluish-violet shade. If the colour were merely due to salt formation, one would expect the colouring matter to give a violet colour with an aluminium mordant. The colour obtained was, however, brown, and since the lake obtained must contain the

grouping, one is inclined to assign another formula to the violet alkaline salts. The only apparent alternative is to assume the formation of an *iso*nitro-group under the influence of alkali, which would, for instance, mean that the potassium salt of *p*-nitrobenzeneazo-α-naphthol must possess the constitution :

EAST LONDON COLLEGE.

V.—*The Action of Water on Diazo-salts.*

By JOHN CANNELL CAIN and GEORGE MARSHALL NORMAN.

DURING an investigation carried out by one of us (J. C. C.) and Nicoll on the rate of decomposition of diazo-compounds, it was noticed (Trans., 1902, 81, 1440) that the products of decomposition of the tetrazo-salts prepared from *oo*-dichlorobenzidine and dianisidine were not the corresponding dihydroxy-compounds, as in the case of benzidine and tolidine, although the nitrogen was evolved as usual. It was accordingly of much interest to investigate the nature of the substances formed, and the result of this work was to show that in each case the chief, if not the only product formed was of a quinonoid character (Trans., 1903, 83, 688). The very great difference in behaviour of these two tetrazo-salts as compared with those from benzidine and tolidine was apparently to be attributed to the presence of a chlorine and a methoxyl group respectively in the ortho-position to the diazonium group.

With the object of testing this view a search was made in the literature for any abnormal cases of decomposition of ortho-substituted

diazo-salts belonging to the benzene series, and, curiously enough, all
the instances found answered this description. It was therefore
decided to make a thorough examination of these compounds with the
object of discovering whether the alleged non-formation of phenols
was to be explained by the presence of an ortho-substituent or
otherwise.

Methods of carrying out the Decomposition.

What may be regarded as the normal method of decomposition of
such diazo-salts is to diazotise in moderately strong mineral acid solution (preferably hydrochloric or sulphuric acid) by adding a solution
of sodium nitrite and then either heat to boiling the diazo-solution
thus obtained or pass a current of steam through it. When, however,
no phenol was found by this means, the method adopted by Heinichen
(Annalen, 1889, **253**, 281) was tried. This consists in heating the
strong diazo-solution with strong sulphuric acid, whereby the boiling
point becomes raised to 150°. Heinichen, by this method, was successful in obtaining oo-dibromophenol from the corresponding diazo-salt,
after the usual method had failed. Heinichen's method has been
applied by various other workers in attacking similar problems, but
often without success; thus Orton (Proc., 1905, **21**, 170) failed to
obtain s-tribromophenol from the diazo-salt of tribromoaniline. This
method of decomposition is, however, not a rational one owing to the
fact that by using more concentrated sulphuric acid in order to reach
a higher temperature a very considerable retarding influence is introduced. This retarding influence can be measured by determining the
"coefficient of decomposition C" according to the method described by
one of us and Nicoll (loc. cit., p. 1412). Thus, in the case of benzenediazonium chloride the value of C is 0·0298 (loc. cit. p. 1420). A
corresponding experiment with benzenediazonium sulphate (solution
containing 1 per cent. H_2SO_4) gave nearly the same number, namely,
0·0302. When, however, the amount of sulphuric acid is increased
until the solution contains about 35 per cent. of the acid, the value of
C diminishes to 0·0197 (for details of this work, see Cain, Ber., 1905,
38, 2511). It follows, therefore, that an increase in the quantity of
sulphuric acid produces an apparent increase in the stability of the
diazo-compound. This is thought to be due (loc. cit.) to the withdrawal
of water from the reaction by the sulphuric acid. A third method,
which yields by far the best result with refractory substances, is that
described in the English Patent No. 7233 of 1897 (Kalle and Co.
D.R.-P. No. 95339), which consists in dropping the diazo-solution into
a mixture of dilute sulphuric acid and sodium sulphate heated to
135—145° and allowing any volatile products to distil over.

This method, applied in the Patent Specification especially to the production of guaiacol from the diazo-salt of o-anisidine, is successful where others have failed.

Although by using this method we have been successful in obtaining phenols where previous observers have failed, many of the substances examined have given good yields of the corresponding phenols under the ordinary conditions. This applies particularly to the cases of the diazo-salts from dibromoaniline, dibromo-p-toluidine, and bromo- and chloro-p-toluidines, which were described by Wroblewski in 1874 as yielding no trace whatever of phenols, but only the substituted hydrocarbon. More recently (in 1884), he attributes this abnormality to the presence of a minute quantity of alcohol in the diazo-derivative, left in during the preparation. This explanation is possibly correct, although it is evident that the quantity of alcohol contained in the diazo-salt must have been considerable.

In every case which has been examined, we have been able to obtain the corresponding phenols, and therefore prove that there is, so far as we know, no case in the benzene series where an ortho-substituent hinders or diverts the course of the reaction, as is, apparently, the case in the diphenyl series. In addition to this we have been able to throw some light on the course of the diazo-reaction as applied to the substances here described.

EXPERIMENTAL.

o-Anisidine.

Limpach (Ber., 1891, 24, 4136) prepared the diazo-salt from this substance and passed steam through the solution, but "es trat vollständige Verharzung ein" and no guaiacol could be detected. Gattermann (Ber., 1899, 32, 1136) attempted to prepare the hydroxy-compound, and remarks "dass auch dieses (namely, o-anisidine) eine ungewöhnlich beständige Diazoverbindung bildet, die selbst nach dem Erhitzen in einer Bombe auf über 100° noch nicht zersetzt war." As we have been able to isolate guaiacol in the products of decomposition as carried out in the usual way, and apparently by the same method as Limpach used, we give the full details of the experiment.

o-Anisidine (24·6 grams) was dissolved in water and 60 c.c. of concentrated hydrochloric acid, and diazotised with addition of 200 c.c. of normal sodium nitrite at 20—25°. Concentrated sulphuric acid (60 c.c.) was now added and steam passed into the diazo-solution in a large flask arranged for steam distillation. The decomposition proceeded very slowly and there was no violent evolution of nitrogen as in the case of more unstable diazonium salts. The colourless solution gradually turned pink, which was obviously due to the formation of an azo-colouring matter, this being found to dye wool direct from an acid bath.

The colouring matter is therefore most probably produced by the combination of the diazonium salt with the guaiacol formed, and consequently possesses the formula $\overset{1}{C}H_3 \cdot O \cdot C_6H_4 \cdot \overset{2}{N} \colon \overset{5'}{N} \cdot C_6H_3(\overset{2'}{O}H) \cdot O \cdot \overset{1'}{C}H_3$. A tar gradually collected in the flask and an oil was seen to distil with the steam. The distillation was carried on for four to five hours, and the contents of the flask, when tested the following day, gave no colour with "R salt," showing that the diazonium salt had been completely decomposed.

Examination of Residue.—The solidified tar was filtered, dried, powdered, and in one experiment boiled out with water, which dissolved the foregoing azo-colouring matter. In other experiments this treatment with hot water was omitted. The dry powder was then extracted with ether, leaving a tarry residue. The ethereal solution was dried over calcium chloride, filtered, and the ether distilled off; a viscid oil was left which distilled above 300°. The distillate was left for some days in a desiccator over sulphuric acid, when large, flat, square tables had crystallised; these were dried on porous porcelain and recrystallised from dilute acetic acid, when fern-like needles separated (m. p. 88—89°). The boiling point was a little above 310°. The substance was soluble in alkalis, being reprecipitated by acids. The high boiling point suggested the formation of a condensation product, and analysis showed that probably the mono-methyl ether of o-dihydroxydiphenyl, $OH(2)C_6H_4 \cdot C_6H_4(2)O \cdot CH_3$, had been formed. The yield was very small.

0.1074 gave 0.3057 CO_2 and 0.0609 H_2O. $C = 77.63$; $H = 6.30$.

$C_{13}H_{12}O_2$ requires $C = 77.96$; $H = 6.05$ per cent.

Examination of the Steam Distillate.—The aqueous distillate containing drops of oil was saturated with salt and extracted with ether. After evaporating off the ether from the dried solution, an oil was left which was distilled. The temperature rose gradually to nearly 300°. On rectifying, a large fraction was collected at 195—205°, when the distillation was stopped. This oil was found to be guaiacol (b. p. 205°, m. p. 33°); it solidified in a desiccator, gave a green colour with ferric chloride, and was identical with a sample of guaiacol prepared according to the above-mentioned English Patent. The residue in the distilling flask was extracted with ether and the oil left on evaporating off the ether solidified to large, square, tabular crystals. These were recrystallised from dilute acetic acid and melted at 88—89°. A further quantity of the methyl ether of o-dihydroxydiphenyl had thus distilled over with the guaiacol during the steam distillation. The diazo-salt from o-anisidine was also decomposed exactly as described in the foregoing English Patent and a satisfactory yield of guaiacol was obtained.

Dichloroaniline $(NH_2 : Cl : Cl = 1 : 2 : 5)$.

Schlieper (*Ber.*, 1893, **26**, 2465) was unable to prepare the diazo-salt of this substance, obtaining only the diazoamino-compound. This was also confirmed by Zettel (*Ber.*, 1893, **26**, 2471). As no mineral acid was used by these chemists, but the diazotisation attempted by the use of amyl nitrite, we were confident of being able to prepare the diazo-solution in the usual way, and a clear solution was easily obtained. In one experiment the base was dissolved in hydrochloric acid, sulphuric acid added, and sodium nitrite solution dropped slowly into the boiling solution. A red solid formed in the flask was dried and recrystallised from benzene, the solution yielding shining, bronze-coloured plates (m. p. 195°).

0·1330 gave 14·8 c.c. nitrogen at 18·5° and 755 mm. $N = 13·0$.
$C_{12}H_7N_3Cl_4$ requires $N = 12·54$ per cent.

By heating with glacial acetic acid and acetic anhydride, an acetyl compound was obtained, melting at 226°. The substance formed in the reaction was therefore the aminoazo-compound having the above formula; it dissolved with a red colour in concentrated sulphuric acid. During the progress of these experiments, a paper appeared by Noelting and Kopp (*Ber.*, 1905, **38**, 3506) describing a number of derivatives of this dichloroaniline and the preparation of the above aminoazo-compound by heating a mixture of the hydrochloride of the base, the diazoamino-compound, and the base itself. In this way they obtained the aminoazo-compound corresponding exactly with the substance above described. They describe also the preparation of the dichlorophenol by decomposition of the diazo-salt, and therefore there was no necessity for us to carry on our work further in this direction.

Dibromoaniline $(NH_2 : Br : Br = 1 : 2 : 4)$.

Wroblewski (*Ber.*, 1874, **7**, 1061) obtained only dibromobenzene by the decomposition of the diazo-salt and detected no trace of the phenol. By using the method of the above-mentioned English Patent, the dibromophenol (m. p. 39°) was isolated.

0·224 gave 0·3332 AgBr. $Br = 63·0$.
$C_6H_4OBr_2$ requires $Br = 63·5$ per cent.

s-*Trichloroaniline.*

Hantzsch (*Ber.*, 1895, **28**, 685) prepared a solution of the diazo-chloride of this substance, and was unable to detect the formation of the corresponding phenol by the action of heat. He says " Trichlordiazobenzolchlorid lässt sich mit Wasser, ja selbst mit salpetersalzsaüre kochen, ohne Stickstoff zu entwickeln oder sich überhaupt zu verändern." Under the usual conditions we were also unable to isolate any trichlorophenol, but by using the patented method a small quantity of this substance distilled over with the steam, and melted correctly, namely, at 68°.

0·2511 gave 0·5441 AgCl.   Cl = 53·4.

$C_6H_8OCl_3$ requires Cl = 53·8 per cent.

s-*Tribromoaniline*.

Silberstein (*J. pr. Chem.*, 1883, **27**, 98) was unable to obtain any phenolic derivative by boiling s-tribromobenzenediazonium sulphate or nitrate with dilute acids.   Hantzsch also (*Ber.*, 1900, **33**, 2517) obtained " gar kein Tribromphenol."   Orton (*Proc.*, 1905, **21**, 170), by using Heinichen's method, was unsuccessful in his attempt to prepare this substance.   We can confirm the work of these chemists, as under ordinary conditions no phenol can be isolated.   When, however, the patented method is applied, a small quantity of s-tribromophenol (m. p. 92°) can be obtained.

0·2113 gave 0·3590 AgBr.   Br = 72·31.

$C_6H_3OBr_3$ requires Br = 72·51 per cent.

Chloro-p-toluidine $(NH_2 : CH_3 : Cl = 1 : 4 : 2)$.

Wroblewski (*Ber.*, 1874, **7**, 1061 ; *Ann.*, 1873, **168**, 147) states that the diazonium salt of this substance on decomposition by boiling yields m-chlorotoluene and no phenol.   By decomposing the diazo-solution according to the patented method and by extraction of the distillate with ether, an oil was obtained boiling at 191°.

0·1621 gave 0·1607 AgCl.   Cl = 24·47.

C_7H_7OCl requires Cl = 24·86 per cent.

The substance is therefore chlorocresol $(OH : CH_3 : Cl = 1 : 4 : 2)$.

Bromo-p-toluidine $(NH_2 : CH_3 : Br = 1 : 4 : 2)$.

Wroblewski in this case also (*loc. cit.*) failed to obtain the bromo-cresol, but described the production of the m-bromotoluene ; due, no doubt, as indicated above, to the presence of alcohol.   When steam is passed through the solution containing the diazonium salt, an oil (b. p. 214°) distils over, which is identical in every way with the bromocresol described by Schall and Dralle (*Ber.*, 1884, **17**, 2530).

0·2209 gave 0·22053 AgBr.   Br = 42·48.

C_7H_7OBr requires Br = 42·78 per cent.

Dibromo-p-*toluidine* $(NH_2 : CH_3 : Br : Br = 1 : 4 : 2 : 6)$.

Here again Wroblewski obtained "nicht die geringste Spur von Kresol," but only dibromotoluene. By the patented method we have had no difficulty in obtaining the expected dibromocresol (m. p. 48°).

0·211 gave 0·297 AgBr.   Br = 59·90.

$C_7H_8OBr_2$ requires Br = 60·15 per cent.

It is to be concluded from the above experiments that the supposed cases of abnormal behaviour on boiling certain ortho-substituted diazonium salts with water have no foundation in fact. We do not claim to have examined every alleged exception to the general rule, but think that the number of examples shown in this paper is sufficient to indicate that any remaining instances which may possibly have been overlooked by us are to be regarded with suspicion.

In conclusion, we wish to express our grateful thanks to the Chemical Society for the grant from the Research Fund by means of which the cost of this work was partly defrayed..

NOTE BY JOHN CANNELL CAIN.—With the completion of the foregoing work it may be of interest to summarise the general results of a series of researches on the diazo-reaction carried out by me during the past four years. The action of water on diazonium salts (mostly in presence of a mineral acid) has been investigated both quantitatively and qualitatively, and the main conclusions are as follows:

1. Diazonium salts of the benzene and naphthalene series decompose according to the equation expressing a unimolecular reaction, namely:

$$\frac{1}{t}\log\frac{A}{A-x} = C \text{ (a constant)}$$

(Trans., 1902, **81**, 1412; 1903, **83**, 206). Only one tetrazo-salt (from dichlorobenzidine) was found to conform to this rule. By the determination of the value of "C" in the above equation, the relative stability of diazonium salts is obtained.

2. The rate of decomposition increases rapidly with the temperature, the values of "C" obtained being in accordance with Arrhenius' formula for the temperature-coefficient, namely:

$$C_{t_1} = C_{t_0}.e^{A(T_1 - T_0):T_1T_0}$$

(Trans., 1903, **83**, 470).

3. The rate of decomposition (in the case of benzenediazonium salts) is independent of the quantity of mineral acid present (except sulphuric acid, which tends to withdraw water from the sphere of action), and

is independent of the nature of the acid. Equivalent solutions of benzenediazonium chloride, sulphate, nitrate, and oxalate decompose at the same rate (*Ber.*, 1905, **38**, 2511).

4. The primary action in the decomposition (except possibly in the diphenyl series) is that phenols are formed. A number of apparent exceptions to this rule have been proved to be groundless (this paper). Unless special precautions are taken in the decomposition of very stable diazonium salts, the unchanged diazo-salt condenses with the phenol formed (probably yielding either a diazo-oxy-compound or an azo-colouring matter), and consequently the latter is not isolated (*loc cit.*).

In the diphenyl series, the tetrazo-salts from dianisidine and dichlorobenzidine yield as chief products quinones and not phenols whereas benzidine, tolidine, &c., behave normally (Trans., 1903, **83** 688). The tetrazo-salt from dichlorobenzidine behaves normally in the various other diazo-reactions (Trans., 1904, **85**, 7).

5. The influence of a substituent in the ortho-position with respect to one diazonium group contained in a tetrazo-salt is very great, and it has been possible in such a case to decompose one diazonium group completely and leave the other intact, the hydroxydiazonium salt thus being obtained in the crystalline condition from boiling water (Trans. 1905, **87**, 5).

MUNICIPAL TECHNICAL SCHOOL,
    BURY, LANCASHIRE.

VI.—*The supposed identity of Dihydrolaurolene and Dihydroisolaurolene with* 1:1-*Dimethylhexahydrobenzene*

By ARTHUR WILLIAM CROSSLEY and NORA RENOUF, Salters' Research Fellow.

IT was stated in a previous communication (Trans., 1905, **87**, 1487 that the main object the authors had in view when preparing 1:1-di methylhexahydrobenzene was a comparison of its properties with those of dihydrolaurolene and dihydroisolaurolene, with which hydrocarbon it has been supposed by Zelinsky and Lepeschkin (*Annalen*, 1901, **318** 303) to be identical. As these authors based their conclusions on the physical properties of the hydrocarbons and gave no details of their chemical properties, such as oxidation products, it became necessary to prepare dihydrolaurolene and dihydroisolaurolene, and to investigate them from the chemical standpoint.

The following is a list of the more important papers dealing with laurolene and *iso*laurolene, and all references in this communication are to these papers unless otherwise stated.

L a u r o l e n e.—Wreden, *Annalen*, 1877, **187**, 171 ; Reyher, *Inaug. Dissertation, Leipzig,* 1891, 51 ; Aschan, *Annalen,* 1896, **290**, 185 ; Noyes, *Amer. Chem. J.,* 1895, **17**, 432 ; Walker and Henderson, Trans., 1896, **69**, 750 ; Zelinsky and Lepeschkin, *Annalen,* 1901, **319**, 311.

i s o *L a u r o l e n e.*—Moitessier, *Jahresber.,* 1866 ; Wreden, *ibid.* ; Damsky, *Ber.,* 1887, **20**, 2959 ; Koenigs and Meyer, *Ber.,* 1894, **27**, 3470 ; Blanc., *Bull. Soc. chim.,* 1898, [iii], **19**, 699 ; Zelinsky and Lepeschkin, *ibid.,* p. 307.

D i h y d r o l a u r o l e n e.

Laurolene was first prepared by the distillation of camphanic acid, and is a colourless, highly refractive liquid boiling at 119—122°, by far the largest portion distilling at 119·5—120·5°. It possesses the properties previously attributed to it except as regards optical rotation, which was never found to be as high as $+23°$ (see p. 38), a point which is being further investigated.

Laurolene is not identical with 1:1-dimethyl-Δ^3-tetrahydrobenzene (Trans., 1905, **87**, 1500), with which it is isomeric.

The conversion of laurolene into its hydriodide is an operation attended with great difficulty, which information would not be gathered from the description given by Zelinsky and Lepeschkin, who state that they obtained more than 70 per cent. of the theoretical quantity, boiling at 69°/15 mm., by heating the hydrocarbon with fuming hydriodic acid for five hours in a water-bath. The present authors, although varying the amount of hydriodic acid used and the length of time of heating, could never obtain more than 25—30 per cent. of the theoretical quantity of hydriodide, boiling much higher than as above stated, namely, 101—106°/33 mm., and a certain amount of unchanged hydrocarbon was always recovered.

Zelinsky and Lepeschkin do not quote an analysis of the hydriodide, nor was it found practicable on the present occasion to carry out an iodine estimation, as the substance cannot be distilled even in a vacuum without some decomposition and always contains free iodine.

The agent employed by Zelinsky and Lepeschkin for the conversion of the hydriodide into dihydrolaurolene was zinc-palladium, but as experiment showed that reduction of the hydriodide by means of zinc dust and aqueous alcohol leads to the same result the latter method was employed, as it is much less troublesome to carry out. The process is again a wasteful one, as only 30 per cent. of the theoretical quantity of the pure saturated hydrocarbon is obtained, which is largely due to the fact that during reduction the elements of hydrogen iodide are to some extent removed from the hydriodide, giving rise to an unsaturated hydrocarbon, which is destroyed on treating the raw reduction product with potassium permanganate.

That the dihydrolaurolene obtained by the above reactions differs from 1 : 1-dimethylhexahydrobenzene will be seen from the following comparison :

	b. p.	sp. gr.	odour.	oxidation product.
1 : 1-Dimethylhexa- hydrobenzene ...	120°	0·7864	resembling geranium	$\beta\beta$-dimethyl- adipic acid
Dihydrolaurolene	111·5—114°	0·7633	camphoraceous	oxalic acid

Regarding the probable constitution of dihydrolaurolene, there is not much to be said on the present occasion. Zelinsky and Lepeschkin have shown that, when laurolene hydriodide is treated with di-methylaniline, the elements of hydrogen iodide are removed, and laurolene is recovered unchanged except that it is optically inactive. If, therefore, laurolene gave oxidation products containing the carbon complex present in the hydrocarbon, as is the case with *iso*laurolene, the determination of its constitution would not be a difficult matter, but it does not, and it has been found, in accordance with the observations of previous experimenters, that the only definite oxidation products obtainable from laurolene are oxalic and acetic acids.

It seems probable that, as Zelinsky and Lepeschkin point out, laurolene (*ibid.*, p. 312) is a mixture, for its boiling point is not particularly constant, its behaviour towards hydrogen iodide is not that of a homogeneous substance, and, further, although prepared from pure camphanic acid under precisely similar conditions, its optical activity varies (see p. 38). Moreover, this view is supported by a consideration of its magnetic rotation (see p. 36) and of the theory of its formation from camphanic acid (I), which takes place as here represented :

$$\begin{array}{c} \text{CH}_2\cdot\overset{|}{\text{C}}\cdot\text{CO}_2\text{H} \\ | \quad \text{C(CH}_3)_2 \quad \text{O} \\ \text{CH}_2\cdot\text{C}\!\!-\!\!-\!\!-\!\!-\text{CO} \\ \overset{|}{\text{CH}_3} \\ \text{I.} \end{array} \quad = \quad 2\text{CO}_2 \;+\; \begin{array}{c} \text{CH}_2\cdot\text{CH}- \\ | \quad \text{C(CH}_3)_2 \\ \text{CH}_2\cdot\text{C}- \\ \overset{|}{\text{CH}_3} \\ \text{II.} \end{array} .$$

It is obvious that the bonds in a substance having formula II must at once undergo rearrangement to give a stable compound, and this may occur in a variety of ways, giving rise to pentamethylene or hexamethylene derivatives :

$$\begin{matrix} CH_2 \cdot CH \\ | \qquad \!\!\!\!> C(CH_3)_2 \\ CH_2 \cdot C \cdot CH_3 \end{matrix}$$
III.

$$\begin{matrix} CH \cdot CH_2 \\ || \quad\; C(CH_3)_2 \\ CH \cdot CH \cdot CH_3 \end{matrix}$$
IV.

$$\begin{matrix} CH_2 \cdot CH_2 \cdot CH_2 \\ | \qquad\quad C \cdot CH_3 \\ CH_2 \!\!-\!\!\!-\!\!\!-\!\! C \cdot CH_3 \end{matrix}$$
V.

$$\begin{matrix} CH_2 \!\!-\!\!\!-\!\! CH \\ | \qquad\quad C \cdot CH_3 \\ CH_2 \cdot CH \!-\! CH_2 \\ \qquad CH_3 \end{matrix}$$
VI.

Experimental evidence is not yet sufficiently complete to allow a definite expression of opinion as to how the reaction takes place, but it may be pointed out that, although dihydrolaurolene is not identical with dihydro*iso*laurolene, there is, on theoretical grounds, no reason why the former should not contain a proportion of the latter, for, supposing that the intermediate product represented by formula II rearranges itself to form substances having either formula III or IV, then on treatment with hydriodic acid and reduction of the hydriodide formed there would be produced 1 : 1 : 2-trimethyl*cyclo*pentane identical with dihydro*iso*laurolene. Under these conditions, however, some *aa*-dimethylglutaric acid should result from the oxidation of dihydrolaurolene, but up to the present stage of the inquiry it has not been found possible to isolate even traces of this acid.

Dihydroisolaurolene.

*iso*Laurolene has usually been prepared by heating *iso*lauronolic acid in sealed tubes for eight hours at a temperature of 300—340° (Blanc, p. 700 ; Zelinsky and Lepeschkin, p. 307), a process which, as pointed out, frequently means considerable loss on account of the bursting of tubes. A new method was therefore sought, and it was found that if *iso*lauronolic acid is heated with one and a half times its weight of pure anthracene somewhat above the melting point of the mixture, a reaction sets in and *iso*laurolene slowly distils over, the end of the reaction being indicated by the fact that anthracene sublimes into the neck of the distillation flask. The reaction takes three to four hours for completion, but when once the temperature has been regulated no further attention is required, and the yield of *iso*laurolene is almost quantitative. When treated with fuming hydriodic acid, *iso*laurolene is readily converted into the liquid hydriodide (yield 75 per cent. of the theoretical), which, on treatment with zinc dust in aqueous alcoholic solution, gives from 60 to 62 per cent. of the theoretical amount of dihydro*iso*laurolene, which hydrocarbon is not identical with 1 : 1-dimethylhexahydrobenzene, as seen from the following comparison :

	b. p.	sp. gr.	odour.	oxidation product.
1 : 1-Dimethylhexa- hydrobenzene ...	120°	0·7864	resembling geranium	$\beta\beta$-dimethyl-adipic acid
Dihydro*iso*laurolene	113—113·5°	0·7762	sweet cam-phoraceous	$\alpha\alpha$-dimethyl-glutaric acid

Constitution of Dihydroisolaurolene.

As the formula to be assigned to dihydro*iso*laurolene naturally depends on the constitution of *iso*laurolene, the evidence in favour of the latter substance being 1 : 1 : 2-trimethyl-Δ^2-*cyclo*pentene must be very briefly reviewed.

Damsky (p. 2959) was the first experimenter to investigate the properties of *iso*laurolene at all fully ; his oxidation experiments " did not, however, give rise to any solid product, but only to oily fatty acids." This oil must have been the ketonic acid, $C_8H_{14}O_3$, described on p. 46, which, on being extracted from the oxidation liquid, is accompanied by small amounts of acetic acid, and does not show any signs of solidification until it has been distilled.

In 1898, Blanc definitely established the constitution of *iso*laurolene in the following manner. Accepting his formula for *iso*lauronolic acid (VII) as correct, he believed *iso*laurolene to be represented by formula VIII :

$$\begin{array}{c} CH_2 \cdot C(CH_3)_2 \\ |\quad C \cdot CH_3 \\ CH_2 \cdot C \cdot CO_2H \\ \text{VII.} \end{array} \quad = \quad CO_2 \quad + \quad \begin{array}{c} CH_2 \cdot C(CH_3)_2 \\ |\quad C \cdot CH_3 \\ CH_2 \cdot CH \\ \text{VIII.} \end{array}$$

that is to say, that during the loss of carbon dioxide no change in the structure of the ring takes place. This was proved by the fact that *iso*lauronolic chloride (IX), when treated with zinc methide, gave rise to the same ketone (X) as is produced by the action of acetyl chloride on *iso*laurolene in presence of aluminium chloride :

$$\begin{array}{c} CH_2 \cdot C(CH_3)_2 \\ |\quad C \cdot CH_3 \\ CH_2 \cdot C \cdot COCl \\ \text{IX.} \end{array} + Zn(CH_3)_2$$

$$\begin{array}{c} CH_2 \cdot C(CH_3)_2 \\ |\quad C \cdot CH_3 \\ CH_2 \cdot CH \end{array} + CH_3 \cdot COCl$$

$$\begin{array}{c} CH_2 \cdot C(CH_3)_2 \\ |\quad C \cdot CH_3 \\ CH_2 \cdot C \cdot CO \cdot CH_3 \\ \text{X.} \end{array}$$

Blanc further showed that when *iso*laurolene is oxidised with potassium permanganate there is obtained γ-acetyldimethylbutyric acid (XI), previously obtained by him from the oxidation of *iso*lauronolic acid (*Bull. Soc. chim.*, 1898, [iii], **19**, 533):

$$
\begin{array}{ccccc}
CH_2 \cdot \underset{\ }{C}(CH_3)_2 & & CH_2 \cdot \underset{\ }{C}(CH_3)_2 & & CH_2 \cdot \underset{\ }{C}(CH_3)_2 \\
\mid \quad \underset{\ }{C} \cdot CH_3 & \longrightarrow & \mid \quad CO \cdot CH_3 & \longrightarrow & \mid \quad CO_2H \\
CH_2 \cdot CH & & CH_2 \cdot CO_2H & & CH_2 \cdot CO_2H \\
& & XI. & & XII.
\end{array}
$$

and this ketonic acid, on further oxidation, gave αα-dimethylglutaric acid (XII). As Blanc says, the formation of γ-acetyldimethylbutyric acid by the oxidation of *iso*laurolene shows that it can only have the constitution represented by formula VIII and no other. "Aucune ambiguité ici n'est possible." Yet Zelinsky and Lepeschkin in 1901 did not accept this conclusion, but regarded *iso*laurolene as a six-ring compound.

Accepting then the foregoing formula (VIII) for *iso*laurolene, the next point is to prove that when *iso*laurolene is treated with fuming hydriodic acid at a temperature of 120—125° no change in the nature of the ring is produced. This might seem probable, because *iso*laurolene, when brought into contact with hydriodic acid at the ordinary temperature, gives a solid and very unstable hydriodide; whereas at 120—125°, a comparatively stable and liquid hydriodide is produced. Nevertheless, it is easily demonstrated that this liquid hydriodide contains the same carbon complex as *iso*laurolene itself.

For this purpose, the hydriodide was treated with diethylaniline, when it readily lost the elements of hydrogen iodide, giving an unsaturated hydrocarbon, C_8H_{14}, boiling at 108—108·5°, and possessing properties identical with those of *iso*laurolene. In order that there should be no doubt on this point, the hydrocarbon was oxidised with potassium permanganate, when it yielded γ-acetyldimethylbutyric acid, and this, on further oxidation with sodium hypobromite, gave αα-dimethylglutaric acid. These are the same products as Blanc obtained by the oxidation of *iso*laurolene (see above), and conclusively prove that no isomeric change takes place during the production of the hydriodide, which must therefore have one of the following formulæ:

$$
\begin{array}{ccc}
CH_2 \cdot \underset{\ }{C}(CH_3)_2 & & CH_2 \cdot \underset{\ }{C}(CH_3)_2 \\
\mid \quad \underset{\ }{C}H \cdot CH_3 & \text{or} & \mid \quad \underset{\ }{C}I \cdot CH_3 \;. \\
CH_2 \cdot CHI & & CH_2 \cdot CH_2
\end{array}
$$

There is, moreover, no reason to suppose that heating this hydriodide with zinc dust in aqueous alcoholic solution would produce a change in the construction of the ring, and this is proved by the fact that when the resulting hydrocarbon (dihydro*iso*laurolene) is oxidised with diluted nitric acid it gives rise to $\alpha\alpha$-dimethylglutaric acid:

$$
\begin{array}{ccc}
\begin{array}{l}
CH_2 \cdot \underset{|}{C}(CH_3)_2 \\
\;\;\; \underset{|}{C}H \cdot CH_3 \\
CH_2 \cdot CH_2
\end{array}
& \longrightarrow &
\begin{array}{l}
CH_2 \cdot \underset{|}{C}(CH_3)_2 \\
\;\;\; \underset{|}{C}O_2H \\
CH_2 \cdot CO_2H
\end{array}
\end{array}
\qquad \longleftarrow \qquad
\begin{array}{l}
CH_2 \cdot \underset{|}{C}(CH_3) \\
\;\;\; \underset{\|}{C} \cdot CH_3 \\
CH_2 \cdot C \cdot CO_2H
\end{array}
$$

that is, to the same oxidation product as *iso*lauronolic acid yields when treated with diluted nitric acid (Blanc, *Bull. Soc. chim.*, 1898, [iii] **19**, 284).

These experiments prove conclusively that dihydro*iso*laurolene is a pentamethylene derivative and is 1 : 1 : 2-*trimethyl*cyclo*pentane*, a deduction which receives striking confirmation from the magnetic rotation of dihydro*iso*laurolene, and to which Dr. Perkin makes allusion in his report (see p. 36).

Zelinsky and Lepeschkin concluded, as a result of their experiments that dihydrolaurolene and dihydro*iso*laurolene were identical, a conclusion which does not seem to be warranted by the results obtained by the present authors. Further, from a consideration purely of the physical properties of dihydro*iso*laurolene, these authors supposed that it was a hexamethylene derivative, and, since laurolene hydriodide and *iso*laurolene hydriodide, both of which substances contain the same carbon ring as dihydro*iso*laurolene, when treated with diethylaniline regenerate laurolene and *iso*laurolene respectively, that therefore the latter substances contain a six-membered carbon ring, which " very probably is also present in *iso*lauronolic and camphanic acids." " Man konnte weiter gehen und schon voraussetzen, dass die Kamphersäure . . . einen Hexamethylenring besitzt." Komppa's synthesis of camphoric acid (*Ber.*, 1903, **36**, 4332) is a sufficient answer to the latter suggestion.

Zelinsky and Lepeschkin then argue that dihydro*iso*laurolene, C_8H_{16}, must be a dimethylhexamethylene, and since it was not identical with 1 : 2-, 1 : 3-, or 1 : 4-dimethylhexahydrobenzene, all of which substances have been described by Zelinsky, it must be the only remaining possibility, namely, 1 : 1-dimethylhexahydrobenzene, a conclusion which is certainly wrong. They attempt to explain the fact that the boiling point of their supposed 1 : 1-dimethylhexahydrobenzene (114° is lower than the boiling points of the 1 : 2-, 1 : 3-, and 1 : 4-isomeride because the two methyl groups are bound to one and the same carbon atom, and quote in support of this the fact that 1 : 3 : 3-trimethyl hexahydrobenzene, which also contains two methyl groups attached to the same carbon atom, boils 4° to 5° lower than the isomeric 1 : 2 : 5-tri methylhexahydrobenzene. We now know that the boiling point of 1 : 1-dimethylhexahydrobenzene is almost identical with those of the isomeric hydrocarbons, as will be seen from the following table, and therefore the presence of the *gem*-dimethyl group does not in this case give a compound of lower boiling point than its isomerides :

	b. p.	sp. gr.
Observers.		
1 : 2-Dimethylhexahydrobenzene (Zelinsky and Lepesçhkin, *Annalen*, 1901, **319**, 319)	116—118°	0·7733
1 : 3-Dimethylhexahydrobenzene (Zelinsky and Naumoff, *Ber.*, 1895, **28**, 781)	119·5°	0·7688
1 : 4-Dimethylhexahydrobenzene (Zelinsky and Reformatsky, *Ber.*, 1898, **31**, 3207)	119·5—120°	0·7690
1 : 1-Dimethylhexahydrobenzene (Crossley and Renouf, *Trans.*, 1905, **87**, 1498)	120°	0·7864

Zelinsky and Lepeschkin further state that, if dihydro*iso*laurolene were a pentamethylene derivative, it would be 1 : 1 : 2-trimethyl*cyclo*pentane, and its boiling point would not therefore be higher than 111°, because the difference in boiling point for the homologues of this series is about 20° ; but as it contains two methyl groups attached to the same carbon atom, so its boiling point would probably be below 111°.

The boiling point of dihydro*iso*laurolene (1 : 1 : 2-trimethyl*cyclo*pentane) is now shown to be 113—113·5°, or 22° higher than the boiling point of dimethyl*cyclo*pentane (91—91·4°) as given by Zelinsky and Rudsky (*Jour. Russ. Chem. Soc.*, 1899, **31**, 408), which is an almost identical difference as that found between the boiling points of *cyclo*pentane, 50·3—50·7° (Wislicenus, *Annalen*, 1893, **275**, 329), and methyl*cyclo*pentane (Zelinsky, *J. Russ. Chem. Soc.*, **31**, 408) 72—72·2°, namely, 21·5°.

The authors desire to express their warmest thanks to Dr. W. H. Perkin, sen., for the interest he has taken in this work, and for kindly determining the physical constants of the hydrocarbons, on which he reports as follows :

Densities, Magnetic Rotations, and Refractive Powers of Laurolene, Dihydrolaurolene, isoLaurolene, and Dihydroisolaurolene.

Laurolene.

Density : $d4°/4° = 0·8097$; $d10°/10° = 0·8048$; $d15°/15° = 0·8010$;
  $d20°/20° = 0·7974$; $d25°/25° = 0·7939$.

Magnetic rotation :

t.	Sp. rot.	Mol. rot.
19·1°	1·1737	8·987

Refractive power :

$$t = 19{\cdot}5° ; \quad d19{\cdot}5°/4° = 0{\cdot}79650.$$

	Index of refraction. $\mu.$	Sp. refraction. $\dfrac{\mu-1}{d}.$	Mol. refraction. $\dfrac{\mu-1}{d}p.$	Calculated.
H	1·44253	0·55559	61·114	60·5
H	1·45246	0·56806	62·486	—
H	1·45845	0·57558	63·314	—

Dispersion $H_\alpha - H_\gamma = 2{\cdot}20.$

Dihydrolaurolene.

Density : $d4°/4° = 0{\cdot}7718 ; \ d10°/10° = 0{\cdot}7670 ; \ d15°/15° = 0{\cdot}7633 ;$
$d20°/20° = 0{\cdot}7596 ; \ d25°/25° = 0{\cdot}7567.$

Magnetic rotation :

$t.$	Sp. rot.	Mol. rot.
19·6°	1·0181	8·332

Refractive power :

$$t = 19{\cdot}8° ; \quad d19{\cdot}8°/4° = 0{\cdot}7588.$$

	Index of refraction. $\mu.$	Sp. refraction. $\dfrac{\mu-1}{d}.$	Mol. refraction. $\dfrac{\mu-1}{d}p.$	Calculated.
H	1·41424	0·54588	61·138	60·8
H	1·42162	0·55561	62·228	—
H	1·42591	0·56126	62·861	—

Dispersion $H_\alpha - H_\gamma = 1{\cdot}723.$

isoLaurolene.

Density : $d4°/4° = 0{\cdot}7953 ; \ d10°/10° = 0{\cdot}7907 ; \ d15°/15° = 0{\cdot}7867 ;$
$d20°/20° = 0{\cdot}7830 ; \ d25°/25° = 0{\cdot}7795.$

Magnetic rotation :

$t.$	Sp. rot.	Mol. rot
14·3°	1·1270	8·749

Refractive power :

$$t = 16{\cdot}1° ; \quad d16{\cdot}1°/4° = 0{\cdot}78510.$$

	Index of refraction. μ.	Sp. refraction. $\dfrac{\mu-1}{d}$.	Mol. refraction. $\dfrac{\mu-1}{d}p$.	Calculated.
H	1·43227	0·55059	60·565	60·5
H	1·44136	0·56216	61·847	—
H	1·44690	0·56923	62·615	—

Dispersion $H_\alpha - H_\gamma = 2\cdot050$.

Dihydroisolaurolene.

Density : $d4°/4° = 0\cdot7847$; $d10°/10° = 0\cdot7800$; $d15°/15° = 0\cdot7762$; $d20°/20° = 0\cdot7727$; $d25°/25° = 0\cdot7694$.

Magnetic rotation :

t.	Sp. rot.	Mol. rot.
14·3°	1·0298	8·249

Refractive power :

$$t = 16\cdot2° \; ; \; d16\cdot2°/4° = 0\cdot77463.$$

	Index of refraction. μ.	Sp. refraction. $\dfrac{\mu-1}{d}$.	Mol. refraction. $\dfrac{\mu-1}{d}p$.	Calculated.
H	1·42244	0·54534	61·078	60·8
H	1·42998	0·55508	62·169	—
H	1·43398	0·56024	62·815	—

Dispersion $H_\alpha - H_\gamma = 1\cdot737$.

On comparing the densities and magnetic rotations of dihydro-isolaurolene and isolaurolene, also of dihydrolaurolene and laurolene, respectively with those of 1 : 1-dimethylhexahydrobenzene and 1 : 1-dimethyl-Δ^3-tetrahydrobenzene (Trans., 1905, **87**, 1491), considerable differences are noticed :

	Densities.	Difference.	Magnetic rotations.	Difference.
1 : 1-Dimethylhexahydrobenzene	0·7864	−0·0102	8·150	+0·099
Dihydroisolaurolene............................	0·7762		8·249	
1 : 1-Dimethyl-Δ^3-tetrahydrobenzene ...	0·8040	−0·0173	8·903	−0·154
isoLaurolene	0·7867		8·749	
1 : 1-Dimethylhexahydrobenzene	0·7864	−0·0231	8·150	+0·182
Dihydrolaurolene...............................	0·7633		8·332	
1 : 1-Dimethyl-Δ^3-tetrahydrobenzene ...	0·8040	−0·0030	8·903	+0·084
Laurolene	0·8010		8·987	

In the cases of dihydroisolaurolene and isolaurolene, we have large and irregular differences, which are dissimilar to those of dihydrolaurolene and laurolene. These iso-compounds are evidently different from all the others of the same composition, and one especial peculiarity is that the difference between the magnetic rotations of the saturated and unsaturated products is remarkably small, thus :

	Mol. rot.	Difference.
*iso*Laurolene	8·749	0·500.
Dihydro*iso*laurolene	8·249	

Now it has been shown that the average influence of unsaturation, caused by the loss of H_2 in the paraffin series, results in a rise of rotation of $+0·720$ (Trans., 1902, **81**, 292), and that, when these unsaturated chain compounds are joined up by loss of H_2 so as to form ring compounds, the result of unsaturation remains practically the same. This has been found to be the case with dihydrobenzene and with 1 : 1-dimethyl-Δ^3-tetrahydrobenzene. This is, however, a much larger effect than that found in the case of dihydro*iso*laurolene and *iso*-laurolene ; but it has been shown that in the lower members of the aliphatic series the differences for unsaturation are exceptionally small, as seen in the halogen derivatives of ethylene and propylene (Trans., 1884, **45**, 568). If the individual rotations of the unsaturated hydrocarbons be examined (Trans., 1895, **67**, 261) it will be observed that in the case of amylene the difference amounts to only 0·578, and therefore we may assume that this would be about the influence of unsaturation in the corresponding ring compounds. It is very interesting to note that this is near to that found in the case of *iso*laurolene, indicating that this substance and its dihydro-derivative are trimethyl five carbon ring compounds. This is quite in agreement with the view of Blanc as regards *iso*laurolene and the results obtained by the authors of the present communication regarding dihydro*iso*laurolene.

The differences observed in the cases of · dihydrolaurolene and laurolene are difficult to interpret ; for if these hydrocarbons are related to each other in the same way as dimethylhexahydrobenzene is related to dimethyltetrahydrobenzene, these differences should be similar for each of the properties, whereas they are much larger in the case of dihydrolaurolene. It is, however, worth while calling attention to the difference between the magnetic rotations of laurolene and dihydrolaurolene, which is as follows :

		Difference.
Laurolene...............	8·987	+ 0·655.
Dihydrolaurolene ...	8·332	

This, it will be seen, lies between the effect of unsaturation of a five and a six carbon ring and might result if laurolene and dihydro-laurolene were mixtures of such compounds.

Zelinsky and Lepeschkin (page 322) have proposed the two bridged ring formulæ for *iso*laurolene and laurolene :

$$(CH_3)_2C \underset{CH_2 \cdot CH}{\overset{CH \cdot CH_2}{<}} CH_2 \qquad\qquad (CH_3)_2C \underset{CH_2 \cdot CH_2}{\overset{CH-CH_2}{<}} CH$$

*iso*Laurolene.   Laurolene.

but the magnetic rotations of such compounds would be very much smaller than those found, because the effect of the bridged ring is quite different to that of an ordinary double linking (Trans., 1902, **81**, 266).

A comparison of the refractive values of these hydrocarbons, made on the same lines as that of densities or magnetic rotations, does not appear to afford much light in reference to their structure.

EXPERIMENTAL.

Preparation of Laurolene.

Bromocamphoric anhydride was prepared according to the directions given by Zelinsky and Lepeschkin (p. 310), except that the raw product was crystallised from glacial acetic acid instead of chloroform. The yield was about 75—80 grams from 100 grams of camphoric acid. This anhydride was then converted into camphanic acid by heating with a solution of sodium carbonate (Aschan, *Ber.*, 1894, **27**, 3506). The yield after crystallising from benzene is 55—60 grams from 100 grams of the anhydride.

In preparing laurolene from camphanic acid, the very precise details given by Aschan (p. 187) were followed. The yield of laurolene was 33 grams from 150 grams of camphanic acid, and its purity was proved by analysis :

0·1149 gave 0·3665 CO_2 and 0·1326 H_2O.   C = 87·00 ; H = 12·82.

C_8H_{14} requires C = 87·27 ; H = 12·73 per cent.

Properties of the Hydrocarbon.—Laurolene is a clear, colourless, highly refractive liquid boiling at 119—122°/760 mm., by far the major portion distilling at 119·5—120·5°. It possesses an odour resembling both camphor and turpentine, but much less sweet than that of *iso*laurolene. With alcoholic sulphuric acid (carried out as previously described, Trans., 1905, **87**, 1494), the hydrocarbon gives a green colour turning to bronze-green, and when treated with concentrated sulphuric acid in acetic anhydride solution there is produced a dark bronze-green coloration, changing to deep grass-green.

Having observed that when laurolene was treated with hydriodic acid (see p. 40) some of the hydrocarbon remained unattacked and appeared to be unaltered, except as regards optical activity, it was decided to examine the rotations of some of the preparations of laurolene. This seemed to be more desirable, as the observations of previous workers regarding this point do not agree, as will be seen from the following table :

Reference.	Source.	Rotation.
Aschan (p. 189)	Camphanic acid	$[a]_j - 23 \cdot 0°$
Walker and Henderson (p. 752)	Potassium allo-ethyl camphorate	$[a]_D - 29 \cdot 2°$
Tiemaunn (*Ber.*, 1900, **33**, 2949)	Aminolauronolic acid	$[a]$ $+ 19 \cdot 9°$
Zelinsky and Lepeschkin (p. 311)	Camphanic acid	$[a]_D + 22 \cdot 9°$

The last-named authors, after partial oxidation with potassium permanganate, obtained laurolene with a rotation $[a]_D + 16 \cdot 2°$, and express the opinion that on this account laurolene prepared from camphanic acid may possibly be a mixture of isomeric hydrocarbons.

Four separately prepared specimens of laurolene have now been examined, and the following numbers obtained :

1. $[a]_D + 11 \cdot 4°$.   2. Inactive.   3. $[a]_D + 6 \cdot 6°$.   4. $[a]_D + 4 \cdot 1°$.

This irregular rotation might be put down to a difference in the method of distilling camphanic acid, but Aschan gives such very definite instructions as to temperature and the number of drops of the distillate per minute, that it may certainly be said each preparation was carried out, as near as it is possible to do so, under identical conditions. The camphanic acid employed was always in the same state of purity, and, moreover, specimens 3 and 4 were prepared from the same bulk of camphanic acid. It was then thought that two hours' heating of laurolene with metallic sodium might cause some change to take place, especially as the liquid becomes dark brown during the heating. Specimens 3 and 4 were therefore examined after drying over calcium chloride, and then at intervals during the heating with sodium, but no change in the initial rotation was observed, and, so far, no satisfactory explanation of this behaviour has been found.

Action of Bromine on Laurolene.— When a solution of bromine in chloroform is added to a solution of laurolene in the same solvent, cooled in ice-water, a green colour is at once produced, changing to brown, which is destroyed on further addition of bromine, returning as the bromine is used up, and ultimately the solution assumes a violet colour. If then an excess of bromine is added, the violet disappears as the bromine is absorbed, and the green, and finally violet, colorations return. During the whole operation, clouds of hydrogen bromide are evolved, and on evaporating the chloroform a green resin remains.

Having found it impossible to get any idea of the quantity of bromine used up in chloroform solution, carbon tetrachloride was tried in its place, as recommended by Aschan (p. 190). Under these conditions, bromine is gradually absorbed and hydrogen bromide evolved, but none of the colour changes noticed when using chloroform was observed. When 1·0641 grams of the hydrocarbon were taken, the first indication of a permanent bromine colour was observed when 2·2156 grams of bromine had been added, the required amount for absorption of Br_2 being 1·5477 grams. Nor was the absorption then quite complete, as further amounts of bromine were slowly used up on standing. This does not agree with Aschan's statement that under the above conditions laurolene absorbs exactly two atoms of bromine and no more. On evaporating the carbon tetrachloride solution, hydrogen bromide was evolved and a green resin remained.

Oxidation of Laurolene. (1) *With Nitric Acid.*—Six grams of laurolene were heated in a flask, attached to a condenser, with 80 c.c. of one part fuming nitric acid and two parts water. The residue obtained from this liquid by working it up in the usual way was proved to consist for the most part of oxalic acid, together with a minute quantity of a dark oily liquid, smelling of burnt sugar, from which no definite chemical compound could be isolated.

(2) *With Potassium Permanganate.*—Twenty grams of the hydro-carbon were suspended in 1 litre of water, and powdered potassium permanganate was gradually added until the coloration became permanent, a result which required 50 grams and occupied seventy hours. The filtered liquid was evaporated to a small bulk, acidified with sulphuric acid, and distilled in steam. The distillate was neutralised with caustic soda, evaporated to complete dryness, and treated with concentrated sulphuric acid, when a volatile liquid passed over, which was proved to consist of acetic acid by its boiling point, and analysis of a silver salt prepared from it.

0·2110 gave 0·1363 Ag.   Ag = 64·60.

$C_2H_3O_2Ag$ requires Ag = 64·67 per cent.

No product of a definite nature could be isolated from the residue of the above steam distillation, nor were any substances other than acetic and oxalic acids obtained by oxidising laurolene with potassium permanganate in acetone solution, or with a mixture of potassium dichromate and sulphuric acid (compare Aschan, p. 193).

The action of a nitrating mixture on laurolene was found to give results coinciding with those recorded by Walker and Henderson (p. 752) and Aschan.

The action of nitrosyl chloride on laurolene was tried in the hope of obtaining a solid derivative, but only dark green resinous substances were obtained, similar in appearance to the products of the action of bromine on laurolene.

Laurolene Hydriodide.

Laurolene, in quantities of 10 c.c. at a time, was heated, as directed by Zelinsky and Lepeschkin, with 50 c.c. of fuming hydriodic acid (sp. gr. 1·97) in an ordinary stoppered bottle in a water-bath for six hours. The resulting liquid was poured into water and the whole extracted with ether, the ethereal solution washed successively with water, aqueous sodium bicarbonate, a solution of sodium thiosulphate, and finally with water, dried over calcium chloride, and the residue obtained on evaporating the ether distilled under diminished pressure. After working up 33 grams of laurolene in this manner, the following fractions were collected at 33 mm. Below 90° = 10·5 grams; 101—106° = 18·7 grams; 106—150° = 5·4 grams.

The fraction below 90° was repeatedly distilled over metallic sodium, when it passed over as a clear, colourless, highly refractive liquid boiling at 119·5—121°.

0·1166 gave 0·3719 CO_2 and 0·1352 H_2O.　C = 87·00; H = 12·88.

C_8H_{14} requires C = 87·27; H = 12·73 per cent.

This liquid possessed properties identical with those of laurolene, except that it was optically inactive. It is, however, impossible to state, in view of the data given on p. 38, whether racemisation had taken place or not, as the optical activity of the laurolene used in this experiment was not tested before treatment with hydriodic acid.

The fraction 101—106° was a greenish-brown liquid and consisted of laurolene hydriodide, which is very much more unstable than *iso*-laurolene hydriodide. Zelinsky and Lepeschkin (p. 313) give the boiling point of this substance as 69°/15 mm., and state that they obtained 15·5 grams from 10 grams of laurolene, but do not mention the quantity of hydriodic acid used or whether they recovered any unaltered hydrocarbon. Although the experiment was repeated under very varied conditions, no better yield of the hydriodide could be obtained, and there was always recovered a certain quantity of unchanged hydrocarbon which, when again heated with fuming hydriodic acid for six hours, was only partially converted into the hydriodide.

The fraction 106—150° was not further investigated.

Dihydrolaurolene.

Thirty-nine grams of laurolene hydriodide were dissolved in 210 c.c. of 90 per cent. alcohol, and 78 grams of zinc dust mixed with an equal volume of sand added, and the whole heated on the water-bath for ten hours and then worked up as previously described (Trans., 1905, **87**, 1497). The hydrocarbon was suspended in water and treated with potassium permanganate until no further oxidation took place and the whole distilled in steam, when the hydrocarbon slowly passed over. It was separated from the water, dried with calcium chloride, fractionated over sodium, and analysed :

0·1016 gave 0·3198 CO_2 and 0·1318 H_2O.   C $= 85·83$; H $= 14·41$.

C_8H_{16} requires C $= 85·71$; H $= 14·29$ per cent.

Dihydrolaurolene is a clear, colourless, refractive liquid boiling at 111·5—114°/760 mm. and possessing a sweet camphoraceous odour ; it does not absorb bromine nor is it acted on by potassium permanganate, and gives no evidence of the formation of a nitro-derivative on treatment with a mixture of nitric and sulphuric acids. The yield is very much smaller than in the case of dihydro*iso*laurolene, amounting to about 35 per cent. of the theoretical quantity from the hydriodide employed.

Oxidation with Nitric Acid.—Two grams of the hydrocarbon were treated in the usual way with 30 c.c. of fuming nitric acid. Large needle-shaped crystals separated from the residual liquid which were proved to consist of oxalic acid. On evaporating to complete dryness and heating with excess of acetyl chloride, only a minute syrupy residue resulted, which was too small to permit of further examination.

After finding that dihydro*iso*laurolene gave αα-dimethylglutaric acid on oxidation with diluted nitric acid, dihydrolaurolene was again oxidised exactly as described on p. 44. It was not found possible to establish the presence of even the minutest quantities of αα-dimethylglutaric acid in the oxidation products obtained.

Preparation of iso*Laurolene.*

*iso*Lauronolic acid was prepared according to the directions given by Lees and Perkin (Trans., 1901, **79**, 341), and the yield obtained was as these authors state, namely, from 45—50 per cent. of the weight of camphoric anhydride used, but only when the purest form of aluminium chloride was employed.

*iso*Lauronolic acid, mixed with one and a half times its weight of pure anthracene, was placed in a double-necked (Claisen) distillation flask connected with a condenser and carefully heated, when a reaction began almost as soon as the mixture of substances became molten, and a colourless liquid slowly distilled. At the end of three hours, anthracene began to sublime into the neck of the distillation flask, which denoted that the reaction was almost complete, as on heating for two hours longer only one or two grams of hydrocarbon passed over. In the first experiment, 10 grams of *iso*lauronolic acid were used, but the reaction occurs equally readily when 30 or 40 grams are heated at one time. The hydrocarbon was dried over calcium chloride and dis- tilled from metallic sodium, when the whole passed over at 108—109°. For the purpose of analysis, it was again distilled over metallic sodium in an atmosphere of carbon dioxide, when it boiled quite constantly at 108—108·2°.

0·1332 gave 0·4260 CO_2 and 0·1532 H_2O.   C = 87·22 ; H = 12·77.

C_8H_{14} requires C = 87·27 ; H = 12·73 per cent.

*iso*Laurolene (1 : 1 : 2-trimethyl-Δ^2-*cyclo*pentene),
$$\begin{matrix} CH_2{\cdot}\underset{|}{C}(CH_3)_2 \\ \underset{|}{C}{\cdot}CH_3 \\ CH_2{\cdot}CH \end{matrix}$$
, the

yield of which is 85—90 per cent. of the theoretical, is a colourless mobile, highly refractive liquid boiling at 108—108·2°/742 mm., and possessing a sweet odour resembling both camphor and turpentine With sulphuric acid in alcoholic or acetic anhydride solution, it gives only a pale straw colour, and on adding a few drops of the hydrocarbon to a cold saturated solution of mercuric chloride and shaking, a cloudi ness appears, and a sticky, amorphous mass separates, which, after standing for some days, becomes pinkish-brown (compare Zelinsky and Lepeschkin, p. 308).

When a chloroform solution of bromine is added to a solution of the hydrocarbon in the same solvent, cooled in ice-water, the bromine is absorbed without any evolution of hydrogen bromide, the reaction being a quantitative one for the absorption of two atoms of bromine

1·1809 absorbed 1·7609 Br.   Molecular absorption, Br = 164.

C_8H_{14} requires Br_2 = 160.

On careful evaporation of the chloroform solution, a practically colourless, crystalline solid was obtained which, when spread on porou plate, set to a waxy mass somewhat resembling camphor, and having a strong odour of camphor and turpentine. It was twice crystallised from methyl alcohol, in which solvent it is very readily soluble, dried as rapidly as possible, and the bromine estimated.

0·1031 gave 0·1429 AgBr.   Br = 58·98.

$C_8H_{14}Br_2$ requires Br = 59·25 per cent.

2 : 3-*Dibromo*-1 : 1 : 2-*trimethylcyclopentane*,
$$\begin{matrix} CH_2{\cdot}\underset{|}{C}(CH_3)_2 \\ \underset{|}{C}Br{\cdot}CH_3, \\ CH_2{\cdot}CHBr \end{matrix}$$
is ver.

readily soluble in the cold in the usual organic media, but can be crystallised from absolute methyl alcohol, when it separates in fern like aggregates of needles melting at 80—85°. On standing, it decomposes with evolution of hydrogen bromide and gradually resinifies.

Damsky (p. 2961) has described the formation of this compound by the direct action of bromine on isolaurolene, but gave to it the formula $C_8H_{12}Br_2$, and in support of this quotes a bromine estimation 60·30, whereas the calculated value for $C_8H_{12}Br_2$ is 59·70. Damsky did not in any way purify his compound, and it would probably contain some higher brominated products, which are formed, as Damsky points out, when bromine acts directly on the hydrocarbon. There can be no doubt that this substance is really the dibromo-additive compound of isolaurolene, for when prepared as above described, no hydrogen bromide is evolved. It seemed, however, useless to estimate the carbon and hydrogen, as the calculated numbers for $C_8H_{12}Br_2$ and $C_8H_{14}Br_2$ are so close to one another as to prevent any accurate conclusions from being drawn.

isoLaurolene Hydriodide.

Quantities of isolaurolene were worked up in the following manner. Ten c.c. of the hydrocarbon and 40 c.c. of fuming hydriodic acid * (sp. gr. = 1·96) were placed in an ordinary narrow-necked, stoppered bottle of 120 c.c. capacity, the stopper wired down, and the whole heated in a glycerol bath for six hours at 120—125°. The contents of the bottle were then poured into water, the heavy oil which separated extracted with ether, the ethereal solution washed successively with water, dilute aqueous sodium carbonate to remove acid, dilute sodium thiosulphate solution to remove free iodine, and, finally, with water. It was then dried over calcium chloride, and, after evaporation of the ether, distilled under diminished pressure. At first, a very small amount of a volatile liquid passed over, which was proved to consist of some hydriodide and unchanged hydrocarbon ; then the thermometer rose rapidly, and the pure hydriodide distilled constantly at 101·5°/33 mm. (Zelinsky and Lepeschkin give the boiling point of this liquid as 75—80°/15—17 mm.) as a practically colourless, oily liquid with a pungent camphoraceous odour. The yield is about 75 per cent. of the theoretical amount.

Dihydroisolaurolene.

Thirty-six grams of the hydriodide were dissolved in 192 c.c. of 90 per cent. alcohol, and 72 grams of zinc dust, mixed with an equal

---

* During the addition of hydriodic acid to the hydrocarbon, the formation of the solid hydriodide was observed (compare Damsky, p. 2961, and Zelinsky and Lepeschkin, p.1808).

volume of sand, added, and the whole heated on the water-bath for twelve hours and then worked up as previously described (Trans., 1905, **87**, 1497). A second quantity of 36 grams of the hydriodide was treated in exactly the same manner. The hydrocarbon obtained from both experiments was then suspended in 150 c.c. of water and 5 grams of powdered potassium permanganate gradually added, with constant shaking, when the colour of the oxidising agent remained permanent, even on heating to the temperature of the water-bath for several hours. The whole was then distilled in steam, when the hydrocarbon passed over very readily, being notably more volatile with steam than dihydrolaurolene (see p. 41). It was then dried over calcium chloride, twice distilled from metallic sodium, and analysed.

0·1205 gave 0·3792 CO_2 and 0·1554 H_2O.   C = 85·82 ; H = 14·33.

C_8H_{16} requires C = 85·71 ; H = 14·29 per cent.

Dihydro*iso*laurolene (1 : 1 : 2-trimethyl*cyclo*pentane),

$$\begin{array}{l} CH_2 \cdot C(CH_3)_2 \\ \qquad \quad \; CH \cdot CH_3, \\ CH_2 \cdot CH_2 \end{array}$$

is a clear, colourless, refractive liquid boiling at 113—113·5°/750 mm. and possessing a sweet camphoraceous odour. It does not decolorise a chloroform solution of bromine, nor is it acted on by potassium permanganate. The yield of pure fractionated hydrocarbon is from 60—62 per cent. of the theoretical amount.

Oxidation with Fuming Nitric Acid.—Two grams of the hydrocarbon were added to 30 c.c. of fuming nitric acid, when, on warming slightly, a reaction started which gradually became more vigorous. Heating was therefore discontinued until the action had completed itself, and then the whole was heated for half an hour. After removing the nitric acid in the usual manner and evaporating, the residue solidified. It was spread on porous plate, when a white solid was obtained, which was proved to consist entirely of oxalic acid.

Oxidation with Diluted Nitric Acid.—Five grams of dihydro*iso*laurolene and a mixture of 26 c.c. of fuming nitric acid and 14 c.c. of water were heated to boiling on a sand-bath in a reflux apparatus with a ground glass attachment, when oxidation took place slowly. After six hours, the unattacked hydrocarbon was removed and heated with a fresh quantity of nitric acid, and this process repeated until all the hydrocarbon had been oxidised. On evaporating the nitric acid liquors to dryness, an oily residue was obtained which slowly solidified. It was heated for two hours with excess of acetyl chloride, the solvent evaporated, and the residue (1·5 grams) distilled, when it boiled for the most part at 260—265° (boiling point of *aa*-dimethylglutaric anhydride = 265° ; Blanc, *Bull. Soc. chim.*, 1898, [iii], 19, 285). A portion was converted into the anilic acid, which crystallised from dilute alcohol in lustrous plates melting at 141° ; nor was this melting point altered on mixing the substance with pure aa-dimethylglutar-anilic acid. The remainder of the anhydride was dissolved in boiling water, the solution evaporated to dryness, and the solid residue crystallised from a mixture of benzene and light petroleum, when it separated in bunches of minute needles melting at 83°. It was not considered necessary to analyse this substance, as the melting point was unaltered on mixing with the analysed aa-dimethylglutaric acid obtained by the oxidation of γ-acetyldimethylbutyric acid (see p. 46).

Action of Diethylaniline on isoLaurolene Hydriodide.

Thirty-three grams of the hydriodide and 50 grams of freshly-distilled diethylaniline were heated in a long-necked distillation flask attached to a condenser, a thermometer being inserted in the liquid. The first signs of a reaction commenced at 150°, when the source of heat was removed. The temperature gradually rose to 154°, when the whole suddenly became turbid and a rather vigorous reaction set in, which maintained the temperature of the mixture at 157°. When complete, the thermometer was raised out of the liquid and the whole heated until everything boiling below 160° had passed over. The distillate was then suspended in water, hydrochloric acid added, and distilled in steam, and the separated hydrocarbon washed with water, dried over calcium chloride, and fractionated. It contained halogen, which could not be completely removed by a second treatment with diethylaniline, and it was therefore heated for two hours with 100 c.c. of a boiling saturated solution of alcoholic potassium hydroxide and distilled in steam, the distillate poured into a large volume of water, and the separated hydrocarbon dried over calcium chloride and distilled over sodium, when all but a few drops passed over between 108° and 109°. This was again distilled over sodium in an atmosphere of carbon dioxide, when it boiled constantly at 108—108·5° and gave the following numbers on analysis :

0·1111 gave 0·3558 CO_2 and 0·1278 H_2O.   C = 87·34 ; H = 12·78.

C_8H_{14} requires C = 87·27 ; H = 12·73 per cent.

This liquid, which was obtained in almost theoretical amount, had a sp. gr. 15°/15° = 0·7857, gave a solid dibromide melting at 80—85°, and in other properties was identical with *iso*laurolene.

Oxidation with Potassium Permanganate.—Fifteen grams of this hydrocarbon were suspended in 375 c.c. of water, and a 4 per cent. solution of potassium permanganate added during constant shaking.

volume
twelve
1905, 8
was tre⸱
from bo⸱
5 grams
constant
permanen
several h⸱
hydrocarb
with stea⸱
over calci⸱
analysed.

  0·1205 g⸱

Dihydrois⸱

is a clear, col⸱
and possessing
a⸱chloroform
permanganate.
60—62 per cen
  Oxidation w⸱
carbon were add
slightly, a react
Heating was the
itself, and then t
ing the nitric ac
solidified. It wa⸱
obtained, which w
  Oxidation with
laurolene and a mi
of water were beat⸱
with a ground gla
After six hours, th⸱
with a fresh quanti
all the hydrocarbon
liquors to dryness, ▋▋
It was heated for to
evaporated, and th⸱ ⸱
the most part at 2⸱

TABLE **II**

			$[M]_D^{20°}$.		
				Chlorobromo-	
		Dibromo-	ester,	Dichloro-
▋▋re dis- **After** dis-		ester,	Cl : Br,	ester,
▋▋tion. **tillation.**		Br : Br.	Br : Cl.	Cl : Cl.
⸱1 ⸱1°	$-$**⸱9·79**°°	$-173·2°$	$\left\{ \begin{array}{l} -172·5° \\ -193·5 \end{array} \right\}$	$-172·9°$
5⸱ ⸱⸱	$-51·62$	$-215·8$	$\left\{ \begin{array}{l} -203·9 \\ -219·2 \end{array} \right\}$	$-209·6$
	$-$**50·96**	$-213·1$	$\left\{ \begin{array}{l} -190·2 \\ -207·8 \end{array} \right\}$	$-199·4$
$-$**⸱⸱⸱** $\left\{ \begin{array}{l} \text{in benzene} \\ \text{solution} \end{array} \right\}$		$-19·5$	$-32·5$	$-113·0$
	$-55·13°$	$-$**230·7**	$\left\{ \begin{array}{l} -207·9 \\ -220·9 \end{array} \right\}$	$-227·5$
⸱⸱7	$-53·79$ °	$-228·2$	$-235·9$	$-233·2$
⸱⸱⸱⸱ $\left\{ \begin{array}{l} \text{in benzene} \\ \text{solution} \end{array} \right\}$		$-$**236·3**	—	—

⸱ ⸱⸱⸱, a little acid separated on distillation,
⸱ ⸱ ⸱⸱ th⸱ undistilled material. The rotation
⸱ ⸱ ⸱⸱ ⸱h⸱ undistilled ester. In all other cases,
⸱ ⸱ ⸱ which is assumed to be purer.

⸱ ⸱ ⸱ 2 : 6-**ester**, the effect of the substi-
⸱ ⸱⸱⸱ **than that** of the substitution of
⸱ ⸱ ⸱ ⸱ **bromine** atom. This result was
⸱ ⸱⸱ ⸱ b⸱⸱⸱ between the monochloro-
⸱ ⸱⸱⸱, 1⸱03, 83, 121
⸱⸱⸱ ⸱ ⸱⸱⸱i⸱ning with the ester of smallest

⸱ ⸱ ⸱ 4; 3 : 5-; 3 : 4-; phenyl.

⸱ ⸱ ⸱ of the dichloro- and chlorobromo-
⸱⸱ ⸱ and 3 : 5-esters, the positions of
⸱ ⸱ ⸱⸱⸱. The influence of the ortho-
⸱ ⸱ r ⸱⸱tion is very clearly indicated in
⸱ ⸱ that of t⸱ ⸱ ⸱dily
⸱ ⸱⸱⸱ ⸱f

EXPERIMENTA.

the Dibromobenzoic Acids, Acid Chlorides, and Menthyl Esters.

employed for obtaining the ibromobenzoic acids is pre-
in the former preparations f the dihalogen compounds
eds no general descriptio. The acid chlorides were
ified in the manner alreav described. As in former
tho-acid chlorides, the 2 6-dibromobenzoyl chloride
higher temperature for esterification (175—180°)
acid chlorides, all of whh react rapidly below 130°
he esters, with the excepion of the 2 : 6-compound,
er diminished pressure. The rotations before and
varied slightly. In the two cases already referred
of the 2 : 3- and 3 : 5-tters, some decomposition
e free acid separated in the distillate. The effect
tation slightly. It is vident that even at this
distillation no racemission occurs.

omobenzoate.—The dibromobenzoic acid was pre-
-o-toluidine (obtained rom aceto-o-toluidide by
x's method) (*Ber.*, 190, **33**, 2498). The base
n erted into the bromnitro-compound, reduced
d again diazotised as llows :

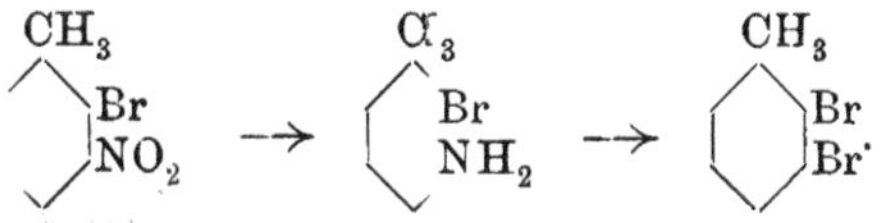

ate the series of rections usually adopted in
her dibromotoluen. Sixty grams of nitro-
grams of 2 : 3-dibrotoluene. The oxidation
as effected by heag in sealed tubes with
f HN to 2 vols. of water) at
h was separated
idation of

The permanganate was used up fairy rapidly at first, but for the completion of the reaction, which reuired 1030 c.c. of the oxidising agent, it was necessary to heat on th water-bath to a temperature of 60—65°. The solution was then fi.red from manganese dioxide, which was washed with hot water, nd the combined filtrate and washings evaporated to about 150 c.c acidified with dilute sulphuric acid, and extracted ten times with eier. The ethereal solution was dried over calcium chloride, the eth' evaporated, and the residue distilled under diminished pressure. At 47 mm., a few drops of a liquid smelling strongly of acetic aci first passed over, but the main portion (5 grams) distilled at 180–95° and solidified almost completely after standing in a cool place r thirty-six hours.

It was spread on porous plate an purified by crystallisation from water, in which it is very soluble, nd from which it separated in transparent, four-sided prisms meltig at 48·5° ; nor was this melting point lowered on mixing the substace with dimethylhexanonic acid, kindly sent to us by M. G. Blane. It gave the iodoform reaction, characteristic of ketonic acids containg the group $CH_3 \cdot CO-$, and on analysis the following numbers were btained :

0·1081 gave 0·2400 CO_2 and 0·086 H_2O. C = 60·55 ; H = 8·92.
$C_8H_{14}O_3$ requires C = 60·7 ; H = 8·86 per cent.

These data prove conclusively tha this substance is identical with the γ-acetyldimethylbutyric acid (diethylhexanonic acid) obtained b Blanc (p. 702) by the oxidation of islaurolene.

This acid was then further oxiised with sodium hyj (Blanc, ibid.). The oily oxidation roduct was heated with acetyl chloride for two hours, the ahydride thus obtained and the portion boiling at 260—26° further examined. was dissolved in boiling water, th solution evaporated when it at once solidified, and after ci isation from benzene and light petroleum meltl a nor point altered on mixing with pure ualil lutaric a to us by M. G. Blanc.

0·1019 gave 0·1960 CO_2 and 0·064   = 52·
$C_7H_{12}O_4$ requires C = 52 0

Another portion of the distilled anilic acid, which crystallised fron melting at 140—141 , the melting anilic acid being 141 .

RESEARCH LABORATORY, PHARMACE
17, BLOOMSBURY SQUARE,

VII.—*The Relation of Position Isomerism to Optical Activity. V. The Relation of the Menthyl Esters of the Isomeric Dibromobenzoic Acids.*

By Julius Berend Cohen and Sarah Emily Jackman.

This present paper is a continuation of previous investigations on this subject (Trans., 1905, 83, 1215; 190., 85, 1262; 1906, 19.6, 67, 1190) and contains an account of certain physical constants of the menthyl esters of the six isomeric dibromobenzoic acids and the products formed in their preparation.

The esters were prepared by methods already described for ... namely, by oxidising the six dibromotoluenes to the corresponding acids, which were then converted into the acid chlorides, and the latter into the esters by heating with menthol. The following tables contain the physical constants of the acids, the chlorides, and esters. The details of their preparation are given in the experimental part.

TABLE I

between the melting points of the acids in the case of the isomeric- and chloro-esters, namely, the lower the melting point of the ester. This ... contains the isomeric menthyl ester together with the molecular of dibromo esters.

The permanganate was used up fairly rapidly at first, but for the completion of the reaction, which required 1030 c.c. of the oxidising agent, it was necessary to heat on the water-bath to a temperature of 60—65°. The solution was then filtered from manganese dioxide, which was washed with hot water, and the combined filtrate and washings evaporated to about 150 c.c., acidified with dilute sulphuric acid, and extracted ten times with ether. The ethereal solution was dried over calcium chloride, the ether evaporated, and the residue distilled under diminished pressure. At 47 mm., a few drops of a liquid smelling strongly of acetic acid first passed over, but the main portion (5 grams) distilled at 180—195° and solidified almost completely after standing in a cool place for thirty-six hours.

It was spread on porous plate and purified by crystallisation from water, in which it is very soluble, and from which it separated in transparent, four-sided prisms melting at 48·5° ; nor was this melting point lowered on mixing the substance with dimethylhexanonic acid, kindly sent to us by M. G. Blanc. It gave the iodoform reaction, characteristic of ketonic acids containing the group $CH_3 \cdot CO-$, and on analysis the following numbers were obtained :

0·1081 gave 0·2400 CO_2 and 0·0868 H_2O.   C = 60·55 ; H = 8·92.

$C_8H_{14}O_3$ requires C = 60·75 ; H = 8·86 per cent.

These data prove conclusively that this substance is identical with the γ-acetyldimethylbutyric acid (dimethylhexanonic acid) obtained by Blanc (p. 702) by the oxidation of isolaurolene.

This acid was then further oxidised with sodium hypobromite (Blanc, ibid.). The oily oxidation product was heated with excess of acetyl chloride for two hours, the anhydride thus obtained distilled, and the portion boiling at 260—265° further examined. A portion was dissolved in boiling water, the solution evaporated to dryness, when it at once solidified, and after recrystallisation from a mixture of benzene and light petroleum melted at 83°; nor was this melting point altered on mixing with pure αα-dimethylglutaric acid, kindly sent to us by M. G. Blanc.

0·1019 gave 0·1960 CO_2 and 0·0604 H_2O.   C = 52·45 ; H = 7·24.

$C_7H_{12}O_4$ requires C = 52·50 ; H = 7·50 per cent.

Another portion of the distilled anhydride was converted into the anilic acid, which crystallised from dilute alcohol in lustrous plates melting at 140—141°, the melting point of pure αα-dimethylglutaranilic acid being 141°.

RESEARCH LABORATORY, PHARMACEUTICAL SOCIETY,
17, BLOOMSBURY SQUARE, W.C.

VII.—*The Relation of Position Isomerism to Optical Activity. V. The Rotation of the Menthyl Esters of the Isomeric Dibromobenzoic Acids.*

By Julius Berend Cohen and Israel Hyman Zortman.

The present paper is a continuation of previous investigations on this subject (Trans., 1903, **83**, 1213 ; 1904, **85**, 1262, 1271 ; 1905, **87**, 1190) and contains an account of certain physical constants of the menthyl esters of the six isomeric dibromobenzoic acids and the products formed in their preparation.

The esters were prepared by methods already described (*loc. cit.*), namely, by oxidising the six dibromotoluenes to the corresponding acids, which were then converted into the acid chlorides, and the latter into the esters by heating with menthol. The following tables contain the physical constants of the acids, acid chlorides, and esters. The details of their preparation are given in the experimental part.

TABLE I.

Substance.	M. p. of pure acid, C. and Z.	Lowest and highest m. p. recorded by previous observers.	M. p. of acid chloride.	M. p. of ester.	B. p. of ester.	Pressure in mm.
2 : 3-	149—150°	147° (Hubner) / 153 (Claus and Lade)	60—62°	52—53°	—	—
2 : 4-	168—169	166·5 (Claus and Weil) / 169 (Miller)	47—49	—	242—245	16
2 : 5-	151—152	153 (Hubner) / 153 (Claus and Weil)	39—41	43—44	238—240	16
2 : 6-	146—147	189 (Claus and Weil) / 136—137° (Meyer and Sudborough)	39—42	151—152	—	—
3 : 4-	229—230	229—230° (Hubner) / 232—233° (Halberstadt)	64—66	41—43	241—244	15
3 : 5-	213—214	209° (Claus and Weil) / 223—227° (Angerstein)	41—42	—	245—250	20

The same relation subsists between the melting points of the acids and esters as was observed in the case of the dichloro- and chloro-bromo-benzoic acids and esters, namely, the lower the melting point of the acid the higher that of the ester. Table II contains the densities and specific rotations of the menthyl esters, together with the molecular rotations of the three series of dihalogen esters.

TABLE II

Menthyl ester of dibromo-benzoic acid.	Density at 20°.		$[a]_D^{20°}$.		$[M]_D^{20°}$.		
					Dibromo-ester, Br : Br.	Chlorobromo-ester, Cl : Br, Br : Cl.	Dichloro-ester, Cl : Cl.
	Before distillation.	After distillation.	Before distillation.	After distillation.			
2 : 3-	1·4189	1·4170	−41·41°	−39·79°*	−173·2°	$\left\{ \begin{array}{c} -172·5° \\ -193·5 \end{array} \right\}$	−172·9°
2 : 4-	1·4007	1·4060	−51·49	−51·62	−215·8	$\left\{ \begin{array}{c} -203·9 \\ -219·2 \end{array} \right\}$	−209·6
2 : 5-	1·3809	1·3821	—	−50·96	−213·1	$\left\{ \begin{array}{c} -190·2 \\ -207·8 \end{array} \right\}$	−199·4
2 : 6-	—	—	−4·68 $\left\{ \begin{array}{l} \text{in benzene} \\ \text{solution} \end{array} \right\}$		−19·5	−32·5	−113·0
3 : 4-	—	1·4258	—	−55·18°	−230·7	$\left\{ \begin{array}{c} -207·9 \\ -220·9 \end{array} \right\}$	−227·5
3 : 5-	1·4114	1·4159	−54·57	−53·79 *	−228·2	−235·9	−233·2
Menthyl benzoate	—	—	−90·92 $\left\{ \begin{array}{l} \text{in benzene} \\ \text{solution} \end{array} \right\}$		−236·3	—	—

* In the case of the 2 : 3- and 3 : 5-esters, a little acid separated on distillation, and the product was probably less pure than the undistilled material. The rotation is therefore calculated from observations on the undistilled ester. In all other cases, the rotation is that of the distilled product, which is assumed to be purer.

With the single exception of the 2 : 6-ester, the effect of the substitution of two bromine atoms is less than that of the substitution of two chlorine or one chlorine and one bromine atom. This result was anticipated from the relation which subsists between the monochloro- and monobromo-derivatives (Trans., 1903, **83**, 1216).

The magnitude of the deviation beginning with the ester of smallest rotation is as follows :

2 : 6- ; 2 : 3- ; 2 : 5- ; 2 : 4 ; 3 : 5- ; 3 : 4- ; phenyl.

The order is the same as that of the dichloro- and chlorobromo-esters with the exception of the 3 : 4- and 3 : 5-esters, the positions of which are reversed in the present case. The influence of the ortho-bromine atom in depressing the rotation is very clearly indicated in all cases, but most strikingly in that of the 2 : 6-ester. The steadily and rapidly decreasing rotation with the increasing atomic weight of the halogens in the di-ortho-positions induces one to predict the almost complete annihilation of activity in compounds which contain iodine in place of the other halogens.

The remarkably high melting point of the diortho-ester corresponds closely with the constants of the other 2 : 6-dihalogen compounds ; the dichloro-ester melts at 134—135°, the chlorobromo-ester at 144—145°, and the dibromo-ester at 151—152°.

EXPERIMENTAL.

Preparation of the Dibromobenzoic Acids, Acid Chlorides, and Menthyl Esters.

The method employed for obtaining the dibromobenzoic acids is precisely that used in the former preparations of the dihalogen compounds (*loc. cit.*) and needs no general description. The acid chlorides were prepared and purified in the manner already described. As in former examples of diortho-acid chlorides, the 2 : 6-dibromobenzoyl chloride required a much higher temperature for esterification (175—180°) than the isomeric acid chlorides, all of which react rapidly below 130° with menthol. The esters, with the exception of the 2 : 6-compound, were distilled under diminished pressure. The rotations before and after distillation varied slightly. In the two cases already referred to, namely, those of the 2 : 3- and 3 : 5-esters, some decomposition occurred and a little free acid separated in the distillate. The effect was to lower the rotation slightly. It is evident that even at this high temperature of distillation no racemisation occurs.

Menthyl 2 : 3-*Dibromobenzoate.*—The dibromobenzoic acid was prepared from 3-nitro-*o*-toluidine (obtained from aceto-*o*-toluidide by Reverdin and Crépieux's method) (*Ber.*, 1900, **33**, 2498). The base was diazotised and converted into the bromonitro-compound, reduced to the bromoamine, and again diazotised as follows :

$$
\underset{NO_2}{\overset{CH_3}{\bigcirc}NH_2} \;\rightarrow\; \underset{NO_2}{\overset{CH_3}{\bigcirc}Br} \;\rightarrow\; \underset{NH_2}{\overset{CH_3}{\bigcirc}Br} \;\rightarrow\; \underset{Br}{\overset{CH_3}{\bigcirc}Br}
$$

This may serve to illustrate the series of reactions usually adopted in the preparation of the other dibromotoluenes. Sixty grams of nitrotoluidine gave nearly 20 grams of 2 : 3-dibromotoluene. The oxidation of the dibromotoluene was effected by heating in sealed tubes with dilute nitric acid (1 vol. of HNO_3, sp. gr. 1·4, to 2 vols. of water) at 130—135° for five and a half hours. The acid which was separated from the product was very impure and, like the product of oxidation of the 3-chloro-2-bromotoluene (Trans., 1904, **85**, 1266), contained a large quantity of a compound melting above 200° and also a substance melting below 135°.

The crude material, amounting to 16 grams, was purified by repeated crystallisation from ligroin, in which the less fusible acid is nearly insoluble and the more fusible compound readily soluble, whilst the 2 : 3-acid dissolves in the boiling liquid, but is nearly insoluble in the cold. Eventually 4·2 grams of acid were obtained, which seemed to

soften slightly a few degrees below 148° and then melted sharply.
Recrystallisation did not seem to alter the melting point, but a small
quantity of acid obtained at a later stage on distilling the menthyl
ester melted sharply at 149—150°, and may be taken as the true
melting point. The acid was heated on the water-bath with an equal
weight of phosphorus pentachloride and the product distilled under
diminished pressure at 100° to remove most of the phosphorus oxychloride. The acid chloride, which solidified on cooling, was dissolved
in benzene to separate the insoluble phosphorus pentachloride and the
benzene distilled off, the last traces being removed by heating to 100°
in a vacuum. The residual solid cake was powdered and pressed on a
porous plate and left in a vacuum desiccator to remove any remaining
phosphorus oxychloride. The crude acid chloride melted at 55—60°
and after one crystallisation from ligroin at 60—62°. The product,
amounting to 3·5 grams, was heated in an oil-bath with 2·5 grams of
pure menthol (Kahlbaum), the specific rotation of which had been
ascertained, and gave the correct number. The reaction began at
100—105°. The mixture was maintained at 130° for an hour, water
and a little sodium carbonate solution were then added, and the ester
submitted to distillation in steam until all trace of free menthol had
been removed.

The residue was extracted with ether, dehydrated with fused calcium
chloride, and the ether removed by distillation. The product solidified
on cooling and melted sharply at 52—53°. The fused substance had
an amber colour, which, however, did not interfere with the polarimeter
readings : $l = 0·302$ dcm., $d/20° = 1·4189$, $a_D - 17·74°$.

The ester, after distilling under reduced pressure, was colourless. A
little free acid which separated in the process was removed by dissolving in light petroleum, in which the acid is insoluble, and filtering.
The distilled product melted at 49—52°, which is lower than the
original material, and was therefore less pure. A redetermination
of the rotation and density confirmed this : $l = 0·302$ dcm.,
$d/20° = 1·4170$, $a_D - 17·03°$.

The analyses of this and the other esters are recorded in a table at
the end of the paper.

Menthyl 2 : 4-*Dibromobenzoate.*—The dibromotoluene which served
for the preparation of the acid was obtained from 2 : 4-nitrotoluidine by
the usual series of operations. Sixty grams of the base gave
35 grams of dibromotoluene boiling at 152—158° under 80 mm.
pressure. Twenty grams gave, on oxidation at 120—125° for five
hours, 16 grams of crude acid (m. p. 145—152°), which were subjected
to steam-distillation to remove a small quantity of less fusible
substance which was volatile in steam. After recrystallisation from
benzene, 6 grams of acid were obtained melting at 164—166°. This

was successively converted in the usual way into the acid chloride (m. p. 43—45°) and the ester, which gave the following polarimeter readings : $l = 0.302$ dcm., $d/20° = 1.4023$, $a_D - 21.94°$.

A second preparation was made in a similar way. The dibromo-toluene gave in succession 15 grams of acid (m. p. 165—167°), 13 grams of acid chloride (m. p. 47—49°), and 13.5 grams of ester. The polarimeter readings (a) before and (b) after distillation were as follows :

    a. $l = 0.302$ dcm. ; $d/20° = 1.4007$; $a_D - 21.78°$.
    b. $l = 0.302$ dcm. ; $d/20° = 1.4060$; $a_D - 21.92°$.

Menthyl 2 : 5-Dibromobenzoate.—The 2 : 5-dibromotoluene was obtained from 2 : 5-nitrotoluidine. Sixty grams of base gave 35 grams of dibromotoluene (b. p. 155—170°/120 mm.). The crude acid was crystallised repeatedly from alcohol and benzene, and finally from water after boiling with animal charcoal to remove colouring matter.

About 26 grams of dibromotoluene yielded 9 grams of pure acid (m. p. 150—151°). The latter gave 6 grams of acid chloride (m. p. 39—41°) and finally 7.3 grams of ester (m. p. 42—44°). The following are the polarimeter readings : (a) before and (b) after distillation :

    a. $l = 0.302$ dcm. ; $d/20° = 1.3809$; $a_D - 21.7°$.
    b. $l = 0.302$ dcm. ; $d/20° = 1.3821$; $a_D - 21.27°$.

Menthyl 2 : 6-Dibromobenzoate.—The 2 : 6-dibromotoluene used for oxidation was obtained from 2 : 6-dinitroluene by a series of alternate reductions and diazotisations. The yield from 150 grams of dinitro-compound amounted to 50 grams of dibromotoluene (b. p. 110—130°/30 mm.). The 29 grams of crude acid obtained on oxidation were first purified by esterifying with methyl alcohol and hydrochloric acid. The 2 : 6-acid is not attacked, whereas other acids which may be present as by-products are esterified, and can be removed by shaking out the alkaline solution with ether. The crude acid was then recrystallised from benzene, when 11 grams of pure 2 : 6-acid were obtained, which crystallised in hexagonal plates and melted at 146—147°. It was converted into the acid chloride (m. p. 35—39°), which, on crystallising from light petroleum, formed colourless needles (m. p. 39—42°). Seven grams of acid chloride gave, after crystallisation from alcohol, 5 grams of pure menthyl ester in colourless needles (m. p. 151—152°).

The rotation was determined in benzene solution as follows :

3.9104 grams in 25.07 c.c. of benzene ; $l = 2$ dcm. ; $a_D - 1.46°$.

10 c.c. of above solution and 10 c.c. of benzene ; $l = 2$ dcm. ; $a_D - 0.64°$.

2·3760 grams in 13 c.c. of benzene ; $l = 2$ dcm. ; $a_D - 1·98°$.

1·2798 grams in 13 c.c. of benzene ; $l = 2$ dcm. ; $a_D - 0·80°$.

Menthyl 3 : 4-*Dibromobenzoate.*—The aceto-*p*-toluidide used in the preparation of 3 : 4-dibromotoluene was brominated and then hydrolysed, and the 3-bromo-*p*-toluidide thus obtained was then diazotised in the usual way. Thirty-four grams of bromotoluidine gave 23 grams of 3 : 4-dibromotoluene (b. p. 160—165°/65 mm.). On oxidation in sealed tubes, 10 grams of acid (m. p. 215—225°) were obtained which were recrystallised from dilute alcohol with the addition of animal charcoal and then gave a colourless product (m. p. 227—229°). About 7 grams of pure acid gave an equal weight of acid chloride (m. p. 64—66°). The ester amounted to 7·8 grams of a brown, viscid liquid which showed a rotation of − 23·5°, but as the colour interfered with the reading, the ester was distilled under reduced pressure, and the colourless distillate then gave the following result: $l = 0·302$ dcm., $d/20° = 1·4258$, $a_D - 23·76°$. The liquid solidified on standing for some weeks and then melted at 41·43°.

Menthyl 3:5-*dibromobenzoate.*—The 3:5-dibromotoluene was obtained from 3-bromo-*p*-toluidine (b. p. 166°/50 mm.), described above, which was brominated and the amino-group replaced by hydrogen in the usual way. Thirty-seven grams of bromotoluidine gave 34 grams of 3 : 5-dibromotoluene (m. p. 38—39°). Thirty grams of dibromotoluene gave 26 grams of crude acid, which, after recrystallisation from benzene, gave 18 grams of pure substance (m. p. 213—214°). From the acid were obtained 15·5 grams of pure acid chloride (m. p. 41—42°), and finally 20 grams of menthyl ester. The following polarimeter readings were made : (*a*) before and (*b*) after distillation :

(*a*) $l = 0·302$ dcm. ; $d/20° = 1·4114$; $a_D - 23·26°$.

(*b*) $l = 0·302$ dcm. ; $d/20° = 1·4159$; $a_D - 23·00°$.

As a little acid separated in the process of distillation, the undistilled product probably gives the more trustworthy value.

It should be added that although the melting points of several of the esters lie much above 20°, at which the densities and polarimeter readings were taken, they are easily supercooled and remain in a fused state at 20° without showing any signs of solidification. The tendency to supercooling is rather remarkable, and it has been frequently observed that when the fused liquid has been cooled to the ordinary temperature in the pyknometer and crystallisation has begun at the capillary end, the main portion will remain liquid for a considerable time.

TABLE III.

Analyses of the Dibromobenzoic Esters.

Menthyl ester.	Substance taken.	AgBr.	Per cent.
2 : 3-	0·1476	0·1338	38·58
2 : 4-	0·2154	0·1920	37·93
2 : 5-	0·1338	0·1185	37·70
2 : 6-	0·1236	0·1114	38·34
3 : 4-	0·1208	0·1087	38·30
3 : 5-	0·1552	0·1392	38·16

$C_{17}H_{22}O_2Br_2$ requires $Br = 38\cdot27$ per cent.

THE UNIVERSITY,
LEEDS.

VIII.—*Caro's Permonosulphuric Acid.*

By THOMAS SLATER PRICE, D.SC.

THE results of the various investigations which have been made to determine the composition of permonosulphuric acid (Caro's acid) have led to the conclusion that the formula is either H_2SO_5 or $H_2S_2O_9$, the acid being monobasic in the former case and dibasic in the latter (compare Price, Trans., 1903, **83**, 543 ; also Mugdan, *Zeit. Elektrochem.*, 1903, **9**, 719 ; Price and Friend, Trans., 1904, **85**, 1526).

The decision between these two formulæ could readily be made if a pure salt, such as the potassium derivative, could be obtained. The simplest method would be to heat a weighed quantity of the salt and determine the weight of potassium sulphate left ; the weight of residue obtained would vary according as to whether the formula was $KHSO_5$ or $K_2S_2O_9$.

Although the author has not yet succeeded in overcoming the difficulties attending the preparation of the pure potassium salt, a mixture has been obtained, from the analysis of which it has been possible to show that the formula of the acid is H_2SO_5. The mixture consisted of the potassium salts of sulphuric, permonosulphuric, and perdisulphuric acids and of potassium hydrogen sulphate. The proportion of each constituent was found as follows: the permonosulphate, calculated as either $KHSO_5$ or $K_2S_2O_9$, was determined by the liberation of iodine from a solution of potassium iodide and titration with thiosulphate ; the perdisulphate by measuring the total oxidising power of the mixture by means of ferrous sulphate and potassium permanganate and allowing for that due to the permonosulphate. The amount of potassium hydrogen sulphate was estimated by a method which depends on the results obtained by Mugdan (*Zeit. Elektrochem.*, 1903, 9, 719). Use was made of the fact that when permonosulphuric acid is decomposed by potassium iodide a diminution in acidity takes place, as shown by the equations

$$H_2S_2O_9 + 4KI = 2K_2SO_4 + H_2O + 2I_2$$
$$\text{or} \quad H_2SO_5 + 2KI = K_2SO_4 + H_2O + I_2.$$

The liberated iodine was titrated with thiosulphate and then the acidity of the solution determined by means of standard caustic potash (free from carbonate). Knowing the amount of permonosulphate present, the diminution in acidity which should take place could be calculated from the above equations; the difference between the observed and calculated results gave the acidity due to the potassium hydrogen sulphate present.

The amount of potassium sulphate was then determined by difference, knowing the weight of the mixture and of the other constituents. All that was then necessary was to find the amount of potassium sulphate obtained by heating a known weight of the mixture, then compare the experimental with the theoretical result.

The following table gives a summary of the results obtained:

Weight of mixture.	Weight of KHSO₅.	Weight of KHSO₄.	Weight of K₂S₂O₈.	Weight of K₂SO₄ in mixture.	Weight of residue = a (observed).	Weight of residue if the salt is K₂S₂O₉ = b (calc.).	Difference (b − a).	Weight of residue if the salt is KHSO₅ = c (calc.).	Difference (c − a).
grams	grams	gram	gram	grams	grams	grams	gram	grams	gram
6·5058	4·2359	0·5328	0·1046	1·6325	4·3780	4·7177	{ 0·3397 = 7·8% }	4·4670	{ 0·0890 = 2·0% }
6·6272	3·6874	0·6153	0·1571	2·1674	4·6960	4·9926	{ 0·2966 = 6·3% }	4·7743	{ 0·0783 = 1·7% }
3·6015	2·3727	0·4050	0·0053	0·8185	2·3800	2·5804	{ 0·2004 = 8·4% }	2·4400	{ 0·0600 = 2·5% }
5·9990	3·2589	0·3686	0·1129	2·2586	4·3470	4·6267	{ 0·2797 6·4% }	4·4338	{ 0·0868 = 2·0% }

The results of the above experiments thus point to the formula $KHSO_5$ in preference to $K_2S_2O_9$ for potassium permonosulphate, and hence to the formula H_2SO_5 for permonosulphuric acid. The difference of about 2 per cent. between the experimental and calculated values for the weight of the residue of potassium sulphate, assuming the formula H_2SO_5 for permonosulphuric acid, seems, at first sight, to be rather large. As will be seen from the experimental details, however, the differences in weight in the last column have been multiplied by 10, the actual differences varying from 6 to 9 milligrams; taking into consideration the number of different determinations which have to be made in order to arrive at the above results, it is very probable that the errors of experiment will account for these differences. All the apparatus used, including the weights, had been carefully calibrated.

It was thought at one time that the error might be due to imperfect drying of the mixture; this was found not to be the case, however, since the errors in all four experiments are about the same, whereas the mixtures in the first two cases were dried for three days in an exhausted desiccator over sulphuric acid, whilst the last two mixtures were dried for five weeks in a vacuum over phosphoric oxide. It is possible that the error may be to some extent explained by the occlusion of some of the mother liquor in the crystals as they are slowly deposited during the concentration of the solution under diminished pressure (compare Richards, *Zeit. physikal. Chem.*, 1903, **46**, 189). This would account for the fact that in all cases the calculated weight of the residue is greater than the experimental weight.

Since the results of previous researches have shown conclusively that the formula for permonosulphuric acid is either H_2SO_5 or $H_2S_2O_9$, the present investigation may be used to decide the question in the same way as the determination of the specific heat of a metal is used to decide the atomic weight. Hence, the first formula must be regarded as the correct one. The constitution of the acid would then be represented graphically by $O \!\!> \!\! S \!\! <^{O \cdot H}_{O \cdot O \cdot H}$, the hydrogen in the hydroxyl group directly attached to the sulphur being presumably the one replaced by metals in the formation of salts. The remaining hydrogen atom would possess only very weak acid properties, if any at all, as it is contained in the group $-O \cdot O \cdot H$, derived from hydrogen peroxide, which is only a very weak acid. Also, from analogy with other dibasic acids, the acidity would be very much diminished by the presence of the much more strongly acid hydrogen ion in the $-OH$ group directly attached to the sulphur.

Since perdisulphuric acid ($H_2S_2O_8$) is dibasic, readily forming the salt $K_2S_2O_8$, it would follow that the constitutional formula

$$O_2S \!\! <^{O \cdot H}_{O} \!\!\!\!-\!\!\!\!-\!\!\!\!-^{H \cdot O}_{O} \!\! > \!\! SO_2,$$

usually assigned to it, is the correct one, in which both the acid hydrogens are contained in $-OH$ groups directly attached to the sulphur atoms, and not $O_5S_2 \!\! <^{OH}_{O_2H}$, as given by Kastle and Loewenhart (*Amer. Chem. J.*, 1903' **29**, 563; see also Price and

Denning, *Zeit. physikal. Chem.*, 1903, **46**, 101), since in the latter case the acid would probably be monobasic, judging by analogy with permonosulphuric acid.

EXPERIMENTAL.

A solution of permonosulphuric acid was made in the usual manner by the action of 25 c.c. of concentrated sulphuric acid on 25 grams of potassium perdisulphate (compare Price and Friend, Trans., 1904, **85**, 1526); potassium perdisulphate was chosen because it is so easily purified. The solution so obtained was diluted and neutralised at the same time by being run very slowly into a solution of potassium carbonate (containing enough carbonate to neutralise the sulphuric acid taken), in which was ice freshly made from distilled water. The temperature of the aqueous potassium carbonate, which was $-7°$ before the solution of permonosulphuric acid was run in, gradually rose to about $0°$ during the process of dilution. The potassium sulphate which separated out was filtered off, the mother liquor being drained away as completely as possible. The filtrate generally contained an amount of persulphuric oxygen which was equivalent to about 18 grams of H_2SO_5 per litre. If the mixture containing the acid was first diluted by being poured on to broken ice, and then neutralised with potassium carbonate, the filtrate was only about half the above strength.

The filtrate was generally acid, and was neutralised by the further addition of anhydrous potassium carbonate, which was added carefully until no further effervescence took place. This method was accurate enough, since it was not essential that the solution should be exactly neutral (see p. 57). The neutral solution was then further concentrated by freezing; the chief advantage of this process lay in the deposition of the potassium sulphate, and not in the increase in concentration of the permonosulphate, since the residue of ice and potassium sulphate always contained an appreciable amount of adherent permonosulphate. Only one freezing was carried out; a second was not found to be advisable, since the increase in concentration of the permonosulphate thus obtained was very small.

It is generally supposed that permonosulphuric acid and its salts are not very stable in solution, but it was found possible to evaporate the solution to dryness by concentration in a vacuum desiccator over concentrated sulphuric acid. The process of evaporation lasted between a week and a fortnight, and the potassium sulphate, which separated out continuously, was filtered off from time to time until the solution became so concentrated that it could not be filtered without undergoing appreciable loss; it was then allowed to evaporate to dryness. During the evaporation of the solution, the concentration of the permonosulphate continually increases, but at the same time some of the salt decomposes with the formation of potassium hydrogen sulphate.   This is the reason why it is immaterial whether the neutralisation with potassium carbonate be exact or not. The mixture so obtained was powdered as finely as possible, and finally dried in a vacuum desiccator.

It should be mentioned that the solutions containing the permonosulphate generally had a strong odour resembling that of bleaching powder (Baeyer and Villiger, *Ber.*, 1901, **34**, 853); this generally disappeared during the concentration under diminished pressure, and the solid substance possessed no odour ; on exposure to the air, the solution acquired this characteristic odour after a short time, but its presence or absence did not affect the analytical results.   No hydrogen peroxide could be detected by the titanium sulphate test, either in the solutions before concentration or in the solutions made from the solid residue obtained.

Analysis of the Mixture.—A weighed quantity of the dry mixture was dissolved in air-free water and the solution diluted to 100 c.c. Aliquot portions of this solution were. taken and the various ingredients determined.   For all the titrations, 2 c.c. of the solution were used and the measurements repeated several times.   In no case did the individual titrations vary more than 0·03 c.c. (on 10 to 25 c.c.).

As has already been mentioned, the permonosulphate was estimated by the liberation of iodine from a solution of potassium iodide, the precautions observed being the same as those detailed in a previous paper (Price, Trans., 1903, **83**, 543).   As is shown by the equations on page 54, a decrease in acidity takes place during this reaction.   For this reason a known amount of dilute hydrochloric acid was added to the aqueous potassium iodide before the permonosulphate solution was introduced. After the iodine liberated had been exactly titrated with thiosulphate, the acidity of the resulting solution was determined by means of standard caustic potash.   From the diminution in acidity which had taken place, the amount of potassium hydrogen sulphate present was calculated in the way already indicated.

This method of determining the acidity of the mixture is indirect, but no direct method could be found which gave trustworthy results. The solution could not be directly titrated with caustic potash, using either methyl-orange or phenolphthalein as indicators.   In the former case, the indicator was rapidly oxidised, and in the latter the colour change was not sharp.   This has been noticed by the author in previous investigations on Caro's acid, but the difficulty could not then be overcome ; the action with phenolphthalein is probably due to the slight acidity of the hydrogen ion in the –OOH group.   Nor could satisfactory results be obtained by the method used by Armstrong and Lowry (*Proc. Roy. Soc.*, 1902, **70**, 94). On heating an aliquot portion of the solution to decompose the persulphates and then titrating with caustic potash, the titrations varied greatly among themselves and were quite untrustworthy; this has also been noticed by Mugdan (*Zeit. Elektrochem.*, 1903, **9**, 719).

In the determination of the total oxidising power of the solution, concordant results could be obtained only in the following manner: a solution of sulphuric acid was boiled to expel all air and then cooled; 5 c.c. of ferrous sulphate solution and 2 c.c. of the solution to be analysed were then added in the order given, and the whole heated to boiling again in order to reduce the persulphate completely; the solution was then cooled and titrated with permanganate.

The only other estimation necessary, which was the determination of the amount of potassium sulphate obtained from an aliquot portion of the solution, was performed by evaporating 10 c.c. of the solution to dryness in a platinum crucible over the water-bath. The crucible was covered with a watch-glass until all the permonosulphate had decomposed, since there was a vigorous effervescence as soon as the solution became hot. The residue in the crucible was then carefully heated to decompose the potassium hydrogen sulphate and any persulphate remaining, and weighed as sulphate in the usual manner. Four different lots of 10 c.c. each were usually treated in this manner. Obviously the simplest method would have been to heat a known weight of the solid mixture directly; there was the uncertainty, however, as to whether the ingredients were thoroughly mixed or not. It remains to be pointed out that the methods given above can only be used to discriminate between the formulæ $KHSO_5$ and $K_2S_2O_9$, which, indeed, is all that is necessary. A short calculation will show that it is not possible to distinguish between the formulæ $K_2S_2O_9$ and KHS_2O_9, or between K_2SO_5 and $KHSO_5$, because of the method used to estimate the proportion of potassium hydrogen sulphate.

CHEMICAL DEPARTMENT,
  MUNICIPAL TECHNICAL SCHOOL,
    BIRMINGHAM.

IX.—*Contributions to the Chemistry of the Amidines. 2-Aminothiazoles and 2-Imino-2 : 3-dihydrothiazoles. 2-Iminotetrahydrothiazoles and 2-Amino-4 : 5-dihydrothiazoles.*

By GEORGE YOUNG and SAMUEL IRWIN CROOKES.

THIS investigation was undertaken with the purpose of studying the constitutions of the derivatives of such amidines as have one of the nitrogen atoms and the carbon atom of the group $-N\dot{:}C\cdot N\dot{-}$ forming part of a closed chain, whilst the second nitrogen atom lies outside of the cyclic nucleus; these are referred to in the following pages as "partially cyclic" amidines. In the present paper, we describe the alkylation of some amidines belonging to the thiazole group, and the determination of the constitutions of the alkyl bases formed.

Traumann (*Annalen*, 1888, **249**, 31) found that the base formed by the action of chloroacetone on thiocarbamide, and to which he ascribed the constitution $\overset{\text{CH}—\text{S}}{\underset{\text{CMe}\cdot\text{N}}{\Big\|}}{>}\text{C}\cdot\text{NH}_2$, yielded on methylation

the derivative $\overset{\text{CH}——\text{S}}{\underset{\text{CMe}\cdot\text{NMe}}{\Big\|}}{>}\text{C}\dot{:}\text{NH}$, and that similarly the base,

$\overset{\text{CH}—\text{S}}{\underset{\text{CPh}\cdot\text{N}}{\Big\|}}{>}\text{C}\cdot\text{NH}_2$, obtained by heating thiocarbamide with phenacyl

bromide, yielded the methyl derivative,

$$\overset{\text{CH}——\text{S}}{\underset{\text{CPh}\cdot\text{NMe}}{\Big\|}}{>}\text{C}\dot{:}\text{NH}.$$

We have methylated the bases obtained by heating methylthiocarbamide, allylthiocarbamide, and phenylthiocarbamide with chloro-acetone. Traumann (*loc. cit.*) has shown that the bases formed in this manner from monosubstituted thiocarbamides, have the constitution $\overset{\text{CH}—\text{S}}{\underset{\text{CMe}\cdot\text{N}}{\Big\|}}{>}\text{C}\cdot\text{NHR}$. The position of the hydrogen atom, which is

substituted on alkylation, is discussed later.

The methyl derivative, obtained from the base with $R = CH_3$, is identical with 2-methylimino-3 : 4-dimethyl-2 : 3-dihydrothiazole, $\overset{\text{CH}——\text{S}}{\underset{\text{CMe}\cdot\text{NMe}}{\Big\|}}{>}\text{C}\dot{:}\text{NMe}$, formed by the action of chloroacetone on *s*-di-

methylthiocarbamide. To determine the constitution of the alkyl derivatives of the bases with $R = C_3H_5$ and C_6H_5, we heated the methyl bases with concentrated hydrochloric acid at 250° in the expectation of obtaining a primary or secondary amine and the oxydihydrothiazole,

$$\underset{CMe \cdot NMe}{\overset{CH----S}{\Big>}} C:NR \quad \longrightarrow \quad NH_2R + \underset{CMe \cdot NMe}{\overset{CH-\quad S}{\Big>}} CO \quad \text{or}$$

$$\underset{CMe \cdot N}{\overset{CH-S}{\Big>}} C \cdot NRMe \quad \longrightarrow \quad NHRMe + \underset{CMe \cdot NH}{\overset{CH----S}{\Big>}} CO.$$

The high pressure in the tube after cooling, the evolution of hydrogen sulphide on evaporation of the mixture, and the separation of methylamine together with a less volatile primary amine (allyl-amine (?) and aniline respectively) from the products showed that the methyl bases had the constitutions

$$\underset{CMe \cdot NMe}{\overset{CH----S}{\Big>}} C:N \cdot C_3H_5 \quad \text{and} \quad \underset{CMe \cdot NMe}{\overset{CH----S}{\Big>}} C:NPh,$$

and that the hydrolysis had proceeded beyond the formation of the oxydihydrothiazole to, at least, partial disruption of the thiazole nucleus. In confirmation of this, it was found that 2-phenylaminothiazole, $\underset{CMe \cdot N}{\overset{CH-S}{\Big>}} C \cdot NHPh$, when hydrolysed with concentrated hydrochloric acid at 250°, yielded aniline and small quantities of ammonia, hydrogen sulphide, and carbon dioxide.

The methylation of the "partially cyclic" amidines,

$$\text{A.} \quad \underset{CH_2-N}{\overset{CHR \cdot S}{\Big>}} C \cdot NHR' \quad \text{or} \quad \underset{CH_2 \cdot NH}{\overset{CHR--S}{\Big>}} C:NR',$$

has been studied by Gabriel, who methylated the bases with $R = H$ and CH_3, and $R' = H$, and obtained the derivatives

$$\underset{CH_2 \cdot NMe}{\overset{CH_2----S}{\Big>}} C:NH \quad \text{and} \quad \underset{CH_2 \cdot NMe}{\overset{CHMe-S}{\Big>}} C:NH$$

(*Ber.*, 1889, **22**, 1142, 2984), whilst Prager, who methylated the bases of this type with $R = CH_3$ and $R' = C_6H_5$ and o-C_6H_4Me (*Ber.*, 1889, **22**, 2998), obtained the derivatives

$$\underset{CH_2-N}{\overset{CHMe \cdot S}{\Big>}} C \cdot NMePh \quad \text{and} \quad \underset{CH_2-N}{\overset{CHMe \cdot S}{\Big>}} C \cdot NMe \cdot C_6H_4Me.$$

We have repeated Prager's preparations, have confirmed his results, and have extended the investigation to the methylation and ethylation of three bases in all, with $R = CH_3$ and $R' = C_6H_5$, o-C_6H_4Me, and p-C_6H_4Me. On successive oxidation and hydrolysis of the alkylated bases, we obtained in each case β-methyltaurine and the secondary base: methyl- or ethyl-aniline, o-toluidine, and p-toluidine respectively, from which follows the constitution of the alkylated bases:

$$\text{B.} \quad \underset{CH_2-N}{\overset{CHMe \cdot S}{\Big>}} C \cdot NR'Alk \quad \longrightarrow \quad NHR'Alk + \underset{CH_2 \cdot NH_2}{\overset{CHMe \cdot SO_3H}{\Big|}}.$$

Our results, together with those of Traumann, Gabriel, and Prager, show that "partially cyclic" amidines of the thiazole series, having a hydrogen atom which may be substituted directly by an alkyl group, yield on alkylation derivatives in which the alkyl is attached to the nitrogen atom of the nucleus, except when the ring is already partially reduced and R' in the formula A is an aryl group, the alkyl group in this case going to the side-chain nitrogen atom, as in B. This rule is probably of general application to the alkylation of "partially cyclic" amidines of any carbo-nitrogen heterocyclic series.

The results of alkylating "partially cyclic" amidines are in accord with those obtained by von Pechmann on methylating "mixed" amidines, if the rule of alkylation be stated in the following form :— on alkylation of an amidine, the alkyl group goes to the more negative nitrogen atom.

Von Pechmann (Ber., 1895, 28, 2362 ; 1897, 30, 1780) found that on methylation of a "mixed" amidine, which might be $-C{\displaystyle \mathop{<}^{NHR}_{NR'}}$ or $-C{\displaystyle \mathop{<}^{NR}_{NHR'}}$, with an R, an alkyl group, or a hydrogen atom, and R', an aryl group, and in which the two nitrogen atoms must vary widely in basicity, there is obtained only one methyl derivative :

$$-C{\displaystyle \mathop{<}^{NR}_{NR' \cdot Me'}}$$

in which the methyl is attached to the same nitrogen atom as in the aryl group, that is to say, to the less basic or more negative nitrogen atom, but on methylation of a "mixed" amidine with R and R', two similar groups, as phenyl and o-tolyl (Ber., 1895, 88, 869) or phenyl and β-naphthyl (Ber., 1897, 30, 1783), in which the difference of the basicity of the two nitrogen atoms can be only small, a mixture of the two possible methyl derivatives is obtained.

The negative nature of a nitrogen atom in an unsaturated heterocyclic nucleus, as exemplified by the ease with which the hydrogen atom of the group –NH– is substituted by metals, is well known, as is also the increase of the basicity of cyclic compounds on reduction. . To quote one example of this : diphenyltriazole, $CPh{\displaystyle \mathop{<}^{N \cdot N}_{NH}}CPh$ (Pinner, Annalen, 1897, 299, 255), has hardly any basic properties, being soluble in dilute alkali hydroxides, but insoluble in dilute acids ; whereas diphenyldihydrotriazole, $CPh{\displaystyle \mathop{<}^{N \cdot NH}_{NH}}CHPh$ (Pinner, loc. cit., p. 266), is a strong base, forming with hydrochloric, nitric, and acetic acids stable salts which are not decomposed by water.

Of the two possible "partially cyclic" amidines,

I. $\begin{matrix} CH{-}S \\ Cme{\cdot}N \end{matrix}{>}C{\cdot}NR{\cdot}Alk$ and   II. $\begin{matrix} CH{-}{-}S \\ CMe{\cdot}NAlk \end{matrix}{>}C{:}NR,$

the alkyl group is in the more negative position in II, and therefore, 2-phenylimino-3 : 4-dimethyl-2 : 3-dihydrothiazole is obtained on
methylation of the base

III. $\begin{matrix} CH{-}S \\ Cme{\cdot}N \end{matrix}{>}C{\cdot}NHR$ or   IV. $\begin{matrix} CH{-}{-}S \\ Cme{\cdot}NH \end{matrix}{>}C{:}NR,$

which has $R = C_6H_5$.

On the other hand, of the alkyl derivatives,

V. $\begin{matrix} CHMe{\cdot}S \\ CH_2{-}N \end{matrix}{>}C{\cdot}NR{\cdot}Alk$ or   VI. $\begin{matrix} CHMe{-}{-}S \\ CH_2{\cdot}NAlk \end{matrix}{>}C{:}NR,$

derived from Gabriel and Prager's bases,

VII. $\begin{matrix} CHMe{\cdot}S \\ CH_2{-}N \end{matrix}{>}C{\cdot}NHR$ or   VIII. $\begin{matrix} CHMe{-}S \\ CH_2{\cdot}NH \end{matrix}{>}C{:}NR,$

the alkyl is probably in the more negative position in V if $R = $ an
aryl group, but in VI if R is a hydrogen atom or an alkyl group;
hence Gabriel's methylated bases are of the latter, but Prager's of
the former type.

The constitution of the derivative obtained on alkylation of an
amidine having been determined, it is assumed usually that the
constitution of the amidine is to be represented by placing a hydrogen
atom in the position taken up by the alkyl. Thus, according to
von Pechmann (*Ber.*, 1897, **30**, 1781), benzphenylamidine must be

$C_6H_5{\cdot}C{<}\begin{matrix} NH \\ NHPh \end{matrix}$, because on methylation it yields the derivative

$C_6H_5{\cdot}C{<}\begin{matrix} NH \\ NMePh \end{matrix}$. If, however, methylation of an amidine takes

place not by direct substitution, but by addition of methyl iodide and
subsequent elimination of hydrogen iodide, the intermediate additive
product in the formation of the methyl derivative of benzphenyl-

amidine must be $C_6H_5{\cdot}CI{<}\begin{matrix} NH_2 \\ NMePh \end{matrix}$ or $C_6H_5{\cdot}C{<}\begin{matrix} NH_2 \\ NMePhI \end{matrix}$ (compare

Beckmann and Fellrath, *Annalen*, 1893, **273**, 24), which could be
formed by addition of methyl iodide to an amidine only of the

constitution $C_6H_5{\cdot}C{<}\begin{matrix} NH_2 \\ NPh \end{matrix}$.

This view of the mechanism of the alkylation of amidines is in
agreement with the behaviour of ethylene-ψ-thiocarbamide and of
propylene-ψ-thiocarbamide. Gabriel, having determined the constitution of their methyl derivatives (see p. 60), represented these bases by
the formulæ

$$\begin{array}{c}\text{CH}_2\text{---S}\\ \text{CH}_2\text{·NH}\end{array}\!\!\!>\!\!\text{C:NH} \quad \text{and} \quad \begin{array}{c}\text{CHMe·S}\\ \text{CH}_2\text{·NH}\end{array}\!\!\!>\!\!\text{C:NH}.$$

Later, Gabriel and Leupold (*Ber.*, 1898, **31**, 2832) treated these two bases with nitrous acid in benzene solution and obtained the derivatives

$$\begin{array}{c}\text{CH}_2\text{-S}\\ \text{CH}_2\text{·N}\end{array}\!\!\!>\!\!\text{C·C}_6\text{H}_5 \quad \text{and} \quad \begin{array}{c}\text{CH}_2\text{-S}\\ \text{CH}_2\text{·N}\end{array}\!\!\!>\!\!\text{C·NO}_2,$$

and

$$\begin{array}{c}\text{CHMe·S}\\ \text{CH}_2\text{---N}\end{array}\!\!\!>\!\!\text{C·C}_6\text{H}_5 \quad \text{and} \quad \begin{array}{c}\text{CHMe·S}\\ \text{CH}_2\text{---N}\end{array}\!\!\!>\!\!\text{C·NO}_2,$$

respectively, probably owing to the intermediate formation of diazo-compounds, pointing to the constitutions $\begin{array}{c}\text{CH}_2\text{-S}\\ \text{CH}_2\text{·N}\end{array}\!\!\!>\!\!\text{C·NH}_2$ and $\begin{array}{c}\text{CHMe·S}\\ \text{CH}_2\text{---N}\end{array}\!\!\!>\!\!\text{C·NH}_2$ for the two bases. These constitutions are in agreement with the imino-formulæ for the methyl derivatives if the methylation takes place through the intermediate formation of the additive products

$$\begin{array}{c}\text{CH}_2\text{---S}\\ \text{CH}_2\text{·NMe}\end{array}\!\!\!>\!\!\text{Cl·NH}_2 \quad \text{and} \quad \begin{array}{c}\text{CHMe---S}\\ \text{CH}_2\text{·NMe}\end{array}\!\!\!>\!\!\text{Cl·NH}_2$$

or

$$\begin{array}{c}\text{CH}_2\text{---S}\\ \text{CH}_2\text{·N(MeI)}\end{array}\!\!\!>\!\!\text{C·NH}_2 \quad \text{and} \quad \begin{array}{c}\text{CHMe---S}\\ \text{CH}_2\text{·N(MeI)}\end{array}\!\!\!>\!\!\text{C·NH}_2.$$

Bamberger and Lorenzen (*Annalen*, 1893, **273**, 274) have pointed out that the addition of methyl iodide takes place to a tertiary, in preference to a primary or secondary, nitrogen atom. Owing to these considerations, we have ascribed formulæ of the type III to the amidines which yield the alkyl derivatives II, VII to those which form the alkyl derivatives VI, and VIII to the parent bases of the compounds V.

The foregoing remarks apply only to the direct alkylation of amidines by the action of alkyl haloids at the laboratory or higher temperature with or without a solvent, and not to indirect alkylation by the action of alkyl haloids on a metallic derivative of the amidine. With the idea that methylation by the latter method might lead to the formation of the isomeric series of alkyl derivatives, we digested the silver derivatives of the 2-arylimino-5-methyltetrahydrothiazoles VIII with one molecule of methyl iodide, but without obtaining more than traces of oily products which might be the alkylated amidines, the original base being recovered almost entirely in each case.

Similarly, Bamberger and Lorenzen (*loc. cit.*, p. 282) found that only very small amounts of 1 : 2 : 5-trimethylbenzimidazole were

formed by the action of methyl iodide on the silver derivative of
2 : 5-dimethylbenzimidazole, whilst Meldola, Eyre, and Lane (Trans.,
1903, **83**, 1185) obtained the same alkyl derivatives from ethenyldi-
aminonaphthalenes by the direct and the indirect methods of alkylation, but in much the poorer yields by the latter process. We suggest
that in reality no alkylation of these amidines takes place by the
indirect method, the small amounts of alkyl derivatives formed being
due to interaction of the alkyl iodide with the liberated amidine,
especially as the best yields by the indirect method appear to be
obtained when, as in the experiments of Meldola, Eyre, and Lane (*loc.
cit.*, p. 1193), an excess of the alkyl haloid is employed.

On the other hand, 2-acetylimino-3 : 4-dimethyl-2 : 3-dihydrothi-

azole, $\overset{\text{CH}\text{---}\text{S}}{\underset{\text{CMe·NMe}}{}}\!\!\!>\!\!\text{C:NAc}$, is formed in good yield by the indirect

method from the acetyl derivative of 2-amino-4-methylthiazole, as was
2-acetylimino-5-phenyl-3-methyl-2 : 3-dihydrothiodiazole,

$$\text{C}_6\text{H}_5\!\cdot\!\text{C}\!\!\ll\!\!\overset{\text{N·NMe}}{\underset{\text{S}}{}}\!\!\!>\!\!\text{C:NAc,}$$

from 2-amino 5-phenylthiodiazole, studied by Young and Eyre (Trans.,
1901, **79**, 54), whereas attempts to methylate the last substance by
the direct method met with no success. It may be that the greatly
diminished basicity of the nitrogen atom of the side-chain consequent
on the introduction of the acetyl group causes isomeric change to take
place :

$$\overset{\text{CH---S}}{\underset{\text{CMe·N}}{}}\!\!\!>\!\!\text{C·NH}_2 \;\rightarrow\; \overset{\text{CH---S}}{\underset{\text{CMe·N}}{}}\!\!\!>\!\!\text{C·NHAc} \;\rightarrow\; \overset{\text{CH---S}}{\underset{\text{CMe·NH}}{}}\!\!\!>\!\!\text{C:NAc,}$$

the hydrogen atom being less labile when attached to the less negative
atom, the subsequent direct substitution, on treatment of the silver
derivative with methyl iodide,

$$\overset{\text{CH---S}}{\underset{\text{CMe·NAg}}{}}\!\!\!>\!\!\text{C:NAc} \;\rightarrow\; \overset{\text{CH---S}}{\underset{\text{CMe·NMe}}{}}\!\!\!>\!\!\text{C:NAc,}$$

being due to the increased acidity of the molecule.

EXPERIMENTAL.

2-Aminothiazoles and 2-Imino-2 : 3-dihydrothiazoles.

2-Anilino-4-methylthiazole was prepared by Hantzsch and Weber
(*Ber.*, 1887, **20**, 3130) by the action of aniline on 2-hydroxy-4-methyl-
thiazole, and was found by these authors to melt at 117°. Traumann
(*loc. cit.*), who prepared the base by acting on phenylthiocarbamide
with chloroacetone, found it to melt at 115°. On repeating the

preparation by Traumann's method, we obtained a substance which crystallised from dilute alcohol in long needles and melted at 117—118°. When boiled with acetic anhydride and sodium acetate, the base formed an *acetyl* derivative, which was readily soluble in alcohol or benzene, but only moderately so in light petroleum ; from the last solvent it crystallised, on slow evaporation, in clusters of soft, white, silky, matted needles which melted at 114·5°.

0·1655 gave 17·0 c.c. moist nitrogen at 16° and 755 mm. N = 11·95.

$C_{10}H_9N_2S\cdot C_2H_3O$ requires N = 12·07 per cent.

2-*Phenylimino*-3 : 4-*dimethyl*-2 : 3-*dihydrothiazole,* $\begin{array}{c} CH\text{----}S \\ CMe\cdot NMe \end{array}\!\!>\!C\!:\!NPh.$

2-Anilino-4-methylthiazole was heated with methyl iodide and methyl alcohol in a closed vessel for one hour in the water-bath, and the product, after being diluted with water and boiled to expel the methyl alcohol, made alkaline and extracted with benzene ; the oily residue obtained on drying and distilling the benzene residue was allowed to solidify slowly in a desiccator over soda-lime. The base prepared in this manner was moderately soluble in warm light petroleum, from its solution in which it separated in small, white crystals melting at 65—66°.

0·1458 gave 17·4 c.c. moist nitrogen at 14·5° and 755 mm. N = 13·93.

$C_{11}H_{12}N_2S$ requires N = 13·72 per cent.

The *platinichloride,* formed by adding platinic chloride to the solution of the base in warm dilute hydrochloric acid, crystallised from the solution on cooling, and after being dried at 105° melted at 189—190°.

0·2446 gave 0·0584 Pt. Pt = 23·88.

$(C_{11}H_{12}N_2S)_2,H_2PtCl_6$ requires Pt = 23·83 per cent.

Hydrolysis of 2-*Phenylimino*-3 : 4-*dimethyl*-2 : 3-*dihydrothiazole.*

The methylated base was heated with concentrated hydrochloric acid in a sealed tube at 245—250° for four and a half hours. After concentration on the water-bath, during which operation considerable quantities of hydrogen sulphide were evolved, an excess of potassium hydroxide was added and the product distilled in a current of steam, any alkaline vapours which passed through the receiver being retained by an acid trap. The distillate, after acidification and concentration, was again made alkaline and boiled in a reflux apparatus until no alkaline vapours could be detected on removal of the hydrochloric acid trap. The residual liquid in the boiling flask gave the *isonitrile* reaction for primary amines and the bleaching powder reaction for aniline, whilst the solution from the hydrochloric acid trap gave only the first of these two reactions, but when made alkaline smelt strongly of methylamine. In one experiment, a small amount of a white, solid substance separated from the distillate; it melted at 78° and was possibly 2-oxy-3 : 4-dimethyl-2 : 3-dihydrothiazole.

2-*Allylamino*-4-*methylthiazole*, $\begin{matrix} CH—S \\ \vert\vert \quad \Large{>} \\ CMe·N \end{matrix}$C·NH·C₃H₅.

The hydrochloride of this base was formed by gradually adding 1 mol. of chloroacetone to 1 mol. of allylthiocarbamide; as the reaction takes place with considerable development of heat, it was necessary to cool the mixture with water. The cold product was treated with aqueous potassium hydroxide, and the solid thus obtained recrystallised from light petroleum, when it formed long, white, silky needles. The base, which was observed to have a slight odour resembling that of thyme, melted at 40—41°.

0·2572 gave 40·9 c.c. moist nitrogen at 15° and 753 mm.   N = 18·44.
0·1780 „   0·2640 BaSO₄.   S = 20·66.
C₇H₁₀N₂S requires N = 18·18 ; S = 20·78 per cent.

The *acetyl* derivative, formed by boiling 2-allylamino-4-methyl-thiazole with acetic anhydride, was obtained as an oil which solidified slowly to a crystalline mass ; it was readily soluble in benzene, but only moderately so in light petroleum, from which it crystallised in thin, square plates. It melted at 36—37° and had, especially before recrystallisation, an odour resembling that of impure acetamide.

0·2590 gave 0·3040 BaSO₄.   S = 16·12.
C₇H₉N₂S·C₂H₃O requires S = 16·33 per cent.

2-*Allylimino*-3 : 4-*dimethyl*-2 : 3-*dihydrothiazole*, $\begin{matrix} CH——S \\ \vert\vert \quad \Large{>} \\ CMe·NMe \end{matrix}$C:N·C₃H₅.

The hydriodide formed by heating 2-allylamino-4-methylthiazole with methyl iodide and methyl alcohol in a closed vessel at 100° separated from the mixture after some days in almost colourless, large, prismatic plates. When dissolved in a small quantity of alcohol, in which it was readily soluble, and precipitated by addition of ether, the salt was obtained as a white, crystalline meal, which melted at 116—117°.

0·2035 gave 0·1616 AgI.   I = 42·94.
C₈H₁₂N₂S,HI requires I = 42·89 per cent.

On treatment of the hydriodide with potassium hydroxide in concentrated aqueous solution, extraction of the product with benzene, and evaporation of the extract, the free base was obtained as a slightly red, viscid oil possessing a characteristic odour; it was readily soluble in dilute acids, and formed a crystalline *platinichloride*, which dissolved to only a slight extent in hot dilute hydrochloric acid.

Hydrolysis of 2-Allylimino-3 : 4-dimethyl-2 : 3-dihydrothiazole.

The methylated base was heated with concentrated hydrochloric acid in a sealed tube at 230—240° for five hours. The product was concentrated, made alkaline, and distilled in a current of steam; the distillate was treated in the same manner as that obtained by the hydrolysis of 2-phenylimino-3 : 4-dimethyl-2 : 3-dihydrothiazole. Both the residual liquid from the reflux apparatus and the solution from the hydrochloric acid trap gave the *iso*nitrile reaction, showing that the hydrolysis had resulted in the formation of two primary amines.

2-*Acetylimino*-4-*methyl*-2 : 3-*dihydrothiazole*,

$$\begin{array}{c} CH \!-\!\!-\! S \\ \| \quad\quad\quad \\ CMe \cdot NH \end{array} \!\!\! > \!\! C\dot{:}NAc.$$

2-Amino-4-methylthiazole hydrochloride was prepared by the action of chloroacetone on thiocarbamide (Traumann, *loc. cit.*), and the base liberated by addition of concentrated aqueous potassium hydroxide and extracted with benzene. On evaporation of the extract, there was obtained a slightly pink oil, which, when cooled over calcium chloride in a desiccator, solidified to a white, slightly hygroscopic solid melting at 42°. The acetyl derivative, formed by boiling the anhydrous base with 1 mol. of acetic anhydride in a reflux apparatus, is moderately soluble in warm water, from which it separates on cooling in shining, colourless crystals. The melting point was found to be 134°, as given by Traumann.

On addition of 1 mol. of silver nitrate and 1 mol. of ammonia in aqueous solution to its solution in alcohol, the acetyl compound formed a *silver* derivative as a dense, white precipitate, which, after being well washed and dried at 100°, was fairly stable to light.

0·1962 gave 0·0809 Ag. Ag = 41·23.

$C_4H_4N_2SAg \cdot C_2H_3O$ requires Ag = 41·03 per cent.

When heated with 1 mol. of methyl iodide in methyl-alcoholic solution in a closed vessel at 100°, the silver derivative forms 2-*acetylimino*-3 : 4-*dimethyl*-2 : 3-*dihydrothiazole*,

$$\begin{array}{c} CH \!-\!\!-\! S \\ \| \quad\quad\quad \\ CMe \cdot NMe \end{array} \!\!\! > \!\! C\dot{:}NAc,$$

F 2

which, on evaporation of the methyl-alcoholic solution and re-crystallisation of the residue, was obtained in white, nodular, crystalline aggregates which melted at about 80°, or sharply at 113°, after being dried over sulphuric acid in a vacuum (Hantzsch and Weber, *Ber.*, 1887, **20**, 3124).

Methylation of 2-Methylamino-4-methylthiazole.

The hydrochloride of 2-methylamino-4-methylthiazole was prepared by slowly mixing chloroacetone and methylthiocarbamide in molecular proportions; it was found necessary to keep the reacting mixture cooled by means of ice-water. The base was liberated by treatment of its hydrochloride with aqueous potassium hydroxide, and recrystal-lised from benzene, from which it separated as a slightly pink oil; this solidified slowly, forming long, white needles, which melted at 64° and dissolved readily in methyl alcohol or light petroleum. Traumann (*loc. cit.*) gives the melting point of this base as 42°.

The result obtained from the sulphur determination, together with the formation of 2-methylimino-3 : 4-dimethyl-2 : 3-dihydrothiazole, is sufficient proof of the constitution of our base.

0·1962 gave 0·3642 $BaSO_4$.   S = 25·52.

$C_5H_8N_2S$ requires S = 25·06 per cent.

When heated with 1 mol. of methyl iodide in methyl-alcoholic solution in a closed vessel at 100°, 2-methylamino-4-methylthiazole yielded 2-methylimino-3 : 4-dimethyl-2 : 3-dihydrothiazole hydriodide,

$$\begin{matrix} CH\text{------}S \\ \| \qquad\quad > C\text{:}NMe,HI, \\ CMe\cdot NMe \end{matrix}$$ which separated in stellate aggregates of

crystals melting at 54°, or at 164° after drying at 105°.

Hantzsch and Weber, who prepared this salt by heating 2-amino-thiazole with 2 mols. of methyl iodide (*Ber.*, 1887, **20**, 3123), found it to melt at 54°, or when anhydrous at 155°, whilst Traumann (*loc. cit.*), who obtained the base by treating *s*-dimethylthiocarbamide with chloroacetone, gives the melting point of the anhydrous hydriodide as 164°.

2-Iminotetrahydrothiazoles and 2-Amino-4 : 5-dihydrothiazoles.

2-Phenylimino-5-methyltetrahydrothiazole,

$$\begin{matrix} CHMe\text{---}S \\ \| \qquad\quad > C\text{:}NPh. \\ CH_2\cdot NH \end{matrix}$$

The hydrochloride, prepared by heating *s*-phenylallylthiocarbamide with concentrated hydrochloric acid in a closed vessel at 100°, was treated with concentrated aqueous potassium hydroxide, and the product recrystallised from hot dilute alcohol. The base obtained in this manner crystallised in needles and melted at 117° (Prager, *loc. cit.*).

0·1673 gave 0·2044 $BaSO_4$. $S = 16·79$.

0·1577 ,, 19·6 c.c. moist nitrogen at 11·5° and 755 mm. $N = 14·71$.

$C_{10}H_{12}N_2S$ requires $S = 16·67$; $N = 14·58$ per cent.

The base formed a picrate, which crystallised from dilute alcohol in rough, yellow needles and melted at 154°.

0·1378 gave 20·7 c.c. moist nitrogen at 19° and 725 mm. $N = 16·46$.

$C_{10}H_{12}N_2S,C_6H_3O_7N_3$ requires $N = 16·63$ per cent.

When shaken with acetic anhydride, the base dissolved with slight development of heat, forming a clear solution; after some time, the mixture was dissolved in ether, and the ethereal solution was washed with aqueous potassium carbonate, dried, and distilled on the water-bath. The residual oil gradually solidified to a crystalline mass which melted at 47°.

0·1808 gave 0·4069 CO_2 and 0·0993 H_2O. $C = 61·38$; $H = 6·10$.

0·1552 ,, 0·1557 $BaSO_4$. $S = 13·79$.

$C_{10}H_{11}N_2S·C_2H_3O$ requires $C = 61·54$; $H = 5·98$; $S = 13·68$ per cent.

The acetyl derivative was readily soluble in alcohol, ether, or benzene, less so in light petroleum, and separated from its solutions as an oil, which rapidly solidified on addition of a small quantity of the crystalline substance.

When dissolved in alcohol and poured into an alcoholic solution of silver nitrate, the base formed the *silver* derivative, $C_{10}H_{11}N_2SAg$, which was obtained as a white precipitate. After being washed with dilute alcohol and dried at 90°, the silver derivative detonated when heated on a piece of porcelain over the Bunsen flame, whilst when heated in a capillary tube it blackened and commenced to melt at 130°.

0·2473 gave 0·0890 Ag. $Ag = 35·99$.

$C_{10}H_{11}N_2SAg$ requires $Ag = 36·08$ per cent.

With the object of obtaining possibly a methyl base isomeric with that prepared by Prager, the silver derivative was warmed with 1 mol. of methyl iodide in methyl-alcoholic solution and the product evaporated and shaken with benzene. The base obtained on evaporation of the benzene solution was converted into its picrate, which crystallised in rough, yellow needles, melted at 154°, and gave analytical results agreeing with those required by the picrate of 2-phenylimino-5-methyltetrahydrothiazole.

0·1279 gave 18·7 c.c. moist nitrogen at 17° and 756 mm. $N = 16·86$.
$C_{10}H_{12}N_2S,C_6H_3O_7N_3$ requires $N = 16·63$ per cent.
$C_{10}H_{11}N_2SMe,C_6H_3O_7N_3$ „ $N = 16·09$ „ „

2-*Phenylmethylamino-5-methyl-4 : 5-dihydrothiazole,*

$$\begin{array}{c} \text{CHMe·S} \\ | \qquad\qquad \diagdown \\ \text{CH}_2\text{—N} \end{array}\!\!\!> \text{C·NMePh.}$$

As the action of methyl iodide on the silver derivative did not lead to the formation of a methyl base, 2-phenylimino-5-methyltetrahydrothiazole was heated with 1 mol. of methyl iodide in methyl-alcoholic solution in a closed vessel at 100°. The base was mixed also with an excess of methyl iodide, and the mixture was allowed to stand overnight in a flask cooled by ice-water (Prager, *loc. cit.*, p. 2997). By these two methods, the same methyl base was obtained as an oil, readily soluble in alcohol, ether, benzene, or dilute acids.

The *platinichloride* of the methyl base crystallised in salmon-coloured, nodular aggregates of plates and melted at 184°.

A.* 0·4758 gave 0·1132 Pt. $Pt = 23·79$.
B.   0·3254 „   0·0775 Pt. $Pt = 23·81$.
$(C_{11}H_{14}N_2S)_2,H_2PtCl_6$ requires $Pt = 23·69$ per cent.

The *picrate* of the methyl base melted under boiling water, in which it was slightly soluble, crystallising on cooling in soft, yellow needles; it dissolved readily in alcohol, from which, on dilution, it crystallised in clusters of yellow needles. It melted at 114—115°; Prager gives its melting point as 125°.

A. 0·1180 gave 17·1 c.c. moist nitrogen at 18° and 740 mm. $N = 16·28$.
B. 0·1525 „ 21·8 „   „   „   19° „ 751 mm. $N = 16·23$.
$C_{11}H_{14}N_2S,C_6H_3O_7N_3$ requires $N = 16·09$ per cent.

2-*Phenylethylamino-5-methyl-4 : 5-dihydrothiazole,*

$$\begin{array}{c} \text{CHMe·S} \\ | \qquad\qquad \diagdown \\ \text{CH}_2\text{—N} \end{array}\!\!\!> \text{C·NEtPh.}$$

2-Phenylimino-5-methyltetrahydrothiazole was heated with 1 mol. of ethyl iodide in ethyl-alcoholic solution in a closed vessel at 100° and the product boiled with a small quantity of water, made alkaline, and extracted with benzene. On evaporation of the benzene, the *ethyl* base was obtained as an oil, which was readily soluble in alcohol or ether, less so in light petroleum, and only sparingly so in water, and had the peculiar odour characteristic of the group of compounds. On addition

* The analyses A and B given for the platinichloride and picrate are for the salts of the base prepared by the first and second methods of methylation respectively.

of platinic chloride to its solution in dilute hydrochloric acid, the base formed the *platinichloride*, which separated as a reddish-yellow, crystalline powder and melted and decomposed at 156°.

0·6488, dried at 105°, gave 0·1488 Pt.   Pt = 22·93.

$(C_{12}H_{16}N_2S)_2,H_2PtCl_6$ requires Pt = 22·94 per cent.

Formation of β-Methyltaurine from 2-Phenylethylamino-5-methyl-4 : 5-dihydrothiazole.

The ethyl base was oxidised with potassium chlorate in slightly warm hydrochloric acid, and the product evaporated to dryness and extracted with a mixture of alcohol and ether.   The residue from this extract was heated with concentrated hydrochloric acid in a sealed tube at 150—200°, and the solution filtered from tarry matters and evaporated to dryness.   On extracting the dried product with absolute alcohol, a white residue was obtained, which, after recrystallisation from 90 per cent. alcohol, separated from a small quantity of warm water in transparent, rhombic plates and agreed in its other properties with those of β-methyltaurine as described by Gabriel (*Ber.*, 1889, **22**, 2984).   When powdered and heated in a capillary tube, β-methyltaurine melts at 284—285°.

0·1732 gave 0·1638 CO_2 and 0·1015 H_2O.   C = 25·79 ; H = 6·51.

0·1845 „   0·3143 $BaSO_4$.   S = 23·39.

0·1936 „   17·5 c.c. moist nitrogen at 17° and 742 mm. N = 10·24.

$NH_2·CH_2·CHMe·SO_3H$ requires C = 25·90 ; H = 6·48 ; N = 10·07 ;
S = 23·02 per cent.

s-p-*Tolylallylthiocarbamide*, $C_6H_4Me:NH·CS·NH·C_3H_5$.

This substance was prepared by boiling *p*-toluidine with a slight excess of allylthiocarbimide in alcoholic solution in a reflux apparatus. On removal of the alcohol and excess of the thiocarbimide by distillation in a current of steam, the product separated as a heavy oil, which solidified on cooling and, after purification by repeated crystallisation from dilute alcohol, was obtained in nodular aggregates of almost colourless needles which melted at 98°.

0·2011 gave 0·2304 $BaSO_4$.   S = 15·75.

$C_{11}H_{14}N_2S$ requires S = 15·53 per cent.

2-p-*Tolylimino-5-methyltetrahydrothiazole,*

$$\begin{matrix} CHMe\cdot S \\ | \\ CH_2\cdot NH \end{matrix}\Big> C\!:\!N\cdot C_7H_7.$$

s-*p*-Tolylallylthiocarbamide was converted into the corresponding tetrahydrothiazole by the action of concentrated hydrochloric acid at 100° under pressure. The clear solution thus obtained was evaporated to dryness, the residue treated with aqueous potassium hydroxide, and the liberated base purified by recrystallisation from alcohol.

2-*p*-Tolylimino-5-methyltetrahydrothiazole crystallised from dilute alcohol in small, shining plates or broad needles, or from ether in diamond-shaped plates, and melted at 106°.

0·1562 gave 0·3664 CO_2 and 0·0970 H_2O.   C = 63·97; H = 6·90.

0·1887   „   22·2 c.c. moist nitrogen at 16° and 750 mm.   N = 13·49.

0·1759   „   0·2001 $BaSO_4$.   S = 15·64.

$C_{11}H_{14}N_2S$ requires C = 64·08; H = 6·80; N = 13·59; S = 15·53 per cent.

The base dissolved readily in dilute hydrochloric acid, and on addition of platinic chloride formed the *platinichloride,* which crystallised in small, rough, thick, yellow needles, and after being dried at 110° melted and decomposed at 204°.

0·5061 gave 0·1199 Pt.   Pt = 23·69.

$(C_{11}H_{14}N_2S)_2,H_2PtCl_6$ requires Pt = 23·74 per cent.

The *acetyl* derivative, $C_{11}H_{13}N_2S\cdot C_2H_3O$, was formed by gently warming the base with acetic anhydride; it crystallised from light petroleum in white prisms and melted at 61°.

0·1689 gave 0·1606 $BaSO_4$.   S = 13·08.

$C_{13}H_{16}ON_2S$ requires S = 12·90 per cent.

When boiled with dilute hydrochloric acid, the acetyl derivative gradually dissolved, and on addition of potassium hydroxide to the solution the parent base, melting at 106°, was precipitated.

2-p-*Tolylmethylamino-5-methyl-*4 : 5-*dihydrothiazole,*

$$\begin{matrix} CHMe\cdot S \\ | \\ CH_2 - N \end{matrix}\Big> C\cdot NMe\cdot C_7H_7.$$

On shaking 2-*p*-tolylimino-5-methyltetrahydrothiazole with slightly more than one mol. of methyl iodide, the mixture became warm and changed to a clear liquid which gradually solidified. The product was boiled with a small quantity of water to free it from the excess of methyl iodide, made alkaline, and extracted with ether. The methyl base, which was obtained as an oil on evaporating the ethereal extract, dissolved readily in dilute hydrochloric acid, alcohol, ether, or benzene, and had the characteristic odour of the 2-amino-4 : 5-dihydrothiazoles.

The *platinichloride* melted and decomposed at 110°.

0·2213 gave 0·0513 Pt.　Pt = 23·18.

$(C_{12}H_{16}N_2S)_2,H_2PtCl_6$ requires Pt = 22·94 per cent.

Oxidation of 2-p-*Tolylmethylamino*-5-*methyl*-4 : 5-*dihydrothiazole.*

The *p*-tolylmethylamino-base was oxidised with potassium chlorate and hydrochloric acid, and the product treated as described under the oxidation of the phenylethylamino-base. The final residue, after extraction with absolute alcohol, crystallised from water in transparent plates, melted at 284—285°, and on analysis gave. numbers agreeing with those required for β-methyltaurine.

0·1821 gave 0·3088 $BaSO_4$.　S = 23·32.

$C_3H_9O_3NS$ requires S = 23·02 per cent.

The absolute alcoholic extract was evaporated with hydrochloric acid, treated with an excess of aqueous potassium hydroxide, and distilled in a current of steam. On addition of hydrochloric acid and platinic chloride to the distillate, the *platinichloride* of methyl-*p*-toluidine separated as a red, crystalline powder, which was dried at 110°.

0·4109 gave 0·1222 Pt.　Pt = 29·74.

$(C_6H_4Me\cdot NHMe)_2,H_2PtCl_6$ requires Pt = 29·91.

2-p-*Tolylethylamino*-5-*methyl*-4 : 5-*dihydrothiazole,*

$$\begin{matrix} CHMe\cdot S \\ | \\ CH_2—N \end{matrix}\!\!>\!C\cdot NEt\cdot C_7H_7.$$

2-p-Tolylimino-5-methyltetrahydrothiazole was heated with slightly more than one mol. of ethyl iodide in ethyl-alcoholic solution in a closed vessel for one hour at 100°, and the product boiled with water, made alkaline and extracted with ether. The ethyl base was obtained, on evaporation of the ethereal solution, as an oil which was easily soluble in alcohol, benzene, or dilute acids.

The *platinichloride* separated from its solution in warm dilute hydrochloric acid in orange-coloured crystals, which melted and decomposed at 189—190°.

0·6832 gave 0·1505 Pt.　Pt = 22·03.

$(C_{13}H_{18}N_2S)_2,H_2PtCl_6$ requires Pt = 22·21 per cent.

Oxidation of 2-p-Tolylethylamino-5-methyl-4 : 5-dihydrothiazole.

The *p*-tolylethylamino-base was oxidised with potassium chlorate and dilute hydrochloric acid, and the product hydrolysed with concentrated hydrochloric acid at 160—200°. The residue obtained on evaporation was extracted with absolute alcohol and recrystallised from water, when it formed transparent plates and melted at 284—285° (β-methyltaurine).

0·1932 gave 0·3286 BaSO$_4$.   S = 23·38.

$C_3H_9O_3NS$ requires S = 23·02 per cent.

The alcoholic extract yielded the *platinichloride* of ethyl-*p*-toluidine.

0·2336 gave 0·0666 Pt.   Pt = 28·51.

$(C_6H_4Me \cdot NHEt)_2, H_2PtCl_6$ requires Pt = 28·68 per cent.

2-o-*Tolylimino-5-methyltetrahydrothiazole*, $\begin{matrix} CHMe \cdot S \\ | \\ CH_2 \cdot NH \end{matrix} > C:N \cdot C_7H_7$.

This base was prepared according to Prager's directions (*loc. cit.*, p. 2999) from *s-o*-tolylallylthiocarbamide; it crystallised in small plates and melted at 126°.

0·1699 gave 0·1941 BaSO$_4$.   S = 15·71.

$C_{11}H_{14}N_2S$ requires S = 15·53 per cent.

On addition of silver nitrate to its solution in alcohol, the base formed the *silver* derivative as a white precipitate, which was washed and dried at 105°.

0·9561 gave 0·3289 Ag.   Ag = 34·40.

$C_{11}H_{13}N_2SAg$ requires Ag = 34·49 per cent.

The *acetyl* derivative, $C_{11}H_{13}N_2S \cdot C_2H_3O$, was prepared by boiling the base with acetic anhydride and shaking the product with water ; it crystallised from light petroleum in stout prisms, melted at 58°, and was easily soluble in benzene, alcohol, ether, or acetone, but only sparingly so in light petroleum.

0·1529 gave 0·3522 CO$_2$ and 0·0897 H$_2$O.   C = 62·82 ; H = 6·52.

0·2137   ,,   0·2029 BaSO$_4$.   S = 13·05.

$C_{13}H_{16}ON_2S$ requires C = 62·90 ; H = 6·45 ; S = 12·90 per cent.

The acetyl derivative dissolved unchanged in cold dilute hydrochloric acid from its solution, in which it was precipitated unchanged on addition of aqueous sodium hydroxide, but when boiled with the dilute acid it was hydrolysed, and the precipitate obtained on addition of the alkali hydroxide consisted of the parent base melting at 126°.

2-o-*Tolylmethylamino-5-methyl-4 : 5-dihydrothiazole,*

$$\begin{array}{c}\text{CHMe·S} \\ | \qquad\quad\searrow \\ \text{CH}_2\text{—N}\end{array}\!\!>\!\text{C·NMe·C}_7\text{H}_7.$$

This substance was obtained as an oil by shaking 2-o-tolylimino-5-methyltetrahydrothiazole with methyl iodide and liberating the methyl base with aqueous potassium hydroxide. The *platinichloride* melted and decomposed at about 200°.

0·5318 gave 0·1217 Pt. Pt = 22·88.

$(C_{12}H_{16}N_2S)_2,H_2PtCl_6$ requires Pt = 22·94 per cent.

On oxidising the base with potassium chlorate and hydrochloric acid and hydrolysing the product with concentrated hydrochloric acid at 150—200°, we obtained, as did Prager, β-methyltaurine melting at 284—285°.

0·3044 gave 0·5106 BaSO₄. S = 23·31.

$C_3H_9O_3NS$ requires S = 23·02 per cent.

2-o-*Tolylethylamino-5-methyl-4 : 5-dihydrothiazole,*

$$\begin{array}{c}\text{CHMe·S} \\ | \qquad\quad\searrow \\ \text{CH}_2\text{—N}\end{array}\!\!>\!\text{C·NEt·C}_7\text{H}_7.$$

The hydriodide of this base was formed by heating 2-o-tolylimino-5-methyltetrahydrothiazole with ethyl iodide and ethyl alcohol in a closed vessel at 100°. When liberated by means of potassium hydroxide, the base was obtained as an oil which had the characteristic odour and dissolved readily in alcohol, ether, benzene, or dilute acids. The *platinichloride* formed orange-red crystals which melted and decomposed at 203°.

0·7959 gave 0·1764 Pt. Pt = 22·17.

$(C_{13}H_{18}N_2S)_2,H_2PtCl_6$ requires Pt = 22·21 per cent.

The base was oxidised with potassium chlorate and hydrochloric acid, the oxidation product hydrolysed with concentrated hydrochloric acid, and the residue obtained by evaporation extracted with absolute alcohol. The part insoluble in absolute alcohol was β-methyltaurine, as after recrystallisation from 90 per cent. alcohol it melted at 284—285°.

0·1587 gave 0·1499 CO₂ and 0·0945 H₂O. C = 25·76 ; H = 6·62.

0·2173 ,, 19·2 c.c. moist nitrogen at 16° and 754 mm. N = 10·22.

$C_3H_9O_3NS$ requires C = 25·90 ; H = 6·48 ; N = 10·07 per cent.

The absolute alcoholic extract yielded the *platinichloride* of ethyl-*o*-toluidine.

0·2786 gave 0·0800 Pt.   Pt = 28·70.

$(C_6H_4Me\cdot NHEt)_2,H_2PtCl_6$ requires Pt = 28·68 per cent.

We take this opportunity of thanking the Research Fund Committee of the Chemical Society for a grant by which the expenses of this investigation were partly defrayed.

<hr>

X.—*The Influence of Certain Amphoteric Electrolytes on Amylolytic Action.*

By JOHN SIMPSON FORD and JOHN MONTEATH GUTHRIE.

IN a recent publication on Lintner's soluble starch and the estimation of diastatic power (*J. Soc. Chem. Ind.*, 1904, **23**, 414), it was pointed out by one of us that under certain conditions the addition of asparagine to starch solutions undergoing hydrolysis by malt diastase gave rise to an increased production of maltose. From the experimental results obtained by working with starch preparations of varying degrees of purification, it was concluded that this augmentation of the hydrolysis was due, not to a specific action of the amide on the amylase, but to an indirect action in preventing or lessening the inhibitive influence of certain impurities in the starch solutions. It was established that the addition of asparagine to starches containing alkaline impurities increased the maltose production at temperatures above 40°, whereas addition to purer starches decreased the maltose production. It was also noted that asparagine was able to lessen the inhibitory influence of traces of copper, which were found to have a very destructive effect on amylolytic action. As these observations are of considerable interest and physiological import, we have further investigated this action of asparagine and also the influence of certain amino-acids on amylolytic hydrolysis.

Preparation of the Starch.

It was pointed out by one of us (*loc. cit.*) that soluble starch, prepared by Lintner's method or otherwise, is extremely difficult to purify ; it obstinately retains traces of phosphorus compounds which prolonged washing with water does not remove, and which are not readily eliminated by solution and precipitation of the starch.

Certain of the preparations of soluble starch used in this investigation were prepared by Lintner's process as usual, then further purified by solution in water and repeated precipitation by means of alcohol, first in the presence of hydrochloric acid and then without addition of acid. The method is tedious and from twenty to thirty precipitations may be necessary before a neutral product is obtained. We have now found that prolonged digestion and extraction of ordinary preparations from maize with dilute acid (HCl) removes the phosphorus compounds completely. After this treatment and washing with water, a few precipitations with alcohol yield an equally pure starch. The criteria of purity we employ are neutrality to rosolic acid and phenolphthalein, and absence of indications of phosphoric acid to molybdate solution in the ash of 5 grams ignited with sodium carbonate and nitrate. The latter is a severe test and it is not often that a preparation is obtained which does not show a faint coloration to this reagent. The specific conductivity of 2 per cent. solutions of such purified starches runs about 5×10^{-6} reciprocal ohms per c.c. at 25°, so that, although not pure, a close approximation to purity is evident. In connection with the drying of alcohol-precipitated preparations of starch, we have noticed on several occasions that starches which were neutral when tested immediately after filtration, showed faint acidity after drying. This acidity is probably due to slight oxidation of the alcohol ; Duchemin and Dourlen (*Compt. rend.*, 1905, **140**, 1466) have recently shown that oxidation of alcohol takes place more readily than is generally supposed. Whatever the origin of the acidity may be, its formation renders the attainment of neutral preparations somewhat difficult if ordinary methods of drying in air be employed. We have found it convenient to dispense with the final drying in many cases, and also to supplement the method of alcohol precipitation by that of freezing out, the separated starch being sucked free from mother liquor on a Buchner funnel, then dissolved at once in boiling water. The strength of the solution is readily deduced from its specific gravity. Working in this manner we have obtained much purer preparations, the specific conductivity of 2 per cent. solutions being reduced to $1·5 \times 10^{-6}$ at 25°. The dried alcohol-precipitated specimens, when moistened with water, form a jelly-like mass which, on heating, gives a solution of specific rotation $[a]_{D4·0u} = 200—202°$. On hydrolysis with malt extract at 55°, the transformation products are the same as those yielded by ordinary starch paste, having the constants $[a]_D$ 150°, R. 80. Therefore, if ordinary starch is a mixture of amylocellulose and amylopectin (Maquenne and Roux, *Compt. rend.*, 1905, **140**, 1303), our preparations must evidently have retained the same proportion of acid modified amylopectin, notwithstanding the prolonged purification and separation into fractions by alcohol precipitation and freezing out.

Preparation of the Amylase.

Ordinary preparations of malt diastase by Lintner's method (*J. pr. Chem.*, 1886, [ii], **34**, 378) are exceedingly impure, and are as a rule strongly alkaline in reaction. We have endeavoured to prepare purer specimens by modifying Lintner's alcohol 'method, the crude product being dissolved in water, containing potassium dihydrogen phosphate, and reprecipitated by addition of excess· of ammonium sulphate. The precipitate so obtained was dialysed for some days and again precipitated, this time with alcohol. The diastase so obtained had only feeble amylolytic properties, and was still far from pure as regards freedom from mineral substances. As our object was to obtain an enzyme of considerable activity and relative freedom from saline impurity, we did not further pursue this method of preparation. Osborne and Campbell (*J. Amer. Chem. Soc.*, 1895, **17**, 503 ; 1896, **18**, 536) have made elaborate investigations on the chemical nature of amylase, and have prepared specimens of great activity and purity by the methods of " salting out " and dialysis. We therefore employed their methods, and from a quantity of highly active malt extract, kindly presented to us by the Distillers' Co., Ltd., Edinburgh, obtained a small yield of a preparation (F_2) suitable for our purpose. This had a diastatic power of fully 300° Lintner. The specific conductivity of a 2 per cent. solution was $7·0 \times 10^{-4}$ at 25°. As only 1 c.c. of a solution of 5 to 15 milligrams per 100 c.c. was taken for the experiments to be described, the amount of impurity contributed by the enzyme preparation was less than that of the distilled water used.

In addition to the experiments with the purified starch and diastase, we have to record several made with ordinary preparations of Lintner's soluble starch and malt extract, the results of which we will consider first, as this is rendered necessary by a recent publication by J. Effront (*Moniteur Scientifique*, 1904, **61**, 561), in which he traverses the conclusions deduced by one of us (*loc. cit.*), and reiterates his opinion that the accelerating influence of asparagine and certain amino-acids on amylolytic action is a specific one, is independent of the temperature and alkalinity of the medium, and is exercised with all natural starches of whatever origin. The results given in his memoir do not, however, justify this conclusion, as the four starches which he employed are by his own showing obviously very impure. The titration values he records indicate that all the starches were alkaline, whilst the fact that different amounts of maltose were yielded by each starch on treatment with equal amounts of malt extract is conclusive proof that the starches contained varying amounts of impurity. It may be pointed out that his purest preparation (D) is, as regards titration value to rosolic acid, ten times more impure than the most impure starch used as an example in the experiments recorded by one of us (loo. cit.), and, further, it is this starch which gives the smallest maltose production, which result is in our experience indicative of metallic contamination. Notwithstanding this, J. Effront, without making any effort to repeat his work on the lines suggested (loo. cit.) with purer preparations, or to verify or disprove our contention as to the significance of these titration values, or the influence of metallic impurity, seeks to extend his generalisations. We do not for a moment doubt the accuracy of his observations, but the mere repetition of experiments under the same conditions does not add additional value to the conclusions he has formed. In order to elucidate further the important and varying influence of the impurities in starches on amylolytic action, we have prepared several impure, as well as purified specimens, and, as will be seen from the results obtained with the more impure, it is possible to transcend the tenfold increase of maltose production mentioned by J. Effront and other workers. A normal preparation of Lintner's soluble starch was shaken with a natural water containing much calcium carbonate (0·4 gram per litre) and traces of iron salts. The starch (Pa) was then filtered, washed with distilled water, and dried. It contained 0·006 per cent. of iron. With colour indicators, the values per 100 grams were as under :

$$\text{Rosolic acid} \quad \ldots\ldots\ldots\ldots\ldots \quad 7\cdot2 \text{ c.c. } N/10H_2SO_4.$$
$$\text{Phenolphthalein} \quad \ldots\ldots\ldots\ldots \quad 8\cdot0 \text{ c.c. } N/10NaOH.$$

One c.c. of malt extract, 1·2 grams of malt (d. p. 38° L.) to 100 c.c., was added to 70 c.c. of a 1·5 per cent. solution of this starch, one hour at 60·3°.

						Milligrams of maltose per 100 c.c.
Starch and malt extract without addition		....				3·0
,,	,,	plus 15 milligrams of asparagine...				73·0
,,	,,	,, 30	,,	,,	...	73·0
,,	,,	,, 50	,,	,,	...	102·2
,,	,,	,, 70	,,	,,	...	111·0
,,	,,	,, 100	,,	,,	...	114·0

The original starch (P) under like conditions gave a decreased maltose formation in presence of asparagine.

We give below several experiments made with a number of starches of varying degrees of purification.

Influence of Asparagine on "Early" Maltose Production.

Series I.—One c.c. of malt extract, 0·6 gram of malt (d. p. 40°) per 100 c.c., to 70 c.c. of 1·5 per cent. solution of starch, one hour at 59·5°.

					Milligrams of maltose formed.			
					Pβ.	N.	M.	Mz.
Starch solution and malt extract without asparagine					29·2	64·2	55·5	49·6
,,	,,	,,	*plus* 10 milligrams of asparagine		—	46·7	—	12·3
,,	,,	,,	,, 30	,,	,,	64·2	46·7	67·1	12·3
,,	,,	,,	,, 50	,,	,,	67·1	32·1	55·5	12·3
,,	,,	,,	,, 100	,,	,,	55·5	—	29·2	—

The titration values of these starches per 100 grams were as under:

	Rosolic acid.	Phenolphthalein.
Pβ.	10·0 c.c. $N/10H_2SO_4$	8·0N/10NaOH
N.	0·2 ,; N/10NaOH	18·0N/10NaOH
M.	0·3 ,, $N/10H_2SO_4$	14·6N/10NaOH
Mz.	0·1 ,, N/10NaOH	0·2N/10NaOH

Series II.—One c.c. of malt extract, 0·4 gram of malt (d. p. 36°) per 100 c.c., to 70 c.c. of 1·5 per cent. starch solution, one hour at 59°.

					Mz_2.	P.	M.	N.
Starch solution and malt extract without asparagine					29·0	29·2	14·4	29·6
,,	,,	;,	*plus* 35 milligrams of asparagine		17·6	20·4	29·0	23·8
,,	,,	,,	,, 50	,,	,,	14·6	14·6	29·0	18·0

The titration values of these starches were, per 100 grams:

	Rosolic acid.	Phenolphthalein.
Mz_2.	neutral	neutral
P.	2·0 c.c. N/10NaOH	19·0N/10NaOH
M.	0·3 ,, $N/10H_2SO_4$	14·6N/10NaOH
N.	0·2 ,, N/10NaOH	18·0N/10NaOH

The foregoing starches, with the exception of Pa, were free from metallic impurities such as iron or copper. We give below some results with starches containing metallic impurities.

L Starch.

Rosolic acid ....3·0 c.c. N/10NaOH per 100 grams.
Phenolphthalein...10·0 c.c. N/10NaOH ,, ,, ,,
Copper ....0·04 per cent.

One c.c. of malt extract, 4 grams of malt (d. p. 30°) per 100 c.c., to 70 c.c. of 1·5 per cent. solution, one hour at 40°.

	Milligrams of maltose formed.
Starch and malt extract without asparagine	79
,, ,, plus 15 milligrams of asparagine ...	324
,, ,, 30 ,, ,, ...	333
,, ,, 50 ,, ,, ...	330
,, ,, 100 ,, ,, ...	324

Drosten's Starch.

Rosolic acid ....3·0 c.c. $N/10H_2SO_4$ per 100 grams.
Phenolphthalein... 12·0 c.c. $N/10NaOH$,, ,, ,,
Copper ....0·0075 per cent.

One c.c. of malt extract, 4 grams per 100 c.c., to 70 c.c. of 1·5 per cent. solution, thirty-five minutes at 40°.

	Milligrams of maltose formed.
Starch and malt extract without asparagine	84·6
,, ,, plus 75 milligrams of asparagine ...	128·4

Numerous other experiments have yielded similar results, and in conjunction with those already published (loc. cit.) confirm conclusively the opinion expressed there, that when augmentation of diastatic action of malt extract is obtained on the addition of asparagine, the augmentation is due to the influence of the asparagine in lessening the inhibitory effect of alkaline or other impurities present in the starch solutions, and not to a specific action in the amylase under such conditions. The conclusion we arrive at as to the influence of asparagine may be extended to the other substances which J. Effront (loc. cit. and Bull. Soc. chim., 1904, [iii], 31, 1230) states stimulate amylolytic action. He concludes that the amino-group, and not the amide, accelerates the action, because he obtained augmentation with aspartic acid, sarcosine, glycine, alanine, leucine (asparagine), glutaminic acid, and hippuric acid, whereas succinamide, acetamide and its homologues, benzamide, the amines, hydroxylamine, and hydrazine exhibit a retarding influence. It is to be observed that the substances he enumerates in the favouring group are either weak acids or amphoteric compounds.

In the paper already referred to, it was pointed out that asparagine, under the conditions of hydrolysis in question, was able to overcome or lessen the inhibitory effect of traces of copper on amylolytic action. We give in the following table some additional experiments as to its influence on metallic impurities.

Influence of Asparagine on Certain Metallic Impurities.

One c.c. malt extract, 4 grams per 100 c.c., to 70 c.c. of 3 per cent. starch, P solution, one hour at 40°.

Grams of
maltose formed.

Starch solution and malt extract without addition				0 32
,,	plus 0·1 milligram of copper as sulphate			0·06
,,	,, 0·1	,,	,, plus 0·1 gram of asparagine	0·29
,,	,, 0·1	,,	mercury as chloride	0·02
,,	,, 0·1	,,	,, plus 0·1 gram of asparagine	0·01
,,	,, 0·1	,,	mercury as cyanide	0·08
,,	,, 0·1	,,	,, plus 0·1 gram of asparagine	0·03
,,	,, 10·0	,,	copper as aminosuccinamate	0·08
,,	,, 10·0	,,	,, plus 0·1 gram of asparagine	0·27
,,	,, 100·0	,,	asparagine	0·31

These results indicate that the protective influence of asparagine in the case of copper is due to the formation of copper aminosuccinamate and its lessened dissociation in presence of excess of asparagine and its salts, the free copper ions being so reduced in quantity as not to interfere greatly with normal amylolytic action. We have here an explanation of the fact already recorded by us (*J. Soc. Chem. Ind.*, 1905, 24, 605), that traces of copper, which greatly inhibit the amylolytic action of precipitated malt diastase (Lintner's), have much less effect on the activity of malt extract.

It was pointed out (*loc. cit.*) that asparagine at temperatures above 40° reacted more strongly acid to colour indicators; we now supply evidence to show that this acidity is really exhibited by ordinary recrystallised specimens of the amide. This is clearly shown by their action on sucrose solutions at temperatures of 40° and 60°.

Acidic Function of Asparagine ($\mu = 0.50$).

Grams of invert
sugar per 100 c.c.
20 hours at

20 per cent. sucrose solution.			40°.	60°.
50 c.c. of sucrose solution, water to 100 c.c.			0·013	0·019
,,	,,	plus 0·5 gram of asparagine, plus water to 100 c.c.	0·019	0·285
	,, 1·0 ,,	,, ,, ,,	0·020	0·343
..	..	,, 1·2 milligrams of hydrochloric acid, plus water to 100 c.c.	0·146	—
,,	,,	,, 0·6 ,, ,, ,, ,,	—	0·971

The asparagine used in the preceding experiments was purified by recrystallisation from alcohol. The molecular conductivity at 25° for $v = 16$ was 0·50, a value in close agreement with that given by Walden (*Zeit. physikal. Chem.*, 1891, 8, 483). As we now know that the active acidity of such recrystallised asparagine is mainly due to the presence of impurity, we will return to this subject subsequently.

We can infer that ordinary specimens of asparagine in virtue of their acid function overcome the inhibitive effect of hydroxyl ions present in our alkaline starches. We have not, however, so far offered any proof that starches with titration values indicating alkalinity, that is, requiring the addition of acid to bring about neutrality to rosolic acid, are really alkaline. The following experiment shows that such starches at least possess a potential, if not an actual, alkalinity.

To 2 grams of each starch (in 70 c.c. water), 20 c.c. of 10 per cent. sucrose were added and 10 c.c. of dilute hydrochloric acid, equal to 1·2 milligrams of hydrogen chloride. Twenty c.c. of sucrose solution *plus* 10 c.c. of the acid were also made to a similar volume. The solutions (in Jena flasks) were kept for seventeen hours at 60°.

The invert sugar produced was as under :

Starch.	Grams of invert sugar.
Mz2	0·293
Pβ	0·147
M	0·250
Aqueous sucrose *plus* 1·2 milligrams of hydrogen chloride	0·350

The starch Mz2 was neutral to rosolic acid and phenolphthalein ; the values of the other starches have already been given. As the viscosity effect would be the same in each case, the reduction of sugar inversion may be regarded as due to alkalinity or to a lessened dissociation of the acid caused by the salts present. The starches were free from chlorides, the salts being phosphates ; whether at such dilutions double decomposition may be looked for is doubtful. In any case this reduction of the number of free hydrogen ions is for our purpose tantamount to alkalinity. Experiments were made to see if these starches increased the rate of mutarotation of freshly dissolved glucose. The results were negative, the rate of change of rotation being the same with each starch. Possibly this indicates that no free hydroxyl ions are present in the so-called alkaline starches ; something, however, is present which is capable of reducing the amount of free hydrogen ions of the added acid.

We have already shown that ordinary preparations of asparagine exhibit distinctly acid properties at 60°, whereas, allowing for reduced velocity of reaction, at 40° it has practically no acidic function. As amylolytic action is admittedly greatly influenced by the degree of alkalinity or acidity of the medium, J. Effront's contention that the favouring influence of asparagine on amylolytic action is independent of the temperature and degree of alkalinity of the starch becomes untenable. Apart from the question of the active acidity of ordinary asparagine at higher temperatures, this amide, glycine and other amino-acids are amphoteric electrolytes, having potential acid and basic functions, capable of neutralising acid or alkali to an extent dependent on the relative proportions of acting substances and the temperature. This is shown by their influence on amylolytic action, to be described later, and also by the following results obtained by

G 2

methods used by Walker (*Zeit. physikal. Chem.*, 1889, **4**, 389), Winkelblech (*ibid.*, 1901, **36**, 546), and others. The diminution in concentration of $H^{\cdot}$ and OH' ions in solutions of hydrochloric acid and caustic soda (due to salt formation) was observed by measurements of electrical conductivity.

Asparagine and Hydrochloric Acid at 25°.

Concentration.			Molecular conduc- tivity.
N/10 hydrochloric acid			381·6
plus 0·012 mol. of asparagine			345·1
,, 0·025	,,	,,	313·4
,, 0·05	,,	,,	248·4
,, 0·10	,,	,,	150·8
,, 0·20	,,	,,	100·6
,, 0·30	,,	,,	86·3

Asparagine and Caustic Soda at 25°.

Concentration.			Molecular conduc- tivity.
N/10 caustic soda			204·6
plus 0·025 mol. of asparagine			166·6
,, 0·05	,,	,,	133·3
,, 0·10	,,	,,	62·7
,, 0·20	,,	,,	56·7
,, 0·40	,,	,,	53·6

Asparagine and Potassium Chloride at 25°.

Concentration.			Molecular conduc- tivity.
N/10 potassium chloride			128·5
plus 0·05 mol. of asparagine			127·5
,, 0·10	,,	,,	127·0
,, 0·20	,,	,,	125·4
,, 0·40	,,	,,	120·1

Glycine and Hydrochloric Acid at 25°.

Concentration.			Molecular conduc- tivity.
N/10 hydrochloric acid			381·7
plus 0·025 mol. of glycine			303·4
,, 0·05	,,	,,	259·5
,, 0·10	,,	,,	153·3
,, 0·20	,,	,,	102·7
,, 0·40	,,	,,	93·6
,, 0·80	,,	,,	87·6

Glycine and Caustic Soda at 25°.

Concentration.			Molecular conduc- tivity.
N/10 caustic soda			204·6
plus 0·025 mol. of glycine			168·7
,, 0·05	,,	,,	136·0
,, 0·10	,,	,,	66·0
,, 0·20	,,	,,	65·0
,, 0·40	,,	,:	63·5
,, 0·80	,,	,,	62·4

Glycine and Potassium Chloride at 25°.

Concentration.			Molecular conduc- tivity.
N/10 potassium chloride			128·5
plus 0·1 mol. of glycine			127·2

a-Alanine and Hydrochloric Acid at 25°.

Concentration.			
N/10 hydrochloric acid			381·7
plus 0·025 mol. of a-alanine			303·4
,, 0·05	,,	,,	236·5
,, 0·10	,,	,,	138·8
,, 0·20	,,	,,	96·7
,, 0·40	,,	,,	86·1

a-Alanine and Caustic Soda at 25°.

Concentration.	Molecular conductivity.
$N/10$ caustic soda	204·4
plus 0·025 mol. of a-alanine	161·3
,, 0·05 ,, ,,	124·4
,, 0·10 ,, ,,	62·2
,, 0·20 ,, ,,	61·4
,, 0·40 ,, ,,	59·2

a-Alanine and Potassium Chloride at 25°.

Concentration.	Molecular conductivity.
$N/10$ potassium chloride	128·5
plus 0·1 mol. of a-alanine...	127·2

It is not necessary to enter into any general discussion of the above results, which we record simply to show in a qualitative manner the amphoteric nature of these substances. The theoretical and quantitative aspect of the subject is fully developed by Winkelblech (*loc. cit.*) and Walker (*Proc. Roy. Soc.*, 1904, **73**, 155, and **74**, 271).

Purification of Asparagine.

We stated previously that the active acidity exhibited by ordinary specimens of recrystallised asparagine is due to the presence of impurity. This impurity we find is present in all preparations we have examined which have been "purified" by the customary methods of recrystallisation. Walker has recently shown (*loc. cit.*) from theoretical calculations that pure asparagine should have a molecular conductivity of 0·087 at $v = 16$. By crystallisation from water twenty-four times he has prepared a specimen with $\mu = 0·096$, using water of $k = 0·7 \times 10^{-6}$ at 18°. We have prepared asparagine of a similar degree of purity, and find by reducing the duration and temperature of dissolution that it is possible to obtain this purity after about twelve recrystallisations. We dissolve the finely-ground amide in a minimum amount of "conductivity" water at 60—65°, cool rapidly, stirring vigorously so as to obtain a crop of very small crystals. These are freed from the mother liquor and washed with a small quantity of ice-cold water, the treatment being repeated until a product of constant conductivity is obtained. The yield, from the nature of the process, is very small. Asparagine of this purity exhibits practically no active acidity; this is evident from its slight inversion of sucrose in the experiments recorded below, the small fall of angle being due to the production of aspartic acid. We find, under like conditions of heating, an obvious increase in the conductivity of aqueous solutions of asparagine; heating on a water-bath for a short time is also sufficient to cause slight decomposition. This provides an explanation of the difficulty of obtaining pure asparagine by simple recrystallisation; when the substance is dissolved in hot or boiling water, aspartic acid is produced, and probably some of the ammonia also formed is driven off, as ammonium aspartate solutions lose ammonia on boiling. On cooling to crystallise, or on addition of alcohol, it is probable that the aspartic acid forms a salt with the amide, the presence of which, or of the ammonium salt, in small quantity is competent to account for the apparent increase of acidity observed on heating solutions of such preparations of asparagine. Even in the presence of an excess of asparagine, the salt undergoes hydrolytic dissociation when the solutions are heated, giving rise to the presence of free hydrogen ions. Apart from this, if we regard asparagine as an internal ammonium salt, it is possible that its hydrolytic dissociation gives rise to the presence of the traces of free H' ions observed in the solutions of our pure asparagine. At the same time we consider that the marked acid function observed by ourselves and by Degener (*Chem. Centr.*, 1897, **2**, 936) is due mainly to the presence of saline impurity in our preparations.

These observations, whilst they force us to modify somewhat our views as to the influence of pure asparagine on pure amylase and starch, do not invalidate our deductions from the foregoing experiments with more or less impure starches. It is perfectly certain that no one has hitherto worked with such pure asparagine, and opinions as to the influence of this amide on amylolytic action have been deduced from experiments made with ordinary preparations which contain an approximately constant amount of impurity. Further, the potential acid function of the pure substance is capable of neutralising impurities, so we need only modify our views to the extent that pure asparagine added to pure amylase and starch will have little influence, whereas ordinary specimens inhibit the action. The inhibition of action brought about by the addition of asparagine to the transformations with the purer starches recorded in the preceding part of this paper is due to the acid-forming impurity in the amide. Pure asparagine does not retard the hydrolysis, nor does it augment this reaction unless the starch contains certain impurities.

Action of Asparagine, Glycine, and a-Alanine on Sucrose.

Rotation of solutions in a 2-dcm. tube at 16°.

	After 20 hours at	
	40°.	60°.
50 c.c. of 10 per cent. sucrose solution, *plus* 0·375 gram of glycine, diluted to 100 c.c.	6·62	6·61
50 c.c. of 10 per cent. sucrose solution, *plus* 0·445 gram of a-alanine, diluted to 100 c.c.	6·60	6·58
50 c.c. of 10 per cent. sucrose solution, *plus* 0·375 gram of asparagine,* $\mu = 0·10$, diluted to 100 c.c.	6·60	6·50

Action of Asparagine, Glycine, and α-Alanine on Sucrose (continued).

Rotation of solutions in a 2-dcm. tube at 16°.

	After 20 hours at	
	40°.	60°.
50 c.c. of 10 per cent. sucrose solution, *plus* 0·375 gram of asparagine,* $\mu = 0·20$, diluted to 100 c.c.	6·57	6·47
50 c.c. of 10 per cent. sucrose solution, *plus* 0·375 gram of asparagine,* $\mu = 0·50$, diluted to 100 c.c.	6·50	5·83
50 c.c. of 10 per cent. sucrose solution, *plus* 1·2 milligrams of hydrochloric acid, diluted to 100 c.c.	6·45	4·80
50 c.c. of 10 per cent. sucrose solution, *plus* 0·6 milligram of hydrochloric acid, diluted to 100 c.c.	—	5·09
50 c.c. of 10 per cent. sucrose solution, *plus* water only, diluted to 100 c.c.	6·62	6·60

Rotation of each solution before heating $= 6·63 \pm 0·03°$.

* At $v = 16$.

The potential basic function of the compounds is well illustrated by the manner in which they decrease the inversion of sucrose by acid. The salts formed undergo very considerable hydrolytic dissociation in dilute aqueous solution, hence a large excess of the base must be added to reduce this. As for our purpose we have only to consider the influence of the substances on minute traces of acid, the experiments recorded below were carried out with acid (HCl) in presence of a distinct excess of the base.

Basic Function of Asparagine, Glycine, and α-Alanine.

Rotation of solutions in a 2-dcm. tube at 16°.

	After 20 hours at	
	40°.	60°.
50 c.c. of 10 per cent. sucrose solution, *plus* 1·2 milligrams of hydrochloric acid, diluted to 100 c.c.	6·50	4·80
50 c.c. of 10 per cent. sucrose solution, *plus* 1·2 milligrams of hydrochloric acid, *plus* 150 milligrams of asparagine, $\mu = 0·10$, diluted to 100 c.c.	6·58	5·85
50 c.c. of 10 per cent. sucrose solution, *plus* 1·2 milligrams of hydrochloric acid, *plus* 150 milligrams of asparagine, $\mu = 0·20$, diluted to 100 c.c.	6·55	5·82
50 c.c. of 10 per cent. sucrose solution, *plus* 1·2 milligrams of hydrochloric acid, *plus* 150 milligrams of asparagine, $\mu = 0·50$, diluted to 100 c.c.	6·53	5·60
50 c.c. of 10 per cent. sucrose solution, *plus* 1·2 milligrams of hydrochloric acid, *plus* 5 milligrams of asparagine, $\mu = 0·10$, diluted to 100 c.c.	—	4·95
50 c.c. of 10 per cent. sucrose solution, *plus* 1·2 milligrams of hydrochloric acid, *plus* 75 milligrams of glycine, diluted to 100 c.c. ...	6·62	6·06
50 c.c. of 10 per cent. sucrose solution, *plus* 1·2 milligrams of hydrochloric acid, *plus* 89 milligrams of α-alanine, diluted to 100 c.c...	6·62	6·04

Rotation of each solution before heating $= 6·68 \pm 0·03°$.

*Influence of Asparagine, Glycine, and a-Alanine on the Hydrolysis of
Purified Amylase and Starch.*

As these compounds are practically neutral substances, it might be
presumed that they would have little influence on amylolytic action
when added to the purified starch and amylase described at the beginning of this communication. As a matter of fact, however, it was
found that they slightly augmented the action, more maltose being
formed in their presence than in the aqueous solution without such
addition. The first interpretation we made of this slight augmentation was that the substances in virtue of their amphoteric properties
neutralised such minute traces of acidity or alkalinity as were
accidentally present in our solutions.

Amylase in this relatively pure state is extremely sensitive to
minute traces of impurity (compare Osborne and Campbell, *loc. cit.*),
so much so that we have found it somewhat difficult to obtain
concordant results in duplicate determinations. Normal amylolytic
action does not take place under such conditions of laboratory
experiment. In the plant or natural product in which the enzyme
works, the media contain mixed phosphates, other salts, amides, and
amino-acids, which ensure the degree of neutrality most suitable for
amylolytic action. This point has as yet received insufficient attention
from biologists, mainly through the misleading values obtained by
ordinary titration methods when applied to the examination of animal
and vegetable fluids or extracts. Foa (*Compt. rend. Soc. Biol.*, 1905,
58, 865), by measurements of electromotive force with hydrogen
electrodes, has lately given examples of this in the case of various
animal fluids. In continuing our work and by using greater precautions
to exclude accidental contamination, we came to the conclusion that
such traces of acidity as might be incidental to our methods of working
were insufficient to provide an explanation for certain apparently
anomalous results obtained. It occurred to us that starch itself might
possibly possess some feeble acid properties, and that if so it might be
possible to obtain some evidence of salt formation by the observation
of the conductivity of caustic soda solutions in presence of starch. A
substance of such a feebly acid nature would form salts readily
hydrolysable in aqueous solution. But, in accordance with the law of
mass action, if sufficient starch were added the hydrolytic decomposition
would be prevented. Such experiments cannot be carried out fully
under the conditions available to us owing to the comparatively slight
solubility of soluble starch. We have, however, made the determinations tabulated below, which are sufficient to prove that soluble starch
of very great purity has feebly acid properties, which, feeble although

they are, are adequate to explain many of the apparently peculiar results we obtained in our experiments with the purified soluble starch and amylase.

Soluble Starch and Caustic Soda at 25°.

Concentration.	Molecular conductivity.
$N/25$ caustic soda...	209·5
,, ,, plus 0·42 gram of starch per 100 c.c.	191·6
,, ,, ,, 0·85 ,, ,, ,, ...	179·6
,, ,, ,, 1·70 ,, ,, ,, ...	151·2
,, ,, ,, 3·40 ,, ,, ,, ...	114·5

Soluble Starch and Hydrochloric Acid at 25°.

Concentration.	Molecular conductivity.
$N/25$ hydrochloric acid	390·3
,, ,, plus 0·42 gram of starch per 100 c.c.	388·7
,, ,, 0·84 ,, ,, ,, ...	385·6
,, ,, 1·70 ,, ,, ,, ...	381·3
,, ,, 3·40 ,, ,, ,, ...	372·0

Soluble Starch and Potassium Chloride at 25°.

Concentration.	Molecular conductivity.
$N/25$ potassium chloride...	132·5
,, ,, plus 0·42 gram of starch per 100 c.c. ...	132·2
,, 0·84 ,, ,, ,, ...	131·7
,, 1·70 ,, ,, ,, ...	129·6
,, 3·40 ,, ,, ,, ...	126·0

These experiments were made with a starch preparation which, in 2 per cent. solution, had a specific conductivity of $2·5 \times 10^{-6}$ at 25°, and so could not contain sufficient impurity to influence greatly the results tabulated. Observations made with other preparations yielded similar values. It is obvious from the results with caustic soda that starch forms compounds with the alkali, the conductivity being reduced fully 45 per cent. by the addition of 3·4 grams of starch per 100 c.c., whereas with hydrochloric acid and potassium chloride the reductions are 4·7 and 5·0 per cent. respectively. The further bearing of these and other results on the nature and constitution of starch we reserve for a subsequent communication.

Experiments with Purified Amylase and Starch.

Influence of Glycine, a-Alanine, and Asparagine.

Conditions of Experiment.—Starch: 70 c.c. of a 1·5 per cent. solution taken. Amylase: 5 milligrams of the preparation F_2 already described were dissolved in 100 c.c. of water; 1 c.c. of this added to each starch solution. The action was allowed to proceed for one hour, when it was stopped by the addition of caustic soda.

I.   59·5°.

Milligrams of
maltose formed.

Starch solution and amylase without addition				95
,,	,,	,,	plus 75 milligrams of glycine	100
,,	,,	,,	,, 89 , ,, a-alanine	103
	,,	,,	,, 150 ,, asparagine	99

In this experiment, a dried alcohol-precipitated specimen of soluble starch was used. A 2 per cent. solution had $k^* = 5 \times 10^{-6}$ at 25°.

II.   45 minutes at 54·5°.

Milligrams of maltose formed.

Starch and amylase without addition				50
,,	,,	plus 0·01 milligram of caustic soda		42
,,	,,	,, 0·01 ,, hydrogen chloride		42
,,	,,	,, 50 milligrams of potassium chloride		64
,,	,,	,, 75 ,, glycine		45
,,	,,	,, 89 ,, a-alanine		64
,,	,,	,, 150 ,, asparagine "A"		55
,,	,,	,, 150 ,, asparagine "B"		8
,,	,,	,, 1·0 milligram of caustic soda		nil
,,	,,	,, 1·0 ,, ,, ,, plus 75 milligrams of glycine		64
,,	,,	,, 1·0 ,, ,, ,, plus 89 milligrams of a-alanine		64
,,	,,	,, 1·0 ,, ,, ,, plus 150 milligrams of asparagine "A"		60

The starch used was purified by "freezing out" in the manner described. A 2 per cent. solution gave $k = 2·0 \times 10^{-6}$ at 25°.

III.   1 hour at 52·2°.

Milligrams of maltose formed.

Starch and amylase without addition			224
,,	,,	plus 0·01 milligram of caustic soda	235
,,	,,	,, 0·01 ,, hydrogen chloride	207
,,	,,	,, 10 0 milligrams of potassium chloride	230
,,	,,	,, 37·5 ,, glycine	257
,,	,,	,, 44·5 ,, a-alanine	269
,,	,,	,, 75·0 ,, asparagine "A"	232
,,	,,	,, 75·0 ,, asparagine "B"	190

A two per cent. solution of the starch gave $k = 2·8 \times 10^{-6}$ at 25°. Amylase, F_2, 15 milligrams to 250 c.c., 5 c.c. to each solution. The asparagine marked "A" in series II and III was a highly purified specimen of $\mu = 0·094$. "B" was an ordinary laboratory "pure"

* No correction has been made, in any of these values of k, for the conductivity of the solvent water for which $k = 1$ to $1·5 \times 10^{-6}$.

specimen recrystallised five times from water, $\mu = 0.50$, $v = 16$. The glycine was a specially purified specimen, $\mu = 0.05$. The α-alanine was crystallised several times, but was not of the same degree of purity as the "A" asparagine or glycine, $\mu = 0.25$, $v = 16$. The potassium chloride was a specially purified preparation used for conductivity values. The water used in the experiments detailed here had $k = 1$ to 1.5×10^{-6} at $25°$. We may mention incidentally that with the still and spray traps described by us (*J. Fed. Inst. Brew.*, 1905, **11**, 3, 218) water of this purity is easily obtained in quantity by a second distillation, a little calcium hydroxide being added to the contents of the still.

Other experiments which we need not detail gave similar results. We, however, record a series of experiments where the amylase used was slightly acid. The acidity of 100 milligrams being equivalent to 0.86 c.c. of N/100NaOH with rosolic acid, 1.17 milligrams of this amylase was added in each case. The starch was the same as in No. I. The asparagine was the impure "B" specimen.

IV.   1 hour at 58.5°.

Milligrams of maltose formed.

Starch and amylase without addition					210
,,	,,	*plus* 75 milligrams of glycine...			215
,,	,,	,, 89	,,	α-alanine	236
,,	,,	,, 150	,,	asparagine "B"	45
,,	,,	. ,, 0.01	,,	caustic soda	252
,,	,,	,, 0.01	,,	,, *plus* 75 milligrams of glycine	254
,,	,,	,, 0.01	,,	,, *plus* 89 milligrams of α-alanine	276
,,	,,	,, 0.01	,, .	,, *plus* 150 milligrams of asparagine "B" ...	50
,,	,,	,, 5.0	,,	,, ...	nil
,,	,,	,, 5.0	,,	,, *plus* 500 milligrams of glycine	160

The maltose formed was determined by gravimetric copper reduction, the only method in our opinion which yields sufficiently accurate results for such work. In connection with the values for maltose formed it must be noted that glycine and α-alanine are not without slight influence on the copper reduction. We have applied experimentally determined corrections in obtaining the above maltose values. Further, owing to the extremely sensitive nature of amylase, even when taking the greatest precautions to exclude atmospheric and other impurities, there is considerable risk of discrepant results through adventitious contamination. Although it is possible that in all the highly purified starches with which we worked some unrecognised impurity may have been present, we feel nevertheless justified in concluding that isolated amylase cannot bring about a normal hydrolysis in starch solutions which are free from saline substances. Even the addition of a neutral salt such as potassium chloride increases the velocity of the reaction (compare Osborne and Campbell, *loc. cit.*). The addition of certain other salts has a like effect; for example, addition of a few milligrams of a mixture of $10KH_2PO_4 + 1Na_2HPO_4$ greatly increases the speed of hydrolysis. In this case, we can conceive the increased action as being due, at least partly, to the attainment of the degree of neutrality most suitable for amylolytic action. The increase with the mixed phosphates is greater than that with potassium chloride, hence we may assume that the influence of the neutral salt depends only on the change it produces in the osmotic pressure of the starch and enzyme solution.

As we are unable to continue this line of investigation, we now record our results, which, although in themselves possibly not conclusive, are at least suggestive, and may be of some assistance to other workers who are in a position to prosecute such investigations under more favourable conditions than obtain in an industrial laboratory. We consider that our results are sufficient to establish that:

(1) Asparagine and the amino-acids mentioned have no specific influence in augmenting the action of amylase: the apparent augmentation of action sometimes obtained by the addition of these amphoteric compounds (or of feeble acids) is due to their neutralising alkaline (or other) impurity in the starch or enzyme solution.

(2) Normal amylolytic action takes place in neutral solution. In the plant substance, this neutrality is brought about by equilibrium between the basic and acid compounds present.

(3) Until the conditions influencing the action of enzymes are more fully established, it is inadvisable to formulate mathematical laws as to the kinetics of enzymic hydrolyses.

(4) Purified soluble starch has the properties of an extremely feeble acid; it is capable of yielding negative ions under the influence of strongly positive ones.

XI.—Studies on Optically Active Carbimides. Part II. The Reactions between l-Menthylcarbimide and Alcohols.

By ROBERT HOWSON PICKARD, WILLIAM OSWALD LITTLEBURY, A.I.C., and ALLEN NEVILLE, B.Sc.

l-MENTHYLCARBIMIDE reacts fairly readily with alcohols, and the resulting l-menthylcarbamates are prepared in the manner already described in Part I (Trans., 1904, 85, 685). These esters are stable substances, which boil under reduced pressure without decomposition, and are easily soluble in the common organic media. Comparing the compounds prepared from the alcohols of the paraffin series, it will be seen that as the molecular weight increases, the melting point and the volatility in steam of the esters decrease.

The rotation of these l-menthylcarbamates exhibits a striking regularity. The molecular rotation of these esters in solution has an approximately constant value for each solvent independent of the nature of the alcohol from which the ester is prepared. Thus, in chloroform, the molecular rotations of these esters approximate to 160°, in benzene to 140°, and in pyridine to 175°. It may be pointed out that there are several isolated exceptions to these values among the molecular rotations of the fourteen esters investigated, but every ester has the approximate value for the molecular rotation in at least one of the three solvents.

The reaction between l-menthylcarbimide and ethyl alcohol has been studied in detail and is easily followed by polarimetric observations. It is a bimolecular reaction, the velocity being proportional to the concentration of the two reacting substances. If the reaction is carried out in ethyl-alcoholic solution, the alcohol concentration remains constant throughout the reaction, and the velocity is dependent only on the concentration of the carbimide and proportional to it. The reaction in this case may be considered as being unimolecular. If, however, the reaction is carried out in a neutral solvent where the concentration of the alcohol is comparable with that of the carbimide, the velocity is dependent on the two concentrations, both of which vary throughout the range of the reaction, which in this case is bimolecular.

The velocity constant of this reaction varies greatly with the nature of the solvent employed. It is noteworthy how the tertiary amines accelerate the velocity of the reaction. Primary and secondary amines cannot be used as solvents, since they react with l-menthylcarbimide.

The following is a table in which the velocity constants obtained under the same conditions with various solvents are compared with that obtained in tripropylamine solution:

Dimethylaniline (about) ...	400	Benzene	5·4
Tripropylamine	100	Chloroform	3·2
Pyridine	30·2	Acetone	2·9
Carbon tetrachloride	9·0	Chlorobenzene	1·7
Nitrobenzene	8·9	Toluene	1·2
		Xylene (from commercial "xylol")	0·9

The influence of these solvents on the velocity of this reaction seems to bear no relationship to their dielectric constants.

Reference to the numerous interesting researches of Menschutkin on the influence of solvents on chemical reactions shows that, as was perhaps to be expected, a solvent has a specific influence on each chemical reaction, although in a series of analogous reactions between allied substances the comparative influence of a number of solvents may be the same. This is a point to which we shall return later when describing the action of *l*-menthylcarbimide on primary and secondary amines.

The velocity of the reaction in ethyl-alcoholic solution has been compared at various temperatures between 20° and 50°. The change in the velocity appears to be approximately proportional to the alteration in temperature, the curve expressing this relationship being a straight line. This result is a very remarkable one, since generally the change in the velocity of a reaction due to an alteration in temperature follows a logarithmic curve (compare J. Plotnikoff, *Zeit. physikal. Chem.*, 1905, 51, 603). It should, however, be remembered that the reaction measured is one between two deliquescent substances, and one, there-fore, in which mere traces of water have a great influence.

The rates of reaction between *l*-menthylcarbimide and various alcohols in pyridine solution have been compared. Little regularity is apparent among the velocity constants obtained. For the lower alcohols of the normal paraffin series, the velocity constants are approximately proportional to the molecular weights of the alcohols, so that when these two variables are plotted, the curve approximates closely to a straight line. The *iso*-alcohols give lower velocity constants than those given by the corresponding normal alcohols, while the unsaturated alcohols give lower constants than the corresponding saturated alcohols. Results somewhat analogous to these have been obtained by Menschutkin (*J. Russ. Phys. Chem. Soc.*, 1893, 23, 263) in a paper on the "Influence of the Constitution of the Alcohols on the Velocity of Etherification." In this paper, the velocity constants of esterification of alcohols mixed in molecular proportion with acetic anhydride dissolved in benzene are given.

In the following table, our results are compared with those of Menschutkin ; in each case the value for ethyl alcohol has been put at 100 for the purpose of comparison :

Alcohol.	Velocity constants of alcohols with l-menthyl-carbimide.	With acetic anhydride.	Alcohol.	Velocity constants of alcohols with l-menthyl-carbimide.	With acetic anhydride.
Methyl ..	106·3	206·6	isoButyl	65·7	74·1
Ethyl	100·0	100·0	isoPropyl......	68·4	27·3
Propyl.. ...	90·5	88·6	Allyl	83·4	53·5
n-Butyl ...	81·3	85·9	Benzyl.........	54·8	51·6
n-Heptyl...	58·8	72·5	Phenylethyl..	10·4	—
n-Octyl ...	49·4	69·6	Phenylpropyl	50·5	
Cetyl	56·1	49·5	Cinnamyl ...	48·7	

The most striking difference in these two series of results is with methyl alcohol. As is generally found in comparing the members of the paraffin series, the first member differs somewhat markedly from those following in its physical and often in some of its chemical properties. Thus, Menschutkin finds that the velocity constant of the reaction between methyl alcohol and acetic anhydride is nearly double that of the reaction between ethyl alcohol and the anhydride, whilst we find that the velocity constant of the reaction between methyl alcohol and l-menthylcarbimide is only slightly greater than that between ethyl alcohol and the carbimide. The differences between the two series of results are to be ascribed not only to the influence of hydrocarbon radicles on the reactivity of the alcohols, but also, of course, to the use of the two specific reagents, acetic anhydride in the one series and l-menthylcarbimide in the other. The comparative regularity of both series of results serves to emphasise the irregular influence of alkyl groups on the reactions of alkyl iodides, the knowledge of which has been extended and summarised by Donnan (Trans., 1904, 85, 555).

Esters of l-*Menthylcarbamic Acid,* $C_{10}H_{19} \cdot NH \cdot CO_2R$.

The esters of l-menthylcarbamic acid are readily prepared (as described in Part I, *loc. cit.*) by heating molecular proportions of the respective alcohols with l-menthylcarbimide. They are easily purified by distillation either with steam or under reduced pressure. They possess a pleasant, characteristic odour, and are soluble in the common organic media except where otherwise stated.

Methyl Ester.—The melting point of this compound is 63°, not 53° as printed in Part I.

The n-*butyl* ester is volatile with steam and solidifies in clusters of colourless, stellate needles which melt at 37°.

0·2266 gave 11·4 c.c. moist nitrogen at 16° and 756 mm.   N = 5·8
$C_{15}H_{29}O_2N$ requires N = 5·51 per cent.

The n-*heptyl* ester was obtained as a pale yellow, refractive oil, which boiled at 215°/22 mm.; when cooled to the temperature of a mixture of ice and salt, it slowly crystallised in long needles, which melted indefinitely at 22—25°. It is only very slowly volatilised by steam.

0·3337 gave 15·3 c.c. moist nitrogen at 24° and 758 mm.   N = 5·1.
$C_{18}H_{35}O_2N$ requires N = 4·71 per cent.

The n-*octyl* ester is a pale yellow liquid which boils at 220°/24 mm.; it is very viscous, strongly refractive, and not volatile with steam.

0·3764 gave 16·8 c.c. moist nitrogen at 22° and 754 mm.   N = 5·0.
$C_{19}H_{37}O_2N$ requires N = 4·50 per cent.

The *cetyl* ester is only slightly soluble in light petroleum, from which it crystallises in colourless needles; these melt at 52·5° and are not volatile with steam.

0·3050 gave 9·1 c.c. moist nitrogen at 15° and 752 mm.   N = 3·4.
$C_{27}H_{53}O_2N$ requires N = 3·31 per cent.

The *allyl* ester, after distillation with steam, solidifies in pearly, colourless plates which melt at 40°.

0·3010 gave 16·6 c.c. moist nitrogen at 15° and 737 mm.   N = 6·2.
$C_{14}H_{25}O_2N$ requires N = 5·86 per cent.

The iso*propyl* ester solidifies, after distillation with steam, in white, lustrous plates which melt at 70°.

0·2502 gave 13·0 c.c. moist nitrogen at 15° and 750 mm.   N = 6·0.
$C_{14}H_{27}O_2N$ requires N = 5·81 per cent.

The iso*butyl* ester solidifies, after distillation with steam, in masses of colourless, stellate needles which melt at 38—40°.

0·2942 gave 15·2 c.c. moist nitrogen at 26° and 758 mm.   N = 5·7.
$C_{15}H_{29}O_2N$ requires N = 5·49 per cent.

The *benzyl* ester is a pale yellow, refractive, viscous oil, which boils at 235°/25 mm. and is not volatile with steam.

0·3408 gave 15·2 c.c. moist nitrogen at 22° and 754 mm.   N = 5·0.
$C_{18}H_{27}O_2N$ requires N = 4·84 per cent.

The *phenylethyl* ester is a pale yellow, refractive, viscous oil, which boils at 240°/25 mm. and is not volatile with steam.

0·3800 gave 15·8 c.c. moist nitrogen at 23° and 752 mm. N = 4·6.
$C_{19}H_{29}O_2N$ requires N = 4·62 per cent.

The *cinnamyl* ester crystallises from petroleum (b. p. 120—130°) in white needles which melt at 68—70°.

0·2250 gave 9·3 c.c. moist nitrogen at 18° and 739 mm. N = 4·6.
$C_{20}H_{29}O_2N$ requires N = 4·41 per cent.

The *phenylpropyl* ester crystallises from light petroleum in white, lustrous plates which melt at 64°.

0·2511 gave 10·5 c.c. moist nitrogen at 17° and 740 mm. N = 4·7.
$C_{20}H_{31}O_2N$ requires N = 4·41 per cent.

Rotations of the l-Menthylcarbamates.

The following rotations were made in a 2-dcm. tube (unless otherwise stated) at the temperature of the laboratory. Preliminary experiments showed that a small variation in temperature between 18° and 25° or a variation in the concentration of the solution from 2 to 20 parts per 100 of solution made practically no difference in the values obtained for the specific rotations.

Rotations in Chloroform.

Ester.	Weight in grams.	Vol. of solution in c.c.	Observed rotation.	$[\alpha]_D$.	$[M]_D$.
Methyl*	1·5720	20·0	− 12·21°	− 77·67°	165·5
Ethyl*	1·5351	20·0	− 11·05	− 71·98	163·4
Propyl............	0·5889	25·0	− 3·23	− 68·56	165·2
n-Butyl	0·9066	20·0	− 5·88	− 64·85	165·4
n-Heptyl.........	1·3556	20·0	− 7·48	− 55·18	163·9
n-Octyl	1·8214	20·0	− 9·49	− 52·10	162·0
Cetyl	0·9754	20·0	− 3·60	− 36·90	156·1
Allyl	1·6045	20·0	− 10·95	− 68·24	163·1
isoPropyl	0·3849	19·7	− 2·57	− 65·77	158·5
isoButyl	0·3678	19·9	− 2·45	− 66·28	169·0
Benzyl	1·5113	20·0	− 7·98	− 52·80	152·6
Phenylethyl ...	0·9328	20·0	− 5·23	− 56·06	169·8
Cinnamyl	0·8846	20·0	− 4·39	− 49·62	156·3
Phenylpropyl...	0·8568	25·0	− 1·65 †	− 48·14	152·6

* The specific rotations of these esters are somewhat higher than given in Part I.

† Observation made in a 1-dcm. tube.

Rotations in Benzene.

Ester.	Weight in grams.	Vol. of solution in c.c.	Observed rotation.	$[a]_D.$	$[M]_D.$
Methyl............	1·0707	20·0	− 6·83°	− 63·79°	135·9
Ethyl	1·1146	20·1	− 6·61	− 59·60	135·3
Propyl............	1·3438	20·0	− 7·73	− 57·52	138·6
n-Butyl	0·5780	19·8	− 3·14	− 53·91	137·5
n-Heptyl.........	0·9028	20·0	− 4·42	− 48·96	145·4
n-Octyl	1·4140	20·0	− 6·56	− 46·38	143·3
Cetyl	1·0809	24·7	− 2·89	− 33·02	139·7
Allyl	1·2383	19·8	− 7·29	− 58·28	139·3
isoPropyl.........	0·7840	20·0	− 4·39	− 55·99	134·9
isoButyl	0·2894	19·9	− 1·56	− 53·63	136·8
Benzyl............	1·1062	19·9	− 5·54	− 49·83	144·0
Phenylethyl ...	1·1949	19·8	− 6·53	− 54·10	163·9
Cinnamyl	0·2309	19·9	− 1·09	− 46·97	147·9

Rotations in Pyridine.

Ester.	Weight in grams.	Vol. of solution in c.c.	Observed rotation.	$[a]_D.$	$[M]_D.$
Methyl............	0·9727	19·7	− 8·00°	− 81·01°	172·5
Ethyl	1·5185	19·9	−11·53	− 75·55	171·5
Propyl............	1·1863	19·9	− 8·59	− 72·05	173·6
n-Butyl	0·4181	19·9	− 2·94	− 69·96	178·4
n-Heptyl.........	0·8628	20·0	− 5·15	− 59·69	177·3
n-Octyl	1·5317	20·0	− 8·70	− 56·80	176·7
Cetyl	1·0119	19·9	− 4·12	− 40·51	171·4
Allyl	0·5002	20·0	− 3·74	− 74·77	178·7
isoPropyl.........	0·4507	20·0	− 3·23	− 71·66	172·7
isoButyl	0·9349	19·7	− 6·59	− 69·43	177·0
Benzyl............	1·3892	19·9	− 8·64	− 61·88	178·8
Phenylethyl ...	0·5713	20·0	− 3·84	− 67·21	203·6
Cinnamyl	0·9757	20·1	− 5·76	− 59·33	186·8
Phenylpropyl...	1·3542	25·0	− 3·15 *	− 58·15	184·3

* Observation made in a 1-dcm. tube.

Reaction between Ethyl Alcohol and 1-Menthylcarbimide.

That the velocity of the reaction between ethyl alcohol and
l-menthylcarbimide is proportional not only to the carbimide concentration, but also to the alcohol concentration, is shown by the determination of the velocity-constants recorded in Tables I and II. The
experiments recorded in Tables I and III to XI were performed with
equivalent quantities in order to determine the influence of the solvent
on the velocity of the reaction, and the " K " values have been
calculated from the formula $K\varDelta = \dfrac{1}{t} \cdot \dfrac{x}{a-x}$.

To obtain further confirmation of the result recorded in Table I,
the reaction was carried out with twice the equivalent quantity of

alcohol (Table II), and the velocity-constant calculated according to the formula for a bimolecular reaction with non-equivalent quantities :

$$K \times C_\infty \times 0\cdot4343 = \frac{1}{t}\log\frac{C_1(C_0 - C_\infty)}{C_0(C_1 - C_\infty)},$$

where C_∞ is the excess of one reagent over the other, that is, in this case one equivalent or equal to the concentration of the carbimide. The "K" value thus obtained and recorded in Table II is in fair accordance with the "K" value obtained with equivalent quantities recorded in Table I.

The ethyl alcohol used in the experiments detailed in this paper had been heated with silver oxide in a reflux apparatus and then dried over barium oxide until a crystal of potassium permanganate produced only a very faint coloration. The other solvents were carefully dried and had constant boiling points.

The method which has been employed to determine all the velocity-constants recorded in this paper (with the exception of those in the section on the influence of temperature) is as follows : an accurately weighed quantity of the carbimide (about 2 grams) is mixed in a graduated flask (20 to 25 c.c.) with the solvent and the calculated quantity of the alcohol. After the rotation has been observed in the polarimeter, the flask is placed in a thermostat at 50°. At intervals, the flask is withdrawn, quickly cooled to the laboratory temperature, and the rotation again observed.

Throughout the paper, the times are expressed in hours, and the concentrations (A) in gram-molecules per litre. The infinity values are in some cases the results of actual observations, and in others are calculated from the specific rotation of ethyl l-menthylcarbamate in the various solvents. Both methods when compared gave identical values.

TABLE I.—*In Toluene.*

Time.	Rotation.	KA.
0	9·06	—
18·25	9·26	0·00548
25·16	9·33	0·00558
90·84	9·80	0·00558
97·66	9·84	0·00562
121·00	9·95	0·00561
138·83	10·03	0·00568
144·40	10·08	0·00598
∞	11·26	—

Mean $KA = 0\cdot00565$.
$A = 0\cdot455$.
$K = 0\cdot0124$.

TABLE II.—*In Toluene.*

Time.	Rotation.	$K \times C_\infty \times 0\cdot4343$
0	9·01	—
71·1	10·29	0·00244
78·2	10·35	0·00240
95·5	10·54	0·00248
101·6	10·56	0·00239
119·4	10·67	0·00234
125·4	10·70	0·00231
∞	11·59	—

Mean = 0·00239.
$C_\infty = 0\cdot462$.
$K = 0\cdot0119$.

TABLE III.—*In Chloroform.*

Time.	Rotation.	KA.
0	8·71	—
22·00	10·36	0·0146
27·41	10·69	0·0151
45·58	11·55	0·0158
99·83	12·85	0·0158
117·83	13·30	0·0178
∞	15·48	—

Mean $KA = 0·0158$.
$A = 0·487$.
$K = 0·0324$.

TABLE IV.—*In Carbon Tetra-chloride.*

Time.	Rotation.	KA.
0	8·10	—
1·66	8·26	0·0392
5·50	8·56	0·0387
23·42	9·36	0·0395
30·33	9 56	0·0415
47·50	9 81	0·0396
53·92	9·90	0·0407
∞	10·72	—

Mean $KA = 0·0398$
$A = 0·442$.
$K = 0·0902$.

TABLE V.—*In Pyridine.*

Time.	Rotation.	KA.
0	8·98	—
1·80	9·69	0·111
2·50	9·92	0·113
4·64	10·44	0·112
5·40	10·66	0·120
6·08	10·79	0·121
7·15	11·01	0·126
∞	13 26	—

Mean $KA = 0·117$.
$A = 0·886$.
$K = 0·302$.

A second experiment with $A = 0·497$
gave $K = 0·301$.

TABLE VI.—*In Tripropylamine.*

Time.	Rotation.	KA.
0	9·58	—
0·33	9·80	0·447
1·08	10·14	0·451
2·58	10·50	0·451
4·91	10·76	0·453
9·66	10·97	0·450
24·16	11·15	0·464
∞	11·29	—

Mean $KA = 0·453$.
$A = 0·452$.
$K = 1·00$.

TABLE VII.—*In Benzene.*

Time.	Rotation.	KA.
0	9·21	—
1·33	9·26	0·0244
5·50	9·40	0·0246
25·08	9·80	0·0235
29 66	9·89	0·0251
49·16	10 07	0·0240
55·99	10·12	0·0239
	10·80	—

Mean $KA = 0·0242$·
$A = 0·449$.
$K = 0·0540$.

TABLE VIII.—*In Xylene.**

Time.	Rotation.	KA.
0	9·50	—
19·75	9·80	0·00390
43·08	10·13	0·00410
67·75	10·42	0·00415
91·67	10·62	0·00398
116·08	10·77	0·00375
163·89	11·11	0·00381
188·09	11·30	0·00400
236·10	11·53	0·00398
∞	13·69	—

Mean $KA = 0·00396$·
$A = 0·451$·
$K = 0·00878$.

* Commercial "xylol" which had been dried and fractionated ; it boiled at 140°.

TABLE IX.—*In Acetone.*

Time.	Rotation.	KA.
0	9·42	—
1·58	9·51	0·0138
·19·66	10·25	0·0125
25·91	10·53	0·0138
43·25	11·06	0·0148
50·08	11·08	0·0130
67·16	11·53	0·0150
∞	13·62	—

Mean $KA = 0·0138$·
$A = 0·467$.
$K = 0·0295$.

TABLE X.—*In Nitrobenzene.*

Time.	Rotation.	KA.
0	9·30	—
2·83	9·57	0·0416
21·66	10·47	0·0388
69·82	11·22	0·0429
94·16	11·35	0·0429
∞	11·86	—

Mean $KA = 0·0415$.
$A = 0·443$.
$K = 0·0937$.

TABLE XL—*In Chlorobenzene.*

Time.	Rotation.	KA.
0	10·40	—
10·75	10·64	0·00759
73·75	13·04	0·00748
80·50	13·13	0·00754
∞	13·58	—

Mean $KA = 0·00753$. $A = 0·451$.
$K = 0 0167$.

The reaction between ethyl alcohol and the carbimide in dimethylaniline solution under the conditions of the above experiments proceeds too rapidly for readings in the polarimeter to be observed.

Influence of Temperature on the Rate of Reaction between Ethyl Alcohol and l-Menthylcarbimide.

The velocity constants recorded in this section were determined in a jacketed polarimeter tube round which circulated water previously passed through a long copper coil immersed in a thermostat kept at the required temperature. These experiments were carried out in ethyl-alcoholic solution and the concentration of the carbimide varied from 0·50 to 0·15, preliminary experiments having proved that this variation did not affect the results.

The temperatures at which the observations were made were read on a standard thermometer immersed in the reaction mixture. The maximum variation in temperature was less than 0·25⁰.

The " K " values are calculated from the formula $K = 1/t \log. C_0/C_t$ and the infinity values from the specific rotations of ethyl l-menthylcarbamate in ethyl-alcoholic solution at the various temperatures.

TABLE XII.—At 24°.

Time.	Rotation.	K.
0	2·92	—
6·08	3·23	0·0135
18·16	3·77	0·0152
23·50	3·90	0·0141
28·08	4·01	0·0143
34·05	4·13	0·0141
∞	4·72	—

Mean $K = 0·0142$.

A second experiment made at 24·5° gave $K = 0·0143$.

TABLE XIII.—At 31°.

Time.	Rotation.	K.
0	10·44	—
2·92	11·36	0·0274
3·58	11·64	0·0300
4·56	11·83	0·0279
5·00	12·04	0·0301
5·66	12·22	0·0302
6·16	12·35	0·0303
6·75	12·46	0·0297
9·33	13·01	0·0296
10·25	13·13	0·0287
11·3	13·29	0·0282
12·60	13·37	0·0278
∞	15·90	—

Mean $K = 0·0290$.

TABLE XIV.—At 39°.

Time.	Rotation.	K.
0	4·36	—
2·25	4·92	0·0489
3·25	5·11	0·0476
4·33	5·29	0·0466
5·75	5·55	0·0488
7·91	5·85	0·0497
9·75	6·06	0·0507
∞	6·86	—

Mean $K = 0·0470$.

TABLE XV.—At 43°.

Time.	Rotation.	K.
0	9·01	—
1·00	9·56	0·0531
2·00	10·03	0·0521
3·08	10·40	0·0484
4·16	10·86	0·0510
5·33	11·23	0·0509
6·15	11·43	0·0498
10·83	12·37	0·0486
∞	13·79	—

Mean $K = 0·0505$.

TABLE XVI.—At 50°.

Time.	Rotation.	K.
0	4·87	—
0·50	5·03	0·0678
1·58	5·33	0·0688
2·75	5·63	0·0697
4·66	5·97	0·0677
6·16	6·15	0·0647
18·25	6·84	0·0680
∞	7·00	—

Mean $K = 0·0674$.

Rate of Reaction between 1-*Menthylcarbimide and Various Alcohols.*

The rates of reaction between l·menthylcarbimide and various alcohols have been measured in pyridine solution. The alcohols, which, with the exception of the phenylpropyl alcohol, were all obtained from Kahlbaum, had constant boiling points and were inactive. The phenylpropyl alcohol, which was prepared from Kahlbaum's cinnamyl alcohol, had a constant boiling point and did not decolorise bromine. The maximum difference in the " K " values obtained when the determinations were repeated was under 10 per cent.

TABLE XVII.—*With Methyl Alcohol.*

Time.	Rotation.	KA.
0	10·11	—
1·16	10·76	0·134
1·84	11·05	0·132
2·40	11·32	0·139
3·00	11·56	0·143
3·50	11·64	0·132
4·40	11·96	0·141
5·00	12·15	0·146
5·66	12·32	0·149
6·56	12·65	0·169
∞	14·94	—

Mean $KA = 0·143$.
$A = 0·446$.
$K = 0·320$.

TABLE XVIII.—*With n-Propyl Alcohol.*

Time.	Rotation.	KA.
0	7·39	—
1·16	7·74	0·0874
1·88	7·94	0·0900
3·08	8·19	0·0866
4·50	8·44	0·0848
5·32	8·61	0·0889
6·32	8·70	0·0832
8·88	9·14	0·0961
24·40	10·01	0·0910
∞	11·19	—

Mean $KA = 0·0885$.
$A = 0·321$.
$K = 0·275$.

TABLE XIX.—*With n-Butyl Alcohol.*

Time.	Rotation.	KA.
0	9·54	—
0·58	9·85	0·0972
1·21	10·15	0·0969
1·88	10·48	0·1063
3·08	10·88	0·0973
4·75	11·45	0·1031
5·91	11·77	0·1054
7·08	12·11	0·1120
11·30	12·82	0·1147
∞	15·35	—

Mean $KA = 0·1041$.
$A = 0·4304$.
$K = 0·242$.

TABLE XX.—*With n-Heptyl Alcohol.*

Time.	Rotation.	KA.
0	10·85	—
1·16	11·43	0·0892
1·66	11·65	0·0896
2·16	11·80	0·0841
3·25	12·14	0·0812
4·16	12·45	0·0839
5·00	12·72	0·0868
5·83	12·99	0·0908
6·91	13·14	0·0852
∞	17·03	—

Mean $KA = 0·0863$.
$A = 0·480$.
$K = 0·179$.

Other values obtained were $K = 0·171$ and $0·182$.

TABLE XXI.—*With n-Octyl Alcohol.*

Time.	Rotation.	KA.
0	13·83	—
1·55	14·81	0·0894
2·25	15·17	0·0887
3·00	15·59	0·0933
4·33	16·15	0·0935
5·55	16·63	0·0961
6·76	16·97	0·0946
8·38	17·34	0·0922
∞	21·88	—

Mean $KA = 0·0925$.
$A = 0·619$.
$K = 0·149$.

TABLE XXII.—*With Cetyl Alcohol.*

Time.	Rotation.	KA.
0	10·02	—
1·92	10·75	0·0802
3·00	11·05	0·0773
3·75	11·25	0·0774
4·50	11·35	0·0714
5·33	11·57	0·0742
6·33	11·85	0·0794
25·75	13·62	0·0748
∞	15·49	—

Mean $KA = 0·0764$.
$A = 0·452$.
$K = 0·169$.

TABLE XXIII.—*With Allyl Alcohol.*

Time.	Rotation.	KA.
0	7·34	—
1·25	7·68	0·0751
1·88	7·83	0·0751
2·50	7·99	0·0785
3·00	8·07	0·0753
3·83	8·27	0·0801
4·40	8·38	0·0809
5·00	8·48	0·0808
25·24	10·02	0·0829
∞	11·30	—

Mean $KA = 0·0786$.
$A = 0·316$.
$K = 0·249$.

TABLE XXIV.—*With isoPropyl Alcohol.*

Time.	Rotation.	KA.
0	12·56	—
0·60	12·93	0·105
1·46	13·44	0·112
2·28	13·83	0·112
5·28	14·96	0·119
6·95	15·30	0·113
∞	18·80	—

Mean $KA = 0·112$.
$A = 0·544$.
$K = 0·206$.

TABLE XXV.—*With isoButyl Alcohol.*

Time.	Rotation.	KA.
0	8·88	—
1·50	9·36	0·0734
2·41	9·59	0·0711
4·41	10·09	0·0755
5·16	10·24	0·0756
6·66	10·51	0·0762
7·41	10·65	0·0777
9·25	10·94	0·0801
∞	13·72	—

Mean $KA = 0·0756$.
$A = 0·387$.
$K = 0·195$.

A second experiment gave $K = 0·202$.

TABLE XXVI.—*With Benzyl Alcohol.*

Time.	Rotation.	KA.
0	11·39	—
0·83	11·87	0·0924
1·66	12·15	0·0766
2·41	12·51	0·0827
3·91	13·03	0·0822
4·75	13·29	0·0827
5·66	13·56	0·0839
6·83	13·84	0·0836
8·40	14·22	0·0862
∞	18·13	—

Mean $KA = 0·0838$.
$A = 0·507$.
$K = 0·165$.

TABLE XXVII.—*With Phenylethyl Alcohol.*

Time.	Rotation.	KA.
0	10·50	—
0·63	10·58	0·0148
1·95	10·74	0·0146
4·61	11·04	0·0145
6·00	11·19	0·0145
7·50	11·37	0·0149
9·00	11·52	0·0149
10·50	11·67	0·0149
∞	19·14	—

Mean $KA = 0·0147$.
$A = 0·470$.
$K = 0·0312$.

TABLE XXVIII.—*With Phenylpropyl Alcohol.*

Time.	Rotation.	KA.
0	10·81	—
1·00	11·25	0·0718
1·83	11·55	0·0694
2·66	11·89	0·0739
3·66	12·11	0·0674
4·50	12·38	0·0698
5·33	12·63	0·0719
6·50	12·95	0·0743
9·17	13·48	0·0746
10·75	13·75	0·0753
11·92	13·79	0·0696
∞	17·38	—

Mean $KA = 0·0718$.
$A = 0·531$.
$K = 0·152$.

TABLE XXIX.—*With Cinnamyl Alcohol.*

Time.	Rotation.	KA.
0	12·08	—
0·70	12·46	0·0724
1·46	12·87	0·0763
2·36	13·33	0·0799
3·00	13·53	0·0752
4·58	14·14	0·0773

Time.	Rotation.	KA.
5·41	14·43	0·0785
6·16	14·69	0·0804
7·00	14·93	0·0813
8·55	15·31	0·0812
∞	19·96	—

Mean $KA = 0·0781$.
$A = 0·531$.
$K = 0·147$.

We desire to express our thanks to Mr. H. L. Leech for valuable assistance in some of the preparative and analytical work for this paper and also to the Research Fund Committee of the Chemical Society for a grant defraying much of the cost of this work.

MUNICIPAL TECHNICAL SCHOOL,
BLACKBURN.

XII.—a-Chlorocinnamic Acids.

By JOHN JOSEPH SUDBOROUGH and THOMAS CAMPBELL JAMES.

IN previous communications (Sudborough and Thompson, Trans., 1903, **83**, 666 and 1154), attention has been drawn to the action of various alkalis on cinnamic acid dibromide and its esters. By means of the reactions :—olefinic acid + bromine – hydrogen bromide, it is usual to obtain the bromo-derivative of an acid stereoisomeric with the original acid. Our previous results have shown that with cinnamic acid itself this is largely true, the final product being mainly a-bromo*allo*cinnamic acid, but that when the acid is replaced by one of its esters the product obtained by the elimination of hydrogen bromide is a mixture of a-bromocinnamic and a-bromo*allo*cinnamic acids, the a-bromo-acid, however, preponderating.

In this paper, we give an account of similar experiments which have been made with cinnamic acid dichloride (αβ-dichloro-β-phenylpropionic acid) and its esters. The results indicate that the reaction with the chloro-acid is quite different from that with the dibromide, as the product obtained was always a mixture of a-chloro- and a-chloro*allo*cinnamic acids, the a-chloro-acid predominating. The yields of the two acids are affected to only a slight extent by altering the alkali or the temperature, or even by substituting the methyl or ethyl ester for the acid dichloride. Experiments in which organic bases were employed in place of potassium hydroxide did not give satisfactory results.

The reaction between cinnamic acid dichloride and alcoholic potash has been previously studied by Tutz (*Ber.*, 1882, **15**, 788), Michael and Pendleton (*J. pr. Chem.*, 1889, [ii], **40**, 65), and Mulliken (*Dissertation, Leipzig*, 1890). All obtained a mixture of two chloro-acids, which were separated by taking advantage of the considerable difference in solubility of their potassium salts in absolute alcohol. The acid melting at 136—137° was termed a-chlorocinnamic acid by Michael and Pendleton, and the acid melting at 110—111° a-chloro-*allo*cinnamic acid.

The numbers given by Mulliken (*loc. cit.*) indicate that the two acids are formed in almost equal quantities; thus from 5 grams of cinnamic acid dichloride the following weights of potassium salts were obtained in different experiments :—

Experiment.	1.	2.	3.	4.	5.	6.	7.
a-Chloro-	2·64	2·63	2·30	2·31	2·48	2·51	2·29 grams
a-Chloro*allo*-:...	2·26	2·29	2·55	2·47	2·38	2·37	2·52 ,,

Michael and Pendleton, on the other hand, state that the amount of a-chloro*allo*-acid is always much less than that of the a-chloro-acid. These same chemists also show that the acid synthesised by Plöchl (*Ber.*, 1882, **15**, 1945) from benzaldehyde, sodium chloroacetate, and acetic anhydride is identical with the acid melting at 136—137°, and thus the a-position of the chlorine in the latter is established. As the acid melting at 110° is readily transformed into this acid, it also presumably contains the chlorine atom in the a-position. This is all the more probable since Michael and Pendleton have prepared two stereoisomeric β-chlorocinnamic acids by the addition of hydrogen chloride to phenylpropiolic acid.

I. *Preparation of $a\beta$-Dichloro-β-phenylpropionic Acid and its Esters.*

Very good yields of cinnamic acid dichloride can be obtained by passing chlorine for several hours into a suspension of finely-divided cinnamic acid in five times its weight of dry carbon disulphide (Erlenmeyer, *Ber.*, 1881, **14**, 1867 ; Mulliken, *loc. cit.*).

In our different experiments, we have obtained 134, 138, 127, 140, 120, 133, and 140 grams from 100 grams of cinnamic acid by removing the crystals and washing with a little carbon disulphide. We have attempted to chlorinate methyl cinnamate in a similar manner and also in chloroform solution, but the yields were poor. Good yields of methyl cinnamate dichloride may be obtained by esterifying cinnamic acid dichloride by Fischer and Speyer's method (*Ber.*, 1895, **28**, 3252). The quantities we used were one of acid to two of pure methyl alcohol containing 4 per cent. of hydrogen chloride, and the mixture was boiled for two hours. When cold, a considerable quantity of the ester had separated. This was removed, more hydrogen chloride passed into the filtrate, and the solution again heated for two hours. We obtained on an average a 92 per cent. yield of ester melting at 101° (Finkenbeiner, *Ber.*, 1894, **27**, 890).

Ethyl cinnamate dichloride has been prepared by Finkenbeiner (*loc. cit.*), who describes it as an oil. We have prepared the same ester by the Fischer-Speyer method from cinnamic acid dichloride. After pouring into water, extracting with ether, and shaking out the ethereal solution with dilute sodium carbonate solution, we dried the solution with calcium chloride and removed the ether. The oil which was left (yield, 92 per cent.) solidified to a crystalline mass melting at 30°. It is readily soluble in all organic solvents, and crystallises from light petroleum in well-developed, colourless prisms melting at 30—31°.

II. *Separation of* a-*Chloro- and* a-*Chloro*allo-*cinnamic Acids.*

As the method adopted by Michael (*J. pr. Chem.*, 1889, [iii], **40**, 63) and Mulliken (*loc. cit.*) for separating the acids by means of the different solubilities of the two potassium salts in alcohol was somewhat tedious, we attempted the separation by means of the barium salts, as described in the separation of a-bromo- and a-bromo*allo*-cinnamic acids (Trans., 1903, **83**, 673), and the results proved that complete separation is readily effected by this method, although loss of acid, especially of the *allo*-acid, occurs.

Expt. 1.—Two grams of each acid, when mixed and separated as the barium salts, gave 1·80 grams of *allo*-acid melting at 110—111° and 1·91 grams of a-chloro-acid melting at 136°.

Expt. 2.—Two grams of each acid gave 1·80 grams of *allo*- and 1·92 grams of a-chloro-acid melting at 137°.

The losses in these experiments are rather larger than with the corresponding bromo-acids (*loc. cit.*).

III. *Action of Alkalis on Cinnamic Acid Dichloride and its Esters.*

We have made a number of experiments on the action of alkalis on the dichloride and its esters, using the same general method as described in the case of the bromo-derivatives (*loc. cit.*, p. 674), with the object of determining the influence of the following factors : (*a*) temperature, (*b*) the alkali, (*c*) replacing the acid by its esters, (*d*) solvent. The results we have obtained are given in the following table. The numbers given in each case are the mean of several experiments.

TABLE I.

Cinnamic Acid Dichloride, 10 *grams*.

Alkali.	Weight of α-chloro-acid.	Weight of allo-acid.	Total.	Conditions.
Alcoholic potassium hydroxide ...	4·52	2·88	7·40	Standing overnight at 0°.
,, ,, ...	4·70	2·75	7·45	Ordinary temperature.
,, ,, ...	4·82	2·83	7·65	Boiling for 10 minutes.
Aqueous potassium hydroxide, *N*.	6·35	1·27	7·62	Standing overnight at 0°.
,, ,, ...	5·75	1·80	7·55	15°.
...	5·25	2·14	7·28	Ordinary temperature in July.
...	4·32	1·35	5·67	At 50—60° for 6 hours.
,, ,, ...	4·0	0·87	4·87	On boiling water-bath for 40 minutes.
Aqueous sodium hydroxide.........	5·86	1·71	7·57	Ordinary temperature.
Sodium ethoxide	4·62	2·84	7·46	Boiling for 15 minutes.
,, ,, ...	4·46	2·85	7·31	Overnight at ordinary temperature.
Alcoholic sodium hydroxide	5·77	1·82	7·59	15°.
Aqueous potassium hydroxide, 3*N*	6·0	1·68	7·68	15°.
,, ,, *N*/3	5·35	1·88	7·25	15°.

Methyl Cinnamate Dichloride, 10·64 *grams*.

Alcoholic potassium hydroxide ...	6·64	1·30	7·94	Both at 0° and on boiling.

Ethyl Cinnamate Dichloride, 11·28 *grams*.

Alcoholic potassium hydroxide ...	6·49	1·40	7·89	Both at 0° and on boiling.

The theoretical amount of mixed α-chloro-acids is 8·33 grams.

The conclusions to be drawn from these experiments are :

(1) The alkali used evidently affects the relative amounts of the two acids formed. Aqueous potassium hydroxide gives a better yield of the α-chloro-acid than does alcoholic potash, as do also aqueous and alcoholic sodium hydroxide.

(2) The temperature factor is small and, in the case of aqueous potassium hydroxide, tends slightly to increase the relative yield of *allo*-acid and, at the same time, to form neutral products. At the higher temperatures, chlorocinnamene is undoubtedly formed and may be recognised by its odour.

(3) The effect of alteration of concentration is slight.

(4) The substitution of an ester for the cinnamic acid dichloride has not the same marked effect as with the dibromide. The yield of α-chloro-acid is increased somewhat and that of the *allo*-acid correspondingly decreased.

(5) It will be noticed that in all the experiments the yield of α-chloro-acid is much greater than the yield of α-bromo-acid from the dibromide and its esters under similar conditions (*loc. cit.*, p. 680).

The action of the tertiary amines, dimethylaniline, and quinoline on cinnamic acid dichloride has also been studied. Ten grams of the dichloride were boiled on the water-bath for four hours with the requisite amount of dimethylaniline (2 mols.) in methyl-alcoholic solution. The alcohol was removed and the residue treated with hydrochloric acid. From the residue, 0·9 gram of a chloro-acid was obtained, melting at 138° after recrystallisation from benzene. This was undoubtedly α-chlorocinnamic acid; the other product was an oil which only slowly solidified. When 10 grams of the acid were boiled in a similar manner for two hours with a methyl-alcoholic solution of quinoline, the products were 6·3 grams of unchanged acid dichloride and a small amount of oil. When the heating was continued for eight hours, a product was obtained from which were isolated, by means of the barium salts, 3·1 grams of α-chloro-acid melting at 138° after recrystallisation, 1·5 grams of unaltered dichloride melting at 164°, and 1 gram of neutral oil.

IV. *Transformation of α-Chloroallocinnamic Acid into its Isomeride.*

(a) *Influence of Light. Expt.* 1.—Two grams of powdered solid α-chloro-acid were exposed to bright sunlight from July to November, 1904. The melting point at the end of this time was still 137°, indicating that no change had occurred.

Expt. 2.—Two grams of α-chloro*allo*cinnamic acid were exposed under exactly similar conditions, and the melting point fell from 111° to 90°. The exposed acid was separated into α-chloro- and α-chloro-*allo*-acids by means of the barium salts, and gave 1·1 grams of unaltered *allo*-acid melting at 108° and 0·65 gram of α-chloro-acid melting at 136°.

Expt. 3.—Similar to 2, but exposed from June to August. The melting point was 90—95°, and the product gave 0·9 gram of α-chloro-acid melting at 137° and only 0·65 gram of unaltered *allo*-acid melting at 111°.

Expt. 4.—Some light petroleum mother liquors obtained from crystallising α-chloro*allo*cinnamic acid were exposed to bright sunlight during four summer months, and gave a considerable quantity of well-developed prisms of the α-chloro-acid melting at 137°.

Expt. 5.—Two grams of α-chloro*allo*cinnamic acid were dissolved in benzene and allo̊wed to remain for three months in an open tube in ordinary daylight. As the benzene evaporated, crystals were deposited. These melted at 111° and proved to be unchanged α-chloro*allo*-acid.

(*b*) *Influence of Temperature.*—The following series of experiments shows the effect of heating small quantities of the α-chloro*allo*-acid at a temperature of 155° for the stated periods.

TABLE II.

Weight of α-allo-acid.	Time in hours.	Weight of α-chloro-acid.	M. p.
2·0 grams	0·5	0·081	137—138°
2·0 ,,	1·0	0·132	137—138
2·0	3·0	0·320	138

These numbers, compared with those obtained for the conversion of α-bromo*allo*cinnamic acid into α-bromocinnamic acid (*loc. cit.*, p. 687.), indicate that the transformation of the α-chloro*allo*-acid proceeds more slowly than that of the α-bromo*allo*-acid at the same temperature.

V. *Elimination of Hydrogen Chloride and Hydrogen Bromide from the α-Chloro- and α-Bromo-cinnamic Acids.*

Experiments have been made by heating given weights of the respective acids with known volumes of standard potassium hydroxide solution (2 mols.) for given periods of time in a boiling water-bath and titrating the excess of alkali with standard oxalic acid, using phenolphthalein as indicator.

TABLE III.

α-Bromocinnamic Acid.—In each experiment 1·244 grams of acid were dissolved in 54·41 c.c. of 0·2015 N-potassium hydroxide. When $t = 0$, the number of c.c. of oxalic acid required $= A = 27·40$.

Time in hours.	$A - x$.	x.	$1/tx/A(A-x)$.
1·0	11·4	16·00	0·0512
1·5	9·0	18·40	0·0497
1·75	8·4	19·00	0 0472
3·0	6·4	21·00	0·0399

Mean = 0·0470. Calculated for N-potassium hydroxide = 1·16.

TABLE IV.

a-*Chlorocinnamic Acid.*—In each experiment, 1 gram was dissolved in 54·41 c.c. of 0·2015 N-potassium hydroxide. $A = 27·4$.

Time in hours.	$A - x$.	x.	$1/tx/A(A - x)$.
4	24·0	3·40	0·0001297
7	22·4	5·00	0·0001164
8	21·7	5·70	0·0001198
9	20·8	6·60	0·0001286

Mean = 0·001236. Calculated for N-potassium hydroxide = 0·0309.

TABLE V.

a-*Chlorocinnamic Acid.*—In each experiment, 1 gram of the acid was dissolved in 10·96 c.c. of N-potassium hydroxide. The number of c.c. of oxalic acid required when $t = 0$ was $5·48 = A$.

Time in hours.	$A - x$.	x.	$1/tx/A(A - x)$.
1	4·6	0·88	0·0350
2	3·81	1·67	0·0399
4	3·07	2·41	0·0358
6	2·42	3·06	0·0384

Mean = 0·0373.

TABLE VI.

a-*Bromo*allocinnamic *Acid.*—Similar to experiments in Table V, except that 1·244 grams of acid were used in each experiment. $A = 5·48$.

Time in hours.	$A - x$.	x.	$1/tx/A(A - x)$.
1	4·94	0·54	0·0199
2	4·37	1·11	0·0230
4	3·90	1·58	0·0185
6	3·15	2·33	0·0225

Mean = 0·0210.

TABLE VII.

a-*Chloro*allocinnamic *Acid.*—Same as for experiments in Table V. $A = 5·48$.

Time in hours.	$A - x$.	x.	$1/tx/A(A - x)$.
6·0	4·75	0·73	0·00468
8·0	4·50	0·98	0·00496
10·5	4·41	1·07	0·00422
24·0 *	4·16	—	—
24·0 *	4·32		

* In these experiments, there was a strong odour of chlorocinnamene.

TABLE VIII.

The constants of the four acids calculated for normal solutions are :

	$K.$		$K.$
a-Bromocinnamic	1·16	a-Brom$allo$cinnamic	0·0210
a-Chlorocinnamic	0·0373 and 0·0309	a-Chloro$allo$cinnamic	0·00462

These numbers clearly show that the hydrogen haloid is eliminated much more readily from the a-acids than from the $allo$-isomerides, and, further, that hydrogen bromide is eliminated much more readily than hydrogen chloride.

The readiness with which hydrogen haloids are eliminated from the a-halogen acids as compared with the stereoisomerides cannot be used as an argument in favour of the cis-positions of hydrogen and halogen in the a-acids, since it is now known that $trans$-addition of halogen and $trans$-elimination of halogen hydracid frequently occur (compare Werner, $Stereochemie$, p. 225). The relative rates of esterification of the a-acids and their $allo$-isomerides (compare Sudborough and Lloyd, Trans., 1898, **73**, 91 ; Sudborough and Roberts, Trans., 1905, **87**, 1840) harmonise best with the view that in the $allo$-acids the hydrogen and halogen are in the cis-positions.

$$\begin{array}{cc} \text{Ph·C·H} & \text{Ph·C·H} \\ \text{CO}_2\text{HC·X} & \text{X·C·CO}_2\text{H} \\ a\text{-}allo\text{-Acid.} & a\text{-Acid.} \end{array}$$

where $X = Cl$ or Br.

Preparation of Phenylpropiolic Acid from Cinnamic Acid Dichloride.

The numbers given in Table VIII show that in attempting to prepare phenylpropiolic acid from cinnamic acid dichloride it is advisable to proceed in two stages. The cinnamic acid dichloride is decomposed with aqueous potassium hydroxide at 0°, or with aqueous sodium hydroxide at the ordinary temperature, or the methyl ester is decomposed with alcoholic potash. In each case, the a-chloro- and a-chloro$allo$-acids are separated by means of their barium salts, and the pure a-chloro-acid is then heated with 2·5 mols. of 20 per cent. aqueous potassium hydroxide for eight hours on the water-bath. The mixture is allowed to cool—generally overnight—and any crystals of potassium a-chlorocinnamate removed. The clear solution is gradually acidified with concentrated hydrochloric acid, and the phenylpropiolic acid which separates as an oil quickly solidifies, and when dry is crystallised from carbon disulphide or from a mixture of chloroform and light petroleum (boiling at 80—90°).

From 60 grams of a-chlorocinnamic acid, we have obtained 25 grams of crude phenylpropiolic acid melting at 110—120°, and after crystallisation 20 grams of pure acid melting at 136°, which corresponds with a 42 per cent. yield of pure acid.

VI. *Derivatives of a-Chlorocinnamic Acids.*

We have prepared a number of derivatives of the a-chloro- and the a-chloro*allo*-acid from the acids obtained in the experiments recorded.

Methyl a-chlorocinnamate, $C_6H_5 \cdot CH\!:\!CCl \cdot CO_2Me$, is readily obtained by Fischer and Speyer's method of esterification, using a 4 per cent. solution of hydrogen chloride in methyl alcohol. It is readily soluble in all organic solvents, even in light petroleum (b. p. 40—50°), and crystallises from the latter in colourless prisms melting at 33—33·5°. The same ester has been prepared by Mulliken (*loc. cit.*, p. 27) from the silver salt.

a-Chlorocinnamyl chloride, $C_6H_5 \cdot CH\!:\!CCl \cdot COCl$, obtained in the usual manner by the action of phosphorus pentachloride (18 grams) on the acid (15 grams), is a yellow oil with a penetrating odour, and distils at 156° under 22 mm. pressure. During cold weather it sets to a mass of long, flat needles and may be crystallised from light petroleum, in which it is somewhat readily soluble, in the form of long, colourless, flat needles some 2 inches long, melting at 32·5°; 0·6436 gram was gently warmed with an excess of pure potassium hydroxide solution, then acidified with nitric acid, filtered from the a-chlorocinnamic acid when cold, and the filtrate precipitated with silver nitrate.

AgCl = 0·4606. Cl = 17·70. Theory requires 17·64 per cent.
When heated with lime, 0·4766 gave 0·6750 AgCl. Cl = 35·02.
$C_9H_6OCl_2$ requires 35·29 per cent.

a-Chlorocinnamide, $C_6H_5 \cdot CH\!:\!CCl \cdot CO \cdot NH_2$, obtained by adding the chloride to concentrated ammonium hydroxide, crystallises from dilute alcohol in long, flat, glistening plates melting at 121—122°; it also crystallises from benzene in colourless plates with a mother-of-pearl lustre.

0·4 gave 28·5 c.c. of moist nitrogen at 17° and 763 mm. N = 8·3.
C_9H_8ONCl requires N = 7·7 per cent.

When distilled with caustic potash, 0·6016 gram evolved ammonia which neutralised 31·93 c.c. of 0·1039 N sulphuric acid. NH_2 = 8·82.
C_9H_8ONCl requires NH_2 = 8·82 per cent.

The *anilide*, $C_6H_5 \cdot CH\!:\!CCl \cdot CO \cdot NH \cdot C_6H_5$, obtained in a similar manner, crystallises from alcohol in compact, colourless needles melting at 116—116·5°.

0·5 gave 24·5 c.c. of moist nitrogen at 18° and 748 mm.   N = 5·57.
$C_{15}H_{12}ONCl$ requires N = 5·44 per cent.

The p-*toluidide*, $C_6H_5 \cdot CH{:}CCl \cdot CO \cdot NH \cdot C_6H_4 \cdot CH_3$, crystallises from alcohol in well-developed, flat prisms or from benzene in snow-white, glistening plates melting at 116°.   A mixture of the anilide and p-toluidide begins to melt at 90° and is almost completely molten at 98°.

0·5 gave 22·7 c.c. of moist nitrogen at 17° and 760 mm.   N = 5·27.
$C_{16}H_{14}ONCl$ requires N = 5·16 per cent.

The isomeric o-*toluidide* crystallises from dilute alcohol in compact, colourless prisms, melts at 78°, and is much more readily soluble in most organic solvents than its isomeride.

0·526^6 gave 0·2667 AgCl.   Cl = 12·52.
$C_{16}H_{14}ONCl$ requires Cl = 13·06 per cent.

The a-*naphthalide*, $C_6H_5 \cdot CH{:}CCl \cdot CO \cdot NH \cdot C_{10}H_7$, crystallises from benzene or alcohol in small, colourless, felted needles melting at 134°. In the preparation of the a-naphthalide, the acid chloride does not react with the a-naphthylamine at all readily, but the reaction becomes vigorous when the mixture is heated.

0·5 gave 21·2 c.c. of moist nitrogen at 17° and 754 mm.   N = 4·88·
$C_{19}H_{14}ONCl$ requires N = 4·56 per cent.

The β-*naphthalide*, obtained in a similar manner, crystallises from benzene in compact prisms and from alcohol in flat, glistening plates or in needles melting at 139°.

0·506 gave 0·2348 AgCl.   Cl = 11·47.
$C_{19}H_{14}ONCl$ requires Cl = 11·53 per cent.

*Derivatives of a-Chloro*allocinnamic Acid.—The chloride was obtained by mixing together chloroform solutions of the *allo*-acid and phosphorus pentachloride, leaving the mixture for twenty-four hours, and then removing the chloroform and the oxychloride by distillation under reduced pressure.   The chloride itself was not distilled and was obtained as a yellow oil.

The *amide* crystallises from benzene in slender, white needles melting at 134°.

0·3230, when distilled with potassium hydroxide, evolved ammonia which neutralised 18·62 c.c. of 0·0958 N sulphuric acid. $NH_2 = 8·84$.
C_9H_8ONCl requires $NH_2 = 8·82$ per cent.

The *anilide* crystallises from dilute alcohol in slender, felted needles and melts at 138—139°.

0·5019 gave 0·2735 AgCl. Cl = 13·48.

$C_{15}H_{12}ONCl$ requires Cl = 13·77 per cent.

The p-*toluidide* crystallises from dilute alcohol in snow-white, prismatic needles melting at 132°.

0·5830 gave 0·3094 AgCl. Cl = 13·12.

$C_{16}H_{14}ONCl$ requires Cl = 13·06 per cent.

In conclusion, we desire to express our thanks to Mr. S. H. Beard for assistance in the preparation and analysis of certain of the compounds and to the Research Fund Committee of the Chemical Society for a grant which has assisted in meeting the expenses involved in this investigation.

UNIVERSITY COLLEGE OF WALES,
ABERYSTWYTH.

XIII.—*Some Derivatives of Naphthoylbenzoic Acid and of Naphthacenequinone.*

By JAN QUILLER ORCHARDSON and CHARLES WEIZMANN.

Naphthacene,

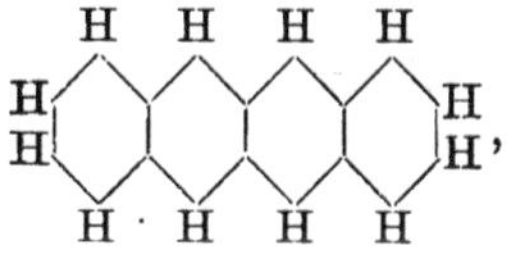

and *naphthacenequinone,*

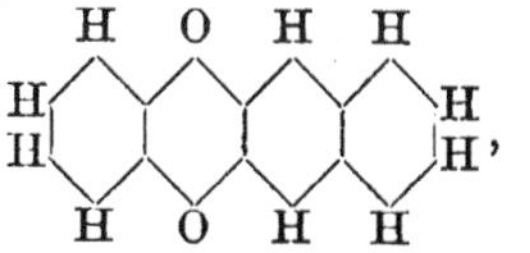

were first prepared from ethindiphthalide by Gabriel and Leupold (*Ber.*, 1898, 31, 1279), and subsequently Deichlers and Weizmann (*Ber.*, 1903, 36, 547) obtained (1)-*hydroxynaphthacenequinone* by the action of sulphuric and boric acids on a mixture of phthalic anhydride and a-naphthol :

1 2

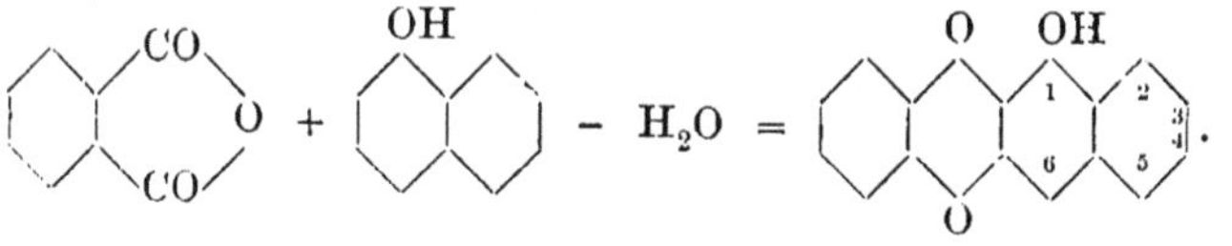

By melting phthalic anhydride with α-naphthol and boric acid alone they obtained (1)-hydroxynaphthoylbenzoic acid,

and found that this acid is evidently an intermediate product in the formation of (1)-hydroxynaphthacenequinone by the above-mentioned process, since when heated with sulphuric and boric acids it is converted into this substance.

In the following experiments, we have sought to make various derivatives of (1)-hydroxynaphthoylbenzoic acid, and with their aid to obtain the corresponding derivatives of naphthacenequinone, this indirect method being necessary on account of the difficulty of obtaining pure products by direct substitution in naphthacenequinone itself. We met, however, in the course of the experiments, with greater difficulty than we had anticipated, owing to the facility with which most of the substituent groups are eliminated when the attempt is made to condense the derivatives of naphthoylbenzoic acid to the corresponding naphthacenequinones by means of concentrated sulphuric acid.

During the preparation of (1)-chloronaphthoylbenzoic acid by the action of phosphorus pentachloride on (1)-hydroxynaphthoylbenzoic acid (Proc., 1904, 20, 220), we had previously noticed the formation of a red substance insoluble in caustic potash. By a modification of the experimental conditions this was ultimately obtained as the chief product of the reaction, and we find that it is evidently a monochloronaphthacenequinone, $C_6H_4{\displaystyle <}_{CO}^{CO}{\displaystyle >}C_{10}H_5Cl$, and isomeric with that prepared by Pickles and Weizmann (Proc., 1904, 20, 220).

The new monochloronaphthacenequinone, when digested with aniline, is converted into phenylaminonaphthacenequinone,

$$C_6H_4{\displaystyle <}_{CO}^{CO}{\displaystyle >}C_{10}H_5 \cdot NH \cdot C_6H_5.$$

We next found that (1)-hydroxybromonaphthoylbenzoic acid,

$$C_6H_4{\displaystyle <}_{CO_2H}^{CO-}{\displaystyle >}C_{10}H_5{\displaystyle <}_{Br(6?)}^{OH(1)},$$

was very readily prepared by the action of bromine on (1)-hydroxynaphthoylbenzoic acid, and the corresponding (1)-hydroxybromonaphthacenequinone, $C_6H_4{\displaystyle <}_{CO}^{CO}{\displaystyle >}C_{10}H_4{\displaystyle <}_{Bz(6?)}^{OH(1)}$, was obtained from this by the action of concentrated sulphuric acid, although considerable decomposition occurred during the reaction. The position of the bromine is not yet known with certainty, but as the substance does not form an azo-compound with diazobenzene chloride, whereas the (1)-hydroxynaphthoylbenzoic acid does, it is probable that the bromine occupies the position (6) indicated in the above formula.

Considerable difficulty was experienced in obtaining a monochloro-monobromonaphthoylbenzoic acid, $C_6H_4 <{}^{CO}_{CO_2H} C_{10}H_5 <{}^{Cl}_{Br}$, but we ultimately obtained a good yield of the acid by treating (1)-hydroxy-bromonaphthoylbenzoic acid with phosphorus pentachloride in the presence of benzene.

In attempting to prepare nitro- and amino-derivatives of naphthacenequinone, we found that the action of nitric acid on naphthacenequinone did not give satisfactory results owing to the difficulty of separating the various products of the reaction. Again, (1)-hydroxy-naphthoylbenzoic acid is decomposed by the action of nitric acid ; we therefore first methylated it, and on treating the methyl ester of the methoxynaphthoylbenzoic acid thus formed with nitric acid, a good yield of the mononitro-compound was obtained, which, on hydrolysis, yielded 1-hydroxynitronaphthoylbenzoic acid :

$$C_6H_4 <{}^{CO}_{CO_2H} C_{10}H_6 \cdot OH \;\rightarrow\; C_6H_4 <{}^{CO}_{CO_2CH_3} C_{10}H_6 \cdot O \cdot CH_3 \;\rightarrow$$

$$C_6H_4 <{}^{CO}_{CO_2CH_3} C_{10}H_5 <{}^{NO_2}_{O \cdot CH_3} \;\rightarrow\; C_6H_4 <{}^{CO}_{CO_2H} C_{10}H_5 <{}^{NO_2}_{OH} .$$

We were unable to obtain the hydroxynitronaphthacenequinone corresponding to this acid by the action of sulphuric acid, and therefore proceeded to reduce it with zinc and acetic acid. It was interesting to find that by this means, instead of obtaining 1-hydroxyamino-naphthoylbenzoic acid, condensation occurred simultaneously with reduction and yielded 1-hydroxyaminonaphthacenequinone,

$$C_6H_4 <{}^{CO}_{CO}> C_{10}H_4 <{}^{OH}_{NH_2} .$$

Experiments are at present in progress which we hope will clearly demonstrate the positions of the nitro- and amino-groups in the compounds mentioned above, and the results of which will be published shortly.

EXPERIMENTAL.

Monochloronaphthacenequinone.

In order to prepare this substance perfectly dry, (1)-hydroxy-naphthoylbenzoic acid (40 grams) was mixed with benzene (200 c.c.) and a considerable excess of phosphorus pentachloride (80 grams) was added gradually in such a way that the reaction was kept well under control. The whole was then heated on the boiling water-bath with a reflux condenser until no further evolution of hydrogen chloride occurred. The benzene was next distilled off over a free flame (not from the water-bath), and when most of the solvent had passed over, a brisk reaction again set in, with further evolution of hydrogen chloride, and the contents of the flask assumed a deep red colour. Water was added to decompose phosphorus oxychloride and pentachloride, the solid was collected at the pump, digested with caustic potash, washed, and dried. It was then recrystallised from nitrobenzene, from which it separated in bright red needles, which retained their colour after repeated crystallisation in presence of animal charcoal.

0·1790 gave 0·4810 CO_2 and 0·0582 H_2O.   C = 73·3; H = 3·6.

0·1698   ,,   0·0768 AgCl.   Cl = 11·2.

$C_{18}H_9O_2Cl$ requires C = 73·3; H = 3·1; Cl = 12·1 per cent.

This monochloronaphthacenequinone melts at 254°, and is therefore isomeric with the yellow compound obtained by the action of sulphuric acid on 1-chloronaphthoylbenzoic acid (Pickles and Weizmann, Proc., 1904, 20, 220). *Phenylaminonaphthacenequinone* was obtained from the above chloronaphthacenequinone by boiling it for two hours with just sufficient aniline to dissolve it. On cooling, the substance separated in red leaves melting at 245°.

0·1120 gave 0·3379 CO_2 and 0·0472 H_2O.   C = 82·2; H = 4·7.

0·2108   ,,   7·6 c.c. nitrogen at 18° and 756 mm.   N = 4·12.

$C_{24}H_{15}O_2N$ requires C = 82·2; H = 4·3; N = 4·00 per cent.

When warmed with sulphuric acid, this phenylaminonaphthacene-quinone yields a strongly fluorescent solution, indicating that condensation has taken place with formation of the corresponding acridine derivative, but the latter substance has not yet been obtained in a state sufficiently pure for analysis.

(1)-*Hydroxybromonaphthoylbenzoic Acid and* (1)-*Hydroxybromo-naphthacenequinone.*

This acid was readily prepared by slowly adding bromine (40 grams) to (1)-hydroxynaphthoylbenzoic acid (60 grams) suspended in carbon disulphide. When the initial vigorous reaction had subsided, the whole was boiled on the water-bath for four hours and until no further evolution of hydrogen bromide was observed. The carbon disulphide was then distilled off and the residue, which consisted of the almost pure acid, was recrystallised from glacial acetic acid, a small quantity of sodium bisulphite being added to remove free bromine. The new bromo-acid separated in pale yellow crystals melting at 236°.

0·1822 gave 0·3875 CO_2 and 0·0455 H_2O. C = 57·9 ; H = 2·8.

0·1724 „ 0·0862 AgBr. Br = 21·5.

$C_{18}H_4BrO_4$ requires C = 58·2 ; H = 3·0 ; Br = 21·5.

(1)-*Hydroxybromonaphthoylbenzoic* acid dissolves in sulphuric acid, yielding a brown solution which, on warming, becomes green, then blue, and finally deep red. If this solution is heated to about 140°, at which temperature bromine begins to be evolved, and is then immediately poured into water, a red precipitate is deposited, which consists mainly of (1)-*hydroxybromonaphthacenequinone*. It is collected on a filter, washed with hot sodium carbonate solution, and crystallised from nitrobenzene, from which it separates in red needles, which do not melt at 300°. Owing to some elimination of bromine during the preparation, a specimen sufficiently pure to give good results on analysis was not obtained.

(1)-*Chlorobromonaphthoylbenzoic Acid and* (1)-*Chlorobromonaphthacenequinone.*

Twenty grams of (1)-hydroxybromonaphthoylbenzoic acid (see last section) were mixed with a small quantity of benzene, and 23 grams of phosphorus pentachloride were added gradually, the whole being then warmed on the water-bath until no further evolution of hydrogen chloride occurred. The time required for the completion of the reaction was about four hours. Only a small quantity of the solvent' and no excess of phosphorus pentachloride should be employed, otherwise a pale yellow, crystalline compound containing phosphorus is obtained, which, owing to its similar appearance and behaviour, is readily mistaken for the chlorobromonaphthoylbenzoic acid. This phosphorus compound gave considerable trouble, and various expedients were tried to avoid its formation, such as varying the solvent used and heating phosphorus pentachloride with hydroxybromonaphthoylbenzoic acid in the dry state. The former procedure always gave the phosphorus compound, and the latter variously halogenated mixtures. It was found by using benzene as the solvent and taking special precautions that an excellent yield of the chlorobromo-acid could be obtained. At the end of the reaction, no solid matter should have separated, but the product should be a rather viscid, dark brown oil. From this as much benzene is distilled off as possible at the temperature of the boiling water-bath. The residue is then treated with water and allowed to stand for some hours with frequent shaking, and until the acid chloride is completely decomposed. The white solid which separates is collected on a filter, washed, and recrystallised from glacial acetic acid, from which it separates in almost colourless crystals melting at 180°.

0·2313 gave 0·4666 CO_2 and 0·0531 H_2O.   C = 55·0 ; H = 2·5.

0·2296   ,,   0·1542 AgCl and AgBr.   Cl = 8·9 ; Br = 20·0.

$C_{18}H_{10}O_3ClBr$ requires C = 55·4 ; H = 2·6 ; Cl = 8·7 ; Br = 20·0 per cent.

(1)-*Chlorobromonaphthoylbenzoic acid* undergoes similar colour changes to (1)-hydroxybromonaphthoylbenzoic acid when treated with concentrated sulphuric acid, giving finally a chlorobromonaphthacenequinone, which, however, on account of the occurrence of partial decomposition during the reaction, was not obtained in a sufficiently pure state for analysis.

Methyl   (1)-Methoxynaphthoylbenzoate,   Methyl   (1)-Methoxy-6-nitro-naphthoylbenzoate, and (1)-Hydroxy-6-nitronaphthoylbenzoic Acid.

In order to prepare the first mentioned of these substances, (1)-hydroxynaphthoyl benzoic acid (40 grams) was dissolved in an excess of caustic potash (25 per cent.) containing about 40 grams of this alkali. To this solution, when quite cold, methyl sulphate (30 grams) was added in very small portions at a time with frequent shaking, cooling being resorted to when necessary. A still greater excess of methyl sulphate may often be used with advantage, but the solution must always remain alkaline. During the operation, a yellow substance separates out, either in the solid state or as an oil, according to the temperature at which the reaction is carried out. When cold, the solid is collected on a filter, washed, and recrystallised from glacial acetic acid.

Methyl (1)-*methoxynaphthoylbenzoate* separates from acetic acid in colourless, transparent crystals, which become opaque on exposure to air and melt at 110°.

0·2198 gave 0·6038 CO_2 and 0·1068 H_2O.   C = 74·9 ; H = 5·3.

$C_{20}H_{16}O_4$ requires C = 75·0 ; H = 5·0 per cent.

The yield of this substance varies and a considerable quantity of partially methylated acid remains in the alkaline solution.

We have several times attempted to obtain (1)-methoxynaphthacene-quinone by first partially saponifying the above-described methyl ester with caustic potash, and then acting on the free acid with concentrated sulphuric acid. Complete hydrolysis, however, occurs under these conditions and the product obtained is (1)-hydroxynaphthacene-quinone.

Nitration.—The finely-powdered methyl (1)-methoxynaphthoyl-benzoate was treated with nitric acid (sp. gr. 1·42) in the cold, the vigorous reaction being kept under control by immersion in ice when necessary. A small portion of the ester passed into solution, the remainder changing first to a soft, and finally to a brittle mass, which floated on the surface of the liquid. After diluting with water, the solid matter was collected, washed well, and recrystallised from glacial acetic acid, from which it separated in bright yellow crystals melting at 136°.

0·1715 gave 0·4305 CO_2 and 0·0640 H_2O. $C = 68·4$; $H = 4·1$.
0·2272 ,, 8·6 c.c. N at 18° and 762 mm. $N = 4·37$.
$C_{20}H_{15}O_6N$ requires $C = 68·3$; $H = 4·4$; $N = 3·82$ per cent.

From the *methyl* (1)-*methoxy-6-nitronaphthoylbenzoate*, by boiling for five hours with strong caustic potash and then precipitating with hydrochloric acid, (1)-*hydroxy-6-nitronaphthoylbenzoic acid* was obtained. It crystallises from glacial acetic acid in slender, lemon-yellow needles.

0·2126 gave 0·4920 CO_2 and 0·0460 H_2O. $C = 63·5$; $H = 3·3$.
0·2310 ,, 8·3 c.c. N at 17° and 762 mm. $N = 4·18$.
$C_{18}H_{11}O_6N$ requires $C = 64·1$; $H = 3·0$; $N = 4·15$ per cent.

(1)-*Hydroxy-6-nitronaphthoylbenzoic acid* melts at 220° and dissolves in caustic potash, forming a deep orange-red solution. When treated with concentrated sulphuric acid, it does not yield hydroxynitronaphthacenequinone, because decomposition takes place with evolution of oxides of nitrogen.

We next attempted to prepare (1)-hydroxy-6-aminonaphthoylbenzoic acid by reducing the above-described hydroxynitro-acid, but found that the product of the reaction consisted of (1)-*hydroxy-6-aminonaphthacenequinone*.

The experiment was conducted as follows : (1)-hydroxy-6-nitronaphthoylbenzoic acid was dissolved in hot glacial acetic acid, and then zinc dust gradually added in small quantities, when the solution became deep red and finally deposited minute, dark red, glistening crystals. After gently boiling for half an hour, the crystalline precipitate was collected on a filter while hot, washed with acetic acid and water, dried well, and extracted with nitrobenzene, from which solvent deep red crystals, insoluble in sodium carbonate, separated.

0·2131 gave 0·5896 CO_2 and 0·0847 H_2O. $C = 75·4$; $H = 4·40$.
0·2495 ,, 10·3 c.c. N at 15° and 762 mm. $N = 4·9$.
$C_{18}H_{11}O_3N$ requires $C = 74·7$; $H = 3·8$; $N = 4·7$ per cent.

(1)-*Hydroxy-6-aminonaphthacenequinone* melts above 300°. It dissolves in caustic potash with a deep red colour, and its solution in concentrated sulphuric acid exhibits a beautiful and very strong green fluorescence.

THE VICTORIA UNIVERSITY OF MANCHESTER.

XIV.—*Ethyl β-Naphthoylacetate.*

By Charles Weizmann and Ernest Basil Falkner.

During the course of his researches on ethyl acetoacetate, Claisen (*Annalen*, 1896, 291, 67) showed that this reagent may be used with great advantage in the synthesis of other ketonic esters, and it thus became possible to prepare a considerable number of important compounds which were either previously unknown or had only been obtained with difficulty. Ethyl benzoylacetate, for example, is readily prepared by acting on ethyl sodioacetoacetate (2 mols.) with benzoyl chloride (1 mol.) and then decomposing the sodium compound of ethyl benzoylacetoacetate, which is produced by digesting with ammonia and ammonium chloride.

$$2CH_3 \cdot CO \cdot CHNa \cdot CO_2Et + C_6H_5 \cdot COCl = \genfrac{}{}{0pt}{}{C_6H_5 \cdot CO}{CH_3 \cdot CO}\!\!\!>\!CNa \cdot CO_2Et +$$
$$CH_3 \cdot CO \cdot CH_2 \cdot CO_2Et + NaCl.$$

$$\genfrac{}{}{0pt}{}{C_6H_5 \cdot CO}{CH_3 \cdot CO}\!\!\!>\!CNa \cdot CO_2Et + H_2O = C_6H_5 \cdot CO \cdot CH_2 \cdot CO_2Et + CH_3 \cdot CO_2Na.$$

At a later date, Needham and Perkin (Trans., 1904, 85, 150) used a similar decomposition, namely, the interaction of *o*-nitrobenzoyl chloride with ethyl sodioacetoacetate for the preparation of ethyl *o*-nitrobenzoylacetate.

Since the compounds in the naphthalene series corresponding to ethyl benzoylacetate are unknown and should be of considerable value as synthetical agents, we undertook the present research, the object of which was to prepare and investigate ethyl β-naphthoylacetate and some of its derivatives.

Pure β-naphthoic acid was converted, by the action of phosphorus pentachloride, into β-naphthoyl chloride, and this was then allowed to react with the sodium derivative of ethyl acetoacetate, when a sparingly soluble sodium derivative was obtained, which on treatment with acid yielded a solid mass of ethyl β-naphthoylacetoacetate,

$$\text{C}_{10}\text{H}_7\!\!-\!\genfrac{}{}{0pt}{}{CO \cdot CH \cdot CO_2Et}{CO \cdot CH_3}$$

The substance melts at 57°, and when digested with ammonia and ammonium chloride is partially hydrolysed with elimination of the acetyl group and formation of ethyl β-naphthoylacetate,

$$\text{C}_{10}\text{H}_7\!\!-\!CO \cdot CH_2 \cdot CO_2Et,$$

an interesting substance which melts at 34°, gives with ferric chloride a red coloration, is not readily soluble in dilute sodium hydroxide, and in general shows properties similar to those of ethyl benzoylacetate. When treated with phenylhydrazine, it is readily converted into a hydrazone which crystallises in yellow needles, melts at 95°, and probably possesses the formula

$$\text{—C·CH}_2\text{·CO}_2\text{Et}$$
$$\text{N·NH·C}_6\text{H}_5$$

Preliminary experiments indicate that exactly similar substances are produced when α-naphthoic acid is substituted for the β-acid in the above experiment, and it is proposed to submit both series of substances to a detailed investigation.

Ethyl β-Naphthoylacetoacetate.

β-Naphthoyl Chloride.—After several experiments, it was found that the following process gives a good yield of this acid chloride. β-Naphthoic acid (35 grams) and phosphorus pentachloride (45 grams) are mixed in small quantities at a time in a distilling flask, which is warmed gently in a water-bath to start the reaction. The decomposition takes place rapidly, and at the end of half an hour the acid chloride is ready for distillation. After the phosphorus oxychloride had passed over, the β-naphthoyl chloride distilled at 208° under the ordinary pressure and solidified to a light lemon-yellow, crystalline mass melting at about 40° and possessing a sweet and rather nauseating smell. The yield obtained was 40 grams.

Condensation of β-Naphthoyl Chloride with the Sodium Derivative of Ethyl Acetoacetate.

The sodium ethoxide required was first made by dissolving sodium (5 grams) in absolute alcohol (90 c.c.), and the solution was allowed to cool. Ethyl acetoacetate (28 grams) was weighed into a dry wide-necked bottle, and to this half the quantity of sodium ethoxide was added and the mixture well shaken. The bottle was now fitted with a mechanical stirring apparatus and the products cooled to 5° by means of ice water.

β-Naphthoyl chloride (20 grams) was then dissolved in pure dry ether (75 c.c.) and half this quantity added, drop by drop, through a burette, to the sodium derivative of ethyl acetoacetate, the whole being well stirred during the addition. The process of adding the acid chloride took from ten to fifteen minutes, and at the end of this time the mixture, which had assumed a bright yellow colour, was allowed to stand for half an hour, the temperature still being kept below 5°. Half the remaining quantity of the sodium ethoxide was then slowly added with stirring, and then half the remaining quantity of acid chloride solution as before. The mixture, which had now become viscid, was again allowed to stand for half an hour. This process was continued until all the ethoxide and acid chloride solution had been added, the bottle was then removed from the ice water, a little dry ether added, and the whole allowed to stand overnight in a cool place.

The semi-solid mass was filtered at the pump and the bright yellow sodium derivative of ethyl β-naphthoylacetoacetate washed with ether and dried on a porous plate. The crude sodium derivative, which was obtained in a yield of 51 grams, was dissolved in aqueous alcohol (10 per cent.) and mixed with an excess of dilute acetic acid, when an oil separated which, when cooled with ice, solidified to a pale pink, brittle mass. This was collected and recrystallised from alcohol with the aid of animal charcoal. The melting point was found to be 57°, and the yield obtained was 17 grams.

0·1942 gave 0·507 CO_2 and 0·099 H_2O.   C = 71·2 ; H = 5·7.
$C_{17}H_{16}O_4$ requires C = 71·83 ; H = 5·63.

Ethyl β-naphthoylacetoacetate melts at 57° and is readily soluble in ether, alcohol, and sodium carbonate solution, and its alcoholic solution gives a reddish-violet coloration with ferric chloride.

Ethyl β-Naphthoylacetate.

Ethyl β-naphthoylacetoacetate (10 grams) was finely powdered, placed in a large beaker, and dissolved in aqueous ammonia (100 c.c.), made by diluting concentrated ammonia solution with an equal volume of water. When the ester had completely dissolved, ammonium chloride (12 grams) dissolved in a small quantity of water was added, and the liquid was then well stirred and gently heated in a water-bath. The clear yellow solution gradually became milky and, after an interval of fifteen minutes, an oil was deposited which, on cooling with ice, solidified. This solid was collected at the pump, washed with water, and recrystallised from alcohol, when very pale pink, opaque crystals were obtained which melted at 34° and gave the following results on analysis.

0·268 gave 0·728 CO_2 and 0·136 H_2O.   C = 74·08 ; H = 5·63.
$C_{15}H_{14}O_3$ requires C = 74·3 ; H = 5·7.

Ethyl β-naphthoylacetate is sparingly soluble in caustic soda solution ; its alcoholic solution gives a greenish-blue precipitate with copper sulphate and a red coloration with ferric chloride. When the ester is added to a solution of phenylhydrazine in acetic acid, a crystalline solid soon separates which, after recrystallising from acetic acid, from which it separates in golden-yellow needles, melts at 95°.

0·198 gave 0·552 CO_2 and 0·114 H_2O.   C = 76·0 ; H = 6·39.

0·157 ,, 11·8 c.c. N at 17° and 755 mm.   N = 8·8.

$C_{21}H_{20}O_2N_2$ requires C = 75·9 ; H = 6·02 ; N = 8·4.

This substance is therefore the *hydrazone* of ethyl β-naphthoyl-acetate.

THE VICTORIA UNIVERSITY OF MANCHESTER.

XV.—*Some New Platinocyanides.*

By LEONARD ANGELO LEVY and HENRY ARNOTT SISSON.

CONSIDERING the comparatively advanced state of our knowledge of the constitution of organic substances possessing the property of fluorescence, it is remarkable that so little should be known about this phenomenon as exhibited by platinocyanides. We were accordingly led to investigate the effect of the basic radicle and also such conditions as hydration, purity, state of division, &c., on the character of the fluorescence. For example, the lithium salt has a red fluorescence, whilst the sodium and barium salts exhibit yellow and green colorations respectively, these being the colours of the characteristic lines in the spectra of these metals.

In the course of our researches, we have prepared several platinocyanides, including the hydrazine and hydroxylamine salts, of which we can find no account in the literature. These substances showed remarkable colour changes on a very slight alteration of temperature, and as these properties appeared interesting and uncommon we have investigated them further.

Hydrazine platinocyanide, $N_2H_4,H_2Pt(CN)_4,3H_2O$, is prepared by double decomposition between equivalent quantities of hydrazine sulphate and barium platinocyanide ; the solution thus obtained, when allowed to evaporate spontaneously, deposits red crystals, showing blue and purple colours by reflected light.

These crystals are unstable under ordinary atmospheric conditions, and, when air-dried, become partially or completely light yellow and opaque, according to the hygrometric state of the atmosphere. The transformation can always be completed by passing dry nitrogen over them. We analysed this light yellow modification, this being the most stable in air.

0·1300 gave 24·43 c.c. nitrogen at 16° and 748 mm.   N = 21·56.

0·2471 ,, 0·1246 Pt.   Pt = 50·42.

0·3807 ,, 0·1730 CO_2 and 0·1049 H_2O.   C = 12·39 ; H = 3·07.

0·2269, after heating for several hours at 100°, left a dull olive-green powder, the loss of weight being 0·0309. H_2O = 13·6 per cent.

$C_4H_6N_6Pt,3H_2O$ requires C = 12·40 ; H = 3·12 ; N = 21·77 ; Pt = 50·34 ; H_2O = 13·9 per cent.

Hydrates.—The damp red crystals become white when slightly warmed, as when a tube containing them is held in the hand, and they recover their original colour on cooling. This transformation occurs at about 28° and appears to be due to loss of water.

We arrived at this conclusion from the following facts :

(i) The red crystals, when partly exposed to air, pass through the white form before becoming yellow.

(ii) The red modification becomes white when dry nitrogen is passed over it.

(iii) Methyl alcohol turns the red crystals white before dissolving them.

This white modification is very unstable at the ordinary temperature, and rapidly becomes red and yellow on exposure to air. No determination of the percentage of water is possible owing to the rapidity with which this change occurs. The state of hydration of the red salt cannot be accurately estimated, as it is very unstable when dry. This red salt may be kept in a damp atmosphere, but the amount of adherent water would vitiate any analysis.

When pure dry nitrogen is passed over the red salt, its colour changes in succession to white, then light yellow, dark yellow, brown, and olive-green. An approximate estimation of the water in the red modification was obtained by passing dry nitrogen over the salt until it became yellow.

0·1500 gram, when so treated, gave 0·1409 gram of yellow salt.

The loss of one and two molecules of H_2O require respectively 0·1432 and 0·1366 gram of the yellow salt.

Hence this hydrate probably contains four molecules of water. The slightly greater loss of water is probably due to dampness of the red salt. The latter fluoresces faintly under radium, but no more brightly than does glass under similar conditions.

All modifications are readily soluble in methyl alcohol, from which they are wholly precipitated by ether. The colour of the precipitate varies according to the amount of water present and may be red, purple, orange, yellow, or white. If the process is repeated once or twice on the precipitate, a pure white, crystalline salt can always be produced. When thus obtained, it is even more unstable than when prepared by the method already described.

Action of Light.—Hydrazine platinocyanide is affected by light. A print can be obtained from a negative by placing it over a sheet of paper soaked in the methyl-alcoholic solution and allowed to dry. The print thus obtained is yellow and grey and shows details fairly well. We have not yet been able to fix this image, which is destroyed by damp, but can be reprinted when dry. The paper is also attacked and becomes brittle.

If hydrazine platinocyanide is prepared with excess of barium platinocyanide, fine red crystals are obtained together with some normal salt. These are permanent in air, contain barium, and, on heating to about 60°, or when placed in a vacuum desiccator, become opaque and assume a lustrous, beetle-green colour. We intend to pursue the investigation of these substances.

Hydroxylamine Platinocyanide, $(NH_2OH)_2H_2Pt(CN)_4,2H_2O$.—This salt is prepared by double decomposition between barium platinocyanide and hydroxylamine sulphate. The solution thus obtained, when allowed to evaporate spontaneously, leaves very soluble red crystals stable in air at the ordinary temperature.

0·1475 gave 26·1 c.c. nitrogen at 17° and 751 mm.   N = 20·3.

0·2338 ,, 0·1132 Pt.   Pt = 48·42.

0·2387, after remaining in a vacuum desiccator over sulphuric acid, gave a black powder, the loss of weight being 0·0217 ; $H_2O = 9·09$ per cent.

$C_4H_8N_6Pt,2H_2O$ requires N = 20·8 ; Pt = 48·39 ; $H_2O = 8·93$ per cent.

Hydrates.—The red crystals become bright yellow when slightly warmed, and this is exactly analogous to the colour variations occurring with the hydrazine salt and takes place quite as readily. These yellow crystals again become red on cooling ; the salt is soluble in methyl alcohol, but is not satisfactorily precipitated by ether. The change from red to yellow is accompanied by loss of weight.

The foregoing salts are the first members of a series of platinocyanides which we hope shortly to prepare in the course of our investigations on the effect of the molecular weight of the base on the character of the fluorescence. We propose to prepare platinocyanides of alkyl-substituted hydrazines, hydroxylamines, and other bases, thus obtaining a series of platinocyanides the molecular weights of which differ by equal or known increments. The more common aromatic bases, such as phenylhydrazine, do not yield the well-crystallised salts so essential for the purpose of comparing their fluorescence.

Platinocyanides of fluorescent bases or radioactive substances should prove of special interest. The radium salt should be self-luminous, and an investigation of its fluorescent properties might throw some light on the origin of the fluorescence conferred by the platinocyanide group on its salts.

Most of the materials necessary for the foregoing researches have been purchased by means of funds kindly supplied by the Government Grant Committee of the Royal Society.

Our thanks are due to Dr. H. J. H. Fenton, F.R S., for valuable advice.

Chemical Laboratory,
Cambridge University.

XVI.—*Studies in Fermentation. I. The Chemical Dynamics of Alcoholic Fermentation by Yeast.*

By Arthur Slator, Ph.D.

The study of the velocity of chemical reactions consists to a large extent of the investigation of the dependence of the velocity on the concentrations of the reacting substances and certain accelerating and inhibiting agents. The concentrations of the reagents are varied and the corresponding velocities measured. If the chemical change in question is unaccompanied by disturbing side reactions, the simplest method of changing the concentrations of the reagents is to allow these substances to be used up in the reaction. In such cases, an integrated formula is employed to calculate the results, which are usually expressed in the form of "constants" or values of K. If these values of K remain constant throughout the reaction, this is probably the best method of investigation. If through any disturbing influences the values do not remain constant, the results expressed in this way are sometimes difficult to interpret and may even be misleading. In these cases the more direct method of investigation is desirable. The velocity of the chemical change is measured over as short a range of the reaction as is consistent with accuracy, and the concentrations of the reagents altered by dilution with the solvent. This method of considering only initial velocities is the one best employed in the investigation of the reaction which forms the subject of the present communication. The dynamics of the fermentation of sugars under the influence of yeast and various preparations from yeast has been studied by a number of investigators, and many methods have been used to follow the reaction. If we consider the fermentation of dextrose $C_6H_{12}O_6 = 2C_2H_5 \cdot OH + 2CO_2$, it is evident that the reaction-velocity can be measured by observing the rate of decrease of dextrose, or the rate of formation of alcohol or carbon dioxide. Dumas (*Ann. Chim. Phys.*, 1874, [iii], 81) uses the copper reagent test for dextrose and observes the time for complete fermentation. A. J. Brown (Trans., 1892, 61, 369) and J. O'Sullivan (*J. Soc. Chem. Ind.*, 1898, 17,559; *J. Inst. Brewing*, 1899, 5, 161) estimate in some experiments the alcohol formed, in other cases the change in optical activity owing to disappearance of the sugar. The latter method is also used by A. L. Stern (Trans., 1899, 75, 201) in his experiments on yeast growth during fermentation, and also by J. H. Aberson (*Rec. trav. chim.*, 1903, 22, 78) in his measurements of the velocity of fermentation throughout the whole range of the reaction. Buchner (*Die Zymase-gärung*, 1903, E. Buchner, H. Buchner, M. Hahn) working with yeast-juice, R. O. Herzog working with " zymin "-yeast treated with acetone (*Zeit. physiol. Chem.*, 1902, 37, 149), and H. Euler also using yeast-juice, estimate volumetrically or gravimetrically the amount of carbon dioxide evolved. The methods devised by these investigators require the fermentation to proceed for some time before the velocity can be accurately measured. Although many interesting and important facts have been discovered in these researches, it is probable that a more sensitive method of investigation would lead to a greater knowledge of the reaction.

A modification of the method of estimating the carbon dioxide evolved seemed to offer a degree of sensitiveness greater than that obtained by other workers. If a sugar solution undergoing fermentation is placed in a closed vessel, the amount of carbon dioxide liberated can be estimated by the pressure produced. Thus 50 c.c. of a 10 per cent. dextrose solution fermenting in a flask of 150 c.c. capacity at a temperature of 15° gives a change in pressure of 1 cm. of mercury for a fermentation of 7·4 milligrams of dextrose.* An apparatus constructed on the principle of measuring the rate of fermentation by change in pressure due to the gas evolved was found to be workable and to give concordant results. This method differs from those mentioned above in that the time of the experiment extends over only a few minutes, only small quantities of sugar are fermented, and only a very small range of the reaction is considered.

* In the calculation, it is assumed that the solution is well shaken to overcome supersaturation. At 15°, the concentration of carbon dioxide is approximately the same in the gaseous as in the liquid phase.

The Apparatus and Method of Investigation.

Details of the apparatus for estimating the rate of fermentation are as follows : an ordinary glass bottle of about 150 c.c. capacity, with a fairly narrow neck, is connected by a piece of pressure tubing to a manometer as represented in the figure. The side-tube is connected to the pump in order to exhaust the apparatus, which is

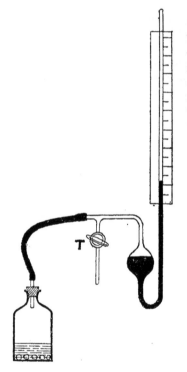

then kept air-tight by means of the tap *T*. The reacting solution, usually 50 c.c., is placed in the bottle, which also contains twenty to thirty fairly large glass beads in order to assist in the thorough shaking of the solution. The bottle rests in a thermostat and during the experiment is taken out at intervals and shaken vigorously by hand, so that the carbon dioxide in the solution is in equilibrium with that in the space above. The pressure at the beginning of the experiment is usually 3—4 cm. of mercury. The change in pressure registered on the metal scale of the manometer is a measure of the amount of fermentation. Throughout this investigation, only relative amounts are measured by a direct comparison of the manometer readings in parallel experiments.

To the solution containing sugar and yeast was usually added a small quantity of nutrient salts (3 grams of asparagine, 1 gram of K$_2$HPO$_4$, 0·5 gram of MgSO$_4$,7H$_2$O in 1 litre), although it was found later that these salts had little influence on the initial rate of fermentation.

The Influence of the Amount of Yeast on the Velocity of Fermentation.

The influence of quantity of yeast on the velocity of alcoholic fermentation has been examined by J. O'Sullivan (*loc. cit.*) ; the experiments are, however, probably complicated by change in activity of the yeast. If such complications are eliminated, the influence of the concentration of the yeast can be predicted, for if each cell acts individually then the velocity of fermentation must be proportional to the concentration of the yeast. The following tests show that this result can be experimentally realised. Some brewery yeast was shaken up with water containing a small quantity of nutrient salts. The rates of fermentation of six solutions of five per cent. dextrose containing respectively 1, 3, 5, 10, 20, and 25 c.c. of the yeast mixture were measured and shown to be almost exactly in the ratio of the concentrations of the yeast. The experiments were carried out in the manner described at a temperature of 30°, for at this temperature the activity of the yeast is unaltered during the time of the experiment. Details of the measurements are given in Table I, where $t =$ time in minutes, $P =$ manometer reading in cm., $D =$ differences in equal time intervals.

TABLE I.—*The Influence of the Concentration of Yeast on the Velocity of Fermentation.*

Temperature = 30°.   Vol. of solution = 50 c.c.

1 .c.c.

t.	P.	D.
0 minutes	6·5 cm.	—
30 ,,	8·05 ,,	1·55
60 ,,	9·7 ,,	1·65

Velocity = 0·533 cm. per 10 minutes.

3 c.c.

t.	P.	D.
0 minutes	6·9 cm.	—
10 ,,	8·5 ,,	1·6
20 ,,	10·15 ,,	1·65
30 ,,	11·8 ,,	1·65

Velocity = 1·63 cm. per 10 minutes.

5 c.c.

t.	P.	D.
0 minutes	6·6 cm.	—
10 ,,	9·25 ,,	2·65
20 ,,	11·9 ,,	2·65
30 ,,	14·55 ,,	2·65

Velocity = 2·65 cm. per 10 minutes.

10 c.c.

t.	P.	D.
0 minutes	7·15 cm.	—
5 ,,	9·8 ,,	2·65
10 ,,	12·5 ,,	2·7
15 ,,	15·2 ,,	2·7

Velocity = 5·36 cm. per 10 minutes.

20 c.c.

t.	P.	D.
0 minutes	8·45 cm.	—
5 ,,	13·85 ,,	5·4
10 ,,	19·05 ,,	5·2

Velocity = 10·6 cm. per 10 minutes.

25 c.c.

t.	P.	D.
0 minutes	6·8 cm.	—
2 ,,	9·45 ,,	2·65
6 ,,	14·85 ,,	2 × 2·7
8 ,,	17·45 ,,	2·6
10 ,,	20·05 ,,	2·6

Velocity = 13·25 cm. per 10 minutes.

Yeast concentrations = 1 : 3 : 5 : 10 : 20 : 25
Velocities = 0·99 : 3·04 : 4·94 : 10 : 19·8 : 24·7

The concordance in the values of D in the single experiments and the agreement in the ratios of the concentrations and velocities give a

K 2

satisfactory proof of the reliability of the method of investigation. These numbers give us no indication of the change in velocity corresponding to the change in the concentration of the enzyme, as the enzyme is contained in the yeast cell.

Euler working with yeast juice and Herzog working with "zymin" give values of $n = 1\cdot29$—$1\cdot67$ and $2\cdot0$ respectively, where n is calculated from the formula $K_1/K_2 = (C_1/C_2)^n$, K_1 and K_2 being the velocity constants corresponding to the concentrations of ferment C_1 and C_2. If C_1 and C_2 are concentrations of yeast, $n = 1\cdot00$ from Table I. The velocity of fermentation is therefore a measure of the quantity of active yeast present. With ordinary brewery yeast, 10 cells per 1/4000 c.mm. at 30° gave an average velocity of about 4·5 cm. per ten minutes on the manometer scale of the apparatus.

The Influence of the Concentration of Dextrose on the Rate of Fermentation.

It has been shown by Dumas (loc. cit.), Tammann (Zeit. physikal. Chem., 1889, 3, 25), A. J. Brown (loc. cit.), and J. O'Sullivan (loc. cit.) that the rate of fermentation is practically independent of the concentration of sugar. This is also the case in the fermentation by yeast-juice. H. Euler (loc. cit.) has shown that the values of K calculated for a unimolecular reaction vary with the initial concentrations of dextrose and are numbers proportional to

$$\frac{\text{the velocity of fermentation}}{\text{concentration of sugar}},$$

and, as he shows that they are approximately inversely proportional to the concentration of the sugar, it is evident that the fermentation velocity is independent of the concentration of dextrose. The constancy of the values $K = \frac{1}{t}.\log\frac{a}{a-x}$ for a small part of the reaction is due probably entirely to the enzyme being slowly destroyed.

J. H. Aberson (loc. cit.) also considers the reaction to be unimolecular, whilst Herzog (loc. cit.) uses both the unimolecular formula and Henri's empirical formula (Zeit. physikal. Chem., 1901, 39, 194) to calculate his results. On examining their data, it is, however, clear that the reaction is approximately independent of the concentration of the sugar. This result has been confirmed and extended by this method of investigation. Solutions containing the same amount of yeast, but different amounts of dextrose (0·2—20 grams of dextrose per 100 c.c.), were tested, and the results are given in Table II. From this table and the accompanying curve, it is seen that with this concentration of yeast and at this temperature a maximum velocity is reached with about 5 grams of dextrose per 100 c.c.; the change in the velocity of fermentation between 0·5 gram and 10 grams per 100 c.c. is, however, only slight. Below 0·5 per cent., the concentration has an influence, and above 10 per cent. the excess of sugar has a distinct retarding influence.

TABLE II.—*Influence of the Concentration of Dextrose.*
Temperature = 30°.

Grams of dextrose per 100 c.c.	Velocity in cm. per 10 minutes.	Grams of dextrose per 100 c.c.	Velocity in cm. per 10 minutes.
0·16	2·9	4 0	5·5
0·28	4·05	5·0	5·4
0·52	4·7	8·0	5·05
0·66	5·1	12·0	5·05
1·0	4·8	20·0	4·4
2·0	5·2		

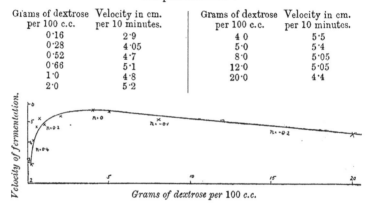

Grams of dextrose per 100 *c.c.*

In the study of the velocity of enzyme actions, it is often found that when the concentration of the reagent in question is large compared to that of the enzyme, the change proceeds as a linear function of the time; with relatively larger concentrations of enzyme the logarithmic law holds, as, for example, in hydrolysis by invertase (C. O'Sullivan and F. W. Tompson, Trans., 1890, 57, 843; A. J. Brown, Trans., 1902, 81, 373), hydrolysis by diastase (Horace Brown and Glendinning, Trans., 1902, 81, 388; Victor Henri, " Lois générales des diastases," Paris, 1903), and in hydrolysis by sucroclastic enzymes (E. F. Armstrong, *Proc. Roy. Soc.*, 1904, 73, 500, 516, 526; 74, 188, 195). It is therefore of interest to examine these data more closely and see whether these two stages of the reaction can be traced.

The numbers can best be studied by calculating " n " the order of the reaction with regard to the sugar according to the equation $V_1/V_2 = (C_1/C_2)^n$, where V_1 and V_2 are velocities corresponding with concentrations of dextrose C_1 and C_2.

If $n = 0$, V is independent of C, and the change proceeds as a linear function of the time.

If $n = 1$, V is proportional to C, and the logarithmic law holds. The results are summarised in Table III, where n is calculated from a number of experiments with different concentrations of sugar. The experiments at 20° are carried out with about twice the quantity of yeast used in those at 30°.

TABLE III.

Temp.	$C_1 : C_2$ in grams of dextrose per 100 c.c.	$V_1 : V_2$.	n.
30°	0·16 : 0·52	1 : 1·62	+0·41
(from Table II).	0 28 : 0·66	1 : 1·26	+0·27
	1 : 2	1 : 1·08	+0·11
	2 : 4	1 : 1·06	+0·08
	4 : 8	1 : 0·92	−0·12
	8 : 20	1 : 0·87	−0·15
20°	0·09 : 0·34	1 : 1·60	+0·35
	0·6 : 1·02	1 : 1·17	+0·30
	1·12 : 2·53	1 : 1·24	+0·26
	2·85 : 8·55	1 : 1·13	+0·11
	10 : 20	1 : 0·90	−0·15

The value of n for higher concentrations is approximately zero and increases as the concentration becomes less, but never reaches 1. The influence of the concentration of dextrose is therefore never so great that the velocity is proportional to this concentration. This enzyme action thus differs from those already studied in that the logarithmic part of the curve has not been realised with these concentrations. The results obtained in these experiments may be summarised by the differential equation:

$$- \frac{d(\text{dextrose})}{dt} = + \frac{d(\text{carbon dioxide})}{dt} = K.(\text{yeast})^m (\text{dextrose})^n,$$

where $m = 1$ using the same yeast, n varies with the concentration of dextrose and yeast, but for the main part of the reaction is approximately 0. With dilute solutions of sugar, values up to 0·5 have been obtained.

The fact that the velocity of fermentation is independent of the concentration of the sugar is most simply explained by the assumption of the formation of a compound between the enzyme and the sugar (compare Horace T. Brown and Glendinning, *loc. cit.*; E. F. Armstrong, *Proc. Roy. Soc.*, 1904, **73**, 502). If with a certain concentration of sugar the main part of the enzyme is combined with sugar, a further addition of the latter reagent would not appreciably alter the amount of this compound. If the velocity which is experimentally measured is the rate of decomposition of the compound or the rate of a reaction involving this compound, then the velocity under these conditions would be independent of the concentration of dextrose. The sugar concentration would have an influence when the concentrations are such that an appreciable quantity of enzyme is left uncombined. If the mechanism of fermentation is worked out on this hypothesis, we have the sugar diffusing into the yeast cell and combining with the enzyme. This compound decomposes either directly or indirectly into alcohol and carbon dioxide with regeneration of the enzyme and immediate formation of more compound. Of this series of processes, diffusion, combination, and decomposition of the compound, the latter is the reaction which proceeds slowest, and is therefore the important one in determining the reaction-velocity. The reactions which precede this serve to bring the reagents which take part in the slow reaction up to a certain concentration, and any subsequent reaction serves to remove the primary products, forming alcohol and carbon dioxide. In dealing with living yeast cells, it must not be forgotten that the reaction takes place within the cell, where little is known of the concentration of the reagents, and this mechanism of fermentation can only be accepted as a working hypothesis. Some other results are not easy to explain by this mechanism of reaction. Thus, if the enzyme is completely combined with the sugar, the amount of compound formed must be proportional to the amount of enzyme present, and proportionality should exist between the concentration of the enzyme and the velocity of reaction. This is difficult to bring into harmony with Herzog's experiments showing that the velocity is proportional to the square of the concentration of the zymin (*loc. cit.*, p. 159). It is also a question whether the enzyme is entirely regenerated. The values of K calculated from the formula for a unimolecular reaction in Herzog's series of experiments show a fair degree of constancy through a large range of the reaction. This is evidently due to the enzyme being decomposed during fermentation, and not to the disappearance of sugar, for the values do not agree in different experiments with varying initial concentrations of sugar.

This is shown in Tables Ia and II (*loc. cit.*, pp. 153, 154), where the same quantity of zymin is used, but different initial concentrations of sugar.

Temperature $= 24 \cdot 5°$.

Ia.			II.		
Dextrose concentration, $a=1 = 20\cdot45$ grams per 100 c.c.			$a = 0\cdot5$.		
1·2 grams of zymin per 10 c.c.			1·2 grams of zymin per 10 c.c.		
t.	$a-x$.	$0\cdot4343K=$ $1/t\log.a/a-x$.	t.	$a-x$.	$0\cdot4343K$.
120	0·961	0·000144	* 240	0·409	0·000363
* 240	0·922	0·000147	420	0·349	0·000374
1200	0·673	0·000143	1440	0·142	0·000379
2690	0·396	0·000149	1740	0·119	0·000359
3000	0·359	0·000150			

The initial velocity in the concentrated solution (calculated from the values in 240 minutes) is slightly less than in the dilute instead of being double, as would be the case with a true unimolecular reaction with regard to the sugar.

The Influence of Temperature.

It is well known that the rate of fermentation is greatly influenced by temperature. Aberson (loc. cit., p. 105), working between 12° and 33°, gives a mean temperature quotient for 10° (K_{t+10}/K_t) 2·72. Herzog (loc. cit., p. 160) gives values of the velocities from 14·5° to 28·5°, $K_{24·5}/K_{14·5} = 2·88$. Some preliminary experiments with the apparatus described confirmed these results, but showed that the temperature quotient varied with the temperature. An investigation was therefore undertaken over as large a range of temperature as possible. Below 5°, the reaction proceeded too slowly to be measured, whilst above 40° the enzyme was destroyed. Experiments were therefore carried out between these temperatures. The method of working was as follows: the velocity was measured at a given temperature in the manner described, the temperature was then quickly raised 5° and the velocity again measured. The apparatus was then cooled to the original temperature and a third measurement of the velocity taken. The mean value of the velocities in the first and third experiments was compared with that in the second, and in this way change in activity of the yeast was eliminated. An example will make this clear.

Temperature.	Time for pressure to change by 4 cm. of mercury.
20°	18·2 minutes
25	9·7 ,,
20	14·2 ,,

Ratio of observed velocities = 16·2/9·7 = 1·67 : 1.

A small correction, amounting in this case to 5 per cent., has to be made for the decreased solubility of the carbon dioxide at the higher temperature, and for the influence of temperature on the pressure. Applying this correction, the ratio of the velocities $V_{25}/V_{20} = 1·59$.

Other experiments gave 1·52, 1·56, 1·60; mean 1·57.

Table IV gives the result of a series of experiments between the temperatures 5° and 40°. The value V_{t+5}/V_t is obtained from the observed ratio by applying the solubility correction given in the third column. As temperature-coefficients are usually given as quotients for 10°, another column is added, giving the values V_{t+10}/V_t at various temperatures.

TABLE IV.

Temperature.	Observed ratio.	Percentage correction.*	V_{t+5}/V_t.	V_{t+10}/V_t.	Herzog.	Aberson.
5°						
10	2·94	10	2·65			
15	2·29	8	2·11	5·6		
20	1·89	5	1·80	3·8		
25	1·65	5	1·57	2·8	2·88	2·72
30	1·50	5	1·43	2·25		
35	1·42	5	1·35	1·95		
40	1·27	5	1·20	1·6		

* These corrections are calculated for 50 c.c. of solution fermenting in an apparatus of 140 c.c. capacity.

It may be now pointed out that if the chemical action brought about by yeast is due to a definite enzyme in the cell and that the enzyme is the same in different kinds of yeast, then certain characteristics of the reaction will remain the same independently of the class of yeast used to excite fermentation. The temperature-coefficient would probably be one of these constant factors. The experiments were therefore extended to an examination of other yeasts. The temperature quotients for 5° of the fermentation reaction with some "distillery" yeast and "wine" yeast were found to be practically identical with those obtained with "brewery" yeast, and we may therefore conclude that the enzyme present in the three kinds of yeast is the same.

TABLE V.—*Temperature Quotients with Different Yeasts.*

V_{t+5}/V_t.	"Brewery" yeast.	"Distillery" yeast.	"Wine" yeast.	Mean.
V_{10}/V_5	2·65	2·50	2·30	2·50
V_{15}/V_{10}	2·11	1·97	1·85	1·98
V_{20}/V_{15}	1·80	1·98	1·96	1·91
V_{25}/V_{20}	1·57	1·62	1·62	1·60
V_{30}/V_{25}	1·43	1·47	1·41	1·44
V_{35}/V_{30}	1·35	1·36	1·33	1·35
V_{40}/V_{35}	1·20	1·26	1·24	1·23

In the study of enzymes it is very seldom found that these substances can be characterised in any but a qualitative manner. They cannot be isolated and analysed, and the velocity of the reaction which they bring about is usually very sensitive to inhibiting agents, and cannot be used directly as a means of identification. We have, however, in the temperature-coefficient of the reaction numbers which seem to be characteristic of the enzyme zymase and may serve this purpose.

These numbers were found to be independent of the concentration of yeast and dextrose, the class of yeast, presence or absence of nutrient salts, and the same when inhibiting agents are present (Table VII). This enzyme occurs not only in yeast and in various preparations from yeast, but also in different animal and vegetable tissues, and these numbers may be of value in proving the identity or otherwise of zymase obtained from various sources.

The Rate of Fermentation of Different Sugars.

It is known that the various fermentable sugars undergo fermentation at almost the same velocity, and the few experiments which are given below confirm this result. If this is true for dextrose and lævulose, then it would also be probably correct for sucrose and maltose, as enzymes are present in the yeast which hydrolyse these disaccharides, giving the fermentable monoses. The numbers given in Table VI show that lævulose is fermented somewhat more slowly than dextrose, sucrose a little faster, and maltose at the higher temperature with almost the same velocity as dextrose, and at a lower temperature somewhat slower.

TABLE VI.

Velocities of fermentation.

Temp.	Grams of sugar in 100 c.c.	Dextrose. In cm. per 10 mins.		Lævulose.	Sucrose.	Maltose.
30°	5	2·5	100	91	103	101
30	5	5·25	100	90	—	—
25	5	2·95	100	—	106	—
25	4	2·0	100	94	—	—
25	10	6·2	100	91	—	84

The velocities are all referred to dextrose as 100, and the actual velocities with this sugar are given in order to indicate the quantity of yeast present. It is remarkable that constant values of the velocity of fermentation of maltose and sucrose are obtained in a few minutes showing that enough sugar is almost instantaneously hydrolysed for the fermentation reaction to attain its maximum velocity.

The Action of Inhibiting Agents.

Enzyme actions are peculiarly sensitive to inhibiting agents or "poisons," and a study of the action of such poisons affords some insight into the nature of the enzyme. Thus Senter, on measuring the retarding influence of acids on the rate of decomposition of hydrogen peroxide by hæmase, an enzyme isolated from blood, shows that this effect is approximately proportional to the $H^{\cdot}$-ion, and concludes that the enzyme is probably a weak base (*Proc. Roy. Soc.*, 1904, **74**, 204 ; *Zeit. physikal. Chem.*, 1905, **51**, 680).   E. F. Armstrong (*loc. cit.*) has been able to show, by the inhibiting action of certain sugars on sucroclastic changes, a close correlation in configuration between enzyme and hydrolyte.   Many "poisons" have been discovered which inhibit alcoholic fermentation, but some preliminary experiments with this apparatus showed in many cases the great influence of the "incubation" time on the activity of the poison.   Moreover, as one is dealing with a mixture of a great many substances, there are present so many disturbing influences that a quantitative estimation of the action of small quantities of poisons cannot be carried out as Senter has done for hæmase.   The investigation was continued on somewhat different lines in the hope of throwing some light on the mechanism of the reaction.   From considerations already discussed, the following steps in fermentation will be assumed.

A. Diffusion of the sugar into the cell.

B. Combination of dextrose and enzyme.

C. Decomposition of the compound forming an intermediate compound (lactic acid ?).

D. Decomposition of this compound forming carbon dioxide and alcohol.

E. Diffusion of the products from the cell into the solution.

Of this series of reactions, *C* is the one which proceeds slowly, and the velocity of this reaction is measured in these experiments.   If an intermediate compound exists—the supposed formation of lactic acid has attracted some attention lately (Buchner and Meisenheimer, *Ber.*, 1904, **37**, 417 ; McKenzie, *Trans.*, 1905, **65**, 1378)—then reaction *D* must proceed rapidly to prevent accumulation of this compound.   The action of a poison on this system may be that it retards any one of the five reactions.   If, for example the poison prevented the diffusion of sugar into the cell sufficiently to influence largely the rate of formation of the end-products, then reaction *A* would be the slow reaction of the series and would be the important one in velocity measurements.   The characteristics of the inhibited reaction would then be the characteristics of the diffusion reaction *A* and would probably be very different to those of reaction *C* which is primarily measured.   To put the matter generally, whichever reaction is retarded by the poison, this reaction becomes the important one in determining the velocity of formation of alcohol and carbon dioxide.   It seemed of interest, therefore, to investigate whether the temperature-coefficient of the inhibited reaction is the same as that of the original reaction.   This characteristic was chosen as being easily and rapidly measured.   Table VII gives a summary of the results obtained, and it is evident that in all these cases the temperature quotient for 5° is practically identical with that of the original reaction, showing that the "poison" is inhibiting the reaction which determines the velocity in these experiments.

TABLE VII.—*Temperature-coefficient of the Reaction inhibited by Sulphuric Acid.*

50 c.c. of solution.	Temp.	Time to fall 3 cm.	
Without acid	25°	6·7 minutes	
+5 c.c. $N/5$ H_2SO_4	25	6·2 ,,	
After 4 hours	25	9·4 ,,	
,, 7 ,, ...	25	18·4 ,,	
	30	12·9 ,,	(calcd. for 25° = 20·4)
	25	22·4 ,,	

Ratio of times = 1·58 : 1. $V_{30}/V_{25} = 1·50$.   From Table IV = 1·43.

50 c.c. of solution.	Temp.	Time to fall 6 cm.	
Without acid	15°	14·8 minutes	
+5 c.c. $N/5$ H_2SO_4	15	14·4 ,,	
After 6 hours	15	19·1 ,,	
	20	10·3 ,,	(calcd. for 15° = 20·1)
	15	21·0 ,,	

Ratio of times = 1·95 : 1. $V_{20}/V_{15} = 1·85$.   From Table IV = 1·80.

Summary.

Concentration of "poison."	Percentage reduction on the velocity.	V_{t+5}/V_t.	For original reaction from Table IV.
0·02N H_2SO_4	67	V_{30}/V_{25} = 1·50	1·43
0·02N H_2SO_4	26	V_{20}/V_{15} = 1·85	1·80
0·035N oxalic acid	67	V_{30}/V_{25} = 1·40 ·	1·43
8 per cent. alcohol.....................	40	V_{30}/V_{25} = 1·39	1·43
0·004 per cent. mercuric chloride...	50	V_{30}/V_{25} = 1·40	1·43

It is interesting to note that in the case of sulphuric acid the first effect is a slight raising of the velocity and then a gradual fall in the activity of the yeast. The enzyme is in some way rendered inactive, and is probably destroyed, for on neutralising the acid the activity is not regained. It was found that the sugar to a certain extent protects the enzyme against the acid, a point in favour of the view of a combination between the enzyme and sugar. Thus, yeast which has been allowed to stand three hours at 25° with sulphuric acid (0·02N) lost 53 per cent. of its activity, whilst when dextrose was present the loss was only 29 per cent. The addition of lactose to a fermenting solution has practically no influence on the rate of evolution of carbon dioxide. A 5 per cent. dextrose solution fermenting at the rate of 5·7 cm. per ten minutes on the addition of the same quantity of lactose gave a velocity of 5·95 cm. per ten minutes.

Lactic Acid as an Intermediate Compound in Fermentation.

It has been suggested that lactic acid is an intermediate compound in the fermentation of dextrose, and that two enzymes take part in the reaction: zymase, which converts dextrose into lactic acid, and lactacidase, which converts the lactic acid into alcohol and carbon dioxide (Buchner and Meisenheimer, *Ber.*, 1905, **38**, 620). Velocity experiments do not, however, confirm this supposition. On adding small quantities of lactic acid to a fermenting solution, no very appre-, ciable change in the velocity is noticed. As the second reaction (Reaction *D*, p. 139) must proceed more quickly than the first (Reaction *C*) in order to prevent a large accumulation of lactic acid, we should expect a very considerable increase in the velocity of evolution of carbon dioxide in the presence of lactic acid.

Temp. 30°.	5 per cent. dextrose solution.
Without lactic acid............	4·9 cm. per ten minutes.
0·007N ,, ,, ...	4·65 ,, ,, ,,
0·05N ,, ,, ...	4·5 ,, ,, ,,

Buchner and Meisenheimer's conclusions are based on experiments which show the appearance and disappearance of lactic acid in certain fermentation experiments. The results are perhaps more easily explained on the supposition that lactic acid is formed by some side reaction and not in an intermediate reaction. A small quantity of sugar may be converted into lactic acid, which is subsequently converted into alcohol, but it is improbable that all the sugar goes through this intermediate step. If an intermediate compound exists, it is probably much less stable than lactic acid, and would be difficult to isolate.

Conclusions.

The chief results obtained in this paper may be summarised as follows:

1. In the study of the rate of alcoholic fermentation, many complications are eliminated by measuring the velocity over very small ranges of the reaction, and changing the concentrations by dilution.

2. The change of pressure due to evolution of carbon dioxide is a convenient and sensitive method of measuring this velocity.

3. The rate of fermentation of dextrose is proportional to the concentration of yeast over a wide range of concentrations.

4. The rate is almost independent of the concentration of the sugar except in very dilute solutions. The influence of this concentration is never so great that the velocity is proportional to the concentration

of the sugar; the reaction is therefore never one of the first order with regard to the sugar.

5. The temperature-coefficient of the reaction is large and varies with the temperature. $V_{15}/V_5 = 5\cdot6$, $V_{40}/V_{30} = 1\cdot6$, and intermediate values are obtained between these temperatures. The temperature quotient for 5° from 5° to 40° forms a series of numbers which seems to be characteristic of the enzyme zymase.

6. The initial rates of fermentation of dextrose, lævulose, sucrose, and maltose are in the ratio $1 : 0\cdot92 : 1\cdot05 : 0\cdot9$.

7. The temperature-coefficient of the reaction inhibited by "poisons" is the same as that of the original reaction.

8. It is improbable that in fermentation any but small. quantities of sugar go through the intermediate step of lactic acid.

9. These results indicate that the reaction which is measured in these experiments is the slow decomposition of a compound between the enzyme and the sugar.

In conclusion, the author wishes to acknowledge his indebtedness to Mr. C. O'Sullivan, F.R.S., and Dr. A. L. Stern, with whom he had the advantage of discussing the matters treated in this paper.

XVII.—The Slow Combustion of Carbon-Disulphide.

By Norman Smith.

In 1890 (*Brit. Assoc. Reports*, p. 776), G. S. Turpin showed that carbon disulphide undergoes a "slow combustion" at temperatures as low as 130° with the formation of a dark reddish-brown substance. He states that "this powder contains both carbon and sulphur, but its composition has not yet been thoroughly made out." Dixon and Russell (Trans., 1899, **75**, 603), in their investigation of the combustion of carbon disulphide, also noticed the formation of this reddish-brown substance at temperatures below that at which explosion of the mixture of carbon disulphide and oxygen takes place.

During the last four years, the author has been engaged in the investigation of this reddish-brown deposit. The chief difficulty has been the small amounts of substance which could be obtained. In the earlier experiments, a stream of air mixed with a small quantity of oxygen was drawn over the surface of some carbon disulphide and then through a long tube heated at temperatures varying from 130—190°. A reddish-brown film gradually formed on the sides of the tube. If boiling water was poured into the tube, the film peeled off, and on analysis was found to contain carbon, sulphur, hydrogen, and oxygen, the hydrogen and oxygen being approximately in the same proportion as in water.

Experiments were next carried out to determine whether moisture was necessary for the formation of the deposit. The carbon disulphide was dried by calcium chloride, and the mixture of air and oxygen passed through sulphuric acid before use. In all the experiments where moisture was excluded, no deposit could be obtained. On the other hand, it was found that, if the gases were quite moist, the action took place much more readily.

Various methods have been tried to increase the yield, but with little success. The introduction of platinum gauze into the heated tube did not appear to cause any increase in the rate of formation of the substance. Finally, the method adopted as giving the best results was the following: a mixture of carbon disulphide and water vapour, obtained by bubbling an inert gas (carbon dioxide or nitrogen) through a tube containing pure carbon disulphide covered with a layer of water, was passed through glass tubes about 90 cm. long and 3 cm. in diameter. These tubes were packed with lengths of glass tubing of 3 to 4 mm. bore, and heated in a large Lothar-Meyer air-oven, kept at 175—180°. By means of a T-piece, oxygen was mixed with the carbon disulphide and water vapour just before the gases entered the heated tube. The most favourable proportion was obtained when slightly less oxygen than that which would cause explosion was introduced into the mixture. The heated tube became slowly covered with the reddish-brown film; in the cooler portions, a very light black powder, resembling soot, was deposited, although in extremely small quantities, whilst from the end of the tube a faint smoke was emitted. The issuing gases contained a considerable quantity of sulphur dioxide, but only a very small amount of carbon dioxide could be detected.

The film peeled off much more readily when a hot solution of sodium carbonate was used instead of boiling water. After drying, the substance was very light; it was dark brown and had a bright lustre. It decomposed when heated strongly, yielding dark yellow fumes with a smell resembling the mercaptans, and a hard black mass insoluble in alkali was left. The substance was insoluble in the usual organic solvents, such as alcohol, ether, benzene, aniline, &c. Water and carbon disulphide dissolved a small quantity, but only sufficiently to colour the liquid faintly. With the exception of the small amount of black powder, all dissolved readily in a hot solution of caustic alkali forming a dark brown solution. On acidifying this solution, a reddish-brown, flocculent precipitate, resembling ferric hydroxide, was deposited. This, on drying, changed to a hard, black solid.

The deposit taken from the tubes generally contained a little free sulphur. After removal of this by repeated digestion with carbon disulphide, the slightly varying analyses gave as a mean :

C = 33·9 ; S = 49·9 ; H = 0·9 ; O = 15·3 per cent.

The substance reprecipitated from caustic potash also gave numbers which varied somewhat.

It was found later that a separation of a substance of constant composition could be effected by means of sodium carbonate. The deposit from the tubes was boiled with a solution of sodium carbonate, when the greater portion dissolved, giving a dark brown solution. This was separated by filtration and acidified with hydrochloric acid. The reddish-brown precipitate formed was collected, washed thoroughly, and dried at 100°. After repeated digestion with carbon disulphide, the substance was kept at 110° until the weight was constant. In the two analyses given, one sample was kept for a period of six months before digestion with carbon disulphide, whilst the other was treated immediately after preparation.

(a) 0·1913 gave 0·2625 CO_2 and 0·0202 H_2O. C = 37·4 ; H = 1·17.
   0·2062 „ 0·7362 $BaSO_4$. S = 49·04.
(b) 0·2106 „ 0·2898 CO_2 and 0·0219 H_2O. C = 37·5 ; H = 1·15.
   0·1843 „ 0·649 $BaSO_4$. S = 48·4.

$C_{16}H_6O_4S_8$ requires C = 37·06 ; S = 49·4 ; H = 1·16 ; O = 12·38 per cent.

The small portion insoluble in sodium carbonate is almost completely dissolved by caustic soda. Analyses of the reprecipitated product from this solution did not give constant results, but in all cases the percentage of carbon was less and that of the sulphur more than in either the original deposit or the substance precipitated from sodium carbonate solution.

Preparation of the Silver Compound.

The substance obtained by reprecipitation from sodium carbonate was carefully purified from any free sulphur and then dissolved in caustic soda. The precipitate formed on acidifying with hydrochloric acid was collected, washed thoroughly with distilled water, and then dissolved in a mixture of equal parts of ammonia (sp. gr. 0·88) and water. Silver nitrate solution, with which ammonia had been mixed until the precipitate which first formed redissolved, was now added, and the dark brown precipitate which formed was filtered off after some time and washed with a dilute solution of ammonia. The salt was then dried in a vacuum over sulphuric acid until the weight was constant. The silver was estimated by heating with nitric and hydrochloric acids in a sealed tube.

0·1926 gave 0·1279 AgCl.   Ag = 50·01.

0·2820 ,, 0·1866 CO_2 and 0·0084 H_2O.   C = 18·04 ; H = 0·32.

$C_{16}HO_4S_8Ag_5$ requires C = 18·24 ; H = 0·09 ; Ag = 51·2 per cent.

Preparation of the Ammonium Compound.

The substance obtained by precipitation from the alkaline solution was dissolved in equal parts of ammonia and water and evaporated in a vacuum over sulphuric acid.   No crystals separated, but a brownish-black solid was left.   This substance had no smell of ammonia and dissolved readily in water.   On treatment with alkali, ammonia was evolved.   The solid ammonium compound on heating gave off ammonia and then decomposed into a yellow oil and a black solid, as was the case with the original deposit.

These experiments lead to the conclusion that, in the slow combustion of carbon disulphide and oxygen, the reddish-brown substance deposited consists chiefly of a compound having the composition $C_{16}H_6O_4S_8$, along with small quantities of another acid substance or substances containing less carbon and more sulphur, and also very small quantities of free carbon and sulphur.

THE UNIVERSITY,
    MANCHESTER.

XVIII.—*The Liberation of Tyrosine during Tryptic Proteolysis.*

By ADRIAN JOHN BROWN and EDMUND THEODORE MILLAR.

WHEN studying the various methods suggested for the purpose of measuring the activity of proteolytic enzymes, it occurred to us that a method of directly estimating tyrosine by bromination recently described by James H. Millar (*Trans. Guinness Research Laboratory*, 1903, 1, Part I) might furnish a means of determining the course of proteolytic change in those cases in which tyrosine is liberated during the breaking down of the protein molecule.   An investigation in this direction was therefore commenced and the results so far obtained are described in this paper.

The paper may be summarised as follows :

(1) J. H. Millar's method of estimating tyrosine by bromination is applicable to the estimation of tyrosine in the presence of proteins and their earlier cleavage products due to enzyme action, if suitable control experiments are employed.

(2) Tyrosine is not a late product of tryptic proteolysis, as is usually supposed ; on the contrary, the tyrosine nucleus of a protein is attacked and the whole of the tyrosine liberated during the first stage of tryptic digestion.

(3) The resistance of the protein tyrosine nucleus to peptic hydrolysis is confirmed.

(4) Attention is called to the similarity of Emil Fischer and E. Abderhalden's recent observations on the actions of tryptic and peptic enzymes on polypeptides containing a tyrosine nucleus (*Zeit. physiol. Chem.*, 1905, **46**, 52) to the authors' observations on the actions of the same enzymes on proteins containing a tyrosine nucleus.

(5) The authors' investigations appear to indicate a reliable means of differentiating enzymes of a peptic from those of a tryptic nature, and may assist in throwing some light on the confused state of knowledge with regard to the existence of a tyrosine nucleus in the different albumoses resulting from peptic and tryptic proteolysis.

EXPERIMENTAL.

J. H. Millar's method of directly estimating tyrosine (*loc. cit.*) is based on its reaction with free bromine, by which a bromine compound of tyrosine is formed.

Tyrosine is dissolved in hydrochloric acid to which potassium bromide is added. The solution is then titrated with a $N/5$ sodium bromate solution. The liberated bromine resulting from the interaction of the sodium bromate and bromide in acid solution is rapidly absorbed by the tyrosine present, and the end of the reaction determined by employing starch and potassium iodide as an indicator for free bromine.

J. H. Millar's experiments with pure tyrosine show that the reaction results in the formation of dibromotyrosine according to the following equation :

$$C_6H_4(HO){\cdot}CH_2{\cdot}CH(NH_2){\cdot}CO_2H + 4Br =$$
$$C_6H_2Br_2(HO){\cdot}CH_2{\cdot}CH(NH_2){\cdot}CO_2H + 2HBr.$$

The method of estimating tyrosine is shown by J. H. Millar to be applicable not only to the accurate estimation of the pure substance, but also to the substance when it exists in intermixture with ammonium salts and amides and amino-acids such as asparagine, aspartic acid, leucine, and phenylalanine, which result from complete acid proteolysis.

J. H. Millar's work does not, however, show whether his method is applicable to the determination of tyrosine in the presence of proteins or their primary cleavage products, such as albumoses or peptones ; it was necessary, therefore, for us to investigate this point as preliminary to an attempt to employ the method for the estimation of tyrosine when present among the products of enzyme proteolysis. Preliminary experiments with solutions of egg-albumin, edestin, and gelatin indicated that they possessed to some extent the property of absorbing bromine under the conditions employed by J. H. Millar to estimate tyrosine.

Following on this observation, a solution of edestin was prepared and divided into two equal volumes. One part was titrated direct with $N/5$ bromate solution and its power of absorbing bromine noted. A known amount of tyrosine was dissolved in the second volume and it was also titrated with bromate solution. It was then found, after correcting the result of the second titration for the amount of bromine absorbed by the edestin alone, indicated by the first titration, that an accurate measure was obtained of the amount of tyrosine introduced. Similar results were also obtained when gelatin and egg-albumin were employed in the place of edestin.

Our preliminary experiments therefore showed that it was possible to determine tyrosine in the presence of proteins if control experiments were made in order to correct for the bromine absorbed by the proteins.

A series of experiments were then made in which edestin* was digested with pancreatic extract and the products of change examined by the bromine method as follows :

A 1 per cent. solution of edestin was prepared by dissolving 2 grams of the dry substance in 200 c.c. of a 0·5 per cent. sodium carbonate solution, 50 c.c of this solution being placed in each of four flasks. To each of three of these flasks, 5 c.c. of active pancreatic extract (Benger) were added, and to the fourth flask, employed as a control, 5 c.c. of pancreatic extract were added which had been previously heated to 100° to render it inactive. All the flasks were placed in a water-bath kept at 32°. After 24 hours, the control and the contents of one of the flasks containing active pancreatic extract were titrated with bromate solution after the addition of 20 c.c. of 20 per cent. hydrochloric acid and 10 c.c. of a 20 per cent. solution of sodium bromide ; and after 72 and 144 hours respectively the contents of the second and third flasks containing active pancreatic extract were titrated in a similar manner.

The results obtained are given in the following table :

* Edestin was employed in this and many of the following experiments, as it contains a tyrosine nucleus and can also be readily prepared in a comparatively pure state.

	Time of digestion.	$N/5$-Bromate solution used.	$N/5$-Bromate solution after deducting control.	Calculated per cent. of tyrosine formed from edestin during proteolysis.
Active digestion...	24 hours	0·80 c.c.	0·38 c.c.	4·06
,, ,,	72 ,,	0·80 ,,	0·38 ,,	4·06
,, ,,	144 ,,	0·80 ,,	0·38 ,,	4·06
Control	24 ,,	0·42 ,,	—	—

It appeared from the above experiments, if the method of estimating tyrosine adopted was reliable, that 4·06 per cent. of tyrosine resulted from the tryptic digestion of edestin during the periods of 24, 72, and 144 hours. On the supposition that proteolysis had proceeded far enough during 24 hours—the shortest period employed—to liberate the whole of the tyrosine from its containing nucleus in the edestin molecule, the results appeared quite reasonable, but on other grounds they were open to question.

In the first place, it was questionable whether the small volumes of bromate solution consumed in the above experiments measured the tyrosine present with any approach to accuracy. It was found, however, on experimenting with known amounts of tyrosine, comparable with those measured in the preceding experiments, that very accurate results were obtained considering the small volumes of bromate solution employed.

A second more difficult objection to meet questioned the accuracy of the correction obtained from the control experiment. The control indicated the amount of bromine absorbed by the pancreatic extract and the undigested protein, and was subtracted as a correction from the total amount of bromine absorbed by an intermixture of digested products and pancreatic extract in order to arrive at the amount of bromine absorbed by the tyrosine liberated. It was open to doubt whether the correction remained constant under these conditions. When solutions of edestin and pancreatic extract of similar concentration to those employed in the foregoing experiments were titrated separately, it was found that each absorbed bromine to some extent.* For instance, when 0·5 gram of edestin and 2·5 c.c. of pancreatic extract were titrated separately with bromate solution, bromine equal to 0·3 c.c. of bromate solution was absorbed by the edestin and to 0·1 c.c. by the pancreatic extract. It was possible to digest pancreatic extract alone as a control in order to ascertain whether its original power of absorbing bromine underwent any change during digestion, and it was found on doing so that no alteration took place. The accuracy of the control experiment so far as it concerned the pancreatic extract was therefore established. But with

* Tyrosine has been found in all pancreatic extracts examined.

regard to the part of the correction applying to the bromine absorbed by edestin previous to digestion, it still remained open to doubt whether it could be employed with accuracy after digestion had taken place and the edestin molecule had been broken down to a greater or less extent.

There appeared to be no way of obtaining an answer to this question by means of experiments with the tryptic digestion products of edestin, but experiments with the tryptic digestion products of gelatin pointed to the conclusion that no change takes place.

Gelatin, unlike a typical albumin or globulin, does not contain a tyrosine nucleus ; it appeared, therefore, that if an examination of its digestion products by the bromine method were made, the complicating presence of tyrosine would be avoided and some light might be thrown on the constancy of the control referred to above.

A series of digestion experiments with a 1 per cent. solution of gelatin and pancreatic extract were made under similar conditions to those with edestin (p. 148) with the following result :

50 c.c. of gelatin solution and 2 5 c.c. pancreatic extract.	Time of digestion.	$N/5$-Bromate solution used.	$N/5$-Bromate solution used after deduction of control.	Calculated percentage of tyrosine formed during proteolysis.
1. Active digestion...	48 hours	0·35 c.c.	none.	none.
2. ,, ,,	100 ,,	0·35 ,,	,,	,,
3. ,, ,,	148 ,,	0·35 ,,	,,	,,
4. Control	—	0·35 ,,	—	—

It will be seen from the foregoing results that the amounts of bromine ·absorbed by the different digestions of the gelatin do not vary from the amount originally absorbed by the gelatin prior to digestion, which shows that for gelatin at least the cleavage products of its molecule absorb the same amount of bromine as the original molecule prior to hydrolysis. It seemed probable, therefore, that the same conditions might obtain with edestin. More convincing evidence of this was, however, obtained by an examination of the products of a peptic digestion of edestin itself.

There was good reason to anticipate from the results obtained by previous investigators that it would be found that the tyrosine nucleus of a protein such as edestin was not attacked during peptic digestion, and consequently that free tyrosine would not be present among the products of peptic proteolysis. On this assumption, the following experiment on the digestion of edestin by peptase was made : 3 grams of edestin were dissolved in 300 c.c. of a solution containing 0·27 per cent. hydrochloric acid, and 50 c.c. of liquor pepticus (Benger) were then added. Immediately after intermixture, 50 c.c. of the solution were withdrawn and titrated with $N/5$ bromate solution in order to ascertain the amount of bromine absorbed by the original mixture of edestin and liquor pepticus prior to digestion. The rest of the solution was kept in a water-bath at 32°, and during digestion portions were withdrawn and titrated with bromate solution at successive intervals of time. The results obtained are given below:

	$N/5$-Bromate solution employed to titrate 50 c.c. of edestin solution.
Control, prior to commencement of digestion ...	0·50 c.c.
After 24 hours' digestion	0·55 ,,
,, 48 ,, ,, ...	0·50 ,.
,, 72 ,, ,, ...	0·60 ,,
,, 96 ,, ,, ...	0·55 ,,
,, 192 ,, ,, ...	0·55 ,,

During the course of the prolonged peptic digestion of edestin in the above experiment, it will be noticed that—within errors of experiment—the original edestin and its digestion products absorbed equal amounts of bromine—a result markedly different from that which was obtained when digesting edestin with pancreatic extract. The experiment therefore confirmed the impression that tyrosine is not liberated during peptic digestion, and further strengthened the view that our method of employing a control for the bromine absorbed by proteins in digestion experiments was reliable.

Before proceeding to make use of J. H. Millar's bromine method for a further investigation of the conditions governing the liberation of tyrosine during tryptic proteolysis, it seemed desirable, however, to inquire as to the existence of another possible source of error. It is known that tryptophane (scatoleaminoacetic acid) is very generally found among the products of proteolysis, and that it readily forms derivatives with free bromine. The presence of tryptophane was, moreover, recognised by us among the tryptic digestion products of edestin. It seemed possible, therefore, that the accuracy of the estimation of tyrosine by means of bromine might be influenced by the presence of tryptophane, although the results obtained in the experiments with edestin previously described appeared to render this unlikely. The investigations of S. Vines indicate that tryptophane is liberated in gradually increasing quantities during tryptic proteolysis; however, the amounts of bromine absorbed in our experiments with edestin (p. 148) remained constant during digestion for very varying intervals of time, a result not likely to be obtained if tryptophane takes part in absorbing the bromine.

In order, however, to settle this point definitely, tryptophane was prepared by Hopkins and Coles' method (*Journ. Physiol.*, 1901, **27**; 418), and subjected to the test of direct experiment.

During protein digestion, tyrosine and tryptophane are said to be

liberated in the approximate proportions of $4 : 1\cdot5$. In order, therefore, to obtain conditions in some degree parallel with the digestion
experiments with edestin (p 148), tyrosine and tryptophane in the
proportion of $4 : 1\cdot5$.—that is, $0\cdot5$ gram of tyrosine and $0\cdot018$ gram of
tryptophane—were dissolved in 50 c.c. of water and titrated with
bromate solution : bromine absorbed $= 0\cdot9$ c.c. $N/5$ bromate solution.

A solution of $0\cdot5$ tyrosine alone in 50 c.c. of water was also
titrated : bromine absorbed $= 0\cdot9$ c.c. $N/5$ bromate solution.

The experiments, therefore, indicated that no bromine was absorbed
by the tryptophane employed.

A second series of experiments in which more tryptophane was
employed than in the foregoing series led to a similar conclusion.

In a further experiment in which $0\cdot037$ gram of tryptophane alone
was titrated with bromate solution, no bromine was absorbed.*

As the evidence obtained from the experiments described above
appeared to show that the bromine method, when employed with a
control, was capable of estimating tyrosine when present among the
products of tryptic proteolysis, further experiments on the tryptic
digestion of edestin were made.

The first series of experiments with edestin (p. 148) indicated that
the maximum amount of tyrosine was liberated within the first twenty-
four hours of tryptic digestion, and this appeared to show that the
tyrosine nucleus of the protein was attacked at a much earlier stage of
digestion than is usually supposed.

Five hundred c.c. of a 1 per cent. solution of edestin, rendered
alkaline by the addition of $2\cdot5$ grams of sodium carbonate, were
digested with 25 c.c. of pancreatic extract (Benger) in the presence of
a little toluene at $32°$. A control experiment in which the pancreatic
extract was rendered inactive by heat was also prepared.

The following results were obtained :

	$N/5$-Bromate solution used.	Percentage of tyrosine calculated after deducting control.
Active digestion after 30 minutes ...	$0\cdot80$ c.c.	$2\cdot2$
,, ,, ,, 1 hour	$0\cdot95$,,	$3\cdot8$
,, ,, ,, 1½ hours ...	$0\cdot90$,,	$3\cdot3$
,, ,, ,, 2 ,, ...	$0\cdot95$,,	$3\cdot8$
,, ,, ,, 20 ,, ...	$0\cdot95$,,	$3\cdot8$
Control	$0\cdot60$,,	—

The above experiments show that the tyrosine nucleus of the edestin
was attacked during a very early stage of proteolysis. Although the
digestion was carried on under conditions which did not favour very

---

* The non-absorption of bromine by tryptophane under the conditions of J. H.
Millar's method of estimating tyrosine is probably due to the hydrogen chloride
which is present.

rapid hydrolysis, within thirty minutes more than half the tyrosine was liberated, and within one hour the whole of it was set free.

Other experiments with edestin, which it is not considered necessary to describe in detail, also indicated that the whole of the contained tyrosine is liberated within a remarkably short period of time after digestion commences. Egg-albumin subjected to tryptic digestion under similar conditions to edestin also appeared to yield the whole of its tyrosine within three hours.

From the above experiments with edestin and egg-albumin, it appeared, therefore, that the tyrosine nucleus of these proteins was one of the first constituent parts of their molecule to be attacked and hydrolysed during pancreatic digestion.

As it appeared desirable to confirm the presence of tyrosine among the first products of tryptic digestion by some means other than the bromine one hitherto employed, 100 c.c. of a solution containing 1 gram of edestin and 0·5 gram of sodium carbonate were digested with 5 c.c. of pancreatic extract for one hour at 32° and at once precipitated with trichloroacetic acid. The dense white precipitate of protein matter was filtered off and the filtrate evaporated to a small volume. After standing, crystals of tyrosine of characteristic appearance were obtained. Following on this, the crystals, together with the mother liquor, were titrated with sodium bromate solution, with the result that 4·3 per cent. of tyrosine, calculated on the original edestin employed, was found.

In a control experiment carried on under similar conditions to the above, but in which the pancreatic extract was rendered inactive by heat, no crystals of tyrosine were obtained.

In a second experiment, 25 grams of edestin dissolved in 2500 c.c. of 0·5 per cent. sodium carbonate solution were digested with 125 c.c. of pancreatic extract for forty-five minutes at 32° and the digestion products precipitated with phosphotungstic acid in presence of dilute sulphuric acid. The precipitate was filtered off and the filtrate treated with barium hydroxide. The solution was again filtered to remove barium sulphate and the filtrate concentrated by evaporation to a small volume. On standing, tyrosine crystallised out freely. The tyrosine after separation was redissolved in dilute hydrochloric acid and again recrystallised from the solution after the addition of ammonium hydroxide. By this means, tyrosine was obtained in apparently a pure state. The above experiments therefore confirmed our original impression that the tyrosine nucleus of proteins is attacked and hydrolysed during a very early stage of tryptic digestion.

An attempt was then made to gain some knowledge regarding the extent of protein degradation accompanying the breaking down of the tyrosine nucleus and the liberation of tyrosine. A method employed by Weiss (*Compt. rend. Trav. Laboratoire de Carlsberg*, 1903, 5, II, 133) when investigating the proteolytic enzymes of germinating barley was used for this purpose. This investigator showed if a solution of a protein such as edestin is precipitated· by tannic acid in the presence of sodium acetate, that almost the whole of the protein is thrown out of solution, and that the filtrate contains a mere trace of nitrogen. If, on the other hand, the protein in solution is subjected to the action of a proteolytic enzyme previous to precipitation by tannic acid, varying amounts of the original protein nitrogen are found in solution, and the amounts found provide to some extent a measure of the proteolytic change which has taken place. As the primary cleavage products of proteolysis, such as albumoses and peptones, are precipitated by tannic acid, it appears from Weiss's work that the soluble nitrogen found after proteolysis and precipitation with tannic acid is the nitrogen of amino-acids and other substances of simpler constitution than albumoses or peptones.

A solution containing approximately 2 per cent. of edestin and 0·5 per cent. of sodium carbonate was prepared: ·

(1) A nitrogen determination by Kjeldahl's method on 50 c.c. of the solution indicated that it contained 0·1518 gram of nitrogen.

(2) Fifty c.c. of the same solution were digested with 5 c c. of pancreatic extract until a control digestion of the same volume of the extract showed by the bromine method of titration that the maximum quantity of tyrosine was liberated. The time taken for digestion to this point was forty-five minutes. Digestion was then stopped by precipitating the solution with tannic acid according to Weiss's method. After filtration, it was found that a volume of the filtrate equal to 50 c.c. of the original solution contained 0·0548 gram of nitrogen.

(3) A control experiment, which is required when employing Weiss's method, was made by digesting 50 c.c. of the original solution of edestin with 5 c.c. of pancreatic extract, the activity of which·was previously destroyed by heat, and treating the solution in an exactly similar manner to Expt. 2. The nitrogen in 100 c.c. of the filtrate, equal to 50 c.c. of the original solution of edestin, was 0·0416 gram. This amount represents the nitrogen of the pancreatic extract employed which has not been precipitated by tannic acid, and also that of a very small amount of edestin not precipitated by tannic acid. The total amount of nitrogen in the control experiment, 0·0416 gram, must therefore be subtracted from the total nitrogen found in Expt. 2 in order to ascertain the amount of nitrogen which has been rendered soluble during the digestion of the original edestin. The amount found was 0·0132.

The total amount of nitrogen in the edestin present in 50 c.c. of the original solution was 0·1518 gram, therefore only 8·7 per cent. of this nitrogen was present in such form as to remain in solution unprecipitated by tannic acid after proteolytic digestion had proceeded sufficiently to liberate the whole of the tyrosine. In a second experiment with edestin, it was found that 9·1 per cent. of its contained nitrogen was rendered soluble.

About one-third of the amount of soluble nitrogen found in the above experiments can be accounted for as being present in the tyrosine liberated; the condition in which the remaining two-thirds exists is at present unknown.

The results of the above experiments should be regarded as only roughly indicating the maximum amount of decomposition of edestin into substances not precipitated by tannic acid during the liberation of tyrosine; it appears very probable that further investigation will show that the amount is less. However this may be, the experiments confirmed our previous conclusion that liberation of tyrosine takes place during the first stage of the tryptic hydrolysis of edestin.

E. Fischer and E. Abderhalden have recently shown (loc. cit.), when synthetically prepared polypeptides containing a tyrosine nucleus are submitted to tryptic digestion, that they are hydrolysed and tyrosine is liberated. On the contrary, the same polypeptides are shown to resist the action of peptic digestion, and consequently no tyrosine is liberated.

It appears interesting to compare these results with our observations on the actions of peptic and tryptic enzymes on proteins containing a tyrosine nucleus. Although the protein molecule is of far greater complexity than that of the polypeptides referred to, the behaviour of the two enzymes with regard to it has the appearance of being the same. This suggests that the tyrosine nucleus of both protein and polypeptide constitutes a point of attack for the tryptic enzyme. But, on the other hand, E. Fischer and E. Abderhalden (loc. cit.) show that some peptides which do not contain a tyrosine nucleus, such as alanylglycin, are hydrolysed by the tryptic enzyme, whilst others of somewhat similar constitution, such as glycyl-alanin, are not decomposed. In those cases in which hydrolysis takes place, it cannot here be associated with a tyrosine nucleus, and it appears desirable to bear this in mind when considering the mode of action of tryptase on the protein molecule. The liberation of tyrosine may be merely a secondary effect accompanying the cleavage of the molecule at some other point than the tyrosine nucleus.

A reliable means of differentiating enzymes of a peptic from those of a tryptic nature is required, for attempted classification from the behaviour of these enzymes in acid or alkaline solution has proved insufficient, particularly with regard to vegetable proteolytic enzymes,

It appears that the rapid and complete liberation of tyrosine during tryptic digestion may furnish a satisfactory means of differentiating enzymes of a tryptic from those of a peptic nature.

At present, the state of knowledge with regard to the existence of a tyrosine nucleus in the different albumoses resulting from tryptic and peptic proteolysis is in a somewhat confused state, and contradictory statements are frequently met with regarding this question. From the results of our experiments, presumably no albumose resulting from tryptic digestion contains a tyrosine nucleus, since the whole of the tyrosine appears to be liberated in the free state in the earliest stage of digestion. On the other hand, one or more of the albumoses or other of the earlier cleavage products of peptic digestion should contain the whole of the protein tyrosine. At present we have not experimented with albumoses formed during tryptic digestion, but some preliminary experiments with albumoses formed during peptic digestion indicate the presence of a tyrosine nucleus in some and not in others.

We have some reason to believe that tyrosine is liberated from edestin during a very early stage of acid proteolysis as well as during tryptic proteolysis.

SCHOOL OF BREWING,
UNIVERSITY OF BIRMINGHAM.

XIX.—Halogen Derivatives of Substituted Oxamides.

By FREDERICK DANIEL CHATTAWAY and WILLIAM HENRY LEWIS.

THE action of the halogens on substituted oxamides has been little studied, and the description of the substances formed is not satisfactory, inasmuch as the crude material was never subjected to any process of purification. By passing chlorine for different periods through a solution of oxanilide in boiling glacial acetic acid, Dyer and Mixter (*Amer. Chem. J.*, 1886, 8, 349) obtained two products : one, the melting point of which is not given, they regard as being possibly a trichloro-oxanilide ; the other, melting at 255°, they show to be somewhat impure tetrachloro-oxanilide.

In the course of the authors' study of substituted nitrogen chlorides, the action of chlorine on a boiling acetic acid solution of oxanilide has been investigated. In this action, a mixture of chloro-oxanilides is formed from which it is difficult to isolate any pure substance in quantity, although both s-di-p-chloro- and s-di-2 : 4-dichloro-oxanilides can be separated in sufficient amount for identification. For purposes of comparison, these two compounds and some closely related derivatives have been prepared from pure specimens of the anilines. The symmetrical disubstituted oxanilides or the ethyl esters of the corresponding substituted oxanilic acids are formed almost quantitatively when the substituted aniline is heated with ethyl oxalate, the product varying according as the aniline or the ethyl oxalate is present in excess.

$$\begin{matrix} CO\cdot O\cdot C_2H_5 \\ CO\cdot O\cdot C_2H_5 \end{matrix} + C_6H_4Cl\cdot NH_2 = \begin{matrix} CO\cdot NH\cdot C_6H_4Cl \\ CO\cdot O\cdot C_2H_5 \end{matrix} + C_2H_5\cdot OH.$$

$$\begin{matrix} CO\cdot NH\cdot C_6H_4Cl \\ CO\cdot O\cdot C_2H_5 \end{matrix} + C_6H_4Cl\cdot NH_2 = \begin{matrix} CO\cdot NH\cdot C_6H_4Cl \\ CO\cdot NH\cdot C_6H_4Cl \end{matrix} + C_2H_5\cdot OH.$$

Should the reaction yield a mixture of the two compounds, these can easily be separated from one another by dissolving out the oxanilic esters with alcohol, in which the disubstituted oxanilides are almost insoluble. When treated in alcoholic solution with ammonia, the oxanilic esters yield mono-substituted oxamides, thus:

$$\begin{matrix} CO\cdot NH\cdot C_6H_4Cl \\ CO\cdot O\cdot C_2H_5 \end{matrix} + NH_3 = \begin{matrix} CO\cdot NH\cdot C_6H_4Cl \\ CO\cdot NH_2 \end{matrix} + C_2H_5\cdot OH,$$

and when heated in alcoholic solution with the equivalent quantity of potassium hydroxide they yield the potassium salts of the substituted oxanilic acids, from which the acids are liberated on the addition of acetic acid:

$$\begin{matrix} CO\cdot NH\cdot C_6H_4Cl \\ CO\cdot NH\cdot C_6H_4Cl \end{matrix} + KOH = \begin{matrix} CO\cdot OK \\ CO\cdot NH\cdot C_6H_4Cl \end{matrix} + C_6H_4Cl\cdot NH_2.$$

If to a boiling glacial acetic acid solution of oxanilide a saturated solution of bleaching powder is added, a white solid is deposited consisting mainly of a mixture of s-dichloro-oxanilide and its dichloro-amino-derivative. The latter can be separated easily owing to its ready solubility in chloroform. Unlike most nitrogen chlorides containing phenyl residues with both ortho-positions to the nitrogen unoccupied, it is transformed into the isomeric oxanilide with the greatest difficulty, and its solution in acetic acid can be boiled until the whole is hydrolysed with regeneration of s-di-p-chloro-oxanilide; if any s-2:4-dichloro-oxanilide results, it is produced in too small a quantity to be recognised. Related to this reaction is the circumstance that if a saturated solution of bleaching powder is added to a boiling solution of s-2:4-dichloro-oxanilide in glacial acetic acid, the substituted oxamide is deposited from solution unchanged, no recognisable quantity of its chloroamino-derivative being formed. This behaviour is probably due to the hindrance offered to the addition of hypochlorous acid to the nitrogen by the spatial arrangement of the atoms forming the large molecule.   It is less probable that it is due to the practical insolubility of the substituted oxanilides in even slightly diluted acetic acid.

That symmetrically disubstituted oxamides containing groups of less complexity can readily yield nitrogen chlorides and bromides is shown by the behaviour of *s*-dimethyloxamide and *s*-diethyloxamide, which are readily converted by hypochlorous or hypobromous acid into their *s*-dichloroamino- or *s*-dibromoamino-derivatives.

s-Di-p-*chlorophenyloxodichloroamide,*

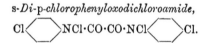

Cl⟨   ⟩NCl·CO·CO·NCl⟨   ⟩Cl.

Oxanilide is so slightly soluble in water that an aqueous solution of hypochlorous acid has practically no action on it, and its dichloro-amino-derivative has not up to the present been obtained. If to a solution of the anilide in boiling glacial acetic acid a saturated solution of bleaching powder is slowly added, a white solid is thrown out of solution, which consists of a mixture of *s*-di-*p*-chloro-oxanilide and its dichloroamino-derivative. If chloroform is added, the latter dissolves, and can be thus separated from the substituted anilide. To ensure complete conversion, the solution in chloroform is ·best shaken with a further quantity of bleaching powder solution acidified with acetic acid, and on separating the chloroform solution, drying, and evaporating off the solvent, the dichloroamino-derivative is left as a white, crystalline mass, which, after several crystallisations from a mixture of chloroform and petroleum, is obtained pure. It is readily soluble in boiling chloroform and sparingly so in petroleum ; it crystallises from a mixture of the two in colourless, transparent rhombs (m. p. 169°).

0·3042 liberated $I = 32·3$ c.c. $N/10$ I.   Cl as $NCl = 18·82$.

$C_{14}H_8O_2N_2Cl_4$ requires Cl as $NCl = 18·76$ per cent.

It is a stable substance, which undergoes transformation very slowly, if at all.   On heating for some hours with boiling glacial acetic acid, hypochlorous acid or chlorine is gradually given off and *s*-di-*p*-chloro-oxanilide regenerated.   *s*-Di-*p*-chlorophenyloxodichloroamide can be prepared also in a similar manner from *s*-di-*p*-chloro-oxanilide itself.

Ethyl p-*Chloro-oxanilate*, $Cl\langle\ \rangle NH \cdot CO \cdot CO_2 \cdot C_2H_5$.

This compound is formed almost exclusively when ethyl oxalate (1 mol.) is heated to 180—200° for several hours with slightly less than the equivalent quantity (1 mol.) of p-chloroaniline. Ethyl alcohol is evolved and a clear, brown liquid obtained, which, on cooling, solidifies to a mass of crystals. On rubbing this to a paste with a little alcohol and pressing it on a porous plate, the ester is obtained as a white, soft, crystalline powder. It is moderately soluble in boiling alcohol, from which it crystallises well in thin, transparent, colourless plates (m. p. 155°), which have a pearly appearance when pressed together.

0·1886 yielded 0·1212 AgCl.   Cl = 15·89.
$C_{10}H_{10}O_3NCl$ requires Cl = 15·58 per cent.

p-*Chloro-oxanilamide*, $Cl\langle\ \rangle NH \cdot CO \cdot CO \cdot NH_2$.

This compound, which is produced when a warm alcoholic solution of ethyl p-chloro-oxanilate is mixed with an alcoholic solution of ammonia, separates on cooling the liquid as a mass of fine needles and melts at 241°; it crystallises from boiling alcohol, in which it is sparingly soluble, in small, colourless needles, which form a felted mass from which the mother liquor can only be removed by considerable pressure.

0·1956 yielded 0·1435 AgCl.   Cl = 18·14.
$C_8H_7O_2N_2Cl$ requires Cl = 17·85 per cent.

s-*Di*-p-*chloro-oxanilide*, $Cl\langle\ \rangle NH \cdot CO \cdot CO \cdot NH \langle\ \rangle Cl$.

This compound, which is produced when ethyl p-chloro-oxanilate is heated with p-chloroaniline, is most easily made by heating p-chloroaniline in slight excess ($2\frac{1}{2}$ mols.) with ethyl oxalate (1 mol.) to 180—200° for three to four hours, the alcohol formed being allowed to escape. On cooling and extracting the crystalline mass repeatedly with boiling alcohol, the substituted oxanilide is left as a white, crystalline residue, scarcely soluble in any ordinary solvent. It can with some difficulty be recrystallised from boiling glacial acetic acid, in which, however, it is only slightly soluble and from which it separates in glittering, four-sided, rhombic plates. The best solvent for this and similarly substituted oxamides is hot nitrobenzene, in which they and oxanilide itself readily dissolve and from which they crystallise exceedingly well on cooling. The adhering nitrobenzene can be removed by boiling the crystals for a short time with alcohol. s-Di-*p*-chloro-oxanilide crystallises from hot nitrobenzene in glittering, long, colourless, transparent, thin plates, which are probably much flattened rhombic prisms (m. p. 288°).

0·4583 yielded 0·4189 AgCl.   Cl = 22·60.

$C_{14}H_{10}O_2N_2Cl_2$ requires Cl = 22·94 per cent.

This compound can also be obtained by the direct chlorination of oxanilide dissolved in boiling glacial acetic acid, and is formed when *s*-di-*p*-chlorophenyloxodichloroamide is decomposed by a solution of hydriodic acid or by boiling alcohol.

Ethyl 2 : 4-*Dichloro-oxanilate*, $Cl\underset{Cl}{\underline{\big\langle\ \ \big\rangle}}NH\cdot CO\cdot CO_2\cdot C_2H_5$.

This compound is prepared by heating 2 : 4-dichloroaniline (1 mol.) with rather more than the theoretical amount (1 mol.) of ethyl oxalate at 180—200° for two to three hours. On adding a little alcohol to the hot product and cooling, the ester separates as a felted mass of white needles. These, after washing with a little alcohol, can be recrystallised from boiling alcohol, in which the oxamate is moderately soluble ; it crystallises from alcohol in colourless, transparent, long, hair-like crystals, seen under the microscope to be slender prisms (m. p. 119°).

0·2318 yielded 0·2556 AgCl.   Cl = 27·26.

$C_{10}H_9O_3NCl_2$ requires Cl = 27·06 per cent.

2 : 4-*Dichloro-oxanilamide*, $Cl\underset{Cl}{\underline{\big\langle\ \ \big\rangle}}NH\cdot CO\cdot CO\cdot NH_2$.

This compound separates as a mass of slender, white needles when a hot alcoholic solution of ethyl 2 : 4-dichloro-oxanilate is mixed with an alcoholic solution of ammonia. It is sparingly soluble in boiling alcohol, from which it crystallises as a network of colourless, branched, hair-like crystals (m. p. 234°).

0·1800 yielded 0·2228 AgCl.   Cl = 30·60.

$C_8H_6O_2N_2Cl_2$ requires Cl = 30·43 per cent.

s-*Di*-2 : 4-*dichloro-oxanilide*, Cl⟨◯⟩NH·CO·CO·NH⟨◯⟩Cl.
          Cl   Cl

This compound is produced when ethyl oxalate (1 mol.) is heated at 180—200° for three to four hours with a slight excess ($2\frac{1}{2}$ mols.) of 2 : 4-dichloroaniline. On cooling, a dark, semi-crystalline mass is left, from which s-di-2 : 4-dichloro-oxanilide is obtained as an insoluble crystalline powder on extracting with boiling alcohol. It is practically insoluble in all ordinary solvents, but can be recrystallised from boiling nitrobenzene, in which it is readily soluble, and from which it crystallises in colourless, glittering, transparent, very slender, flattened prisms (m. p. 276°).

0·2512 yielded 0·3800 AgCl.   Cl = 37·40.
$C_{14}H_8O_2N_2Cl_4$ requires Cl = 37·52 per cent.

Action of Chlorine on Oxanilide.—It is not easy to obtain a pure product by the direct chlorination of oxanilide. If the latter compound is dissolved in a large excess of boiling glacial acetic acid and chlorine passed in to saturation, crystals resembling those of s-di-*p*-chloro-oxanilide separate on cooling. These are not pure, however, but contain some s-di-2 : 4-dichloro-oxanilide, and on repeatedly crystallising from nitrobenzene, the melting point of the substance, which at first, as a rule, melted not very sharply at about 245°, can be raised to about 280°. On passing chlorine for a long time into a glacial acetic acid solution of oxanilide, a little s-di-2 : 4-dichloro-oxanilide can be obtained, but the yield is poor and the process not a convenient one for preparing the compound.

s-*Dimethyloxodichloroamide*, $CH_3·NCl·CO·CO·NCl·CH_3$.

This compound is easily prepared by suspending finely-divided s-di-methyloxamide in a solution of hypochlorous acid made by adding potassium hydrogen carbonate in excess to a solution of sodium hypochlorite. On adding a little chloroform and shaking, the dichloro-amino-derivative is formed and dissolves in the chloroform. To ensure the complete conversion of the oxamide, the chloroform solution is again shaken with a fresh quantity of hypochlorous acid. On separating the chloroform solution, drying with calcium chloride, and driving off the solvent, the dichloroamino-derivative is left as a very pale yellow liquid, which solidifies on cooling and stirring with a little light petroleum. It forms a white, crystalline powder, easily soluble in light petroleum, from which it can be crystallised with some difficulty and from which it slowly separates in clusters of long, colourless, slender prisms (m. p. 37°).   This substance and the other nitrogen chlorides described in this paper were analysed in the usual way by titrating with thiosulphate the iodine liberated by a weighed amount dissolved in acetic acid and mixed with excess of a solution of potassium iodide.

0·4070 liberated I $= 87$·9 c.c. $N/10$ I.   Cl as NCl $= 38$·28.

$C_4H_6O_2N_2Cl_2$ requires Cl as NCl $= 38$·32 per cent.

s-*Diethyloxodichloroamide*, $C_2H_5 \cdot NCl \cdot CO \cdot CO \cdot NCl \cdot C_2H_5$.

This compound was prepared from s-diethyloxamide and isolated exactly as the compound previously described ; it is a very pale yellow, viscid liquid, which, even after some months, shows no sign of crystallising.   On strongly heating, it decomposes with evolution of gas.

0·4832 liberated I $= 91$·2 c.c. $N/10$ I.   Cl as NCl $= 33$·45.

$C_6H_{10}O_2N_2Cl_2$ requires Cl as NCl $= 33$·28 per cent.

s-*Dimethyloxodibromoamide*, $CH_3 \cdot NBr \cdot CO \cdot CO \cdot NBr \cdot CH_3$.

This was prepared from s-dimethyloxamide in the same manner as the corresponding chlorine compound, using a solution of hypobromous acid made from mercuric oxide and bromine.   A little free mercuric oxide was added to the solution to prevent the development of free bromine.   On filtering off the chloroform solution, drying, and expelling the solvent, the dibromoamino-derivative was left as a very pale yellow, crystalline mass.   It crystallises well from chloroform in long, flattened prisms or plates of a very pale yellow colour (m. p. 95°).

0·3373 liberated I $= 49$·2 c.c. $N/10$ I.   Br as NBr $= 58$·32.

$C_4H_6O_2N_2Br_2$ requires Br as NBr $= 58$·35 per cent.

s-*Diethyloxodibromoamide*, $C_2H_5 \cdot NBr \cdot CO \cdot CO \cdot NBr \cdot C_2H_5$.

This compound was prepared from s-diethyloxamide and isolated exactly as the preceding compound ; it crystallises from chloroform in brilliant, glittering plates having a very pale yellow colour (m. p. 82°).

0·2794 liberated I $= 36$·8 c.c. $N/10$ I.   Br as NBr $= 52$·66.

$C_6H_{10}O_2N_2Br_2$ requires Br as NBr $= 52$·94 per cent.

St. Bartholomew's Hospital and College,
London, E.C.

XX.—The Osmotic Pressure of Solutions of Sugar in Mixtures of Ethyl Alcohol and Water.

By PERCIVAL SMITH BARLOW.

WHEN a copper ferrocyanide membrane is used with alcoholic solutions, no osmotic pressure is set up. This has been long known as the result of Tammann's work (*Ann. Phys. Chem. Neue Folge*, 1888, **34**, 309), and has been further confirmed by the author. Other work (*Phil. Mag.*, 1905, [vi], **10**, 1) has also shown that this membrane is very sensitive to the presence of water. These results, obtained almost simultaneously, suggested the inquiry as to how far the presence of alcohol might modify the osmotic pressure of an aqueous solution of sugar if the experiment was arranged so that there were equal concentrations of water and alcohol on opposite sides of the membrane.

Cane sugar was used as the dissolved substance, since it is fairly soluble in a mixture of alcohol and water, and, being the substance employed in Pfeffer's experiments, which were the basis of the van't Hoff "gas" theory, it seemed the most suitable compound for the following work. Special conditions of experiment are employed. With the copper ferrocyanide membrane, the water of the solvent can alone produce an osmotic current, and of the pure liquids is the one which alone dissolves the sugar.

In preparing for each experiment, the clean cell was soaked for from one to two days in the solvent, this (unless otherwise stated) being a mixture of equal volumes of ethyl alcohol ("absolute") and water. The cell was then filled with the solution to be used in the experiment, and was again left for two or three days in the solvent. Fresh solution and solvent were used for the experiment itself. In order to avoid as far as possible any osmotic pressure which might arise from any inequality in the liquids themselves apart from the dissolved sugar, the solvent was always prepared beforehand in large quantity. The same liquid could then be used for soaking the cell, for making the solution, and for the outer liquid. Except in Experiment I, the cell stood in a glass bottle which was well corked. By means of a small mercury manometer through the cork, atmospheric pressure could be maintained above the outer liquid.

Experiment I. Strength of solution, 0·0075 normal. Theoretical pressure, 128 mm. Initial pressure shown by the gauge was 39·5 mm. After ten days this was 35 mm., and was followed by a steady fall.

Experiment II. Strength of solution, 0·0117 normal. Theoretical pressure, 210 mm. Initial pressure, 24 mm. After thirteen days the pressure was 6·8 mm. below atmospheric pressure.

This result indicates a small outflow. It may be due to greater concentration of water inside the cell, in spite of the precautions taken. The membrane is certainly sensitive to small differences of concentration of water, but the difference in this case can only be very small, and should be more than counterbalanced by the sugar present.

Experiment III. Strength of solution, 0·027 normal. Theoretical pressure, 458 mm. The cell was closed for five weeks. After that time a rise in pressure of 9 mm. was shown.

If one can assume steady values after so long a time, this experiment indicates that as the strength of the sugar solution increases, the possibility of the osmotic pressure being shown also increases, and that there is a strength of solution the osmotic pressure of which is nullified by the alcohol present. Solutions below this strength show no osmotic pressure.

In Experiment IV, dextrose was the dissolved substance. The previous history of the cell used in this case is important, and is shortly as follows. The cell was washed in several changes of freshly-boiled distilled water for nine days. It was then used with an aqueous solution of dextrose. The cell was closed and under pressure for eleven days. After that time, the outer water was examined for dextrose by Fehling's solution ; there was no reaction. Test cases showed that the solution was working well. This absence of sugar in the outer water is in agreement with the general behaviour of a copper ferrocyanide membrane, but is important evidence when one considers the result of the following experiment. Finally, the cell was washed for twelve days in distilled water, and then used in the ordinary way with the mixed solvent.

Experiment IV. Solution of dextrose : strength, 0·019 normal. Theoretical pressure, 325 mm. The cell was closed for five weeks and then showed a pressure of 6 mm.

The outer liquid was examined for dextrose, and a small but distinct precipitate of copper oxide was obtained. The membrane, however, had been proved (as above) to be impermeable to dextrose under greater internal pressure ; it therefore appears as though the membrane ceased to be impermeable to the dextrose in the presence of alcohol. It will be necessary to consider this part of the experiment later. The actual osmotic effect is seen to be in agreement with those of the first three experiments.

Having found that, as the strengths of the solutions increased, the possibility of demonstrating the osmotic pressure also increased, it was natural to conclude that the greatest osmotic effect would be shown with saturated solutions. Whatever theoretical considerations we may apply to the grouping of the different molecules in these solutions, it will be found that there should be an inflow of water. If one considers the two liquids as aqueous solutions of alcohol, the inner solution also contains dissolved sugar. This, on the generally accepted theory, will necessitate an inflow, given a suitable membrane. The osmotic current always tends to set up conditions which oppose it ; in this case, this opposition is twofold. There is the increase of internal pressure and the diminution in the concentration of the water outside. Hence, with saturated solutions of sugar in this solvent, no very large pressure can be expected.

In the following experiments, the solutions were saturated. Excess of sugar was placed in the mixed solvent and left for three or four days, the mixture being shaken occasionally.

Experiment V. Solvent, 3 vols. of water to 1 vol. of alcohol. Saturated solution of cane sugar. After eight days, the osmotic pressure was 164 mm. The cell was opened and closed several times ; there was a steady rise in each case, showing altogether a large inflow. The gauge was an open one, and the highest pressure which could be shown on it was not large.

Experiment VI. Saturated solution of cane sugar in equal volumes of alcohol and water. After eight days, the pressure was 62 mm. ; after a month, it was 198 mm.

Traube, as a result of his work on capillary constants of solutions (*Phil. Mag.*, 1904, [vi], 8, 704), has put forward the theory that the osmotic current is caused by the difference of the surface tensions of the liquids separated by the membrane. In another connection, I have had occasion to show how previous work bears on this theory (*Phil. Mag.*, 1905, [vi], 10, 11). The theory seems to demand the passage of the whole liquid ; with a membrane of copper ferrocyanide, the current is caused by the water alone. The theory, therefore, does not include a simple case. Moreover, the surface-tension theory neglects the part played by the membrane, and therefore does not take into account all the conditions of work.

In considering the function of the membrane, the author's work emphasises the importance of Nernst's researches on the absorptive action of the membrane. The part played by the membrane is itself a phenomenon of solution. One of the necessary conditions for an osmotic current is that the solvent (or one of the liquids) can be absorbed by the membrane.

Traube further argues, on lines difficult to follow, that in the presence of alcohol the membrane may be rendered permeable to the sugar or dissolved substance, to which it is impermeable in aqueous solution. The latter part of Experiment IV is in favour of this view. This breaking down of the impermeability of the membrane to sugar is especially difficult to understand in the light of the fact that the cells remain sound for further use. This isolated case affords no basis for inference, but shows that the point is worthy of further examination. If this temporary permeability does occur, it is probable that the alcohol forms a true solution with the sugar.

So far, the membrane used has been permeable to water only. Bladder is the only membrane known to the author which is permeable, in an osmotic sense, to alcohol and water ; this was used with a brass cell consisting of two parts fitted together by flanges and screws, the prepared bladder being fixed between the flanges and two india-rubber rings. With such a cell, a cane-sugar solution in equal volumes of alcohol and water (theoretical pressure = 1240 mm.), and far from being saturated, gave a rise in pressure of 140 mm. in five days, a rate of increase greater than that in experiments with saturated solutions, even although bladder is imperfectly semipermeable. This result suggests that both parts of the solvent formed the osmotic current, as the previous knowledge of the membrane would lead one to expect. The selective action of the bladder would cause a larger inflow of water than of alcohol.

The last case is where the membrane allows the alcohol only to form the osmotic current. Such a membrane is obtained with gutta-percha tissue. This was used in the same way as the bladder in the type of cell just described. Three cells were used for these experiments, but two, after considerable care and time, gave no result. There seems reason for thinking that this is due to deterioration of the membrane, and not to the solutions being too weak. The last experiment is thought to be of interest because of the length of time over which it extends.

Experiment VII. Solvent : equal volumes of alcohol and water. Strength of solution, 0·227 normal. Theoretical pressure = 3850 mm. Gutta-percha membrane.

After being closed for two months, the osmotic pressure was 48 mm. In this experiment, the movement of the gauge can be no indication of the volume of the liquid which has crossed the membrane ; the outward sag of the membrane necessitates a greater inflow than the same gauge-movement would indicate in the case of a copper ferro-cyanide membrane. Alcohol is the inflowing liquid.

The ultimate object of osmotic research must be further knowledge of the internal conditions of solution. Incidentally there also arises the question of the cause of the current and the part played by the membrane. Previous reference to this makes further remark necessary. The cause of the current is found in the mutual potential energy of solution : that a current should flow depends on the ability of the membrane to dissolve the liquid. For mixed solvents, the liquid of the current depends on the selective action of the membrane. This method of explanation, in addition to being in agreement with experiment, brings osmotic phenomena into line with the principle that potential energy tends to a minimum.

On the matter of internal grouping of the molecules of the solution, not much can be said here. Recent work is all in support of "hydration" of the dissolved substance (whether ionised or not), and, more especially, work on non-aqueous solutions. It is still to be discovered why water dissolves sugar and alcohol does not, although one can quite truly say that there is some kind of attraction between sugar and water which does not exist between sugar and alcohol. In this paper, the question arises as to whether the alcohol has some of this "solution attraction" for the sugar after this has been brought into solution by the water. In the light of the last experiment, it is probable that there is "true" solution between the sugar and the alcohol. Here the osmotic current is one of alcohol, and in the usual osmotic sense the alcohol can be regarded as the solvent. If there is no true solution between the alcohol and the sugar, then the aggregation of the water and sugar molecules causes a greater concentration of free alcohol in the liquid containing the sugar than in the mixed solvent. Hence the flow of alcohol in the last experiment should have been outwards and not inwards. On the same assumption the van't Hoff "gas" theory would give the same result. Experiment is therefore in favour of the "gas" particles of the solution being aggregates of the three molecules: in other words, the sugar is "hydrated" by the water *and* the alcohol.

The difficulties of finding the osmotic pressures of these solutions have been mentioned. The selective action of the membrane causes inequalities in the concentrations of the liquids themselves. This might be avoided by adopting a method of applying pressure from the outside (see Lord Berkeley's paper, *Proc. Roy. Soc.*, 1904, **73**, 436), and so finishing the experiment before there could be any appreciable exchange of liquid. Without some such method, it appears to be futile to look for van't Hoff values for mixed solvents and their solutions.

These experiments were carried out at the Cavendish Laboratory; and, in conclusion, the writer thanks Professor J. J. Thomson for the kindness shown to him during the progress of the work.

XXI.—*The Action of Ammonia and Amines on Diazobenzene Picrate.*

By Oswald Silberrad, Ph.D., and Godfrey Rotter, B.Sc.

It was observed that when diazobenzene picrate was exposed to ammonia vapours a vigorous reaction took place, the product having a dark red colour and no longer possessing the characteristic properties of the diazo-compounds. It appeared of interest to investigate the products, since two distinct possibilities presented themselves as to the course of the reaction.

First, the intense red colour suggested that the diazobenzene picrate might have undergone a transformation into the isomeric trinitro-benzeneazophenol, a process somewhat analogous to the change from diazoaminobenzene to aminoazobenzene, thus :

$$C_6H_5 \cdot N_2 \cdot O \cdot C_6H_2(NO_2)_3 \longrightarrow HO \cdot C_6H_4 \cdot N \vdots N \cdot C_6H_2(NO_2)_3.$$

Secondly, the ammonia might set free diazobenzene hydroxide and enter into reaction therewith. In this case, ammonium picrate would be simultaneously formed. The latter assumption seemed more probable owing to the para- and ortho-positions of the phenol being blocked, and has indeed proved to be in accordance with the true course of the reaction.

Diazobenzene picrate, prepared according to Baeyer and Jaeger (*Ber.*, 1875, **8**, 893), forms a lemon-yellow precipitate, which is insoluble in most organic solvents and explodes on heating at 95—100°. The action of ammonia is best examined by placing the solid picrate in a desiccator over a 10 per cent. ammonia solution. Concentrated ammonia gives rise to a violent reaction and leads to explosions of considerable violence when the quantities under investigation are large.

The dark red product was first examined for the presence of azo-compounds, but these were found to be entirely absent. Thus the transformation to trinitrobenzeneazophenol was disproved.

The product was, relatively speaking, devoid of explosive properties ; it burnt quietly on heating with evolution of the characteristic fumes of picric acid. It dissolved, for the most part, in water, but the solution did not give the diazo-reaction with alkaline β-naphthol. Alcohol and benzene effected a partial extraction, but left a considerable amount of insoluble residue. The whole passed readily into solution in acetone, from which thick, yellow prisms were obtained on evaporation. These on recrystallisation from water were obtained as fine, yellow needles which were readily identified as ammonium picrate : analysis gave $N = 22 \cdot 69$; $C_6H_6O_7N_4$ requires $N = 22 \cdot 81$ per cent.

For the further examination of the nature of the reaction, a large quantity of the product was warmed with water rendered alkaline with caustic soda and extracted with ether. The aqueous residue, when acidified and distilled with steam, yielded a turbid distillate containing phenol, the presence of which was confirmed by the preparation of tribromophenol bromide. The ethereal extract was then shaken with a small quantity of dilute hydrochloric acid. On neutralising this solution and testing for aniline with bleaching powder, the characteristic coloration was obtained. The presence of aniline was further confirmed by the preparation of sodium benzene-azo-β-naphthol-6-sulphonate. The ethereal solution was then shaken with concentrated hydrochloric acid in excess, which again extracted a part of the dissolved matter. On addition of ammonia to the hydro-chloric acid extract, extraction with benzene, and treatment of the ben-zene solution with dry hydrogen chloride, a precipitate of diphenylamine hydrochloride was obtained. This was further purified by dissolving in water, precipitating the base with ammonia, extracting with benzene, and finally reprecipitating from the benzene with dry hydrogen chloride. The purified salt melted at $166°$, whilst diphènylamine hydrochloride melts at $167°$.

$0 \cdot 1065$ gave $6 \cdot 50$ c.c. moist nitrogen at $20 \cdot 5°$ and 765 mm. $N = 6 \cdot 92$. $C_{12}H_{12}NCl$ requires $N = 6 \cdot 83$ per cent.

On warming its aqueous solution, the free base separated, which is in accord with the behaviour of diphenylamine hydrochloride.

The above compounds formed practically the whole of the product obtained from ammonia and diazobenzene picrate. Traces of a compound melting at $142°$ were precipitated by the addition of light petroleum to the benzene solution of the crude product, and a small quantity ($0 \cdot 2$ c.c.) of a yellow oil boiling at $135—137°$ resulted on evaporation of the filtrate and distillation of the residue with steam.

The main products are therefore ammonium picrate, diphenylamine, aniline, and phenol, the last being formed in smaller quantities, probably as a decomposition product of bisdiazobenzeneamide which results from the action of ammonia on diazo-salts :

$$2C_6H_5 \cdot N_2 \cdot O \cdot C_6H_2(NO_2)_3 + 3NH_3 =$$
$$(C_6H_5 \cdot N_2)_2NH + 2NH_4 \cdot O \cdot C_6H_2(NO_2)_3.$$
$$(C_6H_5 \cdot N_2)NH = (C_6H_5)_2NH + 2N_2.$$
$$(C_6H_5 \cdot N_2)_2NH + H_2O = C_6H_5 \cdot NH_2 + C_6H_5 \cdot OH + 2N_2.$$

Action of Aniline on Diazobenzene Picrate.—The action of ammonia on diazobenzene picrate having been thus established, that of aniline could be deduced with considerable probability. Aniline picrate is obviously to be expected as one of the products ; the remaining products will be determined by the interaction of aniline with the liberated diazo-complex, thus aminoazobenzene was also to be expected. Experiment confirmed these deductions.

The reaction between diazobenzene picrate and aniline was carried out in benzene. The diazo-salt passed at first into solution, but a crystalline precipitate slowly separated. After washing with benzene and recrystallisation, this was found to agree in its properties with aniline picrate.

0·1150 gave 17·1 c.c. moist nitrogen at 14·8° and 757·7 mm. N = 17·37. $C_{12}H_{10}O_7N_4$ requires N = 17·43 per cent.

On heating, the salt darkened at 168° and blackened and melted at 181°. Pure aniline picrate, prepared directly from aniline and picric acid, was found to behave similarly ; it melts at 181°, and not at 165° as stated by Smolka (*Monatsh.*, 1885, 6, 921).

The benzene solution was shaken with hydrochloric acid, which gave rise to a crystalline precipitate. On filtration and recrystallisation from alcohol, this was found to consist of aminoazobenzene hydrochloride. By treatment with ammonia and extraction with alcohol, the free base melting at 126° was obtained; found N = 21·42, $C_{12}H_{11}N_3$ requires 21·36 per cent.

The above deductions are thus confirmed, and the reaction between aniline and diazobenzene picrate may be represented as follows :

$$C_6H_5 \cdot N_2 \cdot O \cdot C_6H_2(NO_2)_3 + 2C_6H_5 \cdot NH_2 =$$
$$C_6H_5 \cdot N_2 \cdot NH \cdot C_6H_5 + C_6H_5 \cdot NH_3 \cdot O \cdot C_6H_2(NO_2)_3.$$

Hence it may be concluded in general that the action of amino-compounds on diazobenzene picrate is to form the picrate of the base as the main product. Other compounds result from secondary reactions between the free base and the diazobenzene hydroxide.

Our thanks are due to the Explosives Committee for permission to publish these results.

RESEARCH LABORATORIES,
   ROYAL ARSENAL, WOOLWICH.

XXII.—*The Preparation of* p-*Bistriazobenzene.*

By Oswald Silberrad, Ph.D., and Bertram James Smart.

In consequence of the explosive properties of bistriazobenzene, the identity of this compound has never been confirmed by analysis. For this reason a revision of its preparation and the analysis of the product have been undertaken. For the preparation, the best results were obtained by taking acetanilide as the starting point and carrying it through the successive stages indicated in the following scheme :

$$CH_3 \cdot CO \cdot NH \cdot C_6H_5 \ \rightarrow\ CH_3 \cdot CO \cdot \overset{1}{N}H \cdot C_6H_4 \cdot \overset{4}{N}O_2 \ \rightarrow$$
$$CH_3 \cdot CO \cdot NH \cdot C_6H_4 \cdot NH_2 \ \rightarrow\ CH_3 \cdot CO \cdot NH \cdot C_6H_4 \cdot N_2Br_3 \ \rightarrow$$
$$CH_3 \cdot CO \cdot NH \cdot C_6H_4 \cdot N_3 \ \rightarrow\ NH_2 \cdot C_6H_4 \cdot N_3 \ \rightarrow$$
$$Br_3N_2 \cdot C_6H_4 \cdot N_3 \ \rightarrow\ N_3 \cdot C_6H_4 \cdot N_3 .$$

Griess (*Ber.*, 1888, **21**, 1559) proceeded from *p*-phenylenediamine, converting it first into the corresponding oxamic acid and then replacing the remaining amino-group by the triazo-group through the diazo-perbromide; from this he removed the oxalic residue by hydrolysis, leaving *p*-triazoaniline, which was then converted into the bistriazo-compound as above. Experience has shown, however, that the diazotisation of the oxamic acid derivative of *p*-phenylenediamine gives rise to by-products which are more difficult to remove than those formed with the corresponding acetyl derivative.

*Acetyl-*p*-aminodiazobenzene   perbromide,* $CH_3 \cdot CO \cdot NH \cdot C_6H_4 \cdot N_2Br_3$, has not hitherto been described ; it is readily obtained as follows :

Seventy-eight grams of *p*-aminoacetanilide, 1620 c.c. of water, and 108 c.c. of concentrated hydrochloric acid are brought together in a large vessel surrounded with ice and fitted with stirring appliances. Diazotisation is carried out in the usual way with a strong solution of 64·8 grams of potassium nitrite, the temperature of the mixture being maintained between 0° and 4°. When all the nitrite has been added, a solution of 42 c.c. of bromine in 540 c.c. of hydrobromic acid (sp. gr. 1·49) is slowly added to the mixture with vigorous stirring. A yellow, flocculent precipitate separates, but this is very readily converted to a brown oil if any excess of bromine be added. By using the above quantities, the product is obtained almost free from admixed bromine and can be readily dried at the pump. It is then spread out on a porous plate overnight and recrystallised from warm alcohol, from which it separates in orange-coloured needles melting at 126° with immediate decomposition. The yield amounts to 95 per cent.

*Acetyl-*p*-aminotriazobenzene.*—Eighty grams of the crude, dry substance are mixed with 420 c.c. of water and placed in a vessel fitted with a stirring apparatus cooled to 0°. Strong ammonia (70 c c.) is then very carefully added, the temperature being maintained below 2°. The mixture is left for ten minutes in ice, and then quickly filtered, washed with water, and dried on a porous plate. The product is best purified by recrystallising from dilute acetic acid and then from alcohol, and when pure melts at 122·5°; the yield is 60 per cent. of the theoretical. This compound has been previously prepared by Rupe and von Majewski (*Ber.*, 1900, 33, 3406), who were, however, unable to saponify it, and thus failed to obtain the amino-compound.

p-*Aminotriazobenzene*, $NH_2 \cdot C_6H_4 \cdot N_3$.—It has been found possible, under suitable conditions, to remove the acetyl group from acetyl-*p*-aminotriazobenzene by hydrolysis. If the following procedure be adhered to, a yield of 30 per cent. of the theoretical may be obtained.

Fifteen grams of the crude acetyl-*p*-aminotriazobenzene were covered with 150 c.c. of a 40 per cent. solution of potassium hydroxide in a distilling flask. The liquid was slowly distilled and as soon as an oily liquid made its appearance in the condenser the product was removed continuously from the hot solution by distillation with steam. The oil soon began to crystallise in the condenser, and a further quantity was obtained by adding caustic potash to the clear aqueous distillate. The product melted at 62° and was identical with that obtained by Griess from the oxamic acid derivative (*Ber.*, 1888, 21, 558).

Bistriazobenzene.—This was prepared from the *p*-aminotriazobenzene by diazotisation, conversion to the diazoperbromide, and thence to the bistriazo-compound. When crystallised from ether, the product separates in clear yellow tablets melting at 83° and identical in all respects with that described by Griess (*loc. cit.*).

The highly explosive nature of the compound has hitherto precluded the verification of its composition by analysis. This has now been successfully accomplished by making use of the vacuum method, the following results being obtained in this way.

I. 0·06335 gave 28·12 c.c. moist nitrogen at 17° and 767·3 mm. and 53·62 c.c. carbon dioxide at 0° and 760 mm.

II. 0·06385 gave 54·03 c.c. carbon dioxide at 0° and 760 mm.

III. 0·0843 gave 36·05 cc. moist nitrogen at 8·2° and 770 mm.

	Found.			
	I.	II.	III.	$C_6H_4N_6$ requires
Carbon	45·39	45·75	—	44·92
Nitrogen	51·95	—	52·22	52·56
Hydrogen (by difference)	2·66	—	—	2·52

Our thanks are due to the Explosives Committee for permission to publish these results.

RESEARCH LABORATORIES,
ROYAL ARSENAL, WOOLWICH.

XXIII.—Studies on Nitrogen Iodide. III. The Action of Methyl and Benzyl Iodides.

By Oswald Silberrad, Ph.D., and Bertram James Smart.

Action of Methyl Iodide.—Previous experiments have been carried out by Stahlschmidt (*Poggendorff's Ann.*, 1863, **119**, 421) on the reaction between nitrogen iodide and methyl iodide under water. He believed that the reaction could be best explained on the assumption of the formula NI_3 for nitrogen iodide. Since Stahlschmidt's experiments, Chattaway has shown the formula to be $N_2H_3I_3$ (Proc., 1899, **17**, 20; *Amer. Chem. J.*, 1900, **23**, 363, &c.), and more recently the constitution $NH_3 \cdot NI_3$ has been established (Silberrad, Trans., 1905, **89**, 55 and 66), and thus Stahlschmidt's formulation of the reaction becomes inapplicable. It appeared desirable, therefore, to repeat Stahlschmidt's experiments, firstly because this author omitted certain precautions which should obviously be taken, and secondly in order to ascertain quantitatively the molecular proportions of the products formed, and thus to establish the nature of the reaction. In Stahlschmidt's experiments, the nitrogen iodide was prepared by the addition of an alcoholic solution of iodine to concentrated aqueous ammonia. As has been pointed out by Chattaway (*loc. cit.*, p. 363), considerable quantities of iodoform are always produced when alcohol is used in the preparation of nitrogen iodide. Thus the iodoform, which Stahlschmidt believed to be a product of the action of methyl iodide on nitrogen iodide, was simply a by-product in the preparation of the nitrogen iodide, and had no bearing on the reaction. Further, Stahlschmidt evidently did not take sufficient precautions to exclude light, for he found that nitrogen was evolved during the reaction, whereas a repetition of his experiments shows that when the reaction is carried on at a low temperature in the dark no such evolution of nitrogen occurs.

The products isolated by Stahlschmidt, apart from nitrogen and unchanged methyl iodide, were hydriodic acid, ammonium iodide, tetramethylammonium penta-iodide, iodoform, iodine, and a brick-red insoluble compound which was not further investigated. The residual solution evolved ammonia on addition of caustic potash and was then found by him to contain di-iodomethylamine, potassium iodide and iodate, and traces of potassium butyrate. For the formation of iodoform and tetramethylammonium pentaiodide, Stahlschmidt gave the following explanation, assuming the formula of nitrogen iodide to be NI_3:

$$2NI_3 + 6CH_3I = N(CH_3)_4I_5 + NH_4I + 2CHI_3.$$

This equation is based on the presence of iodoform amongst the products, and as this compound is not due to the methyl iodide, but to the presence of alcohol, Stahlschmidt's deductions as to the constitution of nitrogen iodide are unfounded.

In the present work, nitrogen iodide was allowed to react with a slight excess of methyl iodide under water at a low temperature in the dark. After the removal of the unchanged methyl iodide, the products found were tetramethylammonium penta-iodide, ammonium iodide and iodate, and free iodine; neither iodoform nor di-iodomethylamine could be detected. As is shown quantitatively in the experimental part, the formation of these compounds takes place in accordance with the following equation :

$$6N_2H_3I_3 + 24CH_3I + 3H_2O = 6N(CH_3)_4I_5 + 5NH_4I + 3I_2 + NH_4 \cdot IO_3.$$

Thus, the action of methyl iodide on nitrogen iodide does not furnish evidence in favour of the formula NI_3, as supposed by Stahlschmidt, but is completely in accord with the formula $NH_3 \cdot NI_3$.

Action of Benzyl Iodide.—Similar experiments carried out with benzyl iodide have given rise to two interesting compounds, namely, tribenzylammonium penta-iodide and tribenzylammonium di-iodide. No compound was obtained with benzyl iodide analogous to the tetramethylammonium penta-iodide which resulted from the action of methyl iodide on nitrogen iodide. The two benzyl derivatives both result from the action of benzyl iodide on nitrogen iodide under water, the composition of the resultant compound being dependent on the relative quantities of the reagents.

EXPERIMENTAL.

Action of Methyl Iodide.—The reaction was first carried out under the conditions indicated by Stahlschmidt (*loc. cit.*). A quantity of about 2 grams of nitrogen iodide was prepared by adding a solution of iodine in alcohol to a concentrated aqueous solution of ammonia. After washing well with water, an excess of methyl iodide in alcohol was added, and the mixture left for twenty-four hours in a cold dark room. A brown liquid was obtained, which on evaporation smelt strongly of iodoform and gave a quantity of green crystals, the properties of which agreed precisely with those of the tetramethylammonium penta-iodide described by Stahlschmidt. The crystals were purified by crystallisation from alcohol, in which they are sparingly soluble. Of the five atoms of iodine present in the molecule, four are loosely combined and can be readily estimated by allowing the substance to remain in contact with a slight excess of sodium thiosulphate until it passed completely

N 2

into solution, and then titrating back the excess of thiosulphate with decinormal iodine solution.

(i) 0·2033 required 11·46 c.c. of decinormal sodium thiosulphate.
(ii) 0·0921　　,,　　5·19 ,,　　　　,,　　　　　,,　　　　　,,
Found (i) 71·57, (ii) 71·55 per cent. of loosely combined iodine.
$N(CH_3)_4I \cdot I_4$ requires 71·63 per cent.

The green crystalline compound can readily be obtained without the use of alcohol ; thus, if iodine chloride is allowed to react with aqueous ammonia and the resultant nitrogen iodide treated with methyl iodide under water, a good yield of the compound is obtained.

In order to determine the nature of the reaction between nitrogen iodide and methyl iodide, and to estimate the proportions in which the various products were formed, it was necessary to proceed from a known quantity of nitrogen iodide, and since the drying of the nitrogen iodide in the necessary quantity is impracticable, experiments were first carried out to ascertain the yield of nitrogen iodide from a known quantity of iodine chloride.

Fifty c.c. of a standard solution of iodine chloride (16 per cent. of iodine) were added drop by drop to 1200 c.c. of 10 per cent. aqueous ammonia cooled to 2°. The nitrogen iodide produced was repeatedly washed with water by decantation, the washings being passed through a filter. The whole of the nitrogen iodide, including the small quantity collected on the filter paper, was then decomposed with a measured quantity of a 20 per cent. solution of sodium sulphite and the excess of sulphite titrated with decinormal iodine solution. From the results thus obtained, the quantity of nitrogen iodide present was calculated. As a mean of three experiments, it was found that 50 c.c. of iodine chloride solution gave rise to 8·270 grams of nitrogen iodide. The values obtained in the individual experiments were (i) 8·265, (ii) 8·270, (iii) 8·274.

Having thus ascertained the quantity of nitrogen iodide which results from a known quantity of iodine chloride, the action of methyl iodide was examined with a quantity of nitrogen iodide prepared under precisely similar conditions. The nitrogen iodide was washed, as before, by decantation, the washings being poured through a filter. That which collected on the filter was estimated separately and deducted from the total quantity. When thoroughly washed, the nitrogen iodide was covered with about 250 c.c. of water and treated with 16·47 grams of methyl iodide. The mixture was left for three days at a low temperature in the dark. At the end of this time, crystals of tetramethylammonium penta-iodide had separated, and the liquid in the flask was brown. No evolution of nitrogen occurred at any stage of the operation.

The crystals were ground and allowed to remain in contact with the liquid for another day, after which the flask was heated to about 60° on a water-bath. The distillate which collected consisted of methyl iodide and a small quantity of iodine (distillate A). This distillate was shaken with sodium thiosulphate, and the methyl iodide was separated and weighed. The thiosulphate was titrated back for the estimation of the iodine which had distilled over.

The residue from the distillation was filtered at the pump and washed thoroughly with water (washings B). The crystals were then dried, ground under light petroleum, filtered, and washed (washings C); they were finally dried over sulphuric acid in a vacuum desiccator until their weight was constant.

The aqueous liquid (washings B) was then treated with chloroform for the extraction of the free iodine. When no more free iodine remained in the aqueous solution, the chloroform extract was removed and titrated.

The residual solution (D), which contained ammonium iodide and iodate, was diluted to 500 c.c. with water; a measured portion was treated with sulphurous acid to reduce the iodate to iodide, and the whole of the iodine was liberated with hydrogen peroxide and titrated.

A separate measured portion was treated with excess of potassium iodide and acetic acid in order to estimate the quantity of iodate present, whilst a third portion was taken for the estimation of the ammonia. Thus all the products of the reaction were quantitatively determined. The following data show the quantities found :

Quantity of nitrogen iodide which took part in the reaction.

Total quantity, calculated from iodine
chloride...................................... = 8·27 grams.
Quantity deposited on filter paper............ = 0·36 „
Quantity which entered into reaction ...... = 7·91 „

Quantity of methyl iodide which took part in the reaction.

Quantity taken (moist, free from iodine)... = 16·47 grams.
Quantity recovered unchanged (distillate A) = 5·32 „
Quantity which entered into reaction ...... = 11·15 „

Free iodine.

(i) Quantity found in distillate A .... = 0·245 gram.
(ii) From washings C by titration with
thiosulphate .... = 0·221 „

(iii) From chloroform extract by titration
with thiosulphate........................ = 1·965 grams.
Total free iodine .... = 2·431 „

Total combined iodine: 25 c.c. of solution D were treated with
sulphurous acid and then with hydrogen peroxide in acid solution to
liberate the whole of the iodine. The iodine was extracted with
chloroform and titrated with thiosulphate.

Iodine found in 25 c.c. .... = 0·1234 gram.
Total combined iodine in solution D ...... = 2·468 grams.

Iodine present as iodate: 25 c.c. of solution D were treated with
excess of potassium iodide and acetic acid, and the free iodine was
titrated with sodium thiosulphate.

Iodine found by titration..................... = 0·1242 gram.
Hence iodine present as iodate .... = 0·0207 „
Total iodine as iodate in solution D......... = 0·414 „
Total iodine as iodide in solution D (by
difference) .... = 2·054 grams.

Ammonia: 25 c.c. of solution D were distilled with caustic soda
and the ammonia collected in $N/10$ hydrochloric acid and estimated.

Ammonia obtained from 25 c.c. .... = 0·0161 gram.
Total ammonia in solution D................. = 0·322 „

The following summary shows the quantities of all the different
products which result from nitrogen iodide by the action of methyl
iodide. When these are reduced to equivalents, it is seen that they
agree closely with the quantities required by the equation:

$$6N_2H_3I_3 + 24CH_3I + 3H_2O = 6N(CH_3)_4I_5 + 5NH_4I + 3I_2 + NH_4 \cdot IO_3.$$

	Quantity.	Divided by equivalent weight.	Equivalents.	Equivalents required by above equation.
Nitrogen iodide	7·91	1·92	5·89	6
Methyl iodide	11·15	7·85	24·08	24
Free iodine	2·431	1·91	5·85	6
Combined iodine	2·468	1·94	5·95	6
Iodine as iodate	0·414	0·326	1·00	1
Iodine as iodide	2·054	1·62	4·97	5
Ammonia	0·322	1·89	5·80	6
Tetramethylammonium penta-iodide	13·96	1·97	6·04	6

Thus the reaction takes place practically quantitatively in accordance with the above equation and also affords a ready method of preparing tetramethylammonium iodide, and is an additional proof of the formula $NH_3 \cdot NI_3$ for nitrogen iodide. The equations given by Stahlschmidt, in which iodoform is assumed to be a product of the reaction, are seen to be erroneous, and the formula NI_3 is entirely unsupported by experimental evidence.

Preparation of Tetramethylammonium Penta-iodide direct from the Base.—In order to prepare the periodide direct from the base, 2 grams of the iodide dissolved in the least possible quantity of dilute alcohol were added to an alcoholic solution of 5·2 grams of iodine ; on cooling, the characteristic green crystals of the periodide separated (m. p. 126—127°).

Action of Benzyl Iodide, Formation of Tribenzylammonium Penta-iodide, $N(C_7H_7)_3,HI_5$.—Forty-two c.c. of an aqueous solution of iodine chloride containing 0·36 gram of iodine per c.c. were added to a slight excess of ammonia. The nitrogen iodide formed was well washed by decantation, and the supernatant liquid was poured off as completely as possible. Twenty-six grams of benzyl iodide were added and the mixture allowed to stand at 10—15° in the dark. Dark green crystals were gradually deposited ; after twenty-four hours,.these were thoroughly washed with water and then with light petroleum. They still contained a moderate amount of impurity, but were obtained pure by recrystallising rapidly from hot alcohol. It is somewhat difficult to remove the last traces of free iodine from the penta-iodide, since a slight decomposition always occurs during recrystallisation with liberation of iodine, which contaminates the product to some extent. In the pure state, it is sparingly soluble in hot alcohol, but its solubility is considerably increased by the presence of impurities. It melts at 121—122°. For analysis, the product was recrystallised rapidly from hot alcohol and dried quickly over sulphuric acid in a vacuum desiccator.

0·1033 gave 0·1323 AgI. $I = 69·23$.
0·1665 ,, 0·2122 AgI. $I = 68·95$.

$$N(C_7H_7)_3,HI_5 \text{ requires } 68·78 \text{ per cent.}$$

On liberating the free base with alkali and extracting with ether, an oil was obtained which gradually crystallised when the ether was removed. The base melted at 90·5°, which agrees with the melting point of tribenzylamine (90°).

If the tribenzylamine penta-iodide is left with a slight excess of sodium thiosulphate and a little chloroform, the loosely combined iodine enters into reaction with the thiosulphate, and may be estimated by titration.

0·0842 gave 0·0468 iodine.   I = 55·65.

0·1538   „   0·0849 iodine.   I = 55·20.

$N(C_7H_7)_3HI_5$ requires $4I = 55·02$ per cent.

Tribenzylammonium Di-iodide, $N(C_7H_7)_3,HI_2$.—This compound is obtained similarly from nitrogen iodide and benzyl iodide, the only difference being in the ratio in which the reagents are brought together. The di-iodide was obtained by bringing together 50 c.c. of iodine chloride solution (1 c.c. = 0·36 gram) and an excess of ammonia, thoroughly washing the precipitated nitrogen iodide with water by decantation, and adding 42 grams of benzyl iodide. On allowing the mixture to stand, garnet-red crystals were obtained, the yield amounting to 16·5 grams. The crystals dissolved with moderate ease in alcohol and could be readily crystallised from this solvent. This treatment did not, however, effect a complete purification, as the following analyses of the crystallised product show:

0·1523 gave 0·1290 AgI.   I = 45·78.

0·5661   ,,   0·4759 AgI.   I = 45·44.

$N(C_7H_7)_3,HI_2$ requires $I = 46·84$ per cent.

Probably the crystals contained small quantities of tribenzylamine iodide. A much purer product was obtained by grinding the recrystallised compound with water and then drying it in a vacuum desiccator.

0·1204 gave 0·1039 AgI.   I = 46·65.

0·1573   ,,   0·1360 AgI.   I = 46·74.

$N(C_7H_7)_3,HI_2$ requires 46·84 per cent.

The compound melted at 115—116° and gave rise to tribenzylamine when treated with caustic potash. If the tribenzylamine di-iodide is left with a slight excess of sodium thiosulphate, the loosely combined iodine enters into reaction with the thiosulphate and may be estimated by titration.

0·2307 required 4·25 c.c. $N/10$-sodium thiosulphate.   I = 23·39.

$N(C_7H_7)_3,HI_2$ requires $I = 23·42$ per cent.

Interconversion of the Pentaiodide and Di-iodide.—The formula of the red compound is further proved by its formation from the green penta-iodide by the partial abstraction of the iodine by potassium iodide. One gram of the pentaiodide was boiled with 3 grams of potassium iodide in 25 c.c. of water for half an hour. On filtering the red liquid, garnet-red crystals melting at 115—116° were obtained. Conversely, the green compound may be regenerated from the red crystals by warming with the theoretical quantity of iodine in a little alcohol, and both compounds may be prepared directly from the base by the addition of the requisite quantity of iodine dissolved in hot chloroform to the hydriodide of the base also dissolved in the same solvent. For the preparation of the di-iodide, the best results are obtained by the addition of 16·8 grams of iodine dissolved in 170 c.c. of hot chloroform to 56 grams of tribenzylamine hydriodide in 120 c.c. of chloroform, and for the preparation of the penta-iodide by the addition of 35 grams of tribenzylamine hydriodide dissolved in 200 c.c. of chloroform to a hot solution of 43 grams of iodine in 550 c.c. of chloroform. In both cases, the compounds crystallise out on cooling, and are identical with those prepared by the action of benzyl iodide on nitrogen iodide.

In conclusion we wish to express our thanks to the Explosives Committee for permission to publish these results.

RESEARCH LABORATORIES,
ROYAL ARSENAL, WOOLWICH.

XXIV.—Gradual Decomposition of Ethyl Diazoacetate.

By OSWALD SILBERRAD, Ph.D., and CHARLES SMART ROY.

IN previous papers by one of the authors (Hantzsch and Silberrad, *Ber.*, 1900, 33, 58, and Silberrad, Trans., 1902, 81, 598), the products of the polymerisation of ethyl diazoacetate have been investigated. The present paper deals with the spontaneous decomposition of ethyl diazoacetate.

Ethyl 4 : 5-*Dihydropyrazole*-3 : 4 : 5-*tricarboxylate*,

$$CO_2(C_2H_5)\cdot CH-C\cdot CO_2\cdot C_2H_5$$
$$CO_2(C_2H_5)\cdot CH\quad N$$
$$\diagdown\diagup$$
$$NH$$

When ethyl diazoacetate is exposed to light at the ordinary temperature, a slow evolution of nitrogen occurs and, after several years, thick colourless rhombic crystals resembling cane sugar gradually make their appearance. The formation of the compound is assisted by sunlight and may be further accelerated by a preliminary heating on the water-bath. If, however, the latter procedure is adopted, the heating should not exceed three days, as otherwise dark syrupy products are formed which refuse to crystallise.

The crystalline product which had separated was collected, recrystallised from alcohol, and analysed with the following result:

0·1116 gave 0·2056 CO_2 and 0·0642 H_2O.   C = 50·24; H = 6·39.

0·2051 „   0·3780 CO_2 „   0·1192 H_2O.   C = 50·26; H = 6·42.

0·1136 „   10·05 c.c. moist nitrogen at 20·4° and 739·6 mm.   N = 9·78.

0·1097 „   9·85 c.c.   „   „   19·2° „ 739·6 mm.   N = 9·78.

$C_{12}H_{18}O_6N_2$ requires C = 50·30; H = 6·35; N = 9·81 per cent.

The molecular weight was determined by the lowering of the freezing point in benzene.

Solvent.	Substance.	$\Delta t.$	M.W.
17·35	0·2357	0·2325	286
16·12	0·1267	0·1325	291

$C_{12}H_{18}O_6N_2$ requires M.W. = 286.

The compound is slightly soluble in water, but readily so in ether, alcohol, acetone, and chloroform; from alcohol it separates in colourless needles melting at 97·5° (uncorr.).

From this it becomes evident that the compound is the ethyl ester of 4 : 5-dihydropyrazole-3 : 4 : 5-tricarboxylic acid, a compound first obtained by Buchner and von der Heide (*Ber.*, 1901, **34**, 345) by the action of ethyl αα-dimethylacrylate on ethyl diazoacetate at 100°. Its identity was further confirmed by the analysis of its silver salt, which is precipitated by the addition of silver nitrate to a neutral solution of the sodium salt, the latter being obtained by saponifying the ester with alcoholic soda. The silver salt ignites with explosive violence on heating, hence it was found necessary to carry out the analysis under special precautions. This was done by heating a weighed quantity in a tube, one end of which had been previously drawn out and closed with glass wool, the other being securely plugged with an india-rubber stopper. The silver salt was then decomposed by cautious heating and subsequently ignited in a current of pure air until the weight became constant.

0·2291 gave 0·1418 Ag.   Ag = 61·89.

0·2530 „   11·5 c.c. moist nitrogen at 18° and 772·4 mm.   N = 5·31.

$C_6H_3O_6N_2Ag_3$ requires Ag = 61·92.   N = 5·37 per cent.

On the addition of calcium chloride to the neutral solution, the calcium salt is produced as a white precipitate, soluble in acetic acid. Mercuric chloride slowly forms a white, crystalline precipitate insoluble in dilute hydrochloric acid.

Action of Copper Dust on Ethyl Diazoacetate.—It seemed probable that in Buchner and von der Heide's experiments in which they obtained ethyl dihydropyrazoletricarboxylate by heating a mixture of ethyldiazoacetate and ethyl aa-dimethylacrylate that the latter ester acted in some way catalytically. It therefore became of interest to ascertain whether the same acceleration could not be brought about by mere contact action, and to this end the action of copper dust on ethyl diazoacetate was investigated.

When the diazoacetate is added to copper dust, no reaction appears to take place below 80°, but above that temperature the addition of the first drop of ester is accompanied by an explosion of sufficient violence to shatter the flask. In order to moderate the violence of the reaction, the ester was diluted with absolute ether and after several experiments the following procedure was found to give satisfactory results. Two hundred grams of copper powder are placed in a flask heated in a water-bath and fitted with a dropping funnel and a tube connecting the flask with two wash-bottles filled with cold alcohol for the collection of the volatile products of the reaction. Fifty grams of ethyl diazoacetate, diluted with an equal volume of absolute ether are then cautiously added. The first addition is accompanied by a lurid green flash, which, however, may be avoided by previously displacing the air in the flask by means of carbon dioxide. The reaction is accompanied by a stormy evolution of thick white fumes, which condense readily in the alcohol wash-bottles. After allowing the flask to cool down to the ordinary temperature, the alcoholic solutions are added, the whole filtered, the alcohol removed, and the residue distilled under diminished pressure. In this manner, three fractions are obtained, of which the last, boiling at 220—254°/2 mm., consists chiefly of ethyl 4 : 5-dihydro-3 : 4 : 5-tricarboxylate, which slowly solidifies to a crystalline mass which on recrystallisation melts at 97·5° (uncorr.).

0·1134 gave 9·95 c.c. moist nitrogen at 21° and 760·5 mm. $N = 9·95$.
$C_{12}H_{18}O_6N_2$ requires $N = 9·81$ per cent.

The lower fractions consisted chiefly of unaltered ethyl diazoacetate. The reaction is evidently due to the intermediate formation of ethyl fumarate (or maleate) and may be expressed as follows :

$$CO_2(C_2H_5)\cdot CH{<}^{N}_{N} \atop {}^{N}_{N}{>}CH\cdot CO_2\cdot C_2H_5 \ = \ {CO_2(C_2H_5)\cdot CH \atop CH\cdot CO_2\cdot C_2H_5} \ + 2N_2$$

$$CO_2(C_2H_5)\cdot \underset{CH\cdot CO_2\cdot C_2H_5}{\overset{\displaystyle CH}{|}} \quad + \quad \underset{N \diagdown\!\!\diagup N}{\overset{\displaystyle CH\cdot CO_2\cdot C_2H_5}{}} \quad =$$

$$\underset{CO_2(C_2H_5)\cdot CH}{\overset{CO_2(C_2H_5)\cdot CH-C\cdot CO_2\cdot C_2H_5}{}}\underset{\diagdown\!\!\diagup}{\overset{N}{|}}$$
$$NH$$

This view is further confirmed by the work of Buchner and Witter, who prepared the methyl ester by the action of the corresponding ester of fumaric or maleic acid on methyl diazoacetate (*Annalen*, 1893, **273**, 239).

XXV.—*Action of Bromine on Benzeneazo-o-nitrophenol.*

By John Theodore Hewitt and Norman Walker.

In several earlier communications it has been shown that when benzeneazophenol and allied compounds are treated with dilute nitric acid or with bromine in presence of sodium acetate, substitution takes place in the ortho-position to the . phenolic hydroxyl. Typical cases are the production of benzeneazo-*o*-nitrophenol (Hewitt, Trans., 1900, **77**, 49) and of benzeneazo-*o-o*-dibromophenol (Hewitt and Aston, Trans., 1900, **77**, 712, 810) under the conditions mentioned. It is, however, noticeable that whereas only one nitro-group is introduced, it has been found impossible to isolate such a compound as benzeneazo-*o*-bromophenol, even when only one molecular proportion of bromine has been employed, a mixture of the unaltered azophenol and its dibromo-derivative resulting.

Whilst it appeared very probable that, if benzeneazo-*o*-nitrophenol were subsequently brominated in presence of sodium acetate, a mono-bromo-derivative with the following substituents,

$$OH : NO_2 : Br_2 : C_6H_5\cdot N_2 = 1 : 2 : 6 : 4,$$

would be produced, no direct proof had been furnished, and the result appeared to be even somewhat doubtful when, on trying the experiment, a compound was obtained melting at 154·5° (corr.), practically the same melting point as that obtained by H. V. Mitchell with a product obtained from diazotised *p*-bromoaniline and *o*-nitrophenol. The two substances were, however, found to depress one another's melting points, and we are now in a position to give the definite proof that bromination really took place in the other free ortho-position (relatively to the phenolic hydroxyl).

Preparation of 4-Benzeneazo-2-bromo-6-nitrophenol.

The benzeneazo-*o*-nitrophenol was prepared according to the method previously described (Trans., 1900, **77**, 49), but the purification was considerably facilitated by taking advantage of the insolubility of the barium salt. The crude nitration product was dissolved in excess of hot dilute ammonia and barium chloride solution added, the red precipitate was collected at the pump and well washed with hot water until the filtrates were of a light shade. By decomposition with dilute hydrochloric acid and recrystallisation from glacial acetic acid, the benzeneazo-*o*-nitrophenol can be obtained with the correct melting point. In fact, the well-washed barium salt furnishes directly on decomposition with acid a substance melting within two or three degrees of the correct temperature, and usually this product has been employed in further experiments.

Benzeneazo-*o*-nitrophenol is dissolved in four times its weight of boiling glacial acetic acid, together with half its weight of fused sodium acetate. The solution is cooled down to the ordinary temperature under continual stirring, whereby a thin paste is produced. The molecular proportion of bromine diluted with twice its weight of glacial acetic acid is then added with continual stirring and external cooling. Usually rather more acetic acid has to be added after two-thirds of the bromine has been introduced owing to the thickening of the mixture. The resulting mass is treated with hot water and collected at the pump, the crude product usually melting at about 140—145°. By two recrystallisations from boiling glacial acetic acid, the substance is obtained in small yellow needles melting at 154·5—155°.

0·1379 gave 0·2259 CO_2 and 0·0368 H_2O. C = 44·68 ; H = 2·42.

0·1254 ,,   14·4 c.c. nitrogen at 19° and 763 mm. N = 13·47.

0·0899 ,,   0·0520 AgBr. Br = 24·39.

$C_{12}H_8O_3N_3Br$ requires C = 44·70 ; H = 2·51 ; N = 13·07 ; Br = 24·82 per cent.

The substance is readily soluble in acetone and boiling glacial acetic acid, but only moderately in alcohols and aromatic hydrocarbons. The solubility in pyridine is accompanied by salt formation, the solution leaving on evaporation an orange powder melting at 112°. This azophenol gives no hydrochloride. If hydrogen chloride is led into its benzene solution, a slight separation of the unchanged azophenol occurs, but no salt formation can be detected.

Constitution of the Benzeneazobromonitrophenol.

Many experiments were initiated having for their object the determination of the position of the bromine atom. Attempts at reduction
with isolation of the resulting diaminobromophenol as tribenzoyl derivative failed on account of the elimination of the bromine.

The appearance of the recent paper by Otto Schmidt (*Ber.*, 1905,
38, 3021 ; compare Meldola and Morgan, Trans., 1889, **55**, 603) led to
the trial of strong nitric acid as an agent of fission. Two grams of
the azophenol were added in small quantities at a time to 20 grams of
nitric acid (sp. gr. 1·5) cooled externally by ice. The dark red
solution was poured off after five to seven minutes on to crushed ice,
the precipitate collected at the pump, and the filtrate added to an ice-
cold solution of 1·5 grams of β-naphthol and 15 grams of sodium
hydroxide in 200 c.c. of water. A dark red precipitate was formed,
which was collected, washed with boiling water, and dried.

The first precipitate obtained by pouring the nitric acid solution on
to ice weighed 1·5 grams, and, without recrystallisation, melted at
116° (uncorr.). When mixed with 2-bromo-4 : 6-dinitrophenol (m. p.
118°, Körner, *Jahresber.*, 1875, 337), the melting point was not
depressed. Theoretically, 1·6 grams of bromodinitrophenol should
have been obtained.

The product obtained by coupling the simultaneously produced
diazonium salt with β-naphthol, which weighed 1·2 grams, melted in
the crude condition at 237° (uncorr.) and at 239° (= 246° corrected)
after recrystallisation from toluene. Its identification with *p*-nitro-
benzeneazo-β-naphthol was completed by a nitrogen estimation and a
determination of melting point after mixture with synthesised *p*-nitro-
benzeneazo-β-naphthol.

0·1345 gave 16·3 c.c. N at 12° and 759 mm. N = 14·48.

$C_{16}H_{11}O_3N_3$ requires N = 14·37 per cent.

The result leaves no doubt as to the position of the bromine atom,
which must have entered into the ortho-position to the phenolic
hydroxyl.

The *sodium* salt separates from hot solutions in very small crystals
fairly soluble in water; the salt was analysed after drying over sulphuric acid in a vacuum for three days.

0·2079 gave 0·0453 Na_2SO_4. Na = 7·06.

$C_{12}H_7O_3N_3BrNa$ requires Na = 6·81 per cent.

The *potassium* salt forms small, extremely dark-coloured prisms.

0·0810 gave 0·0182 K_2SO_4. $K = 10·07$.

$C_{12}H_7O_3N_3BrK$ requires $K = 10·88$ per cent.

The *ammonium* salt, which has not been isolated for analysis, is apparently only sparingly soluble in cold water.

An ammoniacal solution of the azophenol gives precipitates with many metallic salt solutions.

Metallic salt.	Precipitate.	Metallic salt.	Precipitate.
Silver nitrate	Orange-red, insoluble.	Stannous chloride..	Brown precipitate, soluble in hot water.
Copper sulphate ...	Yellow.		
Barium chloride ...	Orange-red, soluble in boiling water.	Lead acetate.........	Reddish-brown precipitate, insoluble on boiling.
Magnesium sulphate	Orange, soluble in hot water.	Manganous chloride	Brick-red, soluble on boiling.
Zinc sulphate	Orange.		
Mercuric chloride...	Yellow.	Ferrous sulphate ...	Yellow.
Ammonia alum......	Yellow cloudiness in cold solution, yellowish-brown precipitate on boiling.	,, chloride ...	Colloidal.
		Cobalt nitrate	Light brown.
		Nickel sulphate ...	Brick-red, insoluble in hot water.

The *acetyl* derivative, $C_6H_5 \cdot N \colon N \cdot C_6H_2Br(NO_2) \cdot O \cdot CO \cdot CH_3$, was prepared by boiling the substance with its own weight of fused sodium acetate and four times its weight of acetic anhydride for five hours in a reflux apparatus. The product was isolated in the usual manner and recrystallised from glacial acetic acid, when dark red needles (m. p. 137°) were obtained.

0·1679 gave 0·2862 CO_2 and 0·0434 H_2O. $C = 46·49$; $H = 2·86$.

0·2329 ,, $H \cdot C_2H_3O_2$ requiring 6·2 $N/10$ NaOH. $C_2H_3O = 11·44$.

$C_{14}H_{10}O_4N_3Br$ requires $C = 46·13$; $H = 2·77$; $C_2H_3O = 11·82$ per cent.

The *benzoyl* derivative, $C_6H_5 \cdot N \colon N \cdot C_6H_2Br(NO_2) \cdot O \cdot CO \cdot C_6H_5$, was produced by gently boiling the substance with three times its weight of benzoyl chloride for two hours in a reflux apparatus. The product was poured into 80 per cent. alcohol, and the precipitate, when hard, collected and recrystallised from toluene. Radiating brown aggregates of needles were obtained which melted at 131°.

0·1447 gave 0·2849 CO_2 and 0·0417 H_2O. $C = 53·7$; $H = 3·2$.

$C_{19}H_{12}O_4N_3Br$ requires $C = 53·5$; $H = 2·8$ per cent.

4-p-*Tolueneazo-2-bromo-6-nitrophenol.*—To control the foregoing result, the action of bromine on *p*-tolueneazo-*o*-nitrophenol (Hewitt and Lindfield, Trans., 1901, 79, 156) was also studied. Again it was found that the free ortho-position to the phenolic hydroxyl group was attacked with formation of the above-mentioned substance. We discovered that a slip had occurred in the earlier paper, the melting point of *p*-tolueneazo-*o*-nitrophenol being given as 147°, when, as a matter of fact, it is 174°. In these circumstances, the purity of our product was controlled by analysis.

0·1578 gave 0·3521 CO_2 and 0·0624 H_2O. C = 60·9 ; H = 4·4.
$C_{13}H_{11}O_3N_3$ requires C = 60·7 ; H = 4·3 per cent.

The azonitrophenol and half its weight of fused sodium acetate were dissolved in eight parts by weight of glacial acetic acid. A molecular proportion of bromine dissolved in about twice its weight of acetic acid was added under previously mentioned precautions ; after dilution with water, the product was collected and washed with hot water. Acetic acid is not suitable for purification, as any unattacked azonitrophenol is not removed by recrystallisation from this solvent. Two recrystallisations from ethyl acetate furnish fine orange needles melting sharply at 161°.

0·0975 gave 0·1666 CO_2 and 0·0294 H_2O. C = 46·6 ; H = 3·4.
$C_{13}H_{10}O_3N_3Br$ requires C = 46·4 ; H = 3·0 per cent.

The position of the bromine atom was determined by decomposition with fuming nitric acid, when 2-bromo-4 : 6-dinitrophenol melting at 118° was obtained.

The *acetyl* derivative, $C_7H_7 \cdot N_2 \cdot C_6H_2Br(NO_2) \cdot O \cdot CO \cdot CH_3$, was prepared in the usual manner, and after recrystallisation from ethyl acetate formed orange needles melting at 124°.

0·0934 gave 0·1646 CO_2 and 0·0279 H_2O. C = 48·1 ; H = 3·3.
$C_{15}H_{12}O_4N_3Br$ requires C = 47·7 ; H = 3·2 per cent.

The *benzoyl* derivative, obtained from the azophenol by boiling in a reflux apparatus with five times its weight of benzoyl chloride, excess of which was destroyed by alcohol, crystallised from ethyl acetate in small, yellowish-orange crystals melting at 129°.

0·1102 gave 0·2203 CO_2 and 0·0338 H_2O. C = 54·5 ; H = 3·4.
$C_{20}H_{14}O_4N_3Br$ requires C = 54·5 ; H = 3·2 per cent.

East London College.

XXVI.—*The Condensation of Dimethyldihydroresorcin and of Chloroketodimethyltetrahydrobenzene with Primary Amines. Part 1. Monamines.—Ammonia, Aniline, and p-Toluidine.*

By PAUL HAAS, B.Sc., Ph.D.

THE constitution of dimethyldihydroresorcin may be expressed by either of the two formulæ:

$$CMe_2 \!<\!\begin{smallmatrix} CH_2 \cdot CO \\ CH_2 \cdot CO \end{smallmatrix}\!>\! CH_2, \qquad CMe_2 \!<\!\begin{smallmatrix} CH_2 \cdot C(OH) \\ CH_2 \!-\!\!-\!\! CO \end{smallmatrix}\!>\! CH.$$

I.   II.

Although dimethyldihydroresorcin behaves towards hydroxylamine (Vorländer and Erig, *Annalen*, 1897, **294**, 316) and semicarbazide (compare p. 198) as a diketone, the fact of its giving with ferric chloride in aqueous solution a violet-red colour and of its reacting with phosphorus trichloride (Crossley and Le Sueur, Trans., 1903, **83**, 110) to give 5-chloro-3-keto-1 : 1-dimethyl-Δ^4-tetrahydrobenzene points to the presence of a hydroxyl group as indicated by formula II. It is this hydroxyl group which is responsible for the acid reaction of the substance, as well as its solubility in sodium or potassium hydroxide and in sodium hydrogen carbonate ; so acidic is this hydroxyl that it may even be esterified by means of alcohol and strong sulphuric acid (Vorländer, *loc. cit.*) after the manner of a carboxylic acid ; whereas, however, carboxylic acids produce in a naphthalene solution a bimolecular depression of freezing point, dimethyldihydroresorcin exhibits no such abnormality, although its molecular conductivity $K = 0\cdot00071$ is no less than one-third of that of γ-acetylbutyric acid, which has the value $K = 0\cdot0022$ (v. Schilling and Vorländer, *Annalen*, 1899, **308**, 184).

The replacement of the hydroxyl group by chlorine gives rise to chloroketodimethyltetrahydrobenzene ; this substance forms a semicarbazone, showing that it still contains a carbonyl oxygen, and its constitution must therefore be represented by the following formula :

$$CMe_2 \!<\!\begin{smallmatrix} CH_2 \cdot CCl \\ CH_2 \!-\! CO \end{smallmatrix}\!>\! CH.$$

The disappearance of the hydroxyl group is accompanied by a loss of acid properties, the substance being insoluble in cold potassium hydroxide and dissolving only on warming with the formation of the potassium salt of dimethyldihydroresorcin. On the whole, then, this hydroxyl group is more phenolic than alcoholic in nature, and it was therefore thought to be of interest to determine whether the chlorine atom in chloroketodimethyltetrahydrobenzene resembles the chlorine in an aromatic or an aliphatic compound. With this object in view, the behaviour of chloroketodimethyltetrahydrobenzene towards amines has been studied. In every case investigated, it was found to react in one molecular proportion with two molecules of the base, giving rise to a type of compound which Vorländer (*loc. cit.*, p. 305) states he was unable to prepare by the direct condensation of alkyldihydroresorcins with *p*-toluidine; it became necessary, therefore, to include in this investigation the condensation of dimethyldihydroresorcin with amines, and it was found that here, in every case, the two constituents reacted together in molecular preportions only, although the condensation products so obtained could be made to react with a second molecule of the base in the presence of zinc chloride.

Condensations of Dimethyldihydroresorcin.

(a) *Ammonia.* 5-*Hydroxy-3-amino*-1 : 1-*dimethyl*-Δ⁴-*tetrahydrobenzene* (*Monoamine*).—Dimethyldihydroresorcin dissolves in aqueous ammonia to a colourless solution which probably contains the ammonium salt, since it gives with silver nitrate a precipitate of the corresponding silver salt ; when, however, the solution is evaporated to dryness, the resulting solid is not an ammonium salt, but an amine derived from this substance by the loss of water ; this amine, which may be referred to briefly as the monoamine, gives no precipitate with silver nitrate in aqueous solution.

(b) p-*Toluidine.* 5-*Hydroxy-3-p-tolylamino*-1 : 1-*dimethyl*-Δ³ : ⁵-*dihydrobenzene* (*Monotoluidide*), page 196, has already been described by Vorländer (*loc. cit.*).

(c) *Aniline.* 5-*Hydroxy-3-phenylamino*-1 : 1-*dimethyl*-Δ³ : ⁵-*dihydrobenzene* (*Monoanilide*), page 202.

Condensations of Chloroketodimethyltetrahydrobenzene.

(a) *Ammonia.* 5-*Imino-3-amino*-1 : 1-*dimethyl*-Δ³-*tetrahydrobenzene* (*Diamine*) *Hydrochloride.*—In spite of repeated attempts, it was not found possible to isolate the free base in a state of purity, since it breaks up at once, evolving ammonia.

(b) p-*Toluidine.* 5-p-*Tolylimino-3-p-toiylamino*-1 : 1-*dimethyl*-Δ³-*tetrahydrobenzene* (*Ditoluidide*) *Hydrochloride* and

(c) *Aniline.* 5-*Phenylimino-3-phenylamino*-1 : 1-*dimethyl*-Δ³-*tetrahydrobenzene* (*Dianilide*) *Hydrochloride.*—The two latter substances, when dissolved in a large volume of boiling water, give on treatment with caustic alkalis the corresponding bases, which are very stable substances ; they are not hydrolysed by heating with concentrated hydrochloric acid in sealed tubes at 125°; if, however, they are heated with the same acid for three hours at 150—200°, they undergo a curious change, yielding p-toluidine hydrochloride and $\beta\beta$-dimethylglutaric acid.

The formation of this acid was explained by a separate experiment, in which it was found that a 35 per cent. yield of $\beta\beta$-dimethylglutaric acid may be obtained from dimethyldihydroresorcin by heating it under pressure with concentrated hydrochloric acid.

The formation of the monoamine from dimethyldihydroresorcin should be expressed by the equation:

$$
\begin{array}{ccc}
\text{CMe}_2 & & \text{CMe}_2 \\
\diagup\diagdown & & \diagup\diagdown \\
\text{CH}_2 \quad \text{CH} & & \text{CH}_2 \quad \text{CH}_2 \\
| \quad\quad | & + \text{NH}_3 \; = & | \quad\quad | \\
\text{CO} \quad \text{C·OH} & & \text{CO} \quad \text{C·NH}_2 \\
\diagdown\!\!\diagup & & \diagdown\!\!\diagup \\
\text{CH} & & \text{CH}
\end{array}
+ \text{H}_2\text{O},
$$

III.

but the chemical behaviour of the amine is not in agreement with formula III; no evidence of a carbonyl group can be obtained by means of either hydroxylamine, semicarbazide, or p-bromophenylhydrazine, whilst the substance gives a deep cherry-red colour with ferric chloride; from these facts, it would appear that the constitution is better represented by the formula

$$
\begin{array}{c}
\text{CMe}_2 \\
\diagup\diagdown \\
\text{CH} \quad \text{CH}_2 \\
|| \quad\quad | \\
\text{COH} \quad \text{C·NH}_2 \\
\diagdown\!\!\diagup \\
\text{CH}
\end{array} .
$$

When this monoamine is heated with an excess of acetic anhydride, a monoacetyl derivative is formed, which no longer gives any colour with ferric chloride; it must therefore be concluded either that the introduction of the acetyl group into the molecule has caused the hydroxylic oxygen to become ketonic or that the hydroxyl group has been acetylated; seeing, however, that the substance gives a semicarbazone, the former alternative is the more likely and the formula accordingly becomes

$$
\begin{array}{c}
\text{CMe}_2 \\
\diagup\diagdown \\
\text{CH}_2 \quad \text{CH}_2 \\
| \quad\quad | \\
\text{CO} \quad \text{C·NH·CO·CH}_3 \\
\diagdown\!\!\diagup \\
\text{CH}
\end{array}
$$

The monoanilide and monotoluidide, just like the monoamine, also produce colour reactions with ferric chloride and do not give oximes or semicarbazones ; when, however, they are heated with semicarbazide in alcoholic solution, they undergo hydrolysis, liberating aniline and toluidine respectively, and both give the disemicarbazone of dimethyl-dihydroresorcin. On acetylation, both the monotoluidide and the monoanilide give rise. to substances which no longer give colour reactions with ferric chloride, but which condense readily with semicarbazide, giving compounds having the general formula

$$\mathrm{NH_2 \cdot CO \cdot NH \cdot N{:}C} \underset{\underset{\displaystyle CH}{\diagdown\!\!\diagup}}{\overset{\overset{\displaystyle CMe_2}{\diagup\!\!\diagdown}}{\overset{CH_2 \ CH_2}{|\quad\ |}}} \mathrm{CN}{<}^{\mathrm{R}}_{\mathrm{CO \cdot CH_3}}$$

With a view to obtaining evidence of the hydroxyl group in the monotoluidide, this substance was treated in chloroform solution with phosphorus trichloride ; a violent reaction at once set in, resulting in the formation of the above-mentioned ditoluidide hydrochloride. Phosphorus pentachloride produced a similar decomposition, giving the ditoluidide hydrochloride and chloroketodimethyltetrahydro-benzene ; this reaction may be explained by assuming the hydroxyl group in the monotoluidide to be replaced by chlorine with the formation of an intermediate compound,

$$\mathrm{CMe_2}{<}^{\mathrm{CH_2 \cdot C(NHC_7H_7)}}_{\mathrm{CH{=\!=\!=\!=\!=}CCl}}{>}\mathrm{CH,}$$

which then condenses with a second molecule of p-toluidine, set free from the monotoluidide by the hydrolytic action of the liberated hydrochloric acid ; the dimethyldihydroresorcin which is thereby set free is then converted by the remaining phosphorus haloid into chloro-ketodimethyltetrahydrobenzene. The formulæ for the dianilide and ditoluidide are taken to be of the type

$$\mathrm{RN{:}C} \underset{\underset{\displaystyle CH}{\diagdown\!\!\diagup}}{\overset{\overset{\displaystyle CMe_2}{\diagup\!\!\diagdown}}{\overset{CH_2 \ CH_2}{|\quad\ |}}} \mathrm{C \cdot NHR} \qquad \text{and not} \qquad \mathrm{RNH \cdot C} \underset{\underset{\displaystyle CH}{\diagdown\!\!\diagup}}{\overset{\overset{\displaystyle CMe_2}{\diagup\!\!\diagdown}}{\overset{CH \ CH_2}{|\quad\ |}}} \mathrm{C \cdot NHR'}$$

inasmuch as both substances are mono-acid bases, giving rise to monohydrochlorides only ; moreover, the dianilide gives a monoacetyl, and the ditoluidide a monobenzoyl derivative ; the formula of the diamine hydrochloride would accordingly by analogy be

$$CMe_2$$

$$HN\!:\!C \quad C\!\cdot\!NH_2,HCl\ ^{\cdot}$$

$$CH$$

The action of nitrous acid on these substances is still under investigation.

Summary.

1. 5-Chloro-3-keto-1 : 1-dimethyl-Δ^4-tetrahydrobenzene reacts directly with two molecules of a primary base, yielding a compound of the type $RN\!:\!C_8H_{12}\!\cdot\!NHR,HCl$.

2. Dimethyldihydroresorcin condenses with one molecule only of a primary base, yielding a compound of the type $C_8H_{10}NHR$, the elements of water being eliminated between the hydroxyl group of the resorcin and a hydrogen from the amino-group.

3. Neither chloroketodimethyltetrahydrobenzene nor dimethyldihydroresorcin reacts with secondary amines.

4. The replacement of the hydroxyl group in dimethyldihydroresorcin by a basic group such as $-NH_2$, $-NH\!\cdot\!C_6H_5$, or $-NH\!\cdot\!C_7H_7$, causes the remaining ketonic oxygen atom of the resorcin to become hydroxylic, as shown by the behaviour of the resulting substance towards ferric chloride and phosphorus trichloride. If, however, the substituting group is rendered more acidic by the introduction of an acetyl group, this second oxygen atom becomes ketonic and is able to condense with semicarbazide.

5. The mono-derivatives obtained from dimethyldihydroresorcin can be converted into di-derivatives either by condensation with a second molecule of the primary base in presence of zinc chloride or by the action of the phosphorus haloids.

6. An attempt to prepare an unsymmetrical di-derivative from the monoamine and p-toluidine in the presence of zinc chloride resulted in the displacement of ammonia by p-toluidine and the formation of the monotoluidide.

EXPERIMENTAL.

Action of Ammonia on Dimethyldihydroresorcin.

A solution of 20 grams of dimethyldihydroresorcin in 15 c.c. of concentrated ammonia was evaporated to small bulk over a water-bath ; after drying in a vacuum, there remained a light yellow solid residue, which was purified by crystallisation from chloroform. A further quantity of material was obtained from the chloroform mother liquors by evaporating them to dryness and crystallising the residue from water, when some unchanged dimethyldihydroresorcin separated ; the aqueous solution on evaporation left almost pure material.

0·1260 gave 0·3184 CO_2 and 0·1112 H_2O. C = 68·93 ; H = 9·86.

0·1304 ,, 11·6 c.c. moist nitrogen at 18° and 760·5 mm. N = 10·28.

$C_8H_{13}ON$ requires C = 69·06 ; H = 9·36 ; N = 10·07 per cent.

5-*Hydroxy-3-amino*-1 : 1-*dimethyl*-$\Delta^{3:5}$-*dihydrobenzene* (monoamine), $CMe_2{<}^{CH_2 \cdot C(NH_2)}_{CH=C(OH)}{>}CH_2$, crystallises from chloroform or benzene in colourless, flattened needles melting at 163·5—164°; it is readily soluble in cold water, acetone, ether, or alcohol, is very sparingly soluble in cold chloroform or benzene, and is insoluble in light petroleum ; it dissolves readily in hydrochloric acid, giving a hydrochloride, and is insoluble in strong caustic potash. Its solution in water which is neutral to litmus gives no precipitate with silver nitrate, showing that it is not an ammonium salt. Dissolved in alcohol and boiled with strong potassium hydroxide solution and a few drops of chloroform, it develops a faint but distinct odour of carbylamine. It does not yield an oxime, a *p*-bromophenylhydrazone, or a semicarbazone, and its aqueous solution gives a cherry-red colour with ferric chloride.

The *hydrochloride*, $C_8H_{13}ON,HCl$, is obtained by evaporating a hydrochloric acid solution of the base to small bulk over a water-bath ; on standing in a desiccator over caustic potash, the solution deposits large, transparent, lozenge-shaped slabs which melt at 186—188° ; the crystals are readily soluble in alcohol or water, giving an acid solution.

0·1176 required 6·75 c.c. $N/10$ NaOH = 0·0246 HCl. HCl = 20·92.

$C_8H_{13}ON,HCl$ requires HCl = 20·77 per cent.

The *platinichloride*, $(C_8H_{13}ON)_2,H_2PtCl_6$, was obtained by dissolving the base in the least quantity of alcohol, acidifying the solution with a little hydrochloric acid, and adding to it the calculated amount of platinic chloride dissolved in alcohol. It separates from the solution in orange-coloured plates which melt with decomposition at 197—198° and are readily soluble in alcohol and in water.

0·1978 gave 0·0551 Pt. Pt = 27·86·

$C_{16}H_{28}O_2N_2Cl_6Pt$ requires Pt = 28·32 per cent.

• The *picrate*, $C_8H_{13}ON,C_6H_2(NO_2)_3 \cdot OH$, separates from a chloroform solution of the two constituents as a yellow oil which, after warming on the water-bath, solidifies to a mass of deep canary-yellow crystals which melt at 135°.

0·1217 gave 16 c.c. moist nitrogen at 16° and 757 mm. N = 15·26.
$C_{14}H_{16}O_8N_4$ requires N = 15·21 per cent.

The *acetyl* derivative was prepared by heating 10 grams of the monoamine with 15 grams of acetic anhydride for half an hour on the water-bath. After removing the excess of the anhydride by boiling the solution with methyl alcohol, the resulting solid was crystallised from benzene.

0·1207 gave 8·5 c.c. moist nitrogen at 17° and 758 mm. N = 7·76.
$C_{10}H_{15}O_2N$ requires N = 7·73 per cent.

5-*Keto-3-acetylamino*-1 : 1-*dimethyl*-Δ³-*tetrahydrobenzene*,

$$CMe_2 \begin{matrix} CH_2 \cdot C \cdot NH \cdot Ac \\ CH_2 \overline{} CO \end{matrix} CH,$$

crystallises from benzene in oblong plates and melts at 157° ; it is readily soluble in cold chloroform, alcohol, or water, sparingly so in benzene, and insoluble in ligroin. Treated in aqueous solution with bromine water, it is converted into dibromodimethyldihydroresorcin. The substance gives no colour with ferric chloride in aqueous solution. When evaporated on the water-bath with dilute hydrochloric acid, it loses acetic acid, giving the hydrochloride of the monoamine.

The *semicarbazone*, $NH_2 \cdot CO \cdot NH \cdot N:C_8H_{11} \cdot NH \cdot Ac$, of the above acetyl derivative was obtained by dissolving 1 gram of this substance in an alcoholic solution of 1·2 grams of semicarbazide hydrochloride and 1·2 grams of potassium acetate, from which the precipitated potassium chloride had been removed by filtration. After heating for one and a half hours on the water-bath, the major portion of the alcohol was evaporated off ; the solution on standing deposited 0·8 gram of a white, crystalline solid, which was analysed after recrystallisation from water.

0·1275 gave 25·6 c.c. moist nitrogen at 20° and 762 mm. N = 22·93.
$C_{11}H_{18}O_2N_4$ requires N = 23·53 per cent.

The substance is readily soluble in cold methyl or ethyl alcohol and is fairly soluble in warm ethyl acetate or acetone, but is insoluble in light petroleum ; it crystallises from water or from a mixture of alcohol and light petroleum in rhombic plates melting at 208·5—209·5°.

Action of Bromine Water on the Monoamine.

Bromine water added to an aqueous solution of the monoamine produced at once a white, crystalline precipitate of dibromodimethyldihydroresorcin, which crystallised from dilute alcohol in flattened needles melting at 144—146°.

0·1028 gave 0·1295 AgBr.   Br = 53·62.

$C_8H_{10}O_2Br_2$ requires Br = 53·68 per cent.

The aqueous mother liquors on evaporation yielded ammonium bromide. An aqueous solution of dimethyldihydroresorcin treated as above with bromine water yielded the same dibromide (m. p. 144—146°).

Action of Potassium Hydroxide on the Monoamine.

A solution of 1 gram of the monoamine and 1 gram of potassium hydroxide in 6 grams of water was boiled for six hours; on acidifying the solution with hydrochloric acid, 0·6 gram of dimethyldihydro-resorcin was precipitated.

Action of p-Toluidine on the Monoamine.

A solution of 4 grams o the monoamine and 3 grams of p-toluidine in 10 grams of alcohol was heated with 1 gram of zinc chloride in a sealed tube for six hours at 150—160°. The contents of the tube were then filtered and evaporated to dryness; the residue, after extracting with boiling water, was crystallised from acetic acid, from which it separated in hexagonal plates melting at 199—200°. This substance had all the properties of the monotoluidide described on page 196, and when mixed with it did not depress its melting point.

Action of Ammonia on Chloroketodimethyltetrahydrobenzene.

A mixture of 8 grams of chloroketodimethyltetrahydrobenzene with 35 c.c. of alcoholic ammonia was heated in a sealed tube for three hours at 100°; the yellow alcoholic solution was then filtered from some crystals of ammonium chloride formed during the reaction, and evaporated to dryness. The yellowish-brown, solid residue, after decolorising by extraction with chloroform, was purified by crystal-lisation from a mixture of alcohol and ether.

0·1237 gave 0·2611 CO_2 and 0·1064 H_2O.   C = 55·33 ; H = 9·18.

0·1494 „   21 c.c. moist nitrogen at 764 mm. and 21·5°. N = 16·05.

0·1256 „   0·1059 AgCl.   Cl = 20·85.

$C_8H_{14}N_2,HCl$ requires C = 55·01 ; H = 8·60 ; N = 16·05 ; Cl = 20·34 per cent.

5-Imino-3-amino-1 : 1-dimethyl-Δ³-tetrahydrobenzene (diamine) hydro-

chloride, $CH_2 <^{CMe_2-CH_2}_{C(:NH)\cdot CH}> C\cdot NH_2, HCl$, is readily soluble in cold alcohol, giving a greenish-yellow solution, from which it is precipitated by ether in fine, silken-white needles which, on standing, change to stout prisms melting at 257—258°; it is very soluble in methyl alcohol, formic acid, or water, but is insoluble in acetone, chloroform, ether, or light petroleum. The aqueous solution has a blue fluorescence which is changed to greenish-yellow by the addition of caustic potash in the cold; on gently warming, however, ammonia is readily evolved.

The *platinichloride*, $(C_8H_{14}N_2)_2, H_2PtCl_6$ separates from the solution in orange-yellow, hexagonal plates on mixing an alcoholic solution of the hydrochloride with one of platinic chloride; it melts with decompositon at 215°.

0·1765 gave 0·0499 Pt.   Pt = 28·27.

$C_{16}H_{30}N_4Cl_6Pt$ requires Pt = 28·40 per cent.

Attempts to prepare 5-Imino-3-amino-1 : 1-dimethyl-Δ^3-tetrahydrobenzene (the Diamine Base) from its Hydrochloride.

(1) *With Silver Oxide.*—An aqueous solution of the hydrochloride was thoroughly shaken with freshly precipitated silver oxide and filtered ; the strongly alkaline filtrate, after evaporation to dryness over a water-bath, left a syrupy residue ; the latter was extracted with benzene, but during the process there was a considerable evolution of ammonia, and the benzene solution on cooling deposited crystals of the monoamine.

(2) *With Silver Sulphate and Barium Hydroxide.*—The diamine hydrochloride dissolved in water was treated with a solution of silver sulphate until no further precipitate of silver chloride was formed ; the filtered solution was then warmed with a slight excess of barium hydroxide, the excess being subsequently precipitated by means of a current of carbon dioxide. After filtering off the precipitated barium salts, the aqueous solution was evaporated to dryness ; the syrupy alkaline residue so obtained was then boiled with benzene, but a considerable evolution of ammonia was noticed during the process ; the benzene solution on cooling deposited silken needles which melted between 95° and 105°. This substance, which appeared only to be decomposed by further crystallisation, gave numbers on analysis which showed it to be a mixture of the diamine base with the monoamine.

The *picrate* of the diamine, $C_8H_{14}N_2, C_6H_2(NO_2)_3\cdot OH$, was prepared by heating some of the syrupy residue, obtained from the diamine hydrochloride as described in the previous experiment, with a chloroform solution of picric acid ; after heating for an hour, the precipitated yellow solid was filtered off and crystallised from a mixture of acetone and light petroleum ; it forms orange-yellow, flattened needles and melts at 175°.

0·1128 gave 18·5 c.c. moist nitrogen at 18° and 768·5 mm. N = 19·14.

$C_{14}H_{17}O_7N_5$ requires N = 19·07 per cent.

Conversion of the Diamine into the Monoamine.

Two grams of the diamine hydrochloride dissolved in a small amount of water were warmed on the water-bath for half an hour with a solution of 1·2 grams (2 mols.) of caustic potash in 10 c.c. of water. On cooling the mixture, a mass of yellow needles separated out which, when purified by recrystallisation from benzene, melted at 164—165° and in every way resembled the monoamine ; the substance was further characterised by conversion into its hydrochloride melting at 183—185°.

Conversion of the Monoamine into the Diamine Hydrochloride.

Two [and a half grams of the monoamine dissolved in 10 c.c. of alcoholic ammonia were heated in a sealed tube with 2 grams of fused and powdered zinc chloride for three hours at 160° ; the tube now contained an alcoholic solution and some well-formed prismatic crystals attached to the sides of the tube ; the crystals collected from a number of experiments were found to be insoluble in all organic solvents ; when treated with water, they were decomposed with the liberation of zinc hydroxide, giving a green, fluorescent solution ; the latter on evaporation yielded a solid residue which crystallised from a mixture of alcohol and ether in tabular prisms melting at 258° and had all the properties of the hydrochloride of the diamine described on page 195.

Action of p-Toluidine on Dimethyldihydroresorcin.

The monotoluidide which results from the condensation in alcoholic solution of *p*-toluidine with dimethyldihydroresorcin has already been described by Vorländer and Erig (*Annalen*, 1897, **294**, 315).

4-*Hydroxy-3-tolylamino*-1 : 1-*dimethyl-*$\Delta^{3\cdot5}$-*dihydrobenzene*   (mono-toluidide), $CMe_2 \begin{smallmatrix} CH_2 \cdot C(NH \cdot C_7H_7) \\ CH \underline{\qquad\qquad} C(OH) \end{smallmatrix} CH$, when perfectly pure is colourless and melts at 202° and not at 200°, as stated by the above-mentioned authors ; it gives with ferric chloride in alcoholic solution a cherry-red colour and does not yield a semicarbazone.

The *hydrochloride* of the monotoluidide, $C_{15}H_{19}ON,HCl$, was obtained by saturating a chloroform solution of the base with dry hydrogen chloride. The residue left on evaporating the solution to dryness in a vacuum crystallised from a mixture of alcohol and ether in prisms which melted with decomposition at 208—212° after contracting some degrees below that temperature. The substance dissolves readily in water, giving an acid solution.

0·1269 required 4·7 c.c. $N/10$ NaOH = 0·0171 HCl. HCl = 13·52.
$C_{15}H_{19}ON$,HCl requires HCl = 13·74 per cent.

The *acetyl* derivative was prepared by boiling a solution of 10 grams of the monotoluidide with 12 grams of acetic anhydride for two hours over an iron gauze ; after evaporating down the mixture several times with methyl alcohol, it was placed in a vacuum over caustic potash, when it solidified after about a week. The brown solid was then purified by boiling in chloroform solution with animal charcoal and crystallising from a mixture of chloroform and light petroleum (b. p. 60—80°).

' 0·1247 gave 0·3426 CO_2 and 0·0902 H_2O. C = 74·93 ; H = 8·03.
0·134 ,, 6·2 c.c. moist nitrogen at 22·5° and 774 mm. N = 5·32.
$C_{17}H_{21}O_2N$ requires C = 75·27 ; H = 7·75 ; N = 5·15 per cent.

5-*Keto*-3-*acetyl*-p-*tolylamino*-1 : 1-*dimethyl*-Δ^3-*tetrahydrobenzene*,

$$CMe_2 <^{CH_2 \cdot C(NAc \cdot C_7H_7)}_{CH_2 \underline{} CO} > CH,$$

crystallises from light petroleum in long, colourless, transparent prisms which melt at 95—97°; it is readily soluble in cold chloroform, alcohol, or ether, and fairly soluble in boiling light petroleum (b. p. 40—60°); its solution in alcohol gives no colour with ferric chloride.

The *semicarbazone*, $NH_2 \cdot CO \cdot NH \cdot N:C_8H_{11}NAc \cdot C_7H_7$, of the acetyl derivative of the monotoluidide, was prepared without the application of heat by allowing the acetyl derivative to stand with two molecular proportions of semicarbazide acetate ; it crystallises from alcohol in stellar aggregates of oblong plates and melts with evolution of gas at 216° or 221° according to the rate at which it is heated ; it is very slightly soluble in cold alcohol and is insoluble in acetone.

0·1102 gave 16 c.c. moist nitrogen at 19° and 769 mm. N = 16·84.
$C_{18}H_{24}O_2N_4$ requires N = 17·07 per cent.

Action of Semicarbazide on the Monotoluidide.

One gram of the monotoluidide was heated on the water-bath for two hours with an alcoholic solution of semicarbazide acetate obtained by mixing 1 gram of semicarbazide hydrochloride, dissolved in the least quantity of water, with an alcoholic solution of 1 gram of potassium acetate and filtering off the precipitated potassium chloride. On cooling the mixture, an odour of p-toluidine was observed, and the solution deposited 0·4 gram of a crystalline solid which, when washed with alcohol, melted with decomposition at 213—216°; this substance was identified as the semicarbazone of dimethyldihydroresorcin both by its melting point and by its insolubility in ordinary organic solvents.

0·1014 gave 27 c.c. moist nitrogen at 10·5° and 784 mm. N = 32·88. $C_{10}H_{18}O_2N_6$ requires N = 33·07 per cent.

The alcoholic mother liquors on further concentration yielded unchanged monotoluidide.

Semicarbazone of Dimethyldihydroresorcin.

One gram of dimethyldihydroresorcin dissolved in alcohol was added to an alcoholic solution containing two molecular proportions of semicarbazide acetate obtained in the manner described above. On warming the mixture for five minutes on the water-bath, a copious white precipitate was formed, which, after filtering, was thoroughly washed with hot alcohol and dried in a vacuum.

0·1059 gave 29·8 c.c. moist nitrogen at 16° and 771 mm. N = 33·31. $C_{10}H_{18}O_2N_6$ requires N = 33·07 per cent.

Dimethyldihydroresorcin disemicarbazone,
$$NH_2 \cdot CO \cdot NH \cdot N \colon C_8H_{12} \colon N \cdot NH \cdot CO \cdot NH_2,$$
melts with evolution of gas at 213—216°; it is very slightly soluble in boiling ethyl alcohol and is insoluble in all ordinary organic solvents.

Action of p-Toluidine on Chloroketodimethyltetrahydrobenzene.

When chloroketodimethyltetrahydrobenzene (1 mol.) and p-toluidine (2 mols.) are heated together, they react with considerable violence; in alcohol or benzene solution, the reaction takes place quite gently in the cold on standing for some hours, or more rapidly on heating, with the formation of a mass of yellow crystals; the latter, dissolved in a large volume of boiling water and treated with dilute potassium hydroxide, gave a light yellow, curdy precipitate which, after washing with water, was purified by crystallisation from alcohol.

0·1257 gave 0·3826 CO_2 and 0·0966 H_2O. C = 83·01 ; H = 8·54.
0·1262 „ 0·3842 CO_2 „ 0·0967 H_2O. C = 83·10; H = 8·51.
0·1346 „ 10·4 c.c. moist nitrogen at 21·5° and 770 mm. N = 8·89.
$C_{22}H_{26}N_2$ requires C = 83·02 ; H = 8·17 ; N = 8·80 per cent.

5-p-*Tolylimino*-3-p-*tolylamino*-1 : 1-*dimethyl*-Δ^3-*tetrahydrobenzene* (di-toluidide), $CMe_2\!\!<^{CH_2 \cdot C(NH \cdot C_7H_7)}_{CH_2-C(:N \cdot C_7H_7)}\!\!>CH$, is a light yellow solid which persistently retains its colour even after boiling in alcoholic solution with animal charcoal ; it crystallises from alcohol in light yellow plates melting at 208—210° ; it is readily soluble in cold chloroform and slightly so in hot ether ; it dissolves readily in glacial acetic acid, but with difficulty in hot dilute hydrochloric or sulphuric acid, and is insoluble in water. A solution of the base in alcohol has a strongly alkaline reaction to litmus. The substance is unaffected by boiling with alcoholic potash.

The *hydrochloride*, $C_{22}H_{26}N_2,HCl$, is formed directly in practically theoretical yield by the condensation of *p*-toluidine with chloroketo-dimethyltetrahydrobenzene, as described above ; it crystallises from formic acid in canary-yellow, hexagonal plates which contain a molecule of formic acid and melt with decomposition at 320°.

0·1248 gave 0·3170 CO_2 and 0·0838 H_2O. C = 69·27 ; H = 7·46.

0·1307 „ 8·2 c.c. moist nitrogen at 21° and 775·6 mm. N = 7·28.

0·1283 „ 0·0463 AgCl. Cl = 8·92.

0·2660 on heating to 130° lost 0·0276. H_2CO_2 = 11·58.

$C_{22}H_{27}N_2Cl,H_2CO_2$ requires C = 68·91 ; H = 7·24 ; N = 6·99 ; Cl = 8·86.

H_2CO_2 = 11·48 per cent.

It is sparingly soluble in hot alcohol and dissolves only in a very large volume of boiling water, giving a neutral solution ; it is insoluble in all ordinary organic solvents except formic and acetic acids. It may also be prepared by dissolving the base in alcohol and adding hydrochloric acid to the boiling solution ; on cooling, yellow, hexagonal plates separate out, which do not contain a molecule of the solvent.

0·1208 gave 0·0500 AgCl. Cl = 10·23.

$C_{22}H_{26}N_2,HCl$ requires Cl = 10·02 per cent.

The *platinichloride*, $(C_{22}H_{26}N_2)_2,H_2PtCl_6$, is formed as an immediate yellow precipitate on mixing an alcoholic solution of the base acidified with hydrochloric acid with the calculated amount of platinic chloride dissolved in alcohol ; it decomposes at 253°.

0·2016 gave 0·0377 Pt. Pt = 18·70.

$C_{44}H_{54}N_4Cl_6Pt$ requires Pt = 18·62 per cent.

The *benzoyl* derivative, $C_{22}H_{25}N_2 \cdot CO \cdot C_6H_5$, was prepared by vigorously shaking a suspension of the base in aqueous caustic potash with an ethereal solution of benzoyl chloride ; a white, crystalline solid was formed, which was purified by crystallisation from dilute alcohol.

0·1222 gave 0·3702 CO_2 and 0·0845 H_2O.   C = 82·62 ; H = 7·68.
0·1269 „   7·5 c.c. moist nitrogen at 18·5° and 760·7 mm.   N = 6·82.
$C_{20}H_{30}ON_2$ requires C = 82·46 ; H = 7·11 ; N = 6·63 per cent.

It crystallises from dilute alcohol in needles which melt with decomposition between 144° and 149°; it is very soluble in cold alcohol, chloroform, acetone, or ether.

Action of Fuming Hydrochloric Acid on the Ditoluidide.

Two and a half grams of the ditoluidide hydrochloride heated with 12 grams of fuming hydrochloric acid in a sealed tube at 125° for two hours remained practically unchanged.

The same proportions as above were therefore heated for two hours at 250°; the contents of the tube, consisting of a brown liquid in which were suspended crystals of p-toluidine hydrochloride, were poured into water, and the acid solution A thus obtained was extracted with ether. The ethereal extract on evaporation yielded 0·8 gram of a brown, viscous oil, which solidified after some days ; the solid was purified by crystallisation from a mixture of alcohol and light petroleum, when a well-defined, crystalline substance was obtained which melted at 101°. This was identified as $\beta\beta$-dimethylglutaric acid by conversion into the anhydride (m. p. 124·5°), anilic acid (m. p. 134—135°), and anil (m. p. 156°), all these melting points agreeing closely with those given by Perkin (Trans., 1896, **69**, 1473). The acid solution A, when rendered alkaline and extracted with ether, gave 1 gram of p-toluidine, which was identified by its crystalline form and characteristic odour and melting point (45°).

Action of Fuming Hydrochloric Acid on Dimethyldihydroresorcin.

Two grams of dimethyldihydroresorcin and 8 grams (16 mols.) of fuming hydrochloric acid were heated in a sealed tube fo two hours at 230°; the brown, resinous mass contained in the tube was then extracted with water ; on evaporating the aqueous solution to dryness, there remained 0·8 gram of white crystals which melted at 94—100°; they were converted without further purification into the anhydride, melting after crystallisation at 125°, and the latter into the anilic acid (m. p. 134·5°), which was analysed.

0·1236 gave 6·8 c.c. moist nitrogen at 9·5° and 776 mm.   N = 6·72.
$C_{11}H_{17}O_3N$ requires N = 6·63 per cent.

The amount of $\beta\beta$-dimethylglutaric acid obtained in this experiment corresponds to a yield of about 35 per cent.

Conversion of the Monotoluidide into the Ditoluidide.

To a mixture of 5·5 grams of the monotoluidide with 2·5 grams of *p*-toluidine ' heated to 200° were added, in portions, 1·5 grams of freshly fused and powdered zinc chloride, the heating being continued for about one hour. On cooling, the contents of the flask solidified; they were broken up, extracted with dilute hydrochloric acid to remove any unchanged *p*-toluidine, filtered, washed, and dried, 7·5 grams of solid being thus obtained. On dissolving this substance in a large volume of boiling water and pouring the filtered solution into dilute caustic potash, a yellow precipitate was at once formed which, after crystallising from alcohol, had the melting point of the ditoluidide base.

0·1223 gave 0·3720 CO_2 and 0·0950 H_2O.   C = 82·95 ; H = 8·63.

0·1271 ,,   10·6 c.c. moist nitrogen at 22° and 761 mm.   N = 9·46.

$C_{22}H_{26}N_2$ requires C = 83·02 ; H = 8·17 ; N = 8·80 per cent.

Action of Phosphorus Trichloride on the Monotoluidide.

Fifteen grams (3 mols.) of the monotoluidide, dissolved in 70 grams of chloroform, were treated with 3 grams (1 mol.) of phosphorus trichloride ; a fairly violent reaction at once set in, and the solution assumed a claret-red colour. After heating for three hours over a water-bath, the chloroform was distilled off ; on extracting the red resin so obtained with ether, a yellow, solid residue was obtained, which was identified as the hydrochloride of the ditoluidide by crystallisation from formic acid, from which it separated in yellow, hexagonal plates.

0·1253 gave 0·0458 AgCl.   Cl = 9·04.

$C_{22}H_{27}N_2Cl,H_2CO_2$ requires Cl = 8·86 per cent.

Action of Phosphorus Pentachloride on the Monotoluidide.

To a solution of 10 grams of the monotoluidide dissolved in 50 grams of chloroform were added, in portions, 10 grams of phosphorus pentachloride, the whole being heated on the water-bath for one and a half hours, by which time the evolution of hydrochloric acid had considerably diminished ; the brown chloroform solution was then poured into water and extracted with ether. After some minutes, a yellow solid began to separate from the ethereal solution ; this substance, which weighed 3·6 grams, was identified as the hydrochloride of the ditoluidide ; when dissolved in a large volume of boiling water and precipitated with caustic potash, it gave the free base melting at 206—208°. The ethereal solution, on evaporation, gave 6·5 grams of a brown liquid, which, on distillation in steam, yielded 2·2 grams of a light yellow oil; the latter was identified as chloroketodimethyltetrahydrobenzene by conversion into its semi-carbazone, which melted at 196—197° with evolution of gas; another portion of the oil was converted into the ditoluidide hydrochloride by condensing it with p-toluidine; the hydrochloride was then converted into the base and found to give the melting point characteristic of this substance. The residue left in the flask after submitting the above-mentioned brown oil to steam distillation was crystallised from benzene; it separated from this solvent in short, flat needles melting at 183—184°.

$$0.1042 \text{ gave } 0.1360 \text{ AgCl. } Cl = 32.29.$$

This substance, which was soluble in ether, alcohol, or chloroform, but insoluble in light petroleum, was not further investigated.

Action of Aniline on Dimethyldihydroresorcin.

Four grams of dimethyldihydroresorcin and 3 grams of aniline were heated in alcoholic solution for three hours over a water-bath. The yellow solid remaining after evaporation of the alcohol was crystallised once from acetic acid to remove any free aniline and then from benzene until it was obtained colourless.

0.1302 gave 0.3722 CO_2 and 0.1004 H_2O. $C = 77.96$; $H = 8.57$.

0.1391 ,, 8.2 c.c. moist nitrogen at $21°$ and 758 mm. $N = 6.79$.

$C_{14}H_{17}ON$ requires $C = 78.14$; $H = 7.90$; $N = 6.51$ per cent.

5-*Hydroxy*-3-*phenylamino*-1 : 1-*dimethyl*-$\Delta^{3:5}$-*dihydrobenzene* (mono-anilide), $CMe_2 \begin{smallmatrix} CH_2 \cdot C(NH \cdot C_6H_5) \\ CH \underline{\quad\quad} C(OH) \end{smallmatrix} CH$, crystallises from acetic acid in rhombic plates and from benzene in clusters of oblong plates which melt at 180°. It is readily soluble in cold alcohol or chloroform, but is insoluble in cold ether. Its alcoholic solution is neutral to litmus and gives with ferric chloride a cherry-red colour.

The *hydrochloride*, $C_{14}H_{17}ON,HCl$, was prepared by saturating a chloroform solution of the base with dry hydrogen chloride. The solution on standing deposited silken-white needles which were recrystallised from a mixture of alcohol and ether.

0.1327 required 5.24 c.c. $N/10$ NaOH $= 0.0191$ HCl. HCl $= 14.41$.

$C_{14}H_{18}ONCl$ requires HCl $= 14.51$ per cent.

The substance melts with decomposition at 184—186°.

The acetyl derivative was prepared by boiling 4 grams of the mono-anilide with 7 grams of acetic anhydride for one and a half hours over a wire gauze. The mixture was then evaporated with methyl alcohol and left in a vacuum for some days; as it still showed no signs of solidifying, it was dissolved in a mixture of chloroform and light petroleum; the solution, on being allowed to evaporate slowly, deposited well-formed amber-coloured plates with truncated angles.

0·1226 gave 0·3374 CO_2 and 0·0845 H_2O.   C = 75·05; H = 7·65.

0·1750 ,,   9 c.c. moist nitrogen at 25° and 767 mm.   N = 5·79.

$C_{16}H_{19}O_2N$ requires C = 74·70; H = 7·39; N = 5·45 per cent.

5-*Keto-3-acetylphenylamino*-1 : 1-*dimethyl-Δ³-tetrahydrobenzene*,

$$CMe_2\begin{matrix}CH_2 \cdot C(NAc \cdot C_6H_5)\\ \overline{CO}\\ CH_2\end{matrix}CH,$$

melts at 65·5—66·5°; it is readily soluble in cold alcohol, ether, chloroform, ethyl acetate, or benzene, but is only slightly soluble in boiling light petroleum. An alcoholic solution of the substance gives no colour with ferric chloride.

The *semicarbazone* of this acetyl derivative,

$$NH_2 \cdot CO \cdot NH \cdot N{:}C_8H_{11}NAc \cdot C_6H_5,$$

which was prepared in a manner similar to the one described for the corresponding toluidine derivative on p. 197, crystallises from alcohol in hexagonal plates and melts with evolution of gas at 210·5°. It is fairly soluble in hot alcohol, but only slightly so in hot acetone or ethyl acetate.

0·1046 gave 16·2 c.c. moist nitrogen at 15·5° and 762 mm. N = 18·15.

$C_{17}H_{22}O_2N_4$ requires N = 17·83 per cent.

Action of Aniline on Chloroketodimethyltetrahydrobenzene.

As in the case of *p*-toluidine, when aniline is heated with chloro-ketodimethyltetrahydrobenzene, a violent reaction takes place with the formation of a yellow solid. This substance is, however, best prepared by heating one molecular proportion of the chloro-compound with two of aniline in benzene solution. The yellow crystals so obtained are the hydrochloride of the dianilide; they are dissolved in a large volume of boiling water and filtered hot into a solution of caustic potash; the light yellow, curdy precipitate which is at once formed is filtered, washed free from alkali, and crystallised from dilute alcohol.

0·1248 gave 0·3782 CO_2 and 0·0910 H_2O.   C = 82·65; H = 7·97.

$C_{20}H_{22}N_2$ requires C = 82·76; H = 7·60 per cent.

5-*Phenylimino-3-phenylamino*-1 : 1-*dimethyl-Δ³-tetrahydrobenzene*,

$$CMe_2\begin{matrix}CH_2 \cdot C(NH \cdot C_6H_5)\\ CH_2 \overline{}C({:}N \cdot C_6H_5)\end{matrix}CH,$$ crystallises from dilute alcohol in

light yellow, oblong plates and melts at 193—195°; it is readily soluble in cold chloroform, alcohol, or glacial acetic acid, and in hot benzene, but it is only sparingly so in ether, and is insoluble in water. The solution in alcohol has a strongly alkaline reaction. Boiling alcoholic potash has no action on the substance.

The *hydrochloride* may be obtained either as described above, by condensation of aniline with chloroketodimethyltetrahydrobenzene, or by crystallising the base from alcohol acidified with hydrochloric acid. Prepared in the former way by condensation and recrystallised from formic acid, the substance gave on analysis the following numbers :

0·1273 gave 0·0518 AgCl.   Cl = 10·06.

0·1600 heated to 150° lost 0·0186. $H_2CO_2 = 11·55$.

$C_{20}H_{23}N_2Cl,H_2CO_2$ requires Cl = 9·79 ; $H_2CO_2 = 12·34$ per cent.

It crystallises from formic acid in stellar aggregates of stout, flat needles, which contain 1 molecule of the solvent ; ·the crystals slowly lose this formic acid and change their form to stout, rhombic slabs resembling sulphur.

The *platinichloride*, $(C_{20}H_{22}N_2)_2,H_2PtCl_6$, was prepared by adding to an alcoholic solution of the hydrochloride the calculated amount of platinic chloride dissolved in alcohol. A crystalline precipitate of orange-yellow needles was at once formed, which, after washing with alcohol and drying, gave the following numbers on analysis :

0·2029 gave 0·0396 Pt.   Pt = 19·51.

$C_{40}H_{46}N_4Cl_6Pt$ requires Pt = 19·68 per cent.

The salt is insoluble in alcohol and in water.

The *acetyl* derivative, $C_{20}H_{21}N_2·CO·CH_3$, was prepared from two grams of the dianilide base by heating it over a water-bath for two hours with excess of acetic anhydride ; after evaporating the solution several times with methyl alcohol in order to remove the excess of the anhydride, the brown, syrupy residue was crystallised from a mixture of chloroform and light petroleum and subsequently from dilute alcohol.

0·1341 gave 9 c.c. moist nitrogen at 13° and 767·5 mm.   N = 8·01.

$C_{22}H_{24}ON_2$ requires N = 8·43 per cent.

The substance separates from dilute alcohol in stellar aggregates of prisms melting at 161—162° ; it is readily soluble in cold alcohol, acetone, ethyl acetate, or chloroform, but is insoluble in ether, light petroleum, or water.

Conversion of the Monoanilide into the Dianilide.

Two and a half grams of the monoanilide, when heated with 2 grams of aniline and 1 gram of zinc chloride for one hour at 200—220°, gave an olive-brown, viscous liquid, which, after extraction with dilute hydrochloric acid, solidified to a yellow mass ; the latter dissolved in a large volume of boiling water to a clear solution, which, on treatment with caustic potash, gave a canary-yellow precipitate ; this substance, which crystallised from dilute alcohol in plates melting at 193—195°, had all the properties of the dianilide.

Behaviour of Chloroketodimethyltetrahydrobenzene and of Dimethyl-dihydroresorcin towards Methylaniline.

Three grams (1 mol.) of chloroketodimethyltetrahydrobenzene and 2 grams (1 mol.) of methylaniline were heated together in alcoholic solution for four hours ; the solution was then acidified and distilled in steam, when some unchanged chloro-compound was recovered ; the major portion of this substance had, however, been hydrolysed by the acid solution and was recovered therefrom in the form of a crystalline precipitate of dimethyldihydroresorcin weighing 2 grams. The acid mother liquors, when made alkaline and extracted with ether, yielded 2 grams of dimethylaniline unchanged.

Molecular proportions of dimethyldihydroresorcin and methylaniline heated together in alcoholic solution for three hours also did not react, the unchanged materials being recovered from the mixture.

CHEMICAL LABORATORY,
ST. THOMAS'S HOSPITAL,
LONDON, S.E.

XXVII.—*The Determination of Available Plant Food in Soil by the Use of Weak Acid Solvents. Part II.*

By ALFRED DANIEL HALL, M.A., and ARTHUR AMOS, B.A.

THE use of some weak acid solvent to extract that portion of the mineral plant food in soil, particularly phosphoric acid and potash, which may be regarded as immediately "available" for the use of the crop, has become an established part of soil analysis. It has been shown to yield results in accord with field trials and to be of value in determining the manurial requirements of the soil, although opinions still differ as to the best method to follow and as to the interpretation of the analytical figures. Of the methods employed, the use of a 1 per cent. solution of citric acid, as suggested by Dyer (Trans., 1894, 65, 115), has become most general; on the other hand, the American chemists have adopted in preference a $N/200$ solution of hydrochloric acid. Dyer's method was based on the idea of obtaining a solvent approximating in composition and strength to the acid sap which is to be found in the roots of most plants, which sap was supposed to have a direct action on the mineral particles with which the roots were in contact. This view, that the root excretes an acid other than carbon dioxide, has now been generally abandoned (see Czapek, Prings. Jahrb. wiss. Bot., 1896, 29, 321 ; Kossowitsch, Ann. de la Sci. Agron., 2nd S., 1, 1903, 220 ; Hall, Proc. Roy. Soc., 1905, 77, Series B, 1), and the method should be taken as an empirical one to be judged by its agreement with the evidence afforded by the crop. The theoretical basis for the use of $N/200$ hydrochloric acid is that it extracts from the soil quantities of mineral plant food approximately equal to those removed by an ordinary crop from the same soil. It would be difficult, however, to justify this theory in view of the fact that such extraneous factors as cultivation, supply of water or of nitrogen, and the nature of the crop will make radical variations in the amounts of the constituents in question assimilated by the crop.

To both theories, however, there is one further objection : they regard the material extracted by the solvent as differing essentially from that which is left behind ; the one is "available" for the crop, the other not, and when the "available" portion has been removed from the soil there should be nothing left for the crop until weathering, &c., has brought a fresh portion into a more soluble condition (see Ingle, Trans., 1905, 87, 43). In such cases, however, both in the laboratory and in nature, the process of solution must be considered dynamically ; there is no fixed point when all the material soluble in the medium employed will have gone into solution, the extraction proceeds until an equilibrium is established between the material in the solid state and that in solution, and if the original material be homogeneous in nature, its mass will not affect the concentration attained by the solution.

Whitney (United States Department of Agriculture, Bureau of Soils, Bulletin 22, 1903) has advanced this argument in support of the idea that the soil water, which must be regarded as the culture medium on which all plants feed, possesses a constant composition for all soils, because it is always in equilibrium with the same slightly soluble soil phosphates. Therefore, the addition of more plant food in the form of fertilisers does not supply the plant with further nutrient material, because an equivalent amount of the constituent added will be thrown out of solution and the original position of equilibrium established. But even assuming that the soil gives rise to a solution of constant concentration, the mass of the constituent in the soil would still come into play by regulating the ease or the frequency with which the solution could be renewed. If, for example, the phosphates in the soil give rise to a solution of phosphoric acid in the soil water, the strength of which is independent of the mass of phosphate present, yet as soon as the crop withdraws phosphoric acid from the solution the equilibrium will be disturbed and more phosphate will be attacked. But the rapidity with which the phosphate will pass into solution, and in its turn to the plant, will be conditioned by the mass of it present, and if the amount is near the limit required for saturation of the soil water, then the solution might not be replenished so often as the plant requires. Thus, even if a particular substance establishes a solution of constant composition in the soil water, its mass will still affect the supply of nutrient to the plant, because with it is bound up the renewal of the solution as it becomes depleted by the growth of the plant.

Again, a soil solution of constant composition would necessitate the identity in all soils of the state of combination of the constituent in question. If, for example, all soils contained a similar tricalcium phosphate and no other compound of phosphoric acid, then the soil water in equilibrium with the soil would attain the same concentration of phosphoric acid. whatever the amount of phosphate in the soil. This identity, however, of the compounds of phosphoric acid in all soils has not been demonstrated.

It was with the view of obtaining more light on the conditions of solution of soil phosphates and kindred substances, both in the soil itself and in the laboratory processes for soil analysis, that the following investigation was undertaken. The method adopted was to attack the soil continuously with the solvent. After equilibrium had been attained with the first portion of solvent, it was removed and renewed to see to what extent similar solutions could be obtained with fresh portions of solvent. The earlier experiments were undertaken with carbon dioxide and water, with the idea of realising thereby as nearly as possible the conditions prevailing in the field, where water charged with carbon dioxide is the great natural solvent. For reasons which will be given, this method of attack was abandoned in favour of repeated extractions with a 1 per cent. solution of citric acid. The investigation was limited to a consideration of the phosphoric acid, since in its progress it became clear that the potash in the soil would be likely to behave in a similar fashion.

The extractions were all made in 1 litre (half Winchester) green glass bottles containing 100 grams of soil, 10 grams of citric acid, and

1 litre of distilled water. The bottles were placed in a shaker and kept in continual end-over-end rotation for twenty hours, this, as will be seen later, being a sufficient period to establish equilibrium. After settling and filtering off the bulk of the liquid for analysis, the remaining soil was washed free from acid and returned to the bottle together with 10 grams of citric acid and water to the level which had been marked after the first filling. The shaking was then renewed and the process repeated as often as need be.

Solutions in Water charged with Carbon Dioxide.

Preliminary experiments showed that the amount of carbon dioxide in the solution was a factor in determining how much phosphoric acid was dissolved from the soil, hence it was necessary to ensure a solution of approximately constant composition. Distilled water was charged with carbon dioxide in an ordinary selzogene, and the required volume was drawn off and added to the soil in the bottle; the mixture was then shaken for a minute or two and the stopper withdrawn to allow the escape of any excess of carbon dioxide shaken out of the super-saturated solution. In this way, the liquid in the bottle would be approximately saturated under atmospheric pressure and contain, within negligible limits, always the same quantity of carbon dioxide. After shaking for twenty to twenty-four hours and filtration, an aliquot portion of the extract was evaporated and ignited, the residue was dissolved in hydrochloric acid, again evaporated, heated in an air-bath at 120—150° for an hour to render the silica insoluble, and finally taken up with dilute hydrochloric acid before being precipitated with molybdic acid. As the amounts of phosphoric acid which go into solution in the water charged with carbon dioxide are very small, they were estimated by a colorimetric method devised by Pagnoul (*Ann. Agron.*, 1899, **25**, 5), which depends on the depth of the brown colour produced when a solution of the phosphomolybdic acid precipitate in ammonia is added to an acid solution containing potassium ferrocyanide. The tint is compared by a method of trial and error with that similarly produced by successive quantities of a standardised solution of phosphoric acid.

The experiments were begun on some of the soils from the Broadbalk Field, Rothamsted, which had been growing wheat under known conditions of manuring since 1843. The treatment of the various plots has repeatedly been described (Dyer, *Phil. Trans.*, 1901, **194**, Series *B*, 235; Hall and Plymen, Trans., 1902, **81**, 117); it will be sufficient here to summarise very briefly the important features as regards the annual supply of phosphoric acid.

TABLE I.—*Description of Soils.*

Plot.	Date at which treatment began.	Annual manuring.	Phosphoric acid. Lbs. per acre in manure.	Total produce. Average lbs. per acre.

Broadbalk Wheat Field.

Plot.	Date	Annual manuring.	Phos.	Total
2b	1843	Dung	78(?)	6049
3	1843	Unmanured	—	2009
5	1849	Superphosphate; potassium, sodium, and magnesium sulphates	64	2308
7	1849	Superphosphate; alkaline sulphates and ammonium salts	64	5775
8	1852	Superphosphate; alkaline sulphates and ammonium salts............................	64	6913
10	1845	Ammonium salts only......................	—	3408

Hoos Barley Field.

Plot.	Date	Annual manuring.	Phos.	Total
1	1852	Sodium nitrate only	—	3743
2	1852	Superphosphate; sodium nitrate	64	5431
3	1852	Alkaline sulphates; sodium nitrate......	—	3987
4	1852	Superphosphate; alkaline sulphates and sodium nitrate	64	· 5529

Thus, Plots 2, 5, 7, and 8 receive an excess of phosphoric acid every year; on Plot 2 it is applied as dung, on the others as "super-phosphate." In the absence of nitrogen, the crop and the loss of phosphoric acid are very small on Plot 5; they are much increased on Plot 7, and still further on Plot 8. Plot 3 receives no phosphoric acid, nor does Plot 10, but in the latter case the loss of phosphoric acid has been greater, because of the use of 86 lbs. of nitrogen every year.

TABLE II. – *Phosphoric Acid dissolved by Water saturated with Carbon Dioxide.*

Milligrams per 100 grams of soil.

Expt.	Plot.	Extractions.							
		1st.	2nd.	3rd.	4th.	5th.	6th.	7th.	8th.
1	Broadbalk, 2b ...	3·9	2·6	3·2	3·3	4·7	3·4	3·4	2·4
2	,, 2b ..	3·6	2·6	3·0	3·3	2·8	2·6	3·4	2·7
3	,, 3 ...	0·25	0·25	0·16	0·15	0·22	0·28	—	—
4	,, 7 ...	2·8	1·9	1·9	4·6	2·2	3·3	—	—
5	,, 7 ...	2·6	2·1	2·3	4·6	—	—	—	—
6	,, 10 ...	0·26	0·27	0·25	0·18	0·58	0 55	—	—

Calcium carbonate replaced after each extraction.

7	Broadbalk, 5 ...	5·0	2·8	2·2	1·8	1·4	1·6	1·3	1·2
8	,, 8 ...	1·7	1·2	1·0	0·6	0·9	0·7	0·7	0·6

Table II gives some of the results obtained, which are also set out in graphic form in **Fig. 1.**

As the extractions proceeded, the process became disturbed by the difficulty of filtration; with the first and second extractions it was easy, the finer particles of the soil being kept in a flocculated condition by the presence of calcium bicarbonate and other salts in the solution. But as soon as all the calcium carbonate in the soil had been removed by the carbonated water, filtration became exceedingly slow, and it was almost impossible to obtain a clear filtrate. The presence of even a small quantity of fine soil in the extract would add a quantity of

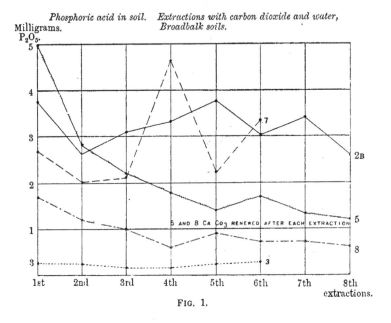

Phosphoric acid in soil.   *Extractions with carbon dioxide and water,*
Milligrams.   *Broadbalk soils.*
P_2O_5.

FIG. 1.

phosphoric acid too great to be neglected, and the introduction of even soluble flocculators, such as calcium chloride or magnesium sulphate, seemed to precipitate some of the soluble phosphoric acid with the sediment. Putting aside the expenditure of time the process began to involve, such a fertile source of errors was thus introduced in the determination of the very small quantities of phosphoric acid that in view of the trend of such results as could be trusted it was decided to abandon the process.

A consideration of the results and curves shows that in most cases less phosphoric acid is extracted by the second than by the first extraction; a steady equilibrium is not immediately attained, indicating that soil phosphate which is first attacked is not combined in the same way as the rest. After a falling off in the second extraction, a rise in the amount of phosphoric acid going into solution is generally observed, and this rise coincides with the complete removal of calcium carbonate from the soil. Evidently at each extraction a somewhat complex condition of equilibrium is attained, in which the free carbon dioxide, the calcium carbonate, and the soil phosphates all are variable factors. This is made more clear by a consideration of experiments 7 and 8, in which, after each extraction, an amount of calcium carbonate was returned to the soil equivalent to that removed by the previous extraction. It will be seen that after the initial decline in the first two or three extractions practically a constant condition is attained, and the state of equilibrium between the soil phosphates, the calcium carbonate, and the free carbon dioxide is repeated indefinitely.

Looking at the results as a whole, it may be concluded that the soil produces with carbon dioxide and water a characteristic solution of phosphoric acid which does not vary greatly in concentration when the solution is renewed. This position of equilibrium is, however, different for each soil, and since it does fall off considerably from the first to the second extraction, and as again it bears no particular relation to the total amount of phosphoric acid in the soil, it would seem to be determined by the nature rather than the mass of the soil phosphates. The concentration of the solution, however, reflects the manurial treatment the soil has received, so that if we may take these laboratory solutions as a guide to the composition of the soil water, the fertilisers applied in the past, so far from being without effect, condition wholly the concentration in phosphoric acid of the soil water, and in consequence the supply of this constituent to the roots of the plant.

During the progress of these experiments, the tile drains beneath the plots yielding the soils under investigation began to run for the first time after a long period of dry weather. Samples of the water were taken for the determination of phosphoric acid, the concentration of which, however, was far below that obtained in the carbonated water in the laboratory, being in most cases only about one-tenth as much. Plot 5 yielded 0·00044 per cent. of phosphoric acid, Plot 7, 0·0003 per cent., and Plot 3 a trace too small for estimation. The samples were taken at hourly intervals during the day when the drains were running and then mixed ; probably the more concentrated early runnings, which represent the true soil solution, had become much diluted by the later water running rapidly through the soil. The composition of these samples, then, settles nothing as to whether the carbon dioxide solutions obtained in the laboratory can be taken to represent the soil water. These experiments with carbon dioxide and water as a solvent may be taken to demonstrate that a hard and fast line cannot be drawn between the phosphoric acid going into solution and that which remains behind. The solution can be renewed repeatedly, and although it will diminish somewhat in concentration with successive extractions, the change is such as would indicate the presence in the soil of an indefinite series of phosphoric acid compounds, akin to one another and only varying slightly in composition. For practical purposes, the average position of equilibrium obtained in the first six extractions would supply a very fair index to the character of the soil as judged from its past history, but there is no indication that any beyond empirical conclusions can be drawn from the results. Since the difficulties of manipulation put the process out of question as a working method of analysis, the further use of carbon dioxide was abandoned in favour of the 1 per cent. solution of citric acid.

Extractions with 1 per cent. Solution of Citric Acid.

The process adopted was exactly similar to that previously described. One hundred grams of the air-dried soil were shaken for twenty hours with 10 grams of citric acid and 1 litre of water, the solution being filtered and the phosphoric acid determined in the usual way. The washed soil was then returned to the bottle and the process repeated. No attempt was made to restore the original calcium carbonate, most of which was removed by the first extraction. It has been shown that the presence of calcium carbonate does diminish the amount of phosphoric acid, &c., dissolved by the citric acid solution (Cousins and Hammond, *Analyst*, 1903, 28, 238), but to add a further amount of citric acid equivalent to the calcium carbonate in the soil seems to introduce a fresh source of error. The Rothamsted soils only contain about 3 per cent. of calcium carbonate, and the amount does not vary greatly in the particular plots under examination, so that it was not considered necessary to attempt any correction on this score.

It has already been stated that an extraction of from twenty to twenty-four hours with continual shaking has been adopted instead of digestion for one week with occasional shaking, as originally recommended by Dyer (*loc. cit.*, p. 142). In experiment 11, each extraction proceeded for five days with continual shaking; in 10, the same soil was extracted for twenty hours only. Other experiments with carbon dioxide led to the same conclusions. The results are practically the same whether the extraction proceeds for one or five days; by twenty hours, the solution has come into equilibrium with the soil, and will dissolve no more phosphoric acid however long the contact continues. It does not, however, follow, as subsequent extractions show, that all the phosphoric acid soluble in the citric acid has been removed in the first extraction. In view of the fact that phosphoric acid continues to be dissolved with each separate extraction, it was not considered necessary to study the time factor further.

The soils already described from the Broadbalk Field were examined, and, in addition, four soils from the Hoos Field, which has carried barley under the same system of manuring since 1852. Four other

TABLE III.—*Phosphoric Acid dissolved by* 1 *per cent. Solution of Citric Acid.*

Milligrams per 100 *grams of soil.*

Expt.	Field.	Period.	Extractions.						Sum of first 5 extractions only.	Ratio of first to last col.
			1st.	2nd.	3rd.	4th.	5th.	6th.		
9	Broadbalk, 2b ...	5 days	—	1·6	6·2	6·0	—	—	—	—
10	,, 2b'	20 hours	47·7	14·2	7·5	6·1	4·4	—	—	—
11	,, 2b ...	5 days	51·0	15·8	8·8	—	—	—	—	—
		Mean...	49·3	15·3	7·5	6·0	4·4	—	82·5	59·8
12	Broadbalk, 7 ...	20 hours	56·4	22·0	8·9	6·5	4·4	4·4	—	—
13	,, 7 ...	20 ,,	55·8	23·6	—	—	—	—	—	—
		Mean...	56·1	22·8	8·9	6·5	4·4	4·4	98·7	56·8
14	Broadbalk, 3 ...	20 hours	6·6	6·6	4·0	3·5	2·4	—	—	—
15	,, 3 ...	20 ,,	6·2	7·0	3·8	2·6	2·7	—	—	—
		Mean...	6·4	6·8	3·9	3·0	2·5	—	22·6	28·3
16	Broadbalk, 5 ...	20 hours	69·0	28·0	11·3	7·3	4·5	2·3	120·1	57·5
17	,, 8 ...	20 ,,	46·3	18·9	7·8	5·3	4·0	3·0	82·3	56·4
18	,, 10 ...	20 ,,	7·7	5·2	3·3	2·7	2·7	2·7	21·6	35·7
19	Hoos, 1AA	20 ,,	6·3	3·5	2·2	1·9	2·0	1·2	15·9	39·6
20	,, 2AA	20 ,,	52·2	21·2	8·9	6·5	3·8	2·9	92·6	56·4
21	,, 3AA	20 ,,	6·3	2·7	2·3	2·1	1·9	1·5	15·3	41·2
22	,, 4AA	20 ,,	53·5	10·6	6·4	4·9	4·5	3·8	79·9	67·0
23	Shenington	20 ,,	7·4	2·6	2·2	1·8	1·3	1·2	15·3	48·4
24	Saxmundham ...	20 ,,	7·2	5·8	5·3	4·1	3·1	2·1	25·5	28·2
25	Cockle Park	20 ,,	14·3	8·0	7·4	5·2	4·3	3·8	39·2	36·5
26	Bramford	20 ,,	72·6	28·4	19·5	—	5·2	3·2	(125·7)	(57·8)

soils were used, two of which, Bramford and Saxmundham, have long been under experiment by the Cambridge University Department of Agriculture (*Report on Experiments, East Suffolk*, 1904). They are in sharp contrast, Saxmundham giving a very pronounced return-for applications of phosphoric acid, which on Bramford show no return in the crop. Cockle Park is another soil on which field experiments have been repeatedly tried (Northumberland Education Committee, *Seventh*

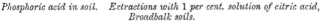

Phosphoric acid in soil. Extractions with 1 *per cent. solution of citric acid, Broadbalk soils.*

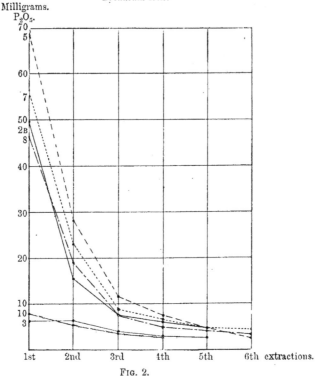

Fig. 2.

Report on Experiments, 1903); it shows a normal, but no marked, response to phosphatic manures. Shenington is again a very exceptional soil, for although it contains as much as 0·84 per cent. of phosphoric acid, crops grown on it stand in great need of phosphatic manuring, probably because the soil contains the wholly exceptional amount of 28·16 per cent. of ferric oxide (Hall, *Jour. Agri. Sci.*, 1905, 1, 85). The results obtained are set out ˙in Table III and in Fig. 2.

It will at once be seen that all the results partake of the same general character; in the second extraction, there is obtained something less than half the phosphoric acid dissolved by the first extraction, the third extraction yields about half of the second, in the fourth the amount dissolved does not fall so much, whilst eventually, about the sixth extraction, the amount going into solution shows a tendency to become constant.   When the amount of phosphoric acid dissolved is plotted against the number of extractions, the curve shows at first a steep descent, then the fall becomes less

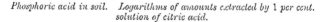

Phosphoric acid in soil.   Logarithms of amounts extracted by 1 per cent. solution of citric acid.

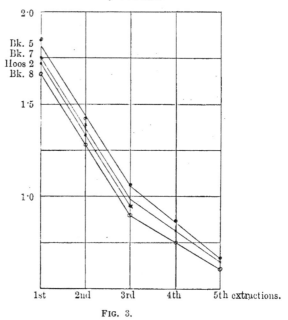

Bk. 5
Bk. 7
Hoos 2
Bk. 8

2·0

1·5

1·0

1st   2nd   3rd   4th   5th extractions.

FIG. 3.

and less until the curve becomes practically parallel to the axis.   The position of the curve throughout differs for each soil, and the relative character of the curves agrees very well with what is known of the cropping and manuring of the plots from which the soils were drawn.

New light is thrown on the character of the curves if, instead of the amounts dissolved at each extraction, their logarithms are plotted, Fig. 3, whereupon it is seen that a series of straight lines is obtained from the Broadbalk and Hoos plots which have been similarly manured with superphosphate every year. The first three extractions yield points which fall in a straight line, and the straight lines obtained for each plot are parallel. After the third extraction, there is a sudden change of direction, but straight lines are again obtained; at this stage, however, the quantities dealt with are so small that the experimental error forbids any detailed discussion of the results. The logarithmic decrement of the amount of phosphoric acid extracted from Plots 5, 7, and 8 indicates that in each case there exists a particular quantity of phosphoric acid of which a certain fraction goes into solution at the first extraction; of the remainder, the same fraction goes into solution at the second extraction, and again the same fraction of what is still left dissolves at the third extraction. The fraction of the whole which goes into solution approximates very closely to 0.6; by taking this coefficient, the following comparison between the calculated and observed amounts of phosphoric acid going into solution is obtained.

TABLE IV.—*Phosphoric Acid dissolved at Successive Extractions. Milligrams per* 100 *grams of soil.*

| | Plot. | Calculated. | | | Found. | | |
		First.	Second.	Third.	First.	Second.	Third.
Broadbalk	5	70·2	28·0	11·3	69·0	28·0	11·3
,,	7	55·8	22·3	8·9	56·4	22·0	8·9
,,	8	46·8	18·7	7·5	46·3	18·9	7·8
Hoos	2	52·2	20·7	8·5	52·2	21·2	8·9

Only the plots which have been continually receiving the same superphosphate manure show this logarithmic rate of decrement in the amounts of phosphoric acid going into solution; with Plot 2b, which receives the variable phosphates contained in dung, Plot 3, which is unmanured, and Bramford, which is farmed in the ordinary way, the results when plotted do not fall into straight lines. But as regards the four Rothamsted soils receiving superphosphate, the strength of the solution of phosphoric acid obtained is the product of the active mass of a particular phosphoric acid compound present and a coefficient which is the same in each case.

The exact significance of the observed regular partition between the solvent and the soil of one portion of the phosphoric acid of the soil on these plots is not as yet apparent; it does not fall under any of the recognised laws of chemical dynamics of a solid going partially into solution. There is doubtless a somewhat complex condition of equilibrium set up between the dissolving acid, the phosphoric acid, and the various bases in the soil, which can only be interpreted by a study, now in hand, of the similar action of the solvent on various pure phosphates in the presence of such bases as are found in the soil. It can hardly be doubted, however, that the logarithmic portions of the curves represent a single phosphoric acid compound in the soil, which has been almost wholly removed after the third extraction, and that the non-logarithmic character of the curves yielded by the other soils indicates that their phosphoric acid is present in a more varied and irregular state of combination, as, indeed, would be expected from the history of the treatment of the plots.

We are indebted to Mr. R. D. Watt, B.Sc., Carnegie Scholar of the University of Glasgow, for some further experiments on the conditions of equilibrium set up between phosphoric acid, the citric acid solution, and soil.

One possible factor is the action of the surface of the soil particles, which is very large, from 10 to 40 square metres for 100 grams of the soils employed. Accordingly, 1 gram of basic slag was shaken with the citric acid solution alone and with the citric acid solution to which had been added 100 grams of "slimes," or kaolin, or soil. The "slimes" consists of the finest portions of the crushed gold-bearing rock from the Rand ; the particles vary in size from 0·04 to 0·002 mm. in diameter, and are mainly quartz with a little felspathic material. This crushed rock had previously been treated with strong hot hydrochloric acid.

The following results show the amounts of phosphoric acid going into solution in 1 litre of 1 per cent. citric acid solution.

TABLE V.—*Phosphoric Acid dissolved by* 1 *litre of* 1 *per cent. Citric Acid Solution.*

Expt.		Phosphoric acid.
27	1 gram of basic slag alone....................................	0·2117 gram
28	,, ,, *plus* 100 grams of "slimes "......	0·2085 ,,
29	,, ,, ,, ,, kaolin	0·2080 ,,
30	,, ,, ,, ,, soil (Hoos 1)...	0·1428 ,,
31	100 grams of soil (Hoos 1) alone	0·0059 ,,

The amount retained by the "slimes" and the kaolin is inappreciable, so that surface actions may be dismissed from consideration, but the presence of the soil withdraws about 35 per cent. of the phosphoric acid which would otherwise have gone into solution. In a second series of experiments, pure dicalcic phosphate was employed as being wholly soluble in the citric acid solution, and the following results were obtained :

TABLE VI.—*Phosphoric Acid dissolved by* 1 *litre of* 1 *per cent. Citric Acid Solution.*

Expt.				Phosphoric acid.
32	1 gram of dicalcic phosphate alone			0·486 gram
33	,,	,,	,, with 0·64 per cent. of citric acid ...	0·470 ,,
34	,,	,,	,, *plus* 100 grams of Woburn soil (R)*.	0·435 ,,
35	,,	,,	,, ,, ,, ,, (2A).	0·429 ,,
36	,,	,,	,, ,, 100 grams of Shenington soil...	0·252 ,,
37	Shenington soil alone			0·0128 ,,

* A light sandy soil from the Stackyard Field of the Woburn Experimental Farm.

The Woburn and Shenington soils contain no calcium carbonate to neutralise the citric acid solution, although in the latter case the oxide of iron reduced the acidity of the solution. But on acting on 1 gram of the dicalcic phosphate alone with a citric acid solution of this reduced strength, there was but little fall in the amount of phosphoric acid dissolved.

These results show very clearly that when the citric acid solvent is in contact with the soil, the solution of phosphoric acid which is attained represents a state of equilibrium between that which goes into the acid solution and that which remains combined with the soil bases. The action of the citric acid is not simply to dissolve the whole of one or more phosphoric acid compounds· present in the soil ; in the cases just cited, experiments 30 and 36, phosphoric acid already seen to be soluble in the acid is retained in a solid state by the soil.

Mr. Watt also attempted to ascertain the nature of the soil phosphates attacked by the successive citric acid solutions by determining the bases which were simultaneously dissolved, with the following results :

TABLE VII.—*Bases dissolved by* 1 *per cent. Citric Acid Solution.*

	Broadbalk, Plot 3.			Broadbalk, Plot 5.		
	CaO.	Fe$_2$O$_3$.	Al$_2$O$_3$.	CaO.	Fe$_2$O$_3$.	Al$_2$O$_3$.
First	1·92	0 090	0·398	1·74	0·080	0·343
Second	0·127	0·056	0·074	0·129	0·064	0·059
Third	0·042	0·068	0·055	—	0 054	0·041

These experiments, however, lead to nothing ; there is always more of each base dissolved than would be necessary to combine with the whole of the phosphoric acid in solution at the same time.

Reviewing the experiments as a whole, it may be concluded that the soil contains compounds of phosphoric acid of varying solubility ; the more easily dissolved compounds removed in the first few extractions are present in comparatively small amounts, so that the amounts going into solution fall rapidly with each extraction. After the fourth or fifth extraction, a point is reached when the compound remaining in the soil seems to be uniform, as shown by the approximately constant concentration in phosphoric acid which the later solutions now attain. This constant equilibrium point is not, however, the same for all soils; it is evidently determined by certain differences in the nature of the phosphoric acid compounds which are characteristic of each soil.

The results lend no support to Whitney's theory, that all soils will form in nature solutions of approximately constant composition, representing an equilibrium between the soil phosphoric acid and the solvent water which is independent of the mass of the former, for although the 1 per cent. solution is a much more drastic solvent than the natural soil water, there is no reason to suppose that its action will be essentially different in kind; indeed one of us has already attempted to show that all weak acid solvents attack the soil in a very similar fashion (Hall and Plymen, *loc. cit.*). The present results indicate that there are essentially different compounds of phosphoric acid in different soils, possessing initially different points of equilibrium with solvents, and the more soluble of these compounds are present in such comparatively small amounts that their mass becomes a factor in the maintenance of the strength of the solution. In other words, the soil water will be of varying concentration in phosphoric acid in different soils, and when the crops remove this phosphoric acid by their growth, the soils will again differ in their power of renewing the solution, because of variations in the mass as well as the nature of the phosphate which gave rise to the original solution.

There remains the practical question of how far these results obtained by repeated extractions of phosphoric acid from a soil with the dilute citric acid solution bear on the practice of soil analysis, in which one extraction only is attempted. It may be assumed that all the "available" or readily soluble phosphoric acid will be represented by the sum of the phosphoric acid going into solution in the first four or five extractions, until the low, but fairly constant, equilibrium is attained between the solvent and the large mass of "dormant" soil phosphates. In the case of the Rothamsted soils manured with superphosphate, it is also easy to determine from the curve the original amount of phosphoric acid which divided itself between solvent and soil bases according to the ascertained coefficient. In Table VIII (p. 220) these two sets of quantities are set out as percentages and as pounds per acre in the first nine inches of soil, below which depth Dyer has shown (*Phil. Trans.*, *loc. cit.*) the manurial applications of phosphoric acid do not sink, even when soluble in water. Against them are set

TABLE VIII.—*Phosphoric Acid dissolved by 1 per cent. Citric Acid compared with that supplied by Manure.*

Plot.	Total of five extractions.		Amount calculated from the first extraction and the coefficient.		Supplied in manure.	Removed in crop.	Surplus.
	Per cent.	Lbs. per acre.	Per cent.	Lbs. per acre.			
Broadbalk, 3 ...	0·0226	565	—	—	0	· 550	− 550
,, 5 ...	0·1201	3000	0·117	2925	3960	790	3170
,, 7 ...	0·0987	2470	0·093	2325	3810	1370	2440
,, 8 ...	0·0823	2055	0·078	1950	3810	1520*	2290*
,, 2b ...	0·0825	2060	—	—	4780	1650	3130
Hoos, 1......... ...	0·0159	400	—	—	0	555	− 555
,, 2	0·0926	2315	0·087	2175	3390	1200	2190
,, 4............	0·0799	2000	—	—	3390	1240	2150

* Approximate estimate. The crop of Plot 8 has seldom been analysed.

the total applications of phosphoric acid in the manure since the beginning of the experiments together with the amounts removed in the crop, so as to show the surplus accumulated in the soil. These surpluses agree very closely with the total amounts of phosphoric acid soluble in citric acid. On the one hand might be deducted the amount of soluble phosphoric acid supplied by the unmanured soil itself to the citric acid solution, and on the other the amount contained in the crop grown by the unmanured plot without any applications of manure, but as these two quantities of phosphoric acid happen to coincide very closely it is unnecessary to attempt the correction.

For the dunged plot, the agreement between the phosphoric acid dissolved by the citric acid and the surplus of the manurial phosphoric acid over that removed by the crops is by no means so close, but in this case the amount of phosphoric acid supplied in the manure is but an approximate estimate, and some of it is, without doubt, combined in forms less attackable by the citric acid than are the compounds formed when superphosphate is precipitated in the soil.

It may be concluded that the repeated extractions in citric acid do eventually dissolve all the phosphoric acid which has been added to the soil in the shape of soluble phosphates for the previous fifty years or so. But the total amount of such readily soluble phosphoric acid bears a fairly constant ratio to that which is dissolved in the first extraction, as may be seen from the last column in Table III. Since also the numbers obtained for citric acid soluble phosphoric acid can only be interpreted empirically by comparison with the results yielded by soils the response of which to phosphatic manures has been determined by field trials, then for ordinary analytical purposes the single extraction is likely to give as much information as repeated attacks with the same solvent.

Particularly on the majority of soils under ordinary cultivation which have not been regularly manured year after year in the same manner, the attack of the citric acid is not a sharply defined action, bringing into solution one or two definite compounds of phosphoric acid which can be termed "available," but is, instead, a complicated process in which equilibrium is established between the solvent and a large number of compounds of varying grades of solubility, so that for practical purposes it is useless to repeat the extractions until more knowledge is attained of the separate actions which go to make up the observed results. For the present, the process of extraction with citric acid must still be regarded and interpreted empirically. Probably the solvent process is parallel to the actions going on in nature, which render the soil phosphoric acid available for the plant, but this cannot be decided *a priori*; it must be determined by the agreement between the results yielded by the solvent and the response of the crop on the same soil to phosphatic fertilisers.

General Conclusions.

The conclusions reached in this examination of the process of attacking a soil with weak acid solvents may be summarised as follows.

1. The solvent does not at once remove all the phosphoric acid capable of going into solution in the particular solvent employed ; instead an equilibrium is established between the phosphoric acid in the solvent and in the soil.

2. The concentration of the solution in equilibrium with the soil falls with each successive attack of the soil by the same solvent. This indicates the presence in the soil of several compounds of varying solubility, the mass of the more soluble being small and of the same order as the amounts going into solution in the earlier extracts. When these more soluble compounds have been removed, an approximate constant equilibrium is attained between the phosphoric acid remaining in the soil and that going into solution at each extraction, indicating that after the more soluble compounds have been removed there remains a phosphate in each soil of such low solubility that the amount going into solution at each extraction is independent of the mass present in the soil.

3. With soils which have been for many years manured with a

Q 2

particular phosphate, the amounts of phosphoric acid going into solution in successive extractions with 1 per cent. citric acid solution follow a logarithmic law of decrement, indicating the presence of one particular phosphate which dissolves in proportion to the mass of it present in the soil. This law does not, however, hold for ordinary soils which have been variously manured.

4. In the case of the Rothamsted soils, the sum of the phosphoric acid dissolved out by the first five extractions with citric acid approximates very closely to the known surplus of phosphoric acid accumulated by the additions of manure to the soil.

5. Assuming that the solvent actions of the soil-water and of the weak acid solvents employed in the laboratory are comparable, the evidence lends no support to the theory that all soils give rise to a natural soil solution of approximately constant composition, which is not disturbed by the use of fertilisers containing phosphoric acid.

6. For the practical purposes of soil analysis, the evidence afforded by a single extraction of the soil for twenty hours with continual shaking is very similar to that obtained from a series of successive extractions by the same solvent and leads to the same conclusions as to the manurial requirements of the soil.

XXVIII.—*Studies in the Camphane Series. Part XXI. Benzenediazo - ψ - semicarbazinocamphor and its Derivatives.*

By Martin Onslow Forster.

Although it is well known that diazonium salts produce diazoamino-compounds with aliphatic bases, these mixed fatty-aromatic derivatives have not been studied so extensively as the corresponding substances obtained from benzenoid amines. The unusual chemical activity displayed by camphoryl-ψ-semicarbazide (Forster and Fierz, Trans., 1905, **87**, 722, 826), and the stability conferred on its derivatives by the presence of the camphor nucleus, suggested the possibility of coupling the substance with the diazonium complex, and I find that, in general, the ψ-semicarbazide nitrate dissolved in water yields a precipitate immediately on treatment with the diazonium salt prepared from a benzenoid amine. The resulting compounds are derivatives of *benzenediazo-ψ-semicarbazinocamphor,*

$$C_8H_{14} \Big\langle \begin{array}{l} CH\text{———}N\cdot NH\cdot N\colon N\cdot C_6H_5 \\ C(OH)\cdot NH\cdot CO \end{array} \Big. ,$$

the substance obtained from aniline itself, and in this communication they are referred to by the class name *diazo-ψ-semicarbazine*, in order to signalise their general analogy to diazoamines ; the members of the new series are well-defined, stable compounds, which, in the undissolved state, may be preserved for an indefinite period.

In spite of the fact that the diazo-ψ-semicarbazines are prepared by a general method precisely similar to that which furnishes the diazo-amines, they distinguish themselves from the older group of compounds in several fundamental particulars. One of the best-known peculiarities of the latter class is the behaviour of its members in the guise of a solid solution composed of the diazoamine, $X \cdot N \vdots N \cdot NHY$, and the iso-meric $Y \cdot N \vdots N \cdot NHX$, and unless the formation of such a system depends on the presence of two aromatic nuclei, it might be ex-pected that benzenediazo-ψ-semicarbazinocamphor would display the properties of the isomerides

$$C_8H_{14} \Big\langle {}^{CH \underline{\qquad} N \cdot NH \cdot N \vdots N \cdot C_6H_5}_{C(OH) \cdot NH \cdot CO} \quad \text{and}$$

$$C_8H_{14} \Big\langle {}^{CH \underline{\qquad} N \cdot N \vdots N \cdot NH \cdot C_6H_5}_{C(OH) \cdot NH \cdot CO}$$

simultaneously. This, however, is not the case. The diazo-ψ-semi-carbazine in question appears to be an individual, because dilute alkalis resolve it quantitatively into camphoryl-ψ-carbamide and phenylazo-imide :

$$C_{17}H_{23}O_2N_5 = C_{11}H_{18}O_2N_2 + C_6H_5N_3,$$

and the same change occurs when an alcoholic solution is exposed to light during several days. Moreover, it has not been found possible to separate benzenediazonium salts from it by acids, or to obtain from it on reduction either aniline or phenylhydrazine by a process which does not first liberate phenylazoimide. Furthermore, the circum-stances in which benzenoid diazoamines are transformed into aminoazo-compounds fail to bring about the corresponding change in the benzenediazo-ψ-semicarbazino.

From these observations, it is clear that the diazo-ψ-semicarbazines resemble more closely the buzylene derivatives than the diazoamines. The foregoing expression for benzenediazo-ψ-semicarbazinocamphor contains the substituted group, $NH_2 \cdot NH \cdot N \vdots NH$, characteristic of the compound first obtained by Curtius (*Ber.*, 1893, **26**, 1263), and further investigated by Wohl and Schiff (*Ber.*, 1900, **33**, 2471). The probable formation of such derivatives was suggested by the work of Griess (*Ber.*, 1876, **9**, 1659), who found that the products from hydrazino-benzoic acid and benzenediazonium nitrate are identical with those from diazotised aminobenzoic acid and phenylhydrazine, namely, aminobenzoic acid, phenylazoimide, aniline, and triazobenzoic acid ;

similarly, Emil Fischer (*Annalen*, 1878, **190**, 94) observed that benzenediazonium nitrate or sulphate acting on phenylhydrazine gives phenylazoimide and aniline. Curtius (*loc. cit.*) isolated hippurylphenyl-buzylene, $C_6H_5 \cdot CO \cdot NH \cdot CH_2 \cdot CO \cdot NH \cdot NH \cdot N \vdots N \cdot C_6H_5$, from benzene-diazonium sulphate and hippurylhydrazine, whilst Wohl (*Ber.*, 1893, **26**, 1587; compare also Wohl and Schiff, *loc. cit.*) prepared unsymmetrical diphenylbuzylene or benzenediazophenylhydrazine from benzenediazonium nitrate and phenylhydrazine. Hippuryl-phenylbuzylene, however, is a very unstable substance, even in the solid state; solutions rapidly undergo decomposition, whilst aqueous alkalis, alkali carbonates, and dilute mineral acids attack the com-pound at once. This distinction from benzenediazo-ψ-semicarbazino-camphor is doubtless due to the fact that the terminal nitrogen atom, to which is attached the hippuryl group, is still associated with an atom of hydrogen, this being displaced in the camphor derivative. A still more important difference, however, is the fact that whilst the sole decomposition products of the diazo-ψ-semicarbazine are camphoryl-ψ-carbamide and phenylazoimide,

$$C_8H_{14} \Big\langle \begin{matrix} CH \text{————} N \text{—} NH \cdot N \vdots N \cdot C_6H_5 \\ C(OH) \cdot NH \cdot CO \end{matrix} =$$

$$C_8H_{14} \Big\langle \begin{matrix} CH \text{————} NH \\ C(OH) \cdot NH \cdot CO \end{matrix} + C_6H_5 \cdot N_3,$$

hippurylphenylbuzylene yields not only hippuramide and phenylazo-imide, which spring from the normal modification already formulated, but also aniline and hippurylazoimide, which may be regarded as arising from the isomeric form $NHBz \cdot CH_2 \cdot CO \cdot N \vdots N \cdot NH \cdot NHPh$ or $NHBz \cdot CH_2 \cdot CO \cdot NH \cdot N \vdots N \cdot NHPh$. From this point of view, hippuryl-phenylbuzylene, like the diazoamines, would appear to be a solid solution comprising two tautomeric substances.

Benzenediazo-ψ-semicarbazinocamphor being clearly an individual, and not a solid solution, an attempt was made to prepare the isomeric form from camphorylnitroso-ψ-carbamide and phenylhydrazine. Action between the substances takes place readily enough, but not in the direction indicated by the equation

$$C_8H_{14} \Big\langle \begin{matrix} CH \text{————} N \cdot NO \\ C(OH) \cdot NH \cdot CO \end{matrix} + C_6H_5 \cdot NH \cdot NH_2 = H_2O +$$

$$C_8H_{14} \Big\langle \begin{matrix} CH \text{————} N \cdot N \vdots N \cdot NH \cdot C_6H_5 \\ C(OH) \cdot NH \cdot CO \end{matrix}.$$

On the contrary, a quantitative yield of phenylsemicarbazide is obtained, the elements of cyanic acid being removed by phenylhydr-azine in a manner recalling the elimination of the same substance when camphorylazoimide is produced by the action of nitrous acid on the nitrate of camphoryl-ψ-semicarbazide (Forster and Fierz, Trans., 1905, **87**, 826).

In reviewing the general properties of the diazo-ψ-semicarbazines, it is noteworthy that whilst the o-, m-, and p-nitrobenzenediazo-compounds are intensely coloured, and benzenediazo-ψ-semicarbazino-camphor is distinctly, although faintly, yellow, the toluene, p-chloro-benzene, and p-bromobenzene derivatives are snow-white. This observation appears to have some bearing on the value of colour as a clue to structure, to which attention has been directed for many years. There can be no reasonable doubt that the diazo-ψ-semicarbazines described in this paper have a common structure, not merely because the method of preparation is a general one, but because the quantita-tive resolution into camphoryl-ψ-carbamide and the substituted phenylazoimide is undergone by each one. If colour is to be regarded as establishing the presence in a molecule of some particular form of linkage, which appears to be ₁the view of H. E. Armstrong, the members of any group in which the compounds have a common structure should be either all coloured or all colourless. Here is a case in which that is not so. A still more striking instance has been brought to light quite recently by Stobbe (*Ber.*, 1905, **38**, 3673), in working with certain carboxylic acids of the butadiene series, derived from the unknown butadiene-$\beta\gamma$-dicarboxylic acid,

$$\mathrm{CH_2{:}\underset{|}{C}{\cdot}CO_2H}$$
$$\mathrm{CH_2{:}C{\cdot}CO_2H'}$$

by the introduction of hydrocarbon radicles in the place of methylene hydrogen atoms. These are called fulgenic acids, and Stobbe has found that of this series only the tetraphenyl- and triphenyl-fulgenic acids are coloured, whilst the tetramethyl-, trimethylphenyl-, dimethyl-diphenyl-, and diphenyl-fulgenic acids are colourless; similarly, amongst the anhydrides, the tetramethylfulgide alone is colourless, the fulgides of the remaining acids displaying a colour which becomes intensified from sulphur-yellow to blood-red as the accumulation of benzenoid substituents proceeds. Each of these compounds contains four ethenoid linkages, which are thus powerless to produce colour until stimulated by benzene nuclei. They show clearly enough that the presence of ethenoid linkages may predispose to colour without actually causing selective absorption in the visible part of the spectrum, this being finally developed by the introduction of some exciting group.

There appears, therefore, to be no justification for the assertion made in a recent paper by H. E. Armstrong and W. Robertson (Trans., 1905, **87**, 1283) that of the four camphordioximes " not one can be regarded as a compound of the formula $C_8H_{14}(C{:}NOH)_2$," since they are not only colourless, but form colourless solutions in alkali."

Armstrong and Robertson ignore the fact that the monoimine of camphorquinone and the derivative it yields with bornylcarbimide,

$$C_8H_{14}\begin{array}{c}C:NH\\|\\C:O\end{array} \quad\text{and}\quad C_8H_{14}\begin{array}{c}C:N\cdot CO\cdot NH\cdot O_{10}H_{17}\\|\\C:O\end{array},$$

are colourless (Forster and Fierz, *loc. cit.*), and maintain that the formulæ

$$C_8H_{14}\begin{array}{c}C:N\cdot OH\\|\\C:O\end{array} \quad\text{and}\quad C_8H_{14}\begin{array}{c}C:N\cdot NHPh\\|\\C:O\end{array}$$

for *iso*nitrosocamphor and camphorquinonephenylhydrazone respectively are precluded by the absence of colour in these substances.

The danger of drawing hasty conclusions from a physical property before determining the exact relation between that property and constitution can be illustrated from another remark by Armstrong and Robertson (*loc. cit.*, p. 1279). In discussing the condensation product of camphorquinone and unsymmetrical diphenylhydrazine, which they admit has the old-fashioned type of constitution,

$$C_8H_{14}\begin{array}{c}C:N\cdot NPh_2\\|\\C:O\end{array},$$

they say, "moreover, notwithstanding that it contains two phenyl groups next the ethenoid junction, it has but a moderately high molecular rotatory power (420°), thus confirming our assumption that the C:N *azethenoid* junction is not likely to exert so powerful an influence as the C:C (ordinary ethenoid) junction." Surely this argument is fallacious? Comparison is drawn between benzylidene-camphor and camphorquinonediphenylhydrazone:

$$C_8H_{14}\begin{array}{c}C:CHPh\\|\\C:O\end{array} \qquad\qquad C_8H_{14}\begin{array}{c}C:N\cdot NPh_2\\|\\C:O\end{array}$$
$$[M]_D\ 1010°. \qquad\qquad\qquad [M]_D\ 420°$$

and it is claimed that the molecular rotatory power reveals a greater influence exerted by the C:C junction than by the C:N junction. This superior influence may or may not exist, but for the following reasons I submit that the case under discussion does not justify the conclusion drawn. In the first place, the two phenyl groups in the hydrazone are not "next the ethenoid junction," as stated by Armstrong and Robertson, the nitrogen atom with which they are associated being separated by another atom of nitrogen from the double linkage, whilst benzylidenecamphor contains the phenyl group directly combined with the carbon atom which participates in the junction; the optical effect might well be greater in benzylidene-camphor from this cause. In the second place, the displacement of hydrogen by phenyl is symmetrical in the case of the hydrazone, unsymmetrical in benzylidenecamphor, and I have shown (Trans., 1899, **75**, 937) that the latter condition can exert a marked effect on rotatory power; comparing bornylamine with methylbornylamine and dimethylbornylamine,

$$C_8H_{14}\!<\!\begin{matrix}CH_2\\CH\cdot NH_2\end{matrix} \qquad C_8H_{14}\!<\!\begin{matrix}CH_2\\CH\cdot NHMe\end{matrix} \qquad C_8H_{14}\!<\!\begin{matrix}CH_2\\CH\cdot NMe_2\end{matrix},$$

$$[a]_D\ 57\cdot1^\circ. \qquad\qquad [a]_D\ 95\cdot9^\circ. \qquad\qquad [a]_D\ 59\cdot6^\circ.$$

it is found that symmetrical displacement of hydrogen leaves the rotatory power almost unchanged, whilst unsymmetrical displacement produces a very considerable difference. The argument of Armstrong and Robertson regarding the optical value of the azethenoid junction seems to me, therefore, hardly justified by facts. But, even if sound, it might well be used against their general proposition as to the dioximes, because if the optical influence of the C:N group is inferior to that of the C:C group, the chromogenic effects of the two forms of linkage might stand in the same relation. It is to be expected that the further study of absorption spectra will go far towards explaining the difficulties presented by these compounds.

In their survey of the hydrazones derived from camphorquinone, Armstrong and Robertson state (*loc. cit.*, p. 1295) that the. phenylhydrazone, the methylphenylhydrazone, and the benzylphenylhydrazone " must be regarded as compounds altogether different in structure from the diphenylhydrazone on account of their colourless character and their great optical activity." Recalling the fact that the method by which these compounds are prepared is a general one, and that Armstrong and Robertson have brought forward no single distinction of a chemical nature, I venture to think that this conclusion is without foundation, particularly as one of the materials upon which their demonstration depends is variously described as " colourless " (*loc. cit.*, p. 1295), "almost colourless" (*loc. cit.*, p. 1294), and "very pale yellow" (*loc. cit.*, p. 1290). The substances I describe in this paper show clearly enough that there may be a group of compounds prepared by a general method and alike in chemical behaviour, some members of which are intensely coloured, whilst others are snow-white. Equally definite in its bearing on the hydrazones of camphorquinone is the latest work of Stobbe, who shows that not only does the substitution of phenyl for a methyl group in tetramethylfulgide lead from a colourless to a coloured substance, but that a similar change occurs in the trisubstituted series, for whilst *iso*propyldimethylfulgide is colourless, dimethylphenylfulgide, dimethyl-*p*-tolylfulgide, and dimethylcumylfulgide are yellow (*Ber.*, 1905, **38**, 3893), dimethylfurylfulgide being orange and difurylfulgide reddish-brown (*loc. cit.*, p. 4075). Unless an alternative representation of these coloured compounds is forthcoming, for which at present there is no chemical demand whatever, the obvious conclusion is that the appearance or absence of colour in a series of substances constituted alike is due to the character of the group introduced into the potentially chromogenic nucleus. It is always possible that more delicate experiments and a clearer insight into the geometrical requirements will invalidate the Hantzsch-Werner hypothesis, but the investigation of Armstrong and Robertson leaves it unshaken.

Turning now from the question of colour, there remain only one or two points to be mentioned in connection with the diazo-ψ-semicarbazines. During some experiments with mercury acetamide (Trans., 1898, **73**, 783), I found that diazoaminobenzene combines with that substance, forming a pale yellow, sparingly soluble compound having the composition $C_{12}H_{11}N_3,Hg(NHAc)_2$, and so stable in character that, unlike diazoaminobenzene itself, it has remained unchanged in appearance during the interval of eight years, although exposed to light throughout that period. Mercury acetamide combines also with every one of the diazo-ψ-semicarbazines described in this paper, forming substances which, with the single exception of that derived from the orthonitrobenzene compound, are insoluble in cold alcohol; they are all darker in colour than the original diazo-ψ-semicarbazines, and those obtained from the colourless members of the series are yellow.

With regard to the colourless diazo-ψ-semicarbazines themselves, it has not been found possible to remove every trace of colour by crystallisation only, but the toluene, parachlorobenzene, and parabromobenzene compounds were obtained snow-white by suspending them in glacial acetic acid, with which all the diazo-ψ-semicarbazines form sparingly soluble salts, and adding a small proportion of zinc dust; in each case, the rotation of the decolorised substance has been compared with that of the original material to make sure that no fundamental change had taken place. It might be argued that possibly the coloured members of the series also could be rendered colourless by this process. In the case of the benzenediazo-ψ-semicarbazine, faintly yellow crystals are obtained always, and the nitrobenzene and paramethoxybenzene compounds undergo some profound change which precludes application of the method; but it is certain that the paranitrobenzene derivative is red, because it has been prepared independently from the pale yellow benzenediazo-ψ-semicarbazine by adding solid sodium nitrite to a suspension of the acetate in glacial acetic acid.

The regularity with which the new substances are resolved into camphoryl-ψ-carbamide and the corresponding derivative of phenylazoimide suggests this as a method for preparing substituted azoimides from such bases as will not withstand the action of bromine. In this connection, it is interesting to notice that the osmophoric value of the triazo-group is that of anise. Phenylazoimide has not that smell, but the metanitro- and paranitro-derivatives have the odour faintly, p-tolyl-azoimide and chlorophenylazoimide displaying it to a marked extent, whilst p-methoxyphenyl-(anisyl)azoimide smells almost as strongly as anethole itself, although the anise perfume is less fragrant and more pungent than with that substance. In view of the superior osmophoric effect of ortho-substituted derivatives of benzenoid compounds when compared with the meta- and para-isomerides, it is remarkable that orthonitrophenylazoimide should be odourless.

EXPERIMENTAL.

Benzenediazo-ψ-semicarbazinocamphor,

$$C_8H_{14} {<} {\begin{matrix} CH \text{------} N \cdot NH \cdot N \vdots N \cdot C_6H_5 \\ | \\ C(OH) \cdot NH \cdot CO \end{matrix}} .$$

Five grams of aniline in 75 c.c. of 10 per cent. hydrochloric acid were diazotised with 20 c.c. of 20 per cent. sodium nitrite solution and added to 15 grams of camphoryl-ψ-semicarbazide nitrate in 100 c.c. of water. The colourless precipitate which separated almost immediately was considerably augmented by sodium acetate, and, becoming too bulky to be filtered conveniently with the pump, was drained on an ordinary filter and roughly dried on porous earthenware. In this condition, the substance is freely soluble in cold alcohol, but if left in the desiccator during twelve hours it becomes yellow, granular, and quite wet, dissolving much less freely in spirit. The separation of water, accompanied by the change in colour, suggests that the material precipitated initially consists of a hydrate of the diazo-ψ-semicarbazine.

It is best purified by dissolving in cold alcohol and precipitating with water, when minute, dark yellow needles are formed. The colour is largely due to some staining impurity, which is removed by suspension with zinc dust in glacial acetic acid ; on adding much water, extracting the filtered mixture of zinc and diazo-ψ-semicarbazine with alcohol, and diluting the filtrate with water, minute, prismatic needles are obtained, which are very faintly yellow and appear colourless when powdered. The melting point, at which the substance also undergoes profound decomposition, is very indefinite, and, as might be expected, depends largely upon the rate at which the temperature is raised ; the highest observed is 191°, but several specimens have melted a few degrees lower.

0·1598 gave 0·3613 CO_2 and 0·1012 H_2O. $C = 61·66$; $H = 7·04$.

0·1770 ,, 33·1 c.c. nitrogen at 21° and 764 mm. $N = 21·40$.

$C_{17}H_{23}O_2N_5$ requires $C = 62·01$; $H = 7·00$; $N = 21·27$ per cent.

Benzenediazo-ψ-semicarbazinocamphor appears to be quite stable in the solid state when protected from light, but it gradually becomes deep yellow when exposed to sunlight ; the same agency effects in the dissolved substance the resolution into camphoryl-ψ-carbamide _and phenylazoimide already mentioned as being brought about more quickly by alkalis. This appears from the polarimetric examination. A solution containing 0·1800 gram dissolved in 25 c.c. of absolute alcohol was very faintly yellow when freshly prepared, and gave a_D 3°21′ in a 2-dcm. tube, whence $[a]_D$ 232·6° ; the colour gradually became intensified during twenty days, at the end of which period the solution was slightly lævorotatory (camphoryl-ψ-carbamide has $[a]_D$ − 13·5°). On allowing the solvent to evaporate, the crystalline residue had a distinct odour of phenylazoimide, and on recrystallisation from hot water was found to consist of camphoryl-ψ-carbamide, melting at 194° with decomposition. In order to be sure that the process of purification by means of zinc and acetic acid has no effect on the substance itself, the specific rotatory power of the original material was determined in alcohol and found to be $[a]_D$ 235·8° for a 1·2 per cent. solution.

The diazo-ψ-semicarbazine is insoluble in boiling benzene and petroleum, and dissolves sparingly in cold chloroform ; alcohol and ethyl acetate dissolve it readily. When treated with an ammoniacal solution of silver oxide, a yellow coloration is developed, and slight reduction takes place on continued boiling. Concentrated nitric acid dissolves the substance with effervescence, developing an intense blue coloration which rapidly changes to dark green ; concentrated sulphuric acid also dissolves it with gas evolution, producing a red colour, which changes to green on dilution. There is no change of colour when the solution in pyridine is warmed with β-naphthol, and there is no precipitate in the concentrated alcoholic solution with picric acid or platinic chloride ; Fehling's solution is not reduced on boiling.

Attempts to prepare benzoyl and phenylcarbamide derivatives have been unsuccessful, but a nitroso-compound is produced when a solution of sodium nitrite is added to a suspension of the diazo-ψ-semicarbazine in ice-cold dilute acetic acid. If, however, the solid salt is allowed to act on the substance in glacial acetic acid, no steps being taken to reduce the temperature, an intense red coloration is developed, and, on diluting the liquid with water, p-nitrobenzenediazo-ψ-semicarbazino-camphor is precipitated, identical with the product from diazotised p-nitroaniline and camphoryl-ψ-semicarbazide nitrate.

Cold acetic acid converts the benzenediazo-ψ-semicarbazine into an unstable acetate which is dissociated by water, but crystallises from a warm solution in the glacial acid, forming slender, lustrous, snow-white needles ; if the solution is heated to about 60°, no crystals separate on cooling, and, on neutralising the liquid with sodium carbonate, phenylazoimide is precipitated. A compound is formed also with pyridine, stable only in excess of the base ; when the diazo-ψ-semicarbazine is covered with that substance, a slight rise of temperature occurs, and a clear solution is first formed, from which minute, deep yellow crystals separate on agitation ; but when the product is spread on earthenware, it changes to a yellow, resinous material from which nothing crystalline can be obtained.

It has not been found possible to subject the diazo-ψ-semicarbazine to a change analogous to the scission which diazoamines undergo when treated with concentrated acids. Ice-cold dilute hydrochloric acid is without effect, but the concentrated acid changes the colour of the substance to chocolate-brown ; the filtrate, however, does not yield the benzeneazo-derivative with β-naphthol, and the colour of the solid appears to be due to stain only, being removed by ether, after which treatment the residue has all properties of the original material. Hot dilute hydrochloric acid gives rise to phenylazoimide.

Action of Mercury Acetamide.—The benzenediazo-ψ-semicarbazine was dissolved in cold absolute alcohol and treated with a solution of mercury acetamide in 50 per cent. alcohol ; the pale yellow colour changed immediately to very dark yellow, and after a few seconds a dull yellow precipitate separated suddenly. This was filtered, washed with absolute alcohol, dried in the desiccator, redissolved in chloroform, and precipitated with light petroleum.

The substance has no definite melting point, and is insoluble in boiling petroleum, ether, and absolute alcohol ; it dissolves sparingly in hot ethyl acetate, readily in hot benzene, and freely in cold chloroform. It does not yield phenylazoimide with boiling alkali, and hot dilute sulphuric acid liberates gas and produces camphor. A minute quantity of the substance develops a pure copper sulphate blue coloration with concentrated nitric acid, quite distinct from that of the diazo-ψ-semicarbazine itself ; the colour with concentrated sulphuric acid is dark brown, and does not become green on dilution.

It will be worth while to investigate this compound more closely, because it is clearly something more than an additive compound of the type furnished by a mercuric salt with a base. Two determinations of nitrogen gave 11·9 per cent., instead of 15·2 per cent. required by the formula $C_{17}H_{23}O_2N_5 + Hg(NHAc)_2$, and the chloroform solution displayed pronounced mutarotation, the initial observation giving $[a]_D$ 323°, which changed in thirty-six hours to $[a]_D$ 474°.

Decomposition by Alkali.—When the diazo-ψ-semicarbazine is warmed with 10 per cent. aqueous potassium hydroxide, it melts to a heavy, red oil, and a sweet, intensely pungent odour becomes perceptible immediately. Five grams of the compound were suspended in 20 c.c. of the alkali and submitted to a current of steam, which carried over 1·5 c.c. of a heavy, pale yellow oil, insoluble in cold dilute acids, indifferent to hot Fehling's solution and to sodium hypochlorite. The liquid which remained in the distilling flask became pasty on cooling, from separation of lustrous crystals consisting of camphoryl-ψ-carbamide; this product melted and decomposed at 194° and furnished the characteristic nitroso-derivative, which, after crystallisation from benzene, melted at 158°.

Before the volatile oil was recognised as phenylazoimide, 10 c.c. were distilled, boiling at 161—162°, with slight decomposition under 754 mm.; no explosion occurred, but the operation has not been repeated. A specimen of the paranitro-derivative, prepared by the action of warm nitric acid, crystallised from petroleum in straw-coloured leaflets melting at 71° and containing 34·1 per cent. of nitrogen ($C_6H_4O_2N_4$ requires $N = 34·14$ per cent.).

o-*Nitrobenzenediazo-ψ-semicarbazinocamphor,*

$$C_8H_{14} \begin{cases} CH\text{------}N \cdot NH \cdot N\dot{:}N \cdot C_6H_4 \cdot NO_2 \ (o) \\ C(OH) \cdot NH \cdot CO \end{cases}$$

On adding 2 grams of diazotised o-nitroaniline to 4 grams of the ψ-semicarbazide nitrate dissolved in water, a bulky, pale red, gelatinous precipitate was formed immediately; on completing the separation by means of sodium acetate, the substance was filtered without the pump, roughly dried, and made into a paste with cold absolute alcohol. When spread on porous earthenware, the product became dry very slowly, and was then crystallised from boiling chloroform, which deposited minute, bright red, silky needles; it melts and decomposes at 160°.

0·1381 gave 26·4 c.c. nitrogen at 17° and 774 mm. $N = 22·60$.

$C_{17}H_{22}O_4N_6$ requires $N = 22·46$ per cent.

The substance is insoluble in boiling light petroleum and only very slightly soluble in warm ether and boiling benzene; it dissolves readily in hot ethyl acetate, from which it crystallises in minute, red needles. When covered with cold chloroform, the compound remains at first suspended, quickly changing to a gelatinous pulp which dissolves only in the boiling solvent. The diazo-ψ-semicarbazine dissolves in concentrated nitric acid without effervescence, forming a red solution which remains clear on dilution, concentrated sulphuric acid also dissolves it, destroying the colour and liberating gas. When the alcoholic solution is treated with ammoniacal silver oxide, a deep red coloration is developed, and on continued boiling reduction takes place, but Fehling's solution is left unaltered. The orthonitro-compound is the only diazo-ψ-semicarbazine yet examined which gives no precipitate with mercury acetamide in 50 per cent. alcohol; a deep red colour is developed, but the alcoholic liquid remains clear until much water is added.

Decomposition by Alkali.—Owing to the extremely bulky character of the orthonitro-compound, the decomposition by alkali is attended with mechanical difficulties. On passing steam through a suspension of the substance in very dilute aqueous caustic potash, a very pale yellow odourless oil distilled slowly; it solidified in melting ice, and, when recrystallised from light petroleum, furnished hard, pale yellow prisms fusing at 53°. The melting point of *o*-nitrophenylazoimide is stated to be 51—52°.

m-*Nitrobenzenediazo-ψ-semicarbazinocamphor*,

$$C_8H_{14} \begin{cases} CH \text{———} N \cdot NH \cdot N{:}N \cdot C_6H_4 \cdot NO_2 \ (m) \\ C(OH) \cdot NH \cdot CO \end{cases}.$$

The metanitro-derivative was precipitated immediately on adding the diazotised base to the ψ-semicarbazide salt, and required but little sodium acetate to complete the separation. The deep yellow substance was filtered, drained, and crystallised from hot alcohol, in which it is moderately soluble, forming minute, orange-yellow needles which soften at about 135° and melt at 170° with decomposition.

0·0974 gave 18·8 c.c. nitrogen at 15° and 758 mm. N = 22·55.

$C_{17}H_{22}O_4N_6$ requires N = 22·46 per cent.

This compound is also insoluble in boiling light petroleum, and sparingly soluble in warm ether, boiling benzene, and boiling chloroform, crystallising from the last named in silky, yellow needles; it is moderately soluble in ethyl acetate. Dissolution in concentrated nitric acid is attended with vigorous effervescence, and from the brown solution water precipitates a deep red solid; concentrated sulphuric acid liberates gas and forms a pale pink solution which remains clear when diluted. The behaviour of the metanitro-compound towards Fehling's solution and ammoniacal silver oxide is the same as that of the foregoing substance, but with mercury acetamide only a slight darkening takes place, followed almost immediately by the separation of an orange precipitate.

Decomposition by Alkali.—Steam was passed into hot water containing a few drops of 10 per cent. aqueous caustic potash and having the metanitro-derivative in suspension; the pale yellow solid which passed over had a faint odour of anise, and crystallised from petroleum in long, lustrous, straw-coloured needles melting at 55°. This is the temperature at which *m*-nitrophenylazoimide melts according to Nölting and Grandmougin.

p-*Nitrobenzenediazo-ψ-semicarbazinocamphor*,

$$C_8H_{14} \big\langle{}^{CH\underline{\quad\quad}N \cdot NH \cdot N \vdots N \cdot C_6H_4 \cdot NO_2 \; (p)}_{C(OH) \cdot NH \cdot CO}$$

On mixing the diazonium salt with ψ-semicarbazide nitrate, the diazo-ψ-semicarbazine separated in the form of a pale red precipitate; by crystallisation from dilute alcohol, it was obtained in aggregates of small, pale red needles, becoming pale yellow at 145° and melting with intumescence somewhat indefinitely at 170°.

0.1435 gave 27.2 c.c. nitrogen at 20° and 766 mm. $N = 21.85$.

$C_{17}H_{22}O_4N_6$ requires $N = 22.46$ per cent.

The paranitro-derivative behaves just like the isomeric substances towards light petroleum, ethyl acetate, and ether; it combines with benzene, which dissolves it fairly readily at first, quickly depositing bulky, orange flocks, which require a considerable quantity of the hot hydrocarbon for complete dissolution, finally separating in silky, yellow needles. When dissolved in concentrated sulphuric acid, there is vigorous effervescence, but no characteristic coloration is developed; concentrated nitric acid forms a pale brown solution, and alcoholic potash yields a rich magenta-coloured solution, which remains clear when diluted with water, regenerating the yellow nitro-derivative with acids. Fehling's solution is unaffected by the paranitro-compound, which develops a deep red coloration with ammoniacal silver oxide; there is no reduction, however, even on boiling, but a red, crystalline precipitate separates slowly. With mercury acetamide, a pale red colour is developed, quickly followed by a red precipitate.

Decomposition by Alkali.—When the paranitrobenzenediazo-ψ-semi-carbazine is covered with 40 per cent. aqueous caustic potash, its colour changes to that of iodine, but it does not dissolve until water is added, when an intense permanganate-coloured solution is formed. In order to resolve the substance into camphoryl-ψ-carbamide and *p*-nitro-phenylazoimide, however, a trace of alkali is sufficient; the product of steam distillation with alkali crystallises from petroleum in lustrous, straw-coloured leaflets having a faint odour of anise, melting at 71°, and becoming red on exposure to light. A determination of nitrogen gave 34.0 per cent., $C_6H_4O_2N_4$ requiring $N = 34.14$ per cent. Bamberger and Renauld have recorded the same melting point, which was not depressed by admixture with a specimen of *p*-nitrophenylazoimide prepared by nitrating phenylazoimide.

p-*Toluenediazo-ψ-semicarbazinocamphor*,

$$C_8H_{14}\!\!<_{\substack{CH \text{———} N \cdot NH \cdot N \vdots N \cdot C_6H_4 \cdot CH_3 \, (p) \\ C(OH) \cdot NH \cdot CO}}$$

Six grams of p-toluidine were diazotised and added to 16 grams of camphoryl-ψ-semicarbazide nitrate dissolved in water. The very faintly yellow precipitate was drained and crystallised first from dilute spirit and then from absolute alcohol, with which it combines, forming pale brown, lustrous, transparent, hexagonal prisms; when freshly withdrawn from the mother liquor, the substance dissolves in its alcohol of crystallisation at 80°, when it loses the solvent, solidifies, and melts at about 163° with decomposition. After several weeks in the desiccator, the crystals are found to have become opaque, but retain their lustre; the substance then undergoes no preliminary fusion at 80°, but intumesces and melts indefinitely at about 145°.

0·1715 gave 30·1 c.c. nitrogen at 17° and 768 mm. N = 20·58.

$C_{18}H_{25}O_2N_5$ requires N = 20·41 per cent.

By the method of purification with zinc and acetic acid already described in dealing with the benzenediazo-ψ-semicarbazine, the toluene compound has been obtained snow-white; a solution containing 0·3468 gram in 25 c.c. of absolute alcohol gave a_D 6°10′ in a 2-dcm. tube, whence $[a]_D$ 222·2°, whilst the undecolorised substance gave $[a]_D$ 221·9°. Vivid colours are developed by concentrated sulphuric and nitric acids, the former being purple, changing to greenish-blue on dilution, whilst nitric acid produces a transient green colour which changes to brown, and remains brown on dilution.

The toluene compound dissolves sparingly in boiling petroleum, from which minute needles crystallise on cooling; it is freely soluble in cold ether, benzene, chloroform, and ethyl acetate. It does not reduce Fehling's solution, but ammoniacal silver oxide produces a yellow precipitate and is reduced vigorously on boiling. Mercury acetamide dissolved in 50 per cent. alcohol develops a deep yellow colour in the alcoholic solution, and after an interval of a few seconds a bright yellow precipitate separates quite suddenly.

Decomposition by Alkali.—The recrystallised diazo-ψ-semicarbazine was suspended in very dilute alkali and steamed, when a heavy, pale yellow oil passed over, having an agreeable perfume of anise. p-Tolylazoimide does not appear to have been described, and it will be necessary to characterise the product more fully; in the meantime, it may be stated that when distilled under atmospheric pressure it undergoes considerable decomposition at about 180°.

p-*Chlorobenzenediazo-ψ-semicarbazinocamphor,*

$$. C_8H_{14} \Big\langle \begin{matrix} CH \text{————} N \cdot NH \cdot N \dot{:} N \cdot C_6H_4 \cdot Cl \ (p) \\ C(OH) \cdot NH \cdot CO \end{matrix} \Big.$$

Five grams of p-chloroaniline were diazotised and added to 11·4 grams
of the ψ-semicarbazide nitrate dissolved in water ; a cream-coloured
turbidity was produced, increasing to a bulky precipitate when sodium
acetate was added. On crystallising the drained substance twice from
dilute alcohol, it was obtained in silky needles which could be scarcely
called coloured, and yet were not snow-white, but by suspending the
substance in glacial acetic acid, which converts it first into a sparingly
soluble acetate, and then leaving it with zinc dust during twenty-four
hours, recrystallisation from dilute alcohol gave snow-white, silky
needles which melt and decompose at 157°, shrinking several degrees
beforehand.

0·2115 gave 35·8 c.c. nitrogen at 20° and 753 mm.   N = 19·18.
0·1873 gave 0·0754 AgCl.   Cl = 9·94.
$C_{17}H_{22}O_2N_5Cl$ requires N = 19·25 ; Cl = 9·76 per cent.

The substance is moderately soluble in boiling petroleum, separating
as a crystalline meal on cooling ; it dissolves freely in cold ether,
chloroform, ethyl acetate, and benzene. A solution containing 0·0972
gram in 25 c.c. of absolute alcohol resembled plain alcohol in appearance, there being not the faintest suggestion of colour when viewed in
the 2-dcm. tube ; it gave a_D 1°39′, whence $[a]_D$ 212·2°. To show that
the treatment with zinc and acetic acid produces no effect beyond
purification, the original substance was examined in the polarimeter, a
1·1 per cent. solution in absolute alcohol giving $[a]_D$ 214·3°, and a
nitrogen determination 19·76 per cent.

The diazo-ψ-semicarbazine has no effect on Fehling's solution, but
yields an orange precipitate with ammoniacal silver oxide which undergoes complete reduction on boiling ; with mercury acetamide, the
colourless, alcoholic solution becomes pale yellow, and a sulphur-
yellow precipitate follows almost immediately. Concentrated sulphuric
acid forms an intense, magenta-coloured solution, which becomes
greenish-blue on dilution ; concentrated nitric acid also dissolves it
with effervescence, the solution being intense carmine. When very
dilute boiling alkali acts on the substance, a heavy oil distils over,
having a pungent odour of anise ; this is doubtless an impure specimen
of p-chlorophenylazoimide, which Griess describes as crystalline, without giving a melting point.

p-*Bromobenzenediazo-ψ-semicarbazinocamphor*,

$$C_8H_{14} \Big\langle \begin{matrix} CH \text{———} N \cdot NH \cdot N \vdots N \cdot C_6H_4 \cdot Br \, (p) \\ C(OH) \cdot NH \cdot CO \end{matrix}$$

The diazo-ψ-semicarbazine was obtained quantitatively in the form of a very pale yellow powder on adding 4 grams of diazotised p-bromo-aniline to 6·8 grams of camphoryl-ψ-semicarbazide dissolved in water. Recrystallisation from dilute alcohol furnished the substance in minute, very faintly yellow needles, but it was obtained snow-white by applying the method of purification employed in previous cases; the melting point is very indefinite, for the substance intumesces at 155—160°. The tendency to combine with alcohol, already noticed in the foregoing diazo-ψ-semicarbazines, is very marked in the case of the p-bromo-compound, so much so that the analyses have yielded low results even with specimens which have remained in the desiccator during several weeks.

0·2290 gave 32·7 c.c. nitrogen at 18° and 770 mm. $N = 16·71$.
0·3410 ,, 0·1506 AgBr. $Br = 18·80$.
$C_{17}H_{22}O_2N_5Br$ requires $N = 17·16$; $Br = 19·60$ per cent.
$C_{17}H_{22}O_2N_5Br + \frac{1}{2}C_2H_6O$ requires $N = 16·24$; $Br = 18·56$ per cent.

A solution containing 0·2738 gram in 25 c.c. of absolute alcohol gave a_D 3°57′ in a 2-dcm. tube, whence $[a]_D$ 180·3°. The substance is sparingly soluble in boiling petroleum, freely, however, in cold ether and ethyl acetate; it dissolves readily also in chloroform and in benzene, crystallising from the latter in white needles. It develops an intense magenta coloration with concentrated nitric acid, and a deep purple with concentrated sulphuric acid; it does not reduce Fehling's solution, and the orange precipitate formed with ammoniacal silver oxide is completely reduced on boiling. With mercury acetamide, the colourless solution becomes dark yellow, and a canary-yellow precipitate separates quickly.

Hot dilute caustic potash resolves the substance into camphoryl-ψ-carbamide and p-bromophenylazoimide, which was obtained as a heavy, pale yellow oil with a pronounced odour of dibromobenzene; the substance solidified on cooling and melted at 20°, the temperature recorded by Griess.

p-*Methoxybenzenediazo-ψ-semicarbazinocamphor*,

$$C_8H_{14} \Big\langle \begin{matrix} CH \text{———} N \cdot NH \cdot N \vdots N \cdot C_6H_4 \cdot OCH_3 \, (p) \\ C(OH) \cdot NH \cdot CO \end{matrix}$$

Five grams of p-anisidine were diazotised and added to an aqueous solution of the ψ-semicarbazide nitrate (11·7 grams). The diazo-ψ-semicarbazine separated very slowly, even after adding sodium acetate, and this was the only case in which the yield was not quantitative. The substance, when dry, was washed with a mixture of ether and petroleum (1:2) until the filtrate was colourless, the undissolved material being then extracted with hot absolute alcohol; on diluting the filtrate, minute, brown needles were obtained, melting at 166° with decomposition.

0·1180 gave 19·5 c.c. nitrogen at 14° and 770 mm. $N = 19·70$.

$C_{18}H_{25}O_3N_5$ requires $N = 19·50$ per cent.

The substance could not be obtained colourless by treatment with zinc dust and acetic acid, because some profound alteration took place instead, an intense brown liquid being produced.

It is insoluble in boiling light petroleum and in benzene, being only very slightly soluble in ether; it is freely soluble in ethyl acetate and moderately in chloroform. It has no action on Fehling's solution, but reduces ammoniacal silver oxide. Concentrated nitric acid forms a deep brown solution, whilst concentrated sulphuric acid develops a purple coloration which changes to Prussian blue on dilution with water. When warmed with aqueous alkali, a heavy oil is produced which has a powerful odour of anise; this is doubtless p-methoxy-phenylazoimide, which has not yet been characterised.

Since the foregoing observations were made, Mr. E. C. C. Baly has been good enough to examine decinormal solutions of the *para*toluene (colourless) and *ortho*nitrobenzene (red) derivatives in the spectroscope. His report is in the following terms :—" The absorption spectra of *para*toluenediazo-ψ-semicarbazinocamphor and *ortho*nitrobenzenediazo-ψ-semicarbazinocamphor are shown in the figure.

" It will be seen that a well-marked absorption band in the visible blue region is developed in the second case. This band is the origin of the colour of this substance, and is the result of a process of isorropesis such as occurs in the nitroanilines and nitrophenols. In the case of the *para*toluene compound, no such band is developed, showing that the process of isorropesis is absent. The difference in colour of these compounds can in no way be considered as evidence that they differ fundamentally in constitution; the conditions for isorropesis may exist perfectly well in each substance, while the influence necessary to start the oscillation may be absent in the *para*toluene compound. A similar pair of compounds is met with in diacetyl and diacetyldioxime. It may be pointed out that the band in the ultra-violet given by *para*toluenediazo-ψ-semicarbazinocamphor may be traced to the *para*toluene nucleus; this band is absent from the spectrum of *ortho*nitrobenzenediazo-ψ-semicarbazinocamphor, owing, no doubt, to the benzene nucleus being restrained in its motions by the nitro-group."

Oscillation frequencies.

18 2000 22 24 26 28 3000 32 34 36 38 4000 42 44

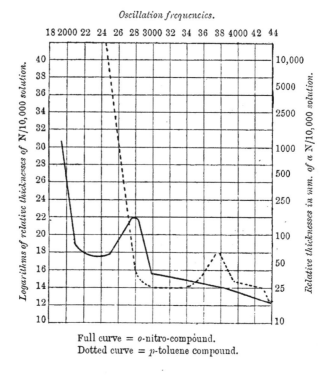

Full curve = *o*-nitro-compound.
Dotted curve = *p*-toluene compound.

I am greatly indebted to Mr. Baly for making this examination, the result of which appears to substantiate the views expressed in the introductory portion of this paper.

ROYAL COLLEGE OF SCIENCE, LONDON,
    SOUTH KENSINGTON, S.W.

XXIX.—*Hydroxylamine-αβ-disulphonates* (*Structural Isomerides of Hydroximinosulphates or Hydroxylamine-ββ-disulphonates*).

By TAMEMASA HAGA.

IN a previous communication to the Society (Trans., 1904, **85**, 78), it was shown that Fremy's *metasulphazilate*, which until then had been considered to be constituted as an amine oxide, $O\overset{..}{:}N(SO_3K)_3$, is in reality a *hydroxylaminetrisulphonate*. The nature of the products of the proximate hydrolysis of the metasulphazilates seems to afford the strongest additional evidence that these salts are hydroxylaminetrisulphonates, and, as such, mixed anhydrides of an acid-sulphate and a hydroxylaminedisulphonate.

The ultimate hydrolysis of a hydroxylaminetrisulphonate, through intermediate stages, into hydroxylamine and an acid-sulphate is difficult to carry to completion (Trans., 1904, **85**, 97), but the first stage of it, into a hydroxylaminedisulphonate and one-third of the quantity of acid-sulphate which is produced in the ultimate hydrolysis, is very easily accomplished. The hydroxylaminedisulphonate thus obtained proves to be an entirely new salt, structurally isomeric with the corresponding *hydroxylaminedisulphonate* (Fremy's *sulphazotate*), from which it differs greatly in essential physical and chemical properties. The one is a *ββ*-derivative, the other an *αβ*-derivative of hydroxylamine. Which, is determined by the behaviour of the two salts towards sodium amalgam. A salt of the series already known remains unaffected, whilst one of the new series decomposes (p. 247) into sulphate and aminemonosulphonate (sulphamate, aminosulphate), just as its parent salt, hydroxylaminetrisulphonate, decomposes, in like circumstances, into sulphate and aminedisulphonate (iminosulphate). It must therefore have the αβ-constitution, as shown in the following equation, framed to express its decomposition by sodium :

$$(SO_3Na) \cdot O \cdot NH(SO_3Na) + H_2O + 2Na =$$
$$(SO_3Na) \cdot ONa + NH_2(SO_3Na) + NaOH,$$

and the other salt must have the *ββ*-constitution, $HON(SO_3Na)_2$, which it has always been assumed to have. The activity of the *αβ*-salts towards sodium amalgam serves also to demonstrate their sulphatic constitution (derived from that of the hydroxylaminetrisulphonates) as the mixed anhydrides * of an acid-sulphate and a hydroxylamine-monosulphonate.

* Two other examples of such mixed sulphatic anhydrides are any hydroxyl-aminetrisulphonate and any hyponitrososulphate. Raschig has recently adduced evidence (*Zeit. angew. Chem.*, 1905, **18**, 1309) that nitrosyl sulphate is, after all, not a sulphate, but a nitrosulphonate, $O_2N \cdot SO_3H$.

The action of potassium hydroxide on the two salts seems clearly to establish the difference there is in their constitution. The old salt reverts to nitrite and sulphite when it is left, even in the cold, in a concentrated solution of the alkali (Trans., 1894, **65**, 539), whereas the new salt is only incompletely decomposed into sulphate, aminemono-sulphonate, and nitrogen (p. 248) after several hours' digestion at 100—125° with the alkali. These results are exhibited by the follow-ing equations, the upper for the old or $\beta\beta$-salt and the lower for the new or $\alpha\beta$-salt :

$$\mathrm{HON(SO_3K)_2 + 3KOH = 2(SO_3K)K + NO_2K + 2H_2O} \; ;$$
$$\mathrm{3(SO_3K)ONH(SO_3K) + 5KOH = 5(SO_3K)OK +}$$
$$\mathrm{NH_2(SO_3K) + N_2 + 3H_2O.}$$

The lower of these equations recalls the action of heated alkalis on hydroxylamine (Lossen) and on hydroxylaminemonosulphonate (Claus). The upper equation brings out strongly the oximidic constitution of the $\beta\beta$-salt.

It is an interesting fact that the course of the hydrolysis of the two salts in acidified solution is widely different, but it is a fact which cannot apparently be used to establish the nature of the difference in their constitution. The $\beta\beta$- or long-known salt, by losing one of its two sulphonate groups, readily passes into the hydroxylamine-β-mono-sulphonate, $\mathrm{HONH(SO_3K)}$, whereas the new or $\alpha\beta$-disulphonate passes (p. 249), much less easily, into acid-sulphate and hydroxylamine itself (and products of its well-known decomposition), without ever affording evidence of the production of a monosulphonate, which in this case should have the α-constitution, expressed by $\mathrm{(SO_3K)ONH_2}$, and, as an amidogenium salt, be perhaps incapable of existence. The slowness with which an $\alpha\beta$-salt begins to hydrolyse is illustrated by the fact that a solution of the potassium salt will remain clear for five minutes at the common temperature after it has been mixed with hydrochloric acid and barium chloride, whereas a solution of the $\beta\beta$-salt will almost at once begin to show turbidity. When the hydrolysis of the $\alpha\beta$-salt proceeds in the absence of much or any hydrochloric acid, nitrogen and ammonia very largely take the place of the hydroxylamine, which is obtained in nearly the theoretical quantity when the salt hydrolyses in presence of a sufficiently concentrated hydrochloric acid solution. The following equation expresses what principally happens in the absence of hydrochloric acid :

$$\mathrm{3(SO_3K)ON(SO_3K) + 3H_2O = 5(SO_3K)OH + N_2 + (SO_3K)ONH_4.}$$

Silver oxide and lead peroxide seem to be without action on the $\alpha\beta$-salts in solution ; they certainly do not produce the deeply coloured peroxylaminesulphonate which so strikingly results from their action on the $\beta\beta$-salts. This difference accords with that indicated in the constitution of the two series of salts. A hydroxyl-amine-$\alpha\beta$-disulphonate also unexpectedly agrees with a hydroxyl-amine-$\beta\beta$-disulphonate in not reducing copper or silver oxide in alkaline solution. It is also inactive on a solution of iodine in presence of sodium acid-carbonate. A hydroxylaminemono-sulphonate, $HONH(SO_3K)$, which, like it, has an aminic hydrogen atom in its constitution, has the activity of hydroxylamine on both alkaline copper solution and on alkali bicarbonate iodine solution.

Like the $\beta\beta$-salts, the hydroxylamine-$\alpha\beta$-disulphonates decompose with gentle explosion when heated. Also like the $\beta\beta$-series, that of the $\alpha\beta$-disulphonates includes highly alkaline normal salts, such as $(SO_3K)ONK(SO_3K)$ and $(SO_3Na)ONNa(SO_3Na)$.

A concentrated solution of the disodium salt is not precipitated by silver nitrate, mercuric nitrate, lead nitrate, or barium chloride. A concentrated solution of basic lead acetate precipitates from it an oil which becomes crystalline on standing. A concentrated solution of a potassium salt precipitates from it the very much less soluble potassium salt. Barium hydroxide gives a voluminous, apparently amorphous, precipitate, probably of a sodium barium salt.

The molecular magnitude of the normal or trisodium hydroxyl-amine-$\alpha\beta$-disulphonate (p. 246), as determined cryoscopically by means of melted Glauber's salt (Löwenherz), is that expressed by $O_7NS_2Na_3$, the same, therefore, as that found for the normal sodium $\beta\beta$-salt (Trans., 1904, 85, 100—101).* Remarkably, however (and it is a unique experience with this method, so far as has been ascertainable), the depression of the solidifying point of the sodium sulphate is at first much less than that which corresponds with the simple molecular weight, the number for which it only reaches, and remains steady at, in the course of an hour or two and after several repetitions of remelt-ing and solidifying. It would seem from this that the solid salt consists of associated simple molecules which require time to separate from each other after dissolution in melted Glauber's salt.

The discovery of this new series of salts, establishing as it does the existence of significant structural isomerism in other than carbon compounds, should prove to be of very special interest, there being hardly any other instance known, except that of nitramine with hypo-nitrous acid, the existence of which is disputed by Hantzsch (*Zeit. anorg. Chem.*, 1898, 19, 106) just because it would be the only case known in inorganic chemistry.

The subjoined scheme of equations may serve to show at a glance

* There is an error in the calculated number given there ; it should read 259·3 instead of 239·3. Other errors occurring there are to be found in the list of errata since published.

the relation by derivation of the new series of salts to the old series. A hydroxylamine-$\beta\beta$-disulphonate, a salt formed by the union of nitrous acid with a metasulphite, is oxidisable wholly into a peroxylamine sulphonate. This, by hydrolysis in presence of an alkali, becomes, to the extent of half its nitrogen, hydroxylaminetrisulphonate ; to the extent of a fourth of its nitrogen, the $\beta\beta$-salt again ; and, to the extent of the remaining fourth of its nitrogen, nitrous acid (nitrite) again. Lastly, by acid hydrolysis, the hydroxylaminetrisulphonate becomes sulphate and a hydroxylamine-$a\beta$-disulphonate. From this it will be seen that, at most, only half of the $\beta\beta$-salt comes out as the $a\beta$-salt, one-fourth of it being regenerated and the remaining fourth reverting to its parent salts, of which the sulphite has suffered oxidation to sulphate.

$$4HNO_2 + 4S_2O_5K_2 = 4HON(SO_3K)_2 \ (\beta\beta\text{-salt}) ;$$
$$4HON(SO_3K)_2 \ (\beta\beta\text{-salt}) \ + 2PbO_2 = 2[ON(SO_3K)_2]_2 + 2Pb(OH)_2 ;$$
$$2[ON(SO_3K)_2]_2 + H_2O = 2(SO_3K)ON(SO_3K)_2 + HON(SO_3K)_2 \ (\beta\beta\text{-salt}) + NO_2H ;$$
$$2(SO_3K)ON(SO_3K)_2 + 2H_2O = 2(SO_3K)ONH(SO_3K) \ (a\beta\text{-salt}) + 2HO(SO_3K).$$

The return of an $a\beta$-salt to the state of its $\beta\beta$-isomeride can hardly be regarded as possible, its own production having been due to oxidation of a fourth of the sulphonate groups into acid-sulphate.

Salts.

Dipotassium Hydroxylamine-$a\beta$-disulphonate, $(SO_3K)ONH(SO_3K)$.— Potassium hydroxylaminetrisulphonate, dissolved in ten times its weight of warm water (it is much less soluble in cold water, Trans., 1904, **85**, 83), soon begins to hydrolyse when its solution is quickly cooled and mixed with a drop of dilute sulphuric acid just before it would otherwise crystallise out again. The hydrolysis, to the end of its first stage, is complete in about four days. When the solution is deprived of sulphate and neutralised by the addition of barium carbonate or hydroxide, filtered, and evaporated, the new disulphonate is obtained in crystals to the extent of at least two-thirds of the calculated yield. That hydrolysis of the trisulphonate proceeds to the extent shown in the equation (see above) in about four days, and then proceeds much more slowly, has been ascertained both acidimetrically and by estimation of the sulphuric acid produced.

The dipotassium salt is an anhydrous salt, about twice as soluble in water as the corresponding (but hydrated) $\beta\beta$-salt. Of it, 6·44 parts at 16·4°, 7·18 parts at 17·8°, and 8·05 parts at 20° dissolve in 100 parts of water. Its solution is neutral to litmus, to methyl-orange, and to phenolphthalein.

The salt forms hard, monoclinic, prismatic crystals, which are sometimes short, thick prisms, sometimes flattened tables, and sometimes long, slender prisms or needles. Crystals of the one or other habit generally recrystallise in that habit, but the salt is not dimorphous. A saturated solution of one form of the salt is saturated also towards another form of it, whilst the two forms will lie side by side unchanged for a length of time in the same mother liquor. Prof. Jimbo has kindly supplied the following account of his examination of a short, thick prism : a monoclinic crystal, developed perfectly on one end of the clinodiagonal, about 6 mm. long, was measured by means of a contact goniometer, only two angles, ce and ed, having been measured by reflection. The faces a and g were depressed ; the other faces also did not give good images. Seven faces were recognised ; one other could not be determined.

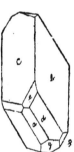

$a.\quad \infty P\overline{\infty}\qquad bd\ =\ 133°25'.$
$b.\quad \infty P\overset{\diagdown}{\infty}\qquad ce\ =\ 144°27'.$
$c.\quad 0P\qquad cf\ =\ 127°53'.$
$d.\quad \infty P\qquad ed\ =\ 141°25'.$
$e.\quad -P\qquad df\ =\ 126°15'.$
$f.\quad P(?)\qquad cg\ =\ 117°.$
$g.\quad \tfrac{1}{4}P\overline{\infty}(?).$

The results of analyses of (A) the tabular form and of (B) the acicular form of the salt are as follows :

A. 0·2863 gave 0·1845 K_2SO_4.   K = 28·98.
    0·2105 ,, 0·3673 $BaSO_4$.   S = 23·96.
    0·2739 ,, 12·58 c.c. moist nitrogen at 19·5° and 756·1 mm.
    N = 5·22.

B. 0·2679 gave 0·1715 K_2SO_4.   K = 28·74.
    0·2684 ,, 11·9 c.c. moist nitrogen at 16·7° and 765·5 mm·
    N = 5·19.

    $HO_7NS_2K_2$ requires K = 29·06 ; S = 23·79 ; N = 5·21 per cent.

In all analyses given in this paper, sulphur was determined by heating the salt with hydrochloric acid in a sealed tube at 180° for five hours or at 200° for two hours.

Tripotassium Hydroxylamine-αβ-disulphonate.—This hydrated salt, $(SO_3K)ONK(SO_3K),2H_2O$, is precipitated, at first as an oil, when alcohol is added to its concentrated aqueous solution, prepared by dissolving the disulphonate in a little hot water and adding to it the calculated quantity of potassium hydroxide solution. The oily salt slowly solidifies into lumps of microscopic, crystalline plates. The quantity obtained should be about equal in weight to that of the disulphonate used, the calculated quantity being five parts from four. It is very soluble in water. Its solution is not precipitated by barium chloride, in this respect being unlike a solution of the corresponding $\beta\beta$-salt. It is caustic in taste and explodes suddenly when heated.

0·1696 gave 0·1286 K_2SO_4. $K = 34·05$.

0·2588 ,, 0·1944 K_2SO_4. $K = 33·73$.

0·1568 ,, 0·2343 $BaSO_4$. $S = 20·52$.

$H_4O_9NS_2K_3$ requires $K = 34·18$; $S = 18·66$ per cent.

The *disodium* salt, $(SO_3Na)ONH(SO_3Na)$, is, like the $\beta\beta$-salt, anhydrous. The very soluble sodium hydroxylaminetrisulphonate is dissolved in five times its weight of water and acidified with dilute sulphuric acid. In two or three days at the ordinary temperature, it will have all hydrolysed, and the solution is then to be neutralised with sodium carbonate. On exposure for a night in the ice-chamber, almost all the sodium sulphate will crystallise out and then the mother liquor can be evaporated to get the new salt. Like the corresponding $\beta\beta$-salt, this salt forms hard masses firmly adhering to the sides of the vessel. These masses are stellar or warty groups of microscopic, thick, rhombic plates. The salt is exceedingly soluble in water, from which it can be nearly all precipitated by alcohol.

Two preparations of the salt were analysed (I and II):

I. 0·2209 gave 0·1313 Na_2SO_4. $Na = 19·28$.

0·1059 ,, 0·2096 $BaSO_4$. $S = 27·17$.

0·5151 ,, 26·45 c.c. moist nitrogen at 17° and 756·8 mm. $N = 5·94$.

II. 0·2509 gave 0·1502 Na_2SO_4. $Na = 19·41$.

0·0959 ,, 0·1910 $BaSO_4$. $S = 27·35$.

$HO_7NS_2Na_2$ requires $Na = 19·43$; $S = 27·02$; $N = 5·92$ per cent.

The *trisodium* salt, $(SO_3Na)ONNa(SO_3Na),2H_2O$, is prepared just in the same way as the tripotassium salt. It is obtained as a crystalline powder. For determination of its molecular magnitude, see p. 246.

0·1450 gave 0·1054 Na_2SO_4. $Na = 23·57$.

0·2269 required 7·56 c.c. $N/10$ HCl with methyl-orange. Alkalinity as $Na = 7·66$.

0·3594 gave 0·5698 $BaSO_4$. $S = 21·75$.

$H_4O_9NS_2Na_3$ requires $Na = 23·41$; Na (alkalinity) $= 7·8$; $S = 21·71$ per cent.

Diammonium Salt, $(SO_3NH_4)ONH(SO_3NH_4)$.—Ammonium hydroxyl-aminetrisulphonate (Trans., 1904, 85, 84) hydrolyses in the same way as the sodium salt. The ammonium acid-sulphate is got rid of by adding just enough barium hydroxide solution, filtering, and evaporating, at first at a gentle heat and then in the cold, over sulphuric acid under reduced pressure. It occurs as small, thick plates, which are somewhat hard, and as nodules composed of minute, tabular crystals. It is a very soluble salt, three parts dissolving normally in just two parts of water at 18°, but it is very apt to form supersaturated solutions. It is a more stable salt than the corresponding $\beta\beta$-compound. Its crystals are probably anhydrous, but those analysed showed the presence of 0·25 H_2O per molecule.

0·3559 gave 0·7189 $BaSO_4$.   S = 27·74·

0·2820 „ 35·5 c.c. moist nitrogen at 16·2° and 754·5 mm. N = 18·18.

$H_4O_7N_2S_2$ requires S = 28·21; N = 18·53. With $\frac{1}{4}H_2O$ added, it requires S = 27·66; N = 18·17 per cent.

Barium Salts.—Barium salts have not been prepared in a state suited for satisfactory determination of their nature. Evaporation of a solution of the ammonium salt with excess of barium hydroxide in a vacuum over sulphuric acid to a small volume removed all ammonia. After removal of the excess of barium hydroxide by carbon dioxide, the filtered solution was further evaporated in the desiccator. First a viscid liquid and then a bulky, friable, porous mass, devoid of crystalline character, were obtained. The latter was not quantitatively analysed, but it yielded, when hydrolysed, barium sulphate and hydroxylamine sulphate in crystals which were further identified by a very satisfactory sulphuric acid determination. The product was therefore undoubtedly a barium hydroxylamine-$\alpha\beta$-disulphonate. By using less barium hydroxide, crystallised ammonium barium salts of varying composition may be obtained. One of these, when quantitatively examined, had a composition corresponding fairly well with that of a compound of 6 mols. of the diammonium salt with 1 mol. of the 2·3-normal barium salt.

Molecular Magnitude of Trisodium Hydroxylamine-$\alpha\beta$-disulphonate.

By Löwenherz's method, the molecular magnitude of the trisodium hydroxylamine-$\beta\beta$-disulphonate has been found to be (anhydrous) 253 and 259·6, whilst $O.NS_2Na_3$ requires 259·35 (Trans., 1904, 85, 101). By the same method, the following approximations to the same number have been obtained for the $\alpha\beta$-salt, namely, 269·6, 279·3, and 256·4, using the constant, 32·6, for sodium sulphate found by Löwenherz. The details of these determinations of the molecular magnitude of the $\alpha\beta$-salt are specially interesting (p. 242). Of this salt, 0·8371 gram dissolved in 40·48 grams of melted Glauber's salt produced a depression of $0.22°$ in the crystallising point, corresponding with the molecular weight 306.2. After solidifying and melting three times, the depression reached $0.24°$ and remained at that, which corresponds with 269.6. There was now added 0.5848 gram more of the salt. and the depression due to this addition was at first only $0.005°$. corresponding with a molecular weight of 9416, that is, about forty times the normal magnitude. On allowing the mixture to solidify and remelt, the depression grew in amount until at the sixth repetition it reached its maximum, corresponding with the simplest formula of the salt. The whole quantity, 1.4219, now caused a depression which gave the molecular magnitude as 279.3. Adding now 0.9026. the additional depression was at first only $0.04°$, corresponding with a molecular weight of 1817, which is about eight times the simple molecule. But, again as before, after several times repeated melting and solidifying, the normal depression was reached and remained constant, and the total quantity of the salt, namely, 2.3245, gave a depression of $0.73°$, indicating the molecule magnitude 256.4.

Reduction of the Disodium Salt by Sodium Amalgam.

Disodium hydroxylamine-$\alpha\beta$-disulphonate (0.4811 gram) was shaken with 12 grams of 3 per cent. sodium amalgam (which acts very slowly on it) and left with it, with occasional shaking, for three days. Much sodium remained unconsumed. The solution was rendered neutral to phenolphthalein by acetic acid and the sulphate present precipitated as barium sulphate. Of the sulphur in the quantity of salt taken, 13.73 per cent. was thus found as sulphate, instead of 13.51, the calculated third. The aminemonosulphonate in the filtrate from this sulphate was hydrolysed by heating the solution at $150°$ for three hours in a sealed tube with hydrochloric acid. It thus yielded the rest of the sulphur as sulphate, a result which, when effected at such a temperature, showed that that sulphur was all in direct union with nitrogen and that none of the hydroxylaminedisulphonate had been left undecomposed by the sodium. As a check, the ammonia, the other product of the hydrolysis of the aminemonosulphonate, was also determined and found to be equal to 5.5 per cent. instead of 5.94, the full amount. The production of aminemonosulphonate by the reduction of the hydroxylamine-$\alpha\beta$-disulphonate was further established quantitatively in a separate experiment, in which, after the reduction, the aminesulphonate was precipitated characteristically by mercuric nitrate, and the acid itself, after recovery from its mercury salt, crystallised out and otherwise tested.

way as the sodium salt. The ammonium acid-sulphateis got rid of by adding just enough barium hydroxide solution, filtering, and evaporating, at first at a gentle heat and then in he cold, over sulphuric acid under reduced pressure. It occurs s small, thick plates, which are somewhat hard, and as nodules compced of minute, tabular crystals. It is a very soluble salt, three prts dissolving normally in just two parts of water at 18°, but it is ve7 apt to form supersaturated solutions. It is a more stable salt tan the corresponding $\beta\beta$-compound. Its crystals are probably nhydrous, but those analysed showed the presence of 0·25 H_2O per mcecule.

0·3559 gave 0·7189 $BaSO_4$. S = 27·74.

0·2320 ,, 35·5 c.c. moist nitrogen at 16·2° an 754·5 mm. N = 18·18.

$H_9O_7N_3S_2$ requires S = 28·21 ; N = 18·53. With $\frac{1}{4}H_2O$ added, it requires S = 27·66 ; N = 18·17 per cent.

Barium Salts.—Barium salts have not been prepad in a state suited for satisfactory determination of their nature Evaporation of a solution of the ammonium salt with excess of barim hydroxide in a vacuum over sulphuric acid to a small volue removed all ammonia. After removal of the excess of barium ıydroxide by carbon dioxide, the filtered solution was further evoorated in the desiccator. First a viscid liquid and then a bulky, .iable, porous mass, devoid of crystalline character, were obtained. 'ıe latter was not quantitatively analysed, but it yielded, when hydılysed, barium sulphate and hydroxylamine sulphate in crystals whic were further identified by a very satisfactory sulphuric acid determnation. The product was therefore undoubtedly a barium hydroylamine-$a\beta$-disulphonate. By using less barium hydroxide, crystallhd ammonium barium salts of varying composition may be obtained. One of these, when quantitatively examined, had a composition corresoonding fairly well with that of a compound of 6 mols. of the diammoium salt with 1 mol. of the 2/3-normal barium salt.

Molecular Magnitude of Trisodium Hydroxylamine-a[disulphonate.

By Löwenherz's method, the molecular magnitue of the trisodium hydroxylamine-$\beta\beta$-disulphonate has been foun to be (anhydrous) 233 and 239·6, whilst $O_7NS_2Na_3$ requires 259·3 (Trans., 1904, 85, 101). By the same method, the following approxnations to the same number have been obtained for the $a\beta$-salt, amely, 269·6, 279·3, and 256·4, using the constant, 32·6, for sodium ılphate found by Löwenherz. The details of these determinations o the molecular magnitude of the $a\beta$-salt are specially interesting (p. 42). Of this salt, 0·8371 gram dissolved in 40·48 grams of melted łlauber's salt produced a depreẟou of 0·22° in the crystallising point, corresponding with the molecur weight 306·2. After solidifying and melting three times, the dpression reached 0·24° and remained at that, which corresponds with 69·6. There was now added 0·5848 gram more of the salt, and thedepression due to this addition was at first only 0·005°, corresponng with a molecular weight of 9416, that is, about forty times the ormal magnitude. On allowing the mixture to solidify and remel the depression grew in amount until at the sixth repetition it reacbd its maximum, corresponding with the simplest formula of the ılt. The whole quantity, 1·4219, now caused a depression which ave the molecular magnitude as 279·3. Adding now 0·9026, the aditional depression was at first only 0·04°, corresponding with a mlecular weight of 1817, which is about eight times the simple molecie. But, again as before, after several times repeated melting an solidifying, the normal depression was reached and remained constant. nd the total quantity of the salt, namely, 2·3245, gave a depression ť 0·73°, indicating the molecule magnitude 256·4.

Reductio of the Disodium Salt by Sodium Amalyam.

Disodium hydrovlamine-$\alpha\beta$-disulphonate (0·4811 gram) was shaken with 12 grams of per cent. sodium amalgam (which acts very slowly on it) and left w,h it, with occasional shaking, for three days. Much sodium reṁined unconsumed. The solution was rendered neutral to phenolphthalein by acetic acid and the sulphate present precipitated as baıum sulphate. Of the sulphur in the quantity of salt taken, 13 7 per cent. was thus found as sulphate, instead of 13·51, the calculatd third. The aminemonosulphonate in the filtrate from this sulphate̦as hydrolysed by heating the solution at 150° for three hours in a seled tube with hydrochloric acid. It thus yielded the rest of the sulpur as sulphate, a result which, when effected at such a temperaturḙshowed that that sulphur was all in direct union with nitrogen and hat none of the hydroxylaminedisulphonate had been left undecompẟed by the sodium. As a check, the ammonia, the other product of th hydrolysis of the aminemonosulphonate, was also determined and fond to be equal to 5·5 per cent. instead of 5·94, the full amount. Th production of aminemonosulphonate by the reduction of the hdroxylamine-$\alpha\beta$-disulphonate was further established quantitativẉ in a separate experiment, in which, after the reduction, the amesulphonate was precipitated characteristically by mercuric nitrṅ, and the acid itself, after recovery from its mercury salt, crystḍised out and otherwise tested.

way as the sodium salt. The ammonium acid-sulphate is got rid of by adding just enough barium hydroxide solution, filtering, and evaporating, at first at a gentle heat and then in the cold, over sulphuric acid under reduced pressure. It occurs as small, thick plates, which are somewhat hard, and as nodules composed of minute, tabular crystals. It is a very soluble salt, three parts dissolving normally in just two parts of water at 18°, but it is very apt to form supersaturated solutions. It is a more stable salt than the corresponding $\beta\beta$-compound. Its crystals are probably anhydrous, but those analysed showed the presence of 0·25 H_2O per molecule.

0·3559 gave 0·7189 $BaSO_4$. S = 27·74.

0·2320 ,, 35·5 c.c. moist nitrogen at 16·2° and 754·5 mm. N = 18·18.

$H_9O_7N_3S_2$ requires S = 28·21 ; N = 18·53. With $\frac{1}{4}H_2O$ added, it requires S = 27·66 ; N = 18·17 per cent.

Barium Salts.—Barium salts have not been prepared in a state suited for satisfactory determination of their nature. Evaporation of a solution of the ammonium salt with excess of barium hydroxide in a vacuum over sulphuric acid to a small volume removed all ammonia. After removal of the excess of barium hydroxide by carbon dioxide, the filtered solution was further evaporated in the desiccator. First a viscid liquid and then a bulky, friable, porous mass, devoid of crystalline character, were obtained. The latter was not quantitatively analysed, but it yielded, when hydrolysed, barium sulphate and hydroxylamine sulphate in crystals which were further identified by a very satisfactory sulphuric acid determination. The product was therefore undoubtedly a barium hydroxylamine-$\alpha\beta$-disulphonate. By using less barium hydroxide, crystallised ammonium barium salts of varying composition may be obtained. One of these, when quantitatively examined, had a composition corresponding fairly well with that of a compound of 6 mols. of the diammonium salt with 1 mol. of the 2/3-normal barium salt.

Molecular Magnitude of Trisodium Hydroxylamine-$\alpha\beta$-disulphonate.

By Löwenherz's method, the molecular magnitude of the trisodium hydroxylamine-$\beta\beta$-disulphonate has been found to be (anhydrous) 233 and 239·6, whilst $O_7NS_2Na_3$ requires 259·35 (Trans., 1904, **85**, 101). By the same method, the following approximations to the same number have been obtained for the $\alpha\beta$-salt, namely, 269·6, 279·3, and 256·4, using the constant, 32·6, for sodium sulphate found by Löwenherz. The details of these determinations of the molecular magnitude of the $\alpha\beta$-salt are specially interesting (p. 242). Of this salt, 0·8371 gram dissolved in 40·48 grams of melted Glauber's salt produced a depression of $0.22°$ in the crystallising point, corresponding with the molecular weight 306.2. After solidifying and melting three times, the depression reached $0.24°$ and remained at that, which corresponds with 269.6. There was now added 0.5848 gram more of the salt, and the depression due to this addition was at first only $0.005°$, corresponding with a molecular weight of 9416, that is, about forty times the normal magnitude. On allowing the mixture to solidify and remelt, the depression grew in amount until at the sixth repetition it reached its maximum, corresponding with the simplest formula of the salt. The whole quantity, 1.4219, now caused a depression which gave the molecular magnitude as 279.3. Adding now 0.9026, the additional depression was at first only $0.04°$, corresponding with a molecular weight of 1817, which is about eight times the simple molecule. But, again as before, after several times repeated melting and solidifying, the normal depression was reached and remained constant, and the total quantity of the salt, namely, 2.3245, gave a depression of $0.73°$, indicating the molecule magnitude 256.4.

Reduction of the Disodium Salt by Sodium Amalgam.

Disodium hydroxylamine-$\alpha\beta$-disulphonate (0.4811 gram) was shaken with 12 grams of 3 per cent. sodium amalgam (which acts very slowly on it) and left with it, with occasional shaking, for three days. Much sodium remained unconsumed. The solution was rendered neutral to phenolphthalein by acetic acid and the sulphate present precipitated as barium sulphate. Of the sulphur in the quantity of salt taken, 13 73 per cent. was thus found as sulphate, instead of 13·51, the calculated third. The aminemonosulphonate in the filtrate from this sulphate was hydrolysed by heating the solution at 150° for three hours in a sealed tube with hydrochloric acid. It thus yielded the rest of the sulphur as sulphate, a result which, when effected at such a temperature, showed that that sulphur was all in direct union with nitrogen and that none of the hydroxylaminedisulphonate had been left undecomposed by the sodium. As a check, the ammonia, the other product of the hydrolysis of the aminemonosulphonate, was also determined and found to be equal to 5·5 per cent. instead of 5·94, the full amount. The production of aminemonosulphonate by the reduction of the hydroxylamine-$\alpha\beta$-disulphonate was further established quantitatively in a separate experiment, in which, after the reduction, the aminesulphonate was precipitated characteristically by mercuric nitrate, and the acid itself, after recovery from its mercury salt, crystallised out and otherwise tested.

Decomposition of the Potassium Salt by Potassium Hydroxide.

Whether potassium hydroxylamine-$\alpha\beta$-disulphonate is decomposed solely into acid-sulphate, aminemonosulphonate, and nitrogen when heated with a concentrated solution of potassium hydroxide (p. 241), or whether it is, to a small extent, decomposed into nitrous oxide, according to the equation $4(SO_3K)ONH(SO_3K) + 6KOH = 6(SO_3K)OK + 2NH_2(SO_3K) + N_2O + 3H_2O$, is not certain. It is experimentally difficult to get sufficiently trustworthy quantitative results. Even qualitatively, there is the uncertainty to deal with as to the entire absence of nitrous oxide from the nitrogen obtained. In these experiments, the gas given off extinguished a flaming match and refused, when mixed with hydrogen, to explode by the electric spark. The occasional production of some nitrous oxide is, perhaps, to be in-ferred from the ratio of the quantity of sulphur as sulphate to that as aminemonosulphonate found in one experiment, although its produc-tion was not thus indicated in another experiment. But, as accuracy in determining this ratio is affected by the fact that the analytic separation of sulphate from aminemonosulphonate is only approximate (Trans., 1896, **69**, 1613—1615 ; 1900, **77**, 982—983), the production of any nitrous oxide still remains uncertain. Another difficulty in the quantitative examination of the products of the decomposition is that the decomposition is far from complete after several hours' heating at 120°. The presence of still undecomposed salt is shown by the production of hydroxylamine when the products are hydrolysed and by the fact that a temperature of 180°, instead of 150°, is necessary to ensure complete hydrolysis of the products. No ammonia is generated, the only products being those already mentioned.

When the gases were to be collected, the salt was heated with one and a half to twice its weight of potassium hydroxide and about four times its weight of water for six hours at 100°, or for four hours at 120° in a tube retort connected with a Sprengel pump. Needless to say, explosive ebullition and corrosion of the hard glass tube (rendering it opaque) had to be encountered as difficulties. When the gases were to be allowed to escape, the mixture was heated in a platinum dish on the water-bath., For analysis, the residue in either case was made faintly acid to phenolphthalein by nitric acid, and sulphate then precipitated by barium nitrate. The thoroughly washed precipitate was purified in the usual way by fusion with alkali [carbonate before weighing. The aminesulphonate was pre-cipitated by mercuric nitrate, the mercuric salt was hydrolysed, and the sulphuric acid and sometimes the resulting ammonia were determined. The filtrate from the mercury precipitate always showed the presence of hydroxylaminedisulphonate.

In an experiment with 1·057 gram of salt, the sulphur found as sulphate was 66·82 per cent., and that as aminesulphonate 15·73 per cent., of the total sulphur, leaving 17·45 to be accounted for as undecom· posed salt. Of the total nitrogen, 29·64 per cent. was obtained as ammonia from the aminesulphonate and 48·89 per cent. as gas (27·85 c.c. of moist nitrogen at 18° and 639·8 mm. = 0·002692), leaving 21·47 per cent. as undecomposed salt. The difference between 17·45 and 21·47, perhaps due to slight leakage of air into the apparatus during the six hours' heating, is not at all so significant as it is made to appear by the way of stating the results, the total percentage of nitrogen in the salt being only 5·2 per cent. On the assumption that 65 per cent. of the salt decomposed so as to give nitrogen (p. 241) and 16·7 so as to give nitrous oxide (p. 248), and that 18·3 per cent. remained undecomposed, the numbers should be sulphur as sulphate, 66·68, and as aminesulphonate, 15·01 per cent., nitrogen as aminesulphonate, 30·02, and as free nitrogen and nitrous oxide, 48·30 per cent. of the total.

In another experiment, in which 0·9792 gram of salt was heated at 120—125° for four hours, the indications of the analysis were that 19·9 per cent. of the salt had resisted decomposition and that no nitrous oxide had been formed. The numbers obtained were 66·69 per cent. of sulphur as sulphate, instead of 66·75 calculated; and 12·94 per cent. of sulphur as aminesulphonate, instead of 13·35 calculated. Some of the gas was lost, so that the distribution of the nitrogen could not be sufficiently tested.

In an experiment in which the salt was heated at 100° with potassium hydroxide for many hours in a platinum dish with occasional renewal of the water, the sulphur of the sulphate produced amounted to 77·24 per cent. of the total. Decomposition of all the salt with production of nitrogen would give 83·33 per cent.

Products of Hydrolysis of Hydroxylamine-$\alpha\beta$-disulphonates.

Potassium hydroxylaminetrisulphonate, which for these experiments could be substituted for the disulphonate, the product of its hydrolysis, was moistened with dilute sulphuric acid and heated in a vacuum at 100° in order to effect its hydrolysis and collect the gas evolved. The gas had no action on ferrous sulphate and was not appreciably soluble in alcohol. It therefore contained no nitric oxide and apparently no nitrous oxide.

Sodium hydroxylaminetrisulphonate in solution in water, just acid with sulphuric acid, was left for several days to hydrolyse slowly, principally to the disulphonate.   It was then kept for fifty minutes at 95°, during which brisk effervescence of nitrogen occurred, slackening only when near the end of the time.   The acidity was then found to be somewhat more than that indicated by either of the following equations, the one for hydroxylamine sulphate, the other for the unknown hydroxylamine-α-monosulphonate :

$$(SO_3Na)ONH(SO_3Na) + 2H_2O = (SO_3H)ONH_3OH + Na_2SO_4.$$
$$(SO_3Na)ONH(SO_3Na) + H_2O = (SO_3Na)ONH_2 + NaHSO_4.$$

The additional acidity and the escape of nitrogen indicated the occurrence, to some extent, of the decomposition expressed by

$$3(SO_3Na)ONH(SO_3Na) + 3H_2O = N_2 + NH_4HSO_4 + 4NaHSO_4 + Na_2SO_4.$$

The solution gave only a moderate reaction for hydroxylamine with the copper test, and on evaporation, with or without previous neutralisation, gave nothing but sodium and ammonium sulphates.

In presence of sufficient hydrochloric acid, say, one volume of the fuming solution to ten volumes of solution of the salt, the production during hydrolysis of nitrogen and ammonia is very slight.   The disodium salt in such a solution, after it had been kept heated for five minutes or so by immersing the vessel in boiling water, gave evidence, on titrating with iodine, of the presence of hydroxylamine equivalent to 95 per cent. of the salt.   In another experiment, in which the solution was left for fifty days at the ordinary temperature, 76·5 per cent. of the salt had then yielded hydroxylamine.

In another similar experiment, the solution, when left in the cold, was also evaporated in the cold under reduced pressure until the salts crystallised out, and still nothing else but sodium and hydroxylamine sulphates, except a very little ammonium sulphate, was obtained.

In other experiments, using in these cases the dipotassium salt, the hydrolysis was allowed to proceed either in the cold or at 60°, and portions of the solution occasionally tested to see whether some indication could be got of the production of hydroxylamine-α-monosulphonate at any stage of the hydrolysis.   In making the test, the acid-sulphate was removed by barium chloride and the solution then tested with iodine in presence of sodium acid-carbonate.   Since the consumption of the iodine caused no precipitation of barium sulphate, evidence was thus obtained that the reducing substance was hydroxylamine only, and not its sulphonate derivative.   Ultimately, by evaporating the solution, when sufficiently hydrolysed, hydroxylamine sulphate was crystallised out, along with the sodium sulphate.

XXX.—Some Oxidation Products of the Hydroxybenzoic Acids. Part II.

By Arthur George Perkin, F.R.S.

IN the previous communication (Trans., 1905, **87**, 1412), it was shown that ellagic acid is formed when gallic acid dissolved in a mixture of acetic and sulphuric acids is oxidised by means of potassium persulphate. Experiments carried out in a similar way with protocatechuic, *p*-hydroxybenzoic, and *m*-hydroxybenzoic acids gave only minute traces of oxidation products insoluble in water, but, on the other hand, when concentrated sulphuric acid was employed as the solvent, moderate yields of interesting compounds were produced, which in constitution bear a simple relationship to ellagic acid. The present paper deals with a description of the oxidation products which are obtained from gallic acid itself when dissolved in both dilute and concentrated sulphuric acid, and which have been already referred to in a preliminary note (*loc. cit.*, p. 429).

EXPERIMENTAL.

The Oxidation of Gallic Acid in the Presence of Dilute Sulphuric Acid.

Twenty grams of finely-powdered gallic acid were treated with 160 c.c. of 96 per cent. sulphuric acid, 66 c.c. of water added, and the hot solution thus obtained was cooled to 50° and maintained at this temperature during the gradual addition of 40 grams of potassium persulphate. The resulting deep brown liquid, from which minute crystals of an oxidation product had separated, was poured into water, causing the deposition of a pale brown precipitate, which was collected and washed with water. The dry substance weighed 15·1 grams, and this yield, it was subsequently ascertained, was not materially affected by a slight variation of the temperature or the dilution of the sulphuric acid employed in the reaction. This compound, which closely resembled ellagic acid, and at first appeared to consist of this colouring matter, was purified by crystallisation from pyridine, which is probably the only solvent suitable for this purpose. It was soon noted, however, that its solubility in this respect was somewhat less than that of ellagic acid, and a fractional crystallisation was accordingly resorted to, as it was only likely from previous results (*loc. cit.*) that some quantity of the latter substance might be present. Thus obtained, the pure compound consisted of small, pale yellow, prismatic needles, which contain pyridine, and this was partially removed by means of hot alcohol, and the final traces by drying at 160°.

Found, $C = 53\cdot08$; $H = 2\cdot22$. $C_{14}H_6O_9$ requires $C = 52\cdot82$; $H = 1\cdot89$ per cent.

It does not melt below 360°, and, like ellagic acid, is almost insoluble in the usual solvents, but it is distinguished from this colouring matter by its somewhat yellower appearance and by the fact that its solution in dilute alkali has a well-marked green tint. With nitric acid containing nitrous acid and subsequent dilution, it gives a blood-red coloration (Griessmayer reaction) which has long been considered as characteristic of ellagic acid. By distillation with zinc dust, a small quantity of a hydrocarbon was obtained ; this, after crystallisation from alcohol, formed colourless leaflets, melting at 109°, and evidently consisted of *fluorene*. This compound, for which the name "*flavellagic acid*" is suggested, dyes mordanted woollen cloth in shades somewhat resembling, but stronger than, those produced by ellagic acid. The following results were obtained :

Chromium.	Aluminium.	Tin.	Iron.
Yellowish-olive	Pale greenish-yellow	Pale yellow	Dark olive-brown

Acetylflavellagic acid is prepared by digesting the substance with a large excess of boiling acetic anhydride for some hours, or more readily in the presence of two or three drops of sulphuric acid. On cooling, a colourless, crystalline precipitate separates, which is purified by recrystallisation from acetic anhydride.

Found, $C = 54\cdot24$, $54\cdot29$; $H = 3\cdot28$, $3\cdot26$. $C_{14}HO_9(C_2H_3O)_5$ requires $C = 54\cdot54$; $H = 3\cdot03$ per cent.

It formed glistening needles melting at 317—319°, very sparingly soluble in the usual solvents.

Acetyl determinations were carried out by the indirect method, employing 25 c.c. of acetic acid, 3 c.c. of sulphuric acid, and 0·6 gram (approximately) of substance.

Found, $C_{14}H_6O_9 = 59\cdot68$, $59\cdot78$. $C_{14}HO_9(C_2H_3O)_5$ requires $C_{14}H_6O_9 = 60\cdot22$ per cent.

It was therefore a *penta*-acetyl compound.

Obtained in this manner, the regenerated flavellagic acid is a pale yellow, microcrystalline powder, but if the acetyl compound (1 gram approximately) is dissolved by digestion with 12 c.c. of pyridine, the solution treated with 15 c.c. of acetic acid, and 12 c.c. of hydrochloric acid is then gradually added to the boiling liquid, the colouring matter is deposited in the form of long, glistening, hair-like needles.

Benzoylflavellagic acid may be produced by treating the substance with a large excess of benzoic anhydride at 200—210° until a clear solution is formed, and subsequently digesting at 170—180° for three hours. The product is poured into alcohol, and the precipitate which separates on standing is collected and purified by crystallisation from nitrobenzene with the aid of animal charcoal.

Found, $C = 70.20$; $H = 3.44$. $C_{14}HO_9(C_7H_5O)_5$ requires $C = 70.16$; $H = 3.10$ per cent.

It forms prismatic needles melting at 287—289°.

A molecular weight determination was carried out by the cryoscopic method, employing naphthalene as a solvent.

0.5190 in 14.19 naphthalene gave $\Delta t - 0.3125°$. M.W. $= 819$. $C_{49}H_{26}O_{14}$ requires M.W. $= 838$.

These results therefore indicate that flavellagic acid has the formula $C_{14}H_6O_9$.

Action of Boiling Potassium Hydroxide Solution.

Twelve grams of flavellagic acid dissolved in 60 c.c. of boiling 50 per cent. aqueous potassium hydroxide were digested at the boiling point until the solution possessed a pale brown colour and a drop of the liquid on exposure to air developed a bluish-violet tint. The product was treated with twice its volume of water and at once neutralised with dilute sulphuric acid, causing the formation of a clear solution, which on standing became semi-solid owing to the separation of almost colourless needles. These were collected, and when dry weighed 8.71 grams (72.5 per cent.), but in some operations, owing to the difficulty in determining the end of this reaction, as little as 5.5 grams (45.8 per cent.) could be obtained.

This compound is best purified by conversion into its acetyl derivative, which after recrystallisation is hydrolysed by acid. In the dry condition it is somewhat sparingly soluble in alcohol, but in the moist, freshly-precipitated state it dissolves much more readily, and it is possible by treating such a solution with animal charcoal and subsequently evaporating to obtain a fairly pure product. Air-dried, it contains 1 molecule of water of crystallisation, which is slowly removed by heating at 160°.

Found, $H_2O = 5.63$. $C_{13}H_8O_8,H_2O$ requires $H_2O = 5.80$ per cent.

The anhydrous substance thus obtained is exceedingly hygroscopic, and on standing in air soon reassumes its original weight. Owing to this behaviour, the analyses of the product dried at 160° (A) were not very satisfactory, and in order to corroborate these an air-dried sample (B) was also examined.

s 2

A. Found, $C = 53 \cdot 03$; $H = 3 \cdot 03$. $C_{13}H_8O_8$ requires $C = 53 \cdot 42$; $H = 2 \cdot 74$ per cent.

B. Found, $C = 50 \cdot 32$; $H = 3 \cdot 39$. $C_{13}H_8O_8,H_2O$ requires $C = 50 \cdot 32$; $H = 3 \cdot 22$ per cent.

This compound, to which the formula $C_{13}H_8O_8$ was therefore assigned, forms almost colourless needles, which on exposure to light become somewhat greyer in appearance. When heated, it does not melt below 300°, and probably not below 360°, but this could not be satisfactorily determined owing to the discoloration of the tube at this temperature. It is sparingly soluble in most solvents with the exception of pyridine, in which it readily dissolves, but attempted purification in this manner did not give a satisfactory result. Solutions of the alkaline hydroxides dissolve it with a dull orange-yellow tint, and these on dilution with water and exposure to air develop a strong bluish-violet coloration. With alcoholic lead acetate, it gives an orange-yellow precipitate which rapidly passes to olive-green, with alcoholic ferric chloride a bluish-green coloration, and its solution in sulphuric acid is orange-yellow. Nitric acid dissolves it with an orange-red coloration, which on dilution with water becomes redder, but is devoid of the blood-like tint produced in this manner from both ellagic and flavellagic acids. It dyes mordanted calico rather feebly, but with mordanted woollen cloth the following some-what better results were obtained.

Chromium.	Aluminium.	Tin.	Iron.
Brown	Pale brownish-yellow	Pale yellow	Brownish-purple

This substance was acetylated by a three hours' digestion with boiling acetic anhydride. The solution was concentrated, and the crystalline product, which separated out on cooling, was collected and recrystallised from the same solvent, employing animal charcoal.

Found, $C = 55 \cdot 09$; $H = 3 \cdot 85$. $C_{13}H_2O_8(C_2H_3O)_6$ requires $C = 55 \cdot 15$; $H = 3 \cdot 67$ per cent.

Thus obtained, it forms colourless, prismatic needles melting at 232—234°, somewhat sparingly soluble in the usual solvents.

The determination of the acetyl groups in this substance was, as will be seen later, a matter of considerable importance with regard to the constitution of flavellagic acid, and analyses were therefore carried out both by the direct and indirect methods. In the former case, the ethyl acetate method was adopted, and in the latter the finely-powdered acetyl derivative (1 gram approximately in 20 c.c. of acetic acid) was hydrolysed by the gradual addition of 10 c.c. of hydro-chloric acid. The digestion was continued for at least fifteen minutes after the reaction had apparently terminated, for, owing to the insoluble nature of the original compound, special care was necessary in case some portion remained unattacked.

Found, acetic acid $= 67\cdot01$, $65\cdot30$, $66\cdot12$, $67\cdot30$. $C_{13}H_2O_8(C_2H_3O)_6$ requires $C_2H_4O_2 = 66\cdot17$ per cent.

Found, $C_{13}H_8O_8 = 53\cdot43$. $C_{13}H_2O_8(C_2H_3O)_6$ requires $C_{13}H_8O_8 = 53\cdot67$ per cent.

Found, $C_{13}H_8O_8,H_2O = 56\cdot64$, $56\cdot93$. $C_{13}H_2O_8(C_2H_3O)_6$ requires $C_{13}H_8O_8,H_2O = 56\cdot98$ per cent.

These results indicate that the substance $C_{13}H_8O_8$ contains six hydroxyl groups.

For the purpose of a molecular weight determination, a *benzoyl* derivative was prepared by treating the substance (1 gram) dissolved in pyridine (15 grams) with benzoyl chloride ($11\cdot5$ grams). The new compound was readily isolated by the usual methods and was purified by crystallisation from a mixture of benzene and alcohol. It formed colourless prisms melting at 261—263°, and when heated with benzoic anhydride at 170° for two hours suffered no change.

$0\cdot6877$ in $14\cdot21$ of naphthalene gave $\Delta t - 0\cdot38°$. M.W. $= 892$.

$C_{55}H_{32}O_{14}$ requires M.W. $= 916$.

It thus appeared certain that the formula assigned to the substance $C_{13}H_8O_8$ is correct.

To obtain, if possible, some further insight into its constitution, it was heated with ten times its weight of potassium hydroxide at 220—240° for twenty minutes. The product of the fusion, on solution in water and neutralisation with acid, yielded to ether a mixture of compounds, some quantity of which consisted of unattacked substance. That portion which was soluble in water contained *gallic acid* (found, $C = 49\cdot89$; $H = 3\cdot91$ per cent.), which was identified by the usual tests, and on evaporating the mother liquors obtained during the isolation of this compound a semi-crystalline but very soluble residue remained, the nature of which could unfortunately not be ascertained. The presence of a small quantity of pyrogallol derived from the gallic acid was to be anticipated, but there were indications that it did not constitute the main bulk of this product.

Summary of Results.

In considering the constitution of flavellagic acid, one cannot help observing that not only its composition, $C_{14}H_6O_9$, but its general reactions, notably the Griessmayer coloration and the production of fluorene from it by means of zinc dust, indicate a close relationship to ellagic acid. The analogy between these colouring matters is also strikingly brought out by their behaviour towards boiling potassium hydroxide solution, for whereas in this manner ellagic acid gives pentahydroxydiphenylmethylolid, $C_{13}H_8O_7$, from flavellagic acid the similar compound $C_{13}H_8O_8$ is thus formed. These facts, together with the other results described above, can only be interpreted to mean that *flavellagic acid* is *hydroxyellagic acid*, and the substance $C_{13}H_8O_8$ a *hexahydroxydiphenylmethylolid*, and constitutions (*loc. cit.*) similar to those previously assigned to ellagic acid and its decomposition product are accordingly suggested for these compounds.

Ellagic acid.

Pentahydroxydiphenylmethylolid.

Flavellagic acid.

Hexahydroxydiphenylmethylolid.

Such a formula for flavellagic acid indicates that during the reaction, either before or after the diphenyl condensation, a fourth hydroxyl is introduced into one of the pyrogallol nuclei, a type of oxidation which, as is well known, is frequently effected by this method. The production, for instance, of catellagic acid from *p*-hydroxybenzoic acid, described in the former communication, may be cited as an illustration of the case in point (*loc. cit.*).

In the earlier stages of the work, and having regard to the fact that flavellagic acid differs from gallic acid by possessing a well-marked yellow tint, it appeared just possible that the former colouring matter was constituted as follows :

this being apparently the only other possible formula for this substance.

Apart from those changes occurring during the formation of such a substance from gallic acid in the presence of concentrated sulphuric acid due to oxidation, and which have been discussed above, the elimination of a carboxyl group and its subsequent reintroduction in another portion of the molecule are points which, although necessary for an explanation of this formula, do not appear to be very likely to occur.   Again, if this were the case, the substance $C_{13}H_8O_8$ must of necessity contain seven hydroxyls, whereas according to the analytical methods employed the presence of only six such groups could be detected.

An examination of the more soluble fractions obtained during the fractional crystallisation of the crude flavellagic acid from pyridine indicated that these occasionally contained a minute trace of *ellagic acid*.   It was noted, for instance, on exhausting the crude dry hexahydroxydiphenylmethylolid, obtained by the action of boiling potassium hydroxide solution on these residues, that a trace of a readily soluble substance could be isolated.   This, which when pure crystallised in long, hair-like needles, had the properties of *penta-hydroxydiphenylmethylolid*, the characteristic decomposition product of ellagic acid.

Oxidation of Gallic Acid in the Presence of 96 *per cent. Sulphuric Acid.*

When gallic acid (10 grams) dissolved in 50 c.c. of 96 per cent. sulphuric acid is treated with 20 grams of potassium persulphate at 50°, the features of the oxidation are of a similar nature to those occurring when a more dilute acid · is employed.   The mixture, on treatment with water, deposits a pale brown, crystalline powder, which after washing and drying weighs on the average 6·6 grams.   In certain cases, the product of the experiment when poured into much water gave at first a clear liquid, from which, on standing, crystals somewhat slowly separated, but this peculiar effect appears to be due to some accidental circumstance not as yet ascertained, and ˙recent attempts ᴛo reproduce it have, curiously enough, been unsuccessful. Owing to the sparing solubility of this compound in the usual solvents, it was at first purified by means of its acetyl derivative, which after recrystallisation was hydrolysed by acid.   The product thus obtained formed groups of microscopic needles of a very pale yellow colour resembling ellagic acid in appearance, and was soluble in alkaline solutions with a greenish-yellow tint.   With nitric acid, it gave the Griessmayer reaction, and when distilled with zinc dust gave a small quantity of a hydrocarbon which was found to consist of *fluorene.* Again, its dyeing properties were almost identical with those yielded by ellagic acid, so that in many respects there was a close resemblance between these substances.   The analytical numbers (found, $C = 54·08$, 53·55 ; $H = 2·36$, 2·48 per cent.) obtained with distinct preparations did not, however, accord with one another so closely as was to be desired, and, again, could not be made to harmonise with any simple formula. The *acetyl* derivative, prepared by boiling with acetic anhydride or more rapidly in the presence of sulphuric acid, separated as very minute needles, but these on recrystallisation from the same solvent assumed a peculiar gelatinous appearance, and were in this condition difficult to manipulate. This product melted at 224—226°, gave on analysis $C = 55\cdot04$; $H = 3\cdot20$, and on examination by the indirect method of acetyl determination gave 64·22 and 64·06 per cent. of regenerated colouring matter. These numbers suggested as possible that this compound might consist of an anhydride of flavellagic acid, $(C_{14}H_6O_9)_2—H_2O$, but this was negatived by the failure to obtain such a substance from flavellagic acid by means of sulphuric acid. Large quantities of the oxidation product were now fused with potassium hydroxide at 210—220° in the usual manner, and the fused product, after solution with water, was neutralised with acid. On cooling, some quantity of a crystalline precipitate (*A*) separated, which was removed by filtration and the filtrate extracted with ether. The residue obtained by the evaporation of this extract was digested with boiling water, in which a portion dissolved, and the clear solution was allowed to stand for twelve hours. A quantity of fine hair-like needles (*B*) had then separated, which were collected and purified by crystallisation from dilute alcohol.

Found, $C = 56\cdot31$; $H = 3\cdot12$. $C_{13}H_8O_7$ requires $C = 56\cdot52$; $H = 2\cdot89$ per cent.

The *acetyl* derivative of this substance melted at 224—226°, and by the ethyl acetate method gave acetic acid = 62·45 per cent.; again, the benzoyl compound (found, $C = 72\cdot39$; $H = 3\cdot87$ per cent.) melted at 258—259°, so that it was evident that this compound was identical with the *pentahydroxydiphenylmethylolid* previously obtained from ellagic acid. The aqueous mother liquors arising from the purification of this substance (*B*) contained among other products some quantity of *gallic acid*, which was identified by means of its acetyl derivative (m. p. 172—174°).

The more sparingly soluble substance (*A*), present in but small amount, on solution in alkali and exposure to air yielded a bluish-violet-coloured liquid, and appeared likely to consist of the *hexahydroxy-diphenylmethylolid* previously described in this paper. To ascertain if this was correct, the purified product was acetylated and the acetyl derivative isolated by the former method.

Found, $C = 55\cdot17$; $H = 3\cdot94$. $C_{13}H_2O_8(C_2H_3O)_6$ requires $C = 55\cdot15$; $H = 3\cdot67$ per cent.

It melted at 232—234°, and no doubt as to its identity could there-

fore exist. More recent experiments have shown that by digesting
the gallic acid oxidation product with boiling 50 per cent. potassium
hydroxide solution, the same compounds are obtained, so that the use
of alkali at so high a temperature as that given above is not
necessary.

The formation in this manner of two substances respectively
obtainable by the action of alkali on ellagic and flavellagic acids
suggested that the oxidation of gallic acid by the method employed
had yielded a mixture of the latter substances. Knowing that
flavellagic acid is less soluble than ellagic acid in pyridine, a repeated
fractional crystallisation from this solvent was carried out, but
without effecting the desired separation, and as the acetyl derivative
in various solvents did not give a better result, it appeared at one
time that the product in question was after all a distinct substance.
An interesting point was, however, subsequently obtained in support
of the mixture theory, for it was found that in the presence of a
small quantity of flavellagic acid, ellagic acid gave an ill-defined, semi-
gelatinous acetyl compound, corresponding in appearance with that
given by the oxidation product itself. The difficulty was eventually
surmounted by a study of the benzoyl derivative.

The crude product of the oxidation was heated at 210° with a large
excess of benzoic anhydride until a clear liquid resulted, and the
temperature was then kept at 190° for three hours. The hot solution
was cautiously treated with a few drops of alcohol, causing the
almost immediate separation of a colourless crystalline precipitate,
which was collected on a warm funnel by means of the pump and
washed with hot benzoic anhydride containing a trace of alcohol. It
was again dissolved in benzoic anhydride at about 200°, reprecipitated
as before, collected, and finally purified by crystallisation from nitrobenzene.

Found, $C = 70\cdot37$; $H = 3\cdot29$. $C_{14}H_2O_8(C_7H_5O)_4$ requires $C = 70\cdot19$;
$H = 3\cdot06$ per cent. .

This substance consisted of prismatic needles melting at 332—333°,
and had therefore the composition of a *tetrabenzoylellagic acid*. Such
a compound has been prepared by Goldschmidt and Jahoda (*Monatsh.*,
1892, **13**, 51), but as they do not give a melting point, a sample was
prepared and found to be identical with the product isolated as above
(found, $C = 70\cdot19$; $H = 3\cdot21$ per cent.). This compound, it was
incidentally noted, is prepared more readily by the agency of benzoic
anhydride than by means of benzoyl chloride, which was the reagent
employed by the above authors.

It was now interesting to obtain, if possible, ellagic acid itself from
the benzoyl compound, and for this purpose a modification of the

indirect acetyl method described in a former paper (Trans., 1905, **87**, 107) was employed; 0·65 gram of the finely-powdered substance suspended in 25 c.c. of acetic acid was treated with 8 c.c. of sulphuric acid and digested at the boiling temperature for three hours. The colourless benzoyl derivative had then disappeared, and was replaced by a pale greenish-yellow deposit consisting of fine naedles. The mixture was diluted with half its volume of water and the crystals collected and weighed (0·2682 gram).

Found, $C_{14}H_6O_8 = 41·26$.   Theory requires $C_{14}H_6O_8 = 42·06$ per cent.

That this product was indeed ellagic acid was determined by analysis :

Found, $C = 55·55$; $H = 2·29$. $C_{14}H_6O_8$ requires $C = 55·62$; $H = 1·98$ per cent. ;

and also by an examination of its acetyl compound, which melted at 343—346°.

Experiments were now carried out in order to isolate the more soluble compound produced during the benzoylation, and which was by far the minor constituent of the mixture. The alcoholic benzoic anhydride mother liquors obtained as above described were treated with some quantity of alcohol, and the crystalline deposit which separated on standing was collected and freed from a small quantity of benzoylellagic acid still present by fractionally precipitating its solution in benzoic anhydride with alcohol. The more soluble product was finally crystallised from a mixture of nitrobenzene and alcohol.

Found, $C = 70·21$; $H = 3·50$. $C_{14}HO_9(C_7H_5O)_5$ requires $C = 70·16$; $H = 3·10$ per cent.

This substance consisted of prismatic needles melting at 287—289°, and was identical with the *benzoylflavellagic acid* previously described.

A portion of this substance was now hydrolysed with sulphuric acid according to the method employed for benzoylellagic acid.

Found, $C_{14}H_6O_9 = 37·13$.   Theory requires $C_{14}H_6O_9 = 37·94$ per cent.

An analysis gave :

Found, $C = 52·85$; $H = 2·21$. $C_{14}H_6O_9$ requires $C = 52·82$; $H = 1·89$ per cent.

It dissolved in alkaline solutions with the greenish-yellow tint characteristic of *flavellagic acid,* and as its acetyl derivative melted at 317—319° there could be no doubt as to its identity with this substance.

The product of the oxidation of gallic acid in the presence of 96 per cent. sulphuric acid is therefore a mixture of ellagic and flavellagic acids. It was not possible to determine the exact proportion of each which was present, but it was ascertained that the former colouring matter is without doubt the main substance produced by this reaction. Again, it has been observed that, due probably to some slight variation in the conditions of the experiment, the quantities of these two constituents also vary somewhat in distinct preparations, and on at least one occasion, owing to some cause which unfortunately could not be accounted for, a product which consisted almost entirely of ellagic acid was found to have been formed. As, however, the necessary features of the reaction have now been explained, it was not worth while to devote more time to this point. If gallic acid is oxidised in the presence of 100 per cent. sulphuric acid, the yield is somewhat poorer than when the ordinary commercial acid is employed, and the crude substance does not appear to differ in any marked manner from that described above.

The molecular weight of ellagic acid, although most probably represented as $C_{14}H_6O_8$, has not been determined by cryoscopic methods, and experiments were carried out in the hope that benzoylellagic acid would be sufficiently soluble in naphthalene for this purpose. It was observed, however, that a trace of this substance crystallised out at or about the freezing point of the mixture.

0·3057 in 15·13 naphthalene gave $\Delta t - 0·1525$. M.W = 927.

$C_{42}H_{22}O_{12}$ requires M.W = 718.

There can be little doubt, however, that, although this result is not satisfactory enough to indicate a definite formula, ellagic acid must be represented as C_{14} rather than C_{21} or C_{28}, which is practically all the information that is necessary. If this was not the case, it is obvious that under the imperfect conditions of the experiment a much higher number would have been obtained.

An investigation is in progress with the view of obtaining some insight into the constitution of ellagitannic acid, a substance of much importance in the tanning industry, and from which the natural ellagic acid appears to be almost entirely derived.

CLOTHWORKERS' RESEARCH LABORATORY,
LEATHER INDUSTRIES' RESEARCH LABORATORY,
THE UNIVERSITY,
LEEDS.

XXXI.—*Contributions to the Chemistry of Oxygen Compounds. I. The Compounds of Tertiary Phosphine Oxides with Acids and Salts.*

BY ROBERT HOWSON PICKARD and JOSEPH KENYON.

MANY new conceptions of valency have been described during recent years. That the valencies commonly ascribed to the elements do not serve to explain all experimental results is readily recognised from a consideration of many of the so-called molecular compounds. It is obvious that the affinities to which the formation of molecular compounds is due are weaker than those by reason of which atomic compounds exist, and many names have been given to these weaker valencies: for example, krypto, complex, neutral, partial, residual, supplementary. That these subsidiary valencies differ from the main valencies only in degree has been expressed by Werner (*Annalen*, 1902, **322**, 261) in his theory of principal ("Haupt") and supplementary ("Neben") valencies, according to which the constitution of a compound is determined by the principal and supplementary valencies of the elements along with the co-ordination constant.

The discovery of the basic properties of oxygen compounds (Collie and Tickle, Trans., 1899, **75**, 710; Baeyer and Villiger, *inter alia*, *Ber.*, 1901, **34**, 2679) gave a stimulus to the speculations as to valency, and many attempts have been made to determine the constitution of these "oxonium" compounds. We are making a detailed study of the "molecular" compounds formed by substances containing oxygen and some other element in columns 4—8 of the periodic table. In the present communication are described a number of such compounds formed by the tertiary phosphine oxides with acids and metallic salts. These are all well-defined compounds which have been recrystallised from water or alcohol, and have the general formula

$$2R_3P\!:\!O,HX \text{ or } 2R_3P\!:\!O,M''X_2.$$

Thus the compounds which tri-methyl-, -ethyl-, -propyl-, -phenyl-, and -benzyl-phosphine oxides form with the following acids: ferrocyanic, cobalticyanic, dichromic, chloroauric, camphoric, iodobismuthic, and iodomercuric, and the following salts: zinc chloride and iodide, cadmium iodide, and mercuric and cobalt chlorides, conform to these general formulæ, whilst those with hydrochloric, trichloroacetic, pyruvic, and chloroplatinic acids and with cupric and ferric chlorides are exceptions.

The constitution of one of these compounds, namely, triphenylphosphine oxide hydrochloride, $(C_6H_5)_3PO,HCl$, could be represented by either of the following formulæ :

(i) $(C_6H_5)_3P{<}{OH \atop Cl}$ 　　　　(ii) $(C_6H_5)_3P{:}O{<}{H \atop Cl}.$

The former formula is very improbable for the following reasons : (i) there is no evidence that a phosphine oxide behaves in solution as a basic hydroxide; an aqueous solution of triethylphosphine oxide, for example, has no effect on the birotation of dextrose, and may therefore be regarded as free from hydroxyl ions, and (ii) it offers no explanation (among other things) of the analogy between the compounds formed by the oxides with acids and those with metallic salts.

The second formula is based on the assumption of the quadrivalency of oxygen. This assumption, whilst it may explain the constitution of this hydrochloride, does not satisfactorily explain the constitution of such compounds as $2(\dot{C}H_3)_3PO,H_4Fe(CN)_6$, $4(CH_3)_3PO,H_2PtCl_6$, and so on. For the addition of R_3PO to HX is assumed to form a saturated compound, and therefore the addition of a second molecule of R_3PO to HX seems improbable. A formula such as

$$(CH_3)_3P{:}O{\diagdown}{-}OH{:}P(CH_3)_2$$
$$\diagdown PtCl_6$$
$$(CH_3)_3P{:}O{\diagup}OH{:}P(CH_3)_2$$

appears highly improbable, as it is opposed to the known constitution of chloroplatinic acid. This assumption of the quadrivalency of oxygen is rendered less likely by the close analogy existing between the compounds of the oxides with acids and those with metallic salts, thus, $2(CH_3)_3PO,ZnI_2$, $(C_2H_5)_3PO,CuCl_2$, and the series of compounds of the type of $3(C_7H_7)_3PO,Fe_2Cl_6$, prepared by Fleissner (*Ber.*, 1880, **13**, 1665).

In fact we regard the results described in this communication as a confirmation of the theory of " oxonium " compounds put forward by Werner (*Annalen*, 1902, **322**, 296), and therefore prefer to formulate these compounds on the assumption of supplementary valencies on the part of the oxygen atom in the oxide, as $(R_3PO) \ldots . HX$ and $(R_3PO) \ldots . MeX$. Thus the combination in the first case is supposed to occur through the agency of the supplementary valencies of the oxygen of the oxide and the hydrogen of the respective acids, and in the second case through the agency of the supplementary valencies of the oxygen of the oxide and of the metallic radicles of the salts. The platinichlorides then would have a co-ordination formula,

$$Cl_4Pt{\diagup}^{ClH{\cdots}{:}OP(CH_3)_3}_{\diagdown ClH{\cdots}{:}OP(CH_3)_3}{\cdots}^{OP(CH_3)_3}_{OP(CH_3)_3}.$$

What influence (if any) the phosphorus has on the constitution of these molecular compounds can be only brought out by further studies on other oxygen compounds, the results of which we hope shortly to communicate to the Society. In the present communication we have been able to show that the relative mass of R in compounds R_3PO has no effect within the limits tried, namely, methyl, ethyl, propyl, phenyl, to benzyl, on the constitution of these molecular compounds.

The constitution of "Grignard" reagents has often been represented by formulæ of the type $\mathrm{\frac{CH_3}{IMg}}\!\!>\!\!O\!\!<\!\!\frac{C_2H_5}{C_2H_5}$ (compare Baeyer and Villiger, *loc. cit.*), but can be better represented, according to Werner's theory, by $(C_2H_5)_2O$ . . . . CH_3MgI. We have discovered that the ether may be replaced in these reagents by compounds containing oxygen other than ethers; thus, magnesium dissolves in benzene containing methyl iodide and tribenzylphosphine oxide giving a compound of the probable formula $\{(C_7H_7)_3PO\}_2$ . . CH_3MgI, assuming its formation to be due to the agency of supplementary valencies of the oxygen atoms and the carbon of the organo-magnesium iodide.

These phosphine oxides are only weak bases. Thus, trimethylphosphine oxide hydrogen trichloroacetate is hydrolysed in $N/6$ aqueous solution at 30° to the extent of about 89 per cent. as measured by the inversion of cane sugar method.

Preparation of Tertiary Phosphine Oxides.—The old method of preparing the tertiary phosphine oxides by treating with potassium hydroxide the mixture of iodides obtained when phosphorus is heated with an alkyl iodide in a sealed tube is a very tedious and unpleasant operation. It is far easier to prepare them from the mixture obtained by treating a "Grignard" reagent with phosphorus oxychloride. During the course of our experiments, in a paper by Sauvage (*Compt. rend.*, 1904, 139, 674) it was briefly shown that phosphorus oxychloride reacts with aromatic organo-magnesium compounds to form the substances R_3PO and $R_2PO\cdot OH$. As the method of preparation in the case of the aromatic series presents no difficulty, it will be sufficient to give the details of the preparation by the latter method of tripropylphosphine oxide, a trialkyl phosphine oxide, which has not previously been described.

Tri-n-*propylphosphine Oxide.*—A dilute ethereal solution of magnesium *n*-propyl bromide is placed in a large flask, provided with a reflux condenser, and well cooled with a mixture of ice and salt. This solution is then very slowly, with constant shaking, mixed with a dry ethereal solution of the calculated quantity of phosphorus oxychloride. After the very vigorous reaction is over, water is added, along with sufficient hydrochloric acid to dissolve any excess of magnesium, and the ether distilled off. The solution is then mixed with a large excess of sodium hydroxide and the resulting paste slowly distilled from a copper flask. When the aqueous distillate is fractionated, the oxide is obtained in the fraction boiling at 260—265°/759 mm. *Tri-n-propylphosphine oxide*, when freshly distilled, solidifies in colourless, lustrous, silky needles, which melt at 38°, have no smell, and are very hygroscopic.

Compounds with Ferrocyanic Acid.

These compounds may be prepared by mixing aqueous solutions of the acid and the phosphine oxide. Small, colourless needles separate and soon turn green on exposure to light or moist air. Owing to the solubility of these compounds in water, they are more conveniently prepared from alcoholic solutions of the acid and oxide. These solutions, when mixed, deposit white or pale green, microcrystalline precipitates, which after washing with alcohol are sufficiently pure for analysis.

Trimethylphosphine oxide hydrogen ferrocyanide is very soluble in water, but insoluble in cold alcohol; it crystallises from either of these media in small, colourless needles. Analysis gave $N = 21.03$; $H_4Fe(CN)_6 = 53.39$; $2(CH_3)_3PO,H_4Fe(CN)_6$ requires $N = 21.00$; $H_4Fe(CN)_6 = 54.00$ per cent.

Triethylphosphine oxide hydrogen ferrocyanide was only obtained from the mixed aqueous solutions of the acid and oxide after the addition of concentrated hydrochloric acid, which precipitates it in the form of a microcrystalline powder. This quickly turns green in moist air. Analysis gave $C = 44.65$; $H = 7.38$; $Fe = 11.55$; $H_4Fe(CN)_6 = 44.53$; $2(C_2H_5)_3PO,H_4Fe(CN)_6$ requires $C = 44.62$; $H = 7.02$; $Fe = 11.60$; $H_4Fe(CN)_6 = 44.62$ per cent.

Tripropylphosphine oxide hydrogen ferrocyanide is much less soluble than the analogous methyl compound. Analysis gave $C = 50.37$; $H = 8.27$; $H_4Fe(CN)_6 = 38.41$; $2(C_3H_7)_3PO,H_4Fe(CN)_6$ requires $C = 50.70$; $H = 8.10$; $H_4Fe(CN)_6 = 38.03$ per cent.

Triphenylphosphine oxide hydrogen ferrocyanide is insoluble in water and only slightly soluble in warm alcohol. Analysis gave $C = 65.08$; $H = 4.63$; $N = 11.38$; $H_4Fe(CN)_6 = 28.33$;
$$2(C_6H_5)_3PO,H_4Fe(CN)_6$$
requires $C = 65.28$; $H = 4.40$; $N = 10.88$; $H_4Fe(CN)_6 = 27.98$ per cent.

Compounds with Cobalticyanic Acid.

These can all be prepared by mixing hot alcoholic solutions of the acid and the tertiary phosphine oxide. On cooling the mixture, the compounds separate out in a crystalline condition.

Trimethylphosphine oxide hydrogen cobalticyanide is very soluble in warm water or alcohol, from both of which it crystallises in colourless, glistening needles. When heated, it changes to a light green colour at about 200°. Analysis gave C = 33·05 ; H = 5·88 ; N = 19·61 ; $H_3Co(CN)_6$ = 50·35 ; $2(CH_3)_3PO,H_3Co(CN)_6,1\frac{1}{2}H_2O$ requires C = 33·59 ; H = 5·83 ; N = 19·59 ; $H_2Co(CN)_6$ = 50·77 per cent.

The water of crystallisation cannot be accurately determined as the compound does not lose this below 115°, and prolonged heating above this temperature partially decomposes the substance.

Triethylphosphine oxide hydrogen cobalticyanide was obtained in the form of white, prismatic crystals which, when recrystallised from water, melted at 174°. Analysis gave C = 38·66 ; H = 7·37 ; $H_3Co(CN)_6$ = 38·85 ; $2(C_2H_5)_3PO,H_3Co(CN)_6,4H_2O$ requires C = 38·69 ; H = 7·34 and $H_3Co(CN)_6$ = 39·06 per cent.

Tripropylphosphine Oxide Hydrogen Cobalticyanide.— When concentrated aqueous solutions of tripropylphosphine oxide and cobalticyanic acid are mixed, pale yellow, cubical crystals separate out quite readily. These, when recrystallised from water, still retain a faint yellow colour and melt at 119°. Analysis gave C = 42·27 ; H = 8·10 ; N = 12·40 ; $H_3Co(CN)_6$ = 32·33 ; $2(C_3H_7)_3PO,H_3Co(CN)_6,6H_2O$ requires C = 42·48 ; H = 8·41 ; N = 12·39 ; $H_3Co(CN)_6$ = 32·15 per cent.

Triphenylphosphine oxide hydrogen cobalticyanide is a very pale yellow substance soluble in water and alcohol. From the latter it separates out in transparent, indefinite, prismatic crystals, which melt at 172°. Analysis gave C = 60·57 ; H = 5·05 ; N = 10·11 ; $H_3Co(CN)_6$ = 25·71 ; $2(C_6H_5)_3PO,H_3Co(CN)_6,3H_2O$ requires C = 60·87 ; H = 4·71 ; N = 10·14 ; $H_3Co(CN)_6$ = 26·33 per cent.

Compounds with Chloroauric Acid.

These compounds are prepared by mixing hot concentrated aqueous or alcoholic solutions of the phosphine oxide and chloroauric acid. On cooling, the solution deposits beautiful, yellow, flat, hexagonal plates, which after washing with cold water are sufficiently pure for analysis.

Trimethylphosphine oxide hydrogen aurichloride is very soluble in hot water, from which it readily crystallises ; it melts at 94·5°. Analysis gave C = 13·52 ; H = 3·58 ; Au = 37·8 ; $2(CH_3)_3PO,HAuCl_4$ requires C = 13·74 ; H = 3·63 ; Au = 37·6 per cent.

Triethylphosphine oxide hydrogen aurichloride is less soluble in water than the corresponding methyl compound and melts at 54—56°. Analysis gave Au = 32·8 ; $2(C_2H_5)_3PO,HAuCl_4$ requires Au = 32·3 per cent.

Tripropylphosphine oxide hydrogen aurichloride crystallises from hot water and melts at 67—69°. Analysis gave C = 30·90 ; H = 6·35 ;

$Au = 28\cdot51$; $2(C_3H_7)_3PO,HAuCl_4$ requires $C = 31\cdot20$; $H = 6\cdot22$; $Au = 28\cdot47$ per cent.

Triphenylphosphine oxide hydrogen aurichloride melts at 179°. Analysis gave $Au = 22\cdot09$; $2(C_6H_5)_3PO,HAuCl_4$ requires $Au = 21\cdot99$ per cent.

Tribenzylphosphine oxide hydrogen aurichloride melts at 222·5°. Analysis gave $Au = 20\cdot3$; $2(C_7H_7)_3PO,HAuCl_4$ requires $Au = 20\cdot1$ per cent.

Compounds with Dichromic Acid.

Solutions of the phosphine oxides when mixed with saturated solutions of chromic anhydride in dilute sulphuric or nitric acid readily deposit crystals of these molecular compounds, which are very similar in appearance to potassium dichromate. On exposure to light, they turn black and then become insoluble in water.

Trimethylphosphine oxide hydrogen dichromate crystallises from water or dilute nitric acid in bright red, hard, glistening prisms which darken at 200° and begin to melt at 204°. Analysis gave $H_2Cr_2O_7 = 54\cdot5$; $CrO_3 = 49\cdot8$; $2(CH_3)_3PO,H_2Cr_2O_7$ requires $H_2Cr_2O_7 = 54\cdot2$; $CrO_3 = 49\cdot7$ per cent.

Triethylphosphine oxide hydrogen dichromate crystallises in hair-like needles and melts at 100—102°. Analysis gave $C = 29\cdot64$; $H = 6\cdot20$; $H_2Cr_2O_7 = 45\cdot14$; $2(C_2H_5)_3PO,H_2Cr_2O_7$ requires $C = 29\cdot63$; $H = 6\cdot59$; $H_2Cr_2O_7 = 44\cdot86$ per cent.

Tripropylphosphine oxide hydrogen dichromate melts and decomposes at 164°. It is only slightly soluble in water. Analysis gave $C = 38\cdot09$; $H = 7\cdot38$; $H_2Cr_2O_7 = 44\cdot53$; $2(C_3H_7)_3PO,H_2Cr_2O_7$ requires $C = 37\cdot89$; $H = 7\cdot72$; $H_2Cr_2O_7 = 44\cdot86$ per cent.

Compounds with Iodobismuthic Acid.

Trimethylphosphine oxide hydrogen bismuthic iodide was prepared by mixing concentrated aqueous solutions of the oxide and potassium bismuthic iodide. On adding hydriodic acid, brilliant red crystals separated out, which crystallised from warm dilute hydriodic acid in clusters of bright red, prismatic needles. It is decomposed by excess of water, giving a dirty brown powder (probably of BiI_3) which dissolves in hydriodic acid. When heated, it blackens and evolves dark vapours, chiefly iodine. Analysis gave $C = 8\cdot10$; $H = 2\cdot34$; $Bi = 22\cdot93$; $2(CH_3)_3PO,HBiI_4$ requires $C = 7\cdot99$; $H = 2\cdot11$; $Bi = 23\cdot08$ per cent.

Triethylphosphine oxide hydrogen bismuthic iodide was prepared in a similar manner to the methyl compound, which it closely resembles in properties. When heated, it blackens and decomposes at 130°.

Analysis gave $C = 14\cdot45$; $H = 3\cdot12$; $Bi = 21\cdot33$; $2(C_2H_5)_3PO,HBiI_4$
requires $C = 14\cdot62$; $H = 3\cdot14$; $Bi = 21\cdot21$ per cent.

Compounds with Iodomercuric Acid.

Triethylphosphine oxide hydrogen mercuric iodide was prepared by
mixing concentrated aqueous solutions of the oxide and potassium
mercuric iodide. To this solution, cooled in a freezing mixture, was
added hydriodic acid, the reddish-brown oil which separated was
treated with a dilute solution of sodium thiosulphate and desiccated.
After a short time, the oil solidified to a mass of beautiful, yellow,
crystalline needles, which were pressed and dried on a porous plate.
On heating, the crystals soften at $31°$ and melt at $32—33°$. Analysis
gave $C = 16\cdot81$; $H = 3\cdot66$; $HI = 16\cdot2$; $2(C_2H_5)_3PO,HHgI_3$ requires
$C = 16\cdot94$; $H = 3\cdot65$; $HI = 15\cdot1$ per cent.

Tripropylphosphine oxide hydrogen mercuric iodide was prepared in a
similar manner to the corresponding ethyl compound, which it closely
resembles in properties; it melts at $52—54°$. Analysis gave $C = 22\cdot51$;
$H = 4\cdot70$; $HI = 12\cdot6$; $2(C_3H_7)_3PO,HHgI_3$ requires $C = 23\cdot12$; $H = 4\cdot60$;
$HI = 13\cdot7$ per cent.

Compounds with Chloroplatinic Acid.

Compounds of the oxides and chloroplatinic acid are readily obtained
by mixing warm concentrated aqueous solutions of the two components.
When the solutions cool they separate out in beautiful, large, flaky
crystals, which seem to vary in colour with the amount of acid in the
molecule.

Trimethylphosphine oxide hydrogen platinichloride is insoluble in
alcohol, but very soluble in warm water, from which it crystallises in
beautiful, deep red, pyramidal crystals which melt at $126°$. Analysis
gave $C = 18\cdot49$; $H = 5\cdot21$; $H_2PtCl_6 = 52\cdot73$; $4(CH_3)_3PO,H_2PtCl_6$
requires $C = 18\cdot51$; $H = 4\cdot88$; $H_2PtCl_6 = 52\cdot67$ per cent.

A similar compound has been described by Hofmann and Collie
(Trans., 1888, **53**, 636), but they give the formula

$$3(CH_3)_3PO,H_2PtCl_6,H_2O.$$

Triethylphosphine oxide hydrogen platinichloride was recrystallised
from dilute hydrochloric acid and melts at $150°$. Analysis gave
$C = 30\cdot34$; $H = 6\cdot44$; $H_2PtCl_6 = 43\cdot53$; $4(C_2H_5)_3PO,H_2PtCl_6$ requires
$C = 30\cdot44$; $H = 6\cdot55$; $H_2PtCl_6 = 43\cdot34$ per cent.

Tripropylphosphine oxide hydrogen platinichloride crystallises in light
brown flakes, which are soft and are quite different in appearance from
the methyl and ethyl compounds; it is very soluble in water and melts
at $92—93°$. Analysis gave $C = 44\cdot18$; $H = 9\cdot19$; $6(C_7H_7)_3PO,H_2PtCl_6$
requires $C = 44\cdot17$; $H = 8\cdot73$ per cent.

Tribenzylphosphine oxide hydrogen platinichloride has been described by Fleissner (*loc. cit.*) and also by Letts and Collie (*Trans. Roy. Soc. Edin.*, 1888, **30**, Pt. I, 181). We have repeated their experiments and confirmed the formula $4(C_7H_7)_3PO,H_2PtCl_6$ given by the latter. We could not obtain a compound having the formula $3(C_7H_7)_3PO,PtCl_4$, as described by Fleissner. The compound melts and decomposes at 240—241°.

Compounds with Organic Acids.

Trimethylphosphine oxide hydrogen camphorate was prepared by mixing concentrated alcoholic solutions of the oxide and the acid. On standing in a vacuum, beautiful, clear, hard crystals were formed, which, after recrystallisation from alcohol, melted at 91—93°. Analysis gave $C = 50.06$; $H = 9.50$; $C_8H_{14}(CO_2H)_2 = 52.58$;
$$2(CH_3)PO,C_8H_{14}(CO_2H)_2$$
requires $C = 50.00$; $H = 8.86$; $C_8H_{14}(CO_2H)_2 = 52.09$ per cent.

Triethylphosphine oxide hydrogen pyruvate was prepared by mixing concentrated alcoholic solutions of the oxide and the acid. On standing, small, colourless, prismatic needles separated out. After recrystallisation from alcohol, they melted at 75—77°. Analysis gave $C = 46.88$; $H = 6.96$; $CH_3 \cdot CO \cdot CO_2H = 57.7$;
$$(C_2H_5)_3PO,2CH_3 \cdot CO \cdot CO_2H$$
requires $C = 46.45$; $H = 7.42$; $CH_3 \cdot CO \cdot CO_2H = 58.8$ per cent.

Trimethylphosphine oxide hydrogen trichloroacetate was prepared by mixing concentrated aqueous solutions of the oxide and the acid. On standing, small, colourless crystals separated out. They are quite odourless, are not deliquescent, and melt at 67°. Analysis gave $C = 23.29$; $H = 4.10$; $CCl_3 \cdot CO_2H = 64.2$; $(CH_3)_3PO,CCl_3 \cdot CO_2H$ requires $C = 23.48$; $H = 3.91$; $CCl_3 \cdot CO_2H = 64.0$ per cent.

Triphenylphosphine oxide hydrogen trichloroacetate was prepared by mixing concentrated alcoholic solutions of the oxide and the acid. On standing for some time, small crystals separated out, which melted at 95—97° ; it was recrystallised from alcohol, from which it separates in colourless rhombs which melt at 97—99°. Analysis gave $C = 54.17$; $H = 3.80$; $CCl_3 \cdot CO_2H = 37.09$; $(C_6H_5)_3PO,CCl_3 \cdot CO_2H$ requires $C = 54.35$; $H = 3.62$; $CCl_3 \cdot CO_2H = 37.03$ per cent.

Compounds with Hydrochloric Acid.

These derivatives, of which we have only succeeded in preparing the phenyl and benzyl compounds, are obtained by dissolving the phosphine oxide in a warm alcoholic solution of hydrochloric acid. On standing, white, cubical crystals separate out.

Triphenylphosphine oxide hydrogen chloride melts at 183—185°.

Analysis gave $HCl = 11\cdot56$; $(C_6H_5)_3PO,HCl$ requires $HCl = 11\cdot61$ per cent.

Tribenzylphosphine oxide hydrogen chloride separates from alcoholic hydrochloric acid in clusters of small, white crystals, which, on heating, melt at 169° and evolve bubbles of gas ; then they solidify and remelt at 208—210° (the free oxide melts at 214°). On adding an aqueous solution of sodium carbonate to a warm alcoholic solution of the hydrochloride, a faint effervescence is seen, and on cooling crystals of the oxide separate. On analysis, the crystals gave $HCl = 10\cdot43$ (by acidimetry) and $10\cdot25$ by titration with silver nitrate ; $(C_7H_7)_3PO,HCl$ requires $HCl = 10\cdot24$ per cent.

Compounds of the Phosphine Oxides with Metallic Salts.

These compounds, which are prepared by mixing a concentrated aqueous or alcoholic solution of the phosphine oxide with an aqueous or alcoholic solution of the particular salt, generally separate quite readily, and were recrystallised and analysed.

Trimethylphosphine oxide zinc iodide was obtained in small, white prisms which are not deliquescent ; they are very soluble in water and in hot alcohol, and on recrystallisation from the latter solvent melt at 168°. Analysis gave $C = 14\cdot40$; $H = 3\cdot78$; $2(CH_3)_3PO,ZnI_2$ requires $C = 14\cdot31$; $H = 3\cdot58$ per cent.

Triethylphosphine oxide cupric chloride separates from alcohol in light brown prisms which melt at 233° to a dark liquid ; it is decomposed by water. Analysis gave $C = 26\cdot92$; $H = 5\cdot48$; $(C_2H_5)_3PO,CuCl_2$ requires $C = 26\cdot87$; $H = 5\cdot60$ per cent.

Hofmann (*Annalen*, 1861, Suppl. I, 2) described the compound $2(C_2H_5)_3PO,ZnI_2$, and Pebal (*Annalen*, 1862, **120**, 194) the compound $2(C_2H_5)_3PO,CuSO_4$.

Triphenylphosphine oxide cadmium iodide crystallises from hot alcohol in white, stout prisms which melt at 192·5° and are insoluble in water. The analysis of this compound gave $C = 46\cdot80$; $H = 3\cdot64$, whereas $2(C_6H_5)_3PO,CdI_2$ requires $C = 46\cdot85$; $H = 3\cdot25$ per cent.

Triphenylphosphine oxide zinc chloride crystallises from alcohol in hard, colourless, small prisms, which are insoluble in water and melt at 229—230°. Analysis gave $C = 62\cdot82$; $H = 4\cdot43$; $2(C_6H_5)_3PO,ZnCl_2$ requires $C = 62\cdot43$; $H = 4\cdot34$ per cent.

Triphenylphosphine oxide mercuric chloride crystallises from alcohol in white prisms which melt at 120—122°. Analysis gave $C = 52\cdot37$; $H = 3\cdot79$; $2(C_6H_5)_3PO,HgCl_2$ requires $C = 52\cdot24$; $H = 3\cdot63$ per cent.

Triphenylphosphine oxide cobalt chloride separates from warm alcohol in deep blue, cubical crystals which melt at 233°. Analysis gave $C = 62\cdot92$; $H = 4\cdot61$; $2(C_6H_5)_3PO,CoCl_2$ requires $C = 62\cdot98$; $H = 4\cdot38$ per cent.

Triphenylphosphine oxide zinc iodide crystallises from alcohol in small, colourless rhombs and melts at 223—226°. Analysis gave C = 49·28 ; H = 3·44 ; $2(C_6H_5)_3PO,ZnI_2$ requires C = 49·37 ; H = 3·43 per cent.

Compound of Methyl Magnesium Iodide and Tribenzylphosphine Oxide.

A mixture of 250 c.c. of dry benzene (free from thiophen), methyl iodide (1·5 grams), and magnesium powder (0·3 gram) was boiled for several hours ; the solution remained quite clear and no magnesium dissolved. Then to this mixture was added tribenzylphosphine oxide (3·0 grams) dissolved in benzene. After a short time, the clear liquid became cloudy and the magnesium began to dissolve. The mixture was boiled for seven hours and then filtered hot. On cooling, small, colourless, prismatic needles separated ; these were collected, washed with the mother liquor, and dried in a desiccator. When heated, they shrivel up and melt at 163—166°. When treated with dilute hydrochloric acid, they are decomposed, a colourless, inflammable gas is evolved, the free oxide is formed, and the solution contains magnesium chloride and iodide. The compound undergoes partial decomposition when recrystallised from benzene. An estimation of the metallic constituent gave Mg = 2·41 ; $2(C_7H_7)_3PO,CH_3MgI$ requires Mg = 2·98 per cent.

Experiments to show that Triethylphosphine Oxide is a very weak base.

The birotation of a 5 per cent. solution of dextrose was compared with that of a similar solution containing also 1 per cent. of triethylphosphine oxide. The rate of change in the rotation of the dextrose was found to be the same in both solutions, showing that the oxide is a very weak base and that its aqueous solution contains no hydroxyl ions.

TABLE I.—Dextrose alone at 25°.			TABLE II.—Dextrose with the Oxide at 25°.		
Time in minutes.	Rotation.	$\frac{1}{t}\log\frac{a_0 - a_\infty}{a_t - a_\infty}$.	Time in minutes.	Rotation.	$\frac{1}{t}\log\frac{a_0 - a_\infty}{a_t - a_\infty}$.
0	11·53°	—	0	11·00°	—
8	10·89	0·0107	5	10·66	0·0108
18	10·23	0·0108	18	10·00	0·0108
38	9·32	0·0109	30	9·47	0·0108
45	9·16	0·0104	46	9·06	0·0105
57	8·84	0·0105	54	8·86	0·0108
62	8·64	0·0114	71	8·65	0·0103
∞	7·94	—	∞	8·10	—

Mean K = 0·0107.   Mean K = 0·0107.

Experiments to determine the amount of Hydrolysis of the Compounds of Trimethylphosphine Oxide with Trichloroacetic and Cobalticyanic Acids.

A 25 per cent. solution of cane sugar was inverted with the acids and the rate of inversion compared with that of a similar solution containing the compound of the oxide and the acid. The experiments were performed at 30° and the strength of the solutions taken so that they each contained the same amount (N = 0·158) of acid, free or combined.

TABLE III.—*Cane Sugar and Trichloroacetic Acid.*

Time in minutes.	Rotation.	$1/t \log \frac{a_0 - a_\infty}{a_t - a_\infty}$.
0	+15·93°	—
112	13 16	0·000573
155	12·20	0·000573
205	11·10	0·000579
243	10·30	0·000585
307	8·98	0 000597
437	6·65	0·000612
2840	−1·11	0·000577
∞	−4·24	—

Mean $K = 0·000585$·

TABLE IV.—*Cane Sugar and the Compound of Oxide and Cobalticyanic Acid.*

Time in minutes.	Rotation.	$1/t \log \frac{a_0 - a_\infty}{a_t - a_\infty}$.
0	+16·01°	—
77	·14·25	0·000510
103	13·67	0·000515
123	13·20	0·000525
144	12·82	0·000514
232	11·08	0·000519
298	9·96	0·000514
328	9·36	0·000524
∞	−4·23	—

Mean $K = 0·000517$.

Whence the approximate hydrolysis $\frac{517}{585} \times 100 = 88·3$ per cent.

TABLE V.—*Cane Sugar and Cobalticyanic Acid.*

Time in minutes.	Rotation.	$1/t \log \frac{a_0 - a_\infty}{a_t - a_\infty}$.
0	+15·71°	—
100	13·11	0·000596
165	11·46	0·000621
285	8·67	0·000652
340	7·80	0·000635
348	7·76	0·000622
441	6·52	0·000641
1810	−2·89	0·000607
∞	−4·50	—

Mean $K = 0·000625$.

TABLE VI.—*Cane Sugar and the Compound of Oxide and Cobalticyanic Acid.*

Time in minutes.	Rotation.	$1/t \log \frac{a_0 - a_\infty}{a_t - a_\infty}$.
0	+15·87°	—
55	14·52	0·000544
88	13·71	0·000555
152	12·23	0·000565
330	8·93	0·000551
402	7·77	0·000551
1515	−1·52	0·000558
1900	−2·55	0·000545
∞	−4·52	—

Mean $K = 0·000553$.

Whence the approximate hydrolysis $\frac{553}{625} \times 100 = 88·5$ per cent.

We wish to acknowledge with grateful thanks the patient assistance given to us by Mr. Charles Allen in some of the preliminary work for this paper.

MUNICIPAL TECHNICAL SCHOOL,
BLACKBURN.

XXXII.—A Synthesis of Aldehydes by Grignard's Reaction.

By GORDON WICKHAM MONIER-WILLIAMS.

WHEN two molecules of an organo-magnesium compound are allowed to react with one molecule of ethyl formate, the chief product of the reaction is a secondary alcohol, but if the formic ester is in excess, an aldehyde is formed according to the following equation (Gattermann and Maffezzoli, *Ber.*, 1903, **36**, 4152):

$$HCO \cdot OC_2H_5 + IMgX = C_2H_5 \cdot O \cdot MgI + X \cdot CHO.$$

In addition to this, the first synthesis of aldehydes by Grignard's reaction, numerous other methods have been proposed by various investigators, which all depend on the action of organo-magnesium compounds on derivatives of ethyl formate. Bromoform, iodoform, ethyl orthoformate, disubstituted formamides, and isonitriles have all yielded aldehydes when allowed to react with organo-magnesium compounds. It might appear superfluous to look for a further synthesis of aldehydes along these lines, were it not for the fact that most of the above methods, although giving very good results with certain substances, are found to fail completely in other cases, or to give at most an exceedingly small yield. On this account it seemed desirable to find some other method by which better yields might be obtained.

The following synthesis depends on the action of organo-magnesium compounds on ethoxymethyleneaniline, $C_6H_5N{:}CH \cdot O \cdot C_2H_5$, which leads in the first place not to the aldehydes themselves, but to the anhydro-compounds of these with aniline:

$$C_6H_5N{:}CH \cdot O \cdot C_2H_5 + IMgX = IMgO \cdot C_2H_5 + X \cdot CH{:}N \cdot C_6H_5.$$

By the action of mineral acids, the anhydro-compounds are hydrolysed with formation of free aldehyde and aniline :

$$X \cdot CH{:}NC_6H_5 + H_2O + HCl = X \cdot CHO + C_6H_5 \cdot NH_2,HCl.$$

To determine the most favourable conditions of reaction, experi ments were first carried out with bromobenzene and o-bromotoluene, from which the yields of the respective aldehydes were more than 50 per cent. The method was then applied to the synthesis of a- and β-naphthaldehydes from a- and β-bromonaphthalenes, and the yields in both cases indicated a considerable improvement on the older methods. The latter of these naphthaldehydes has been comparatively little investigated, and several derivatives of it were therefore prepared.

My chief object was the preparation of a new class of aldehydes of the aromatic series, namely, those containing a thio-ether group attached to the benzene ring. The starting point for this synthesis was p-chloronitrobenzene, the chlorine atom of which is easily exchanged for the SH group. From the resulting p-nitrothiophenol, the ethyl ether was obtained, and this on reduction yielded the hitherto unknown p-thiophenetidine, the amino-group of which ·was diazotised and replaced by iodine. The resulting p-iodothiophenetole was used for the synthesis of p-thiophenetylaldehyde, $C_2H_5 \cdot S \cdot C_6H_4 \cdot CHO$.

For the preparation of ethoxymethyleneaniline, I made use at first of Claisen's method (Annalen, 1895, 287, 362), that is, the condensation of aniline with ethyl orthoformate, but this was afterwards abandoned in favour of the method given by Comstock and Kleeberg (Amer. Chem. J., 1890, 12, 497), which consists in treating with ethyl iodide the silver derivative, $C_6H_5N:CH \cdot OAg$, of formanilide. The dry silver derivative is mixed with the calculated quantity of the iodide in absolute ether, the mixture left for twenty-four hours, after which the silver iodide formed is filtered off, and the ether evaporated. It is somewhat difficult to obtain the silver formanilide perfectly pure and dry when operating with large quantities. If any moisture is present, the resulting ethoxymethyleneaniline undergoes more or less complete decomposition into diphenylformamidine and alcohol, and cannot be used for the aldehyde synthesis. The silver salt, precipitated according to Comstock and Kleeberg's method, forms a finely-divided white powder, which decomposes on heating and gives up its moisture with difficulty under reduced pressure over calcium chloride. It was, however, found that the moist substance could be quickly and completely dried by treating it in a porcelain dish with light petroleum, which has the effect of expelling the water and causing it to collect in drops on the surface, whence it can be easily removed by pouring off the main portion and taking up the rest with pieces of filter paper. This action of the petroleum is evidently due to surface tension ; the substance is found to have completely exchanged its water for petroleum, and if spread out on a porous plate may be obtained perfectly dry in a few minutes. It is very sensitive to the action of light. An analysis of the salt dried in this way gave the following result :

0·5122 gave 0·2426 Ag.   Ag = 47·36.

C_7H_6ONAg requires Ag = 47·34 per cent.

The ethoxymethyleneaniline, obtained from the dried silver form-anilide, is a light yellow oil. When perfectly dry it can be kept for several days in a well-closed vessel, and need not be distilled before use.

Preparation of Benzaldehyde, o-Tolualdehyde, and α- and β-Naphth-aldehydes.

Numerous experiments were made to ascertain the conditions under which the best results could be obtained. The method will therefore be given in the form in which it was eventually applied to the pre-paration of β-naphthaldehyde.

Five grams of β-bromonaphthalene dissolved in a small quantity of anhydrous ether are heated with 0·6 gram of magnesium filings in a reflux apparatus until almost all the magnesium is dissolved. The addi-tion of traces of iodine and ethyl iodide helps the reaction considerably. Anhydrous ether is added until the volume is about 90 c.c., and the mixture heated again to boiling for a few minutes. A solution of 3·6 grams of ethoxymethyleneaniline in a small quantity of anhydrous ether is then added gradually from a dropping funnel to the warm ethereal solution of the magnesium naphthyl bromide, the liquid being well stirred during the addition. The product separates on the bottom and sides of the flask as a viscid, reddish-brown mass, soluble with difficulty in ether. The mixture is then carefully treated with well-cooled, moderately dilute hydrochloric acid in a large flask, the ether evaporated, and the β-naphthaldehyde together with naphthalene distilled over with steam. The aldehyde is separated from the naphthalene by means of a fairly-concentrated solution of sodium hydrogen sulphite, and is obtained pure by a second steam distillation. Thus obtained, it is a white, crystalline substance melting at 61°, and is only very slowly oxidised by the air. The yield amounted to 36 per cent., and in the case of benzaldehyde, o-tolualdehyde, and α-naphth-aldehyde the yields were 46, 54, and 48 per cent. respectively.

It will be seen from the accompanying tables how large is the influence on the yield of (1) the concentration of the ethereal solution and (2) the temperature at which the reaction is effected.

(i) *Influence of Concentration.*

Temperature = 35°. Volume of ether = 50—60 c.c.

Amount of bromonaphthalene.	Yield of aldehyde.
2·5 grams	23 per cent.
5 ,,	27 ,,
10	18 ,,
14	14 ,,
30	5 ,,

(ii) *Influence of Temperature.*

Amount of bromonaphthalene = 5 grams. Volume of ether = 90 c.c.

Temperature.	Yield.
+ 35°	36 per cent.
− 15°	3 ,,

It is therefore evident that the reaction proceeds best at considerable dilution and at a fairly high temperature. An attempt was made to improve the yield still further by using boiling toluene as a solvent, as recommended by Bodroux in his synthesis with ethyl orthoformate (*Compt. rend.*, 1904, **138**, 700), but in this case unsatisfactory results were obtained, the temperature being evidently too high.

The *bisulphite* compound of β-naphthaldehyde,

$$C_{10}H_7 \cdot CH(OH) \cdot O \cdot SO_2Na,$$

was obtained by dissolving the aldehyde in an excess of sodium hydrogen sulphite solution, and crystallises on cooling in lustrous, white plates. It is only sparingly soluble, 100 c.c. of moderately dilute sodium hydrogen sulphite solution dissolving 0·5 gram at the ordinary temperature.

0·1009 gave 0·0936 $BaSO_4$. S = 12·77.

$C_{11}H_9O_4NaS$ requires S = 12·31 per cent.

The *phenylhydrazone*, $C_{10}H_7 \cdot CH:N \cdot NH \cdot C_6H_5$, crystallises in colourless leaflets which melt at 205—206° with partial decomposition and become reddish-brown when exposed to light.

0·1126 gave 11·35 c.c. nitrogen at 16·2° and 750 mm. N = 11·58.

$C_{17}H_{14}N_2$ requires N = 11·40 per cent.

The *semicarbazone*, $C_{10}H_7 \cdot CH:N \cdot NH \cdot CO \cdot NH_2$, forms long, white needles, soluble with difficulty in water and alcohol (m. p. 245°).

0·1389 gave 24·75 c.c. nitrogen at 18·4° and 744 mm. N = 19·89.

$C_{12}H_{11}ON_3$ requires N = 19·74 per cent.

The *azine*, $C_{10}H_7 \cdot CH:N \cdot N:CH \cdot C_{10}H_7$, crystallises in microscopic, yellow needles, melting at 232°.

0·0808 gave 6·50 c.c. nitrogen at 12° and 747 mm. N = 9·38.

$C_{22}H_{16}N_2$ requires N = 9·11 per cent.

β-Naphthylideneaniline, $C_{10}H_7 \cdot CH:N \cdot C_6H_5$, is easily obtained by warming together β-naphthaldehyde and aniline; it forms microscopic needles of a light yellowish-green colour and melts at 113°.

0·1997 gave 10·4 c.c. nitrogen at 16° and 750 mm. N = 5·99.

$C_{17}H_{13}N$ requires N = 6·06 per cent.

β-*Naphthylacrylic acid*, $C_{10}H_7\text{·}CH\text{:}CH\text{·}CO_2H$, is best prepared by the condensation of β-naphthaldehyde with aniline or ammonia and malonic acid (Knoevenagel, *Ber.*, 1898, **31**, 2585); when thus obtained, it forms small, white needles melting at 203°, which are fairly soluble in hot water.

0·1075 gave 0·3113 CO_2 and 0·0535 H_2O.   C = 78·96 ; H = 5·57.

$C_{13}H_{10}O_2$ requires C = 78·75 ; H = 5·10 per cent.

β-*Naphthylpropionic acid*, $C_{10}H_7\text{·}CH_2\text{·}CH_2\text{·}CO_2H$, is obtained by reducing β-naphthylacrylic acid with sodium amalgam. It forms brilliant, white leaflets melting at 129—130°, and is much more easily soluble in hot water than the corresponding acrylic acid.

0·1961 gave 0·5619 CO_2 and 0·1104 H_2O.   C = 78·14 ; H = 6·31.

$C_{13}H_{12}O_2$ requires C = 78·00 ; H = 6·00 per cent.

The nitration of the aldehyde was effected by adding the finely-powdered substance gradually to a well-cooled solution of potassium nitrate (1 mol.) in concentrated sulphuric acid. After standing for three hours at 0°, the liquid was poured into ice and the precipitate collected. By repeated extraction with xylene, treatment with animal charcoal, and precipitation with ligroin, a small quantity of a crystalline substance was obtained which, on recrystallisation from xylene, formed silky, white needles melting at 206·5°. On analysis, it proved to be a dinitro-β-naphthaldehyde, $C_{10}H_5(NO_2)_2\text{·}CHO$.

0·1372 gave 14·2 c.c. nitrogen at 23° and 744 mm. N = 11·36.

$C_{11}H_6O_5N_2$ requires N = 11·41 per cent.

From the xylene and ligroin filtrates, a very small quantity of a substance crystallising in long, yellow needles melting at 139—140° was obtained, together with much unchanged aldehyde. The amount was too small for analysis, but there can be little doubt that the compound is a mononitro-aldehyde. Both substances give azines of high melting point, that of the dinitro-aldehyde being slightly soluble in hot water. They are both decomposed on warming with caustic soda, giving a yellow coloration, and with acetone and caustic soda they give a dark red coloration, but no indigo. This seems to show that neither of them is an o-nitro-aldehyde.

Preparation of p-*Thiophenetylaldehyde.*

Willgerodt showed in 1885 (*Ber.*, 1885, **18**, 331) that *p*-nitrothio-phenol, $NO_2 \cdot C_6H_4 \cdot SH$, may be obtained by the action of alcoholic potassium hydrosulphide on *p*-chloronitrobenzene, and by the action of sodium and methyl or ethyl iodide on *p*-nitrothiophenol, Blanksma (*Rec. Trav. chim.*, 1901, **20**, 403) has obtained *p*-nitrothioanisole, $NO_2 \cdot C_6H_4 \cdot S \cdot CH_3$, and *p*-nitrothiophenetole, $NO_2 \cdot C_6H_4 \cdot S \cdot C_2H_5$. *p*-Nitrothiophenol, prepared by Willgerodt's method, was heated with the calculated quantity of sodium ethoxide and ethyl iodide in a reflux apparatus until the dark red colour of the liquid changed to light yellow. On filtering from the sodium iodide and allowing to cool, long, yellow needles of the thio-ether separated out, and on adding water to the mother liquor the remainder was obtained as an amorphous or semi-crystalline mass. The pure substance melts at 44°.

The *p*-nitrothiophenetole was reduced by warming gently with tin and hydrochloric acid on the water-bath. A violet, crystalline substance was formed which is probably a compound of the amine hydrochloride with stannous chloride. The liquid was made alkaline with sodium hydroxide and the *p*-thiophenetidine, $NH_2 \cdot C_6H_4 \cdot S \cdot C_2H_5$, distilled over with steam. On distillation it was obtained as a faintly yellow oil boiling at 280—281° with a somewhat unpleasant odour. The crude oil, however, before it has been distilled, has hardly any odour. The yield is 93 per cent.

0·2058 gave 0·3139 $BaSO_4$.   S = 20·96.

$C_8H_{11}NS$ requires S = 20·91 per cent.

The acetyl compound of the amine, thiophenacetin,

$$C_2H_5 \cdot S \cdot C_6H_4 \cdot NH \cdot CO \cdot CH_3,$$

prepared by warming the amine with acetic anhydride and precipitating with water, crystallises from dilute alcohol in flat, colourless needles several centimetres in length. On account of its insolubility in cold water, this substance does not exert the physiological action of phenacetin.

0·1323 gave 0·1587 $BaSO_4$.   S = 16·50.

$C_{10}H_{13}ONS$ requires S = 16·41 per cent.

The amine was dissolved in dilute hydrochloric acid and diazotised with a solution of sodium nitrite. The clear diazo-solution was treated with the calculated quantity of potassium iodide in water, and the *p*-iodothiophenetole, $I \cdot C_6H_4 \cdot S \cdot C_2H_5$, which separated out was obtained on distilling with steam as a reddish-yellow oil. When distilled under atmospheric pressure, the substance was almost completely decomposed, but by distilling in a partial vacuum it was obtained as a faintly yellow oil of pleasant odour boiling at 146—147° under 11 mm. pressure. The yield was 85 per cent.

0·2234 gave 0·1938 $BaSO_4$ and 0·1994 AgI. S = 11·93 ; I = 48·23.

C_8H_9IS requires S = 12·12 ; I = 48·10 per cent.

p-*Thiophenetylaldehyde*, $C_2H_5 \cdot S \cdot C_6H_4 \cdot CHO$, was prepared from the iodo-compound by the method given above for β-naphthaldehyde, the best results being obtained with 5 grams of the iodo-compound and 70 c.c. of ether. It is a yellow liquid of very characteristic and not unpleasant odour boiling at 244—245° ; the yield is 32 per cent. It is remarkable that the boiling point is not very different from that of the corresponding p-phenetylaldehyde ; from the analogy of other nearly related oxy- and thio-compounds, a higher boiling point would be expected.

The *azine* of p-thiophenetylaldehyde,

$$C_2H_5 \cdot S \cdot C_6H_4 \cdot CH\text{:}N \cdot N\text{:}CH \cdot C_6H_4 \cdot S \cdot C_2H_5,$$

is very characteristic, crystallising from glacial acetic acid in well-defined, golden-yellow leaflets which melt at 152° ; it gives with concentrated sulphuric acid an intense red coloration, a reaction which is also shown by other compounds of this series.

0·1154 gave 0·1653 $BaSO_4$. S = 19·71.

0·1779 , 14·3 c.c. nitrogen at 26° and 740·3 mm. N = 8·63.

$C_{18}H_{20}N_2S_2$ requires S = 19·51 ; N = 8·54 per cent.

The *phenylhydrazone*, $C_2H_5 \cdot S \cdot C_6H_4 \cdot CH\text{:}N \cdot NH \cdot C_6H_5$, crystallises from dilute alcohol in colourless plates which melt at 115° and are coloured reddish-brown by exposure to light.

0·1120 gave 0·1033 $BaSO_4$. S = 12·68·

$C_{15}H_{16}N_2S$ requires S = 12·50 per cent.

The *semicarbazone*, $C_2H_5 \cdot S \cdot C_6H_4 \cdot CH\text{:}N \cdot NH \cdot CO \cdot NH_2$, forms long, white needles melting at 193° ; it is only sparingly soluble in dilute alcohol.

0·1717 gave 0·1819 $BaSO_4$. S = 14 55.

$C_{10}H_{13}ON_3S$ requires S = 14·36 per cent.

The aldehyde combines even in the cold with p-aminothiophenetole to form a solid condensation product, $C_2H_5 \cdot S \cdot C_6H_4 \cdot CH\text{:}N \cdot C_6H_4 \cdot S \cdot C_2H_5$; it crystallises from alcohol in large leaflets of a faint golden-yellow colour (m. p. 114—115°).

0·2263 gave 0·3522 $BaSO_4$. S = 21·37.

$C_{17}H_{19}NS$ requires S = 21·26 per cent.

The oxidation of the aldehyde was carried out with alkaline permanganate, the mixture· being gently warmed on the water-bath.

Contrary to expectation, the acid $C_2H_5 \cdot S \cdot C_6H_4 \cdot CO_2H$ was not obtained, but a further oxidation took place giving phenylethylsulphonecarboxylic acid, $C_2H_5 \cdot SO_2 \cdot C_6H_4 \cdot CO_2H$. This acid forms white leaflets soluble in water and melting at 211°; an analysis of the silver salt gave the following result:

0·1975 gave 0·0664 Ag. Ag = 33·62.

$C_9H_9O_4SAg$ requires Ag = 33·64 per cent.

In conclusion, I wish to express my sincere thanks for the assistance received during the progress of the work from Professor Ludwig Gattermann, Director of the Chemical Laboratory at Freiburg University. Further investigations on the thioanisyl and thiophenetyl compounds are being carried out in the above laboratory.

XXXIII.—The Effect of Constitution on the Rotatory Power of Optically Active Nitrogen Compounds. Part I.

By Mary Beatrice Thomas and Humphrey Owen Jones.

THE investigation of the effect of the constitution of carbon compounds on their rotatory power has occupied a large number of workers for some years past. The work received a great stimulus in 1890, when the hypotheses of Crum Brown and Guye were suggested, and seemed to offer a possibility of establishing some relation between the molecular structure of a compound and its optical rotatory power.

The hypothesis of Guye, which assumes that the masses of the groups attached to the carbon atom are the chief factors which determine the optical rotatory power of a molecule, is capable of being tested by experiment, and during the past fifteen years an enormous mass of data has been accumulated with that object in view. At first, the results obtained seemed to be in fairly good agreement with the predictions of the hypothesis; but latterly this has not been so, and it has been found that the optical rotatory power of the compounds examined, which are almost invariably liquid or solid non-electrolytes, is affected to such a large extent by molecular association, the nature of the solvent, and by temperature, that it is difficult to apply a real test to the hypothesis.

Electrolytes seem to be free from some of these disturbing causes. In dilute aqueous solution, the Oudemans-Landolt law holds within very narrow limits for strong electrolytes and especially for salts, so that complications due to the effect of solvents and of molecular association are avoided, and there remains only the effect of temperature, which, in cases where it neither alters the electrolytic dissociation to any extent nor causes hydrolytic dissociation, might be expected to be fairly regular if not small. The rotatory power of an ion may be regarded as a constant independent of external conditions except temperature.

It is therefore probable that any regularities in the rotatory power of compounds would be more evident in the case of ions than of undissociated molecules. The examination of the rotatory power of a series of ions, the constitution of which could be varied, would therefore be of interest as affording a test of the applicability of Guye's hypothesis, and might possibly indicate in what direction a further connection between constitution and rotatory power is to be sought.

Suitable carbon compounds into which a number of similar groups of different masses can be introduced are difficult to obtain. Patterson (Trans., 1904, 85, 1116) examined potassium methyl, ethyl, and n-propyl tartrates, and found that the rotatory power of these compounds was dependent to a great extent on concentration and temperature, and that each of the compounds exhibited a maximum rotatory power at a different temperature. These substances, therefore, seem to be affected largely by external conditions, and some part of the effect may be due to hydrolysis.

The simplest compounds available for this purpose are the active sulphur and selenium salts, but in these compounds there is little scope for substitution; the salts might in some cases be very difficult to resolve into their optical antimers and, judging from the compounds hitherto examined, they exert only a feeble rotatory power.

The substituted ammonium compounds containing an asymmetric nitrogen atom seem to offer a better field for investigation. These compounds can be resolved into their optical antimers with moderate ease, their salts are extensively ionised in solution and are not hydrolysed, and the groups present, being all alkyl groups, might therefore be expected to produce similar effects on the rotatory power. Further, in the salts of the two bases hitherto examined, namely, phenylbenzylmethylallylammonium salts and phenylbenzylmethylethylammonium salts, there is such a great difference in their rotatory powers ($[M]_D = 166°$ and $[M]_D = 19°$ respectively for the ions) that the effect of constitution is sufficiently well marked to be readily examined.

The investigation of compounds with similar substituting radicles is important, since here the effect of mass would not be so likely to be masked by the very considerable effect of constitution noticed in many cases, such, for instance, as $-CO \cdot C_6H_5$ and $-CO \cdot CH_2 \cdot C_6H_5$.

Here again the choice of compounds is limited to some extent, since in practice it is only convenient that the last group introduced shall be either methyl, allyl, or benzyl : the iodides of these radicles are the only ones which, as a rule, react extensively with tertiary amines when these contain aromatic and other heavy alkyl groups. Two series of compounds were chosen and examined ; these all contain the phenyl and methyl groups, each series consisting of five compounds, one set containing the benzyl group together with ethyl, *n*- or *iso*propyl, *iso*butyl, and *iso*amyl, the other containing the allyl group together with the same five aliphatic groups.

The methyl compounds of both series have already been shown by one of us to be inactive. The ethyl compound of the benzyl series had been previously examined and a preliminary examination of the ethyl compound of the allyl series had also been made (Trans., 1903, **83**, 1418 ; 1904, **85**, 223). The examination of the *n*-propyl and *iso*butyl compounds of the benzyl series has been undertaken by Wedekind, and his results have recently appeared (Wedekind and Fröhlich, *Ber.*, 1905, **38**, 3438 and 3933).* Thus, including the phenylbenzylmethyl-allylammonium compounds first resolved by Pope and Peachey, there are now eleven optically active quinquevalent nitrogen compounds known, which fall naturally into two series of six, both including the last-named compound.

During the course of the work, a number of interesting observations have been made on the methods of resolving these compounds and on their properties, which will be mentioned in due course.

Method of Resolving the Compounds.

The method used for the resolution of the substituted ammonium compounds was that of Pope and Peachey, and consisted in fractional crystallisation of the *d*-camphorsulphonate or *d*-bromocamphorsulphonate of the base from a suitable solvent. In the preparation of the camphorsulphonate and bromocamphorsulphonate, some trouble was frequently occasioned by the formation of a crystalline double compound of silver iodide and the substituted ammonium iodide, which was partially soluble in organic solvents. This is often avoided by the addition of a little methyl alcohol or water, or by the use of excess of silver camphorsulphonate or bromocamphorsulphonate, but best by adding the ammonium iodide in successive small quantities to the mixture of silver salt and solvent. The choice of solvent is in some cases of the utmost importance, whilst in others almost every

---

* We have delayed the publication of our results until now, in order that Wedekind and Frohlich might publish their work on these compounds. Their experiments will be discussed later.

possible solvent effects resolution. To avoid the possibility of race-misation, which appears to take place at higher temperatures in some cases, it is well, as a rule, to use a solvent of low boiling point, and acetone is the one which has been found most generally useful. The other solvents which have been found useful are ethyl acetate, ethylal, methylal, chloroform, and water. Pope and Peachey (Trans., 1899, **75**, 1127) laid some stress on the necessity for using a dry non-hydroxylic solvent to avoid hydrolytic dissociation ; this we have not found to be necessary ; we have frequently used moist acetone and ethyl acetate, and have in two cases found that the bromocamphor-sulphonate was readily resolved by crystallising from boiling water.

Some cases were found where resolution of the camphorsulphonate could not be effected, whilst the bromocamphorsulphonate was easily resolved, and *vice versâ* ; in other cases, it was possible to resolve both, though the resolution of one of them was usually difficult. Since the object of this investigation was merely to determine the value of the molecular rotatory power of the ion, we have been content to solate the salt of one of the *d*- or *l*-isomerides only, the less soluble of the salts derived from *d*-camphorsulphonic or *d*-bromocamphor-sulphonic acid, to determine its rotatory power in aqueous solution, and thus to obtain the rotatory power of the substituted ammonium ion by difference.

Effect of Concentration on the Rotatory Power of Camphorsulphonates.

In order to determine the extent to which the rotatory power of the salts of these bases in aqueous solution was affected by dilution, *d*-phenylbenzylmethylallylammonium *d*-camphorsulphonate, already described by Pope and Harvey (Trans., 1901, **79**, 828), was isolated by repeated crystallisation from acetone, and its rotatory power in aqueous solutions of different concentration determined. The salt melted at 172—173° ; 1·0044 grams were dissolved in water and the solution weighed 16·932 grams. This solution was examined in a 2-dcm. tube,* its density taken, then it was diluted and the operation repeated at a temperature of 16°.

Strength of solution in grams per 100 grams of solution.	a_D in 2-dcm. tube.	Specific gravity.	$[a]_D$.	$[M]_D$.
5·93	5·47°	1·016	45·4°	214·0°
3·02	2·78	1·004	45·8	216·0
1·54	1·42	1·002	46·0	216·7
1·064	1·00	1·001	46·9	221·0
0·783	0·73	1·000	46·6	219·5
0·553	0·52	1·000	47·0	221·0

* Unless otherwise stated, all the determinations of rotatory power given in this paper have been carried out in 2-decimetre tubes in a Landolt-Lippich triple field polarimeter supplied by Schmidt and Haensch.

The above numbers show that dilution affects the rotation to a slight extent, but that in this particular case the variation is almost within the limits of experimental error when the concentration of the solution is not more than 1 per cent.

The values obtained by Frye and Harvey for a 0·5 per cent. solution were about 21·5°.

A similar experiment was made with l-phenylbenzylmethylamyl-ammonium l-camphorsulphonate $[M_D]$ 343°, see p. 331, and this also showed that the effect of dilution was quite small and negligible for solutions more dilute than 1·5 per cent.

The bromocamphorsulphonates were examined in very dilute solution, so that the effect of concentration was in this way eliminated.

The Rotatory Power of Ions at Different Temperatures.

In order to determine the effect of change of temperature on the rotatory power of the ions under examination, some solutions of the active salts were examined at intervals of about 10° between 10° and 50°: the results are given for each compound. The effect of temperature change on the ammonium salts of l-camphorsulphonic acid and l-bromocamphorsulphonic acid and also on phenylbenzylmethylyl-ammonium l-camphorsulphonate was examined with the following results. The densities of the solutions used were between 1·001 and 1·000 at 15°, and for the other temperatures have been taken equal to those of water at the same temperature, since the effect of density is in these cases negligible.

Ammonium l-Camphorsulphonate.

Three solutions, 1·5 per cent., 1 per cent., and 0·5 per cent., were used. The values given are obtained from curves plotted with the means of the three experiments, which were very concordant. The value of c_D measured was about 15°.

t	$[\alpha]_D$	$[M]_D$
10	19·3	47·7
20	20·7	52·3
30	21·4	54·3
40	22·4	55·4
50	23·2	57·4

Frye and Feacher Trans., 1906, 75, 1185 give $[\alpha]_D$ 20·7° and $[M]_D$ 51·7° at 15°.

Ammonium l-Bromocamphorsulphonate.

The value of $[M]_D$ for the active ion of this salt has long been taken as 273° but Rippins Trans., 1905, 87, 633 has shown that the true value is more nearly 289°. The specimen examined had been recrystallised several times from water, and $[M]_D$ 275° at 15°. The determination of the rotatory power of these solutions at different temperatures gave as a mean the following results.

The concentration of the solutions was about 1 per cent., and the observed rotation was therefore about 1·7°

t	$[\alpha]_D$	$[M]_D$
10	39·7	272
20	45·7	277
30	50·2	292
40	55·16	284
50	59·0	302

d-Phenylbenzylmethylallylammonium d-Camphorsulphonate.

Two experiments were made with solutions of about 1·4 per cent., and gave as a mean the following results: the observed rotation at 20° was 1·27°.

t	$[\alpha]_D$	$[M]_D$	$[M]_D$ for basic ion.
10	46·7	217·0	147·5
17	46·8	218·1	148·1
30	47·4	220·5	150·5
40	47·2	222·0	152·2
50	47·8	220·5	153·5

From the observations recorded above, it was expected that the effect of change of temperature on the rotatory power of the series of ions under examination would be small. This, however, was not found to be the case; the rotatory power of some of the compounds was found to be affected to a considerable extent by temperature change, whilst that of others altered but little. When the effect of the acidic ion was taken into account, it was found that the rotatory power of the basic ion in all cases diminished with increase of temperature.

In many of the measurements recorded here, the value of α_D is so small that great accuracy cannot be claimed for the results, which are, however, quite accurate enough to leave no doubt as to the relative magnitudes of the rotatory powers of all the compounds.

Rotatory Power of the Substituted Ammonium Iodides.

In all cases where the solubility of the iodide in water was small enough, it was precipitated from solutions of the camphorsulphonate or bromocamphorsulphonate by the addition of potassium iodide solution. The rotatory power of the iodide so recovered was examined in solution in alcohol and in chloroform. The rotatory power in chloroform was always greater than that of the same iodide in alcohol. The iodide

284 THOMAS AND JONES: EFFECT OF CONSTITUTION ON ROTATORY

The above numbers show that dilution affects the rotation to a slight extent, but that in this particular case the variation is almost within the limits of experimental error when the concentration of the solution is not more than 1 per cent.

The values obtained by Pope and Harvey for a 0·8 per cent. solution were about 218·6°.

A similar experiment was made with *l*-phenylbenzylmethyl*iso*amyl-ammonium *d*-camphorsulphonate ($[M_D]$ 235°, see p. 296), and this also showed that the effect of dilution was quite small and negligible for solutions more dilute than 1·5 per cent.

The bromocamphorsulphonates were examined in very dilute solution, so that the effect of concentration was in this way eliminated.

The Rotatory Power of Ions at Different Temperatures.

In order to determine the effect of change of temperature on the rotatory power of the ions under examination, some solutions of the active salts were examined at intervals of about 10° between 10° and 50°: the results are given for each compound. The effect of temperature change on the ammonium salts of *d*-camphorsulphonic acid and *d*-bromocamphorsulphonic acid, and also on phenylbenzylmethylallyl-ammonium *d*-camphorsulphonate, was examined with the following results. The densities of the solutions used were between 1·001 and 1·000 at 15°, and for the other temperatures have been taken equal to those of water at the same temperature, since the effect of density is in these cases negligible.

Ammonium d-Camphorsulphonate.

Three solutions, 1·6 per cent., 2 per cent., and 1·8 per cent., were used. The values given are obtained from curves plotted with the means of the three experiments, which were very concordant. The value of a_D measured was about 0·8°.

t.	$[a]_D$.	$[M]_D$.
10°	19·9°	49·5°
20	20·7	51·6
30	21·8	53·6
40	22·3	55·6
50	23·1	57·6

Pope and Peachey (Trans., 1899, **75**, 1085) give $[a]_D$ 20·7° and $[M]_D$ 51·7° at 16°.

Ammonium d-Bromocamphorsulphonate.

The value of $[M]_D$ for the acidic ion of this salt has long been taken as 270°, but Kipping (Trans., 1905, **87**, 628) has shown that the true value is more nearly 280°. The specimen examined had been recrystallised several times from water, and $[M]_D$ 275° at 15°. The determination of the rotatory power of three solutions at different temperatures gave as a mean the following results.

The concentration of the solutions was about 1 per cent., and the observed rotation was therefore about 1·7°.

t.	$[\alpha]_D$.	$[M]_D$.
10	83·5°	273°
20	84·7	277
30	86·25	282
40	87·46	286
50	89·0	291

d-*Phenylbenzylmethylallylammonium* d-*Camphorsulphonate*.

Two experiments were made with solutions of about 1·4 per cent., and gave as a mean the following results; the observed rotation at 20° was 1·27°.

t.	$[\alpha]_D$.	$[M]_D$.	$[M]_D$ for basic ion.
10°	46·0°	217·0°	167·5°
17	46·3	218·1	167·1
30	46·8	220·5	166·9
40	47·2	222·0	166·4
50	47·3	222·5	164·4

From the observations recorded above, it was expected that the effect of change of temperature on the rotatory power of the series of ions under examination would be small. This, however, was not found to be the case ; the rotatory power of some of the compounds was found to be affected to a considerable extent by temperature change, whilst that of others altered but little. When the effect of the acidic ion was taken into account, it was found that the rotatory power of the basic ion in all cases diminished with increase of temperature.

In many of the measurements recorded later, the value of a_D is so small that great accuracy cannot be claimed for the results, which are, however, quite accurate enough to leave no doubt as to the relative magnitudes of the rotatory powers of all the compounds.

Rotatory Power of the Substituted Ammonium Iodides.

In all cases where the solubility of the iodide in water was small enough, it was precipitated from solutions of the camphorsulphonate or bromocamphorsulphonate by the addition of potassium iodide solution ; the rotatory power of the iodide so recovered was examined in solution in alcohol and in chloroform. The rotatory power in chloroform was always greater than that of the same iodide in alcohol. The iodides retained their rotatory power in alcoholic solution in the cold, but racemised in chloroform solution at the ordinary temperature with varying velocity ; this is attributed to partial dissociation into benzyl or allyl iodide and tertiary amine and subsequent recombination, resulting eventually in equilibrium with equal quantities of the d- and l-isomerides. Observations of the velocity of racemisation at different concentrations show that the reaction is one of the first order which was to be expected on the above view of the process. The phenomenon of autoracemisation will probably be found to be exhibited only by compounds containing the allyl and benzyl groups, since they appear to have other properties in common ; these groups are usually those the iodides of which react most readily with tertiary amines ; they are replaced by the methyl group by the action of methyl iodide, and quaternary compounds containing them undergo a gradual change in molecular weight in chloroform solution. These peculiarities are still under investigation, and other compounds are now being resolved, some of which contain neither the benzyl nor the allyl group, in order, if possible, to confirm the above view that the allyl and benzyl groups alone cause racemisation, and also to extend further the study of the relation between optical rotatory power and constitution in these compounds.

The Benzyl Series.*

d-*Phenylbenzylmethylethylammonium* d-*Camphorsulphonate.*

This compound has already been described by one of us (Trans., 1904, 85, 224) ; the dB, dA, and lB, lA salts were found to have $[M]_D \pm 71 \cdot 1°$. And hence the value of $[M]_D$ for the basic ion was given as $\pm 19 \cdot 4°$.

The effect of temperature change on the rotatory power of the dB, dA salt was now examined with the following results.

0·448 gram in 13·638 grams of solution (density = 1·004 at 20°).

t.	a_D.	$[a]_D$.	$[M]_D$.	$[M]_D$ for basic ion.
10°	1·00°	15·15°	69·2°	19·7°
20	1·01	15·3	69·9	18·3
30	1·05	15·9	72·8	19·2
40	1·06	16·1	73·5	17·9
50	1·08	16·4	75·1	17·5

The values of $[M]_D$ for the ions here obtained are smaller than the actual values, since the solution used was comparatively concentrated (3·27 per cent.). The real value is probably about 2° higher in each case.

---

* A note on the *iso*propyl and *iso*amyl compounds of this series has already been published (*Proc. Camb. Phil. Soc.*, 1904, 13, 33).

The *n*-propyl and *iso*butyl compounds have been prepared and resolved (the *l*-component was isolated) by Wedekind and Fröhlich (*loc. cit.*). Their experiments on the *n*-propyl compound have been repeated, and the *iso*butyl compound has been resolved by a different method. These experiments are here described, as there are some discrepancies between the results of these chemists and ours as regards the melting point and the hygroscopic nature of some of the salts.

*Phenylbenzylmethyl-*n-*propylammonium iodide* was prepared by mixing equivalent quantities of methylpropylaniline and benzyl iodide. The crystalline salt began to separate out almost immediately, and on recrystallisation from alcohol was deposited in colourless prisms which melted with decomposition at 167°. Wedekind and Fröhlich give 147° as the melting point.

0·1872 gave 0·1020 H_2O and 0·3805 CO_2.   C = 55·43 ; H = 6·06.

$C_{17}H_{22}NI$ requires C = 55·58 ; H = 5·99 per cent.

*Phenylbenzylmethyl-*n-*propylammonium* d-*camphorsulphonate* was prepared by mixing the calculated quantities of the iodide and silver *d*-camphorsulphonate, boiling with ethyl acetate, and filtering. On concentrating and cooling the filtrate, the salt was deposited in prisms, which after four recrystallisations from ethyl acetate, melted at 188°. The crude substance gave $[\alpha]_D$ 14·59° and $[M]_D$ 68·72° ; this high rotation was probably due to the presence of some silver camphorsulphonate as impurity, for on further recrystallisation the value for $[M]_D$ became 50·8°, and the iodide precipitated from it was inactive in alcoholic solution. Acetone was also used as a solvent, but did not effect resolution.

*Phenylbenzylmethyl-*n-*propylammonium* d-*bromocamphorsulphonate* was prepared as described by Wedekind and Fröhlich (*loc. cit.*). It was purified by repeated crystallisation from ethyl acetate and separated in colourless needles which were stable in air and melted with decomposition at 148°. These authors state that the compound is too hygroscopic to permit of a melting point determination. We found, however, that the purified salt was not visibly affected when left on the laboratory bench overnight, and that it gave quite a sharp melting point. After crystallising six times from ethyl acetate, the rotatory power was found to be practically constant, the following result being obtained :

0·295 gram in 15·118 grams of solution gave $\alpha_D - 0\cdot17°$, whence $[\alpha]_D - 4\cdot36°$ and $[M]_D - 23\cdot06°$. Hence $[M]_D$ for the basic ion is $- 299°$, a value slightly higher than, but agreeing fairly well with, that given by Wedekind and Fröhlich.

Methylisopropylaniline was prepared by heating together methylaniline and *iso*propyl bromide in molecular proportions for eight hours on a water-bath in a reflux apparatus. When left overnight, the contents of the flask solidified to a mass of the crystalline hydrobromide; this was decomposed with aqueous caustic potash, the oil dried over solid caustic potash, and distilled. The fraction boiling at 211—214° was colourless and almost pure. It gave a crystalline platinichloride which separated from hot alcohol in small, yellow needles melting at 193—194° with decomposition.

*Phenylbenzylmethyliso*propyl*ammonium iodide* was deposited in the crystalline form on mixing equivalent quantities of methyl*iso*propyl-aniline and benzyl iodide. The crystals began to separate within half an hour after mixing, and in twenty-four hours a solid cake was obtained. The substance was recrystallised from methylated spirit, and was deposited in long, colourless prisms belonging to the oblique system and melting at 133°.

0·2545 gave 0·5141 CO_2 and 0·1291 H_2O.   C = 55·09 ; H = 6·07.

$C_{17}H_{22}NI$ requires C = 55·58 ; H = 5·99 per cent.

*Phenylbenzylmethyliso*propyl*ammonium* d-*camphorsulphonate* was pre-pared by boiling equivalent quantities of the silver salt of the acid and the substituted ammonium· iodide with moist ethyl acetate. After filtering from silver iodide, the salt was deposited on cooling in large tables melting at 174—175°. Attempts were made to resolve it into its *d*- and *l*-components by fractional crystallisation from ethyl acetate, acetone, or ethylal. After repeated crystallisations from ethyl acetate, however, $[M]_D$ 49·7°, from ethylal $[M]_D$ 50·8°, and from acetone $[M]_D$ 50·0°, showing that no resolution had been effected. The iodide recovered by addition of potassium iodide was quite inactive in alcohol solution.

*Phenylbenzylmethyliso*propyl*ammonium* d-*bromocamphorsulphonate* was prepared in a similar manner, the solvent used being ethyl acetate. The salt was deposited on cooling the filtrate from the silver iodide in colourless, lustrous prisms, which, on recrystallising, melted with decom-position at 184°.

0·2255 gave 0·4840 CO_2 and 0·1291 H_2O.   C = 58·54 ; H = 6·36.

$C_{27}H_{36}O_4NBrS$ requires C = 58·9 ; H = 6·54 per cent.

The mother liquors deposited more crystals, and afterwards on evaporation left a gummy residue.

The salt was resolved into its *d*- and *l*-constituents by fractional crystallisation from chloroform, acetone, ethyl acetate, or water, the salt of the lævorotatory base being in each case the less soluble.

After three recrystallisations from ethyl acetate the rotatory power was examined at 15°: 0·116 gram in 16·798 grams of aqueous solution (density = 1·002) gave a_D − 0·31°, whence $[a]_D$ − 22·4° and $[M]_D$ − 123°.

After two additional crystallisations, 0·227 gram in 15·401 grams solution gave a_D − 0·65°, whence $[a]_D$ − 22·0° and $[M]_D$ − 121°.

0·183 gram in 12·49 grams of solution gave a_D − 0·66°, whence $[a]_D$ − 22·5° and $[M]_D$ − 124°.

The value of $[M]_D$ may therefore be taken as − 123° at 15°, and the value of $[M]_D$ for the basic ion as 398°.

A nearly saturated solution of concentration 1·431 per cent. (density = 1·002 at 17°) gave the following results at temperatures between 7° and 50° :

t.	a_D.	$[a]_D$.	$[M]_D$.	$[M]_D$ for basic ion.
7°	− 0·65°	− 22·6°	− 124·0°	396°
17	0·62	21·6	119·0	395
30	0·53	18·5	102·0	384
50	0·44	15·5	85·2	376

1-*Phenylbenzylmethylisopropylammonium iodide* was precipitated from the solution of the *d*-bromocamphorsulphonate by the addition of potassium iodide solution and purified by recrystallisation from alcohol.

The crystals were prisms belonging to the tetragonal system, and differ therefore from those of the inactive iodide, which belong to the oblique system (see below). The inactive salt is therefore a racemic compound. The melting point, 132°, was slightly lower than that of the inactive compound, 133°. Both compounds, however, decompose at the melting point, and the melting point of mixtures of active and inactive iodide is only slightly (0·5—1°) lower than that of the former.

The following determinations of rotatory power were made :

0·102 gram in 10·464 grams of ethyl alcohol solution gave a_D − 1·83° (density = 0·811), hence $[a]_D$ − 116° and $[M]_D$ − 425°.

0·103 gram in 12·933 grams of solution (density = 0·794) gave a_D − 1·47°, whence $[a]_D$ − 116·2° and $[M]_D$ − 427°.

The rotatory power of the solutions of the iodide in alcohol did not change when left for three days in the tube.

The rotatory power of the substance in chloroform was also determined. 0·109 gram in 25·069 grams of solution (density = 1·50) gave a_D − 1·80°, whence $[a]_D$ − 138·0° and $[M]_D$ − 506°.

0·110 gram in 21·192 grams of solution (density = 1·496) gave a_D − 2·15°, whence $[a]_D$ − 138·2° and $[M]_D$ − 507°.

This solution slowly racemised on standing.

Time in hours after first observation.	a_D.
0	$-2\cdot15°$
2	$1\cdot86$
4	$1\cdot61$
6	$1\cdot31$
24	$0\cdot4$

After two days, the solution was practically inactive.

Crystalline Form of d- and i-Phenylbenzylmethylisopropylammonium Iodides.

The examination of the crystals of the active and inactive iodides was considered important, since the crystals seemed to differ, and the inactive iodide was probably a racemic compound. Mr. G. R. Dain, of Clare College, kindly undertook their examination, and we are indebted to him for the following account of them.

i-Phenylbenzylmethylisopropylammonium Iodide.

The crystals were invariably long prisms which were found to belong to the oblique system. The pinacoids $a\{100\}$ and $c\{001\}$ were

FIG. 1.

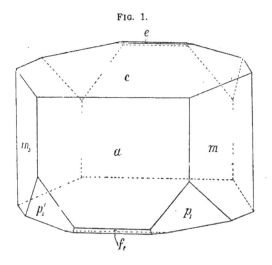

dominant. The prisms $m\{110\}$ were well developed and gave good reflections. The other forms which were nearly always present gave bad images on the goniometer, and the domes $e\{\bar{1}01\}$ and $f\{\bar{2}01\}$ were mere lines. No general forms were observed, and there was no evidence of hemihedrism; in fact, certain vicinal faces indicated the existence of a plane of symmetry.

Crystalline system. Oblique : $a:b:c = 1\cdot613 : 1 : 1\cdot476$, $\beta = 76°4'$.

Forms observed : $a\{100\}$, $c\{001\}$, $m\{110\}$, $e\{\bar{1}01\}$, $f\{\bar{2}01\}$, and $p\{\bar{2}11\}$.

The angles $ac = (100):(001)$ 76°4', $mm_1 = (110):(\bar{1}10)$ 65°8', and $p_1p' = (21\bar{1}):(\bar{2}11)$ 103°54' were taken as parametral angles, and from fifty to seventy measurements of each were taken.

Calculation and measurement gave the following table of angles :

ce	$=$	$001 : \bar{1}01$	45°13'	am	$= 100 : 110$	57°26'
ef	$=$	$\bar{1}01 : \bar{2}01$	27 18	em	$= \bar{1}01 : 110$	106 11
fa'	$=$	$\bar{2}01 : \bar{1}00$	31 25	mp_1	$= 110 : 21\bar{1}$	28 14
ap_1	$=$	$100 : 21\bar{1}$	47 47	p_1e_1	$= 21\bar{1} : 10\bar{1}$	45 35
ap	$=$	$100 : \bar{2}11$	132 13	p_1f_1	$= 21\bar{1} : 20\bar{1}$	38 3

d-Phenylbenzylmethylisopropylammonium Iodide.

The crystals (Fig. 2), which belong to the tetragonal system, were found to have two distinct habits; the commoner were short, squat crystals

FIG. 2.

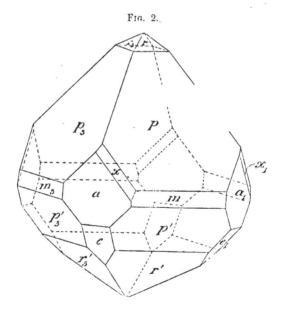

with the pyramid faces of the form $p\{111\}$ well developed; the other type were long, distorted prisms closely resembling the crystals of the inactive iodide; the prismatic form was due to the development of the faces (111), (100), (111), and (100). The similarity in appearance between these crystals of the inactive and active iodides is extended to the magnitudes of some of the principal angles.

There is little doubt that the only element of symmetry is a single tetrad axis, and that the crystals belong to the acleistous tetragonal class (Class III, Lewis), members of which are very rare. Proof of this could not be obtained by demonstrating that the axis was pyro-electric, nor could definite evidence be obtained from etched figures ; but the two ends of the crystals were invariably developed differently, at one end $\{111\}$ is dominant, while at the other $\{11\bar{2}\}$ is the dominant form. Again, although no definite faces of any general form $\{hkl\}$ were observed, indications of the existence of vicinal faces were obtained which were so disposed as to exclude the existence of diad axes perpendicular to the tetrad axis.

Crystalline system. Tetragonal. Acleistous tetragonal class.
$$c = 1\cdot0676.$$

Forms observed : $a\{100\}$, $m\{110\}$, $p = \tau\{111\}$, $p' = \tau\{11\bar{1}\}$, $r = \tau\{112\}$, $r = \tau\{11\bar{2}\}$, $e = \tau\{10\bar{1}\}$, and $x = \tau\{322\}$.

The angles $(111):(\bar{1}\bar{1}1)$ and $(11\bar{1}):(\bar{1}\bar{1}\bar{1})$ were taken as para-metral. Measurement and calculation gave the following table of angles :

mp	=	110 : 111	33°31′	ax	=	100 : 322	11°28′
pr	=	111 : 112	19 26	xp	=	322 : 111	42 25
rr_2	=	112 : $\bar{1}\bar{1}2$	74 6	ep'	=	10$\bar{1}$: 11$\bar{1}$	36 7
pp_2	=	111 : $\bar{1}\bar{1}1$	112 58	pp_1	=	111 : $\bar{1}11$	72 14

All the other angles are readily obtained from the above table. The angles in the zone am are naturally all 45°.

d-*Phenylbenzylmethylisopropylammonium iodide* was precipitated by addition of potassium iodide solution to a solution of the gummy residue containing the bromocamphorsulphonate of the d-base, which was left by evaporation of the mother liquors when no further separation of crystals took place. The rotatory power of this was examined in alcoholic solution and gave $[M]_D$ 314°. The d-iodide is therefore still contaminated with some of the l-iodide.

Methylisobutylaniline was prepared in the manner previously described by one of us (Trans., 1903, 83, 1408). The fraction collected boiled at 225—228°. The base gave a crystalline platinichloride which melted at 180—184°.

Phenylbenzylmethylisobutylammonium iodide was prepared by mixing equivalent quantities of the base and benzyl iodide. The deposition of crystals began almost immediately, and after twenty-four hours a solid mass was obtained. The salt was recrystallised from alcohol, from which solvent it was deposited in stellate aggregates of prisms which melted at 148°. Wedekind and Fröhlich (*loc. cit.*) give the melting point as 130—131°.

0·2095 gave 0·4345 CO_2 and 0·122 H_2O.   C = 56·56 ; H = 6·47.

$C_{18}H_{24}NI$ requires C = 56·69 ; H = 6·30 per cent.

Phenylbenzylmethylisobutylammonium d-*camphorsulphonate* was prepared in the usual manner, and, after recrystallising from acetone, was obtained in colourless crystals which melted at 181°. No resolution was effected by crystallising from acetone, ethyl acetate, or a mixture of methylal or chloroform and light petroleum. After three recrystallisations from ethyl acetate, $[M_D]$ 50·3°, and the iodide recovered by the addition of potassium iodide to the aqueous solution was found to be inactive in alcoholic solution.

The d-*bromocamphorsulphonate* was therefore prepared by extracting the calculated quantities of silver salt and substituted ammonium iodide with boiling acetone, evaporating off the excess of solvent, and allowing to stand. The salt is very soluble in acetone and does not crystallise easily from this solvent. It was found that better crystals were obtained when the salt was dissolved in hot water and the solution allowed to cool slowly. The mixture of the two isomerides is first converted into an oil on treatment with hot water and subsequently dissolves ; the crystals of the salt of the lævorotatory base, which is only slightly soluble in water, separate first, and on concentrating the mother liquor the salt of the dextrorotatory base is precipitated as a gummy residue. The salt of the lævorotatory base was deposited in groups of lustrous prisms, which were unchanged on standing in air and melted at 180°. Wedekind and Fröhlich (*loc. cit.*) state that the salt is strongly hygroscopic and melts at 165°. We found that when purified it could be left exposed on the laboratory bench for weeks without change.

0·2015 gave 0·4390 CO_2 and 0·1171 H_2O.   C = 59·42 ; H = 6·46.

$C_{28}H_{38}O_4NBrS$ requires C = 59·56 ; H = 6·74 per cent.

The rotatory power became constant after crystallising seven times from water or acetone.

0·134 gram in 18·526 grams of aqueous solution gave a_D - 0·13°, whence $[a]_D$ - 8·99° and $[M]_D$ - 50·68°.

0·1125 gram in 15·354 grams of solution gave a_D - 0·12°, whence $a]_D$ - 8·19° and $[M]_D$ - 46·18°.

0·082 gram in 15·128 grams of solution gave a_D - 0·09°, whence $[a]_D$ - 8·3° and $[M]_D$ - 46·82°. The density of these solutions is practically that of water.

Hence the mean value for $[M]_D$ for the basic ion at 15° is - 323°.

l-*Phenylbenzylmethylisobutylammonium iodide* was precipitated from the solution of the bromocamphorsulphonate by addition of potassium iodide solution. Determination of its rotatory power in alcohol gave the following results :

0·063 gram in 11·404 grams of solution (density = 0·80) gave a_D − 0·82°, whence $[a]_D$ − 97·3° and $[M]_D$ − 370·7°.

0·062 gram in 12·00 grams of solution (density = 0·80) gave a_D − 0·82°, whence $[a]_D$ − 99·2° and $[M]_D$ − 377·9°.

The active iodide dissolved easily in chloroform, and a determination of its rotatory power showed that it racemised rapidly in chloroform solution in the dark.

Thus, 0·1165 gram in 22·078 grams of solution (density = 1·5) gave a_D − 1·52°, whence $[a]_D$ − 96·0° and $[M]_D$ − 366°; after one and a half hours, a_D − 1·09°; after three and a half hours, a_D − 0·71°; after six hours, a_D − 0·20°, and after seven hours the solution was inactive.

Another determination taken within ten minutes of making up the solution gave a_D − 1·09° for 0·083 gram in 22·5 grams of solution, whence $[a]_D$ − 98·5° and $[M]_D$ − 375°. This solution also became inactive after seven hours. It is therefore probable that $[M]_D$ in chloroform is really about 390° and greater than in alcohol.

Methylisoamylaniline was obtained by mixing equivalent quantities of methylaniline and *iso*amyl bromide and heating on a water-bath for sixteen hours or until the mixture had solidified to a crystalline mass. The hydrobromide was treated with aqueous caustic potash, the oil separated, dried over solid caustic potash, and fractionated. The oil distilling from 246° to 248°, which was almost colourless, was collected and used for the experiments. The platinichloride was precipitated as a pale yellow powder from solutions of the hydrochloride, but on dissolving it in alcohol it underwent some decomposition and could not be purified.

Phenylbenzylmethylisoamylammonium iodide was prepared by mixing methyl*iso*amylaniline and benzyl iodide in molecular proportions. Crystals began to be formed almost immediately, and after an hour a solid cake was obtained. The solid was crystallised several times from alcohol; colourless, fine needles were deposited, which melted sharply at 156° when heated rapidly.

0·2043 gave 0·4325 CO_2 and 0·1266 H_2O.   C = 57·7 ; H = 6·8.

$C_{19}H_{26}NI$ requires C = 57·72 ; H = 6·58 per cent.

Phenylbenzylmethylisoamylammonium d-camphorsulphonate was prepared by boiling equivalent quantities of phenylbenzylmethyl*iso*amylammonium iodide and silver *d*-camphorsulphonate with acetone. In order to avoid the formation of the double compound of the ammonium iodide and silver iodide, it was found advisable to get the silver salt finely powdered, to cover it with acetone, and to add the iodide in successive small quantities, shaking and boiling after each addition. The silver iodide was filtered off and the filtrate heated on the water-bath until all the acetone was driven off; on allowing the residue to cool in a desiccator, it crystallised readily. The salt is fairly soluble in ethyl acetate and in acetone. It was found that after repeated crystallisation from acetone the melting point of the salt rose gradually and the rotatory power fell, became zero, and then the lævorotation increased.

It is clear that the salt of the l-base is being separated. In several experiments with quantities varying from 30 to 100 grams of the camphorsulphonate, the material was exhausted after some twenty or twenty-five crystallisations, when the value of $[M]_D$ was about $-220°$ * and not constant. Finally, about 270 grams of the salt were prepared and repeatedly crystallised from hot acetone; after sixteen· crystallisations, 10 grams of the salt were left, melting at 170—171° and having $[M]_D - 219°$.

This was crystallised twice, when the melting point rose to 176—177° and $[M]_D$ to $-230°$. The salt was again crystallised twice; the melting point became 179—180° and $[M]_D - 235°$. Further re-crystallisation did not change either the melting point or the rotatory power, so that the pure salt of the l-base has been isolated.

0·1163 gave 0·2975 CO_2 and 0·0875 H_2O. C = 69·76 ; H = 8·36.

$C_{20}H_{41}O_4NS$ requires C = 69·74 ; H = 8·22 per cent.

The following determinations of rotatory power were made in aqueous solution. at 15°.

0·120 gram in 12·585 grams of solution gave $a_D - 0·9°$ (density = 1·000), hence $[a]_D - 47·2°$ and $[M]_D - 235·4°$.

0·131 gram in 12·387 grams gave $a_D - 1·00°$ (density = 1·000), hence $[a] - 47·5°$ and $[M]_D - 235·8°$.

0·131 gram in 12·330 grams gave $a_D - 1·01°$ (density = 1·001), hence $[a]_D - 47·5°$ and $[M]_D - 237·0°$.

0·133 gram in 12·588 grams gave $a_D - 0·99°$ (density = 1·000), hence $[a]_D - 46·90°$ and $[M]_D - 233·8°$.

The effect of concentration on the rotatory power of the salt at 18° was examined as the salt is particularly suitable for the purpose on account of its large rotatory power and its solubility in water.

---

* The values of $[M]_D$ for the camphorsulphonate and the active iodide previously given (*loc. cit.*) were erroneous.

Weight of salt in 100 grams of solution.	a_D.	Specific gravity.	$[a]_D$.	$[M]_D$.
2·75	− 2·54°	1·004	− 46·0°	− 229·7°
1·50	1·40	1·002	46·6	232·4
1·02	0 96	1·001	47·1	235 2
0·633	0·59	1·000	46·6	232·6

It is evident from the above numbers that, as in the case of the phenylbenzylmethylallylammonium salt, the value of the rotatory power is constant within the limits of experimental error when the concentration is about 1·5 per cent. or lower.

The mean value of $[M]_D$ for l-phenylbenzylmethyl*iso*amyl d-camphorsulphonate at 15° is therefore − 235·5°, and the value of $[M]_D$ for the basic ion is − 287·2°

The effect of temperature change on the rotatory power of the 1·5 per cent. solution above was also investigated (density = 1·002 at 18°).

t.	a_D.	$[a]_D$.	$[M]_D$.	$[M]_D$ for basic ion.
7°	− 1·41°	− 46·8°	− 233·5°	− 282·4°
13	1·40	46·6	232·5	282·7
18	1·40	46·6	232 4	283·6
30	1·36	45·3	226·0	279·6
40	1·31	43·8	218·6	274 2
50	1·29	43·1	215·0	272·4

Phenylbenzylmethyliso amylammonium   d-*bromocamphorsulphonate* crystallises readily, is very soluble in acetone, and was therefore crystallised from a mixture of acetone and ethylal. After recrystallising several times, the colourless plates melted at 179—180° and gave $[M]_D$ 195° ; the value of $[M]_D$ fell slowly on further crystallisation, but, as it was clear that resolution was not being effected rapidly, the further examination was abandoned.

l-*Phenylbenzylmethyliso amylammonium iodide* was immediately precipitated as fine needle-shaped crystals when aqueous solutions of the camphorsulphonate and potassium iodide were mixed. The precipitated iodide melts at 151—152° and after recrystallisation from cold alcohol melts at 155—156° when heated rapidly ; a mixture of this with the inactive iodide melts practically at the same temperature.

Determinations of the rotatory power of the iodide were made in solution in alcohol and in chloroform.

In alcohol, 0·1440 gram in 10·394 grams gave a_D − 1·90° (density = 0·811), hence $[a]_D$ − 84·5° and $[M]_D$ − 333·8°.

After recrystallising from cold alcohol, 0·1470 gram in 11·067 grams gave a_D − 1·87° (density = 0·811), hence $[a]_D$ − 87·0° and $[M]_D$ − 344°.

0·1140 gram in 10·358 grams gave a_D − 1·55° (density = 0·810), hence $[a]_D$ − 86·9° and $[M]_D$ − 343°.

In chloroform, 0·1340 gram in 18·519 grams gave a_D − 2·16° (density = 1·492), hence $[a]_D$ − 100° and $[M]_D$ − 395°.

The solution racemised slowly in the dark at the ordinary temperature; after twenty-eight hours, a_D − 1·02°; after forty-eight hours, a_D − 0·50°; after seventy-two hours, a_D − 0·2°; and after ninety-six hours the solution was completely inactive.

0·1030 gram in 17·259 grams gave a_D − 1·75° (density = 1·49), hence $[a]_D$ − 98·6° and $[M]_D$ − 390°.

After one hour, a_D − 1·67°; after eighteen hours, a_D − 0·87°; after twenty-four hours, a_D − 0·56°; after forty-eight hours, a_D − 0·16°; and after seventy-two hours the solution was inactive.

The Allyl Series.*

d-*Phenylmethylethylallylammonium* d-*bromocamphorsulphonate* was prepared in the usual way from silver d-bromocamphorsulphonate and the iodide, $N(C_6H_5)(CH_3)(C_2H_5)(C_3H_5)I + CHCl_3$, prepared by Wedekind, in acetone solution. After filtering off the silver iodide and evaporating most of the acetone on the water-bath, the residue crystallised easily. The salt could be crystallised from acetone, ethyl acetate, or ethylal; it was crystallised several times from a mixture of acetone and light petroleum and then from acetone alone, when it was obtained in short, colourless prisms melting at 138—139°.

0·1582 gave 0·3090 CO_2 and 0·0895 H_2O. C = 53·2; H = 6·28.
$C_{22}H_{37}O_4NBrS$ requires C = 53·1; H = 6·59 per cent.

The salt was recrystallised from acetone until the rotatory power became constant when examined during three successive crystallisations.

0·4576 gram in 25 c.c. solution gave a_D 2·20°, $[a]_D$ 60·1°, and $[M]_D$ 292°.

0·168 gram in 12·457 grams solution gave a_D 1·80°, $[a]_D$ 59·9°, and $[M]_D$ 291·2°.

0·157 gram in 12·317 grams solution gave a_D 1·52°, $[a]_D$ 59·6°, and $[M]_D$ 289·6°.

The value of $[M]_D$ for the salt is therefore 291°, which agrees closely with the number given in a preliminary notice (Trans., 1903, 83, 1420); the value of $[M]_D$ for the basic ion is therefore about 16° at the ordinary temperature (15°).

* A note on the results obtained with this series has already been published (*Proc. Camb. Phil. Soc.*, 1905, 13, 190).

Two determinations of the effect of temperature change were made with another sample of salt resolved separately. The following result served to show that the change in rotatory power is extremely small.

0·171 gram in 13·25 grams of solution (density = 1·001) at 20°.

l.	a_D.	$[a]_D$.	$[M]_D$.	$[M]_D$ for basic ion.
10°	1·54°	59·6°	290·0°	17·0°
19	1·55	60·0	291·6	14·6
31	1·55	60·0	291·6	9·6
44	1·55	60·0	291·6	3·6
50	1·56	60·3	293·4	2·4

It is curious that the rotatory power of the basic ion should vanish at some temperature above 50° and then become lævorotatory, as it probably would.

Phenylmethyl-n-propylallylammonium Iodide.

A mixture of methyl-n-propylaniline and allyl iodide in molecular proportions rapidly sets to a solid, crystalline mass of the quaternary salt ; this is very soluble in alcohol, acetone, and ethyl acetate, and was purified by dissolving in alcohol, adding ether until a turbidity was produced, and allowing to stand, when colourless, rectangular tables were deposited. After repeating the process several times, the crystals melted sharply at 109—110° and gave the following results on analysis :

0·2116 gave 0·3815 CO_2 and 0·1217 H_2O.   C = 49·1 ; H = 6·4.

$C_{13}H_{20}NI$ requires C = 49·2 ; H = 6·3 per cent.

Phenylmethyl-n-propylallylammonium d-*camphorsulphonate*, prepared in the usual way, crystallised readily from a mixture of acetone and benzene in rhombic plates which melted at 167—168°.

After three crystallisations from acetone, 0·354 gram of the salt dissolved in 12·663 grams of solution gave a_D 0·66° (density = 1·001), hence $[a]_D$ 11·9° and $[M]_D$ 49·8°.

The d-camphorsulphonate is therefore not easily resolved.

Phenylmethyl- n -*propylallylammonium*   d-*bromocamphorsulphonate* crystallised easily from a mixture of acetone and benzene and was then crystallised repeatedly from acetone, in which it becomes less soluble as this process goes on. The salt is thus obtained in lustrous prisms melting at 169—170°.

0·2298 gave 0·466 CO_2 and 0·1475 H_2O.   C = 55·3 ; H = 7·12.

$C_{23}H_{34}O_4NBrS$ requires C = 55·2 ; H = 6·8 per cent.

The rotatory power of the salt in aqueous solution was found to increase rapidly at first and then more slowly after recrystallising several times.

0·150 gram in 12·37 grams of solution gave a_D 1·85° in a 2-dcm. tube, hence $[a]_D$ 76·3°, and $[M]_D$ 381·5°.

The salt was then recrystallised twice more from acetone and the first and second crops examined at 15°.

0·106 gram in 12·39 grams of solution gave a_D 1·30°, $[a]_D$ 76·1°; $[M]_D$ 380·5°.

0·106 gram in 12·45 grams of solution gave a_D 1·30°, $[a]_D$ 76·5°; $[M]_D$ 382·5° (density = 1 001 in all cases).

Hence $[M]_D$ for d-phenylmethyl-n-propylallylammonium d-bromocamphorsulphonate is 381·5° and the value of $[M]_D$ for the basic ion is 106·5° at 15°.

The effect of temperature change on a solution of this salt was examined and found to be negligible.

A 1·08 per cent solution at 15° gave a_D 1·65° between 4° and 48°, so that the value of $[a]_D$ remains practically constant at 76·4° throughout this range of temperature.

The values of $[M]_D$ for the basic ion may therefore be taken as 108·5° at 10°, 104·5° at 20°, 99·5° at 30°, 95·5° at 40°, and 90·5° at 50°.

The active iodide is not precipitated from a strong solution of the bromocamphorsulphonate by the addition of excess of potassium iodide.

The platinichloride is precipitated on the addition of platinic chloride as a pale yellow, crystalline powder extremely sparingly soluble in all solvents, so that its rotatory power could not be determined.

Phenylmethylisopropylallylammonium Iodide.

A mixture of methylisopropylaniline and allyl iodide in molecular proportions rapidly becomes turbid, and in the course of twenty-four hours sets almost completely to a crystalline mass.

This was crystallised several times from hot alcohol and formed colourless prisms melting at 171—172° which were analysed with the following result :

0·1620 gave 0·2905 CO_2 and 0·0955 H_2O.   C = 48·9 ; H = 6·5.

$C_{13}H_{20}NI$ requires C = 49·2 ; H = 6·3 per cent.

l-Phenylmethylisopropylallylammonium d-Camphorsulphonate.

The above iodide was converted into the d-camphorsulphonate in the usual way, and this crystallised partially on standing. The salt is very soluble in acetone, and separates from its solutions sometimes in short, thick prisms and tables rather resembling cane sugar crystals in appearance, sometimes as long, needle-shaped crystals, and often as a mixture of the two; as the crystallisation proceeded, the acicular form became the more common. It was found most convenient to dissolve the salt in hot acetone, to add to this solution about one-tenth its own volume of toluene, and then to set this solution in a desiccator over sulphuric acid, when well-formed crystals separated, from which the mother liquor could be much more easily removed than when acetone was used alone.

After several crystallisations, the crystals melted at 167—168° and gave the following result on analysis :

0·2000 gave 0·4802 CO_2 and 0·1477 H_2O. $C = 65·4$; $H = 8·2$.

$C_{23}H_{35}O_4NS$ requires $C = 65·5$; $H = 8·3$ per cent.

Determinations of the rotatory power showed that this fell slowly on recrystallisation, became zero, and then the lævorotation increased.

Thus, after five crystallisations, $[M]_D - 11·3°$, and after eight $[M]_D - 23·2°$.

The salt after recrystallising five or six times always crystallised entirely in needles and long prisms. After the thirteenth crystallisation, it was found that the rotatory power of the salt had become constant, and further repetition of the process three times did not alter the value of $[M]_D$, but it was found that the rotatory power of this salt was affected to a greater extent by temperature than that of any of the other salts examined.

The melting point of the salt was now 168—169°.

The following determinations of the rotatory power were made.

0·258 gram in 12·609 gram solution gave $a_D - 0·48°$ at 20° (density = 1·002), hence $[a]_D - 11·7°$ and $[M]_D - 49·3°$.

0·260 gram in 12·637 grams of solution (density = 1·002 at 20°).

t.	a_D.	$[a]_D$.	$[M]_D$.
13°	− 0·52°	− 12·62°	− 53·1°
20	0·49	11·9	50·0
26	0·48	11·65	49·1
40	0·36	8·74	36·8
45	0·33	8·02	33·8

0·286 gram in 13·367 grams of solution (density = 1·002 at 20°).

t.	a_D.	$[a]_D$.	$[M]_D$.
9°	− 0·55°	− 12·83°	− 54·0°
13	0·53	12·4	52·05
25	0·46	10·84	45·27
32	0·40	9·35	39·38
50	0·28	6·40	27·58

It is extremely difficult to get really concordant results with a salt

·of this kind with its small rotatory power and large temperature effect, but by taking a mean of the above results we get the following values of $[M]_D$ for the basic ion, which are probably not far from the truth.

103·5° at 10°, 100·6° at 20°, 96·6° at 30°, 91·6° at 40°, and 86·6° at 50°.

The value of $[M]_D$ for the salt at 15° is taken to be 52°, and for the basic ion 102·6°; this value may be slightly smaller than the true one, and on account of the great experimental error it is impossible to decide from these data whether the *n*- or the *iso*-propyl compound has the greater rotatory power; it may be assumed that they are very nearly identical.

Phenylmethylisopropylallylammonium d-*bromocamphorsulphonate* was prepared and found to crystallise readily; it was very soluble in water, sparingly soluble in methylal, but more so in acetone and ethyl acetate. Specimens were recrystallised several times from acetone and others from ethyl acetate and the rotatory powers of the different fractions examined, and were found to vary between 274° and 277° and not to be altered by further crystallisation. This salt is therefore not resolved by recrystallisation from these solvents. The salt melted at 169—171° and gave the following result on analysis:

0·1599 gave 0·3222 CO_2 and 0·0987 H_2O. C = 54·9; H = 6·86.

$C_{23}H_{34}O_4NBrS$ requires C = 55·2; H = 6·8 per cent.

l-*Phenylmethylisopropylallylammonium iodide* was slowly precipitated in beautiful, lustrous prisms when solutions of the camphorsulphonate and potassium iodide were mixed. The iodide is so soluble that small quantities only could be recovered. The crystals melted at the same temperature, 171—172°, as the inactive iodide, and were rather sparingly soluble in cold alcohol and less soluble in chloroform, so that accurate determinations of the rotatory power were difficult, as they were necessarily affected largely by experimental error.

ı The following results were obtained:

In alcohol:

0·158 gram in 10·604 grams of solution gave a_D - 0·50° (density = 0·810), hence $[a]_D$ − 20·7° and $[M]_D$ − 65·7°.

0·165 gram in 9·101 grams of solution gave a_D − 0·50° (density = 0·811), hence $[a]_D$ − 20·0° and $[M]_D$ − 63·4°.

In chloroform:

0·125 gram in 21·805 grams of solution practically saturated gave a_D − 0·52° (density = 1·499), hence a_D − 30·3° and $[M]$ − 96°.

The solution racemised at a very slow rate.

After two, nine, and fourteen days, a_D was $-0.47°$, $-0.26°$, and $-0.15°$ respectively, and after standing about a month the solution was inactive.

Phenylmethylallylisobutylammonium iodide was prepared by mixing the calculated quantities of methyl*iso*butylaniline and allyl iodide. An oil was deposited at first, which on standing for twenty-four hours was converted into a solid, crystalline mass. The crystals were very soluble in alcohol, chloroform, or acetone, from any of which solvents they are deposited as an oil which slowly crystallises. They were recrystallised from hot ethyl acetate, in which they are easily soluble, and were deposited on cooling in long prisms which melt at 143°.

0.1986 gave 0.3707 CO_2 and 0.1195 H_2O. C = 50.9 ; H = 6.70.

$C_{14}H_{22}NI$ requires C = 50.75 ; H = 6.65 per cent.

Phenylmethylallylisobutylammonium d-*camphorsulphonate* was prepared by extracting a mixture of the substituted ammonium iodide and silver d-camphorsulphonate with boiling acetone to which a few drops of methyl alcohol had been added, and filtering from the precipitated iodide. On allowing the filtrate to stand over sulphuric acid in a desiccator, the salt was deposited in rhombic prisms. These were purified by recrystallisation from benzene or acetone and melted at 173° with decomposition.

0.1604 gave 0.3895 CO_2 and 0.1190 H_2O. C = 66.5 ; H = 8.24.

$C_{24}H_{37}NSO_4$ requires C = 66.21 ; H = 8.50 per cent.

The salt was resolved into its d- and l-constituents by fractional crystallisation from acetone. After four recrystallisations from this solvent, 0.579 gram in 14.56 grams of aqueous solution gave a_D 0.07°, whence $[a]_D$ 0.88° and $[M]_D$ 3.83°. After two more recrystallisations, 0.302 gram in 14.494 grams of solution gave a_D $-0.06°$, whence $[a]_D$ $-1.44°$ and $[M]_D$ $-6.26°$.

On further recrystallisation, a solution containing 0.2625 gram of the salt in 15.122 grams of solution (1.785 per cent.) gave the following results (density = 1.002 at 15°) :

l.	a_D.	$[a]_D$.	$[M]_D$.	$[M]_D$ for basic ion.
7°	$-0.04°$	$-1.13°$	$-4.93°$	$-54.46°$
13	-0.01	-0.28	-1.24	-51.8
30	$+0.06$	$+1.71$	$+7.41$	-46.16
50	$+0.1$	$+2.87$	$+12.5$	-45.1

Repetition of the experiments gave similar results.

In the determination of the rotatory power of this compound, a large experimental error is necessarily involved, seeing that the observed rotation is very small ; the values obtained for the molecular rotation of the basic ion at 15° vary between $-52°$ and $-57°$, so that the value for $[M]_D$ may be taken as $-55°$. Although great accuracy cannot therefore be claimed for the numbers given, the experimental error is not great enough to affect the place of the ion in the series, which lies between those of the ions containing the *iso*amyl and *iso*propyl groups respectively.

l-*Phenylmethylallyl*isobutylammonium iodide was precipitated in small, colourless prisms, which melted at 143°, by addition of potassium iodide to aqueous solutions of the camphorsulphonate. Their rotatory power in alcohol and chloroform was determined and gave the following results :

0·114 gram in 15 c.c. of alcohol gave a_D $-0·29°$ (density $= 0·800$), whence $[a]_D$ $-19·08°$ and $[M]_D$ $-63·15°$.

The iodide was easily soluble in chloroform, and the solution slowly racemised. 0·086 gram in 15 c.c. of chloroform gave a_D $-0·27°$, whence $[a]_D$ $-23·55°$ and $[M]_D$ $-77·95°$; after four hours, $a_D - 0·21°$, after twenty-four hours, $a_D - 0·17°$, after forty-eight hours, $a_D - 0·08°$, and after three days the solution was inactive.

*Phenylmethylallyl*isoamylammonium iodide was prepared by mixing the calculated quantities of methyl*iso*amylaniline and allyl iodide and allowing to stand. After twenty-four hours, the mixture, which at first deposited an oil, had solidified to a crystalline mass. This was recrystallised from a mixture of alcohol and ether or from ethyl acetate and separated as small, colourless prisms which melt at 135°.

0·2052 gave 0·3920 CO_2 and 0·1255 H_2O. C $= 52·10$; H $= 6·80$.
$C_{15}H_{24}NI$ requires C $= 52·17$; H $= 6·96$ per cent.

l-*Phenylmethylallyl*isoamylammonium d-*camphorsulphonate* was prepared by mixing together the calculated quantities of the substituted ammonium iodide and silver d-camphorsulphonate and boiling with ethyl acetate, or acetone (in which the salt is very soluble), or benzene, in which it is moderately soluble, and crystallises out on cooling the saturated solution. After twice recrystallising from methylal, the melting point was 155°.

0·2000 gave 0·4385 CO_2 and 0·1567 H_2O. C $= 66·61$; H $= 8·71$.
$C_{25}H_{39}O_4NS$ requires C $= 66·81$; H $= 8·68$ per cent.

The salt was too soluble in acetone, ethyl acetate, and methyl formate to allow of the use of these solvents for the purpose of recrystallising, and when a mixture of any of them with light petroleum was used the salt separated as an oil ; it crystallised easily from benzene in needles, but the aqueous solution of the crystals smelt strongly of benzene even after they had been dried and kept in a

vacuum desiccator over sulphuric acid for some days ; probably, therefore, they contain benzene of crystallisation. Methylal which had
been freed from all traces of methyl alcohol by distillation over caustic
potash was found to be the most convenient solvent ; the l-salt is
sparingly soluble in the cold, and is deposited in long needles on
cooling the hot saturated solution. After recrystallising five times
from this solvent, the salt appeared to be completely resolved.

0·6225 gram in 15·753 grams of solution gave a_D +0·58° (density =
1·004), whence $[a]_D$ 7·31° and $[M]_D$ 32·8°.

0·4655 gram in 17·157 grams of solution (density = 1·003 at 15°)
gave the following results :

$t°$	$a_D.$	$[a]_D.$	$[M]_D.$	$[M]_D$ for basic ion.
7°	+0·37°	+6·80°	+30·5°	−18·5°
15	0·39	7·16	32·2	18·4
30	0·42	7·72	34·7	18·9
50	0·48	8·84	39·7	18·9

The mean of these experiments gives $[M]_D$ −18° for the basic ion
at 15°.

The following tables, which contain the melting points and the
values of the molecular rotatory powers of the compounds described
in this paper and of those previously described, have been prepared
for convenience of reference.

Phenyl-methyl-benzyl Series.

	M. p. of d-camphor-iodide.	M. p. of d-camphor-sulphonate.	M. p. of d-bromo-camphor-sulphonate.	$[M]_D$ of ion at 15°.	$[M]_D$ of iodide in alcohol.	$[M]_D$ of iodide in chloroform.
Ethyl	147·8°	181°	—	+19·4°	30°	33·8°
n-Propyl ...	167·0	188	148°	−299·0	−354	−374·0
isoPropyl ...	133·0	175	184	−398·0	−428	−507·0
isoButyl ...	148·0	181	180	−323·0	−374	−390·0 ?
isoAmyl ...	156·0	178	180	−287·0	−343	−395·0
Allyl	140—142	171	—	+167·0	—	—

Phenyl-methyl-allyl Series.

	M. p. of d-camphor-iodide.	M. p. of d-camphor-sulphonate.	M. p. of d-bromo-camphor-sulphonate.	$[M]_D$ of ion at 15°.	$[M]_D$ of iodide in alcohol.	$[M]_D$ of iodide in chloroform.
Ethyl	—	149—150°	138—139°	16·0°	—	—
n-Propyl ...	109—110°	167—168	169—170	106·5	—	—
isoPropyl...	171—172	168—169	169—171	−102·6	−64·5°	−96°
isoButyl ...	143	173	—	−55·0	−63·0	−78
isoAmyl ...	135	155	—	−18·0	—	—·-

No regularities in the melting points are to be observed. It must
be remembered, however, that fusion is almost invariably accompanied

by decomposition, and that the melting point depends somewhat on the rate of heating in the case of the iodides. The melting point of the d-camphorsulphonate is usually between that of the iodide and the d-bromocamphorsulphonate, but there are some exceptions. The melting points of the phenylmethyl*iso*propylallylammonium salts are surprisingly alike.

The following curves show the effect of temperature is of the same kind in all the compounds we have examined.

FIG. 3.—*Phenyl-methyl-benzyl series.*

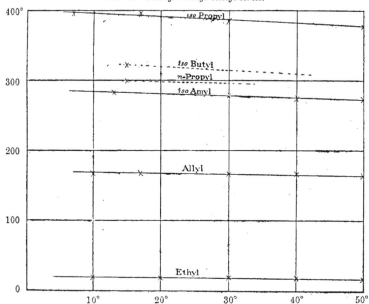

FIG. 4.—*Phenyl-methyl-allyl series.*

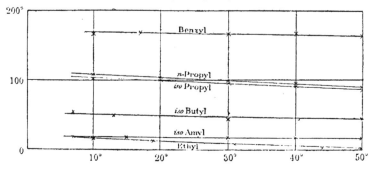

Wedekind has expressed his intention of examining the effect of temperature change on the n-propyl and isobutyl compounds of the benzyl series. There is no doubt that they will behave in a very similar manner to the other salts which we have examined.

The curves show very clearly that the temperature effect is very regular for all the compounds, and is small compared with that for non-electrolytes or for the compounds examined by Patterson (*loc. cit.*).

The Product of Asymmetry for the Nitrogen Atom.

In the two series of optically active nitrogen compounds described above, the large effect of constitution on the molecular rotation is very striking. The existence of a maximum in each case at the second member and the decline of rotatory power in the higher members is clearly indicated. It was explained in the early part of the paper that any connection which might exist between the rotatory power and the constitution of compounds ought to become evident in this simple case of an ion in which the substituting groups were of similar character. Now, although mass cannot be the only factor which affects the rotatory power, it must be one factor and probably also an important one, and in the present state of our knowledge, while we await a development of the connection between light as an electromagnetic phenomenon and chemical atoms and molecules as electric structures, it is the only one which we can take into account.

Assuming, therefore, that optical rotatory power is determined by the mass of the groups attached to the nitrogen atom, and, further, that these masses are concentrated at the angular points of a regular pyramid on a square base, we can develop an expression for the product of asymmetry which will represent the same facts for nitrogen as the function deduced by Guye (*Compt. rend.*, 1893, **116**, 1378) does for carbon. The pyramidal configuration has been shown to be the most probable one for substituted ammonium compounds (Trans., 1905, **87**, 1729).

The pyramid has four planes of symmetry, numbered 1 to 4 in the plane projection of the pyramid on its base in the figure.

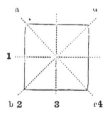

Let d_1, d_2, d_3, d_4 be the distances of the centre of gravity of the molecule from each of these planes.

These are given by the expressions :

$$d_1 = \frac{\{(a+d)-(b+c)\}l\sin\alpha}{(a+b+c+d+x)},$$

$$d_2 = \frac{(a-c)l}{(a+b+c+d+x)},$$

$$d_3 = \frac{\{(a+b)-(c+d)\}l\sin\alpha}{(a+b+c+d+x)},$$

$$d_4 = \frac{(b-d)l}{(a+b+c+d+x)},$$

where a, b, c, and d represent the masses of the four alkyl groups at the base of the pyramid, x the mass of the acidic radicle at its apex, l half the length of the diagonal of the base, and α the angle between two of the planes, 45°.

There is also d_5, the distance of the centre of gravity from the basal plane, which may be expressed as

$$d_5 = \frac{\{(a+b+c+d)-x\}h}{(a+b+c+d+x)},$$

where h is the height of the pyramid.

We have therefore the product of asymmetry :

$$P = d_1 \times d_2 \times d_3 \times d_4 \times d_5 \text{ or } P =$$

$$\frac{\{(a+d)-(b+c)\}\{(a+b)-(c+d)\}(a-c)(b-d)\{(a+b+c+d)-x\}l^4\sin^2\alpha h}{(a+b+c+d+x)^5},$$

or, since l, h, and must be assumed to be constant, otherwise the problem becomes too complicated, we get

$$P' = \frac{\{(a+d)-(b+c)\}\{(a+b)-(c+d)\}(a-c)(b-d)\{(a+b+c+d)-x\}}{(a+b+c+d+x)^5},$$

an expression which should measure (to a very rough approximation only) the magnitude of the rotatory power of an optically active nitrogen compound. It satisfies the necessary conditions, namely, that when $a = c$ the compound is inactive, that is, for a compound with the configuration :

but that if $a = b$ or d the compound is not inactive, that is, a compound with the configuration:

is active, a theoretically possible case which has not been realised experimentally. The above function predicts some curious results, for instance, if $a + b = c + d$ or $a + d = b + c$, the compound should be inactive; this contingency, which would probably not occur, is analogous to the inactivity predicted by Guye's expression for carbon where any two groups are of equal mass, and shows clearly that mass is not the only factor to be considered, a fact which is already sufficiently well recognised.

The above expression, however, refers to the complete molecule, whereas in this paper attention has been confined chiefly to the ions, where the rotatory power is so little affected by external conditions, the solvent is always the same, and the temperature effect is small. In this case, the acidic group x is removed, and its nature does not affect the rotatory power of the basic ion, so that we may put $x = 0$, and we have

$$P'' = \frac{\{(a+d) - (b+c)\}\{(a+b) - (c+d)\}(a-c)(b-d)}{(a+b+c+d)^4}.$$

This function exhibits the same properties as the foregoing more complicated expression for the product of asymmetry of the whole molecule; three maxima and three minima are possible, and $P = 0$, as before, when $a = c$, and also when $a + b = c + d$ or $a + d = b + c$.

There is a practical difficulty in the application of this formula, namely, that there are always three possible configurations for a compound of the type $N\,a\,b\,c\,d\,x$ which may be represented for one case thus:

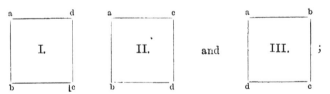

the value of the product of asymmetry is in general different for each of the three arrangements. One of these arrangements is more stable than the others, and that one is produced during the changes which take place in the formation of the compound from a tertiary amine and alkyl halogen compound.

Until we have some means of determining the configuration of these compounds, it seems legitimate, when possible, to choose those configurations for the compounds in a series, which give values for the products of asymmetry in the same numerical order as the rotatory powers actually found for these compounds.

The values of P'' for the three possible arrangements are given in the table below, together with $[M]_D$ for the basic ion at 15°.

Phenyl-methyl-benzyl Series. $a = C_6H_5(77)$, $b = CH_3(15)$, $c = C_7H_7(91)$, d variable.

d.	$[M]_D$ for basic ion.	I. $P'' \times 10^3$.	II. $P'' \times 10^2$.	III. $P'' \times 10^3$.
Methyl	0·0°	0·0	+5·98	+5·98
Ethyl	19·4	0·0	+6·27	0·0
n-Propyl	299·0 }	−0·08	+4·6 ·	−1·76
isoPropyl	398·0 }			
isoButyl	323·0	−0·26	+2·46	−1·71
isoAmyl	287·0	−0·54	+0·61	−0·99
Allyl	167·0	−0·07	+4·87	−1·66

Phenyl-methyl-allyl Series. $a = C_6H_5$, $b = CH_3$, $c = C_3H_5(41)$, d variable.

d.	$[M]_D$ for basic ion.	I. $P'' \times 10^3$.	II. $P'' \times 10^3$.	III. $P'' \times 10^3$.
Methyl	0·0°	0·0	−10·6	−10·6
Ethyl	16·0	−0·8	−2·95	−4·0
n-Propyl	106·5 }	−0·588	−0·44	−0·5
isoPropyl	102·6 }			
isoButyl	55·0	+0·54	+0·11	+2·73
isoAmyl	18·0	+2·14	+0·057	+3·16

The sign of P'' has no significance here as it has in the case of carbon compounds, where changes of constitution are produced in a molecule the configuration of which is always the same, but the sign of its rotation may nevertheless be altered by the change of constitution.

It is at once clear on comparing the above values of the product of asymmetry with the experimental values of $[M]_D$ that there is no simple connection between them. The application of Guye's hypothesis to the nitrogen atom does not therefore give a satisfactory expression connecting the masses of the alkyl groups with the rotatory power of the molecule.

The configuration of the methyl compounds of both series is almost certainly I, since they are inactive. Considering first the benzyl series, there is no reason to suspect that the change from a methyl to an ethyl group would necessitate a change of configuration, so that the ethyl compound might be expected to have a very small rotatory power since $P = 0$. Now if we assume that the *iso*propyl, *iso*butyl, and *iso*amyl compounds have configuration I, we get a maximum value for P'' at the *iso*amyl compound, but if we assume that they have configuration II and that the *n*-propyl and allyl compounds have configuration III, we get a fair approximation to a relation between P'' and $[M]_D$ for the benzyl series. There is as yet no experimental evidence for these changes of configuration, but that there is a difference of configuration between the *n*-propyl and *iso*propyl compounds of the benzyl series seems extremely probable, as we know of no other case where the change from the normal to the *iso* grouping produces any great change in rotatory power, and in the allyl series the *n*- and *iso*-compounds have practically identical rotatory powers. The allyl group in this case might be expected to behave as a group of smaller weight, since, owing to the double linkage, the centre of gravity of the group is nearer its point of attachment to the nitrogen atom than that of a normal grouping.

In the allyl series, it is still more difficult to see a connection between the values of P'' and $[M]_D$, even with the aid of such arbitrary assumptions about configuration as the above, since the ethyl compound is so clearly out of its place. Now the fact that the rotatory power of the ion in this case vanishes above $50°$ might lead to some doubt as to its having been resolved at all, but this doubt is removed by the facts that the observed rotation at the ordinary temperature is outside the limits of experimental error, and, that the platinichloride gives an active solution on treatment with sulphuretted hydrogen in water. With the exception of the ethyl compound, the values of P'' given by configuration II are in the same order as the values of $[M]_D$.

Before arriving at any definite conclusion as to the applicability of Guye's hypothesis to the case of nitrogen, it will be necessary to await the results of our further experiments on the four series of compounds corresponding to the above in which the methyl group is replaced by ethyl and then by *iso*propyl.

The expenses of this investigation have been largely defrayed by grants placed at our disposal by the Government Grant Committee of the Royal Society, for which we are glad to make this grateful acknowledgement.

UNIVERSITY CHEMICAL LABORATORY,
CAMBRIDGE.

XXXIV.—*The Critical Temperature and Value of* $\dfrac{ML}{\Theta}$ *of some Carbon Compounds.*

By James Campbell Brown, D.Sc.

In two former papers (Trans., 1903, **83**, 991, and 1905, **87**, 269) on the determination of latent heat, the author was not able to calculate the value of $\dfrac{ML}{\Theta}$ for some of the compounds because the critical temperature had not been determined, or at least was not recorded in any available journal.

Having a sufficient supply of very pure substances, it seemed desirable to make the determination of critical temperature, as well for the purpose of recording this constant itself as for the purpose of calculating the value $\dfrac{ML}{\Theta}$.

*Method.—*Tubes of soda glass were prepared about 4·5 cm. long having between 4 and 5 mm. internal diameter, the walls being 2 mm. thick. The tubes were sealed at one end and at the other they were drawn out to a long capillary bent at right angles to the tube. To fill a tube with the liquid, it was heated, the end of the capillary dipped into the liquid, and the tube was allowed to cool. The liquid was then boiled off so as to fill the tube with vapour ; on again cooling, the liquid completely filled the tube ; if not, the operation was repeated. Finally the liquid was boiled until only the requisite amount of liquid remained, the tube was sealed at the capillary end, which was then thickened and bent in the form of a hook, the other end of the tube being also provided with a hook to facilitate suspension.

The tube containing the liquid was heated in a three-chambered air-bath of the form usually employed for a similar purpose. This air-bath was made of tinned iron, and two parallel sides of each of the three concentric chambers were provided with mica windows ; behind these windows was placed a light in the line of sight. The middle chamber had four false bottoms of wire gauze to distribute the heat and each chamber was closed by an asbestos board above, perforated to admit the thermometer, which was filled with nitrogen under 20 atmospheres pressure. The heat was obtained from one of Fletcher Russell & Co.'s special high-power burners. The bulb was heated gradually until the meniscus disappeared and striations appeared in the vapour ; the thermometer was then read. The bulb was next allowed to cool slowly and the temperature at which the meniscus reappeared was noted. The heating and cooling were repeated a second time and the mean of the four observations taken.

Each liquid was subjected to at least two such experiments, one with the tube about half full of liquid and the second with the tube one-third full, so that the final result is the mean of at least eight readings. Sometimes an experiment was made with smaller quantities of liquid when satisfactory results were not obtained with the half filled tube.

Results of Experiments.—The mean of the first and second experiments is given to show how far the results are concordant.

Alcohols.

iso*Propyl Alcohol.*—Three experiments were made, namely, with the tube about one-half full, one-third full, and one-fourth full of liquid. Mean value of θ for *iso*propyl alcohol = 243·47°·

The result is probably too low, owing to the supercooling of the vapour delaying the reappearance of the meniscus.

iso*Butyl Alcohol.*—Two experiments were made giving eight observations of the temperature with the stem of the thermometer not wholly immersed in the air-bath. The mean value of θ was 279·07°.

This was not satisfactory, and two experiments were then made with the stem of the thermometer wholly immersed in the air-bath, when the mean value of θ for *iso*butyl alcohol was found to = 277·63°·

The latter was taken as correct, but it may be too low owing to supercooling delaying the reappearance of the meniscus.

sec.-*Butyl Alcohol* :—1st Experiment.—The tube, which was one-third full of the pure dry alcohol, was heated gradually in the air-bath until the meniscus disappeared and striations appeared in the vapour; the thermometer reading was taken at this point. The tube was then allowed to cool slowly and the thermometer again read when the meniscus reappeared. The heating and cooling were repeated a second time.

	Meniscus disappeared.	Meniscus reappeared.
(1)	265·3°	265·5°
(2)	265·6	265·6

2nd Experiment.—Tube about half full of the alcohol.

	Meniscus disappeared. ·	Meniscus reappeared.
(1)	264·8°	264·8°
(2)	265	265·1

Mean critical temperature of *sec.*-butyl alcohol = 265·19°·

iso*Amyl Alcohol* (chiefly active, but contains some inactive).—The boiling point was constant at 130·1°. Two experiments were made with very concordant results, the mean value of θ for *iso*amyl alcohol being 309·77°.

tert.-*Amyl Alcohol.*—Two experiments were made with concordant results, the mean value of θ for *tert.*-amyl alcohol being 271·77°.

n-*Heptyl Alcohol.*—Two experiments were made ; eight observations, maximum •366·5°, minimum 363·25°, the mean value of θ for *n*-heptyl alcohol being 365·3°.

n-*Octyl Alcohol.*—Two experiments were made ; eight observations, maximum 387·25°, minimum 383·5°, the mean value of θ for *n*-octyl alcohol being 385·46°.

sec.-*Octyl Alcohol.*—Maximum 364·5°, minimum 363·75°, the mean value of θ being 364·12°.

Acids.

Several attempts were made to determine the critical temperature of formic acid, using very pure and carefully dried specimens, but in every case the tubes burst owing to decomposition before a critical temperature was reached. There is therefore no critical temperature for formic acid.

In each of the following cases, two experiments with eight observations were made, one with the tube one-third full, one with the tube half full.

n-*Butyric Acid.*—Maximum 355°, minimum 354·5°, the mean value of θ being 354·74°.

iso*Butyric Acid.*—Maximum 336·5°, minimum 336°, the mean value of θ being 336·25°.

n-*Valeric Acid.*—(i) The tube half full of liquid : maximum 378°, minimum 377·5°. (ii) The tube one-third full of liquid gave a higher result, maximum 380°, minimum 379·5°, the mean value of θ being 378·87°.

iso*Valeric Acid.*—Maximum 361°, minimum 360·5°, the mean value of θ being 360·68°.

Esters.

Two experiments were made in each case ; eight observations.

Ethyl iso*Valerate.*—Maximum 315·5°, minimum 314·5°, mean value of θ being 314·87°.

iso*Amyl Acetate.*—Maximum 327°, minimum 325·5°, mean value of θ being 326·18°.

Propyl iso*Valerate.*—Maximum 336°, minimum 335·5°, mean value of θ being 335·93°.

iso*Butyl* iso*Butyrate.*—Maximum 329°, minimum 328·5°, mean value of θ being 328·74°.

iso*Butyl Butyrate.*—Maximum 338·5°, minimum 338°, mean value of θ being 338·25°.

iso*Amyl Propionate*.—Maximum 338·5°, minimum 338°, mean value of θ being 338·24°.

iso*Butyl* iso*Valerate*.—Maximum 348·5°, minimum 348°, mean value of θ being 348·25°.

iso*Amyl Butyrate* (tube half full).—Maximum 347°, minimum 346°; tube one-third full gave a lower result, maximum 345·5°, minimum 344·5°, mean value of θ being 345·68°.

Ethyl Caprylate.—Maximum 386·5°, minimum 384·5°, mean value of θ being 385·56°.

Ethyl Nonylate.—(i) Tube half full of liquid: maximum 402°, minimum 402°; (ii) tube one-third full gave a lower result, maximum 400°, minimum 399·5°, mean value of θ being 400·81°.

Aromatic Hydrocarbons.

Two experiments were made in each case; eight observations.

	Maximum.	Minimum.	Mean value of θ.
o-Xylene	364°	362°	362·95°
m-Xylene	349·5	348·5	349
p-Xylene	349	348	348·5
Mesitylene	371	369·75	370·5
Cymene	387	384	385·15

The values of $\dfrac{ML}{\Theta}$ can now be calculated from the latent heat and critical temperature determinations.

Alcohols (Trans., 1903, 83, 991).

B. p.		θ.	L.	$\dfrac{ML}{\Theta}$.
82·85°	iso Propyl	243·47	161·1	18·71
108·1	iso Butyl	277·63	138·4	18·59
100·0	sec.-Butyl	265·19	136·2	18·72
130·1	iso Amyl (active)	309·77	124·7	18·83
101·8	tert.-Amyl	271·77	115·65*	18·68
176·0	n-Heptyl	365·3	105·0	19·08
195·0	n-Octyl	385·46	97·46	19·24
179·5	sec.-Octyl	364·12	94·48	19·27

* Trans., 1905, 87, 269.

The *iso*propyl and *iso*butyl alcohol numbers for critical temperature are probably nearly a degree too low.

Acids (Trans., 1903, 83, 992).

B. p.		θ.	L.	$\frac{ML}{\Theta}$.
	Formic...............	Decomposes below its critical temperature.		
	Acetic	321·5 (Pawlewski)	97·05	9·78
	Propionic...........	339·9 ,,	128·93	15·55
164°	n-Butyric	354·75	113·96	15·9
153	isoButyric	336·25	111·5	16·1
185	n-Valeric	378·87	103·1	16·13
175	isoValeric	360·68	101·03	16·26

Esters (Trans., 1903, 83, 994).

B. p.			θ.	L.	$\frac{ML}{\Theta}$.
134·3°	C_7	Ethyl isovalerate	314·87	67·84	15·00
143·0		isoAmyl acetate	326·18	69·00	14·97
158·4		Propyl isovalerate	335·93	64·37*	15·22
148·4	C_8	isoButyl isobutyrate	328·74	63·4	15·17
157·0		isoButyl butyrate	338·25	64·59	15·21
160·5		isoAmyl propionate	338·24	65·31	15·38
170·0	C_9	isoButyl isovalerate	348·25	60·41	15·36
179·6		isoAmyl butyrate	345·68	61·79	15·78
207·0	C_{10}	Ethyl caprylate	385·56	60·46	15·79
225·0	C_{11}	Ethyl nonylate	400·81	58·08	16·03

* Trans., 1905, 87, 269.

In consequence of the higher values of $\frac{ML}{\Theta}$ thus obtained for the denser esters, the determinations of the critical temperature of some and of the latent heat of the whole of the esters were repeated with carefully repurified specimens, dried over phosphoric oxide and distilled. The results remained the same, and the values are therefore considered correct except in the case of ethyl nonylate, of which the quantity employed was smaller, and it may have been less perfectly purified and dried. The critical point may be a little too low. The lowest reading was 399·5° repeated three times and the highest 402° repeated four times out of eight readings.

Aromatic Hydrocarbons.

B. p.		θ.	L.	$\frac{ML}{\Theta}$.
144·0°	o Xylene	362·95	82·47	13·74
139·0	m-Xylene	349·0	81·34	13·86
138·5	p-Xylene	348·5	80·98	13·81
165·4	Mesitylene	370·5	74·42	13·87
177·0	Cymene	385·15	67·64	13·77

Cumene was not sufficiently pure.

I have to acknowledge my indebtedness to Dr. A. Rule for much assistance in carrying out the determination of critical temperatures.

THE UNIVERSITY OF LIVERPOOL.

XXXV.—Some Reactions and New Compounds of Fluorine.

By Edmund Brydges Rudhall Prideaux, M.A., B.Sc.

Part I. *Preparation of the Fluorine and Halogen Fluorides.*

THE electrolysis of hydrogen fluoride was carried out according to the well-known procedure due to Moissan. Some modifications related chiefly to details of manipulation and do not call for mention here. It is, however, a noteworthy fact that fluorine thus produced contains, even after prolonged electrolysis, an appreciable amount of oxygen. The presence of oxygen, clearly proved by two experiments especially designed for the purpose, should be taken into account by future investigators, since oxygen generated under such conditions necessarily contains ozone. Now many of the substances which combine easily with fluorine are also acted on to some extent by ozone, whilst the instability of ozone and our scanty knowledge of its physical constants render it a most undesirable impurity of fluorine required for physicochemical investigations. The ozone must therefore be converted into oxygen, and the conversion was effected by heating a section of the metal tubes through which the gases passed before being used. The apparatus has already been described by Mr. Cuthbertson n the author (*Phil. Trans.*, 1905, Series A, **205**, 325).

Iodine Fluoride.

Fruitless attempts were made to prepare a higher fluoride of iodine. The result was in every case the characteristic colourless pentafluoride described by Moissan (*Compt. rend.*, 1902, **135**, 563), who published the result of two analyses carried out by a gravimetric method. In the present instance, many analyses were carried out by an entirely different, volumetric method, and the results are worth recording both on this account and because of the rarity of the compound. The volumetric method proved rapid and convenient and was briefly as follows. The tube containing the colourless fluoride was sealed and plunged into standard alkali, the excess of which was afterwards found by titration. The solution was then made decidedly acid and solution of potassium iodide was added. The iodine liberated according to the equation :

$$5HI + HIO_3 = 3I_2 + 3H_2O,$$

was then titrated in the usual way. One-sixth of the iodine thus found was calculated as cubic centimetres of normal acid and subtracted from the total amount of normal acid equivalent to the compound IF_5. This had been determined by titration. The difference gives the standard acid equivalent to the fluorine and hence the weight of fluorine in the compound.

The results are tabulated below.

Iodine.	Fluorine.	Iodine fluoride.	$\dfrac{I \times 100}{IF_5}$.	
0·112	0·080	0·192	58·11	
0·2195	0·1485	0·368	59·1	Iodine found was confirmed by $AgIO_3 \longrightarrow AgI$ from an aliquot part of the solution.
0·077	0·0564	0·1334	57·7	
0·0365	0·028	0·0645	56·6	

Mean percentage of iodine = 57·9.
IF_5 requires = 57·2.

The specific gravity of the compound roughly determined gave values the average of which is 3·5.

Moissan (*loc. cit.*) has not confirmed the observation of McIvor (*Chem. News*, 1875, **22**, 229) that the pentafluoride is violently acted on by water with a hissing sound and great development of heat. This difference of opinion is probably attributable to different conditions. The former experimenter poured some pentafluoride into water and observed only a slight warming, while the latter presumably allowed the water to come into contact first with an excess of the substance with the effect mentioned above.

Liquid Fluorine and Iodine.—A fractionating bulb containing iodine was immersed in liquid air and fluorine was then liquefied in the same bulb. There is no chemical action; the pale yellow liquid fluorine also dissolves no iodine, since it may be distilled into another bulb and back again without altering in appearance. When the tube is sealed off and removed from the liquid air, a dark colour soon appears in the layer of fluorine next to the iodine, the whole rapidly liquefies, and an energetic action sets in with projection of white fumes some way up the tube ; finally, a pale green flame appears for a few seconds.

Bromine Fluoride.

The fact that fluorine acts on bromine with the formation of some definite fluoride has been put on record by Moissan (*Le Fluor et Ses Composés*, p. 123). As, however, the study of halogen fluorides seemed to have been abandoned since his note on iodine pentafluoride, IF_5 (1902), it seemed desirable to fill this gap in our knowledge. A compound was prepared by the action of fluorine on bromine and analysed for the first time. The formula of the new compound and some of its

Y 2

properties have been published in a brief note to the Chemical
Society (*Proc.*, 1906, **22**, 19). The following is a more detailed
account of the results attained. The formula has since been confirmed by an independent research of M. Lebeau (*Compt. rend.*,
1905, **141**, 1015), who has described additional properties of this
interesting compound.

Comparison of Halogen Fluorides.—A study of the relations between
iodine and fluorine leads to the conclusion that the only fluoride which
gives concordant results on analysis is that which exists when the
change of appearance and properties has become complete and shows a
definite end-point. Further, it was found that this compound
would combine with no more fluorine and was therefore the highest
fluoride. A similarity was to be expected in the relations between
bromine and fluorine. Experiment showed that there was indeed a
general similarity, but that the two reactions differed in one respect.
Iodine liquefies almost immediately, forming a homogeneous dark
liquid which, on further passage of fluorine, separates only incompletely
into an upper colourless and lower dark layer. On the other hand,
the surface of bromine is at once covered by a distinct layer, having
the appearance of the saturated fluoride, and this layer grows at the
expense of the bromine, the line of demarcation remaining until all
the bromine has disappeared. This seems to point to a probability
that no fluoride of bromine can exist containing less than three atoms
of fluorine to one of bromine. The non-existence of a higher fluoride
was proved by experiments in which a large excess of fluorine was
passed over the compound, which is therefore probably the only
compound of fluorine with bromine. Liquid fluorine behaves
in the same way towards solid bromine as it does towards iodine.
The liquid remains of a pale yellow colour and when allowed to boil
off the solid bromine there is no difference of appearance between the
first fraction and the last.

Several attempts were made to determine the composition by the
volumetric method described above in the case of iodine pentafluoride.
Bromine was in this case set free on merely acidifying the alkaline
solution and it soon became evident from this and other considerations
that the fluoride in question was not BrF_5. The following method of
analysis was therefore adopted. The tube containing the compound
was broken under standard ·alkali. The solution having the
appearance and smell of hypobromite was warmed to decompose this
into bromate and bromide. It was then neutralised with standard
nitric acid.

The bromine, being present partly as bromate and partly as bromide,
was estimated by a method recently described. Jannasch and Jahn
(*Ber.*, 1905, **38**, 1576) state that bromates are completely reduced to

bromides when heated with excess of fuming nitric acid. The bromine obtained in this manner from a known mixture of bromide and bromate was within one per cent. of the theoretical value.

To the solution prepared as above was added a little sodium carbonate and excess of pure calcium nitrate, the calcium fluoride being weighed as usual. The filtrate was concentrated by evaporation and treated with excess of silver nitrate and nitric acid. The silver bromide was filtered off, the process repeated with the filtrate, and the whole of the silver bromide dried and heated to constant weight.

The mean of two experiments carried out in this way gave:

$$Br = 57\cdot5 \quad \text{Theoretical} \left\} \begin{array}{l} Br = 58\cdot4. \\ F = 41\cdot6\cdot \end{array} \right.$$
$$F = 42\cdot5 \quad \text{for } BrF_3$$

It was mentioned above that the excess of alkali used for decomposing the bromine trifluoride was titrated with standard nitric acid. The result furnishes a useful check on the gravimetric values, for the latter give the weight of bromine and fluorine, which may of course be expressed as cubic centimetres of normal acid. Their sum represents the total acidity produced by the compound on decomposition with water, and may be compared with the total acidity actually found by titration. These comparative figures are given in the following table :

Cubic Centimetres of Normal Acid.

Acidity from the sum of Br and F.	Acidity by titration.
(1) 2·72	2·62
(2) 10·02	9·96

General Conclusions.—Attempts to set forth regularities in the maximum valencies of elements depending on their position in the periodic table have always led to a distinction between two kinds of valency : that shown towards elements more electro-positive and that towards elements more electro-negative than the element in question. The former kind of valency has only one value in the case of the halogens; chlorine is only univalent towards hydrogen and also towards iodine (ICl_3 has been proved to possess the simplest formula). When we come to consider the compounds of halogens with more electro-negative elements, we are necessarily confined to their compounds with one another, since oxygen compounds cannot be considered as furnishing unimpeachable evidence in questions of valency.

The preparation of iodine pentafluoride for the first time fixed the maximum valency of iodine as not less than five. The existence of iodine trichloride shows that iodine cannot be more than tervalent towards chlorine, and by analogy it cannot be more than univalent towards bromine, a fact established long since by Bornemann.

Similarly, the preparation of bromine trifluoride has fixed the valency of bromine as certainly not less than three. In this case also, the compound with chlorine exhibits a valency two less than that with fluorine. Bromine and chlorine will only combine to form the compound BrCl.

	F	Cl	Br	I
	F	Cl	Br	I
F	F_2	(ClF)	BrF_3	IF_5
Cl		Cl_2	BrCl	ICl_3, ICl
Br			Br_2	IBr
I				I_2

The symmetry shown by these compounds as tabulated is only partial, since a complete symmetry requires also the existence of ClF, BrF, IF_3, and IF. It is by no means improbable that the last two are present in the dark liquid produced in the early stages when fluorine is allowed to react with iodine. Experiments, however, especially designed by other investigators to prepare ClF have always failed, whilst considerations mentioned above tend to show that BrF_3 is the only compound formed by fluorine and bromine.

PART II. *Fluorides of Selenium and Tellurium.*

Information as to fluorides of these elements is scanty and only qualitative. The fluorides described below are, as will be seen from the evidence, undoubtedly the characteristic and important and possibly the only fluorides of selenium and tellurium.

Moissan (*Le Fluor et Ses Composés*, p. 123) has described the appearance of the reactions between fluorine and these elements : "Selenium is attacked in the cold, there are abundant white fumes and the selenium presently melts and takes fire. Around it there is condensed a white, crystalline compound which is decomposed by water and dissolved by hydrofluoric acid. Powdered tellurium put in contact with fluorine combines with incandescence, giving abundant white fumes. The whole mass is quickly covered with a solid crystallised fluoride, easily volatile and very hygroscopic, having the aspect and properties of the fluoride of tellurium described by Berzelius."

Evidently these compounds needed much further study. The selenium used was Kahlbaum's best preparation, and the tellurium (kindly furnished by Dr. L. F. Guttmann) had been purified by redistillation in a vacuum. The solids were contained in the horizontal part of a glass tube bent twice at right angles. One end fitted accurately over the copper tube delivering fluorine, the other was furnished with a guard tube to prevent the entry of moist air. During the action of the fluorine, the appearances were on the whole as noted by Moissan. The incandescence of the tellurium takes the form of vivid blue sparks and the tube soon becomes heated. An important difference is that the white substance obtained was neither hygroscopic nor easily volatile.

The Solid Fluorides.—Although these solids did not appear to be very well-defined compounds, nevertheless some analyses were made of the white substance obtained by the action of fluorine on tellurium. The indefinite nature of this substance was borne out by the results, for whilst the first analyses gave 52·3 and 56·8 per cent. of tellurium, subsequent values were as far from this as 66·9 per cent. The method employed was to warm the substance contained in a glass tube with dilute caustic potash until all the white compound had dissolved, the tube was then washed out and the unchanged tellurium filtered off, dried, and weighed. The filtrate was then evaporated down with strong hydrochloric acid, diluted, and treated with sulphur dioxide until all the tellurium was precipitated. The weight of this gave the amount of combined tellurium, and the weight of white compound *plus* uncombined tellurium being known, the percentage of tellurium in the former could be calculated. This percentage varied widely, as stated above. Moreover, from other considerations it seemed likely that these fluorides would be not solids, but gases, or at least easily volatile liquids, for the fluoride of sulphur is described by Moissan as an extremely stable gas not easily liquefied (melting point, according to Moissan, $-55°$). It was to be expected that selenium and tellurium hexafluorides corresponding to sulphur hexafluoride would at any rate be stable enough to be isolated, and also that they would probably be gases or volatile liquids. For fluorine has a great tendency to form gaseous molecules with other elements. One need only compare silica with silicon fluoride. These considerations, together with the proved indefiniteness of the white compound, made me think that the real fluorides were still to be isolated.

Preparation of the Hexafluorides of Selenium and Tellurium.—A glass tube, of the form described above, containing the tellurium was joined to another which was kept at $-78°$. Immediately after the experiment, the second tube was isolated by sealing off while still in the freezing mixture. It was seen to contain a white, crystalline solid. When this was allowed to warm up, it changed first of all into a clear, mobile liquid, and then completely to the gaseous condition.

Density of Tellurium Fluoride.—As this gas was much compressed, it seemed worth while to attempt a rough measurement of the density. The sealed tube containing the gas was weighed, then 'attached to a Töpler pump, exhausted, and the gas collected over mercury. The gas contained a little air. To determine the amount, it was allowed to stand over water (which completely decomposes tellurium fluoride), and the residual air afterwards measured. Neglecting the weight of that small volume of air and correcting all gaseous volumes to normal temperature and pressure, the density is obtained from the expression :

$$\frac{\text{weight of gas}}{(\text{total vol. of gas} - \text{vol. of air}) \times 0\cdot0000896}.$$

Two density determinations carried out by this method gave molecular weights of 257 and 240, which correspond very fairly to the theoretical 241·6 for tellurium hexafluoride. The density of the gas was afterwards determined accurately by Sir William Ramsay. The gas was first purified as follows (Fig. 1). The tap T was connected with a gas-holder in which the gas was stored, and T_1 with a Töpler pump. T being closed, the apparatus was completely exhausted through T_1. T and T_1 were then closed, and T opened gradually while the level of the mercury in the reservoir attached to the gas-holders was kept about half a barometer height below the level of that in the gas-holder. The gas slowly passed through F, which was immersed in liquid air. This process was repeated until all the gas had passed from the gas-holder through F. The air was finally pumped off until the mercury in M rose to the barometric height. The liquid air was then taken from the bulb F, and the hexafluoride pumped off and collected. This sample was then transferred to a gas burette and its density determined in the manner described by Ramsay and Travers (*Phil. Trans.*, 1901, Series A, **197**, 53 and 54). The density $(O = 16)$ was found to be 119·5, the molecular weight therefore 239, corresponding to the formula TeF_6.

Properties and Analysis of the Hexafluoride.—The gas has a very unpleasant odour, recalling that of tellurium hydride and also ozone. It is decomposed by water slowly but completely. If a sample is confined over water, scarcely any change is noticeable in half an hour, but in the course of a night the gas has been completely absorbed. On evaporating the solution in a platinum dish, a yellow residue of tellurium trioxide is left. By reducing this in a current of hydrogen and dissolving in concentrated sulphuric acid, the crimson colour characteristic of tellurium was produced. This reaction with water affords a method of analysing the gas. A known volume, implying a known weight of the gas, was allowed to stand over water for twenty-four hours, and the solution was then carefully evaporated to dryness and heated until its weight became constant at 150°. The percentage of tellurium was calculated from the weight of the trioxide.

$$TeF_6 + 4H_2O = H_2TeO_4 + 6HF \; ; \; H_2TeO_4 = TeO_3 + H_2O.$$

Two analyses gave 50·06 and 51·9 per cent. of tellurium, the theoretical percentage of tellurium in TeF_6 being 52·6.

Selenium Fluoride.

Selenium fluoride was prepared in a similar way. In this case, the white solid in the reaction tube was not further examined. In the second tube, there appeared a white, crystalline mass. It was sealed off and then allowed to attain room temperature. The gas disengaged was collected over mercury. The density (O = 16), determined in the same manner as that of TeF_6, was found to be 97·23, giving a molecular weight of 194·46, the molecular weight for the simple molecule SeF_6 being 193.

This gas is more stable than tellurium fluoride; it decomposes water with extreme slowness, if at all. To a sample contained over mercury, a little water was added, and the position of the meniscus marked. After more than fifteen hours, there was no noticeable change in the position of the meniscus. The volumes of gas before and after treatment with water (corrected for the presence of water vapour and reduced to normal temperature and pressure) were 7·74 and 7·35 respectively. The gas does not combine with sulphur dioxide in the cold. A volume of selenium hexafluoride mixed with about five times its volume of sulphur dioxide shows no change in volume after mixing. On prolonged sparking, there is a slight contraction and a very small quantity of red solid, probably selenium, is deposited on the walls of the tube.

Physical Constants of the Hexafluorides of Sulphur, Selenium, and Tellurium.

This series of compounds of exactly the same type presents well-marked regularities, some of which have not been observed before, and will probably have some value in connecting the physical properties of compounds with those of the elements which they contain. This series is quite unique in possessing the following properties.

(1) All the compounds are stable gases and fairly easily manipulated.

(2) They are all of the same type and exhibit the maximum valency of the elements which they contain, so that there are no disturbing influences due to "residual affinity."

These gases do not attack glass nor decompose spontaneously. They may be said, therefore, to be on the whole easily manipulated, but with one reservation. Both selenium and tellurium hexafluorides (but not the sulphur analogue) possess the annoying property of making mercury stick to glass in much the same way that ozone does, with this difference, that there is no noticeable diminution in the volume of the fluorides. The mirror of mercury formed on the glass may render it difficult or even impossible to read the volume at some parts of the graduated burette. Also, when a thread of mercury is expelled from a capillary tube, the mirrors and even pools of mercury left sticking to the sides greatly interfere with such measurements as are carried out by bringing a uniform thread of mercury to some constant volume mark.

FIG. 1.

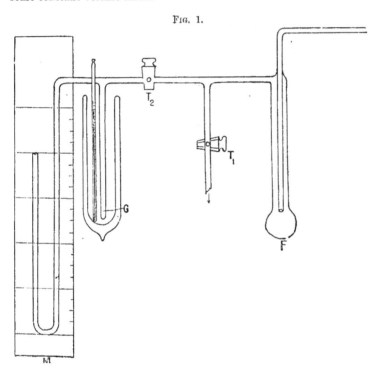

Measurements of Vapour Pressure.—The vapour pressures at various temperatures were measured in the apparatus indicated in Fig. 1.

The manometer of 3 mm. bore was fastened against a glass scale divided in millimetres. This form of manometer was rendered necessary by the fact that the volumes of gas condensed were small. Consequently the volume into which they expanded had also to be small in order that the vapour should remain saturated through a sufficient range of temperature. The whole apparatus was exhausted through T_1 by means of a Töpler pump. Then T_1 and T_2 were closed, the fractionating bulb F was immersed in liquid air, and the gas gradually admitted from a gas-holder. The fluorides condensed in F, and any air was removed by opening T_1 and again exhausting; T_1' was then closed, T_2 opened, and the vessel of liquid air was transferred from F to the side-tube G. The gas quickly condensed into G and the tap T_1 was again opened for a moment. The manometer then showing a vacuum, T_2 was closed and the side-tube allowed to remain for five minutes in the liquid air. The pressure did not alter, showing that the vapour pressures of these compounds are nothing at the temperature of liquid air contained in an open vessel. The liquid air was then replaced by ether or ligroin (preferably the former) cooled with liquid air to about $-110°$. The temperature rose quite slowly and was registered by means of a pentane thermometer immersed in the bath. The manometer was constantly tapped before the reading, which was taken as nearly as possible at the same time as that of the temperature. In the case of selenium and tellurium hexafluorides, the tube soon became coated with mercury as described above, so that readings of the closed limb had to be abandoned. The heights of mercury in the closed limb corresponding to heights in the open limb were in these cases read off from a line which had been plotted by finding many corresponding values of readings on the two limbs. The smoothed curves are given in Fig. 3. The curve very quickly becomes steep in the case of sulphur hexafluoride, but alters its slope more slowly in the case of selenium and tellurium hexafluorides, the vapour pressure curves of which resemble one another very closely and are separated by only a narrow temperature interval ($3·5°$ at the boiling point). The interval between the curves for the sulphur and selenium compounds is on an average $23°$. The points where the pressures of the solids become 760 mm. are respectively $211°$, $234°$, $237·5°$. This temperature was found by a short extrapolation in the case of the sulphur compound, of which there was not a sufficient quantity to give a saturated vapour above $209°$.

Melting Points.—The gases were next sealed up in capillary tubes as follows (Fig. 2). The gas was measured and transferred to the gas-holder A. The whole apparatus up to T_1' (closed) was then exhausted through D. B was then placed in liquid air, T_1' opened to B, and the gas allowed to pass slowly through B until the mercury was beyond T_1'. T_2' was then opened and the solid fluoride completely freed from any trace of air by exhausting through D. T_2' was then opened to C, which was a capillary tube graduated in millimetres and calibrated by running in and weighing a thread of mercury. The liquid air was now transferred from B to C and the vessel was gradually raised. The gases condensed as a white snow, further up the tube the solids appeared as crystalline spangles. The capillary tube was placed for a moment in connection with the vacuum and then sealed off. Those tubes then contain known volumes of the gases existing as liquids under high pressures.   The melting points were determined by immersing the tubes in ether cooled by liquid air to some temperature below the solidifying points.   The temperature of such a freezing mixture contained in a vacuum cylinder rises quite gradually, and the intervals between the first appearance of liquid and the slipping down of the solid were on an average 1°.   The latter temperature was chosen in each case.   The melting points are given in a table, and are also indicated by crosses on the P, T, curves (Fig. 3).*

It will be noticed that the line joining the melting points of sulphur

FIG. 2.

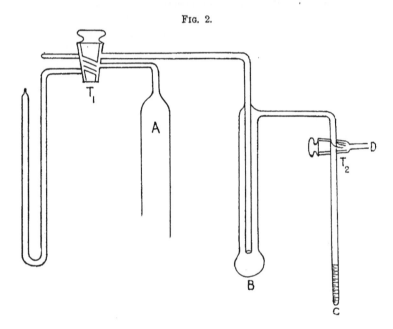

and selenium hexafluorides slants slightly upwards and, if continued, cuts the curve for the tellurium compound at about 241°.   The melting point of the last substance is abnormally low, this being almost the only irregularity exhibited by this series of compounds.

Critical Temperatures.—The tubes were gradually heated in a beaker of water.   Since they were from one-third to one-half full of liquid at the ordinary temperature and the critical volume is usually taken as three times the volume of the liquid, the capacity of the tube was approximately equal to the critical volume of the gas.   Temperatures

* Moissan and Lebeau have found the melting point of sulphur hexafluoride to be − 55°.

were read when the extremely fine line of flattened meniscus vanished and when it reappeared from a band of mist. These temperatures were not more than 1° apart and their mean was taken as the critical temperature. The points are: for SF_6, 327° abs. (54°); SeF_6, 345·35° abs. (72·35°); TeF_6, 356·25° abs. (83·25°).

	Difference of boiling points.	Difference of critical temperatures.
$SeF_6 - SF_6$	23°	18·35°
$TeF_6 - SeF_6$...............	3·5	10·9
$TeF_6 - SF_6$	26·5	29·25

FIG. 3.

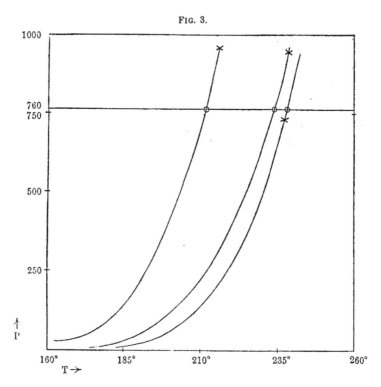

If, then, the critical pressures are approximately equal (that is, within 50 mm., which is the alteration of pressure caused by rise of 1°), the P, T, lines of tellurium and selenium hexafluorides have much the same separation at the critical temperatures as at the temperatures recorded on the curves, but the selenium hexafluoride line bends considerably from that of the tellurium towards that of the sulphur analogue.

In the following table are given the melting points, T, boiling points,

T_1, and critical points, T_2, of these gases, and also T and T_1 for the series SH$_2$, SeH$_2$, and TeH$_2$ as determined by Forcrand and Fonzes-Diacon (*Compt. rend.*, 1902, **134**, 1209). It will be noticed that the differences between T and T_1 are approximately equal in the case of sulphur and selenium hydrides as well as the corresponding fluorides, but that whilst $T - T_1$ is much greater in the case of tellurium hydride, it is much less in the case of tellurium hexafluoride, the melting point of which is very nearly equal to its boiling point. Also the melting points of the hydrides are below their boiling points, whilst the melting points of the fluorides are above the temperatures at which the vapour above the solid attains a pressure of 760 mm.

Compound.	T.	T_1.	T_2.	$T - T_1$.
SF$_6$	$-56°$	$-62°$	$+54°$	$6°$
SeF$_6$	$-34·5$	-39	$+72$	$4·5$
TeF$_6$	-36	$-35·5$	$+83$	$-0·5$
SH$_2$	-86	-62	—	-24
SeH$_2$	-64	-42	—	-22
TeH$_2$	-48	0	—	-48

Coefficients of Expansion.—The above-mentioned calibrated tubes were placed in baths at the required temperature and the volumes of the liquids calculated from the positions of the meniscus on the scale. The expansion is therefore that of the liquid under pressures equal to its own vapour pressures at the temperatures under consideration. It has been abundantly shown that the expansion of a liquid is scarcely affected by pressures of this order. The temperature intervals were chosen such that the lower temperatures were about equal distances above the respective melting points, and the highest temperatures were so far below the critical points as to avoid the abnormal expansion which liquids display near those points.

	Temperature interval.	Coefficient of expansion.
SF$_6$................	$-18·5°$ to $+30°$	$0·027$
SeF$_6$	$-3·5$,, $+51$	$0·030$
TeF$_6$	$-3·5$,, $+51·5$	$0·032$

Specific Gravities of the Liquids.—The volumes of the gases sealed in the tubes being known, as well as the volumes of the liquids formed by their condensation, the specific gravities could be determined. Two series of experiments were made; one in which the capillary tubes were closed by a tap during the observation, and another in the above-mentioned sealed tubes. The absolute values' found in each set of observations differ somewhat, but, taking each set separately, the same relation holds between the molecular volumes of the three liquid compounds. I have chosen the second set because they were more carefully carried out and the method is more accurate. The volumes in the immediate neighbourhood of the melting points being rather irregular, the specific gravities were necessarily taken at some distance above these, the temperature differences being 12°, 11°, and 11·5° respectively.

Boiling point.	Temperature for specific gravity.	Specific gravity.	Molecular volume.
− 62°	− 50°	1·91	76·5
− 39	− 28	2·51	77 2
− 35·5	− 24	3·025	79·9

The close agreement of these molecular volumes accords well with a hypothesis that molecular volumes of binary compounds are conditioned solely by the type of the compound and the nature of the element of smaller valency. On examining other series to see if the like held in their case, it was evident that care had seldom been taken to choose corresponding temperatures for the determination of specific gravity. In the following table, the numbers in brackets indicate the number of degrees between the temperature of the observation and the boiling point of the liquid concerned. In cases where the temperature has not been given, the specific gravity was probably determined at about 15°.

OH_2.	SH_2.	SeH_2.	TeH_2.
18·8 (0°)	39·53 (−)	38·2 (−)	49·75 (− 20°)

PCl_3.	$AsCl_3$.	$SbCl_3$.
93·5 (0°)	84 (− 110°)	85 (− 150°)

$CHCl_3$.	$SiHCl_3$.
85·5 (0°)	94 (−)

It is to be regretted that the data are so scanty. The numbers at least point to a probability that this regularity will be found to hold in other series, since there is no wide difference between the numbers except in the case of water. The densities of arsenious and antimonious chlorides determined at 110° and 150° respectively below their boiling points are hardly comparable with that of phosphorus trichloride determined at its boiling point. Evidently the molecular volumes of the two former chlorides will approach that of the latter as the temperatures of the liquids approach their respective boiling points. The fluorides of sulphur, selenium, and tellurium form the only series of three which shows a very close agreement, but they are also the only series of three the molecular volumes of which have been measured at corresponding temperatures. Denoting by the word "similar" a series of compounds in which the atoms of higher valency belong to the same vertical groups in the periodic classification by long periods (double octaves), and the atoms of lower valency are the same throughout the series, it may be provisionally stated that "equal"

volumes of "similar" liquid compounds at corresponding temperatures contain the same numbers of molecules, or, in other words, compounds belonging to "similar" series are always associated to the
same extent in the liquid state when the vapour pressures of the
liquids are equal.

Refractivities of the Gases.

The refractivities were kindly determined for me by Mr. Cuthbertson and Mr. Metcalfe, using the method of interference due to
Jamin. Interference bands were counted for differences of pressure
amounting to 70 and 80 mm. The refractivities were calculated by
the formula $\mu - 1 = \dfrac{5893 \times \text{No. of bands} \times T \times 760}{\text{Length of tube} \times 273 \times P}$. The values found
were most remarkable when compared with those obtained from
compounds formerly examined. These have usually shown a rough
approximation to an additive law, although in some cases they vary
more than 20 per cent. from the value required by such a law—a variation too great to be accounted for by experimental error. In the case
of sulphur, selenium, and tellurium hexafluorides, there is no approach
to an additive law, the refractivities found being 783, 895, and 991,
while those calculated from $3F_2 = 576$, $S = 540$, $Se = 810$ $(3/2 \times 540)$,
$Te = 1350$ $(5/2 \times 540)$ (according to Cuthbertson's law connecting
the refractivities of the elements) are 1116, 1386, and 1926 respectively. The refractivity of tellurium hexafluoride is actually less
than that calculated for one atom of gaseous tellurium. There is,
however, a regularity connecting these refractivities with the densities
of the gases, which is shown in Fig. 4. The refractivities plotted
against the densities fall very close to a straight line. The theoretical
densities calculated from the atomic weights were chosen, since these
depending on the results of analysis are probably more accurate than
the densities obtained by weighing the gases.

If the points for sulphur and tellurium hexafluorides are joined by
a straight line, this cuts the ordinate from 96·5 at 884, eleven units
away from the observed refractivity 895. Thus, if the refractivity of
the selenium analogue were unknown, it could be calculated within
2 per cent. from those of the sulphur and tellurium compounds. The
experimental error in determining these refractivities being about
2 per cent., it is quite certain that the refractivity minus any number
greater than 500 (where the line cuts the ordinate from the origin
of densities) is proportional to the density. The reason that this
relationship has never been observed before, is that no such series has
ever been examined before in this way. The only series of three compound gases the refractivities of which have been determined, is that of
the halogen hydrides. In their case, $\mu - 1$ for hydrogen bromide found

by joining the points for the chloride and iodide lies about 10 per cent. below $\mu - 1$ actually observed. The indices of refraction for the series, hydrogen chloride, &c., are difficult to determine accurately on account of the hygroscopic and corrosive nature of the gases. It seems extremely probable that when the refractivities of other comparable series have been accurately determined the same relation will

FIG. 4.

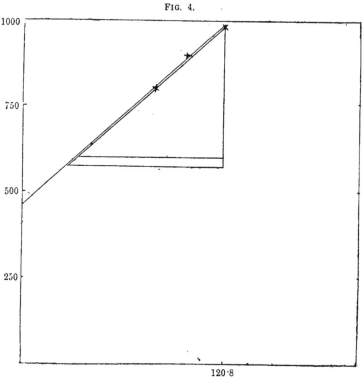

120·8

Ordinates $= \mu - 1$ for SF_6, SeF_6, TeF_6.
Abscissæ $=$ Densities of SF_6, SeF_6, TeF_6.
$SF_6 = 73, 783$; $SeF_6 = 96·5, 895$; $TeF_6 = 120·8, 991$.

Refractivities and densities of the fluorides of sulphur, selenium, and tellurium.

be found to hold, although no doubt the characteristic slope of the line would be different in each case.

The points have been marked where the line cuts the ordinate from F_6 $(d = 57)$ and F_2 $(d = 19)$. The latter point gives an index of 552, which is within 5 per cent. of the index for $3F_2$ (576). At present this can only be regarded as a coincidence.

If the assumption is made that $\mu - 1$ for F_6 in such a combination remains constant throughout the series, this value subtracted from the refractivities will give the retardation due to sulphur, selenium, and tellurium in combination. These values, 207, 319, and 415, are very nearly in the ratio 4, 6, and 8. The ratios of $\mu - 1$ for the elements according to Cuthbertson's law (*Phil. Trans.*, 1905, Series *A*, **204**, 323) of simple integers should be 4, 6, and 10. There is no explanation at present of the variation in the case of tellurium, but the simplicity of the relation is noteworthy.

This law connecting the refractivities of compounds with the alteration of density caused by the substitution of an element of higher atomic weight may be compared with the well-known law connecting refractivities with alterations of density caused by compression of the same gas. This relation, $\dfrac{\mu - 1}{d} = $ a constant, is universally employed in order to render comparable the results of refractivities determined at different pressures.

The above-mentioned regularity shows that if $(\mu - 1)_c = $ refractivity of the compound, $K = $ any number greater than 450:

$$\frac{(\mu - 1)_c - K}{d} = \text{a constant.}$$

In conclusion, I desire to express my thanks to the Royal Society for a grant in aid of this research, to Sir William Ramsay for his valuable advice and assistance, and to members of the Staff at University College for the kind interest they have shown in this investigation.

HERIOT-WATT COLLEGE,
EDINBURGH.

XXXVI —*Menthyl Benzenesulphonate and Menthyl Naphthalene-β-sulphonate.*

By THOMAS STEWART PATTERSON and JOHN FREW, M.A.

THE experiments detailed in the following paper were undertaken with the object of obtaining some data, since none seems to be available, regarding the influence of the radicles of benzenesulphonic and naphthalene-β-sulphonic acids in modifying the rotation of optically active compounds into which they may be introduced. The results are also of interest in connection with the work of Tschúgaeff and of J. B. Cohen. The former has shown (*Ber.*, 1898, **31**, 364) that the molecular rotations of the menthyl esters of the aliphatic acids have very nearly the same value (about $-157°$), and that a somewhat similar behaviour is shown by the menthyl esters of benzoic and substituted benzoic acids (*Ber.*, 1898, **31**, 1778; Abstr., 1903, ii, 2). The rotations of the aromatic derivatives are, however, much higher than those of the aliphatic esters, and this seems to be primarily caused by the presence of the benzene nucleus in the former.

The work of Cohen and Briggs (Trans., 1903, **83**, 1213), Cohen and Raper (Trans., 1904, **85**, 1262, 1272), Cohen and Armes (Trans., 1905, **87**, 1190), and Cohen and Zortman (this vol., p. 47) on menthyl esters of substituted benzoic acids shows that the molecular rotations of these compounds are affected considerably—in a few cases very considerably—both by the nature of the substituents in the benzene nucleus and by their position. It nevertheless seems to be the case that the presence of the benzene nucleus is chiefly responsible for the high rotations observed, which for the large majority of compounds examined lie between $-190°$ and $-250°$. The question whether the benzene nucleus would still exert a dominant effect in a menthyl benzenesulphonate or whether the introduction of the sulphur atom would produce a marked change seemed therefore of interest.

Menthyl Benzenesulphonate.

Menthol (18·7 grams) was dissolved in pyridine (60 grams), the solution cooled in a freezing mixture, and benzenesulphonic chloride (21·1 grams) very slowly added with frequent stirring. After some time, a crystalline substance collected at the bottom of the beaker and eventually a mass of small, white crystals was produced. A large quantity of water was then added, and after the mixture had stood for a short time with occasional stirring the solid was filtered off at the pump and dried on porous plate.

Menthyl benzenesulphonate dissolves readily in warm light petroleum or alcohol, and on cooling, even when a considerable quantity of the solvent is used, the mass becomes almost solid, and the mother liquor must be removed by pressing and draining. A second crystallisation is usually sufficient to free the substance from uncombined menthol. The compound thus prepared consists of very fine needle-shaped crystals melting at 80°. In the above preparation, 10·5 grams were obtained.

0·2789 gave 0·2227 BaSO$_4$.   S $= 10·96$.

C$_{16}$H$_{24}$O$_3$S requires S $= 10·81$ per cent.

An attempt was made to distil the compound under reduced pressure, but decomposition took place. A clear oil passed over at a temperature a few degrees above the melting point of the substance and did not solidify on cooling. On heating the menthyl benzenesulphonate to 85—90° and keeping it at that temperature for a short time, decomposition took place. The decomposition products formed two layers, a clear oil above a brown, viscous, deliquescent substance. Both these compounds will be further investigated.

An attempt was made to determine the rotation of the compound in the molten condition. The result was as follows, but since decomposition had commenced it is not very reliable: $a_D^{84°}$ ($l = 40$ mm.) $- 36·2°$. Assuming a density of 1·2, this value gives for $[a]_D^{84°} - 75·4°$, and for $[M]_D^{84°} - 223·2°$ which is not far removed from the value given below for solution in alcohol.

The rotation of the menthyl benzenesulphonate was then examined in solution in ethyl alcohol, benzene, and pure nitrobenzene, with the following results :

Solvent : *Ethyl Alcohol.* Sp. gr. $20°/4° = 0·7918$.

$p = 4·72$.

$t.$	$a_D^{t°}$ (170 mm.).	$d.$	$[a]_D^{t°}.$	$[M]_D^{t°}.$
18·8°	− 4·696°	0·8016	− 73·02°	− 216·1°

*$p = 3·416$.

19·7°	− 3·344°	0·8000	− 71·97°	− 213·0°
51·8	3·071	0·7720	68·50	202·7

Solvent : *Benzene.* Sp. gr. $20°/4° = 0·8786$.

$p = 5·014$.

$t.$	$a_D^{t°}$ (170 mm.).	$d.$	$[a]_D^{t°}.$	$[M]_D^{t°}.$
19·2°	− 4·855°	0·8867	− 64·17°	− 189·9°

*$p = 5·376$. Sp. gr. of benzene $20°/4° = 0·8798$.

18·2°	− 5·153°	0·8911	− 63·27°	− 187·3°
50·1	4·824	0·8579	61·53	182·1

* These values were obtained later than the others for the purpose of determining the temperature-coefficient. The specimen examined was not quite so pure as that previously used.

Solvent: *Nitrobenzene (Distilled under Reduced Pressure).*

$p = 7\cdot857.$

$t.$	a_D^c (100 mm.).	$d.$	$[a]_D^c.$	$[M]_D^c.$
15·7°	−6·232°	1·1975	−66·23°	−196·0°
16·5	6·210	1·1970	66·03	195·5
30·7	6·013	1·1830	64·69	191·5
70·5	5·540	1·1442	61·62	182·4
16·5	6·212	1·1970	66·03	195·5

$p = 18\cdot769.$

	(30·48 mm.).			
16·0°	−4·438°	1·1865	−65·38°	−193·5°
39·9	4·232	1·1638	63·55	188·1
63·5	4·029	1·1420	61·66	182·5
78·4	3·800	1·1270	58·94	174·5

The data for the more dilute solution in nitrobenzene show that no decomposition had occurred on heating at 70·5°, since on again cooling to 16·5° the rotation was found to be unaltered. On further heating of the solution, however, to about 110° for half an hour in order to determine the rotation at this higher temperature, decomposition occurred, the solution becoming very dark. Decomposition thus takes place in solution at about the same temperature as in the homogeneous condition.

Menthyl Naphthalene-β-sulphonate.

Menthol (18·7 grams) was dissolved in pyridine (90 grams) in a large beaker surrounded by a freezing mixture; naphthalene-β-sulphonic chloride (27·2 grams) was then added, and on stirring it completely dissolved. In a short time, crystals appeared in the solution, and became more plentiful on stirring. After three to four hours, a large quantity of water was added. This produced a bulky, crystalline precipitate, which was filtered and drained.

The substance thus prepared is moderately soluble in hot ethyl alcohol, and crystallises from it in well-formed, opaque crystals totally different in appearance from those of the corresponding benzene compound. After recrystallisation, the melting point was found to be 114—114·5°.

0·3277 gave 0·2286 BaSO$_4$. S = 9·58.

$C_{20}H_{14}O_3S$ requires S = 9·26 per cent.

The rotation of this compound was also examined in ethyl alcohol, benzene, and nitrobenzene with the following results:

Solvent : *Ethyl Alcohol.* Sp. gr. $20°/4° = 0·7918.$

$p = 1·8403.$

$t.$	$a_D^{t°}$ (170 mm.).	$d.$	$[a]_D^{t°}.$	$[M]_D^{t°}.$
14·7°	$-1·47°$	0·8010	$-58·66°$	$-203·0°$
49·1	1·372	0·7715	56·86	196·7

Solvent : *Benzene.* Sp. gr. $20°/4° = 0·8786.$

$p = 1·0185.$

$t.$	$a_D^{t°}$ (170 mm.).	$d.$	$[a]_D^{t°}.$	$[M]_D^{t°}.$
18·3°	$-0·772°$	0·8823	$-50·52°$	$-174·8°$
42·4	0·745	0·8564	50·24	173·8

Solvent : *Nitrobenzene.*

$p = 2·946.$

$t.$	$a_D^{t°}$ (170 mm.).	$d.$	$[a]_D^{t°}.$	$[M]_D^{t°}.$
12·5°	$-2·490°$	1·2060	$-41·24°$	$-142·7°$
36·0	2·521	1·1840	42·52	147·1
61·7	2·540	1·1598	43·73	151·3

$p = 8·643.$

12·7°	$-7·284°$	1·2030	$-41·20°$	$-142·6°$
41·0	7·435	1·1756	43·04	148·9
61·4	7·440	1·1563	43·80	151·6

On heating the nitrobenzene solutions to a temperature of about 100°, decomposition took place. The solutions became dark, and a crystalline substance separated out on cooling. The latter, when collected and recrystallised from chloroform, was obtained in white, pearly scales melting at 83° ; it will be further investigated.

Collecting now the data relative to the two compounds examined, we obtain the following table :

	Menthyl benzenesulphonate.		Menthyl naphthalene-β-sulphonate.	
Solvent.	$p.$	$[M]_D^{19°}.$	$p.$	$[M]_D^{19°}.$
Ethyl alcohol......	4·72	$-216·1°$	1·84	$-202·2°$
Benzene	5·014	189·9	1·019	174·8
Nitrobenzene	7·857	194·7	2·946	144·0
,, ...	18·769	192·8	8·643	144·2

With regard to these numbers, the following points may be noticed :

(1) In both cases, there is considerable variation of rotation with change of solvent, the variation being especially great with the naphthalene-β-sulphonate.

(2) Comparing the values for the two active compounds in each solvent separately, it appears that the rotations are consistently lower for the β-naphthalene derivative. The difference between the rotations in ethyl-alcoholic solution is 13·9°, in benzene 14·1°, and in nitrobenzene, taking the numbers for the more dilute solutions in each case, it is 50·7°.

(3) The solvents do not, however, exert proportionate influences on the rotations of the two active compounds. The highest rotation occurs with both substances in alcohol, but whilst the lowest rotation for the benzenesulphonate is found in benzene, for the naphthalene-β-sulphonate it occurs in nitrobenzene.

(4) The results obtained with our two preparations in benzene solution may be compared with the values given by Tschúgaeff for menthyl benzoate (*Ber.*, 1898, **31**, 1778) and for menthyl β-naphthoate (*Abstr.*, 1903, ii, 2), which were also obtained in benzene solution, although the strengths of the solutions examined are not recorded.

	$p.$	$t°.$	$[M]_D^{t°}.$
Menthyl benzoate	[?]	20°	$-236·3°$
,, β-naphthoate	[?]	20 [?]	287·6
Menthyl benzenesulphonate	5·014	19	189·9
,, naphthalene-β-sulphonate	1·019	19	174·8

The sulphonates thus appear to have distinctly smaller rotations than the carboxylates, and it is very noticeable that whereas the β-naphthoate has a greater rotation than the benzoate, the naphthalene-β-sulphonate has a less rotation than the benzenesulphonate.

Menthyl naphthalene-β-sulphonate may be regarded as a 2 : 3-di-substitution product of menthyl benzenesulphonate, so that the behaviour of the sulphonates is in agreement with the rule observed by Cohen and Briggs (Trans., 1903, **83**, 1213), Cohen and Raper (Trans., 1904, **85**, 1262), and Cohen and Zortman (this vol., p. 47), that 2 : 3-disubstituted benzoyl groups cause a lower rotation in the menthyl ester than the simple benzoyl radicle.

Tschúgaeff's observations on the benzoate and β-naphthoate are, of course, in direct opposition to this.

(5) As regards the influence of temperature change on the rotations of these compounds in solution, a regularity which is not brought out very distinctly by the numerical data is rendered somewhat clearer by the diagram, and what appears as a remarkable contradiction is reconciled with the other results.

The rotation data given above show that whilst rise of temperature causes increase in the rotation of menthyl naphthalene-β-sulphonate in nitrobenzene solution, the same cause brings with it diminution of rotation in all the other solutions examined, both of the benzene-sulphonate and the naphthalene-β-sulphonate.

In the diagram, the data for the solutions of the benzenesulphonate

Molecular rotation of menthyl benzenesulphonate (full lines) *and menthyl naphthalene-β-sulphonate* (broken lines) *in alcohol, benzene, and nitrobenzene.*

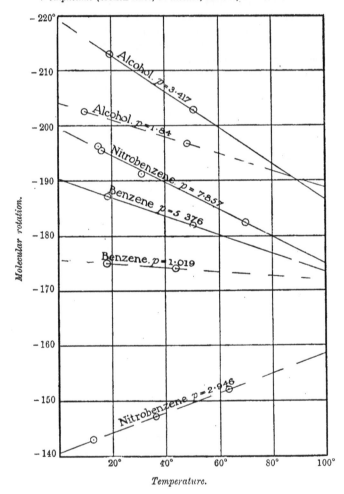

Temperature.

are represented by full lines, whilst those for the β-naphthalene derivative are represented by broken lines.

Taking the three curves for the benzenesulphonate first, it is to be observed that the diminution in rotation with increase of temperature is greatest in ethyl-alcoholic solution; in nitrobenzene solution, it is distinctly less, and it is least in benzene. Thus, the diminution of rotation with increase of temperature is greatest in the solvent which produces the highest rotation.

A similar behaviour is shown by the solutions of the naphthalene-β-sulphonate. The diminution of rotation with rise of temperature is most rapid in the alcoholic solution. In benzene, the change is only slight—the molecular rotation of the solution is almost insensitive to temperature change—whilst in nitrobenzene, where the rotation is much lower, the temperature-coefficient becomes opposite in sign and the rotation increases with increase of temperature. The behaviour of the two substances in these solvents thus seems to be similar—where the rotation is highest its rate of diminution on heating is greatest.

It may be noticed finally that in consequence of this the molecular rotations of these dilute solutions tend at higher temperatures towards a common value lying at about $-170°$.

Part of the cost of this investigation was defrayed by a grant from the Chemical Society's Research Fund, for which the authors have pleasure in expressing their indebtedness.

UNIVERSITY OF GLASGOW.

XXXVII.—*The Attractive Force of Crystals for Like Molecules in Saturated Solutions.*

By EDWARD SONSTADT.

ONE of the earliest experiments made with the object of tracing the effect of crystals on the strength of a solution was performed on considerably more than half a litre of solution of potassium platini-chloride standing in a beaker over a quantity of crystals of the salt which had been deposited from it by cooling. More than half of the solution (A) was poured into a separate beaker, and a trace of crystals of the same salt was added to it. The remainder of the solution (B) was left standing over the crystals. The two beakers, loosely covered with paper, were set in a cellar side by side, where the temperature varied but very slowly, and on the whole diminished. The first trial was made when the cellar temperature was $13·3°$. A portion was placed in a tared porcelain dish, which was weighed, and set on the water-bath. In this and in all similar experiments, a certain fixed interval of time intervened after the drying before removing the dish from the bath, and a definite number of minutes was allowed to elapse before weighing, the conditions in all respects being strictly comparative. The platinichloride remaining was calculated as parts dissolved in 100 parts of water. The specimens were taken in immediate succession for each pair, as in all other cases.

	Temperature.	Pt salt to 100 aq. (A).	Pt salt to 100 aq. (B).
Sept. 16, 1897......	13·3°	0·786 gram	0·775 gram

The experiment was started on the 14th September, 1897, and the lowest temperature attained in the two days' interval was 12·8°.

	Temperature.	Pt salt to 100 aq. (A).	Pt salt to 100 aq (B).
Oct. 1, 1897	13·3°	0·790 gram	0·740 gram
„ 8, „	9·4	0·788 „	0·712 „
„ 15, „	11·1	0·759	0·696 ,
Nov.17, „	10	0·691	0·664 ,

Crystals in beaker, 0·465 gram.   Crystals, 3·74 grams.

Thus, the differences between the quantity of the salt held in solution by 100 grams of water in the two beakers was: for two days, 0·011 gram; for sixteen days, 0·050 gram; for twenty-four days, 0·076 gram; for thirty-one days, 0·063 gram, and for forty-eight days 0·027 gram, the maximum difference being after twenty-four days. Obviously, the gradual accumulation of crystals in solution A, as time went on, would tend to diminish the difference.

In an experiment started November 3, 1897, about 50 c.c. of a solution of the platinum salt saturated at about 10° was put into each of two small flasks. To one flask (I), some minute particles of the solid salt were added, and about one gram of crystals of the salt was added to the second flask. Both flasks were stoppered lightly with cotton wool, and they were set side by side in the cellar.

	Temperature.	Pt salt to 100 aq. (I).	Pt salt to 100 aq. (II).
Nov. 9, 1897......	10°	0·757 gram	0·672 gram
„ 15, „	8·8	0·706 „	0·649 „

Another experiment was made on a specimen of platinichloride specially crystallised in such manner as to produce crystals of comparatively large size. A saturated solution (167 c.c.) of these crystals was formed, a fourth part of which was put into each of four flasks. To one flask, 1·49 grams of platinum salt crystals were added; this flask was marked II. To the other three flasks, traces of the salt were added, and into flask III were put 10·45 grams of glass beads (1494 beads having an estimated total surface of slightly more than 23 sq. inch). Flask IV was set on a layer of the platinum salt. All were plugged as usual and set in the cellar side by side. When tested after twenty-two days, at 5·5°, the results were :

Flask I	0·698	Flask IV	0·692
Flask III	0·667	Flask II	**0·589**

these numbers representing parts of platinum salt in 100 parts of water.

In this experiment, the difference between the solution contained in flask II, including the platinum salt crystals and the others, is well marked. But the influence of the beads is also noticeable, tending to show that the mere surface of indifferent substances exercises an influence on crystallisation. Hence, in such experiments, the vessels used for comparison of results should be as nearly alike in form and capacity as possible.

Test for Errors of Experiment.

From a solution of the platinichloride four portions were successively taken for the trials. The portion first taken was weighed in the dish, evaporated, dried, and the residue weighed with the dish, precisely under the conditions observed in the previous experiments, and so in like manner with the three succeeding portions. The quantities taken varied from about 13 c.c. to 18 c.c. The results were, when calculated for each 100 parts of water, 0·5771, 0·5788, 0·5777, and 0·5783, showing a maximum difference, due to experimental errors, of 0·0017, or a mean error of 0·0009.

Potassium Chloride.—The potassium chloride had been several times crystallised. A part of a solution of the salt standing in a beaker over a considerable quantity of potassium chloride crystals was transferred to another similar beaker and some minute crystals were added to it. The two beakers, loosely covered with paper, were set side by side in the cellar. The next morning, the density of the solution in the first beaker (containing the crystals) was **1·1637** at 7·8°, and that of the second solution 1·1646, both uncorrected, the conditions being exactly the same in both cases.

Potassium Sulphate.—The salt employed was that commercially known as " pure." A saturated solution of the salt was divided between two beakers, into the second (*B*) of which 10 grams of the crystallised salt were put, a minute portion being introduced into the first beaker (*A*). The beakers, covered by paper, were set side by side in the collar. After four days (temperature 7·7°), the density of the solution in beaker *A* was 1·0690, and that of the solution in the second beaker (*B*) was

1·0709, that is, a higher instead of a lower density, as should have been the case had the salt been pure. To test this further, the solution in flask B was heated sufficiently to dissolve much of the salt contained in it, and the solution was again set to crystallise beside flask A. On testing two days afterwards, solution A gave, at 7·8°, the density 1·0687, and solution B 1·0716, showing a still greater difference on the wrong side, obviously due to the presence of impurities, probably in this case of a small proportion of sodium sulphate.

The crystals of potassium sulphate were then twice crystallised, and from the crystals thus obtained a saturated solution was formed and divided into two portions, A and B, to one of which, B, 11·17 grams of the crystals were added. After the usual procedure (temperature 6·1°), the trials were made after three days (the temperature having in the interval fallen to 5·5°). The solution in flask A (no added crystals, except the usual traces) had a density of 1·06850, whilst that in flask B (containing the added crystals) had a density of 1·06596. The result in this case is therefore normal, showing that the salt used was, at least, approximately pure. The crystals which formed in flask A during the experiment were collected and dried, and weighed 0·637 gram. Of course, the crystals thus deposited also exercised attractive force in proportion to their quantity ; and the ideal conditions for conducting such experiments would involve a minimum separation of crystals during their progress.

Boric Acid.—Equal parts of a solution of twice-crystallised boric acid (saturated at 17·2°) were placed in two small glass flasks, with addition to the solution in flask I of 2 grams of boric acid crystals. The second flask (II) was primed, other conditions being as usual (temperature, 13·9°). After eighteen days, with cellar and balance room temperatures of 14·4° and 16·1° respectively, the density for flask I (with the crystals) was 1·01494, whilst that of flask II (no added crystals) was 1·01553. As the conditions were strictly comparable in both cases, no corrections were made.

Oxalic Acid.—This acid, after being twice crystallised from nitric acid solution and washed, was dissolved to form a saturated solution, of which equal portions were taken at 16·1° and placed in two similar flasks, to one of which (I) much crystallised oxalic acid was added, the solution in the other flask (II) remaining without addition beyond the priming. Portions were taken two days later with cellar and balance room temperatures of 13·9° and 15° respectively, when the density of the solution in flask I (with much solid acid) was 1·03055, whilst that of the solution in flask II (solution only) was 1·0314.

Crystallised Copper Sulphate.—Equal portions of a saturated solution of this salt at 16·6° were placed in two small flasks, which were set in the cellar under the usual conditions (temp. 14·4°). To the solution in the second of the flasks, 2 grams of the cupric sulphate crystals were added. After two days at 14·4°, the density of the solution in flask I was 1·1859, and that of the solution in flask II was 1·1832. A diminution of temperature from 16·6° to 14·4° caused considerable crystallisation in the solutions, which tended to diminish the difference which appeared. When the experiment was repeated under similar conditions (except as to temperature) at 11·1°, the densities of the solutions in flasks I and II were 1·1788 and 1·1740 respectively.

Potassium Ferrocyanide.—Thirty c.c. of potassium ferrocyanide solution saturated at 12·2° were put into each of three small flasks, together with a quantity of the crystals in flask I. The other two flasks were only primed. The trials were made eighteen days later, when the temperature both in cellar and balance room was 8·9°. The density of the solution in flask I (with crystals) was 1·1191, whilst those of the solutions in flasks II and III (solution only) were 1·1265 and 1·12615.

In the course of this experiment, the temperature increased a little after the commencement, thus necessitating slight addition of crystals to flasks II and III and frequent shaking of flask I to ensure saturation. The final temperature, however, was relatively low.

Ferrous Sulphate.—A saturated solution (25 c.c.) of the commercial "pure" salt at 9·4° was put into each of three small flasks, into the first of which 2·3 grams of crystals were added, the other flasks containing only the (primed) solution. The trials were made after two days at 8·9°. The density of the solution in flask I (with crystals) was 1·1949, whilst those of solutions II and III (solution only) were 1·1969 and 1·1964.

Borax.—A saturated solution of this salt was placed in each of three small flasks, into the first of which about 2 grams of the crystals were introduced (cellar temperature 7·2°, and varying subsequently between 9·4° and 4·4°). The trials were made forty-two days later, the temperatures in cellar and balance room being 4·4° and 5° respectively. The density of the solution in flask I (with crystals) was 1·0140, whilst those of solutions in flasks II and III were both 1·0155.

Potassium Iodate.—Thirty c.c. of a saturated solution of a highly purified specimen of this salt at 6·6° were placed in each of three small flasks, into the first of which about 2 grams of the crystals were introduced. The cellar temperature was 6·6° and during the experiment the temperature varied from 4·4° to 7·2°. The trials were made after four days at 7·2°. The density of the solution in flask I (with crystals) was 1·0317, whilst the densities of the other two solutions (without crystals) were 1·0338 and 1·0335.

Potassium Bromide.—Thirty c.c. of a solution of the commercial salt saturated at about 7·7° were placed in each of three flasks, into the first of which about 2 grams of the crystals were placed, the solutions in the other flasks being primed only with traces of the crystals. The cellar temperature was 7·2°. The trials were made after six days, the temperature being 5·5°. The density of the solution in flask I (with the crystals) was 1·3431 whilst the densities of the two primed solutions were 1·3499 and 1·3502.

Thirty c.c. of a solution of recrystallised potassium bromide, saturated at about 5·5°, were placed in each of three flasks, into the first of which about 2 grams of the crystals of this salt were introduced, the solution in the other flasks being primed only; but in flask III about 2 grams of washed and ignited sand were placed. The trials were made after two days, the temperature being 5°. The density of solution with the crystals was 1·33655, whilst the density of the solution in flask II (solution) was 1·3441, and that of the solution of flask III (solution and sand) was 1·3443.

It will be noted that the difference between the densities of the solutions with and without added crystals is sensibly greater with the recrystallised than with the ordinary bromide.

A number of experiments of a similar kind was made with potassium hydrogen and with sodium hydrogen carbonate, only to find that, notwithstanding the greatest care taken to work with the pure salts, and to ensure that these should be fully carbonated at the moment of use, the behaviour was always that of an impure salt. Either the portion of solution containing the added crystals proved to be of approximately the same density as that to which no crystals had been added, or the density of the former was even greater than that of the latter. I was forced to the conclusion that, under the conditions of the experiments, potassium hydrogen or sodium hydrogen carbonate was not stable; that, under ordinary atmospheric conditions, it parted with carbon dioxide, so as to become mixed with a small proportion of the neutral carbonate, a conclusion which I find has been independently arrived at by another investigator.

These experiments throw some light on the curious discrepancies existing between the solubilities of some well-known salts as determined by experimenters of unimpeachable skill and accuracy. While attending with the minutest care to the conditions which might be supposed to influence the results, there is no reason to suppose that attention was paid to the presence or otherwise in the solutions tested of a body of crystals. It is probable, also, that in exact determinations it may be advisable to work under certain standard conditions, as to the surface, relatively to the mass of the solution, exposed to the air, and the surface of the containing vessel in contact with the liquid. The experiment already recorded, in which a known surface of glass beads was exposed to a solution, renders it at least probable that surfaces chemically inert may yet exercise a certain influence.

There is an obvious practical application of the property of crystals here described, in furnishing a very ready and reliable means of determining the purity of any given crystallisable salt. The test, if properly carried out, is capable of detecting a very small proportion of impurity, and the differences between saturated solutions depend-ing on the presence or absence of a foreign salt are so considerable that very ordinary means suffice for their detection. In most cases, an experiment of the kind described need not be continued beyond two or three days.

XXXVIII.—Cuprous Formate.

By ANDREA ANGEL.

IN the course of an investigation by the author of the decomposition by heat of cupric formate, the question of the existence of cuprous formate arose, and attempts were therefore made to prepare this substance by different methods, but although definite indications of its existence were obtained, all endeavours to prepare it in quantity failed. The preparation by Péchard (Compt. rend., 1903, 136, 504) of cuprous acetate by a new method, namely, the acidification with acetic acid of an ammoniacal solution of cupric acetate which had been reduced with an aqueous solution of hydroxylamine sulphate, seemed to indicate a possible way of preparing cuprous formate. Working in a similar way, using an ammoniacal solution of cupric formate decolorised by hydroxylamine sulphate and acidifying with formic acid, a white precipitate was momentarily produced. This, however, could not be separated from the liquid, as it decomposed almost imme-diately with the precipitation of metallic copper and the formation of the cupric salt. No modification of the conditions of the experiment could be found which would render the precipitate stable. It seemed, however, probable that cuprous formate could exist, but that in the presence of water or formic acid it underwent decomposition. It was thought, therefore, that if the precipitation could be brought about in a solvent other than water the decomposition might be prevented or delayed. Alcohol seemed a suitable liquid, but then the reduction method became inapplicable, since cuprous chloride was precipitated if hydroxylamine hydrochloride were used for the reduction; or if the sulphate were used, ammonio-cuprous sulphate, which is insoluble in alcohol, as Péchard found, was precipitated. Hydrazine sulphate gave

A A 2

equally unsatisfactory results. The reduction method was therefore abandoned, and eventually the process described below was devised.

In the meantime, however, Joannis (*Compt. rend.*, 1904, **138**, 1498), by the action of liquefied ammonia on ammonium formate and cuprous oxide in a sealed glass apparatus previously exhausted with air, obtained a compound having the formula $Cu_2(O_2CH)_2,4NH_3,\frac{1}{2}H_2O$, which he described as " cuprous formate." The present author, however, considers that this substance would more appropriately be described as " ammonio-cuprous formate," since it differs considerably in properties from the compound which he has obtained by the method described below, and which is the free cuprous formate corresponding to the formula $Cu_2(O_2CH)_2$.

Instead of reducing the cupric salt, cuprous oxide was made the starting point. When an aqueous solution of ammonium formate is heated with cuprous oxide and a little strong ammonia solution under a layer of light petroleum to minimise oxidation, a blue solution is obtained (since some oxidation does occur) which contains a cuprous salt, probably an ammonio-cuprous formate. If this liquid is diluted with alcohol, no precipitation occurs, but on adding formic acid to this alcoholic solution and cooling under the tap, colourless crystals are precipitated which have all the appearance of a cuprous salt and which are fairly stable, the alcohol arresting the decomposition which occurs in an entirely aqueous solution. The decomposition, however, is only delayed and not entirely prevented by the alcohol, for if the crystals are allowed to remain for a long time in the liquid they decompose as before into copper and the cupric salt.

It was therefore necessary to find some means of rapidly separating the crystals from the mother liquor in an air-free atmosphere and also of finding a liquid with which they might be washed without decomposition occurring. Here, again, much difficulty was encountered. Various forms of filtering apparatus were devised, but the only one which proved satisfactory was a method of reverse filtration ; for in all cases of direct filtration, the filter soon became blocked, and the crystals on prolonged contact with the mother liquor decomposed. Further, the ease with which the crystals decompose renders the washing a difficult matter. Water immediately hydrolyses them to cuprous oxide. Ether, chloroform, benzene, petroleum, and absolute methyl and ethyl alcohols were tried with unsatisfactory results; absolute alcohol causes hydrolysis. Formic acid decomposes the crystals with precipitation of copper. Partial success was obtained by using as a washing liquid absolute alcohol containing a little formic acid. Any attempt, however, to prepare more than a small quantity of the crystals resulted in failure, for if the washing liquid was too acid, copper was precipitated, whilst if insufficient acid were used the crystals gradually became yellowish-red during the drying owing to the hydrolysing action of the alcohol. These results suggested that possibly ethyl formate might cause neither of these two decompositions. This was found to be the case ; further, the greater volatility of ethyl formate hastens the drying, and its use as a washing liquid has enabled the author to prepare cuprous formate as a comparatively stable substance.

Details of the Method of Preparation.

A strong solution of ammonium formate is made by dissolving 136 grams of the salt in 100 c.c. of distilled water. A separate solution of 30 grams of ammonium formate in 10 c.c. of water is then prepared, and in this 9 grams of pure cuprous oxide are dissolved (under a layer of light petroleum) with the aid of heat and the addition of just sufficient strong ammonia solution (sp. gr. 0·880) to enable the cuprous oxide to dissolve, forming a blue solution. For the above quantities, the amount of ammonia solution required should not exceed 11 c.c.

The blue liquid so obtained is then diluted with half its own volume of the first solution of ammonium formate, transferred by reverse filtration through a pad of glass wool enclosed between silk (to remove any fine particles of cuprous oxide) to a tube similar to the tube *A* in Fig. 1, and preserved under petroleum.

For the precipitation, this blue liquid is diluted with ten times its volume of alcohol, and more light petroleum added, if necessary, to form a layer on top. Aqueous formic acid of sp. gr. 1·15 is then gradually introduced from a burette. The deep blue colour is discharged, the liquid becomes green, and, on cooling under the tap, colourless crystals of cuprous formate are deposited. About 15 c.c. of the blue liquid is a convenient quantity to take for a preparation, and after dilution with alcohol the formic acid required is about 8 c.c.

Much difficulty was encountered in finding the right conditions for the precipitation. If too much ammonia is used in preparing the blue solution, the heat evolved in the neutralisation causes local decomposition, and small particles of copper become mixed with the crystals and cannot afterwards be separated from them. If the blue solution is too dilute, the products of the hydrolysis of the salt, namely, cuprous oxide and formic acid, are obtained instead. If too much formic acid is used in the precipitation, the preservative influence of the alcohol and ammonium formate is counteracted and decomposition again occurs, whilst if the blue liquid is largely diluted with alcohol, the salt is found to be contaminated with ammonia and cupric salt

owing to the small solubility of ammonium formate and cupric
· formate in alcohol.

For the separation of the crystals, the apparatus shown in **Fig. 1**
was employed. It consists of a vessel, D, made from part of a large
test-tube about $3\frac{1}{2}$ cm. in diameter, into which is fitted an india-rubber
cork, E, with four holes, coated with a very thin layer of paraffin wax.
Through two of the holes pass delivery tubes, A and B, fitted with
taps; a third allows hydrogen to enter through a tube, and through
the fourth and central hole passes the filter tube, C. This consists of
a piece of quill tubing expanded at the bottom; over this is tied
a single layer of silk. The upper end of C is connected by stout india-
rubber tubing to a thick-walled flask

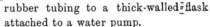
TO PUMP »»

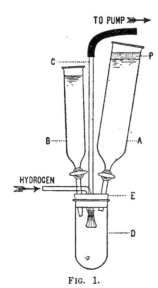
FIG. 1.

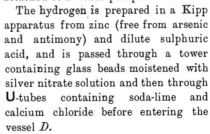

attached to a water pump.

The hydrogen is prepared in a Kipp
apparatus from zinc (free from arsenic
and antimony) and dilute sulphuric
acid, and is passed through a tower
containing glass beads moistened with
silver nitrate solution and then through
U-tubes containing soda-lime and
calcium chloride before entering the
vessel D.

After the whole apparatus has been
filled with hydrogen, the liquid containing the precipitated crystals is
transferred, still under a layer of light
petroleum, to the tube A. It is then
run into D, excluding the petroleum
layer P, the liquid drawn off, and the
crystals repeatedly washed (lowering
C when necessary and eventually inverting the whole apparatus to drain
away the washing liquid), first with a mixture of four volumes of
absolute alcohol with one volume of absolute ethyl formate, and
finally with absolute ethyl formate delivered through the tube B.
When the crystals are quite clean, any remaining liquid is poured
out of A and B, and hydrogen is passed through the inverted
apparatus for one and a half to two hours. The crystals form
a cake on the round end of D and adhere until they are dry.
They may then readily be detached from the glass by a slight
tap and quickly transferred through a dry funnel to carefully dried
bottles with ground tubular stoppers, which are then exhausted of air,
left in connection with phosphoric oxide on the mercury pump, and
then sealed off (Fig. 2),

Analysis.—The salt was quickly weighed out into a porcelain dish and decomposed with sodium carbonate solution. The dish and its contents were heated to boiling and then the liquid decanted from the precipitated cuprous oxide on to a filter, the filtrate being collected in a measuring flask. The oxide was repeatedly washed with distilled water, and the washings added to the filtrate and made up to 500 c.c. The cuprous oxide was dissolved in dilute nitric acid and the copper estimated as cupric oxide by precipitation with sodium hydroxide. The formic acid (as sodium formate) in the filtrate was estimated by Lieben's method, that is, by titration with potassium permanganate in alkaline solution, 50 c.c. of the filtrate being used for an estimation.

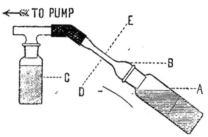

A = Bottle containing cuprous formate.
B = Tubular stopper, originally open at both ends, finally sealed off at DE.
C = Bottle containing phosphoric anhydride.

FIG. 2.

Analyses of three different preparations gave the following results :

Weight of substance taken.	Percentage of copper found.	Percentage of (O_2CH) found.
0·4085 gram	58·54	41·31
0·4051 ,,	58·34	40·79
0·4068 ,,	58·32	41·22

$Cu_2(O_2CH)_2$ requires $Cu = 58·55$; $(O_2CH) = 41·45$ per cent.

Properties of Cuprous Formate.—The compound described by Joannis under this name is said to consist of pale blue crystals, which become intensely blue in moist air and brownish-black in dry air, and when thrown into a dilute acid give immediately a precipitate of cuprous oxide.

Cuprous formate obtained as above consists of colourless, needle-like crystals, which are very light and appear white with a very faint pink tinge in a compact mass. The appearance of the crystals is very characteristic ; they almost invariably occur associated in double or quadruple groups, as shown in the diagram in Fig. 3, whilst in some cases they appear to be grouped about the three axes of a cube, with the result that the whole complex has a globular appearance. This generally occurs when the precipitation is slow and in a solution which has been largely diluted with alcohol.

In moist air, the substance is rapidly decomposed, and assume;

a brilliant orange-red colour owing to the formation of cuprous oxide. When treated with water, it is also hydrolysed to cuprous oxide and the liquid has an acid reaction. When dropped into strong ammonia solution, it is instantly decomposed with a slight hissing noise, and if sodium carbonate solution is poured on the dry substance it decomposes it with effervescence. When thoroughly dried, it is practically permanent in dry air, as it may be kept over sulphuric acid for weeks without any apparent change. Aqueous formic acid rapidly decomposes it into metallic copper and the cupric salt, and when it is thrown into dilute sulphuric acid metallic copper is immediately precipitated.

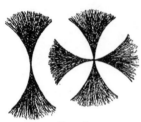

FIG. 3.

The difficulties met with in the preparation of cuprous formate have been chiefly due to the fact that formic acid decomposes it. In the case of the higher members of the same series of acids, this decomposition does not occur. Péchard showed that cuprous acetate is stable in presence of acetic acid. Preliminary experiments made by the author seem to show that this holds for other members of the series, since by a similar method, omitting, however, the addition of alcohol, white precipitates of cuprous salts were obtained with chloroacetic, propionic, and butyric acids.

In conclusion, the author wishes to express his thanks to Mr. Vernon Harcourt and Dr. Brereton Baker for their helpful interest in the work, and to the latter especially also for his kindness in allowing the author every facility in the use of the new research laboratory at Christ Church.

DR. LEE'S LABORATORY,
CHRIST CHURCH,
OXFORD.

XXXIX.—o-Cyanobenzenesulphonic Acid and its Derivatives.

By ANDREW JAMIESON WALKER and ELIZABETH SMITH, B.Sc.

o-CYANOBENZENESULPHONIC ACID is first mentioned by Jesurun (Ber., 1893, 26, 2288), who states that it is formed when o-cyanobenzene-sulphonic chloride, $CN \cdot C_6H_4 \cdot SO_2Cl$, is boiled with water. He did not, however, isolate the acid, but described its sodium salt. When heated under pressure at 95—100° with a small quantity of water for a long time, the chloride yields o-benzoic sulphonamide, $CO_2H\cdot C_6H_4\cdot SO_2\cdot NH_2$, this being converted at 180° into ammonium o-benzoic sulphonate, $CO_2H\cdot C_6H_4\cdot SO_3\cdot NH_4$.

Remsen (Amer. Chem. J., 1895, 17, 309) and Remsen and Saunders (ibid., 347) describe the formation of ammonium o-cyanobenzenesulphonate by the action of aqueous ammonia on the unsymmetrical dichloride of o-sulphobenzoic acid, and Remsen and McKee (ibid., 1896, 18, 794) also obtained this ammonium salt from the corresponding symmetrical chloride, whilst Remsen and Karslake (ibid., 819) prepared it from a mixture of the symmetrical and unsymmetrical chlorides, and obtained the sodium, potassium, and barium salts by treating the ammonium salts with the corresponding hydroxides. They could not isolate the free acid, the barium salt yielding with dilute sulphuric acid o-benzaminosulphonic acid, $CO\cdot NH_2\cdot C_6H_4\cdot SO_3H$, and the ammonium salt with dilute hydrochloric acid, ammonium o-benzaminosulphonate, $NH_2\cdot CO\cdot C_6H_4\cdot SO_3\cdot NH_4$. On heating ammonium o-cyanobenzenesulphonate with hydrochloric acid in a sealed tube, they obtained acid ammonium o-sulphobenzoate,

$$CO_2(NH_4)\cdot C_6H_4\cdot SO_3H.$$

List and Stein (Ber., 1898, 31, 1648) showed that the two dichlorides of o-sulphobenzoic acid are tautomeric, and called Remsen's symmetrical and unsymmetrical forms "stable" and "labile" respectively. They found that ammonia rapidly converts the labile chloride in ethereal solution into Remsen and Saunders' ammonium o-cyanobenzenesulphonate,

$$C_6H_4{<}^{CCl_2}_{SO_2}{>}O \ + \ 4NH_3 \ = \ C_6H_4{<}^{CN}_{SO_3\cdot NH_4} \ + \ 2NH_4Cl,$$

but acts slowly on the stable chloride to form o-benzoic sulphinide,

$$C_6H_4{<}^{CO}_{SO_2}{>}NH.$$

Jesurun prepared o-cyanobenzenesulphonic chloride (loc. cit.) by heating "saccharin" (1 mol.) with phosphorus pentachloride (2 mols.) in a sealed tube for two hours at 70—75°. We have found it better to heat the mixture under the atmospheric pressure for about two hours at 130°. Ether is a more suitable solvent than benzene, a fact noted independently by List and Stein (loc. cit.), who obtained the chloride by treating ammonium o-cyanobenzenesulphonate with phosphorus pentachloride.

By concentration of the aqueous mother liquor obtained in the preparation of o-cyanobenzenesulphonic chloride by a modification of Jesurun's method, we have isolated o-cyanobenzenesulphonic acid,

$CN \cdot C_6H_4 \cdot SO_3H$; it crystallises from water in white needles melting at 279—279·5°. The silver salt was prepared by dissolving the acid in an aqueous solution of the calculated quantity of sodium carbonate and adding the requisite amount of silver nitrate dissolved in water. When heated on the water-bath with fuming nitric acid, o-cyano-benzenesulphonic acid yields two substances, white needles melting at 255—255·5° and a smaller quantity of yellow, prismatic crystals melting between 140° and 150°. The action of bromine on the acid has been investigated, and it has also been shown that it is not reduced by zinc dust and boiling water.

On reduction of o-cyanobenzenesulphonic chloride with zinc dust, there is formed *zinc o-cyanobenzenesulphinate*, $(CN \cdot C_6H_4 \cdot SO_2)_2Zn$. Heating the crude reaction product with a solution of sodium carbonate converts this salt into the corresponding *sodium o-cyano-benzenesulphinate*, $CN \cdot C_6H_4 \cdot SO_2Na$, which forms a green solution. This sodium salt was not isolated. When its solution is acidified with dilute sulphuric acid, *o-cyanobenzenesulphinic acid*, $CN \cdot C_6H_4 \cdot SO_2H$, is precipitated; it crystallises from glacial acetic acid in small, white needles melting at 226·5—228°. When heated on the water-bath with 2*N*-sodium hydroxide, o-cyanobenzenesulphonic chloride yields a solution from which hydrochloric acid does not precipitate the sulphonic acid, but a substance, probably o-benzoic sulphinide, melting at 221·5—223° and possessing the sweet taste of "saccharin." With bromine, o-cyanobenzenesulphinic acid yields two compounds melting at 156—156·5° and 172·5—173° respectively. Neither of these substances contains bromine. When treated with nitrous acid by König's method (*Ber.*, 1878, 11, 615), o-cyanobenzenesulphinic acid is converted into a yellow solid, probably *o-cyanodibenzsulphohydroxamic acid*, $(CN \cdot C_6H_4 \cdot SO_2)_2NOH$.

The further reduction products of o-cyanobenzenesulphonic acid and o-cyanobenzenesulphinic acid by chemical and electrolytic methods are now under investigation. A future communication will treat of the results of these experiments, of the substances obtained by the action of nitric acid on o-cyanobenzenesulphinic acid, and of bromine and nitrous acid on o-cyanobenzenesulphinic acid, and of amino-derivatives resulting from the interaction of o-cyanobenzenesulphonic chloride and various amino-compounds.

EXPERIMENTAL.

o-Cyanobenzenesulphonic Chloride, $CN \cdot C_6H_4 \cdot SO_2Cl$.

Jesurun (*Ber.*, 1893, 26, 2228) prepared this substance by heating "saccharin" (1 mol.) with phosphorus pentachloride (2 mols.) in a sealed tube for two hours at 75°. The product was poured on to ice, the chloride separating as an oil, which solidified, and, after drying in the air, was crystallised from benzene and obtained in prismatic crystals melting at 69—70°. The following procedure gives more satisfactory results.

Thirty grams of commercial "insoluble saccharin" (or "soluble saccharin") and 70 grams of phosphorus pentachloride were heated in an Erlenmeyer flask with an air condenser for an hour and a half in a paraffin bath at 120—140°, much hydrochloric acid being evolved. The hot mixture was poured on to ice, and, after solidification of the oil, the crude chloride was collected and pressed on a porous plate, the yield being 32 grams. This was extracted twice with ether dried over sodium and the chloride obtained by concentrating the solution. The residue after extraction consisted of 5 grams of unchanged "saccharin," and was worked up again for the chloride. After trituration with a 1 per cent. solution of sodium hydrogen carbonate to remove "saccharin," and repeated crystallisation from dry ether, the o-cyano-benzenesulphonic chloride was obtained in prismatic crystals melting constantly at 67·5°. Light petroleum was a less satisfactory solvent, but yielded small needles melting at 67—68°.

Jesurun (loc. cit.) gives the melting point of the crystals from benzene as 69—70°. List and Stein (Ber., 1898, 31, 1648) used ether as a solvent and mention the same melting point. The chloride is readily soluble in cold ether, chloroform, acetone, benzene, and xylene, and in hot light petroleum and carbon disulphide; it has a slightly sweetish taste, but much less pronounced than that of "saccharin," and also has an acid flavour distinguishing it from this substance.

Action of Nitric Acid.—The chloride is dissolved by heating with fuming nitric acid on the water-bath, but not changed chemically.

Action of Sodium Hydroxide.—To 15 grams of the chloride, 65 c.c. of 2N-sodium hydroxide were added gradually, heat being evolved during the reaction. After the mixture had been heated for an hour and a half on the water-bath, solution was complete. While still hot, the liquid was acidified with 2N-hydrochloric acid, crystallisation being facilitated by stirring with a glass rod. The precipitate was collected and dried in the steam-oven. Five grams of product were obtained which melted without decomposition to a yellow liquid at 221·5—223°, proving that it was not o-cyanobenzenesulphonic acid, which melts at 279—279·5°. It had the sweet taste of "saccharin," and was probably o-benzoic sulphinide, although the analyses differ slightly from those required by $C_7H_5O_3NS$. The molecular weight obtained by the cryoscopic method with glacial acetic acid as solvent was 197, $C_7H_5O_3NS$ requiring 183.

o-*Cyanobenzenesulphonic Acid*, $CN \cdot C_6H_4 \cdot SO_3H$.

On evaporation of the aqueous mother liquor from the preparation of o-cyanobenzenesulphonic chloride, a white precipitate was obtained, which melted at 228—250° to a colourless liquid and was soluble in dilute solutions of sodium hydroxide and sodium carbonate. The crude product was dissolved in dilute caustic soda and the solution extracted twice with ether. After removal of the ethereal layer, the acid was precipitated from the alkaline aqueous solution by acidification with dilute sulphuric acid, filtered off at the pump, thoroughly washed with water, and dried by pressing on a porous plate. By repeated crystallisation from water, the acid was obtained in the form of white needles melting at 279—279·5°, soluble in a large quantity of boiling water and in alcohol, and slightly soluble in boiling ether and chloroform.

0·2575 gave 17·5 c.c. moist nitrogen at 26° and 769 mm. $N = 7·63$.

0·19305 ,, 0·2442 $BaSO_4$. $S = 17·34$.

$C_7H_5O_3NS$ requires $N = 7·65$; $S = 17·48$ per cent.

Silver Salt.—Silver o-*cyanobenzenesulphonate*, $CN \cdot C_6H_4 \cdot SO_3Ag$, was prepared by the following method : 0·5 gram of the acid was dissolved in an aqueous solution of 0·1448 gram of anhydrous sodium carbonate and two drops of a concentrated aqueous solution of 0·4644 gram of silver nitrate added. The slight precipitate was filtered off, and on addition of the rest of the silver nitrate solution the silver salt was precipitated as a white solid. This was collected, washed with alcohol and then with ether, pressed out on a dry plate, and dried in an air-oven at 120—130°. The light was excluded as far as possible, but nevertheless the salt became somewhat dark in colour.

0·2520 gave 0·0929 Ag. $Ag = 36·86$.

$C_7H_4O_3NSAg$ requires $Ag = 36·90$ per cent.

Action of Bromine and Nitric Acid.—When heated with bromine, o-cyanobenzenesulphonic acid yielded a brownish-tinted, crude product, very soluble in alcohol, somewhat soluble in glacial acetic acid, and almost insoluble in benzene and light petroleum. Heating on the water-bath with fuming nitric acid converted o-cyanobenzenesulphonic acid into a yellowish-white, very hygroscopic substance, the crude product melting at 255° to a dark brown liquid. On crystallisation from glacial acetic acid, white, prismatic needles melting at 255—255·5° were obtained, accompanied by a small proportion of yellow, prismatic crystals melting at 140—150° with evolution of gas. On raising the temperature, the liquid resolidified, and the dark product melted again at about 230° to a black, viscid liquid.

o-*Cyanobenzenesulphinic* *acid*, $CN \cdot C_6H_4 \cdot SO_2H$, was prepared as follows from o-cyanobenzenesulphonic chloride, $CN \cdot C_6H_4 \cdot SO_2Cl$: 40 c.c. of water were heated to boiling in a 300 c.c. Erlenmeyer flask, 10 grams of zinc dust added, and then, without further heating, 15 grams of finely-powdered o-cyanobenzenesulphonic chloride in small portions. The mixture was stirred frequently with a glass rod. At each addition of the chloride, an energetic reaction with evolution of heat ensued, and when all the chloride had been added the contents of the flask were heated for ten minutes over a small flame, allowed to cool, and filtered at the pump. To decompose the zinc sulphinate, the residue on the filter was heated for ten minutes with a solution of 20 grams of anhydrous sodium carbonate in 200 c.c. of water, filtered, and the residue washed with water. The filtrate was evaporated to half its volume, cooled, and acidified with dilute sulphuric acid. o-Cyano-benzenesulphinic acid was precipitated as a white solid, which was filtered off, a further quantity of the sulphinic acid being obtained by concentrating the filtrate. The acid was dried on a porous plate, the yield of crude product being about one-third of the weight of the chloride used; it melted at 217—225° to a dark liquid and was crystal-lised by dissolving in hot glacial acetic acid, adding hot water until the solution became cloudy, clarifying by the application of heat, and finally allowing the liquid to cool. From this solvent, the sulphinic acid separated in clusters of small, white needles, which, when purified by repeated crystallisation, became moist at 220° and melted to a yellow liquid at 226·5—228°. The yield was somewhat better when a large quantity of the chloride was employed.

0·1767 gave 13 c.c. moist nitrogen at 17° and 756 mm. $N = 8·49$.
0·1618 ,, 0·22725 $BaSO_4$. $S = 19·29$.
$C_7H_5O_2NS$ requires $N = 8·38$; $S = 19·16$ per cent.

In the nitrogen estimation, it was found better not to mix the sulphinic acid with copper oxide in the boat.

o-Cyanobenzenesulphinic acid is very slightly soluble in water, absolute alcohol, ether, benzene, light petroleum, or chloroform; it dissolves readily in hot glacial acetic acid. The *sodium* salt was obtained by treating the acid with the equivalent weight of aqueous sodium carbonate. On concentrating to small bulk, it separated in minute, white needles, which were very soluble in water and became dull in a desiccator through loss of water of crystallisation.

In preparing the acid, the o-cyanobenzenesulphonic chloride must not be mixed with the zinc dust in a mortar, since a vigorous reaction soon ensues, leaving a charred mass from which no sulphinic acid can be extracted.

An attempt was made to prepare o-cyanobenzenesulphinic acid by reducing an aqueous solution of o-cyanobenzenesulphonic acid in an analogous manner, but the sulphonic acid underwent no change.

Action of Bromine.—2·6 grams of o-cyanobenzenesulphinic acid were treated at the ordinary temperature with a solution of 2·6 grams of bromine in chloroform. The red colour of the bromine solution was almost immediately discharged, and excess of the bromine solution was then added until the colour persisted for an hour. The yellow solid was then filtered off, the crude product melting partially at 165—167° and completely at 178°. Its weight was equal to that of the sulphinic acid taken. It was dissolved in dilute caustic soda and reprecipitated by addition of dilute hydrochloric acid. On crystallising several times from water, it was obtained in white crystals melting at 172·5—173°. These dissolved fairly readily in water and alcohol, and very easily in ether and glacial acetic acid : they were insoluble in chloroform, light petroleum, and benzene. The substance contained no bromine, and its aqueous solution was strongly acid to litmus. In estimating the percentage of nitrogen, it was necessary to mix the compound in the boat with fine copper oxide.

0·1933 gave 13·3 c.c. moist nitrogen at 15° and 758 mm.   N = 8·04.
0·2037 ,,   13·8 ,,   ,,   ,,   17° ,,   760 ,:   N = 7·87.
0·1585 ,,   0·2001 BaSO₄.   S = 17·33.

On modifying the conditions, a different product resulted, also containing no bromine: 0·8 gram of o-cyanobenzenesulphinic acid was treated with 0·8 gram of bromine dissolved in chloroform. The bromine solution was almost instantly decolorised, with absorption of considerable heat and evolution of hydrobromic acid. The chloroform was volatilised by heating on the water-bath : the residue weighed 1·12 grams ; 0·5 gram of bromine in solution was then added and evaporated off, but there was no further increase in weight. The crude product was red in colour ; it was dried on a porous plate, and after several crystallisations from alcohol white needles melting at 156—156·5°, soluble in water and alcohol, and insoluble in chloroform, were obtained. On addition of water, the red crude product turns darker red, then white, and finally dissolves.

0·16055 gave 11·2 c.c. moist nitrogen at 18° and 764 mm.   N = 8·10.
0·13735 ,,   0·1929 BaSO₄.   S = 19·28.

The correct interpretation of the analyses of the two compounds obtained by the action of bromine on the sulphinic acid is still a matter of doubt. The percentages of nitrogen and sulphur required by certain formulæ are appended.

(a) $(CN \cdot C_6H_4 \cdot SO_3)_2$ requires $N = 7.69$; $S = 17.58$ per cent.
(b) $(CN \cdot C_6H_4 \cdot SO_2)_2$,, $N = 8.43$; $S = 19.28$. ,, ,,
(c) $CN \cdot C_6H_4 \cdot SO_3H$,, $N = 7.65$; $S = 17.48$,, ,,
(d) $CN \cdot C_6H_4 \cdot SO_2H$,, $N = 8.38$; $S = 19.16$,, ,,

Substances with formulæ (a) and (b) have not been prepared previously. (c) and (d) represent o-cyanobenzenesulphonic acid and o-cyanobenzenesulphinic acid respectively. The melting points and other properties of the substances melting at 156—156·5° and 172·5—173° exclude the possibility of the identity of either with the sulphonic or sulphinic acid described in this paper.

We are indebted to Mr. R. S. Bowman, B.Sc., for analysing and determining the molecular weight of the o-benzoic sulphinide obtained from o-cyanobenzenesulphonic chloride by the action of dilute caustic soda.

TECHNICAL COLLEGE,
DERBY.

XL.—Preparation and Properties of some New Tropeines.

By HOOPER ALBERT DICKINSON JOWETT and ARCHIE CECIL OSBORN HANN.

IN the course of an investigation on the chemistry and pharmacology of the jaborandi alkaloids by one of us and Professor C. R. Marshall, it was shown (Trans., 1900, 77, 481) that, when caustic alkali is added to an aqueous solution of pilocarpine or isopilocarpine, the specific rotatory power is diminished and that the minimum value is obtained when a molecular proportion of alkali has been added. The fact has also been observed (Marshall, J. Physiol., 1904, 31, 153) that an aqueous solution of pilocarpine to which a molecular amount of caustic alkali has been added does not possess the characteristic physiological action of pilocarpine. Furthermore, it has been found that the specific rotatory power of pilocarpine in aqueous solution dropped from 100·5° to 77·5° simply on standing for three weeks ; it has also been shown (Albertoni, Arch. exp. Path. Pharm., 1879, 11, 415 ; Marshall, loc. cit., 144) that, when instilled into the eye, aqueous solutions of the base are less active than solutions of a salt of similar strength. The explanation of these facts would appear to be that, under the above conditions, the lactone ring opens and the corresponding hydroxy-acid

or its salt is formed, and that the hydroxy-acid has a lower specific
rotation and is less active physiologically than the lactone :

$$C_2H_5 \cdot \underset{\underset{O}{\overset{|}{CO}}}{CH}—\underset{\underset{\diagdown\diagup}{\overset{|}{CH_2}}}{CH} \cdot CH_2 \cdot \underset{\underset{CH—N}{\overset{||}{C} \cdot N(CH_3)}}{} {\Large>}CH \rightarrow$$

$$C_2H_5 \cdot \underset{\overset{|}{CO_2H}}{CH}——\underset{\overset{|}{CH_2 \cdot OH}}{CH} \cdot CH_2 \cdot \underset{\underset{CH——N}{\overset{||}{C} \cdot N(CH_3)}}{} {\Large>}CH.$$

This connection between chemical constitution and physiological
action derives additional importance from the fact that pilocarpine acts
specifically on the so-called nerve-endings in the heart.   It was therefore thought of interest to determine whether this difference .in
physiological action between a lactone and its corresponding hydroxy-
acid could be observed in the case of other physiologically active bases.
It has not been found possible, so far, to obtain a glyoxaline derivative
similar to pilocarpine, and as atropine also acts on the nerve-endings
of the heart, though antagonistically to pilocarpine, the tropeines were
selected as suitable substances with which to investigate the difference
in action between a lactone and its corresponding hydroxy-acid.   For
this purpose, it was decided to attempt the preparation of tropeines
containing an acyl group similar to that existing in pilocarpine.
Methylparaconyl- and terebyl-tropeines were therefore prepared, and
the relation between these bases and pilocarpine is shown by the
following formulæ, where P is the pharmacophore or nitrogen-containing
complex :

$$CH_3 \cdot \underset{\underset{\underset{CO}{\diagdown\diagup}}{\overset{|}{O}}}{CH} \cdot \underset{\overset{|}{CH_2}}{CH} \cdot CO \cdot P \quad : \quad (CH_3)_2 \cdot \underset{\underset{\underset{CO}{\diagdown\diagup}}{\overset{|}{O}}}{C}—\underset{\overset{|}{CH_2}}{CH} \cdot CO \cdot P \quad : \quad C_2H_5 \cdot \underset{\underset{\underset{O}{\diagdown\diagup}}{\overset{|}{CO}}}{CH} \cdot \underset{\overset{|}{CH_2}}{CH} \cdot CH_2 \cdot P$$

Methylparaconyltropeine.   Terebyltropeine.   Pilocarpine.

In case these bases should prove to be physiologically inactive, an
aromatic tropeine, similar to homatropine, but containing a lactone
group, was also prepared.   The relation between these bases is shown
as under :

Phthalidecarboxyltropeine.   Homatropine (mandelyltropeine).

At the same time it was thought of interest to investigate the
physiological action of these and other tropeines and to determine to

what extent Ladenburg's generalisation applies. As the result of an investigation of the physiological action of a number of tropeines, this chemist has stated that the characteristic action of a tropeine, namely, its mydriatic action, depends not only on the presence of a tropine complex, but on the nature of the acyl group attached to it, which must contain (1) a benzene residue, (2) an aliphatic hydroxyl in the side-chain containing the carboxyl group. Thus, acetyl-, benzoyl-, and salicyl-tropeines do not produce mydriasis, but mandelyltropeine does. To test this generalisation further, we prepared glycollyl- and pro-tocatechyl-tropeines in addition to those already mentioned. The present paper contains an account of the preparation and properties of these substances, as well as those of their principal salts. The physiological experiments conducted by Professor C. R. Marshall will be described in detail elsewhere, but a brief account of his results may now be given.

First, as regards the difference in action between a lactone and its corresponding hydroxy-acid, it has been found that both terebyl- and phthalidecarboxyl-tropeines, which produce an atropine-like effect on the heart, lose this action after a molecular proportion of alkali has been added to the base. They thus show, in aqueous and alkaline solution, a difference in action analogous to that observed in the case of pilocarpine.

As regards Ladenburg's generalisation that, to obtain a tropeine possessing mydriatic action, it is necessary to have the tropine complex attached to an acyl group containing an aromatic nucleus, it is found that this does not apply in the case of terebyltropeine. Whilst glycollyltropeine may be said to be inactive physiologically, terebyl-tropeine has a distinct mydriatic action, though very much weaker than atropine or homatropine.

Of the five tropeines examined, all were found, when tested on the vagus endings in the heart, to have an action similar in kind to that of atropine but—especially in the case of glycollyltropeine—very much weaker. The order of activity was as follows :

(1) Phthalidecarboxyltropeine,　　(4) Methylparaconyltropeine,
(2) Terebyltropeine,　　　　　　　(5) Glycollyltropeine,
(3) Protocatechyltropeine,

the last on the list being very feebly active and much weaker than the rest. When applied to the eye in 1 per cent. solution (as a salt), the phthalidecarboxyl- and terebyl-tropeines both produced marked dilatation of the pupil, but both were very much weaker than either atropine or homatropine. The other three tropeines had no distinct mydriatic action.

The general results of this inquiry have proved, therefore :

(1) That the peculiar difference in physiological action between a lactone and its corresponding hydroxy-acid, as exemplified by pilocarpine and pilocarpic acid, also occurs in the case of a tropeine having a haptophore group similar to that in pilocarpine, namely, terebyltropeine, and also in the case of phthalidecarboxyltropeine.

(2) That Ladenburg's generalisation, so far as it refers to the necessity for a mydriatic tropeine to contain a benzene nucleus, does not strictly hold, since terebyltropeine possesses a distinct mydriatic action.

It would appear, however, that the conditions most favourable for the development of the mydriatic action in a tropeine are those stated by Ladenburg, namely, that the acyl group should contain a benzene nucleus and an aliphatic hydroxyl in the side-chain containing the carboxyl group.

EXPERIMENTAL.

Glycollyltropeine, $CH_2(OH) \cdot CO \cdot C_8H_{14}ON$.

This base was made by the general method devised by Ladenburg for the preparation of tropeines (*Annalen*, 1883, 217, 82), namely, by neutralising tropine with glycollic acid and digesting the resulting solution with dilute hydrochloric acid (1 : 40) for twenty-four hours on a water-bath.

The crude base was purified by its conversion into the hydriodide and recrystallisation of this salt from methyl alcohol until pure. On regeneration, the base was obtained crystalline and was recrystallised from benzene until the melting point was constant ; it formed laminar crystals melting at 113—114°, which are readily soluble in alcohol, moderately so in water, but dissolve only sparingly in ether.

0·11 gave 0·2436 CO_2 and 0·0864 H_2O.   C = 60·4 ;   H = 8·7.

$C_{10}H_{17}O_3N$ requires C = 60·3 ;   H = 8·5 per cent.

The *hydrochloride* formed exceedingly deliquescent crystals, which, after drying at 110°, melted at 171—172°.

0·3644 gave 0·22 AgCl.   Cl = 14·9.

$C_{10}H_{17}O_3N$,HCl requires Cl = 15·0 per cent.

The *hydriodide* separated from methyl-alcoholic solution in stout, acicular crystals which melted at 187—188° ; it is easily soluble in water, sparingly so in alcohol, and insoluble in ether. The salt contained half a molecule of water of crystallisation, which was not lost after five hours' heating at 110°, and at a higher temperature it became decomposed.

0·199 gave 0·1389 AgI. I = 37·7.

0·1987 gave 0·138 AgI. I = 37·5.

$(C_{10}H_{17}O_3N,HI)_2,H_2O$ requires I = 37·8 per cent.

The *nitrate* was obtained by the decomposition of the pure hydriodide with the requisite quantity of silver nitrate. It separated from its aqueous solution, on evaporation in a vacuum over sulphuric acid, as a viscid oil which gradually solidified. After recrystallisation from absolute alcohol, it was obtained in oblong, laminar crystals, which, after drying at 100°, melted at 120—121°.

0·2038 gave 0·3444 CO_2 and 0·1273 H_2O. C = 46·1 ; H = 6·9.

$C_{10}H_{17}O_3N,HNO_3$ requires C = 45·8 ; H = 6·9 per cent.

The *aurichloride* formed yellow, acicular crystals, which, after recrystallisation from hot water, melted at 186—187°.

0·2080 gave 0·0762 Au· Au = 36·5.

$C_{10}H_{17}O_3N,HAuCl_4$ requires Au = 36·6 per cent.

The *platinichloride* was not precipitated on the addition of the reagent to an aqueous solution of the hydrochloride. On evaporation, however, orange crystals were obtained, which separated from hot water in short, stout needles. After drying over sulphuric acid, the crystals melted and decomposed at 225—226°.

0·0595 gave 0·0145 Pt. Pt = 24·4.

$(C_{10}H_{17}O_3N)_2,H_2PtCl_6$ requires Pt = 24·2 per cent.

Methylparaconyltropeine, $\begin{array}{c} CH_3 \cdot CH - CH \cdot CO \cdot C_8H_{14}ON \\ O \cdot CO \cdot CH_2 \end{array}$.

This base, as well as the remaining tropeines described in this paper, was prepared by passing hydrogen chloride through a solution of tropine neutralised with the acid in question and maintained at a temperature of 120—125° for two to three hours (Täuber, D.R.-P. 95853). The dark brown gum thus obtained was decomposed by ammonia and the base extracted with chloroform ; the crude tropeine was purified by conversion into the hydriodide. The pure base, regenerated from the purified hydriodide, was obtained as a colourless oil which refused to crystallise.

The *hydrobromide* separated from strong alcohol in square, laminar crystals which melted at 196—197° ; this salt is anhydrous, and is easily soluble in water, but moderately so in absolute alcohol.

0·152 gave 0·2695 CO_2 and 0·0868 H_2O. C = 48·4 ; H = 6·3.

0·1942 ,, 0·1057 AgBr. Br = 23·2.

$C_{14}H_{21}O_4N,HBr$ requires C = 48·3 ; H = 6·3 ; Br = 23·0 per cent

B B

The *hydriodide* crystallised from alcohol in triangular groups of crystals, which, after drying in the air, melted at 177—178°; it is easily soluble in water, sparingly so in alcohol, and insoluble in ether.

0·178 gave 0·1054 AgI.   I = 32·0.

$C_{14}H_{21}O_4N,HI$ requires $I = 32\cdot2$ per cent.

The *aurichloride* was precipitated as a yellow oil, which solidified on rubbing with a glass rod. It was recrystallised from dilute hydrochloric acid containing a little alcohol, and thus obtained in the form of yellow, silky leaflets, which, after drying in the air, melted at 64—65°; this salt is moderately soluble in water and in alcohol.

0·1604, after drying at 100°, lost 0·0051 and gave 0·0504 Au. $H_2O = 3\cdot2$; Au = 31·4.

$C_{14}H_{21}O_4N,HAuCl_4,H_2O$ requires $H_2O = 2\cdot9$; Au = 31·5 per cent.

The *platinichloride* was obtained as an amorphous precipitate, and it separated from solution in dilute hydrochloric acid as a yellow powder which melted at 233—234°.

0·1002 gave 0·0206 Pt.   Pt = 20·6.

$(C_{14}H_{21}O_4N)_2,H_2PtCl_6$ requires $Pt = 20\cdot7$ per cent.

The *picrate*, after recrystallisation from alcohol, formed yellow, laminar crystals which melted at 190—191°.

$$Terebyltropeine, \quad \begin{matrix} CMe_2\text{-}CH\cdot CO\cdot C_8H_{14}ON \\ | \qquad | \\ O\cdot CO\cdot CH_2 \end{matrix} .$$

This tropeine was prepared by a method similar to that employed in the case of methylparaconyltropeine, but the mixture was maintained at a temperature of 130—135° for two hours. The pure base, obtained through the hydriodide, solidified on standing, was dried on porous earthenware over sulphuric acid and recrystallised from acetone by the gradual evaporation of the solvent in a vacuous desiccator, and separated in small, diamond-shaped crystals which melted at 66—67°; it is very soluble in water or alcohol.

0·2086 gave 0·4892 CO_2 and 0·1542 H_2O.   C = 64·0; H = 8·2.

$C_{15}H_{23}O_4N$ requires C = 64·1; H = 8·2 per cent.

The *hydrochloride*, which separated from its concentrated aqueous solution as a soft, crystalline mass, was drained on porous earthenware, and after two recrystallisations from acetone was obtained in the form of leaflets which softened at 80° and melted at 82°; it is very soluble in water or alcohol, but is insoluble in ether.

0·194 dried, first at 60° and then at 110°, lost 0·0196. $H_2O = 10·1·$
0·1722 dried at 110° gave 0·077 AgCl.   Cl = 11·0.

$C_{15}H_{23}O_4N,HCl,2H_2O$ requires $H_2O = 10·2$.
$C_{15}H_{23}O_4N,HCl$ requires Cl = 11·2 per cent.

The *hydrobromide* separated from strong alcohol in small, laminar crystals which melted at 230—231°; it is easily soluble in water and moderately so in alcohol.

0·1927, dried at 110°, gave 0·1017 AgBr.   Br = 22·5.
$C_{15}H_{23}O_4N,HBr$ requires Br = 22·1 per cent.

The *hydriodide*, after recrystallisation from alcohol, was obtained in the form of laminar crystals which melted at 213—214°; the salt is moderately soluble in water.

0·1986 gave 0·114 AgI.   I = 31·1.
$C_{15}H_{23}O_4N,HI$ requires I = 31·1 per cent.

The *aurichloride* was precipitated as an oil which solidified after standing for several days; it was recrystallised from hot dilute hydrochloric acid and separated in imperfect crystals which melted indefinitely at 85—86°. The air-dried salt contains a molecule of water of crystallisation which is lost at 100°.

0·1324 air dried lost 0·0032 and gave 0·041 Au. $H_2O = 2·4$; Au = 31·0.
$C_{15}H_{23}O_4N,HAuCl_4,H_2O$ requires $H_2O = 2·8$; Au = 30·8 per cent.

The *platinichloride* separated as a gelatinous precipitate which could not be obtained crystalline. The *picrate* crystallised from dilute alcohol in yellow, matted leaflets which melted at 198—199°.

Phthalidecarboxyltropeine,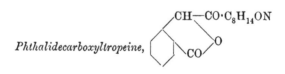

This tropeine was prepared by a method similar to that employed in the case of methylparaconyltropeine, but the crude base was purified through the hydrobromide; it was recrystallised from ethyl acetate, and separated in square, laminar crystals which melted at 79—80°. It is very soluble in alcohol and moderately so in water or ether.

0·2028 gave 0·5037 CO_2 and 0·1195 H_2O.   C = 67·7; H = 6·5.
$C_{17}H_{19}O_4N$ requires C = 67·8; H = 6·3 per cent.

The *hydrochloride* separated from absolute alcohol in laminar crystals which melted and decomposed at 242—244°.

0·1564 gave 0·0661 AgCl.   Cl = 10·4.

$C_{17}H_{19}O_4N,HCl$ requires Cl = 10·5 per cent.

The *hydrobromide* was obtained in the form of glistening leaflets from alcohol, and, after drying over sulphuric acid, melted at 128—129°. The salt is readily soluble in water and contains a molecule of water of crystallisation.

0·1538 gave 0·2884 CO_2 and 0·0774 H_2O.   C = 51·1 ; H = 5·6.

0·2014 ,,   0·0948 AgBr.   Br = 20·0.

0·2028 ,,   0·0955 AgBr.   Br = 20·0.

$C_{17}H_{19}O_4N,HBr,H_2O$ requires C = 51·0 ; H = 5·5 ; Br = 20·0 per cent.

The *nitrate* crystallises from water in square plates or in tufts of acicular crystals ; the air-dried salt contains water of crystallisation, and, after drying at 110°, melted at 169—171°. It is easily soluble in water or alcohol, but it is insoluble in ether.

0·1822 air dried lost 0·0087 H_2O. H_2O = 4·8.

0·1998 dried at 110° gave 0·4090 CO_2 and 0·1004 H_2O.   C = 55·8 ; H = 5·6.

$C_{17}H_{19}O_4N,HNO_3,H_2O$ requires H_2O = 4·7 per cent.

$C_{17}H_{19}O_4N,HNO_3$ requires C = 56·0 ; H = 5·5 per cent.

The *aurichloride* crystallised from alcohol in golden-yellow leaflets which melted at 184—185°.

0·1776 gave 0·0546 Au.   Au = 30·7.

$C_{17}H_{19}O_4N,HAuCl_4$ requires Au = 30·8 per cent.

The *platinichloride* was obtained as a yellow, amorphous powder which melted at 234—235°.

0·208 gave 0·0396 Pt.   Pt = 19·0.

$(C_{17}H_{19}O_4N)_2,H_2PtCl_6$ requires Pt = 19·3 per cent.

Protocatechyltropeine,

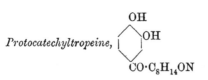

This tropeine was prepared by a method similar to that employed in the case of methylparaconyltropeine, but the crude base was purified by recrystallisation from absolute alcohol ; it separated in stout, acicular crystals which melted at 253—254° with decomposition. The base is sparingly soluble in water or alcohol.

0·1829 gave 0·4338 CO_2 and 0·1172 H_2O.   C = 64·7 ; H = 7·1.

$C_{15}H_{19}O_4N$ requires C = 65·0 ; H = 6·9 per cent.

The *hydrochloride* crystallised from water in small, glistening plates or needles which did not melt below 300° ; it is moderately soluble in water, sparingly so in alcohol, and insoluble in ether.

0·2027 gave 0·0944 AgCl.   Cl = 11·5.

$C_{15}H_{19}O_4N,HCl$ requires Cl = 11·3 per cent.

The *nitrate* was so rapidly oxidised, either in solution or when exposed to the air, that it was not further investigated.   The *aurichloride* separated as an amorphous precipitate which rapidly underwent reduction.   The *platinichloride* separated from a hot solution in small, laminar crystals which melted at· 228—229° with effervescence ; it is sparingly soluble in water and almost insoluble in alcohol.

0·2019 gave 0·0404 Pt.   Pt = 20·0.

$(C_{15}H_{19}O_4N)_2,H_2PtCl_6$ requires Pt = 20·2 per cent.

The *picrate* crystallised from alcohol in yellow plates which darkened at 255° and decomposed at 260—262°.

In conclusion, we wish to express our thanks to Professor A. W. Crossley for kindly supplying us with a quantity of pimelic acid, which was utilised for the preparation of terebic acid by the method described by Lawrence (Trans., 1899, **75**, 527).

THE WELLCOME CHEMICAL RESEARCH LABORATORIES,
LONDON, E.C.

XLI.—*Studies in Asymmetric Synthesis. IV. The Application of Grignard's Reaction for Asymmetric Syntheses.*

By ALEXANDER MCKENZIE.

IT has been shown in a former paper (Trans., 1904, **85**, 1249) that, when *l*-menthyl benzoylformate is acted on by magnesium methyl iodide, the experimental conditions may be so chosen that the attack of the Grignard reagent takes place exclusively at the carbonyl grouping of the ketonic ester, whilst the carboxyaryl grouping is not attacked.   When the product of this action is decomposed by ice and mineral acid, a mixture of unequal amounts of *l*-menthyl *d*-phenyl· methylglycollate and *l*-menthyl *l*-phenylmethylglycollate is produced, and, when this mixture is saponified by an excess of alkali and the resulting menthol completely removed, the potassium salt formed is lævorotatory, as also is the acid obtained from it. The asymmetric synthesis of an optically active phenylmethylglycollic acid (atrolactinic acid) was therefore accomplished in accordance with the scheme:

$$C_6H_5 \cdot CO \cdot CO_2H \text{ (inactive)} \longrightarrow C_6H_5 \cdot CO \cdot CO_2 \cdot C_{10}H_{19} \text{ (active)} \longrightarrow$$

$$\underset{\overset{|}{OH}}{\overset{CH_3}{\underset{|}{C_6H_5 \cdot C \cdot CO_2 \cdot C_{10}H_{19}}}} \text{ (active)} \longrightarrow \underset{\overset{|}{OH}}{\overset{CH_3}{\underset{|}{C_6H_5 \cdot C \cdot CO_2H}}} \text{ (active)}.$$

An active phenylethylglycollic acid can be synthesised in an analogous manner.

The only other examples of the genesis of optically active compounds by asymmetric synthesis are the formation of a lævorotatory valeric acid from methylethylmalonic acid by the agency of brucine (Marck-wald, *Ber.*, 1904, **37**, 349) and the formation of a lævorotatory lactic acid by the reduction of *l*-menthyl pyruvate (McKenzie, *Trans.*, 1905, **87**, 1373).*

It was accordingly of interest to extend the application of Grignard's reaction for the purpose of obtaining other cases of asymmetric synthesis, and the present paper contains a description of the results so far attained in this direction.

The action of magnesium propyl iodide, magnesium *iso*butyl iodide, magnesium *tert.*-butyl iodide, and magnesium α-naphthyl bromide respectively on *l*-menthyl benzoylformate was investigated. In each case, the asymmetric synthesis of a substituted glycollic acid was effected. The influence of the increase of weight of the hydrocarbon group introduced into the molecule of *l*-menthyl benzoylformate appears to be to diminish the value for the specific rotation of the mixture of unequal amounts of the *d*- and *l*-acids obtained; for example, the mixture of *d*- and *l*-phenylmethylglycollic acids resulting from the action of magnesium methyl iodide on *l*-menthyl benzoylformate is very much more lævorotatory than either the mixture of phenyl-*tert.*-butylglycollic acids resulting from the action of magnesium *tert.*-butyl iodide or the mixture of phenyl-α-naphthylglycollic acids resulting from the action of magnesium α-naphthyl bromide. It is not, however, permissible to conclude from this result that the ratio of *l*- to *d*-acid produced in the latter cases is less than in the former, since neither of those substituted glycollic acids has yet been prepared in the pure optically active form, and it is accordingly impossible to state whether *l*-phenylmethylglycollic acid is a more active substance than *l*-phenyl-*tert.*-butylglycollic acid or *l*-phenyl-α-naphthylglycollic acid. Again, the mode of formation of the Grignard reagent and the conditions under

* No asymmetric synthesis of a sulphur compound is recorded by Smiles in his paper entitled "An Asymmetric Synthesis of Quadrivalent Sulphur" (*Trans.* 1905, **87**, 450).

which its action on the ketonic ester is conducted doubtless in many cases influence the extent to which an asymmetric synthesis takes place. Whilst the activity of the mixture of unequal amounts of d- and l-phenylmethylglycollic acids obtained from l-menthyl benzoylformate does not very appreciably vary with variation of the conditions under which the Grignard reagent is employed, the case is different when magnesium *iso*butyl iodide acts on l-menthyl benzoylformate, since, when the Grignard reagent is added in the proportion of $2\frac{1}{2}$ mols. of iodide to 1 mol. of ester, the extent of the asymmetric synthesis is much less marked than when the reagent is added in the proportion of $1\frac{1}{4}$ mols. of iodide to 1 mol. of ester. Besides, in the preparation of magnesium *iso*butyl iodide, it was found that the reaction was always incomplete when equimolecular weights of magnesium and iodide interacted in presence of anhydrous ether, a certain amount of metal remaining undissolved. In the experiments quoted in this paper, accordingly, where magnesium *iso*butyl iodide was added in the proportion of $1\frac{1}{4}$ mols. to 1 mol. of ester, the action on the ketonic ester was probably incomplete, so that one might expect to get very varying results with each individual experiment. It is likely that the formation of magnesium *iso*butyl iodide is accompanied by side reactions such as Tschelinzeff (*J. Russ. Phys. Chem. Soc.*, 1904, **36**, 549) has described as happening during the formation of magnesium *iso*propyl iodide in ethereal solution. In the latter case, 2 mols. of the iodide are requisite for the solution of 1 mol. of magnesium and, in addition to the normal formation of the organo-magnesium compound, both propane and propylene are produced, thus: $2CHMe_2I + Mg = CH_2Me_2 + CH_2:CHMe + MgI_2$; as much as 41 per cent. of the iodide is used for this reaction and 1·47 per cent. of it undergoes the change represented by $2CHMe_2I + Mg = 2CH_2:CHMe + H_2 + MgI_2$.

In an asymmetric synthesis, an optically active substance is generated, without any analytical separation, from a non-asymmetric one, in which a carbon atom becomes asymmetric under the influence of an optically active asymmetric grouping, which is introduced and then eliminated. The question as to whether the extent of a synthesis of this kind, as measured by the optical activity of the resulting product, depends on the numerical value of the optical activity of the active group introduced and then eliminated was tested by acting on l-bornyl benzoylformate with magnesium alkyl (or aryl) halides. l-Borneol is not so strongly lævorotatory as l-menthol, the specific rotations in ethyl-alcoholic solution being respectively $-39\cdot7°$ and $-49\cdot4°$ under similar conditions of temperature and concentration. It was thought likely that the influence of the l-bornyl group would be less marked than that of the l-menthyl group, and experiment showed this to be the case. When magnesium methyl iodide

($1\frac{1}{4}$ mols.) acted on l-bornyl benzoylformate (1 mol.) under the conditions described in the experimental portion of this paper, the mixture of atrolactinic acids obtained had $[a]_D^{15\cdot5°} - 1\cdot9°$ ($c = 7\cdot57$) in ethyl-alcoholic solution, a value considerably less than that obtained from l-menthyl benzoylformate under similar conditions, namely, $[a]_D^{18°} - 9\cdot5°$ ($c = 9\cdot0704$). The interaction of magnesium ethyl iodide and l-bornyl benzoylformate also gave a lævorotatory mixture of phenylethylglycollic acids, having $[a]_D^{17°} - 4\cdot2°$ ($c = 17\cdot052$) in ethyl-alcoholic solution, but, as the specific rotation had not been determined with the product obtained from l-menthyl benzoylformate (Trans., 1904, 85, 1249), the results are not comparable. A striking observation as to the effect of the bornyl as contrasted with the menthyl group was made with the products resulting from the action of magnesium *iso*butyl iodide and magnesium a-naphthyl bromide respectively on l-bornyl benzoylformate. A dextrorotatory acid mixture was obtained in both cases, whereas the corresponding acid mixtures obtained from l-menthyl benzoylformate were lævorotatory.

The asymmetric synthesis of phenylmethylglycollic acid can be accomplished not only by the action of magnesium methyl iodide on l-menthyl benzoylformate, but also by the action of magnesium phenyl bromide on l-menthyl pyruvate, thus :

$$CH_3 \cdot CO \cdot CO_2H \text{ (inactive)} \longrightarrow CH_3 \cdot CO \cdot CO_2 \cdot C_{10}H_{19} \text{ (active)} \longrightarrow$$

$$\underset{\underset{OH}{|}}{CH_3 \cdot \overset{\overset{C_6H_5}{|}}{C} \cdot CO_2 \cdot C_{10}H_{19}} \text{ (active)} \longrightarrow \underset{\underset{OH}{|}}{CH_3 \cdot \overset{\overset{C_6H_5}{|}}{C} \cdot CO_2H} \text{ (active).}$$

Whilst the first method, however, gives an asymmetric synthesis of the l-acid, the latter gives an asymmetric synthesis of the d-acid, the acid mixture obtained having $[a]_D^{16°} + 5\cdot5°$ ($c = 14\cdot7$) in ethereal solution. The mixture of d- and l-phenylethylglycollic acids obtained from the interaction of magnesium ethyl bromide and l-menthyl pyruvate was also dextrorotatory, although the rotation was not nearly so pronounced as with the mixture of phenylmethylglycollic acids referred to.

The application of Grignard's reaction to menthyl esters of the acetoacetic ester type was also studied. Grignard, who examined the action of magnesium methyl iodide on ethyl acetoacetate, found that the latter compound acts in this case in accordance with its enolic structure (*Compt. rend.*, 1902, 134, 849 ; *Ann. Chim. Phys.*, 1902, [vii], 27, 548). Now, since ethyl acetoacetate is generally regarded as consisting at the ordinary temperature of an allelotropic mixture of the ketonic and enolic forms, and since Lapworth and Hann (Trans., 1902, 81, 1499) conclude that l-menthyl acetoacetate, which can readily be obtained as a crystalline solid by heating ethyl acetoacetate with l-menthol, probably possesses the ketonic structure, it appeared possible that the latter compound might lend itself for purposes of asymmetric synthesis. If it reacted with magnesium phenyl bromide, for instance, in the ketonic form, then the following changes might be effected :

$$CH_3 \cdot CO \cdot CH_2 \cdot CO_2 \cdot C_2H_5 \text{ (inactive)} \longrightarrow CH_3 \cdot CO \cdot CH_2 \cdot CO_2 \cdot C_{10}H_{19}$$

$$\text{(active)} \longrightarrow CH_3 \cdot \underset{\overset{|}{OH}}{\overset{C_6H_5}{C}} \cdot CH_2 \cdot CO_2 \cdot C_{10}H_{19} \text{ (active)} \longrightarrow$$

$$CH_3 \cdot \underset{\overset{|}{OH}}{\overset{C_6H_5}{C}} \cdot CH_2 \cdot CO_2H \text{ (active)}.$$

It was found, however, that l-menthyl acetoacetate reacts in accordance with its enolic structure. When the product of its action with magnesium ethyl bromide was decomposed by ice and mineral acid, the original ester was regenerated. The following equations may accordingly be taken as representing the action :

(1) $CH_3 \cdot C(OH) : CH \cdot CO_2 \cdot C_{10}H_{19} + C_2H_5 \cdot MgBr = C_2H_6 +$
$$CH_3 \cdot C(O \cdot MgBr) : CH \cdot CO_2 \cdot C_{10}H_{19}.$$

(2) $CH_3 \cdot C(O \cdot MgBr) : CH \cdot CO_2 \cdot C_{10}H_{19} + H_2O = MgBr \cdot OH +$
$$CH_3 \cdot C(OH) : CH \cdot CO_2 \cdot C_{10}H_{19}.$$

This result was not altogether unexpected, since it has been found that substances which in solution form an allelotropic mixture, such as amides, thioamides, &c., act in accordance with their hydroxylic structure (Tschugaeff, *Ber.*, 1902, **35**, 3912. Compare also Béis, *Compt. rend.*, 1903, **137**, 575).

Grignard (*loc. cit.*) has further shown that the products of the action of magnesium methyl iodide on ethyl ethylacetoacetate are the unchanged ester and the ester $OH \cdot CMe_2 \cdot CHEt \cdot CO_2Et$, the latter on hydrolysis yielding β-hydroxy-β-methyl-a-ethylbutyric acid,

$$OH \cdot CMe_2 \cdot CHEt \cdot CO_2H.$$

When the action of magnesium methyl iodide on ethyl ethylacetoacetate was conducted at a higher temperature than in the case just quoted, the unchanged ester and the glycol, $OH \cdot CMe_2 \cdot CHEt \cdot CMe_2 \cdot OH$, were formed. Those results showed that ethyl ethylacetoacetate reacted both in its ketonic and enolic forms.

By the action of magnesium ethyl bromide on l-menthyl ethylacetoacetate, which is easily prepared by heating ethyl ethylacetoacetate with l-menthol, the formation of an optically active β-hydroxy-$a\beta$-diethylbutyric acid was not observed, although there was evidence that a mixture of unequal amounts of menthyl esters had actually been produced.

When l-menthyl diethylacetoacetate was acted on by magnesium ethyl bromide or by magnesium phenyl bromide, no evidence of an asymmetric synthesis was obtained, although in this case, where the ester contains no hydroxyl group, the reaction must have taken place at the ketonic group. Disruption of the molecule apparently took place most probably during the saponification by alkali (compare Grignard, *loc. cit.*).

A slight asymmetric synthesis was detected when magnesium phenyl bromide acted on l-menthyl lævulate, the lævorotation observed being presumably due to the formation of the asymmetric lactone,

$$\text{CMePh}\underset{\text{O·CO}}{\overset{\text{CH}_2^-}{<}\!>}\text{CH}_2.$$

EXPERIMENTAL.

The asymmetric synthesis of l-phenylmethylglycollic acid (atrolactinic acid), recorded in a former paper (*loc. cit.*), can also be accomplished when a large excess of alkali is used for the saponification of the mixture of unequal amounts of l-menthyl l-phenylmethylglycollate and l-menthyl d-phenylmethylglycollate resulting from the action of magnesium methyl iodide on l-menthyl benzoylformate. In other words, the alkali did not exert a racemising influence such as the author has already observed in the saponification of unequal amounts of l-menthyl d- and l-mandelates. It is necessary, however, to ensure the complete saponification of the ester mixture resulting from the Grignard reaction in question in order to prove that the rotations observed were actually due to an asymmetric synthesis and not to a resolution of a partially racemic ester by the fractional hydrolysis method, first used by Marckwald and McKenzie. Thus, if it were the case that the product of the action is l-menthyl dl-phenylmethylglycollate and not a mixture of unequal amounts of the d- and l-isomerides, saponification by an insufficiency of potassium hydroxide would yield an optically active potassium salt (compare McKenzie and Thompson, Trans., 1905, **87**, 1004). In the experiments formerly quoted, due regard was paid to this consideration, precautions being taken that the ester mixture was completely saponified; the following may, however, be submitted as corroborative evidence. The action of magnesium methyl iodide (1 mol.) on 5 grams of l-menthyl benzoylformate (1 mol.) was conducted as in the case of the second experiment formerly described (Trans., 1904, **85**, 1260), and the ester mixture was boiled for one and a half hours under a reflux condenser with 50 c.c. of alcoholic potassium hydroxide containing 3·1 grams of alkali. After sixteen hours, the mixture was again boiled for one hour and the alcohol and menthol entirely removed. The aqueous solution of potassium salt (20 c.c.) obtained was lævorotatory, 14 c.c.

in a 1-dcm. tube giving a_D − 1·50°, whilst the resulting acid (2·1 grams) in ethyl-alcoholic solution gave $l = 2$, $c = 13·266$, a_D − 2·25°, $[a]_D$ − 8·5°. Although the saponification was in this case conducted with a large excess of alkali, the rotation of the acid was practically the same as in the previous case, when the specific rotation of the acid was − 8·3°.

Whilst the mixture of phenylmethylglycollates is readily saponified by alcoholic potassium hydroxide, the saponification is slow when aqueous baryta is used. A solution of magnesium methyl iodide ($1\frac{1}{4}$ mols.) in 25 c.c. of ether was added within a minute to 5 grams (1 mol.) of l-menthyl benzoylformate dissolved in 25 c.c. of ether, and, when the vigorous action had subsided, the product was boiled for a quarter of an hour. The ester mixture obtained after the addition of ice and hydrochloric acid was freed from ether and then heated on the water-bath for twenty-four hours with a solution of 10 grams of barium hydroxide in 100 c.c. of water and the product distilled in a current of steam for several hours until all the menthol had been removed. The excess of barium hydroxide was separated by means of carbon dioxide and the aqueous solution of barium salt concentrated to 52 c.c., of which 28 c.c., when examined in a 4-dcm. tube, gave $a_D^{10°}$ − 1·41°· The concentration of the solution ($c = 3·634$)·was estimated by evaporating 10 c.c. to dryness and then drying the residue at 140°; whence $[a]_D^{10°}$ − 9·7°.

0·3634 gave 0·1802 BaSO$_4$. Ba = 29·2.

$C_{18}H_{18}O_6$Ba requires Ba = 29·4 per cent.

Action of Magnesium Propyl Iodide on l-Menthyl Benzoylformate.

When a large excess of magnesium propyl iodide was added to a solution of 2·8 grams of ester in 15 c.c. of ether, the substituted glycollic acid obtained was found to contain some benzoylformic acid in admixture with it. The acid was lævorotatory, 1 gram dissolved in 14 c.c. of ethyl alcohol giving a_D − 0·28° ($l = 2$).

Action of Magnesium isoButyl Iodide on l-Menthyl Benzoylformate.

A solution of magnesium ($1\frac{1}{4}$ mols.) in isobutyl iodide ($1\frac{1}{4}$ mols.) and 30 c.c. of ether was added by means of the siphon apparatus previously described (loc. cit.) within an interval of forty minutes to a solution of 5 grams of ester (1 mol.) in 20 c.c. of ether. The action was accompanied by the separation of iodine, the tint of which disappeared as the addition of the magnesium compound proceeded. After twelve hours, the product was decomposed by the successive addition of crushed ice and dilute sulphuric acid and the ethereal solution decolorised by sulphurous acid. The product obtained after removal of the ether was then saponified by a solution of 5 grams of potassium hydroxide in a mixture of 50 c.c. of ethyl alcohol and 20 c.c. of water. After the solution had remained for five days at the ordinary temperature, it was boiled for two hours, the ethyl alcohol was removed by evaporation, water was added, and the precipitated menthol drained off. The aqueous solution of potassium salt, from which the menthol was entirely removed, was decolorised by charcoal, acidified by sulphuric acid, and extracted with ether. A deliquescent acid was obtained, which proved to be lævorotatory, its rotation in ethyl alcohol being $l = 2$, $c = 12\cdot1$, $a_D^{14°} - 4\cdot09°$, $[a]_D^{14°} - 16\cdot9°$. The concentration was determined by withdrawing an aliquot portion, evaporating off the solvent, and weighing the residue, dried at 100°. An analysis of the silver salt of this acid showed that it contained some benzoylformic acid, since the percentage of silver was 35·8, whereas $C_{12}H_{15}O_3Ag$ requires $Ag = 34\cdot3$ per cent.

The acid resulting from other experiments on the action of magnesium *iso*butyl iodide on *l*-menthyl benzoylformate had not such a marked rotation as in the instance quoted. The experimental conditions were, however, purposely varied in each case and the solutions submitted to polarimetric examination had, in each separate experiment, to be heated under varying conditions with charcoal in order to obtain results on which reliance could be placed. The experiment just quoted was repeated with the following modifications.

The Grignard reagent ($1\frac{1}{4}$ mols.) was added within an interval of thirty minutes to an ethereal solution of 5 grams of the ester (1 mol.) and the product boiled for one hour. The ester mixture was saponified by boiling with a solution of 4·3 grams of potassium hydroxide in 50 c.c. of ethyl alcohol. The acid obtained was crystallised from water; the small crop which separated was inactive, whilst the filtrate (28 c.c.) gave only $a_D - 0\cdot40°$ ($l = 4$).

In another experiment, the Grignard reagent ($1\frac{1}{4}$ mols.) was added to an ethereal solution of the ester (5 grams, 1 mol.) within forty-five minutes and the solution then allowed to remain at the ordinary temperature for eighteen hours. The ester mixture, obtained as usual, was dissolved in a solution of 5·2 grams of potassium hydroxide in 50 c.c. of ethyl alcohol and after twelve hours was boiled for one hour; 2 grams of acid were obtained which, when made up to 14 c.c. with ethyl alcohol, gave $l = 2$, $a_D^{11°} - 1\cdot42°$, and $[a]_D^{11°} - 5°$.

When the action was conducted with a larger excess of the Grignard reagent, the asymmetric synthesis was less marked. A solution of magnesium *iso*butyl iodide ($2\frac{1}{2}$ mols.) in 20 c.c. of ether was added within one minute to a solution of 5 grams of ester (1 mol.) in 32 c.c. of ether; the product of the vigorous action was then boiled for half an hour. The saponification of the ester mixture was conducted by dissolving it in a solution of 4·6 grams of potassium hydroxide in 50 c.c. of ether, and, after twelve hours, boiling the mixture for one and a half hours : 2·1 grams of acid were obtained, which, when made to 14 c.c. with ethyl alcohol, gave only $l = 2$, $a_D - 0·35°$.

Action of Magnesium tert.-Butyl Iodide on 1-Menthyl Benzoylformate.

Magnesium (0·53 gram, 1½ mols.) was added to a mixture of *tert.*-butyl iodide (4 grams, 1½ mols.) and 20 c.c. of ether. The action began at the ordinary temperature, but, after heating for four and a half hours, it was found that 0·16 gram of metal was not dissolved. The solution was quickly added to a solution of 4 grams of ester in 20 c.c. of ether. After twenty-four hours, the product was heated for a quarter of an hour and manipulated in the usual manner, the saponification of the ester mixture having been conducted by boiling for forty-five minutes with a solution of 3·3 grams of potassium hydroxide in 25 c.c. of ethyl alcohol. Only 0·85 gram of acid was obtained and this, when dissolved in ethyl alcohol and then examined in a 2-dcm. tube (14 c.c.), gave $a_D - 0·07°$, a reading which, although feeble, was quite distinct with the polarimeter used.

In a second experiment, where 2½ mols. of the Grignard reagent were used, 0·93 gram of acid was obtained from 4 grams of ester and this, when dissolved in ethyl alcohol as before, gave $a_l^{16°} - 0·09°$.

Action of Magnesium a-Naphthyl Bromide on 1-Menthyl Benzoylformate.

A solution of magnesium a-naphthyl bromide (2¼ mols.) in 25 c.c. of ether was gradually added to a solution of 5 grams of *l*-menthyl benzoylformate (1 mol.) in 25 c.c. of ether. At first a red coloration was imparted to the solution, but this disappeared as the addition of the Grignard reagent proceeded. The action was moderated in such a manner that the ether was gently boiling during the addition. After two days at the ordinary temperature, the product was decomposed first by crushed ice and then by dilute sulphuric acid, the ether was removed from the ethereal solution, and the residual oil submitted to distillation in steam until all the naphthalene had been separated. The remaining oil was then extracted with ether, the ether expelled, and the saponification conducted by boiling for two hours with an alcoholic solution of potassium hydroxide (5 grams). The alcohol was expelled, water added, and the precipitated menthol drained off. After the aqueous solution of potassium salt had been completely freed from menthol and after attempts to decolorise it had been unsuccessful, the addition of dilute sulphuric acid brought down a voluminous precipitate of the substituted glycollic acid, which was quantitatively extracted with ether. The acid, which is sparingly soluble in water, was finally decolorised with difficulty by heating with charcoal its solution in a mixture of ethyl alcohol and water.

A mixture of i- and l-acids (3·3 grams) was withdrawn, whilst the filtrate, when evaporated to dryness and then dissolved in acetone (0·17 gram of acid in 14 c.c. of solution), gave a_D $-0·25°$ ($l = 2$). The whole of the acid obtained was then converted into barium salt, which was crystallised from water, in which it is sparingly soluble. The crop which separated gave, on analysis, Ba = 19·2, whereas $C_{36}H_{26}O_6Ba$ requires Ba = 19·7 per cent.; the filtrate (14 c.c.) gave $a_D^{12°}$ $-0·35°$ ($l = 2$).

A second experiment was more successful. The saponification of the product obtained after the removal of the naphthalene was conducted by boiling with a solution of 5·2 grams of potassium hydroxide in 50 c.c. of ethyl alcohol for one and a half hours. The barium salt obtained was crystallised from water. *Barium phenyl-α-naphthyl-glycollate* separates from a mixture of ethyl alcohol and water in glassy prisms.

0·3748 (dried at 130—140°) gave 0·1244 $BaSO_4$.

Ba = 19·54 ; $C_{36}H_{26}O_6Ba$ requires Ba = 19·87 per cent.

The filtrate (28 c.c.) gave $l = 4$, $c = 1·022$, a_D $-0·45°$, $[a]_D$ $-11·0°$.

Action of Magnesium Methyl Iodide on l-*Bornyl Benzoylformate.*

l-*Bornyl benzoylformate* was conveniently prepared as follows. Benzoylformic acid was heated on a boiling water-bath for ten hours with three times its weight of l-borneol, a current of dry hydrogen chloride having been passed at intervals into the mixture. The ethereal solution of the product was washed successively with water and dilute sodium carbonate, the ether removed, and the product distilled in a current of steam until the borneol was practically all removed. The residue, which solidified on cooling, was purified by crystallisation from ethyl alcohol, from which it separates in colourless, glassy prisms with pyramidal ends ; it melts at 42—43°.

0·1897 gave 0·5252 CO_2 and 0·1326 H_2O. C = 75·5 ; H = 7·8.

$C_{18}H_{22}O_3$ requires C = 75·5 ; H = 7·7 per cent.

A determination of its specific rotation in ethyl-alcoholic solution gave the result : $l = 2$, $c = 10·819$, $a_D^{20°}$ $-5·76°$, $[a]_D^{20°}$ $-26·6°$.

The ester is readily soluble in hot ethyl alcohol and easily soluble in cold chloroform, acetone, benzene, carbon tetrachloride, or light petroleum.

When exposed to bright sunlight, the ester assumed an ochreous tint iu the course of two minutes, and after five minutes the tint had attained its maximum; when placed in the dark, the crystals again became colourless after a lapse of twelve hours. When the colourless crystals were exposed for five minutes to the light from an arc lamp, they also assumed an ochreous tint.

A solution of magnesium methyl iodide ($1\frac{1}{4}$ mols.) in 20 c.c. of ether was quickly dropped by means of a siphon into a solution of 5 grams of ester (1 mol.) in 20 c.c. of ether cooled at 0°. The violent action was accompanied by the separation of iodine. After the product had been boiled for one and a quarter hours, ice and dilute acetic acid were successively added, and the ethereal solution, from which the acetic acid was removed, evaporated. Thirty c.c. of alcoholic potassium hydroxide (1 c.c. = 0·0565 gram of KOH) were added, and after eighteen hours the solution was boiled for one hour. The alcohol was evaporated, water added, and the precipitated borneol drained off. The filtrate, which was extracted with ether and then evaporated for several hours to ensure the complete removal of the borneol; was found to be lævorotatory. After acidification by sulphuric acid and decolorisation by charcoal, the acid was extracted with ether; the ethereal solution, dried by sodium sulphate, yielded 2·12 grams of an acid crystallising in leaflets. This, when dried at 100°, gave the following rotation in ethyl-alcoholic solution: $l = 4$, $c = 7·57$, $a_D^{15\cdot5°} - 0·57°$, $[a]_D^{15\cdot5°} - 1·9°$. After removal of the alcohol, the acid was dried at 100° and analysed by titration with standard sodium hydroxide:

1·5516 required 18·9 c.c. alkali (0·4964N) for neutralisation, the amount calculated for $C_9H_{10}O_3$ being 18·8 c.c.

In a second experiment, where the addition of the Grignard reagent ($1\frac{1}{4}$ mols.) to the ester (1 mol., 5 grams) was conducted within an interval of three-quarters of an hour instead of quickly, as in the experiment just desciibed, and where the saponification of the resulting ester mixture was effected by a solution of 2·5 grams of potassium hydroxide in 30 c.c. of ethyl alcohol, the rotation of the acid obtained was practically the same as before, namely, $l = 2$, $c = 11·18$, $a_D - 0·36°$, $[a]_D - 1·6°$ (in ethyl-alcoholic solution). After the alcohol had been expelled from the latter solution and the residue crystallised from water, a crop of *i*-atrolactinic acid was obtained, whilst the filtrate gave $[a]_D - 4·2°$ ($c = 2·33$).

Action of Magnesium Ethyl Iodide on l-*Bornyl Benzoylformate.*

Magnesium ethyl iodide ($1\frac{1}{4}$ mols.) in ethereal solution (30 c.c.) was added within half an hour to a solution of 12 grams (1 mol.) of *l*-bornyl benzoylformate in 30 c.c. of ether. After the mixture had been allowed to remain at the ordinary temperature for one hour, it was boiled for one hour and decomposed in the usual manner by ice and mineral acid. The ester mixture resulting from the·ethereal extract was a yellow, viscid oil (13·25 grams), which was saponified by being heated for one hour with a solution of 2·95 grams of potassium hydroxide in 100 c.c. of ethyl alcohol. The alkaline aqueous solution, resulting after the removal of the ethyl alcohol and the borneol, was neutralised by hydrochloric acid, decolorised by charcoal, and concentrated to 25 c.c., of which 14 c.c., examined in a 2-dcm. tube, gave $a_D^{18°} - 1·70°$. The phenylethylglycollic acid (4·3 grams) obtained by acidifying the potassium salt and then extracting with ether crystallised in long, fine needles which were dried at 100° and then polarimetrically examined in ethyl-alcoholic solution : $l = 1$, $c = 17·052$, $a_D^{17°} - 0·72°$, $[a]_D^{17°} - 4·2°$.

After the expulsion of the alcohol, the residual acid was exactly neutralised by aqueous potassium hydroxide. The aqueous solution of potassium phenylethylglycollate thus obtained gave $l = 1$, $c = 27·71$, $a_D^{18°} - 0·90°$, $[a]_D^{18°} - 3·2°$. The concentration was determined by evaporating off the water from an aliquot portion of the solution and drying the residue at 140° until constant in weight.

0·2266 (dried at 140°) gave 0·0915 K_2SO_4. $K = 18·1$.

$C_{10}H_{11}O_3K$ requires $K = 17·9$ per cent.

Action of Magnesium isoButyl Iodide on l-Bornyl Benzoylformate.

A solution of magnesium ($1\frac{1}{4}$ mols.) in *iso*butyl iodide ($1\frac{1}{4}$ mols.) and 25 c.c. of ether was added within an interval of one hour to a solution of 10·8 grams of *l*-bornyl benzoylformate (1 mol.) in 25 c.c. of ether. After two days, the product was decomposed in the usual manner by ice and mineral acid and the ethereal solution washed with an aqueous solution of sodium hydrogen sulphite. The ester mixture remaining after the removal of the ether was submitted to distillation in steam, when no *iso*butyl iodide was detected in the distillate ; it was then saponified by the addition of a solution of 8·2 grams of potassium hydroxide in a mixture of ethyl alcohol and water, the mixture having been allowed to remain at the ordinary temperature for two days, after which it was boiled for one and a half hours. The ethyl alcohol and borneol were removed as usual and the aqueous solution (50 c.c.), which was decolorised with some difficulty by animal charcoal, proved to be dextrorotatory, 28 c.c. of it in a 4-dcm. tube giving $a_D^{12°} + 0·19°$. The potassium salt, when acidified and extracted with ether, yielded 3·6 grams of a crystalline acid, which in· ethyl-alcoholic solution (14 c.c.) gave $l = 2$, $a_D^{14°} + 0·50°$. The ethyl alcohol was expelled and· the residue crystallised from a mixture of benzene and light petroleum.

The crop, which separated as glistening needles grouped in rosettes, melted at 119—120° after having first been dried at 100°, and was inactive. The filtrate contained 2·2 grams of acid and was dextrorotatory; the silver salt was analysed.

0·4338 gave 0·1517 Ag. Ag = 34·2.

$C_{12}H_{15}O_3Ag$ requires Ag = 34·3 per cent.

Action of Magnesium a-Naphthyl Bromide on l-Bornyl Benzoylformate.

A solution of magnesium a-naphthyl bromide ($2\frac{1}{2}$ mols.) in 32 c.c. of ether was gradually added within an interval of forty minutes to a solution of 10 grams of l-bornyl ̄benzoylformate (1 mol.) in 25 c.c. of ether. After twelve hours, the product was decomposed in the usual manner and the naphthalene removed from it by distillation in steam. The residue in the flask was a yellow, semi-solid mass and the supernatant aqueous solution was neutral. After it had been found that the saponification could not readily be effected by aqueous potassium hydroxide (9·6 grams), ethyl alcohol was added and the saponification completed by boiling for several hours. The ethyl alcohol and borneol were removed, but it was not found possible to decolorise the aqueous solution of potassium salt to an extent necessary for accurate polarimetric observations. The acid obtained by acidifying the potassium salt and extracting with ether was crystallised first from a mixture of ethyl alcohol and water and then from chloroform, after which treatment it was optically inactive. For analysis, the acid was dried at 100°.

0·1975 gave 0·5618 CO_2 and 0·0941 H_2O. C = 77·6; H = 5·3.

$C_{18}H_{14}O_3$ requires C = 77·7; H = 5·1 per cent.

i-Phenyl-a-naphthylylycollic acid, when dehydrated at 100°, melts at 143—144° to a green liquid; it is somewhat sparingly soluble in hot benzene, from which, on cooling, it separates in glassy prisms grouped in rosettes. It is practically insoluble in cold water and sparingly soluble in hot chloroform, from which it separates in needles; it dissolves with difficulty both in hot and cold light petroleum and in carbon tetrachloride; it is easily soluble in cold acetone or cold ethyl alcohol.

The mother liquors, from which the i-acid (4·2 grams) had been removed by filtration, and which contained the optically active product of the asymmetric synthesis, were evaporated, but the dark brown mass thus obtained, when boiled in various organic solvents with animal charcoal, could not be sufficiently decolorised. The following method, however, was successful. The acid was boiled with water, charcoal, and an excess of barium carbonate for several days. A considerable amount of water was necessary to bring all the barium salt into solution. The aqueous solution (28 c.c.) which was filtered off from charcoal, barium carbonate, and some barium r-phenyl-a-naphthyl-glycollate gave the following rotation : $l = 4$, $c = 2·456$, $a_D^{15°} + 0·96°$, $[a]_D^{15°} + 9·8°$.

The concentration of the salt was determined by withdrawing 10 c.c. of the solution, evaporating off the water, and drying the residue at 130° until constant in weight.

Action of Magnesium Ethyl Bromide on l-Menthyl Pyruvate.

The l-menthyl pyruvate used for this action was portion of the product employed for the asymmetric synthesis of l-lactic acid (Trans., 1905, **87**, 1373); it had $[a]_D^{19·6°} - 92·8°$. A solution of magnesium ethyl bromide ($2\frac{1}{2}$ mols.) in 25 c.c. of ether was added within an interval of thirty minutes to a solution of 11 grams of l-menthyl pyruvate (1 mol.) in 25 c.c. of ether. The action, which was very vigorous, was accompanied in the initial stages by the separation of a white solid which gradually dissolved, whilst the solution assumed a green tint. After twelve hours, the product was decomposed as usual and the ester mixture obtained as a dark green oil, which became brown on the addition of a solution of 4·4 grams of potassium hydroxide in 40 c.c. of methyl alcohol, the solution having been boiled for one and a half hours. After removal of the methyl alcohol and menthol, the solution of potassium salt was decomposed by dilute sulphuric acid and extracted with ether. The resulting acid was converted into barium salt, the aqueous solution of which was decolorised by charcoal and a crop withdrawn, whilst the filtrate (18 c.c.) gave $l = 1$, $c = 15·8$, $a_D + 0·20°$. The acid obtained from this solution of barium salt gradually crystallised in the form of silky needles, and a portion, when sublimed, had the properties of i-methylethylglycollic acid already described by E. Frankland and Duppa (*Annalen*, 1865, **135**, 37) ; it melted at 67°.

Action of Magnesium Phenyl Bromide on l-Menthyl Pyruvate.

A solution of magnesium phenyl bromide ($1\frac{1}{4}$ mols.) in 30 c.c. of ether was added within an interval of thirty minutes to a solution of 18 grams of l-menthyl pyruvate (1 mol.) in 50 c.c. of ether. After eighteen hours, the product was decomposed as usual and the ester mixture saponified by being boiled for one hour with a solution of 6·2 grams of potassium hydroxide in 100 c.c. of ethyl alcohol. Attempts to decolorise the aqueous solution of potassium salt, which had been freed from menthol, were unsuccessful. The free acid, however, was decolorised by boiling its aqueous solution with charcoal. A crop of an optically inactive acid was removed, which analysis and melting point determinations showed to be i-atrolactinic acid. The filtrate (30 c.c.) proved to be dextrorotatory, 28 c.c. in a 4-dcm. tube giving $a_D + 2\cdot14^\circ$. The concentration of the solution, as determined by titrating an aliquot portion of it with standard alkali, was $c = 2\cdot9784$, whence $[a]_D + 18\cdot0^\circ$.

A second experiment was performed in order to confirm this dextrorotation. A solution of the Grignard reagent ($1\frac{1}{4}$ mols.) in 25 c.c. of ether was added within an interval of thirty-five minutes to a solution of 10 grams of ester (1 mol.) in 25 c.c. of ether. The ester mixture, obtained as usual, was in this case saponified by dissolving it in 50 c.c. of ethyl alcohol, adding 3·6 grams of solid potassium hydroxide, and then allowing the mixture to remain overnight, and finally boiling for one hour after the addition of a few c.c. of water. The ethereal solution of the acid, decolorised in the manner indicated in the previous experiment, measured 24 c.c., of which 14 c.c. in a 2-dcm. tube gave $a_D^{16^\circ} + 1\cdot6^\circ$; its concentration, as determined by titration against standard baryta, was $c = 14\cdot7$, whence $[a]_D^{16^\circ} + 5\cdot4^\circ$. The whole of the acid obtained was converted into barium salt (50 c.c.), 28 c.c. of the aqueous solution giving $a_D^{13^\circ} + 1\cdot89^\circ$ in a 4-dcm. tube. The concentration, as determined by withdrawing an aliquot portion and drying it at 130° until constant in weight, was $c = 5\cdot034$, whence $[a]_D^{14^\circ} + 9\cdot4^\circ$.

The acid obtained by the action of magnesium methyl iodide ($1\frac{1}{4}$ mols.) dissolved in 20 c.c. of ether on 5 grams of l-menthyl pyruvate (1 mol.) dissolved in 20 c.c. of ether was optically inactive, a result which was expected, as no asymmetric synthesis was possible under these conditions.

Action of Magnesium Alkyl Halides on l-Menthyl Acetoacetate.

l-Menthyl acetoacetate has already been described by Cohn (*Monatsh.*, 1900, 21, 200), Cohn and Tauss (*Ber.*, 1900, 33, 731), and Lapworth and Hann (Trans., 1902, 81, 1499). The method of preparation described by those authors was slightly modified. A mixture of l-menthol ($1\frac{1}{2}$ mols.) and ethyl acetoacetate (1 mol.) was heated in a paraffin-bath at 140—150° for six hours. The bulk of residual menthol was then removed by distillation in steam, and the product remaining in the flask extracted with ether. The ethereal solution was dried, the ether removed, and the product fractionated under diminished pressure. l-Menthyl acetoacetate, obtained in this manner, boiled at 146—147° under 9 mm. pressure (Cohn gives 145° under 9 mm. pressure), readily solidified, and had the following rotation in ethyl-alcoholic solution :
$l = 2$, $c = 5\cdot24$, $a_D^{20^\circ} - 7\cdot35^\circ$, $[a]_D^{20^\circ} - 70\cdot1^\circ$.

Lapworth and Hann give $[a]_D - 68 \cdot 5°$ for $c = 1 \cdot 5$ (temperature not quoted). No multirotation in ethyl-alcoholic solution was observed.

A solution of magnesium phenyl bromide ($1\frac{1}{4}$ mols.) in 20 c.c. of ether was gradually added to a solution of 10 grams of l-menthyl acetoacetate (1 mol.). The aqueous solution of potassium salt, obtained by the methods previously indicated, was decolorised and proved to be inactive.

A solution of magnesium ethyl bromide ($1\frac{1}{4}$ mols.) in 33 c.c. of ether was gradually added within an interval of forty minutes to a solution of $8 \cdot 7$ grams of menthyl acetoacetate (1 mol.) in 21 c.c. of ether. The action was vigorous, the white precipitate at first formed gradually disappearing. After eighteen hours at the ordinary temperature, ice and dilute hydrochloric acid were successively added, and the ethereal solution removed, and the aqueous solution extracted twice with ether. The ethereal extracts were united, washed once with a little water, and dried over anhydrous sodium sulphate. After the expulsion of the ether, an oil was obtained which quickly crystallised to a solid mass when nucleated with menthyl acetoacetate. A determination of the specific rotation of the product ($7 \cdot 6$ grams) in ethyl-alcoholic solution proved it to be l-menthyl acetoacetate: $l = 2$, $c = 5 \cdot 333$, $a_D^{20°}$ $- 7 \cdot 46°$, $[a]_D^{20°} - 69 \cdot 9°$.

Action of Magnesium Ethyl Bromide on 1-Menthyl Ethylacetoacetate.

1-Menthyl ethylacetoacetate was prepared as follows: equimolecular weights of ethyl ethylacetoacetate and l-menthol were heated for five hours in an oil-bath at 145—155°. The bulk of the menthol was removed by distillation of the product in a current of steam, and the remaining aqueous liquid, which was neutral to litmus, extracted with ether together with the oil. The ethereal extract was dried by calcium chloride, the ether distilled off, and the residual menthol readily separated by fractionation under diminished pressure. l-Menthyl ethylacetoacetate was obtained as a colourless oil, which did not solidify when immersed for several hours in a freezing mixture; it boils at 159—161° under 9—10 mm. pressure.

$0 \cdot 3244$ gave $0 \cdot 8482$ CO_2 and $0 \cdot 3109$ H_2O. $C = 71 \cdot 3$; $H = 10 \cdot 7$.

$C_{16}H_{28}O_3$ requires $C = 71 \cdot 6$; $H = 10 \cdot 5$ per cent.

The following polarimetric results were obtained: $l = 0 \cdot 5$, $a_D^{20°} - 30 \cdot 40°$, $d20°/4°$ $0 \cdot 9653$, $[a]_D^{20°} - 63 \cdot 0°$.

In ethyl-alcoholic solution: $l = 2$, $c = 4 \cdot 442$, $a_D^{20 \cdot 3°} - 6 \cdot 03°$, $[a]_D^{20 \cdot 3°} - 67 \cdot 9°$.

When a few drops of an anhydrous ethereal solution of ferric chloride were added to a solution of the ester in anhydrous ether, a coloration was not observed until the mixture had remained for about an hour at the ordinary temperature; ,it then gradually intensified, but was never very marked. Under the same conditions, the ethyl ethylacetoacetate used in the preparation of the menthyl ester gave a pronounced violet coloration at once.

A solution of magnesium ethyl bromide ($1\frac{1}{4}$ mols.) in 20 c.c. of ether was added to a solution of 10 grams of the ester (1 mol.) in 20 c.c. of ether. No perceptible action took place until about two-thirds of the Grignard reagent had been added, when the ether boiled and continued to do so until the addition was complete. After remaining overnight, ice and mineral acid were added and the ethereal solution dried with sodium sulphate. The residue, after evaporation of the ether and drying over sulphuric acid in a partial vacuum, was polarimetrically examined in a 5 per cent. ethyl-alcoholic solution, when the value $[\alpha]_D^{20°} - 64·6°$ was obtained. This result indicates that menthyl ethylacetoacetate had not acted towards the Grignard reagent exclusively in accordance with the structure

$$CH_3 \cdot C(OH) \dot{:} CEt \cdot CO_2 \cdot C_{10}H_{19},$$

since the variation of the value $-64·6°$ from that of menthyl ethylacetoacetate itself is beyond the limit of experimental error.

The product (8·5 grams) was saponified by heating it with a solution of 2·2 grams of potassium hydroxide in 50 c.c. of ethyl alcohol for one and a half hours, and the solution of potassium salt, obtained in the usual manner, was practically inactive.

Action of Magnesium Alkyl Halides on l-Menthyl Diethylacetoacetate.

l-Menthyl diethylacetoacetate is not formed in any appreciable amount when ethyl diethylacetoacetate is heated with an excess of l-menthol even at 190°, nor can it be conveniently obtained from the latter substances by heating them in the presence of hydrogen chloride according to the method which Patterson and Dickinson (Trans., 1901, 79, 280) have successfully devised for the interconversion of methyl and ethyl tartrates. The following method was employed : a solution of sodium ethoxide, prepared from sodium ($2\frac{1}{2}$ grams) and ethyl alcohol (35 grams), was added to a mixture of l-menthyl ethylacetoacetate (33 grams) and ethyl iodide (20 grams). After having been gently boiled for two hours, the liquid was neutral to litmus. The alcohol was expelled, water added to the residue, and the whole extracted with ether. The ethereal solution was dried with calcium chloride, the ether distilled off, and the resulting oil submitted to fractional distillation under diminished pressure.

l-Menthyl diethylacetoacetate, $CH_3 \cdot CO \cdot CEt_2 \cdot CO_2 \cdot C_{10}H_{19}$, was obtained as a colourless oil boiling at 180—182·5° under 13 mm. pressure, the yield being 15 grams.

0·1926 gave 0·5150 CO_2 and 0·1926 H_2O. C = 72·9 ; H = 11·2.

$C_{18}H_{32}O_3$ requires C = 72·9 ; H = 10·9 per cent.

A determination of the specific rotation of the freshly-prepared ester gave the result: $l = 0·5$, $d20°/4°$ 0·9605, $a_D^{20°}$ − 26·30°, $[a]_D^{20°}$ − 54·8°.

A solution of magnesium ethyl bromide ($1\frac{1}{4}$ mols.) in 20 c.c. of ether was added within an interval of twenty minutes to a solution of 5·5 grams of l-menthyl diethylacetoacetate (1 mol.) in 20 c.c. of ether. A reaction, evidenced by the boiling of the ether, took place. The usual treatment was followed, but the potassium salt obtained proved to be practically inactive.

In a similar manner, when magnesium phenyl bromide was substituted for magnesium ethyl bromide, the generation of an additional asymmetric carbon atom was not accompanied by an asymmetric synthesis, since the aqueous solution of potassium salt, prepared in a manner analogous to that usually employed, was optically inactive.

Action of Magnesium Alkyl Halides on l-*Menthyl Lævulate.*

For the preparation of l-menthyl lævulate, a mixture of lævulic acid (50 grams) and l-menthol (160 grams) was heated for eighteen hours in an oil-bath at 95—105°, whilst a current of dry hydrogen chloride was passed occasionally through the liquid ; the product was dissolved in about its own volume of ether, and the solution washed first with water and then with sodium hydrogen carbonate solution until it no longer gave an acid reaction. The ether was distilled off and the residual oil submitted to distillation in steam in order to remove the bulk of the menthol, after which process the ester was extracted with ether from the liquid in the distilling flask. The ethereal solution was dried over potassium carbonate, the ether expelled, and the residue fractionated under diminished pressure.

l-*Menthyl lævulate*, $CH_3 \cdot CO \cdot CH_2 \cdot CH_2 \cdot CO_2 \cdot C_{10}H_{19}$, was obtained as a colourless oil which boiled at 169° under 12 mm. pressure, the yield being 80 grams.

0·1679 gave 0·4377 CO_2 and 0·1620 H_2O. C = 71·1 ; H = 10·8.

$C_{15}H_{26}O_3$ requires C = 70·8 ; H = 10·3 per cent.

A determination of its specific rotation gave the following result : $l = 0·5$, $d19·8°/4°$ 0·9773, $a_D^{19·5°}$ − 29·59°, $[a]_D^{19·5°}$ − 60·6°.

No asymmetric synthesis was detected when magnesium ethyl bromide interacted with l-menthyl lævulate, but the solution of potassium salt, obtained in the usual manner, was so highly coloured, as also was the acid (or lactone) derived from it, that much reliance could not be placed on this result. A similar difficulty was also encountered when magnesium α-naphthyl bromide was substituted for magnesium ethyl bromide ; attempts to decolorise the aqueous solution of potassium salt and of the free acid respectively failed to give solutions sufficiently colourless for accurate polarimetric observations, whilst even the barium salt was unsuitable, as it was found to be practically insoluble in water. On the other hand, a positive result was obtained when magnesium phenyl bromide was used. A solution of magnesium phenyl bromide ($1\frac{1}{4}$ mols.) in 20 c.c. of ether was added within an interval of thirty minutes to a solution of 9·5 grams (1 mol.) of l-menthyl lævulate in 20 c.c. of ether. After twenty-four hours, the product was treated as usual, the saponification being conducted by boiling with a solution of 4 grams of potassium hydroxide in methyl alcohol. All attempts to decolorise the potassium salt and the free acid having failed, the latter was converted into barium salt, the aqueous solution of which was decolorised and found to be distinctly lævorotatory.

My thanks are due to Mr. R. V. Stanford, B.Sc., for able assistance rendered in the experiments with menthyl acetoacetate and its mono- and di-ethyl derivatives. I am also indebted to the Research Fund Committee of the Society for a grant which has defrayed most of the expense of this work.

THE UNIVERSITY,
BIRMINGHAM.

XLII.—*The Resolution of* 2 : 3-*Dihydro-3-methylindene-2-carboxylic Acid into its Optically Active Isomerides.*

By ALLEN NEVILLE, B.Sc.

BY the action of concentrated sulphuric acid on ethyl benzylacetoacetate, v. Pechmann obtained a crystalline acid, which from various reactions he concluded was identical with the dihydronaphthoic acid prepared by Berthelot. However, Roser (*Ber.*, 1887, **20**, 1574 ; *Annalen*, 1888, **247**, 165) prepared the same substance and showed it to be 3-methylindene-2-carboxylic acid, $C_6H_4{<}^{CMe}_{CH_2}{>}C{\cdot}CO_2H$. This acid is readily reduced by sodium amalgam in alkaline solution,

giving a dihydro-derivative, $C_6H_4\langle^{CHMe}_{-CH_2}\rangle CH \cdot CO_2H$, which contains two asymmetric carbon atoms.

The preparation of d-dihydronaphthoic acid by Pickard and Neville (Trans., 1905, **87**, 1763) led to the anticipation that probably this 2 : 3-dihydro-3-methylindene-2-carboxylic acid could be resolved in a similar manner and that a determination of its molecular rotation might be of some interest. This has been done by the fractional crystallisation of the l-menthylamine salt. Since the dihydromethylindenecarboxylic acid contains two asymmetric carbon atoms with different groups attached to each, four optically active isomerides are theoretically possible, but from the yield obtained of a pure dextrorotatory acid it would seem that only two of these are formed in any quantity during the reduction of the methylindenecarboxylic acid. The fact that acids of this type give one pair of isomerides in much larger quantity than the other has also been noticed in the case of phenylparaconic acid, details of the resolution of which it is hoped will be published shortly. The molecular rotation $[M]_D$ 118·41° might belong to either of the theoretically possible dextrorotatory isomerides, but since only one was separated it cannot be said to which it should be ascribed.

It may be pointed out that only in one or two cases has an optically active primary amine proved of service for the resolution of inactive acids. l-Menthylamine, however, appears to be particularly well adapted for this purpose, as will be gathered from the paper on dihydronaphthoic acid (Pickard and Neville, *loc. cit.*) and the present communication. Experiments directed towards the resolution of other analogous acids are in progress in which this base is used as the resolving agent.

2 : 3-*Dihydro*-3-*methylindene*-2-*carboxylic Acid.*

Ethyl benzylacetoacetate is mixed with eight to ten times its weight of concentrated sulphuric acid and the mixture warmed by the addition of a small quantity of water. After a few hours, the mixture solidifies to a crystalline mass, which is then treated with a large volume of water and filtered off. The acid so obtained is dissolved in sodium carbonate solution and reduced by means of sodium amalgam. The reduction takes place easily and a good yield of dihydro-acid is obtained, which, after crystallisation from water, melts at 82°,

l-*Methylamine* 2 : 3-*Dihydro-3-methylindene-2-carboxylate*,
$$C_{10}H_{11}\cdot CO_2H, C_{10}H_{19}\cdot NH_2.$$

Ethereal solutions of the foregoing dihydro-acid and *l*-menthylamine are mixed and the mixture heated for some time on a water-bath. The ether is then distilled off and the viscous mass desiccated, when in the course of a few days it sets to a hard, dark mass. The salt is insoluble in water, but very soluble in all the common organic media except ethyl acetate, from which it can be crystallised. Owing to the dark colour of the substance and the readiness with which resolution takes place when it is crystallised from ethyl acetate, no polarimetric observations of the racemic salt were obtained. After one crystallisation from ethyl acetate, it melted at 152°.

0·2513 gave 9·4 moist nitrogen at 16° and 752 mm.   N = 4·32.
$C_{21}H_{33}O_2N$ requires N = 4·22 per cent.

l-*Menthylamine* d-2 : 3-*Dihydro-3-methylindene-2-carboxylate.*

When the racemic salt described above is crystallised five or six times from ethyl acetate, the *lBdA*-salt is obtained in a pure state, as is shown by consecutive crystallisations giving fractions with a constant rotation. The pure salt crystallises in long, white needles which melt at 170°; it is soluble in most of the ordinary organic media, sparingly so in ethyl acetate and ether, and insoluble in water. The crystals do not lose weight at 100°.

0·2115 gave 0·5931 CO_2 and 0·2500 H_2O.   C = 76·47 ; H = 13·13.
0·2083 ,,   7·6 c.c. moist nitrogen at 14° and 758 mm.   N = 4·28.
$C_{21}H_{33}O_2N$ requires C = 76·13 ; H = 13·09 ; N = 4·23 per cent.

The following polarimetric observations * on successive fractions were made :

0·3802, made up to 20 c.c. with absolute alcohol, gave a +1·04°, whence $[a]_D$ +27·35° and $[M]_D$ +90·52°.
0·2511, made up to 20 c.c. with absolute alcohol, gave a +0·68°, whence $[a]_D$ +27·08° and $[M]_D$ +89·63°.

d-2 : 3-*Dihydro-3-methylindene-2-carboxylic Acid.*

When the pure *lBdA*-salt is shaken with ether and caustic soda and the alkaline solution separated and acidified, the pure dextrorotatory acid gradually crystallises out in the form of long, flat needles which

* All these observations were made in decimetre tubes.

melt at 84°. The acid is insoluble in water, but very soluble in the ordinary organic media. It does not lose weight when dried at 100°.

0.2431 gave 0.8145 CO_2 and 0.1787 H_2O. C=74.55; H=6.61.
$C_{11}H_{10}O_2$ requires C=75.00: H=5.81 per cent.

The following determinations of rotatory power were made:

0.2731, made up to 20 c.c. with absolute alcohol, gave α +2.53°, whence $[\alpha]_D$ +67.25° and $[M]_D$ +113.41°.

0.2014, made up to 20 c.c. with absolute alcohol, gave α +1.36°, whence $[\alpha]_D$ +67.68° and $[M]_D$ +113.09°.

0.2473, made up to 20 c.c. with benzene, gave α +1.90°, whence $[\alpha]_D$ +76.85° and $[M]_D$ +133.27°.

0.1456, made up to 20 c.c. with toluene, gave α +1.75°, whence $[\alpha]_D$ +89.23° and $[M]_D$ +137.38°.

Boiled with a large excess of caustic soda or dilute sulphuric acid for four hours, practically no racemisation took place, a small alteration in rotatory power being probably due to some slight decomposition, as the solution deepened considerably in colour.

The sodium, potassium, and barium salts are soluble in water, the silver and lead salts insoluble. The barium salt crystallises from alcohol in needles and gave the following rotation:

0.4221, made up to 20 c.c. with water, gave α +0.49°, whence $[\alpha]_D$ +24.04° and $[M]_D$ +115.49°.

The methyl ester, prepared by saturating a methyl-alcoholic solution of the acid with hydrogen chloride, was obtained as a crystalline solid melting at 69°, which gave the following analytical and polarimetric results.

0.1527 gave 0.4143 CO_2 and 0.1013 H_2O. C=75.55; H=7.42.
$C_{12}H_{12}O_2$ requires C=75.78; H=7.35 per cent.
0.2151, made up to 20 c.c. with alcohol, gave α +1.36°, whence $[\alpha]_D$ 63.22° and $[M]_D$ +130.11°.

1:2:3-Dihydro-3-methylindene-2-carboxylic Acid.

When the mother liquors obtained in the crystallisation of the dextrorotatory acid were worked up, they gave, as the most soluble fraction of the menthylamine salt, a viscid, brown syrup, which, even after prolonged desiccating, would not solidify. The salt was therefore treated with ether and caustic soda solution, and the acid precipitated from the alkaline solution. The acid thus obtained has $[\alpha]_D$ −46.0° in alcoholic solution. On crystallising several times from aqueous alcohol, a small quantity of pure l-acid was obtained.

The acid crystallises in long, flat needles and melts at 85°. In general properties, it is exactly similar to the corresponding d-acid.

0·1817 gave 0·4990 CO_2 and 0·1120 H_2O. $C = 74·89$; $H = 6·84$.
$C_{11}H_{12}O_2$ requires $C = 75·00$; $H = 6·81$ per cent.

The following polarimetric observations were made :

0·2565, made up to 20 c.c. with absolute alcohol, gave $\alpha -1·71°$, whence $[\alpha]_D -66·66°$ and $[M]_D -117·32°$.

0·1641, made up to 20 c.c. with benzene, gave $\alpha -1·24°$, whence $[\alpha]_D -75·56°$ and $[M]_D -132·98°$.

A mixture made by dissolving equal quantities of the pure d- and l-acids in alcoholic solution and evaporating the solution to dryness gave, after crystallisation, the racemic acid melting at the same temperature, namely, 82°.

Although indications of the presence of the other theoretically possible isomerides were obtained, sufficient material was not available for their isolation.

COUNTY LABORATORIES,
CHELMSFORD.

XLIII—The Condensation of Dimethyldihydroresorcin and of Chloroketodimethyltetrahydrobenzene with Primary Amines. Part II. Diamines.—m- and p-Phenylenediamine.

By PAUL HAAS, D.Sc., Ph.D.

IN a previous communication (this vol., p. 187), the condensation of dimethyldihydroresorcin and of chloroketodimethyltetrahydrobenzene with primary monamines was described, and it was there shown that these two substances condensed directly with 1 and 2 molecules respectively of a monamine. When molecular proportions of the resorcin and a primary diamine are heated together in alcoholic solution they interact to give an 80 per cent. yield of the simple condensation product (I) formed from 1 molecule of each constituent.

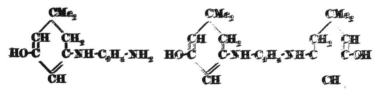

melt at 86°.   The acid is insoluble in water, but very soluble in the ordinary organic media.   It does not lose weight when dried at 100°.

0·2968 gave 0·8146 CO_2 and 0·1767 H_2O.   C = 74·85 ; H = 6·61.
$C_{11}H_{12}O_2$ requires C = 75·00 ; H = 6·81 per cent.

The following determinations of rotatory power were made :

0·3760, made up to 20 c.c. with absolute alcohol, gave a +2·53°, whence $[a]_D$ +67·28° and $[M]_D$ +118·41°.

0·2010, made up to 20 c.c. with absolute alcohol, gave a +1·36°, whence $[a]_D$ +67·66° and $[M]_D$ +119·08°.

0·2472, made up to 20 c.c. with benzene, gave a +1·90°, whence $[a]_D$ +76·86° and $[M]_D$ +135·27°.

0·1959, made up to 20 c.c. with toluene, gave a +1 75°, whence $[a]_D$ +89·33° and $[M]_D$ +157·22°.

Boiled with a large excess of caustic soda or dilute sulphuric acid for four hours, practically no racemisation took place, a small alteration in rotatory power being probably due to some slight decomposition, as the solution deepened considerably in colour.

The *sodium*, *potassium*, and *barium* salts are soluble in water, the *silver* and *lead* salts insoluble.   The *barium* salt crystallises from alcohol in needles and gave the following rotation :

0·4121, made up to 20 c.c. with water, gave a +0·99°, whence $[a]_D$ +24·02° and $[M]_D$ +125·62°.

The *methyl* ester, prepared by saturating a methyl-alcoholic solution of the acid with hydrogen chloride, was obtained as a crystalline solid melting at 68°, which gave the following analytical and polarimetric results.

0·1527 gave 0·4248 CO_2 and 0·1016 H_2O.   C = 75·86 ; H = 7·39.
$C_{12}H_{14}O_2$ requires C = 75·78 ; H = 7·36 per cent.

0·2151, made up to 20 c.c. with alcohol, gave a +1·36°, whence $[a]_D$ 63·22° and $[M]_D$ +120·11°.

l-2 : 3-Dihydro-3-methylindene-2-carboxylic Acid.

When the mother liquors obtained in the crystallisation of the dextro-rotatory acid were worked up, they gave, as the most soluble fraction of the menthylamine salt, a viscid, brown syrup, which, even after prolonged desiccating, would not solidify.   The salt was therefore treated with ether and caustic soda solution, and the acid precipitated from the alkaline solution.   The acid thus obtained has $[a]_D$ − 46·0° in alcoholic solution.   On crystallising several times from aqueous alcohol, a small quantity of pure *l*-acid was obtained.

The acid crystallises in long, flat needles and melts at 86°.   In general properties, it is exactly similar to the corresponding d-acid.

0·1817 gave 0·4990 CO_2 and 0·1120 H_2O.   C = 74·89 ; H = 6·84.

$C_{11}H_{12}O_2$ requires C = 75·00 ; H = 6·81 per cent.

The following polarimetric observations were made :

0·2565, made up to 20 c.c. with absolute alcohol, gave a − 1·71°, whence $[a]_D$ − 66·66° and $[M]_D$ − 117·32°.

0·1641, made up to 20 c.c. with benzene, gave a − 1·24°, whence $[a]_D$ − 75·56° and $[M]_D$ − 132·98°.

A mixture made by dissolving equal quantities of the pure d- and l-acids in alcoholic solution and evaporating the solution to dryness gave, after crystallisation, the racemic acid melting at the same temperature, namely, 82°.

Although indications of the presence of the other theoretically possible isomerides were obtained, sufficient material was not available for their isolation.

County Laboratories,
    Chelmsford.

XLIII.—*The Condensation of Dimethyldihydroresorcin and of Chloroketodimethyltetrahydrobenzene with Primary Amines. Part II. Diamines.*—m- and p-Phenylenediamine.

By Paul Haas, D.Sc., Ph.D.

In a previous communication (this vol., p. 187), the condensation of dimethyldihydroresorcin and of chloroketodimethyltetrahydrobenzene with primary monamines was described, and it was there shown that these two substances condensed directly with 1 and 2 molecules respectively of a monamine. When molecular proportions of the resorcin and a primary diamine are heated together in alcoholic solution they interact to give an 80 per cent. yield of the simple condensation product (I) formed from 1 molecule of each constituent.

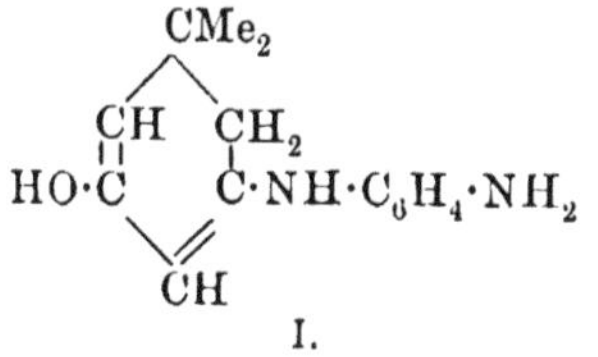

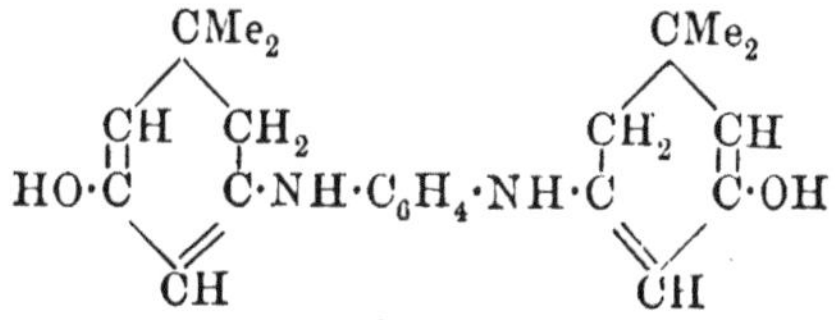

In addition to this there is, however, also formed a small amount of the compound II, resulting from the condensation of 2 molecules of the resorcin with 1 of the diamine. The compounds of the formula I, which will be referred to as the monometa- or monopara-compounds, are, unlike the original diamines, quite stable substances; they are both colourless when pure, but the para-compound slowly assumes a reddish-yellow tint on exposure to daylight; they function as di-acid bases, giving rise to dihydrochlorides which react with one molecular proportion of platinic chloride to form platinum salts. The latter tenaciously retain a certain amount of alcohol even when dried in a vacuum, and only part with it completely when heated to 120°. The bases are insoluble in water, but dissolve in alcohol to form neutral solutions which produce with ferric chloride a reddish-brown coloration; on acetylation by means of acetic anhydride, they yield monoacetyl derivatives only, which still give a colour reaction with ferric chloride. In the previous communication, it was shown that the introduction of an acetyl group into a monoamino-derivative destroyed the ferric chloride colour reaction of the free base by causing the originally hydroxylic oxygen atom in the compound to become ketonic; in the case of these substances, however, the acetyl group is not able to exert the same influence on the oxygen atom, since it is the free amino-nitrogen which is acetylated and not the nitrogen atom directly attached to the resorcin complex; this fact was proved in the case of the monopara-compound by showing that its acetyl derivative could also be prepared by the condensation of dimethyldihydroresorcin with p-aminoacetanilide.

When either the monometa- or the monopara-compound is boiled with hydrochloric acid, it is hydrolysed, giving rise to the hydrochloride of the corresponding m- or p-phenylenediamine and s-bisresorcyl-m- or -p-phenylenediamine (II); the latter substance may also be prepared by the condensation of the mono-derivative with a second molecule of the resorcin. These di-substituted phenylene-diamines are also di-acid bases having a neutral reaction; they are not acetylated by boiling with acetic anhydride and may be recrystallised without change from glacial acetic acid; in alcoholic solution they give with ferric chloride a reddish-yellow colour.

Chloroketodimethyltetrahydrobenzene condenses at once with 2 molecules of m-phenylenediamine to give the hydrochloride of a base having the formula III.

$$\text{H}_2\text{N·C}_6\text{H}_4\text{·N:C}\overset{\displaystyle \overset{\text{CMe}_2}{\diagup\diagdown}}{\underset{\underset{\text{CH}}{\diagdown\diagup}}{\overset{\text{CH}_2 \quad \text{CH}_2}{}}}\text{C·NH·C}_6\text{H}_4\text{·NH}_2$$

III.

This hydrochloride is only very slightly soluble in boiling water, giving a neutral solution from which the free base is precipitated in a crystalline form on the addition of potassium hydroxide; the base, which may be referred to as the dimeta-compound, is insoluble in water, and dissolves readily in alcohol, forming a strongly alkaline solution. On mixing the latter with an alcoholic solution of dimethyldihydroresorcin, a buff-coloured precipitate is at once formed which is a molecular combination of the base with the resorcin; it has the same percentage composition, but twice the molecular weight of the mono-meta-compound; it dissolves readily in glacial acetic acid and on treatment with potassium hydroxide gives a precipitate of the dimeta-base, thus showing that it is a salt; it is insoluble in ordinary organic solvents except boiling methyl alcohol, and even in this case the process of solution is accompanied by decomposition. It was not found possible to convert this salt into a true condensation product by elimination of the elements of water.

EXPERIMENTAL.

Action of Dimethyldihydroresorcin on m-Phenylenediamine.

A solution of 14 grams of dimethyldihydroresorcin (1 mol.) and 10 grams of freshly distilled m-phenylenediamine (1 mol.) in absolute alcohol was boiled for two hours on the water-bath; a portion of the alcohol was then distilled off and the brown solution A set aside to crystallise; 16·5 grams of a yellowish-pink solid separated, which, after washing with cold alcohol and powdering in a mortar, were recrystallised from alcohol.

0·1249 gave 0·3348 CO_2 and 0·0954 H_2O. $C = 73\cdot10$; $H = 8\cdot48$.

0·1302 „ 14·1 c.c. moist nitrogen at 18·5° and 747 mm. $N = 12\cdot28$.

$C_{14}H_{18}ON_2$ requires $C = 73\cdot05$; $H = 7\cdot83$; $N = 12\cdot18$ per cent.

5-$Hydroxy$-3-m-$aminophenylamino$-1 : 1-$dimethyl$-$\Delta^{3:5}$-$dihydrobenzene$,

$CMe_2 {<}^{CH_2 \cdot C(NH \cdot C_6H_4 \cdot NH_2)}_{CH \xrightarrow{\hspace{1cm}} C(OH)} {>} CH$ (monometa-compound), prepared

as above, is a pink substance which crystallises from alcohol in flattened needles, or more commonly in hard, dome-shaped aggregates, and melts at 234—234·5°; it dissolves only with difficulty in hot methyl or ethyl alcohol, although the solubility is somewhat increased by first powdering the crystals; when, however, it is dissolved, it does not separate out again without considerably concentrating the solution; it dissolves also in boiling acetone or pyridine, but is insoluble in ethyl acetate, chloroform, ether, or benzene. The crystals after prolonged boiling in alcoholic solution with animal charcoal lose their pink colour, but still retain a faint yellow coloration, which can only be removed by dissolving the base in hydrochloric acid and precipitating it from solution by the addition of potassium hydroxide. An alcoholic solution of the base, which is neutral to litmus, gives a reddish-brown coloration with ferric chloride.

Mr. A. J. Ewins very kindly undertook the determination of the molecular weight of this substance according to the method recently described by Barger and Ewins (Trans., 1905, **88**, 1756); using pyridine as the solvent and benzil as the standard, he obtained the value 217 ; $C_{14}H_{18}ON_2$ requires M.W. = 230.

The brown alcoholic mother liquors and washings from the pink crystals after boiling with animal charcoal and filtering became somewhat lighter in colour ; on precipitation with water, they yielded 5·5 grams of a yellow solid melting at 210—220°, which proved to be a mixture of the above-mentioned pink crystals with a small amount of the *s*-bisresorcyl-*m*-phenylenediamine (m. p. 268—269·5°) described on p. 392.

The *hydrochloride*, $C_{14}H_{18}ON_2,2HCl$, was prepared by dissolving the base in the least quantity of absolute alcohol and saturating in the cold with dry hydrogen chloride. On addition of ether, a pinkish-white solid was deposited which gave the following numbers on titration of its aqueous solution.

0·1251 required 8·26 c.c. $N/10$ NaOH.   HCl = 24·06.

0·1206 gave 0·1145 AgCl.   HCl = 24·17.

$C_{14}H_{20}ON_2Cl_2$ requires HCl = 24·09 per cent.

The *platinichloride*, $C_{14}H_{18}ON_2,H_2PtCl_6$, obtained by adding an alcoholic solution of platinic chloride to a warm alcoholic solution of the hydrochloride, separates from the solution on cooling in needles. The substance tenaciously retains alcohol, as shown by the low values 26·03 and 26·43 obtained on estimating the platinum in the substance when dried in a vacuum. On heating to constant weight at 120°, the following numbers were obtained :

0·2822 gave 0·0856 Pt.   Pt = 30·33.

$C_{14}H_{20}O_2N_2Cl_6Pt$ requires Pt = 30·66 per cent.

The *acetyl* derivative, $C_8H_{11}O\cdot NH\cdot C_6H_4\cdot NHAc$, was obtained by heating 1 gram of the base for ten minutes over a water-bath with 3 grams of acetic anhydride and 2 grams of glacial acetic acid. On pouring the mixture into water, an oily substance separated ; the latter, on solidifying, was crystallised from a mixture of alcohol and ethyl acetate, from which it separated in faintly yellow plates melting at 210·5—211·5°.

0·1202 gave 10·7 c.c. moist nitrogen at 18° and 758 mm.   N = 10·27.

$C_{16}H_{20}O_2N_2$ requires N = 10·29 per cent.

The substance is readily soluble in cold ethyl alcohol, but is insoluble in ethyl acetate, chloroform, ether, light petroleum, or benzene. In alcoholic solution, it gives with ferric chloride a reddish-brown coloration.

Action of Hydrochloric Acid on the Monometa-compound.

(a) *Under Atmospheric Pressure.*—Four grams of the pink monometa-compound were boiled in a reflux apparatus for twenty minutes with 7 grams of concentrated hydrochloric acid and 3 c.c. of water; the solution was then largely diluted with water and neutralised with potassium hydroxide, and the yellowish-brown precipitate thus obtained was filtered off and repeatedly extracted with hot water until the washings were practically colourless. The residue A, which weighed 2·5 grams and did not give a sharp melting point, was purified by boiling with animal charcoal and recrystallising several times from aqueous alcohol, when it melted at 268—269° and was found by means of a mixed melting point to be identical with the s-diresorcyl-m-phenylenediamine described on p. 392.

The aqueous filtrate from A was made alkaline and extracted with ether; the residue from the ethereal extract was found to distil without decomposition, giving a light yellow oil, which solidified on sowing with a crystal of m-phenylenediamine, and then melted at 60—63°; this substance was further identified as m-phenylenediamine by condensing it in alcoholic solution with chloroketodimethyltetra-hydrobenzene, when it gave the hydrochloride described on p. 393; the latter was further converted into the corresponding base (m. p. 118°, with evolution of gas), as described on p. 393.

(b) *Under Increased Pressure.*—Three grams of the base were heated in a sealed tube with 12 grams of concentrated hydrochloric acid for three hours at 180°; the contents of the tube, consisting of a brown liquid and some prismatic crystals, were dissolved in water and made alkaline with solid sodium carbonate ; a small quantity of a brown oil separated, which was filtered off and rejected. The filtrate was then evaporated to dryness, acidified with hydrochloric acid, and extracted with ether; the ethereal extract on evaporation yielded 0·5 gram of a yellowish-brown oil which slowly solidified, and was identified as $\beta\beta$-dimethylglutaric acid by the fact that when mixed with a pure specimen of this acid it did not depress its melting point.

Condensation of Dimethyldihydroresorcin with the Monometa-compound.

Five grams of the monometa-compound (p. 389) and 4 grams of dimethyldihydroresorcin were heated together in boiling alcoholic solution for twelve hours; on evaporating off the major portion of the alcohol, the solution deposited 7 grams of a yellow, crystalline solid which, after being decolorised by means of animal charcoal, was recrystallised from aqueous alcohol.

0·113 gave 0 3104 CO_2 and 0·0871 H_2O. $C = 74·91$; $H = 8·56$.

0·1126 gave 7·7 c.c. moist nitrogen at 19° and 775 mm. $N = 8·02$.

$C_{22}H_{28}O_2N_2$ requires $C = 75·00$; $H = 7·95$; $N = 7·95$ per cent.

s-*Bisdimethyldihydroresorcyl*-m-*phenylenediamine*,

$$\text{CMe}_2 \Big\langle \begin{array}{c} \text{CH}_2 \\ \text{CH:C(OH)} \end{array} \Big\rangle \text{C} \Big\langle \begin{array}{c} \text{NH·C}_6\text{H}_4\text{·NH} \\ \text{CH} \quad\quad \text{CH} \end{array} \Big\rangle \text{C} \Big\langle \begin{array}{c} \text{CH}_2 \\ \text{C(OH):CH} \end{array} \Big\rangle \text{CMe}_2,$$

separates from aqueous alcohol in colourless plates which melt at 268—269·5°; it dissolves readily in cold glacial acetic or formic acid, but is insoluble in all other organic solvents; its alcoholic solution, which is neutral to litmus, gives a marked reddish-brown colour with ferric chloride. The substance was recovered unchanged after boiling for three hours in alcoholic solution with one molecular proportion of m-phenylenediamine, showing that it would not condense any further.

The *hydrochloride*, $C_{22}H_{28}O_2N_2,2HCl$, was prepared by suspending the base in alcohol and saturating the solution in the cold with dry hydrogen chloride; on evaporating the clear yellow solution so obtained to dryness in a vacuum, a white solid remained which could not be recrystallised without undergoing decomposition; it was therefore analysed without further purification.

0·1256 gave 0·0856 AgCl.   HCl = 17·32.

$C_{22}H_{30}O_2N_2Cl_2$ requires HCl = 17·17 per cent.

The substance is readily soluble in cold alcohol; when this solution is diluted with water, the free base is precipitated, whilst the hydrochloric acid is quantitatively liberated and may be estimated by titration.

0·1303 required 6·08 c.c. N/10 NaOH.·   HCl = 17·09.

0·128   „   5·94 c.c. N/10 NaOH.   HCl = 16·93.

Action of Chloroketodimethyltetrahydrobenzene on m-*Phenylenediamine.*
Seven grams (1 mol.) of chloroketodimethyltetrahydrobenzene and 9 grams (2 mols.) of m-phenylenediamine dissolved in alcohol were heated over a water-bath; a yellow, crystalline precipitate of the hydrochloride (p. 393) began to separate almost immediately; after heating for one and a half hours, the precipitate was filtered off, washed with alcohol, and dried on a porous tile; its weight was 12·7 grams. When dissolved in a large volume of boiling water and poured in a thin stream into a dilute solution of potassium hydroxide,

a buff-coloured, crystalline precipitate was produced, which, after washing free from alkali and drying on a porous tile, was recrystallised from aqueous alcohol. The following numbers were obtained on analysis of the substance heated to constant weight at 105—110°.

0·138 gave 0·378 CO_2 and 0·1012 H_2O. C = 74·70 ; H = 8·14.

0·1122 ,, 17 c.c. moist nitrogen at 18° and 748 mm. N = 17·19.

$C_{20}H_{24}N_4$ requires C = 75·00 ; H = 7·50 ; N = 17·50 per cent.

5-m-*Aminophenylimino*-3-m-*aminophenylamino*-1 : 1-*dimethyl*-Δ^3-*tetrahydrobenzene* (dimeta-compound),

$$CMe_2 \overset{CH_2 \cdot C(NH \cdot C_6H_4 \cdot NH_2)}{\underset{CH_2 \cdot C(:N \cdot C_6H_4 \cdot NH_2)}{\Big\langle}} CH,$$

is an extremely hygroscopic substance, which crystallises from dilute alcohol in buff-coloured, flat needles which melt at 118—120° with evolution of gas. The substance when dried in a vacuum still retained a considerable amount of moisture; for analysis it was weighed in a boat contained in a weighing bottle and heated to constant weight at 105°; when thus dried, the residue melts at 148—150°, but on exposure to air it recovers its original melting point of 118°; it is readily soluble in cold ethyl alcohol, acetone, or ethyl acetate, sparingly so in hot ether, chloroform, or benzene, and is insoluble in water; in alcoholic solution, the substance has a strongly alkaline reaction; it yields an oily picrate and an amorphous platinum salt, neither of which was analysed.

The *hydrochloride*, $C_{20}H_{24}N_4$,HCl, obtained by the condensation of chloroketodimethyltetrahydrobenzene with *m*-phenylenediamine as above described, was crystallised from alcohol.

0·1306 gave 0·0504 AgCl. Cl = 9·49.

$C_{20}H_{25}N_4Cl$ requires Cl = 9·95 per cent.

The salt is very slightly soluble in a large volume of boiling alcohol and separates from this solution in oblong plates; it dissolves with difficulty in a large volume of boiling water to give a neutral solution.

Reaction of the Dimeta-compound with Dimethyldihydroresorcin.

On mixing together alcoholic solutions containing equal weights of the dimeta-compound and dimethyldihydroresorcin, a yellow, crystalline precipitate was at once formed ; this substance, when washed with alcohol, was dried in a vacuum.

0·1164 gave 0·3119 CO_2 and 0·0895 H_2O. C = 73·08 ; H = 8·54.

0·1324 ,, 13·9 c.c. moist nitrogen at 17° and 769 mm. N = 12·33.

$C_{28}H_{36}O_2N_4$ requires C = 73·05 ; H = 7·83 ; N = 12·18 per cent.

5-m-*Aminophenylimino*-3-m-*aminophenylamino*-1 : 1-*dimethyl*-Δ^3-*tetra*-

hydrobenzene   resorcylate, $CMe_2\Big\langle\begin{array}{l}CH_2 \cdot C(NH \cdot C_6H_4 \cdot NH_2), C_8H_{12}O_2\\ \qquad\qquad\qquad\qquad\quad CH\\ CH_2 \cdot C(\colon N \cdot C_6H_4 \cdot NH_2)\end{array}$

prepared as above, forms fine, yellow needles, which darken at 258° and melt with evolution of gas at 260—261°; it is insoluble in ethyl acetate, acetone, chloroform, light petroleum, or benzene, is slightly soluble with decomposition in boiling methyl or ethyl alcohol, and readily dissolves in cold glacial acetic or formic acid. A solution in glacial acetic acid, on making alkaline with potassium hydroxide, gave a precipitate of the dimeta-compound, which was characterised by its melting point (118°) and by the fact that when mixed with a sample of the pure substance it did not depress its melting point.

Action of Dimethyldihydroresorcin on p-Phenylenediamine.

Ten grams of dimethyldihydroresorcin and 7 grams of p-phenylene-diamine were boiled together in alcoholic solution for three hours; after distilling off some of the alcohol, the solution was allowed to cool, when it deposited a mass of light yellow crystals, which, when filtered off and dried, weighed 13 grams; these crystals were boiled with alcohol and filtered from about 1 gram of a yellow, insoluble solid A; the filtrate, on addition of light petroleum, deposited a mass of faintly yellow, flat needles.

0·1253 gave 0·3358 CO_2 and 0·0904 H_2O. C = 73·09; H = 8·01.
0·1096 ,, 11·5 c.c. moist nitrogen at 19° and 758 mm. N = 12·03.
$C_{14}H_{18}ON_2$ requires C = 73·05; H = 7·83; N = 12·18 per cent.

5-Hydroxy-3-p-aminophenylamino-1 : 1-dimethyl-$\Delta^{3 \;:\; 5}$-dihydrobenzene,

$CMe_2\Big\langle\begin{array}{l}CH_2 \cdot C(NH \cdot C_6H_4 \cdot NH_2)\\ CH\!=\!\!=\!\!=\!\!=\!\!=C(OH)\end{array}\Big\rangle CH$ (monopara-compound), crystal-

lises from aqueous alcohol or a mixture of alcohol and ligroin in colourless needles and melts at 209—210°; it dissolves readily in cold alcohol, is fairly soluble in hot acetone, ethyl acetate, or chloroform, and is slightly soluble in boiling water; its solution in alcohol is neutral and gives with ferric chloride a reddish-brown colour. Although colourless when quite pure, the substance slowly assumes a reddish-yellow colour on exposure to light.

A determination of the molecular weight by the freezing point method, using diphenylamine as solvent, gave the following result :

0·304, in 26·3596 diphenylamine, gave $\Delta t - 0·41$. M.W. = 247.
$C_{14}H_{18}ON_2$ requires M.W. = 230.

The yellow, insoluble substance A, mentioned above, proved to be identical with the s-bisresorcyl-p-phenylenediamine described on p. 396.

The *hydrochloride*, $C_{14}H_{18}ON_2,2HCl$, obtained as a white, crystalline precipitate on saturating an alcoholic solution of the base with dry hydrogen chloride, gave the following numbers on titration in aqueous solution :

0·1258 required 8·1 c.c. $N/10$ NaOH.   HCl = 23·50.

$C_{14}H_{20}ON_2Cl_2$ requires HCl = 24·09 per cent.

The *platinichloride*, $C_{14}H_{18}ON_2,H_2PtCl_6$, is obtained in the form of golden-yellow, glistening plates on mixing warm alcoholic solutions of the foregoing hydrochloride and platinic chloride.   The crystals were analysed after heating to constant weight at 110°.

0·1751 gave 0·0524 Pt.   Pt = 29·92.

$C_{14}H_{20}ON_2Cl_6Pt$ requires Pt = 30·66 per cent.

The *acetyl* derivative, $C_8H_{11}O \cdot NH \cdot C_6H_4 \cdot NHAc$, was prepared by heating 2 grams of the base with 6 grams of acetic anhydride and 4 grams of glacial acetic acid for ten minutes over a water-bath.   On pouring into water, a light yellow solid was precipitated, which crystallised from a mixture of alcohol and acetone in oblong plates melting at 255—256°.

0·1219 gave 10·8 c.c. moist nitrogen at 19° and 757 mm. N = 10·15.

$C_{16}H_{20}O_2N_2$ requires N = 10·29 per cent.

The compound is fairly soluble in hot alcohol, acetone, or chloroform, and is insoluble in ethyl acetate, benzene, or ether ; it gives a reddish-yellow colour with ferric chloride in alcoholic solution.   The same substance was also obtained by condensing dimethyldihydroresorcin in alcoholic solution with *p*-aminoacetanilide, showing that in the monopara-compound it was the nitrogen in the para-position to the resorcin complex which had been acetylated.

The monopara-compound did not yield a diacetyl derivative even on boiling with acetic anhydride and acetic acid.

Action of Hydrochloric Acid on the Monopara-compound.

A solution of 4 grams of the monopara-compound in 9 grams of concentrated hydrochloric acid and 10 c.c. of water was boiled in a reflux apparatus for a quarter of an hour ; on diluting the solution and making it faintly alkaline, a yellow solid was precipitated ; this substance, after extraction with boiling water and drying, weighed 2 grams ; it was identified as the *s*-bisresorcyl-*p*-phenylenediamino described on p. 396 by the fact that it did not melt at 300° and by its insolubility in all ordinary organic solvents.

The alkaline filtrate from this substance slowly deposited 0·8 gram of unchanged monopara-base ; the mother liquors were thereupon made strongly alkaline with potassium hydroxide and extracted with ether; the ethereal extract on evaporation yielded 0·3 gram of a solid which melted between 130° and 140° and was proved to be p-phenylenediamine by the following tests for this substance : (a) it dissolved in hydrochloric acid to give a brown solution, which was turned violet by ferric chloride, and (b) on heating with potassium dichromate and sulphuric acid it gave p-benzoquinone.

Condensation of Dimethyldihydroresorcin with the Monopara-compound.

A solution of 2 grams of the monopara-base and 1·5 grams of dimethyldihydroresorcin in alcohol was boiled over a water-bath for twelve hours ; the solution was then filtered from a yellow, crystalline precipitate, which on drying weighed only 0·6 gram ; on further heating, the solution deposited more of this precipitate, but the reaction was very slow. The substance, which was insoluble in nearly all ordinary organic solvents, was purified by dissolving in a large volume of boiling glacial acetic acid and precipitating it by means of water.

0·1100 gave 0·3005 CO_2 and 0·0833 H_2O.   C = 74·50 ; H = 8·41.

0·1112 „   7·6 c.c. moist nitrogen at 17° and 761 mm. N = 7·94.

$C_{22}H_{28}O_2N_2$ requires C = 75·00 ; H = 7·95 ; N = 7·95 per cent.

s-*Bisdimethyldihydroresorcyl-p-phenylenediamine* (for formula, compare the corresponding meta-compound described on p. 392), prepared as above, forms light yellow needles which do not melt below 300° ; it is insoluble in all ordinary organic solvents.

The *hydrochloride*, $C_{22}H_{28}O_2N_2,2HCl$, was obtained in a manner similar to the one employed in preparing the corresponding meta-derivative described on p. 392 ; it cannot be recrystallised without undergoing hydrolysis and yielding the free base. Dissolved in aqueous alcohol and titrated by means of sodium hydroxide, using phenolphthalein as indicator, it gave the following numbers :

0·1274 required 6·02 c.c. $N/10$ NaOH.· HCl = 17·26.

0·1303 „   6·18 c.c. $N/10$ NaOH. HCl = 17·32.

$C_{22}H_{30}O_2N_2Cl_2$ requires HCl = 17·17 per cent.

CHEMICAL LABORATORY,
  ST. THOMAS'S HOSPITAL, LONDON, S.E.

XLIV.—Silicon Researches. Part X. Silicon Thiocyanate, its Properties and Constitution.

By J. Emerson Reynolds.

SILICON thiocyanate was obtained by Miquel in 1877 (*Ann. Chim. Phys.*, [v], II, 343) by heating to about 350° a mixture of lead thiocyanate and silicon tetrachloride, when a brown, oily liquid distilled over which quickly solidified to a crystalline mass. In this process, charring occurs even when small quantities of 10 or 12 grams are distilled, and an impure product results.

As I desired to attempt to settle the constitution of this substance, it was advisable to avoid a high temperature in its preparation, and, finding that silicon thiocyanate is soluble in perfectly dry benzene, it seemed probable that interaction between the materials could be effected in presence of that solvent and the compound be thus obtained at once in a pure state. This turned out to be the case, but it was also found that a much larger proportion of the lead salt must be used than is required by the equation :

$$SiCl_4 + 2Pb(SCN)_2 = 2PbCl_2 + Si(SCN)_4.$$

The reason for this is that the lead salt is not converted into the simple chloride in the first instance, but that there is the intermediate production of the compound $Pb(SCN)Cl$,* which latter is then converted very slowly and partially into the chloride by prolonged treatment ; hence it is better to begin with a considerable excess of lead thiocyanate as in the following case.

One hundred grams of dehydrated and very finely-divided lead thiocyanate were diffused through about 400 c.c. of benzene contained in a flask provided with a reflux condenser ; 17 grams of silicon chloride were then added and after brisk agitation for some time the mixture was heated on a water-bath. As the interaction proceeded, the lead salt assumed a pale yellow colour owing to the formation of the above-mentioned chlorothiocyanate. The treatment was continued until the whole of the silicon chloride had interacted, as evidenced by the cessation of fuming. The hot liquid was rapidly filtered into a distilling flask and much of the benzene thence removed by distillation ; when the residual solution cooled, a fine crop of small, colourless crystals of pure silicon thiocyanate separated.

During this process it is necessary to exclude carefully any moisture

* A similar compound is easily produced by digesting lead thiocyanate with an alkaline chloride.

as the thiocyanate is easily decomposed and speedily becomes yellow owing to the separation of perthiocyanic acid, hence all the benzene is best removed by gentle heating in a current of dry hydrogen.   Once free from benzene, however, the thiocyanate changes very slowly in nearly dry air.

Silicon thiocyanate prepared as above from its benzene solution forms small, apparently trimetric, prisms ; it melts at 143·8° (corr.) and a clear, pale yellow-coloured liquid results.   This liquid when further heated deepens in colour, becoming somewhat brown, and then distils, giving a distillate which quickly crystallises in the receiver.   Miquel states (loc. cit.) that the thiocyanate boils " vers 300° degrés," but I find that the corrected boiling point is 314·2°, and this result was checked by comparison with pure diphenylamine under exactly the same conditions.   The crystals obtained by distillation were identical in composition with those from the benzene solution, as they afforded respectively 10·68 and 10·65 per cent. of silicon : $Si(SCN)_4$ requires $Si = 10·76$.

In the absence of moisture and oxygen, silicon thiocyanate is not so easily decomposed by heat as Miquel supposed.   I passed the vapour from about 2 grams of substance very slowly through a long and narrow Jena glass tube which was heated to redness in a combustion furnace ; nearly the whole of the substance passed through the tube unchanged and condensed at the cool end outside the furnace, a little carbon disulphide was given off, and at the hottest part of the tube some white silicon nitride was left.   Very little further decomposition occurred even after passing the compound four times through the visibly red-hot tube.   At the end of the treatment, the white crystals obtained on cooling the condensed liquid were compared with the original specimen of thiocyanate, and found to be identical in general properties and to melt at the same temperature.

As the thiocyanate was thus proved to be able to withstand a temperature far beyond its boiling point without suffering material decomposition, it appeared probable that its vapour density could be obtained with sufficient accuracy to serve as a control of the molecular weight.

A Victor Meyer tube of Jena glass was filled with dry nitrogen and heated to 413° in a constant temperature air-bath provided with Callendar's excellent electrical thermometer ; the substance used in the determination was that which had been passed in the state of vapour through a strongly heated tube as described above.   The following are the data obtained :

Weight taken, 0·1671 gram.   Gas expelled, 16·2 c.c.   T. 18·5°.   Bar. 763 mm.   The vapour density is therefore 129·5 $(H = 1)$, consequently the molecular weight is 259.   Theory for $Si(SCN)_4 = 260$.

It is obvious that so good an agreement with theory could not have been obtained if any decomposition occurred at 413°, some 99° above the boiling point of the compound.

Silicon thiocyanate is stated by Miquel to be insoluble in ether, carbon disulphide, chloroform, benzene, and petroleum, and that its best solvent is "l'acide sulphocyanique en solution benzénique"; in my experience, benzene alone is a sufficiently good solvent, even in the absence of thiocyanic acid, to render the preparation of silicon thiocyanate described above quite easy and satisfactory.

Ten c.c. of a cold saturated benzene solution from which a crop of crystals of the thiocyanate had separated gave, after decomposition with water, oxidation by nitric acid, and ignition of the residue, 0·288 gram of silica ; 100 c.c. of this benzene solution therefore contained 12·5 grams of silicon thiocyanate.

Again, I find that carbon disulphide, chloroform, and even light petroleum dissolve the compound in small but sensible proportions, contrary to Miquel's experience. It is difficult for anyone not accustomed to work with silicon compounds to realise the care necessary in the removal of all traces of moisture from all solvents used with them ; a very slight hydration leads to superficial decomposition of the solid and the formation of a slight layer of a silicic acid, which much impedes the action of a solvent. In this consideration is probably to be found the explanation of the differences between the two sets of observations, as all my solvents had been specially dehydrated. Ether, free from alcohol as well as moisture, does not dissolve the thiocyanate, but gradually decomposes it, and alcohol acts rapidly. It is scarcely necessary to add that acids and alkalis readily break up the compound.

The constitution of silicon thiocyanate is the question of chief interest connected with this very curious substance, as Augustus E. Dixon has shown (Trans., 1901, 79, 541) that the somewhat analogous phosphorus "trithiocyanate" acts as if it were a tautomeric compound. Dixon's admirable and extensive work in the difficult department of organic chemistry which he has studied so successfully entitles his opinion to great weight, and I therefore readily accept his view that phosphorus trithiocyanate exhibits a marked tendency to act not only as $P(SCN)_3$, but in certain cases as $P(NCS)_3$. It was therefore a matter of considerable interest to ascertain whether silicon thiocyanate acts under favourable conditions as $Si(NCS)_4$ rather than as $Si(SCN)_4$.

The elements silicon and phosphorus, although near to each other in atomic weight, belong to two very distinct periodic groups, and their compounds even when of the same general class do not necessarily undergo similar changes ; therefore, silicon tetrathiocyanate and phosphorus trithiocyanate are not so closely related that the tautomerism which the latter substance exhibits is to be presumed to exist in the

case of the silicon compound. On the contrary, the following considerations leave little, if any, doubt that silicon and sulphur are
directly united in silicon thiocyanate, and that it does not afford any
real evidence of thiocarbimide structure under the conditions to which
it has been subjected.

The fact that silicon thiocyanate which has been prepared at
a temperature not exceeding 80° can be vaporised and heated in that
state to quite 500° without undergoing isomeric change, and without
decomposition, is good evidence that nitrogen is not directly united to
silicon as in $Si(NCS)_4$; for all other silicon compounds which are
known to include four atoms of nitrogen lose two atoms of it at
temperatures below 200°.

The action of aniline on phosphorus thiocyanate is, however, that
which has afforded Dixon the chief evidence in favour of the view
that the compound is tautomeric; consequently I have examined the
action of the same substance on silicon thiocyanate with care in order
to ascertain whether under the influence of the strong base it would
act as $Si(NCS)_4$. The test was carried out in the following manner.

Ten grams of pure silicon thiocyanate dissolved in 85 c.c. of benzene
were mixed with 28 grams (8 mols.) of pure aniline; the mixture
became warm, indicating that chemical action had taken place, but no
solid matter separated, as aniline thiocyanate is freely soluble in
benzene. No external heat was applied, and the solution was allowed
to stand at the ordinary temperature for ten days in order that the
action might be completed; at the end of that time, the liquid was
still perfectly clear, and a drop of it when shaken up with ferric
chloride gave the red thiocyanate reaction very strongly. The
solution was now diluted with about three volumes of benzene and
boiled in a reflux apparatus; after a short time, a little of the liquid
was removed and cooled, but nothing separated. If any material quantity of a phenylthiourea had been formed, it must have separated at
this stage owing to its low solubility in cold benzene. The boiling
was continued and a white substance began to form; the process was
prolonged until no further separation of the white substance took
place; the latter was then quickly filtered off, thoroughly washed
with boiling benzene, and dried. The filtered solution gave a good
crop of crystals on cooling; much of the benzene was distilled off
and two more crops of similar crystals were obtained. The mother
liquor from the last crop contained some free aniline. The crystals,
which were mixed and recrystallised from alcohol and ultimately from
boiling water, presented all the characters of monophenylthiourea;
they gave the lead reaction strongly; when heated in a test-tube, they
gave much ammonia and a thiocarbamine, and the melting point
was 153—154°.

0·3994 gave 0·6036 $BaSO_4$. $S = 21·17$.

$CS(NH·C_6H_5)·NH_2$ requires $S = 21·5$ per cent.

The white substance which separated on prolonged heating included all the silicon and gave a very strong thiocyanic reaction with ferric chloride; it slowly dissolved in caustic alkali with separation of aniline.

0·4478 gave 0·748 SiO_2. $Si = 7·78$.

0·5404 ,, 0·0905 SiO_2 and 0·38 $BaSO_4$. $Si = 7·78$; $S = 9·65$.

$Si(N·C_6H_5)_2, NH_2·C_6H_5, HSCN$ requires $Si = 7·73$; $S = 8·83$ per cent.

The sulphur is high, as usual, in these cases owing to the imidothiocyanate precipitate carrying down with it a little of the perthiocyanic acid which is inevitably formed in small quantity on long-continued heating of the benzene solution.

The interpretation of these facts presents but little difficulty in view of the known habits of certain of the silicon compounds described in former papers of this series. Silicon thiocyanate evidently interacts with excess of aniline very much as the chloride does, although not quite so energetically, and the following equation doubtless represents the first stage :

$$Si(SCN)_4 + 8NH_2·C_6H_5 = Si(NH·C_6H_5)_4 + 4NH_2·C_6H_5, HSCN.$$

Prior to diluting and boiling the solution, no phenylated thiourea was detected, but further changes proceeded *pari passu* on heating the diluted liquid. Probably the first to begin is the gradual molecular rearrangement of aniline thiocyanate * into the more stable isomeric form of monophenylthiourea, which most of it ultimately assumes. Under the same conditions, the silicophenylamide parts with two molecules of aniline and is reduced to the di-imide, a change which has been shown in previous papers to take place easily, and the imide carries down with it a molecule of aniline thiocyanate, with which it appears to form an additive compound similar to those obtained in other cases.

Under all the conditions specified, there is therefore no doubt that silicon thiocyanate is correctly represented by the expression $Si(SCN)_4$.

THE DAVY-FARADAY LABORATORY,
ALBEMARLE STREET, LONDON.

* This change takes place much more easily than is commonly supposed. Strong alcoholic solutions of aniline hydrochloride and ammonium thiocyanate, if mixed in the cold, precipitate ammonium chloride, and aniline thiocyanate is retained in solution and remains unchanged for a considerable time. When the solution is boiled and then poured into cold water, an abundant precipitate of monophenyl-thiourea can be obtained, so that boiling in alcoholic or even aqueous solution determines the molecular change :

$$NH_2·C_6H_5, HSCN = CS{<}{\begin{matrix} NH·C_6H_5 \\ NH_2 \end{matrix}}.$$

XLV.—Studies in the Camphane Series. Part XXII. Nitrogen Halides from Camphoryl-ψ-carbamide.

By Martin Onslow Forster and Hans Grossmann, Ph.D.

The description of camphoryl-ψ-carbamide (Forster and Fierz, Trans., 1905, **87**, 110) made it clear that the substance is one of an unusual type, and although the constitutional formula ascribed to it agrees with subsequent observations, it is desirable to gain further information respecting the changes which it undergoes. We have therefore studied the behaviour of the compound towards potassium hypobromite and hypochlorite, the investigations of Chattaway and Orton having shown that anilides of various types readily yield nitrogen halides with those agents. Up to the present time, these authors have dealt with anilides, both substituted and unsubstituted, and also with symmetrical diphenylcarbamide (*Ber.*, 1901, **34**, 1073 and 1078); Chattaway has investigated diacylammonias and sulphonamides, and in association with Wadmore (Trans., 1902, **81**, 191) has studied cyanuric acid from this point of view. Quite recently, Chattaway and Lewis have prepared halogen derivatives of substituted oxamides (this vol., p. 155).

As in the case of diphenylcarbamide and the substituted oxamides, camphoryl-ψ-carbamide presents two points of attack for the halogen atom, whilst the corresponding derivative prepared from methylamino-camphor affords opportunity for the entrance of one atom only :

$$C_8H_{14}\diagdown \begin{array}{c} CH\text{---}NH \\ C(OH)\cdot NH \end{array} \diagup CO \qquad C_8H_{14}\diagdown \begin{array}{c} CH\text{---}NMe \\ C(OH)\cdot NH \end{array} \diagup CO.$$

Camphoryl-ψ-carbamide.   Camphorylmethyl-ψ-carbamide.

Accordingly we find that when treated with potassium hypobromite these amides yield the compounds :

$$C_8H_{14}\diagdown \begin{array}{c} CH\text{---}NBr \\ C(OH)\cdot NBr \end{array} \diagup CO \quad \text{and} \quad C_8H_{14}\diagdown \begin{array}{c} CH\text{---}NMe \\ C(OH)\cdot NBr \end{array} \diagup CO,$$

respectively, and the corresponding nitrogen chlorides are obtained when sodium hypochlorite is used. The chemical behaviour of these substances places them unquestionably in the same category as the nitrogen halides described by previous workers. They oxidise alcohol to aldehyde and sulphites to sulphates, liberating iodine from potassium iodide, nitrogen from ammonia, sulphur from sulphuretted hydrogen, and oxygen from hydrogen peroxide, the *pseudo*carbamide being regenerated in each case. The action with potassium iodide can be utilised, as Chattaway and Orton have shown, for quantitative determination of the halogen, and the agreement of results obtained in this way with those furnished by the Carius method proves that none of the halogen enters the camphor nucleus.

In one of their later communications (Trans., 1901, **79**, 461), Chattaway and Orton showed that when nitrogen halides act on excess of phenylhydrazine the latter is converted into phenylazoimide, which they suggest is due to the intermediate formation of a hydrazino-halogen compound :

$$2C_6H_5 \cdot NH \cdot NHCl = C_6H_5 \cdot N_3 + C_6H_5 \cdot NH_2 + 2HCl.$$

The dibromide from camphoryl-ψ-carbamide behaves in the same manner, and when acting on less than one molecular proportion of phenylhydrazine it converts the base into bromobenzene, corresponding to the chlorobenzene formed when phenylhydrazine acts on excess of acetylchloroamino-2 : 4-dichlorobenzene (*loc. cit.*, p. 468). There is thus complete agreement between our compounds and those described by Chattaway and Orton.

Having established this point, it became of interest to compare the behaviour of the normal carbamide and methylcarbamide,

$$C_8H_{14} \big\langle \genfrac{}{}{0pt}{}{CH \cdot NH \cdot CO \cdot NH_2}{CO} \quad \text{and} \quad C_8H_{14} \big\langle \genfrac{}{}{0pt}{}{CH \cdot NMe \cdot CO \cdot NH_2}{CO},$$

towards hypobromite with that of the *pseudo*-compounds. We find that action takes place readily in each case; as might be expected, however, from the readiness with which alkalis convert the normal type into the *pseudo*-modification, the product is always identical with the hydrogen halide obtained from the corresponding *pseudo*-carbamide.

The production of these halogen derivatives from camphorylcarbamide furnishes another illustration of the great stability conferred on its derivatives by the camphor nucleus. The decomposition of carbamide itself by alkali hypobromite and hypochlorite has long been known to be rapid and complete; experiments with phenylcarbamide and bornylcarbamide have shown us that although potassium hypobromite acts vigorously on these substances, it is not possible under ordinary conditions to isolate a definite compound of the class under consideration.

In conclusion, it may be mentioned that, as in the case of the compounds obtained by previous workers, the stability of the nitrogen chlorides is greatly superior to that of the nitrogen bromides. The latter compounds rapidly become yellow, which is deep in the case of the dibromide, whilst the chlorides remain snow-white; moreover, owing to the absence of a benzenoid nucleus into which the halogen might wander, these derivatives can be recrystallised from warm glacial acetic acid.

EXPERIMENTAL.

Camphoryldibromo-ψ-carbamide, $C_8H_{14}\begin{matrix}CH\!-\!-\!NBr\\C(OH)\cdot NBr\end{matrix}\!>\!CO.$

Preliminary experiments on the behaviour of camphoryl-ψ-carbamide towards potassium hypobromite showed that varying concentration of the alkaline liquid gave widely different results and furnished products of indefinite composition. For example, when hypobromite of maximum concentration was employed, the carbamide formed a clear solution which became warm spontaneously, then evolved gas, and precipitated a tarry product. On using a dilute solution, however, a crystalline substance separated from the clear liquid, and after precipitation from chloroform by petroleum melted at 111°, the analytical data indicating a mixture. We accordingly adopted the use of an alkaline bicarbonate, as recommended by Chattaway and Orton, a modification which led to satisfactory results.

Forty grams of bromine were added slowly to 40 grams of potassium hydroxide in 50 c.c. of water mixed with 200 grams of crushed ice; 20 grams of camphoryl-ψ-carbamide were suspended in 100 c.c. of water and added to the hypobromite, which was cooled externally with ice. The amide rapidly dissolved, forming a clear, pale yellow solution, and to this was immediately added an ice-cold, saturated solution of sodium hydrogen carbonate, which yielded a bulky, colourless precipitate. This was washed thoroughly, drained on porous earthenware, and dissolved in ice-cold, absolute alcohol, which was then diluted with water; colourless needles separated rapidly, and were filtered immediately, as the substance quickly decomposes when left in contact with the mother liquor. When heated in a capillary tube, the freshly crystallised dibromide becomes yellow at about 100°, and detonates with great violence at 120°, leaving a charred mass; on a platinum spatula, it merely chars suddenly with a slight hissing noise. Specimens which have remained a few days in the desiccator are found to have become deep yellow, and detonate at temperatures between 110° and 115°. The yield is theoretical.

0·0934 gave 0·0949 AgBr. Br = 43·23.

0·1244 liberated 0·0865 I_2 from KI. Br = 43·81.

$C_{11}H_{16}O_2N_2Br_2$ requires Br = 43·48 per cent.

A solution containing 0·2428 gram in 25 c.c. of chloroform gave a_D 0°36′ in a 2-dcm. tube, whence $[a]_D$ 30·9°; the solution darkened rapidly on exposure to sunlight. The substance is insoluble in hot petroleum, but dissolves in boiling benzene, from which it crystallises on cooling in lustrous, transparent, hexagonal plates ; it is moderately soluble in glacial acetic acid, from which it separates in snow-white leaflets on dilution with water, and is freely soluble in cold alcohol, chloroform, ethyl acetate, and ether.

The dibromo-derivative obtained from the normal camphoryl-carbamide is identical with the substance just described. Employing the same proportions as before, it was noticed that before the carbamide dissolved completely in the hypobromite a colourless precipitate began to separate ; on adding a further quantity of ice-water, however, a clear, yellow solution was obtained, yielding immediately a bulky, colourless precipitate with sodium hydrogen carbonate. The product was dissolved in ice-cold, absolute alcohol, from which it separated in minute, lustrous needles on dilution with water ; it detonated at 120° and gave $[a]_D$ 31·8° in chloroform.

The production of the dibromide in alkaline solution is presumptive evidence in favour of its derivation from the *pseudo*carbamide rather than the normal isomeride when it is remembered that the latter is changed by alkalis into the former with great facility. Any doubt on this point is removed by the fact that the carbamide regenerated by potassium iodide, ammonia, hydrogen sulphide, sulphurous acid, and hydrogen peroxide, with all of which the dibromide acts readily, is the *pseudo*-compound, not the normal.

Although camphoryl-ψ-carbamide contains two atoms of hydrogen capable of undergoing replacement by halogens, and situated unsymmetrically as regards the complete molecule, it has not been found possible to displace one at a time ; an experiment in which only half the above proportion of hypobromite was employed led to the dibromo-derivative, the remaining *pseudo*carbamide escaping attack.

Action of Phenylhydrazine.—In the first experiment, the base was in excess, as required by the equation :

$$3C_6H_5 \cdot NH \cdot NH_2 + C_{11}H_{16}O_2N_2Br_2 = C_6H_5 \cdot N_3 + C_6H_5 \cdot NH_2,HBr +$$
$$C_{11}H_{18}O_2N_2 + C_6H_5 \cdot NH \cdot NH_2,HBr.$$

Eighteen grams of the dibromide dissolved in dry chloroform were treated with 21 grams of phenylhydrazine (4 mols.) in the same solvent, the action being moderate and accompanied by precipitation of the crystalline hydrobromides ; the filtered liquid was extracted with dilute hydrochloric acid, the chloroform being then distilled and the residue submitted to a current of steam. The pale yellow oil obtained in this way boiled at 160—170° under 760 mm. pressure, 163° being the boiling point of phenylazoimide, and furnished a nitro-derivative which melted at 71° when crystallised from petroleum, and did not depress the melting point of *p*-nitrophenylazoimide when mixed with that substance.

In the second experiment, which was expected to proceed according to the equation:

$$C_6H_5\cdot NH\cdot NH_2 + C_{11}H_{16}O_2N_2Br_2 = C_6H_5Br + C_{11}H_{18}O_2N_2 + N_2 + HBr,$$

5·4 grams of the base were added to 32 grams of the dibromide, both materials being dissolved in chloroform; a vigorous action ensued and gas was evolved, but no precipitation occurred. On removing the chloroform and passing a current of steam through the residue, bromo-benzene was obtained, and was identified by conversion into p-nitro-bromobenzene, which melted at 123° and did not depress the melting point of the pure substance.

$$\textit{Camphoryldichloro-}\psi\textit{-carbamide, } C_8H_{14} {<} {\overset{\text{CH}-\text{NCl}}{\underset{\text{C(OH)}\cdot\text{NCl}}{}}} {>} \text{CO.}$$

Twenty grams of the *pseudo*carbamide were suspended in 400 c.c. of an aqueous solution of sodium hypochlorite cooled with ice; the substance gradually dissolved, forming a pale yellow solution, and when this process was complete a concentrated solution of sodium hydrogen carbonate was added, yielding a bulky, colourless precipitate of the dichloro-derivative. The product was dissolved in ice-cold alcohol, from which it separated in minute, snow-white needles on dilution with water; it decomposed quite suddenly at 140° without detonation.

0·1588 liberated 0·1449 I_2 from KI.   Cl = 25·50.

0·3120   „   0·2801 I_2 „ KI.   Cl = 25·09.

$C_{11}H_{16}O_2N_2Cl_2$ requires Cl = 25·45 per cent.

A solution containing 0·2987 gram in 50 c.c. of chloroform gave a_D 0°19′ in a 2-dcm. tube, whence $[a]_D$ 26·5°. The superior stability of the substance is indicated not only by the higher melting point, but also by the fact that the chloroform solution may be exposed to direct sunlight during several hours without developing colour; moreover, the compound may be recrystallised from warm glacial acetic acid, which deposits beautiful, lustrous prisms on cooling. The dichloro-compound is insoluble in petroleum, but dissolves sparingly in boiling benzene, from which it crystallises in lustrous, silky needles on cooling; it is moderately soluble in warm ether and freely soluble in cold pyridine and ethyl acetate.

The same dichloro-derivative was obtained on treating Rupe's normal camphorylcarbamide with an ice-cold solution of sodium hypochlorite.

$$Camphorylmethylbromo\text{-}\psi\text{-}carbamide, \quad C_8H_{14} \underset{C(OH)\cdot NBr}{\overset{CH\text{---}NMe}{<}} > CO.$$

On treating the methyl pseudocarbamide with a quantity of hypobromite corresponding to one-half that employed in preparing the dibromo-derivative, the monobromo-compound began to separate before the carbamide had dissolved completely; sodium hydrogen carbonate was added, and the granular precipitate was washed, dried, and dissolved in ice-cold, absolute alcohol, from which it separated in lustrous, silky needles on dilution with water. The freshly-prepared substance is white and decomposes suddenly at 101° with a faint hissing noise, leaving a dark brown mass.

0·1173 gave 0·0721 AgBr. Br = 26·15.

0·1163 ,, 0·0711 AgBr. Br = 26·01.

$C_{12}H_{19}O_2N_2Br$ requires Br = 26·37 per cent.

A solution containing 0·3275 gram in 25 c.c. of chloroform gave a_D 0°15′ in a 2-dcm. tube, whence $[a]_D$ 9·5°. The compound is very sparingly soluble in boiling petroleum and in ether, more readily in hot benzene, from which it separates in silky needles; it is moderately soluble in glacial acetic acid and ethyl acetate, dissolving freely in pyridine and chloroform. .

On treating the normal camphorylmethylcarbamide originally described by Duden and Pritzkow with an ice-cold solution of potassium hypobromite, a monobromo-derivative was obtained identical in all points with the substance just described.

$$Camphorylmethylchloro\text{-}\psi\text{-}carbamide, \quad C_8H_{14} \underset{C(OH)\cdot NCl}{\overset{CH\text{---}NMe}{<}} > CO.$$

Twenty grams of the methyl-ψ-carbamide were finely powdered and suspended in 50 c.c. of the sodium hypochlorite solution, in which the substance partly dissolved; the chloro-derivative began to separate before dissolution was complete, however, and after half an hour at the temperature of the laboratory a concentrated solution of sodium hydrogen carbonate was added, precipitating the remainder in granular form. The substance was only sparingly soluble in cold alcohol, but dissolved on warming gently, and crystallised in minute, lustrous, snow-white needles which decompose suddenly at 147°.

0·1746 gave 0·0954 AgCl. Cl = 13·52.

$C_{12}H_{19}O_2N_2Cl$ requires Cl = 13·71 per cent.

The chloro-derivative is optically inactive, or so feebly rotatory that a 1 per cent. solution in chloroform has no action on polarised light.

The substance is insoluble in boiling petroleum, but dissolves moderately in boiling benzene, from which it crystallises in minute, silky needles; it is only sparingly soluble in warm ether, more readily in boiling ethyl acetate and warm glacial acetic acid, crystallising from both these solvents in long, lustrous, transparent prisms, whilst pyridine dissolves it freely.

On attempting to prepare an isomeric chloro-derivative from normal camphorylmethylcarbamide and sodium hypochlorite, there was obtained a compound identical in every respect with that derived from the *pseudo*carbamide.

ROYAL COLLEGE OF SCIENCE, LONDON,
SOUTH KENSINGTON, S.W.

XLVI.—*Note on the Application of the Electrolytic Method to the Estimation of Arsenic in Wall-papers, Fabrics, &c.*

By THOMAS EDWARD THORPE.

ONE of the questions to be considered at the forthcoming International Congress of Applied Chemistry, to be held at Rome during the spring of this year, relates to the methods which should be employed in the detection and determination of arsenic in a variety of manufactured articles, more especially in paper and woollen fabrics, in view of the fact that certain countries impose restrictions on the importation of goods containing arsenic.

As the electrolytic method which I described to the Society in 1903 (Trans., 1903, 83, 974), and which has been in constant use in this laboratory during the last three years in testing various brewing materials for arsenic, appeared to lend itself to the estimation of arsenic in paper and fabrics, I have caused a series of experiments to be made in order to ascertain whether the larger amounts of arsenic which are known to be occasionally present in such articles can be accurately and expeditiously determined by means of it. The results have shown that the method is readily applicable to the matter in question.

Where a dyed material with a design composed of more than one colour is to be tested, it is advisable to take such a portion that all the colours are represented. The selected portion may then be measured, or preferably weighed. So far as the amount of material actually taken is concerned, this must depend on the material itself and the design and colours it contains, but for fabrics in general, such as woollen goods and wall-papers, 2 grams is a suitable quantity. The weighed portion of the sample, cut into pieces of convenient size, is placed in a platinum dish, about 7·5 cm. in diameter, and moistened with hot water. When the water has been absorbed by the fabric, 20 c.c. of arsenic-free lime-water and 0·5 gram of calcined magnesia are added, the latter being stirred with a glass rod among the pieces of the fabric. The platinum dish is then placed on a hot plate or over a small Bunsen flame and the liquid evaporated. The dried material is then thoroughly charred and heated in a muffle furnace until practically all the carbon is burnt off. When cold, the ash is moistened with water and 20 c.c. of the dilute sulphuric acid (*loc. cit.*, p. 978) added. The dish is warmed and the contents transferred to a flask of about 120 c.c. capacity. Half a gram of potassium meta-bisulphite is added and the solution boiled until free from sulphurous acid. The liquid is cooled and diluted to a bulk of 50 c.c. in a calibrated flask or measuring tube. An aliquot portion may then be taken for the test.

It is found in practice that the density of deposit most suitable for purposes of comparison is equivalent to a quantity of arsenic falling between 0·005 milligram and 0·0125 milligram of arsenious oxide (As_4O_6). In order, therefore, to obtain a deposit coming within this range, 5 c.c. of the solution are added to the apparatus. If at the end of ten to fifteen minutes a deposit is being produced which may be expected to be of a density suitable for comparison with the standard deposit, the experiment is continued for the thirty minutes required for the test. Should, however, the deposit be unsuitable for purposes of comparison by being too dense, a further test is carried out on such smaller aliquot portion as may be considered suitable. On the other hand, if at the expiration of ten to fifteen minutes only a faint appearance of arsenic is produced or none at all, a further 20 or 25 c.c. of the solution may be at once added to the same apparatus and the electrolysis continued for thirty minutes, when the deposit produced will be that from the sum of the aliquot portions added.

In this way a single incineration of the material suffices for several tests. Where a larger amount of material than 2 grams is taken, and the quantity of arsenic may therefore be expected to be considerable, the solution after reduction is made up to a larger volume, say 100 c.c. or 200 c.c., and the test carried out on an aliquot portion of the liquid. The amounts of lime-water and magnesia given above have been proved by direct experiment to retain amounts of arsenic ranging from 0·0025 to 5 milligrams when contained in 2 grams of wool or paper.

The reduction of the solution with sulphurous acid before addition to the apparatus is necessary, since, under the conditions of the experiment, arsenic in the form of an arsenate or arsenic acid doe not yield hydrogen arsenide.

The amount of arsenic deposited by heating the hydrogen arsenides determined by comparison with deposits obtained in precisely the same manner from wool and paper containing known quantities of arsenic.

One gram of arsenic-free wool, or paper, is placed in a platinum dish and warmed with a small quantity of water so that the fabric is wetted through without leaving an excess of water in the dish; 1 c.c. of standard arsenic solution containing 0·01 milligram of arsenious oxide is then added from a narrow burette. The whole is treated in the manner described, and the deposit of arsenic obtained is suitably preserved for purposes of comparison as the standard depot from 0·01 milligram of As_4O_6. Other deposits are obtained similarly for 0·0025, 0·005, 0·0075, and 0·0125 milligram respectively.

In the working out of the electrolytic method for the special purpose of this investigation, a number of experiments have been carried out with the object of ascertaining whether a known amount of arsenic added to an arsenic-free material would be recovered, and what were approximately the limits within which the method as described could be employed.

In the first place, difficulty arose as to obtaining a woollen material free from arsenic. Undyed natural wool flannels and wools, even after repeated washing and after long use as wearing apparel, were found to contain arsenic. Specimens of wool were then secured from young lambs and sheep which had not been treated with an arsenical dip. This raw wool was examined for arsenic as follows : first the clean white ends of the wool next the body of the sheep were tested without washing or treatment of any kind. then the other portions of the wool, generally containing much dirt, were washed with cold and warm water containing a small quantity of ammonia, dried, and taken for the test. These results were obtained :

(1) Wool from lamb six weeks old ; the mother had been treated with arsenical dip six months before lamb was born. Wool entirely free from arsenic.

(2) Wool from lamb ; the mother had probably been dipped six months earlier. The white ends of wool were free from arsenic, the remainder of wool showed a trace—about 0·001 milligram in 4 grams of wool.

(3) Wool from lamb ; the mother had been treated with arsenical dip six months before lamb was born. White ends of wool next the body were free from arsenic ; the remainder of wool contained small quantity—about 0·001 milligram in 2 grams of wool.

(4) Wool from lamb four weeks old ; the mother had been treated with arsenical dip. Wool free from arsenic.

(5) Wool from sheep six months old ; the mother was treated with ·senical dip during the same month in which the lamb was born ; :mb had not been dipped, but had been treated with fly powder. The white ends of wool next the body contained arsenic equal to ₡04 milligram of arsenious oxide in 1 gram of wool ; the remaining ɟrtions of wool, washed, contained arsenic, equal to 0·019 milligram c arsenious oxide in 1 gram of wool.

(6) Wool from sheep nine months old, never dipped. The un-wshed wool contained a trace of arsenic, about 0·001 milligram in 4 rams.

7) Wool from sheep twenty months old, once clipped, and since tnted—three months ago—with carbolic dip. Ends of wool next the bcy contained arsenic equal to 0·047 milligram of arsenious oxide in 1 ram of wool.

3) Wool from ewe ; treated with carbolic dip in October 1904 (fieen months ago), clipped July 1905 (six months ago), not since diped. Ends of wool next the body, washed, contained arsenic equal to ·025 milligram in 1 gram of wool.

onfirmatory results were obtained on all the above specimens of wo with one exception, where the total amount of wool available wainsufficient for more than one experiment. Wool No. 1 on the abce list was used for the purpose of preparing the standard arsenic depsits for wool, and also for the various experiments carried out wit the view of ascertaining to what extent arsenic could be recovered fror woollen materials. The preparation of the standard deposits had shon that a notable deposit of arsenic was obtained from amounts of arseic so low as 0·0025 milligram in 1 gram of wool, and it was not condered necessary to carry the test to smaller quantities than 0·005 milligram. The arsenic so obtained is commensurate with the ɖposit from the same quantity of arsenic added directly to the appaatus without incineration with wool in the prescribed manner. Ther is, therefore, no loss of arsenic where small quantities are concned, and the gradation of the tubes for the standard deposits procds in proper order up to the deposit equal to 0·015 milligram of arserɔus oxide, beyond which amount the deposits are too dense for purpɔes of comparison.

Exɔriments with wool containing larger amounts of arsenic were then ırried out.

(1)ʃwo grams of arsenic-free wool were soaked with water in a platinm dish and 3 c.c. of a solution containing 0·03 milligram of arseniɯs oxide added. The whole was treated in the prescribed mannɛ the solution of the ash after reduction being made up to 50 c.c Three separate quantities of this were electrolysed, namely, 20 c.c. qual to 0·012 milligram of As_4O_6, 15 c.c. equal to 0·009 milli-

experiment, arsenic in the form of an arsenate or arsenic acid does not yield hydrogen arsenide.

The amount of arsenic deposited by heating the hydrogen arsenide is determined by comparison with deposits obtained in precisely the same manner from wool and paper containing known quantities of arsenic.

One gram of arsenic-free wool, or paper, is placed in a platinum dish and warmed with a small quantity of water so that the fabric is wetted through without leaving an excess of water in the dish ; 1 c.c. of standard arsenic solution containing 0·01 milligram of arsenious oxide is then added from a narrow burette. The whole is treated in the manner described, and the deposit of arsenic obtained is suitably preserved for purposes of comparison as the standard deposit from 0·01 milligram of As_4O_6. Other deposits are obtained similarly for 0·0025, 0·005, 0·0075, and 0·0125 milligram respectively.

In the working out of the electrolytic method for the special purpose of this investigation, a number of experiments have been carried out with the object of ascertaining whether a known amount of arsenic added to an arsenic-free material would be recovered, and what were approximately the limits within which the method as described could be employed.

In the first place, difficulty arose as to obtaining a woollen material free from arsenic. Undyed natural wool flannels and wools, even after repeated washing and after long use as wearing apparel, were found to contain arsenic. Specimens of wool were then secured from young lambs and sheep which had not been treated with an arsenical dip. This raw wool was examined for arsenic as follows : first the clean white ends of the wool next the body of the sheep were tested without washing or treatment of any kind. then the other portions of the wool, generally containing much dirt, were washed with cold and warm water containing a small quantity of ammonia, dried, and taken for the test. These results were obtained :

(1) Wool from lamb six weeks old ; the mother had been treated with arsenical dip six months before lamb was born. Wool entirely free from arsenic.

(2) Wool from lamb ; the mother had probably been dipped six months earlier. The white ends of wool were free from arsenic, the remainder of wool showed a trace—about 0·001 milligram in 4 grams of wool.

(3) Wool from lamb ; the mother had been treated with arsenical dip six months before lamb was born. White ends of wool next the body were free from arsenic ; the remainder of wool contained a small quantity—about 0·001 milligram in 2 grams of wool.

(4) Wool from lamb four weeks old ; the mother had been treated with arsenical dip. Wool free from arsenic.

(5) Wool from sheep six months old ; the mother was treated with arsenical dip during the same month in which the lamb was born ; lamb had not been dipped, but had been treated with fly powder. The white ends of wool next the body contained arsenic equal to 0·004 milligram of arsenious oxide in 1 gram of wool ; the remaining portions of wool, washed, contained arsenic, equal to 0·019 milligram of arsenious oxide in 1 gram of wool.

(6) Wool from sheep nine months old, never dipped.   The unwashed wool contained a trace of arsenic, about 0·001 milligram in 4 grams.

(7) Wool from sheep twenty months old, once clipped, and since treated—three months ago—with carbolic dip.   Ends of wool next the body contained arsenic equal to 0·047 milligram of arsenious oxide in 1 gram of wool.

(8) Wool from ewe ; treated with carbolic dip in October 1904 (fifteen months ago), clipped July 1905 (six months ago), not since dipped.   Ends of wool next the body, washed, contained arsenic equal to 0·025 milligram in 1 gram of wool.

Confirmatory results were obtained on all the above specimens of wool with one exception, where the total amount of wool available was insufficient for more than one experiment.   Wool No. 1 on the above list was used for the purpose of preparing the standard arsenic deposits for wool, and also for the various experiments carried out with the view of ascertaining to what extent arsenic could be recovered from woollen materials.   The preparation of the standard deposits had shown that a notable deposit of arsenic was obtained from amounts of arsenic so low as 0·0025 milligram in 1 gram of wool, and it was not considered necessary to carry the test to smaller quantities than 0·0025 milligram.   The arsenic so obtained is commensurate with the deposit from the same quantity of arsenic added directly to the apparatus without incineration with wool in the prescribed manner. There is, therefore, no loss of arsenic where small quantities are concerned, and the gradation of the tubes for the standard deposits proceeds in proper order up to the deposit equal to 0·015 milligram of arsenious oxide, beyond which amount the deposits are too dense for purposes of comparison.

Experiments with wool containing larger amounts of arsenic were then carried out.

(1) Two grams of arsenic-free wool were soaked with water in a platinum dish and 3 c.c. of a solution containing 0·03 milligram of arsenious oxide added.   The whole was treated in the prescribed manner, the solution of the ash after reduction being made up to 50 c.c.   Three separate quantities of this were electrolysed, namely, 20 c.c. equal to 0·012 milligram of As_4O_6, 15 c.c. equal to 0·009 milli-

gram of As_4O_6, and 10 c.c. equal to 0.006 milligram of As_4O_6. The deposit obtained in each case was equal to the proportionate amount of arsenic added.

(2) A solution of arsenic containing 0.1 milligram of As_4O_6 for each c.c. was employed for this test. One gram of wool was taken, and 5 c.c. of the arsenic solution equal to 0.5 milligram were added. The whole was then treated in the prescribed manner, the solution of the ash after reduction being made up to 100 c.c. Four separate quantities, each 2 c.c. equal to 0.01 milligram of the arsenic added, were electrolysed. All the deposits were equal to the standard 0.01 milligram deposit.

(3) A similar experiment was carried out, using, however, 2 grams of wool and 10 c.c. of the arsenic solution, equal to 1.0 milligram of arsenious oxide. The solution of the ash, after reduction, was made up to 200 c.c., and 2 c.c. representing 0.01 milligram of arsenic were used for testing. Six experiments were carried out and all the deposits were of the proper amount compared with the standard 0.01 milligram deposit.

(4) Lastly, two experiments were made, using for each 2 grams of wool and 50 c.c. of the arsenic solution, equal to 5 milligrams of As_4O_6. To neutralise the acidity of the standard arsenic solution, the wool was first treated in the dish with the requisite amount of caustic soda solution and the arsenic solution added at intervals, evaporating to dryness between each addition of the arsenic solution. The whole was then treated in the prescribed manner, using the same quantities of materials as in the previous experiments. The solutions, after reduction, were made up in each case to a litre, and 2 c.c. equal to 0.01 milligram of As_4O_6 were used for the test. The deposits show conclusively that the whole of the arsenic is recovered by the process when present to the extent of 5 milligrams of As_4O_6.

So far, therefore, as the experiments with wool are concerned, the method gives rapidly and accurately the whole of the arsenic when present to so great an amount as 5 milligrams.

Paper.—Experiments with paper were carried out in a similar manner. Of papers tested, the only class of paper found to be free from arsenic was filter paper.

A series of standard deposits was prepared using 1 gram of arsenic-free filter paper to which definite amounts of arsenic were added in the manner described in connection with wool.

The series so formed shows that with quantities of arsenic varying from 0.0025 milligram to 0.015 milligram there is no loss of arsenic when these deposits are compared with the deposits obtained from the corresponding amounts of arsenic added directly to the apparatus.

Experiments with paper containing higher amounts of arsenic were carried out on similar lines to those already described in connection with wool, and in every case the amount of arsenic added was recovered by the process described.

In addition to the raw wools, of which particulars have been given, various specimens of materials have been examined by the electrolytic method as to the quantity of arsenic present in them, and a table is annexed giving the results in those cases where an estimate was made as to the amount.

Table showing the Amounts of Arsenic found by the Electrolytic Method in Various Fabrics and Papers.

Materials.	Mgm. of As$_4$O$_6$ for 1 gram of material.	Materials.	Mgm. of As$_4$O$_6$ for 1 gram of material.
1. Flannel No. 2 (natural wool)	0·005	10. Vest wool (undyed)	0·011
2. ,, 3 ,,	0·009	11. Blotting paper (white)	0·001
3. ,, 3 ,,		12. Writing paper (azure-blue)..	0·024
after 2 washings	0·009	13. Foolscap (white, blue-lined)	0·028
4. Flannel No. 3 (natural wool)		14. Wrapping paper (white) ...	0·024
after 4 washings	0·01	15. Paper (for sugar)	0·003
5. Flannel No. 3 (natural wool)		16. ,, (for butter, grease-	
after 6 washings	0·009	proof)	0·001
6. Old worn flannel (undyed)...	0·004	17. Japanese paper	Free
7. White Berlin wool	0·037	18. Wall paper (white, for	
8. Cream flannel	0·004	lining)	0·018
9. Welsh flannel	0·015	19. Linen (white)	Free
		20. Silk (undyed)........, ...	0·001

I have to thank Mr. George Stubbs and Mr. S. Bell for the assistance they have afforded me in carrying out the experiments described in this communication.

GOVERNMENT LABORATORY,
    LONDON.

XLVII.—*The Refractive Indices of Crystallising Solutions, with Especial Reference to the Passage from the Metastable to the Labile Condition.*

By HENRY ALEXANDER MIERS, D.Sc., F.R.S., and FLORENCE ISAAC.

DURING the course of some experiments on the properties, especially the refractive indices, of solutions in contact with a growing crystal of the solute, it became necessary to investigate the refractive indices of known solutions of sodium nitrate of various strengths at different temperatures. The method pursued was that adopted by one of us in the experiments described in *Phil. Trans.*, 1903, Series A, 202, 459, and the apparatus used is the same. This method renders it possible to trace the changes in the refractive index of a solution while its temperature or its concentration, or both, are changing. A trough, the front surface of which is a parallel-faced plate of glass, contains the solution, and is placed below the inverted goniometer described on pp. 464—466 of the memoir just quoted (Figs. 1 and 16, Plate 13).

In the solution is immersed a glass prism of known refractive index mounted on the crystal holder of the goniometer, and the refractive index of the solution is determined by the method of total reflection within the prism, as described on pp. 491—516 of the same memoir.

Preliminary Experiments on Sodium Nitrate.

In the first experiment, the solution taken contained a large quantity of crystals at about 15°, and was heated above 40° until all the crystals had disappeared; the trough was then filled with the warm, clear solution and the prism immersed. The sodium flame was adjusted and the prism rotated until the edge of the shadow indicating total reflection was visible in the field. The telescope having been set perpendicular to the front face of the trough, the edge of the shadow was adjusted on the vertical cross-wire of the eyepiece, and was watched while the solution cooled from 45·5° to 15·3°. The refractive index (n) of the solution was calculated from the formula $\dfrac{\mu}{n} = \text{Cos} p$,

where $\tan p = \dfrac{2\text{Sin}\left(45 + \dfrac{A-\theta}{2}\right)\text{Sin}\left(45 - \dfrac{A+\theta}{2}\right)}{\text{Sin} A}$; where A = angle of

the prism, μ = refractive index of prism, θ = angle of emergence of the totally reflected light.

At 41·5°, the angle of emergence of the totally reflected light was $-1°33\frac{3}{4}'$, but as the solution cooled, the edge of the shadow gradually moved from right to left in the field, and the angle of emergence became $-1°11\frac{3}{4}'$ at the temperature 19·4°. The refractive index of the solution at 41·5° was 1·388010, and as the solution cooled to 19·4° the index was found to increase gradually until it became 1·392537 at this temperature. As the solution cooled below 19·4°, the edge of the shadow remained stationary for some time, and then slowly moved back again in the opposite direction, that is, from left to right in the field. The angle of emergence of the reflected light changed from $-1°11\frac{3}{4}'$ at 18·35° to $-1°15\frac{3}{4}'$ at 15·3°, and the refractive index of the solution as calculated from these values decreased from 1·392537 at 18·35° to 1·391714 at 15·3°.

The solution having now attained the temperature of the room, further cooling was not possible.

At 27°, crystals were seen forming on the surface of the solution and at the bottom of the trough, and these continued to grow steadily until at 15·3° there was a large quantity of crystals in the trough. At the end of the experiment, a curve was plotted to show the variation of the refractive index of the solution with temperature, the values of index being taken as ordinates and the corresponding temperatures as abscissæ. From 41·5° to 27°, the resulting curve is almost a straight line inclined at about 45° to the axes. At 27°, crystals are first seen in the trough, and a depression appears in the curve at this temperature, followed by a slight difference in the direction of the curve from 27° to 19·4°. From 19·4° to 18·35°, the curve is flat and parallel to the temperature axis, after which it begins to fall gradually until the lowest temperature is reached. From 18·35° to 15·3°, the direction of the curve is somewhat steeper than it was before the maximum point was reached. This curve may be taken as a type of several experiments next to be described in which a supersaturated solution is allowed to cool at rest. The general form of the curves obtained is that of the upper curve in Fig. 1 (p. 416), which relates to experiment 13, and has a maximum at 23° and a depression at 37°.

Another similar sodium nitrate solution was next taken and heated until all the crystals had disappeared. Readings for refractive index were taken as the solution cooled from 37° to 14·5°, and a similar curve was plotted having values of refractive index for ordinates and corresponding temperatures for abscissæ.

From 37° to 22°, the index rose steadily as in the previous experiment. At 22°, crystals are first seen, and a depression occurs in the curve. The index continues to rise until the temperature is 18°, with, however, a slight change in the direction of the curve; after 18°, the index begins to fall again until the temperature of the room is attained. The curve obtained is similar in all respects to that obtained in the first experiment, except that the temperature at which crystals are first seen in the solution, and the temperature at which the maximum point takes place, are both lower in this experiment than in the previous one.

In these first two preliminary experiments, the initial strength of the solution was not known, and in order to ascertain what bearing this might have on the maximum value of the index and the temperature at which it is attained, several experiments were next made with sodium nitrate solutions of known strength. In all the following experiments, the solutions were of approximately known concentration, a known weight of salt being dissolved in a known weight of water. The concentrations given below are always estimated as parts by weight of salt to 100 parts of solution.   The salt was always carefully dried before weighing, but since, in spite of the usual precautions, a certain small amount of water is lost while the solution is being heated to dissolve the crystals and during its transference to the trough, the percentage of salt in the solution can only be regarded as approximately known.

First Series of Experiments.   Solutions Crystallising at Rest.

Expt. 1.—Concentration : $NaNO_3 = 43\cdot23$ per   cent.   Refractive index rose steadily from $1\cdot378206$ at $38\cdot5°$ to $1\cdot382236$ at $21\cdot5°$.   No

FIG. 1.

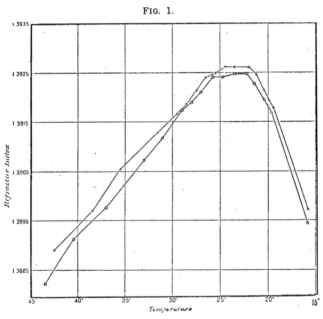

Experiment 13.

crystals appeared in the trough between these temperatures, and no sudden change or maximum value of refractive index.   Solution cooled in two hours.

Expt. 2.—Concentration : $NaNO_3 = 53\cdot10$ per   cent.   Crystals appeared in trough at once before any readings were taken.   Index rose from $1\cdot39341$ at $45\cdot5°$ to $1\cdot394332$ at $39\cdot5°$.   Index then fell gradually from $1\cdot394332$ at $39\cdot5°$ to $1\cdot309430$ at $16°$.

Expt. 3.—Concentration : $NaNO_3 = 46\cdot62$ per   cent.   Index rose from $1\cdot387188$ at $34\cdot5°$ to $1\cdot391253$ at $16\cdot8°$, and remained stationary as solution cooled further to 14·5°. Crystals appeared at 18·8°, and there is a corresponding depression in the curve just below this temperature.

Expt. 4.—Concentration : $NaNO_3 = 49·48$ per cent. Index rose from 1·388062 at 47·5° to 1·393004 at 24·5° and then fell from this value to 1·391253 at 20·8°. Crystals appeared at 30·5° and there is a corresponding depression in the curve at this temperature.

Expt. 5.—Concentration : $NaNO_3 = 48·68$ per cent. Index rose from 1·387391 at 44·5° to 1·392328 at 21·4°, and then fell from this value to 1·390889 at 17·8°. Crystals appeared in the trough at 29·5°, and there is a depression in the curve below this temperature.

Expt. 6.—Concentration : $NaNO_3 = 47·85$ per cent. Index rose from 1·386572 at 44·5° to 1·392073 at 19·4°. Crystals appeared in trough at 24·5° and there is a corresponding depression in the curve, which shows no maximum point between the temperatures taken.

Expt. 7.—Concentration : $NaNO_3 = 45·76$ per cent. Index rose from 1·381876 at 49° to 1·389811 at 15·8°. Crystals and a depression in the curve appeared at 23·5° and no maximum point is reached.

Expt. 8.—Concentration : $NaNO_3 = 47·45$ per cent. Index rose from 1·386054 at 41·5° to 1·391351 at 17·6°, and then fell from this value to 1·390481 at 16·7°. Crystals and a depression in the curve appeared at 25·5°.

Expt. 9.—Concentration : $NaNO_3 = 48·67$ per cent. Index rose from 1·387495 at 45·5° to 1·392020 at 23·45°. A depression appears in curve at 30·5°. Index fell from 1·392020 at 23·45° to 1·390787 at 19·9°. Crystals appeared in the trough at 36 5°.

In some of the preliminary experiments, a cleavage rhombohedron of calcite had been employed as the totally reflecting prism, and it had been noticed that the refractive index of a saturated solution of sodium nitrate appears to be slighly higher when in contact with glass than when in contact with calcite. It appeared that this might be due to some specific action of calcite on a solution of the nitrate, with which it is isomorphous. To inquire further into this point, a compound prism was constructed by Messrs. Hilger, formed of two small prisms cemented together by their bases, so that one is a continuation of the other. The upper half was of glass, and the lower half of calcite cut with the refracting edge parallel to the axis. The angle of both parts of the compound prism was 60°0½', so that the faces of the upper and lower halves of the compound prism were in the same planes. The compound prism was immersed in the nitrate solution in the trough exactly as the glass prism had been in the preceding experiments. The refractive index of the glass part of the prism was 1·650888, and that of the calcite 1·658553 at 19°. When the compound prism is immersed in the solution, the shadows indicating total reflection from both parts of the prism are visible in the field at the same time, the angle of emergence of the internally reflected light from both parts of the prism only differing by about 30′. Only the ordinary ray in the calcite is totally reflected so as to emerge from the prism, and no account need therefore be taken of the extraordinary index of the calcite.

The following are some of the results obtained for the refractive index of sodium nitrate solutions by total reflection from the compound prism.

Expt. 10.—Concentration : $NaNO_3 = 49.53$ per cent. Index rose from 1·387954 at 44° with calcite, and 1·388623 at 41·5° with glass, to 1·392047 at 23·45° with calcite, and 1·392234 at 24·45° with glass, and then fell from these values to 1·389810 at 16·8° with calcite, 1·389904 at 16·8° with glass. Crystals appeared in the trough at 31°, and there were corresponding depressions in both curves.

Expt. 11.—Concentration : $NaNO_3 = 48.594$ per cent. Index rose from 1·386672 at 44·5° with calcite, 1·387520 at 43° with glass, to 1·391567 at 22·4° with calcite, 1·391916 at 22·95° with glass, and then fell to 1·390774 at 19·9° with calcite, 1·390803 at 19·9° with glass. Crystals appeared in the trough at 32·5°.

Expt. 12.—Concentration : $NaNO_3 = 49.414$ per cent. Index rose from 1·388000 at 44·5° with calcite, 1·389109 at 42·5° with glass, to 1·392267 at 26° with calcite, 1·392444 at 25·7° with glass, and then fell to 1·390160 at 17·8° with calcite, 1·390220 at 17·8° with glass. Crystals appeared at 30·5°, a depression occurring in the curves at about 35°.

Expt. 13.—Concentration : $NaNO_3 = 49.25$ per cent. Index rose from 1·388215 at 43·5° with calcite, 1·388897 at 42·5° with glass, to 1·392470 at 23·45° with calcite, 1·392602 at 24·45° with glass, and then fell to 1·389441 at 15·8° with calcite, 1·389691 at 15·8° with glass. Crystals appeared at 37°.

The results of this experiment are shown in Fig. 1, in which the lower curve relates to the calcite prism.

In these thirteen experiments we may now compare the highest values reached by the refractive indices of the solutions, the amount of salt contained in each solution, and the temperature at which the maximum value of the refractive index is reached, corresponding to the highest point of the curve.

Parts of salt contained in 100 parts of solution. Approx.	Maximum value of refractive index reached.	Temperature at highest point.
53·10 per cent.	1·394332	39·5°
49·53 ,,	1·392234	24·5
49·48 ,,	1·393004	24·5
49·414	1·392444	25·7
49·25	1·392602	24·5
48·68	1·392328	21·4
48·67	1·392020	23·45
48·594	1·391916	23·0
47·85	1·392073	19·4
47·45	1·391351	17·6
46·62	1·391253	16·8
45·76 ,,	1·389811	15·8

Deductions from the foregoing Experiments.

(1) In a cooling solution of sodium nitrate, the refractive index generally continues to increase long after crystals have made their appearance, attains a maximum value, and then diminishes.

(2) It appears that the higher the percentage of salt in the solution the higher is the value of the refractive index at the maximum point.

(3) The higher the percentage of salt in the solution, the higher is the temperature at which the maximum point of the curve is reached.

(4) The appearance of crystals in the trough causes an abrupt depression in the temperature-index curve.

(5) When the compound prism is used, the value of the refractive index of the solution obtained by total reflection from the glass part of the prism is slightly greater than the value of the refractive index of the solution obtained by total reflection from the calcite part of the prism.

This result agrees with that obtained in the preliminary experiments with cleavage pieces of calcite.

It should be added that the refractive index of distilled water as determined with this prism was the same whether the glass part or the calcite part of the prism was used.

We propose to return to the subject of the effect of calcite on a solution of sodium nitrate in contact with it at some future time, and for the present confine our attention to the refractive indices as determined by glass prisms.

Second Series of Experiments.   Solutions Crystallising in Motion.

A new set of experiments was now started in which the solution was kept continually stirred during cooling. The stirrer employed was a vane of platinum immersed in the solution and driven by means of a small water motor.

Expt. 14.—A solution of approximately equal parts by weight

of water and sodium nitrate was heated above 40° to dissolve all the crystals, the trough was filled with the solution, and the prism immersed as in the previous experiments. The changes in the refractive index of the solution as it cooled were observed, as before, by means of the shadow denoting total reflection, and a curve was plotted with refractive indices as ordinates and corresponding temperatures as abscissæ.

At the higher temperatures, the curve resembled those obtained from the unstirred solutions. The index at 38·5° was 1·390273. Crystals appeared in the trough at 35° and continued to grow steadily at the top and bottom of the trough and on the vanes of the stirrer, while the index continued to rise to its maximum value, 1·392867, at 27°. At 27°, the refractive index began to decrease rapidly and fell from 1·392867 to 1·390433, while the temperature changed from 27° to 25·5°, a range of only 1½°. From the beginning of the experiment at 38·5° down to 26·25°, the solution appeared quite clear except for the crystals growing to a considerable size on the top and bottom of the trough and on the vanes, but at 26·25° a shower of crystals appeared suddenly throughout the whole solution and gradually sank to the bottom of the trough, where they lay like a white snow. The crystals of this shower were very much smaller than the crystals which had been growing steadily at the top and bottom of the trough during the experiment. The solution became almost clear again at 25·5°, and from this point down to 17° the refractive index decreased much more slowly with the temperature, the resulting curve being much less steep.

This experiment differs from those in which no stirrer was used in the following respects:

(1) A shower of crystals appeared throughout the whole solution soon after the maximum value for the refractive index was reached, whereas no shower was observed in the experiments when the solution was kept at rest as it cooled.

(2) The fall of the refractive index of the solution during the shower was much more sudden than the fall in the unstirred solutions.

(3) After the shower of crystals had fallen to the bottom of the trough and the solution had become clear again, the refractive index continued to decrease slowly as the temperature decreased.

Other experiments were made with the compound glass-calcite prism immersed in sodium nitrate solutions of varying concentration, the solutions being kept continually stirred. The following are some of the results obtained.

Expt. 15.—Concentration: $NaNO_3 = 49·81$ per cent. Index of solution increased from 1·389810 at 40·5° from calcite, 1·390273 at

38·5° from glass, up to 1·392572 at 27·5° from calcite, 1·392867 at 27° from glass. Index then dropped suddenly from these values to 1·390664 at 25·95° from calcite, 1·390433 at 25·45° from glass. From this point, index decreased slowly with temperature to 1·388923 at 17·3° from calcite, 1·389003 at 17·3° from glass. Crystals began to grow steadily at top and bottom of the trough and on the stirrer vanes at 36°. At 26·25°, a shower of small crystals appeared throughout the whole solution.

Expt. 16.—Concentration : $NaNO_3 = 49·02$ per cent. Index of solution increased from 1·388444 at 39·5° from calcite, 1·389215 at 37·5° from glass, up to 1·391943 at 22·75° from calcite, 1·392287 at 22·75° from glass. Index then dropped suddenly from these values to 1·389927 at 21·9° from calcite, 1·389956 at 21·7° from glass, and then continued to decrease slowly with temperature to 1·389388 at 19° from calcite, 1·389479 at 18·85° from glass. Crystals began to grow on vanes of stirrer and at top and bottom of the trough at 31°. At 22·75°, a shower of crystals appeared throughout the solution in the trough.

Expt. 17.—Concentration : $NaNO_3 = 48·67$ per cent. Index of solution increased from 1·386764 at 44° from calcite, 1·387724 at 42·5° from glass, up to 1·391452 at 23·95° from calcite, 1·391588 at 23·45° from glass. The index then dropped suddenly from these values to 1·389743 at 22·95° from calcite, 1·389788 at 22·45° from glass, and then continued to decrease slowly with temperature to 1·389374 at 20·4° from calcite, 1·389426 at 20·4° from glass. Crystals appeared at 33° and a shower took place at 23·5°.

Expt. 18.—Concentration : $NaNO_3 = 46·95$ per cent. Index increased from 1·385590 at 45·5° from calcite, 1·386505 at 43° from glass, up to 1·390871 at 22·4° from calcite, 1·391007 at 22·4° from glass. Then fell suddenly from these values to 1·389850 at 21·9° from calcite, 1·389733 at 21·9° from glass, and continued to decrease with temperature to 1·388856 at 18·3° from calcite, 1·388941 at 18·3° from glass. Crystals began to grow in trough at 28°. A shower of small crystals fell at 21·9°.

Expt. 19.—Concentration : $NaNO_3 = 53·1$ per cent. The refractive index increased from 1·389709 at 58·5° up to 1·393712 at 36·5°, then fell suddenly from this value to 1·391404 at 35·5° and continued to decrease slowly with temperature to 1·389503 at 19·4°. Crystals appeared in the trough at 54°. A cloud of crystals fell at 36·5°. The results of this experiment are shown in Figs. 2 and 3.

The general results of these experiments (14—19) are the same as those obtained from experiments with unstirred solutions, namely :

(1) The greater the percentage of salt in the solution, the greater is the maximum value of refractive index reached.

(2) With one exception, the greater the percentage of salt in the solution, the higher is the temperature at which the maximum point occurs.

(3) The maximum value for the refractive index of the solution obtained by total reflection from the glass part of the prism is greater than the maximum index obtained from the calcite part.

At the end of each experiment, the crystals which grew on the vanes of the stirrer were found to be far larger than the crystals in any other part of the trough, and they were seen on both upper and under sides of the vanes.

At the end of experiment 18, after the index of the solution

FIG. 2.

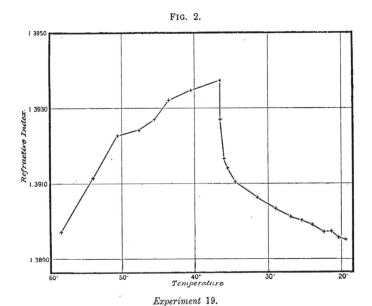

Experiment 19.

had become constant at the temperature of the room (19°) for over half an hour, a large quantity of sodium nitrate crystals was added to the solution in the trough and the stirring continued. This did not affect the last reading for the refractive index in any way.

Expt. 20.—The trough was next filled with a solution containing 48·19 per cent. of sodium nitrate at 50°, so that at this temperature the solution was quite clear and contained no crystals. A glass prism was immersed and the solution was stirred continuously. As it cooled the changes of refractive index were watched by noting the motion of the shadow for total reflection. At 34°, crystals began to appear in the trough, the refractive index of the solution increasing from 1·387461 at 43·5° to 1·390001 at 33°.

At 32°, a large excess of sodium nitrate crystals heated up to 32° was added to the solution in the trough. The temperature continued to fall steadily and the value of the refractive index to rise, just as if no crystals had been added. At 29°, the refractive index reached its maximum value, 1·390529. From this point, the index decreased slowly with the temperature until it reached 1·388512 at 14·5°.

The crystals grew steadily on the top and bottom of the trough and on the vanes of the stirrer. There was no shower of crystals seen at any time, the solution appearing clear throughout. There was no sudden drop in the index, as in the case of the other stirred solutions to which no crystals were added. The maximum value for the refractive index was 1·390529, whereas the maximum value for a stirred solution of equal strength when no crystals were added would be about 1·391300. Hence the effect of adding crystals at 32° is to lower the maximum point reached by the index. Also the fall of index after reaching the maximum point is far more gradual than in any of the preceding experiments, whether the solution was stirred or not.

A curve was drawn with values of refractive index as ordinates and temperatures as abscissæ. At the maximum point, this curve is very blunt, the index varying very little, whilst the solution cooled through a considerable range of temperature (see Fig. 7).

Expt. 21.—Another solution of sodium nitrate containing approximately 49·5 per cent. of the salt was placed in the trough and stirred as before, but an attempt was made to arrest the sudden drop in refractive index and to prevent the shower of crystals which usually takes place after the solution has reached its maximum refractive index. A large plate of copper was therefore inserted just under the glass trough containing the solution and was heated by means of a spirit lamp. In this way, the solution was kept between 32° and 31° for more than an hour, and its refractive index was watched. The index increased quite regularly from 1·386210 at 54° to 1·391630 at 32°. Crystals were seen on the surface of the liquid at 41·5°, and there is a slight depression in the curve at this temperature.

After reaching 1·391630 at 32°, the index remained constant for half an hour, while the temperature fell to 30·5°. Up to this point, the solution was quite clear, with crystals growing on the top and bottom of the trough.

After falling to 30·5°, the heated copper plate under the trough kept the solution at this constant temperature for more than half an hour, and the refractive index of the solution began to fall directly this temperature was attained. The index decreased from 1·391630 to

1·391054 in this time, the temperature being 30·5° throughout. During this fall in index, a very fine, but not sudden, shower of crystals appeared throughout the solution. This fine shower did not compare in density with the sudden showers obtained when a solution of equal strength was stirred and cooled uniformly, and only a small amount of solid material was deposited at the bottom of the trough. When the index reached the value 1·391054, the solution was nearly clear again. The temperature having been kept constant for more than half an hour, the spirit lamp was removed, and from the temperature 30·5° the solution was allowed to cool regularly. As the solution cooled from 30·5° to 28·5°, the refractive index remained constant, its value being 1·391054. On cooling below 28·5°, another fine shower appeared in the solution, similar to the one which formed at 30·5°, and the refractive index fell from 1·391054 at 28·5° to 1·390222 at 26°. The solution was not cooled below this temperature (see Fig. 7).

Expt. 22.—Another experiment was tried with a sodium nitrate solution containing approximately 49·75 per cent. of the salt. In the preceding experiment, an attempt was made to keep the temperature of the solution constant at the point where the refractive index was a maximum. In experiment 22, the temperature was kept constant at an earlier phase, so that the solution was at a constant temperature of 37·5° for one hour and a quarter. The temperature was then allowed to fall, and the index rose to a maximum. As in the preceding experiment, the solution was kept at the constant tempera- ture required by means of a heated copper plate under the glass trough. In this experiment, two spirit lamps were used to heat the copper plate. The refractive index of the solution rose from 1·386158 at 54·5° to 1·390787 at 37·5°, crystals appearing in the trough at 45·5°.

The solution remained at the temperature 37·5° for one hour and a quarter, during which time the refractive index rose from 1·390787 to 1·391490. The solution was quite clear, with a large quantity of crystals at the top and bottom of the trough. Up to this point, the solution was at rest, the stirrer not being used. After keeping the temperature constant for one hour and a quarter, the solution was stirred and the lamps heating the copper plate were removed. The temperature then fell, and the refractive index rose from 1·391490 at 37·5° to 1·391917 at 33·5°. At this point, a very fine shower appeared throughout the solution and the refractive index fell very slightly, its value being 1·391814 at 31°. The shower of crystals then became rather thicker, and the fall in index greater, until it reached 1·390787 at 28·5°. At this point, the solution was almost clear again, and the index remained constant at 1·390787 until the temperature reached 26·5°. On cooling below this temperature, the index fell again, and another fine shower of crystals appeared in the solution. At 20·4°, the value of the index was 1·389657.   In this experiment, as in the preceding one, the showers were very thin and quite unlike the thick showers obtained when a stirred solution is allowed to cool regularly.

The general effects of arresting the cooling of the solution seem to be:

(1) A lowering of the maximum value of the refractive index.   In the solutions cooling regularly, a solution containing 49·4 per cent. of sodium nitrate reached a maximum refractive index of 1·392444, whereas in the two preceding experiments the maximum values for the index are 1·391630 and 1·391917, the solutions containing 49·5 and 49·75 per cent. of the salt respectively.

(2) The solid material is deposited in the trough in a series of very fine showers of crystals instead of in one thick shower, as in the regularly cooled solutions.

(3) After the maximum point is reached, the fall in refractive index is very irregular, and not nearly so sudden as in the regularly cooled solutions.

In experiments 2, 7, 8, 12, 13, 15, 16, 17, 18, 19, crystals appeared to form in the solution at a temperature higher than that corresponding to saturation, as may be seen by comparing the numbers with the solubility curve S in Fig. 3.

It is possible that in these experiments the temperature of the solution was suddenly and temporarily lowered during its transference to the trough.

In experiment 21, crystals appeared on the surface of the liquid at about $41\frac{1}{2}°$ instead of at $32\frac{1}{2}°$; it should be mentioned that in this experiment the stirring was very irregular, so that the temperature of the surface layer may possibly have fallen to $32\frac{1}{2}°$, a temperature at which germs falling into the liquid would continue to grow.

In experiment 22, in which the solution was not stirred until 38°, the crystals appeared at $45\frac{1}{2}°$ instead of at $33\frac{1}{2}°$.

On the other hand, it is certain that in all these experiments, except with very rapid stirring, the temperature of the upper part of the solution was somewhat greater than that of the lower part; in general, both temperature and refractive index were measured somewhat below the middle of the trough.

Criticism of the Preceding Results.

The foregoing experiments prove that during the cooling of a supersaturated solution of sodium nitrate, even while crystallisation is proceeding, a more or less sudden change in the refractive index occurs, corresponding apparently to a more or less sudden weakening of the solution.   This is accompanied or closely followed by a more or less sudden separation of crystals.   If the solution is crystallising at rest, the change is gradual; if the solution is stirred, the change is quite sudden, and is accompanied or followed by a dense shower of crystals.

We may add at once that experiments made with other substances, such as sodium chlorate, alum, sodium thiosulphate, and ammonium oxalate, showed that these substances behave in the same way; the experiments are described below.

The change of index might indicate a physical change in the liquid corresponding perhaps either (a) to a transition from one modification of the solute to a different modification, that is, to a polymorphous change, or (b) to a formation of different hydrates in the solution. These were the explanations which suggested themselves to us when operating with solutions at rest.

An argument, however, against (a) is that similar results are afforded by two substances so different as· sodium nitrate and sodium chlorate.   The first of them both at high and low temperatures, and alike from fusion and from solution, crystallises only in one rhombohedral modification and shows no sign of polymorphism.   The sodium chlorate, on the other hand, certainly crystallises in at least two modifications: when fused and allowed to cool it solidifies in a birefringent modification which soon becomes transformed into the usual isotropic crystals.   A drop of supersaturated solution of sodium chlorate allowed to crystallise on a microscope slide deposits the same birefringent crystals, possessing strong double refraction and having the form of rhombic plates; these subsequently become transformed into the isotropic modification and continue to grow as cubes.   The two modifications may even grow side by side in the same drop; sometimes, indeed, cubes make their appearance before the birefringent plates; when the two come into contact, the latter rapidly become isotropic.   Now sodium nitrate and sodium chlorate behave alike as regards the change of refractive index in the cooling solution, so that any variation due to the polymorphous change must be too slight to be detected in the curves, and the cause for the great change of index must be sought elsewhere.

Against both suggestions (a) and (b) it may be urged that similar curves are obtained with all the substances examined, whether hydrated or anhydrous, and that with each of them crystals of one and the same form are being actually deposited in the solution while the change is taking place: the ordinary rhombohedra from the sodium nitrate solutions, the ordinary octahedra from the alum solutions, &c. We are therefore led to regard the change in index as simply due to an increase in the rate of crystallisation, and this is confirmed by the behaviour of the solutions when they are kept stirred during the observations.

Shortly after the maximum value of the refractive index has been reached during the descent of the curve, a copious shower of crystals is produced which fall to the bottom of the trough like a white snow. This sudden separation of material is quite enough to account for the sudden change of index. In the case of sodium thiosulphate, the heat evolved during this copious crystallisation is sufficient to raise the temperature of the cooling solution considerably, as may be seen by the backward slope of the descending branch of the curve in Fig. 10.

When small crystals, few in number, first begin to make their appearance at an earlier stage of the cooling, a small change of index is produced, and is indicated by a slight indentation on the upward branch of the curve.

The whole series of observations then goes to show that there are two stages in the process of crystallisation. As the supersaturated solution is allowed to cool, small crystals make their appearance at the surface of the solution ; some of these drop to the bottom ; owing to fall of temperature, the index of refraction rises, and the diminution of index owing to the separation of crystals is so slight that it is not sufficient to compensate the effect due to cooling ; this is the stage of slow growth.

Suddenly a copious crystallisation takes place not only at the surface and bottom of the solution, but as a cloud of small crystals throughout the liquid ; the effect is to weaken the solution to such an extent that the fall of index due to this cause is far greater than the rise due to cooling, and the latter is also partially counteracted by the liberation of heat. This is the stage of rapid growth ; the minute crystals increase quickly in size and fall to the bottom, leaving the solution clear again, and it soon tends towards saturation and a constant index at a lower temperature. During the period of slow growth, the crystals are few in number and appear to start at the surface of the solutions ; during the period of rapid growth, they increase enormously in number throughout the liquid, and correspond to what is understood by the spontaneous crystallisation of a supersaturated solution.

We have no hesitation in regarding these two stages as Ostwald's metastable and labile conditions (*Zeit. physikal. Chem.*, 1897, 22, 302) This interpretation was first suggested to us by Mr. H. B. Hartley, of Balliol College, from whom we have received much helpful advice and criticism during the course of these experiments.

During the metastable stage, the solution is not so highly supersaturated, and can only crystallise when in contact with the solid crystals, which consequently grow singly ; on entering the labile stage, the temperature has been lowered so far that the solution is highly supersaturated, and spontaneous crystallisation is set up by the stirring, although the solution already contains rapidly growing crystals.

In a cooling solution at rest, the change takes place so gradually that it escapes ordinary observation, and it is clear that crystallisation may proceed in an unstirred metastable solution for a very long time before either the labile state or a condition of mere saturation is attained. That a solution in which crystals are growing may remain super-saturated for quite a long time may easily be proved by experiments on a metastable solution of sodium nitrate or sodium chlorate. A drop of either solution taken at any time during several hours while crystals are growing in it and placed on a microscope slide behaves exactly like the supersaturated solution before the crystals have begun to separate.

As has been found by many observers, it is necessary to stir a solution vigorously and for a long time in order to reduce it from a supersaturated to a saturated condition.

Interpretation of the Results in Terms of Concentration.
The Supersolubility Curve.

In the preceding observations, it is impossible to say how much solvent has been lost at any given moment by evaporation and how much solute by crystallisation; the composition is only approximately given by the original strength of the solution, and in order to interpret the results in terms of concentration it is necessary to make an independent series of determinations of refractive index at different temperatures on solutions of known strength before crystals begin to separate; these will enable us to ascertain: (1) the change of index due to fall of temperature for solutions of different degrees of concentration, and hence (2) the refractive index of any given solution at any desired temperature.

Solutions of various concentrations were therefore made up in a small flask; they were warmed to about 60° to dissolve the crystals, the flask being loosely corked throughout, so that little evaporation takes place. The solution was then poured into the trough and covered immediately with a thin layer of olive oil raised to the temperature of the solution, so that evaporation did not occur to any appreciable extent as the solution cooled. It was found that the presence of the layer of oil prevented the solution from crystallising until a somewhat lower temperature than in previous experiments.

Solutions of various concentrations were taken, and the refractive index of each observed as the solution cooled down from about 50° until crystals appeared in the trough. The results of all these experiments taken together were then expressed by curves drawn with concentrations as ordinates and temperatures as abscissæ the refractive index being constant for each curve.

Such curves were drawn for the following values of refractive index: 1·3500, 1·3540, 1·3560, 1·3580, 1·3600, 1·3620, and 1·3640. On examining these concentration-temperature curves for different values of index it is seen that between 20° and 50° and concen-

Fig.

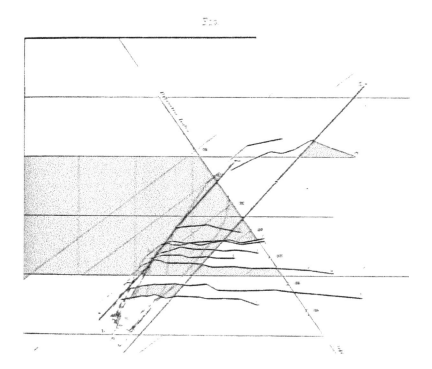

stirring, although the solution already contains rapidly growing crystals.

In a cooling solution at rest, the change takes place so gradually that it escapes ordinary observation, and it is clear that crystallisation may proceed in an unstirred metastable solution for a very long time before either the labile state or a condition of mere saturation is attained. That a solution in which crystals are growing may remain supersaturated for quite a long time may easily be proved by experiments on a metastable solution of sodium nitrate or sodium chlorate. A drop of either solution taken at any time during several hours while crystals are growing in it and placed on a microscope slide behaves exactly like the supersaturated solution before the crystals have begun to separate.

As has been found by many observers, it is necessary to stir a solution vigorously and for a long time in order to reduce it from a supersaturated to a saturated condition.

Interpretation of the Results in Terms of Concentration. The Supersolubility Curve.

In the preceding observations, it is impossible to say how much solvent has been lost at any given moment by evaporation and how much solute by crystallisation; the composition is only approximately given by the original strength of the solution, and in order to interpret the results in terms of concentration it is necessary to make an independent series of determinations of refractive index at different temperatures on solutions of known strength before crystals begin to separate; these will enable us to ascertain: (1) the change of index due to fall of temperature for solutions of different degrees of concentration, and hence (2) the refractive index of any given solution at any desired temperature.

Solutions of various concentrations were therefore made up in a small flask; they were warmed to about .60° to dissolve the crystals, the flask being loosely corked throughout, so that little evaporation takes place. The solution was then poured into the trough and covered immediately with a thin layer of olive oil raised to the temperature of the solution, so that evaporation did not occur to any appreciable extent as the solution cooled. It was found that the presence of the layer of oil prevented the solution from crystallising until a somewhat lower temperature than in previous experiments.

Solutions of various concentrations were taken, and the refractive index of each observed as the solution cooled down from about 50° until crystals appeared in the trough. The results of all these experiments taken together were then expressed by curves drawn with concentrations as ordinates and temperatures as abscissæ, the refractive index being constant for each curve.

· Such curves were drawn for the following values of refractive index: 1·3820, 1·3810, 1·3860, 1·3880, 1·3900, 1·3910, and 1·3920. On examining these concentration-temperature curves for different values of index, it is seen that between 20° and 50°, and concen-

Fig. 3.

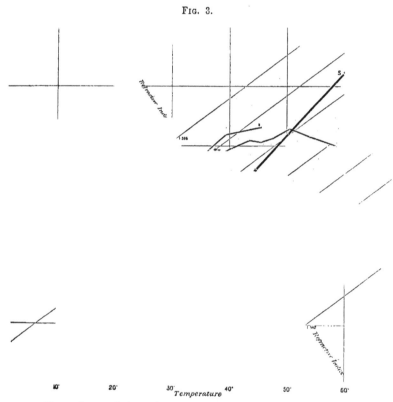

The number attached to each curve is that of the corresponding experiment.

Experiments A, B, 2, 8, 10, 12, 13, 15, 16, 19.

trations of 47 and 51 per cent., the curves are approximately straight lines parallel to each other, and inclined at an angle tan.⁻¹·68 to the temperature axis : the scales being such that a length representing one degree on the temperature axis represents 0·2 per cent. on the concentration axis. Two of these experiments are shown in Fig. 3, and are denoted by the letters *A* and *B* respectively.

Also the perpendicular distance between any two of these curves is approximately proportional to the difference in their refractive indices, and also, therefore, to the difference in their concentrations. Hence if a line be drawn in the temperature-concentration diagram inclined to the temperature axis at an angle cot.$^{-1.68}$, the refractive index of any sodium nitrate solution as it cools may be measured along this line in the same way as the temperature is measured along the horizontal axis. Any point on the diagram will now give not only the temperature and concentration of a solution by its co-ordinates, as measured along the horizontal and vertical axes, but also the refractive index by its co-ordinate measured along the line so constructed.

If now the previous index-temperature observations for any solution are transferred to the new diagram, measuring indices along this new line and temperatures along the horizontal axis, the concentration of the solution at any point is given directly by the corresponding ordinate parallel to the vertical axis (Fig. 3).

S in this diagram is the solubility curve; in the case of sodium nitrate, as constructed from the observations of Lord Berkeley, it appears to be nearly a straight line. The various curves in the diagram are plotted from the refractive indices obtained at different temperatures from the various sodium nitrate solutions as they cooled. *It will be seen that the points on these curves corresponding to the maximum refractive index attained by each solution lie approximately on a straight line nearly parallel to the solubility curve.* This line, T, which may be called the "supersolubility curve," represents the temperature and concentration of each solution when it passes from the metastable to the labile state. The curves for uniformly cooling solutions, stirred or unstirred, first cross the line S, at or about which point crystals begin to appear in the solutions, and the curves drop slightly, showing a decrease in concentration from this point until they touch the line T. After this, the curves drop much more rapidly owing to the sudden formation of a quantity of crystals in the solution, and gradually approximate to the solubility line S. The line S divides the unsaturated from the metastable region.

The line T separates the metastable from the labile region. On reaching the labile state, the solutions usually crystallise out in a shower, and the concentrations then fall suddenly, and continue to fall gradually until the solutions reach the saturated condition. It may, however, be possible by keeping a solution at rest to take it into the labile region, and reduce it to a temperature at which the supersaturation far exceeds that corresponding to the curve T, and yet to prevent crystals from forming.

In considering the crystallisation of a supersaturated solution, we have then to take account not only of the solubility curve S, but also of the supersolubility curve T in the accompanying diagrams.

The following cases may occur:

(1) Let a supersaturated solution made by adding the salt to hot water be allowed to cool slowly while being stirred. The whole process is represented by the line $A\,B\,C\,D$ (Fig. 4).

The solution cools from A to B, and, unless they can be kept out, the first crystals (forming from germs introduced from without) make their appearance at B. From B to C, the liquid is cooling and crystals are growing slowly. At C, it passes into the labile condition, a cloud of crystals is deposited, and the concentration rapidly falls to D on the solubility curve, generally with a slight rise of temperature. This is the case of most of the stirring experiments described above, except

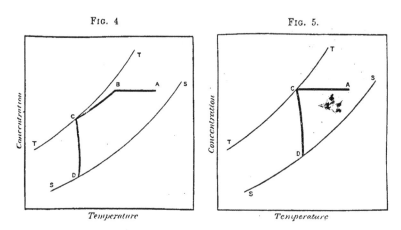

FIG. 4   FIG. 5.

that the rise of temperature is not perceptible in the case of sodium nitrate. If crystals can be kept out of the solution until the labile condition is reached, the process is represented by Fig. 5.

(2) Let a supersaturated solution in the metastable condition have crystals introduced into it, and let the crystals be allowed to grow at nearly constant or slightly diminishing temperature: this is the ordinary case of a not too strong solution crystallising at rest, and is represented by Fig. 6. The supersaturated solution cools from A to B; the crystal is then introduced; from that point, the concentration gradually diminishes by withdrawal of material from the solution, and if the change of temperature be sufficiently slow the labile condition is never reached, and the cloud of crystals is not produced. If the fall of temperature be sufficiently rapid, a cloud, or a sudden increase of crystallisation, is produced at C as before (compare Fig. 4).

It may be objected that as soon as the labile condition has been reached the sudden separation of material at once restores the solution to the metastable condition, in which the growth of crystals should again be gradual, and a repetition of these changes would make the process of crystallisation an oscillatory one. This may actually be the case, and is suggested by some of the curves which we have obtained. In general, however, it may be supposed that the immediate effect of entering the labile state is to produce so many nuclei of crystallisation throughout the solution that they all continue to grow steadily in the now metastable liquid until the saturation point is attained, without any return to the labile state. In fact, the curves show clearly that,

FIG. 6.

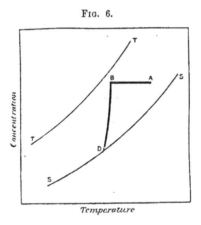

in general, after the shower has been precipitated, the solution goes on diminishing in concentration until the state of saturation, and therefore of equilibrium, is reached.

Effect of Rapid Stirring.

Up to the present the solutions were stirred regularly, but not very rapidly, by means of a water motor.

Expts. 23 and 24.—Experiments were next made on sodium nitrate solutions with more rapid stirring. The solutions used contained approximately 49·87 and 47·616 per cent. of the nitrate respectively, and were stirred very rapidly by means of an electric motor. The variations of refractive index with temperature were plotted as before. The general form of the curves obtained is the same as when the solutions were only moderately stirred, but they differed from the other curves in the following respects :

(1) The maximum values reached by the refractive index are not so great as the maximum values for solutions of equal strength when the stirring was only moderate.

(2) The temperatures at which the maximum values of refractive index are reached are higher than the temperatures for the maximum points of similar solutions which are only moderately stirred.

FIG. 7.

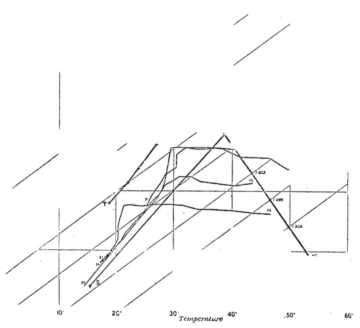

The number attached to each curve is that of the corresponding experiment.

Experiments 20, 21, 23, 24.

The variations of refractive index with temperature for these rapidly stirred solutions are plotted in curves 23 and 24 of Fig. 7, which also shows the concentrations. The concentration rises slightly at first owing to evaporation. After the refractive index has attained its maximum value, the curve falls again somewhat suddenly.

Curve 23 shows that above 37° this solution is unsaturated. At this temperature, it crosses the solubility curve and is saturated. As it cools further, it passes into the metastable region between the curves S and T' of Fig. 7. The curve, however, never reaches the labile region, and begins to fall in the direction of the solubility curve without having touched the supersolubility curve T. The behaviour of the weaker rapidly stirred solution is shown in curve 24, Fig. 7, and is similar to the previous one, since it also never reached the labile condition.

It appears probable that the rapid stirring of a solution which already contains growing crystals has the effect of promoting the crystallisation by bringing the whirling crystals more rapidly into contact with the supersaturated solution, so that the strength of the solution always remains below that of transition to the labile condition; unless, therefore, crystals can be kept out of the liquid, it is not surprising that such solutions do not reach the transition point. The crystals which make their appearance when the curve crosses the line S are almost certainly introduced from without.

Expt. 25.—An experiment in the case of sodium chlorate in which the solution was covered with a sheet of paper as soon as it had been poured into the trough was suggestive. The paper covered the entire trough except for a small slit surrounding the prism holder and the stem of the stirrer. The first crystals which appeared in the trough were seen to form exactly under this slit in the paper cover. It would appear, therefore, that crystallisation was in this case first started in the solution by crystal germs falling into the trough from the air outside, growing on the surface of the solution and then falling to the bottom of the trough, where they continued to grow; it is also possible that crystals may sometimes originate by evaporation of drops near the surface of the liquid at its margin.

The Crystallisation of Supersaturated Solutions of Sodium Nitrate in Sealed Tubes.

Our next experiments were therefore designed to keep out extraneous germs and to prevent evaporation, and so to prevent, if possible, any crystallisation until the labile condition is reached. In these experiments, an attempt was made to determine the "supersolubility" curve T by an independent method. Sodium nitrate solutions of accurately known concentrations were enclosed in sealed glass tubes. The solutions varied in concentration from 52·76 to 47·58 per cent. of the salt in 100 parts of solution. These tubes were boiled in a beaker of amyl alcohol for more than twelve hours and shaken at intervals to ensure the complete solution of the salt. The tubes were then taken out of the alcohol and immediately immersed completely iu a water-bath at about 80° and allowed to cool.

As they cooled in the water, the tubes were kept continually in motion, being fixed on a board which was rocked by means of an electric motor ; and as the water cooled the temperature at which crystals first made their appearance in each tube was noted.

It was found that in all cases where the cooling was slow the solution in the tubes crystallised in a shower at the exact temperature at which they reach the supersolubility curve T (*established above*), *and not before.*

The following table shows the concentration of the solutions in the tubes and the corresponding temperatures at which crystals first appeared in them when they were cooled and shaken in the water-bath.   Each experiment was in general repeated several times.

Experiment.	Percentage of sodium nitrate.	Temperature of crystallisation.
26	52·76	38°, 37½°, 38°
27	51·67	32°, 33°, 33½°
28	50·917	30°, 29°
29	49·57	21·9°
30	47·58	13·5°, 13·5°, 12°, 12°
31	49·226	22°
32	50·69	28½°

It appears, therefore, that with continuous shaking and regular cooling crystals appear in solutions in sealed tubes exactly when the labile state is reached, and the curve plotted with concentrations as ordinates and temperatures of crystallisation as abscissæ coincides almost exactly with the supersolubility curve T obtained by the totally different and more complicated method already described.

The part of the diagram above the curve T represents conditions under which spontaneous crystallisation may take place, whereas in the area below T spontaneous crystallisation cannot occur : if, however, the solution be either kept at rest or insufficiently stirred or shaken there is no reason why it should not remain uncrystallised at temperatures below those corresponding to the curve T.

Expt. 33. *Effect of Rapid Cooling.*—An experiment was made in which the sealed tubes were cooled very rapidly while being shaken continually, and the temperature at which crystals first appeared in the tubes was again noted.   The rapid cooling was effected by immersing a beaker of boiling water containing the sodium nitrate tubes in another vessel through which cold water was circulating.

It was found that the solutions could be cooled in this manner without crystallisation until the temperature was far below that at which they passed into the labile state.   Thus a tube containing 52·76 per cent. of the nitrate was cooled to 27·5° before crystallisation took place.   The same solution passes from the metastable into the labile state at 38°, and with regular cooling and continuous shaking crystals had previously always appeared in the tube at the latter temperature.

Similarly, tubes containing 49·226 and 47·58 per cent. of nitrate respectively, which pass into the labile state at 22° and 13·5°, were cooled down to 16·5° and 7·5° respectively before crystallisation took place.

Expt. 34. *Effect of Slow Cooling.*—An experiment was also tried in which the tubes containing the sodium nitrate solutions were cooled very slowly; they were immersed in a beaker of boiling water and then allowed to cool gradually for a whole day until they reached 49°. They were then kept at this temperature for about eighteen hours, being shaken from time to time, and it was found that in no case did the solutions crystallise out while the temperature was a few degrees above that at which they pass from the metastable to the labile state. As soon as the temperature was allowed to fall below 49°, the various tubes crystallised out in turn at the temperatures already observed in the previous experiments with sealed tubes: that is, the temperatures at which they cross the supersolubility curve T and pass from the metastable to the labile state.

The preceding experiments indicate that sodium nitrate solutions in sealed tubes can in no way be induced to crystallise until the labile state is reached; with continuous shaking and regular cooling, crystals always appear in the tubes at this point, but by means of rapid cooling it is possible to reduce the solutions to a temperature far within the labile region without crystallisation taking place.

Experiments with Sodium Chlorate.

The position of the supersolubility curve T for sodium nitrate separating the labile and metastable regions being thus fixed on the concentration-temperature-index diagram by means of the foregoing experiments, an attempt was made to establish a similar curve for sodium chlorate. The method adopted was the same as with sodium nitrate. Preliminary experiments were first made upon solutions of known strength before the crystals began to separate. These experiments give : (1) the change of index due to the fall of temperature for solutions of different concentrations, and hence (2) the refractive index of any given solution at any desired temperature. Solutions of six different concentrations were then made up containing from 51 to 56 per cent. of sodium chlorate.

Each solution was heated in a closed flask to dissolve the crystals and poured into the trough of the goniometer and covered immediately with a layer of warm olive oil to prevent evaporation as the solution cooled. The solution was stirred slowly in order to avoid inequalities of temperature due to convection. The refractive index was observed as the solution cooled down from about 55° until crystals first appeared in the trough. Curves were then drawn (as previously for sodium nitrate) with concentrations as ordinates and temperatures as abscissæ, the refractive index being constant for each curve. Such curves were drawn for the following values of refractive index : 1·388, 1·389, 1·390, 1·391, 1·392, 1·393, 1·394, 1·395. Between 50° and 20°, these curves, though not actually straight lines, are approximately equidistant and parallel to each other, and for the purposes of the diagram are regarded as straight lines inclined at an angle 38$\frac{1}{4}$° to the temperature axis on the scale employed in Fig. 3.

It may be mentioned that these constant index curves on the concentration-temperature diagram are more irregular than the similar curves obtained for sodium nitrate, and the direction of the line along which we propose to measure refractive indices can therefore only be regarded as an approximation.

This irregularity of the constant index curves may perhaps be due to the polymorphism of sodium chlorate already referred to in this paper.

The preliminary experiments being finished, another series* of experiments was made with solutions of sodium chlorate of various concentrations. The solutions were uncovered and stirred moderately by means of the water motor; they were introduced into the trough at about 50°, and the variations in their refractive indices were traced by means of the glass prism which was used in the sodium nitrate experiments.

The following are the results obtained :

Expt. 35.—Approximate concentration : $NaClO_3 = 54\cdot835$ per cent. The refractive index rose from 1·391253 at 50·55° to 1·394846 at 34·5°, and fell from this value to 1·389606 at 21·5°. Crystals appeared at 44·5°, and a shower occurred at 33·5°.

Expt. 36.—Approximate concentration : $NaClO_3 = 54\cdot054$ per cent. Index rose from 1·391814 at 48·55° to 1·395003 at 33·5°, and then fell suddenly to 1·391148 at 31·5°, after which it fell slowly to 1·389370 at 20·2°. Crystals were in the trough throughout the experiment, and a dense shower occurred at 33°.

Expt. 37.—Approximate concentration : $NaClO_3 = 51\cdot736$ per cent. The refractive index rose from 1·390430 at 35° to 1·392742 at 24·5°, and then fell suddenly from this value to 1·390481 at 24°, after which it decreased slowly to 1·388988 at 19·7°. Crystals appeared in the trough at 29°, and a shower occurred at 24·5°. The results of this experiment are shown in Fig. 8.

Expt. 38.—Approximate concentration : $NaClO_3 = 51\cdot035$ per cent.

The refractive index rose from 1·387083 at 43·5° to 1·392020 at 22·45°, and fell from this value to 1·389709 at 20·7°. Crystals appeared in the trough at 26°, and a slight shower at 22·5°.

Expt. 39.--Approximate concentration: $NaClO_3 = 52·397$ per cent. The refractive index rose from 1·385488 at 55° to 1·392690 at 25°, and then fell suddenly from this value to 1·389657 at 24°. Crystals appeared in the trough at 27°, and a copious shower occurred at 24·9°.

Expt. 40.—Approximate concentration: $NaClO_3 = 56·004$ per cent.

FIG. 8.

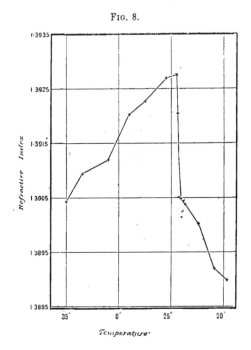

Experiment 37.

The index rose from 1·392874 at 54° to 1·396381 at 35°, and fell from this value to 1·391455 at 24·2°. Crystals appeared in the trough at 44°, and a dense shower occurred at 33°.

Expt. 41.—Approximate concentration: $NaClO_3 = 54·46$ per cent. The refractive index rose from 1·387495 at 57° to 1·394229 at 28·5°, and then fell suddenly from this value with rise of temperature to 1·390430 at 29°. Crystals appeared at 34°, and a dense shower occurred at 28·5° and 29°.

Expt. 42.—Approximate concentration: $NaClO_3 = 55·77$ per cent.

The index rose from 1·391250 at 53·5° to 1·395054 at 35°, then fell suddenly from this value to 1·391043 at 32°, after which it decreased slowly with temperature to 1·389195 at 22·2°. Crystals appeared in the trough at 49°, and a dense shower occurred from 34° to 32°.

Expt. 43.—Approximate concentration : NaClO₃ = 55·037 per cent. The index rose from 1·388577 at 57·5° to 1·394434 at 32°, and then fell suddenly from this value to 1·391455 at 28·8°. Crystals appeared in the trough at 45°, and a shower occurred at 29°.

All the above observations are shown in Fig. 9.

The curves obtained from the experiments with sodium chlorate by plotting refractive indices as ordinates and the corresponding temperatures as abscissæ do not differ at all in their general character from the similar curves obtained from the experiments with stirred sodium nitrate solutions. After reaching the maximum, the fall in refractive index is perhaps rather more sudden than in the sodium nitrate solutions, and the shower appears also to be somewhat more dense. A depression usually appears in the curves at the temperature at which crystals were first observed, as with the sodium nitrate curves.

One of these index-temperature curves obtained from Expt. 37 is shown in Fig. 8.

The index-temperature curves are now transferred to a new diagram, measuring indices along a line obtained in the manner already described, which is inclined at an angle 51¾° to the temperature axis, Fig. 9. The ordinates on this diagram then give the concentrations of the solutions at the various temperatures. S is the solubility curve and is obtained from Comey's *Dictionary of Solubilities*. The points on the index-temperature curves corresponding to the highest refractive index attained by each solution are seen to be approximately on a straight line nearly parallel to the solubility curve S. This is the supersolubility curve T, which separates the metastable and labile regions for sodium chlorate solutions.

This new diagram (Fig. 9) for sodium chlorate solutions corresponds closely to the similar diagram, Fig. 3, obtained for sodium nitrate solutions. It will be seen that, after the maximum values for the refractive indices of sodium chlorate solutions have been reached, the fall in concentration is somewhat more rapid than was the case for the sodium nitrate solutions. After this rapid fall in concentration and index, most of the curves on Fig. 9 coincide almost exactly with the solubility curve S for a considerable range of temperature. For this reason, the curves have not been carried on to the region in which they nearly coincide with the solubility curve, but the complete curves may be traced by means of the tables on pp. 447—452.

The supersolubility curve T may be regarded as approximately

ished by the above experiments on the refractive indices of the
g sodium chlorate solutions.

Experiments with Sodium Chlorate in Sealed Tubes.[*]

following experiments were carried out in order to make an
determination of the supersolubility curve T by enclosing

FIG. 9.

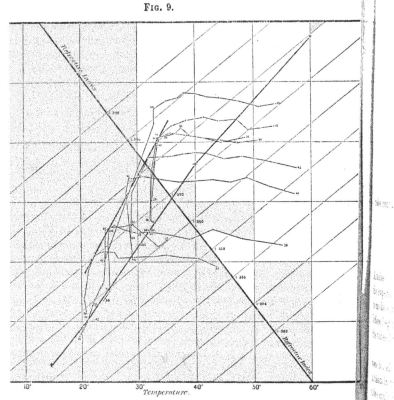

The number attached to each curve is that of the corresponding experiment.

Experiments 35 to 43.

m chlorate solutions of known concentrations in sealed tubes and
g the temperatures at which crystals first appear when the tubes
haken, as was done in the case of sodium nitrate.

hese experiments were made with the object of illustrating the necessity of
n.

Solutions containing 51·55 and 53·578 per cent. of the chlorate were enclosed in sealed tubes. They were then boiled in water for two days and nights and shaken at intervals. The tubes were then immediately immersed in a water-bath at 89° and allowed to cool. The tubes were kept continually in motion during cooling by means of the same apparatus which was used for the experiments with the sealed sodium nitrate tubes (experiments 26—32). The temperature of the water-bath was observed as the tubes cooled. The water-bath cooled finally to the temperature of the room without crystals appearing in either of the sodium chlorate tubes, although the temperature attained was far below that at which the solutions passed into the labile state, as determined by the experiments relating to the refractive index.

Since, therefore, cooling and continual motion fail to bring down crystals in the sealed sodium chlorate tubes, other tubes were made up containing known sodium chlorate solutions and also a fragment of some insoluble substance, such as mica, galena, or fluorspar, in order to ascertain whether friction combined with continual motion will bring crystals down in the tubes at the labile temperature. The cubic substances were chosen in view of the possibility that one or other may be structurally related to sodium chlorate. It was found that a tube containing sodium chlorate solution together with a fragment of one of these cubic substances, when cooled down slowly in the water-bath and shaken continually, crystallised in a shower at almost exactly the temperature at which the solution reached the supersolubility curve T. The same result was obtained with garnet, magnetite or zinc-blende. A tube containing a fragment of glass also crystallised in a shower on the supersolubility curve. In all these experiments, it is most important to make sure that all the solid sodium chlorate in the tube is dissolved before beginning any experiment. To do this it was necessary to boil the tubes for at least two days and nights in water, or for one day and night in amyl alcohol, shaking them at intervals. If the tubes were boiled for less than this time, crystals almost invariably appeared in them in the metastable state, showing that the solid material in the tubes could not have been completely dissolved.

The following table gives the concentrations of the solutions in the tubes, the substances contained in the tube together with the solution, the temperature at which the labile state is reached for each solution taken from the diagram, and the observed temperatures at which crystals appeared in the various tubes.

No. of expt.	Percentage of NaClO₃ in tube.	Substance in the tube with the solution.	Labile temperature taken from Fig. 9.	Observed temperatures at which crystallisation took place.	
				Tubes boiled two days and nights in water or 24 hours in amyl alcohol.	Tubes-boiled one day and night only in water.
44	51·55	Solution only in tube	19·5°	No crystals appeared in tube which was cooled to 16°	—
45	52·668	Galena	23·5	24°, 23·5°	23·1°, 24°, 27°
46	53·578	Solution only in tube	26·0	No crystals appeared in tube which was cooled to 16°	19°
47	54·108	Garnet	28·0	28·8°	30°
48	54·27	Fluorspar	28·5	28·5°, 28·5°, 28·2°	28·5°, 28·5°, 28·5°, 28·2°
49	55·142	Galena, mica, & zinc-blende	31·5	32°, 31·5°, 31·5°	36·8°, 34·7°, 32°, 39°
50	55·772	Fluorspar	33·5	33°, 34·3°, 33·2°	33·2°, 32°
51	56·212	Glass	34·8	34·7°	34·8°

It appears, therefore, that with moderate shaking and friction within the tubes, and regular cooling, crystals appear as a cloud almost exactly when the labile state (indicated by the experiments on refractivity) is reached ; the curve plotted with concentrations as ordinates and temperatures of crystallisation as abscissæ coincides almost exactly with the supersolubility curve T, Fig. 9, obtained by the observations on the refractive indices of sodium chlorate solutions already described. It also appears from these experiments that a fragment of glass is quite as effective in bringing down a shower of crystals at the labile temperature in a closed tube as is a fragment of some cubic substance. We therefore assume that it is the friction within the tube which brings down the shower of crystals, and not any action due to the cubic symmetry of the mineral substances employed.

Experiments with other Salts.

Experiments were also tried with solutions of the following substances : potassium alum, ammonium alum, sodium thiosulphate, ammonium oxalate, potassium sulphate, and sodium chloride.

As in the first experiments described, solutions of approximately known concentrations were examined in the trough of the inverted goniometer. With the exception of sodium chloride, all the salt solutions treated behaved in· a similar manner to sodium nitrate : that is, as the solutions cooled, crystals appeared in the trough, where they continued to grow, while the refractive index first rose to a maximum and then began to fall rapidly as the temperature decreased further. In the case of moderately stirred solutions, shortly after the index reached its maximum a cloud of crystals suddenly appeared throughout the solution, causing a very sudden drop in the refractive index, while in unstirred solutions the index rose to a maximum and then fell more gradually than in the stirred solutions, there being no sudden liberation of crystals in the trough, but only a quantity of large crystals growing quietly at the bottom of the vessel. The following are some of the results obtained for the various salts.

Experiments with Potassium Alum.

Expt. 52.— Concentration : $K_2Al_2(SO_4)_4,24H_2O = 24\cdot48$ per cent. Stirred solution. Index rose from $1\cdot354400$ at 45° to $1\cdot355400$ at 39·5° and then fell from this value to $1\cdot348574$ at 28°, after which it decreased more gradually with temperature until it reached $1\cdot344565$ at 17·3°. Crystals were growing in the trough throughout the experiment. A very thick cloud of crystals appeared throughout the solution at 33° which lasted until the temperature reached 28°.

Expt. 53.—Concentration : $K_2Al_2(SO_4)_4,24H_2O = 20\cdot462$ per cent. Stirred solution. The index rose from $1\cdot350391$ at 46·5° to $1\cdot351121$ at 39·5°, and then fell from this value to $1\cdot346857$ at 32°, after which it decreased more gradually until it reached $1\cdot344617$ at 19·4°. Crystals appeared in the trough at 41·5°. At 40°, a very thick cloud appeared throughout the solution, which lasted until the the temperature reached 32°.

Expt. 54.—Concentration : $K_2Al_2(SO_4)_4,24H_2O = 15\cdot375$ per cent. Stirred solution. The index rose from $1\cdot350755$ at 42·5° to $1\cdot352422$ at 30·5°, and then fell from this value to $1\cdot346909$ at 24·2°, after which it decreased more gradually until it reached $1\cdot345139$ at 20·4°. Crystals appeared in the trough at 30·5°, and there is a depression in the curve at 32·5°. A dense cloud of crystals appeared throughout the solution at 26·5°, which lasted until the temperature reached 24°.

Experiments with Ammonium Alum.

Expt. 55.—Concentration : $(NH_4)_2Al_2(SO_4)_4,24H_2O = 21\cdot065$ per cent. Stirred solution. The index increased from $1\cdot353255$ at 39·45° to $1\cdot355178$ at 24·45°, and then dropped suddenly from this value to $1\cdot343731$ at 15·3°. A dense cloud of crystals appeared throughout the solution between 20° and 19°. Crystals first appeared in the trough at 36·5°.

Expt. 56.—Concentration : $(NH_4)_2Al_2(SO_4)_4,24H_2O = 23\cdot0015$ per cent. Stirred solution. The index rose from $1\cdot353567$ at 47·55° to $1\cdot357311$ at 26·45°. It then fell suddenly from this value to $1\cdot343214$ at 14·85°. Crystals first appeared in the trough at 37·45°, and a very thick shower of crystals appeared at 22·4°.

The refractive-index temperature curves obtained from the potassium and ammonium alums differ from those of sodium nitrate in that the fall of index from the maximum point to the lowest temperature attained is far greater for the alums than for sodium nitrate. Thus, for the potassium alum solution of experiment 52 we have a fall of index from 1·355440 at the maximum point to 1·344565 at 17·3°, or a fall of 0·010875 in refractive index ; and for the ammonium alum of experiment 56 a fall in index from 1·357311 at the maximum point to 1·343214 at 14·85°, or a fall of 0·014097 in refractive index.

The greatest fall in index obtained from sodium nitrate solutions was in experiment 19, when the index fell from 1·393712 at the maximum point to 1·389503 at 19·4°, or a fall of 0·004209 in refractive index.

The cloud of crystals which forms in the alum solutions after the maximum point is reached is usually so dense as to make the liquid nearly opaque, and stirring had to be stopped before a reading for the refractive index of the solution could be taken.

The maximum values of refractive index yielded by the solutions of potassium alum do not appear to lie on a single definite curve T like that established for sodium nitrate and sodium chlorate, but on two such curves separated by an interval of about 5°. This may be due to the undoubted fact that two sorts of crystals may separate from an alum solution, one isotropic and the other birefringent.

We propose to return to this subject in a subsequent investigation.

Experiments with Sodium Thiosulphate.

Expt. 57.—Approximate concentration unknown. As the solution was stirred, the index rose from 1·436878 at 46·5° to 1·441460 at 27·15°, and then fell suddenly from this value to 1·429778 at 22·85°. Crystals appeared in the trough at 32°. The resulting temperature refractive-index curve shows two maxima, the first at 30°, after which the index falls slightly and then rises again to the final maximum value at 27·15°. A shower of crystals appeared throughout the solution at 26·65°, causing a considerable rise in temperature from 26·65° to 29·75°. The index then continued to fall more slowly with temperature in a regular manner.

Expt. 58.—Concentration : $Na_2S_2O_3,5H_2O = 71·965$ per cent. Stirred solution. The index rose from 1·432069 at 39·45° to 1·436135 at 21·9°, and then fell suddenly from this value to 1·426484 at 18·85°. Crystals appeared in the trough at 30·5°. The resulting curve again shows two maxima, the first at 32° and the second at 21·9°. A shower of crystals appeared at 21·4°, causing a rise of temperature of 2°.

Expt. 59.—Concentration : $Na_2S_2O_3,5H_2O = 70$ per cent. Stirred solution. The index rose from 1·434946 at 44·5° to 1·439296 at 24·45°, and then fell suddenly from this value to 1·430773 at 21·9°. Crystals appeared in the trough at 36°. The resulting curve again shows two maxima, the first at 38·45° and the second at 24·45°. A shower of crystals appeared at 24·45°, causing a rise of temperature of 2°. The results of this experiment are plotted in Fig. 10.

Experiments with Ammonium Oxalate.

Expt. 60.—Concentration : $(NH_4)_2C_2O_4 = 8·576$ per cent.   Stirred solution.   The index rose from 1·343266 at 49·55° to 1·346282 at 30°. It then fell suddenly from this value to 1·343318 at 29·5°, after which

FIG. 10.

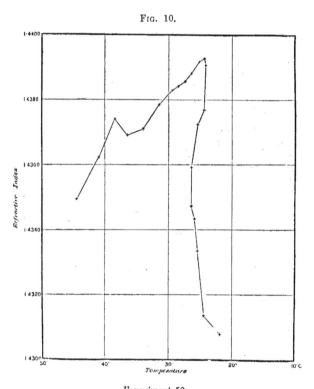

Experiment 59.

it decreased more gradually to 1·340560 at 14·35°.   Crystals appeared in the trough at 33°, and a depression appears in the resulting curve at this temperature.   A very dense shower of crystals appeared throughout the solution at 30°, causing the solution to be absolutely opaque.

Expt. 61.—Concentration : $(NH_4)_2C_2O_4 = 8·2126$ per cent.   The index rose from 1·342049 at 48·55° to 1·345866 at 28°.   It then fell suddenly from this value to 1·342329 at 24·45°, after which it decreased more gradually with temperature to 1·340560 at 13·65°. Crystals appeared in the solution at 31·5°, and there is a slight depression in the curve at this point. At 28°, a very dense cloud of crystals appeared in the solution, causing it to appear almost solid.

Sodium Chloride.

Expt. 62.—Concentration: $NaCl = 26·932$ per cent. The index rose from 1·377484 at 48·55° to 1·381516 at 14·35°· The solution had crystals in it throughout the experiment. The curve shown is almost a straight line, the index increasing regularly as the temperature decreased and showing no maximum point; in this respect, the solution differs from all the other solutions examined, but sodium chloride does not easily form supersaturated solutions.

The Effect of Friction on Sodium Nitrate Solutions in Sealed Tubes.

The action of friction, combined with shaking, in inducing crystallisation within the sealed tubes containing sodium chlorate solutions suggested that experiments should be made to test the action of friction within a tube containing a sodium nitrate solution. Angular fragments of glass were therefore inserted in tubes containing sodium nitrate solutions of known concentrations and the tubes were sealed. They were boiled in water to dissolve all crystal germs, after which they were placed in a water-bath at 80° and continually shaken as they cooled in the manner already described.

The following are the results obtained :

Experiment.	Concentration of NaNO₃ in the tube.	Labile temperature taken from diagram 3.	Temperature at which crystals appeared in tube.
63	54·091 per cent.	44·0°	45·5°, 45°, 44°
64	48·357 ,,	17·3	18·6°, 18·8°
65	51·639 ,,	32·5	34°
66	50·676 ,,	27·7	29°

The friction caused by shaking angular glass fragments within the sodium nitrate tubes appears to cause the solutions to crystallise at a slightly higher temperature than that indicated by the supersolubility curve T on diagram 3, which was fixed from the results of the series of experiments 1—19 on the refractive index, and verified by the experiments 26—32 on solutions in sealed tubes containing nothing but the solution.

These new experiments seem to indicate that the position of the supersolubility-curve T may have been fixed about $1\frac{1}{2}$ degrees too low on the diagram 3.

If by different methods slightly different temperatures are obtained for the transition from the metastable to the labile condition, we must

regard the highest values as being nearest to the true temperature
since, as was indicated by the sodium chlorate experiments, the
solution may go over into the labile state without crystallising unless
appropriate mechanical means are employed to make it crystallise.
It is possible, therefore, that the true supersolubility curve for
sodium nitrate is a line approximately parallel to T in diagram 3, but
slightly to the right of it.

Selected experiments, most of which are represented in Figs. 3, 7, and 9.

Sodium Nitrate.

Experiment 1.—Concentration unknown.

Index of solution.	Temperature.
1·388577	36·95°
1·388936	35·5
1·389607	32·0
1·390015	29·5
1·390634	27·45
1·391105	25·45
1·391544	23·45
1·391969	22·2
1·392378	21·1
1·392484	20·4
1·392643	19·4
1·393004	18·35
1·393004	17·3
1·392914	16·8
1·392690	16·5
1·391302	14·55

Experiment 2.—53·1 per cent. $NaNO_3$.

Index of solution.	Temperature.
1·393410	45·5°
1·394332	39·45
1·394230	36·45
1·393922	32·5
1·393310	29·5
1·393895	27·45
1·392590	25·45
1·392227	23·45
1·391917	22·4
1·390430	16·05

Experiment A.—49·325 per cent. $NaNO_3$.

Index of solution.	Temperature.
1·384816	54·5°
1·385849	50·5
1·386516	47·55
1·387443	43·5
1·388063	40·45
1·388782	37·45
1·389606	33·5
1·390171	31·5
1·390634	29·5
1·391096	27·45
1·391508	25·45
1·391814	24·45
1·392071	23·45
1·392174	22·4
1·392277	21·4
1·392226	20·4
1·392174	19·4
1·391814	17·8
1·391148	16·3
1·390634	16·3

Experiment B.—48·7 per cent. $NaNO_3$.

Index of solution.	Temperature.
1·382545	59·5°
1·384455	51·5
1·385385	47·55
1·386158	44·5
1·387391	38·45
1·388267	34·5
1·388936	32·5
1·388555	29·5
1·390171	26·95
1·390634	24·45
1·390888	23·45
1·391045	22·4
1·391251	21·4
1·391457	19·9
1·391560	19·4
1·390328	14·85

Experiment 8.—Approximate concentration 47·45 per cent. NaNO₃.

Index of solution.	Temperature.
1·386054	41·45°
1·386979	38·45
1·387650	35·5
1·388267	32·5
1·388577	31·5
1·388988	29·5
1·389414	27·45
1·389760	25·45
1·390378	23·45
1·390531	22·45
1·390736	21·4
1·391351	17·6
1·390481	16·7

Experiment 10.—49·53 per cent. NaNO₃.

Index of solution.	Temperature.
1·388623	41·45°
1·389162	39·45
1·389638	37·45
1·389957	35·5
1·390380	34·0
1·390803	32·0
1·391121	30·5
1·391755	28·0
1·392022	25·95
1·392234	24·45
1·392234	23·45
1·392022	22·7
1·391916	21·65
1·391702	21·4
1·390591	17·9
1·389904	16·8

Experiment 12.—49·414 per cent. NaNO₃.

Index of solution.	Temperature.
1·389109	42·5°
1·389744	38·45
1·390327	36·5
1·390909	33·0
1·391121	31·5
1·391650	29·5
1·391863	28·0
1·392234	26·45
1·392444	25·7
1·392444	23·95
1·390220	17·8

Experiment 13.—49·25 per cent. NaNO₃.

Index of solution.	Temperature.
1·388897	42·5°
1·389691	38·45
1·390538	35·45
1·391280	31·5
1·391650	29·5
1·391863	28·5
1·392128	27·45
1·392392	26·45
1·392496	25·45
1·392602	24·45
1·392602	22·0
1·392444	21·2
1·392128	20·4
1·391755	19·4
1·389691	15·8

Experiment 15.—49·81 per cent. NaNO₃.

Index of solution.	Temperature.
1·390273	38·45°
1·391227	35·0
1·391916	32·5
1·392234	30·75
1·392496	29·5
1·392761	28·0
1·392867	26·95
1·392392	26·25
1·390433	25·45
1·390220	24·85
1·390062	24·25
1·389850	22·45
1·389585	21·4
1·389532	20·4
1·389109	18·35
1·389003	17·3

Experiment 16.—49·02 per cent. NaNO₃.

Index of solution.	Temperature.
1·389215	37·45°
1·389850	34·5
1·390327	32·5
1·390909	29·5
1·391386	28·0
1·391702	26·45
1·392022	24·95
1·392128	23·95
1·392234	23·05
1·392287	22·75
1·391386	22·65
1·389956	21·7
1·389904	21·1
1·389691	20·4
1·389532	19·4
1·389479	18·85

Sodium Nitrate.

Experiment 19.—53·1 per cent. NaNO₃.

Index of solution.	Temperature.
1·389709	58·5°
1·391148	54·0
1·392277	50·5
1 392430	47·55
1·392690	45·5
1·393204	43·5
1·393459	40·45
1·393712	36·5
1·392690	36·5
1·391663	36·0
1·391404	35·5
1·391045	34·5
1·390634	31·5
1·390328	29·0
1·390120	26·95
1·390015	25·45
1·389709	22·4
1·389503	19·4

Experiment 20.—48·19 per cent. NaNO₃.

Index of solution.	Temperature.
1·387461	43·5°
1·387985	40·45
1·388728	37·45
1·389471	34·5
1·390001	33·0
1·390423	31·0
1·390476	29·5
1·390529	29·0
1·390529	28·5
1·390476	28·25
1·390264	27·45
1·390212	26·45
1·390104	25·45
1·389896	24·45
1·389788	23·25
1·389630	21·7
1·389418	20·9
1·389148	19·15
1·388512	14·5

Experiment 21.—49·5 per cent. NaNO₃.

Index of solution.	Temperature.
1·386210	53·5°
1·386877	49·5
1·388062	46·55
1·388525	44·5
1·389194	41·45
1·390222	38·45
1·390681	36·5
1·391105	34·5
1·391630	32·0
1·391630	30·5
1·391531	30·5
1·391054	29·5
1·391054	28·5
1·390839	28·0
1·390531	26·45
1·390222	25·45

Experiment 23.—49·87 per cent. NaNO₃.

Index of solution.	Temperature.
1·385022	59·5°
1·386726	52·5
1·387443	49·5
1·388370	46·0
1·388988	43·5
1·389761	40·45
1·390480	37·45
1·390837	35·5
1·391251	33·5
1·391508	32·5
1·391713	31·5
1·391970	30·5
1·392123	29·5
1·391457	29·0
1·390736	28·5
1·390428	28·0
1·390120	26·95
1·389913	25·45
1·389709	23·45
1·389301	20·4
1·389092	17·3

Experiment 24.—47·616 per cent. NaNO$_3$.

Index of solution.	Temperature.	Index of solution.	Temperature.
1·385281	46·5°	1·390325	24·45°
1·385951	43·5	1·390685	23·15
1·386826	40·0	1·390839	22·4
1·387650	36·5	1·390941	21·4
1·388164	34·0	1·390634	20·8
1·388885	31·5	1·390221	20·4
1·389370	29·5	1·389400	19·9
1·389811	27·5	1·389092	18·3
1·390015	26·45	1·388988	17·3

Sodium Chlorate.

Experiment 35.—54·835 per cent. NaClO$_3$.

Time. Hours Mins.	Index of solution.	Temperature.
2 33½	1·391253	50·55°
2 35	1·391864	47·55
2 37	1·392590	44·5
2 42	1·393867	39·45
2 44	1·394332	37·45
2 46	1·394589	36·5
2 49½	1·394846	34·5
2 52	1·394795	33·5
2 54	1·393643	33·0
2 57	1·392378	32·5
3 0	1·391148	32·0
3 5	1·390787	30·5
3 13	1·390481	28·5
3 22	1·390221	26·45
3 30	1·390015	24·45
3 45	1·389709	22·45
3 51	1·389606	21·5

Experiment 36.—54·054 per cent. NaClO$_3$.

Index of solution.	Temperature.
1·391814	48·55°
1·392484	44·5
1·393265	41·45
1·394023	38·45
1·394743	35·5
1·395003	33·5
1·393699	32·7
1·391544	32·3
1·391148	31·5
1·390889	30·5
1·390531	28·5
1·390377	26·25
1·390118	24·45
1·389416	21·1
1·389370	20·2

Experiment 37.—51 736 per cent. NaClO$_3$.

Index of solution.	Temperature.
1·390430	35·0°
1·390941	33·5
1·391200	31·0
1·392020	29·0
1·392277	27·45
1·392690	25·45
1·392742	24·45
1·392020	24·35
1·391200	24·25
1·390481	24·05
1·390377	23·65
1·390015	22·4
1·389301	20·9
1·388988	19·7

Experiment 38.—51·035 per cent. NaClO$_3$.

Index of solution.	Temperature.
1·387083	43·5°
1·388063	40·45
1·388731	37·45
1·389400	35·0
1·389861	32·5
1·390480	29 5
1·391003	27·45
1·391457	25·95
1·391814	24·45
1·391970	23·45
1·392020	22·45
1·391663	21·4
1·391404	21·2
1·391251	21·1
1·390532	20·4
1·389709	20·7

Sodium Chlorate.

Experiment 39.—52·397 per cent. NaClO₃.

Index of solution.	Temperature.
1·385488	55·0°
1·387340	48·0
1·388526	43·0
1·389246	39·0
1·390024	36·0
1·391045	32·0
1·391457	30·0
1·391970	28·0
1·392277	27·0
1·392639	26·0
1·392690	25·0
1·392590	25·0
1·392071	24·9
1·391253	25·0
1·389862	25·0
1·389657	24·0

Experiment 40.—56·004 per cent. NaClO₃.

Index of solution.	Temperature.
1·392174	54·0°
1·392690	51·0
1·393815	47·0
1·394538	44·0
1·395054	42·0
1·395613	40·0
1·396075	38·0
1·396075	37·0
1·396381	35·0
1·396330	34·0
1·395767	33·0
1·391455	24·2

Experiment 41.—54·46 per cent. NaClO₃.

Index of solution.	Temperature.
1·387495	57·0°
1·388782	52·5
1·389606	49·0
1·390325	46·0
1·391043	43·0
1·391662	40·0
1·392277	37·0
1·393047	34·0
1·393815	31·0
1·394126	29·0
1·394229	28·5
1·393712	28·0
1·390839	29·0
1·390430	29·0

Experiment 42.—55·77 per cent. NaClO₃.

Index of solution.	Temperature.
1·391250	53·5°
1·392175	49·0
1·392996	47·0
1·393559	44·0
1·394229	41·5
1·394743	38·5
1·395054	36·0
1·395054	35·0
1·394952	34·0
1·394640	33·5
1·393308	32·5
1·391509	32·3
1·391043	32·0
1·389195	22·2

Experiment 43.—55·037 per cent. NaClO₃.

Index of solution.	Temperature.	Index of solution.	Temperature.
1·388577	57·5°	1·393308	38·0°
1·390015	52·0	1·393661	36·0
1·391043	48·0	1·394023	34·0
1·391611	45·0	1·394434	32·0
1·392484	42·0	1·393460	29·0
1·392944	40·0	1·391455	28·8

Potassium Alum.

Experiment 52.—24·48 per cent. alum.

Index of solution.	Temperature.	Index of solution.	Temperature.
1·354400	45·0°	1·351902	29·5°
1·354762	41·45	1·349509	28·5
1·355440	39·45	1·348574	28·0
1·355334	37·45	1·347481	26·45
1·354917	34·5	1·346857	25·45
1·354502	33·0	1·346387	23·45
1·353723	31·5	1·345347	21·4
1·353255	30·5	1·344565	17·3

Ammonium Alum.

Experiment 55.—21·065 per cent. alum.

Index of solution.	Temperature.
1·353255	39·45°
1·353775	36·5
1·354086	34·5
1·354298	32·5
1·354710	30·5
1·354969	28·5
1·355125	26·45
1·355178	24·45
1·355074	23·45
1·354813	22·4
1·354034	21·4
1·353359	20·9
1·352631	20·4
1·351486	19·9
1·350443	19·4
1·348782	19·4
1·347584	19·4
1·346699	19·1
1·346075	18·6
1·345451	18·3
1·343731	15·3

Sodium Thiosulphate.

Experiment 59.—70 per cent. thiosulphate (approximate).

Index of solution.	Temperature.
1·434946	44·5°
1·436234	40·95
1·437420	38·45
1·436927	36·5
1·437124	34·0
1·437864	31·5
1·438310	29·5
1·438458	28·5
1·438606	27·45
1·438853	26·45
1·439197	25·2
1·439296	24·45
1·439098	24·2
1·437716	24·45
1·437272	25·45
1·435937	26·45
1·434748	26·45
1·434352	25·95
1·483660	25·45
1·431371	24·45
1·430773	21·9

Ammonium Oxalate.

Experiment 60.—8·576 per cent. oxalate.

Index of solution.	Temperature.
1·343266	49·55°
1·344099	44·5
1·344556	41·45
1·345138	38·45
1·345710	36·5
1·345970	33·0
1·346282	30·5
1·346282	30·0
1·345452	30·0
1·343995	30·0
1·343318	29·5
1·342845	28·5
1·342535	26·45
1·342226	24·45
1·342019	23·45
1·340560	14·35

Sodium Chloride.

Experiment 62.—26·932 per cent. NaCl.

Index of solution.	Temperature.
1·377484	48·35°
1·377949	45·5
1·378363	42·5
1·378777	38·95
1·379083	36·5
1·379293	34·5
1·370502	32·5
1·379708	30·0
1·379915	28·5
1·380122	26·45
1·380431	24·45
1·380740	22·4
1·380946	20·4
1·381050	18·85
1·381206	17·3
1·381516	14·35

Summary.

It may seem remarkable that the facts described in this paper have not been established before, but on consideration it will be seen that it would be difficult to ascertain the position of the supersolubility curve by any other method than that of the refractive indices. Crystallisation in sealed tubes would not be sufficient, for, although spontaneous crystallisation (that is, growth of crystals without inoculation) cannot take place until the solution reaches the curve T, it is by no means necessary that crystals should form then. In a still tube, the solution may pass far into the labile state without crystallising, and even vigorous shaking is not enough to promote crystallisation in all solutions. The behaviour of sodium chlorate shows that some other mechanical stimulus, such as friction, may be required.

When, however, the crystals have been brought down by appropriate mechanical means, it is found that the maximum temperatures at which they appear lie on the supersolubility curve which was obtained by optical measurements in an open trough. On the other hand, it would be very difficult to determine the curve T from observations in the open trough otherwise than by measurements of the refractive index. In a still solution, crystals are growing continuously, the change is gradual, and the solution may pass into the labile state without any visible alteration. In a stirred solution, it is true that a cloud comes down near the temperature corresponding to the supersolubility curve, but owing to the presence of crystals in the solution minor clouds may appear before that point is reached, and it is in any case difficult to fix the exact temperature at which the more profuse cloud descends.

Without the knowledge (derived from the refractive index) that a real change takes place in the nature of the solution, one would scarcely expect a transition from the metastable to the labile state to be possible in a solution in which crystals are forming freely, and even the production of a profuse cloud would scarcely be attributed to such a change, but would merely seem to indicate that at a certain temperature crystallisation becomes more rapid. But when observations on the refractive index are combined with experiments with sealed tubes, we may draw the following conclusions from the facts established for the salts which we have examined (with the exception of sodium chloride, which does not readily form supersaturated solutions).

(1) For a cooling supersaturated aqueous solution of certain strength, there is a temperature t^o at which a sudden diminution of refractive index occurs.

(2) This is accompanied or followed by a copious separation of crystals.

(3) The same solution enclosed in a sealed tube cannot be made to crystallise at a lower temperature except by inoculation with a crystal of the solute (or of an isomorphous substance).

(4) But, in general, the solution in a sealed tube does crystallise at or slightly below $t°$ when shaken, especially if friction be employed.

(5) The increase of crystallisation at $t°$ is therefore due to the spontaneous growth of new crystals in addition to those already growing; in other words, at $t°$ the solution passes into the labile state.

In an open solution, the presence of growing crystals in moderate quantities while the solution is in the metastable state, so far from preventing, actually facilitates the spontaneous growth of new crystals at $t°$. If, however, they are present in excess they may prevent the latter, since their growth so far weakens the solution as to prevent it from passing into the labile state. In sealed tubes, although spontaneous growth may take place at $t°$, it by no means necessarily does so; in those containing sodium thiosulphate, we have been unable to produce crystallisation even at temperatures far below $t°$.

If the change in refractive index be simply due to an increase in the velocity of crystallisation on passage into the labile state, as suggested above, an index-time curve constructed with refractive indices as ordinates and times as abscissæ should exhibit much the same form as the index-temperature curve. A number of experiments have established this fact. One of these (Expt. No. 35) is quoted on p. 437.

MINERALOGICAL DEPARTMENT,
UNIVERSITY MUSEUM,
OXFORD.

XLVIII.—*The Relation of Position Isomerism to Optical Activity. VI. The Rotation of the Menthyl Esters of the Isomeric Chloronitrobenzoic Acids.*

By JULIUS BEREND COHEN and HENRY PERCY ARMES.

IT was pointed out in a previous paper on this subject (Trans., 1905, **87**, 1192) that whereas chlorine and bromine in the ortho-position to the active group lower its rotation, the nitro-group has the opposite effect of greatly increasing it. It seemed, therefore, an interesting problem to ascertain the combined effect of halogen and nitro-group on the rotation of a series of menthyl esters. This has been completed as far as practicable. We found ourselves obliged to omit two members of the series of isomeric chloronitrobenzoic esters owing to the difficulty of obtaining the 3-chloro-2-nitrobenzoic acid. According to Holleman (Abstr., 1900, **1**, 638), a small quantity can be separated from the products of nitration of *m*-chlorobenzoic acid, but this method did not appear very satisfactory. In conjunction with Mr. H. G. Bennett we have succeeded in obtaining 3-chloro-2-nitrotoluene, which gives the required acid on oxidation, but the method of preparing the toluene derivative affords too small a yield to be practicable for the present purpose. There seemed little object in preparing the 2-chloro-3-nitro-ester without the corresponding 3-chloro-2-nitro-isomeride, and we therefore decided to omit both from the present investigation.

The details of preparation of acids, acid chlorides, and esters are fully described in the experimental part. Table I contains a list of the physical constants of the compounds in question.

TABLE I.

Substance, Cl : NO₂.	M. p. of acid, C. and A.	M. p. of acid, previous observers.	M. p. of acid chloride.	M. p. of ester.
2 : 4	140—142°	136—137° (Wachendorff)	—	—
4 : 2	140—143	{ 138—139 (Varnholt) { 140—141 (Green, Lawson) } 31—34°		63—66°
2 : 5	164	165 (Hubner)	60	55—57
5 : 2	132—136	137—138 ,,	—	80—82
2 : 6	157—161	161 (Green, Lawson)	32—34	127—129
3 : 4	180—182	185—186 (Claus, Kurz)	—	54—56
4 : 3	177—179	178—180 (Hubner)	51—52	112—113
3 : 5	148—149	147 ,,	—	· 42—44

The relation between the high melting point of the 2 : 6-ester and the low melting points of the 2 : 4- and 3 : 5-esters is in close agreement with that of the dihalogen esters, but the correspondence between the high melting point of the acid and the low melting point of its ester observed among the dihalogen compounds does not hold in the present case, a notable exception being the 4 : 3-compound, which exhibits a high melting point both of the acid and ester. On the other hand, the least fusible acid (3 : 4) forms a very fusible ester.

Table II gives the densities and specific and molecular rotations of the esters fused at 100°, at which temperature, with two exceptions, they are all fusible. The rotation of the 4 : 3-ester (m. p. 112—113°) was determined at 120°, and that of the 2 : 6-ester (m. p. 127—129°), owing to its high melting point, was examined in benzene solution at 20°. As we have shown that a range of even 80° does not affect the rotation in most cases by more than 2 or 3 per cent., the constants for the above two substances will not be subject to any serious error arising from differences of temperature. Moreover, we have found that the solutions of the esters in benzene give very nearly normal values.

TABLE II.

Menthyl ester, Cl : NO$_2$.	Density at 100°.	$[a]_D^{100°}$.	$[M]_D^{100°}$.
2 : 4	1·119	− 51·63°	− 175·39°
4 : 2	1·117	109·30	− 371·30
2 : 5	1·118	63·17	− 214·60
5 : 2	1·116	113·45	− 385·40
2 : 6	In benzene solution 69·77		− 237·00
3 : 4	1·112	61·05	− 207·39
4 : 3	1·091	67·53	− 229·30
3 : 5	1·109	73·01	− 248·00

Perhaps the most instructive way of viewing the above rotations is to calculate them on the supposition that the mononitro- and monochloro-benzoic esters (Trans., 1905, 87, 1192) retain their values unaltered in the chloronitro-esters., The difference produced by an orthonitro-group would then be to increase the negative rotation by 145°, that of the metanitro-group by 15°. The effect of the orthochlorine atom would be to decrease the negative rotation by 41°. We may regard the paranitro-group and the meta- and para-chlorine atoms as having roughly no rotational effect. Calculated in this way, we obtain the following values (Table III):

TABLE III.

Cl : NO$_2$.	Observed.	Calculated.	Cl : NO$_2$.	Observed.	Calculated.
2 : 4	175	195	2 : 6	237	340
4 : 2	371	340	3 : 4	207	236
2 : 5	214	210	4 : 3	229	251
5 : 2	385	381	3 : 5	248	251

Anything like an exact agreement could scarcely be expected, seeing that the rotations of each group of esters has been determined at a different temperature; but, taking this fact into account, there is a remarkably close correspondence between the observed and calculated values for the 2 : 5-, 5 : 2-, and 3 : 5-esters and a rough agreement between the remaining members of the series with the striking exception of the 2 : 6-ester, which falls far below the calculated number. Its low rotation recalls the abnormally low values of the dihalogen esters (Trans., 1906, 89, 48). It will be instructive to see now whether the combined effect of two nitro-groups in the 2 : 6-positions will enhance or depress the rotation to a greater extent than the presence of a single nitro-group. With this object in view, we are at present engaged in preparing a few of the more accessible dinitrobenzoic esters, including the 2 : 6-compound, which we hope to report on shortly.

In the course of the present investigation, we have determined the rotations of some of the esters not only in the fused state, but also in benzene and chloroform solution, with the result that the chloroform values are abnormally high. We propose to investigate this matter in greater detail in order to find whether the difference in the action of the solvents can be traced to any specific group.

EXPERIMENTAL.

Menthyl 2-Chloro-4-nitrobenzoate.—The acid was obtained according to Wachendorff's method (*Annalen*, 1877, **185**, 275), namely, by oxidising 2-chloro-4-nitrotoluene with dilute nitric acid. We obtained the best result by heating with 4 to 5 volumes of nitric acid (1 vol. of nitric acid, sp. gr. 1·4 to 2 vols. of water) at 125° for five hours. Two tubes containing 7 grams each gave 9·5 grams of crude acid (m. p. 139°). The acid is easily purified by recrystallising from benzene, and separates in large, shining prisms containing benzene of crystallisation, which they gradually part with in the air and lose their lustre. The purified crystals free from benzene melted sharply at 140—142°. The acid was heated with rather more than its own weight of phosphorus pentachloride on the water-bath for an hour. The acid chloride could not be frozen in spite of the fact that Grohmann (*Ber.*, 1891, **24**, 3812) states that it gradually solidifies and melts at 115°. It was freed from the greater part of the phosphorus oxychloride by heating under diminished pressure, the remainder being removed by shaking with small quantities of light petroleum, in which the acid chloride is only slightly soluble. A small quantity of phosphorus pentachloride was then separated by dissolving the acid chloride in benzene and filtering.

The product freed from benzene and weighing 9 grams was heated in the oil-bath with an equal weight of menthol. The reaction began at 105—110° and was completed by heating to 130° for an hour, when the evolution of hydrogen chloride had nearly ceased. The product, after being rendered alkaline with sodium carbonate, was distilled in steam until the smell of menthol could no longer be detected in the distillate. The ester was extracted with ether, dehydrated, and the ether removed. The ester obtained in this way is a light brown, viscid liquid which showed no signs of crystallisation when kept for nearly a year. A second preparation made in the same manner gave a similar product.

The following are the polarimetric readings, densities, and specific and molecular rotations of the two specimens I and II ($l = 0·302$ dcm.):

Menthyl 2-Chloro-4-nitrobenzoate.

Temperature.	Reading.		Density.	$[a_D^c]$		$[M]_D^c$	
	I.	II.		I.	II.	I.	II.
100°	−17·44°	−17·45°	1·119	−51·62°	−51·64°	−175·36°	−175·42°
80	17·62	17·61	1·135	51·40	51·37	174·61	174·51
70	17·80	17·79	1·141	51·65	51·63	175·46	175·39
65	17·85	17·85	1·145	51·62	51·62	175·36	175·36
40	17·89	17·88	1·164	50·89	50·87	172·88	172·81
20	17·90	17·90	1·182	50·14	50·14	170·33	170·33

The analyses of this and the other esters are given in Table IV at the end of the paper.

Menthyl 4-Chloro-2-nitrobenzoate.—The acid was prepared by oxidising 4-chloro-2-nitrotoluene, according to Varnholt (*J. pr. Chem.*, 1887, [ii], **36**, 30), the best results being obtained by using 25 c.c. of dilute nitric acid (1 vol. HNO_3, sp. gr. 1·4, to 2 vols. of water) and 5 grams of the mononitrotoluene and heating for twelve hours at 120°.

The yield of crude acid amounted to about half the weight of the original material and melted at 130—139°. After crystallising repeatedly from water and finally from benzene, a colourless product was obtained (m. p. 140—143°). The acid chloride (m. p. 31—34°), prepared in the usual way, was heated with the calculated quantity of menthol in the oil-bath. The reaction started at 110—120° and was completed at this temperature after an hour. The ester, which crystallises from alcohol in needles, melted at 63—66°.

Menthyl 4-Chloro-2-nitrobenzoate.

Temperature.	Reading.	Density.	$[a]_D^c$	$[M]_D^c$
100°	−36·87°	1·117	−109·3°	−371·30°
80	36·91	1·129	108·2	367·56
70	37·01	1·138	107·6	365·52

Menthyl 2-Chloro-5-nitrobenzoate.—The acid (m. p. 164°) was obtained in two ways : (1) by nitrating o-chlorobenzoic acid as described by Hübner (*Annalen*, 1883, **222**, 195) and (2) by oxidising 2-chloro-5-nitrotoluene (m. p. 43—44°) with dilute nitric acid at 125—130°. The products obtained by the two methods were identical. The acid chloride, prepared in the usual way, was a solid, and after crystallisation from ligroin melted at 60°. On heating with menthol, the reaction began at 120° and was completed in an hour at 130—140°. The ester was recrystallised from methyl alcohol, from which it separated in colourless needles (m. p. 55—57°). Fourteen grams of o-chlorobenzoic acid gave 9 grams of ester by the first method, and 16 grams of chloronitrotoluene gave 4·5 grams of ester by the second method. The rotations of the two preparations (I and II) are given in the following table.

Menthyl 2-Chloro-5-nitrobenzoate.

Temperature.	Reading.		Density.	$[a]_D^{e}$.		$[M]_D^{e}$.	
	I.	II.		I.	II.	I.	II.
10)°	− 21·30°	− 21·34°	1·118	− 63·11°	− 63·23°	− 214·39°	− 214 80°
80	21·76	21·77	1·136	63·43	63·46	215·48	215·58
70	21·96	21·98	1·144	63·56	· 63·62	215·92	216·12
65	22·10	22·12	1·148	63·75	63·81	216·56	216·77

Menthyl 5-Chloro-2-nitrobenzoate.—The acid was obtained by Hübner's method by the action of fuming nitric acid on *m*-chlorobenzoic acid (*Annalen*, 1883, **222**, 95). Ten grams of chlorobenzoic acid mixed with 30 c.c. of fuming nitric acid were heated on the water-bath for twenty minutes. The product was poured into 10 volumes of water, concentrated, and the acid which separated extracted with small quantities of water until it no longer melted under water. It still contained an impurity of a bright yellow colour, which was removed by repeated crystallisation from water and from benzene. The melting point (132—136°) remained unchanged on further crystallisation. A second preparation of the acid was made in a similar way, but with the modification that after diluting the nitration product with water and concentrating to about 50 c.c. the acid which then separated was recrystallised from benzene only, in which the chief impurities are insoluble, and the yellow compound remains in solution. In this way, a much better yield (8 grams in the present case) was obtained, although the melting point of the acid was the same as that prepared by the previous method.

Each preparation was converted separately into acid chloride and ester. The acid chloride did not crystallise, but the menthyl ester forms colourless prisms from alcohol (m. p. 80—82°).

Menthyl 5-Chloro-2-nitrobenzoate.

Temperature.	Reading.		Density.	$[a]_D^{e}$.		$[M]_D^{e}$.	
	I.	II.		I.	II.	I.	II.
100°	− 38·23°	− 38·25°	1·116	− 113·4°	− 113·5°	− 385·23°	− 385·57°

Menthyl 2-Chloro-6-nitrobenzoate.—The acid was obtained by oxidising 2-chloro-6-nitrotoluene according to the method of Green and Lawson (Trans., 1891, **59**, 1019), but in place of permanganate the substance was heated in sealed tubes with dilute nitric acid at 120—130°. The reaction proceeds very slowly, 5 grams requiring about twenty hours for complete oxidation. The product was very impure and after repeated crystallisation from benzene melted at 157—161°. The acid chloride, which was purified by crystallisation from light petroleum, melted at 32—34°. The reaction with menthol does not begin until the temperature reaches 150° and then proceeds very slowly at 150—160°. This high temperature causes the substance to darken and yields a product which requires repeated crystallisation from alcohol and animal charcoal before becoming colourless. The purified ester melts at 127—129°. The rotation in benzene solution gave the following readings ($l = 2$ dcm.).

Menthyl 2-Chloro-6-nitrobenzoate.

Concentration per cent.	Rotation.	$[a]_D^{20°}$.	$[M]_D^{20°}$.
16·288	− 22·64	− 69·52°	− 236·16°
8·144	11·37	69·81	237·15
4·072	5·70	70:00	237·80

Menthyl 3-Chloro-4-nitrobenzoate.—The acid used in the preparation of the ester was obtained by nitrating m-chloroacetanilide, which yields the two isomerides, the 3-chloro-4-nitro- and 3-chloro-6-nitro-derivatives. The product was hydrolysed and distilled in steam, whereby the more volatile 3-chloro-6-nitroaniline is removed and the 3-chloro-4-nitro-aniline remains in the form of yellow needles (m. p. 155—156°). The latter was converted into the cyanide by Sandmeyer's reaction and separated and purified by distillation in steam. The cyanide (m. p. 87°) was hydrolysed by boiling with a mixture of equal volumes of strong sulphuric acid and water until the originally fused substance was transformed into a solid mass of crystals. The crude acid (m. p. 170—177°) obtained in this way was crystallised from hot water and melted at 180—182°. It was converted into the acid chloride, which is a viscid liquid, and then into the menthyl ester in the usual way. The ester when recrystallised from alcohol forms colourless needles melting at 54—56°. Seventy-eight grams of m-chloroacetanilide gave 13·5 grams of menthyl ester.

Menthyl 3-Chloro-4-nitrobenzoate.

Temperature.	Reading.	Density.	$[a]_D^c$.	$[M]_D^c$.
100°	− 20·50°	1·112	− 61·05°	− 207·39°
80	20·86	1·130	61·13	207·66
70	21·00	1·137	61·16	207·66
60	21·14	1·142	61·30	208·24

Menthyl 4-Chloro-3-nitrobenzoate.—The acid was prepared by nitrating p-chlorobenzoic acid with fuming nitric acid (Hübner, *Zeit. Chem.*, 1866, 615) and melted at 177—179°. It was converted into the acid chloride (m. p. 51—52°) and into the ester (m. p. 112—113°) in the usual way. The rotation was determined by fusing the substance in the vapour of boiling amyl alcohol (132°) and of ethyl butyrate (120°) and also in benzene solution at 20°.

Menthyl 4-Chloro-3-nitrobenzoate.

Temperature.	Reading.	Density.	$[\alpha]_D^e$.	$[M]_D^e$.
132°	− 22·0°	1·078	− 67·57°	− 229·4°
120	22·25	1·091	67·53	229·3

In Benzene Solution.

Temperature.	Concentration per cent.	Reading.	$[\alpha]_D^{20°}$.	$[M]_D^{20°}$.
20°	9·765	− 13·69°	− 70·11°	− 238·17°
	7·324	10·38	70·86	240·72
	3·662	5·20	71·01	241·23

Menthyl 3-Chloro-5-nitrobenzoate.—The acid in this case was obtained by reducing the methyl ester of 3 : 5-dinitrobenzoic acid with hydrogen sulphide in presence of a little ammonia (Trans., 1905, 87, 1267). The amino-group was then replaced by chlorine according to Sandmeyer's method. The diazotisation does not go smoothly, and although the conditions were varied it was impossible to obtain more than half the theoretical yield. This result was owing to the simultaneous formation of a yellow, insoluble by-product which was not further investigated. The ester was hydrolysed with sulphuric acid (1 vol. of strong sulphuric acid to 2 vols. of water) and crystallised repeatedly from hot water with the addition of a little animal charcoal. The pure and colourless acid melted at 148—149°; it was converted into the acid chloride, which is a viscid liquid, and into the menthyl ester, which gradually solidified, forming colourless prisms which melted at 42—44°. The reaction between the acid chloride and menthol began at a temperature of 95—100°, which is rather lower than that observed in the other cases. From 7·5 grams of acid, 7 grams of acid chloride and 11 grams of ester were obtained.

A second preparation was made in the same manner with the following results :

Menthyl 3-Chloro-5-nitrobenzoate.

First preparation.

Temperature.	Reading.	Density.	$[\alpha]_D^e$.	$[M]_D^e$.
100°	− 24·44°	1·109	− 73·00°	− 247·99°
80	24·95	1·127	73·33	249·11
70	25·25	1·134	73·74	250·50
65	25·40	1·138	73·91	251·03
40	26·00	1·161	74·15	251·89
20	26·53	1·179	74·40	253·05

Second preparation.

100	24·45	1·109	73·03	248·09

Table IV contains the analytical data obtained by estimating the chlorine in the series of esters.

TABLE IV.

Analyses of the Chloronitrobenzoic Esters.

Menthyl ester,

Cl : NO₂.	Substance taken.	AgCl.	Per cent.
2 : 4	0·1972	0·0828	10·39
4 : 2	0·1732	0·0575	10·34
2 : 5	0·2102	0·0821	10·34
5 : 2	0·1770	0·0726	10·15
3 : 4	0·1944	0·0835	10·63
4 : 3	0·1960	0·0851	10·74
2 : 6	0·1920	0·0804	10·36
3 : 5	0·2039	0·0870	10·55

$C_{17}H_{22}O_4ClN$ requires $Cl = 10·39$ per cent.

We desire in conclusion to thank Mr. I. H. Zortman for preparing the 3 : 4- and 4 : 3-esters and Mr. P. W. Chadwick for furnishing the 2 : 5-ester.

THE UNIVERSITY,
LEEDS.

XLIX.—*The Action of Light on Benzaldehydephenylhydrazone.*

By FREDERICK DANIEL CHATTAWAY.

ALTHOUGH light has been observed in many cases to cause profound alteration of the properties of both inorganic and organic substances, in comparatively few have the changes been at all fully investigated or any adequate explanations of the phenomena been put forward. Roloff (*Zeit. physikal. Chem.*, 1898, 26, 335) distinguishes between chemical and physical actions of light, including among the former only those in which different molecules interact ; known instances of isomeric change, however, and others probably due to this which he includes in the latter group are more properly considered as strictly chemical.

Marckwald (*Zeit. physikal. Chem.*, 1899, 30, 140) uses the term "phototropy" to designate a type of change which he regards as purely of a physical nature, where a colour which appears on exposing a compound to the light disappears on heating or on placing the substance again in the dark.

Among organic substances which behave thus, benzaldehydephenylhydrazone is perhaps the one most easily obtained. Fischer, who first prepared the ordinary stable α-modification of this compound, noted (*Annalen*, 1878, **190**, 135) that when exposed to the air it reddened. He appears to have regarded this reddening as due to slight oxidation similar to that which so many amino-compounds undergo in contact with oxygen and dismisses the phenomenon without further comment. Biltz (*Annalen*, 1899, **305**, 171, and *Zeit. physikal. Chem.*, 1899, **30**, 527) recalls this observation and states that the foregoing hydrazone and a number of osazones are sensitive to light in the same way as two compounds mentioned by Marckwald (*loc. cit.*). He states that in all these cases the substances which are greyish-white or yellow become red on exposure to light, and that in the dark or on heating the original colour returns, and also that the red colour shows signs of fading if the exposure to light lasts a week or thereabouts. Reutt and Pawlewski (*Bull. Acad. Sci. Cracow*, 1903, 503) record the same facts with regard to this hydrazone; they appear to have regarded the reddened product as a third modification of the substance, but beyond noting the changes of colour made no further observations.

Benzaldehydephenylhydrazone does not redden at all, even on prolonged exposure to the air, if light is excluded ; a specimen freely exposed on a watch-glass for a year in a wooden box, and thus screened from light, showed no trace of red colour or of decomposition and its melting point remained unaltered. On the other hand, if exposed to the action of light, its colour alters in a remarkable way ; in diffused light, reddening takes place slowly and is only perceptible after some hours, but in sunlight the change becomes very noticeable in a few minutes and the colour quickly deepens until after a few hours a maximum intensity is reached, when the crystals have a brilliant scarlet colour resembling that of azobenzene.

The presence of air is not essential, as the change takes place equally well in an atmosphere of dry hydrogen. Only those parts directly exposed to light are affected, whilst subjacent layers or any portions screened by black paper, sheet copper, or yellow glass remain uncoloured. Seen under the microscope the crystals appear not to have undergone any change in shape or brilliancy ; the surfaces reflect light exactly as before and the colour only is altered.

The light from an electric arc playing between iron poles, which is extremely rich in ultra-violet rays, also brings about the change, the velocity of which falls off in the ordinary way as the distance from the arc is increased ; exposed in a quartz tube at a distance of five centimetres, the change was about as rapid as in direct winter sunlight. The transformation, however, appears not to be due to the ultra-violet rays, as its rate is not affected by interposing a sheet of ordinary crown glass two millimetres, or a sheet of lead glass two centimetres, in thickness, or by both together.

The light from an incandescent mantle, giving a continuous spectrum but deficient in violet rays, also brings about the coloration very slowly. Placed at a distance of about five centimetres from the glass of the vacuum tube and exposed for over an hour to a powerful stream of x rays, benzaldehydephenylhydrazone undergoes no change whatever. Exposed to light under various coloured glasses which absorb different parts of the spectrum, it is found that only the violet, and to a less extent the blue, portions are active, the other portions of the spectrum causing no appreciable colour change.

If a solution of the compound in alcohol or acetic acid is exposed to light, no change of colour occurs; if, however, crystals separate, these are slightly reddened on the side exposed to the light. If, when the colour change has reached its maximum, the scarlet crystals are dissolved in warm alcohol, a pale yellow solution results from which ordinary benzaldehydephenylhydrazone separates on cooling, no appreciable amount of any other substance being produced.

On allowing the scarlet crystals to remain for a long period screened from light at the ordinary temperature, the colour slowly fades; this fading is more rapid if the temperature is increased and takes place in a few minutes if the coloured product is heated at 100°. The pale yellow product thus regenerated is ordinary benzaldehyde-phenylhydrazone and is indistinguishable from the original pure substance. On rapidly heating the coloured product, fading takes place as the temperature rises, and when about 150° is reached the red colour completely disappears; the product is then indistinguishable from the original material, both substances melting at exactly the same temperature (158—160°). On exposing the faded product again to light or on powdering and exposing the melted product, reddening again occurs as before. On powdering benzaldehydephenylhydrazone which has been melted, a smell resembling that of benzaldehyde is noticed; this is also observed on powdering the coloured product after transformation and melting have taken place; it is not noticed on powdering the coloured product which has lost its colour by exposure to a temperature of 100°, and consequently appears not to be connected with the transformation of coloured to colourless material, but to be due to some decomposition of ordinary benzaldehydephenylhydrazone itself at this temperature.

The colour also disappears in a curious way on long exposure to sunlight. Quantities of benzaldehydephenylhydrazone contained in sealed tubes of thin glass filled with air or hydrogen were exposed in the open to direct sunlight; when the colour had deepened to a maximum, it began slowly to fade. After several days of bright sunshine, the loss of colour was unmistakable, and after several weeks all red tint had completely disappeared. Continued exposure during the whole of a summer caused apparently no further change. The crystals exposed in hydrogen, after this treatment, stuck to the sides of the tube somewhat more than at first, but did not seem otherwise altered or decomposed, and melted within 2 degrees of the original melting point. The crystals exposed in air, on the other hand, had evidently undergone slight decomposition, as those sticking to the inner sides of the tube had lost their sharp edges and had a faint brown tint. On opening the tubes, a slight smell of benzaldehyde was noticed in both. On dissolving the contents in warm alcohol, a pale yellow solution was obtained from the product exposed in hydrogen and a slightly brown solution from that exposed in air, but from each solution pure benzaldehydephenylhydrazone crystallised on cooling, only a very slight amount of brown residue being left, apparently not more than is obtained on recrystallising the pure hydrazone from alcohol and evaporating the mother liquor to dryness.

Several of the tubes exposed during the whole summer out-of-doors became cracked, so that air had entered freely. In these, considerable decomposition had taken place under the combined influence of light, air, and moisture, and the inner surface of the tubes had become coated with a soft, adhering, sticky, brown film, but even from this decomposed residue benzaldehydephenylhydrazone was obtained pure in considerable quantity by a few crystallisations from alcohol.

The colour only appears to give any clue as to what takes place under the influence of light, and supplies a possible explanation of the nature of the change, even if we do not accept in its entirety Armstrong and Robertson's recent contention (Trans., 1905, **87**, 1285) that arguments based on colour are already among the most absolute at our command.

It seems probable that in these transformations of benzaldehydephenylhydrazone we have a definite reversible intramolecular rearrangement brought about by the action of light, and if we accept colour as our guide in interpreting it, we are led to the conclusion that it is a change from the hydrazino- to the azo-configuration, thus :

$$C_6H_5 \cdot CH \colon N \cdot NH \cdot C_6H_5 \; \rightleftharpoons \; C_6H_5 \cdot CH_2 \cdot N \colon N \cdot C_6H_5.$$

Benzaldehydephenylhydrazone is not colourless as generally described, but has a distinct although very pale yellow colour which is quite marked in alcoholic solution, the grouping $-CH \colon N-$ being only slightly colour-producing; the grouping $-N \colon N-$, on the other hand, gives rise to strongly coloured compounds, and it is worth again noting that the colour developed by light in benzaldehydephenylhydrazone is almost identical with that of azobenzene itself. The hydrazone configuration of the compound under consideration is under ordinary conditions the more stable, and the coloured product having the azo-configuration, or consisting of an equilibrium mixture of both forms, passes readily and completely again into the former on standing or heating or dissolving in alcohol.

This explanation of the colour change as being due·to a reversible intramolecular rearrangement is supported by the behaviour of benzaldehydediphenylhydrazone, $C_6H_5 \cdot CH \vdots N \cdot N(C_6H_5)_2$, of benzaldehydephenylbenzylhydrazone, $C_6H_5 \cdot CH \vdots N \cdot N(C_6H_5) \cdot CH_2 \cdot C_6H_5$, and of benzaldehydeacetylphenylhydrazone, $C_6H_5 \cdot CH \vdots N \cdot N(CO \cdot CH_3) \cdot C_6H_5$, which resemble benzaldehydephenylhydrazone in structure, but in place of the labile hydrogen atom have the groups C_6H_5, $C_6H_5 \cdot CH_2$, and $CO \cdot CH_3$, which are not labile and consequently would not be likely to pass from the nitrogen to the carbon under the disturbing influence of light. As might be expected, none of these compounds undergoes a colour change on exposure to sunlight.

Further investigation is needed to decide what causes the disappearance of the colour on prolonged exposure to light, but as some slight decomposition undoubtedly takes place, the reverse transformation may be brought about· under the influence of one of the decomposition products, or may be .due to slight superficial decomposition leading to the formation of a slightly orange-coloured film which absorbs the blue and violet rays and so allows the normal reverse change leading to fading to go on. The latter seems the more probable reason, as on powdering the faded crystals the powder was readily coloured by light, showing that only the surface is involved.

As showing that a group linked to nitrogen in the hydrazone configuration exhibits a marked tendency to pass to the carbon, the easy transformation of diphenyldibenzylidenehydrotetrazone into β-benzilosazone, observed by Ingle and Mann (Trans., 1895, 67, 606), may be cited.

It is worth noting that benzophenonephenylhydrazone, contrary to what might be expected, is not at all affected by light; the presence of the two phenyl groups attached to the carbon atom may, however, offer steric hindrance to the transference of the hydrogen from the nitrogen to the carbon atom and so prevent change occurring.

The consideration of this intramolecular rearrangement caused by light and accompanied by marked colour-change leads one to put forward the suggestion that the fading of organic colouring matters in sunlight may, in some cases at least, be due to a similar reversible isomeric change. Light, as is well known, can cause definite intramolecular rearrangement or can cause such vibration or oscillation within the molecular structure as makes such rearrangement possible. In the case described in this paper, the change under the influence of light is from the colourless to the strongly coloured configuration; in others it may be from coloured to colourless.   In this connection, it would be of interest to ascertain whether any faded dyed material has ever been observed to recover after being kept screened from light for a long period.

CHEMICAL LABORATORY,
ST. BARTHOLOMEW'S HOSPITAL, E.C.

L.—*Studies on Optically Active Carbimides. III. The Resolution of α-Phenyl-α'-4-hydroxyphenylethane by means of* l-*Menthylcarbimide.*

By ROBERT HOWSON PICKARD and WILLIAM OSWALD LITTLEBURY, A.I.C.

IN Part I (Trans., 1904, **85**, 685), one of us in conjunction with A. Neville suggested the use of an optically active carbimide as a suitable reagent to effect the resolution of hydroxyl compounds containing an asymmetric carbon atom into its optical isomerides.   Of the carbimides there described, one of them, bornylcarbimide, was shown by Forster and Attwell (Trans., 1904, **85**, 1188) to react very sluggishly with alcohols, and this led Pope (*Ann. Reports*, 1904, **1**, 136) to assume that the other carbimide described by us, namely, l-menthylcarbimide, would also be of little use for the purpose.   We have, however, shown that l-menthylcarbimide reacts readily with various types of alcohols (Part II, this vol., p. 93), and we hope shortly to communicate to the Society a paper describing the phenolic esters of l-menthylcarbamic acid.   Therefore, since this carbimide reacts in general with all hydroxyl groups, we have worked out a method for the resolution of racemic hydroxy-compounds.   This consists (1) in the fractional crystallisation of a mixture of the dextro- and lævo-rotatory esters of l-menthylcarbamic acid and (2) in the hydrolysis of the dextrorotatory (or lævorotatory) ester to obtain the optically active hydroxyl compound. These carbamates in several cases under investigation vary considerably in solubility and can be easily separated.   In most cases they crystallise very readily and are quite stable, although we have been unable to apply the method to sec.-octyl alcohol, as its l-menthylcarbamic ester, like the phenylcarbamic ester (Bloch, *Bull. Soc. chim.*, 1904, [iii], **31**, 50), is an oil.

The carbamates can be prepared in two ways : (1) by direct combination of the hydroxyl compound with l-menthylcarbimide and (2) by conversion of the hydroxyl compound into an ester of chlorocarbonic acid by treatment with phosgene and subsequent conversion of the chlorocarbonate into the l-menthylcarbamate by treatment with l-menthylamine and sodium bicarbonate. The choice of either method depends on the relative cost of the hydroxyl compound and l-menthylcarbimide; in (1), a quantitative yield of the carbamate is usually obtained, but the preparation of the carbimide is somewhat costly and tedious; in (2), the preparation of the chlorocarbonate generally entails some loss of the hydroxyl compound, but the yield of the menthylcarbamate is quantitative.

The limitations of the process are twofold. First, alcohols of the type, for example, $C_6H_5 \cdot CH(OH) \cdot CH_3$ are often dehydrated by a carbimide giving an unsaturated hydrocarbon, and, secondly, the hydrolysis of the carbamates tends in some cases to racemise the active hydroxyl compound. So far, we have been unable to hydrolyse the carbamates with any milder reagent than alcoholic sodium hydroxide.

Preparation of the d-a-*Phenyl-a'-4-hydroxyphenylethane Ester of* l-*Menthylcarbamic Acid,* $C_{10}H_{19} \cdot NH \cdot CO_2 \cdot C_6H_4 \cdot CH(CH_3) \cdot C_6H_5.$

a-Phenyl-a'-4-hydroxyphenylethane was prepared according to the method of Koenigs (*Ber.*, 1890, **23**, 3145) by the condensation of styrene and phenol. We found it advantageous in the purification of the compound to replace the distillation in superheated steam by fractional distillation under reduced pressure of that portion of the mixture which is soluble in sodium hydroxide. In this manner, a colourless, refractive oil is obtained which boils at 200°/20 mm. and slowly crystallises. It is readily recrystallised from light petroleum and melts at 58°.

For the preparation of the $(d + l)$ esters of l-menthylcarbamic acid, the racemic a-phenyl-a'-4-hydroxyphenylethane (7 grams) was mixed in molecular proportion with l-menthylcarbimide (6·4 grams), heated on the water-bath for twenty-four hours, and then in the oil-bath kept at 130° for ten hours. The very viscous mass was then crystallised from light petroleum and yielded 5·6 grams of colourless crystals melting at 109°. These had $[a]_D$ $-43·16°$ in chloroform. After a second crystallisation from light petroleum, 3·8 grams were obtained in the form of lustrous, prismatic needles melting at 117°.

A polarimetric observation * on 1·0060 grams made up to 20 c.c. with chloroform gave a rotation of $-4·21°$, whence $[a]_D$ $-41·8°$ and $[M]_D$ $-158·4°$.

0·1594 gave 5·7 c.c. moist nitrogen at 15° and 756 mm. $N = 4·1$.
$C_{25}H_{33}O_2N$ requires $N = 3·7$ per cent.

* All observations recorded in this paper were made in a 2-dcm. tube.

The dextrorotatory ester is very soluble in ether, benzene, chloroform, and warm petroleum, but is only very sparingly soluble in cold petroleum. Repeated crystallisations of similar preparations did not alter this melting point or specific rotation.

d-a-*Phenyl-a'-4-hydroxyphenylethane.*

To effect the hydrolysis of the carbamate, it was heated with alcoholic sodium hydroxide ($3N/2$) for ten hours in a reflux apparatus, the alcohol and the resulting menthylamine being then removed by distillation in steam. The solution of the sodium salt of the phenol was concentrated over the free flame until the concentration of the free sodium hydroxide in excess was about 15 per cent. On cooling, the sodium salt crystallised in small, white leaflets; these were dissolved in water and the phenol liberated with dilute hydrochloric acid. Precipitated as an oil, it rapidly solidifies in a crystalline condition and melts at 56°. 1·2442 grams of this crude product made up to 20 c.c. with chloroform gave a rotation of +0·70°, whence $[a]_D$ +5·62°.

When recrystallised from light petroleum, it was obtained in a well-defined crystalline condition ; deposited from an undisturbed solution, it crystallises in fan-shaped feathers, four radiating from a point ; otherwise lustrous, prismatic needles are obtained. Both forms melt very sharply at 64°, that is, six degrees higher than the inactive compound.

Polarimetric observations gave the following results :

1·1162 grams made up to 20 c.c. with chloroform gave a rotation of +0·73°, whence $[a]_D$ +6·54°.

0·9897 gram made up to 20 c.c. with benzene gave a rotation of +0·77°, whence $[a]_D$ +7·78°.

Another preparation made from different materials throughout gave $[a]_D$ +6·65° in chloroform and $[a]_D$ +7·86° in benzene. The rotation of the compound is unaffected by prolonged boiling with sodium hydroxide.

The active phenol, when treated with *l*-menthylcarbimide in the manner described above, yields the carbamate melting at 117°.

Benzoyl Derivative.—This compound was prepared by shaking a dilute solution of the phenol in sodium hydroxide with benzoyl chloride. After repeated washing with sodium carbonate solution, it was obtained after one crystallisation from light petroleum in colourless, lustrous, silky needles which melted at 80°. The melting point of the inactive benzoyl compound is 83° (Koenigs, *loc. cit.*).

0·5763 gram made up to 20 c.c. with chloroform gave a rotation of $+0·20°$, whence $[a]_D$ $+3·47°$. Another preparation gave $[a]_D$ $+4·17°$.

1-a-Phenyl-a'-4-hydroxyphenylethane.

The mother liquors from the crystallisation of the dextrorotatory carbamate from light petroleum, when allowed to evaporate, left a viscous oil from which, after some months, a few crystals separated ; these were isolated and proved to be the nearly pure (dl) carbamate melting at 117°. The remaining oil containing the (ll) carbamate was not crystallised, but was hydrolysed in the manner described above with sodium hydroxide and yielded crystals of the phenol which, after one crystallisation from light petroleum, melted at 58°. As was only to be expected, this product was not the pure lævorotatory phenol, as the following polarimetric observation shows :

1·1130 grams made up to 20 c.c. with chloroform gave a rotation of $-0·25°$, whence $[a]_D$ $-2·24°$.

We desire to express our thanks to the Research Fund Committee of the Chemical Society for a grant defraying much of the cost of this work.

MUNICIPAL TECHNICAL SCHOOL,
BLACKBURN.

LI.—A Modification of the Volumetric Estimation of Free Acid in the Presence of Iron Salts.

By C. CHESTER AHLUM.

AN accurate estimation of free acid in chalybeate waters by means of standard sodium hydroxide is found to be impossible because of the inapplicability of indicators under these conditions. Phenolphthalein is destroyed by ferrous salts, whilst in the case of ferric salts the red colour appears after the iron has been entirely precipitated as hydroxide and when the sodium hydroxide is in excess. Methyl-orange is also worthless, as the colour change is gradual and obscured by the precipitate.

From the foregoing facts, it is evident that in order to titrate the free acid, that is, in presence of iron salts, by means of standard alkali, one must remove the metallic base from solution. The following facts must be taken into consideration in arriving at a suitable method.

Ferrous Salts.—These salts are neutral to methyl-orange.

Sodium dihydrogen phosphate gives no precipitate and does not alter the neutrality.

Disodium hydrogen phosphate gives a light green precipitate of ferrous hydrogen phosphate, becoming dark green on exposure to air.

Trisodium phosphate gives a mixed precipitate of ferrous phosphate and hydroxide, rapidly oxidising, and assuming first a dark green and finally a brown tint.

Ferric Salts.—These salts are acid to methyl-orange.

Sodium dihydrogen phosphate gives a cream-coloured precipitate of ferric phosphate, the solution becoming strongly acid.

Disodium hydrogen phosphate gives a yellow precipitate of ferric phosphate with traces of hydroxide.

Trisodium phosphate gives a brownish-red precipitate consisting of a mixture of ferric phosphate and hydroxide.

If a weighed quantity of ferric sulphate is dissolved in water containing a few drops of methyl-orange, a 10 per cent. solution of sodium dihydrogen phosphate then added in excess, the precipitate of ferric phosphate filtered off, and the filtrate titrated with standard sodium hydroxide, the acidity is found to be equivalent to two molecules of sulphuric acid.

$$Fe_2(SO_4)_3 + 2NaH_2PO_4 = 2FePO_4 + Na_2SO_4 + 2H_2SO_4.$$
$$Fe_2Cl_6 + 2NaH_2PO_4 = 2FePO_4 + 2NaCl + 4HCl.$$

If the above principles are applied to the estimation of free acid in the presence of ferric salts, a correction would become necessary. The acid in combination in the ferric salts which has been liberated in the reaction of the latter with the sodium dihydrogen phosphate raises the acidity above its true value.

Referring to the above equations, it will be noticed that 56 parts of ferric iron are equivalent to 98 parts of liberated sulphuric acid or 73 parts of liberated hydrochloric acid. The amount of acid in combination which is capable of being liberated is obtained from these proportions, the iron being previously estimated. The true amount of free acid is obtained by deducting the amount of acid liberated from the ferric salts from the total amount obtained from the titration.

The method of estimating free acid in water based on the foregoing principles is as follows. To 100 c.c. of water is added * in excess a 10 per cent. solution of sodium dihydrogen phosphate. The precipitate is filtered off and the filtrate titrated with $N/10$ sodium hydroxide, using methyl-orange as indicator. The amount of acid liberated from

* An idea of the amount may be obtained from the amounts of iron, calcium, and magnesium found in the regular analysis of the water which should precede this estimation.

the ferric salts is obtained from the determination of the ferric iron in the regular analysis, using the following factor : $Fe : H_2SO_4 = 1\cdot7500$.

Preparation of the Sodium Dihydrogen Phosphate.—One hundred grams of disodium hydrogen phosphate are dissolved in 1000 c.c. of water containing a sufficient quantity of methyl-orange to colour it yellow. Dilute sulphuric acid is added until a yellowish-red colour is obtained. A dilute solution of sodium hydroxide is then added drop by drop until the red colour is discharged and the solution becomes yellow.

Example : 100 c.c. of water treated in the above manner were titrated with $N/10$ sodium hydroxide requiring $21\cdot7$ c.c. of the same. The amount of iron in the ferric state was found to be $0\cdot0215$ gram in 100 c.c. of water.

$21\cdot7 \times 0\cdot0049 = 0\cdot1063$ gram H_2SO_4 from titration.

$0\cdot0215 \times 1\cdot7500 = 0\cdot0376$ „ H_2SO_4 liberated from the ferric salts.

Difference $\quad 0\cdot0687$ „ free sulphuric acid.

If chlorine is present in the water, the free acid may be stated as hydrochloric acid, but this is discretional, as are the other statements of combinations of bases and acids.

A number of mixtures of known quantities of ferric sulphate and sulphuric acid were prepared and the free acid determined with the following results :

Grams of $Fe_2(SO_4)_3$.	Grams of H_2SO_4 added.	Grams of H_2SO_4 found.	Error.
0·1	0·0201	0·0197	− 0·0004
0·2	0·0402	0·0409	+ 0·0007
0·3	0·0603	0·0585	− 0·0018
0·4	0·0804	0·0781	− 0·0023
0·5	0·1215	0·1201	− 0·0014
0·7	0·1614	0·1632	+ 0·0018
0 9	0·1876	0·1847	− 0·0029
1·2	0·2142	0·2109	− 0·0033
1·8	0·2567	0·2573	+ 0·0006
2·5	0·2867	0·2854	− 0·0015

Calcium, magnesium, and ferrous salts do not interfere with the estimations.

The following is an analysis of a chalybeate water heavily charged with sulphuric acid. Waters of this character are particularly abundant in the mining regions of western Pennsylvania and in West Virginia, and are a source of much trouble to operators.

	Grains per gallon.		Grains per gallon.
Sodium sulphate	83·369	Silica	5·210
Calcium ,,	35·738	Organic and undetermined.	15·060
Magnesium sulphate	17·839		
Ferrous ,,	408·121	Total solids	565·337
Ferric ,,	traces		
		Free sulphuric acid	36·671

The foregoing sample was taken in the bituminous coal region of western Pennsylvania.

LABORATORIES OF GEO. W. LORD COMPANY,
PHILADELPHIA, U.S.A.

LII.—*Slow Oxidations in the Presence of Moisture.*

By NORMAN SMITH.

I. *The Oxidation of Ammonia.*

THE oxidation of ammonia at the ordinary temperature is of great interest not only from a scientific, but also from an economic, standpoint. The formation of nitrates both in the air and soil is a question of great importance to the agriculturist. Ammonia does not oxidise at the ordinary temperature when mixed with air or oxygen, but several investigators (Traube and Biltz, *Ber.*, 1904, **37**, 3130; Müller and Spitzer, *Ber.*, 1905, **38**, 778 ; Traube and Schönewald, *Ber.*, 1905, **38**, 828) have recently shown that ammonia may be readily oxidised to nitrite electrolytically. Matignon and Desplantes (*Compt. rend.*, 1905, **140**, 853) have also shown that the oxidation not only of copper but of many other metals is facilitated by the presence of ammonia. In this section are described the results of an investigation made with the object of determining whether the oxidation of ammonia may be brought about by the use of catalysts or by induced oxidation.

Catalytic Oxidation.—The experiments consisted in exposing certain catalysts to a mixture of ammonia and air, the catalysts selected being ferric oxide, stannic oxide, manganese dioxide, platinum, and lead peroxide.

The substance carefully freed from any nitrite or nitrate was placed in a platinum dish and then moistened with a small quantity of dilute ammonia solution prepared by bubbling ammonia gas into distilled water contained in a platinum vessel. A desiccator was filled with air which had been passed through caustic potash and sulphuric acid ; the platinum dish containing the substance was then

1 1 2

put into the desiccator, and finally the whole was covered by a glass bell jar standing in water. After remaining about fourteen days (temperature 13—17°), the substance was treated with a little distilled water, filtered through a filter paper which had been washed free from nitrogen compounds, and carefully tested for nitrite and nitrate. For the nitrite, the very delicate sulphanilic acid—α-naphthylamine test was used.* To test for nitrate, part of the filtrate was evaporated to dryness on a water-bath and tested with phenol and sulphuric acid (Wiley, *Agricultural Analyses*, I, 554). This reaction is characteristic, and in this respect is much superior to the ordinary brucine or diphenylamine test.

Stannic oxide and manganese dioxide gave both nitrite and nitrate; ferric oxide and lead peroxide gave small quantities of nitrite and nitrate, whilst negative results were obtained with platinum.

The experiments were now made quantitative, the amount of nitrate being determined by comparison with a standard solution. The results obtained are indicated in the following table:

Five grams of ferric oxide		gave 0·01 milligram of nitrate.
,,	stannic oxide	,, 0·05 ,, ,,
,,	manganese dioxide	,, 0·05 ,,
,,	lead peroxide	,, 0·01 ,, .
,,	platinum	,, nil ,, ,,

At higher temperatures, the amount of oxidation is of course greater. At the ordinary temperature, platinum has no action, but if a very dilute solution of ammonia is evaporated by dropping into a heated platinum crucible and the steam condensed and tested, nitrite is always found. This experiment is best carried out in the apparatus shown in Fig. 3 and described later in the paper (Section III, p. 479).

Induced Oxidation.—The experiments were carried out in a similar manner to those just described under "catalytic oxidation." Ferrous hydroxide, manganous hydroxide, and the metals copper, zinc, and tin in the form of pure foil were tried. The results are tabulated below:

Ferrous hydroxide	No nitrite and only a trace of nitrate.
Manganous hydroxide ...	Trace of nitrite, no nitrate.
Copper	Large amount of nitrite and nitrate (0·3 gram).
Zinc	No nitrite and no nitrate, but in some experiments performed in glass dishes positive results were obtained.
Tin	Some nitrite and nitrate (about 1 milligram).

* It has been found that the presence of a large quantity of free ammonia interferes with this test. Care must therefore be taken to remove any large excess of ammonia.

The comparatively large amount of oxidation which takes place in the case of copper is noteworthy. So much nitrite and nitrate are produced, that on treatment of the product with sulphuric acid there is a brisk evolution of brown fumes. The formation of a cuprammonium compound may be the cause of the increased action in the case of copper.

In the preceding experiments, it has been shown that the oxidation of ammonia can be brought about both by catalytic action and by induced oxidation. The amounts transformed in the former case are only small, but in the latter, where positive results are obtained, the amount of oxidation is much greater.

II. *The Oxidation of Nitrogen.*

It was now thought to be of interest to determine whether or not the oxidation of nitrogen would take place in a similar manner to that of ammonia.

(a) *Catalytic Oxidation.*—The experiments were performed in a manner similar to that adopted for the oxidation of ammonia with the following modifications. The catalysts were moistened with freshly distilled water instead of a solution of ammonia, whilst the bell jars covering the desiccators stood in sulphuric acid to prevent leakage of ammonia from the atmosphere.

Results.

Ferric oxide ...... No evidence of the formation of nitrite or nitrate.
Stannic oxide ....
Manganese dioxide
Lead peroxide......
Platinum...... .....

The results obtained with platinum at a higher temperature will be discussed later (Section III, p. 479).

(b) *Induced Oxidation.*—In some of the experiments, dishes composed of the metal under investigation and partly filled with water were used, whilst in others the substance was placed in platinum dishes with the distilled water. In the cases of potassium and sodium, a stream of purified air was passed through distilled water and then over the metal contained in a dry U-tube, the gases then passing through a small wash-bottle containing distilled water. The solid remaining was dissolved in alcohol, and both this solution and the water from the wash-bottle tested for nitrite and nitrate.

Results.

Ferrous hydroxide	No indications of nitrite or nitrate.
Manganous hydroxide ...	.. ..
Tin	,, . ,,
Zinc	One or more of the substances, ammonia, nitrite, or nitrate, usually found, but in a few experiments negative results were obtained.
Iron	,, ,,
Magnesium	One or more of the substances always obtained.
Cuprous oxide	Usually a trace of nitrite or nitrate (never more than 0·005 milligram for 5 grams of substance oxidised).
Cuprous chloride	..
Potassium	
Sodium	.. ,,

The action of zinc, iron, and magnesium can be further tested in another way. The litre flask, *A* (Fig. 1), contains distilled water, and

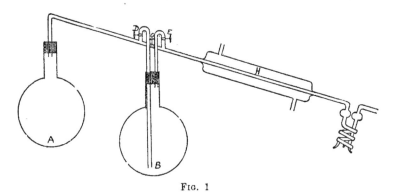

Fig. 1

the metal, zinc or magnesium foil or fine iron wire, is placed in another flask, *B*, both flasks being connected with an ordinary condenser, *H*.

By closing taps *D* and *E* and leaving *C* open, water can be distilled from flask *A* to wash out the apparatus thoroughly. After some time, *D* and *E* are opened and *C* closed. The steam now passes through the flask *B*, which is kept hot by boiling water, and distillation is continued until 50 c.c. of the distillate gives no colour with Nessler reagent. *D* is now closed and some of the water which has collected in the flask B is distilled over. This distillate should also be free from ammonia. The open end of the condenser is then connected with a glass-worm containing sulphuric acid and the distillation stopped. In this way, all the air entering the apparatus is freed from any ammonia by passing through the sulphuric acid. The whole apparatus is now left for ten days, the flask B being shaken at intervals; 50 c.c. of water are then first distilled from flask A by opening C and closing D and E. This distillate will contain any ammonia which might possibly have leaked in, and is kept for comparison. C is now closed and E opened, and 50 c.c. of water distilled from flask B. Nessler solution is added to each distillate and the colours compared with standard solutions.

Metals used.	Ammonia in 50 c.c. of water from flask A.	Ammonia in 50 c.c. of water from flask B.
2·5 grams of pure iron wire	Much less than 0·001 milligram	0·002 milligram
5 ,, zinc foil	Slight trace	0·005 ,,
5 ,, magnesium ribbon.	,, ,,	0·025 ,,

It will be noticed that in all the experiments the metals giving positive results are those which form nitrides fairly readily. These, with the water present, would form ammonia, which might be converted by oxidation into nitrite and nitrate. To decide whether the nitrogen compounds were produced during the oxidation of the metal or were due to small quantities of nitride originally present in the metal, the following experiment was carried out.

Two flasks, B B (Fig. 2), of about 400 c.c. capacity and fitted with ground glass joints, C C, were sealed on to an ordinary distillation flask, A. On the other side they were connected through a sulphuric acid wash-bottle with a pump. The whole apparatus was first thoroughly cleansed with potassium dichromate and sulphuric acid and washed out with hot distilled water. Equal quantities of the metal under investigation were placed in each of the flasks B B and a mixture of "ammonia free" water in which potassium permanganate had been dissolved and concentrated sulphuric acid was drawn into the flask A, which was then sealed off at D. The apparatus was now exhausted and the flask A surrounded by hot water. The first portions of water which distilled over were allowed to pass into the sulphuric acid wash-bottle. After some time, the flasks B B were cooled in a mixture of ice and salt and, after about 10 c.c. of liquid had condensed in each, the taps E E were closed and the flasks were sealed off at F F. One flask was filled with purified air and the other with oxygen prepared from potassium permanganate. Both gases had been kept for some time over sulphuric acid. After about ten days, the water was carefully tested for ammonia and nitrite. With the three metals, iron, zinc, and magnesium, both substances were found to be present and no difference could be observed in the quantities produced in the flasks containing air and oxygen respectively.

From these experiments it appears that the nitrogen must come from the slight impurities in the metal.   This reaction may explain some of the results obtained in the much debated question: "Is ammonium nitrite formed in the evaporation of water?"

Schönbein (*J. pr. Chem.*, 1861, **84**, 215; 1862, **86**, 131) described experiments showing the formation of ammonium nitrite in the evaporation of water.   These results were called in question by Böhlig (*Annalen*, 1863, **125**, 21), who thought that the nitrite obtained was

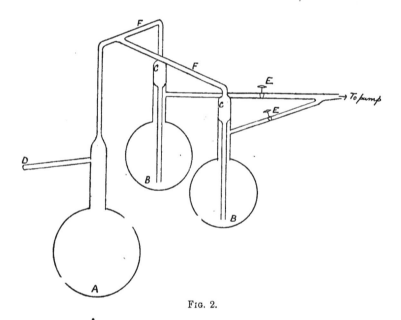

Fig. 2.

already present in the air.   Liebig (*Annalen*, 1863, **125**, 33), however, pointed out that Böhlig's experiments were carried out under different conditions.   Carius (*Annalen*, 1874, **174**, 31) and, later, Leeds (*Chem. News*, 1879, **40**, 70) and Warington (Trans., 1881, **39**, 229) pointed out that Schönbein's results might be explained by the presence of nitrite in the air and in the products of combustion from flames.   A. v. Loesecke (*Arch. Pharm.*, 1879, [iii], **14**, 54) and Scheurer-Kestner (*Bull. Soc. chim.*, 1883, [ii], **39**, 289), however, appear to have obtained evidence of the formation of nitrite in the evaporation of water, and although Baumann (*Landw. Versuchs-Stat.*, 1888, 217, or Abstr., 1889, **56**, 183) in a series of very careful experiments has

obtained exactly the opposite results, further experimental evidence
was thought to be advantageous to the decision of
this question.

III. *Experiments on the Evaporation of Water.*

(*a*) Schönbein found that
if water was evaporated
drop by drop in a platinum
crucible, heated to a temperature just below that
necessary to give the
spheroidal state, and the
steam condensed, nitrite
was always present in the
collected water. This experiment, when repeated by
me, always gave nitrite,
and as this salt may come
from the air, or probably
from the flame used to heat
the crucible, the experiment was modified so as to
avoid these sources of error.

A platinum crucible, *A*,
was carefully fastened with
asbestos packing into a
glass vessel, *B*, of the form
shown in Fig. 3. The
crucible was heated on a
large asbestos sheet and
a current of air, purified
by passing through acid
potassium permanganate,
caustic potash, and sulphuric acid, sent through
the apparatus, and then
through a little distilled
water contained in testtube *C*. After about fifty
minutes, the apparatus was
washed out with the water in *C* and tested for nitrite. If this
experiment gave no result, water was then allowed to fall drop

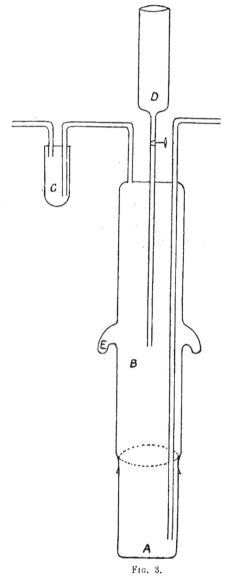

Fig. 3.

by drop from the funnel D on to the hot crucible, each drop being evaporated before the next was run in. The greater portion of the steam was condensed on the sides of the vessel and collected in the receiver E, whilst any not condensing was caught by the water in C. After about 10 c.c. of water had been evaporated in this way, the liquid in C and E was tested for nitrite by the sulphanilic acid— α-naphthylamine solution. In the earlier experiments, traces of nitrite were always found, but when special care was taken to free the water from ammonia and albuminoid matter no nitrite was formed. If a trace of ammonium salt or albuminoid matter was added to the water before evaporation, nitrite was always obtained.

(b) Water when sealed up with air in tubes of various kinds of glass and heated for some time at 170° invariably yielded nitrite, but in all cases the glass was attacked and was probably the source of the nitrite.

(c) A fine stream of purified air was forced through water at temperatures varying from 15° to 80° for periods of about 100 hours, when no trace of nitrite could be detected. This result is in complete agreement with Carius's experiments.

(d) Steam from a boiler indicating a pressure of 40 lbs. to the square inch was allowed to escape through a nozzle into a long wide glass tube. No nitrite was present in the water which condensed on the sides of the tube.

(e) Schönbein exposed to the air a strip of moistened filter paper and a similar strip of dry paper. After a time, he found that the former contained a nitrite, whilst the latter did not. Scheurer-Kestner confirms this result and, so far as I know, it has not been contradicted. On repeating the experiment in purified air in a desiccator, no nitrite was found in either strip. The difference observed by the above-mentioned chemists was probably due to the moistened paper retaining more dust from the ordinary air.

(f) Water and air were sealed in a glass tube and shaken violently for ninety hours. No nitrite was formed.

In the light of the foregoing experiments, it is evident that previous experimenters who obtained positive results fell into one or more of the following errors. (1) The traces of nitrite in the air were not removed. (2) Nitrite was produced in the flames used. (3) Ammonium compounds were present in the water or leaked in from the air, and yielded nitrite by oxidation. Ammonia being lighter than air will of course diffuse more readily. The experience gained while working on this subject shows that ammonia will leak in rapidly through the minutest aperture. (4) In the cases where metal dishes were used, traces of nitride as impurity form ammonia by interaction with the water.

IV. *The Formation of Hydrogen Peroxide.*

The question of the formation of hydrogen peroxide in the evaporation of water has also received attention during the progress of the above investigation. H. B. Dixon (Trans., 1886, **49**, 108) obtained evidence which led him to believe that under certain conditions hydrogen peroxide was formed in the evaporation of water,* and in the same year Ramsay (*Proc.*, 1886, **2**, 225) examined water which was evaporating in air and found a substance capable of decomposing potassium permanganate. The experiments have been carried out in the same manner and at the same time as those described in section III. A solution of titanium dioxide in sulphuric acid was found to be the best test for hydrogen peroxide and was used in all the experiments. It is a delicate test, and is not affected by ozone or other substances, nor is there any danger of oxidation as in the potassium dichromate and ether test. In many of the experiments, the action of hydrogen peroxide on a photographic plate (see W. J. Russell, *Proc. Roy. Soc.*, 1899, **64**, 409) was also used as an additional test.

The results obtained are given in the following table :

(*a*) Evaporation of water in a platinum crucible at a temperature just below that necessary to give the spheroidal state .... No hydrogen peroxide formed.

(*b*) Water and air heated to 170° in glass tubes ....

(*c*) Air forced through water in a fine stream at various temperatures from 15° to 80° ....

(*d*) Steam under high pressure allowed to escape into the air and condensed... ....

(*e*) Water and air violently shaken together for ninety hours ....

(*f*) A sample of water taken from the base of a waterfall and tested ...

(*g*) Evaporation of water contained in metal dishes at the ordinary temperature :

    Zinc .... Hydrogen peroxide always found.

    Iron .... Generally no hydrogen peroxide.

    Tin .... No hydrogen peroxide formed.

    Platinum ....   ,,   ,,

* Professor Dixon informs me that he has several times repeated the experiment without finding hydrogen peroxide,

In a paper on "The Rusting of Iron," recently published (Trans., 1905, **87**, 1559), Dunstan, Jowett, and Goulding have shown that many metals give hydrogen peroxide in presence of water containing a trace of acid. Iron is one of the exceptions. This is explained by these investigators by the assumption that hydrogen peroxide is formed but immediately decomposes. In this connection it is interesting to note that in two of my experiments with iron, indications of hydrogen peroxide were obtained, but on further standing the liquid no longer gave positive results, nor in other experiments could any hydrogen peroxide be detected.

The preceding results lead to the conclusion that hydrogen peroxide is not produced in the evaporation of water in air. If, however, a metal such as zinc is present, some hydrogen peroxide is formed and the metal is largely oxidised at the same time.

THE UNIVERSITY OF MANCHESTER.

LIII.—*Studies in the Acridine Series.   Part III.   The' Methylation   of   Chrysaniline   (2-Amino-5-p-amino- phenylacridine).*

By ALBERT ERNEST DUNSTAN and JOHN THEODORE HEWITT.

IN two previous papers, Hewitt and Fox (Trans., 1904, **85**, 529, and 1905, **87**, 1058) have described the formation of paraquinonoid anhydro-bases resulting from the hydrolysis of the methiodides of 8-acetylamino-3 : 7-dimethylacridine   and   2 : 8-diacetylamino-5-phenyl-3 : 7-dimethyl-acridine (diacetylbenzoflavine) respectively, and subsequent precipitation by an alkali. In the former case, a carbinol base was isolated, which lost the elements of water at 200° ; in the latter case, no such intermediate compound was observed ; the changes occurring may be represented by the following schemes :

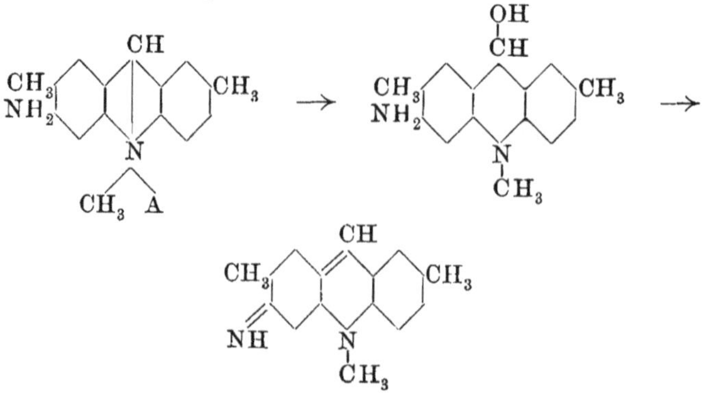

$$\text{CH}_3\,\text{NH}_2\!\!\left(\!\!\begin{array}{c}\text{CPh}\\ \end{array}\!\!\right)\!\!\text{CH}_3\,\text{NH}_2 \quad\rightarrow\quad \left[\text{CH}_3\,\text{NH}_2\!\!\left(\!\!\begin{array}{c}\text{CPh(OH)}\\ \text{N(CH}_3)\end{array}\!\!\right)\!\!\text{CH}_3\,\text{NH}_2\right]$$

$$\text{CH}_3\!\!\overset{}{\underset{\text{N}}{\underset{|}{}}}\!\!\text{A}$$

$$\rightarrow\quad \text{CH}_3\,\overset{}{\underset{\text{NH}}{}}\!\!\left(\!\!\begin{array}{c}-\text{CPh}-\\ \text{N(CH}_3)\end{array}\!\!\right)\!\!\text{CH}_3\,\text{NH}_2\;.$$

The fact that the carbinol base cannot be isolated in the second instance is indicated by the inclusion of the hypothetical formula of such a substance in brackets.

As a further instance of the elimination of water which occurs between the carbinol group and a p-amino-group, the behaviour of chrysaniline in similar circumstances may be cited; it will be seen in the sequel that no carbinol base has been isolated in this case, but simply the product of its dehydration.

Chrysaniline (2-amino-5-p-aminophenylacridine) was the subject of an extended investigation by Hofmann (*Ber.*, 1869, **2**, 379), O. Fischer and G. Körner (*Ber.*, 1884, **17**, 203; *Annalen*, 1886, **226**, 175), and Anschütz (*Ber.*, 1884, **17**, 434). This colouring matter occurs commercially as an impure nitrate or hydrochloride under the name of phosphine, and is used to a limited extent as a basic dye for leather and silk. The chrysaniline used in this investigation was obtained in the following manner. The commercial phosphine was added to hot water in small portions at a time to avoid clotting, constant stirring being employed. After cooling, precipitation of the fine yellow colour base was effected by addition of caustic soda, the precipitate being collected at the pump and well washed, then dried and dissolved in boiling benzene. After boiling some hours in a reflux apparatus, the solution was decanted, the solid matter in the flask being worked up with a fresh batch of material. Crystalline crusts separate from the cooling solution, and these, as observed by Fischer and Körner, contain benzene of crystallisation. By boiling with dilute sulphuric acid, the benzene is driven off, the solution of the sulphate of the base yielding purified chrysaniline on decomposition with caustic soda. Recrystallisation from 50 per cent. alcohol furnishes a product containing two molecules of water; after drying, the base was found to melt at 265° (uncorr.). Fischer and Körner give the melting point as 267—270°.

A solution of the base in hydrochloric acid furnishes a chromate by precipitation.

0·0821 gave 0·0123 Cr_2O_3. $Cr = 12·5$.

$C_{19}H_{15}N_3,H_2CrO_4$ requires $Cr = 12·8$ per cent.

Acetylation of Chrysaniline.

Anschütz (*loc. cit.*) acetylated chrysaniline by heating it with excess of acetic anhydride in sealed tubes at 140°; Fischer and Körner found that acetylation might be effected under the ordinary pressure. The diacetylchrysaniline required in this investigation was obtained by boiling equal weights of the base and fused sodium acetate with five parts by weight of acetic anhydride for four hours in a reflux apparatus. The chrysaniline at first dissolves with a deep red colour, but this gradually changes to an orange shade as acetylation proceeds (compare Hewitt and Fox, Trans., 1905, **87**, 1058). The fused mass was poured into dilute alcohol to destroy excess of the anhydride ; after standing overnight, a brown residue was left which separated on recrystallisation from acetic anhydride in yellowish-brown needles. This substance proved to be the tetra-acetyl derivative.

0·1853 gave 0·4840 CO_2 and 0·0824 H_2O. $C = 71·25$; $H = 4·88$.

0·1491 „ 0·3875 CO_2 „ 0·0757 H_2O. $C = 70·9$; $H = 5·63$.

$C_{27}H_{23}O_4N_3$ requires $C = 71·5$; $H = 5·07$ per cent.

On pouring the filtrate from the tetra-acetyl derivative into dilute ammonia solution, the diacetylchrysaniline, described by Anschütz, was obtained as a pale yellow precipitate ; it was recrystallised by solution in boiling absolute alcohol, water was then added until a faint turbidity was noticed, this was removed by addition of a few drops of alcohol, and the solution allowed to cool slowly. Minute crystals separated which melted at 200° with decomposition.

Diacetylchrysaniline dissolves in dilute hydrochloric acid and behaves as a monacid base, whereas the parent substance forms salts with 2 molecules of a monobasic acid. The chromate, which was prepared by precipitation from the solution in hydrochloric acid, gave the following result on analysis :

0·1893 gave 0·0168 Cr_2O_3. $Cr = 5·96$.

$2C_{23}H_{21}O_2N_3,H_2CrO_4$ requires $Cr = 6·07$ per cent.

This result is in agreement with the observations of Anschütz, who found that diacetylchrysaniline furnishes a mononitrate and monohydrochloride. It is not merely acridine which may be characterised and purified by means of its chromate, but very many of its derivatives also give characteristic chromates.

Action of Methyl Iodide on Diacetylchrysaniline.

Diacetylchrysaniline, one equivalent of methyl iodide, and an excess of methyl alcohol were heated for four hours at 120° under pressure. The magnificent crimson liquid thus obtained deposited small, dark needles, very readily soluble in alcohol and dyeing cotton directly to a deep brown shade. If the alcoholic solution is poured into ammonia, a red precipitate is obtained which darkens considerably on drying. When the base thus produced is boiled with hydrochloric acid, it dissolves with a deep red colour, the solution on cooling depositing a hydrochloride in the form of dark-coloured needles having a green reflex. The percentage of chlorine showed that the base was monacid and that the acetyl groups had been completely hydrolysed.

0·0915 gave 0·0396 AgCl. Cl = 10·7.

$C_{20}H_{13}N_3Cl$ requires Cl = 10·6 per cent.

By precipitation of a solution of the hydrochloride with ammonia and recrystallisation from absolute alcohol, the anhydro-base was obtained.

0·1193 gave 0·34625 CO_2 and 0·0647 H_2O. C = 79·9 ; H = 6·03.

0·1515 „ 0·4421 CO_2 „ 0·0903 H_2O. C = 79·6 ; H = 6·62.

0·2586 „ 0·7580 CO_2 „ 0·1604 H_2O. C = 79·9 ; H = 6·88.

0·0773 „ 8·5 c.c. N at 15° and 770 mm. N = 13·3.

$C_{20}H_{17}N_3$ requires C = 80·25 ; H = 5·69 ; N = 14·04 per. cent.

The low numbers for carbon and nitrogen and the high results for hydrogen are to be ascribed to the fact that the substance being very difficult to burn completely, the combustions were necessarily protracted ; we may point out, however, that the percentages obtained for carbon quite exclude the carbinol formula, $C_{20}H_{19}ON_3$, which requires C = 75·7 .per cent.

The two following formulæ are possible for the anhydro-base :

and

Of these we prefer the first from analogy with benzoflavine, although no definite proof can be assigned.

The methylation of diacetylchrysaniline may also be effected by means of methyl sulphate. Equal weights of the two substances were made into a paste and heated for half an hour at 120°. The clear red product of fusion was readily soluble in water, giving a solution having a colour comparable with that of the methiodide previously described ; to obtain the anhydro-base, boiling for one hour with dilute sulphuric acid and precipitation with ammonia were adopted. The base obtained resembled in all respects that produced by means of methyl iodide ; a portion was converted into a platinichloride by solution in hydrochloric acid and precipitation with hydrogen platinichloride. The brown, amorphous salt obtained was washed with water and alcohol, dried, and analysed.

0·0712 gave 0·0138 Pt.   Pt = 19·63.

$(C_{20}H_{18}N_3)_2PtCl_6$ requires Pt = 19·34 per cent.

The solution of the methylsulphate readily furnishes other salts (for example, chromate, nitrate, &c.), which will be described in another communication, together with further derivatives of chrysaniline.

In his classical research on the acridines, Bernthsen (*Annalen*, 1885, **224**, 13) described a dinitro-derivative of phenylacridine which he considered might yield chrysaniline or an isomeride on reduction. This base has been examined during the past year, and corresponding derivatives (chromate, platinichloride, acetyl derivative, and quaternary methiodide) certainly exhibit a striking resemblance to compounds derived from chrysaniline. Since, however, it is unlikely that the two nitro-groups enter into the molecule unsystematically we consider it not improbable that the positions 2 and 8 are those which are substituted.

Colour and Fluorescence of Acridine Derivatives.

The colours exhibited by acridine, *meso*phenylacridine, chrysaniline, benzoflavine, and various derivatives of the two latter substances call for some comment. Acridine and its phenyl derivative both exhibit a somewhat pale yellow colour, whilst the introduction of the two amino-groups in para-positions to the *meso*-carbon atom, whether as in benzoflavine the 2- and 8-positions be occupied or the 2- and 5-*p*-positions be replaced as in chrysaniline, results in the production of deep yellow colouring matters, the substances in the solid condition exhibiting a deep orange shade. If, now, the amino-group is acetylated, light yellow compounds are formed, the colour of which is strictly comparable with that of the parent phenylacridine.

This behaviour has suggested the possibility that the colour bases may possess a paraquinonoid structure, and it is certainly worth mentioning that the undoubted paraquinonoid methyl derivatives in·

which the chance of tautomerism is excluded are intensely-coloured compounds. The intermediate shade exhibited by the non-methylated bases (chrysaniline and benzoflavine) suggests the possibility that both the quinonoid and non-quinonoid forms may be represented, but speculation on this point would be vain without a careful spectrographic investigation.

The fluorescent phenomena of these substances call for some remark : the fluorescence exhibited by acridine, phenylacridine, and the diacetyl derivatives of benzoflavine and chrysaniline is blue in shade and somewhat resembles that exhibited by anthracene. The diamino-compounds, however, exhibit a strong green fluorescence if the free bases are dissolved in alcohol, this fluorescence being very nearly destroyed by the addition of acids : the fluorescence is probably a property of the free bases. As one of the present authors has already pointed out, the more symmetrical benzoflavine shows a much more marked fluorescence than the less symmetrical chrysaniline (*Zeit. physikal. Chem.*, 1900, **34**, 14). In the case of the non-tautomeric paraquinonoid methyl derivatives the observed fluorescence of the free bases is so faint as to be perhaps attributable to a small amount of admixed non-methylated base.

The connection between fluorescence and constitution of organic compounds has engaged the attention of several chemists. R. Meyer (*Zeit. physikal. Chem.*, 1897, **24**, 468) attributes fluorescence to the occurrence of certain "fluorophore" groups in the molecules of substances exhibiting the property. These groups are mostly six-membered rings, for example, the pyrone ring and its congeners, the middle 6-carbon ring in anthracene, the pyridine ring in acridine compounds, and the paradiazine ring in safranines.

J. T. Hewitt (*Zeit. physikal. Chem.*, 1900, **34**, 1) attributes fluorescence to internal vibrations in the molecule conditioned by symmetrical double tautomerism ; the molecules of acridine and benzoflavine, for example, swinging between the extreme positions indicated schematically as follows :

NH   HN   NH₂   CH₃   CH₃   C   C₆H₅   ⇌   N   NH₂   CH₃   NH₂   CH₃   C   C₆H₅

⇌   NH   NH   NH₂   CH₃   CH₃   C   C₆H₅

Finally, H. Kauffmann (*Ber.*, 1900, **33**, 1731; 1904, **37**, 294; 1905, **38**, 789; *Annalen*, 1906, **344**, 30), as a result of his extensive investigations, has propounded the theory that the molecules of a fluorescent substance contain at least two groups. The first of these groups is the one from which the radiant energy is emitted when the molecules are excited either by light of suitable wave-length, Tesla currents, or exposure to the action of radium. This group Kauffmann terms the luminophore.

The second group, the "fluorogen," has a specific action. Thus, in the case of anthranilic acid, the aniline grouping is considered as the luminophore. Aniline is not a fluorescent substance, but the introduction of a fluorogen (in this case, carboxyl) results in the production of fluorescent anthranilic acid. The precise action of the fluorogen is at present left unsolved, but it is apparently an unsaturated group in all cases.

Whilst the theories of Meyer and Hewitt have not been applied hitherto to an explanation of the fluorescence of such compounds as anthranilic acid, it may be mentioned that the theory connecting double tautomerism with fluorescence agrees very well with the observed facts in the case of acridine, phenylacridine, and its diamino-derivatives. Where only shifting of linkings is possible, as with acridine and the diacetyl derivatives of benzoflavine and chrysaniline, the character of the fluorescence is very similar. But with the introduction of the two amino-groups the fluorescence partakes more of the character of that exhibited by fluorescein, and the possible mechanism of the tautomerism is not unlike that with alkaline solutions of fluorescein.

Finally, if the tautomerism is inhibited to a greater or less extent, either by salt formation or by methylation of the *meso*-nitrogen atom, the fluorescence is strikingly diminished.

In conclusion, we wish to express our thanks to Mr. R. O'F. Oakley for much valuable assistance.

EAST HAM TECHNICAL COLLEGE.   EAST LONDON COLLEGE.

LIV.—*The Relation between Absorption Spectra and Chemical Constitution. Part I. The Chemical Reactivity of the Carbonyl Group.*

By Alfred Walter Stewart (Carnegie Research Fellow)
and Edward Charles Cyril Baly.

It has been shown by many different workers that certain reactions which can be carried out easily with parent substances are much more difficult to bring about when derivatives of these compounds are used. The work of Menschutkin on the esterification of aliphatic acids and the analogous researches of Victor Meyer and Sudborough on the aromatic acids are examples of what is meant.

From the standpoint of stereochemistry, it has been usual to attribute the effects in question to a variation in the amount of free space around the reactive groups of the molecule, since it is obvious that the substitution of a larger for a smaller radicle in the vicinity of the reactive part of the molecule will decrease the possibility of new reagents coming in contact with the active radicle. It may be granted that, if atoms have any volume at all, these premises are correct; but it is not yet proved that the effects attributable to this cause play any very considerable part in the reactions in question. It seems more probable that the free paths of the atoms in their intramolecular vibrations are so large in comparison with the size of the atoms themselves that this heaping up of substituents in the vicinity of the reactive group will have no very marked effect. For the present, however, the hypothesis of steric hindrance forms a convenient mechanical explanation of the non-reactivity of certain compounds.

Stewart (Trans., 1905, **87**, 185 ; Proc., **21**, 78) has shown that when a hydrogen atom near the carbonyl group of a ketone is replaced by a methyl radicle, the result is a decrease in the additive capacity of the carbonyl group. This is what might have been anticipated from the hypothesis of steric hindrance, since the volume of the methyl radicle must be greater than that of a hydrogen atom. A contradiction between theory and practice is to be found in the case where, instead of a methyl radicle, a $-CO_2Et$ group is introduced into the molecule. In the case of the latter group, it is found that instead of decreasing the velocity of addition of sodium hydrogen sulphite, as its bulk might lead one to predict, it has the contrary effect, for some of those ketones which contain a carboxyl group are much more reactive than the corresponding simple ketone, and consequently still more active than the methyl-substituted ketone. The numbers found for

K K 2

acetone, methyl ethyl ketone, and ethyl acetoacetate show this clearly :

	10	20	30	40 minutes.	
Ethyl acetoacetate	37·4	47·0	56·0	60·0 per cent.	bisulphite
Acetone	28·5	39·7	47·0	53·6 ,,	compound
Methyl ethyl ketone	14·5	22·5	25·1	29·1 ,,	formed

It is thus evident that some new influence has come into play which tends to mask or modify the steric hindrance due to the more voluminous group.

The carboxyl group, however, is not in itself sufficient to produce this increased reactivity of the carbonyl radicle, as the rate of addition of sodium hydrogen sulphite to ethyl lævulate was found to be slightly lower than that found in the case of methyl propyl ketone, which contains a carbon chain of the same length, and a like result was observed in the case of the diketone, acetonylacetone. On the other hand, ethyl acetonedicarboxylate has an additive capacity even greater than that of ethyl acetoacetate. Acetone shows very little sign of tautomeric change, whilst, on the contrary, ethyl acetoacetate and ethyl acetonedicarboxylate are tautomeric compounds. Thus, here again theory and practice appear to be opposed to one another : the true carbonyl compound having much less reactive power than the semi-enolised substance. It occurred to us that in this fact was to be found the key of the problem, and the exceptionally great reactivity of the carbonyl group in tautomeric compounds was due to the actual process of tautomeric change. Ethyl acetoacetate, under ordinary conditions, exists as an equilibrium mixture of the two substances (I) and (II), and the conversion of the first into the second and *vice versâ* is going on continuously.

$$CH_3 \cdot \underset{\underset{OH}{|}}{C} : CH \cdot CO_2Et \qquad\qquad CH_3 \cdot \underset{\underset{O}{||}}{C} \cdot CH_2 \cdot CO_2Et$$

$$\text{I.} \qquad\qquad\qquad\qquad \text{II.}$$

Now, when a molecule of (I) changes into a molecule of (II), the result is the formation of a carbonyl group from a hydroxyl group. From analogy with the behaviour of atoms in the nascent state, we must suppose that this " nascent carbonyl group " is endowed with a much greater reactivity than that possessed by the ordinary non-nascent carbonyl radicle. This activity, however, need not be occasioned purely by the actual wandering of the hydrogen atom from the oxygen to the carbon : it may be due to some finer play of forces within the molecule which manifests itself in the production of the character-istic absorption band in the β-diketone spectrum. The condition into which the hydrogen atom is thrown as a result of this play of forces may be termed a condition of " potential tautomerism," and in it the hydrogen atom will possess a reactive power more or less analogous to that acquired by an atom as a consequence of the ionisation process (Baly and Desch, Trans., 1905, **87**, 766). Evidence of the probability of this hypothesis will be adduced later ; for the present it will be sufficient to call attention to the conception of the " nascent carbonyl group."

If we now apply this idea to several cases which have hitherto been classed under the head of steric hindrance, it will be found that they can be satisfactorily explained. Taking the cases of the ketones which have already been dealt with by one of us (Stewart, *loc. cit.*), a marked decrease in reactivity of the carbonyl group is shown when the hydrogen atoms of acetone are successively replaced by methyl radicles.

In the course of their investigations of the spectra of derivatives of ethyl acetoacetate, Baly and Desch (Trans., 1904, **85**, 1039) proved that the change from the enolic into the ketonic form produced an absorption band in the ultra-violet region of the spectrum, and they also showed that this band was not due merely to the shifting of a hydrogen atom, but was rather to be considered as the result of some intra-atomic change. In the hope of finding some analogous process in the simple ketones, we examined the spectra of several, and we found that a similar absorption band exists there as well. We further noticed that the persistence of this band decreases *proportionately to the diminution in the reactivity of the carbonyl group in the ketone.*

For instance, the following numbers show the percentages of oxime formed by various ketones in twenty minutes, and on comparing these amounts with the curves of the absorption spectra shown in Fig. 1, the relation between the two will be apparent.

	Oxime.		Oxime.
Acetone	49·7 per cent.	Methyl *iso*propyl ketone	31·5 per cent.
Methyl ethyl ketone	39·2 ,,	Pinacolin	17·0 ,,
,, propyl ,,	37·3 ,,		

Lapworth (Trans., 1904, **85**, 32) showed that the action of halogens on acetone was preceded by the production of the enolic form of the ketone, and he found, further, that the presence of acids hastened the reaction. Now, he had already shown (Trans., 1902, **81**, 1503, and 1903, **83**, 1121) that the presence of acids brings about a rapid attainment of equilibrium between the tautomeric forms of carbonyl compounds : or, in other words, the addition of acid has a tendency to produce a " nascent carbonyl group." Hence, in the case of acetone itself, not only is there direct spectroscopic evidence in favour of tautomeric change, but the chemical evidence at our disposal is also favourable. Instead of attributing Lapworth's results to the actual formation of the enolic form and an immediate addition of halogen, we prefer to look at them from another point of view. It is obvious that if we consider the change of the group $-CH\overset{*}{:}CH(O\overset{*}{H})-$ into $-CH\overset{*}{H}\cdot CO-$, the hydrogen atom marked with an asterisk must

FIG. 1.

Oscillation frequencies.

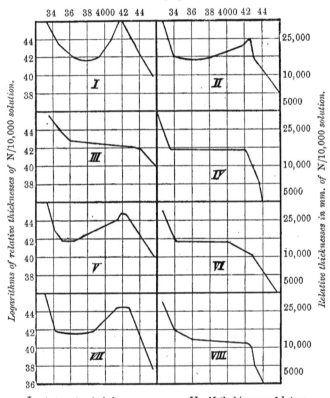

I. *Acetone in alcohol.*   II. *Methyl isopropyl ketone.*
III. *Acetone in water.*   IV. *Pinacoline.*
V. *Methyl ethyl ketone.*   VI. *Methyl hexyl ketone.*
VII. *Methyl propyl ketone*   VIII. *Methyl nonyl ketone.*

become "pseudo-nascent" in the process of change. It would therefore be peculiarly liable to chemical action, and would be easily replaced by halogens. The very great ease with which the methylene hydrogen atoms in acetylacetone are replaceable by halogens lends further support to our hypothesis.

In their second paper (Trans., 1905, **87**, 760), Baly and Desch stated that acetonylacetone and ethyl lævulate were pure ketonic substances ; but on examining their spectra at greater concentrations than were previously employed we have been able to detect at one point a rapid extension of the spectrum, which corresponds to a very

FIG. 2.

Oscillation frequencies.

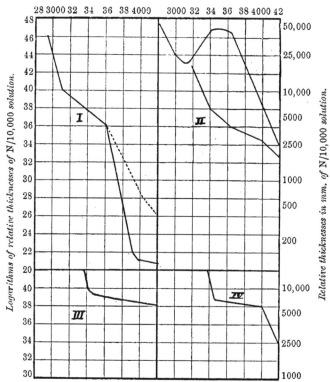

I. *Ethyl acetoacetate in alcohol* (full curve).   II. *Ethyl pyruvate.*
   Ethyl acetoacetate in water (dotted curve).   *Ethyl diethylacetoacetate.*
III. *Ethyl lævulate.*   IV. *Acetonylacetone.*

shallow absorption band (see Fig. 2). The shallowness of the band indicates that the tautomerism in these two compounds is very weak, which agrees well with what has been found with regard to the reactivities of their carbonyl groups. The close agreement between theory and practice in these cases is very noteworthy.

Now, Petrenko-Kritschenko has shown (*J. Russ. Phys. Chem. Soc.*, 1903, **35**, 404) that the speed of phenylhydrazone formation is greatly influenced by the nature of the solvent in which the reaction is carried out. It appeared probable to us that this might be due to the influence of the solvent on the tautomerism process, and to test the matter we examined the spectra of acetone and ethyl acetoacetate in aqueous solution, using as a control in the latter case the spectrum of an alcoholic solution of ethyl diethylacetoacetate, which is much less tautomeric than the parent substance. From the curves for acetone in alcoholic and in aqueous solution (Fig. 1), it will be seen that the influence of water is very marked, the band in the latter case being much shallower than in the former. The three curves shown in Fig. 2 give the absorption spectra of the ethyl acetoacetate series and it is obvious from them that the water has reduced the tautomerism very considerably. It is probable that the greater the unsaturation of the solvent, the less reactivity will be shown by the carbonyl group of the dissolved ketone.

The evidence from simple ketones being so far favourable, we must now examine the case of ketones containing a carbethoxyl group. If tautomeric change alone were the cause of the reactivity of the carbonyl radicle, compounds containing the group $-CO \cdot CH_2 \cdot CO-$ should be more reactive than those which do not contain it, and the reactivity of the carbonyl group in ethyl pyruvate should not be at all abnormal, since the grouping in question does not occur in it; if, however, the reactivity were found to be great, we hoped that some light might be thrown on the problem by a study of the spectrum exhibited by the compound.

We therefore decided to compare the rates of addition of potassium hydrogen sulphite to acetone, ethyl acetoacetate, ethyl acetonedicarboxylate and ethyl pyruvate. At first, experiments were made to find out how rapidly the " bisulphite compound " was decomposed by water, and it was observed that a considerable dissociation occurred, even at zero. Error is thus introduced into the problem, and all that could be done was to reduce this error to a minimum. Since the inverse change increases in the direct ratio to the amount of " bisulphite compound " formed, it is evident that the aim of the experiments must be to estimate the amounts of compound formed before equilibrium is established. Secondly, since the surplus sulphite is destroyed in the course of the estimation, thus bringing about a disturbance in the state of equilibrium of the solution, it is advisable to reduce the relative quantity of potassium hydrogen sulphite as much as possible. Thirdly, no harm could follow from keeping down the temperature, as this would act as a retarding agent on the direct and inverse reactions, while facilitating the measurement of differences in the rates of formation. With these facts in view, the following method of estimation was devised. Fifty c.c. of a $M/20$ solution of potassium hydrogen sulphite were cooled to zero and mixed with an equal volume of a 50 per cent. alcoholic $M/10$ solution of the ketone, which had also been cooled to zero. The liquid was immersed in a freezing mixture and maintained at a temperature of $-10°$. Ten c.c. were withdrawn every five minutes and titrated with iodine. The results obtained by this method are given in the table below :

	5	10	15 minutes.	
Acetone	5	7	9	⎫ Percentage of
Ethyl acetoacetate	12	18	24	⎪ bisulphite
Ethyl acetonedicarboxylate.	30	36	42	⎰ compound
Ethyl pyruvate	52	64	76	⎭ formed.

In spite of the precautions taken, these numbers are probably not quite accurate, owing to various causes which cannot be controlled, but the differences between the numbers themselves are very much larger than any possible experimental error under the conditions employed.

An examination of the numbers shows that the introduction of a carbethoxyl group into acetone increases the additive capacity of the carbonyl group; the introduction of two carbethoxyl groups still further enhances the reactivity of the carbonyl, but the most striking effect is produced when, as in the case of ethyl pyruvate, the carbonyl and carboxyl groups are brought into juxtaposition in the chain. Now, in the case of ethyl pyruvate, although the compound sometimes reacts in the enolic form, it is most improbable that the change from the enolic to the ketonic form and *vice versâ* is going on at a rate at all comparable to that at which it is occurring in ethyl acetoacetate or ethyl acetonedicarboxylate, so that it is not likely that the exceptional additive capacity of the carbonyl group in ethyl pyruvate is due to this kind of tautomerism.

We thought it advisable to examine the spectrum of ethyl pyruvate in the hope that some light might thus be thrown on the problem of the activity of the carbonyl group. We found that ethyl pyruvate gives an absorption band which lies much nearer the red end of the spectrum than the band given by ethyl acetoacetate (Fig. 2). The origin of the band in the ethyl pyruvate spectrum might be looked for in two phenomena : either in the keto-enol change of the group $CH_3 \cdot CO-$ or in the interaction of the carbonyl and carboxyl groups of the radicle $-CO \cdot CO_2Et$. The first explanation is impossible, since if the band were produced by a similar state of intra-atomic vibration in both instances, it would occur in nearly the same place in the spectrum, whilst actually the new band has its head at 3100, while that of the tautomeric β-diketones lies at 3700 ; and, further, since the molecule of ethyl pyruvate is lighter than that of ethyl acetoacetate, we should expect to find the band in the latter case nearer to the red end of the spectrum than in the former, whilst the reverse of this is observed.

In order to make certain that the band in question was actually produced by the proximity of the two true carbonyl groups in the chain, that is, that it was not due to the $-\ddot{C}\cdot OEt$ residue of the carboxylate radicle, we examined the spectra of several diketones and found a similar absorption band in all of them, though in them it was

FIG. 3.

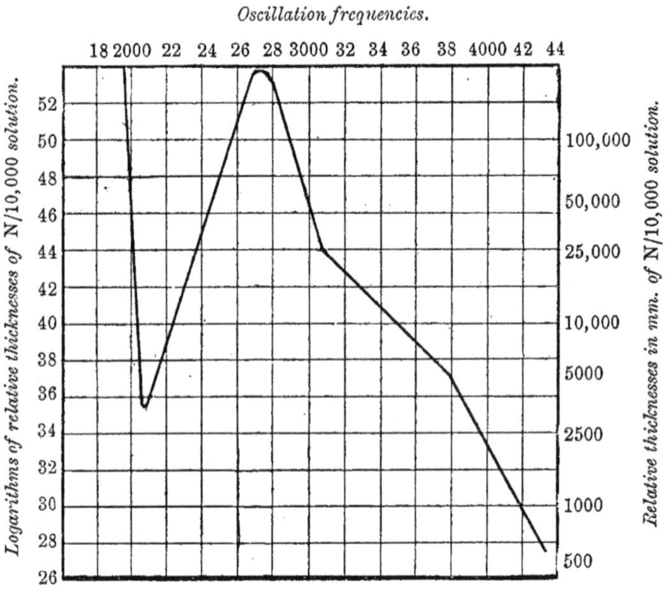

Oscillation frequencies.

Camphorquinone.

situated nearer the red end of the spectrum. For example, in the case of camphorquinone (see Fig. 3) it will be seen that a band is shown of very long persistence, with its head at 2070. The results in the case of the other α-diketones were similar, but as their consideration is reserved for the next paper it is needless to discuss them here.

It was thus proved that the band in question was due to the two carbonyl groups in the α-position to one another. It has already been pointed out that Baly and Desch concluded that, although the absorption band was produced by intra-atomic vibration, it was caused by the change of linking brought about by the oscillation of the hydrogen during the change from the ketonic to the enolic form. From analogy, we should expect to find a somewhat similar state of things in the present case. Now the only possible way in which such a change of linking can be supposed to occur in ethyl pyruvate is by imagining that, like ethyl acetoacetate, it occurs in two forms :

$$CH_3 \cdot \underset{\underset{O}{||}}{C} \cdot \underset{\underset{O}{||}}{C} \cdot OEt \quad \text{and} \quad CH_3 \cdot \underset{\underset{O}{|}}{C} \underset{\underset{O}{|}}{:} C \cdot OEt$$

$$\text{I} \qquad\qquad\qquad \text{II.}$$

It is very hard to indicate exactly what is meant by the aid of the usual structural formulæ, as they only indicate a static condition of the molecule, whilst what we wish to suggest is essentially a dynamic state. We wish to make it perfectly clear that we do not suppose that these two forms actually exist, but, owing to the defect of ordinary structural formulæ, it is impossible to write them otherwise if the usual symbols be employed. We would prefer to indicate the presence of a "nascent carbonyl group" by printing the oxygen symbol in heavy type, thus :

$$CH_3 \cdot C \cdot CO_2 Et$$
$$\underset{\mathbf{O}}{||}$$

Our conception can best be comprehended if it be clearly borne in mind that the two formulæ are not intended to represent actual compounds, but merely two phases of the same compound, just as the two formulæ

represent the same substance in two different phases of vibration. This idea is not new, having been first suggested by Kekulé in 1865, and afterwards extended by Collie (Trans., 1897, 71, 1013). If this conception of phases be understood, it will be apparent that the change of linking is continually going on, and that this change will affect the intra-atomic relations of the molecule very much in the same way as they are affected by the phenomena of tautomerism.

At the same time, it should be noticed that the change of linking from (II) to (I) would produce what we have already defined as a "nascent carbonyl group," which would have great reactivity. Thus we are led to conclude that substances which show these peculiar absorption bands will in general be more active chemically than other compounds which do not exhibit such selective absorption.

The idea which we have put forward cannot be considered as part of the theory of tautomerism, as, owing to its associations, the name tautomerism will always suggest the wandering of a hydrogen atom.

It is unfortunate that the name "desmotropism" has already been employed to denote tautomerism, as it seems well fitted to describe the phenomenon with which we have dealt. We therefore wish to propose the word *Isorropesis* (ἰσορροπία : equipoise) to describe the process.

The arguments in favour of this theory appeared to us to warrant its application to other classes of compounds, and we proceeded to make a further series of investigations, some of which will be dealt with in a later paper. For the present, *p*-benzoquinone is the only compound which need be described. The known close relation to one another in which the two para-positions in a benzene stand seemed to lend probability to the idea that a band somewhat similar to those observed in compounds containing the group –CO·CO– might be found in the spectra of the para-quinones. Our anticipations were again justified, as the *p*-benzoquinone spectrum has a band almost identical with that of camphorquinone, its head being at 2150 (see Fig. 5 in following paper). Now it is known that *p*-benzoquinone can exist in two forms, for both of which chemical evidence has been adduced :

I.   II.

The chief points in favour of the first formula are : the reduction of *p*-benzoquinone yields quinol, a benzenoid derivative ; the action of phosphorus pentachloride produces a *p*-dichlorobenzene ; the oxidising power of benzoquinone, in which it resembles a peroxide. The second formula is supported by the following facts : its reactions with hydroxylamine and phenylhydrazine prove that the quinone carbonyl group resembles that in an aliphatic ketone ; quinone takes up two and four atoms of chlorine or bromine as if it were a tetrahydrobenzene derivative.

Chemical evidence being favourable to both formulæ, it appears not unwarrantable to assume that in this case also the absorption band is caused by the "making and breaking" of contact between the two oxygen atoms. Thus it may be concluded that the actual wandering of a hydrogen atom is not necessary for the production of these absorption bands. Again, since the result of this "making and breaking" would be the production of two nascent carbonyl groups, an explanation is thus given of the great chemical activity of the carbonyl radicles in benzoquinone.

It appears to us that our results throw much light on those reactions which are supposed to be influenced by steric hindrance, for if the tautomerism or isorropesis of any carbonyl group can be influenced by substitution, all the results will be produced which are often attributed to purely spacial relations. We intend to investigate the question more fully later, and at resent need only mention one or two instances.

The action of phosphorus pentachloride on ethyl acetonedicarboxylate produces ethyl monochloroglutaconate. Now, if the tautomerism of the carbonyl group be reduced by substituting alkyl groups for hydrogen atoms in the methylene radicles, the reaction should be more difficult to bring about. Petrenko-Kritschenko (*Annalen*, 1896, **289**, 52) has found that this is actually the case, tri-alkylated derivatives giving very small yields, and no reaction at all being found when tetra-alkylated substances are employed.

Petrenko-Kritschenko and Ephrussi (*Annalen*, 1896, **289**, 59) have shown that substitution produces an analogous effect on hydrazone formation, neither the dimethyl nor diethyl derivatives of ethyl acetonedicarboxylate giving a hydrazone, although the parent substance easily yields one. This might be attributed to purely steric causes, but V. Meyer (*Ber.*, 1896, **29**, 836) has shown that space relations have less effect than purely chemical ones, since, although acetylmesitylene gives no oxime, yet when the more bulky $-CO_2Et$ group is substituted for a hydrogen atom an oxime is easily formed, as in the case of ethyl mesitylglyoxylate. Somewhat similar effects can be found in the difficulty of hydrazone formation in the case of the substituted phloroglucinols (Herzig and Zeisel, *Ber.*, 1888, **21**, 3493).

We have not yet investigated the case of the esterification of substituted acids, although here also the evidence points to the fact that when tautomerism is reduced the esterification constant is lowered. If V. Meyer's view of the esterification process be accepted, it will be seen that the chief factor in the problem is the additive capacity of the carbonyl group, which, according to our view, depends greatly on tautomerism effects.

$$-C\begin{smallmatrix}O\\\\OH\end{smallmatrix} + \text{EtOH} \longrightarrow -C\begin{smallmatrix}OEt\\OH\\OH\end{smallmatrix} \xrightarrow{-H_2O} -C\begin{smallmatrix}OEt\\O\end{smallmatrix}$$

I.   II.   III.

According as the carbonyl group in I is more or less reactive, the yield of II will be greater or less, and hence the esterification process will be more or less rapid.

The question of the bromination or chlorination of fatty acids also throws some light on the point. Here, owing to the proximity of a carbonyl group, the hydrogen atoms attached to the α-carbon atom will be in the condition which we have termed "potential tautomerism," and will therefore be much more reactive than the other hydrogen atoms of the compound. The halogen therefore attacks them first.

We should also like to point out that if our conclusions are accepted they involve the rejection of the deductions drawn by Petrenko-Kritschenko (*J. pr. Chem.*, 1900, [ii], 61, 431 ; 1900, 62, 315 ; *Ber.*, 1901, 34, 1699, 1702 ; *Annalen*, 1905, 341, 150) from his measurements of the reactivity of open chain and cyclic ketones He assumed that if open chain compounds had a more or less cyclic configuration in space, their carbon atoms would exert the same -degree of steric hindrance as those in the corresponding closed chain compound. Hence, if the steric hindrance were different in open and in closed chain compounds, the two substances would have different configurations.

Petrenko-Kritschenko, starting from the above assumption, proceeded to measure the effect of allowing certain open chain and cyclic ketones to react for one hour with potassium hydrogen sulphite, hydroxylamine, and phenylhydrazine, and he found, as he had predicted, that the quantities of product formed by cyclic ketones in that time were greater than those formed by aliphatic ketones. His results are undoubtedly correct and valuable, although it would have been better to measure the rates of formation of the product rather than merely to estimate its gross amount after an arbitrary time had elapsed ; but we venture to point out that the deductions which he draws from his results are not so trustworthy. Tautomerism is much more marked in the case of cyclic compounds than in aliphatic substances, and hence the cyclic carbonyl compound is much more reactive than the fatty one. Consequently, a cyclic ketone might be expected to show much greater activity than an open chain one, quite apart from any purely steric considerations. It seems to us that too much reliance should not be laid on deductions from ketonic reactions as to the space formulæ of carbon chains. Professor Petrenko-Kritschenko has approached the subject from other less debatable standpoints.

All the substances employed in this investigation were obtained in the greatest possible state of purity. As a general rule, $N/10$ solutions in alcohol were made up and the iron spectrum photographed through 35, 30, 25, 20, 17, 15, 12, 10, 8, 6, 5, and 4 mm. This was repeated for the same lengths of $N/100$, $N/1000$, and, if necessary, $N/10,000$ solutions. The curves shown were obtained by plotting the limits of absorption against the logarithms of the relative thicknesses of a $N/10,000$ solution.

Conclusions.

(1) The reactivity of any carbonyl group is not inherent in the group itself, but is produced by the action of neighbouring atoms which render the carbonyl group "nascent."

(2) Such action may take the form of tautomerism or of a modification of tautomerism which does not require the actual transfer of a hydrogen atom from one atom to another, but merely some intra-atomic disturbance in the system $-CH_2 \cdot CO-$.

(3) The action may also take the form of the process which we have termed isorropesis, in which no actual wandering of atoms occurs, but in which a finer play of forces between two carbonyl groups is involved.

(4) Many cases which are at present accounted for on the hypothesis of steric hindrances can be better accounted for either by tautomerism or isorropesis, and some cases which are in direct contradiction to the steric theory can also be explained. It is therefore claimed that the hypothesis of the "nascent carbonyl group" accounts more satisfactorily for the facts, and is preferable to explanations based on the idea of steric hindrance.   ·

(5) When the possibility of the formation of a nascent carbonyl is excluded, the usual ketonic reactions are not observed. The carbonyl group may then be considered an "inactive" carbonyl group in contradistinction to a "nascent" one.

In conclusion, we wish to express our thanks to the Chemical Society for a grant towards the expenses of this research, to Professor Collie and Dr. Smiles for the great interest they have taken in the work, and to Mr. W. B. Tuck, B.Sc., for assistance during the course of the investigation.

ORGANIC AND SPECTROSCOPIC LABORATORIES,
UNIVERSITY COLLEGE,
LONDON.

LV.—*The Relation between Absorption Spectra and Chemical Constitution. Part II. The a-Diketones and Quinones.*

By EDWARD CHARLES CYRIL BALY and ALFRED WALTER STEWART
(Carnegie Research Fellow).

IN the preceding paper, the absorption spectrum of ethyl pyruvate was described, and it was shown how an absorption band with its head at an oscillation frequency of 3100 is developed at from 50 to 23 mm. of a $N/10$ solution. The position of this band and the concentration at which it appears are entirely different from those occurring with the keto-enol tautomeric process (Baly and Desch, Trans., 1904, **85**, 1039, and 1905, **87**, 760), and it was suspected that the new band has its origin in the juxtaposition of the two carbonyl groups because of the exceptional activity of the ketonic group in this compound. That this supposition was entirely justified we proved by observing the absorption spectrum of camphorquinone, in which there are two true ketonic groups in juxtaposition. As was described in the last paper, the new band is strongly exhibited in the case of this substance, showing that the presence of two carbonyl groups adjacent to one another gives rise to a new type of oscillation which causes the absorption of light of a much greater wave-length than that absorbed by the process of keto-enol tautomerism. For this new type of oscillation we have proposed the name "isorropesis," and in the present paper we propose to treat of the phenomenon of isorropesis and show how it is the origin of the colour of the a-diketones and quinones.

Now we have also examined the absorption spectrum of diacetyl, the absorption curve of which is shown in Fig. 1 ; it will be seen that here, too, is present the same absorption band as in the case of camphorquinone. The frequency of the head of the band is a little greater than in the case of camphorquinone, owing to the fact that the molecular weight of diacetyl is considerably less than that of camphorquinone.

The absorption spectra of many β-diketones have already been described (Baly and Desch, *loc. cit.*), and in every case the absorption bands lie in the extreme ultra-violet, and their origin has been traced to the phenomenon of the labile hydrogen or metallic atom. The frequency of these bands is of a mean value of 3800, although they naturally tend to shift towards the longer wave-lengths when the total mass of the molecule is increased. Since the position of this band is far down in the ultra-violet, the compounds showing a simple keto-enol tautomerism are not visibly coloured ; in the compounds such as the α-diketones just described, the oscillation or isorropesis between the residual affinities on the oxygen atoms of the carbonyl groups results in the absorption of light of a mean wave-length of 4200 Ångström units, which is situated in the blue region of the spec rum.

Fig. 1.

Oscillation frequencies.

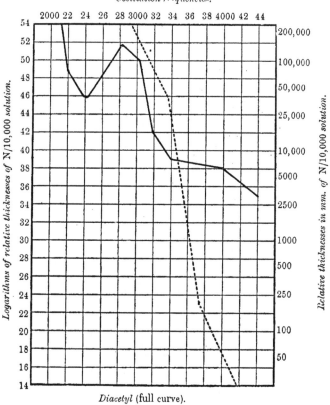

Diacetyl (full curve).
Diacetyldioxime (dotted curve).

These substances are, therefore, strongly yellow, *their colour being d₁ e to the isorropesis between the carbonyl oxygen atoms in juxtaposition.*

We have extended our observations to include other compounds, conta'ning two true carbonyl groups in juxtaposition. For example, we have measured the absorption of acenaphthenequinone and phen-anthraquinone ; in the spectra of both these substances (Fig. 2), the

new absorption band is exhibited, so that here again the isorropesis
between the two oxygen atoms is the origin of the yellow colour.

Isatin is a further example of this type of substance, and, as shown
by Hartley and Dobbie (Trans., 1899, **75**, 640), its absorption spectrum
exhibits a similar band with head at a frequency of 2400, which again

FIG. 2.

Oscillation frequencies.

18 2000 22   24   26   28 3000 32   34   36   38 4000 42   44

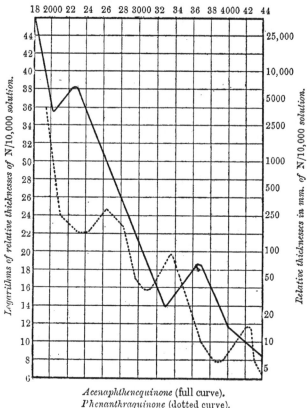

Acenaphthenequinone (full curve).
Phenanthraquinone (dotted curve).

shows the connection between the two adjacent carbonyl groups and
colour.

Another very interesting case of an α-diketone is that of benzil, the
absorption curve of which is shown in Fig. 3.   Now it was shown in a
previous paper, dealing with the absorption spectra of certain monosubstituted derivatives of benzene (Baly and Collie, Trans., 1905, **87**,

1331), that the residual affinity of the oxygen atom in acetophenone modified the absorption spectrum of benzene in a very marked manner. All the absorption bands belonging to benzene have entirely disappeared, showing that the ordinary benzenoid tautomerism has been stopped. This is doubtless owing to the attraction between the residual affinity of the oxygen of the carbonyl group and the atoms of the ring.

Fig. 3.

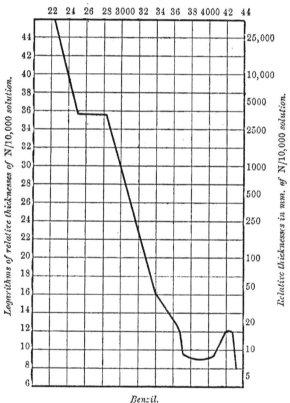

Benzil.

This undoubtedly accounts for the fact that the carbonyl group of acetophenone is unusually inactive towards sodium bisulphite, &c., because the residual affinity of this group is almost entirely occupied with and fixed by the residual affinity of the benzene ring, with the result that the group does not easily exist in the nascent state necessary to the formation of additive compounds. It might be expected, therefore, in

L. L. 2

the case of benzil, that the residual affinities of the two carbonyl groups would each be occupied and fixed by the adjacent phenyl group, and that therefore no isorropesis between the residual affinities would occur. On reference to the absorption curve of this substance, Fig. 3, it will be seen that in the region of least concentration there is an absorption band with head at a frequency of 3900. This band occupies the region of the benzene bands, and its presence in the spectrum of benzil supports the view that the benzenoid tautomerism is undoubtedly present to a certain extent. For this reason, therefore, we may conclude that the residual affinities of the two carbonyl groups are not entirely fixed, and that a small amount of isorropesis between these groups is possible. It is evident that this conclusion is justified from an inspection of the upper portion of the absorption curve of benzil, where a shallow band with head at a frequency of 2650 appears. The existence of this band shows that isorropesis is taking place, and its shallowness proves that it is only present to a small degree (compare Baly and Desch, Trans., 1905, 87, 766). It may be noted that the yellow colour of benzil is not very pronounced, and readily disappears on dilution. The measurements of the additive capacity of the benzil carbonyl groups made by Petrenko-Kritschenko agree very closely with our hypothesis.

The most striking application of this principle is in the case of quinones, for in these compounds we have a type resembling in some respects an a-diketone, and in these compounds, too, the new absorption band is exhibited showing the undoubted existence of the process of isorropesis between the quinonoid oxygen atoms. As regards the absorption spectrum of p-benzoquinone itself, this was observed by Hartley, Dobbie, and Lauder (*Brit. Assoc. Report*, 1903, 126) with an aqueous solution, when they found the presence of two bands with heads at the frequencies of 3400 and 4050. The authors conclude that the yellow colour of p-benzoquinone is only due to the presence of general absorption, since both the absorption bands are in the ultraviolet. Now it is evident that water is an unsatisfactory solvent for organic compounds the absorption spectra of which are to be observed; in every case, we are investigating the influence or the properties of residual affinity, and the use of such solvents as water, which possess strong residual affinity of their own, is clearly inadvisable except in special circumstances (compare previous paper, p. 494). The absorption spectrum of p-benzoquinone in alcoholic solution is entirely different from that observed by Hartley, Dobbie, and Lauder for the aqueous solution. The absorption curve is reproduced in Fig. 4 and shows only one band, with its head at a frequency of 2100. This band is almost identical with the absorption band shown by camphorquinone, and, as stated in the preceding paper, is undoubtedly caused by the isorropesis between the two oxygen atoms in the para-position, exactly in the same way as between those of the α-diketones. Further, it is evident that the new absorption band of *p*-benzoquinone in the visible region of the spectrum (wave-length 480μμ) is the true origin of the colour of this substance; or, in other words, the colour of *p*-benzoquinone is due to the isorropesis between the two oxygens in the para-position.

FIG. 4.

Oscillation frequencies.

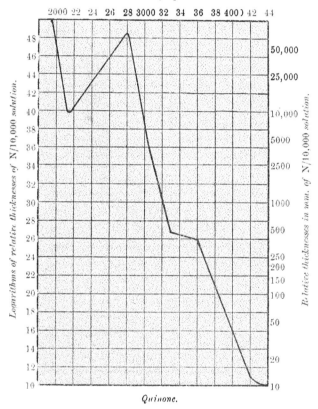

Quinone.

We have also investigated the absorption spectra of the following quinones: toluquinone, *p*-xyloquinone, thymoquinone, α-naphthaquinone, and anthraquinone, and find that the same band is present in each case. The absorption curves are reproduced in Figs. 5, 6, 7, and 8 respectively, and, as can be seen, show the presence of the new band. The process of isorropesis exists in the case of these quinones just as in quinone itself, and, indeed, is the origin of the colour of these compounds.

The importance of these results as regards Armstrong's theory of colour is manife-t; they would seem to supply the key to his generalisations, and at the same time to explain the colours of many

FIG. 5.

Oscillation frequencies.

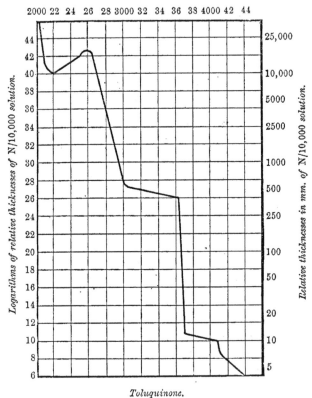

Toluquinone.

substances which are difficult of interpretation by Armstrong's quinonoid linking alone. Armstrong, in stating that colour was due to the quinonoid linking, was, of course, perfectly correct, but this formula gives us no reason why colour is produced. There is no esoteric value in any of the linkings of the formula as light-absorbing centres ; the results given in this paper, however, show that when the quinonoid form exists, a new type of oscillation is set up between the atoms in the para-position, the period of which is equivalent to that of light waves in the visible blue region

FIG. 6.

Oscillation frequencies.

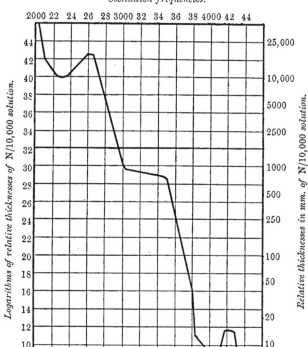

p.-*Xyloquinone.*

of the spectrum ; such a quinonoid linking therefore produces a yellow colour.

The results given above show that the process of isorropesis is common to a-diketones and quinones, so that they may be thus generalised : when two ketonic groups are adjacent to one another in the same molecule, the compound will be coloured owing to the existence of a new type of oscillation which is set up between the two re-idual affinities of the oxygen atoms. There is little doubt th it this generalisation can be extended to include many other types of residual affinity than that of the ketonic ox·gen, and we are at present engaged on a series of investigation⸱ in this direction which we hope to communicate to the Society. In the next paper we deal with compounds of the

FIG. 7.

Oscillation frequencies.

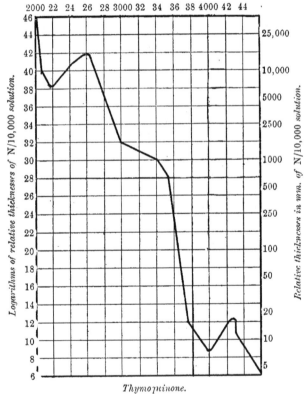

Thymoquinone.

quinone type with one or both atoms of oxygen replaced by nitrogen, and it is shown that the same type of oscillation occurs in these compounds as in the quinones and α-diketones.

In considering the whole question of colour of compounds which may be raised at this point, there is little doubt that the new principle may be extended to include every case; that is to say that isorropesis can take place between any atoms possessing residual affinity. . It must be remembered, however, that in order for the new oscillation to take place, it is absolutely necessary for some exciting or disturbing influence to be present. For example, let us take the group –CO·CO– of the α-diketones ; each oxygen atom possesses a definite amount of residual

FIG. 8.

Oscillation frequencies.

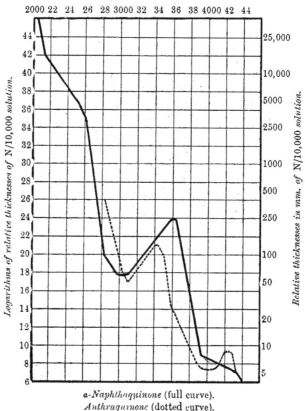

2000 22 24 26 28 3000 32 34 36 38 4000 42 44

α-*Naphthaquinone* (full curve).
Anthraquinone (dotted curve).

affinity, and it is evident that no oscillation can arise between the atoms unless one or both residual affinities are disturbed. Now, in diacetyl, $CH_3 \cdot CO \cdot CO \cdot CH_3$, this influence is furnished by the hydrogens of the methyl groups. In this compound there is an attraction exerted on the hydrogen atoms by the oxygen atoms with the result that the residual affinities on the two oxygen atoms tend to be altered.

This was discussed at length in the preceding paper. Now we have direct evidence of this potential tautomerism in the absorption curve of diacetyl (see Fig. 1), for, as can be seen, the curve shows a sudden extension at the ordinate 38. This extension undoubtedly means the incipient formation of an absorption band which occupies the position of the band due to the tautomerism of a labile hydrogen (Baly and Desch, *loc. cit.*). Clearly, therefore, the residual affinities of the two oxygen atoms are being slightly disturbed, and it is owing to this disturbance that the new oscillation or iso-rropesis takes place. We may now understand why the dioxime of diacetyl is colourless, for in this compound we have apparently the condition for colour, and yet only general absorption is indicated. The residual affinity of the nitrogen atoms exerts no attraction on the hydrogen atoms of the methyl groups and therefore is not disturbed in any way ; thus no isorropesis is set up, and the compound is colourless. The absorption curve of diacetyldioxime is shown in Fig. 1. Perhaps the process may be looked at from another point of view ; diacetyl consists of two $CH_3 \cdot CO$ groups, both of which are potential colour systems, using the word colour in its broadest sense as being the property of any compound which shows an absorption band, whether in the ultra-violet or the visible region. When two or more of these systems are present and mutually dependent, then the new process of oscillation or isorropesis is set up between the two systems. The proviso of mutual dependence is inserted of necessity to account for the new oscillation being started ; two perfectly independent vibrating systems will not combine to give a new note, they must be connected or interdependent to some extent.

It may be pointed out here that these results show that a difference of colour cannot be taken as an argument in favour of a necessary fundamental difference in constitution. Many compounds can and do exist with all the conditions for isorropesis, and yet there is lacking the influence to disturb the equilibrium between the residual affinities and so the compounds are colourless. Other compounds agreeing in every essential detail of constitution are strongly coloured simply owing to their having the disturbing influence present. All assumptions, there-fore, that two compounds must have essential differences in constitution if one is coloured and the other white are untrustworthy.

It is very noteworthy that the wave-length of the light absorbed by the process of isorropesis is about the same as that emitted by the simpler fluorescent substances ($\lambda = 4800$ to $\lambda = 4000$). It may be that there is an intimate connection between fluorescence and isorropesis, and that the former is only a manifestation of the latter. The exist-ence of an absorption band in the spectrum only means that a free period exists within the molecule capable of being excited when the light falls upon it. This is true in the case of isorropesis, the free period being established by the oscillation between the residual affinities. If now the oscillation between the residual affinities were not only able to establish the free period, but also to excite it, then we should have the phenomenon of fluorescence. There is nothing inherently improbable in this idea. In both cases, colour and fluorescence, a free period is produced by the isorropesis; in the former case, the free period is excited by the incident light, and we have absorption; in the latter case, the free period is excited by the isorropesis, and we have emission. An important fact bearing on the connection between isorropesis and fluorescence has recently been recorded by Nichols and Merritt (*Physical Review*, 1904, **18, 447**); these authors have observed that, when the fluorescence of fluorescein and certain other substances is excited by a beam of ultra-violet light, a distinct absorption occurs of light of the same wavelength as that emitted by the substance when fluorescent.

Every possible precaution was taken to obtain the substances in a state of the greatest possible purity, and the absorption curves shown in the figures were drawn by plotting the limits of absorption against the logarithms of the relative thicknesses of a $M/10,000$ solution.

Conclusions.

The following conclusions may be drawn from our experiments :

(1) When two true ketonic groups are in juxtaposition in the molecule, an oscillation or isorropesis occurs between the residual affinities of the oxygen atoms, which results in the absorption of light in the visible region of the spectrum. These substances are therefore coloured.

(2) This isorropesis also occurs between the residual affinities of the oxygen atoms in the quinones, and is the origin of the yellow colour of these substances.

(3) In order to start the oscillation, it is necessary that some influence should be present to disturb the residual affinities on the oxygen atoms.

(4) Subject to the proviso referred to in (3), there is no doubt that this principle may be extended, and that the phenomenon of visible colour is due to the oscillation between the residual affinities on atoms or groups of atoms in juxtaposition.

(5) Any assumption that two compounds must be fundamentally different in constitution if one is coloured and the other white is quite untrustworthy.

. (6) It is possible that colour and fluorescence are evidences of the same phenomenon—isorropesis. In the former case, the isorropesis provides the mechanism, and incident light actuates it; in the latter case, the isorropesis both provides and actuates the mechanism.

Our thanks are due to Mr. W. B. Tuck, B.Sc., for much valued assistance during the carrying out of the experiments. We are again indebted to Professor Collie for the great interest he has taken in this work, and also to the Chemical Society for a grant in aid of this investigation.

SPECTROSCOPIC LABORATORY,
UNIVERSITY COLLEGE, LONDON.

LVI.—*The Relation between Absorption Spectra and Chemical Constitution. Part III. The Nitroanilines and the Nitrophenols*

By EDWARD CHARLES CYRIL BALY, WALTER HENRY EDWARDS, and ALFRED WALTER STEWART (Carnegie Research Fellow).

IN the preceding paper, dealing with the absorption spectra of the quinones and a-diketones, it was shown that the colour of these compounds is due to an absorption band in the visible blue region prod iced by a new type of oscillation which occurs when two ketonic groups are in juxtaposition within the same molecule. To this process we have given the name isorropesis, and we have shown that the yellow colour of the aromatic quinones is also due to this process taking place between oxygen atoms in the para-position. Iu the present paper, the nitroanilines and nitrophenols are discussed, and it is shown how the process of isorropesis is present between the residual affinities of two nitrogen atoms in the one case and of an oxygen and a nitrogen atom in the other.

The spectra of the nitroanilines were examined originally by Hartley and Huntington; we, however, reproduce the curves in Fig. 1 of the meta- and para-compounds (from measurements of our own plates), because the Hartley and Huntington curves are inconvenient for comparison with our curves. It will be noticed that the persistence of the absorption bands is much less in the meta-compound than in the case of the ortho- and para-compounds. There is, however, no possibility of doubt that the absorption band is due to the meta-compound itself, and not to a small quantity of a highly coloured impurity. This can readily be understood from a comparison of the persistences of the ortho- and meta- and the para-compounds; the persistence of the band with the meta-compound is roughly one-third what it is with two isomerids; that is to say, if the colour were due to an impurity present to the amount of 1 per cent., this impurity would necessarily have a colouring power and an absorption band with a persistence more than thirty times that of the o- and p-nitroaı ilines. This is manifestly absurd, as the absorption bands given by the most intensely coloured substances do not show a persistence which in any way approaches this amount. Now we have examined the absorption spectra of solutions of the nitroanilines in hydrochloric acid. These solutions are quite colourless, provided that the concentıation of the

Fıg. 1.

Oscillation frequencies.

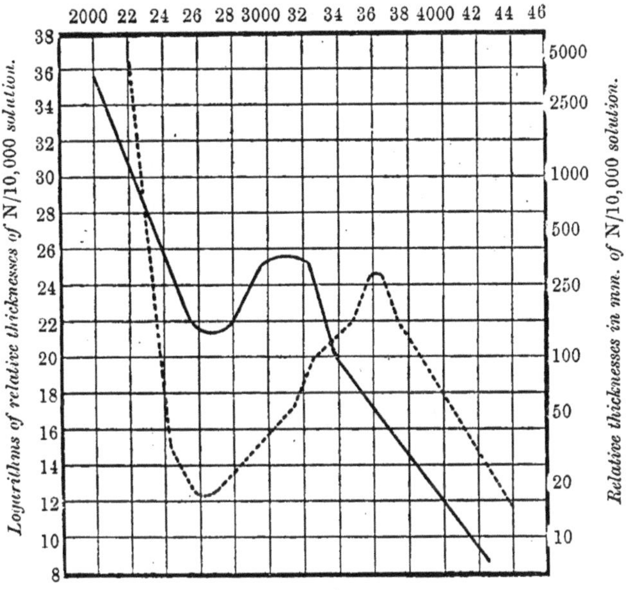

m-*Nitroaniline* (full curve).
p-*Nitroaniline* (dotted curve).

·acid be sufficient. It is very important in view of what follows that whilst the ortho- and para-compounds have to be dissolved in con-centrated acid, the meta-compound requires far less to decolorise its solution. The absorption curve of p-nitroaniline dissolved in hydrochloric acid is shown in Fig. 2, from which it can be seen that theı e is no absorption band ; the curves of the ortho- and meta isomerides are very similar.

From a comparison of the spectra of aniline and its hydrochloride and the monoalkylated benzenes, it has been shown (Baly and Collie,

Trans., 1905, **87,** 1331) that in the hydrochloride the $-NH_2HCl$ group behaves in almost exactly the same way as a single alkyl group ; that is to say, the very striking effect of the residual affinity of the $-NH_2$ group in aniline has entirely disappeared. It was also shown in the same paper that the effect of the nitro-group is to block almost entirely the tautomerism of the benzene ring, thereby introducing a fixed state of strain, with the result that strong general absorption is evidenced. Probably this effect arises from the attraction exerted by the unsaturated oxygen atoms on the atoms in the ring. Arguing from

FIG. 2.

Oscillation frequencies.

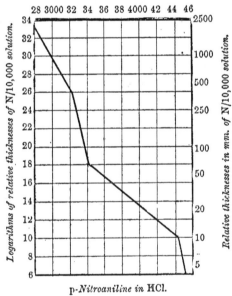

p-*Nitroaniline in* HCl.

these facts, there is little doubt but that the structure of the nitro-anilines in acid solution is that of the true hydrochloride, thus : $C_6H_4(NO_2)NH_2,HCl.$

When, however, the free substances are examined in alcoholic solution, the absorption curves (see Fig. 1) show the presence of a similar absorption band to that present in the quinones and α-diketones. We may therefore conclude at once that the substances have changed into the quinonoid form and that the process of isorropesis is taking place in exactly the same way as in the quinones. The residual affinity of the oxygen atoms of the nitro-group exerts an attraction on the hydrogen atoms of the amino-group, so that the compounds pass over to the quinonoid form :

O   OH
\\   /
N
||
(ring)
||
N
·
H

The two nitrogen atoms then occupy the position of the oxygen atoms of p-benzoquinone and function in the same way. As in the case of p-benzoquinone, this may be expressed graphically by saying that it is possible for the molecule to exist in two phases :

O   OH   O   OH
\\   /   \\   /
N—   N
|| ||
(ring)   ⇄   (ring) .
|| ||
N—   N
H   ·
I.   H
                 II.

Just as in the case of diacetyl and ethyl pyruvate quoted in the preceding papers, the formula (I) represents in all probability only an extreme phase, and therefore we are not justified in attributing this static formula to the nitroanilines. There is no doubt that the residual affinities of the nitrogen atoms as expressed by the formula

are being disturbed by the motions of the benzene nucleus, and that therefore isorropesis is set up between them. Such a process cannot, of course, be represented by any static chemical formula, and that given above (I) is only intended to represent a condition which the isorropesis may tend to bring about. It is not improbable that the possibility of writing both types of formula for a substance may be

·used as a test of the possibility of isorropesis. The nitrophenols and nitrosophenol are very similar to the nitroanilines. ˙The absorption curves of o- and p-nitroanisoles are shown in Fig. 3, and undoubtedly represent the molecular vibration curves of the formulæ

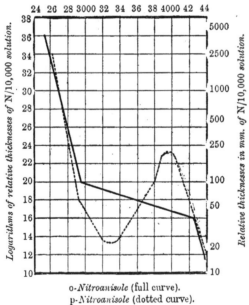

respectively.

Neither of these curves is altered in any way by the addition of sodium ethoxide to the solutions of the nitroanisoles. The absorption

FIG. 3.

Oscillation frequencies.

o-*Nitroanisole* (full curve).
p-*Nitroanisole* (dotted curve).

curves of o- and p-nitrophenols have already been described by Hartley and Huntington (*loc. cit.*); we have reproduced them plotted on logarithmic scale in order to compare them with the curves of the analogous substances. The absorption curve of p-nitrophenol in neutral alcoholic solution is shown in Fig. 5 by the full curve and is identical with that of *p*-nitroanisole; we have no hesitation, therefore, in saying that *p*-nitrophenol in neutral alcoholic solution has the formula

FIG. 4.

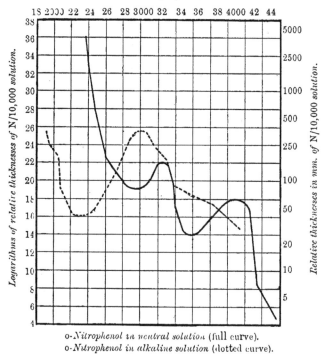

Oscillation frequencies.

o-*Nitrophenol in neutral solution* (full curve).
o-*Nitrophenol in alkaline solution* (dotted curve).

On the addition of sodium ethoxide and the formation of the sodium salt, the absorption spectrum entirely alters, and is shown by the dotted curve in Fig. 6. A similar band has now appeared in the visible region to that in the nitroanilines, and therefore we may conclude that the residual affinity of the oxygen atoms of the nitro-group

exerts insufficient attraction for the hydrogen of the free nitrophenol
to cause the formation of the quinonoid form, but that when the
hydrogen is replaced by the more electro-positive sodium atom, then
the attraction of the oxygen atoms is sufficient to bring the sodium
over, with the formation of the quinonoid form :

Very much the same is the case of o-nitrophenol, the absorption
curves of which are shown in Fig. 4 in neutral (full curve) and alkaline
(dotted curve) solution respectively.

FIG. 5.

Oscillation frequencies.

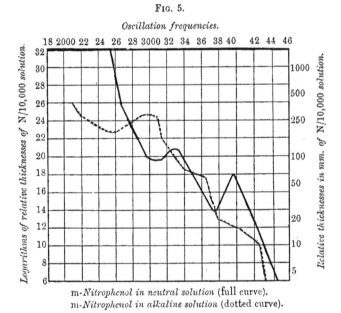

m-*Nitrophenol in neutral solution* (full curve).
m-*Nitrophenol in alkaline solution* (dotted curve).

In the case of m-nitrophenol, the absorption curves of which in
neutral and alkaline solution are shown in Fig. 5 by the full and dotted
curves respectively, the free substance in all probability exists in the
ordinary phenolic form, whilst in alkaline solution the presence of the
isorropesis band shows that the quinonoid form is undoubtedly present.

As the band is shallow, however, it is evident that but a small quantity of this form is possible. That this band is due to the quinonoid form of the *m*-nitrophenol itself and not to the presence of a small quantity of a highly coloured impurity is evident from a comparison of the persistence of the band with those of the ortho- and para-isomerides and with those of any of the colouring matters.

We have also studied the absorption spectrum of *p*-nitrosophenol, and the absorption curves of this substance are shown in Fig. 7, the

FIG. 6.

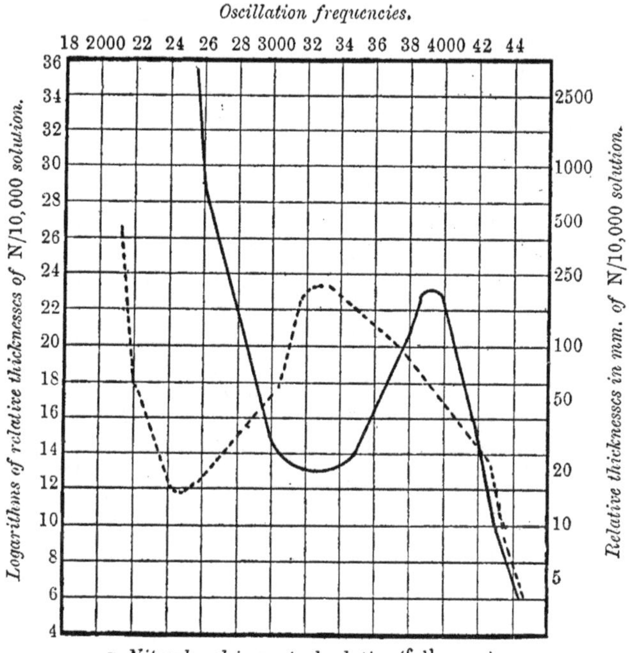

Oscillation frequencies.

p-Nitrophenol in neutral solution (full curve).
p-Nitrophenol in alkaline solution (dotted curve).

full curve being that of the compound in neutral alcoholic solution and the dotted curve being that obtained when the solution is rendered alkaline with sodium ethoxide. The absorption of *p*-nitrosophenol in neutral solution has already been observed by Hartley, Dobbie, and Lauder (*loc. cit.*) and differs considerably from ours ; these authors find two bands whilst we find only one. The principal band observed by Hartley, Dobbie, and Lauder agrees in position with ours. It will be seen that this curve is sufficiently similar to that of nitrosobenzene (Fig. 7, dot and dash curve) to

justify the conclusion that the free substance has the true nitrosophenol formula. The case is absolutely different when alkali has
been added (Fig. 7, dotted curve), and the presence of the band in
the visible region is evidence of isorropesis and the necessary
existence of the quinonemonoxime form. When in the free state, the
residual affinity of the oxygen of the nitroso-group is not sufficient to
attract the hydrogen atom away from the hydroxyl group; when,

FIG. 7.

Oscillation frequencies.

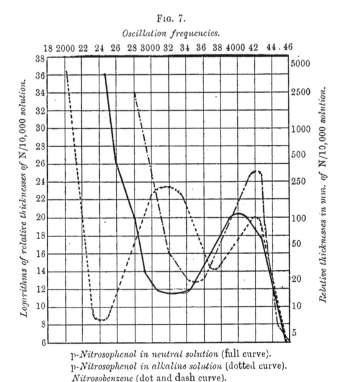

p-*Nitrosophenol in neutral solution* (full curve).
p-*Nitrosophenol in alkaline solution* (dotted curve).
Nitrosobenzene (dot and dash curve).

however, this hydrogen atom has been replaced by sodium, the
attraction of the nitroso-oxygen is powerful enough to bring the
sodium away from the phenolic oxygen, with the result that the
following compound is produced:

From these results we may conclude that the three nitroanilines in neutral solution and the three nitrophenols and nitrosophenol in alkaline solution exist in the quinonoid form and that a process of isorropesis is taking place between the residual affinities of the two nitrogen atoms in the nitroanilines and the nitrogen and oxygen atoms in the nitrophenols and in nitrosophenol, with the result that an absorption band is formed in the visible region and the compounds are coloured.

The cases of *m*-nitroaniline and *m*-nitrophenol require separate consideration, for their absorption spectra show that while the absorption bands are slightly nearer the shorter wave-lengths than in the case of the ortho- and para-compounds, yet there is no doubt that the absorption bands are due to the meta-compounds themselves, and that their structure and the resulting isorropesis must be quite analogous to that occurring in the ortho- and para-compounds. It would thus appear necessary to accept the existence of meta-quinones. Now it is not necessary to insist upon the static existence of a meta-quinonoid linking, and, indeed, the spectroscopic evidence is against this, because in both *m*-nitroaniline and *m*-nitrophenol the per-sistence of the absorption band is much less than in the ortho- and para-compounds, and therefore there is not so much of the quinonoid form present with the meta-compound. No doubt in the ortho- and para-compounds the whole are in the quinonoid form because, in the case of the nitrophenols, no further change is produced after the addition of one equivalent of sodium ethoxide. An increase in the amount of the ethoxide does not increase the amount of quinonoid form ; it can be concluded that the quinonoid and phenolic forms are not in a state of dynamic equilibrium, and that the whole of the molecules are quinonoid in structure. On the other hand, the meta-compound is not wholly quinonoid. We can only assume therefore, that as no change is produced by a further addition of sodium ethoxide, there is some restraining influence acting against the formation of the quinonoid form.

The space formula proposed by Collie (Trans., 1897, **71**, 1013) had the advantage of representing the benzene molecule as a system of atoms in a state of continual vibration, and by this means it was possible to express all the various formulæ which had then been put forward as phases of one formula. We consider that this idea of a system in motion is extremely important, and we wish to emphasise its value by certain considerations which will now be dealt with ; but at the same time it is evident that vibrations of the atoms not expressly described in Collie's original paper must be introduced in order to bring the theory into line with the spectroscopic and chemical evidence now at our disposal. These new motions will now be described.

Now it has been shown (Baly and Collie, *loc. cit.*) that benzene shows seven very similar and closely-situated absorption bands, and it was pointed out that the formation of these can be accounted for by assuming that each band is due to a separate making and breaking of linking between the carbon atoms of the ring. There are seven such makings and breakings possible, as can be seen from the following figures, the asterisks being attached to those atoms which are changing their linking:

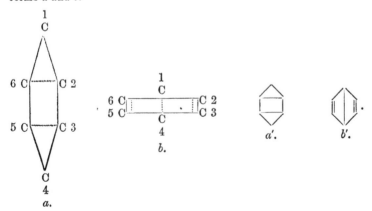

(1.)   (2.)   (3.)   (4.)

(5.)   (6.)   (7.)

It will be seen that in case (2) a single meta-linking is being formed or broken; this throws some light on the possibility of the existence of meta-quinones.

Now in order to bring the seven phases into existence, it is necessary to assume the displacement of the carbon atoms of the ring, and we can do this in the simplest way possible, that is to say, by the ordinary vibration as is accepted by any elastic ring. Thus we may say that the benzene ring is pulsating between the two displaced forms a and b.

Each carbon has residual affinity, and consequently in the condition represented in a, when the atoms 2 and 6 and the atoms 3 and 5 are brought close together, these residual affinities will produce linkings as shown by the dotted lines. The atoms 1 and 4, however, are far removed from one another and from the other atoms, and are therefore unsaturated. On the other hand, when the ring has passed into the other phase b, then the three atoms 2, 1, and 6 come very close to the three atoms 3, 4, and 5 respectively, and linking may be considered to be formed between these pairs of atoms. The linkings existing in phases a and b are shown for greater convenience on the ordinary hexagons in a' and b'. As the ring is pulsating between the forms a and b, many of the seven phases of linking change described above will be obtained. For example, let us consider the ring to have reached the form b; as it starts opening, the first break will occur between the atoms 1 and 4, followed by the breaking of the two ortho-linkings 2 : 3 and 5 : 6. When the ring passes through the half-way stage, that is, the circular form, then we shall have the centric formula, with the result that phase No. 7 is produced. We can in this way account for phases 1, 2, 3, 6, and 7; Nos. 4 and 5 can readily be understood if the motions described above are slightly interfered with by collisions between adjacent molecules. In the above it was assumed that the displacement takes place so that the atoms 1 and 4 are at the ends of the ellipse in the form a, but in general the displacement can take place along any of the three possible axes.

This scheme of displacement of the benzene ring renders it perfectly possible for meta-quinones to have a transitory existence; let us take m-nitroaniline :

a.

and let the displacement take place along the dotted lines, when we shall obtain phase a. When in the form a, then the meta-quinone form can exist, thus :

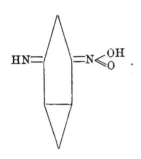

It must be remembered that this meta-quinone can only exist when the displacement occurs along the dotted lines shown on p. 525. It is not, therefore, necessary to conceive of the static existence of a meta-quinone, but it is clearly possible for such a linking to exist during part of the motions of the ring.

The results obtained with m-nitroaniline and with m-nitrophenol in alkaline solution show that only a portion of the substance exists in the quinonoid form. Doubtless the persistence of the absorption bands compared with those of the ortho- and para-compounds will give a measure of the relative number of molecules possessing the quinonoid form, that is to say, the number of molecules vibrating or pulsating in the special way described above. Inasmuch as a special form of vibration is necessary in order that the meta-quinone may exist, we may say that in this fact is to be found the undoubted restraining influence against the formation of the meta-quinone referred to above. In this way we may account for the much greater ease with which m-nitroaniline is decolorised by hydrochloric acid. On these grounds, therefore, we conclude that all three of the nitroanilines and the three nitrophenols in alkaline solution exist in the quinonoid form, and that isorropesis then occurs between the two nitrogen atoms or the nitrogen and oxygen atoms, with the result that an absorption band is formed in the visible region and the substance is coloured.

It may be pointed out that the pulsation of the benzene ring is able to explain very satisfactorily many of the characteristic reactions and properties of benzene and its derivatives. Four of the most striking may be very briefly indicated :

1. It is at once apparent that the carbon atoms in the para-position came very near to one another during the vibration, so that the wandering of atoms or groups of atoms from one carbon atom to that in the para-position is easy of explanation. Furthermore, it has been shown that phenol is a labile substance possessing keto-enol tautomerism (Baly and Ewbank, Trans., 1905, **87**, 134). On account of the near approach of the para-carbon atoms, we should expect the phenolic hydrogen to wander to the para-position, thus:

The action of nitrous acid on phenol is then easily understood, thus:

$$O{=}\langle\times\rangle\begin{matrix}H\\[2pt]+O{=}{=}N{-}OH\\[2pt]H\end{matrix} \;=\; O{=}\langle\rangle{=}NOH \;+\; H_2O.$$

2. The absorption spectra of the disubstituted benzenes show that the para-compound is always more symmetrical than the two isomerides; that is to say, the internal motions of the benzene ring are less disturbed by the para- than by either the ortho- or meta-substitution (Baly and Ewbank, Trans., 1905, **87**, 1355). This fact is clearly accounted for by the theory of a pulsating ring, because it is evident that in a compound such as p-xylene the vibration will take place very readily along the dotted axes shown in a:

$$\begin{matrix} CH_3 \\ \diagup\!\!\diagdown \\ \vert\quad\vert \\ \diagdown\!\!\diagup \\ CH_3 \end{matrix}$$

In the ortho- and meta-compounds the unsymmetrical loading of the ring will to a great extent militate against the vibration of the ring. Further, the meta-substitution will disturb the vibration more than the ortho-substitution.

3. The reduction of the phthalic acids is equally easy of explanation. In terephthalic acid, the vibration is less disturbed than in the case of the ortho- and meta-isomerides. Terephthalic acid must therefore exist to a considerable extent in the phase represented by the following formula. In this phase, the carbon atoms 1 and 4, being far removed from the other atoms of the ring, are unsaturated, with the result that they will each take up a hydrogen atom with ease, with the formation of $\Delta^{2:5}$-dihydroterephthalic acid. Similarly, the reduction of phthalic acid, although with greater difficulty, to $\Delta^{3:5}$-dihydrophthalic acid is explained. Lastly, in the case of isophthalic acid, the motions are very greatly interfered with by the meta-substitution, so that it is doubtful if any of the carbon atoms in the ring reach the unsaturated position readily; isophthalic acid is only reduced with difficulty by nascent hydrogen (Perkin and Pickles, Trans., 1905, **87**, 293), and tetrahydro-acids only are produced.

4. In the chlorination of benzene, the formation of the p-dichloro-compound, to the exclusion of the ortho-

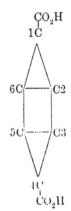

isomeride, is explained. When the benzene ring.exists in phase *a*, the two carbon atoms at the end of the ellipse are unsaturated and take up chlorine, giving the compound shown in *b*. When the opposite

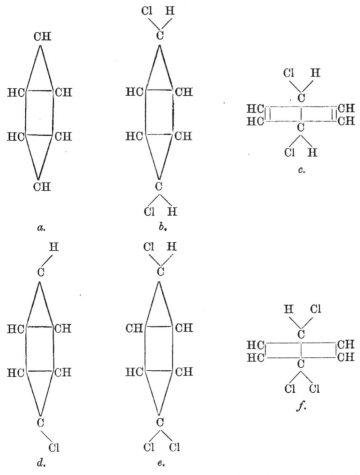

extreme is reached, as in *c*, a molecule of hydrogen chloride is split off, giving chlorobenzene. This, on reaching the first position, as in *d*, again takes up two atoms of chlorine, as in *e*; this additive compound, on reaching the form shown in *f*, again loses a molecule of hydrogen chloride and gives *p*-dichlorobenzene. It will be seen that by these motions of the ring there is no opportunity of the ortho-compound being formed, and in actual experiment none is produced.

Many other examples might be quoted in which the reactions of benzene and its compounds can be at once explained by this conception of the pulsation of the ring. For example, it has been shown (Baly and Collie, Trans., 1905, **87**, 1332) that in nitrobenzene the tautomerism of benzene on the motions of the ring has been stopped, doubtless owing to the attraction of the residual affinities of the oxygens of the NO_2 group. In the chloronitrobenzenes, therefore, there is little or no benzenoid motion, and thus these compounds approximate to the fatty type. It can be thus understood how the chlorine is replaced by hydroxyl on heating with sodium hydroxide.

A strong point in favour of this theory is its simplicity. The motion described is the simplest possible, and is the form of vibration adopted by any elastic ring, as, for example, a bell when struck.

The compounds the absorption spectra of which are described in this paper were all most carefully recrystallised and were undoubtedly pure. The p-nitrosophenol and nitrosobenzene were prepared with the greatest care; it is difficult to account for the difference in our observations of the absorption of the p-nitrosophenol and those recorded by Hartley, Dobbie, and Lauder.

Conclusions.

The following conclusions may be drawn from these observations :

1. The three nitroanilines in neutral solution and the three nitrophenols and p-nitrosophenol in alkaline solution exist in the quinonoid form.

2. The process of isorröpesis then exists between the two nitrogen atoms in the case of the nitroanilines and between the nitrogen and oxygen atoms in the case of the nitrophenols and p-nitrosophenol. This process is the origin of the colour of these substances.

3. It is necessary to assume the transitory existence of a meta-quinonoid linking to account for the phenomena observed with m-nitroaniline and m-nitrophenol.

4. Many of the physical properties of benzene are explained by considering that the ring is elastic and undergoes the same vibrations as are suffered by any elastic ring.

5. The meta-quinone linking is possible during one phase of this displacement of the benzene ring.

6. This simple vibration of the benzene ring accounts for very many of the characteristic reactions and properties of benzene and its compounds—for example, the preparation of p-nitrosophenol by the action of nitrous acid on phenol, the production of only the p-dichloro-compound in the chlorination of benzene, the reduction of the phthalic acids, and also the absorption spectra of the three isomerides in the case of the disubstituted benzenes.

We have again to thank Professor Collie for the great interest he has taken in these experiments and to express our indebtedness to the Chemical Society for a grant in aid of the work.

SPECTROSCOPIC LABORATORY,
UNIVERSITY COLLEGE, LONDON.

LVII.—*The Theory of Alkaline Development, with Notes on the Affinities of Certain Reducing Agents.*

By SAMUEL EDWARD SHEPPARD.

THE following communication deals with the reactions between hydroxylamines, hydrogen peroxide, and certain organic reducing agents on the one hand, and silver salts, particularly the emulsified silver bromide of photographic plates, on the other. It treats, therefore, of the theory of alkaline development, which, according to Abney (*Instruction in Photography*, 8th edition, p. 19), was first introduced in 1862 by Major Russell, and is now universally used in practice. The investigation may be conveniently divided into two parts, the first dealing with the nature and stoichiometry of the chemical reactions involved, the second with the statics and dynamics of development with these agents. Much of the minutiæ of the data and results are chiefly of photographic moment, and will be dealt with elsewhere.

Part I. *The Reactions between Hydroxylamine and Hydrogen Peroxide respectively with Silver Salts, with some Notes on the Reactions of Organic Developers.*

Except for the iron developers, we have but slight knowledge of the stoichiometry of both the inorganic and organic reducers available, and little even of their oxidation products, so that a consistent survey of development is impossible (Abegg, *Eder's Jahrbuch*, 1904, 1—5 ; Eder, *Handbuch*, 1903, III, 307). We have at present the following methods and results : Reeb adds the reducing agent to a solution of silver oxide in alkaline sulphite, and from the silver reduced and the original amount of reducer employed calculates the reducing power. Hurter and Driffield (*Phot. Journ.*, 1892, **22**, 194), by a similar process with ammoniacal silver nitrate, determined the reducing power of pyrogallol. They conclude that one molecule of pyrogallol reduces four atoms of combined silver to the free state, but the reaction with solid silver halide may quite possibly take a different course, and in fact this will be shown to be the case.

Andresen, whose results will be noted later, improves on the foregoing (*Eder's Handbuch*, 1903, III, 312) by adding the alkaline developer to excess of plain precipitated silver bromide and determining the silver reduced, but it is still assumed without independent proof that the reducer is entirely oxidised. On account of the analogy previously mentioned (*Proc. Roy. Soc.*, 1905, Series *A*, **74**, 448) as existing between the structure of organic reducers and the inorganic substances hydrazine, hydroxylamine, and hydrogen peroxide, the reactions of the two latter were quantitatively studied.

Hydroxylamine and Silver Salts.

The hydroxylamine was estimated by the method of Jones and Carpenter (Trans., 1903, **83**, 1394). The hydroxylamine is added with certain precautions to a boiling copper solution, the cuprous oxide formed is treated with a solution of a ferric salt, and the ferrous salt produced titrated with potassium permanganate. The method was found to be both accurate and convenient. Hydroxylamine hydrochloride recrystallised from hot 98 per cent. alcohol was used in all the following experiments, and the analysis both for chlorine and hydroxylamine showed that it could be reckoned as pure. Ordinary precipitated silver bromide, whether exposed to light or not, is rapidly reduced by alkaline hydroxylamine. The estimations were carried out in the red light of the dark room, although the precaution was probably superfluous. The silver bromide was precipitated as a fine suspension by the addition of silver nitrate solution to potassium bromide solution. The following factors were varied : the excess of bromide ions, the concentration of the hydroxylamine and the proportion of alkali, and also the total mass of solid halide. When the evolution of gas had ceased, the alkali was just neutralised with N-sulphuric acid, the solution separated by levigation, and the residue washed with hot water. This liquid was added to the solution and the whole made up to a definite volume. The silver in the residue was dissolved in nitric acid and estimated with $N/10$-potassium thiocyanate. One likely source of error is the spontaneous decomposition of alkaline hydroxylamine solutions (Berthelot, *Ann. Chim. Phys.*, 1877, [v], **10**, 433), especially in the presence of finely-divided silver (Tanatar, *Zeit. physikal. Chem.*, 1902, **40**, 475), but control experiments with reduced silver showed that the discrepancy was less than 0·5 per cent. for the duration and other conditions of the writer's experiments ; the results are as follows :

TABLE I.

Concentration of NH_2OH, HCl.	Concentration of alkali.	Relative amounts of AgBr and Br′.	Grams of silver per gram-molecule of NH_2OH.
0·7020 in 100 c.c. $=0·10N$	25 c.c. of N-NaOH 0·40N	5 c.c. of N-AgNO$_3$ with 10 c.c. of N-KBr	127·0
0·7020 in 100 c.c. $=0·10N$	50 c.c. of N-NaOH 0·80N	3·9 c.c. of N-AgNO$_3$, pptd. with HBr and washed	104·2
0·7028 in 100 c.c. $=0·10N$	0·20N	3·9 c.c. of N-AgNO$_3$ with 5 c.c. of N-KBr	110·0
0·7028 in 100 c.c. $=0·10N$	0·20N	3·9 c.c. of N-AgNO$_3$ with 5 c.c. of N-KBr	109·2
0·7028 in 100 c.c. $=0·10N$	0·20N	2·9 c.c. of N-AgNO$_3$ with 5 c.c. of N-KBr	109·1
0·3503 in 200 c.c. $=0·025N$	0·05N	2·0 c.c. of N-AgNO$_3$ with 3 c.c. of N-KBr	105·8

The results agree very well with the reaction $2NH_2OH + 2AgBr = 2Ag + N_2 + 2HBr + 2H_2O$, which would give 108 grams of silver per gram-molecule of hydroxylamine, but the experiment giving 127 grams of silver, in which the final concentration of the hydroxylamine was only $N/80$, points to a further reaction at great dilution. To test this, hydroxylamine was added to hot ammoniacal silver nitrate, the addition being made from a pipette as in the estimations by the copper process. It was found that 0·0438 gram of hydroxylamine hydrochloride produced 0·1300 gram of silver, hence 206 grams of silver per gram-molecule of reducer, so that the reaction $2NH_2OH + 2Ag_2O$ is approached.

Hydroxylamine and Silver Oxide in Sulphite Solution.

Preliminary experiments showed that silver sulphite solutions were fairly stable below 30° in the presence of reduced silver. The reaction with hydroxylamines is slow, but can be accelerated by adding excess of alkali and a trace of reduced silver. Two estimations gave :

(*a*) 0·236 gram of silver for 0·3515 gram of hydroxylamine hydrochloride, that is, 188 grams per gram-molecule of hydroxylamine.

(*b*) 0·1795 gram of silver for 0·702 gram of hydrochloride, that is, 178 grams per gram-molecule of hydroxylamine.

Owing to the low reaction-velocity, the condition of great dilution might not have been strictly fulfilled. In a further experiment, the hydroxylamine was added in drops to the mechanically-stirred sulphite solution at such a rate that the complete reaction took four hours. This gave :

(*c*) 0·2140 gram of silver for 0·7020 gram of hydroxylamine hydrochloride, that is, 212 grams of silver per gram-molecule of hydroxyl-amine.

Hence it appears that at great dilution the reaction is $2NH_2OH + 2Ag_2O = 4Ag + N_2O + 3H_2O$, and that neither Hurter and Driffield's nor Reeb's method for estimating the reducing power of developers is reliable.

Theory of the Reactions.

The reducing action of free hydroxylamine is greatly intensified by the presence of alkali. It is in fact an amphoteric electrolyte, forming both hydrogen and hydroxyl ions (Winkelblech, Zeit. physikal. Chem., 1901, 36, 550; J. Walker, Proc. Roy. Soc., 1904, Series A, 73, 155), hence acting as a base with acids, its salts being considerably hydrolysed, whilst with strong bases it behaves as a very weak acid. The two stages in which the oxidation of hydroxylamine may take place may be considered as arising from its function as a dibasic acid. At moderate dilutions, the dissociation of the first hydrogen atom gives the ion NH_2O', reacting according to the equation $2NH_2O' + 2Ag^{\cdot} = 2Ag + N_2 + 2H_2O$, whilst at greater dilution the divalent ion NHO'' is formed, giving on oxidation $2NHO'' + 4Ag^{\cdot} = 4Ag + N_2O + H_2O$.

The coupling of the discharged ions $-NH_2O$ and $:NHO$ in the manner outlined above appears to explain satisfactorily the oxidation of hydroxylamine by copper and silver salts, its behaviour being much the same with both.

The Effect of Gelatin.

There is no reason to believe that silver bromide emulsified with gelatin reacts differently with a developer from ordinary precipitated bromide, but control experiments were made with dry plates. Several plates were given a full exposure and developed in alkaline hydroxyl-amine. The hydroxylamine unoxidised was determined, and after photometric measurement of the densities the silver estimated in a manner previously described (Proc. Roy. Soc., 1905, Series A, 74, 451). The results were as follows :

Area of plate in cm.²	Density D (mean value).	Silver in grams.	NH_2OH,HCl required.	Grams of silver per gram-mol. of NH_2OH.	Silver in grams per 100 cm.² for $D = 1.0$.
5 × 87·2	3·01	0·139	0·0935	103·2	0·0103
5 × 87·2	3·21	0·145	0·0972	104·0	0·0104

Hence, in development, one molecule of hydroxylamine reduces one molecule of silver bromide, whilst the "covering-power" of the silver reduced is the same as for ferrous oxalate, for which $P = 0.0103$ was found.

The Reaction between Hydrogen Peroxide in Alkaline Solution and Silver Bromide.

The hydrogen peroxide was estimated by titration with permanganate in acid solution, acetic acid being used, as recommended by Ramsay (Trans., 1901, **74**, 1324), to avoid the discrepancies occurring with sulphuric acid from the formation of persulphuric acid. For the original peroxide solution, the results were: by titration with $N/10$ permanganate in acetic acid solution, 0·0344 gram of peroxide per c.c.; with $N/10$ permanganate in sulphuric acid, 0·0347 gram; and with $N/100$ permanganate in acetic acid, 0·0342 gram per c.c. It was found necessary to add a little manganous sulphate to start the reaction.

The chief difficulty in following the reaction of hydrogen peroxide with silver salts is the well-known catalytic decomposition of peroxide, especially in alkaline solution, by finely-divided silver. To allow as far as possible for this, an exactly similar solution of peroxide to that reacting with the silver bromide was identically treated for the same time with a quantity of silver about one-half that finally reduced. The results are expressed as a " silver correction." In the reaction with silver bromide, the alkaline peroxide was added to a suspension of this and both the silver and the peroxide estimated at the end of the reaction. The results were:

(*a*) 0·0218 gram of peroxide gave 0·0550 gram of silver or 86 grams of silver per gram-molecule of peroxide. With a "silver correction" equivalent to 15·1 grains, this becomes 97 grams of silver per gram-molecule of peroxide.

(*b*) 0·1485 gram of peroxide gave 0·4126 gram of silver. With a "silver correction" equivalent to 9·4 grams, this gives 103·9 grams of silver per gram-molecule of peroxide.

(*c*) 0·162 gram of peroxide gave 0·4830 gram of silver. With a "silver correction" of 9·4 grams, this gives 110·6 grams of silver per gram-molecule of peroxide.

These results are sufficiently near the value 108 grams of silver to indicate that equimolecular quantities react, but the complete reaction-equation remains undecided. In these experiments, the concentration of peroxide was about $N/10$ to $N/5$, with alkali in the proportion of 4 molecules to 1 of peroxide. This substance in solution is a weak acid, forming with strong bases salts which are very considerably hydrolysed (Calvert, *Zeit. physikal. Chem.*, 1901, **38**, 513). Considering the univalent anion O_2H' formed according to the equation $HO_2H + NaOH \rightleftharpoons NaO_2H + H_2O$, the interaction with silver would be $O_2H' + Ag^{\cdot} = Ag + O_2H-$, the complexes $-O_2H$ then coupling according to the scheme $-O_2H + -O_2H = 2O_2 + H_2$, or condensed

$2O_2H' + 2Ag^{\cdot} = 2Ag + 2O_2 + H_2$, that is, hydrogen should be formed in the proportion of one volume of hydrogen to two of oxygen. To test this, the gas evolved from silver bromide in suspension added to alkaline peroxide was collected and analysed. Quantities of hydrogen varying from 2 to 5 per cent., the remainder being oxygen, were obtained, showing at least that hydrogen is formed, but probably the proportion was masked by the oxygen from the catalytic decomposition and the reaction $O_2'' + 2Ag^{\cdot} = O_2 + 2Ag$ (compare hydroxylamine).

Since Thénard's experiments (*Ann. Chim. Phys.*, 1818, **9**, 314), the decomposition of peroxide by silver compounds has been extensively investigated, but with somewhat confused and even contradictory results.

Berthelot (*Compt. rend.*, 1901, **132**, 897) considers that a peroxide of silver is formed, E. Mulder (*Rec. Trav. Chim.*, 1903, **22**, 388) proposes as typical the scheme $2(HO \cdot OH) + Ag_2O = 2(HO \cdot OAg) + H_2O = Ag_2O + H_2O + O_2$, whilst Kastle and Loewenhardt (*Amer. Chem. J.*, 1903, **29**, 397, and 1903, **29**, 563) consider that complex additive products are formed, but all agree that the intermediate substance breaks down into water, oxygen, and the original silver salt. Baeyer and Villiger (*Ber.*, 1901, **34**, 349) disagree with Berthelot's conclusions, finding no intermediate oxide and also that more oxygen is evolved than would be liberated by simple catalytic decomposition of the hydrogen peroxide. No explanation is offered as to the influence of alkalis in reduction with peroxide. The author's experiments with alkaline peroxide acting on suspended silver bromide point to a definite chemical reaction according to the equation : $2H_2O_2 + 2AgBr = 2Ag + 2HBr + H_2 + O_2$ or $2O_2H' + 2Ag^{\cdot} = 2Ag + 2O_2 + H_2$, the silver reduced then catalytically assisting a further decomposition of the peroxide. This result agrees rather with Baeyer and Villiger's view and with our present knowledge of the molecular state in solution of peroxide, according to which it functions as a weak acid, forming first the univalent ion O_2H' (Carrara and Brighenti, *Gazzetta*, 1903, **33**, 362) and at greater dilution the divalent ion O_2''.

The Reactions of Quinol, p-Aminophenol, and Other Organic Reducers with Silver Bromide.

All the organic reducers available for development possess the following configuration. They are substituted aromatic derivatives containing at least two of the groups $-NH_2$, $-OH$, saturated in the ortho- or para-position with respect to one another, the meta-derivatives not functioning as developers (Andresen, *Eder's Jahrbuch Phot.*, 1899, 140), from which it appears probable that the primary oxidation product has a quinonoid structure. The author has dealt elsewhere with the chemistry of the quinol developer (*Zeit. wissent. Phot.*,

1904, **2**, 5) and has arrived at the following conclusions. Assuming quinone to be formed by the reaction $C_6H_4O_2'' + 2Ag^{\cdot} \rightleftharpoons C_6H_4O_2 + 2Ag$, the reverse reaction is greatly lessened by the interaction of quinone with alkali and sulphite. With alkalis, quinol is regenerated and hydrogen peroxide, according to the equation: $C_6H_4{<}^O_O + 2OH' \rightleftharpoons C_6H_4{<}^{O'}_{O'} + H_2O_2$, but atmospheric oxygen leads to the production of tarry substances. Moreover, quinone is reduced by sulphite to quinol, dithionic acid being formed: $C_6H_4{<}^O_O + 2SO_3'' = C_6H_4{<}^{O'}_{O'} + S_2O_6''$. Similar reactions occur for quinonoid compounds formed on mild oxidation of aminophenols (Diepolder, *Ber.*, 1902, **35**, 2816). Owing to these reactions, all attempts to estimate quinol which involve its oxidation to quinone are nullified by the presence of alkali and sulphite. Further, the action of sulphite on quinones gives an explanation of the function of this substance as a preservative and stain preventative in developers, an action hitherto ascribed to a preferential oxidation of the sulphite. On the contrary, organic reducers behave as negative catalysts to the oxidation of sodium sulphite (Bigelow, *Zeit. physikal. Chem.*, 1898, **27**, 585; S. W. Young, *Amer. Chem, J.*, 1902, **28**, 391), so that in the autoxidation of the mixture sulphite *plus* organic reducer we have a "coupled" reaction in which the total velocity is retarded, a result probably due to a cycle of changes.

The views expressed as to the oxidation of hydroxylamine and peroxide may be applied to the organic reducers of similar configuration. Whilst the acid character of the polyhydric phenols is well established and stable monobasic salts are known (Hantzsch, *Ber.*, 1899, **32**, 576), the $-NH_2$ group lowers the acid character, the aminophenols and diamines forming stable salts with strong acids. But their reducing power is intensified by alkali, pointing to their acid character, the true reducing agent being the free anion. The stoichiometry and nature of their oxidation by silver salts are obscured by the cyclic action just described, but we may indicate briefly the probable reactions for the developers investigated. In the following table, the reducing power expresses the number of molecules of silver bromide reduced by one molecule of the developer.

TABLE II.

Developer.	Concentration.	Reducing ion.	Reducing power.	Oxidation product.
Hydroxylamine	To $N/80$	NH_2O'	1	N_2
	Below $N/80$	NHO''	2	N_2O
Hydrogen peroxide	$N/20$	O_2H'	1	$2O_2 + H_2$
	Dilute	O_2''	2	O_2
Quinol	Any	$C_6H_4O_2''$	2	$C_6H_4\Big\langle\begin{smallmatrix}O(1)\\O(4)\end{smallmatrix}$
p-Aminophenol ("metol")	$N/20$	$C_6H_4\big\langle\begin{smallmatrix}O\\NH_2\end{smallmatrix}$	1 *	Azoxindones and other condensed substances
p-Aminophenol ("metol")	Very dilute	$C_6H_4\big\langle\begin{smallmatrix}O'\\NH\end{smallmatrix}$†	2 *	Iminoquinone
p-Phenylenediamine......	,,	$C_6H_4\big\langle\begin{smallmatrix}NH'\\NH\end{smallmatrix}$†	2	Di-iminoquinone

* Andresen finds 1·7 as the reducing power.

† For oxidation with silver oxide, see Willstätter and Pfannensteil (*Ber.*, 1904, **37**, 4605). .

Part II. *The Statics and Dynamics of Development with Alkaline Reducers.*

The methods evolved for the quantitative study of development have been dealt with elsewhere (*Proc. Roy. Soc., loc. cit.*). The velocity is determined by measuring γ, the slope of the exposure or $D\log E$ curve (*Trans.*, 1905, **87**, 1325), which depends only on development, whilst $\log i$, the point where the straight line cuts the exposure axis, may depend on the development, when the free energy of the reaction falls below a certain value (*Trans., loc. cit.*), but is generally independent of it. It has been explained in the foregoing section that all these reducing agents are weak acids, the actual reducer being the anion, the concentration of which is increased by the addition of alkali owing to the greater dissociation of the salts. These are subject to hydrolysis according to the scheme $X' + H_2O \rightleftarrows XH + OH'$, whilst the mass law gives $[OH'[XH = K[X'$ (J. Walker, *Zeit. physikal. Chem.*, 1899, **4**, 319).

We may expect therefore that in the presence of sufficient excess of alkali the velocity will be proportional to the concentration of reducer, with excess of reducer to that of the alkali, whilst with equivalent quantities of acid and alkali the velocity should diminish faster than the dilution owing to increased hydrolysis. If the velocity depends on a dibasic ion the relations will be more complicated.

All the measurements were compared with development for ferrous oxalate, the dynamics of which have already been thoroughly investigated. The values for the first series of plates were :

TABLE III.

t (time in minutes).	γ.	$K=1/t\log\dfrac{\gamma\infty}{\gamma\infty-\gamma}$	t (time in minutes).	γ.	$K=1/t\log\dfrac{\gamma\infty}{\gamma\infty-\gamma}$
2·0	0·56	0·064	6·0	1·18	0·057
3·0	0·63	0·050	8·0	1·50	0·063
4·0	1·02	0·068	10·0	1·62	0·059
5·0	1·15	0·065			
			∞	2·18	K = 0·061 mean

Hydroxylamine.

This base was used in the form of the hydrochloride with excess of caustic soda. The nitrogen evolved first forms a supersaturated solution, which then forms bubbles of gas in the film. This renders it difficult to obtain accurate measurements much above γ = 1, and limits its technical use as a developer. It was found that the density ratios were independent of the time of development and the value of log$i$ the same as for ferrous oxalate. The velocity measurements were as follows:

TABLE IV.

Concentration of NH$_2$OH.	Concentration of NaOH.	Rate of development.				
0·010N	0·087N	t (in minutes)	5 0	10·0	∞	
		γ	0·64	0·95	2·18	
0·010N	0·0335N	t	6·0 10·0	14·0	20·0	
		γ	0·47 0·63	0·67	0·80	
0·010N	0·184N	t		12·0		
		γ	·3	0	1·18	
0·020N	0·077N	t	3·0	5·0		
		γ	0·51	0·75		
0·050N	0·144N	t	0·50	1·0	2·0	
		γ	·35	6	0	
0·050N	0·095N	t	·0		4·0	
		γ	·46	0	9 4	
0·050N	0·047N	t	·0			
		γ	·50.	5	7	2
0·10N	0·046N	t	1·0	2·0	3·0	5·0
		γ	4	0	0·95	6
0·10N	0·016N	t	·0		5·0	6·0
		γ	4	3	·65 · 5	8 1·06

It will be seen from these data and curves that both the course and the velocity of the reaction are modified by the composition of the solution. Beyond a certain value, the velocity is not increased much by the addition of alkali. This corresponds to the diminution of hydrolysis according to the mass law. In the presence of great excess of alkali, we may assume that the concentration of salt is proportional to that of the reducer, and it will be seen that the velocity is proportional to this. Thus with $0.05N$-NH_2OH in $0.144N$-NaOH, the time for $\gamma 0.75$ was 1.3 minutes, and with $0.01N$-NH_2OH in $0.087N$-NaOH, 6.5 minutes. The hydrolysis is evidently very great, as with $N/100$-hydroxylamine some 20 molecules of caustic soda were required

CURVE 1.

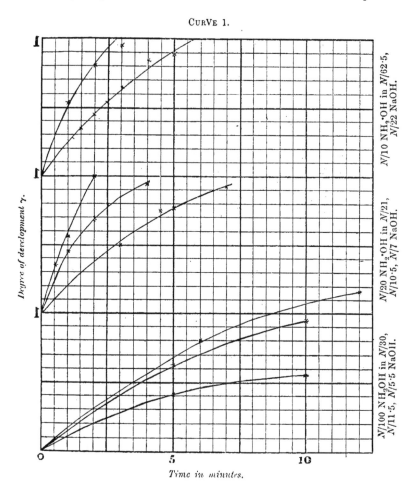

Degree of development γ.

Time in minutes.

to suppress it. Consequently the velocity falls off more rapidly than the total concentration. Thus a solution of $N/20$-hydroxylamine in $N/10$-alkali was diluted to twice the volume and the same densities obtained in 2.0 and 5.0 minutes.

The Mechanism of Development and the Velocity Function.

It has been deduced (*Proc. Roy. Soc.*, 1905, **74**, 448) from the law of constant density ratios and the theory of heterogeneous reactions that the rate of development depends on the micro-diffusion or supply of the reducing ion *in the film* to the affected halide. Under certain conditions, the velocity was given by the equation $dD/dt = K(D_\infty - D)$, or for γ after integration, $1/t \log \gamma_\infty / \gamma_\infty - \gamma = K$, which was verified for iron salts.

This formula was valid for hydroxylamine under certain restricted conditions. Owing to the great velocity of reduction, the invasion of the developer ceases to be very rapid compared with the process in the film, so that concentrated solutions showed some tendency for K to decrease. The most concordant results (Table V) were obtained with sufficient excess of reducer or alkali to check the hydrolysis.

TABLE V.			TABLE VI.		
$0.10N$-NH_2OH in $0.016N$-NaOH.			$0.01N$-NH_2OH in $0.184N$-NaOH.		
t (time in minutes).	γ.	$K = 1/t \log \dfrac{\gamma_\infty}{\gamma_\infty - \gamma}$.	t (time in minutes).	γ.	$K = 1/t \log \dfrac{\gamma_\infty}{\gamma_\infty - \gamma}$.
2.0	0.44	0.0490	5.0	0.63	0.030
2.5	0.53	0.0485	6.0	0.80	0.033
3.0	0.65	0.0510	12.0	1.18	0.028
4.0	0.85	0.0530			
5.0	0.88	0.0450	∞	—	$K = 0.0303$ mean
6.0	1.06	0.0482			
∞	2.18	$K = 0.0491$ mean			

Assuming that the hydrolysis was completely suppressed, the concentration of the reducing ion NH_2O' would be $0.016N$, so that for $N/10$ hydroxylamine the velocity would be 0.306, compared with 0.61 for $N/10$ ferrous oxalate. Another comparison where the hydrolysis was checked by excess of alkali gave the results indicated in Table VI.

Hence K for $N/10$-$NH_2O' = 0.303$, in good agreement with the former value.

The relative velocities of development by the ions NH_2O' and $Fe(C_2O_4)_2''$ are as 0.305 to 0.061 or as 5 to 1.

Viscosity of the Solution and Rate of Development.

The addition of alcohol to the hydroxylamine developer showed that the viscosity of the solution had little or no influence on development. The developer was $N/20$ hydroxylamine in $N/20$ caustic soda.

	(a) No alcohol.		(b) With 25 per cent. ethyl alcohol.	
t	3·0	5·0	4·5	7·0
γ	0·50	0·77	0·75	0·92
	$K = 0·0375$		$K = 0·0375$	

The velocity is the same. According to Dunstan (Trans., 1904, **85**, 817), the viscosity for ethyl alcohol—water mixtures gives $\eta = 0·00891$ for no alcohol and $\eta = 0·01851$ for 24·7 per cent. by weight of alcohol. Hence doubling the viscosity did not influence the rate of development, which agrees with the view that in general the invasion or rate of entry of the developer into the film is much faster than the process of development itself.

The Developing Equivalence of the Alkalis.

Experiments were made on the relative developing values of potassium, sodium, and lithium hydroxides and potassium carbonate. For the plates used, the value of γ_∞ was 2·50. The alkali solutions were titrated against standard sulphuric acid, and for 100 c.c. of developer quantities equivalent to 10 c.c. of N-alkali taken in each case. The developer was 5 c.c. of N-hydroxylamine hydrochloride, 10·0 c.c. of N-alkali to 100 c.c. with water. With potassium carbonate, the hydrochloride was just neutralised with caustic potash, then 10·0 c.c. of molecular carbonate solution added.

For the three caustic alkalis, practically identical values were obtained for the densities for the same time of development, so that the velocity is the same for equimolecular quantities; that is, the developing equivalence of the alkalis is the same as their affinity constants as bases, the effect depending only on the concentration of hydroxyl ions (Nernst, *Theoretical Chem.*, 4th edition, p. 512).

TABLE VII.

NaOH.			KOH.			LiOH.		
$t.$	$\gamma.$	$K.$	$t.$	$\gamma.$	$K.$	$t.$	$\gamma.$	$K.$
2·0	0·43	0·040	2·0	0·54	0·058	2·0	0·56	0·055
3·0	0·83	0·062						
4·0	1·01	0·056	4·0	0·96	0·053	4·0	0·97	0·053
5·0	1·08	0·049						
∞	2·50		∞	2·50		∞	2·50	
	$K = 0·052.$			$K = 0·056.$			$K = 0·054·$	

The values obtained with $N/10$-K_2CO_3 were:

t in minutes	12·5	21·5	25·0	45·0
γ	0·36	0·66	0·81	1·19
K	0·0055	0·0062	0·0067	0·0062

Mean $K = 0·0062.$

From this we have :

$$0.05N\text{-KOH} ; \quad K = 0.054.$$
$$0.10N\text{-K}_2\text{CO}_3 ; \quad K = 0.0062.$$

Assuming the velocity is proportional to the concentration of hydroxyl ions, the hydrolysis of potassium carbonate in decinormal solution is calculated to be 5·4 per cent. For sodium carbonate, Shields (*Zeit. physikal. Chem.*, 1893, **12**, 167), by the saponification of ethyl acetate, found 3·2 per cent., and Koelichen (*ibid*, 1900, **33**, 173), by the velocity of acetone condensation, obtained 2·2 per cent. The alkalinity of free hydroxylamine itself introduces an error in the foregoing method, but the result is sufficient to show that the developing value of carbonates and similar salts simply depends on their hydrolysis and the hydroxyl ions formed thereby.

Quinol.

As a developer, quinol has certain peculiarities ; the image appears slowly, but then gains density with increasing rapidity. The velocity measurements indicate an initial induction, while in some cases, especially with deficiency of alkali, there is a regression of the inertia, the value of $\log i$ not reaching the ferrous oxalate value until an advanced stage of development, an effect also produced by lowering the temperature below the normal 20°. Also the retardation by bromide is great, as will be shown later. The facts point to an initially low chemical velocity, the potential of the system increasing as metallic silver is deposited (Trans., 1905, **87**, 1311) and as the quinone first formed is destroyed by alkali and sulphite (Part I). Hence some difficulty was experienced in obtaining concordant and reliable measurements for low degrees of development, where the law of constant density ratios is not always followed. The following are the data obtained :

TABLE VIII.

The concentration of alkali was constant, that of the quinol varied.

t = time in minutes. t_a = time of appearance of image.

Concentration of quinol.	Concentration of caustic soda.	Rate of development.					t_a.
0·015N	0·0485N	t 5·5	9·0				1·3
		γ 0·45	•0				
0·025N	0·0485N	t 3 0		5·0	8·0		1·0
		γ 0·515	·45	0·735	·15		
0·025N	0·0485N	t 3·5	·	5·0	·0		1·0
		γ 0·50	·8	0·73	·04	·	
0·050N	0·0485N	t 3·0		6·0	·0		1·0
		γ 0·41	·6	0·89	·00		
0·100 V	0·0485N	t 4·0		6·0	·0	9·0	1·3
		γ 0·34	·5	0·61	·76	0·945	
0·200N	0·0485N	t 4 5	· 35				1·5
		γ 0 515					

Table IX.

Concentration of quinol constant at $0.025N$ and $0.100N$.

Concentration of quinol	Concentration of caustic soda.	Rate of development.					$t_{a.}$
$0.025N$	$0.194N$	t	4.0	7.0			0.95
		γ	$.5$	1.00			
$0.025N$	$0.097N$	t	$.$	5.0	6.0	8.0	1.00
		γ	$.0$	0.685	0.87	1.03	
$0.025N$	$0.0485N$	t		4.5	5.0	8.0	1.00
		γ	0	0.68	0.73	1.04	
$0.025N$	$0.0194N$	t	1	17.0	21.0	31.0	3.00
		γ	8	0.73	0.79	1.15	
$0.100N$	$0.0194N$	t	14.0	15.0	20.0	26.0	2.51
		γ	$.44$	0.69	0.875	1.04	
$0.100N$	$0.097N$	t	$.0^4$	3.0	4.0	6.0	0.51
		γ	$.5$	0.85	1.00	1.23	
$0.100N$	$0.0485N$	t	$.0$	4.5	6.0	8.0	9.0	1.3
		γ	$.3$	0.45	0.61	0.75	0.945	

CURVE 2.

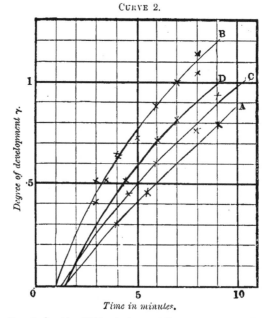

Time in minutes.

A. $N/20$ sodium hydroxide, $N/65$ quinol.　*B.* $N/20$ sodium hydroxide, $N/40$ quinol.
$N/20$,,

C. $N/20$ 　　　$N/10$,,　*D.* $N/20$ 　　　$N/5$

The experiments with constant $N/20$ alkali and increasing quinol give a curious result; up to $N/40$ quinol, the velocity is proportional to the quinol concentration, but at $N/40$ a maximum is reached, the rate for $N/20$ quinol being the same, $N/10$ a lower value, and $N/5$ a lower value than $N/20$, but greater than $N/40$.

Relation of Velocity to Alkali Concentration.—With $N/40$ quinol, increase of alkali up to $N/20$ alkali increased the velocity at a somewhat greater rate than the concentration, but it attains a maximum for 1 molecule of quinol to 2 of alkali ; this agrees with the view that the developing ion is $C_6H_4O_2''$ from the dibasic salt, $C_6H_4(ONa)_2$, but the

CURVE 3.

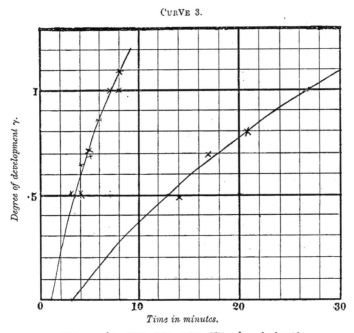

Time in minutes.

$N/40$ quinol in $N/50$, $N/20$, $N/10$, $N/5$ sodium hydroxide.

fact that further addition of alkali has no influence on the velocity would seem to show that the hydrolysis is small. This is surprising, as quinol is a very weak acid (R. Bader, *Zeit. physikal. Chem.*, 1889, **6**, 291). Its behaviour is similar to that of a pseudo-acid (Hantzsch, *Ber.*, 1899, **32**, 576), that is, a weak acid the salts of which show abnormally low hydrolysis. According to Hantzsch, this is characteristic of a tautomeric change in the acid molecule, and it is of interest to note that Baly and Ewbank (Trans., 1905, **87**, 1354), from a study of its absorption spectrum, attribute tautomerism to quinol.

The Velocity Function.

For the steady period, the results agree fairly well with the exponential formula.

TABLE X.

$0.025 N\text{-}C_6H_4(OH)_2$ in $0.0485 N\text{-}NaOH.$

t (time in minutes).	$\gamma.$	$K = 1/t \log \dfrac{\gamma_\infty}{\gamma_\infty - \gamma}.$
3·0	0·51	0·038
3·5	0·50	0·033
4·0	0·65	0·038
4·5	0·68	0·036
5·0	0·73	0·036
8·0	1·10	0·038
∞	2·18	$K = 0.0365$ mean

For $0.10 N\text{-}C_6H_4(OH)_2$ with maximum alkali $K = 0.146.$

TABLE XI.

$0.10 N\text{-}$quinol in $0.097 N\text{-}$alkali.

$t.$	$\gamma.$	$K.$
2·0	0·54	0·062
3·0	0·85	0·071
4·0	1·00	0·066
6·0	1·23	0·060
∞	2·18	$K = 0.065$ mean.

CURVE 4.

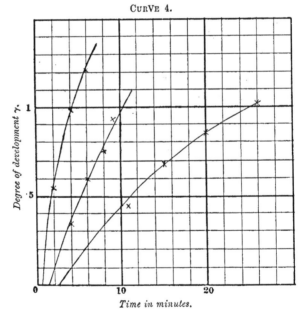

Time in minutes.

$N/10$ quinol in $N/50$, $N/20$, $N/10$ sodium hydroxide.

With two molecules of $N/5$-alkali, $K = 0.130$, in sufficiently good agreement. Compared with ferrous oxalate, for $N/10$ solution we have

K (quinol) $= 0\cdot146$, K (ferrous oxalate) $= 0\cdot061$, or per gram-molecule the quinol velocity is $2\cdot4$.

TABLE XII.—*Influence of Temperature.*

At $11\cdot0°$, $0\cdot05$ N-quinol in $0\cdot485$ N-alkali.

8 minutes	$\gamma = 0\cdot483$	$\log i = 1\cdot55$
13 ,,	$\gamma = 0\cdot750$	$\log i = 1\cdot30$
∞	$\gamma = 2\cdot18$	$\log i = 1\cdot25$

K at $20° = 0\cdot036$, K at $11° = 0\cdot0138$, hence the temperature-coefficient for $10°$, $\dfrac{K+10°}{K} = 2\cdot8$. The marked regression in the inertia at $11°$ indicates a low free energy in the reaction.

p-*Aminophenol* is used photographically as "rodinal." For these experiments, the hydrochloride was used; it was purified by precipitation with pure strong hydrochloric acid and dried in a vacuum over lime. p-Aminophenol is an amphoteric electrolyte similar to hydroxylamine, and its salts with strong acids may be titrated with caustic alkali, using phenolphthalein as an indicator. The characteristics of the aminophenols in development are just the opposite to those of quinol. The image appears rapidly, there being no lag in the lower tones, so that the law of constant density ratios is followed from the start, but the velocity diminishes very rapidly from the commencement, and the course of the reaction varies somewhat with the dilution and composition of the solution, so that comparable velocity measurements on an even scale are not easily obtained.

TABLE XIII.

Effect of Alkali.

Concentration of aminophenol.	Concentration of alkali.		Rate of development.			
$0\cdot020N$	$0\cdot0150N$	t	$1\cdot0$	$2\cdot0$	$3\cdot0$	$5\cdot0$	∞
		γ	$0\cdot297$	$0\cdot43$	$0\cdot64$	$0\cdot865$	$2\cdot18$
$0\cdot020N$	$0\cdot025N$	t	$0\cdot50$	$1\cdot0$	$2\cdot0$	$5\cdot0$
		γ	$0\cdot27$	$0\cdot46$	$0\cdot68$	$1\cdot03$
$0\cdot050N$	$0\cdot050N$	t	$0\cdot50$	$2\cdot0$	$4\cdot0$	$8\cdot0$
		γ	$0\cdot42$	$0\cdot79$	$1\cdot25$	$1\cdot50$
$0\cdot050N$	$0\cdot020N$	t	$0\cdot50$	$1\cdot0$	$2\cdot0$	$4\cdot0$	$6\cdot0$
		γ	$0\cdot32$	$0\cdot33$	$1\cdot00$	$1\cdot10$	$1\cdot32$

Effect of Dilution.

Concentration of aminophenol.	Concentration of alkali.		Rate of development.			
$0 \cdot 10 N$	$0 \cdot 10 N$	t	$1 \cdot 0$	$2 \cdot 0$	$3 \cdot 0$	$6 \cdot 0$
		γ	$0 \cdot 79$	$1 \cdot 25$	$1 \cdot 60$	$1 \cdot 80$
$0 \cdot 05 N$	$0 \cdot 05 N$	t	$0 \cdot 50$	$2 \cdot 0$	$4 \cdot 0$	$8 \cdot 0$
		γ	$0 \cdot 42$	$0 \cdot 79$	$1 \cdot 25$	$1 \cdot 50$
$0 \cdot 025 N$	$0 \cdot 025 N$	t	$0 \cdot 50$	$1 \cdot 0$	$2 \cdot 0$	$5 \cdot 0$
		γ	$0 \cdot 27$	$0 \cdot 50$	$0 \cdot 70$	$1 \cdot 04$

Effect of Temperature.—$0 \cdot 05 N$-aminophenol in $0 \cdot 02 N$-alkali.

At 11°	t	$1 \cdot 5$	$3 \cdot 0$		$\log i = 1 \cdot 20$
	γ	$0 \cdot 62$	$0 \cdot 81$		
At 20°	t	$0 \cdot 50$	$1 \cdot 0$	$2 \cdot 0$	$\log i = 1 \cdot 20$
	γ	$0 \cdot 32$	$0 \cdot 63$	$1 \cdot 00$	

From these data it will be seen that the velocity is approximately proportional to the concentration of the alkali, and nearly inversely proportional to the volume when alkali and reducer are in equimolecular proportions, whilst the temperature-coefficient $K + 10°/K = 1 \cdot 5$.

TABLE XIV.

Time of Appearance and Concentration (see *Proc. Roy. Soc.*, 1905, Series *A*, **76**, 228).

These results are for an equimolecular mixture.

T_a in seconds.	Volume.	T_a/V.	Concentration.
$5 \cdot 0$	$1 \cdot 0$	$5 \cdot 00$	$N/20$
$10 \cdot 3$	$2 \cdot 0$	$5 \cdot 15$	$N/40$
$21 \cdot 0$	$4 \cdot 0$	$5 \cdot 25$	$N/80$
$39 \cdot 6$	$8 \cdot 0$	$4 \cdot 90$	$N/160$
$81 \cdot 0$	$16 \cdot 0$	$5 \cdot 08$	$N/320$
$247 \cdot 4$	$32 \cdot 0$	$7 \cdot 75$	$N/640$

The initial velocity is proportional to the concentration down to $N/400$, and then falls off more rapidly.

The Velocity Function.—The velocity decreases very rapidly with the time, and the exponential formula only holds for a brief early stage. But since the density ratios are constant, it appears that the development is still the same function of the density of the image. The divergence seems to be due to the following cause. The rate of reduction in the film development is so fast that it becomes as rapid or more rapid than the invasion or penetration of the developer, so that the concentration of developer in the film is continually lowered. This makes a comparison with ferrous oxalate somewhat difficult, as the ratio varies. Taking for dynamical purposes an initial or maximum velocity, we have for $\gamma 0 \cdot 5$ (about 20 per cent.) at 20°,

$$K \text{ for } 0 \cdot 02 N \text{-aminophenol} = 0 \cdot 110.$$
$$K \text{ ,, } 0 \cdot 10 N \text{-aminophenol} = 0 \cdot 550.$$
$$K \text{ ,, } 0 \cdot 10 N \text{-ferrous oxalate} = 0 \cdot 061.$$

So that per gram-molecule, p-aminophenol develops nine times as fast as the oxalate.

Methyl-p-*aminophenol* ("*metol*") is used in the form of the sulphate, $(HO \cdot C_6 H_4 \cdot NH \cdot CH_3)_2, H_2 SO_4$.

Measurements at 20° for $0 \cdot 05 N$-"metol" in $0 \cdot 02 N$-alkali gave t $0 \cdot 5$, $1 \cdot 0$, $1 \cdot 5$, $2 \cdot 0$; γ $0 \cdot 33$, $0 \cdot 44$, $0 \cdot 55$, $0 \cdot 72$.

Comparison with "rodinal" showed that the velocity was almost the same, K being $0 \cdot 53$ for $N/10$-"metol." The temperature-coefficient $K + 10°/K$ was $1 \cdot 20$.

Affinity, Reduction-potential, and Bromide Sensitiveness.

The velocity-constant is no measure of the affinity of a reaction and especially so in heterogeneous reactions owing to the question of diffusion. In reductions, the true measure of the affinity of the reducer is its reduction-potential (compare Fredenhagen, *Zeit. anorg. Chem.*, 1902, **29**, 396).

For development we have the general equation $R'' + Ag^{\cdot} \rightleftarrows R' + Ag$, and at equilibrium $[R''[Ag^{\cdot} = K[R'[Ag$ (Trans., 1905, **87**, 1310), the free energy being given by the expression $F = RT \log \dfrac{[R''[Ag^{\cdot}}{[R'[Ag}$. Separating the factors, the potential of an ion R'', which passes to a higher stage of oxidation R', is given by the form $A = RT \log \dfrac{[R''}{[R'}$, which measures its reduction-potential (Peters, *Zeit. physikal. Chem.*, 1898, **26**, 193). If the value $[R''/[R'$ will not naturally reach such a limit that forces $[Ag^{\cdot}/[Ag$ past a certain value such that metallic silver is precipitated from solution, the reducer is too weak to act as a developer. Since $[Ag$ is constant for the same exposure (Trans., 1905, **87**, 1325), this minimum value depends on $[Ag^{\cdot}$ and hence on the concentration of the halide ion, as well as its nature, so that stronger reducers are required for bromide than for chloride plates.

Dependence of Reduction-potential on Concentration and Alkali.

Where the reducer, as in alkaline development, is the anion of a weak acid, its concentration is increased both by increasing the concentration of alkali and acid, and hence also its reduction-potential. But by virtue of certain side reactions, the alkali may further augment the reduction-potential and the reducer-acid diminish it, for the addition of the primary oxidation product should, in accordance with the foregoing passage, decrease the reduction-potential and the free energy. But alkali removes the oxidation product, which is not stable under the conditions of development (Part I). If, therefore, the acid-reducer be in such excess as to lower considerably the concentration of free hydroxyl ions, it may diminish the free energy of the reaction, and so cause a retardation of development. It will be seen that it is not possible to determine electrically the reduction-potentials of organic reducers in alkaline solution, and this was confirmed experimentally. In agreement with Fredenhagen (*loc. cit.*) the potential of a reducing agent is indeterminate unless a definite concentration of its oxidation product be present. Abegg (*Eder's Jahrbuch Phot.*, 1904, p. 1) proposed to characterise the reducing energy of a developer by the concentration of bromide it could just overcome. This static method cannot, however, be used for organic reducers for the reasons just explained. A dynamic method is preferable, since the practical developing energy is a function of its natural potential as defined above and of the stability or time of existence of its oxidation product. The method employed was to determine the depression of density produced at a given degree of development for a given concentration of bromide, and compare the same with that for ferrous oxalate. For this, the numerical relations were known (Trans., *loc. cit.*). ΔD, the constant depression, is proportional to the concentration of bromidion and inversely proportional to the time of development. $\Delta D = a[\mathrm{Br}]/t$ for ferrous oxalate, whilst

$$t = \frac{\log \gamma_\infty - \log(\gamma_\infty - \gamma)}{K}.$$

Hence, if the depression with the other developer be measured, then the concentration of bromide necessary to produce with ferrous oxalate the same depression at the same degree of development γ can be calculated, and this affords a numerical comparison of the developing energy compared with ferrous oxalate; the results are as follows ·

Ferrous Oxalate.—Plates A. $K = 0.061$, $\gamma_\infty = 2.18$.

Depression in three minutes' development: from the curves:

$\Delta D = \gamma(\log i_0 - \log i) = 0.645(1.75 - 1.30) = 0.290$ in $N/100$ bromide.

Hydroxylamine:

$\Delta D = \gamma(\log i_0 - \log i) = 0.645(1.45 - 1.25) = 0.129$ in $N/200$ bromide.

Ferrous Oxalate.—Series B of plates. $K = 0.046$, $\gamma_\infty = 2.50$.

Depression in three minutes' development: $N/100$ bromide:

$\Delta D = \gamma(\log i_0 - \log i) = 0.74(1.80 - 1.25) = 0.410$ at $\gamma = 0.74$.

Quinol.—Four minutes in $N/20$ quinol and $N/100$ bromide :

$$\Delta D = 0.85(2.25 - 1.55) = 0.595 \text{ at } \gamma = 0.85.$$

Six minutes :

$$\Delta D = 0.97(1.60 - 1.25) = 0.340 \text{ at } \gamma = 0.97.$$

p-*Aminophenol.*—Three minutes in $N/20$ p-aminophenol :

$$\Delta D = 0.64(1.40 - 1.20) = 0.128 \text{ at } \gamma = 0.64.$$

Reducing these results by the formulæ given we obtain the value of [Br', which produces the same depression with ferrous oxalate as for the reducer in question. The following are the data. .

Concentrations of bromide producing the same depression as with ferrous oxalate: FeC_2O_4, $0.010N$ potassium bromide; NH_2OH, $0.0113N$ potassium bromide ; $C_6H_4(OH)_2$(p), $0.0052N + 0.0073N$ at 34 and 40 per cent. development respectively ; $NH_2 \cdot C_6H_4 \cdot OH$(p), $0.034N$.

These numbers express the relative reducing energies of the developers compared with ferrous oxalate. It will be seen that "rodinal" is the most energetic, quinol the least.

Summary.

The complete results may be numerically tabulated as follows :

$K =$ Velocity of development in $N/10$ solution at $20°$.

$R =$ Reducing power, that is, the number of molecules of silver bromide reduced by one molecule of reducer.

$E =$ Efficiency, or velocity compared with ferrous oxalate divided by reducing power.

$T.C. =$ Temperature coefficient or $K + 10°/K$.

$F =$ Energy, that is, concentration of bromide producing the same retardation as $N/100$ bromide with ferrous oxalate.

Developer.	R.	K.	E.	T.C.	F.
Ferrous oxalate, $Fe(C_2O_4)_2''$	1	0.061	1.00	1.70	$0.01N = 1.00$
Hydroxylamine, NH_2O'	1	0.305	5.0	2.10	$0.0113N = 1.13$
Quinol, $C_6H_4O_2''$	2	0.146	1.2	2.80	$0.0052N = 0.52$
					$0.0073N = 0.73$
p-Aminophenol, $C_6H_4 <^{O'}_{NH'}$	(2)*	0.550°	4.5	1.5	$0.034N = 3.4$
Methyl-p-aminophenol, $C_6H_4 <^{O'}_{N \cdot CH_3'}$	(2)*	0.500	4.4	1.25	—
Hydrogen peroxide, O_2H'	1	—	—	--	—

* Varies with concentration from 1 to 2.

The author desires to express his sincere thanks to Professor Sir William Ramsay, F.R.S., for his interest in the investigation.

CHEMICAL DEPARTMENT,
UNIVERSITY COLLEGE,
GOWER STREET,
LONDON.

LVIII.—*Fischer's Salt and its Decomposition by Heat.*

By Prafulla Chandra Rây.

To the potassium cobaltinitrite, named after its discoverer, is generally assigned the formula $K_6Co_2(NO_2)_{12},3H_2O$. According to Sadtler, however, "this salt can be formed with $4H_2O$, $3H_2O$, $2H_2O$, H_2O, or anhydrous, according to the degree of concentration of the solution used, passing in colour from a light yellow to dark greenish-yellow, and that in consequence of such dependence, we can, in most preparations, fix no absolute point, but are liable to have a mixture of salts of different degrees of hydration" (*Amer. J. Sci.*, 1870, [ii], 49, 192).

Rosenheim and Koppel, who have recently investigated the subject, find that in order to obtain the salt in a state of purity it is necessary to pass a current of nitrous anhydride—the product of the action of arsenic trioxide on nitric acid—into a solution of potassium nitrite holding in suspension cobalt carbonate. The product thus secured has the formula $K_6Co_2(NO_2)_{12}$ (*Zeit. anorg. Chem.*, 1898, 17, 35).*

Although my immediate object in view has been to study the decomposition products of this compound under the action of heat, the results could scarcely be relied on until the main question of its composition was satisfactorily settled.

The following methods of preparing the salt were therefore examined.

I. *Preparation of the Double Salt.*

Method A.—(*a*) A strong solution of potassium nitrite acidified with acetic acid was added to a fairly concentrated solution of cobalt chloride, care being taken that the whole of the cobalt was not precipitated. After standing overnight, the precipitate, which was of a light yellow colour, was collected on the filter paper and washed, first with a solution of potassium acetate and then with 90 per cent. alcohol, and dried by being allowed to remain under pressure between the folds of blotting paper.

(*b*) The method was much the same as above, the only difference being that a large excess of potassium nitrite was taken. The filtrate was free from cobalt, but had a pale green colour, due to traces of nickel. The mother liquor in both (i) and (ii) was distinctly acid.

(*c, d, e*) A systematic study of the composition of the salt was desirable; with this object in view, three solutions were prepared: (1) of potassium nitrite of sp. gr. of 1·27 ; (2) of cobalt chloride of sp. gr. of 1·16, and (3) of acetic acid containing 25 per cent. of glacial acid and mixed in the following proportions.

* My information is derived from the abstract of the paper (Abstr., 1898, ii, 430).

No.	$CoCl_2$ solution.	KNO_2 solution.	Acid solution.	Remarks.
I.	25 c.c.	25 c.c.	5 c.c.	Colour of the salt: greenish-yellow. Mother liquor: pink, showing the presence of excess of cobalt. Reaction : very feebly acid $=c$.
II.	25 c.c.	50 c.c.	5 c.c.	Mother liquor gave faintly alkaline. reaction ; the sample was consequently rejected.
III.	25 c.c.	25 c.c.	10 c.c.	Colour of the salt : pale yellow. Mother liquor : pink, but less so than in No. I. Reaction distinctly acid $=d$.
IV.	25 c.c.	50 c.c.	10 c.c.	Colour of the salt : dark greenish-yellow. Mother liquor : pale green, free from cobalt, with feebly acid reaction $=c$.

For convenience of reference, the results of analyses of the different preparations are presented below in a tabulated form.

Designation of preparation.	Percentage of			Remarks.
	Co.	K.	H_2O.	
a	13·42	25·12	4·31	Theory for $Co_2(NO_2)_6,6KNO_2,3H_2O$ requires $Co=12·57$; $K=24·99$; $H_2O=$, 5·75 ; $N=17·90$.*
b	12·65	24·85	8·48	Theory for $Co_2(NO_2)_6,6KNO_2,4H_2O$ requires $Co=12·34$; $K=24·52$; $H_2O=$ 7·53 ; $N=17·54$.
c	14·04	25·14	3·08	—
d	13·45	25·2	2·90	Theory for the dihydrated salt requires $Co=12·82$; $K=25·48$; $H_2O=3·91$.
e	13·57	25·18	2·56	Theory for the monohydrated salt requires $Co=13·07$; $K=25·99$; $H_2O=1·99$.

* $(Co=59 ; O=16 ;$ Richards and Baxter.$)$

It will be seen that, judged by the percentage of water alone, preparation a approximates to a salt with 3 molecules of water, and b to one with 4 molecules of water, whilst d and e would lie between a mono- and a di-hydrated salt. Two more preparations according to methods described under a and b gave the percentages of water as 6·35 and 5·64 respectively. As all these preparations were simply air-dried, the percentage of water seemed to be more or less a matter of accident, depending on the hygrometric state of the atmosphere. The percentage of cobalt came out invariably too high, and potassium as a rule too low. As the method of analysis was one which admitted of a high degree of accuracy (see footnote on p. 554), this anomaly could not be explained. Another series of preparations was undertaken, in which special care was taken to obtain as pure a product as possible. Ordinary cobalt chloride, as supplied in the laboratories and labelled "pure," was dissolved in water and treated with sulphuretted hydrogen in order to remove any traces of the metals of the copper group (*Chem. News*, 1898, **77**, 22). The filtrate was evaporated down to a small bulk, from which, on standing overnight, cobalt chloride crystallised out. In some cases, the cobalt solution was fractionated with ammonia, although, as pointed out by Richards and Baxter, this method is of doubtful advantage (*Chem. News*, 1900, **81**, 140).

A concentrated solution of cobalt chloride thus purified was acidified with acetic acid and an insufficient amount of potassium nitrite solution added to it. The precipitate was collected after eighteen hours and washed as before, but dried in the steam oven.[*] The mother liquor, which was distinctly rose-coloured, was now treated with a solution of pure potassium nitrite to which acetic acid had been just previously added. The precipitate collected after twenty-four hours was washed and dried exactly as above. The mother liquor was still decidedly of a rose colour. This time a sufficient excess of potassium nitrite acidified with acetic acid was once more added, and on the next day the precipitate was collected as usual. The mother liquor was now found to be faintly green owing to the presence of nickel. The analyses of the three successive fractions α, β, and γ are given below. It will be noted that although formed under widely varying conditions their composition remains practically constant.

	α.			β.	
	i.	ii.		i.	ii.
Co	13·26	13·19	Co	13·34	13·49
K	25·03	25·02	K	24·17	24·10
H₂O	6·30	6·40	H₂O	—	—

	γ.		Theory for
	i.	ii.	$Co_2(NO_2)_6,6KNO_2,3H_2O$.
Co	13·33	13·32	12·57
K	24·51	24·60	24·99
H₂O	4·75	4·80	5·75

In determining the composition of the salt, the percentages of the two metals and of water have been regarded as sufficient evidence, but several estimations of nitrogen have, however, been made, as will be seen in the second portion of this paper.

The most favourable conclusion which can be drawn from the above investigation seems to be that a salt of the composition of

$$Co_2(NO_2)_6,6KNO_2,3H_2O$$

is uniformly formed, but that it carries down with it traces of an oxide of cobalt, thus increasing the percentage of this metal and

[*] At 100°, this compound does not undergo decomposition.

lowering that of potassium.* Preparations d and e would, to a certain extent, go to support Sadtler's views, namely, that the hydration and colour of the salt depend on the concentration of the solutions used.

B. Rosenheim and Koppel's Method.

The products of this method of preparation had a more attractive colour, with all the gradations of tint varying from orange-yellow to lemon-yellow. The analyses of several distinct preparations are given below :

I.	II.		III.			IV.		
	i.	ii.	i.	ii.	iii.	i.	ii.	iii.
Co... 13·56	Co.. 13·65	13·54	Co... 13·68	13·78	13·71	Co... 13·68	13·60	13·70
K ... 24·00	K... —	—	K ... 24·01	—	—	K ... 24·0	24·01	—
			H₂O. 4·53	—	—			

It will be seen that the cobalt in this series of preparations is even much higher and the potassium correspondingly lower than in those of method A, and the conclusion which has already been drawn is further confirmed.

A modified method was found to give better results. Nitrous fumes were passed into water holding in suspension cobalt carbonate until a perfectly bright, clear, rose-coloured solution was obtained. A solution of potassium nitrite was similarly treated until red fumes began to be evolved. The latter was then added to the former and the mixture set aside until the mother liquor became quite quiescent. The precipitate was washed as usual and dried in the steam chamber.

	Found.		Theory for
	i.	ii.	$Co_2(NO_2)_6,6KNO_2$.
Co	13·09	12·96	13·34
K	24·39	24·37	26·52
H₂O............	4·31	—	—

Judged by the percentage of cobalt alone, one would naturally take this last preparation to be anhydrous, yet not only is the percentage of potassium too low, but the salt is found to be actually hydrated. In fact this is the nearest approximation to a trihydrated salt.

It is thus evident that Rosenheim and Koppel's method can scarcely be regarded as an improvement on the ordinary one, and that it really yields a trihydrated salt contaminated as before with appreciable quantities of some oxide of cobalt.

* The potassium acetate solution with which the salt has to be washed is not completely removed when subsequently treated with alcohol. Hence, as a rule, the percentage of potassium comes out higher rather than lower.

II. *Decomposition by Heat.*   [With ATUL CHANDRA GAÑGULI.]

Erdmann found that Fischer's salt on heating evolves nitrous acid.[*] The method of experiment followed was exactly the same as already described (Trans., 1905, **87**, 180).   As, however, no nitrous fumes were evolved, the interposition of the glass spiral packed with glass beads previously moistened with caustic potash solution was obviated. It was found that the decomposition proceeded according to the following equation :

$$Co_2(NO_2)_6, 6KNO_2, 3H_2O = Co_2O_3 + 6NO + 3KNO_3 + 3KNO_2 + 3H_2O.$$

In other words, half the amount of the total nitrogen in the salt is given off as nitric oxide.[†]   At the same time an internal oxidation takes place.

	Found.	Calculated.
Nitrogen, total	17·15[*]	17·90
,, as nitric oxide	8·85	8·95

[*] The total nitrogen was estimated according to Dumas' method and varied in the several preparations from 16·82 to 17·48 per cent.

This result represents the mean of more than a dozen experiments. After the reaction was over, the residue was exhausted with hot water, and the filtrate on analysis gave the remaining half of nitrogen made up of equal proportions of nitratic and nititric nitrogen.   Traces of nitric oxide begin to be given off below 200°.   At 203°, the evolution proceeded regularly but very slowly, the reaction not being completed in less than eight hours.   The most favourable temperature was found to lie between 210° and 215°.   If the salt is heated in an air-bath instead of in a vacuum, the whole of the potassium nitrite is converted into nitrate.   In this case, the nitric oxide previously set free evidently acts as a carrier of oxygen from the air.

III. *The Analysis of Fischer's Salt.*

Cobalt is often separated from nickel and other metals in the form of this salt.   The estimation of cobalt and of potassium involves a tedious and laborious process.   The latest authoritative work (Moissan, *Chimie Minerale*, vol. iv., pp. 152 and 200) gives a method which is practically the same as that employed by me eighteen years ago in the analysis of nickel, cobalt, and potassium sulphate.[‡]   With

[*] "Es entwickelt dabei saltpetrige säure" (*J. pr. Chem.*, 1866, **97**, 401).

[†] The nitric oxide was invariably found to contain traces of carbon dioxide, no doubt from the decomposition of potassium acetate, which adheres tenaciously to the double salt.

[‡] "Conjugated sulphates of the copper-magnesium group" (*Proc. Roy. Edinburgh*, 1888).

the experience gained in the present investigation, the following expeditious method was devised. The salt is weighed out in a crucible which is placed in an air-bath at a temperature of 215—220° for about a couple of hours or less. The residue in the crucible is then transferred to a porcelain basin and repeatedly washed by decantation with boiling water before being collected on the filter, care being taken not to wash it further at this stage, as otherwise the oxide has a tendency to run through. Traces of oxide which adhere to the basin should be wiped off by means of a moistened filter paper. The cobalt is estimated according to Rose's method. The reduced metal may be washed once more with hot water to remove traces of alkali and again reduced in a current of hydrogen. It has been found, however, that the difference in weight after this treatment seldom exceeds a milligram ; the potassium in the filtrate is weighed as sulphate. The chief recommendation of the method lies in the fact that it dispenses with the use of ammonium sulphide, caustic alkali, or of any reagent whatever. To judge of the degree of accuracy attainable, the reader is referred to the foregoing tables of analyses.

CHEMICAL LABORATORY,
PRESIDENCY COLLEGE.
CALCUTTA.

LIX.—The Constitution and Properties of Acyl Thiocyanates.

By JOHN HAWTHORNE.

IN connection with the work on the thiocarbimides of acid radicles and the thiocarbamides to which they give rise by interaction with bases (Dixon, Trans., 1895, **67**, 1040 ; 1896, **69**, 855, 1593 ; 1897, **71**, 617 ; 1899, **75**, 375, 388 ; 1901, **79**, 541 ; 1903, **83**, 84 ; 1904, **85**, 807 ; Dixon and Doran, Trans., 1895, **67**, 565 ; Doran, 1905, **87**, 331), a paper was published last year (Dixon and Hawthorne, Trans., 1905, **87**, 468) in which the behaviour of acetyl "thiocyanate" (Miquel, *Ann. Chim. Phys.*, 1877, [v], ·11, 295) towards aniline at different temperatures was studied in some detail. It appeared that, at low temperatures, double decomposition occurs, whereby thiocyanic acid is produced, $CH_3 \cdot CO \cdot SCN + PhNH_2 = HSCN + CH_3CO \cdot NHPh$; whilst at high temperatures the substances unite to form chiefly an additive compound, acetylphenylthiocarbamide,

$$CH_3 \cdot CO \cdot NCS + PhNH_2 = CH_3 \cdot CO \cdot NH \cdot CS \cdot NHPh.$$

At intermediate temperatures, the products of both these changes are obtained, the relative amounts depending on the temperature of interaction ; in the neighbourhood of 80°, the sulphur of the " thiocyanate '' is distributed equally, one half appearing as aniline thiocyanate and the remainder as acetylphenylthiocarbamide.

The apparent influence of temperature on the capacity of acetyl "thiocyanate" to react as such or as thiocarbimide being so very marked, the idea suggested itself, to inquire whether this property depends solely on the temperature or to some extent on the nature of the base submitted for interaction, or of the acidic radicle of the "thiocyanate." Below are given some preliminary results on the effect of substituting o-toluidine for aniline and propionyl for the acetyl residue.

Acetyl " Thiocyanate " and o-*Toluidine.*

The methods employed were identical with those mentioned in the former paper (*loc. cit.*), with the following slight differences.

(*a*) The method of determining the combined thiocyanic acid by standard copper sulphate (Barnes and Liddle) was abandoned, titration with $N/10$ caustic alkali, using phenolphthalein as indicator, being found equally satisfactory.

(*b*) Determination of the total yield of products by direct weighing was impossible, owing to the difficulty of crystallising o-toluidine thiocyanate. In all the experiments with this base (except those at high temperatures, where, the yield of thiocyanate being small, this substance probably remained entangled in the thiocarbamide), two distinct products of interaction were obtained, namely, a crystalline solid (a mixture of acetyl-o-tolylthiocarbamide and aceto-o-toluidïde) which remained suspended in the solvents, and a yellow oil which settled on the bottom of the flask, and which consisted of o-toluidine thiocyanate together with a small amount of dissolved acetyl-o-tolylthiocarbamide. The solvents and suspended solids were poured on to a filter and the oil washed as well as possible with ligroin, which dissolves o-toluidine readily, although aniline only with difficulty. The oil was then treated with water, the dissolved thiocarbamide being deposited from solution ; this precipitate was added to the solids already filtered off, and the whole washed with water until free from thiocyanic acid. Filtrate and residue were then analysed as previously described, and the sum of the weights of the constituents, as determined by analysis, was taken to be the total yield obtained. That this is a legitimate procedure is seen by glancing at the table in the previous paper for the aniline—acetyl " thiocyanate " system, where the mean difference between the total by direct weighing and the sum of the weights of the different products found by analysis is 0·15 gram on about 10 grams.

(c) In the former paper, the experimental results were shown by curves in which the percentages of the products were plotted against temperature. Although this method of plotting shows the course of the reaction sufficiently clearly for each individual interaction, it is not adapted for comparing the results of interactions where the combining weight of acid or basic radicle differs from that of the original system. In order to be able to follow the course of the CNS group independently of these combining weights, it is preferable to take into account merely the relative weights of sulphur as calculated from the proportions of thiocarbamide and thiocyanate respectively, and to plot these as percentages of the total sulphur present.

The results of the experiments are summarised in the following table, which shows for each temperature :

(i) The weight of o-toluidine thiocyanate found in the aqueous extract by titration of the thiocyanic acid with $N/10$ caustic potash.

(ii) The total weight of acetyl-o-tolylthiocarbamide : namely, that found in the residue after extraction of the thiocyanate by water, together with the small amount dissolved by the water and the quantity found in the residue from the hydrocarbon solvents.

(iii) The weight of sulphur present as thiocyanate.

(iv) The weight of sulphur present as thiocarbamide.

(v) The total weight of sulphur present (sum of iii and iv).

(vi) The percentage of thiocarbimidic sulphur, calculated from iv and v.

Temperature	12°.	40°.	70°.	85°.
i. Weight of thiocyanate	2·017	1·983	0·817	0·2772
ii. ,, thiocarbamide	7·117	7·357	8·720	8·9500
iii. ,, thiocyanic sulphur	0·389	0·382	0·157	0·0544
iv. ,, thiocarbimidic sulphur...	1·094	1·132	1·341	1·3770
v. Total weight of sulphur	1·483	1·514	1·498	1·4314
vi. Percentage of thiocarbimidic sulphur	73·8	74·7	89·5	96·3

The acetyl "thiocyanate" used in the above experiments was a portion of the pure sample the preparation of which is described below (p. 566). The o-toluidine (Kahlbaum's) boiled at 203—204°/777 mm.

It may be noted, in connection with these results, that they do not accord with the observations of Doran (Trans., 1905, **87,** 338), who found no material difference between the yields of acetyl-o-tolylthiocarbamide at – 3° and at the boiling point of benzene.

On recalculating in the same way the numbers already published for the aniline—acetyl "thiocyanate" system, and plotting them together with those tabulated above for the o-toluidine series, the curves I and II are obtained on the following diagram.

In the system o-toluidine—acetyl "thiocyanate," sets of measurements were made at a few temperatures only, the object being not to determine with precision the outline of the curve, but merely to ascertain whether it would coincide, even approximately, with that derived from the corresponding aniline system. That such is not the case is manifest, for the o-toluidine curve not only lies considerably above that for aniline, but also differs from it in shape.

Propionyl " Thiocyanate " and Aniline.

Amongst other factors which may determine the part played, in given circumstances, by an acidic "thiocyanate" is the nature of the radicle attached to the CNS residue. This subject also is being

Acyl thiocyanates and bases.

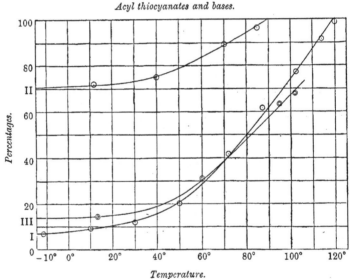

Temperature.

Readings from ${below \atop above}$ measure percentage of total S found as ${thiocarbamide \atop thiocyanate}$.

(I) MeCONCS + PhNH$_2$.   (II) MeCONCS + o-C$_7$H$_7$·NH$_2$.
(III) EtCONCS + PhNH$_2$.

investigated with regard to the effect, if any, of increased combining weight in the fatty group, but so far the propionyl derivative alone has been tested at different temperatures, the base selected being aniline.

The only noteworthy difference between this interaction and the previous ones was due to the much greater solubility in equal parts of benzene or toluene and ligroin of propionylphenylthiocarbamide as compared with acetylphenylthiocarbamide and acetyl-o-tolylthio-carbamide. The residue left on evaporation of the hydrocarbon solvents amounted usually to about 1 gram (as against 0·1 to 0·3 gram in the former cases) and consisted almost entirely of propionylphenyl-thiocarbamide, with scarcely any propionanilide.

The results of the experiments are given in the following table, and the percentages of thiocarbimidic sulphur are plotted as curve III on the preceding diagram.

	Temperature	13°.	60°.	95°.	102°.
i.	Weight of thiocyanate	5·229	4·043	2·363	1·736
ii.	,, thiocarbamide	1·202	2·557	5·680	4·993
iii.	,, thiocyanic sulphur	1·101	0·851	0·497	0·365
iv.	,, thiocarbimidic sulphur	0·185	0·393	0·874	0·768
v.	Total weight of sulphur	1·286	1·244	1·371	1·133
vi.	Percentage of thiocarbimidic sulphur..	14·4	31·0	63·7	68·0

From the diagram it will be seen that, roughly speaking, the acetyl and propionyl curves fall together ; at the ordinary temperature, a little more thiocarbimidic sulphur was found, and at high temperatures a little less, than where acetyl "thiocyanate" was concerned, but approximately the two processes were comparable. Change of the acidic radicle thus appears to have but little effect on the course of the reaction. Even a large increase in the weight of the acid radicle, provided that this is of the fatty series, appears to affect only slightly the distribution of the ;sulphur, so far at least as can be judged from the result of the following single experiment with stearyl "thiocyanate" and aniline.

Stearyl " Thiocyanate " and Aniline.   Stearylphenylthiocarbamide.

The failure of attempts to obtain stearylphenylthiocarbamide having been recorded in an earlier paper (Dixon, Trans., 1896, **69**, 1602), it was decided to repeat the experiments, especially in the light of the foregoing results.

For this purpose, 7·5 grams of pure stearyl chloride were heated on the water-bath with dry lead thiocyanate and toluene until the liquid ceased to show the presence of chlorine. After removal of the lead salts by filtration, the solution was heated to boiling (114°) and mixed quickly with 2·9 grams (1¼ mols.) of aniline, also dissolved in boiling toluene. Vigorous interaction took place, some sulphuretted hydrogen was expelled, and solid matter separated at once ; the mixture was cooled. mixed with an equal volume of ligroin and filtered. The small amount of solid which separated was found to consist of aniline thiocyanate and stearanilide, with practically no thiocarbamide, whilst the hydrocarbon mother liquors left on evaporation in the air 8·78 grams (92 per cent. of the total yield) of a white solid, which contained no thiocyanic acid, and which proved to be practically pure stearylphenylthiocarbamido. After twice crystallising from alcohol, the substance melted at 63—64°.

0·3 gave 0·1688 $BaSO_4$. S = 7·7.

0·836 gave 0·0800 NH_3 (by Kjeldahl's method). N = 6·7.

$C_{25}H_{42}ON_2S$ requires S = 7·6 ; N = 6·7 per cent.

Stearylphenylthiocarbamide, which crystallises from alcohol in long, white, soft needles, is easily soluble in ether, chloroform, benzene, hot ligroin, or hot alcohol, slightly so in cold ligroin or alcohol, and insoluble in water. Silver nitrate desulphurises it quickly in the cold, alkaline lead tartrate only on boiling. Of the total sulphur accounted for in this reaction, 93·5 per cent. appeared as stearyl-phenylthiocarbamide and only 6·5 per cent. as aniline thiocyanate.

Physical Examination of Acetyl " Thiocyanate."

It now remained to be ˹seen whether the isomeric change from thiocyanate to thiocarbimide, which in several cases is known to occur under the influence of heat, might in the case of acidic derivatives be conditioned only temporarily, so that for any given temperature an equilibrium could exist between the proportions of the two constituents. In order to decide this point, it was necessary to examine the substance when not undergoing chemical interaction, that is, by some purely physical means. Of such methods, the most suitable appeared to be the determination of the molecular refraction for light of known wave-length, since this property is nearly independent of the temperature, and the value for a doubly linked sulphur atom in the molecule rises by an amount which can easily be measured.

Before proceeding to the examination of acetyl "thiocyanate," it was considered desirable to determine by direct observation of certain known thiocyanates and thiocarbimides the refraction equivalents of the groups –SCN and –NCS, and, incidentally, to compare the results with those recorded by previous observers.

For determining the indices of refraction, a Pulfrich reflectometer, provided with a heating arrangement, was employed, and the results were plotted on a scale representing single degrees of temperature and indices of refraction to the third decimal place. The densities of the various substances, measured either by means of a specific gravity flask or a Sprengel pyknometer and compared with water at 4°, were, plotted on a similar scale. Within the limits of temperature examined, the results of both series of measurements could be represented as straight lines.

The molecular refraction, M_D, was calculated from the formula $M_D = \dfrac{(\mu_D - 1)m}{d_{t°/4°}}$ (where μ_D = index for D-line, m = molecular weight, and $d_{t°/4°} = $ density)˙ for various temperatures, and the mean taken. It should be noted that the value of M_D, as thus calculated, was found to diminish slightly ·with rising temperature; if the formula $M_D = \dfrac{\mu_D^2 - 1}{\mu_D^2 + 2} \times \dfrac{m}{d_{t°/4°}}$ is used, the converse holds. The calculated refraction value of the hydrocarbon radicle was in each case subtracted from these numbers, the difference representing the refraction-equivalent of –SCN or –NCS.

In order to calculate the refraction equivalents of the various fatty hydrocarbon radicles for sodium light, Cauchy's formula,

$$\mu = A + \frac{B}{\lambda^2},$$

was applied to the numbers for the elements given in the following table for the spectral lines H_α (column II) and H_β (column III) and for $\lambda = \infty$ (column I), the results, R_D, being entered in column IV.

	I. RA. $\lambda = \infty$.	II. R_a. $\lambda = 656·2·$	III. R_β. $\lambda = 486·1.$	IV. R_D. $\lambda = 589·6.$
H	1·29	1·302	1·316	1·306
C′	4·86	5·06	5·17	5·092
C″	5·86	6·075	—	6·126
C:C	11·72	12·15	—	12·252
O″	3·29	3·40	—	3·426
Cl	9·53	9·80	—	9·864

For a reason referred to later, the refraction-equivalents in the case of cyclic or unsaturated hydrocarbon radicles were determined by a different method.

To determine the value of –SCN, experiments were carried out on methyl and ethyl thiocyanates.

(1) Molecular Refraction of Methyl Thiocyanate.

The specimen used boiled at 130—131°/758 mm., and gave the results in Table I. On plotting these results, as mentioned above, the figures in Table II are obtained for the molecular refraction.

Temperature.	μ_D.	$d_{t°/4°}$.	Temperature.	M_D.
12·5°	1·4764	—	20°	32·11
16	—	1·0778	30	32·07
29	1·4670	—	40	32·02
48	1·4562	—	50	31·97
55	—	1·0331	60	31·94
59	1·4450	—	70	31·93
70	—	1·0149		
76	1·4410	—	II.	

I.

The mean value of M_D is 32·01, the maximum variation from this being $+0·1$ (Nasini and Scala found at 23·8°, Gazzetta, 1888, **17**, 70,

$M_D = 31\cdot95$). Now since $CH_3 = 9\cdot01$, it follows that the refraction equivalent of SCN is about $23\cdot0$.

(2) *Molecular Refraction of Ethyl Thiocyanate.*

The sample boiled at $145\cdot2—145\cdot4°/758$ mm., and gave the following results :

Temperature.	μ_D.	$d_{t°/4°}$.	Temperature.	M_D.
15°	1·4684	1·0150	20°	40·22
29	1·4612	—	30	40·15
40	—	0·9885	40	40·07
49	1·4512	—	50	40·05
60	1·4447	—	60	39·99
64	1·4430	—	70	39·96
71	—	0·9557		
75·4	1·4368	—	II.	
78	1·4353	0·9491		

I.

The mean value of M_D is $40\cdot07$, the maximum variation being $+0\cdot15$ [for the same substance, at $23°$, Nasini found (*loc. cit.*) $M_D = 40\cdot2$]. The value of C_2H_5 being $16\cdot71$, it follows that $SCN = 23\cdot36$. From these two series of experiments, the mean refraction-equivalent of the group $-SCN$ for sodium light is $23\cdot18$.

In order to ascertain the refractive value of the $-NCS$ residue, three thiocarbimides were next examined.

(3) *Molecular Refraction of Ethylthiocarbimide.*

As before, the observations are tabulated in I and the molecular refractions in II.

Temperature.	μ_D.	$d_{t°/4°}$.	Temperature.	M_D.
15·4°	—	1·0035	20°	44·73
18	1·5145	—	30	44·71
35·4	1·5050	—	40	44·68
39	1·5028	—	50	44·65
53	1·4946	—	60	44·63
70 6	1·4843	—	70	44·60
73	—	0·9433		
75	1·4818	—	II.	
80	1·4788	—		

I.

The mean value of M_D is $44\cdot67$, the greatest variation from the mean being $\pm 0\cdot07$ [Nasini found at $23°$ (*loc. cit.*) $M_D = 44\cdot66$].

All the preceding observations having shown that M_D is nearly independent of temperature, it was considered sufficient in the next two cases to make determinations at one temperature only.

(4) *Molecular Refraction of Allylthiocarbimide.*

Temperature.	μ_D.	$d_{e/4^\circ}$.	M_D.
15°	1·5298	1·0155	51·65

This agrees with the value, 51·40, found by Nasini (*loc. cit.*), and that given by Berliner, 51·48 (*Inaug. Diss., Breslau,* 1886).

(5) *Molecular Refraction of Benzylthiocarbimide.*

Temperature.	μ_D.	$d_{e/4^\circ}$.	M_D.
15°	1·6049	1·1246	80·14

It has been observed that the refraction-equivalents of benzyl and other cyclic groups are commonly higher than those calculated from the sum of the atomic refractivities, and it was therefore considered that a more trustworthy value would be obtained by deducting the value of chlorine from the observed refractivities of two closely allied compounds, chlorotoluene and benzyl chloride, and taking the difference as representing the actual value of the benzyl group. Adopting for this purpose the careful determinations of Gladstone (Trans., 1884, **45,** 258), the refractivities are 62·36 and 61·97 respectively, the mean being 62·16, which, less 9·86, the value for chlorine, gives 52·3 for the group C_7H_7.

From the data given by Brühl (*Annalen*, 1880, **200,** 139) for allyl chloride, the value of the allyl group is found to be $33·88 - 9·86 = 24·02$. This number, used below, differs but slightly from the sum of the atomic refractivities, namely, 23·87.

Deducting now, as in the case of the thiocyanates, from the observed molecular refraction of each of the above thiocarbimides the refraction-equivalent of the hydrocarbon nucleus, the remainders, representing the effect of the –NCS group, are very nearly equal, amounting in the mean to 27·81, as shown by the following table:

	M_D (observed).	Calculated for hydrocarbon group.	Difference = NCS.
$C_2H_5{\cdot}NCS$ ...	44·67	$2 \times 5·092 + 5 \times 1·306 = 16·71$	27·96
$C_3H_5{\cdot}NCS$ ...	51·65	24·02	27·63
$C_7H_7{\cdot}NCS$ ...	80·14	52·30	27·84
		Mean............	27·81

With regard to the values thus experimentally determined for the –NCS group, Gladstone's observations showed (*loc. cit.*) that the molecular refractivity of ethylthiocarbimide, as measured, is considerably higher than the calculated number, even when allowance has been made for the effect of a doubly linked sulphur atom. The calculated value of the –NCS group for the spectral line *D*, after making the allowance in question and assuming N = about 4·2, would be 4·2 + 5·09 + 16·5 = 25·8.

However, the immediate object was not to measure the separate refraction equivalents of the atoms thus linked, but to learn their total effect, and this it appears may be represented by about 27·8 units for sodium light and reckoning by the $\mu - 1$ formula. It should be added that in the case of phenylthiocarbimide, the molecular refractive power is abnormally high; this case has been discussed elsewhere (Nasini and Scala, *Atti R. Accad. Lincei*, 1886, 617, 623), but hitherto no adequate explanation has been given of its peculiar behaviour.

It remained now to apply the foregoing observations to the case of acetyl "thiocyanate." So far as the acetyl group is concerned, its molecular refractivity, as judged from its combinations with Cl, OH, &c., agrees fairly well with the calculated value, and for the D-line of the spectrum is about 17·8 units. If, therefore, the constitution of acetyl "thiocyanate" be $CH_3 \cdot CO \cdot SCN$, the molecular refractive power should amount to 17·8 + 23·18, or about 41 units, whilst for $CH_3 \cdot CO \cdot NCS$ the value should be 17·8 + 27·8, or about 45·6; this large difference could of course be very readily observed. But if its constitution depended on the temperature, then the molecular refractive power, determined by the joint effect of the two constituents, must rise and fall with this external condition; and since the total variation could amount to nearly 5 units, whilst the probable error of experiment could scarcely reach one-quarter of a unit, a change of composition corresponding to 5 per cent. of thiocyanate converted into thiocarbimide would be appreciable: a number not very different from that attained in the chemical interaction with bases, the results measured in this way being probably not accurate to within 3 per cent.

In the earlier experiments, the specimens of acetyl "thiocyanate" were prepared by following Miquel's instructions (*loc. cit.*), and, without going into details, it may be stated briefly that the molecular refractive power, measured at seven temperatures between 17° and 78°, gave at the former the value 45·37 and at the latter 45·10, the intermediate temperatures giving concordant values, the maximum difference of which was 0·27 unit, the mean value being 45·23. So far, therefore, as these experiments showed, it seemed tolerably certain that no change of constitution occurred within the specified limits of temperature.

But in the rectification of the substance under atmospheric pressure, it was noticed that, even on repeated distillation, and particularly at the commencement, a little carbon disulphide distilled over, and hence the method of preparation was altered so as to avoid exposing the compound to the high temperature—about 134·5°—at which it boils under the atmospheric pressure.

Ninety grams of acetyl chloride, free from phosphorus, were mixed with $1\frac{1}{2}$ molecules of lead thiocyanate moistened with 70 c.c. of benzene dried over sodium; the mixture was digested on the water-bath, in a reflux apparatus fitted with a drying tube, until interaction was complete and the solid residue was filtered off at the pump. After washing with a further 70 c.c. of benzene, the filtrate was distilled under diminished pressure, the solvents passing over mostly at 26—23° under 77—54 mm. pressure. The temperature now rose quickly to 34°/50 mm., whereupon the receiver was changed, and, after distilling a further 10 c.c., the rest was collected apart, practically the whole distilling within two degrees. This portion was again rectified, and, although it boiled constantly, the process was again repeated, the final result being an almost colourless liquid, nearly all of which (some 40 grams) distilled between 42·25° and 43° under 23 mm. pressure. The distillate had a faint straw-yellow colour and differed from other products in that it changed colour only slowly. As usually prepared, acetyl thiocyanate, although nearly colourless at first, changes to deep bromine-red and becomes turbid within twenty-four hours : this sample, after seven days, was still perfectly clear and bright yellowish-red. It was examined at once with the results shown in Table I, the calculated molecular refractions being tabulated in II.

Temperature.	μ_D.	$d_{t°/4°}$.	Temperature.	M_D.
13·25°	—	1·1523	20°	46·08
18·4	1·5231	—	30	45·99
31·8	1·5144	—	40	45·91
33·5	1·5142	—	50	45·86
40	—	1·1230	60	45·80
47	1·5056	—	70	45·73
61	1·4969	1·0953		
69	1·4920	—		II.
71	1·4917	—		
75	—	1·0790		

I.

The values of μ_D and $d_{t°/4°}$, when plotted as above, lie on straight lines.

The mean of these numbers gives 45·89 as the molecular refractivity of acetyl "thiocyanate," the total variation being 0·35 and the maximum difference from the mean being $+0·19$.

So far as can be judged from this purely physical method of investigation, this value, when compared with those calculated above for acetyl thiocyanate (41) and acetylthiocarbimide (45·6), indicates that within the specified limits of temperature, and when not undergoing chemical change, the substance is not *acetyl thiocyanate* but *acetylthiocarbimide*.

If this be so, the hypothesis that acetyl thiocyanate can behave, under certain conditions, as acetylthiocarbimide must now be changed, and the interactions viewed from a converse standpoint. That is, the substance which, in a static condition, is acetylthiocarbimide, may, when undergoing chemical change, behave as acetyl thiocyanate, and in this sense be a tautomeric compound.

From this point of view its behaviour is determined not by its own normal constitution, but by external conditions: for instance, the nature of the substance with which it interacts and the temperature of the experiment. It would be premature, at this stage of the investigation, to venture on any hypothesis to explain the facts observed, since it is now certain that, where a nitrogenous base interacts with an acidic thiocarbimide, three factors, at least, are concerned in the ultimate result: these are (1) the nature of the base, (2) the temperature of interaction, and (3) the nature of the radicle associated with the -NCS group.

I have to thank Professor A. E. Dixon for much assistance given during this research, and I am indebted also to the Research Fund Committee of the Society for a grant which has partly covered the expense of the investigation.

CHEMICAL DEPARTMENT,
QUEEN'S COLLEGE,
CORK.

LX.—Studies on Comparative Cryoscopy. Part IV. The Hydrocarbons and their Halogen Derivatives in Phenol Solution.

By PHILIP WILFRED ROBERTSON.

IN parts I, II, and III of this research (Trans., 1903, **83**, 1425; 1904, **85**, 1617; 1905, **87**, 1574), the cryoscopic behaviour of the fatty and aromatic acids and their esters was investigated in phenol solution, and it is now proposed to give the results obtained with the hydrocarbons. These compounds are characterised by the fact that they give low molecular depressions which decrease slowly with the concentration. This indicates an apparent increase in molecular weight, but the effect must be attributed rather to the action of the associated molecules of the solvent than to the formation of hydrocarbon molecular complexes, especially since in the substituted phenols, which are not so strongly associated, there is less irregularity.

P P

Using the notation of the previous papers, it is found that the " rate " of association (A) is generally small, and for most compounds the actual value is in the neighbourhood of 15.

The experimental results are given in Table I (Δ_1 is the molecular depression for a fall of 1°).

TABLE I.

Compound.	Δ_1.	A.	Compound.	Δ_1.	A.
1. Benzene	68	12	11. Triphenylmethane ...	66	21
2. Chlorobenzene	71	11	12. Naphthalene	68	13
3. Toluene	68·5	14	13. Dibromonaphthalene.	68	15
4. Benzyl chloride	72	14	14. Acenaphthene	70·5	13
5. Ethylbenzene	70	14	15. Retene	69	23
6. m-Xylene	68	17	16. Carbon tetrachloride.	71	24
7. Pseudocumene	73	17	17. Bromoform	72	6
8. Mesitylene	66	18	18. Ethylene dibromide..	69	7
9. Cymene	71	22	19. Caprylene	67	25
10. Diphenyl	67·5	8	20. Hexane	64	42

The data for the following compounds show that the introduction of methyl groups into the benzene nucleus increases the value of A.

$A = 12.$ $A = 14.$ $A = 17.$

$A = 18.$ $A = 22.$

In general, however, constitution appears to have little influence. Thus, benzene and naphthalene give practically the same result, as is also found to be the case for methyl*iso*propylbenzene and methyl*iso*-propylphenanthrene (retene).

The numbers obtained for chlorobenzene, benzyl chloride, and dibromonaphthalene show that the introduction of halogen atoms has little effect on the magnitude of A.

The molecular depressions of the hydrocarbons and their halogen derivatives are much lower than are required by van't Hoff's equation, and, in general, the lower the molecular depression the higher is the value of A. Similarly, it was found that in the case of the esters there was a tendency for the " rate " of negative association to increase with the molecular depression. The following numbers for typical compounds in phenol solution will make this clear :

	Δ_1.	A.
Hexane	64	42
Ethylbenzene	70	14
Ethyl acetate	74	-4
Ethyl oxalate	83	-20
Ethyl tricarballylate	95	-27

With regard to the behaviour of benzene, it is interesting to note that Bruni (*Gazzetta*, 1898, i, **28**, 249), from the fact that this compound gives a molecular depression in phenol solution of 68 instead of 76, as demanded by van't Hoff's equation, concludes that the benzene forms a solid solution with the phenol. In order to prove this, he adds a measured amount of benzil to the solution, allows it to freeze partially, and analyses the crystals which separate. He then makes the assumption that the benzil, phenol, and benzene will adhere to the crystals in the same proportions as they are present in the mother liquor, but, finding that there is always too great a proportion of benzene, he states that this excess (30 per cent.) exists with the phenol in the form of a solid solution.

In addition to this impossible result it should be noted that the mean initial molecular depression of fourteen hydrocarbons is 68·3, so that the behaviour of benzene (with $\Delta_1 = 68$) is by no means abnormal, as Bruni concludes from his observations on this one hydrocarbon.

From these considerations it is possible to draw the following two conclusions, which may be of some importance in the light of the numerous researches published by Bruni and others on solid solutions.

(1) The method of determining the amount of the solute existing in solid solution by the addition of a third normal substance is not reliable.

(2) If the molecular depression of a substance differs from the theoretical by about 10 per cent., it is not justifiable to conclude that the substance forms a solid solution.

As in Part III of this investigation it was shown that the esters tend to become normal in solutions of the substituted phenols, it will be seen from the following results that the abnormality of the hydrocarbons is reduced in a similar manner.

TABLE II.

	Phenol.	o-Cresol.	Thymol.	Guaiacol.	o-Nitro-phenol.
Association factor (Vaubel)...	4·6	3·9	3·5	3·5	3·1
m-Xylene. A	.17	16	13	10	9
Hexane. A	42	34	—	20	—
*iso*Amyl acetate. A	-18	-7	-2	0	$+2$

The values for *iso*amyl acetate are added for the sake of comparison.

In conclusion it may be stated that the hydrocarbons and their halogen derivatives in phenol solution give low molecular depressions, which diminish with the concentration. The abnormality appears to be connected with the association of the phenol molecules, as the substituted phenols, which are less associated, tend to give normal results.

THE VICTORIA COLLEGE LABORATORY,
WELLINGTON, NEW ZEALAND.

LXI.—*The Occurrence of Methane among the Decomposition Products of Certain Nitrogenous Substances as a Source of Error in the Estimation of Nitrogen by the Absolute Method.*

By PAUL HAAS, D.Sc., Ph.D.

DURING the course of the investigation of certain nitrogenous bases already described by the author (this vol., pp. 187 and 387), considerable difficulty was experienced in interpreting the analytical results. The bases, all of which are compounds of one of the three following types,

when analysed in the ordinary way in a tube filled with copper oxide, gave numbers for the percentage of nitrogen which, although agreeing fairly closely amongst themselves (compare table, p. 574), were from 2 to 5 per cent. too high for the calculated values; the hydrochlorides, on the other hand, gave correct values. Although the formulæ calculated from the analytical results were not in agreement with the general properties or reactions of these substances, it was not until in one case an acetyl derivative was found to give a higher percentage of nitrogen than the base from which it had been prepared that the method of analysis began to be suspected ; the gas contained in the nitrometer was accordingly analysed and found to contain methane. On consulting the literature, it was found that an isolated case of the same nature had already been recorded by Dunstan and Carr (*Proc.*, 1896, **12**, 48), who observed that the alkaloid aconitine and its hydrochloride presented exactly the same anomaly as the compounds described above. Although the results contained in the present communication have been to a great extent anticipated by the above-mentioned authors, yet their paper contains no experimental details and refers only to a single substance ; in view of the importance of the question from the point of view of the reliability of the absolute method for the estimation of nitrogen, it was thought worth while to investigate the matter thoroughly, inasmuch as there was here a whole class of compounds, more than a dozen in number, which showed this abnormal behaviour.

In order to establish the presence of methane, the gas contained in the nitrometer was first washed with distilled water until the strong caustic potash had been displaced some way down the tube; it was then transferred into an explosion burette and, after measuring, passed into a Bunte apparatus, where it was washed first with ammoniacal cuprous chloride to remove any carbon monoxide and subsequently with dilute sulphuric acid to remove ammonia ; on bringing the gas back into the explosion burette, no alteration in volume was ever observed, thus proving the absence of carbon monoxide. The oxygen, prepared by heating potassium permanganate, was introduced in requisite amount into the burette, where, after measurement, the gas was fired. After measuring the contraction on explosion, the volume of carbon dioxide was determined by absorption with caustic potash, and, after absorbing the excess of oxygen by means of an alkaline solution of potassium pyrogallate, the residual nitrogen was also measured. The volume of carbon dioxide was always found to be a little less than half the explosion contraction owing to the oxidation of some of the nitrogen ; consequently also the volume of the residual nitrogen was found to be rather less than it should have been.

The results obtained in five different cases are given below :

(1) The monometa-base, $CMe_2 \begin{subarray}{c} CH_2 \cdot C(NH \cdot C_6H_4 \cdot NH_2) \\ CH ===== C(OH) \end{subarray} CH.$

(a) Analysed with cuprous chloride and lead chromate :

0·1302 gave 14·1 c.c. moist nitrogen at 18·5° and 747 mm. N = 12·28. $C_{14}H_{18}ON_2$ requires N = 12·18 per cent.

(b) Analysed with copper oxide only. Time of combustion one and a half hours.

0·1405 gave 19·3 c.c. moist nitrogen at 18° and 748 mm. N = 15·61.

Volume of gas	=19·3 c.c.			
,, ,, mixed with oxygen...	=30·2 ,,			
,, ,, after explosion.........	=21·1 ,,	Contraction on explosion	=9·1 c.c.	
,, ,, ,, absorbing carbon dioxide	=16·8 ,,	Volume of carbon dioxide	=4·3 ,,	
Residual nitrogen	=14·9 ,,			

Taking the volume of methane to be equal to the volume of carbon dioxide formed and deducting this from the volume of gas originally found in the nitrometer, the volume of nitrogen becomes 19·3 – 4·3 = 15 c.c. : hence 0·1405 gave 15·0 c.c. of moist nitrogen at 18° and 748 mm. N = 12·13 per cent.

In the remaining four examples, only the explosion contractions and the volumes of carbon dioxide and nitrogen are recorded.

(2) The monoanilide, $CMe_2 \big< ^{CH_2 \cdot C(NH \cdot C_6H_5)}_{CH====C(OH)} \big> CH.$

(a) Analysed with cuprous chloride and lead chromate :

0·1391 gave 8·2 c.c. moist nitrogen at 21° and 758 mm. N = 6·79. $C_{14}H_{17}ON$ requires N = 6·51 per cent.

(b) Analysed with copper oxide only. Time of combustion one and a half hours.

0·1725 gave 12·7 c.c. moist nitrogen at 19·5° and 758 mm. N = 8·43 per cent.

Explosion contraction......... = 7·6 c.c.
Volume of carbon dioxide ... = 2·8 ,,   Whence N = 6·57.
,, ,, residual nitrogen = 9·4 ,,   ,,   N = 6·24 per cent.

(3) The acetyl derivative of the monotoluidide,

$$CMe_2 \big< ^{CH_2 \cdot C(NAc \cdot C_7H_7)}_{CH====C(OH)} \big> CH.$$

(a) Analysed with cuprous chloride and lead chromate :

0·1340 gave 6·2 c.c. moist nitrogen at 22·5° and 774 mm. N = 5·32. $C_{17}H_{21}O_2N$ requires N = 5·15 per cent.

(b) Analysed with copper oxide only. Time of combustion, two hours.

0·1565 gave 10 c.c. moist nitrogen at 16° and 763 mm. N = 7·47. per cent.

In this ease only an aliquot part, namely, 4·3 c.c. of gas, were analysed, the dilution being arranged to reduce the oxidation of the nitrogen to a minimum, so that the volume of carbon dioxide should be half of the explosion contraction.

Explosion contraction......... = 2·7 c.c.
Volume of carbon dioxide ... = 1·3 ,,   Whence N = 5·23
,,   ,, residual nitrogen = 3·1 ,,   ,,   N = 5·38 per cent.

(4) The ditoluidide, $CMe_2 < ^{CH_2 \cdot C(NH \cdot C_7H_7)}_{CH_2-C(:N-C_7H_7)} > CH$.

(a) Analysed with cuprous chloride and lead chromate :

0·1346 gave 10·4 c.c. moist nitrogen at 21·5° and 770 mm. N = 8·89. $C_{22}H_{26}N_2$ requires N = 8·80 per cent.

(b) Analysed with copper oxide only. Time of combustion one and a half hours.

0·127 gave 13·3 c.c. moist nitrogen at 18° and 755 mm. N = 12·01.

Explosion contraction = 7·5 c.c.; volume of carbon dioxide = 3·6 c.c. N = 8·77 per cent.

(5) s-Bisresorcyl-m-phenylenediamine,

$CMe_2 < ^{CH_2——C}_{CH:C(OH)} > CH$ —NH·C_6H_4·NH——$C < ^{C——CH_2}_{C(OH):CH} > CMe_2$.

(a) Analysed with cuprous chloride and lead chromate :

0·1126 gave 7·7 c.c. moist nitrogen at 19° and 775 mm. N = 8·02. $C_{22}H_{28}O_2N_2$ requires N = 7·95 per cent.

(b) Analysed with copper oxide only. Time of combustion one and a half hours.

0·1376 gave 15·5 c.c. moist nitrogen at 18° and 761 mm. N = 13·04.

Explosion contraction......... = 13·0 c.c.
Volume of carbon dioxide ... = 5·9 ,,   Whence N = 8·06
,,   ,, residual nitrogen = 9·3 ,,   ,,   N = 7·82 per cent.

In regard to the method adopted for preventing the accumulation of methane, it had been observed that the hydrochlorides of these bases behaved normally on combustion, from which it appeared that the presence of halogen had some influence, and it was found, as stated by Dunstan and Carr, that cuprous chloride when mixed with the substance made it possible to obtain correct analyses; moreover, since it was found (p. 576) that lead chromate was a more powerful oxidising agent for methane than copper oxide, the following modified method of analysis was employed. A layer of fused lead chromate was substituted for the usual layer of granular copper oxide, and the substance, before mixing with powdered copper oxide, was mixed with three to four times its bulk of freshly precipitated and washed cuprous chloride. Carried out in this way, perfectly correct results were obtained, the gas collected in the nitrometer being free from methane. In one case, however, that of *s*-bisresorcyl-*m*-phenylene-diamine, the amount of cuprous chloride mentioned above was not enough, and a correct result was only obtained on repeating the analysis with a larger bulk of the copper salt. On the other hand, a mere trace of chloride would appear to suffice in some cases, inasmuch as several of the substances gave correct results on the first analysis owing to the fact that the estimations were carried out in a tube in which hydrochloride had been previously analysed. It is probably unnecessary to use a fresh quantity of cuprous chloride each time, since (see below) cupric chloride can be substituted for the cuprous salt, but no data are available on this point. The lead chromate, on the other hand, appeared not to lose its efficiency even after a dozen or more analyses. The appended table shows the percentage of nitrogen obtained on analysing a number of these compounds when mixed with copper oxide only or when mixed with copper oxide and cuprous chloride, using lead chromate in place of granular copper oxide.

Percentage of nitrogen.

Substance.	Found on analysis with copper oxide only.	Found on analysis when mixed with copper chloride Calculated. and oxide.	
Monoanilide (this vol., p. 202)	8·09, 8·55	6·51	6·79, 6·82
Monoanilide hydrochloride (*ibid.*)...	5·88	5·58	—
Monotoluidide (this vol., p. 196) ...	7·84, 7·87, 8·13	6·11	6·39
Monotoluidide hydrochloride (*ibid.*)	5·27	·5·37	—
Acetyl monotoluidide (this vol., p. 197)	5·86, 6·20	5·15	5·29, 5·32
Ditoluidide (this vol., p. 199)	11·66	8·80	8·89
Monometa-base (this vol., p. 389)...	$\left\{ \begin{array}{l} 13·91, 14·58, 14·74, \\ 14·74, 14·84, 15·32 \end{array} \right\}$	12·18	12·28, 12·30
Monometahydrochloride (this vol., p. 390)	9·24	9·22	—
Monopara-base (this vol., p. 394)...	13·74, 14·60	12·18	12·03
Acetyl derivative of monometa-base	12·68	10·29	10·27
„ „ monopara-base	10·80	10·29	10·15
s-Bisresorcyl-*m*-phenylenediamine..	$\left\{ \begin{array}{l} 10·96, 11·76, 11·99, \\ 13·06, 13·26 \end{array} \right\}$	7·95	8·02
s-Bisresorcyl-*p*-phenylenediamine...	10·64, 10·72, 10·98	7·95	7·94

Subsequent experiments showed that the cuprous chloride could be replaced by cupric or lead chloride, but not by sodium chloride; it was, moreover, found that correct results could be obtained by mixing the substance with coarsely powdered lead chromate and using a tube filled with lumps of the same material.

The following experiments illustrate the results obtained by a variety of methods of analysis. The time of heating varied from one and a half to one and three-quarter hours, being reckoned in each case from the moment of the complete expulsion of air from the tube until the last traces of gas had been collected in the nitrometer.

The Monotoluidide.—(1) Substance mixed with copper oxide and cuprous chloride. Tube filled with lead chromate.

0·1346 gave 7·6 c.c. moist nitrogen at 21° and 757 mm. $N = 6·39$.

(2) Substance mixed with copper oxide and cuprous chloride. Tube filled with copper oxide.

0·1326 gave 8·0 c.c. moist nitrogen at 23° and 758 mm. $N = 6·77$.

$C_{14}H_{17}ON$ requires $N = 6·11$ per cent.

The Monometa-base.—(1) Substance mixed with copper oxide and cuprous chloride. Tube filled with lead chromate.

0·1302 gave 14·1 c.c. moist nitrogen at 18·5° and 747 mm. $N = 12·28$.

(2) Substance mixed with copper oxide and cupric chloride. Tube filled with copper oxide.

0·1251 gave 12·7 c.c. moist nitrogen at 16° and 778 mm. $N = 12·11$.

(3) Substance mixed with copper oxide and lead chloride. Tube filled with copper oxide.

0·1110 gave 11·7 c.c. moist nitrogen at 16·5° and 758 mm. $N = 12·23$.

(4) Substance mixed with copper oxide. Tube filled with copper oxide mixed with cupric chloride.

0·1220 gave 12·4 c.c. moist nitrogen at 14° and 769 mm. $N = 12·10$.

(5) Substance mixed with lead chromate. Tube filled with lead chromate.

0·1114 gave 11·8 c.c. moist nitrogen at 18° and 764 mm. $N = 12·29$.

$C_{14}H_{18}ON_2$ requires $N = 12·18$ per cent.

(6) Substance mixed with copper oxide and sodium chloride. Tube filled with copper oxide.

0·1316 gave 15·4 c.c. moist nitrogen at 16° and 766·5 mm. $N = 13·76$.

With regard to the influence of the cuprous chloride, two possibilities presented themselves; either this salt was in some way able to effect the combustion of methane more completely than copper oxide alone could do or else it combined with the bases to form compounds which did not yield methane or decompose at a temperature more favourable to the combustion of methane. In order to settle this point, methane was passed through red-hot tubes filled with carbon dioxide and containing pure copper oxide and copper oxide mixed with cuprous chloride respectively, when it was found that, whereas in the first case as much as 43 per .cent. of the gas passed over unburnt, complete oxidation was effected in the second. This shows that the action of the cuprous chloride must be attributed to its exerting some destructive influence on the methane rather than to its combining with the base, a view which gains support from the observation that lead chromate, which may be used as a substitute for cuprous chloride, also does not permit any methane passed over it to escape combustion.

The experiments were carried out as follows : an ordinary combustion tube, filled with the particular oxidising agent the action of which was to be tried, was connected in the usual manner with a hard glass tube containing sodium hydrogen carbonate for generating carbon dioxide. Between the two tubes was inserted a T-piece containing a three-way tap at its centre ; the stem of the T was attached to a graduated gas burette, the lower end of which was connected by an india-rubber tube to a reservoir filled with water. From this burette, pure methane, prepared by the action of water on aluminium carbide, was introduced in a slow stream through the T-piece into the furnace, after all the air had been previously displaced by a current of carbon dioxide. When sufficient methane had been introduced into the tube, the three-way tap in the T-piece was turned so that only carbon dioxide, which had been kept bubbling all the time, entered the furnace, the stream being continued until no more gas collected in the nitrometer. By inserting a second three-way tap between the top of the burette and the tap in the T-piece, the dead space between these two could be freed from air by a current of carbon dioxide before the methane was passed through it.

The results of the experiments are given below in tabular form :

	Tube packed with 50 cm. of granular copper oxide.	Tube packed with 46 cm. of granular copper oxide and 10 cm. of powdered copper oxide mixed with 4 grams of cuprous chloride.	Tube packed with 60 cm. of lead chromate.
Time occupied in passing the methane through the tube	40 minutes	45 minutes	35 minutes	40 minutes
Volume of methane introduced into the tube	18 c.c.	17 c.c.	18·8 c.c.	21·7 c.c.
Volume of methane collected in the nitrometer	7·8 c.c.*	7·6 c.c.*	0	0
Percentage of methane escaping combustion	43·3	44·7	0	0

* In each of these cases, the gas was proved to consist of methane only by exploding it with oxygen and absorbing the carbon dioxide formed with caustic potash.

In attempting to account for the formation of methane from this group of substances, it was originally thought that the phenomenon might be connected with the presence in the compounds of the grouping $R \cdot NH \cdot R_2$; to test the correctness of this view, a nitrogen estimation was carried out with p-aminoacetanilide, a substance first prepared by Nietzki (*Ber.*, 1884, **17**, 343), and only analysed by him in the form of its platinum salt; being a substituted diamine, it was thought that it might present the same anomalies as the compounds here described; however, on analysis with copper oxide it was found to behave quite normally.

0·1233 gave 19·8 c.c. moist nitrogen at 15° and 755 mm. $N = 18·67$.
$C_8H_{10}ON_2$ requires $N = 18·66$ per cent.

The cause of the trouble must therefore be sought in the hydro-aromatic complex, more especially as it is found that the compounds of the type III (p. 570), which contain two such complexes, give the largest amounts of methane and the lowest combustions.

In order to determine whether in this case the attachment of two methyl groups to the same carbon atom had not some special influence, it was decided to analyse an analogous substance which did not contain this grouping; accordingly, the m-phenylenediamine condensation product of monomethyldihydroresorcin was prepared and analysed, and was, indeed, found to give quite correct results when analysed in the usual way.

Condensation of Methyldihydroresorcin with m-*Phenylenediamine.*

Three and a half grams of methyldihydroresorcin and 3 grams of m-phenylenediamine dissolved in alcohol were heated over a water-bath for three hours. On evaporating off the alcohol, there remained a brown, syrupy residue, which, on dissolving in a mixture of alcohol and light petroleum, separated from the solution in colourless, slender prisms melting at 178·5—179·5°.

5-*Hydroxy*-3-m-*aminophenylimino*-1-*methyl*-$\Delta^{3:5}$-*dihydrobenzene*,

$$CH_3 \cdot CH \diagdown \begin{array}{c} CH_2 \cdot C(NH \cdot C_6H_4 \cdot NH_2) \\ \overline{CH\underline{}C(OH)} \end{array} \diagup CH,$$

is readily soluble in cold alcohol or acetone, slightly so in het benzene, and insoluble in chloroform or light petroleum.

0·1355 gave 15·1 c.c. moist nitrogen at 15° and 763 mm. $N = 13·35$.
$C_{13}H_{16}ON_2$ requires $N = 12·96$ per cent.

The analysis was performed in the usual way, using copper oxide only, and no methane was formed.

The combustions for carbon and hydrogen in these substances were originally performed in the ordinary way with copper oxide and a spiral of reduced copper gauze in air, but the results obtained by this method were too low ; it was, however, found that, on substituting lead chromate for the copper oxide and carrying out the combustion in an atmosphere of oxygen with a spiral of silver gauze in place of the reduced copper, correct results could be obtained. It was, moreover, observed that the presence of a little chromium sesquioxide in the boat helped to render the combustion complete.

The following table shows the results obtained by the two methods :

Percentage of carbon and hydrogen.

Substance.	Combustions in air with copper oxide.	Calculated percentage.	Combustions in oxygen with lead chromate.
Acetylmonotoluidide............	C = 73·03	75·27	74·93, 75·38
	H = 7·73	7·75	8·03, 8·47
Monometa-base	C = 71·10, 71·67		⎧73·10
	H = 8·14, 8·37	C = 73·05	⎪ 8·48
Monopara-base	C = 71·35, 70·97	H = 7·83	⎨73·09
	H = 7·75, 7·41		⎩ 8·01
s-Bisresorcyl-m-phenylenedi- amine.	C = 71·21, 71·48, 72·79	C = 75·00	⎧74·91
	H = 7·18, 7·23, 7·67		⎪ 8·56
s-Bisresorcyl-p-phenylenedi- amine.	C = 70·64, 71·51	H = 7·95	⎨74·50
	H = 7·51, 7·73		⎩ 7·95

In conclusion, the author desires to express his indebtedness to Dr. H. R. Le Sueur for many useful suggestions during the course of the work.

CHEMICAL LABORATORY,
ST. THOMAS'S HOSPITAL,
LONDON, S.E.

LXII.—*Silver Dioxide and Silver Peroxynitrate.*

By EDWIN ROY WATSON.

IN 1804, Ritter (*Gehlen's Neues J.*, 1804, 3, 561) obtained a black crystalline substance at the anode during the electrolysis of an aqueous solution of silver nitrate, which he regarded as silver dioxide, Ag_2O_2.

Further investigation of this product, however, has shown that it contains nitrogen and that its composition is expressed by the formula $Ag_7O_{11}N$ (Šulc, *Zeit. anorg. Chem.*, 1896, 12, 89 ; 1900, 24, 305; Mulder and Heringa, *Rec. trav. chim.*, 1896, 15, 1 ; Tanatar, *Zeit. anorg. Chem.*, 1901, 28, 331).

This electrolysis of silver nitrate solution is the only method by which a multivalent silver compound can be obtained in any quantity.

Other methods have been described for the preparation of silver dioxide. Wöhler states that he obtained silver dioxide as a black crust on a silver anode during the electrolysis of dilute sulphuric acid (*Annalen*, 1867, 146, 263), but it is difficult to obtain in this way sufficient even for analysis. Schiel has described the preparation of silver dioxide by the action of ozone on silver monoxide (*Annalen*, 1864, 132, 322), and Berthelot has given reasons (*Compt. rend.*, 1880, 90, 572) for the supposition that an oxide, Ag_4O_3, is formed on the addition of alkali to a mixture of hydrogen peroxide and silver nitrate solution, but has never isolated the compound.

The black powders obtained at the anode during the electrolysis of aqueous solutions of other soluble silver salts have been investigated by Mulder and Tanatar, and apparently to these substances also it is necessary to assign quite complicated formulæ.

I have varied the conditions of electrolysis of silver nitrate by altering the current-concentration and density and also the strength of solution, and have found the product to be the same in all cases and identical with the compound described by Šulc. This disposes of the possibility of this definitely crystallised product being a mixture. Šulc has satisfactorily investigated the effect of heat on this peroxynitrate. I have not been able to confirm Šulc's analytical data for the reaction of the peroxynitrate with ammonia.

Silver Dioxide.—I have examined the decomposition of the peroxynitrate by water. Even at the ordinary temperature of the laboratory (27° to 32°), reaction slowly occurs with the evolution of oxygen. This change occurs more readily on boiling, and is complete in less than an hour. Oxygen is evolved, part of the silver goes into solution, and there remains a black substance which I have found to be silver dioxide, Ag_2O_2, probably obtained pure for the first time. The course of the reaction is represented by the equation :

$$Ag_7NO_{11} = AgNO_3 + 3Ag_2O_2 + O_2.$$

I have determined quantitatively the correctness of this expression for the decomposition of the peroxynitrate by boiling, and also the composition of the insoluble substance as being that of silver dioxide. In warm dilute sulphuric acid, the substance dissolves as silver sulphate with the evolution of half of its oxygen.

It is a greyish-black powder, of sp. gr. 7·44, approximately, which may be heated to 100° without change. At a higher temperature, it evolves oxygen and leaves a residue of silver.

The behaviour of the dioxide with ammonia is remarkable ; it dissolves with the evolution of nitrogen, but in the amount required by the equation :

$$6Ag_2O_2 + 2NH_3 = N_2 + 3H_2O + 3Ag_4O_3,$$

and not, as would have been expected, in accordance with the equation:

$$3Ag_2O_2 + 2NH_3 = N_2 + 3H_2O + 3Ag_2O.$$

Soluble Silver Peroxy-salts.—Both the peroxynitrate and the dioxide of silver, also the peroxysulphate produced by the electrolysis of aqueous silver sulphate solution, dissolve in cold strong nitric acid with the production of an intensely brown solution, and in cold strong sulphuric acid with an olive-green colour. These colours are due to the formation of silver peroxy-salts. It can hardly be doubted that the same coloured salt is formed from the peroxynitrate and the peroxysulphate as from the dioxide, because the colours and absorption spectra of the solutions obtained from the three substances are identical. These coloured salts gradually decompose at the ordinary temperature, and more quickly on heating or on adding water, silver sulphate or nitrate remaining in solution. Up to the present, attempts to isolate these peroxy-salts have been unsuccessful. During the decomposition of these solutions, some evolution occurs of what appears to be oxygen, but no hydrogen peroxide is formed.

An attempt was made to study the rate of decomposition of the coloured compound in the nitric acid solution by measuring the depth of colour of the solution from time to time. It appears that the rate of decomposition of the coloured compound is proportional to the concentration of this substance in the solution. Expressed in symbols, $dx/dt \propto x$, where $x = $ concentration of the coloured compound in the solution, $t = $ time, or $t = A\log x + B$, where A and B are constants.

These observations are not in agreement with the supposition that the coloured compound has the simple formula $Ag(NO_3)_2$. The formula $[Ag(NO_3)_2]_4$ or $Ag_4(NO_3)_8$ satisfies the requirement that the substance shall decompose according to a unimolecular reaction, thus: $Ag_4(NO_3)_8 + 2H_2O = 4AgNO_3 + 4HNO_3 + O_2$. This requirement is also satisfied by $Ag_2(NO_4)_2$, decomposing thus: $Ag_2(NO_4)_2 = 2AgNO_3 + O_2$. The question of the composition of the soluble coloured compound is still under investigation.

EXPERIMENTAL.

Preparation of Silver Peroxynitrate.—In Expt. I, the silver nitrate solution was contained in a platinum dish surrounded by ice and water. The dish served as the cathode, whilst the anode was a square piece of platinum foil having an area of 2 sq. cm. In Expts. II, III, and IV, when stronger currents were employed, the peroxynitrate at the anode and the silver at the cathode formed in needles which grew to a great length, and it was necessary to use a porous cell to separate the products of the two electrodes. The silver nitrate was contained in a small beaker surrounded by ice and water and the electrodes were rectangular pieces of platinum foil, 4 cm. × 2 cm., the cathode being surrounded by a porous cell. In Expt. I, the current was continued for two hours; in Expts. II, III, and IV only for half an hour. In all cases the anodic product was easily separated from the platinum foil and was washed with cold distilled water by decantation and dried at the ordinary temperature over soda-lime in a desiccator.

The four preparations of peroxynitrate were each analysed in the same way. A weighed quantity was heated very gently in a small round flask until the first stormy evolution of gas occurred. The operation was performed in a flask, because in a crucible it was difficult to avoid loss when the sudden evolution occurred. The black residue was, after weighing, transferred as completely as possible to a porcelain crucible and gently heated until it was reduced to metallic silver.

Experiment No.	Percentage of silver nitrate in solution.	Current strength. Amperes.	Current density. Amperes per sq. cm.	Percentage of black residue.	Percentage of silver.
I.	15	0·03	0·015	91·35	79·78
II.	15	0·55	0·07	91·54	79·66
III.	15	1·12	0·14	91·40	79·63
IV.	5	0·55	0·07	(a) 91·70	79·98
				(b) 91·88	80·18
$Ag_7O_{11}N$ requires..................				91·56	79·91

These numbers show that the composition of the anodic product is independent of the concentration of the silver nitrate solution and of the strength and density of the current. The product was uniformly crystalline in octahedra. In I, the crystals were separate or in small irregular aggregates. In II, III, and IV, the octahedral crystals were regularly arranged in needle-like aggregates.

Action of Boiling Water on the Peroxynitrate.—For this and subsequent experiments, the peroxynitrate was prepared as in Expt. III summarised in the previous paragraph. With one cell, about 1·8 grams prepared in one operation lasting thirty minutes.

A weighed quantity of the substance was boiled with excess of distilled water in a beaker for one and a half hours, the water being replaced as required. The insoluble portion was filtered off, washed with hot distilled water, and dissolved in hot dilute nitric acid.

The silver in this solution was estimated as silver chloride. The silver in the filtrate was estimated in the same way.

0·6557 gave 0·5968 silver chloride from the insoluble residuê. Insoluble silver = 68·50 per cent.

0·6842 gave 0·6186 silver chloride from the insoluble residue, and 0·1015 silver chloride from the filtrate.

Insoluble silver = 68·05 ; soluble silver = 11·17 per cent.

In decomposing in accordance with the equation $Ag_7NO_{11} = AgNO_3 + 3Ag_2O_2 + O_2$, Ag_7NO_{11} gives insoluble silver = 68·49 and soluble silver = 11·42 per cent.

In another experiment, the gas evolved during the reaction was collected and found to be completely absorbed by alkaline pyrogallol solution. For collecting the gas, the following apparatus was employed : a flask of about 300 c.c. capacity was fitted with a two-holed cork. In the one hole was fitted a delivery-tube with a stop-cock, and in the other a dropping-funnel with a short wide delivery-tube. The flask was half-filled with distilled water, which was boiled vigorously. When all air had been expelled, the flame was withdrawn and the stop-cock on the delivery-tube closed. A quantity of the peroxynitrate covered with water in the dropping-funnel was then introduced into the flask through the funnel, the flask again heated, the stop-cock on the delivery-tube opened, and the oxygen collected over water.

Silver Dioxide, Ag_2O_2.—The insoluble residue which remained after prolonged boiling of the peroxynitrate with water was washed by decantation with hot water and dried either at the ordinary temperature over soda-lime in a desiccator or in the steam-oven. Two determinations of the specific gravity with about 2 grams of the substance in the specific gravity bottle gave 7·46 and 7·42 respectively.

The silver in the compound was estimated by heating a weighed quantity and weighing the residual silver.

1. 0·7447 gave 0·6475 residual Ag. Ag = 86·94 per cent.
2. 0·3612 „ 0·3138 residual Ag. Ag = 86·88 per cent.

The silver in the second sample was also estimated by dissolving in warm dilute nitric acid and weighing the silver as chloride.

0·3663 gave 0·4232 AgCl. Ag = 86·94.

Ag_2O_2 requires Ag = 87·11 per cent.

The total oxygen in the compound was estimated by heating in a combustion-tube in a current of carbon dioxide and collecting the oxygen over a solution of potassium hydroxide. It was completely absorbed by alkaline pyrogallol solution.

0·0842 gave 8·8 c.c. oxygen at 27° and 757·5 mm. O = 13·07.

Ag_2O_2 requires O = 12·89 per cent.

The Solution of Silver Dioxide in Hot Dilute Sulphuric Acid.

The estimation of the oxygen evolved when the dioxide is dissolved in dilute sulphuric acid solution was carried out in the apparatus previously used for examining the gas evolved on boiling silver peroxynitrate with water.

0·2745 gave 13·7 c.c. oxygen at 26° and 757·5 mm. $O = 6·30$.
Ag_2O_2 requires $O = 6·45$ per cent.

The Solution of Silver Dioxide in Aqueous Ammonia.

The nitrogen liberated in this reaction was estimated in an apparatus similar in principle to that described by Šulc (*Zeit. anorg. Chem.*, 1900, **24**, 305). The substance was placed in a flask fitted with delivery-tube and a dropping-funnel with a delivering-tube reaching to the bottom of the flask and ending in a capillary. The whole apparatus was completely filled with water and strong aqueous ammonia gradually introduced through the funnel. The nitrogen liberated was collected over water. At the end of the reaction, any gas remaining in the apparatus was driven out by water. The solution was effected at the ordinary temperature.

0·4158 gave 7·3 c.c. nitrogen at 28° and 762·5 mm. $N = 1·92$ per cent. Two other estimations gave concordant results.

$6Ag_2O_2 + 2NH_3 = 3Ag_4O_3 + 3H_2O + N_2$ requires $N = 1·88$ per cent.

These numbers indicate that only one-quarter of the oxygen contained in the dioxide reacts with ammonia (see p. 579).

THE CHEMICAL LABORATORY,
CIVIL ENGINEERING COLLEGE,
SIBPUR, CALCUTTA.

LXIII.—*Influence of Substituents in the Trinitrobenzene Molecule on the Formation of Additive Compounds with Arylamines.*

By JOHN JOSEPH SUDBOROUGH and NORMAN PICTON.

s-TRINITROBENZENE and various other trinitro-derivatives readily form additive compounds with α- and β-naphthylamines and other arylamines, including primary, secondary, and tertiary bases (Trans.,

1901, **79**, 522; 1903, **83**, 1334); certain of these additive compounds can be alkylated, acetylated, and benzoylated, and are all resolved by mineral acids into their components.

Since the appearance of our last paper (Trans., 1903, **83**, 1334), several communications bearing on the same subject have appeared. Sommerhoff (*Inaug. Dissertation, Zurich,*]1904; Noelting and Sommerhoff, *Ber.*, 1906, **39**, 76) describes additive compounds of trinitrobenzene with various alkyl derivatives of aniline and also with various diamines; the majority of these compounds contain one molecule of nitro-compound combined with one of the base. The additive compounds of trinitrobenzene with diphenylamine, *p*-ditolylamine, quinoline, and 1 : 2-xyloquinoline are exceptions, which contain two molecules of nitro-compound to one of the amine (compare Wedekind, *Ber.*, 1900, **33**, 434). The additive compounds derived from the quinolines are described as colourless. In the theoretical part of his *Dissertation*, Sommerhoff criticises the quinonoid formula which has been suggested for these coloured additive compounds, and suggests that the products are complex compounds according to Werner's conception, in which C_6 is equivalent to a heavy metal. The two arguments used by Sommerhoff against the quinonoid structure, namely, that the compounds have no pronounced acidic properties and that they do not resemble quinones, are not conclusive. Since the compounds are resolved into their components in the presence of solvents (compare the molecular weight determinations, Trans., 1903, **83**, 1338), it is difficult to say anything with regard to their acidic properties. The compounds have never been compared with quinones, but with quinonoid compounds, for example, with the compounds obtained by the addition of arylamines to quinones. Loring Jackson and Clarke (*Ber.*, 1904, **37**, 176, and *Proc.*, 1906, **22**, 83) describe a number of somewhat unstable compounds of mono- and di-methylanilines with various nitro-compounds. Kreemann (*Monatsh.*, 1904, **25**, 1215, 1271, 1311) has proved the formation of a number of additive compounds, for example, *m*-dinitrobenzene—aniline, 2 : 4-dinitrotoluene—naphthalene, which are too unstable to be actually isolated. This method is based on an examination of melting-point curves obtained by taking melting points as ordinates and molecular percentage of one component as abscissæ.

The work summarised in this communication deals mainly with the influence of various substituents in the trinitrobenzene molecule on the formation of such additive compounds. The previous experiments proved that single substituents—such as methyl, methoxyl, ethoxyl, carbethoxyl, amino—do not completely inhibit the formation of additive compounds, although they appear to render the compounds less stable when once formed (Trans., 1901, **79**, 533). We have studied the effect of the introduction of two and three chlorine, bromine, methyl, or methoxyl radicles. The results may be summarised as follows : (1) three methyl, two methoxyl, or three bromine radicles completely inhibit the formation of the additive compounds * with α- or β-naphthylamine. (2) The inhibition probably depends on two factors : (a) the positive or negative nature of the substituents—thus, methyl and methoxyl groups readily prevent the formation of additive compounds, whereas chlorine radicles do not. (b) The comparative sizes of the substituents—thus, although chlorine and bromine are both negative substituents, three bromines completely prevent the formation of additive compounds, whereas three chlorines do not.

This conclusion is confirmed by Loring Jackson and Clarke's results (Ber., 1904, 38, 176). These chemists find that both trichlorotrinitrobenzene and tribromotrinitrobenzene yield unstable additive compounds with dimethylaniline in the absence of solvents, but that the bromo-compound is far less stable than the chloro-analogue.

The compounds we have obtained are similar in properties to those previously described ; they are all dark, and are decomposed into their components by dilute hydrochloric acid. The coloured additive products derived from chloro-derivatives of trinitrobenzene are characterised by the readiness with which they lose hydrogen chloride when their alcoholic solutions are warmed with an excess of base, and compounds of the type of picrylaniline, $(NO_2)_3C_6H_2·NHPh$, are thus formed.

The compounds of this type which have been prepared are picryl-α-naphthylamine, picryl-β-naphthylamine, picrylethylaniline, picryl-methyl-α-naphthylamine, 1:3-dinitro-2:4-di-α-naphthylaminobenzene, and the isomeric β-compound.

All these compounds are bright red, and in many respects recall the additive compounds already described ; they are, however, remarkably stable, and are not decomposed when boiled with dilute or concentrated hydrochloric acid.

The bright red colour has suggested the possibility that these compounds may be o-(or p-)quinonoid derivatives of the type

$$NO_2 \overset{NPh}{\underset{NO_2}{\bigcirc}} :N{<}^{OH}_{O} ,$$

---

* In this paper we use the expression "inhibit the formation of additive compounds" in the sense that no definite crystalline additive compounds could be isolated. It is quite possible that Kremann's method would indicate the formation of extremely unstable compounds in some of these cases.

and not benzenoid derivatives,

$$\underset{\overset{|}{NO_2}}{NO_2 \diagdown \diagup NO_2,} \quad \text{with } NHPh$$

just as it has been suggested that the additive compounds may also be quinonoid derivatives :

$$NO_2 \diagup \diagdown :N\diagdown_O^{OH} \quad \text{with } H \quad NH \cdot C_{10}H_7, \quad NO_2.$$

We have examined a number of compounds of the type of picryl-aniline and have been able to establish the following points.

(1) Compounds of this type form mono-potassium salts which are intensely coloured and explosive. This is in harmony with either the benzenoid or quinonoid structure, as the salts may be either

$$(NO_2)_3C_6H_2 \cdot NKPh \text{ or } (NO_2)_2C_6H_2(:NO \cdot OK):NPh.$$

Against the second formula may be urged the fact that the salts are readily hydrolysed and are thus derived from a feeble acid.

(2) The compounds do not yield acetyl derivatives when boiled for several hours with acetyl chloride or acetic anhydride. This may be due to stereochemical inhibition (compare Auwers, *Ber.*, 1904, **37**, 3890), and is of no assistance in elucidating the constitution.

(3) Some of them form additive compounds with arylamines. We have been able to prepare the following : (*a*) picrylaniline—α-naphthyl-amine, (*b*) picryl-α-naphthylamine—aniline, (*c*) picryl-β-naphthyl-amine—aniline. These ˌare all red, relatively unstable compounds, and are readily decomposed into their components.

(4) Many of the compounds exist in two isomeric forms : (*a*) a bright red or scarlet modification ; (*b*) a sulphur-yellow variety (compare Bamberger, *loc. cit.*, p. 107). The red form can usually be obtained by crystallising the crude product from benzene or glacial acetic acid, and the yellow form by heating the red compound to a suitable tem-perature.

In certain cases the two compounds appear to melt at the same temperature owing to the fact that the red modification is transformed into the yellow during the process of heating.

So far, we have not been able to determine the nature of the isomerism ; that is to say, whether it is a case of structural isomerism between the benzenoid and quinonoid forms or stereoisomerism between two quinonoid forms.

The isomerism is evidently of the same type as that met with by Loring Jackson and Bentley (*Proc. Amer. Acad. Arts Sci.*, 1891, 70 and 82) in the study of ethyl anilinotrinitrotartronate,

$$C_6H(NO_2)_3(NHPh)\cdot C(OH)(CO_2H)_2.$$

This exists in the form of orange-red prisms melting at 143° and of yellow needles melting at 122°. The yellow modification is transformed into the red when heated alone, when its alcoholic solution is crystallised at 60°, or when it is boiled with water; conversely, the red is transformed into the yellow when dissolved in acetic acid and precipitated with water.

These examples of isomerism deserve further study, and we hope to devote more time to the elucidation of the problem at a later date.

Turpin (Trans., 1891, **59**, 716) and Wedekind (*Ber.*, 1900, **33**, 434) have previously prepared condensation products of picryl chloride with secondary amines, namely, picrylmethyl- and picrylethyl-anilines. We have prepared the same compounds, and, in addition, picrylmethyl-α-naphthylamine. All these compounds are coloured, they do not yield metallic derivatives, and so far have not been met with in isomeric forms. They are remarkably stable and are not decomposed when heated with concentrated hydrochloric acid at 180—200°; these experiments were made in order to determine, if possible, whether the alkyl groups are attached to nitrogen (benzenoid formula) or to oxygen (quinonoid formula).

The fact that the additive compounds of picryl chloride and similar halogen nitro-compounds with α- and β-naphthylamines readily lose hydrogen chloride and yield the condensation products already mentioned is in perfect harmony with the quinonoid structure of the additive compounds, thus:

and, in reality, such a constitution affords probably the simplest explanation of the mechanism of the reaction.

Since halogen and also nitro-radicles which are in the ortho-position to a nitro-group can be thus readily exchanged for an arylamino-group, we have attempted the replacement of an arylamino-radicle which is in the ortho-position with respect to a nitro-group by a different arylamino-radicle, thus: ·NHPh by ·NH·C₁₀H₇ and *vice versâ*. We have attempted this (a) by the preparation of additive compounds of the type picrylaniline—α-naphthylamine and subsequent treatment with hydrochloric acid, but in each case the original nitro-derivative was obtained.   The reaction did not proceed in the manner

$$C_6H_2(NO_2)_3 \cdot NHPh \quad \longrightarrow \quad C_6H_2(NO_2)_2 \Big\langle \begin{matrix} NO \cdot OH \\ NHPh \\ NH \cdot C_{10}H_7 \end{matrix} \quad \longrightarrow$$
$$C_6H_2(NO_2)_3 \cdot NH \cdot C_{10}H_7.$$

It is more than probable that the additive compound in this case would not be formed in exactly the same manner as with picryl chloride and α-naphthylamine.   In all probability the second arylamino-radicle would not unite with the same carbon atom to which the first arylamino-group was attached, and the reaction would thus be :

and by the action of acids the original nitro-compound would be regenerated.

(b) By heating the picrylarylamine with a second arylamine.   In this case also no replacement could be observed, the substance remained unaltered or was completely decomposed with deposition of carbon.

EXPERIMENTAL.

1.   *Methyl Substituents.*

It has already (Trans., 1903, 83, 1335) been stated that s-trinitro-xylene and s-trinitromesitylene do not form definite additive compounds with arylamines.   In the case of trinitroxylene it would be more correct to say that no definite compound can be isolated when the nitro-compound and α-naphthylamine are brought together in the presence of a solvent, although a deep red coloration is developed. An experiment was made by mixing together the two components in molecular proportions and fusing the mixture by gently heating it in a sulphuric acid bath.   The mass, when cooled slowly, set to a red, crystalline solid melting at about 100—105°, which did not lose in weight when exposed to the air at the ordinary temperature and hence presumably contained the α-naphthylamine in a state of combination. The whole of the base may be removed by the following methods : (a) heating the red solid in the steam-oven for some forty hours ; (b) by the addition of practically any organic solvent, and (c) by treatment with mineral acid.   The product left in each case is tri-nitroxylene melting at 180°.

2. Chlorine Substituents.

(a) *Picryl Chloride.*—We have prepared the additive compound of picryl chloride and α-naphthylamine, previously described by Bamberger (*loc. cit.*); it crystallises from alcohol in long, brown needles melting at 110·5—111·5°, and is extremely unstable, readily losing hydrogen chloride and forming picryl-α-naphthylamine.

We have not been able to prepare an additive compound of picryl chloride with β-naphthylamine: the product obtained in each experiment was picryl-β-naphthylamine.

Picryl chloride and *ethyl-α-naphthylamine* yield an additive compound, $C_6H_2Cl(NO_2)_3,C_{10}H_7 \cdot NHEt$, when the two components are separately dissolved in ether, the solutions mixed, and a small amount of alcohol added. The product crystallises in slender, silky, brown needles melting at 85°, and is readily soluble in cold ether or benzene, but only sparingly so in cold alcohol. It is far more stable than the additive compound with α-naphthylamine.

0·7442 gave 0·2594 AgCl. Cl = 8·6.

$C_{18}H_{15}O_6N_4Cl$ requires Cl = 8·5 per cent.

Picryl chloride and *diethyl-β-naphthylamine* yield an additive compound, $C_6H_2Cl(NO_2)_3,C_{10}H_7NEt_2$, which crystallises from warm alcohol in dark purple needles melting at 77°; it dissolves readily in cold benzene, yielding a reddish-purple solution, but the colour disappears when the solution is diluted.

0·3714, when decomposed with dilute hydrochloric acid, gave 0·202 of picryl chloride = 54·4.

$C_{20}H_{19}O_6N_4Cl$ requires 55·4 per cent.

A similar *additive compound*, $C_6H_2Cl(NO_2)_3,C_{10}H_7 \cdot NMe_2$, is obtained when dimethyl-α-naphthylamine is used; it crystallises in slender, red needles and melts at 42°.

0·3368 gave 0·1986 of picryl chloride = 58·9.

$C_{18}H_{15}O_6N_4Cl$ requires 59·1 per cent.

(b) 4 : 6-*Dichloro*-1 : 3-*dinitrobenzene* (Nietzki and Schedler, *Ber.*, 1897, 30, 1666).—The additive compound α-*naphthylamine*-4 : 6-*dichloro*-1 : 3-*dinitrobenzene*, $C_6H_2Cl_2(NO_2)_2,C_{10}H_7NH_2$, may be prepared by dissolving molecular proportions of the components in a small amount of ether, adding about twice the volume of alcohol, and allowing the ether to evaporate ; it is thus obtained in the form of slender, brownish-red needles melting at 95°. The same compound is obtained in the form of stout, prismatic needles when the ethereal solution is allowed to evaporate without the addition of alcohol; it dissolves readily in cold ether or benzene and also in hot alcohol, from which it crystallises in the form of long, red needles. When the alcoholic solution is boiled for a short time, hydrogen chloride is eliminated to a certain extent, and, as the solution cools, a mixture is obtained consisting of the above-mentioned red needles and a yellow compound of high melting point (4 : 6-di-α-naphthylamino-1 : 3-dinitrobenzene). The same yellow compound is formed when the mother liquor from the red needles is evaporated. The yield, however, of this condensation product is not good, even when an alcoholic solution of the additive compound is boiled for some time. The additive compound gave the following results on analysis :

0·2006 gave 0·1496 AgCl.   Cl = 18·44.

$C_{16}H_{11}O_4N_3Cl_2$ requires Cl = 18·68 per cent.

0·3368, when decomposed by hot dilute hydrochloric acid, filtered, and washed, gave 0·2224 gram of residue after allowing for the solubility of the dichlorodinitrobenzene in water (residue melted at 100°), this amount corresponding with 66 per cent. $C_{16}H_{11}O_4N_3Cl_2$ requires 62·4 per cent. It was noticed that when the residue obtained by decomposing the additive compound with dilute hydrochloric acid was washed with cold water, the colour rapidly changed to red, due to the regeneration of the additive compound (compare Trans., 1901, **79**, 526).

The additive compound, 4 : 6-*dichloro*-1 : 3-*dinitrobenzene-β-naphthylamine*, is somewhat more difficult to prepare, owing to the readiness with which hydrogen chloride is eliminated and a condensation product formed. It was ultimately prepared by dissolving the nitrocompound (0·5 gram) and the base (0·4 gram) separately in ether, mixing the solutions, allowing the ether to evaporate, and washing the residue with cold alcohol. It forms brownish-red needles, melts at about 67—68°, is decolorised by the addition of benzene, and when left in contact with alcohol for some time loses hydrogen chloride.

0·1470 gave 0·1038 AgCl.   Cl = 17·46.

$C_{16}H_{11}O_4N_3Cl_2$ requires 18·68 per cent.

0·4042, when decomposed with dilute hydrochloric acid, gave 0·2550 of nitro-compound (m. p. 100°) = 63·0. Calculated for $C_6H_2Cl_2(NO_2)_2,C_{10}H_7 \cdot NH_2 = 62·4$ per cent.

No additive compound of dinitrodichlorobenzene and dimethyl-α-naphthylamine has been obtained. When ethereal solutions of the components are mixed, the unaltered nitro-compound separates.

The additive compound, 4 : 6-*dichloro*-1 : 3-*dinitrobenzenediethyl-β-naphthylamine*, was prepared by dissolving the nitro-compound

(0·75 gram) and the base (1·1 grams) in ether, allowing the ether to evaporate, and keeping the residual oil in a desiccator over sulphuric acid. After several days, it solidified to hard, compact, almost black crystals which were washed with ethyl alcohol. It is practically insoluble in cold alcohol, but dissolves on warming to a deep red solution which, as it cools, yields crystals of the additive compound unmixed with the free nitro-compound and melting at 30°.

0·5005, when decomposed with hydrochloric acid, gave 0·1810 of nitro-derivative = 36·2. $C_6H_2Cl_2(NO_2)_2, C_{10}H_7 \cdot NEt_2$ requires 54·35 per cent., and $C_6H_2Cl_2(NO_2)_2, 2C_{10}H_7NEt_2$ requires 37·3 per cent.

(c) 2 : 4-*Dichloro*-1 : 3 : 5-*trinitrobenzene.*—The nitro-compound was prepared by boiling the 4 : 6-dichloro-1 : 3-dinitrobenzene with a mixture of fuming nitric (1 vol.) and fuming sulphuric acids (2 vols.) for three hours in a Jena flask with a reflux condenser ground into the neck. When cold, the product was poured into a large volume of cold water and the precipitate crystallised from alcohol (m. p. 128°), the yield being 5 grams from 6 of the dinitro-compound.

The additive compound, *dichlorotrinitrobenzene-α-naphthylamine*, was obtained by mixing ethereal solutions of the components, adding twice the volume of alcohol, and keeping for some time; it crystallises in pale brownish-red needles, melts at 126—127°, and is readily soluble in ether or benzene.

0·3276, when decomposed with hydrochloric acid, gave 0·2144 of dichlorotrinitrobenzene = 65·5; $C_6HCl_2(NO_2)_3, C_{10}H_7 \cdot NH_2$ requires 66·3 per cent.

0·2031 gave 0·1300 AgCl. Cl = 15·9.

$C_{16}H_{10}O_6N_4Cl_2$ requires 16·7 per cent.

In the attempts to prepare an additive compound of dichlorotrinitrobenzene and β-naphthylamine, even in ethereal solution, hydrogen chloride was formed and β-naphthylamine hydrochloride was deposited.

2 : 4-*Dichloro*-1 : 3 : 5-*trinitrobenzenediethyl-β-naphthylamine* crystallises from a mixture of ether and alcohol in bluish-black needles melting at 72°; it dissolves readily in benzene and ether, but is practically insoluble in cold alcohol. When warmed with alcohol, a pale red solution is obtained; the colour deepens as the solution cools and ultimately crystals of the additive compound together with colourless crystals of dichlorotrinitrobenzene are deposited.

0·4038, when decomposed with dilute hydrochloric acid, gave 0·2352 of dichlorotrinitrobenzene melting at 127° = 58·2;

$C_6HCl_2(NO_2)_3, C_{10}H_7NEt_2$ requires 58·6 per cent.

(d) s-*Trichlorotrinitrobenzene.*—The additive compound, *trinitro-trichlorobenzene-α-naphthylamine*, was obtained by dissolving the nitro-compound (1 gram) in benzene, adding an excess of base (1·5 grams), allowing the benzene to evaporate, and washing with alcohol ; it crystallises from benzene or ether, in which it readily dissolves, in flat, dark brown plates melting at 149—150°. It is readily decomposed by dilute acids or by acetone.

0·1724 gave 0·1921 AgCl = 27·55 Cl.

$2C_6Cl_3(NO_2)_3,C_{10}H_7·NH_2$ requires Cl = 27·42 per cent.

No definite additive compounds with β-naphthylamine, dimethyl-α-naphthylamine, or diethyl-β-naphthylamine could be prepared.

3. *Bromine Substituents.*

2 : 4 : 6-Tribromo-1 : 3-dinitrobenzene (Jackson and Moore, *Proc. Amer. Acad.*, 1889, **24**, 274, and Jackson and Kock, *ibid.*, 1898, **34**, 126) and 2 : 4 : 6-tribromo-1 : 3 : 5-trinitrobenzene (Jackson and Bentley, *ibid.*, 1891, **2**, 71) ·produce pale yellow colorations when their solutions in chloroform or ether are mixed with α-naphthylamine. As the solvent evaporates, crystals of the unaltered nitro-compounds are deposited, and [no additive products can be isolated. Similar results are obtained when diethyl-β-naphthylamine is used.

4. *Methoxy-substituents.*

(a) Methyl Picrate.—The additive compound, methyl picrate-α-naphthylamine, has been described already (Trans., 1901, **79**, 532).

The additive compound, *methyl picrate-ethyl-α-naphthylamine*, may be prepared by dissolving a mixture of equal weights of the components in benzene and adding light petroleum (b. p. 60—90°) ; it forms dark red needles, dissolves readily in benzene, crystallises from alcohol in slender, dull red needles ' melting at 86°, and is not readily decomposed by cold dilute hydrochloric acid.

0·1690 gave 21·2 c.c. of moist nitrogen at 12° and 757 mm. N = 14·92.
0·0934 „ 11·5 c.c. of moist nitrogen at 13° and 753 mm. N = 14·6.

$C_6H_2(NO_2)_3·OMe,C_{10}H_7·NHEt$ requires N = 13·5 and
$2C_6H_2(NO_2)_3·OMe,C_{10}H_7·NHEt$ requires N = 14·9 per cent.

No definite additive compounds could be obtained with either dimethyl-α-naphthylamine or diethyl-β-naphthylamine.

(b) *Trinitroresorcinol Dimethyl Ether* (*Ber.*, 1878, **11**, 1042).—A good yield of this compound can be obtained by adding a mixture of resorcinol dimethyl ether and its own volume of concentrated nitric acid to a large volume of concentrated sulphuric acid, taking care to keep the mixture cold. After some hours, the mixture is poured into water and the solid product crystallised from alcohol. When benzene or alcoholic solutions of the trinitroresorcinol dimethyl ether are mixed with α-naphthylamine, β-naphthylamine, dimethylaniline, or diethyl-β-naphthylamine, pale red colorations are noticed, but no additive compounds can be isolated, as the unaltered nitro-compound crystallises as the solvent evaporates. No additive compound can be isolated when dimethylaniline is used without a solvent.

5. *Condensation Products of the Type of Picrylaniline*,

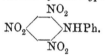

$$NO_2$$
$$NO_2\diagup\diagdown NHPh.$$
$$NO_2$$

(*a*) Picrylaniline (Mertens, *Ber.*, 1878, **11**, 845 ; Bamberger and Müller, *ibid.*, 1900, **33**, 108) crystallises from acetic anhydride in brilliant, scarlet prisms and appears to yield no acetyl derivative. It is not decomposed when heated for several hours at 100° with concentrated hydrochloric acid. A *potassium* salt is obtained when a methyl-alcoholic solution of potassium hydroxide is added to a benzene solution of picrylaniline. It forms black, crystalline needles with an intense metallic lustre, is slowly decomposed when exposed to the air, and also by dilute acids, yielding picrylaniline.

0.1792 gave 0.0438 K_2SO_4.   K = 10.98.

$C_{12}H_7O_6N_4K$ requires K = 11.44 per cent.

An *additive compound* of picrylaniline and α-naphthylamine has been prepared by crystallising a mixture of the nitro-compound (2 grams) and base (7 grams) from benzene ; it forms brownish-red crystals, melts at about 87°, and is decomposed when kept in contact with alcohol or ether ; 0.2106 gave 0.1352 of picrylaniline when decomposed with dilute acid = 64.2 ; $C_6H_2(NO_2)_3\cdot NHPh,C_{10}H_7\cdot NH_2$ requires 68.0 per cent.

(*b*) Picrylmethylaniline and picrylethylaniline have been prepared by Wedekind's method.

The former compound melts at 128—129°, and not at 108—110°, as stated by Turpin (Trans., 1891, **59**, 717) and Wedekind (*Ber.*, 1900, **33**, 434). The two compounds do not appear to yield potassium derivatives ; when their benzene solutions are mixed with methyl-alcoholic potash, very little increase in the depth of the colour is observed, and when the solutions are evaporated the unaltered picryl derivatives are obtained.

(c) *Picryl-a-naphthylamine* (Turpin, *loc. cit.* ; Bamberger and Müller, *loc. cit.*, p. 106) melts at 197—198°, does not yield an acetyl derivative, and is not decomposed when heated with concentrated hydrochloric acid. The *potassium* derivative was prepared in the same manner as the potassium salt of picrylaniline, which it closely resembles.

0·6810 gave 0·1478 K_2SO_4.   K = 9·75.

$C_{16}H_9O_6N_4K$ requires K = 9·98 per cent.

An impure additive compound may be obtained by dissolving the condensation product in aniline, allowing to cool, and pressing the red mass on a porous plate. It melts at about 122—123° and is readily decomposed by dilute hydrochloric acid.

0·464 gave 0·3506 picryl-a-naphthylamine = 75·6.

$C_6H_2(NO_2)_3 \cdot NH \cdot C_{10}H_7, C_6H_5 \cdot NH_2$ requires 79·2 per cent.

An attempt was made to replace the a-naphthylamino-group by the anilino-group by boiling the condensation product with aniline for 1·5 hours, but complete decomposition occurred and carbon separated.

(d) *Picryl-β-naphthylamine* (Bamberger and Müller, *loc. cit.*, p. 107) forms minute, brick-red crystals melting at 231·5°; it yields a *potassium* derivative, which closely resembles the potassium compounds already described, but is somewhat more unstable and readily turns yellow on exposure to the air.

0·4651 gave 0·0994 K_2SO_4.   K = 9·6.

$C_{16}H_9O_6N_4K$ requires 9·98 per cent.

When this potassium salt is decomposed with hydrochloric acid, Bamberger's yellow condensation product is obtained. An unstable additive compound of the red crystals and aniline has been prepared. It melts at about 174° and is readily decomposed into its components.

(e) *Picrylmethyl-a-naphthylamine*, obtained when picryl chloride is boiled for an hour with an alcoholic solution of methyl-a-naphthylamine, crystallises from glacial acetic acid in glistening, black plates melting at 245°; it is only sparingly soluble in hot alcohol, but dissolves readily in hot benzene or acetic acid. It does not appear to form a potassium salt, and cannot be decomposed by hot concentrated hydrochloric acid.

0·1976 gave 26 c.c. moist nitrogen at 13° and 746 mm.   N = 15·3.

$C_{17}H_{12}O_6N_4$ requires 15·2 per cent.

(f) *Di-a-naphthylaminodinitrobenzene*, $C_6H_2(NO_2)_2(NH \cdot C_{10}H_7)_2$, is obtained when 4 : 6-dichloro-1 : 3-dinitrobenzene is boiled in alcoholic solution with a-naphthylamine (4 mols.) and the solution filtered hot. The condensation product is left as a sulphur-yellow, crystalline powder. When crystallised from benzene, a mixture of red and yellow compounds is obtained, but from the mother liquor pure red prisms are deposited. When heated, these change at 180° into the yellow crystals, and then melt at 202—203°. The pure yellow compound may be prepared by heating the red crystals at 180°, by boiling the mixture of red and yellow crystals with a large amount of alcohol, or by crystallising the red modification from a mixture of chloroform and alcohol. No change in weight could be detected when the red compound was converted into the yellow, and exposure of either to sunlight during several months produced no change in colour.

Analyses: (a) *red.*—0·1982 gave 21·7 c.c. moist nitrogen at 15° and 750 mm. N = 12·65.

$C_{26}H_{18}O_4N_4$ requires 12·44 per cent.

(b) *Yellow.*—0·1664 gave 17·5 c.c. moist nitrogen at 13° and 773 mm. N = 12·7 per cent.

The condensation product from dichlorodinitrobenzene and β-naphthylamine also appears to exist in two isomeric forms: an orange-red modification, which becomes transformed into a yellow modification at 180°, and then melts to a red liquid at 183°. The red compound may be obtained by crystallising from glacial acetic acid or from a mixture of benzene and alcohol.

In conclusion, we desire to thank the Research Fund Committee of the Society for grants which have helped to defray the expenses incurred in this investigation, and Professor Loring Jackson for a specimen of s-trichlorotrinitrobenzene.

UNIVERSITY COLLEGE OF WALES,
ABERYSTWYTH.

LXIV.—*The Estimation of Carbon in Soils and Kindred Substances.*

By ALFRED DANIEL HALL, NORMAN HARRY JOHN MILLER, and NUMA MARMU.

THE method of estimating the organic carbon present in soil by oxidising to carbon dioxide with a mixture of chromic and sulphuric acids was originally suggested by Wolff, and was subjected to a critical examination by Warington and Peake (Trans., 1880, **37**, 617), who found that it invariably gave results 10 to 20 per cent. lower than were obtained by direct combustion in oxygen. Cameron and Breazeale (*J. Amer. Chem. Soc.*, 1903, **26**, 29), who recently examined the process, obtained similar results. The deficiency appears to be due, not to any failure of the chromic acid to attack the organic matter in the soil, but to the incompleteness of the resulting oxidation, a certain amount of aldehyde, acetic acid, and other substances less oxidised than carbon dioxide being produced. The authors have re-examined the method and find that by the addition of a short tube containing red-hot copper oxide to complete the combustion the whole of the carbon in the soil can be obtained as carbon dioxide.

The authors also prefer to absorb the evolved carbon dioxide by means of dilute caustic alkali in a Reiset tower, and determine its amount by the method of double titration with phenolphthalein and methyl-orange which was suggested by Hart and worked out more

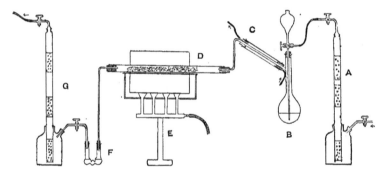

recently by Brown and Escombe (*Phil. Trans.*, 1900, **193**, series *B*, 289). This is by far the most sensitive method for estimating small quantities of carbon dioxide, and is rapid and easy of manipulation when the necessary solutions have been prepared and standardised. A further advantage in the suggested method is that an estimation can first be made of the carbon dioxide combined as carbonate in the soil (see Amos, *Jour. Agric. Sci.*, 1905, **1**, 322), and then, on the same sample and in the same apparatus, after the addition of the copper oxide tube and fresh alkali in the Reiset absorbing tower, the organic carbon can be estimated. The apparatus employed is shown in the figure ; *A* is a Reiset tower containing alkali to remove all traces of carbon dioxide from the incoming gas, *B* is a flask containing the soil, fitted with a side delivery tube and a stoppered funnel, the tap of which makes connection with the Reiset tube, *A*, or with the liquid in the funnel. A small condenser, *C*, is inserted to avoid risks of cracking the combustion tube with condensed water, *D* is a combustion tube

containing about 6—8″ of granular copper oxide and heated by the
large Bunsen burner E; F is a U-tube containing a little silver
sulphate to arrest any chlorine or volatile chlorides, after which the
gas is passed through the Reiset tower, G, to the water-pump. The
procedure is as follows : .

Five to twenty grams of finely-powdered soil are placed in B and
100 c.c. of 4 per cent. sodium hydroxide solution in G ; after allowing
the pump to operate for a little time, G is introduced and sufficient
sulphuric acid diluted with an equal volume of water is poured on to
the soil. As the air current carries the evolved carbon dioxide
forward, the acid is boiled gently to complete the decomposition of the
carbonates and the expulsion of the dissolved carbon dioxide, until in
twenty to thirty minutes the Reiset tower can be detached and the
alkali titrated (see Brown and Escombe, *loc. cit.*). Another 100
c.c. of alkali are then run into the tower, the copper oxide tube is
introduced into the circuit and heated to redness, and 20 to 30 c.c. of
strong sulphuric acid are allowed to flow on to the soil. A drop or two
of water are then passed in to clean the tap of the funnel, followed by
10 c.c. of saturated chromic acid solution. Oxidation now begins and
the process is completed as before. Carbon compounds in the soil
appear to oxidise completely in a short time, as do the carbohydrates
and other pure carbon compounds examined. A few substances,
however—ring compounds containing nitrogen—are very resistant,
betaine, for example, being only oxidised to the extent of 86 per cent.
after one hour's heating. The substitution of phosphoric for sulphuric
acid is of no particular advantage, neither hastening the reaction nor
rendering it more complete in the few cases where the substance
resists attack. The following table shows a comparison between the
results obtained by this modification of the chromic acid wet
combustion and the ordinary process of combustion in a stream of
oxygen.

Carbon, per cent.

Substance.	Found.	Calculated.		Soils.	Chromic acid method.	Combustion in oxygen.
Cellulose	44·35	44·44				
Mannitol	39·46	39·56		Soils.		
Benzoic acid ...	69·03	68·85		Geescroft waste A. 0—9″	1·21	1·13
Quinol	65·87	65·45		,, ,, B. 0—9″	1·58	1·51
Diphenylamine.	85·75	85·21		,, ,, A. 10—18″	0·60	0·64
				,, ,, B. 10—18″	0·46	0·45

Soils.	Chromic acid method.	Combustion in oxygen.		Pasture	0—9″	3·84	3·60
Broadbalk waste 0—9″	1·67	1·57		,,	0—9″	3·75	3·50
,, ,, 10—18″	0·71	0·67		,,	10—18″	2·07	2·09
,, ,, 19—27″	0·50	0·51					

THE ROTHAMSTED EXPERIMENTAL STATION,
LAWES AGRICULTURAL TRUST.

LXV.—*The Electrolysis of Salts of ββ-Dimethylglutaric Acid.*

By JAMES WALKER and JOHN KERFOOT WOOD.

WHEN a solution of sodium ortho-ethyl camphorate is submitted to electrolysis, one of the products is the ester of an unsaturated acid (Walker, Trans., 1893, **63**, 495). This acid, now known as *iso*lauronolic acid, was recognised later as being an *αβ*-unsaturated acid, a view confirmed by its subsequent synthesis (Perkin and Thorpe, Trans., 1904, **85**, 128).

While the complete structure of camphoric acid was in doubt, it was natural to suppose from this electrolytic result that the two carboxyl groups were attached to adjacent carbon atoms. The establishment of Bredt's formula, however, by synthesis shows that camphoric acid is a derivative of glutaric acid, and the genesis of ethyl *iso*-lauronolate from the ester-salt of camphoric acid becomes more difficult of explanation. It has to be assumed that, during the electrolysis, a methyl group breaks off from one carbon atom and becomes attached to an adjacent one. It appeared therefore of interest to ascertain whether the same migration occurred during the electrolysis of acids having a constitution similar to that of camphoric acid. The simplest of such acids is *ββ*-dimethylglutaric acid, a substance which can now be readily obtained from dimethyldihydroresorcinol. By conversion of the dimethylglutaric acid into its anhydride and subsequent addition of sodium ethoxide, a solution of sodium ethyl *ββ*-dimethylglutarate was readily prepared. From this solution there was obtained by electrolysis an oil from which the diethyl ester of *βββ'β'*-tetramethylsuberic acid was isolated. No satisfactory separation of the more volatile esters simultaneously produced could be effected.

By the electrolysis of a solution of potassium glutarate, L. Vanzetti (*Atti R. Accad. Lincei*, 1904, **13**, [ii], 112) obtained a hydrocarbon which was not trimethylene, as might have been expected, but propylene. The following diagram shows that this hydrocarbon is produced by the migration of a hydrogen atom.

$$\begin{array}{ccc}
\text{CH}_2\text{--CO}_2\text{K} & \text{CH}_2\text{--} & \text{CH}_3 \\
| & | & | \\
\text{CH}_2 & \rightarrow \text{CH}_2 & \rightarrow \text{CH} \\
| & | & || \\
\text{CH}_2\text{--CO}_2\text{K} & \text{CH}_2\text{--} & \text{CH}_2
\end{array}$$

When *ββ*-dimethylglutarate is substituted for glutarate in the electrolysis, such a transference of hydrogen is impossible, so that if in this case an open-chain hydrocarbon is formed at all, it can only be by the migration of a hydrocarbon group as a whole. Experiment showed that the hydrocarbon produced is unsymmetrical methylethyl-ethylene.

$$
\begin{array}{ccc}
\underset{\underset{\text{CH}_2\text{-}\vdots\text{-CO}_2\text{Na}}{|}}{\overset{\text{CH}_2\text{-}\vdots\text{-CO}_2\text{Na}}{|}} & \underset{\text{CH}_2\text{-}}{\overset{\text{CH}_2\text{-}}{|}} & \underset{\text{CH}_2}{\overset{\text{CH}_2\cdot\text{CH}_3}{|}} \\
\text{CH}_3\cdot\text{C}\cdot\text{CH}_3 \longrightarrow & \text{CH}_3\cdot\text{C}\cdot\text{CH}_3 \longrightarrow & \text{C}\cdot\text{CH}_3 .
\end{array}
$$

In the original acid, the maximum chain in the hydrocarbon radicle is of three carbon atoms; in the electrolytic hydrocarbon, there is a chain of four carbon atoms. There is thus an undoubted fundamental rearrangement of the carbon nucleus, most easily conceived as a trans-ference of a methyl group, and so a parallel is offered to the produc-tion of an $\alpha\beta$-unsaturated acid by the electrolysis of camphoric acid.

EXPERIMENTAL.

Preparation of $\beta\beta$-Dimethylglutaric Acid.—This acid, first prepared by Perkin and Goodwin (Trans., 1896, **69**, 1473), was shown by Komppa (*Ber.*, 1899, **32**, 1423) to be produced by the action of sodium hypo-bromite on dimethyldihydroresorcinol. We have found that the sodium hypobromite can be replaced with very satisfactory results by sodium hypochlorite, a concentrated solution of which can readily be prepared from bleaching powder and sodium carbonate. A pre-liminary experiment showed that six atoms of available chlorine, or three molecules of sodium hypochlorite, are required to convert one molecule of dimethyldihydroresorcinol into $\beta\beta$-dimethylglutaric acid. The action thus takes place with formation of chloroform, thus :

$$
(\text{CH}_3)_2\text{C}\underset{\text{CH}_2\cdot\text{CO}}{\overset{\text{CH}_2\cdot\text{CO}}{<}}\!\!>\text{CH}_2 + 3\text{NaOCl} = (\text{CH}_3)_2\text{C}\underset{\text{CH}_2\cdot\text{CO}_2\text{Na}}{\overset{\text{CH}_2\cdot\text{CO}_2\text{Na}}{<}} + \\
\text{NaOH} + \text{CHCl}_3.
$$

No appreciable quantity of carbon tetrachloride is produced, and no chloro-derivative of the acid or of the original substance.

One hundred grams of dimethyldihydroresorcinol were dissolved in a solution containing 65 grams of caustic potash in 750 c.c. of water. The resulting solution was slowly added with constant stirring to a slight excess of a concentrated solution of sodium hypochlorite. During the addition, the temperature rose gradually to 40°. On com-pletion of the action, the chloroform was separated and the aqueous liquid acidified with hydrochloric acid, a small quantity of sodium sulphite being simultaneously added to decompose the slight excess of hypochlorite. After evaporation to about one-third of its original bulk, the acid solution was extracted several times with other. The ethereal extract, after drying, left on distillation an oily residue which speedily solidified, the melting point of the solid being 98—100°. From 100 grams of dimethyldihydroresorcinol, 110 grams of the

$\beta\beta$-dimethylglutaric acid were obtained, the theoretical yield being 116 grams.

The acid was converted into its anhydride by means of acetic anhydride after the method of Perkin and Goodwin (*loc. cit.*, p. 1475). From 200 grams of acid, 145 grams of anhydride were obtained melting at 127°.

Preparation of Sodium Ethyl $\beta\beta$-Dimethylglutarate.—Twenty-three grams of sodium were dissolved in 550 c.c. of ethyl alcohol, and to the solution the powdered anhydride (140 grams) was gradually added, the mixture being cooled by means of ice. The liquid became yellow and a small quantity of a crystalline solid was deposited, probably the disodium salt. After remaining several hours at the ordinary temperature, the solution was evaporated on the water-bath until all the alcohol had been expelled. The solid residue was then dissolved in its own weight of water and the solution extracted with ether. The ether on evaporation only yielded a few drops of oil, probably the diethyl ester. The action of sodium ethoxide on the anhydride thus proceeds almost wholly according to the equation:

$$CMe_2{<}^{CH_2 \cdot CO}_{CH_2 \cdot CO}{>}O + NaOEt = CMe_2{<}^{CH_2 \cdot CO_2Na}_{CH_2 \cdot CO_2Et} \; .$$

Electrolysis of Sodium Ethyl $\beta\beta$-Dimethylglutarate.—The aqueous solution containing 45 per cent. by weight of the ester-salt was submitted to electrolysis in portions of about 40 grams. The anode employed was a stout platinum wire, and the current was regulated to 3 amperes. With this current, the temperature during the electrolysis did not exceed 35°. The electrolytic product consisted of a viscid oil which floated on the aqueous solution. From 200 grams of ester-salt, 89 grams of the oil were obtained, of which 7 were extracted from aqueous solution by means of ether.

The electrolytic oil was submitted to distillation with steam in order to separate the less volatile synthetical ester from the more volatile esters produced from one molecule of the original ester-salt. The amount of oil contained in the distillate was at first considerable, but at a certain stage rapidly diminished. At this point, the distillation was stopped, and the oils were then extracted from the distillate and residue respectively with ether. After drying and evaporating off the ether, it was found that the more volatile oil weighed 22 grams and the less volatile oil 59 grams.

$\beta\beta\beta'\beta'$-Tetramethylsuberic Acid,
$$CO_2H \cdot CH_2 \cdot CMe_2 \cdot CH_2 \cdot CH_2 \cdot CMe_2 \cdot CH_2 \cdot CO_2H.$$

The less volatile oil was first distilled under 100 mm. pressure in order to separate a small quantity of more volatile products with which it was contaminated. Nearly the whole of the oil passed over between 210° and 230°, and this portion, when afterwards distilled under the ordinary pressure, boiled without decomposition at 292—293°/755 mm.

0·1828 gave 0·4484 CO_2 and 0·1725 H_2O.   C = 66·89 ; H = 10·48.

$C_{16}H_{30}O_4$ requires C = 67·13 ; H = 10·48 per cent.

A determination of the molecular weight by the cryoscopic method was made : 0·3163 gram of substance dissolved in 18·47 grams of benzene gave a freezing-point depression of 0·322°, corresponding with a molecular weight of 261 ; $C_{16}H_{30}O_4$ requires a molecular weight of 286. The less volatile ester has thus the composition and molecular weight of the normal product of electro-synthesis, namely, *ethyl $\beta\beta\beta'\beta'$-tetramethyl-suberate*, formed according to the equation :

$$2CO_2(C_2H_5)\cdot CH_2\cdot CMe_2\cdot CH_2\cdot CO_2 =$$
$$CO_2(C_2H_5)\cdot CH_2\cdot CMe_2\cdot CH_2\cdot CH_2\cdot CMe_2\cdot CH_2\cdot CO_2\cdot C_2H_5 + 2CO_2.$$

It is a colourless oil of pleasant odour, and has a specific gravity of 0·965 at 16° as compared with water at the same temperature.

The ester was saponified by heating for four hours on the water-bath with a 15 per cent. solution of caustic potash in methyl alcohol. After the alcohol had been removed by distillation, the residue was dissolved in water and acidified with hydrochloric acid ; a precipitate of a solid was obtained immediately, which was recrystallised, first from aqueous alcohol and then from acetone. The colourless crystals melted at 164—165° and gave the following results on analysis :

0·0953 gave 0·2183 CO_2 and 0·0828 H_2O.   C = 62·47 ; H = 9·65.

$C_{12}H_{22}O_4$ requires C = 62·60 ; H = 9·56.

0·0778 required 16·8 c.c. 0·04N NaOH solution for neutralisation. Equivalent = 116. $C_{10}H_{20}(CO_2H)_2$ requires 115.

The acid has thus the composition and equivalent of a tetramethyl-suberic acid ; it is very sparingly soluble in water, the solution saturated at 25° containing only 0·17 gram per litre ; it is only slightly soluble in benzene and in light petroleum, dissolving more readily in ether, and freely in acetone and in alcohol.

During the conversion of the ester into the acid, a part of the *potassium* salt formed by the saponification of the ester crystallised out when the alcoholic solution was cooled. The crystals were separated and recrystallised from hot alcohol.

0·600 yielded 0·331 K_2SO_4.   K = 24·7.

$C_{10}H_{20}(CO_2K)_2$ requires K = 25·5 per cent.

A dilute solution of the ammonium salt gave no precipitate with calcium or zinc chloride ; with copper sulphate it gave a pale green, and with ferric chloride a brownish-yellow precipitate.

The *calcium* salt was prepared by boiling the acid with water and calcium carbonate until the solution became neutral. The filtered liquid when evaporated yielded crystals which gave the following results on analysis:

0·7494 air-dried salt lost 0·1246 at 110° and gave 0·3098 $CaSO_4$. $H_2O = 16·63$; $Ca = 12·16$.

$C_{12}H_{20}O_4Ca,3H_2O$ requires $H_2O = 16·77$; $Ca = 12·43$ per cent.

The *silver* salt was obtained as a white precipitate on the addition of silver nitrate to a solution of the potassium salt. When dried at 110°, it gave the following results on analysis:

0·1976 gave 0·960 Ag. Ag = 48·53.

$C_{12}H_{20}O_4Ag_2$ requires $Ag = 48·64$ per cent.

The more volatile portion of the original electrolytic oil (about 20 grams) was submitted to fractionation under the ordinary pressure. The principal portion boiled between 170° and 175° and contained $C = 65·0$ and $H = 10·0$ per cent. ; its molecular weight, as determined by the elevation of the boiling point of ether, was 143. The molecular weights of the esters $C_8H_{14}O_2$, $C_8H_{16}O_2$, and $C_8H_{16}O_3$, which are respectively the unsaturated, saturated, and hydroxymonobasic esters which by analogy might be expected to result from the electrolytic decomposition of a dimethylglutaric ester-salt, are 142, 144, and 160. The composition of the ester, however, agrees with no single one of these substances, although it is not far removed from the composition of the unsaturated ester, which requires $C = 67·6$ and $H = 9·9$. That the ester contained an unsaturated substance was proved by its at once decolorising alkaline permanganate solution in the cold, and by its power of absorbing bromine in carbon tetrachloride solution. A quantitative saponification of the ester showed it to have the equivalent weight 145, in close agreement with the molecular weight determination.

Several attempts were made to separate the esters, and also the acids produced by their hydrolysis, but we were not successful in isolating any substance in a pure condition, all our analytical results going to show that we obtained only mixtures of saturated and unsaturated substances in varying proportions (compare Brown and Walker, *Annalen*, 1893, **274**, 60—62 ; Bouveault, *Bull. Soc. chim.*, 1903, [iii], **29**, 1043).

Electrolysis of Sodium ββ-Dimethylglutarate.

A solution of the disodium salt, prepared by neutralising a concentrated solution of the acid (one part of the acid in three parts of water) by means of solid sodium carbonate, was electrolysed in a

Hofer apparatus with a current of two amperes, a platinum wire being again employed as anode. Preliminary experiments showed that the gas evolved at the anode consisted principally of carbon dioxide, but that there was also a gas present which decolorised bromine water. The anodic gas was therefore collected in a vessel over caustic potash solution to remove the carbon dioxide, and was found on examination to consist of carbon monoxide and oxygen, together with small quantities of hydrocarbons. During the process of collecting the gas, globules of oil were observed to form on the surface of the caustic potash solution. This oil, which we suspected to be a volatile hydrocarbon, was removed, dried, and distilled over sodium. It was found to boil at 31—33°, and analysis showed it to have the composition of a pentene.

0·0732 gave 0·0940 H_2O and 0·2245 CO_2. C = 83·7 ; H = 14·3.

C_5H_{10} requires C = 85·7 ; H = 14·3 per cent.

The quantity of this hydrocarbon formed was very small,. only two grams being obtained from 200 ‾grams of dimethylglutaric acid. It decolorised bromine and reduced alkaline permanganate solution in the cold, gave no precipitate with ammoniacal silver nitrate solution, and dissolved in a mixture of two volumes of sulphuric acid with one volume of water.

In order to identify the hydrocarbon, a portion was converted into the corresponding iodopentane by direct addition of gaseous hydrogen iodide to the ice-cold liquid. The product obtained, after being dried with anhydrous sodium sulphate, had a boiling point of 125—126° ; this constant, when taken into consideration with the mode of formation of the iodide from an unsaturated hydrocarbon, indicated that the substance was tert.-amyl iodide. On shaking the hydriodide with 1½ parts of water, hydrolysis ensued, and after some time the oily layer rose to the surface, in accordance with the behaviour of tert.-amyl iodide as observed by Bauer (Annalen, 1883, 220, 158).

Now tert.-amyl iodide may be formed by the union of hydriodic acid with any of the following hydrocarbons : trimethylethylene, $(CH_3)_2C{:}CH(CH_3)$; as-methylethylethylene, $(C_2H_5)(CH_3)C{:}CH_2$; 1 : 1-dimethyltrimethylene, $(CH_3)_2C{<}^{CH_2}_{CH_2}$. The last-mentioned compound would be that formed by electrolytic decomposition of $\beta\beta$-dimethylglutarate without atomic rearrangement, and is therefore the hydrocarbon which might be expected. Gustavson and Popper (J. pr. Chem., 1899, [ii], 50, 458), however, give the boiling point of this hydrocarbon as 21°, that is, more than ten degrees lower than that of the electrolytic hydrocarbon. They also give the refractive index as 1·366, that of the electrolytic hydrocarbon being 1·378, a value sufficiently divergent to show that the two substances are not identical.

It therefore remained to decide with which of the other substances our hydrocarbon could be identified, and, in order to effect a direct comparison, both of the isomerides were prepared and their properties compared with those of the electrolytic hydrocarbon under precisely similar conditions. The following table contains the results of the comparison.

	Trimethyl-ethylene.	as-Methyl-ethylethylene.	Electrolytic product.
Boiling point.........	36—38°	32—34°	31—33°
Sp. gr. 0°/0°	0·6775	0·6663	0·6668
μ_D at 16°.........	1·389	1·381	1·378

The comparison leaves no doubt that, in the main at least, the electrolytic hydrocarbon consists of as-methylethylethylene. The slight differences in the numerical values of the constants may readily be accounted for on the supposition that the electrolytic hydrocarbon contains a small quantity of 1 : 1-dimethyltrimethylene.

During the electrolysis, a little comparatively non-volatile oil was produced in the electrolytic vessel. A preliminary investigation showed this oil to be of a complex nature, and the small amount at our disposal precluded a detailed examination. It would appear then that when discharged at the anode, the negative ion of $\beta\beta$-dimethyl-glutaric acid is for the most part oxidised to carbon dioxide, other products being formed in relatively small proportions.

The expenses of the preceding investigation were defrayed by a Government Grant of the Royal Society.

UNIVERSITY COLLEGE,
DUNDEE.

LXVI.—Bromo- and Hydroxy-derivatives of $\beta\beta\beta'\beta'$-Tetramethylsuberic Acid.

By JOHN KERFOOT WOOD.

IT was shown by Perkin and Thorpe (Trans., 1899, **75**, 48) that alcoholic potash has the rather surprising effect of removing hydrogen bromide from the bromo-esters of $\beta\beta$-dimethylglutaric acid, with the formation of the caronic acids. In the light of this result, it appeared to be of interest to study the action of alcoholic potash on the bromo-derivatives of $\beta\beta\beta'\beta'$-tetramethylsuberic acid, the ester of which is the direct synthetical product obtained in the electrolysis of sodium ethyl $\beta\beta$-dimethylglutarate (see Walker and Wood, previous paper).

A trial experiment having shown that the ester of tetramethylsuberic acid was extremely difficult to brominate, experiments were made with the acid itself, and from it, by means of phosphorus pentabromide and bromine, mono- and di-bromo-acids were prepared. If the action of alcoholic potash on α-bromo-$\beta\beta\beta'\beta'$-tetramethylsuberic acid were analogous to its action on α-bromo-$\beta\beta$-dimethylglutaric acid, the product would be either a hexamethylene or a trimethylene derivative, probably the former in view of the position of the carboxyl groups :

$$
\begin{array}{llll}
\text{CHBr·CO}_2\text{H} & & & \text{CH·CO}_2\text{H} \\
\overset{\bullet}{\text{C}}\text{Me}_2 & \diagup\text{CMe}_2\diagdown & & |{>}\text{CMe}_2 \\
\overset{\bullet}{\text{C}}\text{H}_2 & \text{CH}_2 \quad \text{CH·CO}_2\text{H} & & \text{CH} \\
\overset{\bullet}{\text{C}}\text{H}_2 \quad \text{would give} & \text{CH}_2 \quad \text{CH·CO}_2\text{H} & \text{or} & \text{CH}_2 \\
\overset{\bullet}{\text{C}}\text{Me}_2 & \diagdown\text{CMe}_2\diagup & & \text{CMe}_2 \\
\overset{\bullet}{\text{C}}\text{H}_2\text{·CO}_2\text{H} & & & \text{CH}_2\text{·CO}_2\text{H}
\end{array}
$$

A similar action in the case of the $\alpha\alpha'$-dibromo-acid might possibly result in the formation of an acid containing two trimethylene rings :

$$
\begin{array}{lll}
\text{CHBr·CO}_2\text{H} & & \text{CH·CO}_2\text{H} \\
\overset{\bullet}{\text{C}}\text{Me}_2 & & |\ {>}\text{C·Me}_2 \\
\overset{\bullet}{\text{C}}\text{H}_2 & & \text{CH} \\
\overset{\bullet}{\text{C}}\text{H}_2 & \text{giving} & \text{CH} \\
\overset{\bullet}{\text{C}}\text{Me}_2 & & |\ {>}\text{C·Me}_2 \\
\overset{\bullet}{\text{C}}\text{HBr·CO}_2\text{H} & & \text{CH·CO}_2\text{H}
\end{array}
$$

It was found, however, that none of these ring compounds was formed, the only products obtained by the action of the alcoholic potash on the mono- and di-bromo-acids being the corresponding hydroxy-derivatives of $\beta\beta\beta'\beta'$-tetramethylsuberic acid. It seems probable that the different behaviour shown by bromo-$\beta\beta$-dimethylglutaric acid when acted on by alcoholic potash is directly traceable to the constitution of that substance.

EXPERIMENTAL.

Preparation of α-*Bromo-*$\beta\beta\beta'\beta'$-*tetramethylsuberic Acid.*—Thirty grams of $\beta\beta\beta'\beta'$-tetramethylsuberic acid were mixed with 120 grams of phosphorus pentabromide and the mixture heated on the water-bath until evolution of hydrogen bromide ceased; 22 grams of bromine were then added, and after heating for some time the mixture was allowed to stand overnight. The viscous product was then mixed with water ; action at once ensued, an almost solid product being eventually obtained. The aqueous solution was poured off and extracted with ether, the ethereal extract being mixed with a solution of the semi-solid product in the same solvent. After washing with a dilute solution of sodium carbonate, the solution was dried with calcium chloride and the solvent then removed by distillation. The dark oil was then dissolved in benzene and afterwards mixed with light petroleum. In this way, un-changed tetramethylsuberic acid was precipitated and was removed by filtration. The filtrate, after being concentrated, yielded a crystalline deposit, a portion of which was purified for analysis by being crystal-lised several times from benzene. The acid was eventually obtained in the form of clusters of needles (m. p. 107—110°); it was insoluble in water, but very soluble in alcohol, ether, benzene, and chloroform.

0·1340 substance gave 0·08042 AgBr. Br = 25·53. ·

$C_{10}H_{19}Br(CO_2H)_2$ requires Br = 25·89 per cent.

Action of Alcoholic Potash on a-Bromo-βββ'β'-tetramethylsuberic Acid.

The crude bromo-acid, which contained a little unbrominated tetra-methylsuberic acid, was heated for several hours with excess of a concentrated solution of caustic potash in methyl alcohol. After evaporation of the alcohol, the residue was dissolved in water and acidified with hydrochloric acid, when the liquid became turbid and an oil was precipitated. The liquid was extracted several times with ether, and from this solution, after drying with calcium chloride and distilling off the solvent, a brown, viscous oil was obtained. On placing in a desiccator, crystals separated which were filtered from the oily portion and which, on examination, were found to consist of impure tetra-methylsuberic acid. They were crystallised from aqueous alcohol and the product washed with small amounts of benzene and ether, in which solvents tetramethylsuberic acid is only sparingly soluble. The mother liquors and washings were evaporated and the residue mixed with the oil from which the original crystals of crude tetramethylsuberic acid had been separated. This oil slowly solidified on standing; the viscous solid was boiled with calcium carbonate and water until the liquid was neutral. After filtering from excess of calcium carbonate, the solution was acidified with hydrochloric acid and then extracted with ether. The acid so extracted was crystallised from boiling water. A small quantity of tetramethylsuberic acid separated out, and after this had been removed the liquid was evaporated and yielded a viscous oil which slowly crystallised in the desiccator. This product was shown by analysis to be almost pure a-hydroxy-βββ'β'-tetramethylsuberic acid ; it was purified by converting into its benzoyl derivative by the action of benzoyl chloride in alkaline solution. After the smell of benzoyl chloride had disappeared, the liquid was acidified, the precipitated benzoic acid filtered off, and the filtrate distilled with steam. From the residue in the distilling flask, an oil was extracted by means of ether ;

it soon crystallised in the desiccator and, after being spread on tile and crystallised twice from benzene, melted at 102—105°.

0·0579 gave 0·1243 CO_2 and 0·0469 H_2O.   C = 58·54 ; H = 9·00.

$C_{12}H_{22}O_5$ requires C = 58·57 ; H = 8·98 per cent.

0·0425 required for neutralisation 6·78 c.c. $N/20$ NaOH. Equivalent = 125·3. $C_{12}H_{22}O_5$ requires equivalent = 123.

A small quantity of the acid was converted into the ammonium salt, and to a solution of the latter, solutions of various metallic salts were added. White precipitates were obtained with silver nitrate and cadmium chloride, zinc sulphate gave a faint white precipitate, while no precipitate was produced either by calcium chloride or magnesium sulphate. With copper sulphate, a bluish-green precipitate was obtained, which dissolved on the addition of sodium hydroxide, forming a blue solution. This is additional proof that the substance under examination was a hydroxy-compound.

Preparation of aa·'Dibromo-$\beta\beta\beta'\beta'$-tetramethylsuberic Acid.—Ten grams of tetramethylsuberic acid were heated on the water-bath with 40 grams of phosphorus pentabromide until all action ceased. After allowing to cool, 17—18 grams of bromine were added and the mixture heated in a reflux apparatus for about twenty-four hours. The water was then removed from the jacket of the condenser and the heating continued for a short time. The product was mixed with water, the temperature being raised after the preliminary action was over. The pasty solid so produced was separated from the solution and dissolved in ether, the solution also being extracted with ether. The combined ethereal solution was washed successively with solutions of sodium sulphite and sodium carbonate, and was then dried with calcium chloride. After distilling off the ether, an oil remained, which partially solidified when placed in the desiccator. The pasty solid was treated with small quantities of benzene, in which the oily substance present appeared to be soluble, leaving a white solid residue. The solid was crystallised from boiling water and separated out in small, white crystals melting at 178—180°.

0·0995 gave 0·0966 AgBr.   Br = 41·31.

$C_{10}H_{18}Br_2(CO_2H)_2$ requires Br = 41·23 per cent.

Action of Alcoholic Potash on aa'-Dibromo-$\beta\beta\beta'\beta'$-tetramethylsuberic Acid.

The bromo-acid was heated for several hours with a solution of caustic potash in methyl alcohol. After evaporation of the alcohol, the residue was dissolved in water, acidified with hydrochloric acid, and the liquid extracted several times with ether. After drying with calcium chloride, the ether was distilled off, leaving a white solid residue, which melted with decomposition at 210°. This substance was readily soluble in alcohol, moderately so in benzene, but only slightly so in ether and water.

0·0708 gave 0·1418 CO_2 and 0·0556 H_2O.   C = 54·62 ; H = 8·72.
$C_{12}H_{22}O_6$ requires C = 54·96 ; H = 8·39 per cent.

The substance was therefore aa'-dihydroxy-$\beta\beta\beta'\beta'$-tetramethyl-suberic acid.

0·0439 acid required 6·3 c.c. $N/20$ NaOH for neutralisation. Equivalent = 139·3. $C_{12}H_{22}O_6$ requires equivalent = 131.

A solution of the ammonium salt gave white precipitates on the addition of solutions of zinc chloride, cadmium nitrate, and lead nitrate. No precipitates were obtained on the addition of silver nitrate or calcium chloride, while with copper sulphate a pale greenish-blue precipitate was obtained, which dissolved on adding sodium hydroxide, forming a dark blue solution.

The calcium salt of the dihydroxy-acid was prepared by boiling calcium carbonate with water and the acid until the liquid no longer gave an acid reaction ; it was then filtered and the filtrate concentrated until crystallisation occurred. The crystals were filtered off and air-dried.

0·0828 air-dried salt lost 0·0163 on heating at 120° and yielded 0·0318 $CaSO_4$. H_2O = 19·69 ; Ca = 11·30.
$C_{12}H_{20}O_6Ca,4H_2O$ requires H_2O = 19·35 ; Ca = 10·75 per cent.

UNIVERSITY COLLEGE,
DUNDEE.

LXVII.—An Improved Apparatus for Measuring Magnetic Rotations and Obtaining a Sodium Light.

By WILLIAM HENRY PERKIN, sen.

THE magnetic field which is necessary for experiments on magnetic rotations has so far always been produced either by means of an electro-magnet with pierced pole pieces or a long coil or helix, and both methods have their advantages and disadvantages. The electro-magnet has the valuable property of producing a very powerful field, which makes it not only possible to obtain accurate results by the use of comparatively short measuring tubes requiring only small quantities of substance, but also to measure substances which are not perfectly

colourless or absolutely free from turbidity, and therefore impossible
to measure in long tubes, but it has the drawback arising from the
necessity of maintaining the massive pole pieces at the same temperature as the substance under examination, owing to the fact that the
ends of the measuring tube containing the substance are nearly in
contact with the pole pieces, and, as these possess high conducting
power, any variation between the two prevents uniformity of temperature from being obtained. Although means have been devised by
which these difficulties have been overcome, still they are cumbersome
and consume much time (Trans., 1896, 69, 1032, 1035).

In the case of the coil, this difficulty does not exist, but on the
other hand the helical coils employed are usually very long, and
consequently the measuring tubes also must be long, and this necessitates not only the use of much substance, but also requires that the
substance shall be perfectly clear. The coil used by Rodger and
Watson (*Phil. Trans.*, 1895, 186, 623) was 50 cm. long and the
measuring tube 62 cm., or about six times as long as the tubes I
frequently use in my electro-magnet apparatus. In order to overcome
these difficulties, it was desirable to devise, if possible, an arrangement
which would combine the advantages of both the magnet and
coil without their disadvantages, that is to say, to construct an
apparatus without pole pieces which would produce a magnetic field
sufficiently powerful to allow of accurate measurements being made in
comparatively short measuring tubes.

It appeared to me that these conditions could be secured by the use
of a very powerful short coil encased in iron or steel. After discussing
the matter with Prof. Ayrton and giving him the particulars of
my requirements, he very kindly had the necessary details worked out
for its construction. The apparatus was made by Messrs. C. Crompton
& Co. Its construction will be understoood from Figs. 1, 2, 3, 4,
and 5.

Fig. 2 represents the magnet with part of its steel casing removed
so as to show the coil, and Fig. 1 is the cross section.

The coils A A' are wound on a strong gun-metal tube, B, 7·6 cm. in
diameter and 15·6 cm. long, the interior of which constitutes, of
course, the magnetic field. On the ends of this tube are screwed
wrought iron flanges (C C, Fig. 1) 0·5 cm. thick, covered on the
inside with insulating material about 1·5 mm. thick. These flanges
are bolted together in several places by means of iron stay rods (seen
at D and D D D, Figs. 1 and 2) 1·0 cm. thick. The copper wire used
for the coil is square, 3·05 × 2·7 mm., and is doubly covered. There
are 2000 turns of this wire wound in two separate sections A A' of
equal resistance. This arrangement was employed because at present
about twenty-seven or twenty-eight Grove cells have to be used, there

not being any convenient electrical supply at hand, and it was therefore necessary that the two sections should be coupled parallel, but if a current of 100 volts was available they would be connected so as to form a single coil.   The casing of the coil E is made of steel of high permeability, 1·0 cm. in thickness, and is in two halves: these are held together with bolts $F F$, the joint being machine planed ; an opening is left in the centre, of the same diameter as that of the gun-metal tube.   Each coil is provided with two binding screws fixed on an ebonite block and seen at G.   This electro-magnet weighs about 155 kilograms ; it is supported between the polariser and analyser of the polarimeter on a strong pitch-pine stand or stool,· with the legs so arranged that the narrow table carrying the optical parts can go

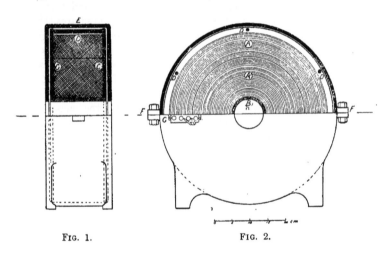

FIG. 1.   FIG. 2.

underneath it.   This polarimeter and its optical parts have been described in detail in a previous communication (Trans., 1896, 69, 1027, 1031).

The new coil gives with the same battery a more powerful field than the ordinary electro-magnet previously used.   Fig. 5 gives the measurements obtained in different parts of this field when using twenty-seven Grove cells, the ammeter in circuit indicating 12·5 amperes.   From this diagram it will be seen that the magnetic field is not uniform, being strongest in the centre and especially near the walls of the gun-metal tube.   As was anticipated, the amount of radiation outside the coil is very small.   The want of uniformity is of no importance, as the measurements are always made relatively to water and in exactly the same position in the field.   In order to support the measuring tubes in the magnetic field and maintain them at any desired temperature, the arrangements shown in end view, Fig. 3, and in section, Fig. 4, are employed. *A* and *AA* consists of a

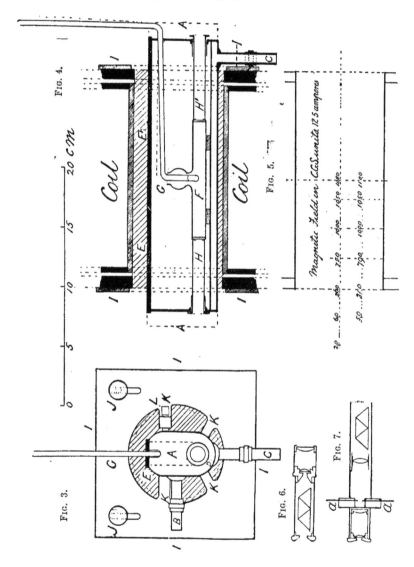

copper trough 23 cm. long, and therefore rather longer than the gun-metal tube. It is made with double walls so that water of different temperatures may circulate between them, and thus the temperature may be regulated as desired. All the joints are silver soldered, so. that the vapour of products of much higher temperatures than even that of boiling water may also be used if desired. The inlet and outlet (B and C, Fig. 3) of this jacket are both at one end of this trough, so that it can easily be drawn out of the gun-metal tube. In order to ensure that the water or hot vapour circulates thoroughly from end to end of the jacket and that the trough is equally heated in all parts, a diaphragm (D, Fig. 3) is inserted along the bottom of the jacket, but not quite throughout its entire length. The trough has a thick copper lid fitted to it by a rebate, which is made in two halves of unequal length, the shorter one, E, being fixed by means of a screw, the longer, E', being loose so that it can be easily removed, and thus allow of the introduction of the measuring-tube and thermometer, F and G. This thick copper cover conducts the heat from the sides of the trough across the top, and thus converts the cavity into an air-bath. The glass measuring-tube, F, for holding the substance is supported in this trough or air-bath by two pieces of hollowed out cork. The inlet of the measuring-tube is enlarged into a bulb to make room for expansion of its contents with rise of temperature; it also receives the bulb of the thermometer, G. The thermometer is bent twice at right angles, as shown, and is graduated to tenths of a degree. Before making the readings, the thermometer bulb is allowed to sink right down into the measuring tube, F, so as to show accurately the temperature of its contents, but when the readings are being made it is raised out of the optical field into the position indicated by the dotted lines. To ensure the glass measuring-tubes being always in exactly the same part of the magnetic field, they are held in their places between sliding brass tubes, H and H' ; one of these, H, when pushed into its place can be firmly fixed, whilst the other, H', which is made slightly longer than necessary, on being pressed in as far as it will go will force the glass tube into its exact position, therefore no matter how often removed and replaced this vessel will always occupy the same part of the magnetic field. With this apparatus, measuring-tubes of different lengths are used according as to whether there is much or little substance at command ; the different lengths used are 100, 150, and 175 mm., and for each of these a pair of brass tubes to place them always in the same part of the magnetic field have been provided. In order to prevent light passing along the walls of the glass tubes, the inner ends of the brass tubes, H H', are provided with stops with apertures about 6 mm., or rather less than the internal diameter of the glass tubes, so that the light can only pass through the centre of the tube and the substance it may contain.

When easily volatile substances are to be examined, the measuring-tube is closed with a cork and the thermometer placed outside and against it ; this arrangement answers well if sufficient time is given for the temperature to become practically constant.

To support the copper trough in the gun-metal tube, two thick square brass plates are provided (*I I I I*, Fig. 3, and *I I I I*, Fig. 4) ; these are fastened on the steel casing of the magnet by screws, as seen at *J J*, Fig. 3, passing through slots to allow of adjustment. Eight cm. of the centre of these plates are cut away with the exception of four narrow pieces (*K K K K*, Fig. 3) which serve as arms to support the trough ; on one of these arms is a set screw, *L*, so that the trough may be fixed quite rigidly if necessary. When using this trough it is pushed in as far as it will go, the outlet water-pipe (*C*, Fig. 4) acting as a stop, so that it and its contents always occupy exactly the same position in the magnetic field, thus rendering all measurements comparable.

When the apparatus is put together it will be seen that there is a free air space between the trough and the gun-metal tube, shown by the lightly shaded parts of Figs. 3 and 4, so that when the trough is heated the coil is but slightly affected, and conduction of heat is further prevented by the use of a tube of asbestos cardboard which lines the gun-metal tube.

With respect to the measurements made with this coil, it may be mentioned that if *absolute values* are required they can very easily be obtained by using the constants for water obtained by Rodgers and Watson (*Phil. Trans.*, 1895, **186**, 650), because all the measurements are relative to water, which is taken as unity. This remark equally applies to all the determinations I have made from the commencement of my study of the subject of magnetic rotations.

The polariser is kept at a distance of 18 cm. from the coil so as to allow room for the withdrawal of the trough. The distance of the analyser from the coil is 9 cm. At these distances it is probable that the small amount of radiating magnetism has no practical disturbing influence on the optical parts : this is, in fact, borne out by observations, but in any case any negative magnetic rotation which may be caused in this way (see Trans., 1884, **45**, 435) will be allowed for during the measurements of the glass ends of the measuring-tubes (see Trans., 1896, **69**, 1038). These glass ends are cut out from one piece of glass, about 1·23 mm. thick, the surfaces of which are worked parallel so that they are all of exactly the same thickness. Their rotatory value in the different positions they occupy in the magnetic field when fixed on the measuring-tubes was carefully determined. For example, when using twenty-six quart Grove cells, the following numbers were obtained.

Two glass ends on empty tube				100 mm. long	......			14′24″
,	,,	,,	,,	,,	,,.	150	,,	,, ...	9′36″
,,	,,	,,	,,	,,	,,	175	,,	,, ...	5′24″

These numbers vary, of course, with the electrical power used.

The optical parts employed in connection with this new coil are the same as previously described (Trans., 1896, **59**, 1027, 1029), and the sugar half shadow cell the same as A', Fig. 3., p. 1030 of the same volume.

The steel-clad electro-magnet, which has now been in use for several years in place of the ordinary electro-magnet previously employed, has been found to be very much more convenient and a great improvement in many respects, and it was satisfactory to find that magnetic rotations made by the old and new apparatus gave identical results.

With this apparatus the following rotations are obtained for water in tubes of the three lengths usually employed, 26 Grove cells being used and the ammeter indicating 12 amperes.

Length of tube.	Rotation.
100 mm.	5°59'
150 mm.	8°14'
175 mm.	9°3'

These readings are double readings, the sum of those obtained when the electric current is passed through the electro-magnet in two opposite directions.

Measurement of the Plane of Polarisation.

It is well known that the accurate measurement of the plane of polarised light by means of the shadow polarimeter and sodium light becomes more and more difficult as the angle of rotation increases on account of the light not being perfectly monochromatic, the two sides of the half-shadow disc forming the field of view being then seen to be differently coloured when at the correct measuring point, whereas they should be of the same tint. To overcome this difficulty, the light is usually filtered through potassium dichromate or some other similarly coloured medium, this being sometimes followed by a green screen, but any such arrangement only improves the light to a small extent and must always result in a considerable loss of intensity.

When measuring the optical activity of substances with large permanent rotations, the difficulty is often overcome by reducing the size of the rotation by dilution with an inactive solvent, or using very short measuring tubes. The measurement of magnetic rotations does not allow of such devices, because the readings are much smaller, and it is therefore necessary to employ the substances either in a pure state or in very concentrated solutions, and for the same reason very short tubes are inadmissible ; in fact, these difficulties make it impossible to measure with any degree of accuracy the magnetic rotations of substances with large permanent rotations, because, although the two properties are independent of each other, the magnetic rotation is additive to the permanent. For example, a tube 100 mm. long filled with *l*-limonene rotates the plane about 91°, and this forms the zero point in the subsequent determination of the magnetic rotation; ordinary sodium light then gives such considerable differences in the colour of the two sides of the shadow disc that accurate readings are impossible.

In attempting to get a pure light very many experiments were made with screens of different kinds, but without success, and in order to reduce the rotations quartz discs of varying thickness and opposite rotations to the substance under examination were tried, but with no really useful result. After many failures, the conclusion was eventually arrived at that satisfactory results would not be obtained until the light was purified by means of a prism, some such arrangement as that used by Sir William Abney (*Phys. Proc., Soc.*, 1885, **7**, 182) being employed, although it was feared that this might lead to some difficulty in obtaining sufficient intensity of light even with the sodium flame.

Whilst experimenting with coloured screens it was often found convenient to use them at the eye-piece of the instrument instead of near the source of light, and this observation led to experiments being made with a direct vision spectroscope in this position. After trying various unsuccessful arrangements, the effect of a direct prism without either slit or lens was investigated, with the result that a well-defined image of the half-shadow disc was obtained, and when at zero, prismatic colours were feebly seen on either side of it. On examining substances with rather large rotations with this arrangement, the image was found to maintain its practically uniform colour over a wide angle; the loss of light was very small, but the prismatic colours on either side of the disc became much brighter as the rotation increased. On introducing lithium and thallium compounds into the source of light, three discs appeared, being respectively red, yellow, and green, so that it was evident that the 6 mm. stop next to the shadow disc acted in the same way as a wide slit. Coloured solutions illuminated with white light of course cannot be used as screens with this arrangement, because, not being monochromatic, a blurred patch only is seen in the place of a sharp image.

As previously mentioned, the colours of the spectrum on either side of the disc become brighter as the rotation increases, and at about 150° they are so bright that the field looks like that of the ordinary spectrum with a yellow moon almost melting into it, and, moreover, the two sides of the disc do not show perfect uniformity of colour, but shade off to the colour of the spectrum, although the centre is nearly uniform, indicating that the 6 mm. stop forms too wide an opening for such large angles. With the ordinary half-shadow polarimeter and a sodium light it is well known that the full shadows of the disc change with large rotations from dark grey or nearly black to bluish-violet, and so foi th, and the colour becomes paler with increase of rotation ; at the same time, it is not possible to get uniformity of colour in the centre of the disc. When using the direct prism, the dark grey shadows also gradually change, becoming coloured when the rotations are very large, and when about 150° and above, the full shadows on the red side of the spectrum become red and almost merge into the spectrum, and the full shadows on the green side become green, but when the measuring position is obtained the centre of the disc is homogeneous, although the sides shade off somewhat to the red and green. In these circumstances it was found advantageous to have an adjustable slit placed just behind the half-shadow disc. Both jaws of this slit have to be adjustable, because it is necessary to keep the centre line of the half-shadow disc in the centre of the slit, which should be just sufficiently open to observe the shadows on either side ; this can be done even if the slit is only one or two millimetres wide. In the case of very large rotations, a grey band comes into the field and lies across the centre line when the correct measuring position is obtained ; this, however, does not interfere with the readings, but is rather helpful. With this slit arrangement, rotations of 360° or more can be read with moderate accuracy, although of course, more care is required and the average of a larger number of readings taken than when smaller rotations are measured.

This addition to the polarimeter commends itself on account of its simplicity, all that is necessary being to have a small direct vision prism screwed on to the eye-piece (see Fig. 6), and, if exceptionally large rotations have to be measured, to have besides this a slit with both sides adjustable placed next to the half-shadow disc. By these arrangements, I have been able to measure the magnetic rotations of the different sugars, camphor and its derivatives, and also the terpenes, which would otherwise have been impossible.

The direct vision prism can be used in other positions besides that in front of the eye-piece ; for example, it can be placed behind the telescope, as in Fig. 7, and if this is done it is easy to cut off the extraneous part of the spectrum by having shutters behind the eye-piece (as shown at a a), but I usually prefer to place it in front (as in Fig. 6), because of the better definition of the image, and one soon gets used to the other colours of the spectrum on each side of the image, although at first they are annoying. With a suitably adjusted outlet, the prism might be used next to the analyser, but experiments in this direction showed a great loss of intensity of light.

In the optical arrangements employed, the half-shadow disc is connected with the *analyser* and not the polariser, as is usually the case.

This arrangement makes it now possible to investigate magnetic rotations from the point of view of dispersion; this was impossible without the use of a prism, because the salts of lithium, thallium, and other usual sources of monochromatic light are not sufficiently pure,

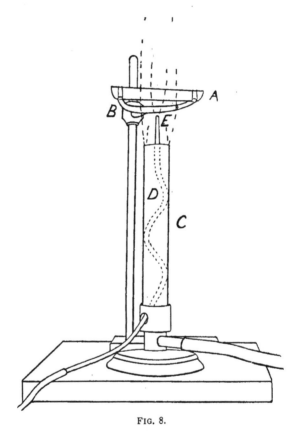

FIG. 8.

this being especially the case with lithium salts, which all contain sodium, even the best preparations.

Sodium Light.

In my previous arrangement for obtaining a monochromatic light the vapour of heated sodium was carried forward by a current of hydrogen and then burnt (Trans., 1884, **45**, 424), but I have now

s s 2

reverted to the use of sodium chloride and heat this to a much higher temperature than is generally employed. The apparatus is shown in · Fig. 8. Fused sodium chloride is placed in a platinum boat, *A*, which is supported by and securely wired to the half of a small retort ring, *B* ; the boat is heated by a large Bunsen burner, *C*, and through one of the air-holes of this and then up the inside of the burner a fine copper tube, *D*, is passed, bent in a zigzag manner so that it may be kept rigidly in its place. The top end of this tube is fitted with a platinum jet, *E* ; this is conveniently and cheaply made by rolling a piece of platinum foil round a slightly conical piece of wire ; it is then gently pressed on to the end of the copper tube and does not require any fastening. The diameter of the jet should be about one millimetre. The copper tube is supplied with oxygen, and the oxygen gas flame produced in the centre of the Bunsen flame is so placed that half of it impinges on the bottom of the platinum boat containing the sodium chloride, the other half passing up in front of the boat and receiving the vapour of the salt as it volatilises. In this way, a very intense yellow flame is obtained. The sodium chloride in the boat on either side of the high temperature flame is kept partially fused by the heat of the Bunsen burner and gradually flows to the centre and replaces that which is volatilised, so that the flame may be kept burning for a considerable time without atten- tion. The air supply of the Bunsen burner should only be moderate, so as to produce a somewhat soft flame, otherwise the light is not so good. This light, of course, like all other sodium flames, is not purely monochromatic, but when the previously described arrangement with the prism is used this is of no consequence.

LXVIII.—*The Relation between Absorption Spectra and Chemical Constitution. Part IV. The Reactivity of the Substituted Quinones.*

By ALFRED WALTER STEWART (Carnegie Research Fellow) and EDWARD CHARLES CYRIL BALY.

IN our previous papers, we have shown the effect of substitution on the absorption spectra of ketonic compounds, and in the case of some ketones we have been able to prove that the persistence of a certain absorption band in their spectra is proportional to the reactivity of their carbonyl groups. From a consideration of our results, we were enabled to put foi ward a chemical explanation which covered all the phenomena hitherto attributed to steric hindrance. The present paper contains an account of similar researches in the quinone series.

Kehrmann (*Ber.*, 1888, **21**, 3315; *J. pr. Chem.*, 1889, **39**, 399; **40**, 257) has shown that when the hydrogen atoms of *p*-benzoquinone are replaced one at a time by methyl radicles or by halogen atoms, a distinct change takes place in the reactivity of the carbonyl groups in the compounds. Although he made no accurate measurements, his results are quite sufficient for the present purpose. His conclusions, which are based on the examination of many substituted quinones, may be summarised as follows :

(1) Monosubstituted quinones, when treated with hydroxylamine, first form a monoxime, the carbonyl group in the ortho-position to the substituent being left unattacked. On further treatment, this monoxime yields a dioxime :

$$
\begin{array}{ccc}
\underset{\displaystyle\underset{O}{\|}}{C} & \underset{\displaystyle\underset{O}{\|}}{C} & \underset{\displaystyle\underset{O}{\|}}{C} \\
\overset{RC{=}CH}{HC{=}CH} \rightarrow & \overset{RC{=}CH}{HC{=}CH} \rightarrow & \overset{RC{=}CH}{HC{=}CH\cdot} \\
\end{array}
$$

(with O / NOH / NOH at carbonyls)

(2) Disubstituted quinones, when both substituents are in the ortho-position to the same carbonyl group, yield only monoximes :

$$
\begin{array}{ccc}
\overset{O}{\underset{RC{=}CR}{\overset{RC{=}CR}{HC{=}CH}}} & \text{gives only} & \overset{O}{\underset{NOH}{\overset{RC{=}CR}{HC{=}CH\cdot}}}
\end{array}
$$

(3) Disubstituted quinones, when the substituents are in the para-position to one another, give mono- and di-oximes, but only with some difficulty :

$$
\overset{O}{\underset{O}{\overset{RC{=}CH}{HC{=}CR}}} \rightarrow \overset{O}{\underset{NOH}{\overset{RC{=}CH}{HC{=}CR}}} \rightarrow \overset{NOH}{\underset{NOH}{\overset{RC{=}CH}{HC{=}CR}}} \cdot
$$

(4) Trisubstituted quinones give only monoximes, that carbonyl group being attacked which has only one ortho-substituent:

gives only

(5) Tetrasubstituted quinones give no oximes.

Up to the present time, it has been usual to attribute the phenomena observed by Kehrmann to the influence of steric hindrance. It was supposed that the chief cause of the non-reactivity of the carbonyl groups was to be found in the occupation of the space around them by the vibrations of the substituents in the ortho-position, this being supposed to be sufficient to prevent the approach of any hydroxyl-amine molecules. This rough and ready mechanical idea has been very useful, as it gave an easily comprehensible explanation of most of the phenomena of hindrance which occur in chemical reactions. It appeared to us, however, that other causes might lie at the root of the matter, and we began to examine the absorption spectra of a series of substitution products of quinone, hoping to find some more probable explanation for the phenomena which Kehrmann indicated.

Having already proved the connection between the persistence of the isorropic absorption band and the reactivity of the carbonyl groups in certain open-chain ketones, we endeavoured to find out whether the same rule holds good in the case of the quinone carbonyl groups. Our results show that it is valid, as can be seen by examining the curves of benzoquinone, toluquinone, and thymoquinone which we have already published (Baly and Stewart, this vol., pp. 507—510).

The measurements of the persistence of the absorption band in each case are as follows:

	Benzoquinone.	Toluquinone.	Thymoquinone.
Absorption band begins at	72·4	21·0	13·8
,, ,, ends at	10·0	10·5	7·2
Change of dilution over which absorption band persists	85·8	50·0	47·8 per cent.

It is evident that the effect of the substitution has been to diminish the persistence of the isorropic band to a considerable degree. But at the same time as its persistence decreases, a new band appears and increases. Benzoquinone in alcoholic solution shows no trace of a benzenoid structure, so far as spectroscopic evidence can be adduced, for it shows no sign in its spectrum of any of the absorption bands which are characteristic of benzene. But already in the spectrum of toluquinone a very shallow band makes its appearance at 3900, and in p-xyloquinone this band deepens and becomes recognisable as that of a benzenoid compound.

The spectrum of benzoquinone in alcohol differs to a great extent from its spectrum in aqueous solution. The curve of its absorption spectrum in alcoholic solution has been published by us (Stewart and Baly, this vol., p. 507), whilst that of an aqueous solution has been published by

FIG. 1.

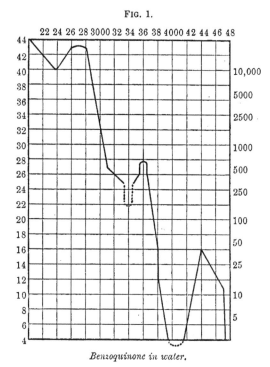

Benzoquinone in water.

Hartley, Dobbie, and Lauder (*Brit. Assoc. Report*, 1902, 99). As these authors did not mention the presence of the isorropic band in benzoquinone, we repeated the examination of an aqueous solution at higher concentrations and found the isorropic band. We give the complete curve in Fig. 1. The effect of the solvent is very marked in this instance. Apparently an additive product is formed; the isorropic band shrinks, and a benzenoid band makes its appearance at 4000. Thus there are three bands in the absorption spectrum of benzoquinone in aqueous solution.

The second point which we wish to mention is the fact that the middle band of the three, the head of which lies at 3400, occurs in the spectrum of an alcoholic solution of benzoquinone, but in that case it appears merely as an extension of the spectrum and not as a true band. The action of the solvent water extends the band considerably. We find that the same band occurs in the spectra of the substituted quinones, hydroquinone, and quinhydrone. We intend at a later date to investigate this point more fully, as it appears likely to throw light on the intramolecular vibration of the quinone system.

FIG. 2.

18 2000 22 24 26 28 3000 32 34 36 38 4000 42 44 46

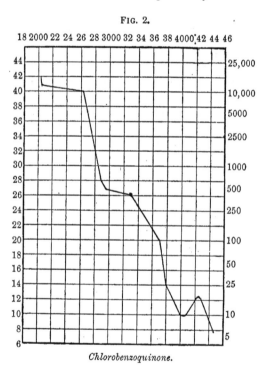

Chlorobenzoquinone.

As most of Kehrmann's investigations were carried out on halogen-substituted quinones, the spectra of several of these compounds were examined. The results obtained confirm what has already been said with regard to the effect of substitution. In chlorobenzoquinone (Fig. 2), the isorropic band becomes merely a slanting line lying between 2000 and 2600 ; in 2 : 6-dichlorobenzoquinone (Fig. 3), the line representing the isorropic band approaches more nearly to the general curve, while in trichlorobenzoquinone (Fig. 4) and trichlorotolu-quinone (Fig. 5) there is no measurable isorropic band. At the same

time as the isorropic band diminishes, the benzenoid band increases
steadily, as can be seen from the following figures :

	Chloro-benzo-quinone.	2 : 6-Di-chloro-benzo-quinone.	Trichloro-benzo-quinone.	Trichloro-tolu-quinone.
Absorption band begins at	17·4	57·5	63·0	87·1
,, ,, ends at '	10·0	15·8	14·5	10·0
Change of dilution over which the absorption band persists	42·0	55·0	77·0	88·0 per cent.

FIG. 3

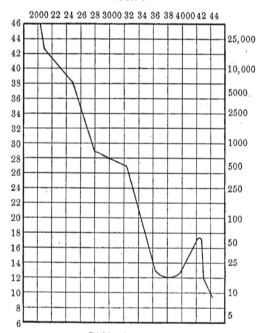

Dichlorobenzoquinone.

Three substances remain to be described. When the curve of
bromobenzoquinone (Fig. 6) is compared with the corresponding
chlorobenzoquinone, a distinct difference is noticeable between the
two. In the former, *both* the isorropic and the benzenoid bands
appear to be less marked than in the chlorine compound. This can
easily be explained. Bromine is more unsaturated than chlorine, and
it has been shown by Baly and Collie (Trans., 1905, **87,** 1332) that
the introduction of an unsaturated group into the benzene nucleus
tends to merge the benzene absorption bands into one another and

interfere with them to a great extent. The effect of introducing bromine into the benzene nucleus is well shown by the absorption spectrum of bromobenzene (Fig. 6), where the seven distinct absorption bands of the benzene spectrum are completely obliterated. The relatively weaker effect of the chlorine atom may be seen by comparing Fig. 6 with the curves of chlorobenzene given by Baly and Collie (*loc. cit.*).

With regard to the second compound, dichlorothymoquinone (Fig. 7),

Fig. 4.

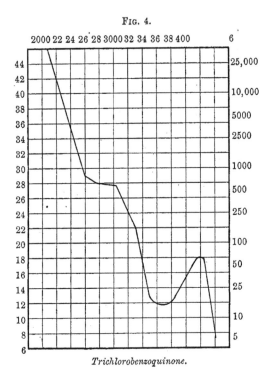

Trichlorobenzoquinone.

it is interesting to compare it with trichlorotoluquinone (Fig. 5), from which it may be derived by the substitution of a propyl group for a chlorine atom. This exchange of an alkyl group for a halogen atom produces a decrease in the persistence of the benzenoid band of the compound, as is shown by the following numbers :

	Trichloro-toluquinone.	Dichloro-thymoquinone.
Absorption band begins at	87·1	66·1
,, ,, ends at	10·0	12·0
Change of dilution over which absorption band persists ....,.....	88·0	82·0 per cent

This decrease in the persistence of the benzenoid band corresponds to the respective influences of the two substituents on the reactivity of the carbonyl group, as observed by Kehrmann, who found that the introduction of a halogen atom had a greater effect than that of an alkyl group. The same difference may be noticed, and is even more strongly marked, in the cases of toluquinone (Baly and Stewart, this vol., p. 508) and chlorobenzoquinone (Fig. 2). The isorropic band, which is clearly marked in the toluquinone spectrum, is almost extinguished when the chlorine atom is substituted for the methyl

FIG. 5.

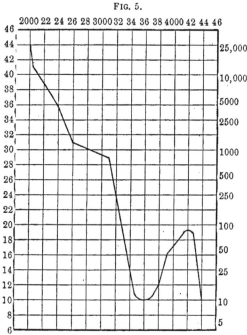

Trichlorotoluquinone.

group, whilst the same exchange extends the benzenoid band from a straight line into a well-marked band.

The third compound, dibromothymoquinone (Fig. 8), when compared with dichlorothymoquinone (Fig. 7), shows the more marked influence exerted by the bromine atoms in comparison with the chlorine substituents.

	Dichloro-thymoquinone.	Dibromo-thymoquinone.
Absorption band begins at	66·1	109·6
,, ,, ends at	12·0	15·0
Change of dilution over which absorp-tion band persists	82·0	85·0 per cent.

It is obvious that the successive substitution of the hydrogen atoms in the quinone nucleus has produced a change in the whole system of the substance. Benzoquinone itself probably exists in the true quinonoid form, but during the course of the substitution it becomes more and more benzenoid in character. What form it eventually takes cannot be determined. It is not improbable that its vibrations approximate more or less closely to those implied in the ordinary peroxide formula for benzoquinone, but no definite conclusion can be

FIG. 6.

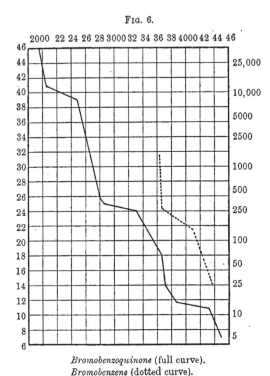

Bromobenzoquinone (full curve).
Bromobenzene (dotted curve).

drawn at present. What is evident is that the process of isorropesis is being gradually diminished, and from this we may infer that less and less of the compound is vibrating in the quinonoid form.

In considering the question of the effect which substitution exerts on the quinone carbonyl group, several factors must be taken into account.

(1) Steric hindrance produced by the vibration of the substituents.

(2) The distortion of the benzene ring consequent on the un

equal distribution of weight in the nucleus which substitution produces.

(3) The possibility of the formation of a nascent carbonyl group.

With regard to the first of these factors, we have already in our previous paper (Stewart and Baly, *loc. cit.*) shown that it need not be assumed to enter into the matter—at least to any measurable extent.

The question of the distortion of the benzene ring owing to its being unequally loaded is of more importance. From the evidence which we

FIG. 7.

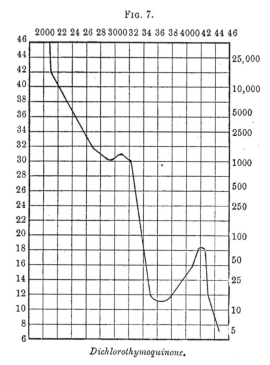

Dichlorothymoquinone.

have published, it seems a fair deduction that in the unsaturated ring system, $CO{<}^{CH=CH}_{CH=CH}{>}CO$, there are two forces at work, which are mutually antagonistic : the isorropic process and the tendency which the system will have to return to the most stable grouping, namely, the benzenoid form. The isorropic process consists of some vibration between the two carbonyl groups, and in order that this process may continue, the compound must exist in the quinonoid form. Undoubtedly, the principal vibrations of the atoms of such a compound would be parallel to the line of symmetry of the molecule, that is, along the line AB :

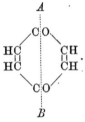

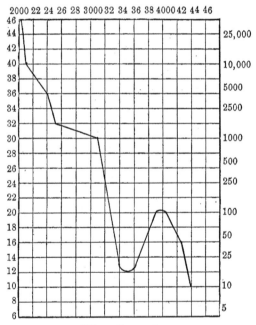

FIG. 8.

Dibromothymoquinone.

If the vibrations were not in this direction, it would hardly be possible for any isorropic process to take place. If one now supposes that the centre of gravity of the system is altered by the replacement of hydrogen atoms by methyl groups, the conditions of stability in the system are completely altered. For instance, if one introduces a single methyl radicle, one would expect that the molecule would vibrate along the new line of symmetry XY :

$$
\begin{array}{c}
\overset{\displaystyle CO}{\diagdown}\qquad\overset{\displaystyle X}{\diagup} \\
\overset{|}{CH}\quad\overset{|}{C\cdot CH_3} \\
\overset{|}{CH}\quad\overset{|}{CH} \\
\overset{Y}{\diagup}\qquad\overset{CO}{\diagdown}
\end{array}
$$

In this case, the isorropic process would not attain the same intensity as in the preceding instance, since the vibration of the molecule would not tend to produce sufficiently frequently the conditions necessary for its existence. But as soon as the isorropic process is diminished, the tendency of the compound will be to assume the most stable benzenoid structure, so that the effect of substitution will be twofold, first, in preventing the isorropesis, and, second, in thus encouraging the formation of the benzenoid type of compound in preference to the quinonoid.

The last of the three factors, namely, the possibility of a nascent carbonyl group being formed in the compound, is probably the determining factor in the problem, although it is to a great extent controlled by the vibratory motions of the ring. We have already shown in our first paper (Stewart and Baly, *loc. cit.*) that when the hydrogen atom of the group $=CH\cdot CO-$ is replaced by a methyl radicle, the tendency to form a nascent carbonyl group is checked, and the reactivity of the carbonyl group in $=C(CH_3)\cdot CO-$ is much less than in the parent substance. We have also shown that the nascent carbonyl group may be formed by a process analogous to tautomerism (potential tautomerism) in which there is no actual transfer of the hydrogen from the carbon to the oxygen, but merely a mutual action between the two atoms, oxygen and hydrogen, which action, if continued, would end in tautomeric change. If one applies the same reasoning to the case of the quinone carbonyl group, one finds that in quinone itself one has the grouping $=CH\cdot CO\cdot CH=$, where two hydrogen atoms could take part in the process of potential tautomerism. In toluquinone, one of these is replaced by methyl, $=CH\cdot CO\cdot C(CH_3)=$, so that the possibility of potential tautomerism is greatly decreased. This alone would suffice to explain the hindering effect of an ortho-substituent, but another and probably more powerful cause exists as well. We pointed out in our second paper (Baly and Stewart, this vol., p. 511) that the isorropic process in diacetyl was not an independent action, but was brought into action by the potential tautomerism in the $CH_3\cdot CO-$ groups. That is to say, if one could destroy this starting mechanism one would prevent the isorropesis and therefore the formation of a nascent carbonyl group in the substance. Now, in the case of toluquinone, half the starting mechanism of the group $=CH\cdot CO\cdot CH=$ has been destroyed by the substitution of a methyl group for one of the hydrogen atoms. This has a marked effect on the isorropic process, as the curves show. When a second substituent is introduced, as in the case of dichlorobenzoquinone, the isorropic process almost ceases, and the second carbonyl group reacts chiefly on account of the potential tautomerism which is still possible.

It appears to us that this purely chemical explanation of these phenomena is more probable than one which depends on an idea of mechanical shocks and collisions between atoms. Such collisions may certainly influence the reaction, but do so, in all probability, only to an immeasurably slight extent. The conception of steric hindrance has never satisfactorily explained several very important cases, such as the ease with which the compound (I) forms an oxime, in contradistinction to the difficulty found in the case of (II), although the methyl radicle is probably much smaller than the carboxyl group :

$$CH_3 \diagdown\diagup \text{·CO·CO}_2\text{H} \qquad CH_3 \diagdown\diagup \text{·CO·CH}_3$$

$$CH_3 \qquad\qquad\qquad\qquad CH_3$$

$$CH_3 \qquad\qquad\qquad\qquad CH_3$$

$$\text{(I.)} \qquad\qquad\qquad\qquad \text{(II.)}$$

The hypothesis which we have put forward in the course of our work on this subject seems much more satisfactory, since it is capable of explaining not only all that the steric hindrance hypothesis can explain, but also those exceptions which cannot be elucidated by any idea of steric hindrance.

To avoid the possibility of misconception, we wish to call attention to the following fact. It is evident that the process of isorropesis is decreased by substitution, but the compounds still remain yellow (although on dilution their colour disappears much more rapidly than is the case with benzoquinone), even when no isorropic band is shown in our curves. There is, however, no contradiction between our present and our previous work. In the photographs which we have taken, the limits of the isorropic band are not always clearly marked, especially in the case of the more highly substituted quinones. In order to avoid inaccuracies due to brighter lines in the spectra, we have smoothed the curves, and in the course of the smoothing the very faintly marked isorropic band has in some cases been practically obliterated. Although this does not matter very much from the point of view of the present paper, we think it advisable to call attention to the fact that our present work in no way invalidates our previous papers.

Conclusions.

The effect of substitution on quinones is as follows :

(1) It tends to diminish the possibility of potential tautomerism and thus indirectly renders isorropesis less frequent.

(2) By unevenly loading the ring, it produces in benzoquinone a greater tendency to assume the benzenoid form ; thus, in another way, diminishing the possibility of isorropesis taking place.

(3) Halogen substituents have more effect on the isorropic process than methyl groups owing to their unsaturated character, which affects the vibrations of the ring.

In conclusion, we wish again to thank Professor Collie for the great interest he has taken in the research during its progress.

THE SPECTROSCOPIC LABORATORY,
    UNIVERSITY COLLEGE, LONDON.

LXIX.—*A Mode of Formation of Aconitic Acid and Citrazinic Acid, and of their Alkyl Derivatives, with Remarks on the Constitution of Aconitic Acid.*

By HAROLD ROGERSON and JOCELYN FIELD THORPE.

WE have already shown (Trans., 1905, **87**, 1686) that ethyl β-methyl-α-cyanoglutaconate, $CN \cdot CH(CO_2Et) \cdot CMe{:}CH \cdot CO_2Et$, could be prepared in considerable quantities by the interaction of ethyl acetoacetate with the sodium derivative of ethyl cyanoacetate, and that either by employing a substituted derivative of ethyl acetoacetate, or by alkylating the product of the condensation, derivatives of ethyl cyanoglutaconate could. be obtained containing two or more substituting groups. The results of this investigation, by which we arrived at the conclusion that glutaconic acid possessed a symmetrical structure and that the same substituting group entering in either the α- or the γ-position gave rise to the same derivative, have led us to investigate another series of acids, closely allied in constitution to glutaconic acid, in order to ascertain whether other compounds of this type exhibit similar phenomena.

Aconitic acid (I) differs from glutaconic acid (II) in containing a carboxyl group attached to the carbon atom :

$$CH_2 \cdot (CO_2H) \cdot C(CO_2H){:}CH(CO_2H) \qquad CH_2(CO_2H) \cdot CH{:}CH(CO_2H)$$
$$\text{(I.)} \qquad\qquad\qquad \text{(II.)}$$

and should, therefore, possess a symmetrical structure similar to that
assigned by us to glutaconic acid :

$$CO_2H\cdot CH \quad \overset{\overset{\displaystyle CO_2H}{|}}{\underset{\underset{\displaystyle H}{}}{C}} \quad CH\cdot CO_2H$$

Aconitic acid (I).

$$CO_2H\cdot CH \quad \overset{\overset{\displaystyle CH}{|}}{\underset{\underset{\displaystyle H}{}}{}} \quad CH\cdot CO_2H$$

Glutaconic acid (II).

The correctness of this view could be readily ascertained by preparing an a- and a γ-alkyl derivative of this acid and comparing their
properties, which in the event of the compound possessing a symmetrical structure would be the same.

Aconitic acid has been known for a considerable time, and has been
prepared by a number of methods, that from citric acid by the action
of hydrochloric acid being probably the most convenient. No alkyl
derivatives of aconitic acid have, however, been prepared, and in the
following process, which is analogous to that used by us in the preparations of the corresponding derivatives of glutaconic acid, we found
a method not only for the preparation of aconitic acid, but also for any
of its alkyl derivatives, excepting those containing two substituting
groups on the a-carbon atom.

The sodium derivative of ethyl cyanoacetate readily condenses with
ethyl oxalacetate 'forming the sodium derivative of ethyl a-cyano-
aconitate (III), according to the equation :

$$\underset{\underset{\displaystyle CO_2Et}{|}}{CN\cdot CNaH} \;+\; \underset{\underset{\displaystyle CO_2Et}{|}}{C(OH)}{=}\underset{\underset{\displaystyle CO_2Et}{|}}{CH} \;\longrightarrow$$

$$\underset{\underset{\displaystyle CO_2Et}{|}}{CN\cdot CNa}{-\!-}\underset{\underset{\displaystyle CO_2Et}{|}}{C}{=\!=\!=}\underset{\underset{\displaystyle CO_2Et}{|}}{CH} \;+\; H_2O$$

$$(III.)$$

a compound which, on treatment with dilute acids, is transformed into
ethyl a-cyanoaconitate, $CN\cdot CH(CO_2Et)\cdot C(CO_2Et){:}CH\cdot CO_2Et$.

Considerable difficulty was experienced in the first instance in discovering the correct conditions for the production of this substance in
sufficient quantities for the purpose of the research. In the preliminary experiments, in which the heating was continued for a short
time only, the yield scarcely reached 15 per cent. of that theoretically
possible, whereas on prolonged heating it was found that the
desired condensation product gradually disappeared. A similar
difficulty had arisen in the condensation of ethyl sodiocyanoacetate
with ethyl acetoacetate, but in this case the formation of an insoluble
sodium derivative of the condensation product, which required only
fifteen minutes for its complete production, rendered the control of the

reaction a much simpler matter and did not necessitate any special precautions being taken to deal with the water formed. This water evidently reacts with the sodium compound of ethyl cyanoacetate, forming sodium cyanoacetate and alcohol, thus :

$$CN \cdot CHNa \cdot CO_2Et + H_2O \longrightarrow CN \cdot CH_2 \cdot CO_2Na + EtOH$$

(compare Thorpe and Young, Trans., 1900, **77**, 936 ; Thorpe, *ibid.*, 925).

A considerable quantity therefore of the ethyl sodiocyanoacetate used is removed from the sphere of action and takes no further part in the condensation. This difficulty was to a great extent overcome by the following method. Molecular quantities of ethyl sodiocyanoacetate and ethyl oxalacetate were heated in alcohol for one hour at 100°, a further half-molecule of the sodium compound in alcohol was then added and the mixture heated for three-quarters of an hour longer, finally another addition of a quarter-molecule of the sodium derivative was made, and the whole heated for half an hour. In this way, we were enabled to obtain the ethyl ester in yields representing 60—70 per cent. of the theory.

Ethyl α-cyanoaconitate is readily converted into aconitic acid on hydrolysis with hydrochloric acid, but if treated with cold concentrated sulphuric acid, is transformed into ethyl 2 : 6-dihydroxypyridine-4 : 5-dicarboxylate (ethyl dihydroxycinchomeronate) (IV), according to the equation :

(IV.)

The latter substance is converted into citrazinic acid (2 : 6-di-hydroxypyridine-4-carboxylic acid), , on hydrolysis with aqueous caustic potash, and the same acid is also formed as the sole product when ethyl α-cyanoaconitate is hydrolysed with alcoholic potash. When ethyl α-cyanoaconitate is treated with sodium ethoxide and methyl iodide, it is converted into ethyl α-cyano-α-methylaconitate, $CN \cdot CMe(CO_2Et) \cdot C(CO_2Et) \colon CH \cdot CO_2Et$, the position of the entering methyl group being proved by the fact that the compound does not react with cold concentrated sulphuric acid (Trans., 1905, **87**, 1675), forming a derivative of $2 : 6$-dihydroxypyridine, and is also insoluble in dilute sodium carbonate solution.

Ethyl α-cyano-α-methylaconitate is hydrolysed by acid hydrolysing agents forming α-methylaconitic acid,

$$CHMe(CO_2H)\cdot C(CO_2H)\!:\!CH\cdot CO_2H,$$

and by alcoholic potash, giving methylcitrazinic acid (2 : 6-dihydroxy-5-methylpyridine-4-carboxylic acid),

$$N\underset{OH}{\overset{OH\ Me}{\diamondsuit}}\cdot CO_2H.$$

Ethyl α-cyano-γ-methylaconitate (V) is formed when ethyl sodio-cyanoacetate is condensed with ethyl methyloxalacetate (prepared from ethyl propionate), the reaction evidently proceeding in accordance with the equation :

$$\underset{CO_2Et}{CN\cdot\underset{|}{C}NaH} \ + \ \underset{CO_2Et\ CO_2Et}{\overset{C(OH)\!:\!CHMe}{\underset{|}{}}} \ \longrightarrow$$

$$\underset{CO_2Et\ CO_2Et\ CO_2Et}{CN\cdot\underset{|}{C}Na\!-\!\underset{|}{C}\!=\!=\!\underset{|}{C}HMe} \ + \ H_2O,$$

(V.)

and yielding a sodium derivative from which the ethyl salt is obtained on treatment with acids.

This ethyl ester, which is soluble in a dilute solution of sodium carbonate, interacts readily with cold concentrated sulphuric acid, forming ethyl 2 : 6-dihydroxy-3-methylpyridine-4 : 5-dicarboxylate (VI) :

$$\begin{array}{ccc}
\underset{CO_2Et\cdot CH\ CMe}{\underset{CN\ CO_2Et}{\overset{C}{\overset{\diagup\diagdown}{}}}} & \underset{CO_2Et\cdot CH\ CMe}{\underset{CO\ CO}{\overset{C}{\overset{\diagup\diagdown}{}}}}\ or & \underset{OH}{\overset{CO_2Et}{\underset{N}{\diamondsuit}}}Me
\end{array}$$

(VI.)

There can therefore be no doubt that the methyl group is in the position indicated by the above formula, and that the two methyl derivatives of ethyl cyanoaconitate prepared have the formulæ

$$CN\cdot CMe(CO_2Et)\cdot C(CO_2Et)\!:\!CH\cdot CO_2Et$$

and $CN\cdot CH(CO_2Et)\cdot C(CO_2Et)\!:\!CMe\cdot CO_2Et$ respectively.

Both these ethyl esters give the same methylaconitic acid on hydrolysis with hydrochloric acid, the identity of the two acids being clearly established not only by their crystalline forms and melting points, but also by the crystalline forms and melting points of their anhydro-acids.

Ruhemann and Orton (*Ber.*, 1894, **27**, 3456) suggest that aconitic acid itself corresponds to fumaric acid in constitution, whereas the isomeric maleinoid form is represented by aceconitic acid :

$$CO_2H \cdot CH \atop CO_2H \cdot CH_2 \cdot C \cdot CO_2H$$
Aconitic acid (fumaroid).

$$CH \cdot CO_2H \atop CO_2H \cdot CH_2 \cdot C \cdot CO_2H$$
Aceconitic acid (maleinoid).

Aceconitic acid was prepared by Baeyer (*Annalen*, 1865, **135**, 306) by the action of sodium on ethyl bromoacetate, but little seems to be known concerning its constitution or properties, since not even its melting point appears to have been recorded.

Ruhemann and Orton do not give their reasons for considering that aceconitic acid represents the maleinoid form of aconitic acid, although they assign the fumaroid structure to aconitic acid owing to its formation from citrazinamide by the action of caustic potash at 150° (Ruhemann, *Ber.*, 1894, **27**, 1271), and also owing to the formation of citrazinamide (VII) by the action of ammonia on ethyl aconitate,

$$NH \cdot CO - CH \atop CO \cdot CH_2 \cdot C \cdot CO \cdot NH_2 \quad \longrightarrow \quad {CO_2H \cdot CH \atop CO_2H \cdot CH_2 \cdot C \cdot CO_2H} \cdot$$
(VII.)

From our experiments, it follows that aconitic acid, like glutaconic acid, is incapable of existing in forms corresponding to maleic and fumaric acids, and that in all probability this property is only exhibited by those derivatives which, like the aa-dialkylglutaconic acids (Trans., 1905, **87**, 1680), contain two substituting groups on the methylene carbon atom ; thus, an aa-dialkylaconitic acid of the formula $CR_2(CO_2H) \cdot C(CO_2H){:}CH(CO_2H)$, in which the double bond must necessarily be fixed, would probably exist in the two forms :

$$CO_2H \cdot CH \atop CO_2H \cdot CR_2 \cdot C \cdot CO_2H \quad \text{And} \quad {CH \cdot CO_2H \atop CO_2H \cdot CR_2 \cdot C \cdot CO_2H}$$

Unfortunately, no dialkylaconitic acids of this constitution have been prepared, and in order to decide this point we have started experiments having for their object the preparation of acids of this type by the elimination of hydrogen bromide from the corresponding dialkylbromotricarballylic acids, the results of which we hope shortly to place before the Society.

Claisen (*Ber.*, 1890, **24**, 126) showed that aconitic acid is transformed on heating at 200° into carbon dioxide and itaconic acid (VIII) :

$$CH - CO_2H \atop C \cdot CO_2H \atop CH_2 \cdot CO_2H \quad \longrightarrow \quad CO_2 + {CH_2 \atop C \cdot CO_2H \atop CH_2 \cdot CO_2H},$$
(VIII.)

and it has been shown by Anschütz and Petri (*Ber.*, 1880, **13**, 1539) that the anhydride of itaconic acid is converted into mesaconic anhydride on distillation under the ordinary pressure.

α-Methylaconitic acid and γ-methylaconitic acid both lose carbon dioxide on heating at 200°, and it is strong evidence in favour of the identity of these two acids that the same products are in each case formed. The corresponding anhydro-acids decompose, however, at a much lower temperature, and here, again, the products formed from the α- and γ-methylanhydro-acids are the same.

Assuming the identity of the two methylaconitic acids, it is apparent that there are two possible isomeric anhydro-acids represented by the formulæ (IX) and (X):

$$\begin{array}{c} \text{Me·C·CO}_2\text{H} \\ | \\ \text{H·C—CO} \\ | \qquad\quad\Large{>}\normalsize\text{O} \\ \text{CH·CO} \end{array} \qquad \begin{array}{c} \text{Me·C·CO} \\ | \qquad\quad\Large{>}\normalsize\text{O.} \\ \text{H·C·CO} \\ | \\ \text{CH·CO}_2\text{H} \end{array}$$

$$\text{(IX.)} \qquad\qquad\qquad \text{(X.)}$$

The anhydro-acid of methylaconitic acid, which is prepared by the action of acetyl chloride on the acid, has a constitution represented by (X), since on heating at 150° it is converted with loss of carbon dioxide into β-methylitaconic anhydride (XI), which, as Fittig showed (*Ber.*,1896,**29**, 1843); is converted into dimethylmaleic anhydride (XII) on distillation:

$$\begin{array}{c} \text{CH}_2\text{:C——CO} \\ | \qquad\qquad\Large{>}\normalsize\text{O} \\ \text{CHMe·CO} \end{array} \qquad \begin{array}{c} \text{Me·C·CO} \\ || \qquad\quad\Large{>}\normalsize\text{O.} \\ \text{Me·C·CO} \end{array}$$

$$\text{(XI.)} \qquad\qquad\qquad \text{(XII.)}$$

It was at first thought that in the decomposition of an anhydride of formula (X) evidence might be obtained of the presence of the mobile hydrogen atom in the molecule by the formation of both β-methylitaconic anhydride (XIII) and dimethylmaleic anhydride (XIV) at the lower temperature, since it might be supposed that when the carboxyl group was eliminated the hydrogen atom would attach itself partly to one carbon atom and partly to the other, in accordance with the equation:

$$\begin{array}{c} \text{Me·CH·CO} \\ \text{CH}_2\text{:C——CO} \end{array}\Large{>}\normalsize\text{O}$$

$$\text{(XIII.)}$$

$$\begin{array}{c} \text{Me——C·CO} \\ | \qquad\quad\Large{>}\normalsize\text{O} \\ \text{H——C·CO} \\ | \\ \text{CH·CO}_2\text{H} \end{array}$$

$$\begin{array}{c} \text{Me·C·CO} \\ || \qquad\quad\Large{>}\normalsize\text{O} \\ \text{Me·C·CO} \end{array}$$

$$\text{(XIV.)}$$

but no trace of dimethylmaleic anhydride could be detected accompanying the β-methylitaconic anhydride, although its presence could have been readily shown owing to the fact that, unlike β-methylitaconic anhydride, it crystallises from hot water as the anhydride and not as the free acid. It is probable, therefore, that in methylaconitic anhydride the double bond is already fixed, and that the formula of this substance should be represented by

$$\begin{array}{l} \text{Me·CH·CO} \\ \underset{||}{\text{C}}\text{—CO} \overset{\displaystyle >}{} O \\ \text{CH·CO}_2\text{H} \end{array},$$

a remark which probably also applies to all other similarly constituted compounds.

When ethyl α-cyano-α-methylaconitate,

$$\text{CN·CMe(CO}_2\text{Et)·C(CO}_2\text{Et):CH·CO}_2\text{Et,}$$

is treated with sodium ethoxide in alcoholic solution, it is converted into the sodium derivative, $\text{CN·CMe(CO}_2\text{Et)·C(CO}_2\text{Et):CNa(CO}_2\text{Et)}$, from which ethyl α-cyano-αγ-dimethylaconitate,

$$\text{CN·CMe(CO}_2\text{Et)·C(CO}_2\text{Et):CMe(CO}_2\text{Et),}$$

can be prepared on treatment with methyl iodide. The same ethyl α-cyano-αγ-dimethylaconitate is also formed when the sodium derivative of ethyl α-cyano-γ-methylaconitate,

$$\text{CN·CNa(CO}_2\text{Et)·C(CO}_2\text{Et):CMe·CO}_2\text{Et,}$$

prepared by the action of sodium ethoxide on ethyl α-cyano-γ-methylaconitate, $\text{CN·CH(CO}_2\text{Et)·C(CO}_2\text{Et):CMe·CO}_2\text{Et,}$ is treated with methyl iodide.

Ethyl α-cyano-αγ-dimethylaconitate is converted by acid hydrolytic agents into αγ-dimethylaconitic acid,

$$\text{CHMe(CO}_2\text{H)·C(CO}_2\text{H):CMe·CO}_2\text{H,}$$

and by methyl-alcoholic potash into dimethylcitrazinic acid (2 : 6-di-hydroxy-3 : 5-dimethylpyridine-4-carboxylic acid), $\text{N}\underset{\substack{| \\ \text{OH Me}}}{\overset{\substack{\text{OH Me} \\ |}}{\diamondsuit}}\text{CO}_2\text{H.}$

An attempt to introduce another methyl group into a molecule of ethyl α-cyano-αγ-dimethylaconitate gave rise to a reaction of the same nature as that noticed in the case of the corresponding derivative of ethyl glutaconate.

In that case (Trans., 1905, **87**, 1681) it was observed that ethyl α-cyano-αβγ-trimethylglutaconate, $\text{CN·CMe(CO}_2\text{Et)·CMe:CMe·CO}_2\text{Et,}$ on warming with sodium ethoxide quantitatively lost a carboxyl group as ethyl carbonate, becoming thereby converted into ethyl γ-cyano-αβγ-trimethylcrotonate, $\text{CN·CHMe·CMe:CMe·CO}_2\text{Et.}$

A precisely similar reaction occurs when ethyl α-cyano-αγ-dimethyl-aconitate (XV) is warmed with sodium ethoxide, a carboxyl group being eliminated as ethyl carbonate and ethyl γ-cyano-αγ-dimethyl-β-carbethoxycrotonate (XVI) being formed.   The constitution of the last-named substance is shown by the fact that on hydrolysis it yields $\alpha\gamma$-dimethylaconitic acid (XVII) and 2:6-dihydroxy-3:5-dimethyl-pyridine-4-carboxylic acid (XVIII):

$$CO_2Et \cdot CMe \overset{\overset{\displaystyle CO_2Et}{\displaystyle |}}{\underset{}{\overset{\displaystyle C}{\diagup\diagdown}}} CMe(CN)\cdot CO_2Et \quad + \ EtOH \ \longrightarrow$$

(XV.)

$$CO_2Et\cdot CMe \overset{\overset{\displaystyle CO_2Et}{\displaystyle |}}{\overset{\displaystyle C}{\diagup\diagdown}} CHMe\cdot CN \quad + \ CO{<}^{OEt}_{OEt}$$

(XVI.)

$$CO_2H\cdot CMe{:}C(CO_2H)\cdot CHMe\cdot CO_2H$$

(XVII.)

$$\underset{N}{\overset{CO_2H}{\underset{HO \qquad OH}{Me \qquad Me}}}$$

(XVIII.)

An attempt to introduce another methyl group into ethyl γ-cyano-$\alpha\gamma$-dimethyl-β-carbethoxycrotonate (XVI) led to no result, and the ethyl ester after treatment with sodium ethoxide and methyl iodide was recovered unchanged.   These experiments, therefore, taken in conjunction with those previously described in dealing with the constitution of glutaconic acid, seem to show conclusively that a compound of this type containing the complex $XCH\cdot C{:}C$ does not react with sodium ethoxide under ordinary conditions, whereas in a compound of the type of ethyl α-cyano-α-methylaconitate,

$$CN\cdot CMe(CO_2Et)\cdot C(CO_2Et){:}C\overset{*}{H}\cdot CO_2Et,$$

the hydrogen atom marked (*) is readily replaced by sodium when it is acted on by sodium ethoxide.

EXPERIMENTAL.

Formation of Ethyl α-Cyanoaconitate,
$CN\cdot CH(CO_2Et)\cdot C(CO_2Et){:}CH\cdot CO_2Et.$

This condensation was effected in the following manner: 23 grams of sodium were dissolved in 270 grams of absolute alcohol and the solution thus formed mixed with 113 grams of ethyl cyanoacetate. To the hot liquid, 188 grams of ethyl oxalacetate were then added, and the whole heated on the water-bath for one hour.   It was noticed that on adding ethyl oxalacetate the insoluble sodium derivative of ethyl cyanoacetate gradually dissolved, the solution becoming deep red.

For the reasons given in the introduction, a further solution of 11·5 grams of sodium in 135 grams of alcohol was mixed with 56·5 grams of ethyl cyanoacetate and added to the condensing mixture, the heating being then continued for forty-five minutes longer, when a final addition of 5·7 grams of sodium in 68 grams of alcohol with 28 grams of ethyl cyanoacetate was made, and the heating again continued for thirty minutes.

The condensation product was isolated by distilling off the alcohol as far as possible on the water-bath, adding water, and extracting the oil which then separated with ether. The ethereal extract was then washed with water to remove alcohol, and subsequently with a dilute aqueous solution of caustic potash. The alkaline washing liquor deposited a considerable quantity of oil on acidifying, which was extracted with ether, the ethereal solution being then dried and evaporated.

Ethyl a-cyanoaconitate obtained in this way is a viscid oil, which can be distilled under diminished pressure if the operation is conducted rapidly, and if only small quantities are employed; the boiling point under these conditions is 215°/25 mm.

0·2157 gave 0·4350 CO_2 and 0·1164 H_2O.   C = 55·00 ; H = 5·99.

$C_{13}H_{17}O_6N$ requires C = 55·1 ; H = 6·0 per cent.

This compound has been prepared by Errera and Perciabosco (*Ber.*, 1901, **34**, 3704) by the action of dilute alkalis on ethyl *iso*imino-dicarboxyaconitate, $CO_2EtC{\displaystyle <}^{CH(CO_2Et)\cdot C:NH}_{C\text{----}CO\text{----}O}$, and described as a heavy, yellow oil, not distillable without decomposition, and having acid properties.

The ethereal solution from which ethyl a-cyanoaconitate had been extracted by caustic alkali gave a quantity of oil on evaporation. On fractionation under diminished pressure, most of this was found to consist of unchanged ethyl oxalacetate, but a small quantity passed over at 191°/20 mm. as a viscid, colourless oil, which gave the following numbers on analysis :

(I) 0·2132 gave 0·4220 CO_2 and 0·1214 H_2O.   C = 53·9 ; H = 6·3.

(II) 0·2085 ,,   0·4088 CO_2 ,,   0·1198 H_2O.   C = 53·5 ; H = 6·3.

This analysis gives no clue as to the identity of the substance, and, since no solid acid could be obtained from it on hydrolysis, it was not further investigated.

Formation of Ethyl 2 : 6-*Dihydroxypyridine*-4 : 5-*dicarboxylate* (*Ethyl Dihydroxycinchomeronate*).

Ethyl α-cyanoaconitate quickly dissolves when mixed with an equal volume of concentrated sulphuric acid, and considerable heat is at the same time generated. After the mixture has been allowed to stand for twelve hours, ethyl 2 : 6-dihydroxypyridine-4 : 5-dicarboxylate (IV) separates on the addition of water as a white solid. This product is recrystallised from alcohol, from which solvent it separates in the form of small leaflets melting at 157°.

0·2109 gave 0·3990 CO_2 and 0·0930 H_2O.   C = 51·60 ; H = 4·91.
$C_{11}H_{13}O_6N$ requires C = 51·8 ; H = 5·1 per cent.

The ethyl ester is insoluble in water and in concentrated hydrochloric acid, but readily dissolves in aqueous caustic alkalis and in solutions of alkaline carbonates. It gives in alcoholic solution an intense reddish-violet coloration with ferric chloride, and a blue colour with a hot solution of sodium nitrite. It is evidently identical with the compound prepared by Errera and Perciabosco (*Ber.*, 1901, **34**, 3713) by the action of hydrochloric acid on ethyl 2 : 6-dihydroxypyridine-3 : 4 : 5-tricarb-oxylate.

Citrazinic acid (2 : 6-*dihydroxypyridine*-4-*carboxylic acid*) is produced when the above ethyl dihydroxypyridinedicarboxylate is boiled with a dilute aqueous solution of caustic soda for six hours, and is precipitated as a white, insoluble compound on acidifying the hydrolysed product with hydrochloric acid ; it can be purified by recrystallisation from concentrated hydrochloric acid, from which solvent, although only sparingly soluble, it separates in the form of microscopic plates. It chars at a high temperature without melting.

0·1978 gave 0·3358 CO_2 and 0·0551 H_2O.   C = 46·30 ; H = 3·09.
$C_6H_5O_4N$ requires C = 46·5 ; H = 3·2 per cent.

The identity of this acid with the compound prepared by Behrmann and Hofmann (*Ber.*, 1888, **21**, 2687) from citramide, and by Schneider (*loc. cit.*, p. 670) from methyl aconitate, was proved by the fact that it gives the characteristic colour with potassium nitrite, and can be converted into aconitic acid on heating in a sealed tube with hydro-chloric acid at 180° (Guthzeit and Dressel, *Annalen*, 1891, **262**, 123). Citrazinic acid is also formed when ethyl α-cyanoaconitate is boiled with a solution of one and a half times the calculated quantity of caustic potash dissolved in methyl alcohol until a test portion on dilution with water remains clear. The acid can be extracted by evaporating the solution until free from alcohol, adding water, and acidifying with hydrochloric acid, when the compound separates as a white, amorphous precipitate. Citrazinic acid appears to be the sole product of this reaction.

Formation of Aconitic Acid, $CO_2H \cdot CH_2 \cdot C(CO_2H) \vdots CH \cdot CO_2H$.

Ethyl α-cyanoaconitate slowly dissolved when boiled in a Geissler flask with concentrated hydrochloric acid and was completely hydrolysed after twelve hours. Since no solid acid separated on cooling, it was necessary to extract the product of hydrolysis with ether, and after evaporating the ether to place the residual gum in an evacuated desiccator for some time; the mass slowly became semi-solid and was purified by mixing with an equal volume of concentrated hydrochloric acid and filtering. The solid acid thus obtained was recrystallised from water, from which solvent it separated in the form of colourless leaflets which melted and decomposed at 187°.

0·1359 gave 0·2076 CO_2 and 0·0443 H_2O. C = 41·61; H = 3·60.

$C_6H_6O_6$ requires C = 41·3; H = 3·4 per cent.

The *silver* salt, $C_6H_3O_6Ag_3$, was formed as a white, flocculent precipitate when the calculated quantity of a solution of silver nitrate was added to a neutral solution of the ammonium salt of the acid.

0·2131 gave 0·1389 Ag. Ag = 65·18.

$C_6H_3O_6Ag_3$ requires Ag = 65·38.

The acid was further identified with ordinary aconitic acid by converting it into the anhydro-acid by means of acetyl chloride. This anhydro-acid crystallised from benzene in the form of cubes which melted at 95°; it was in every way the same as the compound prepared by Easterfield and Sell (Trans., 1892, **61**, 1009). There is apparently no trace of citrazinic acid formed in the hydrolysis of ethyl α-cyano-aconitate with hydrochloric acid.

Formation of Ethyl α-Cyano-α-methylaconitate, $CN \cdot CMe(CO_2Et) \cdot C(CO_2Et) \vdots CH \cdot CO_2Et$.

Twenty-eight grams of ethyl α-cyanoaconitate were mixed with 2·3 grams of sodium dissolved in 16 grams of alcohol and treated in a reflux apparatus with 16 grams of methyl iodide. On mixing the ethyl ester with sodium ethoxide, a considerable amount of heat was generated and the solution darkened; after adding methyl iodide, the reaction was completed by heating on the water-bath for two hours. The methylated ester was isolated in the usual way and was found to consist of a viscid oil which boiled constantly at 210—212°/25 mm.

0·2210 gave 0·4578 CO_2 and 0·1253 H_2O. C = 56·49; H = 6·29.

$C_{14}H_{19}O_6N$ requires C = 56·6; H = 6·4 per cent.

Formation of α-*Methylaconitic Acid,* $CO_2H \cdot CHMe \cdot C(CO_2H) \vdots CH \cdot CO_2H$.

Ethyl α-cyano-α-methylaconitate dissolved on boiling with concentrated hydrochloric acid in a Geissler flask, and usually the hydrolysis was found to be complete when all the oil had passed into solution. Since no crystals separated on cooling, the liquid was extracted with ether and the gummy acid obtained on evaporating the ether placed in an evacuated desiccator. As soon as crystallisation commenced, an equal volume of concentrated hydrochloric acid was added and the mixture allowed to stand. The large quantity of a crystalline acid which separated under these conditions was recrystallised from water and in this way obtained in the form of long needles which melted at 159° and evolved gas at about 200°.

0·2319 gave 0·3790 CO_2 and 0·0841 H_2O.   C = 44·57 ; H = 4·03.
$C_7H_8O_6$ requires C = 44·7 ; H = 4·2 per cent.

The acid instantly decolorises an alkaline solution of permanganate. The *silver* salt, $C_7H_5O_6Ag_3$, is formed as a white, insoluble precipitate when the calculated quantity of silver nitrate solution is added to a neutral solution of the ammonium salt of the acid.

0·2534 gave 0·1607 Ag.   Ag = 63·41.
$C_7H_5O_6Ag_3$ requires Ag = 63·58 per cent.

The *anhydro-acid*, $CO_2H \cdot CH \vdots C{-}CO{>}O$ with $Me \cdot CH \cdot CO$ group, is formed when the acid is heated with acetyl chloride until all has passed into solution and until the evolution of hydrogen chloride has ceased. On evaporating the chloride, the anhydro-acid is left as an oil, which solidifies on scratching and then crystallises from a mixture of chloroform and light petroleum (b. p. 60—80°) in the form of small prisms which melt at 51° and give off gas at about 150°, at the same time becoming brown.

0·2122 gave 0·3829 CO_2 and 0·0643 H_2O.   C = 49·21 ; H = 3·36.
$C_7H_6O_5$ requires C = 49·4 ; H = 3·5 per cent.

The anhydro-acid rapidly dissolves in a solution of sodium hydrogen carbonate with effervescence, and also dissolves on boiling in water, regenerating the same acid as that from which it was derived.

Formation of β-Methylitaconic and Dimethylmaleic Anhydrides.

If anhydro-β-methylaconitic acid is heated to 159°, carbon dioxide is eliminated, and if the compound is kept at this temperature until the evolution has ceased, the product on cooling again becomes solid.

The substance thus obtained is an anhydride which crystallises from carbon disulphide in the form of small leaflets which melt at 63°.

0·2093 gave 0·4375 CO_2 and 0·0868 H_2O. C = 57·01 ; H = 4·60.
$C_6H_6O_3$ requires C = 57·1 ; H = 4·8 per cent.

This anhydride is evidently β-methylitaconic anhydride, since it dissolves on boiling in water and becomes transformed into β-methylitaconic acid melting at 151°, and identical in all respects with the compound prepared by Fittig and Kettner (*Annalen*, 1899, **304**, 166).

0·2110 gave 0·3858 CO_2 and 0·1039 H_2O. C = 49·87 ; H = 5·47.
$C_6H_8O_4$ requires C = 50·0 ; H = 5·6 per cent.

If anhydro-β-methylaconitic acid is slowly heated in a distillation flask until the evolution of carbon dioxide has ceased and then distilled at the ordinary temperature, an oil passes over at 220—225°, which solidifies on cooling. When recrystallised from water, it forms lustrous leaflets which melt at 96°.

0·2078 gave 0·4333 CO_2 and 0·0867 H_2O. C = 56·87 ; H = 4·63.
$C_6H_6O_3$ requires C = 57·1 ; H = 4·8 per cent.

The compound is identical with dimethylmaleic anhydride (compare Fittig, *Ber.*, 1896, **29**, 1843), since on boiling with a 10 per cent. solution of caustic soda it is converted into dimethylfumaric acid, a compound which separates from hot water in long needles which melt at 239—240° (Fittig and Kettner, *loc. cit.*).

Formation of Methylcitrazinic Acid (2 : 6-*Dihydroxy*-5-*methylpyridine-*

4-*carboxylic Acid*),

$$N\underset{\text{OH Me}}{\overset{\text{OH}}{\big\langle \quad \big\rangle}}CO_2H.$$

Ten grams of the ethyl ester were mixed with an alcoholic solution containing one and a half times the calculated quantity of caustic potash and the whole heated on the water-bath for twelve hours. After this time, the alcohol was evaporated on the water-bath and dilute hydrochloric acid added, when a little solid separated ; this was collected and recrystallised from concentrated hydrochloric acid, from which solvent it separated in the form of small prisms which carbonised at a high temperature without melting.

0·2317 gave 0·4216 CO_2 and 0·0836 H_2O. C = 49·63 ; H = 4·09.
$C_7H_7O_4N$ requires C = 49·7 ; H = 4·1 per cent.

The acid is practically insoluble in hot water and is only sparingly soluble in hot concentrated hydrochloric acid ; it is also insoluble in all the usual organic solvents, but dissolves instantly in solutions of alkali carbonates and in caustic alkalis, being reprecipitated from these solutions on the addition of acid.

Unlike citrazinic acid, the salts of methylcitrazinic acid do not show any tendency to become coloured on exposure to the air. Thus the ammonium salt can be boiled in solution for a considerable time without undergoing any alteration; it develops a blue coloration when added to a hot solution of potassium nitrate, and gives, when dissolved in a mixture of alcohol and water, a deep reddish-violet coloration with ferric chloride.

The *diacetyl* derivative, $C_{11}H_{11}O_6N$, is prepared by boiling methyl-citrazinic acid with acetic anhydride until all has passed into solution. The acetic anhydride is then evaporated in an evacuated desiccator over caustic potash and the residue extracted with hot alcohol and filtered.

The alcoholic extract on cooling deposits the diacetyl derivative in the form of small prisms, which melt and decompose at about 165°. The compound does not, however, possess a definite melting point, and can be made to decompose considerably below this temperature if slowly heated.

0·2201 gave 0·4199 CO_2 and 0·0830 H_2O. $C = 52·03$; $H = 4·19$.

$C_{11}H_{11}O_6N$ requires $C = 52·2$; $H = 4·3$ per cent.

We were unable to obtain any characteristic derivative of methyl-citrazinic acid; in the form of its ammonium salt, it reduces silver nitrate solution, and the dibenzoyl derivative, which was prepared by the Schotten-Baumann method, could not be obtained in a sufficiently pure state for analysis.

Cold concentrated sulphuric acid has no action on ethyl α-cyano-α-methylaconitate, and after the ethyl ester had remained in contact with the concentrated acid for twelve hours the greater portion was recovered unchanged on pouring the mixture into water.

Formation of Ethyl α-*Cyano-γ-methylaconitate,*

$$CN·CH(CO_2Et)·C(CO_2Et)\overset{..}{:}CMe·CO_2Et.$$

In order to produce this substance, 23 grams of sodium were dissolved in 270 grams of alcohol and 113 grams of ethyl cyanoacetate were added. To the hot solution containing the sodium derivative in suspension, 202 grams of ethyl methyloxalacetate (prepared from ethyl propionate) were added, when the white sodium derivative rapidly dissolved and the solution became deep red. The mixture was heated for one hour on the water-bath, when an additional quantity of the sodium derivative of ethyl cyanoacetate, made by mixing 56 grams of the ethyl salt with a solution containing 11·5 grams of sodium dissolved in 135 grams of alcohol, was added and the heating continued for forty-five minutes longer. After this time, a final addition of the sodium derivative, made by dissolving 5·7 grams of sodium in 68 grams of alcohol and mixing with 28 grams of ethyl cyanoacetate, was poured into the condensing flask and the whole heated for thirty minutes more. The product of the condensation was isolated by distilling off the alcohol, as far as possible, on the water-bath, adding water, and extracting the separated oil by means of ether. The ethereal solution was then thoroughly washed with water and finally with a dilute aqueous solution of caustic potash.

The potash washings, on acidifying, deposited an oil which was extracted with ether, the ethereal solution was dried and then evaporated. The oil thus obtained was fractionated under diminished pressure, when pure ethyl α-cyano-γ-methylaconitate passed over at 210—212°/25 mm.

0·2109 gave 0·4359 CO_2 and 0·1195 H_2O. C = 56·37 ; H = 6·29·

$C_{14}H_{19}O_6N$ requires C = 56·6 ; H = 6·4 per cent.

The ethyl ester must be rapidly distilled, otherwise decomposition will set in.

Formation of Ethyl 2 : 6-Dihydroxy-3-methylpyridine-4 : 5-dicarboxylate,

OH   Me

N⟨ ⟩CO_2Et.

OH CO_2Et

Concentrated sulphuric acid readily dissolves ethyl α-cyano-γ methylaconitate, considerable heat being at the same time generated, and if the solution be allowed to stand (without cooling) for twelve hours, the above ethyl ester is obtained as a white precipitate on pouring the mixture into water. Recrystallised from alcohol, it is obtained in the form of microscopic needles which melt at 173°.

0·2371 gave 0·4628 CO_2 and 0·1175 H_2O. C = 53·23 ; H = 5·50.

$C_{12}H_{15}O_6N$ requires C = 53·5 ; H = 5·6 per cent.

The ethyl ester is insoluble in water and in concentrated hydrochloric acid, but readily dissolves in solutions of alkaline carbonates and caustic alkalis ; it gives in alcoholic solution a deep reddish-violet coloration with ferric chloride. The same compound is produced to a small extent in the original condensation, and is filtered off from the ethereal solution, in which it is insoluble.

Methylcitrazinic acid (2 : 6-dihydroxy-3-methylpyridine-4-carboxylic

OH Me

acid), $N\langle\overset{\frown}{\underset{\smile}{\quad}}\rangle CO_2H$, is formed when the above ethyl ester is boiled

OH .

with dilute caustic potash solution for six hours; it is in every way the same as the compound prepared by the hydrolysis of the isomeric ethyl ester described on p. 643.

0·2194 gave 0·3984 CO_2 and 0·0793 H_2O. C = 49·53 ; H = 4·02.

$C_7H_7O_4N$ requires C = 49·7 ; H = 4·1 per cent.

Methylcitrazinic acid is formed as the sole product when ethyl a-cyano-γ-methylaconitate is hydrolysed with alcoholic potash.

Formation of γ-Methylaconitic Acid, $CO_2H\cdot CH\cdot C(CO_2H){:}CMe\cdot CO_2H$.

The ethyl ester was boiled with concentrated hydrochloric acid in a Geissler flask until all the oil had passed into solution and the evolution of carbon dioxide had ceased, an operation which takes about six hours. The hydrolysed liquid was extracted with ether and the residual gum placed in a desiccator. As soon as crystals appeared, an equal volume of concentrated hydrochloric acid was added and the mixture allowed to stand; the crystalline acid which separated was collected and recrystallised from water, when it separated in the form of long needles melting at 159°.

0·2022 gave 0·3297 CO_2 and 0·0762 H_2O. C = 44·47 ; H = 4·18.

$C_7H_8O_6$ requires C = 44·7 ; H = 4·2 per cent.

Since the acid gave with acetyl chloride an anhydro-acid crystallising from a mixture of light petroleum in the form of small prisms melting at 51°, which on distillation gave dimethylmaleic anhydride, there can be no doubt that it was identical with methylaconitic acid already described.

The two acids, as well as their anhydrides, are not only similar in crystalline form, melting point, and solubility, but mixtures of them also melt at the same temperatures.

Formation of Ethyl a-Cyano-aγ-dimethylaconitate,
$CN\cdot CMe(CO_2Et){:}C(CO_2Et){:}CMe\cdot CO_2Et$.

Thirty-one grams of ethyl a-cyano-γ-methylaconitate were added to a solution of 2·3 grams of sodium dissolved in 16 grams of alcohol contained in a reflux apparatus and treated with 16 grams of methyl iodide, the mixture being heated on the water-bath for two hours. The methylated ester formed was isolated by adding water and extracting the oil, which was then precipitated by means of ether.

The ethereal solution after being washed with water and [dilute sodium carbonate solution was dried and the ether evaporated, when the residual oil on distillation under diminished pressure was found to consist of pure ethyl α-cyano-αγ-dimethylaconitate boiling at 205—207°/25 mm. The ethyl ester has to be distilled rapidly, since if the process is prolonged it undergoes considerable decomposition.

0·2018 gave 0·4271 CO_2 and 0·1239 H_2O.   C = 57·72 ; H = 6·82.

$C_{15}H_{21}O_6N$ requires C = 57·9 ; H = 6·7 per cent.

Formation of αγ-Dimethylaconitic Acid,
$$CHMe(CO_2H)\cdot C(CO_2H):CMe\cdot CO_2H.$$

The ethyl ester was mixed with twice its volume of concentrated hydrochloric acid and boiled in a Geissler flask until all the oil had passed into solution, when the hydrolysed liquid was poured into an evaporating basin and evaporated to a small bulk. The partially solid residue was then extracted with ether and the ethereal extract, after being dried, freed from ether by evaporation. The gummy residue, which showed no signs of crystallisation, was mixed with an equal volume of concentrated hydrochloric acid and left for some days, at the end of which time a considerable quantity of a crystalline acid had separated. This was collected by filtration and recrystallised from water, being thus obtained in the form of small needles, which melted at 164° and evolved gas at about 180°.

0·1872 gave 0·3244 CO_2 and 0·0844 H_2O.   C = 47·26 ; H = 5·01.

$C_8H_{10}O_6$ requires C = 47·5 ; H = 4·9 per cent.

The *silver* salt, $C_8H_7O_6Ag_3$, is precipitated as a white, amorphous powder on adding a solution containing the calculated quantity of silver nitrate to an aqueous solution of the acid neutralised with ammonia in the cold.

0·2519 gave 0·1555 Ag.   Ag = 61·73.

$C_8H_7O_6Ag_3$ requires Ag = 61·86 per cent.

The *anhydro-acid*, $\begin{array}{l} CMe(CO_2H):C-CO \\ \quad\quad\quad | \quad\quad >O, \\ Me\cdot CH\cdot CO \end{array}$ is best prepared by boiling the acid with acetyl chloride for fifteen minutes or until the evolution of hydrogen chloride has ceased. On evaporating the chloride on the water-bath, the anhydro-acid remains as an oil which solidifies on scratching. When recrystallised from light petroleum (50—60°), it is obtained in the form of microscopic needles which melt at 74°.

0·1976 gave 0·3772 and 0·0747 H_2O.   C = 52·06 ; H = 4·20.

$C_8H_8O_5$ requires C = 52·2 ; H = 4·3 per cent.

The anhydro-acid dissolves on boiling with water, and, on cooling, the acid from which it was derived separates out. Both $\alpha\gamma$-dimethyl-aconitic acid and its anhydro-acid instantly decolorise an alkaline solution of potassium permanganate.

Formation of Dimethylcitrazinic Acid (2 : 6-*Dihydroxy*-3 : 5-*dimethyl-pyridine*-4-*carboxylic Acid*), $\text{N}\langle\!\!\!\begin{array}{c}\text{OH Me}\\ \rule{1.5cm}{0.4pt}\\ \text{OH Me}\end{array}\!\!\!\rangle\text{CO}_2\text{H}.$

Ten grams of the ethyl ester were dissolved in a solution of $1\frac{1}{2}$ times the calculated quantity of caustic potash dissolved in methyl alcohol and the mixture heated on the water-bath for twelve hours. At the end of this time the alcohol was evaporated off and the residue, after diluting with water, acidified with dilute hydrochloric acid. The white precipitate which then separated was recrystallised from glacial acetic acid, being thus obtained in the form of small needles which carbonise at a high temperature without melting.

0·2015 gave 0·3865 CO_2 and 0·0867 H_2O.   C = 52·31 ; H = 4·78.

$\text{C}_8\text{H}_9\text{O}_4\text{N}$ requires C = 52·4 ; H = 4·9 per cent.

The acid closely resembles the corresponding monomethyl derivative in its reactions; it is insoluble in all the usual organic solvents excepting glacial acetic acid, in which, however, it is only sparingly soluble. It is insoluble in mineral acids, but dissolves instantly in dilute alkalis, from which solutions it is reprecipitated on the addition of acid. Its salts are stable in the air and do not become coloured on exposure ; it gives no coloration when added to a hot solution of potassium nitrate, but when boiled in a mixture of equal parts of alcohol and water and filtered, the filtrate, although containing only a small quantity of the acid, gives an intense red coloration with a ferric chloride solution. We were unable to obtain any characteristic derivative of this acid, since neither the acetyl nor the benzoyl derivative could be obtained pure.

Concentrated sulphuric acid is without action on ethyl α-cyano-$\alpha\gamma$-dimethylaconitate, and on mixing the ethyl salt with an equal volume of the concentrated acid and leaving the mixture for some days, the product obtained on pouring the solution into water will be found to consist for the most part of unchanged ethyl ester.

Formation of Ethyl α-Cyano-αγ-dimethylaconitate,
CN·CMe(CO₂Et)·C(CO₂Et)ːCMe·CO₂Et,
by the Methylation of Ethyl α-Cyano-α-methylaconitate,
CN·CMe(CO₂Et)·C(CO₂Et)ːCH·CO₂Et.

This methylation was carried out precisely in the same manner as in the case of the corresponding γ-methyl derivative, 31 grams of the ethyl ester being mixed with a solution of 2·3 grams of sodium dissolved in 16 grams of alcohol and subsequently treated with 16 grams of methyl iodide. The methylated ester was isolated by adding water, extracting the oil thus precipitated with ether, washing the ethereal solution with water and dilute sodium carbonate solution, drying, and evaporating until free from ether. The viscid oil thus obtained boiled at 205—207°/25 mm.

0·2010 gave 0·4268 CO₂ and 0·1243 H₂O. C = 57·91 ; H = 6·87.

C₁₅H₂₁O₆N requires C = 57·9 ; H = 6·7 per cent.

The ethyl ester is evidently the same as that formed by the methylation of ethyl α-cyano-γ-methylaconitate, since on hydrolysis with hydrochloric acid it is converted into the same αγ-dimethylaconitic acid melting at 164°, and on hydrolysis with a methyl-alcoholic solution of caustic potash is transformed into the same dimethylcitrazinic acid.

Formation of Ethyl γ-Cyano-αγ-dimethyl-β-carbethoxycrotonate,
CN·CHMe·C(CO₂Et)ːCMe·CO₂Et, *and Ethyl Carbonate.*

As explained in the introduction, it was found that on treating ethyl α-cyano-αγ-dimethylaconitate with sodium ethoxide (in an attempt to methylate it further) a carboxyl group was quantitatively split off as ethyl carbonate, ethyl γ-cyano-αγ-dimethyl-β-carbethoxycrotonate being formed at the same time. The operation was carried out in the following way : 33 grams of the ethyl ester were mixed with a solution containing 2·3 grams of sodium dissolved in absolute alcohol and the solution warmed on the water-bath for fifteen minutes. Water was then added and the oil which separated out extracted with ether, the ethereal solution being washed with water, dried, and the ether evaporated. The residual oil was fractionated under diminished pressure, care being taken to condense the more volatile portion, when an almost quantitative yield of ethyl γ-cyano-αγ-dimethyl-β-carbethoxycrotonate, boiling at 159°/25 mm., was obtained.

0·1993 gave 0·4387 CO₂ and 0·1297 H₂O. C = 60·03 ; H = 7·23.

C₁₂H₁₇O₄N requires C = 60·2 ; H = 7·1 per cent.

U U 2

The ethyl ester is a colourless, mobile oil, and is converted on hydro-lysis with hydrochloric acid into $a\gamma$-dimethylaconitic acid, an operation which is best effected by boiling in a Geissler flask with concentrated hydrochloric acid until all the oil has disappeared. The acid is extracted by evaporating the solution to a small bulk and leaving the residue to crystallise, when more hydrochloric acid is added and the acid separated by filtration. It is identical with the acid prepared by the methods already described.

The more volatile fraction obtained in the fractionation of the above ethyl ester was rectified under the ordinary pressure, when ethyl carbonate was obtained as a clear, mobile liquid boiling at 126°/753 mm.

0·2115 gave 0·3934 CO_2 and 0·1638 H_2O.   C = 50·73 ; H = 8·61.

$C_5H_{10}O_3$ requires C = 50·8 ; H = 8·5 per cent.

The ethyl ester was identified by its conversion into carbamide on heating in a sealed tube with ammonia for four hours at 180°.

In order to ascertain whether a methyl group could be introduced into ethyl γ-cyano-$a\gamma$-dimethyl-β-carbethoxycrotonate it was treated with the molecular quantity of sodium dissolved in alcohol and the well-cooled solution mixed with the calculated quantity of methyl iodide.

On working up the product in the usual way, it was found that the ethyl ester had remained unchanged, and that practically the whole of the original amount used was recovered unaltered.

Since this paper was written, a communication has appeared by Feist and Beyer (*Annalen*, 1906, **345**, 117) " On the β-Methylglutaconic Acids and $a\beta$-Dimethylglutaconic Acid," in which it is pointed out that two isomeric forms of β-methylglutaconic acid,

$$CH_2(CO_2H)\cdot CMe\dot{:}CH\cdot CO_2H,$$

have been known for a considerable time, but have been included in the literature under other and misleading names.

. These acids, under the name of the acetocrotonic acids, were origin-ally prepared by Genvresse (*Ann. Chim. phys.*, 1891, [i], **24**, 108) by the action of baryta or dilute sodium hydroxide solution on ethyl carbacetoacetate or ethyl *iso*dehydracetate. The *cis*-acid melted at 143° the *trans*-modification at 115—116°, both yielding the same anhydride melting at 86°, from which, on treatment with water, the acid melting at 143° was alone obtained.

The identity of these acids has now been placed beyond question by Feist (*Annalen*, 1906, **345**, 82), who has not only prepared them by the methods described by Genvresse, but has compared the *trans-*

modification (m. p. 115—116°) with the acid prepared by the hydrolysis of the ethyl salt obtained by the action of ethyl tetrolate on ethyl malonate :

$$
\begin{array}{ccc}
\underset{\substack{||| \\ \text{C·CO}_2\text{Et}}}{\text{CMe}} & + & \underset{\substack{| \\ \text{H}}}{\text{NaC(CO}_2\text{Et)}_2} & \longrightarrow & \underset{\substack{|| \\ \text{CO}_2\text{Et·CH}}}{\text{Me·C·CH(CO}_2\text{Et)}_2} \\
& & & & \downarrow \\
& & & & \underset{\substack{|| \\ \text{CO}_2\text{H·CH}}}{\text{Me·C·CH}_2\text{·CO}_2\text{H}}
\end{array}
$$

Although in this reaction only ·the *trans*-modification (m. p. 115—116°) is formed, yet by interaction with acetyl chloride it is readily converted into an anhydride (m. p. 86°) from which the *cis*-acid (m. p. 147°) can be prepared on treatment with water.

There can be no doubt that the acid melting at 147°, which Feist shows is identical with homomesaconic acid prepared by Hantzsch (*Annalen*, 1884, 222, 12) from ethyl *iso*dehydracetate, is identical with the β-methylglutaconic acid prepared by us (Trans., 1905, 87, 1691), although we have been unable to make our acid melt lower than 149°, the temperature given. We are therefore indebted to these chemists for pointing out to us that derivatives of glutaconic acid other than those in which both hydrogen atoms of the methylene group are substituted are capable of existing in *cis*- and *trans*-forms. The main reason, however, for our suggesting a symmetrical formula for glutaconic acid, and later for aconitic acid :

$$
\begin{array}{ccc}
\text{CO}_2\text{H·CH}_4 \underset{\diagdown\text{-H-}\diagup}{\overset{\diagup\text{CH}\diagdown}{\big|}} \text{CH·CO}_2\text{H} & & \text{CO}_2\text{H·CH} \underset{\diagdown\text{H}\diagup}{\overset{\diagup\text{C-}\diagdown}{\big|}} \text{CH·CO}_2\text{H,}
\end{array}
$$

was to explain the identity of the α- and γ-positions, and we expressly state (Trans., 1905, 87, 1680) that the intention of this formula is to express a state of equilibrium between the two α-carbon atoms, in that the mobile hydrogen atom may assume either of the positions

$$
\text{CO}_2\text{H·CH}_2 \overset{\diagup\text{CH}\diagdown}{\diagdown} \text{CH·CO}_2\text{H} \quad \text{or} \quad \text{CO}_2\text{H·CH} \overset{\diagup\text{CH}\diagdown}{\diagup} \text{CH}_2\text{·CO}_2\text{H,}
$$

according to the nature of the reaction involved. It is quite conceivable that the entrance of substituting groups might cause the compound to react as if the bond were fixed, in which case the formation of isomerides corresponding to maleic and fumaric acids would be possible. That this is the case as regards β-methylglutaconic acid is evident, and that it may be shown to, be the case with other substituted glutaconic acids is quite probable. It is, however, exceedingly unlikely that compounds such as the $\alpha\beta$- and $\beta\gamma$-dimethylglutaconic acids and the α- and γ-methylaconitic acids, which have been shown to be tautomeric, will be found to exist in stereoisomeric forms corresponding to maleic and fumaric acids.

The cost of this research has been met by a grant from the Research Fund of the Chemical Society, for which we desire to express our indebtedness.

THE VICTORIA UNIVERSITY OF MANCHESTER.

LXX.—The Interaction of Well-dried Mixtures of Hydrocarbons and Oxygen.

By WILLIAM ARTHUR BONE and GEORGE WILLIAM ANDREW.

THE experiments described in this paper were made to prove whether water vapour has an influence on the combustion of a hydrocarbon at all comparable with that which it exerts in the well-known instances of the combustion of carbon monoxide (H. B. Dixon, *Phil. Trans.*, 1884, 175, 617) and hydrogen (H. B. Baker, Trans., 1902, 81, 400). We may here recall the fact that H. B. Dixon found that a perfectly dry mixture of carbon monoxide and oxygen was rendered explosive by the presence of even a trace of hydrocarbon, and, also, that H. B. Baker, in describing his experiments on the comparative inertness of dry electrolytic gas, emphasised the importance of obtaining it free from hydrocarbon impurity.

Most of our experiments have been made with ethylene and acetylene. In selecting these hydrocarbons for investigation, we were influenced by the fact that no steam is produced during the initial stages of their slow combustion, which has been shown to involve the early formation of formaldehyde, thus:

$$C_2H_4 \longrightarrow C_2H_4O_2 \longrightarrow 2CH_2O, \text{ &c.}$$
$$C_2H_2 \longrightarrow C_2H_2O_2 \longrightarrow CO + CH_2O, \text{ &c.}$$

The experimental method consisted in comparing the amounts of change occurring in two similar tubes containing respectively thoroughly dried, and undried, equimolecular mixtures of one or other of the hydrocarbons with oxygen, when heated under similar conditions for the same length of time.

Comparative experiments were also made iu similar tubes with electrolytic gas, dried and undried, in order to ensure that the degree of dryness attained in the hydrocarbon experiments was such as would practically inhibit the combustion of hydrogen. From the results of

these experiments, we have arrived at the conclusion that the rigid
exclusion of moisture by means of the best-known method of desiccation has little, if any, influence on the rate of oxidation of a hydrocarbon.

Preparation of the Gases.

Electrolytic Gas.—This was prepared by the electrolysis of a solution
of thrice-crystallised barium hydroxide. For the "dry" experiments,
the gas as it left the electrolytic cell was passed through a system of
drying tubes containing (1) solid caustic potash and (2) phosphoric
anhydride,* after which it was collected in a glass holder over dry
mercury. From thence it was finally transferred into the experimental tubes through another series of drying tubes containing phosphoric anhydride. Lastly, traces of moisture were removed by long
standing over phosphoric anhydride in the experimental tubes, in the
manner to be described later. All the glass taps used in the experiments with electrolytic gas were lubricated with glacial phosphoric
acid.

Ethylene.—This was prepared by the interaction of ethyl alcohol and
strong sulphuric acid at 160°. The crude gas was passed through suitable absorption vessels containing (1) ice-cold water, (2) strong sulphuric acid, (3) a strong solution of caustic potash. It was then
collected over a mixture of equal volumes of glycerine and water.

The washed gas was further purified by repeated liquefaction in a
bath of liquid air, followed in each case by a fractionation of the liquid,
in which the first and last thirds were rejected. In the final purification, about a litre of the thrice-fractionated gas was again liquefied,
and on removing the liquid air-bath the first 300 c.c. of gas were
rejected. A middle fraction measuring 250 c.c. was then collected in
a holder over dry mercury, where it was subsequently mixed with an
equal volume of dry oxygen. The mixture was finally passed into the
experimental tubes through a series of drying tubes containing
redistilled phosphoric anhydride.

Acetylene.—The crude gas, prepared from commercial calcium carbide,
was passed through (1) dilute sulphuric acid, (2) a mixture of
bleaching powder with excess of quicklime, (3) towers containing
"kieselguhr" saturated with an acid solution of cupric and ferric
chlorides, and (4) a strong solution of caustic potash. The final
purification was effected by repeated solidification in a small glass
bulb immersed in liquid air and subsequent fractional vaporisation of
the solid, much in the same way as the liquid ethylene had been fractionated. The subsequent admixture with an equal volume of oxygen,

* All the phosphoric anhydride used in the research had been redistilled in a
current of dry oxygen over red-hot platinised asbestos.

and the drying of the mixture before it was filled into the experimental tube, were carried out in precisely the same manner as in the case of ethylene.

Oxygen.—This was prepared by heating recrystallised potassium permanganate. The gas was first passed through a strong solution of caustic potash, and then collected in a glass holder over pure strong sulphuric acid. Before admixture with the hydrocarbons, it was slowly passed through a long tube filled with solid caustic potash.

Cleaning and Drying of the Experimental Tubes.   The Final Drying of the Experimental Mixtures.

The experimental tubes were nearly all made of Jena borosilicate glass (No. 59[III]), which, owing to its remarkably feeble power of retaining surface moisture, is, above all other kinds, suitable for experiments with dried gases. A few experiments were made with tubes of Jena hard combustion glass, which is also admirably adapted to the purpose in view.

Tubes of uniform dimensions (length = 70 cm., internal diameter = 15 mm., capacity = 150 c.c.) were selected, and bent into the form shown in the accompanying figure, at a point B, about 15 or 20 cm. from one end.

To each end of the tube was fused a short length of capillary tubing, D and E, and the tubes to be used for the dry mixtures were also provided with a side-piece, F, in the vertical limb, BC, through which the redistilled phosphoric anhydride was subsequently introduced.

The tubes were carefully cleaned by prolonged treatment with a hot strong solution of chromic anhydride in sulphuric acid, after which they were allowed to drain. They were next thoroughly washed out with several litres of hot distilled water, after which the inside of each tube was steamed for an hour and a half. It was then dried in a current of hot air, previously dried by sulphuric acid and filtered through cotton-wool.

A quantity of redistilled phosphoric anhydride was next introduced into the tubes intended for the diied mixture through the side-piece, F, which was then sealed off along with the capillary tube, D. The operation of finally drying the tube and filling it with the dry experimental mixture was afterwards performed as follows. The horizontal portion of the tube AB was heated in a Lothar Meyer furnace to a temperature considerably higher than that subsequently employed in the actual experiment. The capillary tube, E, was connected, through an arrangement of drying tubes containing phosphoric anhydride, with an automatic Sprengel pump, and the heated tube exhausted during several hours. This, it was thought, would ensure the removal of the film of moisture which often clings so tenaciously to a glass surface. The vacuous tube was allowed to cool slowly, and was then connected, through a drying tube containing redistilled phosphoric anhydride, with the holder containing the experimental mixture. After all the connections had been exhausted, the mixture was very slowly admitted to the tube until the pressure was about 10 mm. below the atmospheric. The capillary, E, was then sealed off, and the tube set aside for several weeks in a dark room, to complete the desiccation of the gases. At the end of this period, the phosphoric anhydride invariably appeared perfectly dry. Meanwhile, other similar tubes, cleaned as above, had been filled with the same experi-

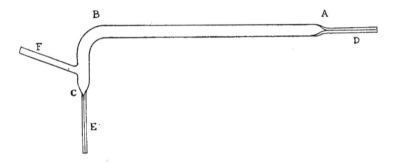

mental mixture, which on its way into each tube had been slowly passed over the surface of distilled water.

Heating of the Tubes.

In an actual experiment, the horizontal portion, AB, of the tube was inserted into a wider tube already heated to the desired experimental temperature in a Lothar Meyer constant temperature furnace. A slow current of hot air, passed through the annular space between the two tubes, ensured the uniform heating of the inner one. The temperature was registered by a recording pyrometer, the thermojunction wires being inserted in the annular space already mentioned.

Throughout the entire duration of the experiment, the vertical limb, BC, containing the phosphoric anhydride, was screened from the furnace and immersed in ice-cold water. The cold portion of the tube was carefully watched for any visible signs of chemical action. After a definite time interval, the tube was withdrawn from the furnace, and, after cooling, was opened either under mercury, or in connection with a manometer of capillary bhore. The gaseous products were afterwards withdrawn and analysed.

In a given experiment, one tube each of the dried and undried gases was heated under precisely similar conditions for the same length of time, and the amounts of combination in each case compared.

Results of the Experiments.

A. *With Electrolytic Gas.*

Altogether five experiments were made with electrolytic gas, the period of drying extending over four weeks in each case. A preliminary trial with a tube containing the undried gases indicated 525° as a suitable experimental temperature, and this was accordingly selected.

In four of the experiments, an enormous difference was observed between the rate of combination in the moist and dried gases, respectively, during ten minutes at 525°. In the case of the undried gases, water invariably appeared in the cold portion of the tube after one and a half minutes, and accumulated as the heating was continued; altogether between 40 and 50 per cent. of the gases combined during any given experiment. In the case of the dried gases, however, there was never any visible sign of combination, and on subsequently opening the cold tubes under mercury, the contractions observed never exceeded 3 per cent. of the original volume, and in two cases were negligibly small. The amounts of combination observed in the four experiments are tabulated below:

Experiment.	Undried gas.	Dried gas.
1	52 per cent.	0:5 per cent.
2	45 ,,	3·0 ,,
3	42 ,,	nil
4	40 ,,	3·0 ,,

In the fifth experiment, a small quantity of the phosphoric anhydride was accidentally projected into the horizontal part of the tube during the introduction of the electrolytic gas. We were therefore not surprised to find a moderate amount of action when the tube was subsequently heated after four weeks' drying. Nevertheless, the percentage combination observed was only 14, as compared with 45 in the corresponding case of the undried gases. In this connection, the fact may be recalled that H. B. Baker had a similar experience with one out of twelve of his tubes of·dried gas (*ibid.*, p. 403).

B. *Experiments with an Equimolecular Mixture of Ethylene and Oxygen.*

An equimolecular mixture of the two gases was selected because previous experiments (Trans., 1904, 85, 1643) had shown that the fastest rates of combination are obtained with these proportions.

1st Experiment.—For this experiment, two tubes of the dried mixture were prepared, the period of drying extending over eleven weeks. With the object of comparing the temperatures at which visible reaction commenced in the dried and undried mixtures, the furnace was lighted after the introduction of the tubes and the temperature slowly raised. The first signs of chemical action (the appearance of a white film of paraformaldehyde in the cold portion of the tube) were noticed at nearly the same temperature (385—395°) in both the case of the dried and the undried mixtures. Moisture first appeared about five minutes later in each case. The heating was continued fifteen minutes longer (maximum temperature = 430—450°), after which the tubes were withdrawn and cooled.

The following numbers show that there was very little difference between the total amounts of change observed in the dried and undried mixtures respectively.

Mixture.	Percentage contraction (corrected) after cooling.	CO_2.	CO.	C_2H_4.	C_2H_6.	H_2.	O_2.
Dried	37·0	10·25	45 75	32·65	2·10	1·05	8·20
,, ...	37·7	10·25	44·50	28·85	6·35	4·00	6·05
Undried	34·8	10·60	48·4	27·6	4·65	2·70	6·05
,, ...	36·1	11·45	50·8	24·85	5·15	3·25	4·50

The heading above the composition columns reads "Percentage composition of gaseous products."

2nd Experiment.—Three tubes, one of undried and the other two of dried mixture (period of drying = twelve weeks), were introduced into the furnace and heated for fifteen minutes in each case. The first visible action was observed after about five minutes in each case. The total amount of change in the undried mixture was somewhat greater than that observed in either of the " dried " tubes.

Percentage composition of gaseous products.

Mixture.	Percentage contraction.	CO_2.	CO.	C_2H_4.	C_2H_6.	H_2.	O_2.
Undried	29·3	7·30	35·35	36·10	1·00	2·30	17·95
Dried	22·9	6·60	24·55	39·75	1·20	0·25	27·65
,, ...	22·9			Not analysed.			

3rd Experiment.—Three tubes, one containing undried and the other two the dried gases (period of drying = twelve weeks), were introduced into the furnace at 457°. A striking difference between the dried and undried gases was observed. For, whereas the combination of the undried gases followed the normal course of slow combustion, that is, without ignition, the dried gases ignited after three minutes, a slow flame traversing the mixtuie and finally dying out a little distance from the cooled end of the tube. There was a slight deposit of carbon in the heated portion of the " dried " tubes, where the flame had started, but none in the case of the " undried " gases. There had, therefore, been a much faster initial rate of combination in the case of the dried gases than with the undried mixture, the ignition following as a result of local heating. This is borne out by the ratio of the pressures in the tubes before and after the experiment (p_2/p_1), and the composition of the gaseous products.

Mixture.	Temperature.	Duration of heating.	p_2/p_1.	Percentage composition of gaseous products.					
				CO_2.	CO.	C_2H_4.	C_2H_6 and CH_4.	H_2.	O_2.
Undried......	457°	15 minutes	0·708	8·0	41·5	33·1	2·5	nil	14·9
Dried.........	457	3 ,,	1·71	0·8	46·6	2·6	1·9	44·0	2·1
,, ...	450	3 ,,	1·75			Not analysed.			

4th Experiment.—Two tubes, the one containing the undried and the other the dried gases (drying period = six weeks), were introduced into the furnace at 434°. In the case of the undried mixture, the first signs of combination were observed after four minutes, after which slow combustion continued during the remainder of the experiment. In the case of the dried gases, however, a film of paraformaldehyde appeared in the cold end of the tube after one minute, and at two and three-quarter minutes the contents of the tube ignited, as in the previous experiment.

5th Experiment.—Two tubes, containing respectively the dried and undried gases (period of drying = six weeks), were introduced into the furnace at 440°. The first signs of combination were observed at the end of five and a half minutes in the case of the dried mixture, and half a minute later with the undried mixture. There was no ignition of the gas in either case, and after fifteen minutes the tubes were withdrawn. The percentage contraction found on opening the tubes under mercury and the composition of the gaseous products in each case are given below.

Mixture.	Percentage contraction.	Percentage composition of gaseous products.					
		CO_2.	CO.	C_2H_4.	C_2H_6.	H_2.	O_2.
Dried	30·0	6·25	36·95	31·85	3·60	4·90	16·45
Undried	28·3	8·55	46·35	30·85	3·20	2·90	8·15

Two other experiments were made with similar results to the above. Altogether ten tubes of the "dried" mixture and ten of the undried gases were examined. Considering the results as a whole, we are inclined to think that the initial rate of oxidation, and consequently the accumulation of aldehyde, was greater in the dried than in the undried gases. It is significant, in this connection, that actual ignition never took place except with the dried gases.

C. *Experiments with Acetylene.*

Two experiments were made with thoroughly dried equimolecular mixtures of acetylene and oxygen, prepared as previously described.

(1) Period of drying = four weeks. Commencing with the furnace cold, the tube was heated to 415° during thirty-five minutes. A white film of paraformaldehyde appeared at 335° after twenty-seven minutes' heating, and four minutes later the formation of water was noticed.

The contraction after cooling, on opening under mercury, was 32·8 per cent., and the following analysis of the gaseous products, expressed in percentages, shows that the oxygen had completely disappeared.

| CO_2 .... | 19·65 | CO .... | 56·65 | C_2H_2 .. .. | 22·5 |
| H_2 .... | 0·85 | C_2H_6 .. ... | 0·35 | O_2 .... | nil |

(2) Period of drying = four weeks. The tube was placed in the furnace previously heated to 520°. Two minutes later, the gases exploded with shattering effect.

D. *Experiments with Ethane.*

The following comparative experiments were also made with tubes containing dried and undried mixtures of ethane and oxygen in equimolecular proportions.

The ethane had been prepared by the action of water on zinc ethyl and purified by repeated liquefaction and fractionation, as in the case of ethylene.

(1) Two tubes, containing respectively the dried and undried gases (period of drying = seven weeks), were placed in the furnace at 415°. The dried gas ignited after nineteen and three-quarter minutes, but after twenty minutes there was no visible sign of reaction in the tube of undried gas.

(2) Two tubes, containing respectively the undried and dried gas (period of drying = seven weeks), were placed in the furnace at 440°. In both cases, the mixture ignited, the dried gas after seven and three-quarter minutes, and the undried gas after twelve and a quarter minutes. Carbon separated and water condensed when the tubes were subsequently cooled. The ratios (p_2/p_1) were 1·503 and 1·501 respectively.

The experiments with both acetylene and ethane thus confirm the conclusions drawn from the more complete investigation in the case of oxidation of ethylene.

In conclusion, the authors have to thank the Government Grant Committee of the Royal Society for grants out of which the expenses of the investigation have been largely met.

FUEL AND METALLURGICAL LABORATORY,
UNIVERSITY OF MANCHESTER.

LXXI.—*The Explosive Combustion of Hydrocarbons.*

By WILLIAM ARTHUR BONE and JULIEN DRUGMAN.

IN the previous papers of the series (Trans., 1902, **81**, 536 ; 1903, **83**, 1074 ; 1904, **85**, 693, 1637 ; 1905, **87**, 1232), it has been shown that the slow combustion of a hydrocarbon may be regarded as a process involving the initial formation of unstable hydroxylated molecules, which subsequently undergo thermal decompositions into simpler products. We shall now endeavour to prove that this theory may be extended to hydrocarbon flames, and that, in so far as the immediate result of encounters between hydrocarbon and oxygen molecules is concerned, there is probably no real difference between slow and rapid combustion.

The only alternative theory worthy of serious consideration in this connection is the one which supposes that, in flames, carbon is burnt preferentially to hydrogen. Those who hold this view believe that whenever the system $C_xH_y + x/2O_2$ is raised to a sufficiently high temperature, it is transformed into $xCO + y/2H_2$, either as the direct result of collisions between single molecules of the hydrocarbon and oxygen (for example, in the case of ethylene or acetylene), or possibly through an intermediate state which may be represented somewhat as follows :

$$C_xH_y + x/2O_2 = [xC + yH + x/O_2] = x/2CO + y/2H_2.$$

This theory was originally put forward by Kersten in 1861 (*J. pr. Chem.*, **84**, 310), who vigorously assailed the prevailing dogma that hydrogen is preferentially burnt in hydrocarbon flames, an idea which would probably never have gained currency had not chemists ignored Dalton's classical experiments on the combustion of methane and ethylene (*New System*, 1808, Vol. I, 437, 444). So firmly rooted was this notion, however, that Kersten's protest remained unheeded for thirty years, when it was reinforced by Smithells and Ingle, in their paper on the "Structure and Chemistry of Flames" (Trans., 1892, **61**, 204), as well as by the work of Lean and Bone on the explosion of mixtures of ethylene and oxygen, in 1892.

The theory of the preferential· combustion of carbon has recently been revived by Wilhelm Misteli (*Journ. Gasbeleuchtung*, 1905, **48**, 802 ; see also Abstr., 1905, **88**, 849), but we shall have no difficulty in proving that, however well it may appear at first sight to explain certain well-known facts, it completely breaks down as a general theory of hydrocarbon combustion.

The present paper contains the results of a systematic study of the explosive combustion * of a number of different gaseous hydrocarbons, including members of the saturated series C_nH_{2n+2}, up to butane, olefines such as ethylene, propylene, and the butylenes, as well as trimethylene and acetylene. Among the new facts brought to light during the research, we need now only refer to those which appear to be crucial as regards the two theories under discussion.

Considering, in the first place, the *end result* obtained when explosive combustion occurs in the system $C_xH_y + x/2O_2$, we have discovered a remarkable difference between the behaviour of an olefine and that of the corresponding paraffin. In the case of an olefine, there is no separation of carbon, and very little (if any) formation of steam, the cooled products consisting chiefly of carbon monoxide and hydrogen, in accordance with the empirical equation:

$$C_nH_{2n} + n/2O_2 = nCO + nH_2.$$

In the case of the corresponding paraffin, however, there is always a considerable separation of carbon and a large formation of steam. The gaseous products contain, besides carbon monoxide and hydrogen, between 8 and 10 per cent. of methane, and fair proportions both of unsaturated hydrocarbons (acetylene and ethylene), and of carbon dioxide. Evidently, therefore, the theory of the preferential combustion of carbon in hydrocarbon flames does not apply to the case of a saturated hydrocarbon.

But it will be shown that the theory also breaks down when the case of an olefine is more closely examined. For whilst it is true that the *end-products* obtained when a mixture $C_nH_{2n} + n/2O_2$ is exploded conform to the requirements of the theory, we find that if the proportion of oxygen in the original mixture is further diminished, much water, as well as carbon, is produced. Indeed the quantity of water formed increases as the supply of oxygen is reduced below the above limit. The same thing also applies in the case of trimethylene. This characteristic behaviour of members of the C_nH_{2n} series obviously does not harmonise with the theory in question, but it is easily explained if the initial formation of hydroxylated products in hydrocarbon flames be admitted.

Comparative experiments on the explosion of paraffins with oxygen in proportions indicated by the general expression $C_nH_{2n+2} + n/2O_2$, and of mixtures of the corresponding olefines with hydrogen and

---

* The terms "*explosive combustion*" and "*explosion*," as applied to gaseous mixtures, are perhaps somewhat ambiguous. They are often used indiscriminately to denote both the propagation of a flame through a combustible mixture under ordinary conditions ("*inflammation*"), and also the conditions existing when an explosion wave is set up ("*detonation*"). It must be clearly understood that, throughout this paper, the terms in question are used in the former sense only.

oxygen in the proportions $C_nH_{2n} + H_2 + n/2O_2$, that is to say, of mixtures initially containing the same relative amounts of gaseous carbon, hydrogen, and oxygen, have revealed a totally different end result in the two cases. The addition of hydrogen, in the above proportion, to a mixture of an olefine and oxygen in the ratio $C_nH_{2n} + n/2O_2$, causes no separation of carbon and very little formation of steam when the gases are ignited, a result which is in striking contrast with the behaviour of a mixture $C_nH_{2n+2} + n/2O_2$ under similar conditions. The same contrast is exhibited by the behaviour of a mixture of acetylene and electrolytic gas in the proportions $C_2H_2 + 2H_2 + O_2$, as compared with that of an equimolecular mixture of ethane and oxygen, $C_2H_6 + O_2$. It is, therefore, impossible to regard the explosive combustion of a hydrocarbon as involving a primary dissolution of the hydrocarbon molecule, followed by a distribution of the oxygen between free carbon and hydrogen.

In many experiments, direct evidence has been obtained of the formation of aldehydes in explosive combustion. We have also succeeded in detecting minute quantities of an aldehyde both in the interconal gases from a Smithells' flame-separator, in which either ethane or ethylene was being burnt, and also among the products of an oxygen flame burning in an atmosphere either of ethylene or coal-gas. We have satisfied ourselves, by actual experiment, that the considerable quantities of aldehydes formed in hydrocarbon flames cannot possibly be attributed to interchanges in the system CO, H_2O, H_2, and CO_2 at high temperatures (see D. L. Chapman and A. Holt, jun., Trans., 1905, **87**, 916). Postponing the further discussion of the mechanism of explosive combustion until the end of the paper, we may now proceed to the detailed consideration of our experiments.

Experimental Method.

The arrangements for carrying out the experiments are shown in the accompanying diagram (Fig. 1). The explosion vessel was a cylindrical bulb, A (capacity = 60 c.c.), made of stout borosilicate glass and provided with platinum firing wires, so arranged that the spark ignited the central portion of the explosive mixture. Each end of the bulb terminated in a glass tube of capillary bore, somewhat constricted at a to facilitate the operation of sealing up the experimental mixture before it was fired.* By means of these capillary ends connections were made, as required, either with a glass gas-holder containing the

---

* The practice of firing the mixtures in the *sealed* bulb was found to be a necessary precaution against the risk of a small escape of gas during the momentary high pressure generated by the explosion.

experimental mixture over mercury, or with the capillary manometer, E, or with a Töpler pump.

Each experiment involved the following operations, in the order given, namely : (1) the sealing up of the dry experimental mixture, at known temperature and pressure, in the explosion bulb, by the application of a small blow-pipe flame at the points a a, with due precautions against any actual handling of the body of the bulb during the process, (2) the ignition of the mixture by a spark passed between the firing wires, (3) the opening of the bulb in connection with the capillary manometer, E, by breaking off one of the capillary ends under a stout india-rubber joint, all connections between the pump and

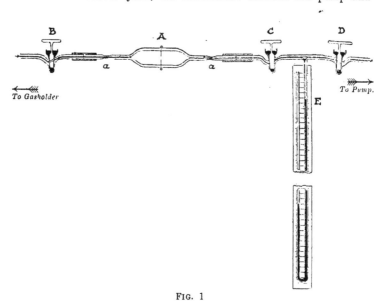

FIG. 1

manometer having been previously exhausted, (4) the determination of the pressure of the *dry* products, after applying the proper correc- tions for "dead space" in the capillary connections between the bulb and the manometer and for any change in the room temperature during the experiment, (5) the removal of the gaseous products through the pump for subsequent analysis, and (6) the rinsing out of the bulb with a small quantity of distilled water, and the testing of the rinsings for aldehydes with a solution of Schiff's reagent capable of indicating one part of formaldehyde in a million parts of water.

. In the case of each particular mixture, experiments were usually carried out under different initial pressures, the upper limit naturally depending on the capability of the bulb to withstand the shock of the explosion. It should be distinctly understood, moreover, that the dimensions of the bulb were too small to allow of the setting up of the " explosion wave" when the mixtures were fired. The experimental conditions, therefore, represented those of ordinary hydrocarbon flames, and not those of " detonation."

Preparation of Gases.

It is perhaps unnecessary to say anything about the preparation of the various hydrocarbons examined during the research beyond stating that every precaution was taken to ensure the highest possible degree of purity in each case. The final purification was always carried out by repeated liquefaction (by means of liquid air, solid carbon dioxide, or other suitable refrigerent) and subsequent fractionation of the liquid. The purity of each hydrocarbon was finally tested by an explosion analysis; the following are examples of the ratios C/A obtained in each case.

	Found.	Calculated.			Found.	Calculated.
Methane	2·00	2·00		Propylene ...	0·841 ⎫	0·833
Ethane	1·25	1·25		Trimethylene..	0·831 ⎭	
Propane	1·00	1·00		Butylenes......	0·755	0·750
n-Butane	0·872	0·875		Acetylene	0·757	0·750
Ethylene	1·005	1·000				

The experimental mixtures were stored over mercury in a glass gas-holder (capacity 500 c.c.), graduated into 5 c.c.

Tabulation of Results.

The experimental results are tabulated so as to show in each case :

(i) $p_1 =$ initial pressure of the dry " nitrogen-free " mixture exploded.

(ii) $p_2 =$ final pressure (corr.) of the dry " nitrogen-free " products.

(iii) The percentage composition of the gaseous products.

(iv) The number of " units " of carbon, hydrogen, and oxygen, respectively, both in the original mixture fired and in the gaseous products.

(v) $\theta =$ the total heat developed during the explosion of 22·4 litres, at N.T.P., of the original mixture.

A. *Experiments with Ethane, Ethylene, and Acetylene.*

The behaviour of each of these hydrocarbons when exploded with its own volume of oxygen will first of all be considered.

In the experiments with either ethylene or acetylene, there was always a sharp explosion without any separation of carbon or condensation of moisture when the products cooled. The ratios p_2/p_1, as well as the composition of the gaseous products, always corresponded to the empirical equations :

$$C_2H_4 + O_2 = 2CO + 2H_2,$$
$$C_2H_2 + O_2 = 2CO + H_2,$$

except for the formation of a small percentage of methane.

The behaviour of an equimolecular mixture of ethane and oxygen was, however, strikingly different. A lurid flame filled the explosion bulb, accompanied by a dense cloud of carbon particles and an abundant formation of steam. The ratio p_2/p_1, instead of being 2·5, as demanded by the empirical equation,

$$C_2H_6 + O_2 = 2CO + 3H_2,$$

was approximately 1·5, whilst the rinsings from the bulb always gave a strong aldehydic reaction. The gaseous products, although largely composed of carbon monoxide and hydrogen, contained nearly 4 per cent. of carbon dioxide, and as much as 10 per cent. of methane, as well as appreciable quantities of both acetylene and ethylene. We estimate that between 12·5 and 16·5 per cent. of the original carbon, and from one-third to two-fifths of the original oxygen, according to the initial pressure, were not accounted for in the gaseous products. The ratio H_2/O_2 in the condensed products was, as a rule, somewhat higher than 2·0, a circumstance possibly connected with the formation of acetaldehyde.

The results of typical experiments with the three hydrocarbons are tabulated on pages 666 and 667 ; in the fourth experiment with the equimolecular mixture of ethane and oxygen, the bulb was kept at 130° in an air-bath when the mixture was fired, in order to prevent any condensation of water during the actual combustion. The end result was not, however, materially different from those obtained in the other three experiments.

The influence of pressure, and therefore of flame temperature, in the above experiments deserves a passing notice. In the case of ethylene, a large difference in the initial pressure had no appreciable influence on the final result. In the case of ethane, however, the quantities of unsaturated hydrocarbons, as well as of steam, in the final products, showed a tendency to increase as the initial pressure diminished. Against the argument that the large formation of steam in the case of ethane, as compared with its entire absence from the ethylene products, may be attributed to a much lower flame temperature, must be set the fact that, *ceteris paribus*,

TABLE IA.—*Experiments with* $C_2H_4 + O_2$ *and* $C_2H_2 + O_2$.

The original mixtures were in each case exactly equimolecular.

Original mixture.	$C_2H_4 + O_2$.		$C_2H_2 + O_2$.	
Experiment ...	1.	2.	1.	2.
p_1	545·3 mm.	280·3 mm.	331 mm.	352 mm.
p_2	1053·5 ,,	546·5 ,,	487 ,,	508 ,,
p_2/p_1	1·93	1·95	1·47	1·44
Percentage composition of gaseous products. $\begin{cases} CO_2 \\ CO \\ H_2 \\ CH_4 \end{cases}$	0·3 50·0 48·0 1·7	0·45 50·00 47·55 2·00	1·05 66·15 31·10 1·70	0·75 67·10 30·60 1·55
Units in original mixture	C. H. O. 545 545 272	C. H. O. 280 280 140	C. H. O. 331 165 165	C. H. O. 352 176 176
Units in gaseous products.........	547 541 267	286 282 139	335 168 166	352 171 175
Mean value of θ...	30		54	

the total evolution of heat during the explosion of the mixture $C_2H_6 + O_2$ was greater than that evolved in the case of the mixture $C_2H_4 + O_2$. The evidence of the research as a whole is, moreover, decidedly against any such view. But whilst the striking contrast between the behaviour of ethane and that of ethylene or acetylene in the above experiments cannot be reconciled with the theory of the preferential combustion of carbon, the following argument will show that it is perfectly consistent with the view that the mechanism of combustion is essentially the same in flames as it is below the ignition point. It has been shown that the slow combustion of ethylene involves the formation of formaldehyde, probably as the result of the thermal decomposition of dihydroxyethylene (Trans., 1904, **85**, 1637), thus:

$$\begin{matrix} CH_2 \\ CH_2 \end{matrix} \rightarrow \begin{matrix} CH \cdot OH \\ CH_2 \end{matrix} \rightarrow \begin{matrix} CH \cdot OH \\ CH \cdot OH \end{matrix} \rightarrow 2CH_2O, \&c.$$

In a similar manner, acetylene yields carbon monoxide and formaldehyde (Trans., 1905, **87**, 1232):

$$\begin{matrix} CH \\ CH \end{matrix} \rightarrow \begin{matrix} C \cdot OH \\ CH \end{matrix} \rightarrow \begin{matrix} C \cdot OH \\ C \cdot OH \end{matrix} \rightarrow CO + CH_2O, \&c.$$

TABLE IB.—*Experiments with* $C_2H_6 + O_2$.
The original mixtures were exactly equimolecular.

Experiment ...	1.	2.	3.	4.
p_1	698·2 mm.	466·6 mm.	270 mm.	664 mm.
p_2	1048·0 ,,	673·8 ,,	382 ,,	1018 ,,
p_2/p_1	1·50	1·44	1·415	1·50

		1.	2.	3.	4.
Percentage composition of gaseous products.	CO_2	4·30	3·90	3·70	3·80
	CO	34·75	34·20	34·10	35·20
	C_2H_2	1·90	3·50	6·65	3·30
	C_2H_4	2·00	2·35		
	CH_4	9·20	9·15	10·70	9·70
	H_2	47·85	46·90	44·85	48·00

	C.	H.	O.	C.	H.	O.	C.	H.	O.	C.	H.	O.
Units in original mixture	698	1047	349	467	700	233	270	405	135	664	996	332
Units in gaseous products.........	587	756	227	397	494	141	236	291	79	563	736	218
Difference	111	291	122	70	206	92	34	114	56	102	260	114

The mean value of θ in the above experiments = 36 approximately.
In the 4th experiment, the bulb was kept at 130° when the mixture was fired.

Now, at high temperatures, formaldehyde decomposes, yielding equal volumes of carbon monoxide and hydrogen, without any separation of carbon or formation of steam (Trans., 1905, **87**, 910). The *apparent* preferential combustion of carbon which is observed when the mixtures $C_2H_4 + O_2$ and $C_2H_2 + O_2$ are exploded is thus easily explained.

The slow combustion of ethane, however, involves the early formation of acetaldehyde and steam :

$$\begin{array}{ccccccc} CH_3 & & CH_3 & & CH_3 & & CH_3 \\ \dot{C}H_3 & \rightarrow & \dot{C}H_2\cdot OH & \rightarrow & \dot{C}H(OH)_2 & \rightarrow & \dot{C}HO \end{array} + H_2O, \&c.$$

But at temperatures somewhat below 400°, acetaldehyde decomposes, yielding equal volumes of methane and carbon monoxide, and at higher temperatures it is rapidly resolved into carbon, hydrogen, methane, and carbon monoxide. The prominence of all these substances, as well as of steam, among the explosion products in the case of the mixture $C_2H_6 + O_2$ is, therefore, in entire accordance with our views.

Further evidence against the preferential combustion of carbon in flames is afforded by the behaviour of ethylene when exploded with *less* than its own volume of oxygen. The results of three experiments with a mixture corresponding to $3C_2H_4 + 2O_2$ are tabulated below. The gases combined with a lurid flame, carbon separated, and water condensed when the products cooled. The rinsings from the bulb gave a strong aldehydic reaction, and the gaseous products contained both methane and unsaturated hydrocarbons. We calculate that between 16 and 24 per cent. of the original oxygen appeared in the condensed products as aldehydes and water. In two of the experiments, the ratio H_2/O_2 in the condensed products was approximately 3·0, indicative of the formation of some such product as C_2H_4O (acetaldehyde, for example).

TABLE II.—*Experiments with Mixtures*, $3C_2H_4 + 2O_2$.

Original mixture	$C_2H_4 = 59\cdot65$; $O_2 = 40\cdot35$ per cent.			$C_2H_4 = 60\cdot5$; $O_2 = 39\cdot5$.		
p_1 p_2 p_2/p_1	524·3 mm. 768·2 ,, 1·465			562·3 mm. 816·4 ,, 1·451	667 mm. 1003 ,, 1·50	
Percentage composition of gaseous products. $\begin{cases} CO_2 \\ CO \\ C_2H_2 + C_2H_4 \\ CH_4 \\ H_2 \end{cases}$	2·45 40·10 5·00 5·45 47·00			2·50 37·20 6·40 6·50 47·4	2·20 39·80 3·90 6·70 47·50	

	C.	H.	O.	C.	H.	O.	C.	H.	O.
Units in original mixture ...	625	625	212	670	670	227	807	807	263
Units in gaseous products...	445	503	173	482	572	172	565	670	221
Difference...............	180	122	39	188	98	55	242	137	42

The mean value of θ in the above experiments is 37·0 approximately.

The formation of so much water in the above experiments is entirely inconsistent with the idea of the preferential combustion of carbon, and it can hardly be attributed to a low flame temperature, since the heat evolved when a given volume of the mixture $3C_2H_4 + 2O_2$ was exploded was actually greater than that evolved in the case of an equal volume of the mixture $C_2H_4 + O_2$ under the same initial pressure. But if hydroxylated molecules are formed and decomposed in hydrocarbon flames, the formation of steam, as the result of the decomposition of monohydroxyethylene, $\dfrac{CH \cdot OH}{\ddot{C}H_2}$, when the supply of oxygen is far short of that required to convert all the hydrocarbon into the dihydroxy-derivative, $\dfrac{CH \cdot OH}{\ddot{C}H \cdot OH}$, is a highly probable event, and may be claimed as an important piece of evidence in favour of our views as applied to flames.

Comparison of the Behaviour of Mixtures corresponding to $C_2H_4 + H_2 + O_2$ *or* $C_2H_2 + 2H_2 + O_2$ *respectively with that of the Mixture* $C_2H_6 + O_2$.

At this stage of the inquiry it seemed important to compare the behaviour of different explosive mixtures, such as $C_2H_6 + O_2$, $C_2H_4 + H_2 + O_2$, and $C_2H_2 + 2H_2 + O_2$, all containing the same relative proportions of gaseous carbon, hydrogen, and oxygen. For if, as Misteli has recently argued (*loc. cit.*), hydrocarbons are not burnt as such in flames, but are first of all resolved into carbon and hydrogen before actual combustion occurs, much the same end result should be obtained in the three cases. The experiments recorded on page 670, however, prove that this view is untenable. For whereas the explosion of the mixture $C_2H_6 + O_2$ was invariably accompanied by the separation of carbon and a large condensation of water (see Table IB, page 667), in neither of the other two cases was there any separation of carbon, and in only one of them (namely, the mixture $C_2H_4 + H_2 + O_2$) was there a visible condensation of steam. Comparing the proportions of the original hydrogen which appeared in the condensed products in the three cases, we find that, whilst for the mixture $C_2H_6 + O_2$ it exceeded 27 per cent., in the case of the mixture $C_2H_4 + H_2 + O_2$ it varied between 5 and 8 per cent., according to the initial pressure, and in the case of the mixture $C_2H_2 + 2H_2 + O_2$ it was negligibly small. So far, indeed, from these experiments supporting the view that hydrocarbons are decomposed into carbon and hydrogen prior to combustion, they afford a striking demonstration of the enormously greater affinity of hydrocarbons, as compared with that of either hydrogen or carbon, for oxygen at temperatures which prevail in ordinary flames.

The complete absence of steam from the explosion products of the mixture of acetylene and electrolytic gas in the above experiments, led us to try whether a further large addition of hydrogen would materially affect the end result. We accordingly fired a mixture corresponding very nearly to $C_2H_2 + 4H_2 + O_2$ ($C_2H_2 = 16 \cdot 15$, $H_2 = 67 \cdot 80$, and $O_2 = 16 \cdot 05$ per cent.) under various initial pressures between 650 and 750 mm. There was, however, never any separation of carbon,

TABLE III.—*Experiments with the Mixtures* $C_2H_4 + H_2 + O_2$ *and*
$C_2H_2 + 2H_2 + O_2$ *respectively.*

Original mixture.	$C_2H_4 + H_2 + O_2$. $C_2H_4 = 33\cdot9$; $H_2 = 32\cdot65$; $O_2 = 33\cdot45$ per cent.			$C_2H_2 + 2H_2 + O_2$. $C_2H_2 = 25\cdot0$; $H_2 = 50\cdot0$; $O_2 = 25\cdot0$ per cent.	
Experiment ..	1.	2.	3.	1.	2.
p_1	370·5 mm.	502·8 mm.	628·2 mm.	286·6 mm.	534·5 mm.
p_2	546·1 ,,	750·3 ,,	927·8 ,,	349·6 ,,	653·6 ,,
p_2/p_1	1·473	1·492	1·476	1·23	1·22
Percentage composition of gaseous products. $\{$ CO_2	0·80	0·35	0·65	0·3	0·2
CO	39·35	39·60	38·95	40·0	39·8
C_2H_2	2·20	1·25	1·30	nil	nil
H_2	54·80	55·15	55·20	59·2	59·8
CH_4	2·25	3·65	3·90	0·5	0·2
Units in original mixture	C. H. O. 251 372 124	C. H. O. 341 505 168	C. H. O. 426 631 210	C. H. O. 143 215 71·5	C. H. O. 267 400 133·5
Units in gaseous products	258 342 111	346 478 151	428 596 187	142 210 71·0	262 394 131·5
θ		23			28

whilst the amount of steam formed was so small that it could only be
detected with difficulty. In one experiment, for example, $p_1 = 753$ mm.,
$p_2 = 850$ mm., and the gaseous products contained

$$CO_2 = 0\cdot2, \quad CO = 28\cdot1, \quad H_2 = 70\cdot2, \text{ and } CH_4 = 1\cdot5 \text{ per cent.}$$

Before leaving the subject of acetylene, we must mention one
respect in which its behaviour differs from that of ethylene, or indeed
any of the other unsaturated hydrocarbons examined, namely, that there
is little or no formation of steam when the supply of oxygen is reduced
below the equimolecular proportion $C_2H_2 + O_2$. Thus, for instance,
we did not detect any condensation of water when a mixture corresponding to $2C_2H_2 + O_2$ was exploded, although much carbon separated.
Part of the acetylene was burnt to carbon monoxide and hydrogen,
the remainder was resolved into carbon and hydrogen *plus* a small
quantity of methane. This circumstance may, we think, be attributed
to the extraordinarily great affinity of this hydrocarbon for oxygen,
and its well-known liability to decompose at high temperatures. It
would appear that, owing to the extreme rapidity of the initial stages

in its combustion, something tantamount to a direct transformation from C_2H_2 to $C_2H_2O_2$ occurs in flames, as the result of collisions between single molecules of the hydrocarbon and oxygen.

B. *Experiments with Higher Members of the* C_nH_{2n+2} *and* C_nH_{2n} *Series.*

The conclusions drawn from the preceding experiments were amply confirmed by a study of the behaviour of the higher members of the C_nH_{2n+2} and C_nH_{2n} series. The results of these experiments, which are set forth in detail in Tables IV to VIII inclusive, may be conveniently discussed together.

The behaviour of propane and n-butane, when exploded with oxygen in the proportion indicated by the expression $C_nH_{2n+2} + n/2O_2$ (see Table IV), closely resembled that of ethane under similar conditions. The gases combined with a lurid flame, carbon separated, and much water condensed when the products cooled. The ratio p_2/p_1 obtained, and the percentages of the original carbon, hydrogen, and oxygen respectively which appeared in the condensed products in each case are given below :

		Percentage of original in condensed product.		
Mixture.	p_2/p_1.	C.	H.	O.
$C_3H_8 + 1\frac{1}{2}O_2$	1·75	9—16	25	33
$C_4H_{10} + 2O_2$	1·97	10	25	27

In each case we had no difficulty in detecting the formation of aldehydes. The gaseous products contained between 8 and 9 per cent. of methane, from 3 to 5 per cent. of unsaturated hydrocarbons (including acetylene), and about 4 per cent. of carbon dioxide, the remainder consisting of carbon monoxide and hydrogen.

The close agreement between the results obtained with ethane, propane, and n-butane warrants the belief that their behaviour is characteristic of the paraffins as a class, and it may be concluded that they do not exhibit any preferential combustion of carbon, when burnt under ordinary conditions.

We found, moreover, a great difference between the behaviour of the above mixture of propane and oxygen, $C_3H_8 + 1\frac{1}{2}O_2$, and that of a mixture of propylene, hydrogen, and oxygen, corresponding very nearly to $C_3H_6 + H_2 + 1\frac{1}{2}O_2$ (see Table IV). In the latter case, there was a barely visible condensation of moisture after the flame had traversed the mixture, and the whole of the original carbon was accounted for in the gaseous products. We estimate that approximately 90 per cent. of the original oxygen reacted with the propylene, forming carbon monoxide and hydrogen ; the remaining tenth appeared in the products as water.

TABLE IV.—*Experiments with Propane and Oxygen*, $C_3H_8 + 1\frac{1}{2}O_2$, *with Propylene, Hydrogen, and Oxygen*, $C_3H_6 + H_2 + 1\frac{1}{2}O_2$, *and with* n-*Butane and Oxygen*, $C_4H_{10} + 2O_2$.

Original mixture.	Propane and oxygen. $C_3H_8=39\cdot9$; $O_2=60\cdot1$ per cent.			Propylene, hydrogen, and oxygen. $C_3H_6=27\cdot5$; $H_2=29\cdot25$; $O_2=43\cdot25$ per cent.		n-Butane and oxygen. $C_4H_{10}+2O_2$.	
p_1	591·9 mm.	417·8 mm.	374·6 mm.	497·0 mm.	357·0 mm.	617·0 mm.	597·5 mm.
p_2	1028·5 ,,	725·5 ,,	656·8 ,,	847·8 ,,	610 1 ,,	1220·7 ,,	1177·3 ,,
p_2/p_1	1·737	1·741	1·753	1·705	1·708	1·978	1·970
Gaseous products. CO_2	4·1	4·4	4·2	110	1·25	3·7	3·8
CO	38·0	37·5	37·8	43·50	41·50	41·7	42·0
$C_2H_2+C_2H_4$	3·8	5·6	5·5	0·80	2·00	3·4	3·0
CH_4	8·0	8·0	9·2	2·40	2·40	8·5	8·7
H_2	46·1	44·4	43·3	52·20	52·81	42·7	42·5
Units in original mixture	C. H. O. 708 945 355	C. H. O. 500 669 251	C. H. O. 448 598 225	C. H. O. 410 555 215	C. H. O. 295 399 155	C. H. O. 823 1028 411	C. H. O. 797 996 398
Units in gaseous products.	593 700 237	444 499 168	408 458 152	412 497 193	300 363 135	741 790 299	712 758 292
Difference......	115 245 118	56 170 83	40 140 73	— 58 22	— 36 20	82 238 112	85 238 106

We next proceeded to examine the behaviour of propylene and trimethylene when each was exploded with proportions of oxygen varying between one and one and a half times its own volume (see Tables V and VI). With neither of the two hydrocarbons was there any separation of carbon when mixtures corresponding to $C_3H_6 + 1\frac{1}{2}O_2$ were fired, although in some experiments we were just able to detect a slight formation of dew when the products cooled. In one experiment with trimethylene there was distinct evidence of the formation of aldehyde. The ratios p_2/p_1, as well as the composition of the gaseous products, in each case indicated that the hydrocarbon had been burnt mainly in accordance with the empirical equation :

$$2C_3H_6 + 3O_2 = 6CO + 6H_2.$$

But that there had been no real preferential combustion of carbon was proved when we exploded each hydrocarbon with its own volume of oxygen. Under these conditions, not only did carbon separate, but there was an abundant formation of water, as much as 25 per cent. of the original oxygen appearing in the condensed products. The average ratio H_2/O_2 in these products was nearly 2·6, a circumstance which points to the formation of something else besides water, and, as a matter of fact, the rinsings from the bulb always gave a strong aldehydic reaction. The gaseous products in each case contained considerable quantities of unsaturated hydrocarbons, among which acetylene was prominent, and methane ; the reduction of the oxygen supply below the limit $C_3H_6 + 1\frac{1}{2}O_2$ also had the effect of increasing materially the formation of carbon dioxide, a result in all probability connected with the large production of steam.

In the case of propylene, we made two experiments with a mixture corresponding to $C_3H_6 + 1\frac{1}{4}O_2$ nearly. There was still a considerable formation of steam, but the proportion of the original oxygen appearing in the condensed products was now only 21 per cent., and the ratio H_2/O_2 exactly 2·0. The great similarity between the behaviour of propylene and trimethylene in these experiments shows that structural differences have but little influence on the mode of combustion of members of the C_nH_{2n} series.

TABLE V.—*Experiments with Propylene and Oxygen.*

	Original mixture.	$C_3H_6+1\frac{1}{2}O_2$.		$C_3H_6+1\frac{1}{4}O_2$. $C_3H_6=45\cdot0$; $O_2=55\cdot0$ per cent.		$C_3H_6+O_2$. Exactly equimolecular.	
p_1	563·0 mm.	383·0 mm.	299·1 mm.	682·0 mm.	598 0 mm.	585·6 mm.	492·3 mm.
p_2	1244·0 „	846·0 „	649·9 „	1213·0 „	1046·0 „	939 0 „	790·0 · „
p_2/p_1	2·20	2 24	2·17	1·78	1·75	1·60	1·60 „
Gaseous products CO_2	0·1	0·8	0·75	3·20	3·10	3·4	3·25
CO	50·9	50·3	50·50	43·05	43·00	39·0	37·70
$C_2H_2+C_2H_4$	1·0	0·8	0·80	4·90	5·00	6·7	8·90
H_2	46·5	46·0	45·35	41·95	42·00	40·4	38·65
CH_4	1·5	2·0	2·60	6·90	6·90	10·0	10·65
O_2	nil	nil	nil	nil	nil	0·5	0·90
Units in original mixture.	C. H. O. 675 675 338	C. H. O. 458 458 230	C. H. O. 359 359 180	C. H. O. 918 918 376	C. H. O. 805 805 329	C. H. O. 878 878 293	C. H. O. 738 738 246
Units in gaseous products.	677 634 318	462 433 219	360 333 169	763 766 300	658 661 257	618 662 219	548 578 182
Difference...	— 41 20	— 25 11	— 26 11	155 152 76	147 144 72	260 216 74	190 160 64
Percentage of original oxygen in condensed products		Between 5 and 6		21		25	

There was nothing in the behaviour of either of the two butylenes examined to call for special comment. When mixtures corresponding to $C_4H_8 + 2O_2$ were exploded, there was practically no separation of carbon, and an only just visible condensation of dew. The ratio p_2/p_1 was 2·43 in the case of n-butylene and 2·20 in the case of *iso*butylene, instead of the 2·66 required by the empirical equation:

$$C_4H_8 + 2O_2 = 4CO + 4H_2.$$

In the case of mixtures corresponding to $C_4H_8 + 1\frac{1}{2}O_2$, however, there was both separation of carbon and a large formation of water. Moreover, the rinsings from the bulb always gave a distinct aldehydic

TABLE VI.—*Experiments with Trimethylene and Oxygen-Mixtures corresponding to* $C_3H_6 + 1\frac{1}{2}O_2$ *and* $C_3H_6 + O_2$ *respectively.*

Original mixture ...	$C_3H_6 + 1\frac{1}{2}O_2$ exactly.		$C_3H_6 = 48\cdot15$; $O_2 = 51\cdot85$ per cent.	
p_1	552·7 mm.	552·2 mm.	563·7 mm.	330·5 mm.
p_2	1186·7 ,,	1219·6 ,,	936·3 ,,	536·8 ,,
p_2/p_1	2·15	2·20	1·66	1·624
Percentage composition of gaseous products. $\begin{cases} CO_2 \\ CO \\ C_2H_2 + C_2H_4 \\ CH_4 \\ H_2 \end{cases}$	0·60 49·30 0·85 0·45 48·80	0·35 48·70 1·15 0·45 49·35	2·8 40·5 5·9 6·3 44·5	3·30 40·25 9·65 6·55 40·25
Units in original mixture...................	C. H. O. 663 663 331	C. H. O. 662 662 331	C. H. O. 814 814 292	C. H. O. 477 477 171
Units in gaseous products	617 610 299	632 626 301	575 618 216	372 364 125
Difference.........	46 53 32	30 36 30	239 196 76	105 113 46

reaction. Between 24 and 28 per cent. of the original oxygen appeared in the condensed products, and the reduction of the oxygen supply below the limiting proportion $C_4H_8 + 2O_2$ materially increased the formation of carbon dioxide.

The results of these experiments are tabulated on p. 675.

C. *Experiments with an Equimolecular Mixture of Methane and Oxygen.*

In order to ascertain whether considerable differences in initial pressure, and therefore in flame temperature, would, in the absence of free carbon, have any marked influence on the quantity of steam in the final products of an explosive mixture, we selected an equimolecular mixture of methane and oxygen, which, as is well known, burns, form- ing much water, but without any separation of carbon. Dalton, who studied the behaviour of this mixture a century ago, writes thus about it: "If 100 measures of carburetted hydrogen be mixed with 100 measures of oxygen (the least that can be used with effect) and a spark passed through the mixture, there is an explosion without any material change of volume; after passing a few times through lime- water, it is reduced a little, manifesting signs of carbonic acid. This residue is found to possess all the characters of a mixture of equal volumes of carbonic oxide and hydrogen. . . . In this case, each atom

TABLE VII.—*Experiments with* n- *or* iso-*Butylene and Oxygen.*

Original mixture.	n-Butylene.			isoButylene.	
	$C_4H_8 + 2O_2$. $C_4H_8 = 32\cdot9$; $O_2 = 67\cdot1$ per cent.	$C_4H_8 + 1\frac{1}{2}O_2$. $C_4H_8 = 39\cdot6$; $O_2 = 60\cdot4$ per cent.	$C_4H_8 + 1\frac{1}{4}O_2$. $C_4H_8 = 46\cdot6$; $O_2 = 53\cdot4$ per cent.	$C_4H_8 + 2O_2$ exactly.	$C_4H_8 + 1\frac{1}{2}O_2$. $C_4H_8 = 39\cdot7$; $O_2 = 60\cdot3$ per cent.
p_1	479·0 mm.	574·0 mm.	487·4 mm.	438·0 mm.	606·5 mm.
p_2	1169·0 ,,	1094·0 ,,	904·0 ,,	982·5 ,,	1107·6 ,,
p_2/p_1	2·43	1·90	1·85	2·20	1·83
Percentage composition of gaseous products. $\{CO_2$	0·90	3·50	1·80	1·40	3·85
CO	50·50	41·40	37·90	50·20	39·85
$C_2H_2 + C_2H_4$	0·65	4·75	8·20	1·90	5·90
CH_4	1·60	7·55	15·2	3·00	8·10
	46·35	42·80	36·9	43·50	42·30

	C. H. O.	C. H. O.	C. H. O.	C. H. O.	C. H. O.
Units in original mixture	630 630 322	899 899 347	909 909 260	584 584 292	958 958 363
Units in gaseous products	634 590 305	678 713 265	644 720 187	574 515 260	702 744 262
Difference......	— 40 17	221 186 82	265 189 73	10 69 32	256 214 101

of the gas requires only two atoms of oxygen ; the one joins to one of hydrogen and forms water, the other joins to the carbone to form carbonic oxide, at the same moment the remaining atom of hydrogen springs off " (*New System*, 1808, vol. I, p. 444).

In our experiments, the results of which are tabulated on page 676, there was never any separation of carbon, and the ratio p_2/p_1 varied between 1·02 and 1·03 only, although p_1 varied between 335 and 651 mm. The products contained from 5·7 to 6·8 per cent. of carbon dioxide, and about 1 per cent. of methane ; the proportion of the original oxygen which appeared in the products as steam varied between 43·5 and 44·8 per cent. only, a variation which is probably within the limits of experimental error.

TABLE VIII.—*Experiments with an Equimolecular Mixture of Methane and Oxygen.*

	651·6 mm.	544·6 mm.	335·3 mm.
p_1 p_2 p_2/p_1	672·2 ,, 1·031	556·3 ,, 1·021	364·5 ,, 1·025

Percentage composition of gaseous products.		651·6 mm.	544·6 mm.	335·3 mm.
	CO_2	6·8	6·3	5·75
	CO	41·3	41·9	42·40
	H_2	50·8	50·6	51·10
	CH_4	1·1	1·2	0·75

	C.	H.	O.	C.	H.	O.	C.	H.	O.
Units in original mixture ...	326	652	326	272	544	272	177·5	355	177·5
Units in gaseous products...	330	356	184	274	295	151	177·0	192	98·0
Difference	—	296	142	—	149	121	—	163	79·5
Percentage of original O_2 as H_2O	43·5			44·5			44·8		

The mean value of θ in the above experiments is approximately 39.

The above results are easily explained by our theory. It has been shown that, below the ignition point, methane burns, forming at an early stage steam and formaldehyde. The process may probably be best expressed as follows :

$$CH_4 \rightarrow CH_3 \cdot OH \rightarrow CH_2(OH)_2 \rightarrow CH_2O + H_2O, \&c.$$

At high temperatures, the formaldehyde would certainly decompose into carbonic oxide and hydrogen, so that in explosive combustion we should obtain

$$CH_4 + O_2 = \overbrace{CO + H_2}^{CH_2O} + H_2O.$$

The 6 per cent. of carbon dioxide formed in our experiments would obviously arise by the secondary interaction of steam and carbonic oxide in the flame.

D. *Further Experiments on the Formation of Aldehydes in Flames.*

Although there had been direct evidence of the formation of aldehydes in several of the experiments just described, it seemed desirable to extend the inquiry to the case of a steady flame burning under ordinary conditions.

We first of all turned our attention to the interconal gases of an ethane or ethylene flame burning in a modified form of the well-known Smithells' separator. The success of an experiment depended on the maintenance of two cones during the withdrawal of a portion of the interconal gases through a small spiral glass condenser ; the water which condensed was afterwards tested for aldehydes with a sensitive Schiff reagent.

In the first experiment, 20 litres of pure ethylene were burnt in air, and about 3 litres of the interconal gases withdrawn. The condensed water gave a distinct aldehyde reaction. A blank experiment, in which the gases were passed unburnt through the apparatus, gave no result.

In the second experiment, 12 litres of pure ethane were burnt with air in half an hour, and between 4½ and 5 litres of the interconal gases were withdrawn. The condensed water gave a strong aldehydic reaction after standing about half a minute with the Schiff reagent. The experiment was repeated with a similar result.

It is, however, much easier to demonstrate the formation of aldehydes in a flame of air or oxygen burning in a hydrocarbon or coal-gas atmosphere. For this purpose we have employed the apparatus shown in Fig. 2. It consists of an inner glass "burner," A, situated within the water-jacketed "combustion chamber," C. The top of this

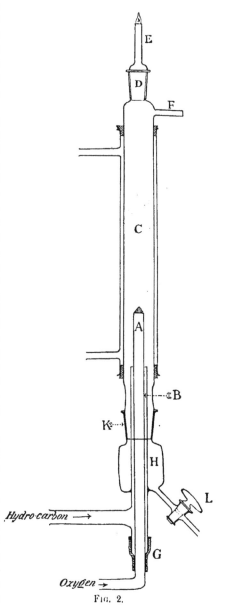

Fig. 2.

chamber is closed by the ground glass joint, D, which carries the narrow vertical tube, E. A small side tube, F, sealed into the combustion chamber just above the water-jacket, allows of the withdrawal of a portion of the gases when required.

The burner of the apparatus consists of two co-axial, vertical, glass tubes, A and B, the inner one of which, A, is made to move freely up and down within the outer, B, by means of the india-rubber joint, G, at the bottom of the apparatus. The outer tube, B, is made in one piece with the glass reservoir, H, which fits into the bottom of the combustion chamber by means of the ground glass joint, K. Into the lower part of this reservoir is sealed the glass stop-cock, L.

In an actual experiment, a steady flame of air (or oxygen) is maintained at A, burning in hydrocarbon or coal-gas, within the water-cooled chamber C, the excess of hydrocarbon being burnt at E, when it comes in contact with the outside atmosphere. The products of the burning oxygen are thus very rapidly cooled, and the steam arising in the flame condenses on the cold walls of the combustion chamber. The resulting water, which carries with it aldehydes formed in the flame, runs down into the glass reservoir, H, from which it is finally withdrawn through the stop-cock, L, for examination.

We have succeeded in proving the formation of aldehydes in the cases of air burning in coal-gas, and also when oxygen burns in ethane or ethylene.

Discussion of Results.

Whilst fully recognising the limitations imposed on the resources of chemical analysis by the conditions prevailing in hydrocarbon flames, we think it may be fairly claimed that the evidence accumulated during this research is conclusive against the theory of the preferential combustion of carbon.

In arguing that there is no essential difference between the mechanism of combustion below and above the ignition point, we do not, of course, infer that the phenomena observed at low temperatures, in slow combustion, are exactly reproduced in flames. We mean rather that the result of the initial encounter between the molecules of hydrocarbon and oxygen is probably much the same in both cases, namely, the formation of an " oxygenated " molecule. At the higher temperatures of flames, secondary thermal decompositions undoubtedly come into operation at an earlier stage, and play a more important rôle than they do below the ignition point.

The affinity of a hydrocarbon for oxygen is so enormously great at high temperatures, that in all probability the initial stage, or stages, of its combustion takes precedence of all other chemical phenomena in the flame. Our experiments with the mixtures $C_2H_2 + 2H_2 + O_2$,

$C_2H_4 + H_2 + O_2$, and $C_3H_6 + H_2 + O_2$ are very significant in this connection. It does not appear that, except in a very limited supply of oxygen, a hydrocarbon is to any great extent decomposed in the flame, much less that it is resolved into its elements, before its actual combustion begins. It is probably not so much the original hydrocarbon as its "oxygenated" molecule which decomposes in the flame. The sudden increase in the intramolecular energy of the hydrocarbon molecule, consequent on its association with oxygen, would render the resulting "oxygenated" molecule extremely unstable; its birth would speedily be followed by dissolution.

If so much be conceded, it still remains to discuss the character of the "oxygenated" molecules formed, and the precise way in which the oxygen is conveyed to the hydrocarbon in flames.

With regard to the first point, we think the facts are, as a whole, best interpreted on the supposition that hydroxylated molecules are formed in flames, as they undoubtedly are at lower temperatures. The case of a saturated hydrocarbon is certainly best explained in this way, for example:

$$CH_4 \longrightarrow CH_3 \cdot OH \longrightarrow CH_2(OH)_2 \longrightarrow \underbrace{CH_2O + H_2O}_{CO + H_2}.$$

Moreover, in the case of ethylene, the formation of so much water when a mixture $3C_2H_4 + 2O_2$ is exploded seems better explained on the supposition that it arises by the decomposition of a monohydroxy-derivative, $\begin{array}{c} CH \cdot OH \\ \| \\ CH_2 \end{array}$, than that it results from such a molecule as $O {\displaystyle <} \begin{array}{c} CH_2 \\ CH_2 \end{array}$. On the other hand, the fact must be recalled that we did not detect any formation of steam by the decomposition of a mono-hydroxy-acetylene when a mixture $2C_2H_2 + O_2$ was exploded.

If it be granted that hydroxylated molecules are formed in hydrocarbon flames, it may be asked what grounds there are for believing that a monohydroxy- rather than a dihydroxy-derivative is initially formed. Two sets of facts have a bearing on this question. The one is the large formation of water when such mixtures as $3C_2H_4 + 2O_2$, $C_3H_6 + O_2$, and $2C_4H_8 + 3O_2$ were exploded. The other is the relative ratio of oxidation observed at low temperatures with mixtures of a given hydrocarbon with varying proportions of oxygen. A comparison of the rates for such mixtures as

(a) $2C_2H_6 + O_2$, $C_2H_6 + O_2$, and $C_2H_6 + 2O_2$,

(b) $2C_2H_4 + O_2$, $C_2H_4 + O_2$, ,, $C_2H_4 + 2O_2$,

(c) $2C_2H_2 + O_2$, $C_2H_2 + O_2$, ,, $C_2H_2 + 2O_2$,

showed that whereas excess of oxygen over and above an equimolecular ratio always greatly retarded the combustion, a corresponding excess of the hydrocarbon had but little, if any, retarding effect. In view of these facts, we are inclined to think that the monohydroxy-derivative is actually formed, although in a sufficient supply of oxygen, and especially at high temperatures, it is very rapidly further oxidised to the dihydroxy-derivative. We can, however, imagine conditions (detonation, for example) under which the transition from the hydrocarbon to the dihydroxy-derivative may be practically direct, that is, the result of a single molecular impact. Possibly, this occurs in the case of acetylene, owing to its extraordinarily great affinity for oxygen at high temperatures.

The facts brought to light in the previous paper warrant the belief that oxygen acts directly on a hydrocarbon, rather than that it is conveyed to it indirectly through the intervention of steam. This view is, of course, opposed to that put forward by H. E. Armstrong, although in most other respects our interpretation of hydrocarbon combustion and his are practically identical. He has drawn attention to a difficulty in the way of accepting our version of the matter, as follows : " If oxygen molecules were directly active as wholes, and the actual, immediate, and sole cause of oxidation, there would seem to be no reason why the dihydroxy-derivative should not be directly produced rather than the mon-hydroxy. . . ." (Proc. Roy. Soc., 1904, series A, 74, 87). But it may be that, in flames, we are not dealing with oxygen entirely in the ordinary molecular condition, and moreover, from the kinetic standpoint, the difficulty urged does not seem any greater than that involved in the view that the formation of a hydroxy-derivative always requires the simultaneous conjugation of hydrocarbon, water, and oxygen. So far as the experiments recorded in the previous paper go, it would hardly appear that the presence of water is necessary for hydrocarbon combustion, and, until we have some definite evidence to the contrary, we prefer to regard the oxygen as directly active.

It now seems possible to arrive at an adequate conception of the mechanism of combustion in hydrocarbon flames burning under ordinary conditions. The picture may not be complete in all its details, but the main outlines are fairly clear. We venture to suggest the following interpretation of the facts established in this paper.

In the case of olefines, we have to account for two very significant facts, namely, (1) that when mixtures corresponding to $C_nH_{2n} + n/2O_2$ are exploded, there is no separation of carbon, and little (if any) formation of steam—the products consisting almost entirely of carbon monoxide and hydrogen, and (2) that with a more limited supply of oxygen, both carbon and steam arise. We imagine that, in an adequate supply of oxygen, the combustion involves successive eliminations of formaldehyde, which at once decomposes into carbon monoxide and hydrogen. The process may be crudely represented as follows, taking propylene as our example :

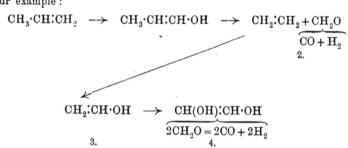

$$CH_3 \cdot CH{:}CH_2 \quad \rightarrow \quad CH_3 \cdot CH{:}CH \cdot OH \quad \rightarrow \quad CH_2{:}CH_2 + CH_2O$$
$$\underbrace{CO + H_2}$$
$$2.$$

$$CH_2{:}CH \cdot OH \quad \rightarrow \quad CH(OH){:}CH \cdot OH$$
$$\underbrace{2CH_2O = 2CO + 2H_2}$$
$$3. \qquad\qquad 4.$$

With a more limited supply of oxygen, say two-thirds of that required by the above scheme, the process would go on as far as Stage 3. There would then be no more oxygen left over to transform

$$CH_2{:}CH \cdot OH$$

into $CH(OH){:}CH \cdot OH$, and it would break down into acetylene and water, or carbon, hydrogen, and water. And since acetylene yields much methane when it undergoes thermal decomposition (Bone and Jerdan, *Proc.*, 1901, **17**, 164), the presence of as much as 10 per cent. of methane in the explosion products of the mixture $C_3H_6 + O_2$ is also explained.

The case of trimethylene may obviously be explained on similar lines. The course of events below the ignition point would probably be somewhat different from the above, owing to the greater stability of the initial oxidation products, and to differences in their mode of decomposition ; acetaldehyde, as well as formaldehyde, might arise in the cases of propylene and trimethylene.

In the case of the paraffins, the guiding fact is that both steam and carbon are produced when a mixture corresponding to $C_nH_{2n+2} + n/2O_2$ is exploded. Two explanations of this seem possible, namely, (1) that, as at lower temperatures, an aldehyde and steam are formed at an early stage, for example :

$$CH_3 \cdot CH_2 \cdot CH_3 \quad \rightarrow \quad CH_3 \cdot CH_2 \cdot CH_2 \cdot OH \quad \rightarrow \quad \genfrac{}{}{0pt}{}{CH_3 \cdot CH_2 \cdot CH(OH)_2}{CH_3 \cdot CH_2 \cdot CHO + H_2O} ;$$

the aldehyde would then probably decompose, yielding carbon monoxide and the lower paraffin, and a similar process would be repeated, until at length acetaldehyde would break down into carbon, hydrogen, methane, and carbonic oxide ; or (2) that the primary oxidation product $C_nH_{2n+1} \cdot OH$ might decompose into $C_nH_{2n} + H_2O$, the olefine subsequently being burnt in its own peculiar way. In either case, the end result with a mixture $C_nH_{2n+2} + n/2O_2$ would be much the same.

In view of the direct bearing of the subject on internal combustion engine practice, one of the authors is making arrangements for an investigation of the explosion of typical hydrocarbons and oxygen under much higher pressures than those which it was possible to employ in this research. It is also our intention to investigate more closely the conditions of equilibrium in hydrocarbon flames when mixtures containing more oxygen than that corresponding to $C_xH_{y+x}/2O_2$ are ignited.

In conclusion, we desire to thank the Government Grant Committee of the Royal Society for grants out of which part of the expenses of this investigation has been defrayed.

FUEL AND METALLURGICAL LABORATORY,
MANCHESTER UNIVERSITY.

LXXII.—*The Action of Phenylpropiolyl Chloride on Ketonic Compounds. Part II.*

By SIEGFRIED RUHEMANN.

IN a recent communication made to the Society in conjunction with R. W. Merriman (Trans., 1905, **87**, 1383), it has been shown that phenylpropiolyl chloride interacts with sodioacetylacetone to form a yellow compound, which was represented by the formula

$$C_6H_5 \cdot CH{:}C \cdot O \cdot C \cdot CH_3$$
$$CO \cdot C \cdot CO \cdot CH_3 \ .$$

This substance, under the influence of piperidine, is transformed into the red isomeride:

$$\begin{array}{c} C(OH){:}C(C_6H_5) \\ CO \cdot C(CO \cdot CH_3) \end{array}\!\!\!>\!C \cdot CH_3,$$

which dissolves in alkalis or their carbonates, as well as in organic bases, to yield blue solutions; on boiling with sodium carbonate, it changes into a colourless acid which has the same molecular composition and which was formulated thus:

$$\begin{array}{c} C(C_6H_5){:}C \cdot CO_2H \\ C(CH_3){:}C \cdot CO \cdot CH_3 \ . \end{array}$$

I have subjected these compounds to a closer study in order to supply further evidence for the correctness of the above formulæ, and, although this task has not been fully accomplished, yet on account of the interesting results obtained up to the present I have thought it advisable to place them on record.

Attention has been especially directed to the examination of the red isomeride, $C_{14}H_{12}O_3$, since its properties, as previously stated, closely resemble the compound obtained from ethyl oxaloacetate (see Ruhemann and Hemmy, Trans., 1897, 71, 34) and oxalyldibenzylketone, which Claisen and Ewan (*Annalen*, 1895, 284, 245) prepared by the action of sodium ethoxide on a mixture of dibenzylketone and ethyl oxalate. The latter substance had been carefully investigated by its discoverers, and the constitution which they derived from the study of this compound closely agrees with the formulæ of the other two substances with similar properties, as indicated by the following symbols:

$$\begin{matrix} C(OH)\!:\!C(C_6H_5) \\ CO\!\!-\!\!-\!\!C(C_6H_5) \end{matrix}\!\!>\!\!C\cdot OH \qquad \begin{matrix} C(OH)\!:\!C(CO_2Et) \\ CO\!\!-\!\!-\!\!C(CO_2Et) \end{matrix}\!\!>\!\!C\cdot CO_2Et$$

Oxalyldibenzylketone.   Ethyl oxalylaconitate.

$$\begin{matrix} C(OH)\!:\!C(C_6H_5) \\ CO\!-\!C(CO\cdot CH_3) \end{matrix}\!\!>\!\!C\cdot CH_3.$$

Acetyloxalylphenylmethylpropene.[*]

Although this fact may be regarded as a sufficient proof for the formula of the red compound $C_{14}H_{12}O_3$, yet I have thought it advisable to support this evidence by an investigation of the substance on lines similar to those followed up by Claisen and Ewan in the case of oxalyldibenzylketone. Like this substance, the red isomeride $C_{14}H_{12}O_3$ interacts with one molecule of phenylhydrazine only to yield a phenyl-hydrazone. The circumstance that it is insoluble in alkalis leads to the conclusion that the compound is to be represented thus :

$$\begin{matrix} C_6H_5\cdot NH\cdot N\!:\!C\!\!-\!\!-\!\!CH(C_6H_5) \\ CO\cdot C(CO\cdot CH_3) \end{matrix}\!\!>\!\!C\cdot CH_3,$$

and that its formation is accompanied by the change of the enolic into the ketonic group. The behaviour of the red substance towards semicarbazide is analogous, as a semicarbazone is produced, which on account of its insolubility in caustic soda must be expressed by the symbol :

$$\begin{matrix} NH_2\cdot CO\cdot NH\cdot N\!:\!C\!\!-\!\!-\!\!CH(C_6H_5) \\ CO\cdot C(CO\cdot CH_3) \end{matrix}\!\!>\!\!C\cdot CH_3.$$

The fact that this compound is colourless is of additional interest, because it indicates that the red colour of the substance $C_{14}H_{12}O_3$ is due to the enolic grouping and disappears with the transformation

---

[*] I have adopted this name instead of the one used before (*loc. cit.*) in order to indicate the relation of this compound to the other two substances.

into the ketonic group. The product of the action of hydroxylamine on the red compound points to the same conclusion, since the colourless oxime,

$$\begin{matrix} C(NOH){\cdot}CH(C_6H_5) \\ | \\ CO \text{------} C(CO{\cdot}CH_3) \end{matrix}\!\!\Big>\!C{\cdot}CH_3,$$

which, indeed, dissolves in sodium carbonate, yields a light yellow solution.

This behaviour of the red compound $C_{14}H_{12}O_3$ allows of a ready explanation of the results at which Claisen and Ewan (loc. cit.) arrived in the course of their investigation of oxalyldibenzylketone. These chemists have found that this substance forms with caustic soda a yellow salt, and blue solutions with an excess of the alkali. The metallic derivatives which are thus produced they represent by the formulæ :

$$\begin{matrix} C(ONa){:}C(C_6H_5) \\ | \\ CO \text{---} CH(C_6\overset{.}{H}_5) \end{matrix}\!\!\Big>\!CO \quad \text{and} \quad \begin{matrix} C(ONa){:}C(C_6H_5) \\ | \\ CO \text{---} CNa(C_6H_5) \end{matrix}\!\!\Big>\!CO.$$

The action of alkalis on the red substance $C_{14}H_{12}O_3$ leads to the conclusion that the blue salt is to be represented thus :

$$\begin{matrix} C(ONa){:}C(C_6H_5) \\ | \\ .\,CO \text{------} C(C_6H_5) \end{matrix}\!\!\Big>\!C{\cdot}ONa,$$

because the grouping $:C(OH){\cdot}CO-$ contained in the compound $C_{14}H_{12}O_3$. as well as in ethyl oxalylaconitate, gives rise to the formation of blue salts. On the other hand, the yellow sodium derivative of oxalyldibenzylketone is to be expressed by the formula :

$$\begin{matrix} CO{\cdot}CH(C_6H_5) \\ | \\ CO \text{---} C(C_6H_5) \end{matrix}\!\!\Big>\!C{\cdot}O\overset{.}{N}a,$$

in which the ketonic group occupies the β-position with respect to the enolic group. It would therefore follow that the former arrangement leads to the formation of blue salts, whilst the other grouping yields yellow salts. At present, there exist only the three above-mentioned compounds which contain the group $:C(OH){\cdot}CO-$ and which form blue salts,* but a number of substances of the type $:C(OH){\cdot}\overset{|}{\underset{|}{C}}{\cdot}CO-$ are known, and they all dissolve in alkalis or their carbonates to yield yellow solutions ; the following may be mentioned as members of this series : benzoylacetylacetone, $_6H_5{\cdot}CO{\cdot}C(CO{\cdot}CH_3){:}C(OH){\cdot}CH_2$; di-

---

* A similar compound is most probably formed by the action of sodium acetate and acetic anhydride on ethyl acetonylacetate, when a violet solution is produced (see Claisen and Stylos, Ber., 1889, 22, 1141). I am engaged in the study of this reaction.

benzoylacetone, $(C_6H_5 \cdot CO)_2C \colon C(OH) \cdot CH_3$, and xanthochelidonic acid, $CO[CH \colon C(OH) \cdot CO_2H]_2$.

The view which I have advanced concerning the constitution of the blue sodium derivative of oxalyldibenzylketone is in complete harmony with the behaviour of the ketonic compound itself. According to Claisen and Ewan, the blue salt interacts with methyl iodide to yield a methyl derivative, which dissolves in alkalis, forming yellow solutions, and, on hydrolysis, furnishes dibenzylmethylketone. It follows, therefore, that this oxalyldibenzylmethylketone has the formula :

$$\begin{array}{l} CO \cdot C(CH_3)(C_6H_5) \\ CO \text{------} C(C_6H_5) \end{array}\!\!\!>\!C \cdot OH,$$

and that its formation is analogous to the production of ethyl methylacetoacetate.

An isomeric methyl derivative has been obtained by Claisen and Ewan (*loc. cit.*) on treatment of the silver derivative of oxalyldibenzylketone with methyl iodide. The facts that this compound gives with alkalis blue solutions and, on hydrolysis, furnishes oxalyldibenzylketone lead to the symbol :

$$\begin{array}{l} C(OH) \cdot C(C_6H_5) \\ CO \text{------} C(C_6H_5) \end{array}\!\!\!>\!CO \cdot CH_3 \ ;$$

a similar structure should be attributed to the acetyloxalyldibenzylketone :

$$\begin{array}{l} C(OH) \colon C(C_6H_5) \\ CO \text{------} C(C_6H_5) \end{array}\!\!\!>\!CO \cdot CO \cdot CH_3,$$

which forms a violet potassium salt.

The properties of the phenylhydrazone and the oxime agree with the formulæ :

$$\begin{array}{l} C(\colon N \cdot NHPh) \cdot CH(C_6H_5) \\ CO \text{--------} C(C_6H_5) \end{array}\!\!\!>\!C \cdot OH \quad \text{and} \quad \begin{array}{l} C(NOH) \cdot CH(C_6H_5) \\ CO \text{------} C(C_6H_5) \end{array}\!\!\!>\!C \cdot OH.$$

The analogous constitution must be assigned to the amino- and anilino-derivatives of the oxalyl compound.

It has previously been pointed out (see Ruhemann and Merriman, . *loc. cit.*) that the properties of the direct product of the action between phenylpropiolyl chloride and acetylacetone differ most markedly from those of its red isomeride. This difference is also manifested by the behaviour towards phenylhydrazine, because an almost colourless substance is formed, the solutions of which are light yellow. The analytical data indicate that two molecules of phenylhydrazine are required for the formation of this substance and that the reaction takes place according to the equation

$$C_{14}H_{12}O_3 + 2NH_2 \cdot NH \cdot C_6H_5 = C_{26}H_{24}ON_4 + 2H_2O.$$

The constitution of this compound shouldmost probably be represented thus : $\begin{array}{l}C(C_6H_5){:}C(NH{\cdot}NHPh){\cdot}C{:}NNHPh \\ C(CH_3){=\!=\!=\!=\!=\!=\!=\!=}C{\cdot}C{\cdot}CH_3\end{array}$. The mode of its formation would therefore be similar to he transformation of the yellow substance $C_{14}H_{12}O_3$ into its red isomeride.

EXPERIMENTAL

The Phenylhydrazone of Acetyloxalylphenylmethylpropene.

On adding phenylhydrazine, dissolved in 50 per cent. acetic acid, to the cold solution of the red compound $C_1H_{12}O_3$ in methyl alcohol, a deep red coloration takes place which is no doubt due to the formation of a salt. The colour, after a short time, canges to yellowish-red and a solid is deposited. This product is very solble in chloroform, sparingly so in ether, but readily dissolves in hot alcohol, and, on cooling, crystallises in yellow plates which melt and decompose indefinitely at 169—171°.

0·2023 gave 0·5595 CO_2 and 0·1033 H_2O. $C = 75·42$; $H = 5·66$.
0·2157 „ 16·2 c.c. moist nitrogen at 20° and 768 mm. $N = 8·69$.
$C_{20}H_{18}O_2N_2$ requires $C = 75·47$: $H = ·66$; $N = 8·80$ per cent.

The phenylhydrazone is insoluble in alkalis or their carbonates; it dissolves in concentrated sulphuric acid yelding a deep red solution.

The Oxime of Acetyloxalylphenylmethylpropene.

This substance is formed on adding a mixture of equivalent quantities of hydroxylamine hydrochloride and sodium carbonate, dissolved in water, to the alcoholic solution of the red compound $C_{14}H_{12}O_3$. The solution develops a deep red colour which rapidly changes to a light yellow. The mixture, after two ays. is extracted with ether, when, after evaporation of the solvent, he oxime is left behind as a slightly coloured oil which shortly sets to a solid. For purification it is dissolved in a little ether and the solution mixed with light petroleum, when colourless prisms gradually separate which melt indefinitely at 129—130°.

0·2009 gave 0·5068 CO_2 and 0·0991 H_2O. $C = 68·80$; $H = 5·48$.
0·2233 „ 11·8 c.c. moist nitrogen at 20° and 755·5 mm. $N = 6·0$.
0·1725 gave 8·8 c.c. moist nitrogen a 20° and 759 mm. $N = 5·83$.
$C_{14}H_{13}O_3N$ requires $C = 69·13$; $H = 5·35$; $N = 5·76$ per cent.

The oxime dissolves freely in ether or alcohol, but only sparingly in boiling water ; it is readily soluble i sodium carbonate, forming a yellow solution from which the oxime is reprecipitated by hydrochloric acid as well a by acetic acid. The alcoholic solution of the oxime yields with ferric chloride an olive-green coloration.

The Semicarbacone of Acetyloxalylphenylmethylpropene.

On adding the mixure of equal weights of semicarbazide hydrochloride and potassium acetate, dissolved in water, to an alcoholic solution of the red subtance $C_{14}H_{12}O_3$, a dark coloration in this case, also, takes place whic quickly becomes light red. After standing overnight, water is adod, when a white solid is precipitated which is insoluble in water, benene, or ether, but dissolves in boiling alcohol, although with great diiculty, and, on cooling, gradually separates in colourless prisms. Tb semicarbazide is readily soluble in warm glacial acetic acid and ı precipitated by water in small crystals which melt and decompose at 08°.

0·2055 gave 0·4760 O_2 and 0·0998 H_2O.   C = 63·17 ; H = 5·39.
0·2160 „ 27·8 c.c. ıoist nitrogen at 20° and 769 mm.   N = 14·89.
$C_{15}H_{15}O_2N_3$ require $C = 63·16$; H = 5·26 ; N = 14·74 per cent.

Action of Phenylhydrzine on Acetylmethylbenzylideneketodihydrofurfuran.

The yellow substance $A_{16}H_{12}O_3$, which is formed by the action of sodioacetylacetone on phnylpropiolyl chloride when treated in alcoholic solution with phenyhydrazine dissolved in acetic acid, yields a deep red coloration whh rapidly changes to yellowish-red. The mixture slowly deposits a olid which is readily soluble in hot alcohol and, on cooling, crystalliss in almost colourless needles which melt and decompose at 161—162°.

0·1240 gave 0·3468 CO and 0·0698 H_2O.   C = 76·27 ; H = 6·25.
0·2013 „ 24 c.c. mois nitrogen at 20° and 769·5 mm.   N = 13·81.
0·2197 „ 26·4 c.c. most nitrogen at 21° and 769·5 mm.   N = 13·85.
$C_{26}N_{24}ON_4$ requires C = 76·47 ; H = 5·88 ; N = 13·72 per cent.

On adding concentrated sulphuric acid to this compound, it first turns yellowish-red, and thn dissolves, yielding a bluish-green solution which gradually changes t yellow. On warming the compound with concentrated hydrochloric acid, it yields a yellow solution which rapidly becomes colourless.

In conclusion, I express ıy thanks to the Grant Committee of the Chemical Society for the help afforded me in carrying out this investigation.

GONVILLE AND CAIUS COLLGE,
CAMBRIDGE.

GONVILLE AND CAIUS COLLEGE,
   CAMBRIDGE.

LXXIII.—*Studies in Asymmetric Synthesis. V. Asymmetric Syntheses from* l-*Bornyl Pyruvate.*

By ALEXANDER MᶜKENZIE and HENRY WREN, B.A., B.Sc., Ph.D.

OPTICALLY active lactic acids are [formed by a variety of methods, which may be sharply differentiated from one another as follows :

I. *Production from Substances Already Optically Active.*

Under this heading are embraced such cases as the formation of "fermentation" lactic acid from carbohydrates by the agency of micro-organisms (compare Trans., 1905, **87**, 1373), the production of *d*-lactic acid from meat extract, the formation of *d*-lactic acid in the organism (Saito and Katsuyama, *Zeit. physiol. Chem.*, 1901, **32**, 214), and the formation of *d*-lactic acid from *d*-alanine (Fischer and Skita, *Zeit. physiol. Chem.*, 1901, **33**, 177).

II. *Production by Resolution of the Inactive Variety into its Optically Active Components.*

(*a*) Biological method by moulds (Lewkowitsch, *Ber.*, 1883, **16**, 2720 ; Linossier, *Bull. Soc. Chim.*, 1891, [iii], **6**, 10 ; McKenzie and Harden, Trans., 1903, **83**, 424) and by bacteria (P. F. Frankland and MacGregor, Trans., 1893, **63**, 1028).'

(*b*) Crystallisation method, by separation of alkaloidal salts (Purdie and Walker, Trans., 1892, **61**, 754 ; Jungfleisch, *Compt. rend.*, 1904, **139**, 56).

(*c*) Crystallisation method, by the addition of a nucleus of active zinc ammonium lactate to a supersaturated solution of zinc ammonium *i*-lactate (Purdie, Trans., 1893, **63**, 1143).

(*d*) Fractional saponification method from *l*-menthyl *dl*-lactate (McKenzie and Thompson, Trans., 1905, **87**, 1004).

III. *Production by Asymmetric Synthesis.*

The asymmetric synthesis of *l*-lactic acid by the reduction of *l*-menthyl pyruvate has recently been described (Trans., 1905, **87**, 1373).

Lactic acid is the only compound containing an asymmetric carbon atom which can be obtained in its optically active forms by such a variety of methods.

The authors have therefore considered it of interest to record in the present paper another instance of the asymmetric synthesis of lactic acid in accordance with the following scheme :

$$CH_3 \cdot CO \cdot CO_2H \text{ (inactive)} \longrightarrow CH_3 \cdot CO \cdot CO_2 \cdot C_{10}H_{17}^{*} \text{ (active)}.$$

$$\longrightarrow CH_3 \cdot C\overset{*}{H}(OH) \cdot CO_2 \cdot C_{10}H_{17}^{*} \text{ (active)}.$$

$$\longrightarrow CH_3 \cdot C\overset{*}{H}(OH) \cdot CO_2H \text{ (active)}.$$

Pyruvic acid is inactive, and does not contain an asymmetric carbon atom. Its l-bornyl ester is active in virtue of the l-bornyl group. When this ester is reduced, a new asymmetric carbon atom is generated, and a mixture of unequal amounts of l-bornyl d-lactate and l-bornyl l-lactate, containing an excess of the latter, is formed. When the latter mixture is saponified by an excess of alkali and the resulting l-borneol completely removed, the aqueous solution of potassium salt is dextrorotatory, and contains a mixture of potassium d- and l-lactates with an excess of the latter, since on acidification by mineral acid it becomes lævorotatory (l-lactic acid forms dextrorotatory salts).   On extracting the lactic acid and converting it into its lithium salt, the latter is found to be dextrorotatory.

In a recent paper by one of us (this vol., p. 365), the action of magnesium alkyl (or aryl) halides on l-menthyl benzoylformate, l-bornyl benzoylformate, and l-menthyl pyruvate respectively was investigated with the view of contrasting the effect, firstly, of the active menthyl and bornyl groups, and, secondly, of the various alkyl (or aryl) halides used, on the extent of the asymmetric syntheses of the resulting substituted glycollic acids.   The direction of rotation of the mixture of d- and l-substituted glycollic acids, resulting from the esters in question by variation of the Grignard reagent, is indicated in the table on p. 690.

From the rotation values quoted (*loc. cit.*) it appears that, in the actions in which l-menthyl benzoylformate is involved, the increase in weight of the hydrocarbon group of the Grignard reagent has the effect of diminishing the lævorotation of the mixture of unequal amounts of the substituted glycollic acid obtained in each case.   With l-bornyl benzoylformate the lævorotation is actually transformed into a dextro one with the increase in weight of the hydrocarbon group of the Grignard reagent, and the mixture of phenyl-α-naphthyl-glycollic acids obtained is more dextrorotatory than the mixture of phenyl*iso*butylglycollic acids.   In the two actions quoted with l-menthyl pyruvate, the substitution of the phenyl for the ethyl group in the Grignard reagent very considerably increases the dextro-rotation of the acid mixture.

If the influence of the bornyl group be contrasted with that of the menthyl group, the atrolactinic acid mixture obtained from l-bornyl

Ester.	Magnesium alkyl halide.	Acid mixture.	Sign of rotation.
$l\text{-}C_6H_5\cdot CO\cdot CO_2\cdot C_{10}H_{19}$	$CH_3\cdot Mg\cdot I$	$\begin{matrix} CH_3 \\ C_6H_5 \end{matrix}\!\!>\!C\!<\!\begin{matrix} OH \\ CO_2H \end{matrix}$	Lævo
,, ,,	$C_2H_5\cdot Mg\cdot Br$	$\begin{matrix} C_2H_5 \\ C_6H_5 \end{matrix}\!\!>\!C\!<\!\begin{matrix} OH \\ CO_2H \end{matrix}$	
,, ,,	$n\text{-}C_3H_7\cdot Mg\cdot I$	$\begin{matrix} n\text{-}C_3H_7 \\ C_6H_5 \end{matrix}\!\!>\!C\!<\!\begin{matrix} OH \\ CO_2H \end{matrix}$	,,
,, ,,	$iso\text{-}C_4H_9\cdot Mg\cdot I$	$\begin{matrix} iso\text{-}C_4H_9 \\ C_6H_5 \end{matrix}\!\!>\!C\!<\!\begin{matrix} OH \\ CO_2H \end{matrix}$	,,
,, ,,	$tert.\text{-}C_4H_9\cdot Mg\cdot I$	$\begin{matrix} tert.\text{-}C_4H_9 \\ C_6H_5 \end{matrix}\!\!>\!C\!<\!\begin{matrix} OH \\ CO_2H \end{matrix}$	,,
,, ,,	$\alpha\text{-}C_{10}H_7\cdot Mg\cdot Br$	$\begin{matrix} \alpha\text{-}C_{10}H_7 \\ C_6H_5 \end{matrix}\!\!>\!C\!<\!\begin{matrix} OH \\ CO_2H \end{matrix}$	,,
$l\text{-}C_6H_5\cdot CO\cdot CO_2\cdot C_{10}H_{17}$	$CH_3\cdot Mg\cdot I$	$\begin{matrix} CH_3 \\ C_6H_5 \end{matrix}\!\!>\!C\!<\!\begin{matrix} OH \\ CO_2H \end{matrix}$	,,
,, ,,	$C_2H_5\cdot Mg\cdot I$	$\begin{matrix} C_2H_5 \\ C_6H_5 \end{matrix}\!\!>\!C\!<\!\begin{matrix} OH \\ CO_2H \end{matrix}$	,,
,, ,,	$iso\text{-}C_4H_9\cdot Mg\cdot I$	$\begin{matrix} iso\text{-}C_4H_9 \\ C_6H_5 \end{matrix}\!\!>\!C\!<\!\begin{matrix} OH \\ CO_2H \end{matrix}$	Dextro
,, ,,	$\alpha\text{-}C_{10}H_7\cdot Mg\cdot Br$	$\begin{matrix} \alpha\text{-}C_{10}H_7 \\ C_6H_5 \end{matrix}\!\!>\!C\!<\!\begin{matrix} OH \\ CO_2H \end{matrix}$	,,
$l\text{-}CH_3\cdot CO\cdot CO_2\cdot C_{10}H_{19}$	$C_2H_5\cdot Mg\cdot Br$	$\begin{matrix} CH_3 \\ C_2H_5 \end{matrix}\!\!>\!C\!<\!\begin{matrix} OH \\ CO_2H \end{matrix}$	,,
,, ,,	$C_6H_5\cdot Mg\cdot Br$	$\begin{matrix} CH_3 \\ C_6H_5 \end{matrix}\!\!>\!C\!<\!\begin{matrix} OH \\ CO_2H \end{matrix}$	,,

benzoylformate is seen to be much less lævorotatory than that obtained from l-menthyl benzoylformate. The effect of the bornyl group is to increase the proportion of the d-acid in the acid mixture.

The present paper contains an account of the action of magnesium ethyl iodide, magnesium isobutyl iodide, magnesium phenyl bromide, and magnesium α-naphthyl bromide respectively on l-bornyl pyruvate, whilst the action of magnesium isobutyl iodide and magnesium α-naphthyl bromide respectively on l-menthyl pyruvate has also been examined. The sign of rotation of the acid mixture obtained in each case is indicated in the table on p. 691.

Whilst the results obtained by the application of the Grignard action to l-bornyl pyruvate are, considered by themselves, not of much interest as typical examples of asymmetric syntheses owing to the feeble optical activity of the resulting substituted glycollic acids, they are in striking uniformity with the results quoted in the former paper (*loc. cit.*), and bear out in a remarkable degree the influences of mass and other factors, to which attention has just been drawn. For example, the mixture of d- and l-atrolactinic acids, resulting from the action of magnesium methyl iodide on l-bornyl benzoylformate, is much less lævorotatory than the mixture resulting from l-menthyl benzoylformate and magnesium methyl iodide : a similar effect of the l-bornyl as contrasted with the l-menthyl group is seen when the

Ester.	Magnesium alkyl halide.	Acid mixture.	Sign of rotation.
l-CH$_3$·CO·CO$_2$·C$_{10}$H$_{17}$	C$_2$H$_5$·Mg·I	$\frac{CH_3}{C_2H_5}{>}C{<}\frac{OH}{CO_2H}$	Dextro
,,	,,	iso-C$_4$H$_9$·Mg·I	$\frac{CH_3}{C_4H_9}{>}C{<}\frac{OH}{CO_2H}$	,,
,,	,,	C$_6$H$_5$·Mg·Br	$\frac{CH_3}{C_6H_5}{>}C{<}\frac{OH}{CO_2H}$	,,
,,	,,	α-C$_{10}$H$_7$·Mg·Br	$\frac{CH_3}{\alpha\text{-}C_{10}H_7}{>}C{<}\frac{OH}{CO_2H}$	Lævo
l-CH$_3$·CO·CO$_2$·C$_{10}$H$_{19}$	iso-C$_4$H$_9$·Mg·I	$\frac{CH_3}{iso\text{-}C_4H_9}{>}C{<}\frac{OH}{CO_2H}$	Dextro
,,	,,	α-C$_{10}$H$_7$·Mg·Br	$\frac{CH_3}{\alpha\text{-}C_{10}H_7}{>}C{<}\frac{OH}{CO_2H}$	,,

action of magnesium phenyl bromide on l-menthyl pyruvate and l-bornyl pyruvate is contrasted. We should expect that the mixture of atrolactinic acids obtained from l-bornyl pyruvate would be much less dextrorotatory than that obtained from l-menthyl [pyruvate, and this on experiment was found to be the case. Again, the effect of increasing the mass of the hydrocarbon group in the magnesium alkyl (or aryl) halide in the actions on l-bornyl benzoylformate is to cause a change in sign from lævo- to dextro-rotation. A similar effect is found with l-bornyl pyruvate; whilst dextrorotatory acid mixtures are obtained by the application of magnesium ethyl iodide, magnesium isobutyl iodide and magnesium phenyl bromide respectively, the sign of the acid mixture changes from dextro to lævo when magnesium α-naphthyl bromide is employed.

The results obtained with l-menthyl pyruvate are also in harmony with the previous observations. The acid mixture obtained by the action of magnesium isobutyl iodide on l-menthyl pyruvate was, as was expected, slightly dextrorotatory, whilst the influence of the heavier α-naphthyl group is seen by the marked dextrorotation of the acid mixture, resulting from the action of magnesium α-naphthyl bromide on l-menthyl pyruvate.

EXPERIMENTAL.

Pyruvic acid was heated with three times its weight of l-borneol for ten hours at 100°, a current of dry hydrogen chloride having been passed at intervals of three hours into the mixture. After the product had been washed with water several times, it was submitted to distillation in steam in order to remove the bulk of the borneol. The residual oil was washed with very dilute sodium carbonate, and its ethereal solution dried with anhydrous sodium sulphate. After expulsion of the ether, the borneol was separated from the ester by allowing the mixture to rise almost to the boiling point under diminished pressure, when the borneol sublimed and was removed from time to time from the side-tube of the distilling flask. The removal of the last traces of borneol in this manner is a tedious process, but presents no difficulty.

l-*Bornyl pyruvate*, $CH_3 \cdot CO \cdot CO_2 \cdot C_{10}H_{17}$, is a colourless oil and boils at 143—144° under 18 mm. pressure.

0·1473 gave 0·3760 CO_2 and 0·1212 H_2O. C = 69·6 ; H = 9·2.

$C_{13}H_{20}O_3$ requires C = 69·6 ; H = 9·0 per cent.

A determination of its specific rotation gave the result : $l = 1$, $d\ 19·9°/4°\ 1·0467$, $a_D^{19·9°} - 54·82°$, $[a]_D^{19·9°} - 52·4°$.

Reduction of l-*Bornyl Pyruvate. The Asymmetric Synthesis of* l-*Lactic Acid.*

A solution of 12·5 grams of *l*-bornyl pyruvate in 50 c.c. of moist ether was added to an excess of aluminium amalgam. A vigorous evolution of hydrogen quickly took place. More ether and a little water were added from day to day during four days. The ethereal solution was drained off and the residue washed with ether. The oil, resulting from the ethereal solution, was dissolved in a solution of 6·5 grams of potassium hydroxide in 70 c.c. of methyl alcohol and allowed to remain overnight at the ordinary temperature. To ensure complete saponification, the liquid was then boiled for one hour under a reflux condenser. The methyl alcohol was expelled, water added, and the precipitated borneol drained off. The filtrate, which was strongly alkaline to litmus, was acidified by the addition of dilute sulphuric acid and made faintly alkaline by potassium hydroxide, since it was desirable that, during the subsequent prolonged evaporation, the solution should not be too strongly alkaline ; the presence of a large excess of alkali might tend to racemise any active potassium lactate present, although this racemising effect of alkali on active potassium lactate is slight (Trans., 1905, **87**, 1373 ; compare also Godchot and Jungfleisch, *Compt. rend.*, 1905, **140**, 719). The aqueous solution of potassium salt was then extracted with éther and decolorised by being heated on the water-bath for several hours with animal charcoal. By this treatment the borneol was completely removed. The solution was then filtered and sterilised. The polarimetric determination, made two days later, showed that the solution was dextrorotatory, 28 c.c. in a 4-dcm. tube giving $a_D^{20°} + 0·44°$. The solution was then concentrated and dilute sulphuric acid added in excess as indicated by Congo red paper ; the precipitated potassium sulphate was removed, and the filtrate, when made up to 28 c.c. and examined in a 4-dcm. tube, proved to be lævorotatory, giving $a_D^{20°}$ $-0.17°$. The lactic acid present in this solution was then extracted with ether by aid of a continuous extraction apparatus, the ether was expelled and the aqueous solution of the resulting syrup neutralised by the exact amount of lithium carbonate. The aqueous solution of lithium salt, after evaporation on the water-bath for one hour, measured 30 c.c. ; it was dextrorotatory, 28 c.c. of it in a 4-dcm. tube giving $a_D^{20°}$ $+ 0.22°$· The concentration of the solution as estimated by withdrawing an aliquot portion, evaporating off the water and drying the residue at 100° until constant in weight was 13·46.

A portion of the solution was evaporated to dryness and the residue on analysis shown to consist of lithium lactate.

0.3892, dried at $120-125°$, gave 0.2206 Li_2SO_4.   Li = 7·24.

$C_3H_5O_3Li$ requires Li = 7·32 per cent.

The specific rotation of the mixture of lithium d- and l-lactates prepared in the manner indicated is $+0.4°$ at 20°, whereas the value for the pure active lithium lactate as deduced from Purdie and Walker's determinations (Trans., 1895, **67**, 616) is 12·0° for an N-solution at about 10°.

In the asymmetric synthesis described, the necessary precautions were taken to ensure that the rotations observed were not due to a resolution of l- bornyl dl-lactate by fractional saponification or to a resolution of potassium r-lactate by the growth of micro-organisms.

Action of Magnesium Alkyl (or Aryl) Halides on 1-Bornyl Pyruvate.

A solution of magnesium ethyl iodide ($1\frac{1}{4}$ mol.) in 25 c.c. of ether was added by means of a siphon within an interval of forty-five minutes to a solution of 10 grams (1 mol.) of l-bornyl pyruvate in 30 c.c. of ether. The saponification of the oil, obtained by decomposing the product resulting from the Grignard action by ice and mineral acid, was conducted with a solution of 5 grams of potassium hydroxide in 100 c.c. of ethyl alcohol. The ethyl alcohol and borneol were removed as previously described (this vol., p. 374), and the acid, obtained from the potassium salt, decolorised in aqueous solution. A crop of inactive acid was separated, and the filtrate proved to be feebly dextrorotatory, giving $a_D + 0.06°$ in a 2-dcm. tube.

A solution of magnesium *iso*butyl iodide ($2\frac{1}{2}$ mol.) in 30 c.c. of ether was added within an interval of twenty minutes to a solution of 10 grams (1 mol.) of l-bornyl pyruvate in 30 c.c. of ether. The saponification of the ester mixture, obtained in the usual manner, was conducted by allowing it to remain for two hours at the ordinary temperature in contact with a solution of 5·4 grams of potassium hydroxide in 100 c.c. of ethyl alcohol and then boiling for one hour with the addition of 5 c.c. of water. The ethyl alcohol and borneol were removed, and the acid, obtained from the potassium salt by acidification and extraction with ether, was converted into barium salt; the aqueous solution of the latter, when decolorised, proved to be dextrorotatory, the observed rotation in a 2-dcm. tube being only $a_D + 0.10°$. The concentration of this solution as estimated by withdrawing an aliquot portion, evaporating to dryness and then heating at 130° until constant in weight, was 5·47. An estimation of barium gave $Ba = 38.2$ per cent., a value which indicated that some barium pyruvate was present in the solution together with the barium methyl-*iso*butylglycollate.

A solution of magnesium phenyl bromide (1¼ mol.) in 25 c.c. of ether was added drop by drop within an interval of seventy minutes to a solution of 10 grams (1 mol.) of *l*-bornyl pyruvate in 30 c.c. of ether. After a night, the product was boiled for fifteen minutes and decomposed by the successive addition of crushed ice and dilute mineral acid. The oil, resulting from the ethereal solution, was allowed to remain for two hours at the ordinary temperature in contact with a solution of 4·9 grams of potassium hydroxide in 100 c.c. of ethyl alcohol, and, after the addition of 5 c.c. of water, the solution was boiled for one hour. After the removal of the ethyl alcohol and borneol, the aqueous solution of potassium salt could not be sufficiently decolorised to permit of accurate polarimetric observation. It was accordingly acidified by mineral acid and the solution extracted with ether. The resulting atrolactinic acid was partially decolorised in aqueous solution and then converted into barium salt by boiling with an excess of barium carbonate. The filtrate (15 c.c.) was slightly but distinctly dextrorotatory, giving $a_D + 0.08°$ in a 2-dcm. tube.

This slight dextrorotation was confirmed by a second experiment, where magnesium phenyl bromide was used in the proportion of 2½ mols. to 1 mol. of ester.

A solution of magnesium α-naphthyl bromide (2½ mol.) in 50 c.c. of ether was added within an interval of twenty minutes to a solution of 10 grams (1 mol.) of *l*-bornyl pyruvate in 30 c.c. of ether. The action was vigorous. After remaining at the laboratory temperature overnight, the product was decomposed by ice and dilute hydrochloric acid, and the oil, resulting from the ethereal solution, submitted to distillation in steam in order to remove naphthalene. The residue in the distilling flask was extracted with ether, the ether expelled and the oil dissolved in a solution of 6 grams of potassium hydroxide in 100 c.c. of ethyl alcohol. After two hours at the ordinary temperature, the solution was boiled for one hour. The ethyl alcohol and borneol were then removed in the customary manner. The aqueous solution of potassium salt, which was rather highly coloured, was decomposed by an excess of sulphuric acid, and the acid, resulting from the extraction with ether, was directly converted into barium salt by the addition of a solution of 15 grams of crystallised barium hydroxide in 100 c.c. of water. The excess of barium hydroxide was removed by carbon dioxide, and, when the solution had been filtered off from barium carbonate, it was decolorised by animal charcoal. This method was found to be a convenient one for obtaining a solution sufficiently colourless for accurate polarimetric observation. It is practically impossible to decolorise the aqueous solution of potassium salt directly resulting from the saponification, nor can the acid, obtained from the potassium salt, be itself conveniently decolorised. The aqueous solution of barium salt was concentrated to 32 c.c., of which 15 c.c. in a 2-dcm. tube gave a_D $-0.14°$. The concentration of this solution, as estimated for anhydrous salt, was 7·54. An estimation of barium gave Ba $= 31.3$ per cent., a result which indicated that the interaction between the magnesium a-naphthyl bromide and l-bornyl pyruvate had been incomplete.

A second experiment yielded a similar result, a lævorotatory acid mixture again being obtained. As a product from this experiment, i-a-naphthylmethylglycollic acid was isolated; when crystallised from benzene and then dried at 100°, it melted at 138—139°, whereas Grignard (*Ann. Chim. Phys.*, 1902, [vii], **27**, 548) gives 143°.

Action of Magnesium isoButyl Iodide on 1-Menthyl Pyruvate.

A solution of magnesium *iso*butyl iodide ($2\frac{1}{2}$ mol.) in 30 c.c. of ether was added within an interval of thirty minutes to a solution of 10 grams (1 mol.) of l-menthyl pyruvate in 30 c.c. of ether. The product was treated as in the case of the corresponding experiment with l-bornyl pyruvate. The barium salt obtained (21 c.c.) was dextrorotatory, 14 c.c. in a 2-dcm. tube giving $a_D + 0.13°$. The concentration of this solution, as estimated for anhydrous salt, was 4·85. The anhydrous salt contained 40·4 per cent. of barium, a result which indicated the presence of barium pyruvate in the solution.

Action of Magnesium a-Naphthyl Bromide on 1-Menthyl Pyruvate.

A solution of magnesium a-naphthyl bromide ($2\frac{1}{2}$ mol.) in 25 c.c. of ether was added within an interval of twenty minutes to a solution of 10 grams (1 mol.) of l-menthyl pyruvate in 30 c.c. of ether. The action was vigorous. After twenty-four hours, the product was decomposed and the naphthalene removed as described for the corresponding experiment with the bornyl ester. The ester mixture was dissolved in a solution of 5·2 grams of potassium hydroxide in 100 c.c. of ethyl alcohol, the solution allowed to remain for two hours at the ordinary temperature and then boiled for one hour. The ethyl alcohol and menthol were then removed by the usual method. The acid, obtained by acidifying the aqueous solution of potassium salt and extracting with ether, was converted into barium salt by the addition of baryta water, the excess of which was removed by carbon dioxide and the resulting barium salt decolorised by charcoal. The aqueous solution of barium salt was concentrated to a bulk of 22 c.c., of which 15 c.c. in a 2-dcm. tube gave $a_D^{15°} + 0·89°$. The concentration of this solution as determined by withdrawing an aliquot portion, evaporating off the water and drying the residue at 130° was 6·43.

The authors desire to thank the Research Fund Committee of the Chemical Society for a grant in aid of this research.

THE UNIVERSITY,   BIRKBECK COLLEGE,
BIRMINGHAM.   LONDON, E.C.

LXXIV.—*Aromatic Sulphonium Bases.*

By SAMUEL SMILES and ROBERT LE ROSSIGNOL.

IT is well known that neither purely aromatic nor mixed fatty aromatic sulphonium bases can be obtained by the usual method which applies in the aliphatic series, namely, the direct union of a sulphide with a halogen derivative. Hence the statement has been made (Meyer and Jacobson, *Lehrbuch der Organische Chemie*, vol. II, p. 130; Kehrmann and Duttenhöfer, *Ber.*, 1905, **38**, 4197) that aromatic sulphonium bases do not exist, but the fact seems to have been overlooked that Michaelis and Godchaux formerly (*Ber.*, 1891, **24**, 757) obtained a substance of this class by the action of thionyl chloride on mercury dimethylaniline. The present paper contains an account of three methods by which purely aromatic sulphonium derivatives may be easily prepared. We have already stated in a preliminary communication to the Society (*Proc.*, 1906, **22**, 24) that whilst our experiments were nearing completion Kehrmann and Duttenhofer published an account (*loc. cit.*) of a research in which they succeeded in obtaining mixed fatty aromatic sulphonium salts by the action of methyl sulphate on aromatic sulphides.

The Action of Thionyl Chloride on Phenetole.

When attempting to prepare phenetyl sulphoxide by adding aluminium chloride to a mixture of thionyl chloride and phenetole, it was noticed that the product of the reaction did not contain the required sulphoxide, but consisted almost entirely of another substance which was found to be triphenetylsulphonium chloride. On making further experiments, it was found that (1) sometimes a small quantity of phenetyl sulphoxide was formed at the same time, and (2) that even when carefully dried materials were employed a considerable amount of sulphur dioxide was evolved, especially towards the end of the reaction. These facts furnished a clue to the course of the reaction.

We have been able to show that phenetyl sulphoxide is formed as the primary product which condenses in presence of the excess of thionyl chloride with another molecule of phenetole, yielding the sulphonium salt. It is questionable, however, whether the sulphoxide condenses directly with the phenetole, thus :

$$(C_6H_4 \cdot O \cdot C_2H_5)_2SO + C_6H_5 \cdot O \cdot C_2H_5 = (C_6H_4 \cdot O \cdot C_2H_5)_3S \cdot OH \ ;$$

probably not, for the aromatic sulphoxides are weak bases, and would exist in the mixture as hydrochlorides owing to the large excess of the mineral acid present. It may be mentioned in passing that phenyl-sulphoxide forms a platinichloride and a hydrochloride, phenetyl sulphoxide also forms a hydrochloride ; the latter salts are not crystalline and are exceedingly unstable ; we therefore have been unable to analyse them, but from analogy with the aliphatic sulphoxides there is little doubt that they are formed from one molecule of the base and one of acid, and may be written as $Ar_2S(OH) \cdot Cl$. The nitrate of methyl sulphoxide has been isolated (Saytzeff, *Annalen*, 1867, **144**, 148), and corresponds to the structure $(CH_3)_2SO, HNO_3$. It is highly probable therefore that the sulphoxide when produced during the reaction forms a hydrochloride, or perhaps unites with the aluminium chloride, forming a double salt. The sulphonium salt must therefore be formed by condensation of the hydrochloride or aluminium chloride double salt with the excess of phenetole, the water which is eliminated decomposing part of the thionyl chloride present. The whole reaction may then be formulated as follows :

I. $2C_2H_5 \cdot O \cdot C_6H_5 + SOCl_2 = 2HCl + SO(C_6H_4 \cdot O \cdot C_2H_5)_2$.

II. $SO(C_6H_4 \cdot O \cdot C_2H_5)_2 + HCl = (C_6H_4 \cdot O \cdot C_2H_5)_2SCl \cdot OH$.

III. $(C_6H_4 \cdot O \cdot C_2H_5)_2SCl \cdot OH + C_6H_4 \cdot O \cdot C_2H_5 + SOCl_2 =$
$$SO_2 + 2HCl + (C_6H_4 \cdot O \cdot C_2H_5)_3SCl.$$

It follows from this explanation of the reaction that a mixture of phenetole and thionyl chloride in the proportion of 3 molecules of the former to 2 molecules of the latter should yield on treatment with aluminium chloride almost exclusively the sulphonium chloride. In practice, we have found that an almost theoretical yield is obtained under these conditions.

Furthermore, by avoiding any excess of the condensing agent, it should be possible to improve the yield of sulphoxide; this also we have succeeded in doing by slowly adding 1 molecular proportion of thionyl chloride to two such proportions of phenetole mixed with the requisite amount of aluminium chloride. With these precautions the yield of sulphoxide is raised to 40 per cent. of the theoretical, and it probably could be improved still further if the acid chloride could be made to react more rapidly with the phenetole; as it is, the action does not seem to be instantaneous, and the sulphoxide at first produced finds itself in presence of excess of thionyl chloride and phenetole, and consequently undergoes further condensation to the sulphonium salt. Experiments have shown that aluminium chloride also can bring about the condensation of sulphoxide with phenetole, but the action is not nearly so rapid or complete as with thionyl chloride.

The action of thionyl chloride and phenetole has already been studied by Loth and Michaelis (Ber., 1894, 27, 2543), but under somewhat different conditions. These investigators added aluminium chloride to an approximately equimolecular mixture of phenetole and thionyl chloride, and when the violent reaction was over they diluted the mixture with dry ether, and finally warmed the solution on the water-bath to complete the reaction. Under these conditions, a moderate yield of thiophenetole was obtained. This result, however, is not surprising, for Loth and Michaelis also showed (loc. cit., p. 2547) that thionyl chloride on being warmed with phenyl sulphoxide reduces that substance to a sulphide; thus, it would seem that any phenetyl sulphoxide produced in the above reaction would also be reduced by the excess of thionyl chloride present.

Similarly, Tassinari (Gazzetta, 1890, 20, 362) obtained hydroxy-phenyl sulphide together with other substances containing sulphur and halogen by the action of thionyl chloride on phenol. In our experiments, where the temperature seems to have been lower and other conditions different to those adopted by Loth and Michaelis, we have only met with very small quantities of thiophenetole.

By oxidising their thiophenetole, Loth and Michaelis (loc. cit.) also prepared phenetyl sulphoxide, and the product, as might be expected, was the same as that obtained directly from thionyl chloride and phenetole.

The phenetyl sulphoxide produced in these reactions must be considered to be the dipara-derivative, for when oxidised it yields a sulphone, $(C_6H_4 \cdot O \cdot C_2H_5)_2SO_2$, which is identical with the substance produced by alkylation of p-dihydroxydiphenylsulphone,

$$SO(C_6H_4 \cdot OH)_2$$

(Annaheim, *Annalen*, 1874, **172**, 36).

When preparing large quantities of the sulphonium chloride we found that sometimes a mixture of two isomeric chlorides was formed : one of these can be obtained crystalline, the other, in spite of much experiment, has remained an oil. We have not been able to determine the exact conditions necessary for the production of the oily chloride ; in most experiments it formed by far the smaller portion of the sulphonium chloride.

As will be seen later, the solid triphenetylsulphonium salt may be prepared from p-phenetyl sulphoxide and phenetole ; it therefore must contain at least two p-phenetyl groups, and we incline to the view that it is the tripara-derivative on account of its relatively high melting point compared to the liquid isomeride, which would be the orthodi-para-compound.

The salts of the aromatic sulphonium bases which we have examined seem to be very much more stable than Kehrmann's mixed fatty aromatic sulphonium derivatives (*loc. cit.*).

Generally speaking, they resemble both in chemical and physical properties the analogous compounds of the aliphatic series. In solution they behave as ordinary salts and are ionised ; for example, silver nitrate precipitates the chloride of that metal from solutions of the chlorides, and barium chloride yields barium sulphate from the sulphonium sulphates. The hydroxides have not been isolated in the free condition, but their solutions may be obtained by treating aqueous solutions of the chlorides with silver oxide. These solutions show a strongly alkaline reaction to litmus.

When dissolved in glacial acetic acid and treated with potassium permanganate at the ordinary temperature, these substances remain unattacked.

When heated, the chlorides readily break up into substances containing chlorine and sulphur, which have not been closely investigated.

The salts of the aromatic sulphonium bases, which contain different hydrocarbon radicles, are not easily obtained crystalline, and are hygroscopic like the similar fatty derivatives. The platinichlorides are crystalline substances, sparingly soluble in most organic media, and are more readily purified than the simple salts ; we have therefore in some cases employed them for analysis and characterisation of the various bases.

Sulphonium Base from a Sulphinic Acid.

The further study of the action of thionyl chloride on phenetole has led to other methods of preparing these aromatic sulphonium bases. The reaction $SOCl_2 + 3C_6H_5 \cdot O \cdot C_2H_5 = HCl + H_2O + (C_2H_5 \cdot O \cdot C_6H_4)_3SCl$, may be regarded as the replacement of three hydroxyl groups in the theoretical ortho-sulphurous acid by aromatic nuclei. This may take place in three distinct stages.

1. The change of sulphurous acid to a sulphinic acid :

$$S(OH)_4 + C_6H_5 \cdot O \cdot C_2H_5 = H_2O + C_2H_5 \cdot O \cdot C_6H_4 \cdot S(OH)_3.$$

2. The change of sulphinic acid to sulphoxide :

$$C_2H_5 \cdot O \cdot C_6H_4 \cdot S(OH)_3{}^* + C_6H_5 \cdot O \cdot C_2H_5 = H_2O + (C_2H_5 \cdot O \cdot C_6H_4)_2S(OH)_3.$$

3. The change of sulphoxide to sulphonium base :

$$(C_2H_5 \cdot O \cdot C_6H_4)_2S(OH)_2{}^* + C_6H_5 \cdot O \cdot C_2H_5 = H_2O + (C_2H_5 \cdot O \cdot C_6H_4)_3S \cdot OH.$$

In the foregoing paragraphs it has been shown that by the aid of thionyl chloride and aluminium chloride it is possible to carry out these three successive reactions in one operation and also that by using certain precautions it is possible to stop the reaction at the end of the second stage. With phenetole we have not been successful in suspending the reaction at the end of the first stage, where ethoxybenzenesulphinic acid should be produced, but it is interesting to notice that Michaelis (*Annalen*, 1900, **310**, 137) has obtained dimethylanilinesulphinic acid from thionyl chloride and dimethylaniline.

We have, however, been able to show that a sulphinic acid may be changed to a sulphoxide or sulphonium base at will according as one or two molecular proportions of phenetole are allowed to enter into the reaction. Thus, when *p*-ethoxybenzenesulphinic acid is mixed with one molecular proportion of phenetole and excess of concentrated sulphuric acid, the *p*-phenetyl sulphoxide mentioned above is produced ; the yield, however, is very poor, for the sulphoxide and sulphinic acid are both easily altered to other condensation products by the strong acid. With two molecules of phenetole an excellent yield of the sulphonium salt is obtained. It has also been shown that benzenesulphinic acid undergoes these changes, but only the final product of the reaction, namely, phenyldiphenctylsulphonium, has been isolated.

It is worth noting that in these reactions, where strong mineral acid is present, the sulphinic acid reacts in the hydroxylic form $R \cdot SO \cdot OH$,

---

* For the sake of clearness, the sulphinic acid and sulphoxide have been written in the theoretical ortho-forms, and not, as in the presence of alkalis, in the sulphone form $R \cdot SO_2 \cdot H$. We thus have further evidence of the desmotropy of the sulphinic acids.

Sulphonium Base from a Sulphoxide

The third stage in the replacement of hydroxyl in sulphurous acid, that is, the change of sulphoxide to sulphonium base, has also been carried out. p-Phenetyl sulphoxide condenses rapidly at the ordinary temperature with phenetole, in presence of strong sulphuric acid, forming triphenetylsulphonium sulphate. Also phenyl sulphoxide, when warmed with phosphoric oxide and phenetole on the water-bath, yields diphenylphenetylsulphonium phosphate. It is most convenient to represent these reactions as taking place between phenetole and a salt of the sulphoxide, thus :

$$(C_6H_5)_2S{<}_X^{OH} + C_6H_5 \cdot O \cdot C_2H_5 = (C_6H_5)_2S{<}_X^{C_6H_4 \cdot O \cdot C_2H_5} + H_2O.$$

It is interesting to note that phenetyl sulphoxide and p-ethoxybenzenesulphinic acid give when dissolved in concentrated sulphuric acid an intense Prussian blue colour, which changes especially rapidly with the sulphinic acid into a deep green.

Since neither phenyl sulphoxide nor benzenesulphinic acid gives any coloration with strong sulphuric acid, it would seem that this reaction is characteristic of the grouping $\overset{-}{\underset{O}{S}} \cdot C_6H_4 \cdot O \cdot C_2H_5(p)$.

The final green colour of the mixture is due to complex sulphonium bases. This colour reaction is especially useful, since it enables the change of sulphoxide to sulphonium base to be easily followed. When phenetole is slowly added to a solution of phenetyl sulphoxide in sulphuric acid, the colour of the latter is gradually discharged and vanishes when molecular proportions of the reagents are present ; the solution then contains triphenetylsulphonium sulphate, a substance which gives colourless solutions in sulphuric acid. Again, if the colourless solution of benzenesulphinic acid in sulphuric acid be gradually mixed with two molecular proportions of phenetole, the liquid at first assumes an intense blue colour, showing the presence of the sulphoxide, $C_6H_5 \cdot SO \cdot C_6H_4 \cdot O \cdot C_2H_5$, and then, when all the phenetole has been added, becomes colourless and contains the diphenetylphenylsulphonium sulphate. In practice it usually requires slightly more than the theoretical amount of phenetole to discharge the colour completely ; this seems probably due to partial sulphonation of the phenetole before condensation has taken place.

In conclusion, it may be of interest to note that triphenol-sulphonium chloride is colourless in distinction to the corresponding compound,* aurin. This relation will be readily understood if it be assumed that aurin contains a quinonoid structure,

$$(C_6H_4 \cdot OH)_2 C \colon C_6H_4 \colon O \; ;$$

the sulphur compound, on the other hand, certainly exists in the sulphonium form, indeed the quinonoid structure in such a substance, $(C_6H_4OH)_2 \cdot S \colon C_6H_4 \colon O$, would undoubtedly be unstable.

To summarise the results of our experiments, it can be said that these aromatic sulphonium bases may be prepared by three methods from phenetole and thionyl chloride, and from a sulphinic acid or a sulphoxide with phenetole. It will be seen that the reactions have only been carried out with phenyl and phenetyl derivatives ; how far they may be extended to other substances has not been determined, but preliminary experiments have given indications that they may be applied to most phenols, their ethers, and to many amino-compounds.

In their general physical and chemical properties these substances resemble the aliphatic sulphonium derivatives, and it is noteworthy that, as in the case of the iodonium bases, the presence of the aromatic groups does not destroy the basic function of the nuclear atom. At present sulphur is the only element known which gives rise to bases of the ammonium type containing entirely either fatty or aromatic groups.

EXPERIMENTAL.

(ppp ?)-*Triphenetylsulphonium*, $(C_6H_4 \cdot O \cdot C_2H_5)_3 S$.

(1) *From Thionyl Chloride and Phenetole.*—Two molecular proportions of thionyl chloride and three of phenetole are mixed together in a flask cooled in ice-water. Powdered aluminium chloride is then added in portions to the mixture, the flask being continually shaken until renewed addition causes no fresh evolution of hydrochloric acid ; the amount of aluminium chloride required is about equal in weight to the thionyl chloride used. To complete the reaction, the mixture is warmed gently on the water-bath. After being set aside to cool, the contents of the flask are decomposed with ice-water, when the impure sulphonium chloride separates as a viscid, oily layer, which is extracted with warm carbon disulphide or ether to remove any phenetyl sulph-oxide or unchanged phenetole. The now purified sulphonium base (usually obtained in 80 per cent. yield) is oily and occasionally very troublesome to crystallise, especially if it contain the isomeric substance described below. The method which generally succeeds the best is to dissolve the chloride in alcohol and gradually to add ether until a faint cloudiness appears ; the solution is then set aside for

* Michaelis and Godchaux found a similar relation between their hexamethyltri-aminotriphenylsulphonium chloride and " methyl-violet " (*loc. cit.*).

some days, when the substance crystallises out in large, colourless prisms on the walls of the vessel. Generally, however, if the oily product is set aside in a cool place it crystallises almost completely. When once obtained in the crystalline state, triphenetylsulphonium chloride can be readily recrystallised from ethyl acetate or water, in both of which media it is sparingly soluble. It is, moreover, very soluble in alcohol and insoluble in ether and carbon disulphide. These crystals contain water of crystallisation, part of which is very readily lost even at the ordinary temperature, as the following numbers show :

0·4317 (air-dried sample) lost 0·0383 H_2O at 110°. $H_2O = 9·8$.

0·4543 (vacuum-dried sample) lost 0·0388 H_2O at 110°. $H_2O = 8·5$.

$C_{24}H_{27}O_3SCl,3H_2O$ requires $H_2O = 11·1$ per cent.

$C_{24}H_{27}O_3SCl,2H_2O$,, $H_2O = 7·7$,,

When the crystalline chloride is dried at 110°, the anhydrous chloride is obtained as a viscous oil, which, on cooling, sets to a brittle glass.

In contact with water, the anhydrous chloride rapidly crystallises. The following analyses of the anhydrous substance were made :

0·2581 gave 0·0762 AgCl. Cl = 7·98.

0·3247 ,, 0·0998 AgCl. Cl = 8·31.

0·2892 ,, 0·1637 $BaSO_4$. S = 7·82.

0·2237 ,, 0·5452 CO_2 and 0·1276 H_2O. C = 66·49 ; H = 6·39.

$C_{24}H_{27}O_3SCl$ requires Cl = 8·24 ; S = 7·43 ; C = 66·89 ; H = 6·27 per cent.

Solutions of the *hydroxide* which are strongly alkaline to litmus may be prepared by shaking the aqueous or alcoholic solutions of the chloride with excess of silver oxide.

The *platinichloride* is deposited as an orange-yellow precipitate by adding platinichloride to solutions of the chloride. To purify the substance, it is dissolved in epichlorohydrin and reprecipitated therefrom by adding alcohol until the solution becomes cloudy. The mixture rapidly sets to a mass of yellow leaflets, which are again collected and finally dried in the air. Triphenetylsulphonium platinichloride when prepared in this way contains 1 molecule of alcohol of crystallisation.

0·2951 lost 0·0097 at 100°. Loss = 3·28.

0·2626 gave 0·0411 Pt. Pt = 15·6.

$(C_{24}H_{27}O_3S)_2PtCl_6,C_2H_6O$ requires $C_2H_6O = 3·70$; Pt = 15·6 per cent.

In this state, the melting point of the substance varies considerably with the rapidity of heating ; when freed from alcohol, the platinichloride melts at 205—206°.

0·2455 gave 0·0400 Pt.   Pt = 16·29.

$(C_{24}H_{27}O_3S)_2PtCl_6$ requires Pt = 16·3 per cent.

Triphenetylsulphonium platinichloride dissolves only sparingly in strong hydrochloric acid or cold alcohol, but is readily soluble in epichlorohydrin.

(2) *From* p-*Ethoxybenzenesulphinic Acid and Phenetole.*—3·4 grams of *p*-ethoxybenzenesulphinic acid are made into a paste with 4·5 grams of phenetole (two molecular proportions) and the mixture gradually added to about 15 c.c. of strong sulphuric acid, which are contained in a flask surrounded with ice. The initial blue colour of the mixture rapidly becomes less intense, and at the end of five minutes a few drops of phenetole are added to complete the reaction, the colour fading to a pale violet. The whole is now mixed with powdered ice and the pale brown oil which is deposited is separated, shaken out with ether, and finally dissolved in alcohol. The solution is now warmed with excess of a concentrated aqueous solution of barium hydroxide, then, after being cooled, is filtered and saturated with carbonic acid to remove excess of barium ; the precipitate is then collected and the filtrate acidified with hydrochloric acid and precipitated with platinic chloride.

The platinichloride was purified as formerly described ; after being dried at 100°, it melted at 205—206°.

0·3210 gave 0·0523 Pt.   Pt = 16·29.

0·2868 „   0·0466 Pt.   Pt = 16·25.

0·1840 „   0·3223 CO_2 and 0·0762 H_2O.   C = 47·8 ; H = 4·6.

$(C_{24}H_{27}O_3S)_2PtCl_6$ requires Pt = 16·3 ; C = 48·1 ; H = 4·6 per cent.

(3) *From* p-*Phenetyl Sulphoxide and Phenetole.*—A mixture of 4 grams of *p*-phenetyl sulphoxide and 2 grams of phenetole is dissolved in about 6 c.c. of ice-cold sulphuric acid. The colour of the liquid, which is at first a deep Prussian blue, fades in about five minutes to a pale violet, indicating the completion of the reaction. If the mixture is allowed to become warm during the reaction it is usually necessary to add a small quantity of phenetole to discharge the colour. The sulphonium salt is then converted into the platinichloride in a precisely similar manner to that described above.

The triphenetylsulphonium platinichloride obtained by this method, after being purified and dried at 100°, melted at 204—205°.

0·2238 gave 0·0368 Pt.   Pt = 16·44.

$(C_{24}H_{27}O_3S)PtCl_6$ requires Pt = 16·3 per cent.

It is worthy of remark that triphenetylsulphonium chloride dissolves in cold sulphuric acid without coloration, but if the solution is gently warmed a pale violet colour appears which gradually deepens to blue ;

on cooling the solution again, the colour does not disappear until a small quantity of phenetole is added. It is therefore probable that the sulphonium base is decomposed by warm sulphuric acid into phenetyl sulphoxide and phenetole, the latter of which is sulphonated, so that when cold the solution retains the blue colour owing to excess of sulphoxide present.

(opp ?)-*Triphenetylsulphonium Chloride*.

In the preparation of the foregoing crystalline chloride, another iso-meric chloride is occasionally met with, which was obtained by setting aside the purified mixture of sulphonium chloride for about two months under moist ether. The crystalline portion was then separated from the oil by filtration and the viscous filtrate converted into the platini-chloride.

After several crystallisations, this platinichloride, when dried in the steam oven, melted at 117°. It is very soluble in epichlorohydrin and crystallises from a mixture of that solvent and alcohol in long needles.

0·2201 gave 0·0355 Pt. Pt = 16·12.

0·2489 gave 0·4355 CO_2 and 0·1008 H_2O. C = 47·7 ; H = 4·5.

$(C_{24}H_{27}O_3S)_2PtCl_6$ requires Pt = 16·3 ; C = 48·1 ; H = 4·5 per cent.

Triphenolsulphonium Chloride, $(HO \cdot C_6H_4)_3SCl$.

This substance was prepared in a similar manner to the triphenetyl derivative by the action of thionyl chloride in phenol and aluminium chloride. The sulphonium chloride could not be obtained crystalline ; when dried under diminished pressure it solidified to a brittle glass which is easily soluble in aqueous solutions of sodium hydroxide, sparingly so in hot water, and insoluble in ether. Tribenzoyloxyphenyl-sulphonium chloride was prepared and converted into the platinichloride. The latter forms a pale yellow, crystalline powder which melts at 168—170°.

0·2255 gave 0·0262 Pt. Pt = 11·54.

0·3121 ,, 0·6452 CO_2 and 0·0935 H_2O. C = 56·4 ; H = 3·3.

$(C_{30}H_{27}O_6S)_2PtCl_6$ requires Pt = 11·79 ; C = 56·6 , H = 3·2 per cent.

Diphenetylphenylsulphonium, $(C_2H_5 \cdot O \cdot C_6H_4)_2S(C_6H_5)$.

Benzenesulphinic acid was mixed with rather more than two molecular proportions of phenetole, the pasty mass was then added in small portions to ice-cold sulphuric acid. When the action was com-plete, powdered ice was added and the precipitated oil, after being extracted with ether, was converted as before into the platinichloride. Diphenetylphenylsulphonium platinichloride forms a yellow, crystalline powder which melts at 135—137° and is insoluble in alcohol or water.

0·2038 gave 0·0362 Pt.   Pt = 17·76.

0·2359 ,,   0·4163 CO_2 and 0·0855 H_2O.   C = 48·0 ; H = 4·0.

$(C_{22}H_{23}O_2S)_2PtCl_6$ requires Pt = 17·57 ; C = 47·56 ; H = 4·1 per cent.

Diphenylphenetylsulphonium, $C_2H_5\cdot O\cdot C_6H_4\cdot S(C_6H_5)_2$.

A mixture of 10 grams of phenyl sulphoxide, 7 grams (a slight excess) of phenetole, and 8 to 10 grams of the phosphoric oxide was heated on the water-bath for an hour and a half ; the red, viscous mass was then boiled with water and the clear solution finally evaporated to dryness. The product, evidently the phosphate of the base mixed with excess of phosphoric acid, was dissolved in alcohol and converted into the hydroxide by boiling with excess of barium hydroxide. The filtered solution, after being freed from barium salts by a current of carbon dioxide, was acidified with hydrochloric acid. By evaporating this solution to dryness, diphenylphenetylsulphonium chloride was obtained in almost theoretical yield as a hygroscopic, crystalline mass. For analysis it was converted into diphenylphenetylsulphonium platinichloride, which melts at 229—230° and forms diamond-shaped plates.

0·3478 gave 0·0676 Pt.   Pt = 19·41.

0·2322 ,,   0·3989 CO_2 and 0·0797 H_2O.   C = 46·8 ; H = 3·8.

$(C_{20}H_{15}OS)_2PtCl_6$ requires Pt = 19·08 ; C = 46·9 ; H = 3·7 per cent.

p-*Phenetyl Sulphoxide*, $(C_2H_5\cdot O\cdot C_6H_4)_2SO$.

(1) *From Thionyl Chloride, Phenetole, and Aluminium Chloride.*—A solution of 25 grams of aluminium chloride in 50 grams of phenetole was placed in a flask surrounded by ice and 25 grams of thionyl chloride were gradually added from a dropping funnel, the mixture being well shaken during the course of the reaction. When all the acid chloride had been added, the whole was set aside for a short time at the ordinary atmospheric temperature and then decomposed with ice-water. Finally the semi-solid product was extracted with warm carbon disulphide, which dissolved the sulphoxide and left the oily sulphonium chloride. In this way 24 grams of sulphoxide were obtained, the yield being approximately 40 per cent.

After recrystallisation from ethyl acetate, phenetyl sulphoxide melts at 115—116°.

0·2456 gave 0·5956 CO_2 and 0·1372 H_2O.   C = 66·1 ; H = 6·28.

0·1702 ,,   0·1388 $BaSO_4$.   S = 11·20.

$C_{16}H_{20}O_3S$ requires C = 66·2 ; H = 6·21 ; S = 11·03 per cent.

(2) *From* p-*Ethoxybenzenesulphinic Acid and Phenetole.*—Seven grams of a mixture containing equimolecular proportions of *p*-ethoxybenzene-

sulphinic acid and phenetole were dissolved in 10 c.c. of cold sulphuric acid and the dark blue solution immediately poured on to pounded ice. The liquid was then extracted with warm ether, and the ethereal layer, after being separated, was shaken with dilute aqueous potassium hydroxide to remove any unchanged sulphinic acid. It was then dried over solid potassium hydroxide and finally distilled; the residue solidified. The product consisted of a mixture of thiophenetole and phenetyl sulphoxide, which were separated by crystallisation from carbon disulphide. The least soluble portion obtained from this solvent was finally crystallised from ethyl acetate; the product melted at 114° and gave the characteristic reaction of phenetyl sulphoxide with sulphuric acid.

0·2424 gave 0·2044 $BaSO_4$. S = 11·58.

$C_{16}H_{18}O_3S$ requires S = 11·03 per cent.

Four grams of the crude substance were obtained from seventeen grams of p-ethoxybenzenesulphinic acid. The portion which was more soluble in carbon disulphide was recrystallised from alcohol; in this way phenetyl sulphide was obtained in broad, colourless prisms which melted at 55° and gave colourless solutions in sulphuric acid.

0·2080 gave 0·184 $BaSO_4$. S = 12·1.

$C_{16}H_{18}O_2S$ requires S = 11·7 per cent.

Diphenetylsulphone, $(C_2H_5 \cdot O \cdot C_6H_4)_2SO_2$.

(1) *From Phenetyl Sulphoxide.*—Michaelis and Loth (*loc. cit.*) have already shown that phenetyl sulphide may be oxidised to the sulphone with potassium permanganate in glacial acetic acid solution. The same reagents were employed for the oxidation of the sulphoxide. The product crystallised from a solution in warm ethyl acetate in thin leaflets which melted at 163°.*

0·1888 gave 0·1418 $BaSO_4$. S = 10·31.

$C_{16}H_{18}O_4S$ requires S = 10·46 per cent.

(2) *From Dihydroxydiphenylsulphone.*—The preparation of diphenetyl-sulphone from the sodium salt of dihydroxydiphenylsulphone and ethyl iodide has been described by Annaheim (*loc. cit.*, p. 52), who stated that the substance melts at 159°. On repeating the experiment, we have found the product to be identical with that resulting from the oxidation of phenetyl sulphoxide; it melts at 163°, and a mixture of the sulphones from the two different sources also melts at that temperature.

---

* Michaelis and Loth (*loc. cit.*) have described this substance as melting at 263°.

0·2535 gave 0·1946 BaSO₄.   S = 10·54.

$C_{16}H_{18}O_4S$ requires S = 10·46 per cent.

In conclusion, the authors desire to express their most hearty thanks to Professor Collie for the kind interest he has taken in these experiments.

THE ORGANIC CHEMISTRY LABORATORY,
UNIVERSITY COLLEGE,
LONDON.

LXXV.—*Acetyl and Benzoyl Derivatives of Phthalimide and Phthalamic Acid.*

By ARTHUR WALSH TITHERLEY and WILLIAM LONGTON HICKS.

In a communication by one of the authors (Trans., 1904, **85,** 1679), in which various methods of preparing acylamides and cyclicimides were discussed, it was shown that succinic anhydride and sodium benzamide readily interact, yielding dibenzoylsuccinamide and benzoylsuccinamic acid. The latter compound, moreover, was obtained by cautious rupture of the cyclic imide ring of benzoylsuccinimide, which, it was shown, could be obtained by the pyridine benzoylation of succinimide.

A further investigation has been carried out on similar lines with phthalic anhydride and phthalimide with similar results, which appear to indicate that (1) the condensation between cyclic anhydrides and sodium acylamides is a general one, and (2) the pyridine benzoylation of cyclic imides proceeds with great ease, although other methods of benzoylation fail.

Phthalic anhydride readily interacts on warming with sodium acetamide or sodium benzamide in presence of benzene, giving rise respectively to acetyl- and benzoyl-phthalamic acids, which were thus isolated, but apparently the diacylphthalamides are not produced in the condensation. It has also been shown that whilst benzoyl chloride in presence of pyridine readily benzoylates phthalimide, it fails to benzoylate phthalamide and benzoylphthalamic acid. Numerous attempts to prepare the diacyl derivatives of phthalamide failed.

Both acetyl- and benzoyl-phthalamic acids were also prepared by rupture of the imide ring of the corresponding acylphthalimides through the absorption of water (compare the formation of benzoylsuccinamic acid from benzoylsuccinimide) thus:

$$C_6H_4{\large\lessgtr}^{CO}_{CO}{\large >}N{\cdot}COR' \quad \xrightarrow{H_2O} \quad C_6H_4{\large\lessgtr}^{CO{\cdot}NH{\cdot}COR}_{CO_2H}.$$

On treatment with dehydrating agents, the resulting acids are readily converted into the corresponding acylphthalimides.

The following scheme sets forth the relationships which were observed :

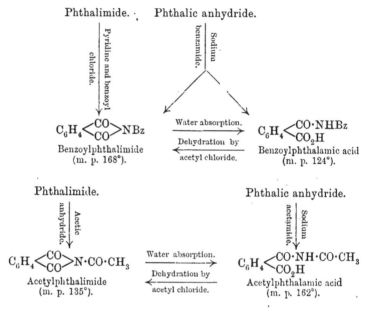

Phthalimide. ·   Phthalic anhydride.

$C_6H_4\!\!<\!\!^{CO}_{CO}\!\!>\!NBz$
Benzoylphthalimide
(m. p. 168°).

Water absorption.
——————————→
Dehydration by
←——————————
acetyl chloride.

$C_6H_4\!\!<\!\!^{CO\cdot NHBz}_{CO_2H}$
Benzoylphthalamic acid
(m. p. 124°).

Phthalimide.   Phthalic anhydride.

$C_6H_4\!\!<\!\!^{CO}_{CO}\!\!>\!N\cdot CO\cdot CH_3$
Acetylphthalimide
(m. p. 135°).

Water absorption.
——————————→
Dehydration by
←——————————
acetyl chloride.

$C_6H_4\!\!<\!\!^{CO\cdot NH\cdot CO\cdot CH_3}_{CO_2H}$
Acetylphthalamic acid
(m. p. 162°).

EXPERIMENTAL.

Benzoylphthalimide, $C_6H_4\!\!<\!\!^{CO}_{CO}\!\!>\!NBz$.

Phthalimide is readily benzoylated by benzoyl chloride in presence of pyridine, with disengagement of heat. Twenty-one grams of phthalimide, partially dissolved and suspended in 70 grams of pyridine, were treated gradually with 20 grams of benzoyl chloride, the mixture being kept well cooled and free from moisture. The liquid, which became dark red, was left for twenty-four hours and treated with 80 c.c. of alcohol, when in the course of ten minutes the solution deposited a white, powdery precipitate of benzoylphthalimide. This product, when washed with cold alcohol, was practically pure and weighed 35 grams (93 per cent. of the calculated amount). On recrystallisation from boiling alcohol it was obtained in colourless, gritty prisms melting at 168°.

0·5172 gave 25·1 c.c. nitrogen at 24° and 761 mm. N = 5·4 per cent.
$C_{15}H_9O_3N$ requires N = 5·5 per cent.

Benzoylphthalimide is sparingly soluble in methyl and ethyl alcohols, benzene, and ether, but readily soluble in acetone and chloroform.

Benzoylphthalamic acid, $CO_2H \cdot C_6H_4 \cdot CO \cdot NHBz$, was obtained both by the interaction of phthalic anhydride and sodium benzamide, and also by the hydrolytic rupture of the benzoylphthalimide ring.

Action of Phthalic Anhydride on Sodium Benzamide.—Thirty-six grams of benzamide and 12 grams of sodamide were converted into sodium benzamide in presence of a little benzene (Trans., 1902, **81**, 1531), the temperature being finally raised to 120° to complete the conversion. The sodium benzamide, representing 2 molecules, after cooling was rubbed to a thin paste with benzene and treated with 22 grams (1 mol.) of powdered phthalic anhydride, the whole being well mixed and heated at 90—95° for two hours. The resulting granular mass contained benzoylphthalamic acid, sodium phthalate, and phthalic acid in quantity, benzamide, and small varying quantities of benzoylphthalimide. The mass was spread on a porous plate, after which it was cautiously treated with a cold freshly-prepared solution of sodium hydrogen carbonate to remove phthalic acid and its sodium salt. The washed insoluble residue was now digested with dilute caustic soda, by which benzoylphthalamic acid was extracted and reprecipitated in an impure state (m. p. 100—110°) by acidifying the filtrate. This product was purified by recrystallisation from hot water and finally by repeatedly dissolving in methyl alcohol and carefully precipitating with water. In this way, benzoylphthalamic acid was obtained as a mass of very light needles melting at 123—124°, the yield amounting to only 8 grams.

0·2415 gave 0·5484 CO_2 and 0·0935 H_2O.   C = 67·2 ; H = 4·3.

0·1823 ,, 8·7 c.c. nitrogen at 23° and 764 mm.   N = 5·4.

0·2259 on titration required 8·4 c.c. $N/10$ NaOH.   M.W. = 269.

$C_{15}H_{11}O_4N$ requires C = 66·9 ; H = 4·1 ; N = 5·2 per cent.   M.W. = 269.

The insoluble residue, after the above treatment with caustic soda, contained benzamide and benzoylphthalimide ; these were separated by treatment with warm alcohol, when the latter compound, being only sparingly soluble in this medium, was isolated as a crystalline powder (m. p. 166—168°) identical with that obtained by the pyridine benzoylation of phthalimide.

0·2234 gave 10·84 c.c. nitrogen at 24° and 760 mm.   N = 5·4·

$C_{15}H_9O_3N$ requires N = 5·5 per cent.

In some experiments, however, no benzoylphthalimide was isolated, and in no case was dibenzoylphthalamide definitely isolated as a result of the condensation (compare Trans., 1904, **85**, 1690), although a small quantity of a substance, probably $C_6H_4(CO \cdot NHBz)_2$, was obtained, melting at 195°, insoluble in sodium hydrogen carbonate and soluble in sodium hydroxide, yet the total quantity isolated from a number of condensations was insufficient for analysis.

Formation of Benzoylphthalamic Acid from Benzoylphthalimide.— The rupture of the cyclic imide ring is easily effected by warming with aqueous sodium carbonate. Five grams of benzoylphthalimide were heated for half an hour on the water-bath with a 5 per cent. solution containing the equivalent quantity of sodium carbonate. The solid dissolved almost completely, and after filtering and cooling benzoylphthalamic acid was deposited as a white precipitate with hydrochloric acid. It was purified and separated from the accompanying benzoic acid by the methyl alcohol method (see p. 710).

0·1987 gave 9·1 c.c. nitrogen at 17° and 759 mm. N = 5·31.

$C_{15}H_{11}O_4N$ requires N = 5·2 per cent.

Benzoylphthalamic acid is very soluble in methyl and ethyl alcohols and in acetone; it is insoluble in cold water, but readily dissolves in hot water, from which it crystallises in a light bulky form on cooling. It is moderately soluble in ether, not very soluble in chloroform, and only slightly so in benzene.

Its lead and mercurous salts are formed as insoluble, curdy-white precipitates, its silver salt as a white, granular precipitate from strong solutions only, and its ferric salt is deposited as a yellowish-buff precipitate.

Benzoylphthalamic acid remains unchanged on heating at 170° for two hours, but on treatment with acetyl chloride even in the cold it rapidly loses water, forming the cyclic imide; 3 grams of the acid were treated with 4·5 grams of acetyl chloride and allowed to stand. The mixture soon liquefied with copious evolution of hydrogen chloride, and, after several hours, large, glistening crystals of nearly pure benzoylphthalimide (m. p. 161—165°) separated. More was obtained from the acetic acid solution, the total amounting to 65 per cent. of the calculated quantity. After crystallisation from alcohol, the benzoylphthalimide was obtained pure and melted at 168°.

0·3966 gave 18·8 c.c. nitrogen at 16·5° and 755 mm. N = 5·48.

$C_{15}H_9O_3N$ requires N = 5·5 per cent.

Benzoylphthalamic acid, on treatment with benzoyl chloride in presence of pyridine, is simply dehydrated, as with acetyl chloride, yielding benzoylphthalimide (m. p. 166°) and not the expected dibenzoylphthalamic acid.

Acetylphthalamic acid, $CO_2H \cdot C_6H_4 \cdot CO \cdot NH \cdot CO \cdot CH_3$, was obtained by the interaction of phthalic anhydride and sodium acetamide, and also by the hydrolytic rupture of acetylphthalimide.

Action of Phthalic Anhydride on Sodium Acetamide.—Twelve grams of acetamide and 8 grams of sodamide were converted in presence of benzene into sodium acetamide, which, in presence of the benzene, was then heated for four hours at 50° with 14 grams of phthalic anhydride. The resulting partly-discoloured solid was separated from the benzene, dried, treated with water at 0°, and filtered into ice-cold dilute hydrochloric acid. The resulting brown precipitate contained acetyl-phthalamic acid, which was purified by digesting with aqueous sodium hydrogen carbonate, filtering, and acidifying the filtrate. On crystallisation from hot water, it separated in shining plates melting at 164°; the yield was 7 grams.

0·0694 gave 3·9 c.c. nitrogen at 15° and 768 mm. N = 6·69.

0·1100 on titration required 5·05 c.c. $N/10$ NaOH. M.W. = 218.

$C_{10}H_9O_4N$ requires N = 6·7 per cent. M.W. = 207.

Formation of Acetylphthalamic Acid from Acetylphthalimide.

Acetylphthalimide, $C_6H_4{<}^{CO}_{CO}{>}N{\cdot}CO{\cdot}CH_3$, was obtained by Aschan (*Ber.*, 1886, **19**, 1400) by heating phthalamide with acetic anhydride. In reference to its behaviour with alkali, he states that it is somewhat unstable and hydrolysed, yielding phthalamide and acetic acid, but this observation is not confirmed by the authors, who, on the contrary, find that the cyclic imide ring is ruptured exactly as in the case of benzoylphthalimide, yielding acetylphthalamic acid, if carefully performed, and, otherwise, acetic and phthalamic acids or phthalic acid and ammonia. The conversion into acetylphthalamic acid was best conducted thus : 5 grams of acetylphthalimide were warmed at about 60° with the theoretical amount of dilute sodium carbonate for twenty minutes, the solution was filtered, cooled, and precipitated by dilute hydrochloric acid. An almost theoretical yield of pure acetylphthalamic acid (5 grams) melting at 162° was deposited. After recrystallisation from hot water, it was obtained in very light pearly-white plates melting at 165°.

0·3216 gave 17·85 c.c. nitrogen at 14° and 756 mm. N = 6·5.

0·2090 on titration required 10·0 c.c. $N/10$ NaOH. M.W. = 209.

$C_{10}H_9O_4N$ requires N = 6·7 per cent. M.W. = 207.

Acetylphthalamic acid (and not, as Aschan states, *loc. cit.*, phthalimide and acetic acid) is also produced by long digestion of acetylphthalimide with hot water.

Acetylphthalamic acid is moderately soluble in warm methyl and ethyl alcohols, from which it slowly crystallises on cooling; it is insoluble in cold, but readily soluble in hot, water, separating in plates on cooling. It is practically insoluble in chloroform, benzene, or ethyl acetate. Its silver and mercurous salts form white precipitates, and its ferric salt is deposited as a light buff precipitate, somewhat darker than that of the corresponding ferric benzoylphthalamate.

Like benzoylphthalamic acid, acetylphthalamic acid readily undergoes internal condensation, yielding acetylphthalimide: 1 gram of the acid was gently warmed with an excess of acetyl chloride and left for two hours. A sparingly soluble crystalline powder which separated, weighing 0·7 gram and melting at 133—135°, consisted of pure acetyl-phthalimide.

0·2910 gave 18·7 c.c. nitrogen at 15° and 756 mm. N = 7·52.

$C_{10}H_7O_3N$ requires N = 7·4 per cent.

Acetylphthalimide is not an " etwas unbeständige Verbindung," as described by Aschan in respect to its behaviour with water and aqueous alcohol. From boiling water it may be recrystallised unchanged, although on prolonged heating it decomposes in the manner already described, whilst it is quite unaffected by boiling alcohol. Acetyl-phthalimide is rather more soluble in methyl than in ethyl alcohol, and dissolves easily in cold chloroform.

ORGANIC LABORATORY,
UNIVERSITY OF LIVERPOOL.

LXXVI.—Some *Thio-* and *Dithio-carbamide Derivatives* of *Ethyleneaniline and the Ethylenetoluidines.*

By OLIVER CHARLES MINTY DAVIS.

ETHYLENEANILINE and the ethylenetoluidines interact with the thio-carbimides to give crystalline compounds. In most cases when the change is effected by heating the two reagents on a water-bath at 100°, a quantitative yield is obtained. With few exceptions the product could be obtained in a state of comparative purity by washing with cold alcohol, and this method was used in each case before recrystallising for analysis.

During the investigation, several instances of steric hindrance were observed, the position of the substituent methyl group in the ethylene-toluidines and the tolylthiocarbimides having an important bearing both on the time of reaction and also on the nature of the product. One molecule of ethyleneaniline unites with two molecules of allyl-, phenyl-, o-tolyl-, or p-tolyl-thiocarbimide to form symmetrical dithio-

3 A 2

carbamide derivatives, but with m-tolylthiocarbimide only one molecule of each unites to form an asymmetrical, substituted thiocarbamide. Ethylene-o-toluidine interacts with difficulty to give asymmetrical derivatives in the case of phenyl-, m-tolyl-, and p-tolyl-thiocarbimides, but with allyl- and o-tolyl-thiocarbimides the yield of product was very small, and the substance could not be purified for analysis.

Ethylene-m-toluidine gives symmetrical derivatives with phenyl- and p-tolyl-thiocarbimides, but with the allyl, o-tolyl, and m-tolyl compounds asymmetrical derivatives result.

Ethylene-p-toluidine gives an asymmetrical derivative with m-tolylthiocarbimide, but with its analogues symmetrical derivatives are formed.

With the ethylenetoluidines, the ortho- and meta-positions therefore seem to modify the reactions, but in the thiocarbimides the meta-position only has this effect. When heated in alcoholic solution with mercuric oxide to about 140°, the thiocarbamide derivatives yield the corresponding oxygen compounds. The product obtained from the interaction of ethyleneaniline and phenylthiocarbimide appears to be identical with a compound synthesised from ethyleneaniline and phenylcarbimide. This fact strengthens the supposition that the compounds described are thio- and dithio-carbamide derivatives, their constitution being represented by the general formulæ

$$\begin{array}{ccc} \text{CH}_2\cdot\text{NR}\cdot\text{CS}\cdot\text{NHR}' & & \text{CH}_2\cdot\text{NR}\cdot\text{CS}\cdot\text{NHR}' \\ \text{CH}_2\cdot\text{NHR} & \text{and} & \text{CH}_2\cdot\text{NR}\cdot\text{CS}\cdot\text{NHR}' \end{array}$$

The table on p. 715 indicates the nature of the compound formed in each case.

EXPERIMENTAL.

Ethyleneaniline and Allylthiocarbimide. — These reagents when heated together for about half an hour at 100° readily interact, yielding a crystalline product. The reaction was carried out under varying conditions, using either excess of ethyleneaniline or excess of the thiocarbimide, but in all cases the chief product is a crystalline compound formed by the union of 1 molecule of ethyleneaniline and 2 of the thiocarbimide. This substance is best recrystallised from absolute alcohol, and is obtained in well-defined needles, very soluble in acetone and benzene, and melts without decomposition at 165°.

0·2764 gave 0·6500 CO_2 and 0·1704 H_2O. C = 64·15 ; H = 6·84.

0·1482 „ 0·1631 $BaSO_4$. S = 15·11.

0·1630 „ 19·6 c.c. moist nitrogen at 23° and 766 mm. N = 13·65.

$C_{22}H_{26}N_4S_2$ requires C = 64·39 ; H = 6·34 ; N = 13·65 ; S = 15 60 per cent.

A molecular weight determination, using the ebullioscopic method

Melting point.

Ethylene compound.	Thiocarbimide.	Thiocarbamide.	Symmetrical.	Asymmetrical.
Ethylene aniline	Allyl	$C_2H_4(NPh\cdot CS\cdot NH\cdot C_3H_5)_2$	165°	—
,, ,,	Phenyl	$C_2H_4(NPh\cdot CS\cdot NHPh)_2$	184	.
,, ,,	o-Tolyl	$C_2H_4[NPh\cdot CS\cdot NH\cdot C_7H_7(o)]_2$	179	—
,, ,,	m-Tolyl	$C_2H_4{<}^{NPh\cdot CS\cdot NH\cdot C_7H_7}_{NHPh}$	—	185°
,, ,,	p-Tolyl	$C_2H_4(NPh\cdot CS\cdot NH\cdot C_7H_7)_2$	190	—
Ethylene-o-toluidine.	Phenyl	$C_2H_4{<}^{N(C_7H_7)\cdot CS\cdot NHPh}_{NH\cdot C_7H_7}$	—	150
,,	m-Tolyl	$C_2H_4{<}^{N(C_7H_7)\cdot CS\cdot NH\cdot C_7H_7}_{NH\cdot C_7H_7}$	—	157
,, ,,	p-Tolyl	$C_2H_4{<}^{N(C_7H_7)\cdot CS\cdot NH\cdot C_7H_7}_{NH\cdot C_7H_7}$	—	118
Ethylene-m-toluidine	Allyl	$C_2H_4{<}^{N(C_7H_7)\cdot S\cdot NH\cdot C_3H_5}_{NH\cdot C_7H_7}$	—	99
,, ,,	Phenyl	$C_2H_4[N(C_7H_7)\cdot C\cdot NHPh]_2$	178	—
,, ,,	o-Tolyl	$C_2H_4{<}^{N(C_7H_7)\cdot CS\cdot NH\cdot C_7H_7}_{NH\cdot C_7H_7}$	—	176
,, ,,	m-Tolyl	$C_2H_4{<}^{N(C_7H_7)\cdot CS\cdot NH\cdot C_7H_7}_{NH\cdot C_7H_7}$	—	170
,, ,,	p-Tolyl	$C_2H_4[N(C_7H_7)\cdot CS\cdot NH\cdot C_7H_7]_2$	187	—
Ethylene-p-toluidine.	Allyl	$C_2H_4[N(C_7H_7)\cdot CS\cdot NH\cdot C_3H_5]_2$	133	—
,, ,,	Phenyl	$C_2H_4[N(C_7H_7)\cdot CS\cdot NH]_2$	195	—
,, ,,	m-Chlorophenyl	$C_2H_4[N(C_7H_7)\cdot CS\cdot NH\cdot C_6H_4Cl]_2$	180	—
,, ,,	o-Tolyl	$C_2H_4[N(C_7H_7)\cdot CS\cdot NH\cdot C_7H_7]_2$	202	—
,, ,,	m-Tolyl	$C_2H_4{<}^{N(C_7H_7)\cdot CS\cdot NH\cdot C_7H_7}_{NH\cdot C_7H_7}$	—	179
,, ,,	p-Tolyl	$C_2H_4[N(C_7H_7)\cdot CS\cdot NH\cdot C_7H_7]_2$	178	—

with benzene as solvent, gave a magnitude of 431, the theoretical value being 410. On treatment for two and a half hours with freshly-precipitated mercuric oxide in a sealed tube at 130—140°, with alcohol as solvent, a small amount of a crystalline substance was obtained, melting at 208°; from analogy with a similar compound, described later, this is in all probability the corresponding dicarbamide derivative.

In addition to the compound melting at 165° a very small amount of a second substance was obtained, melting in the neighbourhood of 128°. This is rather more soluble in solvents than the dithiocarbamide derivative but it was not found possible to isolate it in a state of purity. An analysis shows it to be a mixture of the dithiocarbamide derivative, with an asymmetrical compound resulting from the union of 1 molecule of ethyleneaniline and 1 of the thiocarbimide.

0·2870 gave 0·6985 CO_2 and 0·1766 H_2O. C = 66·39; H = 6·8.

0·3110 ,, 37 c.c. moist nitrogen at 20° and 744 mm. N = 13·32.

0·1596 ,, 0·1480 $BaSO_4$. S = 12·7.

$C_{18}H_{21}N_3S$ requires C = 69·45; H = 6·75; N = 13·5; S = 10·28 per cent.

It is interesting to notice that both the symmetrical and asymmetrical compounds contain almost the same percentage of nitrogen, and consequently the estimation of that element present confirms the supposition that the substance melting at 128° is a mixture of the two derivatives.

Ethyleneaniline and *phenylthiocarbimide* interact together quite readily at 100°, giving a substance only very slightly soluble in boiling alcohol, but moderately in boiling benzene, toluene, or amyl alcohol. Either of these solvents on cooling deposits the dithiocarbamide derivative in needle-shaped crystals melting at 184°.

0·2158 gave 0·5486 CO_2 and 0·1160 H_2O.   C = 69·35 ; H = 6·01.

0·1364   „   13·2 c.c. moist nitrogen at 17° and 764 mm. N = 11·29.

$C_{28}H_{26}N_4S_2$ requires C = 69·70 ; H = 5·39 ; N = 11·61 per cent.

When this compound is heated with alcohol and mercuric oxide in a sealed tube at 140° for three hours, a small amount of a crystalline substance is obtained, which on recrystallisation from alcohol melts sharply at 215°.

0·1608 gave 16·8 c.c. moist nitrogen at 18° and 750 mm. N = 11·91.

$C_{28}H_{26}N_4O_2$ requires N = 12·44 per cent.

That this substance was the substituted dicarbamide corresponding to the dithiocarbamide derivative was confirmed by its synthesis from ethyleneaniline and phenylcarbimide.

These two substances when heated together at 100° readily combined, giving a crystalline compound which, after recrystallisation from alcohol, melted at 215—216°.

0·3100 gave 0·8455 CO_2 and 0·1657 H_2O.   C = 74·37 ; H = 5·93.

0·2230   „   23·4 c.c. moist nitrogen at 15° and 766 mm. N = 12·39.

$C_{28}H_{26}O_2N_4$ requires C = 74·66 ; H = 5·77 ; N = 12·44 per cent.

Ethyleneaniline and o-*Tolylthiocarbimide.*—As in the previous case, the two compounds interact at 100° ; the product, a dithiocarbamide derivative, is very slightly soluble in boiling alcohol, benzene, or acetone, but may be recrystallised from boiling toluene, amyl alcohol, or nitrobenzene. From either of these solvents it separates in needles melting at 179°.

0·2010 gave 0·5168 CO_2 and 0·1196 H_2O.   C = 70·14 ; H = 6 65.

0·2478   „   23 c.c. moist nitrogen at 20° and 766 mm. N = 10·70.

$C_{30}H_{30}N_4S_2$ requires C = 70·58 ; H = 5·88 ; N = 10·97 per cent.

Ethyleneaniline and m-*Tolylthiocarbimide.*—The reaction takes place quite readily at 100°, and the product, after washing with cold alcohol, may be recrystallised from boiling acetone, when needle-shaped crystals melting at 185° are obtained. In this case, the position of the substituent methyl group in the thiocarbimide appears to have modified the reaction, analysis showing the compound to be undoubtedly formed by the union of one molecule of ethylene-aniline and one of the thiocarbimide, although excess of the latter was employed in the preparation.

0·2634 gave 0·7064 CO_2 and 0·1452 H_2O.　$C = 73·13$; $H = 6·12$.

$C_{22}H_{23}N_3S$ requires $C = 73·13$; $H = 6·37$ per cent.

Ethyleneaniline and p-*Tolylthiocarbimide.*—One molecule of ethylene-aniline and two molecules of the thiocarbimide readily unite under the conditions previously mentioned, with the formation of a dithio-carbamide which is only slightly soluble in alcohol, but may be recrystallised from benzene, toluene, or amyl alcohol, and is obtained from either of these solvents in colourless crystals melting at 190°.

0·2182 gave 0·5590 CO_2 and 0·1281 H_2O.　$C = 69·89$; $H = 6·52$.

0·1067 „ 10·4 c.c. moist nitrogen at 19° and 746 mm.. $N = 10·99$.

$C_{30}H_{30}N_4S_2$ requires $C = 70·58$; $H = 5·88$; $N = 10·97$ per cent.

Ethylene-o-toluidine and Thiocarbimides.

The position of the substituent methyl group in the ethylene-toluidine has a very marked effect on the reaction which takes place.

In the case of allyl- and *o*-tolyl-thiocarbimides it was found quite impossible to isolate any compound in a state of purity, the yields obtained being very small and the crystals having very indefinite melting points, but little improved by recrystallisation.

In the case of phenyl-, *m*-tolyl-, and *p*-tolyl-thiocarbimides the reagents had to be heated for several hours before crystals separated, and some difficulty was experienced in their purification. Analysis showed in each instance that one molecule of ethylenetoluidine had united with one of thiocarbimide to form the asymmetrical thio-carbamide derivative.

Ethylene-o-toluidine and Phenylthiocarbimide.—The compound formed is only slightly soluble in benzene, alcohol, or acetone, but more so in toluene and amyl alcohol, from which solvents crystals are obtained melting at 150°.

0·3195 gave 0·8546 CO_2 and 0·1918 H_2O.　$C = 72·94$; $H = 6·6$.

$C_{23}H_{25}N_3S$ requires $C = 73·6$; $H = 6·6$ per cent.

The compound from *ethylene-o-toluidine* and *m-tolylthiocarbimide* yields needle-shaped crystals from acetone, melting at 157°.

0·3368 gave 0·9252 CO_2 and 0·2125 H_2O.　$C = 74·91$; $H = 7·01$.

$C_{24}H_{27}N_3S$ requires $C = 74·03$; $H = 6·91$ per cent.

Ethylene-o-toluidine and p-*Tolylthiocarbimide.*—The crystalline product is readily soluble in alcohol or acetone, and when recrystallised from these solvents melts at 119°.

0.2470 gave 0.6656 CO_2 and 0.1554 H_2O.   C = 73.48 ; H = 6.99.
$C_{24}H_{27}N_3S$ requires C = 74.03 ; H = 6.94 per cent.

*Ethylene-*m*-toluidine and Allylthiocarbimide.*—The reaction between the two substances did not take place at all readily even after prolonged heating. At the end of twelve hours, a viscous mass was obtained which, on treatment with ether, gave a small amount of a crystalline substance.

Since this substance was very soluble in the ordinary organic solvents, considerable difficulty was experienced in its purification. When recrystallised from benzene, it melted at 97—99°, and was obviously not quite pure. Subsequent attempts to obtain it in a purer condition were unsuccessful. The analysis showed that one molecule of ethylenetoluidine had united with one molecule of thiocarbimide, the discrepancy in the proportion of carbon being probably due to contamination with a little ethylenetoluidine.

0.2022 gave 0.5377 CO_2 and 0.1357 H_2O.   C = 72.51 ; H = 7.45.
$C_{20}H_{25}N_3S$ requires C = 70.79 ; H = 7.37 per cent.

*Ethylene-*m*-toluidine and Phenylthiocarbimide.*—Interaction took place readily between the two substances, the product being soluble in boiling acetone. On cooling, needle-shaped crystals of the dithiocarbamide derivative were deposited, melting at 178°.

0.2046 gave 0.5246 CO_2 and 0.1140 H_2O.   C = 69.92 ; H = 6.10.
$C_{30}H_{30}N_4S_2$ requires C = 70.58 ; H = 5.8 per cent.

*Ethylene-*m*-toluidine and* o-*Tolylthiocarbimide.*—The interaction between these two substances only takes place slowly, and eight hours' heating was required for its completion. The crystals obtained were only slightly soluble in benzene, and on recrystallising from acetone melted at 176°. The product was the asymmetrical thiocarbamide derivative.

0.2475 gave 0.6790 CO_2 and 0.1440 H_2O.   C = 74.81 ; H = 6.46.
$C_{24}H_{27}N_3S$ requires C = 74.03 ; H = 6.94 per cent.

*Ethylene-*m*-toluidine and* m-*Tolylthiocarbimide.*—One molecule of each of these substances combines to form the asymmetrical thiocarbamide derivative, which may be recrystallised from boiling acetone, the crystals thus obtained melting sharply at 170°.

0.3007 gave 0.8085 CO_2 and 0.1746 H_2O.   C = 73.32 ; H = 6.45.
$C_{24}H_{27}N_3S$ requires C = 74.03 ; H = 6.94 per cent.

Ethylene-m-*toluidine and* p-*Tolylthiocarbimide.*—This interaction takes place easily and the product is slightly soluble in benzene, but may readily be recrystallised from acetone. The symmetrical dithiocarbamide derivative thus obtained melts at 187°.

0·2614 gave 0·6822 CO_2 and 0·1476 H_2O.   C = 71·16 ; H = 6·25.

$C_{32}H_{34}N_4S_2$ requires C = 71·37 ; H = 6·31 per cent.

Ethylene-p-*toluidine and Allylthiocarbimide.*—The interaction occurs fairly readily, giving a good yield of the dithiocarbamide derivative, which is moderately soluble in alcohol, but is best recrystallised from acetone, this solvent depositing crystals melting at 133°.

0·1797 gave 0·4377 CO_2 and 0·1132 H_2O.   C = 66·42 ; H = 6·9.

$C_{24}H_{30}N_4S_2$ requires C = 65·75 ; H = 6·84 per cent.

Ethylene-p-*toluidine and Phenylthiocarbimide.*—Interaction readily takes place between 1 molecule of the ethylenetoluidine and 2 molecules of phenylthiocarbimide, giving the symmetrical dithiocarbamide which is slightly soluble in benzene and toluene, but more so in acetone ; the crystals obtained from either solvent melt at 195°.

0·2240 gave 0·5740 CO_2 and 0·1174 H_2O.   C = 69·90 ; H = 6·00.

0·2532 ,, 25 c.c. moist nitrogen at 18° and 760 mm.   N = 11·4.

$C_{30}H_{30}N_4S_2$ requires C = 70·58 ; H = 5·8 ; N = 10·98 per cent.

Ethylene-p-*toluidine* and m-*chlorophenylthiocarbimide* interact, giving a small yield of a crystalline compound which, when recrystallised from acetone, melts sharply at 180°.

Unlike the corresponding derivative from ethylene-*p*-toluidine and *m*-tolylthiocarbimide it was found that 2 molecules of thiocarbimide had united with one of the base to form a symmetrical dithiocarbamide derivative, but in spite of the usual criteria of purity it appears to be contaminated with a little of the asymmetrical compound, which is richer in carbon. The nitrogen percentage is approximately the same in both compounds, and an estimation of this element strengthens this supposition.

0·3230 gave 0·7504 CO_2 and 0·1525 H_2O.   C = 63·3 ; H = 5·2.

0·3380 ,, 28·2 c.c. moist nitrogen at 17° and 762 mm.   N = 9·68.

$C_{30}H_{28}N_4S_2Cl_2$ requires C = 62·17 ; H = 4·83 ; N = 9·66.

$C_{23}H_{24}N_3SCl$ requires C = 67·39 ; H = 5·86 ; N = 10·25 per cent.

Ethylene-p-*toluidine* and o-*tolylthiocarbimide* interact readily and furnish a good yield of the dithiocarbamide derivative. This compound is practically insoluble in alcohol, amyl alcohol, benzene, and toluene, but may be crystallised from nitrobenzene and melts at 202°.

0·1971 gave 0·5097 CO_2 and 0·1144 H_2O.   C = 70·52 ; H = 6·4.

0·1934 „ 17·2 c.c. moist nitrogen at 19° and 766 mm.   N = 10·30.

$C_{32}H_{34}N_4S_2$ requires C = 71·37 ; H = 6·31 ; N = 10·40 per cent.

Ethylene-p-*toluidine and* m-*Tolylthiocarbimide.*—The product in this case consists of the asymmetrical thiocarbamide derivative, which may be recrystallised from acetone, when it is obtained in the form of white, silky crystals melting at 179°.

It is interesting to notice the influence of the methyl group in the thiocarbimide, the analogous compounds formed from o- and p-tolyl-thiocarbimides consisting of the dithiocarbamide derivatives.

0·2898 gave 0·7804 CO_2 and 0·1676 H_2O.   C = 73·43 ; H = 6·42.

0·2624 „ 0·7070 CO_2 „ 0·1630 H_2O.   C = 73·47 ; H = 6·90.

$C_{24}H_{27}N_3S$ requires C = 74·03 ; H = 6·94 per cent.

Ethylene-p-*toluidine and* p-*Tolylthiocarbimide.*—Combination of 1 molecule of ethylene-p-toluidine with two of thiocarbimide readily takes place with formation of the symmetrical dithiocarbamide, a substance which crystallises from acetone in feathery needles melting at 178°, but is only slightly soluble in alcohol and benzene.

0·2768 gave 0·7224 CO_2 and 0·1524 H_2O.   C = 71·53 ; H = 6·11.

$C_{32}H_{34}N_4S_2$ requires C = 71·37 ; H = 6·31 per cent.

UNIVERSITY COLLEGE,
BRISTOL

LXXVII.—*The Rusting of Iron.*

By GERALD TATTERSALL MOODY.

IN a recent number of this Journal (Trans., 1905, **87**, 1548) appears a paper by Dunstan, Jowett and Goulding, in which these authors contend that the rusting of iron is a change mainly depending on the direct inter-action of iron, water and oxygen, and that the part played by carbonic acid in the atmospheric corrosion of the metal is merely subsidiary. The conclusions at which they arrived were based mainly on the assumption that they had succeeded in eliminating carbonic acid in an experiment in which they allowed iron, water and oxygen to remain in contact, and in which rusting occurred. From their observations they infer that the formation of hydrogen peroxide is a necessary part of the chemical process of atmospheric rusting. In further support of this explanation they submit analyses of rust which they regard as being fairly re-presented by the formula $Fe_2O_2(OH)_2$ and as having a com-position identical with the red substance formed on immersing iron in

commercial hydrogen peroxide solution.   Dunstan and his co-workers
do not appear to have been aware of the extreme difficulty of completely excluding carbonic acid.   That they in their most important
experiment did not take adequate precautions to ensure the absence of
carbonic acid is clearly shown, for in their description of it they make
the significant admission that on allowing oxygen to pass into the
vessel containing water and iron "action immediately commenced, a
substance of a green colour being produced, which rapidly changed to
the red colour characteristic of rust." No more conclusive evidence
that carbonic acid was present in the materials used could be afforded
than the formation of this green colour, which invariably accompanies
the early stage of attack of carbonic acid on iron in presence of air or
oxygen.

Action of Oxygen and Water on Iron.

The crucial question as to whether or not direct interaction takes
place between oxygen, water and iron has engaged my attention for
several years and the experimental inquiry pursued has led to definite
results which are hereinunder recorded and which are in direct conflict
with the evidence put forward by Dunstan.*   After several partial
failures due to the difficulty of constructing an apparatus capable of
excluding the merest traces of carbonic acid, an arrangement has been
devised by means of which it has been found possible to leave iron,
oxygen and water in contact for many weeks without even a speck of
rust appearing on the surface of the metal.

The apparatus consists of a distilling flask, A, having a capacity of
about one and a half litres, the side tube from which passes through the
long condenser case, B, forming a bend at C in which the metal under
experiment is placed.   The other end of the bend is attached to a
flask, D, which acts as a receiver. The second flask is connected in series
with a large aspirator, G, by means of a U-tube, E, tightly packed with
soda-lime and a caustic potash tower, F. On the posterior side of the
distilling flask, A, is a second U-tube, H, also packed with soda-lime,
which communicates with a caustic potash tower, K, itself in connection
with an air-reservoir, L, containing moistened sticks of caustic potash.
When the aspirator, G, is working, air enters the apparatus through a
minute orifice at the upper part of the tube M, which is filled with soda-
lime.   From and including the stopcock N to the side of the tube joining
E and F all parts of the apparatus are fused together, so that when
reasonable precautions are taken leakage of carbonic acid from the air
into the bend C is entirely prevented.   The other parts of the
apparatus are securely joined together by pressure india-rubber tubing
wired in place and having its surface covered with vaseline.

* *Loc. cit.*   See also *Proceedings of the Royal Artillery Institution*, 1899, No 5,
26 ; and *Report on Atmospheric Corrosion*, Steel Rails Committee, 1900.

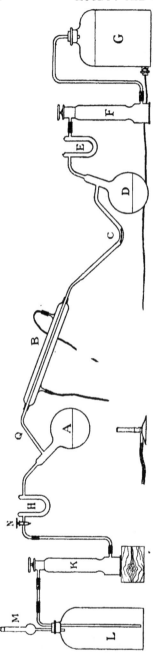

The metal used in the experiments was a soft Swedish iron which was cut from a thick bar into cylindrical pieces about 40 mm. in length and 2 mm. in diameter. These were highly polished and slightly curved so as to lie evenly in the bent tube. The metal contained 99·8 per cent. of iron, and each cylindrical piece weighed a little less than 1 gram.

The experiment consisted in placing one of the pieces of iron in a bent piece of clean, dry glass tube which was as quickly as possible sealed into position at C. A slow current of air was drawn through the apparatus for about three weeks so as to remove all traces of carbonic acid from contact with the iron. Water was distilled from the flask A, which contained a one per cent. solution of barium hydroxide, until more than half a litre had passed over. The cock N was turned off during distillation and, with the object of preventing carbonic acid finding its way into the apparatus through a sudden inrush of air, care was taken to cool the apparatus very slowly after distillation. A current of air was again drawn steadily through the apparatus and the metal kept under observation. The bent tube was of such a size and shape that each bubble of air on passing through the water in it exposed the upper surface of the iron to oxygen, whilst the lower surface remained immersed in water.

When an experiment was made under the conditions described it was found that no general rusting

occurred, but that, however carefully the operator worked, after
an interval varying between a few days and several weeks the
ends of the iron which rested on the glass became discoloured,
and in some cases a few brown specks formed on the surface of
the metal not in contact with the glass. It appeared probable
that this discoloration was due to carbonic acid and moisture shut
in the apparatus during the process of fusing the bent tube in its
place, and accordingly a different mode of procedure was adopted.
Before the bent tube was fused into position, it was partly filled with a
one per cent. solution of chromium trioxide which just covered the iron.
When the apparatus had been swept with air free from carbonic acid
for three weeks, water was distilled through the bend until all the
chromic acid was washed into the receiver, D. After the liquid in the
bent tube became colourless, the distillation was continued until a
volume of water exceeding that already distilled had passed over. Air
was again drawn through the apparatus, and at the end of six weeks
the iron was perfectly bright with the exception of those parts which
rested on the glass and which showed a slight discoloration. As
interaction between glass and iron appeared to take place, a means was
devised whereby in subsequent experiments the contact between these
two substances might be avoided. This was effected by covering the
ends of the iron cylinders with paraffin wax. Commercial paraffin wax
was purified by repeatedly shaking it, in the molten state, with hot
water. The wax was allowed to cool until the temperature fell to
just above the melting point, when each end of the piece of metal was
dipped in and quickly drawn out two or three times. By this means
a blob of wax, covering some 3 mm. of the cylinder, was fixed at
each end. These blobs allowed the metal to lie in the bent tube
without touching the glass. Several experiments were made with
iron protected in this manner, and it was found in every case that,
even after the long-continued passage of air through the water
containing the iron, the metal remained perfectly bright. In one
experiment, after distillation, the current of air was continued for
five weeks. During this time, approximately 56 litres of air
passed over the iron. Since the iron weighed 0·9 gram, the total
weight of oxygen in the air to which it was exposed exceeded by
thirty times the amount required the convert the whole of the metal
into ferric oxide, but not even one speck of rust appeared. In
another experiment, after passing air having a total volume of 32
litres through the apparatus for three weeks, during which time the
metal remained perfectly bright, the glass tube was cut at the point Q
immediately above the distilling flask. Air containing carbonic acid,
but cleaned by its passage through a tower containing pumice-stone
moistened with distilled water, was then drawn through the apparatus.

In the course of six hours, the surface of the metal was distinctly tarnished. After seventy-two hours, during which time approximately 16 litres of air were drawn through, the whole of the surface of the metal was corroded and a considerable quantity of red rust had collected in the bend of the tube.

It must therefore be accepted that when carbonic acid is entirely excluded, no interaction takes place between oxygen and iron in presence of water. Under such conditions, oxygen alone is unable to induce oxidation of the metal, but as soon as air containing its normal quantity of carbonic acid is admitted vigorous rusting results.

Influence of Carbonic Acid on the Absorption of Atmospheric Oxygen by Iron.

Having regard to the remarkable sensitiveness of iron to carbonic acid, to which reference is made later, it appeared of interest to determine the influence of carbonic acid on the absorption of atmospheric oxygen by iron. The apparatus used consisted of a dropping funnel having a globe of 110 c.c. capacity. This funnel was attached by its stem and by a length of pressure india-rubber tubing to a second similar but smaller funnel. The upper opening of the larger funnel could be closed by a caoutchouc stopper into which was fixed a piece of capillary tube bent at a right angle, through which air could be admitted and connection made with a Hempel bulb so as to enable the necessary absorption to be carried out.

In those experiments in which it was desired to exclude carbonic acid, the apparatus and measuring vessels were rinsed and filled with water prepared by distilling one per cent. barium hydroxide solution in an atmosphere which had been exposed to caustic potash and soda-lime.

In estimating the absorption, the dropping funnels were first filled with water, 10 grams of clean, fine iron wire in the form of a spiral were introduced into the globe of the larger funnel, and then 100 c.c. of air admitted and the globe tightly stoppered. After the desired period had been allowed for interaction, the volume of oxygen remaining was directly measured by absorption with pyrogallate.

In the first set of experiments, ordinary air and distilled water which had been exposed to air were used, whilst in the second set these were replaced by air which had been in contact with soda lime for twenty-four hours and water distilled from barium hydroxide solution.

The experiments were made in pairs and the following measurements were recorded :

Percentage of total oxygen in 100 c.c. of
air absorbed by 10 grams of iron.

	Ordinary air and distilled water.	Air and water almost entirely freed from carbonic acid.
After 6 hours' exposure	5·7	none
,, 24 ,, ,,	29·1	none
,, 72 ,, ,,	61·3	0·9
,, 168 ,, ,,	94·3	3·8

In those experiments in which ordinary air and distilled water were used, the iron after six hours' exposure was covered with a coating in part green and in part brown, and the water was slightly turbid. After twenty-four hours, the iron was completely covered with brown rust and the water was very turbid. Precisely similar conditions were observed to prevail in the experiments in which the materials remained together for seventy-two hours and one hundred and sixty-eight hours respectively. In the series of experiments in which air and water almost entirely freed from carbonic acid were used, the iron remained perfectly bright and the water clear after six hours' exposure. In the second experiment, after twenty-four hours' exposure, the surface of the iron was still perfectly bright, but a few small, brown specks were observed in some places where the iron touched the glass. After seventy-two hours' exposure, the iron still remained bright, but the number of brown spots had increased. Specks of rust were to be seen not only where the iron touched the glass, but also in some places where the iron in crossing touched itself. After one hundred and sixty-eight hours' exposure, the appearance of the iron was very much the same as after seventy-two hours' exposure, but the water showed a slight opalescence. The slight attack on the iron and the small absorption of oxygen which took place in this series of experiments were no doubt due to the minute but inevitable leakage of carbonic acid into the apparatus at the time the iron was introduced. The experiments, however, show clearly that the absorption of oxygen by iron when exposed to air and water is almost entirely stopped if the minute quantity of carbonic acid which air and water usually contain is almost entirely removed.

Composition of Iron Rust.

Various analyses of iron rust have been published from time to time, but there is no record that specimens representative of the material in course of comparatively rapid formation have been examined. With the object of obtaining further evidence concerning the composition of rust, samples were obtained from the unpainted interiors of iron flushing tanks in constant use. In such tanks the iron is under conditions specially favourable to rusting, the metal being alternately

and in rapid succession exposed. to air and water. Several of the
tanks from which material was obtained had remained unscraped for
years, and the sides were blistered with masses of rust, brown on the
exterior, but quite black inside. Every sample of this rust when
placed in hydrochloric acid effervesced briskly, yielding a gas which
was completely absorbed by potassium hydroxide. The samples contained an inappreciable quantity of calcium carbonate, and the condition of the iron present in each is given in the following table :

Number of sample	1.	2.	3.	4.	5.	6.
Percentage of iron as ferric oxide	55·73	51·12	64·60	65·13	68·89	67·46
,, ,, ferrous oxide	32·86	36·57	25·74	25·66	23·18	24·40
,, ,, ,, carbonate.	11·40	12·31	9·66	9·21	7·93	8·14

In order to determine the effect of leaving recently-formed rust
fully exposed to air, a portion of one of the samples was roughly
powdered and spread on a porcelain dish. After remaining for eight
days, when the whole of the surface of the material had changed to a
rich brown colour, it was again submitted to analysis. The percentage of iron present as ferrous oxide was found to have been
reduced from 32·86 per cent. to 14·11 per cent. and the percentage of
ferrous carbonate from 11·4 per cent. to 5·62 per cent., or in each
case by slightly more than 50 per cent. The readiness with which
ferrous oxide and ferrous carbonate on exposure to air undergo
oxidation forming ferric oxide accounts for the low percentages of
ferrous iron found in most samples of rust by previous observers.

It is clear that rust kept moist, that is, in contact with what is
really a very dilute solution of carbonic acid, and still adhering to
iron, contains a large proportion of its iron in the ferrous state and
much carbonate. The persistence of ferrous iron under such conditions
is easily understood, for the material is partly surrounded by a
reducing atmosphere of hydrogen, which is constantly being liberated
at the surface of the metal.

The composition of rust in course of formation is altogether out of
harmony with the view that hydrogen peroxide is an effective agent
in its production. In this connection attention must be directed to
the fact that Dunstan and his co-workers assume that by the interaction of iron, oxygen and water twice as much hydrogen peroxide is
liberated as is necessary to oxidise the ferrous oxide simultaneously
formed. If hydrogen peroxide were actually produced in the ratio
stated, ferrous oxide and ferrous carbonate would not be found in rust,
since both these substances undergo immediate oxidation in presence
of hydrogen peroxide. The only rational interpretation of the presence
of a large percentage of ferrous compounds in rust is that the first
stage in the atmospheric corrosion of iron involves the interaction of

carbonic acid and metal, resulting in the formation of ferrous carbonate which gradually becomes basic in character through loss of carbonic acid.

Interaction of Iron and Carbonic Acid.

The remarkable ease with which iron is dissolved by carbonic acid is not generally recognised. The following experiment serves as an illustration of the sensitiveness of the metal to attack. Distilled water which has been shaken with or left in contact with air is poured on a perfectly clean, polished surface of iron. At the end of forty seconds, when the water is seen to be clear and the metal perfectly bright, the water is allowed to run into a porcelain basin containing a drop of a dilute solution of potassium ferricyanide. A marked blue coloration immediately results. If a similar experiment be made with rain water instead of distilled water, the iron is found to be even more readily attacked, ferrous salt in solution being detected after thirty seconds' contact. The ferrous iron found in solution in these experiments is obviously formed by the interaction of a very dilute solution of carbonic acid and iron, for recently-boiled distilled water does not dissolve the metal.

An estimation of the amount of iron dissolved by carbonic acid was made by placing clean iron borings in water kept saturated with carbonic acid at atmospheric temperature and pressure. In these experiments, hydrogen was steadily evolved from the surface of the metal, but, provided air was excluded, the solution remained clear and colourless. In a series of experiments in which 250 c.c. of liquid and 500 grams of clean iron borings were used, the solution was found to contain :

After 20 hours' contact	....	0·2546 gram of FeO per litre	
,, 30 ,,	,, ...	0·3771	,, ,,
,, 56 ,,	,, ...	0·5245	,, ,,
,, 96 ,,	,, ...	0·8172	,, ,,
,, 26 days'	,, ...	2·139	,, ,,

Measurements of hydrogen evolved were also made. In one experiment, 500 grams of clean, fine iron borings were placed in a flask of 2400 c.c. capacity, and the flask and leading tube were subsequently filled with a solution of carbonic acid saturated at 18°. At the end of seven days, 635 c.c. of hydrogen had been collected, and the gas was still being slowly evolved when the experiment was stopped at the end of three weeks. In the absence of air, the solution remained perfectly clear and colourless, but on boiling a green precipitate was formed. This precipitate effervesced with acids, turned red on exposure to air, and gave all the reactions of ferrous carbonate, so that the iron when in solution must have existed as ferrous bicarbonate. The solution was readily decomposed by atmospheric oxygen, yielding a mixture of ferrous carbonate and ferrous and ferric hydroxides, part of the carbonic acid being simultaneously regenerated. On account of the ease with which this change takes place it follows that, in presence of air, a definite weight of carbonic acid will exert a greater corrosive influence on iron than will an equivalent quantity of hydrochloric or sulphuric acid.

Influence of Substances on the Rusting of Iron.

It is not necessary to attribute the inhibiting effects on rusting of such compounds as the alkalis, sodium nitrite, and potassium ferrocyanide to the power they possess of decomposing hydrogen peroxide, for it has been shown that these compounds interact with carbonic acid (*Proc.*, 1903, **19**, 157, 239). Moreover, some substances, such as potassium iodide, which destroy hydrogen peroxide, do not inhibit but actually accelerate rusting. Whilst it is true that a small mass of iron remains bright in a large volume of one per cent. chromium trioxide, nevertheless, iron slowly passes into solution and may be precipitated by the addition of ammonia. If the ratio of iron to chromium trioxide is great, brown ferric hydroxide separates from the solution in the course of a few weeks. Chromic acid, which does not attack iron, appears all the more to exert a protecting influence on the metal because of the ease with which it dissolves ferrous carbonate and ferrous and ferric hydroxides. In solutions exposed to air and containing not more than 0·1 per cent. of chromium trioxide, iron rusts very rapidly, although such solutions react strongly with hydrogen peroxide.

It is assumed by Dunstan that chromic acid and potassium dichromate prevent the rusting of iron because they decompose and therefore interfere with the existence of hydrogen peroxide, which he regards as an intermediate product of rusting. If this form of argument were valid, the inhibiting power of the substances named might equally well be ascribed to their power of decomposing and, therefore, interfering with the existence of ferrous carbonate which, as has been shown, is immediately formed by the interaction of iron and water containing carbonic acid. Like chromic acid, nitric acid may be said to prevent rusting, for a sheet of iron half exposed to air and half immersed in the strong acid remains perfectly bright. A solution of hydrogen peroxide, free from acid, behaves similarly, and it might not unreasonably be suggested that the inhibiting effect results from the power of each of these substances to destroy ferrous carbonate.

Action of Hydrogen Peroxide on Iron.

According to Dunstan, when iron is placed in hydrogen peroxide solution the metal is rapidly oxidised with formation of a substance identical with natural rust, and which, when dried, has the formula $Fe_2O_2(OH)_2$. In his experiments, Dunstan used commercial hydrogen peroxide, but he appears to have been unaware that commercial preparations of hydrogen peroxide are highly impure. I have examined a large number of specimens from various sources and find that every one contains free hydrochloric acid. Many samples also contain phosphoric acid and other impurities. In presence of hydrochloric acid, which energetically attacks iron, and of the powerful oxidising agent, hydrogen peroxide, there is nothing remarkable in the formation of hydrated ferric oxide, but it appeared of interest to compare the behaviour of iron towards hydrogen peroxide solution containing no free acid. Distilled hydrogen peroxide was therefore diluted with unboiled distilled water to 20-volume strength, and a small cylinder of polished iron was introduced. The surface of the metal immediately became covered with bubbles of gas and a slow but steady stream of oxygen escaped from the solution, but no formation of brown oxide could be observed. The experiment was repeated in glass tubes, the open ends of which were drawn out to capillaries bent twice at right angles and terminating under mercury seals. In every case the evolution of oxygen from the surface of the metal continued for from nine to fourteen weeks, at the end of which time the solution failed to decolorise an acid solution of potassium permanganate. During these periods the surface of the metal was exposed to hydrogen peroxide, and was also surface-swept by oxygen in presence of water without the least sign of rust appearing or of the iron undergoing any change in weight or in appearance, thus affording conclusive evidence as to the indifference of iron not only towards hydrogen peroxide, but also towards oxygen in presence of water.

The experiments recorded in this paper conclusively show that oxygen is unable to oxidise iron directly in presence of water, but that when a minute quantity of carbonic acid, such as is contained in air, is present, absorption of oxygen takes place. The explanation of rusting as a process involving the production of hydrogen peroxide, as advanced by Dunstan, is refuted, not only by the complete indifference of iron towards oxygen in presence of water, but also by the composition of rust in actual formation and by the fact that hydrogen peroxide, when free from acid, does not oxidise iron.

On the other hand, the ready interaction of iron and carbonic acid —which exists in all natural waters—resulting in the formation of hydrogen and ferrous salt, affords a satisfactory explanation of the first stage of rusting, which is followed by a more or less complete oxidation of ferrous salt by atmospheric oxygen leading to the production of rust, the composition of which is variable and dependent on the extent to which oxidation of ferrous salt has taken place.

CENTRAL TECHNICAL COLLEGE,
SOUTH KENSINGTON.

LXXVIII.—*The Dynamic Isomerism of Phloroglucinol.*

By EDGAR PERCY HEDLEY, A.R.C.Sc.I.

ON account of the difficulty of deciding from chemical reactions whether phloroglucinol is ketonic or enolic in constitution, Hartley, Dobbie, and Lauder (*Brit. Assoc. Report*, 1902), examined it by means of the spectrograph, and they concluded that it is purely enolic from the close analogy of its absorption curve to that of its trimethyl ether. As further confirmation of this view, they point to the great similarity which exists between the absorption curves of phloroglucinol and pyrogallol. The latter is well known to be enolic in constitution (Baeyer, *Ber.*, 1886, **19**, 163). It would appear that Baly and Ewbank (Trans., 1905, **87**, 1347) join issue with Hartley, Dobbie, and Lauder, on the ground that as tautomerism exists in the mono- and di-phenols themselves, and not in their ethers, the conclusion arrived at regarding the constitution of phloroglucinol was not justified.

Baly and Desch (Trans., 1904, **85**, 1029, and 1905, **87**, 766) have shown that the position of the absorption band of certain aliphatic tautomeric substances undergoes a displacement towards the less refrangible end of the spectrum when sodium hydroxide is added to their solutions, whilst, on the other hand, the addition of acid causes the absorption band to become less persistent. This method, which Baly and Ewbank (*loc. cit.*) applied to the study of the phenols, is also adopted in this investigation.

If the tautomeric process existing in phloroglucinol is of the same type as that existing in the mono- and di-phenols, then on the addition of sodium hydroxide the absorption band ought to be shifted towards the red. Two photographs of phloroglucinol (one milligram-molecule dissolved in 100 c.c. of water) in the presence of the alkali were taken, one with one equivalent and the other with two equivalents of sodium hydroxide. Curve I, Fig. 1, represents the absorption spectrum of the former, the latter being represented by Curve II, Fig. 1. In both cases the band is greatly displaced towards the red. This displacement is considerably greater than that recorded by Baly and Ewbank (*loc. cit.*) in the cases of phenol, quinol, catechol, and resorcinol under the action of five equivalents of sodium hydroxide. It was thought advisable to examine whether the change taking place on the addition of sodium hydroxide to a solution of phloroglucinol was permanent in character. To decide this point, five equivalents of sodium hydroxide were added to an aqueous solution of phloroglucinol (one milligram-molecule in 100 c.c.). The solution, which was deep red, was left for sixty minutes. At the end of this time, five equivalents of hydrochloric acid were added, when the original colourless solution was immediately reproduced. If the change produced by the alkali was a permanent structural change, then, on photographing the solution to which the acid had been added, a series of spectra differing from those of phloroglucinol in neutral solution should result. On the other hand, if the colour and other changes produced by the alkali were of a temporary character, then, on neutralising the alkali, the absorption spectrum of phloroglucinol should be reproduced. These suppositions were tested experimentally and the latter view was definitely confirmed. It would therefore appear that the action of sodium hydroxide is simply to increase the number of ketonic molecules in the solution, and not to produce any other change.

Lowry (*Brit. Assoc. Report*, 1904), in his paper on dynamic isomerism, has pointed that in some cases it has been observed that an ionising solvent influences the equilibrium between two tautomerides in solution. It was therefore thought advisable to examine the absorption spectrum of phloroglucinol in a non-ionising solvent. One milligram-molecule of phloroglucinol was dissolved in 100 c.c. of ether and examined under exactly the same conditions as had existed when phloroglucinol was examined in aqueous solution. The series of spectra obtained were identical with that given by the aqueous solution, and this result shows that, in the present instance, the solvent does not affect the equilibrium between the two forms present in solution.

Finally, the absorption spectrum of phloroglucinol in the presence of hydrochloric acid was examined, with the object of ascertaining whether the persistency of the absorption band would be decreased, as was indicated by Baly and Desch (*loc. cit.*). Such a test applied to phloroglucinol should, if the compound exists in two modifications, decrease the persistency of the absorption band. Ten equivalents of hydrochloric acid were added to a solution of phloroglucinol, which consisted of one milligram-molecule of phloroglucinol dissolved in 100 c.c. of water, and examined through the same thicknesses of liquid as in the cases discussed above. Here again Baly and Desch's theory was supported, as the band was reduced in persistency and became more like phloroglucinol trimethyl ether. Curve I, Fig. 1, represents the band given by one milligram-molecule of phloroglucinol in 100 c.c. of water, and Curve IV that of the same concentration of phloroglucinol in the presence of ten equivalents of hydrochloric acid. The absorption of the trimethyl ether is given in the above-mentioned paper by Hartley, Dobbie, and Lauder.

The spectrum of pyrogallol in alkaline and acid solution has also been examined. Absolutely no difference can be perceived between the absorption of one milligram-molecule of pyrogallol in 100 c.c. of water and that of the same concentration in the presence of ten equivalents of hydrochloric acid, see Curve I, Fig. 2. In the presence of four equivalents of sodium hydroxide, a change is evident. The solution becomes deep brown, due to the absorption of oxygen from the air, and, in consequence, the general absorption is greatly increased, Curve II, Fig. 2. There is, however, no trace of a shift of the absorption band towards the red end of the spectrum, a result which was anticipated for chemical reasons. Great difficulty was experienced in making up the alkaline solution of the pyrogallol to prevent the colour becoming too intense. This was partially obviated by preparing a solution of sodium sulphite in boiling water, thus practically eliminating the dissolved oxygen. On adding four equivalents of sodium hydroxide, the solution assumed a deep amber colour instead of an intense brown. The only effect that this had on the spectrum was to decrease the extent of the general absorption by a very small amount. On adding four equivalents of sodium hydroxide to a solution of pyrogallol (one milligram-molecule dissolved in 100 c.c. of water), excluding the air, allowing to stand for sixty minutes, and then adding four equivalents of hydrochloric acid, the colour of the solution remained unchanged. A photograph of this solution compared with one of the alkaline pyrogallol solution exhibited no difference, thus indicating that the change produced by the alkali is permanent in character.

EXPERIMENTAL.

The method used in photographing the spectra was that described by Hartley (*Phil. Trans.*, 1885, **175**, 325), the lines of cadmium being used as reference. The method of plotting the curves is that employed by Baly and Desch (Trans., 1904, **85**, 1029). In all cases the solutions were prepared by dissolving one milligram-molecule of the substance in 100 c.c. of the solvent, and when necessary these solutions were diluted to one-tenth of this concentration.

The phloroglucinol used was a portion of the specimen recrystallised by Prof. Hartley and described in the British Association Report,

1902. The pyrogallol, which was purified by recrystallisation, melted at 132—134°.

The dot and dash curve, Fig. 1, represents phloroglucinol in aqueous solution in the presence of two equivalents of sodium hydroxide. It will be noticed that the band is broad and persistent, its position being very much nearer the red end of the spectrum than that of phloroglucinol, Curve I, Fig. 1, representing the absorption band of the latter substance in aqueous solution.

The dotted curve, Fig. 1, represents phloroglucinol in aqueous solution in the presence of one equivalent of sodium hydroxide. The absorption band in this case is less persistent than in the former case, a fact to be expected.

FIG. 1.

Oscillation frequencies.

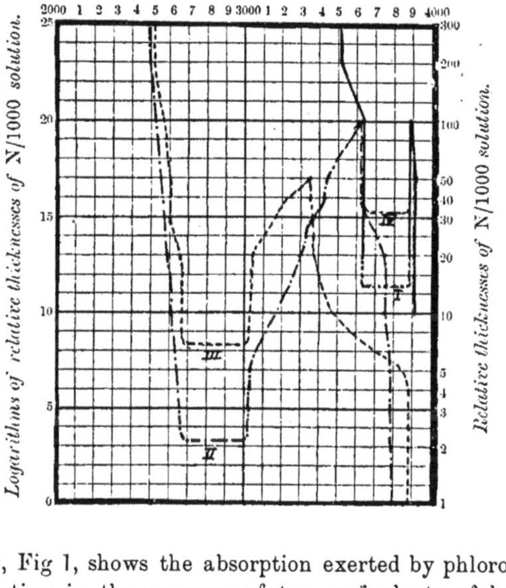

Curve IV, Fig 1, shows the absorption exerted by phloroglucinol in aqueous solution in the presence of ten equivalents of hydrochloric acid.

Curve I, Fig. 2, shows the absorption band characteristic of pyrogallol in aqueous solution. The persistency of the band is not so great as in the case of phloroglucinol; in other respects they are identical. Pyrogallol in aqueous solution and in the presence of ten equivalents of hydrochloric acid produces an absolutely identical absorption spectrum.

The dot and dash curve, Fig. 2, represents pyrogallol in aqueous solution in the presence of two equivalents of sodium hydroxide.

Conclusions.

1. Phloroglucinol exists in neutral solutions in both the ketonic and enolic modifications. That this is the case is shown from the fact that the absorption curve of its trimethyl ether is less persistent than that of phloroglucinol itself, and also because the equilibrium between the two forms is disturbed by the introduction into the solution of alkali and acid.

2. In neutral solution, the enolic form of phloroglucinol is greatly in excess of the ketonic form. This can be seen by comparing the curves of pyrogallol and phloroglucinol, the former being *entirely* enolic.

FIG. 2.

Oscillation frequencies.

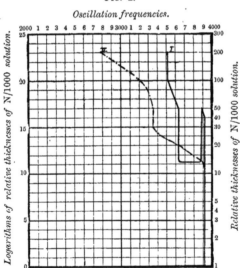

3. The change taking place on the introduction of a "third substance" is not tautomeric but is more correctly described, after Lowry's definition, as a case of dynamic isomerism.

4. The equilibrium between the two forms of phloroglucinol in neutral solution is not affected by the class of solvent used.

In conclusion, I should like to express my warmest thanks to Professor Hartley for affording me facilities for carrying out the experimental work, and for the suggestions and advice with which he has favoured me from time to time.

ROYAL COLLEGE OF SCIENCE,
DUBLIN.

ANNUAL GENERAL MEETING,

MARCH 30TH, 1906.

Professor R. MELDOLA, F.R.S., in the Chair.

THE PRESIDENT declared the ballot open for the election of Officers and Council for the ensuing year, Prof. J. J. SUDBOROUGH and Mr. J. L. BAKER being appointed Scrutators. He then presented the following Report on the state of the Society during the past twelve months :

REPORT OF THE COUNCIL.

THE Council have the satisfaction of being able to report that the past year has been a prosperous one for the Society, the activity of which, as measured by the number of papers read and the number of Fellows on the list, has exceeded that of any previous year.

On the 31st December, 1904, the number of Fellows was 2,711. During 1905, 164 Fellows were elected and 2 reinstated, thus making a gross total of 2,877. The Society has lost 25 Fellows by death, 28 have resigned, the election of 3 has become void, 1 has withdrawn, and 35 have been removed for non-payment of Annual Subscriptions. The total number of Fellows, therefore, on the 31st December, 1905, was 2,785, showing an increase of 74 over the number for the previous year.

The names of the deceased Fellows, with the dates of their election, are :

W. Ackroyd (1897).	F. W. Harrold (1891).
C. M. Blades (1885).	F. M. Mercer (1884).
J. F. Braga (1881).	M. Prasad (1903).
G. B. Buckton (1852).	A. B. Prescott (1876).
J. L. Bullock (1842).	W. T. Rickard (1845).
C. F. Burnard (1849).	R. Roose (1886).
J. H. Calvert (1871).	F. Shapley (1893).
H. S. Carpenter (1875).	C. W. Sutton (1884).
J. Duncan (1863).	C. R. C. Tichborne (1863).
H. S. Elworthy (1886).	L. White (1862).
J. Epps (1885).	W. W. Will (1885).
E. Graham (1895).	R. Yates (1874).
W. H. Greenwood (1873).	

The following Fellows have resigned :

J. Ball.	H. P. Harris.	W. H. Martin.
S. Barlet.	J. H. Hichens.	J. Percival.
G. E. P. Broderick.	S. Hill.	W. S. Rowntree.
C. J. Brooks.	A. Houston.	T. Samuel.
A. J. Carrier.	C. Hunt.	C. E. S. Sherratt.
F. E. Catchpole.	J. T. Johnson.	W. J. Stainer.
C. Childs.	H. W. Kinnersley.	G. R. Tweedie.
C. G. Cresswell.	W. Lang.	T. E. Vasey.
J. Dennant.	H. W. Lawrence.	
H. Goodier.	F. H. Lescher.	

The small number of Fellows still living who were elected in the early days of the Society has been further reduced by the death of Mr. G. B. Buckton, who was elected in March, 1852, and of Mr. J. L. Bullock, elected in 1842. Since the close of 1905, the Society has had to lament the loss of Professor Hermann Johann Philipp Sprengel, who was elected in December, 1864.

The number of Honorary and Foreign Members at the date of the last Annual General Meeting was 35. No names have been added to the list since then, but the Society has sustained a loss in the death of Professor P. T. Cleve, who was elected in February, 1883, and who died on June 18th, 1905. The work of the deceased Honorary and Foreign Member is to be commemorated by a Memorial Lecture, the delivery of which has been undertaken by Professor T. E. Thorpe.

During the year 1905, 233 scientific communications have been made to the Society, 191 of which have been published already in the *Transactions,* and abstracts of all have appeared in the *Proceedings.*

The volume of Transactions for 1905 contains 1,936 pages, of which 1,818 are occupied by 184 memoirs, the remaining 118 pages being devoted to the Wislicenus Memorial Lecture, the Obituary Notices, the Report of the Annual General Meeting, and the Presidential Address ; the volume for the preceding year contains 175 memoirs, which occupy 1,715 pages.

The Journal for 1905 contains also 4,356 abstracts of papers published mainly in foreign journals, which extend to 1,828 pages, whilst the abstracts for 1904 numbered 4,617, and occupied 1,920 pages.

The abstracts for 1905 may be classified as follows :

Part I.

	Pages.	No. of Abstracts.
Organic Chemistry	956	1,727

Part II.

General and Physical Chemistry.......... ..		619
Inorganic Chemistry..........................		562
Mineralogical Chemistry		71
Physiological Chemistry		467
Chemistry of Vegetable Physiology and Agriculture		295
Analytical Chemistry		615
	872	2,629
Total in Parts I. and II.	1,828	4,356

In making comparison with the preceding year, it must be borne in mind that the 1904 volume contained the abstracts for thirteen months. Owing to the change in date of publication, no abstracts appeared between December 1st, 1903, and January 31st, 1904, so that the number of the Journal published on the latter date contained practically the abstracts of two months (see *Trans.*, 1905, **87**, 539).

The Council regrets to announce that Dr. G. T. Morgan has found it necessary to resign the post of Editor of the Society's publications which he has occupied so creditably since the beginning of 1903. Although his resignation was received in September, 1905, Dr. Morgan kindly acted as Editor until the appointment of his successor, Dr. J. C. Cain.

The second part (Subject Index) of the Collective Index for the decade 1893–1902 was issued in December, 1905, to those Fellows who had made application for it in accordance with the printed notices circulated with the monthly parts of the Journal subsequently to July, 1903, and the Council have pleasure in expressing the high appreciation of the ceaseless energy displayed by the Indexer, Mrs. Margaret Dougal, on the completion of this valuable work.

During the vacation of 1905, advantage was taken of the Doctorial Jubilee of Professor Adolph von Baeyer, and of the Professorial Jubilee of Professor Mendeléeff, to address letters of congratulation to these two distinguished Honorary and Foreign Members.

In June, 1905, the Council decided that the Ordinary Meetings of the Society should be held during the ensuing Session on the first and

third Thursdays of the month at 8.30 p.m. The experiment of holding
the meetings on Wednesday afternoons at 5.30 p.m., alternately with
Thursday evenings at 8 p.m., had been in operation since January,
1902, and from the fact that the average attendance of Fellows at the
afternoon meetings had undergone steady diminution during this
period, whilst the number of Fellows attending the evening meetings
had increased, it has been concluded that a majority of the Fellows who
make a practice of attending the meetings find the evening more
convenient than the afternoon.

The Library has received an important addition by the generosity of
Sir Henry E. Roscoe, who has presented to the Society a collection of
136 alchemical and early chemical works of great interest and value.

An increase in the use of the Library is recorded, 1,108 books being
borrowed during 1905, as against 1,057 in the previous year. Additions
to the Library, excluding Sir Henry Roscoe's donation, comprise 165
books, of which 58 were presented, 324 volumes of periodicals, and 48
pamphlets, as against 119 books, 296 volumes of periodicals, and 52
pamphlets last year. On the recommendation of the Library Committee
the Council have made an addition to the Library Rules in the following terms : " No persons other than Fellows of the Society have the
privilege of using the Library, except upon a written introduction from
a Fellow, with whom rests the responsibility for all books consulted by
the person introduced. Such introduction shall be valid for one
occasion only."

Grants amounting in all to £236 have been made during the year
from the Research Fund, and £22 15s. 3d. has been returned. Of the
papers published in the *Transactions* during 1905, 24 were contributed by authors to whom grants had been made from the Research
Fund.

In February of last year the Treasurer was fortunate enough to be
able to increase the invested capital of the Society by almost exactly
£1,500 by the purchase of £1,983 Midland Railway 2½ per cent. Preference Stock for £1,499 14s. 5d. The income from all sources for the
year 1905 exceeds that for 1904 by only £93 8s. 4d., whilst the
expenditure has been abnormally heavy, exceeding the total income
by £305 1s. 5d. This is due to the incidence of several very heavy
accounts, one of which really belongs to several years, and that is the
completion of the Decennial Index for the period 1893–1902, which
has entailed the expenditure this year of no less than £778 10s. 5d.,
or rather more than half the total cost. No further expenditure,
however, on account of Decennial Indexes will be required for
several years to come. As pointed out last year, the continual increase
in the cost of the *Journal* and *Proceedings* is a source of much anxiety,
and this year a further increase has to be recorded on both accounts,

£219 12s. 7d. in the former and £35 18s. 2d. in the latter case, or a net increase over the cost of both in 1904 of £255 10s. 9d., and over that in 1903 of £578 3s. 3d. The Council therefore trust that authors of papers will do their utmost to assist the Publication Committee and the Treasurer in keeping down the cost of printing as far as possible.

The publication of the *Annual Reports on the Progress of Chemistry* has necessarily added materially to the expenditure of the Society ; the cost of Volume I having entailed an expenditure of £445 19s. 6d., of which about £70 has been recovered by their sale. Another item of some magnitude, £52 9s. 6d., is the cost of the printing and circulating of the Bye-laws, together with the changes proposed by the Council, which were submitted to an Extraordinary General Meeting of the Society on February 8th, 1905.

The administrative expenses have been carefully kept in check, and whilst everything has been maintained in a state of thorough efficiency their amount is only £827 16s. 5d., as against £893 1s. 0d. in 1904.

The following facts with regard to the cost of the Decennial Indexes may be of interest. The total cost of Volume IV, 1893-1902 (Parts I and II), exclusive of distribution amounts to £1,454 5s. 6d., of which the cost of printing was £762 6s. 0d. The cost of Volume III, 1883-1892 (issued in one volume) was £1,280 2s. 7d., of which the printing accounted for £585 19s. 7d. The cost of Volume IV was relatively considerably less than that of Volume III, as the former extended to 45½ sheets for the Authors' Index and 108 sheets for the Subject Index, or 153½ sheets in all, whilst Volume III only contained 102¾ sheets. Owing to the way in which the Annual Indexes are now prepared, it is hoped that when Volume V has to be undertaken, the relative cost will be still further reduced and that the chief cost will be that due to the printing.

INCOME AND EXPENDITURE ACCOUNT

Income.

	£ s. d.	£ s d
To Life Compositions		360 0 0
,, Admission Fees		592 0 0
,, Annual Subscriptions—		
Received in advance, on account of 1905	255 0 0	
,, during 1905 on account of 1905	3843 0 0	
,, ,, ,, ,, 1904	284 0 0	
,, ,, ,, ,, 1903	4 0 0	
	4386 0 0	
Less amount included in last year's Income, being valuation of Arrears as per last Balance Sheet	330 0 0	
	4056 0 0	
Add Arrears at date: 1905 £352, 1904 £26, 1903 £4, estimated to realise as per Balance Sheet	250 0 0	4306 0 0
,, Investments :—		
Dividends on £1,050 London and North Western Railway 3 per cent. Debenture Stock	29 18 6	
£6,730 Metropolitan Consolidated 3½ per cent. Stock ...	223 15 8	
£1,520 14s. 3d. Cardiff Corporation 3 per cent. Stock ...	43 6 8	
£1,400 India 2½ per cent. Stock	33 5 0	
£2,400 Bristol Corporation 2½ per cent. Debenture Stock	57 0 0	
£4,341 Midland Railway 2½ per cent. Preference Stock	103 2 0	
£1,200 Leeds Corporation 3 per cent. Stock	34 4 0	
£1,500 Transvaal 3 per cent Guaranteed Stock	42 15 0	
Income Tax Recovered	26 6 11	
Interest on Deposit Account	12 1 10	605 15 7
,, Publications :—.		
Sales :		
Journals	770 3 3	
Proceedings	18 16 9	
General Index	68 5 9	
Memorial Lectures	11 9 6	
Library Catalogue	2 4 9	
Atomic Weight Tables	17 0	
Jubilee Volume	11 0	
Annual Report on the Progress of Chemistry	77 11 0	
	949 19 0	
Less Publishers' Commission	92 9 8	
	857 9 4	
Proceeds of Advertisements in Journal... .. .. £61 15 4		
Less Commission 9 5 3	52 10 1	
		909 19 5
,, Subscriptions from other Societies :—		
Society of Chemical Industry	8 8 0	
Society of Public Analysts	11 11 0	
		19 19 0
,, Balance, being Excess of Expenditure over Income carried to Balance Sheet		305 1 5
		£7098 15 5

I have examined the above Accounts with the Books and Vouchers of the Society, and and the Investments.

W. B. KEEN,

13th March, 1906.   *Chartered Accountant.*

FOR THE YEAR ENDED 31ST DECEMBER, 1905.

Expenditure.

	£	s	d.	£	s.	d.
By Expenses on account of Journal and Proceedings :—						
Salary of Editor	300	0	0			
Salary of Sub-Editor	200	0	0			
Salary of Assistant Sub-Editor	51	5	0			
Salary of Indexer	80	0	0			
Editorial Postages, &c.	23	4	1			
Abstractors' Fees	431	13	6			
Printing of Journal	2313	4	11			
Printing of Advertisements	28	12	0			
Printing of Wrappers	115	16	6			
Distribution of Journal	491	13	9			
Authors' Copies	113	9	6			
Illustrations	22	9	5			
Insurance of Stock at Clay's	10	13	6			
				4182	2	2
Printing of Proceedings	207	7	6			
Distribution of Proceedings	44	4	10			
				251	12	4
,, Annual Report on the Progress of Chemistry				445	19	6
,, List of Fellows				74	0	1
,, Printing and Distribution of Bye-Laws				52	9	"
,, Collective Index :—						
Printing Vol. IV., Part II	592	9	6			
Distribution	38	10	9			
Salaries	147	10	2			
				778	10	5
,, Library Expenses :—						
Salary of Librarian	105	0	0			
Books and Periodicals	232	11	10			
Binding	63	8	0			
				400	19	10
,, Indexing for International Catalogue				30	0	0
,, Balance of Dinner Account				55	5	
,, Administrative Expenses :—						
Tea Expenses	31	5	4			
Salary of Assistant Secretary	170	0	0			
Salary of Office Assistant	28	11	6			
Wages (Commissionaire, Housekeeper, and Charwoman)	148	2	0			
Pension, Mrs. Hall	30	0	0			
Coal and Lighting	43	13	10			
House Expenses and Repairs	90	11	0			
Insurance	6	17	2			
Accountants' Charges	10	10	0			
Accountants' Commission on Recovery of Income Tax	1	16	9			
Legal Charges	21	5	0			
Miscellaneous Printing	83	17	0			
Stationery	37	17	9			
Furniture	5	8	7			
Postages	89	7	5			
Miscellaneous Expenses	28	13	1			
				827	16	5

£7098 15 5

certify them to be in accordance therewith.   I have also verified the Balance at the Bankers

Approved—   E GRANT HOOPER,
HY. FORSTER MORLEY,
H. R. LE SUEUR.

RESEARCH FUND INCOME AND EXPENDITURE ACCOUNT FOR THE YEAR ENDED 31ST DECEMBER, 1905.

Income.	£ s. d.	£ s. d.
To Dividends:—		
On £1000 North British Railway 4 per cent. No. 1 Preference Stock	38 0 0	
On £4400 Metropolitan Consolidated 3½ per cent. Stock	146 6 0	
On £1034 Great Western Railway 2½ per cent. Debenture Stock	24 11 2	
		208 17 2
,, Repayments of Research Grants:—		
G. Dean	1 5 5	
G. Young	1 0 0	
J. C. Cain	7 14 1	
B. D. Steele	9 15 7	
Miss I. Smedley	3 0 2	
		22 15 3
,, Income Tax Recovered		
		£242 1 11

Expenditure.	£ s. d.	£ s
By Grants to—		
W. C. Ball	5 0 0	
E. C. C. Baly	15 0 0	
F. E. Francis	10 0 0	
J. A. N. Friend	3 0 0	
P. Haas	3 0 0	
T. M. Lowry	10 0 0	
R. S. Morrell	10 0 0	
R. H. Pickard	15 0 0	
A. W. Stewart	5 0 0	
J. J. Sudborough	20 0 0	
J. F. Thorpe	15 0 0	
A. E. H. Tutton	20 0 0	
E. F. Armstrong	10 0 0	
P. C. ...	5 0 0	
A. G. Green	10 0 0	
J. Hawthorne	5 0 0	
F. W. Kay	10 0 0	
A. McKenzie	10 0 0	
T. S. Moore	5 0 0	
R. Robinson	10 0 0	
S. Ruhemann	15 0 0	
Miss A. E. Smith	5 0 0	
A. Tattersall	10 0 0	
J. A. Gardner	5 0 0	
		236 0
Balance, being excess of Income over Expenditure, carried to Balance Sheet		6 1
		£242

Liabilities.

	£ s. d.	£ s. d.
To Subscriptions received in advance, 1906 ...	207 18 0	
,, ,, 1907 ...	6 0 0	213 18 0
,, Cash received on Account of Annual Report for 1905 (not published) ...		16 19 0
,, Sundry Creditors ...		1537 12 11
,, Research Fund:—		
As per last Balance Sheet	6784 17 8	
Add Excess of Income over Expenditure for the year	6 1 11	6790 19 7
,, Chemical Society:—		
As per last Balance Sheet	18609 1 6	
Less Excess of Expenditure over Income for the year	305 1 5	18394 0 1
		£26953 9 7

Assets.

	£ s. d.	£ s. d.
By Investments (value when required):—		
£1050 London and North Western Railway 3 per cent. Debenture Stock ...	839 12 0	
£6730 Metropolitan 3½ per cent. Stock	7212 8 0	
£1530 14s. 3d. Cardiff Corporation 3 per cent Stock	1650 0 0	
£1400 2½ per cent. Stock ...	1316 1 0	
£2400 Bristol Corporation 2½ per cent. Debenture Stock ...	2070 2 0	
£2358 Midland Railway 2½ per cent. Preference Stock, as per last Balance Sheet	2072 8 0	
£1983 Midland Railway 2½ per Preference Stock, purchased during year	1499 14 5	
£1200 Leeds Corporation 3 per cent. Stock ...	3572 2 5	
£1143 Transvaal 3 per cent. Stock ...	1143 1 0	
£1500 Stock ...	1460 13 6	10264 0
(Estimated present value of Investments, £170 15s.)		
,, Stock of Publications, &c. (not valued) ...		
Sundry Debtors:—		
Society of Public Analysts	11 11 0	
Advertising Agents	50 9 3	
Messrs. Gurney & Jackson	406 18 11	
Fine Deposit	1 0 0	
,, Subscriptions in Arrear, £382. Estimated to realise...	133 13 0	
,, Insurance paid in advance ...	14 0 4	
,, ,, in hand ...		147 13 4
,, Research Fund:—		
Investments (value when acquired):—		
£1000 North British 4 per cent. No. 1 Preference Stock	1010 0 0	
£4400 Metropolitan Consolidated 3½ per cent. Stock	4587 18 0	
£1084 Great Western Railway 2½ per cent. Debenture Stock ...	1049 15 11	
	6647 13 11	
(Estimated present value of Research Fund Investments, £6594 10s.)		
Cash at Bank...	143 5 8	6790 19 7
		£26953 9 7

L. LXXXIX.

Inc.

To	£ s. d.	£ s. d.
Funds:—		
On £900 North British Railway 4 per cent. No. 1 Stock	38 0 0	
On £4400 Metropolitan Stock	146 6 0	
On £1034 Great rn Railway 2½ per cent. Debenture Stock	24 11 2	208 17 2
,, Repayments of Research Grants:—		
G. Dean	1 5 5	
G. Young	1 0 1	
J. C. Cain	7 14 1	
B. D. Steele	9 15 7	
Miss I. Smedley	3 0 2	22 15 3
,, Income Tax Recovered		10 9 6
		£242 1 11

Expenditure.

	£ s. d.	£ s. d.
By Grants to—		
W. C. Ball	5 10 0	
E. C. C. Baly	15 10 0	
F. E. Francis	10 10 0	
J. A. N. Friend	3 3 0	
P. Haas	3 10 0	
T. M. Lowry	10 10 0	
R. S.	10 10 0	
R. H.	15 10 0	
A. W. Stewart	5 10 0	
J. J. Sudborough	20 10 0	
J. F. Thorpe	15 10 0	
A. E. H. Tutton	20 10 0	
E. F. Armstrong	10 10 0	
J. B. Cohen	5 10 0	
A. Findlay	5 10 0	
A. G. Green	.5 10 0	
J. Hawthorne	10 10 0	
F. W. Kay	10 10 0	
A. McKenzie	5 10 0	
T. S. Moore	10 10 0	
R. Robinson	10 10 0	
S. Ruhemann	15 10 0	
Miss A. E. Smith	5 10 0	
A. Tattersall	10 10 0	
J. A. Gardner	5 10 0	236 0 0
,, Balance, being excess of Income over Expenditure, carried to Balance Sheet		6 1 11
		£242 1 11

I have examined the above Account with the Books and Vouchers of the Society, and certify it to be in accordance therewith. I have also verified the Balance at the Bankers and the Investments.

W. B. KEEN,
Chartered Accountant.

13th March, 1906.

Approved—

E. GRANT HOOPER,
HY. FORSTER MORLEY,
H. R. LE SUEUR.

The adoption of the Report was proposed by Mr. R. J. FRISWELL, seconded by Dr. H. BRERETON BAKER, and carried unanimously.

A vote of thanks to the Treasurer, Secretaries, Foreign Secretary, and Council for their services during the past year was proposed by Dr. T. E. THORPE, seconded by Sir THOMAS STEVENSON, and unanimously adopted. Sir WILLIAM RAMSAY responded.

The PRESIDENT then delivered his Address, which will be found on p. 745.

Sir HENRY E. ROSCOE proposed a vote of thanks to the President, coupled with the request that he would allow his Address to be printed in the Society's *Transactions*. Dr. HORACE BROWN seconded the motion, which was carried by acclamation, and acknowledged by the President.

The Scrutators then presented their Report to the President, who declared the following to have been duly elected as Officers and Council for the ensuing year :—

President : Raphael Meldola, F.R.S.

Vice-Presidents who have filled the office of President: H. E. Armstrong, Ph.D., LL.D., F.R.S. ; A. Crum Brown, D.Sc., LL.D., F.R.S. ; Sir William Crookes, D.Sc., F.R.S. ; Sir James Dewar, M.A., LL.D., F.R.S. ; A. Vernon Harcourt, M.A., D.C.L., F.R.S. ; H. Müller, Ph.D., LL.D., F.R.S. ; W. Odling, M.A., M.B., F.R.S. ; W. H. Perkin, Ph.D., LL.D., F.R.S. ; J. Emerson Reynolds, Sc.D., M.D., F.R.S. ; Sir Henry E. Roscoe. LL.D., F.R.S. ; W. J. Russell, Ph.D., F.R.S. ; T. E. Thorpe, C.B., LL.D., F.R.S. ; W. A. Tilden, D.Sc., F.R.S.

Vice-Presidents : Horace T. Brown, LL.D., F.R.S. ; Harold B. Dixon, M.A., F.R.S. ; Rudolph Messel, Ph.D. ; W. H. Perkin, jun., Ph.D., F.R.S. ; A. Smithells, B.Sc., F.R.S. ; W. P. Wynne, D.Sc., F.R.S.

Treasurer : Alexander Scott, M.A., D.Sc., F.R.S.

Secretaries : M. O. Forster, D.Sc., Ph.D., F.R.S. ; A. W. Crossley, D.Sc., Ph.D.

Foreign Secretary : Sir William Ramsay, K.C.B., LL.D., F.R.S.

Ordinary Members of Council : Edward C. C. Baly ; Bernard Dyer, D.Sc. ; William Gowland ; Alfred D. Hall, M.A. ; H. A. D. Jowett, D.Sc. ; A. Lapworth, D.Sc. ; J. E. Marsh, M.A. ; F. E. Matthews, Ph.D. ; G. T. Moody, D.Sc. ; A. G. Perkin, F.R.S. ; W. J. Sell, M.A., F.R.S. ; John Wade, D.Sc.

PRESIDENTIAL ADDRESS.

Delivered at the ANNUAL GENERAL MEETING, MARCH 30TH, 1906.

By RAPHAEL MELDOLA, F.R.S.

THE first duty which I feel it necessary to discharge at this general meeting of the Fellows is to express my grateful appreciation of the honour which you conferred upon me at the last annual meeting in placing me in the distinguished position of President of the Society. Thirty-five years have elapsed since I joined your ranks, and I find that during no less than twenty years out of that period I have had the privilege of taking part in the administrative affairs of the Society, first as Member of Council, then as Foreign Secretary, as Vice-President from 1902 to 1905, and as President during the past year. The development of the science of chemistry in this country is so closely associated with this Society, of which the presidential chair has been occupied by a succession of such eminent chemists, that any Fellow placed in this honourable position cannot but acknowledge that he has received from his co-workers the highest dignity which it is in the power of British Chemistry to bestow.

You will have gathered from the report of the Council, which is now in your hands, and which, in accordance with the Bye-laws, I have formally presented for your acceptance, that the affairs of the Society are in a flourishing condition. Our activity in every department of work has been well maintained, and the interest in the meetings, as measured by the attendance and the number and quality of the communications and the discussions arising therefrom, indicates that there is no falling off either in zeal or originality on the part of our chemical workers. In fact, if the output of work goes on increasing and we have, as has happened on several occasions during the past year, such crowded programmes that it is impossible to give adequate time to individual authors, it may be necessary for the Council at some future period to take into consideration the advisability of occasionally having extra or overflow meetings so as to give opportunities for the fuller discussion of the more important communications.

From the financial statement of the Treasurer it appears that the prediction of my predecessor in this chair has been fulfilled, and that our expenditure has exceeded our income by more than £300. It may

3 C 2

be pointed out, however, that the past year has been very heavily weighted with special expenses incurred through the publication of the decennial index, as well as by other matters of a non-recurrent character. The Society has also undertaken the new responsibility of publishing the Annual Reports on the progress of chemistry, the second volume of which is now in your hands. Of the immense value of this work there can be no question, but it adds very considerably to our publishing expenses, and some future Council may be confronted with the question whether the finances of the Society will stand this additional burden without some relief from the Fellows in the form of subscriptions for this particular publication. Other societies with which I am connected have, I may add, been compelled to face the very same difficulty, and to come to the conclusion that they were giving their members more than· the value of their annual subscriptions. I personally see no ground for the least uneasiness in the actual financial position, but the present opportunity is a fitting one for pointing out, as has been done from this chair on former occasions, that the main expenditure of the Society being the cost of printing and publishing our Journal, it is to a considerable extent within the power of our contributors to assist in keeping this expenditure down to the lowest reasonable limit. A very long experience in connection with the publications of scientific societies has convinced me that many writers of papers are more or less devoid of the sense of what may be called literary perspective, that is, the faculty of expressing in their true proportions the descriptive details and the general results of their investigations. It is in the nature of our subject that long and laborious experimental work may lead to results which can be stated in a few lines, but that is no reason why the contents of the laboratory note-book should be presented for publication in extenso in our Transactions. I am not now venturing to criticise the literary style of many of our scientific writers ; this subject was dealt with very many years ago by one whose writings have justly placed him in the front rank among scientific authors of the nineteenth century. I refer to the late Prof. Huxley, whose judgment in such matters none will call in question. I do appeal, however, once again to the authors who contribute to our Transactions to express their results with all possible conciseness. It may be pointed out that apart from the question of cost of printing, extreme diffusiveness is injurious to the author himself, as general results of importance are apt to be overlooked if buried under a mass of detail. It must be borne in mind by the writers of chemical papers that they are not addressing a lay public, but a body of experts as familiar as they are themselves with all details of experimental procedure, and that a mere outline statement of method is quite sufficient unless there is some novelty in the apparatus or mode of treatment requiring special description. Another consideration which may be urged in support of this plea for condensation is that the difficulties of the Secretaries, of the Publication Committee, and of the editorial staff are enormously increased by the over-elaboration in which some authors indulge. Delay is often caused by having to place such papers in the hands of referees, and friction may arise between the Editor and author if the criticisms of the referees are not in harmony with the author's views. The readers of the numerous papers which appear in our Transactions do not sufficiently realise that many of these papers in the form in which they are presented to the public represent the end result of much serious deliberation by the Publication Committee, of careful consideration by referees, and often of unpleasant editorial correspondence with the authors. It is the feeling that it is the duty of our Fellows to bear these difficulties in mind when preparing their results for publication that has prompted this appeal from the presidential chair.

Among the events which have recently taken place in the chemical world, attention may be called to the Imperial Chemical Institute which our German colleagues are preposing to establish in Berlin on the lines of their "Physikalische Reichsanstalt." The report of the German Chemical Society on the proposed Institute was published in last month's *Berichte* (1906, **39**, 316), and in that report will be found a number of valuable suggestions concerning the various branches of work which might be carried on in such an establishment. The development of this scheme will be followed with sympathetic interest by all British chemists; its realisation will serve to show other nations that in one European country the national importance of our science to the welfare of the State is fully recognised.

Another event which calls for special mention is the celebration this year of the fiftieth anniversary of the foundation of the coal-tar colour industry by Dr. William Henry Perkin, whose first patent for the production of mauve bears the date Aug. 26th, 1856. Although our Society concerns itself chiefly with the pure science, leaving the special cognisance of the development of chemical industry to its sister Society, yet the relations between the abstract and applied science are in this case so intimate that it may be fairly said that the two aspects of chemistry have in this branch of manufacture become inextricably interwoven. The addresses delivered in Germany on the occasion of the Kekulé celebration are no doubt fresh in your memories, and you will have gathered therefrom how largely the theory of the constitution of the so-called "aromatic" compounds has assisted in the development of the industry and how the latter has in turn reacted on theory through the discoveries arising in the course of manufacturing operations and the supply of new materials placed at the disposal of scientific investigators. In dealing with the history of modern organic chemistry, it is impossible to ignore the great and beneficial influence which the foundation of the coal-tar colour industry has exerted upon this branch of our science. The Fellows of the Chemical Society will, I am sure, desire to convey their sincere congratulations to our esteemed Past-President, Dr. Perkin, to whom it must be a source of great satisfaction to contemplate the enormous developments, both scientific and industrial, which have resulted from his discovery of the first coal-tar colouring matter half a century ago. As you are no doubt aware, an international movement is now being organised for the purpose of celebrating the jubilee of this discovery and of doing honour to the founder of the industry. The long and distinguished association of Dr. Perkin with this Society, which has had, and I rejoice to say still has, the benefit of his scientific activity, makes it only appropriate that the Perkin Memorial scheme should have been organised under our auspices. It is particularly gratifying to me, whose first association with the colour industry began in the year 1870, and with a break of a few years was continued down to 1885, to find myself this year, although only by a pure coincidence, in a position which enables me to take a prominent part in the promotion of the movement. The details of the scheme will shortly be in your hands and I can confidently appeal to all chemists to give it their serious consideration and support.

The completion of one year's active service in the Presidential Chair has made it manifest to me, as it has to my predecessors, how deeply the Society is indebted to its honorary officers for the zeal and judgment with which they discharge the very onerous and ever-increasing duties which they have voluntarily undertaken on our behalf. I desire to take advantage of the present opportunity of expressing my own thanks for the loyal and, indeed, indispensable services which they have rendered during the past year. In parting with Dr. Morgan I am sure also that the Fellows will recognise that during his term of office as Editor the status of our publications has been fully sustained and that in the discharge of his duties he has always had in view the best interests of the Society. His retirement from editorial work, although an immediate loss to us, will, no doubt, be of benefit to the Society in the future, as we may now look to him for increased scientific activity in other directions.

The Living Organism as a Chemical Agency; a Review of some of the Problems of Photosynthesis by Growing Plants.

I PROPOSE on the present occasion offering some remarks on the possibility of extending organic chemistry along certain lines which appear to me to hold but the promise of results of far-reaching importance for the future development of that branch of our subject. The history of Chemistry, as of all other natural and physical sciences, reveals the principle that advances of sufficient importance to be considered as marking epochs have generally resulted from what might be considered the co-opting of the methods and appliances of other departments of science. There is a philosophical basis underlying this principle—the unity of nature ; the existence in reality of one Science only, the subdivision of which into departments is necessitated by the imperfection of our knowledge and is a matter of practical convenience imposed by the limitation of our faculties rather than the expression of any actuality behind the groups of phenomena which we pigeon-hole by our classifications.

It would be setting a dangerous precedent were I to deliver from this chair anything approaching an abstract philosophical disquisition, but with respect to the application of the foregoing principle to our own science of Chemistry, we all know what enormous developments have followed the introduction of the balance, the thermometer and calorimeter, the prism, the polarimeter, and the electric current. Physical methods for the determination of chemical constitution are becoming of greater and greater importance and are destined to play a still greater part in the chemistry of the future. It is only necessary to remind ourselves of the part played by the prism in determining refractivity or specific absorptive power in connection with the structure of molecules, or the use for a similar purpose of the polarising prism in the case of optically active compounds or of compounds rendered temporarily active in the electro-magnetic field. The refined use of the thermometer in determining the depression of freezing point caused by various compounds in solution has resulted in the cryoscopic and ebullioscopic methods of determining molecular weights now in common use in our laboratories. The electric current, utilised at first solely as a decomposing agent by Davy, has in later times been turned to good account in organic synthesis, while the conductivity method of determining the constitution of electrolytes in solution may be regarded as among the recent gains by our science arising from the application of a physical method. Whether the silent electric discharge, the " effluve electrique " of the French physicists, will fulfil the hopes of those who have advocated its use as a synthetical agent must be left for decision by future experimental work.

The modern theory of the constitution of matter, although at present without practical bearing on our purely chemical methods, may yet throw light in the future on such fundamental questions as the structure of the chemical atom and the nature of valency. Whatever may be the subsequent developments of the " electron " theory of the atom in the hands of physicists, it is of interest to remember that this work began by the study of the passage of electricity through gases with the object in the first place of observing and recording their emission spectra. From the old " vacuum tube " we passed to the " high vacuum " and the " ultra-gaseous " form of matter discovered by Crookes, the connection of which with the latest development of the physical conception of the atom is a matter of recent history. From the present point of view, this example illustrates very forcibly the interdependence of physics and chemistry.

So also the discovery of the inert gases of the helium group, which was inaugurated by Lord Rayleigh's observation of the difference in the density of atmospheric nitrogen as compared with nitrogen from chemical sources and by the isolation of argon by Rayleigh and Ramsay, may be regarded as the outcome of the refined use of the balance. The claims of the photographic plate also as an appliance which has been utilised in chemical research with remarkable results cannot be left out of consideration. It is not going too far to say that the new field of radioactivity was opened up by the discovery of the " uranium radiations " by Becquerel through the action of these emanations on the gelatino-bromide film. The hunting down of radium by the Curies was also facilitated—perhaps it may be said even rendered possible—by the use of a physical contrivance, the gold-leaf electroscope, which is capable of detecting and measuring radioactivity by virtue of the " ionising " action of the radiations on the air and the consequent discharge of the electrified leaves.

Examples such as those above mentioned might be multiplied indefinitely from every department of science. The reason for reminding you of these familiar phases in the history of modern chemistry is to emphasise the principle of appealing to other departments of science for the further illumination of the obscure regions of our own subject. If there is any agency in nature of whose teachings we have not sufficiently availed ourselves, it appears to me that in that direction lies the promise of great future developments. There is such an agency which from a very remote period of the earth's history has been solving chemical, and I may perhaps be permitted to add also physical, problems in the most wonderfully mysterious way by methods which we have not yet been able to imitate in our laboratories, and which cannot fail, therefore, when followed up, to lead us into totally new fields of research. The agency to which I refer is the living organism.

Physicists and physiologists are familiar enough with the achievements of the living organism in the way of developing optical and acoustical instruments, in generating and storing an electric charge, in transforming chemical energy into phosphorescent light, in utilising the potential energy (derived originally from solar radiations) of the organic matter of foodstuffs with an efficiency inimitable by any piece of mechanism, in developing structures for the utilisation of gaseous and liquid diffusion, of osmotic pressure and capillarity, or for the production of the most gorgeous colours by diffraction and interference. Nor is it unlikely that certain of the lower animals are in possession of structures for the perception of ethereal disturbances beyond the sensual perceptions of man and the higher animals. It is not for me to ask whether the physicists have exhausted all the possible teachings of vital physics. I can lay claim to being nothing more than an interested onlooker in this field, but I strongly suspect that if a fair account of the present state of knowledge of vital physics were presented, we should find a very large residue of unexplained phenomena. For the purpose of this address, however, the argument is that if these wonderful physical achievements of the living organism are granted, as in face of the facts they must be, no less wonderful and mysterious are its chemical achievements. It is impossible to contemplate the problems of vital chemistry without coming to the conclusion that they are surrounded by mysteries as great as or even greater than those which surround vital physics. I venture to think that the next great step in the domain of organic chemistry is to be looked for in this direction—in the elucidation of the chemical changes going on in living plants and animals as the cause and consequence of their vital activities.

The actual chemical transformations effected by the living organism both in the way of building up and breaking down organic compounds are familiar enough as facts. No less familiar is it that whilst similar transformations can in many cases be carried out in the laboratory, the artificial methods are not the same as the vital methods. By this I mean to say that if it were possible to follow step by step the genesis of any organic compound throughout the course of its vital synthesis— even if the vital cycle set out from the same simple materials, such as carbon dioxide and water—it would be found that the intermediate stages were in most cases, possibly in all cases, quite different in the laboratory and in the living organism. The chemical evolution of organic compounds in the living organism is a branch of investigation beset with great practical difficulties, partly owing to the transitory character of the intermediate stages and partly owing to the want of chemical and microchemical methods of diagnostic value. Such clues to the genetic relationships as have hitherto been obtained have generally come from the physiological side. The chemical realisation of these transformations, quite apart from the question of method, has in most cases eluded the skill of chemists. This fact alone furnishes a strong argument in favour of the contention that the living organism has great and almost unbroached stores of chemical information to contribute to our science when we shall have penetrated further into the mystery of the vital processes. The physiologists are certainly realising the growing importance of this side of their subject. Their methods are being brought more into line with our purely chemical methods, and the modern school comprises workers of sound training as chemists. Chairs of physiological chemistry have been founded in some of our Universities, such as at Liverpool and Glasgow, and Journals devoted to biological chemistry have recently been started. Many of the leading physiologists have insisted on the necessity for co-operation from the chemical side. I need only refer you to Dr. Halliburton's Report on Physiological Chemistry in last year's volume of Annual Reports (*op. cit.*, p. 169). On the occasion of the foundation of the chair of Physiological Chemistry in the University of Glasgow, Prof. McKendrick said :

" I think there can be little doubt that the next great advance in physiology will be from the side of physiological chemistry. . . . During the last sixty years many of the physical phenomena of the living being have been investigated by special methods. It seems to me that we cannot expect much more from the application of the graphic method of registration, nor from the examination of the phenomena of electrical action in living tissues. The microscope and the methods of histological research have left little to be desired as to our knowledge of the structure of the elementary tissues and the structure of organs. A new departure must be made. No method of research seems so inviting or so promising as the rigid and methodical investigation of the chemical phenomena happening in living matter. . . . Hence the extreme importance of the chemist and the physiologist working hand in hand for the future advancement of physiological knowledge " (*Nature*, **70**, 640 ; Oct. 27th, 1904).

In view of this and similar authoritative utterances of late years, I think that the time is opportune for letting the physiologists know that chemistry has just as much to learn from physiology as the latter science from chemistry. An examination of current physiological literature will convince chemists that the developmental stages in the evolution of organic compounds in living plants and animals are either unknown or have been filled in by hypothetical equations which are but speculations, more or less plausible, and based rather on physiological than 'on direct chemical evidence. Such speculative advances are of course legitimate in the absence of real knowledge ; they are useful if properly handled, but they are apt to become dangerous if they are taken without confirmatory evidence as representing the course of the actual biochemical transformations. The fact that when by more thorough investigation or by the discovery of new methods of attack such hypothetical steps are frequently shown to be erroneous, brings out very clearly the necessity for recognising more fully the speculative character of much of the chemistry that has passed into physiological literature. It is the revision of these biochemical hypotheses by the joint labour of physiologists and chemists that is the problem pressing for the attack so strongly advocated by Halliburton, McKendrick, and others. In this country we can point with pride to the results achieved in the domain of phytochemistry by Horace Brown and his colleagues. Results of the highest importance may be looked for also from the systematic study of enzyme action in relation to chemical constitution initiated by Emil Fischer and now being prosecuted so vigorously here and on the continent.

If physiology has forced upon the attention of chemists large classes of chemical changes which have hitherto been unrealisable in the laboratory, and of which the explanation is still in the hypothetical stage, it is no less true that with every advancement of organic chemistry chemists have endeavoured to apply the new knowledge of facts and methods or the new theoretical developments to the elucidation of biochemical phenomena. The invasion of the domain of the physiologist by the chemist has led to splendid results in the hands of such workers as Emil Fischer, whose papers on the sugars and on the purine bases will rank for all time as chemical classics, and whose later advances towards the mysteries of proteid synthesis are being followed with the keenest interest by both physiologists and chemists. It is no disparagement to the labours of this great master of chemical synthesis if I point out that we are still in ignorance of the origin and course of development of an optically active sugar, a purine base, or a proteid in the living organism.

It may be of interest, as an illustration of the small extent to which we have been enabled to penetrate into the vital chemical processes, if I give a brief history of and sum up the present state of knowledge concerning that most fundamental of all biochemical syntheses, the photosynthesis carried on by the green leaves of growing plants. It is somewhat remarkable that the solution of a problem of such enormous importance should have for so long baffled the ingenuity of chemists. The physical aspect of photochemical "assimilation" by the green leaf has been made the subject of exhaustive investigations, notably by Timiriazeff, by Horace Brown and his colleagues, and by F. Blackman. The historical side of the chemistry of this vital process has been dealt with by many writers. I must refer you for full information on this subject to the recognised classics on plant physiology and to certain special writings, among which attention may be called to a lecture by Prof. S. H. Vines, delivered before this Society in 1878 (Trans., 33, 375), to Horace Brown's presidential address to the British Association at Dover in 1899 (B.A. Rep., Dover, 1899, p. 664), and to Mazé's little work, Évolution du Carbone et de l'Azote dans le Monde vivante ("Scientia" series ; No. 6, Carré and Naud, Paris, 1899). The net result of the historical summary is that while distinct advance has been made in our knowledge of the physics of the process, the primary facts of chemical significance which can be regarded as definitely and finally established are that green plants acquire their carbon from carbon dioxide (Priestley, 1771 ; Ingenhouss, 1779 ; Senebier, 1788), that the volume of oxygen exhaled is approximately equal to the volume of carbon dioxide absorbed (De Saussure, 1804 ; Boussingault, Compt. rend., 1861, 53, 862), that in the reverse process of respiration the same relationship between the volumes of carbon dioxide and oxygen is not exactly maintained (Bonnier and Mangin, Compt. rend., 1885, 100, 1303, 1519, and other investigators), that the first visible product of the process is starch (Sachs, Bot. Zeit., 1862, 20, 365), and that the first sugar identifiable by the present known chemical methods is cane sugar (Brown and Morris, Trans., 1893, 63, 604). To these data we may now have to add the observations of Usher and Priestley, communicated to the Royal Society last January, that the first detectable compound of an aldehydic character is formaldehyde (Proc. Roy. Soc., 1906, series B, 77, 369). From these points onwards, in endeavouring to follow the chemical development of the compounds resulting from the photolytic process, we find ourselves in a region more or less hypothetical.

More than thirty-five years have elapsed since v. Baeyer advanced his well-known hypothesis that the first product of assimilation is formaldehyde resulting from the photolysis of carbon dioxide in presence of water with the elimination of free oxygen (Ber., 1870, 3, 63), the aldehyde thus formed undergoing " carbohydrate " polymerisation. Plausible as is this hypothesis from the chemical point of view, there has been no satisfactory physiological evidence in its favour until quite recently. All attempts to prove the presence of formaldehyde in the green parts of plants have hitherto led to inconclusive results.* Experiments undertaken with the object of ascertaining

* For recent work on this subject, see the paper by Plancher and Ravenna, Atti R. Accad. Lincei, 1904, 13, [ii], 459.

whether plants can utilise this aldehyde directly as a source of carbohydrates have only revealed the fact that the compound acts as a
poison, although this conclusion may require modification in view of
the recent statements by Bouilhac (*Compt. rend.*, 1902, **135**, 1369 ;
136, 1155) and by Tréboux (*Flora*, 1903, p. 73), that certain plants
can form starch when grown in very dilute solutions of the aldehyde.
The oft-quoted experiments of Bokorny also (*Ber.*, 1891, **24**, 103), who
succeeded in getting plants to utilise hydroxymethylenesulphonates,
may be considered as evidence in the same direction. In all such
feeding experiments, however, the evidence is obviously indirect, as
there is always a possibility of the aldehyde undergoing previous
saccharisation, and it is known that plants can utilise sugars with the
formation of starch when grown in obscurity. All these points will
be found fully discussed in the standard monographs, and a useful
synoptical summary of the state of knowledge down to 1904 is given
in a paper published in that year by Euler (*Ber.*, **37**, 3412). The
negative evidence, moreover, that is, the absence of proof of the
existence of the free aldehyde in plants, cannot be regarded as fatal to
the hypothesis, since the readiness with which the aldehyde undergoes
condensation with various organic compounds might well account for
its rapid fixation at the photosynthetic centres, as suggested by O. Loew
in 1889 (*Ber.*, **22**, 484).

The latest contribution to the subject of photosynthesis, by Messrs.
Usher and Priestley, to which I have already referred, appears to have
advanced our knowledge by at least one important stage. I say
"appears," because I am given to understand that the authors regard
their paper as of a preliminary character. Their conclusions are,
however, of such great importance that their rigid confirmation will
be anxiously awaited by all who are familiar with the present inconclusive character of the physiological evidence. According to the
results thus far made known, formaldehyde and hydrogen peroxide are
considered to be the normal photolytic products, the former being
rapidly fixed ("condensed") by the living protoplasm and the latter
decomposed by a special enzyme with the liberation of oxygen. The
presence of the aldehyde at the photosynthetic centres was inferred
from the formation of methyleneaniline at these points when the leaves
were soaked in aniline water, the compound being identified by its crystalline form and other properties. But the main point, which appears to
me to be of such far-reaching importance, is that the initial step, the
photolysis of the carbonic acid, is, according to the experiments made
known, not a vital process at all.* Green leaves of *Elodea*, *Ulva*, and

* It is only right to state that the authors do not themselves specifically draw
this conclusion, but it is obviously deducible from their results. There can be
nothing "alive" in the chlorophyll of a leaf which has been immersed in boiling

Enteromorpha, in which both protoplasm and enzymes have been killed by immersion in boiling water, are said to be also capable of producing formaldehyde from carbonic acid in the presence of light. In other words, the efficiency of the chlorophyll apparatus is, according to this view, not destroyed by eliminating the "vital" agencies, but, owing to the non-removal of the photolytic products, the accumulated hydrogen peroxide kills the chlorophyll and the reaction then becomes reversible. Thus for the first time we are led to hope that the extension of work in this direction will enable us to determine with some approach to exactitude how much of the photosynthetic process is purely non-vital and how much due to protoplasmic and enzymic activity. It may be well to point out that the protoplasmic theory of vital synthesis, which postulates the actual combination of the carbonic acid with the living protoplasm as a necessary prelude to photosynthesis, will, for this process, receive its death blow if it is finally established that a dead leaf containing only chlorophyll as an optical sensitiser can produce formaldehyde from carbonic acid. It is for this reason in particular that these recent results require such rigid confirmation. The function of the chlorophyll itself regarded only as an organic pigment has also to be determined.

The hypothesis contributed by chemistry to biology more than thirty-five years ago has thus received more distinct support from the physiological side than has hitherto been furnished. It would appear from the experiments of Messrs. Usher and Priestley that the reason why previous experimenters have not succeeded in obtaining conclusive evidence of the presence of formaldehyde in the green parts of plants is because they failed to remove the active condensing agent, the living protoplasm, before exposure to light. From leaves which had been "killed" these authors claim to have obtained (after exposure to light) sufficient formaldehyde by steam distillation to enable this compound to be detected by the methyleneaniline and tetrabromohexamethylenetetramine tests. From this it follows that the hypothesis of v. Baeyer may now have to be expressed as a reversible equation :

$$H_2CO_3 + 2H_2O \underset{\leftarrow}{\overset{\rightarrow}{\rightleftharpoons}} C\ H_2O + 2H_2O_2.$$

The energy intake insured by the chlorophyll initiates the change from

water. It follows further, if these results are confirmed, that the term "photosynthesis" may have to be abandoned, since the action of light is simply, according to this view, to bring about photolysis of carbonic acid, the carbohydrate synthesis following therefrom being protoplasmic and therefore presumably independent of light. This is in harmony with the fact that green plants can utilise certain sugars and produce starch therefrom in the dark. The enzyme which decomposes hydrogen peroxide may be of the nature of Loew's "catalase" (*Ber.*, 1902, **35**, 2487 ; *Chem. Centr.*, 1903, **1**, 887).

left to right, this change being independent of any " vital " influence. In the absence of such influence, equilibrium is soon reached owing to the reverse change. The functions of the protoplasm and the enzyme are that of catalysts removing both products of photolysis and insuring the *continuous* change from left to right so long as energy is supplied from without.

The position of the "assimilation" problem may be summed up so far as concerns the initial stage by the statements that a possible chemical explanation of carbohydrate synthesis was suggested by v. Baeyer in 1870 and that physiology has, until the present year, furnished but indecisive evidence in favour of this formaldehyde hypothesis. But the problem has been handled from the chemical side by many workers since 1870, and many new hypotheses as well as modifications of the original hypothesis have been suggested. It is unnecessary to go into the details of these later theoretical developments, as they have become a recognised part of the literature of the subject, but it is important to note that the photolysis of carbonic acid with the formation of hydrogen peroxide and formaldehyde was suggested on purely chemical grounds by Erlenmeyer in 1877 (*Ber.*, 10, 634). In view of the very plausible nature of the formaldehyde hypothesis, supported as it is by the earlier and later discoveries that this aldehyde readily undergoes saccharisation, it was naturally realised by many workers that the living plant had probably solved the chemical problem of producing formaldehyde from carbon dioxide and water, and attempts to imitate this process in the laboratory independently of chlorophyll or any "vital" agency have been recorded. Of these, the first positive result was claimed by Bach in 1893 (*Compt. rend.*, 116, 1145). By passing carbon dioxide through a solution of uranium acetate exposed to sunlight, a precipitate of uranous and uranic hydroxides with a trace of uranium peroxide was obtained, and their formation attributed to the photolysis of the carbonic acid with the production of formaldehyde. No distinct proof of the presence of the aldehyde was given. In another communication, published the same year (*loc. cit.*, 1389), dimethylaniline was used as a catalyst, carbon dioxide being passed through a solution of the sulphate of this base exposed to sunlight and the formation of formaldehyde inferred from the colour given by the product (tetramethyldiaminodiphenylmethane) on oxidation (Trillat's test). The details will be found in the original paper. Five years later, Bach further announced that carbonic acid is reduced by hydrogen-palladium with the formation of some formaldehyde, the latter being identified by the methyleneaniline and hexamethylenetetramine tests (*Compt. rend.*, 1898, 126, 479). In this same year, the photolysis of carbonic acid in presence of uranium acetate was repeated in violet light and the formation of formaldehyde again inferred (*Arch. Sci. phys. nat.*, Genève, 1898, [iv], **5**, 401).

These results of Bach have passed into biochemical literature and have remained unchallenged until recently. The importance of the problem has, however, led to the repetition of the work, and the present state of knowledge concerning this fundamental process may be said to be still somewhat indefinite so far as the purely chemical evidence goes. In the first place, the reduction of carbonic acid by all reducing agents which have hitherto been applied has resulted in formic acid only. So far back as 1865, Maly used sodium amalgam for the reduction of carbonates and bicarbonates (*Annalen*, **135**, 119), and his results were confirmed by Ballo (*Ber.*, 1884, **17**, 6) and by Lieben (*Monatsh.*, 1895, **16**, 211 ; 1897, **18**, 582). Bach's experiment with hydrogen-palladium has, so far as I know, not been repeated. The electrolytic reduction of carbonic acid was effected in 1870 by Royer (*Compt. rend.*, 1870, **70**, 731), and has been quite recently made the subject of exhaustive researches by Coehn and Jahn (*Ber.*, 1904, **37**, 2836) and by Löb (*ibid.*, 3593). All these experimenters are unanimous in declaring that formic acid is the sole product ; in other words, with the one exception of Bach's hydrogen-palladium, no reducing agent has yet been found which carries the reduction of carbonic acid to the formaldehyde stage :

$$H_2CO_3 + 2H_2 = CH_2O + 2H_2O.$$

According to Moissan (*Compt. rend.*, 1905, **140**, 1209), such a reducing agent as potassium hydride gives only formic acid.

The synthesis of formaldehyde from carbon dioxide and hydrogen by means of the silent electric discharge was made known more than thirty years ago by Sir Benjamin Brodie (*Proc. Roy. Soc.*, 1874, **22**, 172), and this subject is now undergoing thorough investigation by Walther Löb (see *Ber.*, 1904, **37**, 3593), who promises further information. But even if we admit that there is an analogy between the electric "effluve" and solar radiant energy as an endothermic agency, I do not think that any results obtained by this method are likely to give definite information concerning the process of photosynthesis. It has long been known that under the influence of this discharge dissociation and electrolysis take place. A mixture of carbon dioxide and water vapour might certainly be expected to give carbon monoxide and hydrogen among the products of their decomposition, and it is well known that these give the aldehyde under such conditions (Losanitsch and Jovitschitsch, *Ber.*; 1897, **30**, 136). It may be of interest to note in passing that in its original form the formaldehyde hypothesis of Baeyer postulated the preliminary dissociation of the carbon dioxide into carbon monoxide and oxygen, but this view of the process is no longer held. Only in January last a note was communicated to this Society by Messrs. Chadwick and Ramsbottom (*Proc.*, 1906, 22, 23), in which it was announced that by the action of ultra-violet light carbon dioxide is partially decomposed into carbon monoxide and oxygen. The condition essential for this decomposition is, however, that the gas should be *dry*, which is certainly not the state of affairs with the carbon dioxide undergoing photolysis in the green leaf whether living or dead.

With respect to the photolytic production of formaldehyde independently of "vital" agency, the experiments of Bach have lately been repeated by Euler (*Ber.*, 1904, 37, 3414), who, with the same catalysts, namely, uranium acetate and dimethylaniline, has obtained only negative results. He concludes that no catalyst playing the part of a photochemical reducer of carbonic acid has yet been discovered. On the other hand, Messrs. Usher and Priestley, in their recent paper, state that they have been enabled to confirm the results of Bach both as regards the production of formaldehyde and hydrogen peroxide when uranium acetate is used. There is thus actual conflict of evidence respecting the facts, and we must await the results of further experimental work. Even if positive results are obtained, however, and formaldehyde shown conclusively to be a product of the photolytic decomposition of carbonic acid in presence of uranium acetate, there is still the objection that an organic uranium salt is present and that the aldehyde may arise from the photochemical decomposition of the acetic acid. This objection must have occurred to all chemists who have critically considered Bach's experiments. Messrs. Usher and Priestley have endeavoured to meet this criticism by substituting uranium sulphate for the acetate, but, so far, with this salt formic acid only has been obtained.*

Pending the completion of these researches and the decision of the question from the laboratory side, it is permissible to assume—especially in view of the recent physiological evidence—that the synthesis of carbohydrates in the plant actually sets out from formaldehyde. All Fischer's work on the synthesis of the hexose sugars may be regarded as giving support to this view. But when we attempt to follow out the developmental stages in detail we find ourselves, as I have previously said, in a region of hypothesis. This is tantamount to the admission that the living plant as a chemical agency has some very fundamental principles to contribute to chemical science. Apart

---

* According to a private communication from Mr. Usher, received since writing the above, there is formed, in addition to formic acid, a substance which was obtained as a syrup and which appeared to have the properties of " methylenitan." This substance, however, did not give a compound when treated with phenylhydrazine acetate.

from the fact that synthetical sugars are optically inactive, we are, in the first place, confronted with the questions (1) what causes the polymerisation, that is, the saccharisation of the aldehyde in the plant? and (2) through what stages do the compounds pass in the up-grade course? In attempting to answer questions of this kind it must not be forgotten that we may have enzyme action to deal with at every stage. In the laboratory, no ordinary organic reagent is known as a saccharifier of formaldehyde; the saccharification of this compound has hitherto been effected only by metallic oxides or hydroxides or salts. In this connection I may call attention to recent researches by the Eulers (*Ber.*, 1905, **38**, 2551 ; 1906, **39**, 36, 39 ; compare Auerbach, *Ber.*, 1905, **38**, 2833), who have made a study of the influence of various metallic hydroxides on the aqueous solution of the aldehyde. According to Usher and Priestley, it is the living protoplasm which "condenses" the aldehyde in the assimilating leaf. Their evidence for this as well as for the existence of an enzyme capable of decomposing hydrogen peroxide appears to be sound so far as it goes, but this conclusion, if ultimately found to be true, only shifts the mystery of carbohydrate formation on to the protoplasm. In other words, the production of a sugar from formaldehyde is on this view a case of "protoplasmic synthesis." * It may be that this is the course of events in nature ; if so, the living plant has raised another question for the consideration of chemists, namely, the possibility of finding an organic compound which can saccharify formaldehyde.† It is well known that the aldehyde readily condenses with all kinds of organic compounds, including the proteids (see the *Chemie der Eiweisskörper* by Cohnheim, 1904, p. 126), but no sugar has yet been shown to result from any of these condensation products.

In considering the possible transition stages from formaldehyde to carbohydrate, the plant physiologist will find that the suggestions offered by chemical theory are quite up to, if not far beyond, the existing state of knowledge. From this point of view it may be well to consider, as an illustration of an opposite kind to that furnished by the foregoing invasion of chemistry by physiology, the encroachments of chemical theory on the domain of physiology. The formaldehyde hypothesis of v. Baeyer may be looked upon as the first successful contribution from this purely chemical side. The amended hypothesis of Erlenmeyer in 1877, in which the formation of hydrogen peroxide at

---

* See my address to Section B of British Association, Ipswich, 1895 ; *Reports*, p. 648.

† The only organic saccharifying agent mentioned by Loew is the strongly basic tetraethylammonium hydroxide. This raises the important question whether, in the event of an optically active ammonium base being obtainable, it might not be possible, by this means, to synthesise an optically active sugar.

an intermediate stage was taken into consideration, may be looked on as the second contribution from the chemical side. With respect to the intermediate stages between formaldehyde and the sugars, where we had at first nothing but hypothesis to guide us, it may be useful to call attention to the later discoveries and theoretical suggestions which chemists have given to plant physiologists either for confirmation or refutation. Thus, Emil Fischer suggested in 1890 that it might be worth while looking for the triose, " glycerose," in the green parts of plants (*Ber.*, 1890, **23**, 2138). Glycerose is now known to be a mixture of glyceric aldehyde and dihydroxyacetone, and in 1897 Piloty (*Ber.*, **30**, 3168) indicated certain possible stages in the development of fructose from glycerose, thus :

In the first place through glycollic aldehyde to glyceric aldehyde :

$$\begin{array}{cc} CH_2O \\ CH_2O \end{array} \rightarrow \begin{array}{c} CH_2 \cdot OH \\ CHO \end{array} + CH_2O \rightarrow \begin{array}{c} CH_2 \cdot OH \\ CH \cdot OH \\ CHO \end{array} .$$

In the next place through dihydroxyacetone :

$$\begin{array}{cc} CH_2O & H \\ CH_2O & H \end{array} \!\!\! > \!\! C \colon O \rightarrow \begin{array}{c} CH_2 \cdot OH \\ C \colon O \\ CH_2 \cdot OH \end{array} .$$

The aldehyde and ketone are then supposed to condense to fructose :

$$CH_2(OH) \cdot CH(OH) \cdot CHO$$
$$\qquad CH_2(OH) \cdot CO \cdot CH_2 \cdot OH \quad \rightarrow$$
$$\qquad\qquad CH_2(OH) \cdot [CH(OH)]_3 \cdot CO \cdot CH_2 \cdot OH.$$

For this view there is much chemical, but thus far no biological, evidence. Glyceric aldehyde of biochemical origin is known as a product of the cultivation of certain species of *Bacillus* and *Tyrothrix* in solutions containing mannitol (Péré, *Ann. Inst. Past.*, **10**, 417), whilst dihydroxyacetone is a product of the action of the sorbose *Bacterium* (*B. xylinum*) on glycerol or dextrose (see *Chemical Synthesis of Vital Products*, Vol. I., pp. 242 and 292). But neither glycollic nor glyceric aldehyde nor dihydroxyacetone nor any triose sugar has as yet been found among plant products. This negative evidence may mean that the vital catalyst works so rapidly that the intermediate stages are too transient to be detected, or it may mean that there are no transition stages, and that the formaldehyde is polymerised at once into a polyose sugar. It may also mean that our present chemical methods are incapable of detecting the intermediate stages. In any case here is a suggestive hypothesis concerning the course of events in the living plant, which is worthy of serious consideration from the physiological side, because the chemical evidence is fairly complete all along the line.

It has long been known that the "formose" obtained from formaldehyde by polymerisation with lime is a mixture containing three
or four sugars, among which i-fructose (a-acrose) is present. It is also
known that "glycerose" can be polymerised by the action of alkali
with the formation of a mixture containing i-fructose, and it has been
shown by Fenton and by that author and Jackson that glycollic
aldehyde also readily polymerises into a mixture of a- and β-acrose
(Trans., 1894, **65**, 899 ; 1895, **67**, 48, 774; 1896, **69**, 546; 1897,
71, 375; 1900, **77**, 129). Moreover, some important links in the
chain of chemical evidence were filled in at the beginning of this year
by the Eulers (*Ber.*, 1906, **39**, 45), who have used the more gentle
catalyst, calcium carbonate, instead of lime or alkali, and have thereby
succeeded in producing for the first time both glycollic aldehyde and
dihydroxyacetone directly from formaldehyde. They have also shown
that a pentose sugar, i-arabinoketose, is a product of this polymerisation. Here is another suggestion offered by chemistry to plant
physiology. Natural l-arabinose is an aldose, so also is the synthetical
d-arabinose. Arabinoketose has not yet been found among plant products, but in view of its synthesis by the Eulers it may be worth
looking for.

From formaldehyde to fructose the laboratory evidence is thus fairly
complete. It remains only to connect formaldehyde with carbonic
acid by some photolytic method which may be regarded as above
suspicion—and the discovery of such a method may be looked for
sooner or later—in order to say that the chain of chemical evidence
is quite complete. If, further, glycollic aldehyde or glyceric aldehyde
or dihydroxyacetone could be detected at the photosynthetic centres of
green leaves we should be enabled to say that the laboratory processes
and the vital processes followed the same line of development, the only
differences (confessedly important ones) being the nature of the
catalysts or condensing agents and the ever-present property of
optical activity. But when we attempt to follow the course of development from fructose upwards we find that the living plant soon
leaves all our chemical resources far behind its actual accomplish-
ments. We have to appeal more and more to hypothetical suggestions.
In the first place, we are met with the difficulty that the primary
carbohydrate resulting from the photosynthetic process is, or at any
rate according to Brown and Morris may be, cane sugar. There are
also in the green leaf, besides saccharose and fructose, both dextrose
and maltose.

Now, the transformation of fructose into dextrose (with other
sugars) has been shown by the late Lohry de Bruyn and van Ecken-
stein (*Ber.*, 1895, **28**, 3078 ; *Rec. trav. chim.*, 1897, **16**, 274, 282) to
take place quite readily in the presence of alkali. This chemical

discovery suggests that if fructose is a primary synthetical product. the dextrose may be developed therefrom by isomerisation.* The problem then presented by the living plant for chemists to solve is the discovery of organic transformers playing the part of alkali in the Lobry de Bruyn isomerisation process. From dextrose and fructose, saccharose might arise by condensation and hydration, but this synthesis has as yet eluded all laboratory methods. On the other hand, maltose, which might be expected to arise by the condensation of two molecules of dextrose, has probably been synthesised by the action of hydrochloric acid as well as of certain enzymes on dextrose (E. F. Armstrong, *Proc. Roy. Soc.*, 1905, series *B*, 76, 592). From the twelve carbon atom sugars to carbohydrates such as starch and cellulose, we are practically in the domain of plant physiology ; chemistry has so far dealt with the constitution of these complex compounds only in a tentative way, and no synthesis has as yet been effected.

In considering the questions surrounding this fundamental chemical process, which is generally described as "assimilation," it is absolutely essential that chemists should realise most thoroughly the biological aspects of the problems. The main object of the process is obviously the preparation of food for the immediate or ulterior nourishment of the living protoplasm. Now, it has not yet been proved that either carbonic acid or formaldehyde are "assimilated" as such ; on the contrary, it may be safely asserted that assimilation in the physiological sense is only possible when compounds much higher in the scale of chemical evolution are presented in a suitable form to the protoplasm. For this reason we may have, in accordance with the views of Prof. J. Reynolds Green (*Vegetable Physiology*, 1900, Chaps. X and XI), to reject the term "assimilation" altogether for the initial stages of the process. On the other hand, as I have already indicated, the term "photosynthesis" may be equally inapplicable, unless we agree to regard the photolysis of carbonic acid as a synthesis of formaldehyde. On the whole, I venture to think that it would be safer not to stereotype our nomenclature until the chemical evidence has been strengthened. All that can be said now is that plants can photolyse carbonic acid, and that following this process there is a series of up-grade and down-grade syntheses which have nothing to do with the direct action of light.

From these considerations we are naturally led to the next step in

---

* Of the other sugars which may arise as the result of this process, mannose is a plant product (*Chemical Synthesis of Vital Products*, Vol. I, p. 248), but has not been shown to be generally present among leaf sugars. Glutose, another of the sugars resulting from the isomerisation of fructose, has not yet been found among natural products.   .

the series of physiological processes. It is certain that the protoplasm cannot feed on carbohydrates alone—there must be nitrogenous matter as well, and it is, in fact, generally considered by plant physiologists that the actual food which is assimilated is of a proteid character. The living plant thus furnishes another set of problems for solution by chemistry, namely, the development of proteids from carbohydrates and nitrogenous compounds in some form. Here, of course, we are getting altogether outside the possibilities of the laboratory methods at present at our disposal. The amino-acids are generally, and no doubt correctly, regarded as the simplest compounds which build the nitrogen into the proteid complex (see, for example, J. R. Green, *op. cit.*, pp. 180—181). But the synthesis of any amino-acid from a carbohydrate has not yet been accomplished, excepting through the cyanohydrin and subsequent hydrolysis, a method which can hardly be considered likely to be that followed in the natural course of the synthesis of these acids. By way of physiological data, we have the well-known facts that under normal conditions the majority of plants take in their nitrogen in an inorganic form, and that by cultivation experiments it has been proved that plants can grow when supplied with previously-formed carbohydrates and amino-acids (see for recent experiments on this subject, in addition to the standard works, a paper by Lefèvre, *Compt. rend.*, 1905, **141**, 211).

In the absence of any evidence suggesting a biochemical genetic relationship between carbohydrates and amino-acids, it is permissible to raise the question whether the latter—which are of the same order of importance as carbohydrates from the point of view of plant physiology—may not have some other origin. The only direct relationship between carbohydrates and certain nitrogenous compounds of vegetable origin which recent chemical research has indicated is the interesting discovery by Windaus and Knoop (*Ber.*, 1905, **38**, 1166) that dextrose on treatment with aqueous ammonia in presence of zinc hydroxide breaks down at the ordinary temperature with the formation of methyliminazole, probably through the intermediate formation of glyceric aldehyde, methylglyoxal, and formaldehyde, and the condensation of the two latter with ammonia to form the iminazole ring. The latter is contained, as we know, in some of the natural alkaloids, so that the genesis of these from carbohydrates is at any rate a chemical possibility. But with respect to the amino-acids, all the known synthetical methods are with one exception obviously remote from any biochemical process. The exceptional method is the condensation of unsaturated acids with ammonia, first made known by Engel (*Compt. rend.*, 1887, **104**, 1805 ; 1888, **106**, 1677) and successfully applied by Fischer to the synthesis of diamino-acids (*Ber.*, 1904,

37, 2357 ; *ibid.*, 1905, 38, 3607).* From the biochemical point of view, this method has gained in significance through the recent discovery that ketonic and aldehydic acids can, under suitable conditions, be made to combine with ammonia. Thus, from pyruvic and glyoxylic acids there have been obtained respectively acetylalanine and formylglycine, from which compounds the corresponding amino-acids are obtainable by hydrolysis (Erlenmeyer, jun., and Kunlin, *Ber.*, 1902, 35, 2438 ; also de Jong, as quoted by these authors).

$$\underset{CO_2H}{HC{:}O}\,NH_3\quad \underset{CO_2H}{H{\cdot}C{:}O}\rightarrow \underset{CO_2H}{H_2C{\cdot}NH{\cdot}CHO}\,(+H_2O+CO_2)\rightarrow \underset{CO_2H}{CH_2{\cdot}NH_2}.$$

The presence of unsaturated acids in plants has hitherto received but little confirmation, although, in view of the oxidisability of tartaric to dihydroxymaleic acid, it may be worth while, as suggested by Fenton in 1897 (Trans., 71, 383), making a special search for this acid.

One obvious difficulty in the way of this hypothesis of the formation of amino-acids by plants is the supposed presence of ammonia, if not in the free state, at least in some form of combination. It is generally stated that the higher plants cannot utilise ammonium salts as such for the purpose of proteid formation. The physiological evidence on this point is, however, somewhat conflicting, as will be seen on referring to the standard works.† Admitting that amino-acids are the lower stages in the up-grade synthesis of proteids—an admission which is now warrantable from the chemical side through Fischer's recent work on the polypeptides—no other source of the amino-group than ammonia can in the present state of knowledge be suggested with any degree of probability. This implies the assumption that nitrates can at the synthetic centres of amino-acid production undergo reduction. The assumption is not altogether unwarrantable, and it is accepted as an axiom by many phytochemists. It would, perhaps, be safer to label the supposition as hypothetical and as requiring further confirmation from the living plant. But if we use the hypothesis in the scientific spirit as a suggestive stimulus to further investigation, it raises some very important considerations concerning that fundamental process of photosynthesis from which all phytochemical processes in the higher plants set out.

Speculative explanations of the formation of amino-acids in plants are to be found in the literature of phytochemistry, but on careful

---

* From sorbic acid, a constituent of the juice of mountain-ash berries, Fischer and Schlotterbeck have obtained a new diaminohexoic acid by this method (*Ber.*, 1904, *loc. cit.*).

† See particularly Czapek's *Biochemie der Pflanzen*, 1905, Vol. II, Chaps. XXXIX and XL.

consideration these do not appear very plausible. Those who require details may refer to the papers by Loew (*Chem. Centr.*, 1897, i, 931), suggesting the origination of aspartic acid from formaldehyde and ammonia, by Bach, suggesting the formation of asparagin from formaldehyde and hydroxylamine *via* formaldoxime and formamide (*Chem. Centr.*, 1898, ii, 366), and by Hébert, suggesting the condensation of formaldehyde with hydrogen cyanide with the formation of nitrogenous compounds (*Ann. Agronom.*, 1898, **24**, 416). For these views there is at present practically no evidence either from the physio-logical or chemical side   The discovery of Erlenmeyer and Kunlin, however, directs attention once again to the possibility of certain acids being primary products of photosynthesis. Aldehydic acids of the nature of glyoxylic acid in particular would fulfil the conditions of amino-acid synthesis.

The view that oxalic and other plant acids are primary products of photosynthesis is a very old suggestion, promulgated originally by Liebig in 1843 (*Annalen*, **46**, 66). It has from the time of its initiation been a subject of much controversy and has now been practically abandoned on physiological grounds, the prevailing view being that the plant acids are terminal rather than initial stages of metabolism. The possible transition from acids to amino-acids and the actual discovery in 1886, according to Brunner and Chuard (*Ber.*, 1886, **19**, 595), of glyoxylic acid as a widely distributed acid in the leaves and green parts of plants has, however, resuscitated the early hypothesis of Liebig, at any rate in a modified form. Attention is thus directed to the question whether, after all, carbohydrate synthesis is the sole result of the photolysis of carbonic acid by the living plant. The formaldehyde and carbohydrate hypothesis is very generally assumed to be the only one worthy of consideration, and the reason is quite obvious. In the first place, there is the fact that carbohydrates (formaldehyde, sugars, starch) are the first products detectable microscopically and chemically. In the next place, attention has been concentrated almost exclusively on carbohydrates since the enunciation of the formaldehyde hypothesis by v. Baeyer. Lastly, the carbohydrate hypothesis is in harmony with the fact that the volume of oxygen given off is equal to the volume of carbon dioxide absorbed.

In spite of this accumulation of evidence, and in view of the pressing necessity for accounting for the formation of amino-acids and the absence of any known chemical method, excepting the cyano-hydrin method, genetically connecting the carbohydrates with these acids, it is permissible to direct attention once again to the results of Brunner and Chuard, published twenty years ago. I must confess that in view of the far-reaching importance of their conclusions it is remarkable that no serious steps should have been taken to confirm or refute them.   The paper is of course accessible to all, but it will make this review more complete if I summarise the chief conclusions. These authors consider that they have proved the widespread occurrence of glyoxylic acid, with other well-known plant acids, in the leaves, fruit, &c., of various plants.   They claim also to have discovered among plant products an iodine-absorbing glucoside which is resolvable into a sugar (? dextrose) and succinic acid, and which they term "glucosuccinic acid."   They come to the conclusion, for reasons which are given in detail in the original paper, that glyoxylic and other acids, glucosides, and even starch, are all primary "assimilation" products and not the result of secondary processes.   I propose to speak of this view as the hypothesis of multiple photosynthesis in order to distinguish it from the formaldehyde hypothesis, which sets out from that one compound only.

Before considering the evidence either for or against the hypothesis of multiple photosynthesis, it must, in the first place, be mentioned that the actual occurrence of glyoxylic acid in plants was called in question by Ordonneau in 1891 (*Bull. Soc. chim.*, [iii], 6, 261).   His criticism was, however, met by Brunner and Chuard (*ibid.*, 1895, [iii], 13, 126), and from later independent evidence it must now, I believe, be admitted that both glyoxylic and glycollic acids are entitled to take rank among the vegetable acids.   But in view of the great importance of the issues raised, it does appear that the evidence in favour of the general distribution of glyoxylic acid in the green parts of plants requires strengthening.   The presence of the acid was inferred by Brunner and Chuard mainly from qualitative tests and from one analysis of the calcium salt in which the metal only was determined.   The practical effect of this scantiness of evidence has been that the importance claimed for glyoxylic acid as a primary product of photosynthesis has not been recognised in current litera-ture.   It is only in recent times, and in connection with the discovery of the genetic relationship between this acid and the amino-acids, that it has acquired renewed importance.   One deduction from the glyoxylic acid hypothesis is, however, capable of being submitted to the test of experiment and it is somewhat remarkable that this should not have been done before.   It is only reasonable to expect on this view that plants should find all the materials for normal growth in a culture medium containing carbohydrates, glyoxylic acid, a nitrate, and the other necessary mineral constituents. Such experiments would necessarily have to be made comparative by omitting the glyoxylic acid from some of the cultures.

Turning now to the purely chemical side of the hypothesis, it is considered by the authors named that carbonic acid by continuous photolytic reduction gives rise to certain radicles or residues which, by further reduction accompanied by condensation and polymerisation, give rise to the formation of acids, aldehydes, carbohydrates, &c. The reducing agent is the hydrogen of water, and the liberated oxygen may pass through the hydrogen peroxide stage before elimination. By this process of reduction there would be formed such groups as $CHO,$ $CH_2 \cdot OH$, CO_2H, $CH \cdot OH$, &c., from which the various acids, aldehydes, carbohydrates, &c., could be built up. Thus, by the coalescence of CHO and CO_2H glyoxylic acid would be formed, by CHO and $CH_2 \cdot OH$ glycollic aldehyde, by $(CO_2H)_2$ oxalic acid, and so forth. On a priori grounds this hypothesis of the continuous reduction of carbonic acid, whereby this compound becomes a potential source of all the simpler groups of plant products, has always struck me as being most ingenious. Unfortunately it lacks support from the chemical side. Neither glyoxylic nor any acid other than formic has yet been produced by the direct reduction of carbonic acid. The known synthetical processes for producing oxalic acid from formic acid or from carbon dioxide cannot be considered as having any analogy with the natural processes. Moreover, even if oxalic acid be regarded as a primary photosynthetic product, the reduction of this to glyoxylic acid has not hitherto been realised. So far as I have been able to ascertain, all the reducing agents employed carry the reduction to the glycollic acid stage (Schulze, *Jahresber.*, 1862, 284 ; Church, *Journ. Chem. Soc.*, 1863, **16**, 301 ; Crommydis, *Bull. Soc. chim.*, 1877, [ii], **27**, 3 ; Balbiano and Alessi, *Gazzetta*, 1882, **12**, 190 ; De Forcrand, *Bull. Soc. chim.*, 1883, [ii], **39**, 310 ; Avery and Dales, *Ber.*, 1899, **32**, 2236).

In its initial requirement, the hypothesis of multiple photosynthesis is thus at present in no better position—perhaps it would be more correct to say is in a weaker position—than the formaldehyde hypothesis. It has yet to be shown that glyoxylic and other acids are formed at the photosynthetic centres. It still remains to be proved that glyoxylic acid can be formed by the reduction, photo-chemical or otherwise, of carbonic acid : assuming, of course, that we do not consider the biochemical process to be inimitable. But after we get beyond the initial stage, the hypothesis of Brunner and Chuard has certain advantages which must be mentioned. It accounts for the formation of acids as well as carbohydrates.* Moreover, as will be seen on reference to the original paper, and as is, in fact, now generally well known, the genetic relationships between the various

---

* An ingenious speculation concerning the origin of acids from carbohydrates has been advanced by Emil Fischer (see, for reference, E. O. v. Lippmann's *Chemie der Zuckerarten*, 1904, II, 1766). It is of interest to note that the reverse process, the origin of carbohydrates from acids, was considered a possibility by v. Baeyer in 1870 (*Ber.*, **3**, 68).

plant acids are real, that is, capable of being realised by chemical methods. To these advantages may be added the genetic relationship now shown to exist between aldehydic or ketonic acids and the amino-acids.

One objection which may be urged against the hypothesis is that it is opposed by the facts concerning the ratio between the intake of carbon dioxide and the oxygen eliminated. The formation of glyoxylic acid, for example, accounts for only one-half the volume of oxygen :

$$2CO_2 + H_2O = H_2C_2O_3 + O_2.$$

But if the reduction proceeds to the next (carbohydrate) stage, the remainder of the oxygen would be eliminated, so that this objection then disappears :

$$nH_2C_2O_3 + nH_2O = (C_2H_4O_2)_n + nO_2.$$

Whatever may be the ultimate fate of the glyoxylic acid hypothesis, the broader question raised by Brunner and Chuard's work, whether attention has not been too exclusively concentrated on the carbohydrates as primary photosynthetic products, must, I venture to think, be eventually answered in the affirmative. It is for this reason that I have thought it advisable to bring the hypothesis of multiple photosynthesis once more into prominence. It is well known to those who are familiar with the literature of this subject that, in addition to the modifications of the formaldehyde hypothesis associated with the names of Bach (the percarbonic acid hypothesis, *Compt. rend.*, 1893, **116**, 1145, 1389 ; 1894, **119**, 1218) and of Crato (the cyclic phenol hypothesis, *Ber. deut. bot. Gesell.*, 1892, p. 250), there have not been wanting advocates of the primary photosynthesis of much more complicated products, such as the fats (Mazé, *Évolution du Carbone*, &c., p. 43 ; see also for references E. O. v. Lippmann's *Chemie der Zuckerarten*, I, 1765).

These alternative hypotheses have, it is true, but little evidence, either chemical or physiological, in their favour. Nevertheless, in view of the well-known fact that few organic chemical reactions are " neat," that is, result in the formation of one product only without by-products, it may fairly be doubted whether a biochemical process is likely to be simpler than the majority of laboratory processes. In advocating the claims of multiple photosynthesis to renewed consideration I do not imagine that even the most extreme supporters of this view would desire to displace the carbohydrates as the predominant primary products. The question is, "is there no margin for subsidiary, but not less necessary, products, such as certain acids?" The observed variations in the CO_2/O_2 ratio do certainly give some support to the view that other products may be formed. The

question might be decided if some means could be found for determining the total quantities of carbohydrates formed by the
"assimilating" leaf as the result of the decomposition of a known
quantity of carbon dioxide. I am informed by Dr. Horace Brown—
and a search through the literature has served to confirm his
statement—that the data at present available are insufficient for the
solution of the problem.

The review of the present aspect of the subject of "assimilation"
which I have ventured, I am afraid very imperfectly, to bring under
your notice will at any rate serve to justify the contention that the
living organism as a chemical agency has as yet been made to reveal but
a very small portion of its mysteries, even with reference to what at
first sight may appear one of its simplest chemical achievements. A
general survey of the state of knowledge at which we have arrived
concerning what is unquestionably the most fundamental of all the
biochemical processes going on in the living world must, I imagine,
leave on the minds of chemists the impression that the facts which
have hitherto been wrested from nature are but fragments of the
whole truth and that our observations are really records of the state
of perfection of certain chemical or microchemical methods of detecting and estimating particular compounds or groups of compounds,
rather than the complete story of the chemical processes going on in
the green leaf.

LXXIX.—*Condensation of Benzophenone Chloride with α- and β-Naphthols.*

By George William Clough, B.Sc.

Aldehydes readily form condensation products with phenols in the presence of acids or dehydrating agents.

Baeyer (*Ber.*, 1872, **5**, 25, 280, 1094), for example, obtained products having the general formula $CHR:(C_xH_{y-2}\cdot OH)_2$, and considered that the (CH)''' group was in the para-position to the hydroxyl groups, thus :

$$OH\langle\ \rangle—CH—\langle\ \rangle OH$$
$$\underset{R}{|}$$

Claisen (*Ber.*, 1886, **19**, 3316) prepared similar compounds by the condensation of aldehydes with α-naphthol and assigned to these products the general formula

$$OH\langle\ \rangle—CH—\langle\ \rangle OH$$
$$\underset{R}{|}$$

With β-naphthol, however, he obtained acetals corresponding with the formula $CHR:(O\cdot C_{10}H_7)_2$. This difference in the action of the two naphthols was accounted for by the fact that in α-naphthol the para-position to the hydroxyl group is free, whilst in β-naphthol this position is already occupied. Hewitt and Turner (*Ber.*, 1901, **34**, 202) subsequently obtained benzylidenedi-β-naphthol by the action of benzaldehyde on β-naphthol, and this product, like Claisen's acetal, gave benzylidenedi-β-naphthyleneoxide when heated with acetic acid and a few drops of hydrochloric acid. The formula for this oxide is usually written :

$$\underset{O}{\overset{\overset{\textstyle C_6H_5}{|}}{\underset{|}{CH}}}$$

Consequently the (CH)''' group in benzylidenedi-β-naphthol is supposed to be in the ortho-position to the hydroxyl groups.

Fosse (*Bull. Soc. chim.*, 1900, [iii], **23**, 512) prepared the acetals of phenol and α-naphthol by acting on ethylidene·chloride with sodium phenoxide and sodium α-naphthoxide respectively in aqueous or alcoholic solution :

$$CH_3 \cdot CHCl_2 + 2C_{10}H_7 \cdot ONa = CH_3 \cdot CH(O \cdot C_{10}H_7)_2 + 2NaCl.$$

J. E. Mackenzie (Trans., 1901, **79**, 1216), however, found that on heating phenol with benzylidene chloride the action took place according to the equation :

$$C_6H_5 \cdot CHCl_2 + 2C_6H_5 \cdot OH = C_6H_5 \cdot CH{:}(C_6H_4 \cdot OH)_2 + 2HCl,$$

the condensation product being identical with that obtained by the condensation of benzaldehyde with phenol (Russanoff, *Ber.*, 1889, **22**, 1943). The action of β-naphthol or sodium β-naphthoxide on benzylidene chloride resulted in the production of Claisen's oxide (Mackenzie and Joseph, Trans., 1904, **85**, 793) :

$$C_6H_5 \cdot CHCl_2 + 2C_{10}H_7 \cdot OH = C_6H_5 \cdot CH{:}[C_{10}H_6]_2{:}O + 2HCl + H_2O.$$

The condensations of ketones with phenols have been investigated chiefly by Dianin (*J. Russ. Phys. Chem. Soc.*, 1893, **23**, 488, 523, 601), who obtained compounds of the type $CRR'{:}(C_6H_4 \cdot OH)_2$ by heating phenol in stoppered vessels with aliphatic ketones in the presence of a mixture of acetic and fuming hydrochloric acids. On heating acetone with α-naphthol under similar conditions, the action was found to take place according to the equation :

$$(CH_3)_2CO + 2C_{10}H_7 \cdot OH = (CH_3)_2C{:}[C_{10}H_6]_2{:}O + 2H_2O.$$

Mackenzie (Trans., 1896, **69**, 985) prepared the "ketals" of benzophenone by the action of sodium methoxide and its homologues on benzophenone chloride. The action of sodium phenoxide, however, on the same substance resulted in the formation of a dihydroxy-compound (Trans., 1901, **79**, 1209) :

$$(C_6H_5)_2CCl_2 + 2C_6H_5 \cdot ONa = (C_6H_5)_2C(C_6H_4 \cdot OH)_2 + 2NaCl.$$

In the present paper, the author has studied the action of α- and β-naphthols and also of sodium α- and β-naphthoxides on benzophenone chloride. The condensations of α- and β-naphthols with benzophenone chloride are somewhat similar to those described by Claisen with aldehydes. On heating α-naphthol with benzophenone chloride, di-α-hydroxynaphthyldiphenylmethane was formed according to the equation :

$$(C_6H_5)_2CCl_2 + 2C_{10}H_7 \cdot OH = (C_6H_5)_2C{:}(C_{10}H_6 \cdot OH)_2 + 2HCl.$$

On the other hand, the action of β-naphthol resulted in the formation of the ketal, namely, di-β-naphthoxydiphenylmethane, thus :

$$(C_6H_5)_2CCl_2 + 2C_{10}H_7 \cdot OH = (C_6H_5)_2C{:}(O \cdot C_{10}H_7)_2 + 2HCl.$$

When the sodium naphthoxides in alcoholic solution were allowed to react with benzophenone chloride, the products isolated were the anhydrides of the respective naphthyldiphenylcarbinols :

$$(C_6H_5)_2CCl_2 + C_{10}H_7 \cdot ONa = (C_6H_5)_2C {<}^{C_{10}H_6}_{O} + NaCl + HCl.$$

As these compounds are coloured, perhaps the quinonoid formula,

$$(C_6H_5)_2C{:}C_{10}H_6{:}O,$$

is to be preferred.

A similar condensation occurred when α-naphthol was added to a solution of benzophenone chloride in light petroleum, which recalls Caro and Graebe's synthesis of aurin from p-hydroxybenzophenone chloride and phenol (*Ber.*, 1878, **11**, 1350).

A condensation of α-naphthol with benzophenone has also been effected by heating the substances in the presence of zinc chloride and hydrogen chloride. The reaction took place in accordance with the equation :

$$(C_6H_5)_2CO + 2C_{10}H_7 \cdot OH = (C_6H_5)_2C{:}[C_{10}H_6]_2{:}O + 2H_2O.$$

Experimental.

Di-α-hydroxynaphthyldiphenylmethane.

When benzophenone chloride (23·7 grams) was added in small quantities at a time to α-naphthol (28·8 grams), a deep violet coloration was at once produced, and when the mixture was heated at 50° hydrogen chloride was evolved. When the evolution of gas had almost ceased, the temperature was maintained at 100° for fifteen hours, at the end of which time no more gas was evolved. The diminution in weight amounted to 7·1 grams, whereas the loss calculated for 2 mols. of hydrogen chloride is 7·3 grams. The residue, which was almost black and very viscid, was dissolved in benzene and the solution concentrated. After two days, 4 grams of a dark crystalline substance separated which, when crystallised from benzene, was analysed :

0·1165 gave 0·3750 CO_2 and 0·0588 H_2O.   C = 87·79 ; H = 5·61.

$C_{33}H_{24}O_2$ requires C = 87·61 ; H = 5·31 per cent..

A determination of the molecular weight by the cryoscopic method gave the following results :

0·1540 in 22·50 benzene gave $\Delta t - 0.080°$.   M.W. = 428.

0·3035 ,, 22·50   ,,   ,, $\Delta t - 0.150°$.   M.W. = 450.

$C_{33}H_{24}O_2$ requires M.W. = 452.

Di-α-hydroxynaphthyldiphenylmethane, $(C_6H_5)_2C{:}(C_{10}H_6 \cdot OH)_2$, is a colourless substance which crystallises from benzene in small needles

3 K 2

and melts at 209—210° ; it is easily soluble in ether, acetone, chloroform or benzene, moderately so in alcohol, but only sparingly so in light petroleum. When it is not quite pure it darkens on exposure to air. Its solution in sodium hydroxide is brown and shows a blue fluorescence ; on the addition of an acid, the original substance is precipitated. That it is a dihydroxy-compound is shown by its solubility in sodium hydroxide solution and its conversion into a diacetyl compound by the action of fused sodium acetate and acetic anhydride. When heated with fused sodium acetate (1 part) and acetic anhydride (4 parts), it forms a *diacetyl* derivative which crystallises from acetone in small, white needles and melts at 202·5°.

0·1285 gave 0·3906 CO_2 and 0·0610 H_2O.   C = 82·90 ; H = 5·28.

$C_{37}H_{28}O_4$ requires C = 82·84 ; H = 5·22 per cent.

The *dibenzoyl* derivative, which was obtained by the Schotten-Baumann reaction, crystallises from acetic acid in small, colourless crystals melting at 169°.

0·1416 gave 0·4422 CO_2 and 0·0675 H_2O.   C = 85·17 ; H = 5·30.

$C_{47}H_{32}O_4$ requires C = 85·45 ; H = 4·85 per cent.

Anhydro-a-naphthyldiphenylcarbinol.

The following experiment was carried out in the hope of obtaining a better yield of the above condensation product, but a substance having entirely different properties was produced. Five grams of a-naphthol were added to a solution of 23·7 grams of benzophenone chloride in 300 grams of light petroleum (b. p. 40—45°). A deep blue colour was at once produced and the solution was heated in a water-bath at 45°, when the solution boiled and hydrogen chloride was evolved. When there was no further liberation of gas, a further quantity of a-naphthol was added and the solution again boiled until hydrogen chloride was no longer evolved. This was repeated until the calculated quantity of a-naphthol corresponding with 2 molecules (28·8 grams) had been added. A yellow, crystalline substance separated and the solution was also yellow. The weight of hydrogen chloride evolved was 7·0 grams, whereas the calculated amount corresponding with 2 mols. of the gas is 7·3 grams. The supernatant solution was filtered hot and concentrated ; on cooling, yellow crystals were deposited on the addition of a little ether. These when dried melted at 179—180°.

The residue in the flask darkened when exposed to air, but by repeatedly washing with small quantities of ether the bright yellow colour was restored. The dried substance melted at 179—180°, the total weight obtained being 4·8 grams ; it recrystallises from acetone in small, bright yellow prisms and melts at 180·5—181°.

0·1602 gave 0·5305 CO_2 and 0·0710 H_2O. C = 90·31 ; H = 4·92.

$C_{23}H_{16}O$ requires C = 89·93 ; H = 5·19 per cent.

A determination of the molecular weight by the ebullioscopic method gave the numbers :

0·0998 in 20·03 benzene gave Δt − 0·045°. M.W. = 296.

0·2102 ,, 20·03   ,,   ,, Δt − 0·095°. M.W. = 294.

0·2670 ,, 20·03   ,,   ,, Δt − 0·125°. M.W. = 285.

$C_{23}H_{16}O$ requires M.W. = 308.

Anhydro-a-naphthyldiphenylcarbinol, $(C_6H_5)_2 \!:\! C \!:\! C_{10}H_6 \!:\! O$, is insoluble in water and alkalis, but easily soluble in benzene, alcohol, chloroform, acetic acid, and acetone, less so in ether and light petroleum ; its solution in cold concentrated sulphuric and nitric acids is intensely violet, whilst its warm concentrated hydrochloric acid solution is violet-red. The solution in sulphuric acid turns brown, whilst that in nitric acid becomes yellow on warming. On fusion with caustic potash or on heating with water in a sealed tube at 200°, the compound yields a-naphthol.

As *as*-di-a-naphthoxyethane is produced when an alcoholic solution of sodium a-naphthoxide is heated with ethylidene chloride, the action of sodium a-naphthoxide on benzophenone chloride was investigated. Sodium (2·3 grams) was dissolved in dry ethyl alcohol (200 c.c.) and to this solution was added an alcoholic solution of a-naphthol (14·4 grams). The solution thus obtained was green and showed a blue fluorescence. On the addition of benzophenone chloride (11·8 grams) the liquid became very dark and some sodium chloride separated. The flask containing the mixture was then heated on a water-bath for half an hour, at the end of which time the solution was, neutral to litmus. The sodium chloride filtered off amounted to 5·8 grams, this being the calculated quantity. On distilling off most of the alcohol and allowing the solution to cool, 2·7 grams of a light brown substance were obtained which, when crystallised from acetone, separated in small, yellow crystals melting at 180—180·5° and having properties similar to those of the substance described in the preceding experiment. I hope to give an account of some derivatives of this substance in a subsequent paper.

a-Oxydinaphthyldiphenylmethane.

This condensation product of benzophenone and a-naphthol was obtained in the following experiment. A mixture of 18·2 grams of benzophenone, 28·8 grams of a-naphthol, and 10 grams of fused zinc chloride was heated in an oil-bath to 150° and dry hydrogen chloride passed in for half an hour. The mixture was maintained at 150—160° for three hours, at the end of which time it had become black and very viscid. The product was extracted with boiling benzene and the extract boiled with animal charcoal and concentrated. After cooling, about 3 grams of a dark crystalline substance were obtained, the colour of which was removed by three crystallisations from benzene.

0·1597 gave 0·5360 CO_2 and 0·0702 H_2O.   C = 91·54 ; H = 4·88.

$C_{33}H_{22}O$ requires C = 91·24 ; H = 5·07 per cent.

The substance was not sufficiently soluble in benzene to permit of a cryoscopic determination of the molecular weight in that solvent. A determination carried out by the ebullioscopic method gave the results :

0·0836 in 13·70 benzene gave $\Delta t - 0·040°$.   M.W. = 408.

0·2330 ,, 13·70   ,,   ,, $\Delta t - 0·100°$.   M.W. = 454.

$C_{33}H_{22}O$ requires M.W. = 434.

a-*Oxydinaphthyldiphenylmethane*, $(C_6H_5)_2C{:}(C_{10}H_6)_2{:}O$, is a colourless substance crystallising from alcohol or benzene in small needles and melting at 273°. It is insoluble in water and alkalis, sparingly soluble in ether or alcohol, and more so in benzene. On warming with concentrated sulphuric acid, a yellow solution is obtained which exhibits a green fluorescence. Whether this substance is the anhydride or not of the dihydroxy-compound previously described has not yet been ascertained.

Di-β-naphthoxydiphenylmethane.

Benzophenone chloride (23·7 grams) was added gradually to a boiling solution of β-naphthol (28·8 grams) in dry xylene (300 grams). A steady evolution of hydrogen chloride took place and the boiling was continued for sixteen hours, at the end of which time no more gas was evolved. The solution obtained had a deep red colour. Most of the xylene was distilled off, and, on adding light petroleum to the concentrated solution, a light brown powder was precipitated, the yield amounting to 30 per cent. of the calculated quantity. When crystallised from alcohol, *di-β-naphthoxydiphenylmethane*,

$$(C_6H_5)_2C{:}(O{\cdot}C_{10}H_7)_2,$$

was obtained in small, colourless, prismatic crystals which melted at 137° ; it is insoluble in water and alkalis, easily soluble in ether and benzene, moderately so in hot and slightly so in cold alcohol. By considerably concentrating the mother liquor, 6 grams of a crystalline substance were obtained having the characteristic odour of β-naphthol ; it was soluble in sodium hydroxide solution, and after recrystallisation melted at 121—122°.

Other properties of this substance also agreed with those of β-naphthol. That the substance melting at 137° is the dinaphthoxycompound is shown by the analysis, the determination of the molecular weight, and its hydrolysis into benzophenone and β-naphthol.

0·1900 gave 0·6097 CO_2 and 0·0873 H_2O.   C = 87·51 ; H = 5·11·
$C_{33}H_{24}O_2$ requires C = 87·61 ; H = 5·31 per cent.

A determination of the molecular weight by the cryoscopic method gave the numbers :

0·1610 in 23·25 benzene gave $\Delta t - 0·080$.   M.W. = 433.
0·2470 ,, 23·25   ,,   ,, $\Delta t - 0·120$.   M.W. = 443.
$C_{33}H_{24}O_2$ requires M.W. = 452.

The substance, when heated with sulphuric acid, undergoes the following hydrolysis :

$$(C_6H_5)_2C(O \cdot C_{10}H_7)_2 + H_2O = (C_6H_5)_2CO + 2C_{10}H_7 \cdot OH.$$

The substance was heated in a distilling flask with concentrated sulphuric acid diluted with twice its volume of water. On distillation, a substance which solidified in the receiver passed over with the steam and after crystallising from alcohol it melted at 121°. It was soluble in aqueous sodium hydroxide, had the odour of β-naphthol, and its aqueous solution gave a green coloration with ferric chloride. The residue in the flask was made alkaline with sodium hydroxide and extracted with ether. When the ether had evaporated, a solid remained having the characteristic odour of benzophenone, and after recrystallisation melting at 48°.

$Anhydro\text{-}\beta\text{-}naphthyldiphenylcarbinol.$

To a solution of 2·3 grams of sodium in alcohol was added an alcoholic solution of 14·4 grams of β-naphthol. After the addition of 11·8 grams of benzophenone chloride, the liquid became dark and sodium chloride was formed. The contents of the flask were then heated on a water-bath for half an hour, when the supernatant liquid was found to be neutral. The weight of sodium chloride formed was equal to the calculated quantity. After the solution had been considerably concentrated and allowed to cool, a dark crystalline substance was obtained, which was crystallised from glacial acetic acid and analysed.

0·1204 gave 0·3982 CO_2 and 0·0540 H_2O.   C = 90·20 ; H = 4·98.
$C_{23}H_{16}O$ requires C = 89·93 ; H = 5·19 per cent.

A determination of the molecular weight by the cryoscopic method gave the following results :

0·0785 in 13·80 benzene gave $\Delta t - 0·098°$.   M.W. = 290.
0·1200 ,, 13·80   ,,   ,, $\Delta t - 0·140°$.   M.W. = 310.
$C_{23}H_{16}O$ requires M.W. = 308.

Anhydro-β-naphthyldiphenylcarbinol, $(C_6H_5)_2C\!:\!C_{10}H_6\!:\!O$, crystallises from glacial acetic acid in small, red needles and melts at 194°; it is insoluble in water, but soluble in ether, alcohol, and benzene, giving deep red solutions. It is soluble in concentrated sulphuric, hydrochloric, or nitric acid, giving a deep green solution, the colour disappearing on dilution with water. Its solution in nitric acid becomes yellow on warming. When heated with water in a sealed tube at 200°, it decomposes, β-naphthol being formed.

I desire to express my sincere thanks to Dr. J. E. Mackenzie for suggesting this investigation.

BIRKBECK COLLEGE,
LONDON, E.C.

LXXX.—*Experiments on the Synthesis of Camphoric Acid. Part IV.* The Action of Sodium and Methyl Iodide on Ethyl Dimethylbutanetricarboxylate, $CO_2Et\!\cdot\!CH_2\!\cdot\!CH(CO_2Et)CMe_2\!\cdot\!CH_2\!\cdot\!CO_2Et.$

By WILLIAM HENRY PERKIN, jun., and JOCELYN FIELD THORPE.

DURING the course of a series of experiments on the constitution of *iso*camphoronic acid (Trans., 1899, **75**, 901; compare Trans., 1902, **81**, 246), we gave a description of the synthesis and properties of ββ-dimethylbutane-αγδ-tricarboxylic acid,

$$CO_2H\!\cdot\!\underset{\delta}{C}H_2\!\cdot\!\underset{\gamma}{C}H(CO_2H)\underset{\beta}{C}Me_2\!\cdot\!\underset{\alpha}{C}H_2\!\cdot\!CO_2H.$$

This acid was, at that time, of special interest because Baeyer (*Ber.*, 1896, **29**, 2780) had put forward the suggestion that the constitution of *iso*camphoronic acid was probably represented by this formula. But this acid interested us to an even greater extent for another reason, namely, on account of the possibility that it might prove to be a convenient starting point for the synthesis of camphoric acid.

Dieckmann (*Ber.*, 1894, **27**, 103) had shown that sodium reacts with ethyl adipate with formation of the sodium derivative of ethyl β-ketopentamethylenecarboxylate,

$$\begin{array}{l}CH_2\!\cdot\!CH_2\!\cdot\!CO_2Et \\ | \\ CH_2\!\cdot\!CH_2\!\cdot\!CO_2Et\end{array} + 2Na = \begin{array}{l}CH_2\!\cdot\!CH_2 \\ | \qquad\qquad >\!CO \\ CH_2\!\cdot\!CNa\!\cdot\!CO_2Et\end{array} + NaOEt + H_2.$$

* Part I was published in Trans., 1898, **73**, 45; Part II, Trans., 1899, **75**, 909; Part III, Trans., 1904, **85**, 128.

Since, then, ethyl $\beta\beta$-dimethylbutane-$\alpha\gamma\delta$-tricarboxylate is a substitution derivative of ethyl adipate, it seemed likely that an analogous condensation might take place if it were treated with sodium under similar conditions. The problem would, however, in this case necessarily be a difficult and complicated one owing to the fact that this ester, unlike adipic ester, is unsymmetrical, that is, it contains two CH_2-groups (a and b) which can react with the two carboxyethyl groups (c and d) in such a manner as to give rise to two different condensation products :

$$\overset{d}{CO_2}Et \cdot \overset{a}{CH_2} \cdot CH(CO_2Et)CMe_2 \cdot \overset{b}{CH_2} \cdot \overset{c}{CO_2}Et,$$

$$
\begin{array}{cc}
\begin{array}{c}
CMe_2 \\
\diagup \diagdown \\
CO_2Et \cdot CH \quad CNa \cdot CO_2Et \\
\mid \qquad \mid \\
CH_2 - CO \\
I.
\end{array}
&
\begin{array}{c}
CMe_2 \\
\diagup \diagdown \\
CH_2 \quad CH \cdot CO_2Et \\
\mid \qquad \mid \\
CO - CNa \cdot CO_2Et \\
II.
\end{array}
\end{array}
$$

or

Such sodium derivatives might be expected to react readily with methyl iodide with formation of the two isomeric trimethylketopenta-methylenedicarboxylic esters,

$$
\begin{array}{cc}
\begin{array}{c}
CMe_2 \\
\diagup \diagdown \\
CO_2Et \cdot CH \quad CMe \cdot CO_2Et \\
\mid \qquad \mid \\
CH_2 - CO \\
I.
\end{array}
&
\begin{array}{c}
CMe_2 \\
\diagup \diagdown \\
CH_2 \quad CH \cdot CO_2Et \\
\mid \qquad \mid \\
CO - CMe \cdot CO_2Et \\
II.
\end{array}
\end{array}
$$

or

and the first of these is, obviously, nothing more nor less than the ester of ketocamphoric acid, and ought therefore to yield camphoric acid

$$
\begin{array}{c}
CMe_2 \\
\diagup \diagdown \\
CO_2H \cdot CH \quad CMe \cdot CO_2H \\
\mid \qquad \mid \\
CH_2 - CH_2
\end{array},
$$

if subjected to the action of suitable reducing agents.

When the experiment was instituted, it was found that the product of the action of sodium and then of methyl iodide on ethyl $\beta\beta$-dimethyl-butane-$\alpha\gamma\delta$-tricarboxylate was in fact a *trimethylketopentamethylene-dicarboxylic ester*, but the greatest difficulty was experienced in determining whether its constitution was that represented by formula I or formula II, and for this reason the research had to be set aside on more than one occasion.

Ultimately, however, the investigation of the behaviour of this ester on hydrolysis under certain definite conditions and the subsequent synthesis of the products formed during this hydrolysis furnished the solution of the problem. When the keto-ester is hydrolysed, first with cold alcoholic potash and then by boiling dilute sulphuric acid, it yields three substances, namely, A, a tribasic acid, $C_{10}H_{16}O_6$, melting at 174°, which is readily soluble in water, B, a tribasic acid, $C_{10}H_{16}O_6$, which is formed in much smaller quantity, melts at 204°, and is more sparingly soluble in water, and C, a keto-acid, $C_{10}H_{16}O_3$, of melting point 110°, which is present only in minute quantities, and can be separated from the acids A and B owing to the fact that it is volatile in steam. If we confine our attention in the meantime to the tribasic acid A, of melting point 174°, which constitutes the principal product of the reaction, it is evident that this acid is produced from ethyl trimethylketopentamethylenedicarboxylate according to the equation

$$C_{14}H_{22}O_5 + 3H_2O = C_{10}H_{16}O_6 + 2C_2H_5 \cdot OH, \, .$$

and, further, that the third molecule of water has brought about the disruption of the trimethylketopentamethylene ring.

This disruption will take place (as, for example, in the case of methylacetoacetic ester) at the keto-group, and, since the acid of melting point 174° is tribasic, does not eliminate carbon dioxide on heating, and is derived from either formula I or II (see above), it follows that its constitution must be represented by one of the formulæ $CO_2H \cdot CH_2 \cdot CH(CO_2H)CMe_2 \cdot CHMe \cdot CO_2H$ (from I), or $CO_2H \cdot CH_2 \cdot CMe_2 \cdot CH(CO_2H)CHMe \cdot CO_2H$ (from II).

The only really satisfactory way of deciding between these formulæ was to prepare one of these acids synthetically and, after many failures, this was ultimately acomplished by the following process.

Ethyl $\beta\beta$-dimethylacrylate, $CO_2Et \cdot CH\!:\!CMe_2$, was digested in alcoholic solution with the sodium derivative of ethyl cyanoacetate, $CN \cdot CHNa \cdot CO_2Et$, when direct addition took place, with the result that the sodium derivative of ethyl α-cyano-$\beta\beta$-dimethylglutarate,

$$CO_2Et \cdot CH_2 \cdot CMe_2 \cdot C(CN)Na \cdot CO_2Et,$$

was formed (compare Trans., 1899, **75**, 63). When this sodium derivative was treated with ethyl bromopropionate, it yielded *ethyl aγγ-trimethyl-β-cyanobutane-aβδ-tricarboxylate*,

$$CO_2Et \cdot CH_2 \cdot CMe_2 \cdot C(CN)(CO_2Et)CHMe \cdot CO_2Et.$$

Hydrolysis with caustic potash converted this ester into *aγγ-trimethylbutane-aββδ-tetracarboxylic acid*, which melts at 186° and, when heated at 200°, is decomposed with elimination of carbon dioxide and formation of *aγγ-trimethylbutane-aβδ-tricarboxylic acid*,

$$CO_2H \cdot CH_2 \cdot CMe_2 \cdot CH(CO_2H)CHMe \cdot CO_2H.$$

This acid melted at 174° and was shown, by careful comparison, to be identical with the acid A obtained by the hydrolysis of ethyl trimethylketopentamethylenedicarboxylate. Since, then, this synthesis proves that the acid A has the above constitution, it follows that the keto-ester must be represented by the formula

$$CMe_2$$
$$\begin{array}{c} \diagup\diagdown \\ CH_2 \quad CH \cdot CO_2Et \\ CO\text{---}CMe \cdot CO_2Et \text{ '} \end{array}$$

Ethyl 1 : 1 : 3-*trimethyl*-4-*ketopentamethylene*-2 : 3-*dicarboxylate*.

and it was therefore clearly impossible that camphoric acid could result from the reduction of this ester.

The constitution of the second tribasic acid (B) of melting point $204°$ could not be determined because of the very small quantity of material at our disposal. It is not improbable that it is the *trans*-modification of the acid A. On the other hand, it is possible that the product of the action of sodium and methyl iodide on ethyl dimethyl-butanetricarboxylate may contain small quantities of the isomeric keto-ester represented by formula I (p. 779), in which case the acid B might have the constitution

$$CO_2H \cdot CH_2 \cdot CH(CO_2H)CMe_2 \cdot CHMe \cdot CO_2H.$$

The third substance, C (p. 780), is a keto-monobasic acid of the formula $C_{10}H_{16}O_3$, and yields a characteristic oxime and semicarb-azone. The constitution of this interesting substance (which is isomeric with and closely allied in properties to pinonic acid) was at first very difficult to explain, but the method of its formation is probably a comparatively simple one. If the formula of ethyl trimethyl-ketopentamethylenedicarboxylate (see above) is examined, it will be seen that it contains a methylene group which is adjacent to the keto-group, and therefore capable of undergoing condensation with one of the two carbethoxy-groups.

It is therefore not improbable that, during the action of the sodium and methyl iodide, internal condensation may take place to a small extent with the formation of a diketo-ester, the constitution of which would be represented by one of the formulæ

$$\begin{array}{cc} CMe_2 & CMe_2 \\ \diagup\diagdown & \diagup\diagdown \\ CMe \quad CH \cdot CO_2Et & CMe\text{---}CO\text{---}CH \\ CO & | \qquad | \\ CO\text{---}CMe & CO\text{--------}CMe \cdot CO_2Et \end{array} \quad \text{or} \quad .$$

Such bridged-ring compounds would be decomposed on hydrolysis with disruption of the bridge, and the resulting keto-monobasic acid, $C_{10}H_{16}O_3$, would then possess one of the following formulæ:

CMe2
CHMe CH·CO2H
CO—CHMe

or

CMe2
CHMe CH2
CO—CMe·CO2H

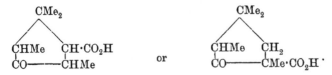

But a substance of the constitution represented by the second of these formulæ would decompose at once, with elimination of carbon dioxide, if it were distilled in a current of steam, and that formula cannot therefore represent the keto-acid of melting point 110°.

For this reason we have adopted the first formula for this acid, and have named it 1 : 1 : 3 : 5-*tetramethyl-4-ketopentamethylene-2-carboxylic acid* (p. 787).

Lastly, we have submitted ethyl trimethylketopentamethylenedicarboxylate to reduction and find that, when treated with sodium amalgam under the conditions described on p. 789, it is reduced with formation of 1 : 1 : 3-*trimethyl-4-hydroxypentamethylene-2 : 3-dicarboxylic acid.*

CMe2
CH2 CH·CO2H
CH(OH)·CMe·CO2H

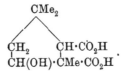

This hydroxy-acid is a syrup which is readily converted into its anhydride (or lactone?) on distillation. It is reduced only with great difficulty, but when treated first with hydriodic acid and then with sodium amalgam in the manner described on p. 790, the hydroxyl group is ultimately replaced by hydrogen and 1 : 1 : 3-*trimethylpentamethylene-2 : 3-dicarboxylic acid,*

CMe2
CH2 CH·CO2H
CH2–CMe·CO2H

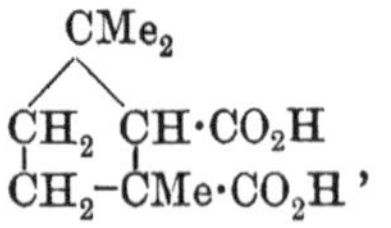

is formed. Although purified as far as possible by conversion into the anhydride and methyl ester, both of which distil without decomposition, the acid has, so far, not been obtained in a crystalline condition. This may be due in part to the syrupy acid consisting of *cis-* and *trans-*modifications, but there is also reason to suspect that slight impurities were present which may have been sufficient to retard or prevent crystallisation.

A careful investigation of the acid and its derivatives would be of considerable interest on account of its close relationship to camphoric acid.

EXPERIMENTAL.

Ethyl 1 : 1 : 3-*Trimethyl-4-ketopentamethylene-2* : 3-*dicarboxylate*.

The ethyl dimethylbutanetricarboxylate required for these experiments was prepared by the process described in detail in a previous communication (Trans., 1899, 75, 902), and the treatment with sodium and methyl iodide was carried out under the following conditions.

Sodium (7 grams) was melted under boiling toluene (50 c.c.) and vigorously shaken in a closed flask so as to bring the sodium into as fine a state of division as possible, the whole being then allowed to cool. The ester (45 grams) was dissolved in toluene (150 c.c.) in a large flask connected with a reflux apparatus, the finely-divided sodium was added all at once, and the flask heated on the sand-bath until action just commenced. The sand-bath was then removed and the violent reaction allowed to proceed; when this slackened, the heating was repeated for about one hour and until the whole of the sodium had disappeared. The product was cooled in ice-water and methyl iodide (50 grams) added through the condenser tube in small quantities at a time, when a vigorous decomposition set in with separation of sodium iodide. After all the iodide had been added, the mass was mixed with methyl alcohol (50 c.c.) and heated on the water-bath for one hour to complete the decomposition.

Water and ether were then added, the oily ethereal layer separated, washed well with water and sodium carbonate,* dried over calcium chloride and evaporated, and the residual oil fractionated under a pressure of 60 mm. As soon as the toluene had passed over, the thermometer rose rapidly to 190°, between which temperature and 220° the product of the reaction passed over.† On refractionation, a

---

* The sodium carbonate extract always yielded a thick oil when it was acidified and extracted with ether. This was esterified with alcohol and sulphuric acid and found to consist of a mixture of ethyl trimethylketopentamethylenedicarboxylate and unchanged ethyl dimethylbutanetricarboxylate, as well as a very thick oil of high boiling point which was not investigated.

† During this operation, a small quantity of oil of lower boiling point is always obtained. This was gradually collected and carefully fractionated under the ordinary pressure, when it distilled constantly at 197—200° and yielded the following results on analysis :

0·1566 gave 0·3248 CO_2 and 0·1260 H_2O. C = 59·7 ; H = 8·9.
$C_{10}H_{18}O_4$ requires C = 59·5 ; H = 8·9 per cent.

Since this ester yielded $\beta\beta$-dimethylglutaric acid on hydrolysis, it is probably the

nearly colourless oil (20 grams) was obtained which distilled at 200—205° (60 mm.).

0·1945 gave 0·4386 CO_2 and 0·144 H_2O.   C = 61·5 ; H = 8·3.
0·2072 „ 0·4701 CO_2 „ 0·1531 H_2O.   C = 61·9 ; H = 8·2.
$C_{14}H_{22}O_5$ requires C = 62·2 ; H = 8·2 per cent.

Ethyl 1 : 1 : 3-*trimethyl*-4-*ketopentamethylene* 2 : 3-*dicarboxylate* is a pale yellow oil which, when pure, may be distilled in small quantities at the ordinary pressure almost without decomposition. Its alcoholic solution gives no coloration with ferric chloride, an important fact, since it shows that the action of the methyl iodide in the above synthesis had been complete and the product did not contain any of the corresponding dimethyl derivative.

Action of Phosphorus Pentachloride on the Keto-ester.—The pure ester (20 grams) was mixed with phosphorus pentachloride (25 grams), when little action took place at the ordinary temperature but, on heating on the water-bath, decomposition set in and a good deal of hydrogen chloride was evolved. As soon as the pentachloride had completely dissolved, the whole was heated to boiling for ten minutes and poured into alcohol. After remaining overnight, water was added, the product was extracted in the usual way and fractionated under reduced pressure, when a considerable quantity of oil was obtained which distilled at 192—194° (35 mm.) and gave the following results on analysis :

0·2107 gave 0·4565 CO_2 and 0·1510 H_2O.   C = 58·7 ; H = 7·9.
0·3085 „ 0·1438 AgCl.   Cl = 11·7.
$C_{14}H_{21}ClO_4$ requires C = 58·2 ; H = 7·3 ; Cl = 12·3 per cent.

This ester is therefore *ethyl* 1 : 1 : 3-*trimethyl*-4-*chlorocyclopentene-dicarboxylate*,

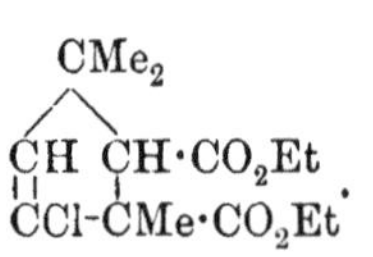

$$CMe_2$$
$$CH \quad CH·CO_2Et$$
$$CCl-CMe·CO_2Et$$

Many attempts were made with the object of reducing this ester to the corresponding trimethyl*cyclo*pentenedicarboxylic acid, but were all unsuccessful, principally because the chlorine atom is very firmly bound and not easily replaced by treatment with reducing agents.

Hydrolysis of Ethyl Trimethylketopentamethylenedicarboxylate.

When this ester was mixed with dilute methyl-alcoholic potash, a deep yellow solution was produced, and, on boiling, potassium carbonate methyl ethyl ester of this acid, namely, $CO_2Me·CH_2·CMe_2·CH_2·CO_2Et$, a substance which must have been produced during the vigorous action described above by a curious decomposition of some of the ethyl dimethylbutanetricarboxylate employed.

separated in quantity. After adding water and filtering from a little neutral oil, the alkaline solution was evaporated until free from alcohol and acidified with hydrochloric acid.

The thick, pale yellow oil which separated was extracted with ether, the ethereal solution washed with water, dried over calcium chloride, and evaporated, when a syrup remained, the aqueous solution of which gave no precipitate with semicarbazide hydrochloride and sodium acetate. Since the oil showed no signs of crystallising, it was dissolved in boiling water, mixed with a slight excess of barium hydroxide, and the excess removed by passing carbon dioxide.

The barium carbonate carried down with it a little coloured impurity and, after filtering and acidifying and extracting with ether, a nearly colourless syrup was obtained which was left for some days over sulphuric acid in an evacuated desiccator and analysed :

0.153 gave 0.3186 CO_2 and 0.1139 H_2O. $C = 56.8$; $H = 8.3$.

$C_9H_{16}O_4$ requires $C = 57.4$; $H = 8.5$ per cent.

The *silver* salt was also prepared and analysed :

0.2172 gave 0.1149 Ag. Ag $= 52.9$.

$C_9H_{14}Ag_2O_4$ requires Ag $= 53.7$ per cent.

The acid which might be expected to result from the hydrolysis of ethyl trimethylketopentamethylenedicarboxylate under the above conditions is α-ethyl ββ-dimethylglutaric acid,

$$CO_2H \cdot CH_2 \cdot CMe_2 \cdot CH(CO_2H) \cdot CH_2Me,$$

one of the acids of the formula $C_9H_{16}O_4$ which does not appear to have been prepared. This acid might, like the isomeric α-ethylpimelic acid (Crossley and Perkin, Trans., 1894, 65, 990), be syrupy, but more probably a slight impurity, indicated by the results of analysis, prevented the syrup from crystallising. A second series of experiments on the hydrolysis of the keto-ester, carried out under quite different conditions, gave much more satisfactory results.

The ester (27 grams) was mixed with methyl-alcoholic potash (17 grams KOH) at 0°, and, after remaining overnight, the product was acidified with a large excess of dilute sulphuric acid (20 per cent.) and distilled in steam until four litres of distillate (A) had collected. The sulphuric acid in the residue was exactly removed by means of baryta water, the filtrate from the barium sulphate evaporated to a small bulk and mixed with an equal volume of strong hydrochloric acid, when a considerable quantity of a crystalline substance gradually separated. This was collected at the pump and several times fractionally crystallised from hydrochloric acid, by which means it was separated into a large proportion of an acid of melting point 172—174° and a small proportion of a much less soluble acid of melting point

204°. As explained in the introduction (p. 780), the acid of melting point 172—174° * is $a\gamma\gamma$-trimethylbutane-$a\beta\delta$-tricarboxylic acid,

$$CO_2H\cdot CH_2\cdot CMe_2\cdot CH(CO_2H)CHMe\cdot CO_2H,$$

and is identical with the acid of this constitution which was subsequently synthesised by the process described on p. 794.

0·1689 gave 0·3216 CO_2 and 0·1078 H_2O.   C = 51·9 ; H = 7·1·
0·1185 „   0·2250 CO_2 „   0·0752 H_2O.   C = 51·8 ; H = 7·0.
$C_{10}H_{16}O_6$ requires C = 51·7 ; H = 6·9 per cent.

When titrated with decinormal caustic soda solution, it was found that 0·2311 required for neutralisation 0·118 NaOH, whereas this amount of a tribasic acid, $C_{10}H_{16}O_6$, should neutralise 0·119 NaOH.

The *silver* salt, $C_{10}H_{13}Ag_3O_6$, was obtained from the ammonium salt by precipitation with silver nitrate as a white, amorphous powder.

0·2441 gave 0·1425 Ag.   Ag = 58·4.
$C_{10}H_{13}Ag_3O_6$ requires Ag = 58·6 per cent.

Anhydrotrimethylbutanetricarboxylic acid,

$$\begin{matrix} Me\cdot CH\cdot CO \\ | \\ CO_2H\cdot CH_2\cdot CMe_2\cdot CH\cdot CO \end{matrix}\!\!>\!\!O,$$

is formed when the acid is boiled with acetyl chloride until solution is complete and the evolution of hydrogen chloride has ceased.

The product is evaporated on the water-bath and the residue left over solid potash until the syrup has crystallised. The solid mass is digested with a little boiling benzene, which dissolves the anhydro-acid, but in which any unchanged acid is very sparingly soluble, and the anhydro-acid then separates from the concentrated filtrate in small prisms which melt at about 105—107°.

0·1504 gave 0·3116 CO_2 and 0·0911 H_2O.   C = 56·5 ; H = 6·7.
.0·1359 „   0·2800 CO_2 „   0·0819 H_2O.   C = 56·2 ; H = 6·7.
$C_{10}H_{14}O_5$ requires C = 56·1 ; H = 6·5 per cent.

Anhydrotrimethylbutanetricarboxylic acid dissolves readily in boiling water and, if the solution is evaporated to a small bulk, mixed with an equal volume of hydrochloric acid, and allowed to cool slowly, small, glistening crystals separate which melt with decomposition at 180°. Although the melting point of this acid is higher than that of the specimen from which the anhydro-acid had been prepared, there can be no doubt that they have the same constitution, since the melting point of the mixed acids was found to be 174—175°.

* This melting point was unchanged by further recrystallisation from hydrochloric acid, and it was not until the anhydro-acid (see below) had been prepared and reconverted into the acid that the melting point rose to 180°.

Examination of the Acid B, of Melting Point 204°.

This acid is formed only in very small quantities when ethyl trimethylketopentamethylenedicarboxylate is hydrolysed under the conditions described on p. 785. Fortunately it is much less soluble in water than the acid of melting point 174°, otherwise it might have escaped detection. It separates from water as a sandy powder which, under the microscope, is seen to consist of warty nodules.

0·1192 gave 0·2256 CO_2 and 0·0756 H_2O. C = 51·6 ; H = 7·1.

0·1492 ,, 0·2817 CO_2 ,, 0·0943 H_2O. C = 51·5 ; H = 7·0.

$C_{10}H_{16}O_6$ requires C = 51·7 ; H = 7·0 per cent.

When rapidly heated, this acid softens at 200° and melts at 204° ; it is almost insoluble in acetyl chloride even on boiling, but it dissolves in hot acetic anhydride, and, if the solution is boiled for a few minutes and evaporated, a thick gum remains which has the properties of a double anhydride of the acid and acetic acid.

It is suggested in the introduction (p. 781) that this acid may be the trans-modification of $a\gamma\gamma$-trimethylbutane-$a\beta\delta$-tricarboxylic acid, but no experimental evidence bearing on this point could be obtained. That the acid is tribasic is indicated by the following titration, which, however, had to be carried out with a specimen of the acid which was not quite pure and melted at 194—197°: 0·1116 required for neutralisation, 0·056 NaOH, whereas this amount of a tribasic acid, $C_{10}H_{16}O_6$, should neutralise 0·058 NaOH.

1 : 1 : 3 : 5-*Tetramethyl-4-ketopentamethylene-2-carboxylic Acid*,

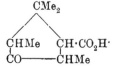

$$\begin{array}{c}
CMe_2 \\
\diagup \quad \diagdown \\
\overset{|}{C}HMe \quad \overset{|}{C}H \cdot CO_2H \cdot \\
\overset{|}{C}O \text{------} \overset{|}{C}HMe
\end{array}$$

During the experiments on the hydrolysis of ethyl trimethylketopentamethylenedicarboxylate, about 20 litres of the steam distillate (A, p. 785) had accumulated. The clear liquid was saturated with salt and extracted three times with ether, the ethereal solution dried over calcium chloride and evaporated, when about 6 grams of a colourless syrup were obtained which, after remaining for several months over sulphuric acid, became semi-solid. In contact with porous porcelain, the oily impurity was gradually absorbed and the dry residue, after digesting in aqueous solution with animal charcoal, crystallised from water in beautiful, colourless prisms of melting point 110°.

0·1584 gave 0·3736 CO_2 and 0·1292 H_2O.   C = 64·7 ; H = 9·1.

0·1630   „   0·3884 CO_2 „   0·1278 H_2O.   C = 65·2 ; H = 8·7.

$C_{10}H_{16}O_3$ requires C = 65·2 ; H = 8·7 per cent.

The method of preparation and properties of this interesting substance point to the probability of its being a tetramethylketopentamethylenecarboxylic acid of the constitution represented at the head of this section (see also p. 782).

The basicity of the acid was determined by titration with decinormal caustic soda, when 0·1890 neutralised 0·0408 NaOH, whereas this amount of a monobasic acid, $C_{10}H_{16}O_3$, should neutralise 0·0411 NaOH.

The *oxime*, $C_8H_{15}(CO_2H)C:N\cdot OH$. In order to prepare this derivative, the keto-acid is dissolved in a considerable excess of caustic potash and mixed with excess of hydroxylamine hydrochloride.

After twenty-four hours, the product is acidified with hydrochloric acid, the viscid mass collected, and washed with water, in which it is sparingly soluble. The most satisfactory way of recrystallising it is to dissolve the substance in methyl alcohol and, after diluting with water, to boil off the methyl alcohol and then stir the solution vigorously as it cools. The oxime then separates as a heavy, granular, crystalline powder which softens at 170° and melts at about 180° with decomposition.

0·1920 gave 11·3 c.c. of nitrogen at 17° and 769 mm.   N = 6·9.

$C_{10}H_{17}O_3N$ requires N = 7·0 per cent.

The *semicarbazone*, $C_8H_{15}(CO_2H)C:N_2H\cdot CO\cdot NH_2$, separates in colourless plates when the aqueous solution of the keto-acid is warmed with semicarbazide hydrochloride and sodium acetate, and may be purified by recrystallisation from much alcohol.

0·1044 gave 15·9 c.c. of nitrogen at 14° and 764 mm.   N = 17·9.

$C_{11}H_{19}O_3N_3$ requires N = 17·4 per cent.

This *semicarbazone* melts at about 215—217° with decomposition, and is very sparingly soluble even in boiling water. It separates from boiling alcohol or dilute acetic acid in well-defined, four-sided plates with bevelled edges.

1 : 1 : 3-*Trimethyl-4-hydroxypentamethylene-2* : 3-*dicarboxylic Acid*,

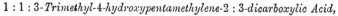
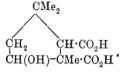

In preparing this acid, pure ethyl trimethylketopentamethylenedi-carboxylate (10 grams) was dissolved in alcohol and mixed with water until a faint turbidity was produced ; the solution was then treated, in a porcelain beaker fitted with a mechanical stirrer, with five times the theoretical quantity of 3 per cent. sodium amalgam, the tempera-ture being kept below 10° and a rapid stream of carbon dioxide passed during the operation. The product which, owing to the separation of sodium bicarbonate, was semi-solid, was extracted with ether, when 8 grams of oil were obtained, showing that little hydrolysis had taken place. This oil, together with that obtained by acidifying the sodium bicarbonate and extracting with ether, was again reduced exactly as before and subsequently hydrolysed by boiling with excess of methyl-alcoholic potash (10 grams KOH); the solution was then diluted with water, evaporated until free from methyl alcohol, acidified, and extracted with ether. After drying over calcium chloride and evaporat-ing, a syrup was obtained which consisted, as analysis showed, of the almost pure hydroxy-acid. This was esterified by dissolving it in alcohol (4 vols.) and sulphuric acid (1 vol.), and after standing for three days water was added, the oil extracted with ether, washed well with sodium carbonate (which extracts a considerable quantity of acid-ester), and dried over calcium chloride.

The ethereal solution was then evaporated and the ester purified by fractionation under reduced pressure, when almost the whole quantity passed over at 194—196° (30 mm.). Two different preparations gave the following results on analysis :

$0 \cdot 1548$ gave $0 \cdot 3451$ CO_2 and $0 \cdot 1244$ H_2O. $C = 60 \cdot 8$; $H = 8 \cdot 9$.
$0 \cdot 1772$ „ $0 \cdot 3974$ CO_2 „ $0 \cdot 1428$ H_2O. $C = 61 \cdot 2$; $H = 8 \cdot 9$.
$C_{14}H_{24}O_5$ requires $C = 61 \cdot 7$; $H = 8 \cdot 8$ per cent.

The ester was now hydrolysed by boiling with excess of methyl-alcoholic potash, the product repeatedly evaporated with water so as to completely remove the methyl alcohol, acidified, and extracted twice with pure ether. The ethereal solution was then dried over calcium chloride, evaporated, and the syrupy hydroxy-acid left over sulphuric acid in an evacuated desiccator for several days.

3 F 2

0·1568 gave 0·3198 CO_2 and 0·1128 H_2O.   C = 55·6 ; H = 7·9.
0·1725 ,,   0·3522 CO_2 ,, 0·1220 H_2O.   C = 55·7 ; H = 7·8.
    $C_{10}H_{16}O_5$ requires C = 55·5 ; H = 7·4 per cent.

Trimethylhydroxypentamethylenedicarboxylic acid is readily soluble in water, and the solution, after neutralising with sodium carbonate, is stable to permanganate. The basicity of the acid was determined by titration with decinormal caustic soda, when 0·4117 required for neutralisation 0·151 NaOH, whereas this amount of a dibasic acid, $C_{10}H_{16}O_5$, should neutralise 0·152 NaOH.

When the acid is distilled under the ordinary pressure, almost the whole quantity passes over at 297—303° as a pale yellow syrup which consists for the most part of the anhydride (or lactone?) of the acid, since the following results were obtained on analysis:

0·1720 gave 0·3818 CO_2 and 0·1178 H_2O.   C = 60·5 ; H = 7·6.
    $C_{10}H_{14}O_4$ requires C = 60·6 ; H = 7·1 per cent.

When, however, the syrup was dissolved in sodium carbonate, it readily reduced permanganate, a behaviour which indicates that, during distillation, some unsaturated acid had been produced.

The hydroxy-acid is not readily attacked by reagents, and, although a large number of experiments were made with the object of converting it into the corresponding chloro- and bromo-acids by the action of phosphorus pentachloride or pentabromide under very varied conditions, no definite products could be isolated.

1 : 1 : 3-*Trimethylpentamethylene*-2 : 3-*dicarboxylic Acid*,

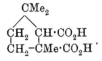

This interesting acid, isomeric with camphoric acid, was obtained when trimethylhydroxypentamethylenedicarboxylic acid was reduced under the following conditions. The pure hydroxy-acid was heated in a sealed tube with fuming hydriodic acid and amorphous phosphorus for four hours at 130° and four hours at 140—145°. The product was diluted with water, extracted with ether, and the syrup, which remained after evaporating the ether and which contained iodine, was dissolved in dilute sodium carbonate and treated with a large excess of sodium amalgam, at first in the cold and then on the water-bath.

Since even after this the reduction was not complete, the whole process was repeated, and the resulting acid dissolved in sodium carbonate and oxidised with permanganate at 0° until the colour remained for ten minutes. The acid was again extracted and distilled, when a considerable quantity of a pale yellow oil was collected at 285—295°, which gave on analysis numbers agreeing approximately with those required for the anhydride of *trimethylpentamethylenedicarboxylic acid*.

0·1471 gave 0·3505 CO_2 and 0·1058 H_2O. C = 65·0 ; H = 8·0.
$C_{10}H_{14}O_3$ requires C = 65·9 ; H = 7·7 per cent.

The anhydride was, however, not pure, since when boiled with water a small quantity of a neutral oil smelling of peppermint remained undissolved. This was removed by filtration through wet paper and the filtrate extracted with pure ether. The ethereal solution was carefully dried over calcium chloride and evaporated, but the almost colourless syrup, even after remaining for fourteen days over sulphuric acid in an evacuated desiccator, showed no signs of crystallisation.

0·2187 gave 0·4793 CO_2 and 0·1591 H_2O. C = 59·8 ; H = 8·1.
0·1693 ,, 0·3708 CO_2 ,, 0·1236 H_2O. C = 59·7 ; H = 8·1.
$C_{10}H_{16}O_4$ requires C = 60·0 ; H = 8·0 per cent.

Although these numbers agree so closely with those required for *trimethylpentamethylenedicarboxylic acid*, the acid is nevertheless not quite pure, as was shown by the results obtained on titration with decinormal sodium hydrate and by the analysis of the silver salt. On titration, 0·2750 of the syrupy acid required for neutralisation 0·105 NaOH, whereas this amount of a dibasic acid, $C_{10}H_{16}O_4$, should have neutralised 0·110 NaOH.

The *silver* salt, prepared by adding silver nitrate to a slightly alkaline solution of the ammonium salt, gave the following results on analysis :

0·3345 gave 0·1721 Ag. Ag = 51·4.
$C_{10}H_{14}O_4Ag_2$ requires Ag = 52·2.

The remainder of the acid (8 grams) was digested with methyl alcohol and sulphuric acid and the ester isolated in the usual manner. The *methyl trimethylpentamethylenedicarboxylate* thus obtained distilled at 145° (13 mm.).

0·1504 gave 0·3465 CO_2 and 0·1215 H_2O. C = 62·9 ; H = 9·0.
$C_{12}H_{20}O_4$ requires C = 63·2 ; H = 8·8 per cent.

This ester was hydrolysed with methyl-alcoholic potash, but although the acid again gave good results on analysis, the syrup did not show any signs of crystallising.*

* Since this was written the acid has commenced to crystallise.

The Action of Ethyl α-Bromopropionate on the Sodium Derivative of Ethyl α-Cyano-ββ-dimethylglutarate.

The sodium derivative of ethyl α-cyano-ββ-dimethylglutarate required for this synthesis was prepared by condensing the sodium derivative of ethyl cyanoacetate with ethyl ββ-dimethylacrylate. Sodium (25 grams) was dissolved in alcohol (300 grams), mixed with ethyl cyanoacetate (113 grams), and, after cooling well, ethyl dimethylacrylate (128 grams) was added and the whole heated on the water-bath for six hours. The condensation between the sodium derivative formed and ethyl α-bromopropionate was effected by cautiously adding 200 grams of the bromo-ester, care being taken that the mixture was kept well cooled during the addition. The mixture was then left for twenty-four hours in running water, and finally heated on the water-bath for two hours. The product was isolated by adding water, extracting with ether, washing the ethereal solution with water and dilute sodium carbonate, drying over calcium chloride, and distilling off the ether. The oily residue was fractionated under diminished pressure and yielded 60 grams of a thick, colourless liquid which boiled constantly at 220—223° (20 mm.).

0·1446 gave 0·3098 CO_2 and 0·1018 H_2O.   C = 58·4 ; H = 7·8.

0·1585 , 0·3415 CO_2 ,, 0·1099 H_2O.   C = 58·7 ; H = 7·8.

0·2941 ,, 11·4 c.c. of nitrogen at 16° and 758 mm. N = 4·6.

$Cl_7H_{27}O_6N$ requires C = 59·8 ; H = 7·8 ; N = 4·1 per cent.

$C_{15}H_{23}O_6N$,,   C = 57·5 ; H = 7·3 ; N = 4·5   ,,

These figures and the subsequent examination of the oil showed that it consisted of a mixture of about equal proportions of two ethyl salts of the formulæ $C_{17}H_{27}O_6N$ and $C_{15}H_{23}O_6N$.

By a process of hydrolysis, described below, it was found to yield both αγγ-trimethylbutane-αβδ-tricarboxylic acid,

$$CO_2H \cdot CH_2 \cdot CMe_2 \cdot CH(CO_2H)CHMe \cdot CO_2H,$$

and the *trans*-modification of αγ-dimethyltricarballylic acid,

$$CHMe(CO_2H)CH(CO_2H)CHMe \cdot CO_2H,$$

and there can therefore be little doubt that it consisted of about equal proportions of *ethyl αγγ-trimethyl-β-cyanobutane-αβδ-tricarboxylate*, $CO_2Et \cdot CH_2 \cdot CMe_2 \cdot C(CN)(CO_2Et)CHMe \cdot CO_2Et$, formed by the direct action of ethyl α-bromopropionate on the sodium compound of derivative of ethyl α-cyano-ββ-dimethylglutarate, and *ethyl αγ-dimethyl-β-cyanopropane-αβγ-tricarboxylate*, formed by the action of two molecules of ethyl α-bromopropionate on the sodium derivative of ethyl cyanoacetate. This latter condensation can be readily understood when it is remembered that the interaction of ethyl dimethylacrylate with the sodium derivative of ethyl cyanoacetate is never complete and that a considerable quantity of ethyl sodiocyanoacetate always remains unchanged.

This then subsequently reacts with ethyl α-bromopropionate and a series of decompositions represented by the following equations takes place (compare Zelinsky and Tschernoswitoff, *Ber.*, 1896, 29, 333):

I. $CHNa(CN)CO_2Et + Me \cdot CHBr \cdot CO_2Et = {\displaystyle \mathop{CH(CN)CO_2Et}_{CHMe \cdot CO_2Et}} + NaBr.$

II. $CHNa(CN)CO_2Et + {\displaystyle \mathop{CH(CN)CO_2Et}_{CHMe \cdot CO_2Et}} = CH_2(CN)CO_2Et +$

$${\displaystyle \mathop{CNa(CN)CO_2Et}_{CHMe \cdot CO_2Et}} \cdot$$

III. $\displaystyle \mathop{CNa(CN)CO_2Et}_{CHMe \cdot CO_2Et} + Me \cdot CHBr \cdot CO_2Et =$

$$\mathop{CO_2Et \cdot CHMe \cdot C(CN)CO_2Et}_{CHMe \cdot CO_2Et} + NaBr.$$

αγγ-Trimethylbutane-αββδ-tetracarboxylic Acid,
$$CO_2H \cdot CH_2 \cdot CMe_2 \cdot C(CO_2H)_2 \cdot CHMe \cdot CO_2H.$$

When the mixed ethyl salts just described (34 grams) were treated with caustic potash (20 grams) dissolved in methyl alcohol, a copious precipitate of a potassium salt (A) soon separated. Since, after standing for fifteen minutes, this separation appeared to be complete, the salt was collected, washed with a little methyl alcohol, and then, without further purification, transferred to a flask and boiled with a solution of caustic potash (15 grams) dissolved in methyl alcohol until the odour of ammonia ceased to be apparent. The acid was then isolated by evaporating until free from methyl alcohol, acidifying, saturating with ammonium sulphate, and extracting at least fifteen times with ether.

After drying over calcium chloride and evaporating, a gum remained which, when stirred with a little concentrated hydrochloric acid, deposited a crystalline acid in small quantity. It was subsequently found that the same acid could be more conveniently isolated, and in much larger quantity, by dissolving the gum in water, neutralising with ammonia, and adding an excess of copper sulphate solution.

The copper salt which was precipitated was collected, washed, suspended in water, and treated with sulphuretted hydrogen until completely decomposed. The filtrate from the copper sulphide yielded, on evaporation on the water-bath, a gum which, when stirred with an

equal quantity of concentrated hydrochloric acid, deposited a considerable amount of a solid acid. This was recrystallised from concentrated
hydrochloric acid, from which it separated in microscopic prisms.

0·1442 gave 0·2495 CO_2 and 0·0806 H_2O.   C = 47·2 ; H = 6·2.

0·1516 „   0·2639 CO_2 „   0·0822 H_2O.   C = 47·5 ; H = 6·0.

$C_{11}H_{16}O_8$ requires C = 47·8 ; H = 5·8 per cent.

αγγ-Trimethylbutane-αββδ-tetracarboxylic acid melts at about 186°
with evolution of gas and is readily soluble in water, but rather
sparingly so in cold concentrated hydrochloric acid.

αγγ-Trimethylbutane-αβδ-tricarboxylic Acid,

$CO_2H·CH_2·CMe_2·CH(CO_2H)CHMe·CO_2H.$

This acid was prepared by heating the tetracarboxylic acid,
described above, in an oil-bath at about 200° until the evolution of
gas had ceased. The gummy residue was dissolved in water, evaporated to a small bulk, mixed with concentrated hydrochloric, acid and
allowed to remain for twenty-four hours, when a considerable amount
of a crystalline acid was found to have been deposited.

This was collected and recrystallised from concentrated hydrochloric acid, from which it separated in small needles which melted at
172—174° with decomposition, due to conversion into the anhydro-
acid (p. 786).

0·1238 gave 0·2358 CO_2 and 0·0784 H_2O.   C = 51·9 ; H = 7·0.

0·1547 „   0·2941 CO_2 „   0·0970 H_2O.   C = 51·8 ; H = 6·9.

$C_{10}H_{16}O_6$ requires C = 51·7 ; H = 7·0 per cent.

The *silver* salt, $C_{10}H_{13}Ag_3O_6$, is precipitated as a white, amorphous
powder when silver nitrate is added to a slightly alkaline solution of
the ammonium salt.

0·2218 gave 0·1294 Ag.   Ag = 58·4.

$C_{10}H_{13}Ag_3O_6$ requires Ag = 58·6 per cent.

It has already been shown that this trimethylbutanetricarboxylic
acid is identical with one of the acids produced by the hydrolysis of
trimethylketopentamethylenedicarboxylic acid with dilute sulphuric
acid (p. 786).

trans-*αγ-Dimethyltricarballylic Acid,*

$CHMe(CO_2H)CH(CO_2H)CHMe·CO_2H.$

The methyl-alcoholic filtrate from the potassium salt (p. 793)
was evaporated on the water-bath until free from methyl alcohol, and
the residue heated with three times its volume of concentrated hydrochloric acid on the sand-bath for twelve hours.

The clear solution was evaporated nearly to dryness on the water-bath and the partially solid residue, which contained large quantities of ammonium chloride, extracted with ether. The ether extract, dried over calcium chloride, yielded, on evaporation, an oil which rapidly crystallised. After remaining in contact with porous porcelain until quite free from oil, the residue was purified by two crystallisations from water, from which it separated in the form of small prisms which melted at 203° with decomposition into water and the anhydro-acid.

0·2365 gave 0·4084 CO_2 and 0·1286 H_2O. C = 47·1 ; H = 6·0.

$C_8H_{12}O_6$ requires C = 47·1 ; H = 5·9 per cent.

The *silver* salt, $C_8H_9Ag_3O_6$, obtained by adding silver nitrate to the ammonium salt, is a white, amorphous powder.

0·2192 gave 0·1348 Ag. Ag = 61·5.

$C_8H_9O_6Ag_3$ requires Ag = 61·6 per cent.

The *anhydro-acid* was prepared by boiling the acid with excess of acetyl chloride for fifteen minutes and then evaporating on the water-bath. It was purified by recrystallisation from a mixture of chloroform and light petroleum, and thus obtained in the form of microscopic needles melting at 112°.

These properties characterise the acid of melting point 203° as trans-α*γ*-*dimethyltricarballylic acid*, which was first obtained by Zelinsky and Tschernoswitoff (*Ber.*, 1896, 29, 334) from dimethylcyanotricarballylic ester by hydrolysis with dilute sulphuric acid.

THE VICTORIA UNIVERSITY OF MANCHESTER.

LXXXI.—*Experiments on the Synthesis of Camphoric Acid. Part V. A Synthesis of Camphoric Acid.*

By WILLIAM HENRY PERKIN, jun., and JOCELYN FIELD THORPE.

WHILE engaged on the problem of the synthesis of camphoric acid, we published, about two years ago (Trans., 1904, 85, 128), a paper in which we described the synthesis of α-campholactone, α-campholytic acid, and β-campholytic acid (*iso*lauronolic acid), the process adopted being briefly as follows.

αα-Dimethylbutane-α*β*δ-tricarboxylic acid,

$$\begin{array}{ll} CH_2 \cdot CO_2H & CMe_2 \cdot CO_2H \\ CH_2 \!\!-\!\!-\!\!-\!\!-\!\!- CH \cdot CO_2H & \end{array},$$

was synthesised and converted into its neutral sodium salt which, when heated with acetic anhydride at 140°, decomposes with evolution of carbonic anhydride and formation of 1-keto-2 : 2-dimethylpentamethylene-3-carboxylic acid :

$$
\begin{array}{l}
\text{CH}_2\text{·CO} \\
\quad | \quad \text{CMe}_2 \\
\text{CH}_2\text{·CH·CO}_2\text{H}
\end{array} \quad .
$$

The ethyl ester of this acid reacts readily with an ethereal solution of magnesium methyl iodide, and one of the products is α-campholactone :

$$
\begin{array}{l}
\text{CH}_2\text{·CMe—O} \\
\quad | \quad \text{CMe}_2 \quad | \\
\text{CH}_2\text{·CH——CO}
\end{array} \quad ,
$$

a substance which is of considerable interest on account of its close relationship to campholactone :

$$
\begin{array}{l}
\text{CH}_2\text{·CMe—CO} \\
\quad | \quad \text{CMe}_2 \quad | \\
\text{CH}_2\text{·CH——O}
\end{array} \quad .
$$

α-Campholactone dissolves in fuming hydrobromic acid and is, in a short time, completely converted into γ-bromotrimethylpentamethylenecarboxylic acid, an acid characterised by the ease with which it loses hydrogen bromide and is converted into α-campholytic acid :

$$
\begin{array}{l}
\text{CH}_2\text{·CMeBr} \\
\quad | \quad \text{CMe}_2 \\
\text{CH}_2\text{·CH·CO}_2\text{H}
\end{array}
\quad \longrightarrow \quad
\begin{array}{l}
\text{CH}= \text{CMe} \\
\quad | \quad \text{CMe}_2 \\
\text{CH}_2\text{·CH·CO}_2\text{H}
\end{array} \quad .
$$

One of the principal reasons for instituting these experiments was, of course, the consideration that the substances synthesised by the processes just described are all very closely allied to camphoric acid. Indeed, it is only necessary to replace the bromine atom in γ-bromotrimethylpentamethylenecarboxylic acid by the carboxyl group in order to accomplish a synthesis of camphoric acid :

$$
\begin{array}{l}
\text{CH}_2\text{·CMe·CO}_2\text{H} \\
\quad | \quad \text{CMe}_2 \\
\text{CH}_2\text{·CH·CO}_2\text{H}
\end{array} \quad .
$$

Many experiments were therefore made with the object, in the first place, of converting the bromo-acid·into the corresponding cyano-acid by the action of cyanides under very varied conditions.

Most of these were completely unsuccessful, and investigation showed that the failure was due to the ease with which the bromo-acid is decomposed, even by the mildest reagents, with elimination of hydrogen bromide. In nearly every case the only product which could be isolated was a-campholytic acid.*

During the course of one experiment, however, in which the bromo-acid was treated with potassium cyanide and hydrocyanic acid under rather unusual conditions (*loc. cit.*, p. 146), a small quantity of an oily acid was obtained, which became semi-solid when left exposed to sulphuric acid for some weeks. After remaining in contact with porous porcelain, a very small quantity of a crystalline acid was obtained which melted between 180° and 190°, and, when treated with acetic anhydride, yielded a crystalline anhydride of melting point 215—217°. The latter was hydrolysed and yielded an acid of melting point 204°. This behaviour is characteristic of *i*-camphoric acid, since this acid melts at 203° and yields an anhydride melting at 221°. After describing these experiments, we added the following comment : " The method of formation of this small quantity of acid and its properties leave scarcely room for doubt that it was *i*-camphoric acid, and experiments were in progress which it was hoped would have yielded enough material for definite identification.

"In the meantime, Komppa (*Ber.*, 1903, **36**, 4332) has published his brilliant synthesis of camphoric acid, which, once for all, establishes the correctness of Bredt's formula, and it is therefore quite unnecessary to investigate our much less satisfactory process further."

The amount of synthetical acid obtained by us was so small that an analysis was quite out of the question, and it was unfortunate that, at that time, the authors were not in possession of a specimen of *i*-camphoric acid. It would otherwise have been possible to demonstrate the identity of the two acids by direct comparison, and especially by the proof that a mixture of the two had the same melting point as the components. Shortly afterwards Prof. W. A. Noyes was kind enough to send us a considerable quantity of *i*-camphoric acid, and we were then at once able to prove that our synthetical acid was identical with this acid. Before publishing this result we thought it wise to repeat the synthesis, and especially to carry out a new series of experiments with the object of improving the yield and of thus making the synthesis more valuable. While we were unsuccessful in this

---

* We encountered a similar difficulty in our synthesis of *i*-camphoronic acid (Trans., 1897, **61**, 1174), during the course of which it was necessary to convert ethyl bromotrimethylglutarate, $CO_2Et \cdot CMe_2 \cdot CBrMe \cdot CH_2 \cdot CO_2Et$, into the corresponding cyano-ester.

When this halogen ester was heated with potassium cyanide, most of it was decomposed with elimination of hydrogen bromide and formation of the ester of trimethylglutaconic acid, $CO_2H \cdot CMe_2 \cdot CMe:CH \cdot CO_2H$, and the yield of cyano-ester was therefore always small, and, indeed, in some experiments only traces were formed.

latter respect, we have, nevertheless, been able to prepare sufficient of the acid to allow of its composition being controlled by analysis, and thus to prove conclusively that the synthetical acid is *i*-camphoric acid.

In the previous paper on this subject, it was mentioned (p. 143) that a considerable quantity of a neutral substance is obtained as a by-product in the preparation of α-campholactone by the action of magnesium methyl iodide on ethyl ketodimethylpentamethylene-carboxylate. As a good deal of this had accumulated during the numerous preparations of the lactone, it was distilled and found to yield a crystalline substance of melting point 80°, which proved to be a *dihydroxyisopropyltrimethylcyclopentane* of the formula

$$CH_2 \cdot C(OH)Me$$
$$| \quad CMe_2$$
$$CH_2 \cdot CH \cdot C(OH) \cdot Me_2$$

When this substance is digested with potassium hydrogen sulphate, two molecules of water are eliminated, and it is converted into the corresponding hydrocarbon:

$$CH{=}C \cdot Me$$
$$| \quad CMe_2 \quad CH_2$$
$$CH_2 \cdot CH \cdot C{<}^{CH_2}_{Me},$$

which distils at 179° and to which we have given the name iso*propenyltrimethyl*cyclo*pentene*. These substances are interesting, not only on account of the fact that very few, if any, terpenes have so far been prepared which contain a 5-carbon ring, but also because of the striking similarity between their properties and those of *trans*-terpin and dipentene, with which they are, indeed, closely related in constitution:

$$CH_2 \cdot C(OH)Me \qquad CH{=}C \cdot Me$$
$$| \quad (CH_2)_2 \qquad | \quad (CH_2)_2 \quad CH_2$$
$$CH_2 \cdot CH \cdot C(OH)Me_2 \qquad CH_2{-}CH \cdot C{<}^{CH_2}_{Me}$$

Terpin.   Dipentene.

The tertiary alcohol, for example, crystallises beautifully from water, and, like terpin, appears to exist in *cis*- and *trans*-modifications (p. 801); the hydrocarbon boils at practically the same temperature as dipentene, and possesses, in a marked degree, the intense odour of lemons so characteristic of that substance.

EXPERIMENTAL.

Synthesis of i-*Camphoric Acid from* γ-*Bromotrimethylpentamethylene-carboxylic Acid.*

The bromo-acid required for these experiments was prepared by the action of fuming hydrobromic acid on a-campholactone, the conditions employed being substantially the same as those described in the previous communication (Trans., 1904, **85**, 145), and the treatment with potassium cyanide and hydrocyanic acid was carried out under the following conditions :

I. Pure potassium cyanide (from Kahlbaum, 10 grams) was dissolved in a little water, mixed with anhydrous hydrocyanic acid (6 c.c.), and then a solution of a-bromotrimethylpentamethylenecarboxylic acid (20 grams) in cold alcohol added. The bottle containing the mixture was securely fastened, shaken on the machine for twenty-four hours, and then allowed to stand at the ordinary temperature for fourteen days.

A further quantity of potassium cyanide (5 grams) and anhydrous hydrocyanic acid (5 c.c.) was then added and the whole heated in the water-bath for three hours. The product, which, owing to the separation of azulmic acid, was quite black, was mixed with concentrated hydrochloric acid, and, after standing for twenty-four hours, heated on the water-bath for five hours. The whole was then distilled in steam until all the β-campholytic acid had passed over and the residue repeatedly extracted with ether. After drying over calcium chloride and evaporating, the ethereal solution deposited a small quantity of a thick, pale yellow oil, which, especially when stirred from time to time with a glass rod, gradually crystallised.

II. In the second series of experiments, ethyl γ-bromotrimethylpentamethylenecarboxylate was employed instead of the acid itself, and some of this ester was prepared by the method already described (*loc. cit.*, p. 145) and some by the following process :

a-Campholactone (20 grams) was dissolved in alcohol (70 c.c.) and saturated with hydrogen bromide, care being taken that the temperature did not rise above 40°. After twenty-four hours, water was added and the oil extracted with ether, when it was found to contain only 15 per cent. of bromine, whereas the formula $C_{11}H_{19}O_2Br$ requires 30·5 per cent.

The treatment with alcoholic hydrobromic acid was therefore repeated as before and once again on the water-bath, after which the ester contained 28·7 per cent. of bromine. This ester was treated with potassium cyanide and hydrocyanic acid in aqueous-alcoholic solution under the conditions described above in the case of the bromo-acid itself, and yielded also small quantities of the pale yellow oil, which gradually crystallised.

The semi-solid mass which had accumulated from six experiments was left in contact with porous porcelain until the oily impurity had been completely absorbed, the residue boiled for a few minutes with a little acetic anhydride, and then left in a cold place for twenty-four hours. The crystals were collected and recrystallised from acetic anhydride, from which the i-*camphoric anhydride* separated in the form of minute, prismatic crystals.

0·1553 gave 0·3751 CO_2 and 0·1090 H_2O.   C = 65·9 ; H = 7·8.

$C_{10}H_{14}O_3$ requires C = 65·9 ; H = 7·7 per cent.

This anhydride melted at 215—217° ; it was dissolved in a little methyl-alcoholic potash, the solution mixed with water, evaporated, acidified, and repeatedly extracted with ether. The ethereal solution was then evaporated, when a solid acid was obtained which crystallised from water in prisms and melted at 202—203°.

These synthetical preparations were directly compared with *i*-camphoric anhydride and *i*-camphoric acid (for a considerable quantity of which the authors are indebted to Prof. W. A. Noyes), with the result that no difference could be observed in their properties. It is especially worthy of note that when about equal quantities of the different specimens were intimately mixed, no alteration in melting point could be detected.

*Dihydroxy*iso*propyltrimethyl*cyclo*pentane   and   *iso*Propenyltrimethyl-cyclopentene.*

During the preparation of a-campholactone (Trans., 1904, **85**, 143), the product of the action of magnesium methyl iodide on ethyl ketodimethylpentamethylenecarboxylate is hydrolysed by means of methyl-alcoholic potash, and there always remains a small quantity of a neutral oil which is unattacked by the potash and is precipitated when the product is subsequently diluted with water. This was collected, and when 65 grams had accumulated it was fractionated under reduced pressure. Rather more than two-thirds distilled at 170—180° (60 mm.) as a colourless oil which, during the winter months, deposited a considerable quantity of crystals. These were collected on the pump, left in contact with porous porcelain until quite free from oil, and then recrystallised from light petroleum.

0·1613 gave 0·4198 CO_2 and 0·1731 H_2O.   C = 70·9 ; H = 11·9.
0·2016 ,, 0·5260 CO_2 ,, 0·2155 H_2O.   C = 71·0 ; H = 11·8.

$C_{11}H_{22}O_2$ requires C = 70·9 ; H = 11·8 per cent.

*Dihydroxy*isopropyl*trimethyl*cyclopentane melts at 79—80° and is readily soluble in alcohol, ether, benzene, or light petroleum ; from the latter solvent it separates in hard crusts. It dissolves in hot water, but does not crystallise readily on cooling ; if, however, the solution is exposed over sulphuric acid, the substance gradually separates in magnificent glistening prisms. When the powdered substance is added to a large excess of fuming hydrobromic acid (saturated at 0°), it liquefies at once and a mobile oil floats on the surface of the acid. After remaining for twenty-four hours, the product was diluted with water, when a heavy oil separated which was apparently a mixture of the bromohydrin, $C_{11}H_{20}(OH)Br$, and the dibromohydrin, $C_{11}H_{20}Br_2$, since it contained 42·2 per cent. of bromine, whereas the former substance contains 32·1 and the latter 51·2 per cent. of bromine.˙ It is curious that a second treatment with the hydrobromic acid did not materially add to the percentage of halogen.

The porous plates which had been used in the purification of the above solid ditertiary alcohol were extracted with ether in a Soxhlet apparatus, the ethereal solution evaporated, mixed with the oily mother liquor which had been removed from the crystals by filtration, and the whole distilled. The oil which passed over at 175—180° (60 mm.) was then left in the ice-chest for four weeks in contact with a crystal of the ditertiary alcohol, when a further slight crystallisation took place. After the crystals had been removed by filtration, the oil was again fractionated and the portion which distilled at 177° (60 mm.) analysed :

0·1630 gave 0·4208 CO_2 and 0·1722 H_2O.   C = 70·4 ; H = 11·7.

$C_{11}H_{22}O_2$ requires C = 70·9 ; H = 11·8 per cent.

This liquid ditertiary alcohol has a penetrating odour of peppermint, and since, when digested with potassium hydrogen sulphate, it yields the same hydrocarbon (see below) as the solid alcohol, it is exceedingly probable that it consists for the most part of the *cis*-modification of *dihydroxy*-isopropyl*trimethyl*cyclopentane. Obviously, from its method of purification, it must contain some of the *trans*-modification, ·but the amount is probably very small, since no further crystallisation took place when the oil was left in the ice-chest for fourteen days.

In order to prepare the corresponding hydrocarbon, the pure solid ditertiary alcohol (10 grams) was digested with powdered potassium hydrogen sulphate (15 grams) in a reflux apparatus for one hour and the product distilled in steam. The distillate was extracted with ether, the ethereal solution washed with sodium carbonate, dried over calcium chloride, and the oil fractionated, when almost the whole quantity passed over at 175—182°.

After refractionation, the fraction 176—180° was twice distilled over sodium and at once analysed :

0·1401 gave 0·4493 CO_2 and 0·1503 H_2O. C = 87·5 ; H = 11·9.

0·1540 ,, 0·4948 CO_2 ,, 0·1659 H_2O. C = 87·7 ; H = 12·0.

$C_{11}H_{18}$ requires C = 88·0 ; H = 12·0 per cent.

iso*Propenyltrimethyl*cyclo*pentene* distils at 177—179° (754 mm.) and possesses in a marked degree the odour of lemons so characteristic of dipentene. It absorbs bromine and combines with hydrogen bromide, but it does not appear to be readily oxidised in contact with air.

The authors wish to thank Messrs. D. T. Jones, S. S. Pickles, and G. B. Stones for valuable assistance in carrying out the experiments described in these communications. They also wish to express their indebtedness to the Research Fund of the Royal Society for several Grants which have defrayed much of the heavy expense which is inseparable from such investigations.

THE VICTORIA UNIVERSITY OF MANCHESTER.

***

LXXXII.—*A Product of the Action of* iso*Amyl Nitrite on Pyrogallol.*

By ARTHUR GEORGE PERKIN, F.R.S., and ALEC BOWRING STEVEN, B.Sc.

DURING experiments with purpurogallin (Trans., 1903, **83**, 192), attempts were made to prepare this substance by treating an alcoholic solution of pyrogallol with acetic acid and *iso*amyl nitrite. Only a trace of a crystalline compound was, however, thus formed, and this possessed reactions which were quite distinct from those of the expected colouring matter. On account of the poor yield, a satisfactory investigation of this product was hardly possible, and a search has been therefore made from time to time, during the last three years, in the hope that some more economical method of preparation could be devised. These experiments have not been successful, and, though they will be continued, it appeared desirable to embody in the present communication some of the characteristic properties of this interesting compound.

EXPERIMENTAL.

Twenty grams of pyrogallol dissolved in 200 c.c. of alcohol were treated with 5 c.c. of acetic acid, the solution placed in a freezing mixture, and 20 c.c. of *iso*amyl nitrite then added a little at a time. The·resulting reddish-brown liquid, after standing for twenty-four hours, had deposited a small quantity of a crystalline substance, which was collected by means of the pump, washed with alcohol until the filtrate was colourless, then with acetic acid, and finally with alcohol. The yield was remarkably regular and in distinct operations varied as little as from 0·2655 to 0·27 gram. In case, however, the mixture was allowed to become warm during the reaction, an amount not larger than 0·1805 to 0·2125 gram could then be obtained, and this product was usually darker in colour and of a less pure nature than that prepared at the lower temperature.

This compound formed a glistening mass of very pale salmon-coloured leaflets, which appeared under the microscope to be perfectly homogeneous, and melted with decomposition at 200—203°. As, however, in one or two operations an almost colourless substance was produced, the presence of a trace of some red-coloured impurity was suspected. To remove this, crystallisation from alcohol and acetic acid was attempted, but as in this manner considerable loss was experienced owing to spontaneous decomposition these solvents were but little employed. A better method consisted in stirring the compound into about twice its bulk of well-cooled pyridine, rapidly collecting the undissolved product by means of the pump, and washing first with a few drops of pyridine and finally with acetic acid. An almost colourless residue was thus obtained which was pure enough for most purposes, but in special cases a portion of this was crystallised from a very large bulk of acetic acid. Examination showed that it contained no nitrogen.

0·1103 gave 0·2349 CO_2 and 0·0347 H_2O. C = 58·08 ; H = 3·49.

$C_6H_4O_3$ requires C = 58·06 ; H = 3·22 per cent.

From acetic acid it is deposited as minute, colourless, prismatic needles, melting with effervescence at 206—208°, very sparingly soluble in the usual solvents. It dissolves in solutions of the alkali hydrates with an orange-brown coloration, which darkens in contact with air, and its solution in sulphuric acid possesses a pale yellowish-brown tint. Nitric acid does not perceptibly attack it in the cold, but on heating a somewhat violent reaction ensues.

As the "crude" salmon-coloured substance differed so little in melting point from the colourless compound isolated from it by re-crystallisation, the former was analysed to determine if this purification had effected any marked change in its composition.

0·1047 gave 0·2235 CO_2 and 0·0331 H_2O.
C = 58·20 ; H = 3·51 per cent.

Evidently this was not the case, and as the general properties of the coloured were found to differ in no respect from those of the colourless compound, it seemed possible that in reality the substance existed in these two modifications. A study of the behaviour of both products towards mordanted fabrics indicated that they possessed well-marked dyeing properties, the shades upon woollen cloth, identical in each case, being as follows :

Chromium.	Aluminium.	Tin.	Iron.
Dull chocolate-brown.	Pale chocolate-brown	Pale dull yellow	Deep olive-brown

Reduction.—The substance (1 gram), suspended in 50 c.c. of boiling acetic acid, was treated with zinc dust, causing the gradual formation of a pale yellow solution. A little dilute sulphuric acid was now added, the clear liquid poured into water, and this, after partial distillation to remove acetic acid, was extracted with ether. On evaporating the extract, a crystalline residue remained, which was purified by crystallisation from benzene.

0·0852 gave 0·1781 CO_2 and 0·0379 H_2O.   C = 57·00 ; H = 4·94.

$C_6H_6O_3$ requires C = 57·14 ; H = 4·76 per cent.

It consisted of colourless needles melting at 130—131°, and had all the reactions of pyrogallol.

Acetylation.—This was readily effected by treating the compound $C_6H_4O_3$ with acetic anhydride containing a drop of sulphuric acid, and digesting at the boiling temperature for two or three minutes. The product, which was sparingly soluble, was collected and recrystallised from a large volume of acetic anhydride.

0·1118 gave 0·2360 CO_2 and 0·0380 H_2O.   C = 57·57 ; H = 3·81.

$C_6H_3O_3(C_2H_3O)$ requires C = 57·84 ; H = 3·61 per cent.

It forms colourless leaflets which, when heated, commence to blacken at about 273° and decompose with effervescence at 283—285°. It is insoluble in cold alkaline liquids, and is very sparingly soluble in the usual solvents.

An attempt to determine the acetyl groups by the acetic ether method was unsuccessful, for the substance was not perceptibly attacked in this manner, and the following modification was therefore adopted. 1·845 grams of the compound were boiled with 50 c.c. of standard alcoholic sodium hydrate solution for thirty minutes, the solution treated with 6 c.c. of sulphuric acid, and made up to 250 c.c. with alcohol. After standing, 50 c.c. of the clear liquid were removed, distilled into alcoholic soda, and treated in the usual manner.

$0·3690 = \dfrac{1·845}{5}$ gave 0·1380 acetic acid. $C_2H_4O_2$ = 37·39.

$C_6H_3O_3(C_2H_3O)$ requires $C_2H_4O_2$ = 36·14 per cent.

This result therefore indicates that the compound is a *monacetyl* derivative, though its stability towards acid hydrolytic agents is remarkable and difficult to understand.

The effect of zinc dust upon the compound $C_6H_4O_3$ during acetylation was now studied, the operation being carried out in the usual manner. The clear liquid was treated with alcohol and evaporated on the steambath, the residue dissolved in a little acetic acid and poured into water. After standing overnight, the crystalline product was collected, distilled in a small test-tube to free it from a trace of amorphous impurity, and the distillate crystallised from alcohol.

0·1081 gave 0·2270 CO_2 and 0·0480 H_2O. C = 57·26 ; H = 4·93.

$C_6H_3O_3(C_2H_3O)_3$ requires C = 57·14 ; H = 4·76 per cent.

This substance melted at 160—162° and was suspected to consist of *triacetylpyrogallol*. As the melting point of this compound did not appear to be on record, a sample was prepared from pyrogallol and found to be identical with the above product.

A similar result was obtained by submitting the acetyl derivative of the substance $C_6H_4O_3$ to the action of zinc dust and acetic anhydride. Addition of water caused the separation of a crystalline precipitate which, after purification, was identified as triacetylpyrogallol.

The Action of Boiling Water.—The substance $C_6H_4O_3$ behaves in a most interesting manner when digested with boiling water. . The colourless crystals obtained by purification with pyridine and acetic acid pass into solution somewhat slowly, and even before this has completely occurred orange-red needles commence to separate out, which dissolve in dilute alkalis with a blue coloration and give an acetyl compound melting at 186—188°. As it was not easy to obtain sufficient of the colourless compound to study this reaction, experiments were carried out with the crude substance to determine if this behaved in a similar manner, and, this being the case, it was accordingly employed.

Two grams were digested with 150 c.c. of boiling water for half an hour and the product was evaporated to dryness on the water-bath. The orange-brown, resinous mass was treated with a little water, the insoluble matter collected, washed with alcohol, and purified by crystallisation from the same solvent, employing animal charcoal.

0·1033 gave 0·2275 CO_2 and 0·0375 H_2O. C = 60·05 ; H = 4·03.

$C_{11}H_8O_5$ requires C = 60·00 ; H = 3·63 per cent.

It formed glistening, orange-red needles, was soluble in dilute alkalis with a blue coloration, and gave an acetyl compound melting as above stated. A further examination proved that it was without doubt *purpurogallin*. On comparing the dyed patterns obtained from the substance $C_6H_4O_3$ with those given by purpurogallin, it was at once evident that they were identical, so that the tinctorial effects produced are not those of the compound itself, but are due to the peculiar change it undergoes in contact with boiling water.

The aqueous filtrate from the purpurogallin was treated with animal charcoal and evaporated to a small bulk. After standing for some days, the pale yellowish-brown liquid had deposited a small amount of a crystalline powder, which was removed and purified by two or three crystallisations from water.

0·1065 gave 0·2115 CO_2 and 0·0367 H_2O.   C = 54·16 ; H = 3·82.

0·1001 ,, 0·1996 CO_2 ,, 0·0365 H_2O.   C = 54·37 ; H = 4·05.

$C_{12}H_{10}O_7$ requires C = 54·14 ; H = 3·76 per cent.

This substance, the nature of which has not yet been determined, consists of small, colourless prisms, sparingly soluble in cold water, which, when heated, commence to darken at 220° and melt with decomposition at 242—243°. Dilute alkaline solutions dissolve it with a brownish-yellow coloration, and with ferric chloride a deep brown liquid is obtained. It does not dye mordanted fabrics. The amount produced was too small for further experiment, and the formula $C_{12}H_{10}O_7$ given above must be considered as merely provisional. The only simpler expression, however, appears to be $C_7H_6O_4$ (C = 54·54 ; H = 3·87), and this does not seem likely to be available.

The mother liquid from this compound on evaporation to dryness gave some quantity of a brittle, brown resin, very easily soluble in water, and which could not be obtained in a crystalline condition. As it was possible that some pyrogallol might be present, distillation was resorted to with a negative result, and a similar attempt with its acetylation product was also unsuccessful. With alkaline solutions, it gave an orange-brown coloration, with ferric chloride a greenish-black liquid, and with lead acetate a pale brown precipitate. The nature of this product must remain unsolved until a readier method of preparing the raw material has been discovered.

Summary.

The compound prepared from pyrogallol by the above method has most probably the formula $C_6H_4O_3$, and is closely allied to the phenol from which it is derived, for it is reconverted into this by reducing agents. By repeated crystallisation, it can be obtained in a colourless condition, but it is suggested that the small quantity of the red coloured substance thus removed may consist of a coloured form of the same compound.

The fact that on boiling with water it is partially converted into purpurogallin is interesting, and points at first sight to a close relationship between these products. On the other hand, this can hardly be the case, because preliminary experiments recently carried out on the behaviour of the latter when acetylated in the presence of zinc dust indicated that no acetylpyrogallol is thus formed, but an ochreous compound of an entirely distinct character. Again, purpuro-gallin, $C_{11}H_8O_5$, when distilled with zinc dust, gives naphthalene, and it thus seems likely that its formation from the substance $C_6H_4O_3$ is of a complex rather than a simple nature. The colourless, crystalline compound $C_{12}H_{10}O_7$? and the accompanying resin require further investigation.

From the information available, the only constitution which appears to be suitable for this product is that of a *hydroxyorthobenzoquinone*,

, though this suggestion is at present made with some reserve. This constitution will explain its behaviour on reduction, and the oxidising action which it seems necessary to presume is involved in the formation of purpurogallin. Again, the instability of the compound in aqueous solution is in harmony with such a formula, for it has been shown by Willstätter and Pfannenstiel (*Ber.*, 1904, **37**, 4744) that *o*-benzoquinone very readily decomposes under these conditions. The fact that attempts to prepare an oxime have as yet failed appears to be accounted for by this peculiar behaviour of its solutions, and though the colourless condition of the main product does not at first sight coincide with the views expressed, this, as is now well known, cannot be employed as an argument against this consti-tution.

On the other hand, such a formula as

appears to be hardly admissible, though it is interesting to note that Wichelhaus (*Ber.*, 1872, **5**, 848) adopted a somewhat similar expression in his suggested constitution for purpurogallin (pyrogallolquinone),

$$HO\cdot C_6H_3{<}^{O\cdot O\cdot C_6H_3(OH)_2}_{O\cdot O\cdot C_6H_3(OH)_2}.$$

Curiously enough, he also predicts that the first product of the oxidation of pyrogallol by his methods is a hydroxyquinone, a point which receives support from the results of this investigation. It is unfortunate that a better method is not yet available for the preparation of this substance, but should such be devised, these experiments will be continued.

The expenses of this research have been partly defrayed by a grant received from the Research Fund of the Chemical Society.

CLOTHWORKERS' RESEARCH LABORATORY,
DYEING DEPARTMENT,
THE UNIVERSITY,
LEEDS.

LXXXIII.—Some New o-Xylene Derivatives.

By GEORGE STALLARD, M.A.

OSCAR JACOBSEN has shown (Ber., 1877, 10, 1011 ; 1878, 11, 22) that on sulphonation o-xylene yields one sulphonic acid only (CH_3:CH_3 : SOH_3 = 1 : 2 : 4). The position of the sulpho-group was determined by fusing the potassium salt with potassium cyanide and treating the nitrile with hydrochloric acid : also by fusion with sodium formate. The acid obtained was p-xylylic acid (1 : 2 : 4), melting at 163°. He also showed (Ber., 1884, 17, 2372) that when o-xylene (1 mol.) is treated in the cold in presence of iodine with bromine (1 mol.) one bromoxylene only is obtained, which he described as a liquid boiling at 214·5° and having a sp. gr. 1·3693 at 15°/15°. The oil froze below 0°, and the solid thus obtained melted at −0·2°.

Since this bromoxylene gave p-xylylic acid when heated with ethyl chlorocarbonate and sodium amalgam, he considered it to be the 1 : 2 : 4-compound (CH_3 : CH_3 : Br = 1 : 2 : 4), and showed also that on sulphonation a monobromosulphonic acid was formed, which on de-bromination gave the 1 : 2 : 4-sulphonic acid, and therefore concluded that it was the 1 : 2 : 4 : 5-compound.

This investigator likewise described the following derivatives : the barium salt (+ $3H_2O$), crystallising in long prisms, the sodium salt (+ $1\frac{1}{2}H_2O$), separating in very long, fine needles, the potassium salt (+ H_2O), obtained in thin, brilliant prisms, and the sulphamide, crystallising in needles melting at 213°.

Having repeated a good deal of Jacobsen's work, I have obtained results identical with his, except that I find a constant sp. gr. of 1·373 at 15°/15° for his bromoxylene, and a boiling point of 215·2°, unaltered by repeated sulphonation and hydrolysis of a salt of the sulphonic acid thus obtained.

Kelbe has studied (Ber., 1886, 19, 2137) the action of bromine (1 mol.) on Jacobson's o-xylenesulphonic acid (1 mol.) and, after removing the mono- and di-bromoxylenes formed, prepared the barium salt, and found that from the hot solution a ·monobromosulphonate, sparingly soluble in water and containing 4 mols. of H_2O, separated on cooling ; the sulphamide melted at 186·5°.

As Jacobsen's monobromosulphonic acid gives a barium salt with $3H_2O$ and a sulphamide melting at 213°, Kelbe concluded that his bromosulphonic acid was isomeric with Jacobsen's, and must therefore have the constitution $CH_3 : CH_3 : Br : SO_3H = 1 : 2 : 3 : 4$ or $1 : 2 : 3 : 5$.

With the object of obtaining the hitherto unknown $(1 : 2 : 3)$ mono-bromo-o-xylene by hydrolysis from Kelbe's sulphonic acid, I have repeated Kelbe's experiments. To a dilute aqueous solution of Jacobsen's $1 : 2 : 4$-o-xylenesulphonic acid (1 mol.) was added a solution of bromine (1 mol.) in hydrochloric acid. At about 40°, as stated by Kelbe, an oil, consisting of mono- and di-bromoxylenes, separates, which partially solidifies on cooling. The aqueous solution was neutralised with sodium carbonate and evaporated to dryness. The residue was taken up with alcohol and the solution evaporated to dryness. The residue was then dissolved in water, and to the boiling concentrated solution a boiling solution of barium chloride was added and the mass left to cool. The solid mass was then collected at the pump, freed from sodium chloride by washing with a little water, and then fractionally crystallised. Only the least soluble portion has been as yet examined. As stated by Kelbe, it consists of a barium bromo-o-xylene sulphonate which crystallises in fine needles, much more soluble in hot than in cold water, but it contains rather less than 8 per cent. of water $(3H_2O = 7·51$ per cent.$)$. For $4H_2O$, as found by Kelbe, 9·8 per cent. of water would be required.

A much better yield of this salt may be obtained by brominating the barium salt of Jacobsen's o-xylenesulphonic acid.

The following derivatives have also been prepared : sodium salt $(+ H_2O)$, crystallising well in needles ; potassium salt (anhydrous) also crystallising well in rather stout needles ; sulphonic chloride, crystallising from ether in stout, six-sided plates, sometimes of considerable size and melting sharply at 85·5° ; sulphamide, crystallising from water in fine needles melting at 191·5° (Kelbe gives 186·5°).

3-*Bromo*-o-*xylene.*—This compound was obtained by steam-distilling the above sodium salt dissolved in excess of sulphuric acid. Hydrolysis began at about 150°, and a colourless, highly refracting liquid was obtained, which, when purified by repeated sulphonation and regeneration from the pure sodium salts of the sulphonic acids formed, boiled constantly at 213·8° and had a sp. gr. of 1·382 at 15°/15°.

Sulphonation of 3-*Bromo*-o-*xylene.*—The bromo-o-xylene was dissolved in slightly fuming sulphuric acid and the liquid left to crystallise. The crystals were then dissolved in water and barium carbonate added in slight excess. A mixture of two barium salts is thus obtained, which can be separated from one another by fractional crystallisation. One of these salts (A) is formed in much larger proportion than the other (B), and is much less soluble in water. A comparison of the properties of derivatives of these two compounds is given below.

	Series A.	Series B.
Barium salt.	(Anhydrous) crystallises in small plates, but sometimes in a powdery form.	(+ 3H$_2$O) crystallises in fine needles.
Sodium salt.	(+ H$_2$O) crystallises in thin, glistening plates.	(+ H$_2$O). The solution prepared from the barium salt first gave a crop of short, thin needles containing ½H$_2$O, whilst from the mother liquor a salt separated in aggregates of thin, transparent plates, which became opaque on losing their water (1 mol.). Both of these salts gave the same sulphonic chloride and sulphamide. The needles (containing ½H$_2$O), on recrystallisation, gave plates identical with the above.
Potassium salt.	(Anhydrous) long, flat needles.	(Anhydrous) crystallises in stout needles.
Sulphonic chloride.	Crystallises from ether in thin, six-sided plates melting sharply at 63·5°.	Crystallises from ether in stout, six-sided plates melting at 85·5°.
Sulphamide.	Crystallises from water in the form of an arborescent growth, melting at 195°.	Crystallises from water in fine needles, melting at 191·5°.

It will be seen that the members of series B are identical with those obtained from Kelbe's bromosulphonic acid, and that therefore when 3-bromo-o-xylene is sulphonated the minor product is an acid identical with that obtained by brominating Jacobsen's o-xylene-4-sulphonic acid. I have not yet ascertained whether this acid is the 1 : 2 : 3 : 4-(CH$_3$: CH$_3$: Br : SO$_3$H) or the 1 : 2 : 3 : 5-derivative. That it cannot be the 1 : 2 : 3 : 6-derivative will be seen from the following considerations.

The isomeric acid, from which the members of the series A are derived and which forms the major part of the sulphonation of the 3-bromo-o-xylene must be the 1 : 2 : 3 : 6-derivative, as its sodium salt yielded, on prolonged treatment with sodium amalgam, the sodium salt of the o-xylene-3-sulphonic acid described by Krüger (*Ber.*, 1885, **18**, 1760) and Moody (*Proc.*, 1892, **8**, 213).

On treatment with phosphorus pentachloride, the last-named sodium salt gave a sulphonic chloride crystallising from ether and melting between 41·5° and 43·5° (Moody gave 47°). This without further crystallisation was converted into a sulphamide, which melted sharply at 166—166·5° (Kruger gave 165° and Moody 167°).

My best thanks are due to Mr. E. C. Camps for help in performing the above-mentioned experiments.

CHEMICAL LABORATORY,
RUGBY SCHOOL.

LXXXIV.—*Note on the Constitution of Cellulose.*

By ARTHUR GEORGE GREEN and ARTHUR GEORGE PERKIN.

THE highest product of the acetylation of cellulose (without concomitant hydration) has hitherto been generally regarded as a tetra-acetyl derivative, $C_6H_6O(OCOCH_3)_4$. The constitutional formula of cellulose proposed by Green (*Zeit. Farb. Text. Ind.*, 1904, 3, 97), namely,

$$CH(OH)\cdot CH-CH(OH)$$
$$\Big| \qquad >O \quad >O \quad ,$$
$$CH(OH)\cdot CH-CH_2$$

contains only three hydroxyl groups, and therefore on acetylation should only give·rise to a *tri*acetyl derivative. In all other respects this formula gives a direct and simple explanation of the reactions of cellulose, especially of its conversion by hydrobromic acid into Fenton's ω-bromomethylfurfural and its general behaviour as a "latent" aldehyde.

It therefore appeared to us desirable to reinvestigate the supposed tetra-acetylcellulose with a view to placing its constitution beyond doubt. The product employed for our experiments was kindly supplied to us by Mr. C. F. Cross, to whom our thanks are due. It had been prepared in the usual manner by acetylation of the cellulose regenerated from the xanthate (Cross, Bevan, and Beadle, Trans., 1895, 67, 447). With this product we have determined the acetyl groups by three different methods, namely : (*a*) by acid saponification according to A. G. Perkin's method (Trans., 1905, 87, 107); (*b*) by alkaline saponification and direct titration ; (*c*) by alkaline saponification followed by treatment as in method (*a*).

Method (*a*) was first employed in the belief that it was less open to possible error than the alkaline saponification. About 0·4 gram of the substance was digested with 30 c.c. of absolute alcohol and 2 c.c. of concentrated sulphuric acid. As soon as the liquid had evaporated to one-half its volume, fresh alcohol was added, and this was repeated three times. The suspended matter gradually passed into solution, until finally a clear liquid was obtained containing only traces of undissolved cuticle. If the operation is properly conducted, this liquid is almost colourless, but if the latter is allowed to evaporate too far, blackening occurs with evolution of sulphurous acid and vitiation of the analysis. The distillate was received in standard alcoholic potash and titrated with acid in the usual manner.

Acetic acid found :

I.	II.	III.	IV.
61·68	61·36	61·50	63·0 per cent.

Theory for triacetylcellulose, $C_6H_7O_2(OCOCH_3)_3$, = 62·5 per cent.

„   „ tetra-acetylcellulose, $C_6H_6O(OCOCH_3)_4$, = 72·72 per cent.

Another sample furnished us by Dr. F. Marsden gave 59·9 per cent. of acetic acid by this method.

To remove any doubt as to the applicability to carbohydrates of this method of estimating acetyl groups, we have applied it to the penta-acetyldextrose and penta-acetyl-lævulose prepared according to Erwig and Koenigs (Ber., 1889, 22, 1464, and 1890, 23, 672).

Penta-acetyldextrose gave 77·68 and 77·25 per cent. acetic acid.

Penta-acetyl-lævulose gave 77·76 per cent. acetic acid.

Theory for each requires 76·92 per cent. acetic acid.

That this method effects a complete saponification of the acetyl-cellulose can scarcely be doubted in view of the fact that at the end of the operation a clear solution is obtained. As, however, it has been suggested that possibly one of the acetyl groups may resist saponification and remain attached to the cellulose residue, we have considered it desirable to repeat the estimation using the method of alkaline saponification. For this purpose, 2 grams of the product were digested for two hours at the boiling point with 100 c.c. of $N/2$-alcoholic caustic soda. After allowing to settle, an aliquot portion of the supernatant liquid was titrated with standard acid.

Acetic acid found :

I.	II.	III.	IV.	V.
59·51	61·24	61·33	59·89	60·92 per cent.

In experiments I, II, and III, ethyl alcohol was used, in experiments IV and V methyl alcohol, the boiling being continued in the last experiment for four hours.

In a third series of determinations, the product obtained by boiling

2 grams of acetylcellulose with 100 c.c. of $N/2$-alcoholic caustic soda was diluted with alcohol and treated with 10 c.c. of concentrated sulphuric acid, making up the volume with alcohol to 250 c.c. in a stoppered cylinder. Then, after allowing the cellulose and precipitated sodium sulphate to settle, 50 c.c. of the clear solution were drawn off and distilled into standard alcoholic caustic soda, afterwards titrating with acid as usual.

Acetic acid found : 59·8 and 59·75 per cent.

Some slight loss would occur in this procedure, which accounts for the rather lower numbers.

From the above results, it appears to be ascertained beyond doubt that the product of the complete acetylation of cellulose (the so-called tetra-acetylcellulose) is in reality the triacetyl compound. It cannot be doubted that higher acetylated products are obtainable if condensing agents such as zinc chloride or sulphuric acid are present during acetylation, but such products are not true compounds of cellulose, but are derivatives of its hydration products. The highest acetate, therefore, like the highest nitrate and benzoate, corresponds to the presence of three hydroxyl groups in the cellulose molecule.

With reference to the objection which has been raised to Green's formula for cellulose on the ground that so simple a structure is at variance with the physical properties, which seem to indicate a body of high molecular weight, it may be pointed out that the suggested formula is only intended to represent cellulose in its simplest or unpolymerised form, in which, for instance, it may be supposed to exist in an ammoniacal cupric solution. The cellulose of fibres may be regarded either as a physical aggregate of these simple molecules, such as we constantly find occurring in the case of colloidal substances, or as a chemical polymer made up of a number of C_6 complexes of the above structure, united together by means of their oxygen atoms. The former hypothesis appears to us the more probable.

DEPARTMENT OF TINCTORIAL CHEMISTRY,
UNIVERSITY OF LEEDS.

LXXXV.—*The Constitution of Salicin. Synthesis of Pentamethyl Salicin.*

By JAMES COLQUHOUN IRVINE, Ph.D., D.Sc. (Carnegie Fellow), and ROBERT EVSTAFIEFF ROSE, Ph.D.

THE method generally adopted in ascertaining the structure of a glucoside is to study the products which it yields when hydrolysed, a change which is effected usually by treatment with mineral acids, by the action of enzymes, or, less frequently, by electrolysis. In cases, however, where the hydrolytic products undergo profound secondary changes when liberated, the results of such decompositions are often perplexing, but in only a few instances has the evidence obtained in this way been supplemented by a direct synthetical preparation of the glucoside itself. The complete structure of a glucoside is determined by (*a*) the identification of the sugar residue and the group with which it has condensed, (*b*) the recognition of the stereoisomeric form (a or *β*) represented by the compound, and (*c*) the proof of the internal linking of the sugar. The first of these factors is determined in the natural and artificial glucosides by hydrolysis and synthesis respectively ; the answer to the second inquiry is found in the selective hydrolytic action of enzymes, but evidence bearing on the linking of the sugar residue can only be obtained by the preparation and hydrolysis of derivatives of the glucoside. Alkylated glucosides are well adapted for such work, owing to the stability of the alkyloxy-groups, and thus the occurrence of secondary changes during hydrolysis is to a large extent precluded.

In previous work it has been shown (Trans., 1903, **83**, 1034) that when the crystalline tetramethyl glucose obtained from either a- or *β*-methylglucoside is oxidised, it is converted into tetramethyl gluconolactone, and this result supports the view that the parent glucosides possess the γ-oxidic linking. The production of the same alkylated glucose from completely methylated sucrose and maltose (Trans., 1905, **87**, 1028) shows that these compounds likewise contain the same internal linking as the artificial glucosides. We considered it advisable to apply similar reactions in order to establish the complete structural formula of a typical natural glucoside, and salicin was selected for experiment. The glucoside in question is one which has been very fully investigated. Its hydrolysis by emulsin shows the compound to be saligenin *β*-glucoside, and its synthetical formation from helicin (*Ber.*, 1881, **14**, 2097) proves that the two parts of the molecule are united through the phenolic hydroxyl group.

In the event of salicin possessing a similar structure to the artificial glucosides, it ought to yield on methylation a pentamethyl salicin, which, on hydrolysis, would be converted into the well-defined tetramethyl glucose already referred to, and the methyl ether of *o*-hydroxybenzyl alcohol. The salicin was alkylated as usual (Trans., 1904, **85**, 1058, and subsequent papers) by the action of silver oxide and methyl iodide, and the pentamethyl derivative was thus obtained in the form of colourless, slender needles (m. p. 62—64°), readily soluble in organic solvents, but practically insoluble in water. Unlike the other alkylated glucosides hitherto examined, the substance underwent decomposition when heated under diminished pressure and could not be purified by distillation. The specific rotation in methyl-alcoholic solution was $-52 \cdot 15°$, a value which does not differ much from the rotatory power of salicin in dilute ethyl alcohol ($[a]_D^{20°} - 50 \cdot 30°$). This result was not unexpected, as previous experience has shown that the methylation of the four hydroxyl groups of either a- or β-methyl glucoside does not affect the values of the optical rotations to any great extent.

The hydrolysis of the compound presented considerable difficulty. Owing, no doubt, to its extreme insolubility in water, emulsin was found to be without hydrolytic action on it, and even when warmed with dilute mineral acids the compound was converted into a resinous mass resembling saliretin, and only a trace of crystalline substance melting at 82—84°, which was doubtless tetramethyl glucose, could be isolated from such reactions. Considering the general stability of alkylated compounds, this behaviour seemed abnormal, but the resinification of salicin derivatives by means of dilute acids seems to extend to compounds in which the aromatic alcoholic group is substituted, as Piria states that even chlorosalicin (*Annalen*, 1845, **56**, 53) undergoes profound decomposition during hydrolysis.

The following synthetical method was therefore devised in order to prove that salicin possesses the γ-oxidic linking. Saligenin and tetramethyl glucose were condensed to saligenin tetramethyl glucoside by heating equimolecular proportions of these substances to 120° in benzene containing 0·25 per cent. of hydrochloric acid. In the course of three successive treatments, the specific rotation of the solution diminished greatly, and the syrupy product was found to behave like a glucoside towards Fehling's solution. In this reaction, the alkylated sugar may condense in two distinct ways with the saligenin, either through the phenolic or the primary alcoholic groups, and in each case both a- and β-modifications of the condensation products might exist.   These possibilities are included in the following alternative formulæ :

I. $CH_2(OH)\cdot C_6H_4\cdot O\cdot CH\cdot CH(OMe)\cdot CH(OMe)\cdot CH\cdot CH(OMe)\cdot CH_2(OMe)$.

$\underbrace{}_{\text{O}}$

α and β

II. $HO\cdot C_6H_4\cdot CH_2\cdot O\cdot CH\cdot CH(OMe)\cdot CH(OMe)\cdot CH\cdot CH(OMe)\cdot CH_2(OMe)$.

$\underbrace{}_{\text{O}}$

α and β

In addition, we found that the internal condensation of the sugar to octamethyl glucosido-glucoside occurred to some extent, as a small quantity of the alkylated disaccharide was isolated from the mixture, but the possible condensation of the saligenin through the primary alcohol group is almost excluded in view of the more acidic character of the phenolic hydroxyl.   We therefore consider that the product of the reaction was essentially a mixture of the α- and β-modifications of tetramethyl salicin (formula I), and the properties of the syrup were comparable with those ascribed by Moitessier (*Jahresber.*, 1866, 676) to tetraethyl salicin, which he obtained by treating the lead derivative of the glucoside with ethyl iodide.

The remaining hydroxyl group in the molecule was now alkylated by the silver oxide method and a compound identical in every respect with the crystalline pentamethyl salicin prepared directly from the glucoside was isolated from the mixture.

This result, besides establishing a new synthetical use for tetramethyl glucose, proves conclusively that salicin, like sucrose, is similar in constitution to methyl glucoside and possesses the γ-oxidic linking.   Moreover, the proof extends to the related glucosides helicin and populin, as these compounds have been prepared from salicin by reactions in which the glucosidic linking is undisturbed.

EXPERIMENTAL.

Alkylation of Salicin.

The methylation of the glucoside was carried out as already described for α-methyl glucoside (*loc. cit.*), methyl alcohol being used as solvent in the early stages of the alkylation.

Two grams of dry salicin were dissolved in 40 c.c. of methyl alcohol containing 30 grams of methyl iodide, and 10 grams of silver oxide were then added.   A moderate reaction occurred on warming, and at the end of each hour an additional quantity of 2 grams of salicin was added, together with 10 grams of silver oxide and 20 grams of methyl iodide.   In this way it was found that the alkylation of 10 grams of the sparingly soluble glucoside could be readily effected without the addition of more solvent alcohol. The total preportions used amounted to salicin (1 mol.), silver oxide (7 mols.), and methyl iodide (14 mols.). In all, 30 grams of the glucoside were treated in this way. The partially methylated product, which was isolated by removing the solvent alcohol from the filtered solution, presented the appearance of a viscid, colourless syrup; this was dried at 100° under diminished pressure. A further alkylation was then carried out, using the same proportions of silver oxide and methyl iodide, but in this case only a small quantity (10 c.c.) of methyl alcohol was necessary to effect complete solution. The third treatment with the alkylating mixture was conducted in acetone solution, after which the product was found to be soluble in methyl iodide, and two successive alkylations in this solvent were then found to be necessary to complete the methylation. The final product, which was extracted from the reaction mixture by means of boiling ether, proved to be a colourless oil practically insoluble in water, but readily soluble in organic solvents. When extracted with boiling light petroleum, the bulk of the syrup passed into solution, and the filtrate, on slow evaporation, deposited a crop of well-formed needles. After drying on a porous plate, and two recrystallisations from light petroleum, the product melted at 62—64°. Contrary to expectation, it was found impossible to purify the substance by distillation under diminished pressure, as at 140° the liquid was suddenly converted into a viscid, brown resin having a strong odour of saliretin. The compound gave a characteristic purple coloration with sulphuric acid.

0·1852 gave 0·4092 CO_2 and 0·1310 H_2O. C = 60·26 ; H = 7·86.

0·1978 ,, by Zeisel's method 0·6590 AgI. CH_3O = 43·55.

Pentamethyl salicin, $C_{13}H_{13}O_2(OCH_3)_5$, requires C = 60·67 ; H = 7·86 ; CH_3O = 43·55 per cent.

The specific rotation was determined in methyl-alcoholic solution, and gave :

$$C = 4·698, \ l = 2, \ a_D^{20°} - 4·90°, \text{ hence } [a]_D^{20°} - 52\ 15°.$$

Action of Hydrolytic Agents on Pentamethyl Salicin.

An attempt was made to hydrolyse the compound by heating with aqueous hydrochloric acid, but without success. When heated to 100° with 10 per cent. acid, the compound only partially dissolved, and was rapidly converted into a viscid oil, which solidified to a resin on cooling. During this change, the liquid became distinctly reactive to Fehling's solution, and the optical activity altered from laevo- to dextro-rotatory, but, nevertheless, the hydrolysis was incomplete. The solution, when filtered from the resinous by-products, was neutralised with barium carbonate and evaporated to dryness. On extracting the dry residue with ether, an oil was obtained which crystallised with difficulty. After drying on porous porcelain and crystallising from light petroleum, the crystals melted at 82—84°, but the quantity obtained was too small to permit of this product being further identified as tetramethyl glucose. The above process was modified by using different concentrations of various mineral acids at lower temperatures, but without improving the result, as, in every case, the alkylated salicin was resinified.

Emulsin was similarly found to be without action on the substance, probably owing to the slight solubility of the compound in water. The finely-powdered glucoside was added to a large excess of emulsin extract and maintained at 37° for ninety hours, the liquid being kept thoroughly mixed by a mechanical stirrer. The optical activity of the solution altered only very slightly, and the hydrolysis therefore did not proceed.

It was afterwards found that the hydrolysis of the alkylated salicin could only be effected in methyl-alcoholic solution, a change which is followed by the formation of tetramethyl methylglucoside.

Synthesis of Pentamethyl Salicin.

In order that the progress of this reaction might be accurately followed by means of polarimetric observations, the tetramethyl glucose used in our experiments was in the first place converted into the equilibrium mixture of permanent rotatory power. This was effected by dissolving the pure α-form of the sugar in benzene, and, after adding a trace of alcoholic ammonia, the solution was left for several days until the permanent value $[\alpha]_D^{20°} + 88.75°$ was recorded. The solvent was then evaporated off under diminished pressure at the ordinary temperature, and the crystalline residue dried, also under diminished pressure.

A preliminary experiment showed that the alkylated sugar underwent partial condensation with saligenin when these substances were dissolved in a large excess of benzene containing 0.25 per cent. of hydrochloric acid, the solution being maintained at 70° for several days. The rotatory power continually decreased, the liquid became turbid owing to the separation of water, and the action on Fehling's solution gradually diminished. The process was, however, extremely slow, and the following modification was finally adopted.

Equimolecular proportions of saligenin and tetramethyl glucose were dissolved in benzene containing 0.25 per cent. of hydrochloric acid, so as to give a 6 per cent. solution of the alkylated sugar, and the mixture was then heated in sealed tubes at 105° for six hours. On opening the tubes, it was evident that reaction had occurred. The solution had become straw-yellow, and was distinctly turbid owing to the separation of water, whilst in addition, a considerable quantity of saliretin had separated. The filtered solution, when clarified by standing over calcium chloride, showed a considerable decrease in rotatory power ($[a]_D + 70 \cdot 1°$), the initial value being $+ 82 \cdot 5°$. The hydrochloric acid was removed from the solution by shaking with barium carbonate, and, on evaporating the filtered liquid to dryness, a yellow oil remained which was practically insoluble in water and which did not crystallise even when a trace of tetramethyl glucose was added. As, however, the substance still reduced Fehling's solution vigorously, the treatment with the acid benzene was repeated. The total product was accordingly dissolved in benzene containing 0·25 per cent. of hydrochloric acid, the molecular proportion of saligenin being again added to compensate for the unavoidable loss of this substance in the form of saliretin. After heating for six hours at 120°, the solution was again found to be turbid, and in addition the specific rotation, calculated on the weight of sugar used, had diminished to $+ 36 \cdot 6°$. Although the product now reduced Fehling's solution very slightly and gave no colour reaction with ferric chloride, it was redissolved in the same excess of the acid benzene, a molecular proportion of saligenin again added, and the solution heated for a further period of ten hours at 110°. Very little water separated during this third treatment, and the specific rotation of the clarified liquid had only diminished to $+ 31 \cdot 7°$.

The product, when isolated as already described, was a pale yellow syrup, slightly soluble in water, and had no action on Fehling's solution. The substance was doubtless a complex mixture, containing, as explained in the introduction, the three forms of octamethyl glucosido-glucoside in addition to the four isomeric saligenin tetramethyl glucosides. Owing to the ready solubility of the former compounds in water, it was found possible to effect a separation by extracting the oil repeatedly with hot water.

Examination of the Aqueous Extract.

On evaporation to dryness over a water-bath, a clear, viscid syrup remained, which behaved like a glucoside towards Fehling's solution. After drying in a vacuum desiccator, the substance showed, in methyl-alcoholic solution, a specific rotation of approximately $+ 116°$. Considering the nature of the purification, this value supports the view that the syrup was essentially a mixture of the three isomeric forms of octamethyl glucosido-glucoside, ($[a]_D + 135 \cdot 9°$), which would certainly be produced under the conditions of the condensation. This supposition was confirmed by a study of the hydrolysis of the compound, which was carried out by the action of 12 per cent. hydrochloric acid at 100°. During this treatment the specific rotation decreased continuously, and the sole product of the reaction was tetramethyl glucose, which after recrystallisation melted at 84—86°.

Examination of the Oil Insoluble in Water.

The oil left undissolved by water was extracted with ether and the solution dried with anhydrous sodium sulphate. On removing the solvent, a straw-yellow syrup remained, which had no action on Fehling's solution, and which gave a brown coloration with ferric chloride. When boiled with aqueous or alcoholic hydrochloric acid, a resin resembling saliretin was produced, and the substance was further characterised by giving a brilliant violet coloration with sulphuric acid. As the compound could not be made to crystallise and underwent decomposition when distilled under diminished pressure, it could not be obtained in a pure state, but the above characteristics correspond with the probable reactions of the different modifications of saligenin tetramethyl glucoside, all of which were doubtless present.

Methylation of the Insoluble Oil.

The insoluble syrup previously described was alkylated by the joint action of methyl iodide (4 mols.) and silver oxide (2 mols.). As the oil did not dissolve completely in the alkyl iodide, a little dry acetone was added to effect solution and thus prevent oxidation. After eight hours' treatment the product was extracted with boiling ether and then worked up as usual. The syrup obtained in this way, after drying at 100° under diminished pressure, was subjected to a second treatment with the same proportions of the alkylating mixture, no acetone, however, being necessary in this instance. The product was a viscid syrup, which crystallised partially after standing for some weeks in a vacuum desiccator. The crystals, when drained on a porous plate and recrystallised from light petroleum, melted at 60—62°. The yield obtained in this way was very small, and consequently the oil absorbed by the tile was recovered and extracted repeatedly with boiling light petroleum. The extract, on slow evaporation, deposited a crop of well-formed needles, which, after drying on porous porcelain and crystallising twice from the solvent previously used, melted sharply at 62—63°. The yield of crystalline product varied considerably in different preparations, and never amounted to more than 30 per cent. of the weight of tetramethyl glucose originally used in the preparation.

0·2241 gave 0·4973 CO_2 and 0·1610 H_2O. $C = 60·52$; $H = 7·98$.
0·1966 ,, by Zeisel's method 0·6228 AgI. $CH_3O = 41·82$.
$C_{13}H_{13}O_2(OCH_3)_5$ requires $C = 60·67$; $H = 7·86$; $CH_3O = 43·55$ per cent.

The identity of the compound with the pentamethyl salicin prepared by the direct method was further established by a determination of the specific rotation in methyl-alcoholic solution.

$C = 5·1400$, $l = 1$, $a_D^{20°} - 2·70°$, $[a]_D^{20°} - 52·53°$.

The constants previously determined for this compound were : m. p. 62—64°, $[a]_D^{20°} - 52·12°$.

Moreover, the product of the synthetical preparation, when boiled with aqueous hydrochloric acid, was also converted into an uncrystallisable resin, and, in addition, it gave the same violet coloration with sulphuric acid.

Simultaneous Hydrolysis and Condensation of Pentamethyl Salicin.

As already explained, the hydrolysis of pentamethyl salicin cannot be effected by treatment with dilute mineral acids, even at moderate temperatures, and hence the only proof of the linking of the sugar residue in the molecule lies in its synthetical formation from tetramethyl glucose. An experiment was subsequently undertaken, in which we endeavoured to effect the joint hydrolysis of the compound and the condensation of the alkylated sugar thus liberated with methyl alcohol, a reaction which would thus be analogous to Fischer's preparation of α-methyl glucoside from starch (*Ber.*, 1895, 28, 1151).

This change ought to yield, in the event of the sugar residue possessing the γ-oxidic linking, a mixture of the α- and β-forms of tetramethyl methylglucoside. The complete reaction is expressed in the following scheme :

Pentamethyl salicin $\nearrow$ o-hydroxybenzyl ether.
$\searrow$ β-tetramethyl glucose → α- and β-tetramethyl methyl-glucosides.

A five per cent. solution of the crystalline pentamethyl salicin (prepared from salicin) was made in methyl alcohol containing 0·25 per cent. of hydrochloric acid, and the liquid heated in the first instance at 50° in a thermostat. The specific rotation, which originally was − 52·6°, decreased steadily. After forty hours' treatment the solution was inactive, and subsequently a continually increasing dextro-rotation was recorded. At the end of ninety hours' treatment the specific rotation was + 25·3°, but examination of the solution showed that it still contained a quantity of an oil insoluble in water, and showing a lævo-rotation in alcoholic solution, so that the hydrolysis was still incomplete.

The solution was accordingly heated for ten hours at 110° to accelerate the reaction, during which process the specific rotation, calculated on the weight of pentamethyl salicin taken, increased to $+81\cdot4°$. As the hydrolysis and condensation were afterwards shown to be complete at this stage, the above value can be recalculated for the concentration of the tetramethyl methylglucoside produced in the reaction. The corrected specific rotation was $[\alpha]_D + 106\cdot5°$, and this agrees exactly with the number found ($[\alpha]_D + 106\cdot6°$) for the equilibrium mixture of α- and β-tetramethyl methylglucosides in the same solvent (Trans., 1905, **87**, 906). The joint hydrolysis and condensation was therefore complete and quantitative. After removing the hydrochloric acid, the solution was evaporated to dryness and the residual oil extracted repeatedly with hot water. The undissolved residue, which contained the aromatic product of the reaction, was quite inactive when examined in alcoholic solution, and was not further investigated. The aqueous extract gave a glucosidic syrup on evaporation, which was hydrolysed as usual by heating with 8 per cent. aqueous hydrochloric acid at 100°. The usual rise and fall of rotation was observed during the hydrolysis, and practically a quantitative yield of tetramethyl glucose melting, after recrystallisation, at 84—86° was obtained. This result affords conclusive evidence of the linking of the sugar residue in alkylated salicin, and therefore also in the parent glucoside.

We take this opportunity of expressing our thanks to Professor Purdie for his interest in our work, and valuable advice, and also to the Executive Committee of the Carnegie Trust for a research grant which defrayed the entire expense of the investigation.

CHEMICAL RESEARCH LABORATORY,
UNITED COLLEGE OF ST. SALVATOR AND ST. LEONARD,
ST. ANDREWS UNIVERSITY.

LXXXVI.—*Reciprocal Displacement of Acids in Heterogeneous Systems.*[*]

By ALFRED FRANCIS JOSEPH, A.R.C.S., B.Sc.

WHEN a salt MA in aqueous solution is evaporated in the presence of an acid HB, the amount of new salt MB which is formed depends on the strengths and volatilities of the acids HA and HB, and also on the relative solubilities of the two salts MA and MB. This last factor —relative solubility—is in many cases of the first importance. Thus, if a fairly soluble silver salt of a slightly volatile acid, such as silver sulphate, be evaporated with hydrochloric acid, a theoretical yield of silver chloride will usually be obtained. Here the relative solubility silver chloride : silver sulphate is very small, and this factor in the transformation overcomes any tendency of the more volatile hydrochloric acid to be expelled.

In a case of this kind, the theoretical quantity of acid is sufficient to transform the whole of the metal from one salt to the other, although sulphuric is far less volatile than hydrochloric acid.

If, however, the solubilities of the two salts concerned be not too different, one acid will displace the other in a varying degree. Prescott (*Chem. News*, 1877, 179) examined the action of a fixed amount of hydrochloric acid on one gram of many different sulphates, the mixtures being evaporated to dryness on the water-bath in each case.

He found, for example, that 100 per cent. of the silver in silver sulphate was transformed into chloride, whereas only 19·3 per cent. of the sodium, and 0·7 per cent. of the potassium, in the corresponding sulphates, were so converted. In the case of calcium, strontium, and barium sulphates, no appreciable transformation into chloride took place.

Prescott's results, however, are not easy to compare, as the proportion of acid he used remained the same although the equivalent weights of the sulphates experimented on varied from about 250 to less than 100 ; the excess of acid would, therefore, also vary to this extent.

Now this excess of acid is of great importance in determining how far transformation will take place. It is well known, for instance, that a nitrate may, by repeated evaporation with hydrochloric acid, be wholly transformed into chloride : a single evaporation with the equivalent quantity of acid being not nearly sufficient, unless the relative solubility of nitrate to chloride be very great.

[*] The title of this paper has been altered from " The Displacement of Acid Ions. Part I." Compare *Proc.*, 1905, **22**, 82.

By taking a considerable excess of hydrochloric acid, much more nitrate may be transformed in a single operation, but experiments tend to show that the process will never be complete.

If, then, a salt MA in aqueous solution is evaporated with the acid HB, the proportion of metal remaining as the new salt depends on :

(a) The amount of acid HB used ;
(b) The relative solubility of the two salts MA and MB ;
(c) The relative volatility of the two acids HA and HB ;
(d) The relative strengths of the two acids.

In the present communication, the first two of these factors are taken into consideration. The acids used were nitric and hydrochloric, and the salts those of potassium, sodium, and strontium.

Method of Experimenting.—In the majority of cases, the following simple method was used :

Normal solutions of the acid and salt were prepared. A measured quantity (usually 10 c.c.) of the salt solution was mixed with the required quantity of the acid solution in a porcelain dish and the mixture evaporated to dryness on the water-bath. The residue was treated in one or more of the following ways : (a) Heated on the water-bath for some time (at least half an hour). (b) A few c.c. of water added, and the solution again evaporated on the water-bath. (c) Heated in an air-bath to about 150° for ten minutes.

The results were apparently independent of the final method used. Thus, in one series of experiments, equivalent quantities of potassium nitrate and hydrochloric solutions being taken :

(a) The residue was heated for half an hour on the water-bath.

(i) 0·275
(ii) 0·290 } of the nitrate was transformed into chloride.
(iii) 0·283

(b) The residue was redissolved in about 5 c.c. of water and again evaporated to dryness on the water-bath.

(i) 0·264
(ii) 0·278 } of the nitrate was transformed.

(c) The residue was heated in the air-bath (about 150°) for ten minutes.

(i) 0·280
(ii) 0·276 } of the nitrate was transformed.
(iii) 0·270

It is therefore evident that this final treatment is of no practical importance in determining the result.

The residue was always neutral, and at the conclusion of the experiment was redissolved and the amount of chloride estimated by titration with $N/10$ silver nitrate solution.

In nearly all cases, at least six similar estimations were made, and the mean of all used in plotting curves. These estimations differed amongst themselves far more than was to be expected when the experimental errors of measurement and titration were taken into consideration. It is thought that this is due to the deposition of salt during evaporation, a circumstance that would hardly be likely to take place with mathematical regularity. Suppose, for example, that a solution of potassium nitrate is being evaporated with hydrochloric acid; at a certain stage, potassium chloride will be deposited. But before this stage, some nitrate may have separated, and so removed from the sphere of action. By taking the mean of a number of estimations, it is hoped that these errors have been considerably reduced.

As previously stated, the evaporations were conducted in open dishes. This method was preferred to others, such as bubbling a current of dry air through the solution, on account of the simplicity of the former, and the fact that it is the method always employed in analytical operations. But more important is the fact that there is practically no possibility of reflux distillation of the volatilised acid when using an open dish, a condition difficult to obtain when air bubbling, or in fact, any other method is used.

When heated over an actively boiling water-bath, the temperature of the mixture was always about $75°$. This was determined by a thermometer when the depth of liquid was sufficiently great, but when the quantity of liquid had diminished to one or two c.c. a few measurements were made calorimetrically. The results showed that the temperature remained practically the same.

For evaporation at the ordinary temperature, the dish was placed over lime in a desiccator, which was then exhausted. When the residue appeared quite dry, it was heated in the air-bath to remove traces of acid.

In what follows, x denotes the number of equivalents of acid taken to one equivalent of salt, and y represents the fraction of the metal transformed from one salt to the other. Thus, in an experiment, 10 c.c. of a normal solution of potassium nitrate were evaporated with 20 c.c. of normal hydrochloric acid, so that $x = 2$; the residue required 42 c.c. $N/10$ silver nitrate on titration, and therefore $y = 0.42$. On graphically representing the values of y obtained with different values of x, it was found that the curves were all convex to the x axis. After trying various empirical relations, it was found that y plotted against log x gave an approximately straight line, which, however, became curved convex to the axis x when y had reached a certain limit (see curves ; common logarithms are plotted). Disregarding this curvature for the time being, it appears that y is a linear function of x

or $x = ae^{by}$ (a and b being constants).

This equation might theoretically be developed as follows :

Suppose x equivalents of hydrochloric acid acting on one equivalent of potassium nitrate in aqueous solution leave on evaporation y equivalents of potassium chloride, then, if it is desired to repeat the operation in order to obtain dy more potassium chloride, the additional quantity dx of hydrochloric acid required will depend (1) on the quantity of

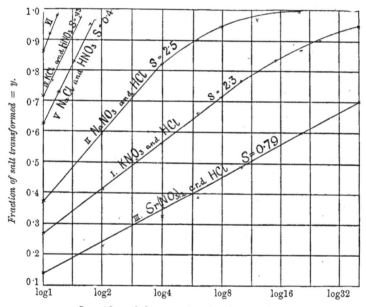

Logarithms of the proportion of acid used $= \log_{10} x.$

S is the ratio of the solubilities of the original and produced salts.

potassium nitrate dy to be changed, and also (2) on the quantity of hydrochloric acid used before ; for if x be large, a larger quantity dx will be required than if x be small. In fact, dy depends on the relative increase of dx. So that dx varies as $x\,dy$, $\dfrac{dy}{dx} = \dfrac{k}{x}$, that is, $y = k\log x + c$, which is the observed form of the relation.

In each case the equation to the line has been deduced from two points on it, and the values of y then calculated by its use for all values of x. These calculated values are shown by the side of the observed ones. The differences are small until the line begins to curve.

It is desired to draw attention to one other point : if x be made less than 1, that is, if 20 c.c. of potassium nitrate solution are evaporated with 10 c.c. of hydrochloric acid, the quantity of potassium chloride formed is the same as if 10 c.c. of potassium nitrate are used. This relation is general. Thus :

No. of c.c. of N-potassium nitrate.	No. of c.c. of N-hydrochloric acid.	Fraction of hydrochloric acid converted into chloride.
10	10	0·273
20	10	0·272
30	10	0·276

This shows that excess of potassium nitrate has no appreciable effect on the action.

No appreciable differences have been observed when the initial concentrations are altered. If, for example, instead of N-hydrochloric acid, the equivalent quantity of $4N$-acid is used, the value of y remains unchanged. Thus, $x = 1$, with N-potassium · nitrate and N-hydrochloric acid, $y = 0·271$, 0·266, 0·264, and 0·275 ; with N-potassium nitrate and $4N$-hydrochloric acid, $y = 0·261$, 0·267, and 0·272.

$x = 4$, with N-potassium nitrate and N-hydrochloric acid, $y = 0·556$, 0·566, and 0·580 ; with N-potassium nitrate and $4N$-hydrochloric acid, $y = 0·578$, 0·576, and 0·582. And if the solution is initially very weak, the result is still the same. For example, 10 c.c. of N-potassium nitrate solution were mixed with 10 c.c. of N-hydrochloric acid and 180 c.c. of water and evaporated : $y = 0·265$, 0·270, 0·268, 0·285, 0·274, and 0·276, this result being practically the same as the foregoing.

The initial concentration has therefore little or no influence on the final result, at any rate within the limits of the above experiments.

I. *Action of Hydrochloric Acid on Potassium Nitrate Solution.*

*Solubility of potassium nitrate at $75° = 1\cdot53$; of potassium chloride, $0\cdot66$.

Solubility ratio : potassium nitrate : potassium chloride $= 2\cdot3$.

Equation to the line, $y = 0\cdot5\log_{10}x + 0\cdot273$.

x.	No. of experiments.	y.			y calculated from above equation.
		Greatest.	Least.	Mean.	
1	11	0·283	0·263	0·273	0·273
2	5	0·431	0·420	0·424	0·423
3	6	0·529	0·512	0·521	0·512
4	12	0·592	0·548	0·573	0·574
6	6	0·680	0·664	0·672	0·662
8	6	0·740	0·712	0·726	0·725
10	4	0·778	0·756	0·768	0·773
15	3	0·860	0·820	0·835	0·861
20	6	0·882	0·853	0·865	0·923
25	3	0·910	0·890	0·895	0·972
40	6	0·946	0·934	0·942	1·074 *
100	6	0·995	0·965	0·970	1·273 *

* These numbers are of no significance, as y can never be greater than 1.

It will thus be seen that in order to transform half the potassium nitrate into potassium chloride, nearly three equivalents of hydrochloric acid are needed ; whilst to transform three-fourths of it, nearly ten are needed, but the transformation is apparently never complete. If the equation above held good throughout, $29\cdot5$ equivalents of hydrochloric acid ought to transform the whole of the potassium nitrate ; actually only about $0\cdot9$ of it is thereby transformed.

* All solubilities referred to in this paper are given as gram equivalents of salt dissolved by 100 grams of water. They are based on data obtained from Comey's *Dictionary of Solubilities.*

II. Action of Hydrochloric Acid on Sodium Nitrate Solution.

Solubility of sodium nitrate at $75° = 1·64$; of sodium chloride, $0·65$. Solubility ratio sodium nitrate : sodium chloride $= 2·52$.

Equation to the line, $y = 0·74\log_{10}x + 0·373$.

x.	No. of experiments.	y.			y calculated from above equation.
		Greatest.	Least.	Mean.	
1	6	0·380	0·369	0·373	0·373
2	5	0·615	0·591	0·599	0·596
3	6	0·742	0·714	0·733	0·726
4	6	0·834	0·800	0·818	0·818
6	6	0·906	0·894	0·900	0·949
8	6	0·948	0·914	0·938	1·041 *
12	5	0·968	0·960	0·961	1·171 *
20	4	1·000	0·984	0·991	1·335 *

* See footnote to previous table.

Here, as before, the agreement between the observed and the equation values is not good after $0·8$ of the sodium nitrate has been transformed. It will be noticed that this line is steeper and starts from a higher point than the former. This is to be expected from the fact that the solubility ratio ($2·52$) is greater than in the former case ($2·3$).

III. Action of Hydrochloric Acid on Strontium Nitrate Solution.

Solubility of strontium nitrate at $75° = 0·9$; of strontium chloride, $1·14$.

Solubility ratio strontium nitrate : strontium chloride $= 0·79$.

Equation to the line, $y = 0·35\log_{10}x + 0·139$.

x.	No. of experiments.	y.			y calculated from above equation.
		Greatest.	Least.	Mean.	
1	6	0·149	0·130	0·139	0·139
2	6	0·214	0·194	0·207	0·244
4	6	0·310	0·300	0·308	0·319
6	5	0·378	0·368	0·374	0·401
10	6	0·490	0·472	0·480	0·489
20	4	0·598	0·586	0·592	0·594
40	4	0·720	0·710	0·716	0·699

The agreement to the logarithmic line is not good for the second three points, the others agree closely. The line is less steep and starts from a lower point than either of the others, the solubility ratio nitrate : chloride being less.

IV. Action of Nitric Acid on Potassium Chloride Solution.

Solubility ratio potassium chloride : potassium nitrate = 0·43.
Equation to line, $y = 1\cdot5\log_{10}x + 0\cdot71$.

x.	No. of experiments.	y.			y calculated from above equation.
		Greatest.	Least.	Mean.	
1·0	9	0·710	0·396	0·706	0·710
1·1	6	0·767	0·752	0·757	0·772
1·2	6	0·835	0·813	0·824	0·828
1·3	6	0·884	0·874	0·878	0·870
1·4	6	0·936	0·916	0·932	0·929
1·5	6	0·980	0·967	0·975	0·974

This transformation is brought about much more easily than the former ; two equivalents of nitric acid transform the whole of the chloride into nitrate.

V. Action of Nitric Acid on Sodium Chloride Solution.

Solubility ratio sodium chloride : sodium nitrate = 0·4.
Equation to the line, $y = 1\cdot3\log_{10}x + 0\cdot63$.

x.	No. of experiments.	y.			y calculated from above equation.
		Greatest.	Least.	Mean.	
1·0	6	0·628	0·619	0·624	0·630
1·25	6	0·756	0·738	0·747	0·756
1·5	5	0·857	0·848	0·852	0·852
1·75	5	0·958	0·949	0·954	0·946

VI. *Action of Nitric Acid on Strontium Chloride Solution.*

Solubility ratio strontium chloride : strontium nitrate $= 1\cdot27$.
Equation to the line, $y = 1\cdot6\log_{10}x + 0\cdot862$.

x.	No. of experiments.	y.			y calculated from above equation.
		Greatest.	Least.	Mean.	
$1\cdot0$	5	$0\cdot868$	$0\cdot856$	$0\cdot862$	—
$1\cdot1$	5	$0\cdot928$	$0\cdot921$	$0\cdot925$	—
$1\cdot2$	3	$0\cdot986$	$0\cdot982$	$0\cdot984$	—

The line rises so rapidly that only three points could be obtained even in the narrow limits above. The middle value is practically the mean of the others (logarithms of quantities so near to 1 are proportional to their numbers).

If the six equations so obtained be considered together, it. is seen that the constants in each have nearly the ratio 2 : 1. If the equation be $y = a\log x + b$,

I. $y = 0\cdot5\log x \quad + 0\cdot273$. $a : b = 1\cdot8$.

II. $y = 0\cdot74\log x \quad + 0\cdot373$. $a : b = 2\cdot0$.

III. $y = 0\cdot35\log x \quad + 0\cdot139$. $a : b = 2\cdot5$.

IV. $y = 1\cdot5\log x \quad + 0\cdot71$. $a : b = 2\cdot1$.

V. $y = 1\cdot3\log x \quad + 0\cdot63$. $a : b = 2\cdot1$.

VI. $y = 1\cdot6\log x \quad + 0\cdot86$. $a : b = 1\cdot9$.

That is, the six equations can all be approximately represented by

$$y = b(2\log_{10}x + 1) \text{ or } y = b\log_{10}x^2,$$

which only contains one constant dependent on the nature of the metallic radicle.

As was pointed out before, the constant b, that is, the quantity of salt changed when equivalent quantities of acid and salt are evaporated ($\log x$ being here $= 0$), increases regularly with the solubility ratio for a given acid acting on a given class of salts.

Thus, in the case of hydrochloric acid acting on nitrates :

For the sodium salts, solubility ratio nitrate : chloride $= 2\cdot52$, $b = 0\cdot373$

„ potassium .. .. „ $= 2\cdot3$, $b = 0\cdot273$

„ strontium „ . „ $= 0\cdot79$, $b = 0\cdot139$

In the case of nitric acid acting on the chlorides, there is a corresponding relation.

This general relation is shown by an examination of the six curves. They may be divided into two groups: Nos. I, II, and III illustrating the displacing power of hydrochloric acid, and Nos. IV, V, and VI that of nitric acid. In each group, the positions are defined by the solubility ratio of the two salts concerned.

Making use of these empirical relations, it becomes possible to calculate roughly the proportion of a nitrate (or chloride) which will be transformed when evaporated with a given quantity of hydrochloric (or nitric) acid. The solubility ratio must be given.

As an example, we may take the case of potassium nitrate and hydrochloric acid, the evaporation being conducted in the cold. At 13°, the solubility ratio potassium nitrate : potassium chloride $= 0.53$. Now, as was pointed out, the solubility ratio strontium nitrate : strontium chloride $= 0.79$ and $b = 0.139$. But when the solubility ratio is 0 (that is, the original salt is insoluble), then b (the amount of new salt produced) is also 0, and by interpolating between these values it is found that with solubility ratio 0.53, b should be 0.09; the constant a should therefore be about $2 \times 0.09 = 0.18$, and the equation: $y = 0.18 \log x + 0.09$.

To test this, evaporations were made at the ordinary laboratory (winter) temperature under diminished pressure over lime. With equivalent quantities of nitrate and acid, 0.085 was transformed; with four equivalents of acid, 0.18 was transformed.

From the above equation, the values of y should be 0.09 and 0.198 respectively, which are not very different from the experimental numbers.

These experiments will be extended to other acids, with a view to determining the importance of the volatility of the acids concerned in the transformation.

The author wishes to express his best thanks to Dr. J. C. Philip for the very valuable advice he has given in connection with this work.

LXXXVII.—*Experiments on the Synthesis of the Terpenes. Part VII. A Synthesis of Tertiary Menthol (p-Menthanol-4) and of Inactive Menthene (Δ^3-p-Menthene).*

By WILLIAM HENRY PERKIN, jun.

Menthene was discovered in the year 1839 by Walter (*Annalen*, 32, 289), who obtained it from menthol by heating with sulphuric acid or phosphorus pentoxide. It is readily prepared, in a yield of 80 per cent., by heating menthol with crystallised oxalic acid (Zelinsky and Zelikoff, *Ber.*, 1901, **34**, 3253), and is described as a colourless oil which distils at 167° and has a faint odour of peppermint.

Menthene has been the subject of repeated investigation, but its constitution was first proved by Wagner, who showed that when oxidised by potassium permanganate it yields the following degradation products :

$$Me \cdot CH \!\!<\!\!\begin{array}{c}CH_2-CH\\CH_2 \cdot CH_2\end{array}\!\!>\!\!C \cdot CHMe_2$$
Δ^3-p-Menthene.

$$Me \cdot CH \!\!<\!\!\begin{array}{c}CH_2 \cdot CH(OH)\\CH_2——CH_2\end{array}\!\!>\!\!C(OH) \cdot CHMe_2$$
Menthandiol (3 : 4).

$$Me \cdot CH \!\!<\!\!\begin{array}{c}CH_2-CO\\CH_2 \cdot CH_2\end{array}\!\!>\!\!C(OH) \cdot CHMe_2 \qquad Me \cdot CH \!\!<\!\!\begin{array}{c}CH_2 \cdot CO_2H\\CH_2 \cdot CH_2 \cdot CO \cdot CHMe_2\end{array}$$
Menthanon(3)-ol(4).   Oxymenthylic acid.

$$Me \cdot CH \!\!<\!\!\begin{array}{c}CH_2 \cdot CO_2H\\CH_2 \cdot CH_2 \cdot CO_2H\end{array}$$
β-Methyladipic acid.

Tertiary menthol was first prepared by Baeyer (*Ber.*, 1893, **26**, 2270 and 2560) by an interesting series of reactions which will be readily understood with the aid of the following formulæ :

$$Me \cdot CH \!\!<\!\!\begin{array}{c}CH_2 \cdot CH(OH)\\CH_2——CH_2\end{array}\!\!>\!\!CH \cdot CHMe_2$$
Menthol.

$$Me \cdot CH \!\!<\!\!\begin{array}{c}CH_2 \cdot CHI\\CH_2-CH_2\end{array}\!\!>\!\!CH \cdot CHMe_2 \qquad Me \cdot CH \!\!<\!\!\begin{array}{c}CH_2-CH\\CH_2 \cdot CH_2\end{array}\!\!>\!\!C \cdot CHMe_2$$
Menthyliodide. Δ^3-p-Menthene.

$$Me \cdot CH \!\!<\!\!\begin{array}{c}CH_2 \cdot CH_2\\CH_2 \cdot CH_2\end{array}\!\!>\!\!CI \cdot CHMe_2 \qquad Me \cdot C_H \!\!<\!\!\begin{array}{c}CH_2 \cdot CH_2\\CH_2 \cdot CH_2\end{array}\!\!>\!\!C(OH) \cdot CHMe_2$$
Tertiary menthyliodide.   Tertiary menthol.

By acting on tertiary menthol with hydrobromic acid, Baeyer obtained tertiary menthyl bromide, and showed that when this bromide is digested with quinoline it yields menthene, and since tertiary menthol is necessarily inactive, it follows that the menthene obtained from it must also be the inactive modification.

Menthene has always taken a leading part in the historical development of the chemistry of the terpenes, and is undoubtedly the most important of the tetrahydro-derivatives of cymene; for this reason it was thought interesting to attempt its synthesis, and ultimately the problem was solved by means of the following series of reactions.

Hexahydro-p-toluic acid is converted into *a-bromohexahydro p-toluic acid* by the action of phosphorus pentachloride and bromine, and the bromo-acid is then treated, in the cold, with caustic potash, when it is readily converted into a mixture of a-*hydroxyhexahydro*-p-*toluic acid* and Δ¹-*tetrahydro*-p-*toluic acid* :

$$Me\cdot CH{<}^{CH_2\cdot CH_2}_{CH_2\cdot CH_2}{>}CBr\cdot CO_2H \quad \longrightarrow$$

$$Me\cdot CH{<}^{CH_2\cdot CH_2}_{CH_2\cdot CH_2}{>}C(OH)\cdot CO_2H \text{ and } Me\cdot CH{<}^{CH_2-CH}_{CH_2\cdot CH_2}{>}C\cdot CO_2H.$$

The hydroxy-acid is readily attacked by cold sulphuric acid with evolution of carbonic oxide and formation of p-*methylcyclohexanone*,

$$Me\cdot CH{<}^{CH_2\cdot CH_2}_{CH_2\cdot CH_2}{>}CO,$$

a colourless, pleasant-smelling oil which distils at 170°, and has already been described by Sabatier and Mailhe (p. 783).

When this ketone is added to an ethereal solution of magnesium *iso*propyl iodide, it is at once converted into *tertiary menthol*,

$$Me\cdot CH{<}^{CH_2\cdot CH_2}_{CH_2\cdot CH_2}{>}C(OH)\cdot CHMe_2.$$

The alcohol obtained in this way had a pleasant odour of peppermint, distilled at 97° (25 mm.), and did not solidify in a freezing mixture, and further comparison showed that its properties agreed exactly with those of the tertiary menthol which Baeyer had obtained from menthol by the process mentioned above. Finally, a quantity of synthetical tertiary menthol was digested with potassium hydrogen sulphate, when it was converted, by elimination of water, into *inactive menthene*,

$$Me\cdot CH{<}^{CH_2-CH}_{CH_2\cdot CH_2}{>}C\cdot CHMe_2.$$

The synthetical hydrocarbon distilled at 168° and yielded the *nitrosochloride* of melting point 128°, which is so characteristic of inactive menthene.

a-*Bromohexahydro*-p-*toluic Acid.*

The method employed in the preparation of this acid is that described in detail in Part II of these researches (Perkin and Pickles, Trans., 1905, **87**, 644), and consists in treating hexahydro-*p*-toluic acid first with phosphorus pentachloride and then with bromine at 100°. The crude bromo-acid is dissolved in boiling light petroleum (b. p. 50—60°), and the solution, after shaking with animal charcoal, filtered and cooled in ice-water, when, especially if the solution is vigorously stirred with a glass rod, the bromo-acid soon crystallises in groups of stars. After two further crystallisations from the same solvent, the acid was analysed with the following results :

0·1988 gave 0·1699 AgBr.   Br = 36·3.

$C_8H_{13}O_2Br$ requires Br = 36·2 per cent.

a-*Bromohexahydro*-p-*toluic acid* melts at 109° and is readily soluble in ether, benzene, alcohol, or chloroform, but rather sparingly so in light petroleum or formic acid in the cold. It crystallises from formic acid in striated leaflets which resemble crystals of benzoic acid.

The bromination of hexahydro-p-toluic acid was first investigated by Einhorn and Willstätter (*Annalen*, 1894, 280, 161), who obtained an a-bromohexahydro-p-toluic acid which, after three crystallisations from formic acid, melted at 71—72°. The considerable discrepancy in melting point may possibly be due to stereoisomerism, that is, the substance obtained by Einhorn and Willstätter may be the *cis*-, and that described in the present paper the *trans*-modification of a-bromohexahydro-p-toluic acid. In order to obtain evidence on this point, the mother liquors of the acid of melting point 109° were very carefully examined. The light petroleum was allowed to evaporate at the ordinary temperature, and the solid residue, which contained only traces of oil, was fractionally crystallised from formic acid.

In this way considerable quantities of the acid of melting point 109° were isolated, but, although there was evidence of the existence of a second acid, it was present in such small quantities that it was impossible to obtain it in a pure state.

a-*Hydroxyhexahydro*-p-*toluic Acid.*

When a-bromohexahydro-p-toluic acid is ground to a paste with water and mixed with an excess of dilute caustic soda, it gradually dissolves, and after standing for three days at the ordinary temperature, and being subsequently heated for half an hour on the water-bath, it will have been completely decomposed with formation of a mixture of the sodium salts of a-hydroxyhexahydro-p-toluic acid and Δ^1-tetrahydro-p-toluic acid. After cooling and acidifying, the solution deposits a precipitate which is sometimes so finely divided that it is difficult to collect, but if the whole is heated for half an hour on the water-bath and then allowed to remain for twenty-four hours this difficulty is overcome. The precipitate is washed with water and crystallised from light petroleum, when colourless prisms are obtained, which melt at 134° and consist of Δ^1-tetrahydro-p-toluic acid (compare Perkin and Pickles, Trans., 1905, 87, 646).

0·1359 gave 0·3408 CO_2 and 0·1047 H_2O.   C = 68·4 ; H = 8·5.

$C_8H_{12}O_2$ requires C = 68·6 ; H = 8·6 per cent.

The aqueous filtrate from this acid was extracted five times with ether, the ethereal solution evaporated, and the residual syrupy mass distilled in steam until the tetrahydro-*p*-toluic acid which it contained had been completely removed. The residue was again extracted with ether and the dried ethereal solution evaporated, when a thick syrup remained which soon solidified, and, after crystallising twice from a mixture of benzene and light petroleum and three times from water, the new hydroxy-acid was obtained pure in the form of pearly plates.

0.2078 gave 0.4632 CO_2 and 0.1668 H_2O.   C = 60.8; H = 8.9.

0.1664 ,,   0.3703 CO_2 ,, 0.1336 H_2O.   C = 60.7; H = 8.9.

$C_8H_{14}O_3$ requires C = 60.8; H = 8.8 per cent.

α-*Hydroxyhexahydro*-p-*toluic acid* melts at 130—132° and is readily soluble in ether or alcohol, but sparingly so in cold water.

The *silver* salt, $C_8H_{13}O_3Ag$, is obtained, on the addition of silver nitrate to a warm slightly alkaline solution of the ammonium salt, as a curdy precipitate which soon breaks up into a satiny, crystalline mass. For analysis, the salt was dried at 100°.

0.4341 gave 0.1698 Ag.   Ag = 40.7.

$C_8H_{13}O_3Ag$ requires Ag = 40.7 per cent.

The slightly alkaline solution of the ammonium salt shows the following behaviour. *Barium chloride* gives a white precipitate which dissolves in boiling water and separates, on slowly cooling, in beautiful colourless needles; *calcium chloride* also gives a similar precipitate which crystallises from water, but it is much more soluble than the barium salt. *Copper sulphate* yields a very pale blue precipitate which is almost insoluble in water, and *lead acetate* and *zinc sulphate* give white, curdy, insoluble precipitates.

p-*Methylcyclohexanone.*

In the introduction to this paper it is stated that α-hydroxyhexahydro-*p*-toluic acid is decomposed by concentrated sulphuric acid with formation of *p*-methylcyclohexanone. Unfortunately, considerable quantities of tetrahydro-*p*-toluic acid are always produced at the same time, and the yield of ketone is therefore unsatisfactory and rarely more than 30 per cent. of that theoretically possible.

The finely-powdered hydroxy-acid is added in small quantities at a time to about ten times its weight of sulphuric acid, which is kept at − 10° by means of ice and salt. As soon as the whole.has been added, the flask is taken from the freezing mixture and shaken until the acid has been completely decomposed, which is the case in about twenty to thirty minutes; the clear brown solution is then poured into ice and water and distilled in steam.

At first the ketone passes over with the condensed water, afterwards the tetrahydro-p-toluic acid separates in the condenser and distillate in the form of crystals. The distillate is saturated with salt and extracted twice with ether; the ethereal solution is repeatedly washed with sodium carbonate,* dried over calcium chloride and evaporated, and the ketone purified by repeated distillation, when, with the exception of a small amount of a condensation product of boiling point 243—245°, almost the whole quantity passes over at 170° (773 mm.).

0·1241 gave 0·3412 CO_2 and 0·1208 H_2O. C = 74·8; H = 10·8.

$C_7H_{12}O$ requires C = 74·9; H = 10·8 per cent.

p-*Methylcyclohexanone* has a pleasant and characteristic odour which somewhat resembles that of peppermint; when it is mixed with a solution of semicarbazide hydrochloride and sodium acetate, combina‑ tion takes place at once, and the very sparingly soluble semicarbazone separates as a crystalline powder. This was collected, crystallised from much alcohol, and the glistening, sandy powder analysed with the following result:

0·1114 gave 23·8 c.c. of nitrogen at 14° and 750 mm. N = 25·0.

$C_8H_{15}ON_3$ requires N = 24·8 per cent.

p-*Methyl*cyclohexanone *semicarbazone* is very sparingly soluble in cold alcohol and melts at about 198° with decomposition.

p-Methylcyclohexanone had been previously obtained by Sabatier and Mailhe (*Compt. rend.*, 1905, **140**, 350—352), who prepared it by reducing p-cresol with hydrogen in the presence of nickel and subsequent oxidation. They give the boiling point of the ketone as 169·5° and the melting point of the semicarbazone as 197°, and these properties agree almost exactly with those found in the case of the p-methylcyclo‑ hexanone produced by the method described above.

Inactive Tertiary Menthol and Inactive Menthene.

In carrying out the synthesis of tertiary menthol, pure p-methyl‑ cyclohexanone (17 grams) was gradually added to an ethereal solution of magnesium *iso*propyl iodide (containing 10 grams of magnesium), care being taken to avoid rise of temperature during the addition.

After remaining for three hours, the product was decomposed by water and dilute hydrochloric acid, the ethereal solution washed well with water, dried over calcium chloride, evaporated, and the residual oil twice fractionated under reduced pressure. In this way, about

---

* When the sodium carbonate extract is acidified, the tetrahydro-p-toluic acid separates at once and, after recrystallising from dilute acetic acid, is pure and melts at 134°.

8 grams of a colourless oil were obtained which contained no unchanged
p-methylcyclohexanone, distilled at 97° (25 mm.), and consisted of
pure *tertiary menthol*.

0·1557 gave 0·4372 CO_2 and 0·1769 H_2O.   C = 76·6 ; H = 12·6.
0·1714 „ 0·4803 CO_2 „ 0·1958 H_2O.   C = 76·5 ; H = 12·7.
$C_{10}H_{20}O$ requires C = 76·9 ; H = 12·8 per cent.

Tertiary menthol is a syrup which possesses in a marked degree the
pungent and pleasant odour of peppermint so characteristic of ordinary
menthol. It decomposes on distillation under the ordinary pressure, but
if the operation is very rapidly carried out a considerable quantity passes
over at 207°, and yields on analysis numbers agreeing approximately
with those given above. Several different specimens of tertiary
menthol were prepared during this investigation, and many attempts
were made to obtain it in a crystalline condition, but without success.
Since, however, ordinary menthol solidifies so readily and is so closely
related to tertiary menthol, it seems likely that the latter will
ultimately also be found to crystallise.* Tertiary menthol was first
prepared by Baeyer (*Ber.*, 1893, **26**, 2270), who obtained it from
menthene by converting it first into menthyliodide, and then, by the
action of silver acetate, into tertiary menthylacetate, which on hydrolysis
yielded tertiary menthol (compare p. 779). Baeyer describes it as an oil,
boiling at 97—101° (20 mm.), which did not crystallise even when
cooled in solid carbonic anhydride.

The synthetical tertiary menthol, obtained as described above, was
digested with twice its weight of powdered potassium hydrogen sulphate for one hour, the mass was then warmed with water until the
salt had dissolved, and the whole distilled in steam. The distillate
was extracted with ether, the ethereal solution dried over calcium
chloride, evaporated, and the residue distilled, when almost the whole
quantity passed over at 158—170°. This was distilled three times
over sodium and the portion boiling at 167—168° analysed.

0·1803 gave 0·5722 CO_2 and 0·2106 H_2O.   C = 86·6 ; H = 12·9.
0·1621 „ 0·5162 CO_2 „ 0·1904 H_2O.   C = 86·8 ; H = 13·1.
$C_{10}H_{18}$ requires C = 87·0 ; H = 13·0 per cent.

In order to prove conclusively that this hydrocarbon was *inactive
menthene*, it was converted into the characteristic *nitrosochloride* in the
following way.

The hydrocarbon (2 c.c.) was dissolved in methyl alcohol (6 c.c.)
and acetic acid (1 c.c.), mixed with freshly-distilled *iso*amyl nitrite
(8 c.c.), and, after cooling to −5°, concentrated hydrochloric acid
(7 c.c.) was added drop by drop from a burette, the whole being well

* Compare Trans., 1905, **87**, 1100, footnote.

shaken during the addition. The product soon began to deposit crystals, and after one hour water was added, the semi-solid mass collected at the pump, and left in contact with porous porcelain until the oily impurity had been completely absorbed.

The colourless residue was then twice crystallised from boiling alcohol, in which this nitrosochloride is rather sparingly soluble.

0·2341 gave 0·1656 AgCl.   Cl = 17·5.

$C_{10}H_{18}ONCl$ requires $Cl = 17·4$ per cent.

This nitrosochloride melted sharply at 128°, which is exactly the melting point of inactive menthene nitrosochloride (Urban and Kremers, *Amer. Chem. J.*, 1894, **16**, 395), so that there can be no doubt that the synthetical hydrocarbon is *inactive menthene*.

THE VICTORIA UNIVERSITY OF MANCHESTER.

LXXXVIII.—*Experiments on the Synthesis of the Terpenes. Part VIII. Synthesis of the Optically Active Modifications of* Δ^3-p-Menthenol(8) *and* $\Delta^{3.8(9)}$-p-Menthadiene.

By FRANCIS WILLIAM KAY and WILLIAM HENRY PERKIN, jun.

THE terpenes and their derivatives which have so far been obtained synthetically (Trans., 1904, **85**, 654 ; 1905, **87**, 639, 655, 661, 1067, and 1083 ; *Proc.*, 1905, **21**, 255) have always been inactive, and, while engaged in further researches on this subject, it was thought that it would be interesting to undertake a series of experiments with the object of synthesising some optically active members of this important group. Our first intention was to resolve dl-Δ^3-tetrahydro-p-toluic acid,

$$Me \cdot C {\displaystyle <_{CH_2 \cdot CH_2}^{CH-CH_2}} > CH \cdot CO_2H,$$

into its active components by the fractional crystallisation of a suitable salt formed by combining it with some active base such as strychnine or brucine. The next step would be to treat the esters of the d- and l-acids with magnesium methyl iodide in ethereal solution, a process which should lead to a synthesis of the d- and l-modifications of terpineol and limonene,

$$Me \cdot C {\displaystyle <_{CH_2 \cdot CH_2}^{CH-CH_2}} > CH \cdot CMe_2 \cdot OH \quad \text{and}$$

$$Me \cdot C {\displaystyle <_{CH_2 \cdot CH_2}^{CH-CH_2}} > CH \cdot C {\displaystyle <_{Me}^{CH_2}}$$

(compare Trans., 1904, **85**, 663—668). Since, however, it is extremely difficult to prepare Δ^3-tetrahydro-p-toluic acid in the quantities
required for a research of this kind, ˚and it moreover appeared to us
that it would be even more interesting to synthesise the optically
active modifications of some terpene which had not been met with in
Nature, we have, in the meantime, postponed our intention of preparing optically active terpineol synthetically, and carried out instead
the following series of experiments.

Hexahydro-p-toluic acid was prepared in quantity from p-toluic
acid by reduction with sodium and *iso*amylic alcohol, and converted
first into α-bromohexahydro-p-toluic acid and then into dl-Δ^1-tetrahydro-
p-toluic acid,

$$\text{Me·CH}\underset{\text{CH}_2\text{·CH}_2}{\overset{\text{CH}_2-\text{CH}}{<}}\hspace{-6pt}>\hspace{-2pt}\text{C·CO}_2\text{H},$$

by a process which has already been described (Perkin and Pickles,
Trans., 1905, **87**, 643—645). A series of experiments on the
behaviour of this dl-acid towards active bases showed that, when
treated with brucine, it yields a salt, $lBlA$, which crystallises well and
is somewhat less soluble than the salt $lBdA$.

By taking advantage of this property, and after many fractional
crystallisations, we were ultimately successful in separating the $lBlA$
salt in a state of purity, and from this we obtained l-*tetrahydro*-p-
toluic acid, the rotation of which was found to be $[a]_D - 100·8°$.

The ester of this acid has $[a]_D - 83·5°$, and reacts readily with an
ethereal solution of magnesium methyl iodide with formation of l-Δ^3-p-
menthenol(8),

$$\text{Me·CH}\underset{\text{CH}_2\text{·CH}_2}{\overset{\text{CH}_2-\text{CH}}{<}}\hspace{-6pt}>\hspace{-2pt}\text{C·CMe}_2\text{·OH},$$

a colourless syrup which distils at 102° (14 mm.) and has a rotation of
$[a]_D - 67·3°$. When this l-menthenol was digested with potassium
hydrogen sulphate, it decomposed in the usual manner with elimination
of water and formation of the corresponding l-$\Delta^{3.8(9)}$-p-*menthadiene*,

$$\text{Me·CH}\underset{\text{CH}_2\text{·CH}_2}{\overset{\text{CH}_2-\text{CH}}{<}}\hspace{-6pt}>\hspace{-2pt}\text{C·C}\underset{\text{Me}}{\overset{\text{CH}_2}{<}},$$

but the rotation of the hydrocarbon obtained in this way was only
$[a]_D - 10·0°$. This result, taken in connection with the high rotation
of the corresponding d-$\Delta^{8\,8(9)}$-p-menthadiene ($+ 98·2°$), clearly proves
that optical inversion must have taken place to a considerable extent
during the heating with potassium hydrogen sulphate, a change which
is somewhat similar to that observed in the case of d-limonene,
which, when strongly heated, is converted into dipentene.

During the course of our experiments on the resolution of dl-tetra-
hydro-p-toluic acid, we made the further fortunate discovery that the

comparative solubilities of the strychnine salts are of the reverse order to those observed in the case of the brucine salts.

It has just been stated that the *dl*-acid combines with brucine and yields a salt, *lBlA*, which is less soluble than the salt *lBdA*.

In the case of strychnine, on the other hand, it is the salt *lBdA* which is the more sparingly soluble. The difference in the solubilities of the strychnine salts was, moreover, sufficient to allow of the ultimate separation of the salt *lBdA* in a state of purity, and from this salt we obtained pure d-Δ^1-*tetrahydro*-p-*toluic acid*.

This acid has $[a]_D + 101 \cdot 1°$, a rotation which compares very favourably with that of the corresponding *l*-acid, for which the value $- 100 \cdot 8°$ had been observed. From this pure *d*-acid, *ethyl* d-Δ^1-*tetrahydro*-p-*toluate* was now prepared in the usual manner and found to have a rotation of $[a]_D + 86 \cdot 5°$, a number slightly higher than that which had been observed in the case of the ester of the *l*-acid ($- 83 \cdot 5°$). When ethyl *d*-tetrahydro-*p*-toluate is treated with magnesium methyl iodide, it is doubtless converted, in the first instance, into d-Δ^3-*p*-menthenol(8), the rotation of which, judging from that observed in the case of the *l*-modification, should be about $[a]_D + 67°$. Since, however, it seemed more important to obtain the d-$\Delta^{3\,8(9)}$-p-*menthadiene* in a pure condition, and our supply of material was limited, we decided not to attempt to prepare the *d*-menthenol in a state of purity, but to proceed at once to the preparation of the hydrocarbon.

The difficulty in obtaining the *d*-hydrocarbon from the *d*-menthenol lies in the fact, mentioned above, that when the usual procedure (digesting with potassium hydrogen sulphate) is adopted optical inversion takes place. It was therefore obvious that, if the pure *d*-hydrocarbon was to be obtained, every care would have to be taken to avoid a high temperature during the elimination of water. Fortunately, during the course of other researches of this series (compare Trans., 1905, **87**, 1100), the observation had been made that the menthenols are converted into the menthadienes when left in contact with an ethereal solution of magnesium methyl iodide at the ordinary temperature.

When this method was applied in the present instance to d-Δ^3-*p*-menthenol(8), the result was most successful, since an almost quantitative yield of d-$\Delta^{3\,8(9)}$-p-*menthadiene* was obtained, which distilled at 184° and had $[a]_D + 98 \cdot 2°$, a rotation which is approximately the same as that of *d*-limonene ($+ 105°$).*

* This interesting result seems to indicate that magnesium methyl iodide may often prove to be a valuable agent for the elimination of water, especially in cases such as the above, where it is necessary for the operation to take place at the ordinary temperature.

l-Δ^1-Tetrahydro-p-toluic Acid.

The resolution of dl-tetrahydro-p-toluic acid may be accomplished, as stated in the introduction to this paper, by the fractional crystallisation of the brucine, quinine, or strychnine salts. In the case of the two first mentioned, the more sparingly soluble salts are the salts of the l-acid; the strychnine salt which is the least soluble is, on the other hand, the salt of the d-acid.

I. *Experiments with Brucine.*—When equivalent quantities of brucine and dl-tetrahydro-p-toluic acid are separately dissolved in pure ethyl acetate and the solutions mixed and allowed to stand, a solid cake of brucine tetrahydro-p-toluate is gradually deposited, and in this the salt of the l-acid predominates. After many fractional crystallisations, this salt may ultimately be obtained in a pure condition.

A convenient amount of acid for one operation is 20 grams, and this is heated on the water-bath with brucine (70 grams) and pure ethyl acetate (500 c.c.) until solution is complete. The liquid is then set aside until crystallisation is complete, which is usually the case in two to three days, care being taken that the solvent does not evaporate during the crystallisation. The mother liquor is decanted and the pale yellow crystals washed with ethyl acetate, the crude salt is then systematically recrystallised from the same solvent until the acid has a constant rotation. As the separation proceeds, the salt loses its original sticky nature, and, especially during the later stages, crystallisation, which at first is always slow, takes place rapidly. Each crop of crystals was tested in the following manner. About one gram was ground to a fine paste with water, warmed on the water-bath with excess of sodium carbonate, and the precipitated base filtered off. The acid obtained by acidifying the filtrate was recrystallised from dilute acetic acid and its rotation determined in ethyl acetate solution in the usual manner. This rather tedious procedure had to be adopted because measurements of the brucine salt itself were not found to be sufficiently reliable.

The following are the rotations of three of the last crystallisations, the acid, in each case, being dissolved in ethyl acetate (20 c.c.) at a temperature of 17° and the measurement made in a 100 mm. tube.

Crop 17.   0·4004 gave a rotation of − 2·02°. $[a]_D$ − 100°
Crop 18.   0·3844   „   „   „   − 1·92°.   „   − 99·9°
Crop 20.   0·3951   „   „   „   − 2·00°.   „   − 100·8°

In the hope of shortening the process of separation, experiments were next carried out on the fractional crystallisation of the brucine salts from water, instead of from ethyl acetate, but, although the salts separated readily, the number of crystallisations necessary before the *l*-acid was obtained pure was about the same as when ethyl acetate was employed as the solvent.

Ultimately, however, by employing a method recommended by Pope and Peachey (Trans., 1899, **75**, 1066), we obtained a much more satisfactory result and were able to prepare the *l*-acid in sufficient quantity for our subsequent synthetical experiments. This important method of separation depends, as its authors state, on the following considerations. The solubilities of the salts (*lBlA* and *lBlA*) of a lævorotatory base (*lB*) with a dextro- and a lævo-acid (*dA* and *lA*) would hardly be expected to differ considerably, because the solubility is partly a function of the chemical nature of the salts.

If, however, the salt *lBlA* is the less soluble, and only sufficient of the active base *lB* necessary for the formation of this salt is added, the balance of base required to dissolve the acid being made up by adding the requisite amount of an optically inactive base, such as caustic soda or ammonia, which forms comparatively soluble salts with the acid, it would be expected that on crystallisation the greater part of the lævo-acid would separate as the sparingly soluble salt, *lBlA*, whilst the mother liquors would retain the very soluble sodium or ammonium salts of the dextro-acid, *dA*.

In applying this method, *dl*-Δ^1-tetrahydro-*p*-toluic acid (14 grams) was exactly neutralised with sodium carbonate and diluted to one litre with boiling water, a boiling concentrated alcoholic solution of brucine (60 grams) and *dl*-acid (14 grams) was slowly poured in, and the liquid boiled until practically all the alcohol had been removed. The solution was then allowed to cool slowly, when a mass of beautiful dagger-shaped crystals rapidly separated and were collected at the pump and washed with a little water. The filtrate and washings were concentrated and precipitated by hydrochloric acid, the acid being collected and again treated with sodium carbonate and brucine exactly as before. The brucine salt was then divided into two equal parts, one of which was ground to an exceedingly fine paste with water and then warmed with an excess of sodium carbonate. The filtrate from the precipitated base was heated to boiling and the excess of alkali exactly neutralised by the cautious addition of dilute hydrochloric acid. The second half of the brucine salt, dissolved in the minimum quantity of boiling alcohol, was then mixed with the aqueous sodium salt and the liquid allowed to stand until crystallisation was complete. After this operation had been repeated five times, the brucine salt of the *l*-acid was obtained quite pure and yielded an acid with a rotation of [*a*]ᴅ −100·8°.

The acid present as sodium salt in the mother liquors from each crystallisation was worked up systematically by the same process, and ultimately more than 25 per cent. of the original dl-acid was obtained in the form of the pure l-acid.

II. *Experiments with Quinine.*—The experiments just described show that when the lævorotatory base brucine is combined with dl-tetra-hydro-p-toluic acid, the more sparingly soluble salt formed is the salt $lBlA$. It seemed possible that quinine (dB), under similar conditions, might yield a salt with the d-acid which was less soluble than the corresponding salt of the l-acid, and that, in this way, a separation of the pure d-acid might prove to be possible. The following experiment showed, however, that the reverse is the case and that the more sparingly soluble quinine salt is the salt $dBlA$.

Quinine (19 grams) was dissolved in alcohol and digested with 14 grams of impure lævo-acid ($[a]_D - 9°$) until solution was complete. A further 14 grams of the same specimen of acid (exactly neutralised with carbonate of soda and dissolved in the minimum amount of water) was added, when a precipitate was produced which was dissolved by the addition of small quantities of alcohol. The copious crystalline precipitate which separated on standing was collected at the pump, washed with water, and decomposed by sodium carbonate, with the result that an acid was obtained which had a rotation of $[a]_D - 20·9°$. This process was not continued as it had no advantages over the separation by means of brucine. In our subsequent experiments on the behaviour of dl-tetrahydro-p-toluic acid towards optically active bases we made the observation that the strychnine salt of the d-acid is less soluble than that of the l-acid, and this led to a process of separation of the pure d-acid which is described on p. 845.

l-Δ^1-Tetrahydro-p-toluic acid melts at 133—134°, or at exactly the same temperature as the dl-acid, a fact which seems to indicate that the latter is a racemic modification and not merely a mixture of the d- and l-components. An analysis of a sample of the l-acid which had a rotation of $[a]_D - 100·1°$ gave the following results :

0·3050 gave 0·7700 CO_2 and 0·2394 H_2O.   C = 68·8 ; H = 8·8.
$C_8H_{12}O_2$ requires C = 68·6 ; H = 8·8 per cent.

This acid separates from dilute acetic acid in slender, prismatic needles and is obtained in large, transparent prisms when its solution in ethyl acetate is allowed to evaporate slowly under ordinary conditions. It dissolves freely in acetic acid, ethyl acetate, or light petroleum, less readily in chloroform, and is rather sparingly soluble in cold alcohol.

Ethyl l-Δ^1-Tetrahydro-p-toluate.—In the preparation of this ester and also of the corresponding ester of the d-acid (p. 847), the use of hydrogen chloride was avoided and every care was taken to prevent any optical inversion by keeping the temperature as low as possible. *l*-Tetrahydro-*p*-toluic acid (15 grams) was dissolved in boiling absolute alcohol (100 c.c.) and the solution cooled quickly under the tap in order that the acid might separate in as finely divided a state as possible. A mixture of concentrated sulphuric acid (10 c.c.) and alcohol (20 c.c.) was gradually added, any rise in temperature being checked by immersing the flask in cold water. The mixture was then set aside and shaken from time to time, when, after five days, all the acid was found to have passed into solution and esterification was practically complete. The oily ester, which separated on the addition of water, was extracted with ether, the ethereal solution washed well with water and dilute sodium carbonate, dried over calcium chloride, and evaporated. The residual oil distilled at 154° (100 mm.) and had $[a]_D - 83·5°$ in a 100 mm. tube at 18°.

d-Δ¹-*Tetrahydro*-p-*toluic Acid.*

When equivalent quantities of strychnine and *dl*-tetrahydro-*p*-toluic acid are digested in ethyl acetate solution, only a portion of the acid combines with the base, and the crystalline deposit which separates contains an excess of the salt of the *d*-acid.

dl-Tetrahydro-*p*-toluic acid (14 grams) and finely-divided strychnine (35 grams) were boiled together with one litre of ethyl acetate in a reflux apparatus for one hour. The hot solution was then filtered at the pump and set aside for two or three days and until crystallisation was complete. The crystalline mass, which consisted of large, transparent, rhomboid plates, was collected at the pump, washed with ethyl acetate, and the filtrate and washings were used to extract the strychnine and salt which had remained undissolved in the first instance. By repeating this operation several times, a considerable additional quantity of the sparingly soluble salt was obtained.

The crops of crystals were combined and recrystallised twice from ethyl acetate, a little chloroform being added to assist solution.

The acid, separated in the usual manner, had a rotation of $[a]_D + 55·9°$. The acetic ester mother liquors were concentrated and allowed to crystallise, when long needles separated, which were found to consist of the free lævo-acid with a rotation of $[a]_D - 32°$. Attempts to purify the crude *d*-salt by further recrystallisation did not lead to a satisfactory result, owing to the ease with which the salt dissociates * in solution, and this method of purification had therefore to be abandoned.

After many unsuccessful experiments, however, we were ultimately able, by employing a slight modification of the method of Pope and

---

* It seems remarkable that the strychnine salts of the *d-* and *l-*tetrahydro-*p*-toluic acids should dissociate so much more readily than the corresponding brucine salts.

Peachey (compare p. 843), to separate the strychnine salt of the d-acid in a state of purity.

The separation could, however, not be carried out in aqueous solution owing to the insolubility of the strychnine salt in water, and, as the sodium salt of the acid is very sparingly soluble in organic solvents, the method of half neutralisation with sodium carbonate could not be employed. When, however, the dl-acid and strychnine, in the proportion of two molecules of the former to one of the latter, were dissolved in chloroform and ethyl acetate, dissociation was largely prevented by the presence of the excess of the acid, and thus a satisfactory separation became possible.

Strychnine (35 grams), dissolved in the least possible quantity of chloroform (1 vol.), was mixed with a hot solution of the dl-acid (28 grams) in acetic ester (4 vols.), and the solvent distilled off until minute crystals began to form. After standing until crystallisation was complete, the salt was collected at the pump and washed once with ethyl acetate. The mother liquors were then treated with half the previous quantity of strychnine, and this operation repeated until no further separation of the crude salt of the d-acid took place.

The various crops of crystals of this salt were combined, divided into two equal parts, one of which was converted into the acid by treatment with sodium carbonate in the usual way; the second half and the free acid were then mixed and crystallised from chloroform and acetic ester as before. After repeating this separation five or six times, the strychnine salt of the d-acid was obtained pure.

If all the mother liquors are carefully collected and systematically treated with strychnine in the way just described, the yield of this salt is at least equal to that of the brucine salt of the l-acid obtained as described on p. 844.

Strychnine d-Δ^1-*tetrahydro*-p-*toluate* crystallises readily from acetic ester, in which it is sparingly soluble, in colourless, transparent, glistening plates frequently half a centimetre in length.

It dissolves sparingly in alcohol, but very readily in chloroform, and separates from the latter solvent in aggregates of large, transparent, hexagonal tablets. This salt shows a marked tendency to dissociate, especially when its solutions are boiled for a considerable time.

d-Δ^1-*Tetrahydro*-p-*toluic acid* was readily obtained by decomposing the pure strychnine salt with sodium carbonate and then recrystallising from dilute acetic acid.

0·2201 gave 0·5561 CO_2 and 0·1757 H_2O.   C = 68·9 ; H = 8·9.

$C_8H_{12}O_2$ requires C = 68·6 ; H = 8·8 per cent.

In order to determine the rotation as accurately as possible, two different specimens of the acid were prepared and examined in the usual manner, with the result that the values $[\alpha]_D$ +101·1° and +100·9° were obtained in the two cases. It will thus bo seen that the rotation of this d-acid compares well with that of the corresponding l-acid, for which the value $[\alpha]_D$ - 100·8° was obtained (p. 842). The other physical properties of the d-acid are identical with those of the l-acid. It melts at 133°, and when it is mixed with an equal quantity of the l-acid (m. p. 133—134°) the mixture melts lower, namely, at about 129°. This seems to confirm the suggestion (p. 844) that the original dl-acid from which the active acids had been prepared is the racemic modification.

Ethyl d-Δ^1-*Tetrahydro*-p-*toluate.*—This ester, which was prepared in exactly the same manner as the corresponding ester of the l-acid (p. 844), distilled at 154° (100 mm.) and had $[\alpha]_D$ +86·5°, a value which is rather higher than that observed in the case of the ester of the l-acid (- 83·5°).

l-Δ^3-p-*Menthenol*(8) *and* 1-$\Delta^{3.8(9)}$-p-*Menthadiene.*

The preparation of l-Δ^3-p-menthenol was conducted under the following conditions. Magnesium (7·5 grams) was covered with dry ether (250 c.c.) and converted into magnesium methyl iodide in the usual manner. The solution was cooled in ice-water and gradually poured into ethyl l-Δ^1-tetrahydro-p-toluate (16 grams) diluted with ether (100 c.c.), any rise of temperature being carefully avoided. After remaining for twenty-four hours, the product was decomposed by water and dilute hydrochloric acid, the ethereal solution washed with a little sodium hydrogen sulphite to remove some iodine, and evaporated. The oil was then freed from any unchanged ester by digesting for half an hour with methyl-alcoholic potash (5 grams KOH), again extracted with ether, and, after washing well and drying over calcium chloride, the ether was evaporated and the residue fractionated under reduced pressure.

0·1091 gave 0·3130 CO_2 and 0·1144 H_2O.   C = 78·3 ; H = 11·7.
$C_{10}H_{18}O$ requires C = 77·9 ; H = 11·7 per cent.

l-Δ^3-p-*Menthenol*(8) distils at 101—102° (14 mm.) and was obtained as a thick, colourless syrup which would doubtless have solidified had a crystal been available with which to start the crystallisation.*

It has a penetrating but pleasant odour which somewhat resembles

---

* Unfortunately the whole of this menthenol had been used up in preparing l-$\Delta^{3.8(9)}$-p-menthadiene before we succeeded in obtaining dl-Δ^3-p-menthenol(8) in a crystalline form (p. 851).

that of terpineol, and its other properties are similar to those of dl-Δ^3-p-menthenol(8), which had previously been prepared by Perkin and Pickles (Trans., 1905, **87**, 647; compare also the present paper, p. 851).

Its rotation, determined in benzene solution in a 100 mm. tube at 18°, was found to be $[\alpha]_D - 67 \cdot 3°$.

The whole of the l-menthenol was next heated to boiling with twice its weight of powdered potassium hydrogen sulphate in a reflux apparatus heated by means of an oil-bath. The product was then mixed with water, distilled in steam, the distillate extracted with ether, and, after drying over calcium chloride, the ether was evaporated and the hydrocarbon fractionated. Almost the whole quantity passed over at 175—185°, and after twice distilling over sodium an oil was obtained which boiled constantly at 182—183° (748 mm.) and had an odour resembling that of dipentene.

0·2353 gave 0·7610 CO_2 and 0·2497 H_2O.   C = 88·1; H = 11·8.

$C_{10}H_{16}$ requires C = 88·2; H = 11·8 per cent.

When this hydrocarbon (0·6111 gram), diluted with benzene (20 c.c.), was examined at 18°, it was found to have a rotation of only $- 0\cdot31°$, or $[\alpha]_D - 10\cdot0°$. Since, however, the rotation of d-$\Delta^{3\ 8(9)}$-p-menthadiene (see below) is $+ 98\cdot2°$, it is clear that the value observed for the above menthadiene from l-Δ^3-p-menthenol(8) is much too low, and that optical inversion must have taken place to a considerable extent during the preparation under the conditions described above.

Conversion of Ethyl d-Δ^1-*Tetrahydro*-p-*toluate into* d-$\Delta^{3\ 8(9)}$-p-*Menthadiene.*

It has just been shown that boiling with potassium hydrogen sulphate had the effect of partially inverting l-$\Delta^{3.8(9)}$-p-menthadiene, and since it seemed probable that the high temperature employed had contributed largely to this result, we were careful to avoid this in the present case by using as the dehydrating agent an excess of magnesium methyl iodide at the ordinary temperature.

A solution of magnesium methyl iodide, prepared from magnesium (9 grams), was gradually poured into ethyl d-Δ^1-tetrahydro-p-toluate (21 grams), and, after remaining for a couple of days, the product was decomposed by dilute hydrochloric acid in the usual manner. After removing any traces of unchanged ester which might be present by hydrolysis with methyl-alcoholic potash, the oil was extracted and distilled under reduced pressure, when most of it passed over at 80° (15 mm.). There was, however, a higher fraction which without doubt consisted of d-Δ^3-p-menthenol(8), but the quantity was small and it was not further examined.

The oil of lower boiling point was distilled at the ordinary pressure and the fraction 178—187° digested twice with sodium and again distilled.

0·1973 gave 0·6343 CO_2 and 0·2110 H_2O. $C = 87·9$; $H = 11·9$. $C_{10}H_{16}$ requires $C = 88·2$; $H = 11·8$ per cent.

d-$\Delta^{3\cdot8(9)}$-p-*Menthadiene* boils at 184° (756 mm.) and has an odour closely resembling that of d-limonene. When examined in benzene solution at 18° in a 100 mm. tube, it had a rotation of $[a]_D + 98·2°$.

Densities, Magnetic Rotations, and Refractive Powers of d-$\Delta^{3\,8(9)}$-p-*Menthadiene, Ethyl* dl-Δ^1-*Tetrahydro-p-toluate,* dl-Δ^3-p-*Menthenol(8),* dl-$\Delta^{3\cdot8(9)}$-p-*Menthadiene, Terpineol, and Dipentene.*

In the present communication, as well as during the course of some of the other researches of this series, terpenes and their derivatives have been described which, so far, have not been met with in Nature. Prof. H. E. Armstrong and others interested in this matter were good enough to suggest to us that valuable results might be expected if it were found possible to prepare some of these new substances in sufficient quantity to allow of their physical properties being carefully examined.

In the first instance, we prepared a very pure specimen of d-$\Delta^{3.8(9)}$-p-menthadiene and sent it to Dr. W. H. Perkin, sen., for examination, but as the results obtained were found to be quite abnormal it was thought necessary, for the sake of comparison, to prepare very pure specimens of the other substances mentioned at the head of this section and examine them. The determination of the whole of the physical constants which are given below was undertaken by Dr. W. H. Perkin, sen., and this laborious investigation has added greatly to the value of the present research.

d-$\Delta^{3.8(9)}$-p-*Menthadiene.*

This hydrocarbon was prepared by the process described on p. 848, and the analysis there given was carried out with the specimen used in the determination of the following physical constants.

Density : $d4°/4° = 0·8712$; $d15°/15° = 0·8634$; $d25°/25° = 0·8574$.

Magnetic rotation :

t.	Sp. rot.	Mol. rot.
13·6°	1·4941	13·061

Refractive power :

$$t = 12\ 7° ; \ d12·7°/4° = 0·86483.$$

	μ.	$\dfrac{\mu - 1}{d}$.	$\dfrac{\mu - 1}{d}p$.
a...............	1·49434	0·57161	77·739
β	1·50913	0·58871	80·064
γ...............	1·51849	0·59953	81·536

Dispersion $\gamma - a = 3\cdot797$.

dl-$\Delta^{-3.8(9)}$-p-*Menthadiene*.

The specimen of this hydrocarbon employed in the following determinations was prepared from crystalline *dl*-Δ^3-*p*-menthenol(8) (see
p. 851) by digesting with potassium hydrogen sulphate. It was
twice distilled over sodium and boiled constantly at 184—185°.

Density: $d10°/10° = 0\cdot8425$; $d15°/15° = 0\cdot8390$; $d20°/20° = 0\cdot83579$.

Magnetic rotation:

t.	Sp. rot.	Mol. rot.
16·2°	1·4356	12·939

Refractive power:

$$t = 17\cdot2°;\ d17\cdot2°/4° = 0\cdot83659.$$

	μ.	$\dfrac{\mu - 1}{d}$.	$\dfrac{\mu - 1}{d}p$.
a...............	1·46945	0·56114	76·315
β	1·48113	0·57510	78·213
γ...............	1·48824	0·58360	79·369

Dispersion $\gamma - a = 3\cdot054$.

Dipentene.

This hydrocarbon was prepared from terpineol (from Schimmel) and
distilled twice over sodium in an atmosphere of carbon dioxide. It
boiled constantly at 180—181°.

Density: $d4°/4° = 0\cdot8627$; $d15°/15° = 0\cdot8548$; $d25°/25° = 0\cdot8486$.

Magnetic rotation:

t.	Sp. rot.	Mol. rot.
15·7°	1·2796	11·315

Refractive power:

$$t = 14\cdot4°;\ d14\cdot4°/4° = 0\cdot85457.$$

	μ.	$\dfrac{\mu - 1}{d}$.	$\dfrac{\mu - 1}{d}p$.
a...............	1·47506	0·55590	75·602
β	1·48629	0·56905	77·391
γ...............	1·49367	0·57768	78·564

Dispersion $\gamma - a = 2\cdot962$.

Terpineol.

The specimen examined was prepared by distilling terpineol (from Schimmel) under reduced pressure. It melted at 35° and boiled at 120—122° (25 mm.).

Density: (surfused) $d10°/10° = 0.9448$; $d15°/15° = 0.9415$; $d25°/25° = 0.9355$.

(Fused) $d40°/40° = 0.9282$; $d45°/45° = 0.9256$.

Magnetic rotation :

t.	Sp. rot.	Mol. rot.
16°	1·1923	10·841
50·5	1·1676	10·823

Refractive power : (a) surfused.

$t = 18.4°$; $d18.4°/4° = 0.93831$.

	μ.	$\dfrac{\mu - 1}{d}$.	$\dfrac{\mu - 1}{d} p.$
a	1·48064	0·51224	78·885
β	1·49028	0·52251	80·466
γ	1·49627	0·52890	81·450

Dispersion $\gamma - a = 2.565$.

(*b*) Fused.

$t = 43.75°$; $d43.75°/4° = 0.91770$.

	μ.	$\dfrac{\mu - 1}{d}$.	$\dfrac{\mu - 1}{d} p.$
a	1·47018	0·51234	78·900
β	1·47966	0·52267	80·491
γ	1·48527	0·52879	81·433

Dispersion $\gamma - a = 2.533$.

dl-Δ^3-p-*Menthenol*(8).

This substance was prepared by the action of magnesium methyl iodide on ethyl $dl\Delta^1$-tetrahydro-p-toluate. When first obtained (Perkin and Pickles, Trans., 1905, **87**, 650), it was described as a syrup, but the specimen prepared on the present occasion solidified during the winter months to a hard, crystalline mass which closely resembled terpineol in appearance.* It melted at 38—40° and distilled at 102° (14 mm.).

---

* Attention has already been directed (Trans., 1905, **87**, 1100, footnote) to the difficulty experienced in obtaining the viscid menthenols in a crystalline condition. This was well instanced in the case of the pure specimens of terpineol and

Density: (surfused) $d10°/10° = 0.9251$; $d15°/15° = 0.9217$; $d25°/25° = 0.9158$; (fused) $d40°/40° = 0.9080$; $d45°/45° = 0.9055$.

Magnetic rotation:

t.	Sp. rot.	Mol. rot.
16·4°	1·2002	11·150
50·5	1·1700	11·085

Refractive power: (a) surfused.

$$t = 16.3°; \quad d16.3°/4° = 0.9200.$$

	μ.	$\dfrac{\mu - 1}{d}$.	$\dfrac{\mu - 1}{d}p$.
a...............	1·47545	0·51679	79·585
β	1·48522	0·52741	81·221
γ...............	1·49105	0·53375	82·197

Dispersion $\gamma - a = 2.612$.

(b) Fused.

$$t = 43.4°; \quad d43.4°/4° = 0.90131.$$

	μ.	$\dfrac{\mu - 1}{d}$.	$\dfrac{\mu - 1}{d}p$.
a...............	1·46341	0·51415	79·179
β	1·47329	0·52511	80·867
γ...............	1·47897	0·53142	81·838

Dispersion $\gamma - a = 2.659$.

Ethyl dl-Δ^1-*Tetrahydro*-p-*toluate.*

This ester was prepared from pure *dl*-Δ^1-tetrahydro-*p*-toluic acid by means of alcohol and sulphuric acid. It distilled at 152—153° under 100 mm. pressure.

Density: $d4°/4° = 0.9877$; $d15°/15° = 0.9792$.; $d25°/25° = 0.9726$.

Magnetic rotation:

t.	Sp. rot.	Mol. rot.
16·3°	1·1576	11·043

Refractive power:

$$t = 15.3°; \quad d15.3°/4° = 0.97813.$$

	μ.	$\dfrac{\mu - 1}{d}$.	$\dfrac{\mu - 1}{d}p$.
a...............	1·46539	0·47580	79·935
β	1·47604	0·48668	81·762
γ...............	1·48259	0·49338	82·887

dl-Δ^3-*p*-menthenol(8) described in the present communication. During the investigation of their physical properties, the specimens were fused and subsequently sent by post, in cold weather, from Sudbury to Manchester without again solidifying. They remained in this condition for several days, and it was not until a crystal of each had been introduced that crystallisation took place.

In discussing the above results it will be convenient to consider, in the first instance, the effect which slight differences in constitution have on the densities of the substances examined. In the case of terpineol (Δ^1-p-*menthenol*-8) and dl-Δ^3-p-menthenol(8),

$$\text{Me·C}\underset{\text{CH}_2\text{·CH}_2}{\overset{\text{CH--CH}_2}{<}}\hspace{-4pt}>\text{CH·CMe}_2\text{·OH} \quad \text{and}$$

$$\text{Me·CH}\underset{\text{CH}_2\text{·CH}_2}{\overset{\text{CH}_2\text{--CH}}{<}}\hspace{-4pt}>\text{C·CMe}_2\text{·OH},$$

we have the following numbers for comparison :

Terpineol..................... $d15°/15° = 0.9415$ ⎱ Difference.
dl-Δ^3-p-Menthenol(8) ...... $d15°/15° = 0.9217$ ⎰ $+0.0198$

It is therefore evident that the change of the double-linking from the Δ^1- to the Δ^3-position has, in the case of these substances, only brought about a slight difference in the density.

Turning next to the hydrocarbons, dipentene (dl-$\Delta^{1\,S(9)}$-p-*menthadiene*) and dl-$\Delta^{3\,S(9)}$-p-menthadiene,

$$\text{Me·C}\underset{\text{CH}_2\text{·CH}_2}{\overset{\text{CH--CH}_2}{<}}\hspace{-4pt}>\text{C}_{\text{H}}\text{·C}\underset{\text{Me}}{\overset{\text{CH}_2}{<}} \quad \text{and} \quad \text{Me·CH}\underset{\text{CH}_2\text{·CH}_2}{\overset{\text{CH}_2\text{--CH}}{<}}\hspace{-4pt}>\text{C·C}\underset{\text{Me}}{\overset{\text{CH}_2}{<}},$$

we obtain the following comparison :

Dipentene .... $d15°/15° = 0.8548$ ⎱ Difference.
dl-$\Delta^{3.S(9)}$-p-Menthadiene...... $d15°/15° = 0.8390$ ⎰ $+0.0158$

Here again the change of density, due to the alteration of the position of the double linking in the ring, is very small, a fact which is interesting when taken in connection with the great difference observed in the magnetic rotation values of these two hydrocarbons. The difference in density is, in fact, less than that between the dl- and d-modifications of $\Delta^{3.8(9)}$-p-menthadiene :

dl-$\Delta^{3.8(9)}$-p-Menthadiene ...... $d15°/15° = 0.8390$ ⎱ Difference.
d- ,, ,, ,, ... $d15°/15° = 0.8634$ ⎰ -0.0244

When these last figures are compared with the corresponding values in the limonene series :

Dipentene .... $d15°/15° = 0.8548$ ⎱ Difference.
d-Limonene* .... $d15°/15° = 0.8498$ ⎰ $+0.005$

it is seen that the difference between the two sets of externally compensated and dextro-modifications is not only considerably larger in the former than in the latter, but also that in the first pair the d-modification has the higher density, whereas the reverse is the case in the limonene series.

* Trans., 1902, **81**, 315.

Taking next into consideration the magnetic rotation values of the menthenols, the following numbers are available for comparison :

Terpineol .... 10·841 ⎱ Difference.
dl-Δ^3-p-Menthenol(8) .... 11·150 ⎰   − 0·309

In other words, the change of the double linking from the Δ^1- to the Δ^3-position has had a distinct but relatively small influence on the magnetic rotation of these substances. It is in the magnetic rotations of the hydrocarbons derived from these menthenols that the most striking differences are to be observed. This was first noticed during a comparison of the values obtained for the magnetic rotations of d-$\Delta^{3\cdot8(9)}$-p-menthadiene and d-limonene (d-$\Delta^{1\cdot3(9)}$-p-menthadiene) :

d-$\Delta^{3\ 8(9)}$-p-Menthadiene............ 12·939 ⎱ Difference.
d-Limonene*...................... 11·246 ⎰   + 1·693

This great difference between the values of two hydrocarbons so closely related is one of the most remarkable observations which has been made during the investigation of the magnetic rotations of organic substances. In order to be certain that this difference was not due to some experimental error, it was thought advisable to prepare pure specimens of dipentene and dl-$\Delta^{3\ 8(9)}$-p-menthadiene and examine their rotations and also to investigate the physical constants of ethyl $dl\Delta^1$-tetrahydro-p-toluate, from which the latter hydrocarbon is derived. The values found for the inactive hydrocarbons are the following :

dl-$\Delta^{3\ 8(9)}$-p-Menthadiene............ 12·939 ⎱ Difference.
Dipentene† .... 11·315 ⎰   + 1·624

This comparison shows that the difference between the rotations of the inactive hydrocarbons is again abnormal and practically the same as that observed in the case of the active modifications. There can therefore be no doubt that this remarkable difference [which is larger than would be required for the introduction of two additional double

---

* The magnetic rotation of l-limonene is 11·162 (Trans., 1902, **81**, 315).

† It may be readily shown that the rotation of dipentene is normal, that is to say, that the value actually observed agrees with that which would be expected as the magnetic rotation of a tetrahydrocymene. The difference between the rotation of benzene (11·284) and hexahydrobenzene (5·664) is 5·620, a number which may be taken as generally representing the effect produced when a benzene hydrocarbon is converted into its hexahydro-derivative.

The magnetic rotation of hexahydrocymene, deduced from that of cymene (15·255) by subtracting the value 5·620, is therefore 9·635.

If to this be added the effect of introducing two double linkings (2 × 0·720 ; compare Trans., 1902, **81**, 300), the calculated rotation of a tetrahydrocymene will be approximately 11·075, a number which agrees well with that actually found in the case of dipentene (11·216).

linkings (2×0.720) into the molecule] is not due to experimental error, and the question naturally arises, what explanation can be offered to account for a difference of this magnitude? It was pointed out on p. 853 that, except in respect of the position of the double linking in the ring, the hydrocarbons $\Delta^{3\,8(9)}$-p-menthadiene and dipentene are exactly similarly constituted, therefore any variation in properties can only be due to a difference in the position of this double linking. Previous researches have, however, shown that the mere alteration of the position of a double linking does not bring about any great change in magnetic rotation.

This fact is well seen from a comparison of the values of dl-Δ^3-p-menthenol(8) and terpineol (see p. 854), the difference being only 0.309. But this small difference is due to precisely the same change in the position of the double linking as the great difference (1.624) between the values of $\Delta^{3\,(9)}$-p-menthadiene and dipentene. Since, then, mere change of the position of the double linking is not sufficient to account for the remarkable alteration in magnetic rotation, the reason must, apparently, be sought in the relative positions of the double linkings.

It has already been pointed out (Perkin, Pickles, and Tattersall, Trans., 1905, 87, 641 ; compare also 1077 and 1101) that dipentene and $\Delta^{3\,8(9)}$-p-menthadiene,

$$\mathrm{Me \cdot C}{\displaystyle <}^{\mathrm{CH-CH_2}}_{\mathrm{CH_2 \cdot CH_2}}{\displaystyle >}\mathrm{CH \cdot C}{\displaystyle <}^{\mathrm{CH_2}}_{\mathrm{Me}} \quad \text{and} \quad \mathrm{Me \cdot CH}{\displaystyle <}^{\mathrm{CH_2-CH}}_{\mathrm{CH_2 \cdot CH_2}}{\displaystyle >}\mathrm{C \cdot C}{\displaystyle <}^{\mathrm{CH_2}}_{\mathrm{Me}},$$

while very similar in most of their properties, exhibit a remarkable difference in their behaviour towards bromine, hydrogen bromide, and hydrogen chloride.

Dipentene, as is well known, combines with *two* molecules of bromine or of hydrogen bromide or chloride to yield the derivatives $C_{10}H_{16}Br_4$, $C_{10}H_{16}2HBr$, or $C_{10}H_{16}2HCl$. $\Delta^{3.8(9)}$-p-Menthadiene, on the other hand, is only capable of combining with *one* molecule of bromine, hydrogen bromide, &c., and when the compounds $C_{10}H_{16}Br_2$ and $C_{10}H_{16}HBr$ have been produced the additive capacity of the hydrocarbon is exhausted. This striking difference in behaviour was attributed to the fact that the formula of $\Delta^{3\,8(9)}$-p-menthadiene has two double linkings in the position $-C\vdots C\cdot C\vdots C-$, whilst the formula of dipentene does not contain this grouping. It is well known that substances containing the above conjugated system of double linkings have unusual properties and are only capable of uniting with one molecule of bromine (or HBr or HCl) to form derivatives in which this grouping becomes $-CBr \cdot C\vdots C \cdot CBr-$. It seems clear that the remarkably high value for the magnetic rotation of $\Delta^{3\,8(0)}$-p-menthadiene must also be ascribed to the presence of this grouping. Future investigation will probably show that other terpenes of similar constitution will yield magnetic rotations higher by about 1·63 than those of hydrocarbons which, whilst in other respects similarly constituted, do not contain this conjugated grouping.

It is rather curious that no correspondingly great difference should have been observed between the permanent rotations of d-$\Delta^{3.8(9)}$-p-menthadiene and d-limonene, and that the values actually observed should be so nearly the same. Again, it is interesting to note that the comparison of the refractive powers of the dl-hydrocarbons :

$$\frac{\mu - 1}{d}p.$$

dl-$\Delta^{3.8(9)}$-p-Menthadiene	76·315 }	Difference.
Dipentene.............,.................	75·602 }	+0·713

exhibits only a very small and not unusual difference, which cannot, of course, be compared with that observed in the case of the magnetic rotations.

For reasons stated on p. 849, the physical constants of ethyl Δ^1-tetrahydro-p-toluate were included in this investigation. The numbers obtained (p. 852) are quite normal and do not call for special comment.

The authors are indebted to the Research Fund Committee of the Society for a grant which has defrayed much of the expense of this investigation.

THE SCHUNCK LABORATORY,
    THE VICTORIA UNIVERSITY OF MANCHESTER.

LXXXIX.—*The Constitution of the Hydroxides and Cyanides obtained from Acridine, Methylacridine, and Phenanthridine Methiodides.*

By CHARLES KENNETH TINKLER, B.Sc.

IN several cases it has been shown that when the methiodides of quinoline and *iso*quinoline compounds are treated with a soluble base there is reason to believe that the following change occurs :

$$-\text{CH:NCH}_3\text{I} \longrightarrow -\text{CH:NCH}_3\text{OH} \longrightarrow -\text{CH(OH)·NCH}_3,$$

and in the case of acridine compounds,

$$R' <\!\!\begin{array}{c} CR \\ | \\ N \end{array}\!\!> R' \;\longrightarrow\; R' <\!\!\begin{array}{c} CR \\ | \\ N- \end{array}\!\!> R' \;\longrightarrow\; R' <\!\!\begin{array}{c} C(OH)R \\ | \\ -N- \end{array}\!\!> R',$$

$$\begin{array}{ccc} \overset{|}{C}H_3 \;\; I & \overset{|}{C}H_3 \;\; OH & \overset{|}{C}H_3 \end{array}$$

the ammonium base first produced being converted into an isomeric carbinol (Decker, *J. pr. Chem.*, 1893, **47**, 222 ; Hantzsch, *Ber.*, 1899, **32**, 3109 ; Dobbie, Lauder, and Tinkler, *Trans.*, 1903, **83**, 598, and Dobbie and Tinkler, *Trans.*, 1904, **85**, 1005, and 1905, **87**, 269). The evidence adduced from the ultra-violet absorption spectra in support of one or other of the two possible formulæ for these hydroxides is obtained by comparing the spectra of the hydroxide under different conditions with those of its various derivatives whose constitution is known with certainty.

By treating the solution of the methiodides with silver oxide instead of with a soluble base, the ammonium form of the hydroxide is, in some cases, obtained in the solution (Hantzsch, *loc. cit.*, Dobbie, Lauder, and Tinkler, and Dobbie and Tinkler, *loc. cit.*), and by the action of water on some of the precipitated carbinols a solution is obtained which contains the substance in the ammonium form, whilst in alcohol the solution contains a mixture of the two forms (Dobbie, Lauder, and Tinkler, and Dobbie and Tinkler, *loc. cit.*). On the addition of a soluble base to the aqueous solutions, the substances are reconverted into the carbinol form, and since this change is probably due to the presence of hydroxyl ions in the solution of the base, which causes a diminution of the dissociation of the ammonium form and reconversion to the carbinol form, a method for comparing the strengths of the bases has been founded on this change (Dobbie, Lauder, and Tinkler, *Trans.*, 1904, **85**, 121).

By treating the methiodides with potassium cyanide, substances are produced in some cases which are constituted in the same way as the hydro-compounds and carbinols (Hantzsch, *Ber.*, 1899, **32**, 3126 ; Dobbie, Lauder, and Tinkler, and Dobbie and Tinkler, *loc. cit.*).

It was on account of the remarkable results obtained by a study of the absorption spectra of cotarnine and its derivatives that the investigation of similar cases of isomerism was undertaken by this method. In the present paper, an account is given of the results obtained by the action of (i) alkalis, (ii) potassium cyanide, on acridine, methylacridine, and phenanthridine methiodides.

The Hydroxides.

(*a*) Acridine combines with methyl iodide when heated with this reagent in a sealed tube. The substance obtained is dark red, and dissolves in water, producing a yellow solution showing a green fluorescence. By the addition of caustic soda to the aqueous solution, the colour disappears, and a white precipitate is slowly formed. If,. however, alcohol be present, no precipitation takes place, but a colourless solution is obtained (Decker, *loc. cit.*). The spectra of the solution of the methiodide before and after the addition of caustic soda are quite distinct : for whilst those of the methiodide agree closely with those of acridine hydrochloride, the spectra of the aqueous-alcoholic solution of the methiodide to which caustic sod ι has been added agree very closely with those of hydroacridine, a substance to which is assigned the formula

$$C_6H_4{<}^{CH_2}_{NH}{>}C_6H_4$$

(curves I and II).

Scale of Oscillation-frequencies.

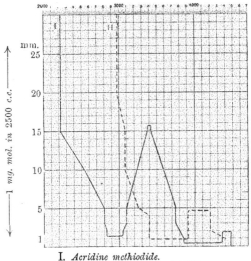

I. *Acridine methiodide.*
II.   ,,   ,,   + NaOH.

The curve of hydroacridine is practically identical with II.

This agreement can only be explained by adopting the carbinol formula for the hydroxide, as suggested by Decker, a change from the ammonium base to the carbinol form taking place as follows :

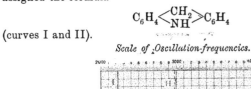

On removing the iodine from the methiodide by means of silver oxide, an exactly similar result is obtained. In this case, as in the case of phenylacridine, the ammonium hydroxide appears to be very readily converted into the carbinol form.

(b) Methylacridine gives results exactly similar to those obtained with acridine and phenylacridine. The addition of caustic soda to the aqueous-alcoholic solution of the methiodide causes a complete change in the spectra, such as would be expected from the conversion of the ammonium form of the hydroxide to the carbinol form, as follows :

$$C_6H_4 \underset{\overset{\displaystyle\wedge}{CH_3\ \ OH}}{\overset{\displaystyle C(CH_3)}{<\underset{N}{\quad}>}} C_6H_4 \ \longrightarrow\ C_6H_4 \underset{\overset{|}{CH_3}}{\overset{\displaystyle C(CH_3)(OH)}{<\underset{N}{\quad}>}} C_6H$$

(c) Phenanthridine, which is isomeric with acridine, is prepared by passing the vapour of benzylidene aniline through an iron tube filled with pumice and heated to bright redness. It is regarded as having the constitution $\dfrac{C_6H_4 \cdot CH}{C_6H_4 \cdot N}$. It yields a methiodide when heated with methyl iodide in a sealed tube, and on treating this compound, which is deep yellow, with caustic soda, a colourless hydroxide is obtained (Pictet and Ankersmit, *Annalen*, 1891, **266**, 138). The absorption spectra of the colourless hydroxide agree very closely with those of the colourless hydro-compound, $\dfrac{C_6H_4 \cdot CH_2}{C_6H_4 \cdot NH}$, obtained by the reduction of phenanthridine (Pictet and Ankersmit, *loc. cit.*), whilst those of the methiodide are quite distinct from and agree closely with the spectra of phenanthridine hydrochloride. It may therefore be concluded that in this case also a change in the linking occurs under the influence of the alkali, with production of a carbinol (curves III and IV):

$$\dfrac{C_6H_4 \cdot CH}{C_6H_4 \cdot N(CH_3)(OH)} \ \longrightarrow\ \dfrac{C_6H_4 \cdot CH(OH)}{C_6H_4 \cdot N \cdot CH_3}.$$

Further evidence in support of this is obtained from a study of the compound produced from the methiodide by the action of potassium cyanide.

On removing the iodine from the methiodide by means of silver oxide, the ammonium form is also converted into the carbinol form as in the case of the acridine compounds.

The Cyanides.

It has been shown by Hantzsch that on treating the salts of cotarnine and phenylmethylacridine, which are strongly-coloured substances, with potassium cyanide, colourless substances are obtained, which are regarded as *pseudo*cyanides (*Ber.*, 1899, **32**, 3126):

$$C_8H_6O_3 \underset{CH_2-CH_2}{\overset{CH=NCH_3 \cdot Cl}{\big<}} + KCN = C_8H_6O_3 \underset{CH_2----CH_2}{\overset{CH(CN) \cdot N(CH_3)}{\big<}} + KCl.$$

$$C_6H_4 \underset{\underset{\underset{CH_3}{\big/ \big\backslash}}{N----}}{\overset{C(C_6H_5)}{\big<}} \!\!\! >\!\! C_6H_4 + KCN = C_6H_4 \underset{----N----}{\overset{C(C_6H_5)CN}{\big<}} \!\!\! >\!\! C_6H_4 + KI.$$
$$\underset{I}{} \qquad\qquad\qquad \underset{CH_3}{}$$

The substances are represented as being constituted similarly to the carbinols and hydro-compounds previously described.

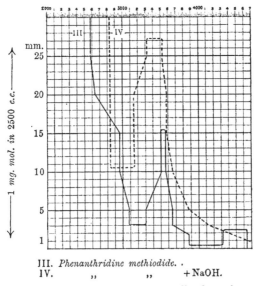

Scale of Oscillation-frequencies.

III. *Phenanthridine methiodide.* ·
IV.   „   „   + NaOH.

The curve of hydrophenanthridine is practically identical with **IV**.

Chemical evidence in support of this structure for these substances was obtained by Freund (*Ber.*, 1900, **33**, 380), who prepared a methiodide from the cotarnine compound, a result most readily explained by regarding the latter as possessing a similar constitution to hydrocotarnine. A study of the absorption spectra also affords confirmation of this view. The results obtained with the cyanogen compounds of cotarnine and hydrastinine have already been published (Dobbie, Lauder, and Tinkler, Trans., 1903, **83**, 595, and Dobbie and Tinkler, Trans., 1904, **85**, 1005).

The cyanide obtained from phenylacridine methiodide, described by Hantzsch (*Ber.*, 1899, **32**, 3126), has been prepared, and its absorption spectra examined. It is found that this substance in ethereal or chloroform solution gives spectra almost identical with those of hydrophenylacridine and phenylmethylacridol (curve V, Trans., 1905, **87**, 270), thus yielding evidence in support of the constitution $C_6H_4\underset{\underset{\overset{|}{CH_3}}{N}}{\overset{C(C_6H_5)CN}{\diagup\diagdown}}C_6H_4$ for this substance. Solutions of the substance in ether and chloroform are colourless, but the alcoholic solution is yellow, and gives spectra which differ from those of the colourless solutions. The spectra shown by the alcoholic solution are those of a mixture of the hydro-compound with the true salts. A similar result was obtained in the case of hydrastinine (Dobbie and Tinkler, *loc. cit.*). It is possible that by solution in alcohol some of the cyanide is changed into a true salt,

$$C_6H_4\underset{\underset{CH_3\ \ CN}{\diagup\diagdown}}{\underset{N}{\overset{C\,(C_6H_5)}{\diagup\diagdown}}}C_6H_4,$$

or that some of the substance is converted into an alcoholate, which assumes the ammonium form. The amount of the ammonium compound formed can be estimated by comparing the spectra of the alcoholic solution with those of the hydro-compound and a true salt, mixed in known proportions, as in the case of cotarnine (Dobbie, Lauder, and Tinkler, *loc. cit.*). On the addition of potassium cyanide to the alcoholic solution of the substance, the colour disappears, and the spectra agree with those of the ethereal and chloroform solutions.

In the case of acridine and methylacridine, the corresponding *pseudo*cyanides have not been isolated. On the addition of potassium cyanide to the alcoholic solutions of the methiodides, the colour disappears, and the spectra given by the solutions are those of mixtures of the carbinol and ammonium forms.

A pseudo*cyanide* has been prepared in the case of phenanthridine by the following method. A solution of potassium cyanide was added to the alcoholic solution of phenanthridine methiodide until the colour of the solution entirely disappeared. The mixture was

then poured into a large volume of water and the turbid liquid so obtained shaken for three hours. In this way the pseúdo*cyanide* was
obtained as a white, granular precipitate. It is readily soluble in ether
and chloroform and melts at 120°.

0·1255 gave 13·6 c.c. moist nitrogen at 12° and 758 mm. $N = 12·85$.
$C_{15}H_{12}N_2$ requires $N = 12·73$ per cent.

The spectra of the ethereal and chloroform solutions of this substance
agree very closely with those of hydrophenanthridine (curves VI
and IV).

Scale of Oscillation-frequencies.

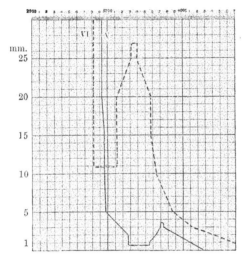

V. *Phenylmethylacridine cyanide* (1 *mg. mol. in* 1000 *c.c.*).
VI. *Methyl phenanthridine cyanide* (1 *mg. mol. in* 2500 *c.c.*).

This agreement can only be explained by adopting the constitution
of a *pseudo*cyanide for this substance, $\begin{array}{l} C_6H_4 \cdot C\,H(CN) \\ C_6H_4 \cdot N \cdot CH_3 \end{array}$.
The spectra of the alcoholic solution are those of a mixture of the
two forms; whether the ammonium form present consists of the
true cyanide or of an alcoholate, it is impossible to decide from the
results of the spectroscopic examination.

THE UNIVERSITY,
    BIRMINGHAM.

XC.—*The Residual Affinity of Coumarin as shown by the Formation of Oxonium Salts.*

By GILBERT THOMAS MORGAN and FRANCES MARY GORE MICKLETHWAIT.

COUMARIN, which is only sparingly soluble in cold water, readily dissolves in concentrated hydrochloric or hydrobromic acid, and from the latter solvent Ebert long ago isolated an extremely unstable hydrobromide which could only be kept at low temperatures and dissociated into its generators before it could be analysed (*Annalen*, 1884, **226**, 347).

The authors, on adding a strong solution of chloroplatinic acid to the coumarin dissolved in concentrated hydrochloric acid, have obtained a well-defined, crystalline substance, which has the composition of a *coumarin platinichloride*, $4C_9H_6O_2,H_2PtCl_6,4H_2O$. This compound, which separates in yellow needles, is quite stable when dried at the ordinary temperature, but is dissociated into its generators by water; it can be recrystallised from concentrated hydrochloric acid, but only in the presence of excess of chloroplatinic acid. The substance was prepared for analysis by washing with cold concentrated hydrochloric acid and drying in a desiccator over quicklime until it ceased to evolve hydrogen chloride or water.

0·2938 gave 0·4304 CO_2 and 0·0911 H_2O. C = 39·96 ; H = 3·44.
0·2880 „ 0·4240 CO_2 „ 0·0850 H_2O. C = 40·15 ; H = 3·28.
0·2568 „ 0·2094 AgCl. Cl = 20·17.
0·2004 „ 0·0367 Pt. Pt = 18·32.
0·3274 „ 0·0597 Pt. Pt = 18·23·
$C_{36}H_{24}O_8,H_2PtCl_6,4H_2O$ requires C = 40·52 ; H = 3·19 ; Cl = 20·00 ; Pt = 18·30 per cent.

It was not found possible to estimate directly the water of crystallisation, as on heating in a current of dry air the substance fused at 95—100°, and from this temperature onward evolved both steam and hydrogen chloride. When treated with water, the double salt readily regenerated coumarin, which gave the correct melting point (69°) and had the other characteristic properties of the lactone.

The preparation of the salt was varied by changing the preportions of coumarin and chloroplatinic acid, and also the concentration of the hydrochloric acid, by passing in gaseous hydrogen chloride, but the composition of the product remained the same. A specimen of the platinichloride recrystallised from concentrated hydrochloric acid containing chloroplatinic acid furnished the following result on analysis:

0·3003 gave 0·0544 Pt.   Pt = 18·11 per cent.

It follows therefore that the compound can be recrystallised under these conditions without changing in composition.

Coumarin aurichloride.—This double salt was produced far less readily than its platinum analogue; when coumarin dissolved in con-ᴊrated hydrochloric acid was treated with a concentrated aqueous solution of chloroauric acid, an amber-coloured oil separated, which under these conditions refused to crystallise when kept for three weeks, but was obtained in a solid form after twenty-four hours by saturating the mixture with dry hydrogen chloride. The yellow salt, being somewhat soluble in concentrated hydrochloric acid, was not easily freed from excess of chloroauric acid, and was probably not quite pure.

0·2613 gave 0·4060 CO_2 and 0·0693 H_2O.   C = 42·38; H = 2·94.

0·2630 ,,   0·1544 AgCl.   Cl = 14·52.

0·3014 ,,   0·0592 Au.   Au = 19·64.

$4C_9H_6O_2,HAuCl_4,4H_2O$ requires C = 43·37; H = 3·41; Au = 19·77;
Cl = 14·25 per cent.

The aurichloride melts indefinitely at 48—51° and evolves steam and hydrogen chloride at 100—120°.

Coumarin platinibromide.—A concentrated solution of bromo-platinic acid, when added to coumarin dissolved in fuming hydro-bromic acid, yielded an orange-coloured precipitate which could not be obtained sufficiently pure for analysis, as it dissociated into its generators on washing with strong hydrobromic acid.

Double Salts of Coumarin Derivatives containing Nitrogen.

The foregoing experiments having shown that coumarin is capable of forming salt-like compounds, its amino-derivatives were examined with the object of ascertaining whether they had any residual affinity beyond that which arises from the quinquevalency of nitrogen.

6-Aminocoumarin, dissolved in concentrated hydrochloric acid and treated with chloroplatinic acid, yielded a yellow, crystalline preci-pitate which, when washed with concentrated hydrochloric acid and dried over quicklime, furnished the following results:

0·2568 gave 0·0686 Pt.   Pt = 26·71 per cent.

The filtrate was then treated with a further excess of platini-chloride, when a second pale yellow precipitate was formed, which was washed with concentrated hydrochloric acid and dried over lime.

0·1652 gave 0·0444 Pt.   Pt = 26·87.

$(C_9H_6O_2\cdot NH_2)_2,H_2PtCl_6$ requires Pt = 26·63 per cent.

This result, which appeared to indicate that in this case the residual affinity of the coumarin complex for chloroplatinic acid was masked by the presence of aminic nitrogen, was confirmed by examining the behaviour of 6-coumaryltrimethylammonium bromide towards bromo-platinic acid. The quaternary salt, when dissolved in hot water and treated with a considerable excess of bromoplatinic acid, yielded a well-defined, sparingly soluble platinibromide, which was collected, washed, and dried over lime.

0·1898 gave 0·1792 CO_2 and 0·0478 H_2O.   C = 25·75 ; H = 2·80.
0·3030 ,,   0·0538 Pt.   Pt = 17·75.
$C_{24}H_{28}O_4N_2Br_6Pt,2H_2O$ requires C = 25·73 ; H = 2·85 ; Pt = 17·42 per cent.

6-Coumaryltrimethylammonium platinibromide,
$$[C_9H_6O_2\cdot N(CH_3)_3]_2PtBr_6,2H_2O,$$
separates in deep orange-red prisms, which evolve steam at about 110° and melt somewhat indefinitely at 218—220° ; the corresponding platinidibromotetrachloride is a yellow, crystalline salt somewhat more soluble than the hexabromide.

As the presence of aminic nitrogen in the coumarin molecule seems to inhibit the formation of the double salts due to the residual affinity of the lactonic complex, the acyl derivatives of 6-aminocoumarin were examined, because in these substances the acyl group to a very large extent, if not entirely, destroys the basic character of the nitrogen. Preliminary experiments showed that the acetyl, propionyl, lactyl, and benzoyl derivatives of 6-aminocoumarin dissolved more or less readily in concentrated hydrochloric acid without undergoing hydrolysis, and the solutions yielded platinichlorides on treatment with chloroplatinic acid. Formyl-6-aminocoumarin under these conditions was very considerably hydrolysed and yielded 6-aminocoumarin platinichloride.

Acetyl-6-aminocoumarin platinichloride,
$$(CH_3\cdot CO\cdot NH\cdot C_9H_5O_2)_4,H_2PtCl_6,4H_2O.$$
—One gram of acetyl-6-aminocoumarin when dissolved in 8 c.c. of slightly warm concentrated hydrochloric acid and treated with excess of chloroplatinic acid in the same solvent yielded a pale yellow, crystalline precipitate, which was washed with concentrated hydrochloric acid and dried on a porous tile over lime.

0·5521 gave 0·8050 CO_2 and 0·1910 H_2O.   C = 39·77 ; H = 3·84.
0·2550 ,,   0·1666 AgCl.   Cl = 16·17.
0·5124 ,,   18·4 c.c. nitrogen at 18° and 769 mm.   N = 4·20.

0·9019 gave 0·1376 Pt.   Pt = 15·25.

$C_{44}H_{36}O_{12}N_4,H_2PtCl_6,4H_2O$ requires C = 40·80 ; H = 3·55 ; Cl = 16·45 ; N = 4·32 ; Pt = 15·07 per cent.

This substance darkened and gave off water at about 115° and decomposed indefinitely at higher temperatures ; it was readily dissociated by water, and the white, insoluble product was shown to be acetyl-6-aminocoumarin (m. p. 216—217°).

Acetyl-6-aminocoumarin aurichloride,

$$(CH_3 \cdot CO \cdot NH \cdot C_9H_5O_2)_2,HAuCl_4,2H_2O,$$

was prepared by adding a concentrated aqueous solution of chloroauric acid to the acetyl derivative dissolved in concentrated hydrochloric acid ; it separated as a pale yellow, pulverulent precipitate, which was washed with this acid only with some difficulty and probably contained a slight excess of auric chloride, which is somewhat sparingly soluble in the strong acid.

0·3150 gave 0·0823 Au. Au = 26·12.

0·2272 ,, 0·1711 AgCl and 0·0593 Au. Cl = 18·63 ; Au = 26·10.

$(CH_3 \cdot CO \cdot NH \cdot C_9H_5O_2)_2,HAuCl_4,2H_2O$ requires Au = 25·19 ; Cl = 18·15 per cent.

Propionyl-6-aminocoumarin, $C_2H_5 \cdot CO \cdot NH \cdot C_9H_5O_2$.—Four grams of 6-aminocoumarin dissolved in warm propionic acid were treated with 4 grams of propionic anhydride and the solution boiled for four to six hours ; the product, when crystallised from water with animal charcoal, separated in colourless needles melting at 186—188°.

0·1729 gave 9·5 c.c. nitrogen at 18° and 769 mm. N = 6·43.

$C_{12}H_{11}O_3N$ requires N = 6·45 per cent.

The platinichloride of propionyl-6-aminocoumarin was so exceedingly soluble in concentrated hydrochloric acid that it could not be washed free from the excess of platinic chloride.

Coumarin hydriodide periodide, $4C_9H_6O_2,HI,I_3$, was prepared by shaking together a warm benzene solution of coumarin with a small amount of concentrated hydriodic acid containing free iodine. As the mixture cooled, dark green needles separated and were collected, washed successively with benzene and light petroleum, and dried over quicklime. The substance may also be prepared by dissolving coumarin in hot fuming hydriodic acid, adding iodine in moderate excess, and allowing the solution to remain for some time ; dark green nodules or needles separate, depending on the concentration.

0·0908 gave 0 0766 AgI. I = 45·51.

0·1862 ,, 0·2775 CO_2 and 0·0500 H_2O. C = 40·64 ; H = 2·98.

0·2466 gave 0·3604 CO_2 and 0·0650 H_2O.   C = 39·86 ; H = 2·92.

0·2474 required 6·6 c.c. $Na_2S_2O_3$ solution (1 c.c. = 0·013102 I). I = 34·95.

0·3424 required 9·0 c.c. $Na_2S_2O_3$ solution (1 c.c. = 0·013102 I). I = 34·43.

$C_{36}H_{24}O_8$,HI,I_3 requires C = 39·60 ; H = 2·29 ; total I = 46·37 ; periodide, I = 34·77 per cent.

A volumetric estimation of the acidity of the substance indicated that one-fourth of the iodine was present as hydrogen iodide.

The hydriodide periodide may be recrystallised from fuming hydriodic acid ; it is, however, readily dissociated by water, regenerating coumarin. This compound may be compared with the periodides $(C_7H_8O_2)_2HI,I_2$ and $(C_9H_{12}O_2)_2HI,I_2$, obtained by Collie and Steele from dimethyl- and tetramethyl-pyrone respectively (Trans., 1900, 77, 1114).

Acetyl-6-aminocoumarin hydriodide periodide,

$$CH_3 \cdot CO \cdot NH \cdot C_9H_5O_2, HI, I_2,$$

was prepared by mixing and shaking together in a stoppered bottle solutions of acetyl-6-aminocoumarin and iodine in concentrated hydriodic acid. The product, which separated in well-defined, dark brown crystals, was washed successively with concentrated hydriodic acid and light petroleum, and dried over lime and caustic potash.

0·4234 gave 0·3546 CO_2 and 0·0782 H_2O. C = 22·84 ; H = 2·05.

0·1548 ,, 0·1835 AgI. I = 64·10.

0·3891 required 13·6 c.c. $Na_2S_2O_3$ (1 c.c. = 0·01247 I). I = 43·57.

0·3754 ,, 13·05 ,, ,, ,, ,, I = 43·35.

$C_{11}H_{10}O_3NI_3$ requires C = 22·56 ; H = 1·74 ; total I = 65·13 ; periodide, I = 43·42 per cent.

A volumetric estimation of the acidity of the substance showed that one-third of the iodine was present as hydrogen iodide.

Coumarin cobalticyanide, $3C_9H_6O_2, H_3Co(CN)_6, 3H_2O$, was readily obtained as a white, crystalline precipitate by adding to a solution of coumarin in concentrated hydrochloric acid an excess of the reagent produced by mixing together 6 grams of potassium cobalticyanide, 18 c.c. of water, and 18 c.c. of concentrated hydrochloric acid, and filtering off the precipitated potassium chloride (compare Baeyer and Villiger, *Ber.*, 1901, 34, 2687). The product was washed repeatedly with cold concentrated hydrochloric acid and dried on a porous tile over lime and caustic potash.

0·2140 gave 0·4327 CO_2 and 0·0816 H_2O. C = 55·14 ; H = 4·24.

0·1912 ,, 20 c.c. nitrogen at 18° and 756 mm. N = 12·02.

0·416 ,, 0·0524 Co_3O_4. Co = 8·34.

$C_{33}H_{27}O_9N_6Co$ requires C = 55·77 ; H = 3·80 ; N = 11·83 ; Co = 8·31 per cent.

This substance, when heated at about 110°, evolved both steam and hydrogen cyanide ; on treatment with water, it was readily dissociated into its generators.

The foregoing platinichlorides of coumarin and acetyl-6-amino-coumarin are compounds of the type $4X\ddot{:}O,H_2PtCl_6$, other members of which have already been described. The phosphine oxides give rise to analogous derivatives ; tribenzylphosphine oxide platinichloride, $4(C_7H_7)_3P\ddot{:}O,H_2PtCl_6$, was discovered in 1880 by Letts and Collie (*Trans. Roy. Soc. Edin.*, 1880, **30**, [i], 181), whilst the corresponding trimethyl and triethyl derivatives have recently been described by Pickard and Kenyon (this vol., p. 268).

Another member of this group, which is even more closely related to the coumarin derivative, is pyrone platinichloride, $4C_5H_4O_2,H_2PtCl_6$, prepared by Werner (*Annalen*, 1902, **322**, 312). Pyrone (I) and coumarin (II), the organic constituents of these two platinichlorides, each contain in their molecules a ring partly composed of an oxidic oxygen atom, a carbonyl group, and two ethylenic carbon atoms :

$$
\begin{array}{cc}
\text{I.} & \text{II.}
\end{array}
$$

The alternative view of the constitution of the pyrones (III) recently put forward by Miss Homfray (Trans., 1905, **87**, 1453) does not destroy this analogy, for coumarin may also be considered to exist in the desmotropic form (IV) :

$$
\begin{array}{cc}
\text{III.} & \text{IV.}
\end{array}
$$

The group of platinichlorides of the type $4X\ddot{:}O,H_2PtCl_6$ is accordingly divisible into two series :

Phosphine oxide series.		Cyclic series.
$4(CH_3)_3P\ddot{:}O,H_2PtCl_6$	Pyrone	$4C_5H_4O_2,H_2PtCl_6$
$4(C_2H_5)_3P\ddot{:}O,H_2PtCl_6$	Coumarin	$4C_9H_6O_2,H_2PtCl_6,4H_2O$
$4(C_7H_7)_3P\ddot{:}O,H_2PtCl_6$		$4CH_3\cdot CO\cdot NH\cdot C_9H_5O_2,H_2PtCl_6,4H_2O$

Whatever view be adopted with regard to the configuration of the foregoing platinichlorides, it is evident that they cannot be considered as normal oxonium salts, since they contain twice as much of the organic constituent as is present in the molecule of dimethylpyrone platinichloride, $2C_7H_8O_2,H_2PtCl_6$ (Collie and Tickle, Trans., 1899, **75,** 712), a substance which may be derived from the normal oxonium hydrochloride, $(C_7H_8O){:}O{<}^H_{Cl}$, by interaction with chloroplatinic acid.

A survey of the literature of oxygen additive compounds shows that there are many of these substances which do not conform to the simple type

$$\frac{R}{R'}{>}O{<}^H_{Cl}$$

V.

Even in the pyrone series it is only the salts of dimethylpyrone which generally correspond with this type ; pyrone and tetramethylpyrone both yield salts in which the amount of organic constituent is double that required by formula V.

In discussing the constitution of pyrone platinichloride, Werner (*loc. cit.*) employs the conception of subsidiary valency ("Nebenvalenz"), according to which the oxidic oxygen of pyrone, in addition to its two principal valencies ("Hauptvalenzen"), has one subsidiary valency, which is satisfied by the subsidiary valencies of the hydrogen atoms of chloroplatinic acid, each of these hydrogens being further assumed to be capable of satisfying the subsidiary valency of two pyronic oxygens. He formulates the compound in the following manner (VI), employing dotted lines to represent the subsidiary valencies :

$$PtCl_6{<}\overset{H{<}\overset{O{:}C_5H_4O}{O{:}C_5H_4O}}{H{<}\overset{O{:}C_5H_4O}{O{:}C_5H_4O}} \qquad PtCl_6{<}\overset{H{<}\overset{O{:}C_9H_6O,H_2O}{O{:}C_9H_6O,H_2O}}{H{<}\overset{O{:}C_9H_6O,H_2O}{O{:}C_9H_6O,H_2O}}$$

VI.   VII.

According to this hypothesis, the coumarin platinichloride might similarly be represented by formula VII.

It is, however, doubtful whether the assumptions made in Werner's theory are necessary in so far as these platinichlorides are concerned. An explanation of the constitution of these compounds based on the mutual affinities of the oxygen and chlorine atoms in the molecules of these salts would involve fewer assumptions, and be more in accordance with the behaviour of these two elements in their simpler combinations (compare the article " Chemistry," *Encyclopœdia Brit.*, vol. 26, p. 717). There is abundant evidence that chlorine can exhibit a valency greater than that of a univalent radicle, whilst the chemical properties of the simplest oxygen compounds—water, the alcohols and ethers, and alkylene oxides—favour the view that their great reactivity is determined by the residual affinity of the combined oxygen. More-over, such a theory would explain without additional assumptions the formation of the anomalous tripropylphosphine oxide platinichloride, $6(C_3H_7)_3P\overset{..}{:}O,H_2PtCl_6$ (Pickard and Kenyon, *loc. cit.*), coumarin aurichloride, and hydriodide periodide, whereas the "Nebenvalenz" theory would, in each instance, involve some alteration in the sub-sidiary valency of hydrogen or oxygen.

Bearing in mind the fact that oxygen chiefly exerts its higher valencies either towards the halogens or towards other oxygen atoms, the anomalous hydrochlorides $2(X\overset{..}{:}O),HCl$ (tetramethylpyrone hydro-chloride, for example) might be formulated as $\begin{array}{c}H-O=X\\Cl-O=X\end{array}$, whilst the platinichlorides of the type $4(X\overset{..}{:}O),H_2PtCl_6$, discussed in this com-munication, would by analogy be regarded as being constituted in the following manner: $\begin{array}{c}H\cdot O\cdot X\\Cl\\ \overset{|}{\underset{|}{Cl}}>Pt<\overset{Cl\overset{..}{:}Cl-O\cdot X}{Cl\overset{..}{:}Cl-O\cdot X\cdot}\\H\cdot O\cdot X\end{array}$ Similar formulæ could be devised for those salts containing even larger proportions of the organic constituents simply by assuming that they contain longer chains of oxygen atoms, and this assumption seems quite justifiable in view of the chemical properties of water and allied oxygen com-pounds and the tendency which these substances have to combine additively to yield more complicated combinations. Moreover, the ring $>Pt<\overset{Cl}{\underset{Cl}{|}}$ might undergo disruption, whereby the higher valencies of the two chlorine atoms would become available for fixing a certain proportion of the organic constituent of the complex molecule.

The foregoing hypothesis, based on the higher valencies of oxygen and chlorine, can, however, only be regarded as a first approximation to a comprehensive theory of the constitution of these additive pro-ducts, inasmuch as it does not take into account the influence of any other unsaturated atoms or groups which may be present in the molecule of the organic component. These so-called "molecular com-pounds" vary very considerably in stability from the additive compounds of the simple alcohols, ethers, &c., which exist only at very low temperatures (Archibald and McIntosh, Trans., 1904, **85**, 219; 1905, **87**, 784), to the crystallisable phosphine oxide and pyrone derivatives, and in all probability this difference depends largely on the degree of unsaturation of the organic component.

The lactonic ring of coumarin (II) contains three centres of residual affinity, the oxidic oxygen, the carbonyl group, and the ethylenic linking, and, moreover, the benzene nucleus offers further possibilities of interaction. It is probably on account of the concurrent action of all these factors that coumarin is capable of combining additively with reagents of such extremely diverse types.

It is interesting to note in this connection that coumarin occurs in *Melilotus officinalis*, not in the free state, but in combination with melilotic acid, the coumarin melilotate having the formula

$$C_9H_6O_2,C_9H_{10}O_3$$

(Zwenger and Bodenbender, *Annalen*, 1863, **126**, 257). In this compound there is one molecular proportion of the coumarin to each hydrogen ion of the acid, in which respect the melilotate corresponds with the coumarin cobalticyanide described in the present communication.

In addition to the salt-like compounds under discussion, coumarin also combines additively with metallic oxides and hydroxides, as was shown long ago by R. Williamson, who obtained the following series : $C_9H_6O_2,2NaOH$, $C_9H_6O_2,Na_2O_2$, $C_9H_6O_2,2KOH$, $C_9H_6O_2,Ba(OH)_2$, $C_9H_6O_2,Ag_2O$, and $C_9H_6O_2,2PbO$ (*Jahresber.*, 1875, 587). In these compounds, the lactone displays a different order of unsaturation to that exhibited in its oxonium salts.

Among the first additive compounds of coumarin discovered were the dichloride and dibromide obtained by W. H. Perkin, sen. (*Annalen*, 1871, **157**, 116), who found that the latter, a well-defined, crystalline substance (m. p. 105°), in alcoholic solution regenerated coumarin, whilst the loosely combined bromine attacked the solvent. Ebert subsequently found that 50 to 70 per cent. of the original lactone could be recovered simply by boiling the dibromide with water (*Annalen*, 1884, **226**, 348). Perkin (*loc. cit.*) expressed the belief that the dibromide was not identical with dibromomelilotic anhydride, $CHBr\diagdown_{CHBr\cdot CO}^{-C_6H_4-}\diagup O$. The foregoing results indicate that in all probability the bromide is an additive compound of the following type, $CH\diagdown_{CH\cdot CO}^{-C_6H_4-}\diagup O\diagdown_{Br}^{Br}$, this formula accounting for the ease with which the bromine is eliminated.

Although graphical formulæ are suggested for the foregoing dibromide, and also for the coumarin platinichloride, the remaining compounds of coumarin and its derivatives have been denoted simply as "molecular compounds," because this formulation accords with our belief that the existence of these additive products is determined not only by the power possessed by the oxidic oxygen of assuming a higher valency, but also to some extent by the presence in the coumarin molecule of other centres of residual affinity, the effect of which cannot be precisely indicated in terms of the theory of integral valencies.

The authors desire to express their thanks to the Government Grant Committee of the Royal Society for a grant which has partly defrayed the expenses of this investigation.

Royal College of Science, London,
South Kensington, S.W.

XCI.—*The Constitution of Ammonium Amalgam.*

By Elizabeth Mary Rich and Morris William Travers, D.Sc., F.R.S.

The study of the so-called ammonium amalgam has formed the subject of numerous researches, which have resulted in throwing a considerable amount of light on its properties, but have led to no very satisfactory conclusions as to its constitution. Recently Moissan (*Compt. rend.*, 1901, **133**, 803) has published an elaborate memoir on this subject in which he describes a convenient method of preparing the amalgam, gives an account of its chemical and physical properties, and in discussing the latter leaves it an open question as to whether it is to be considered to be a solution of "ammonium" in mercury or an emulsion of ammonia, hydrogen, and mercury. With a view to the solution of this problem we decided to determine the freezing points of a series of preparations of the amalgam of different concentrations. If the amalgam were a true solution, that is to say, a homogeneous mixture of ammonium and mercury, its freezing point should decrease linearly with the concentration of the ammonium present in it; if the latter explanation held good, either the freezing point would be identical with that of pure mercury, or a paste of crystals might be formed at some higher temperature.

The amalgam was prepared by Moissan's method (*loc. cit.*). Sodium amalgam (400—600 grams), containing a quantity of sodium equivalent to the concentration of the ammonium amalgam required for the experiment, was cooled in a flask with a long and narrow neck to a temperature just above its freezing point. Anhydrous ammonia was then condensed in the flask, and in this liquid excess of ammonium iodide was dissolved. The mixture was shaken from time to time, the temperature being kept constant by adding small quantities of liquid air to the alcohol in which the flask was cooled. After about an hour, the ammonia was allowed to evaporate, and the liquid amalgam was then filtered several times through an ordinary folded filter paper with a hole pierced in the bottom of it. It was finally transferred to the experimental tube, particular care being taken to keep the temperature as near to the freezing point as possible during the manipulations to which it was subjected.

The freezing point of mercury, and afterwards of the amalgam, was determined by means of an arrangement similar to that commonly employed in molecular weight determinations. The amalgam was contained in a cylindrical tube, 30 cm. long and 3·5 cm. in diameter, rounded at the lower end. It was surrounded for the greater part of its length by an outer tube, the two being kept apart by rings of flannel. The temperature was measured by means of a Callendar resistance thermometer. The resistance, galvanometer, &c., were the same as had already been employed by one of us in the comparison of the platinum and hydrogen scales of temperature (*Proc. Roy. Soc.*, 1905, **74**, 528). The thermometer consisted of a coil of silk-covered platinum wire, soldered at its ends to copper leads. The coil fitted fairly tightly into the bottom of a glass tube, 8 mm. in diameter, which also surrounded the main and "compensating" leads. Its resistance at the ice-point was 17·52 ohms; the coefficient of change of resistance with temperature was 0·0036647, a value which indicates that the metal was not pure platinum. However, as the freezing point of mercury was found to be − 39·40° on the scale of this thermometer, it appears that over the range of our experiments the observed temperatures are so nearly coincident with temperatures on the Centigrade scale as to make the application of a correction unnecessary.

In conducting an experiment, the alcohol-bath surrounding the experimental tube was maintained at a temperature about 5 or 10 degrees below the freezing point of the amalgam. There was not the least difficulty in determining the freezing point, though when a very small trace of free ammonia was present with the amalgam the latter appeared to become slightly pasty at its freezing point. This difficulty was overcome by carefully purifying the amalgam.

After determining the freezing point, the thermometer and stirrer were removed from the tube and a rubber stopper with two holes was fitted to it. Through the holes passed tubes, so that air could be drawn through the experimental tube and afterwards through an absorption apparatus containing a normal solution of hydrochloric acid. After some hours, the tube was warmed to accelerate the evolution of the ammonia, and finally the mercury was itself gently warmed with the acid solution; the slight evolution of gas showed that even the action of heat only slowly decomposed very dilute solutions of ammonium in mercury. The quantity of ammonia evolved was determined by titration, and the solution was always examined for the presence of sodium, which was only found in one experiment which we have not recorded. At the end of each experiment, the mercury was dried and weighed. The following are the results of our experiments:

Freezing Point of Mercury on Platinum Scale, $-39\cdot40°$.

Grams of NH_4 in 100 grams of mercury (a).	Freezing point of amalgam.	$D = \dfrac{\Delta}{a} \times 18$.
0·084	$-41\cdot605°$	470
0·507	$-45\cdot61$	220
0·415	$-44\cdot82$	230
0·079	$-40\cdot81$	320 *
0·027	$-40\cdot01$	410
0·0094	$-39\cdot62$	420
0·0117	$-39\cdot67$	420

* Probably inaccurate.

It appears that with low concentrations the value of "D" is between 410 and 470, and that it first increases and then decreases rapidly as the concentration increases. Tammann (*Zeit. physikal. Chem.*, 1889, **3**, 440) obtained similar results in the case of the amalgams of certain common metals. For sodium he obtained as a mean value of "D" the number 420, corresponding to concentrations between 0·022 and 0·112 gram of sodium in 100 grams of mercury. His results for potassium amalgam are stated below:

a.	D.
0·018	600
0·030	560
0·091	320
0·112	360
0·137	280

The resemblance in the behaviour of sodium amalgam and ammonium amalgam is very marked, and leaves no room for doubt that the latter is really a solution of ammonium in mercury. This view is supported by the fact that the amalgam continues to evolve ammonia and hydrogen after it has been allowed to stand for many hours at the temperature of the experiment. Further, if the temperature of the amalgam is allowed to rise until the volume begins to increase, and the amalgam is then cooled, it appears to return to its original state; hence it is not impossible that "ammonium" can exist transitorily in the free state.

UNIVERSITY COLLEGE,
BRISTOL.

XCII.—*Aromatic Compounds obtained from the Hydroaromatic Series. Part II. The Action of Phosphorus Pentachloride on Trimethyldihydroresorcin.*

By ARTHUR WILLIAM CROSSLEY and JAMES STUART HILLS.

IN 1901, Crossley (Trans., **79**, 144) described a solid melting at 76·5°, obtained by the action of phosphorus pentachloride on trimethyl-dihydroresorcin (I), and suggested that it was 3 : 5-dichloro-1 : 1 : 2-tri-methyl-$\Delta^{2\cdot4}$-dihydrobenzene (II):

$$
\begin{array}{cc}
\text{C(CH}_3)_2 & \text{C(CH}_3)_2 \\
\text{H}_2\text{C} \diagdown \diagup \text{CH·CH}_3 & \cdot\text{H}_2\text{C} \diagdown \diagup \text{C·CH}_3 \\
\text{HO·C} \diagup \diagdown \text{CO} & \text{ClC} \diagup \diagdown \text{CCl} \\
\text{CH} & \text{CH} \\
\text{I.} & \text{·II.}
\end{array}
$$

At a later period (Trans., 1902, **81**, 826) it was shown that phosphorus pentachloride acts on dimethyldihydroresorcin (III) to give 3 : 5-dichloro-1 : 1-dimethyl-$\Delta^{2\cdot4}$-dihydrobenzene (IV), but that a

$$
\begin{array}{ccc}
\text{C(CH}_3)_2 & \text{C(CH}_3)_2 & \text{C·CH}_3 \\
\text{H}_2\text{C} \diagdown \diagup \text{CH}_2 & \text{H}_2\text{C} \diagdown \diagup \text{CH} & \text{HC} \diagdown \diagup \text{C·CH}_3 \\
\text{HO·C} \diagup \diagdown \text{CO} & \text{ClC} \diagup \diagdown \text{CCl} & \text{ClC} \diagup \diagdown \text{CCl} \\
\text{CH} & \text{CH} & \text{CH} \\
\text{III.} & \text{IV.} & \text{V.}
\end{array}
$$

secondary reaction also takes place, consisting in the transformation of this hydroaromatic dichloride into 3 : 5-dichloro-*o*-xylene (V) (Trans., 1902, **81**, 1543). It was therefore thought desirable to investigate further the action of phosphorus pentachloride on trimethyldihydroresorcin, and in 1903 (*Proc.*, **19**, 227) a short notice was published, in which it was indicated that, during the reaction, two substances are produced, a liquid, which is 3 : 5-dichloro-1 : 1 : 2-trimethyl-$\Delta^{2\cdot4}$-di-hydrobenzene (VI), and the solid melting at 76·5°, which belongs to the aromatic series and is 3 : 5-dichloro-1 : 2 : 6-trimethylbenzene (VII).

The former boils at 118—119° (33 mm.), and its hydroaromatic nature is shown by the fact that it absorbs two atoms of bromine to form a very unstable dibromide, which readily loses two molecules of hydrogen bromide to form dichlorotrimethylbenzene (VII):

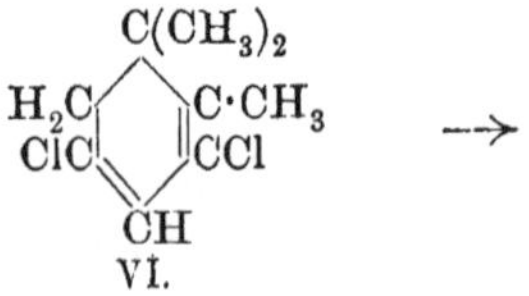

$$C(CH_3)_2$$

$$H_2C \diagdown C\cdot CH_3$$
$$ClC \diagup CCl \qquad \longrightarrow$$
$$CH$$
VI.

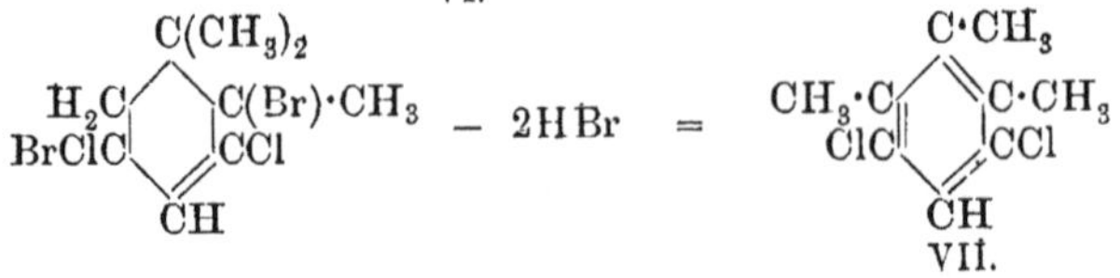

$$C(CH_3)_2$$
$$H_2C \diagdown C(Br)\cdot CH_3$$
$$BrClC \diagup CCl \qquad - 2HBr =$$
$$CH$$

$$C\cdot CH_3$$
$$CH_3\cdot C \diagdown C\cdot CH_3$$
$$ClC \diagup CCl$$
$$CH$$
VII.

If, however, a large excess of bromine be used, then 3 : 5-dichloro-4-bromo-1 : 2 : 6-trimethylbenzene is formed,

$$CH_3\cdot C\diagdown^{C(CH_3)\cdot CCl}_{C(CH_3):CCl}\diagup CBr.$$

This is an exactly similar behaviour to that of 3 : 5-dichloro-1 : 1-dimethyl-$\Delta^{2:4}$-dihydrobenzene (VIII), which under the influence of bromine (Trans., 1904, 85, 264) is transformed into 3 : 5-dichloro-o-xylene (IX) or one of the two possible dichlorobromo-o-xylenes (X or XI) derivable from it :

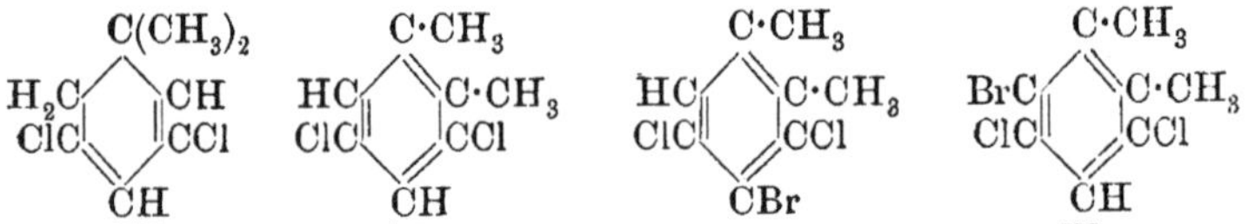

$$C(CH_3)_2$$
$$H_2C \diagdown CH$$
$$ClC \diagup CCl$$
$$CH$$
VIII.

$$C\cdot CH_3$$
$$HC \diagdown C\cdot CH_3$$
$$ClC \diagup CCl$$
$$CH$$
IX.

$$C\cdot CH_3$$
$$HC \diagdown C\cdot CH_3$$
$$ClC \diagup CCl$$
$$CBr$$
X.

$$C\cdot CH_3$$
$$BrC \diagdown C\cdot CH_3$$
$$ClC \diagup CCl$$
$$CH$$
XI.

A new point arises in discussing the constitution of dichlorotrimethyldihydrobenzene, for, unlike the corresponding dichloride obtained from dimethyldihydroresorcin, it is not a symmetrical molecule and may therefore be represented by one of the two following formulæ (XII or XIII):

$$C(CH_3)_2$$
$$H_2C \diagdown C\cdot CH_3$$
$$ClC \diagup CCl$$
$$CH$$
XII.

$$C(CH_3)_2$$
$$HC \diagdown CH\cdot CH_3$$
$$ClC \diagup CCl$$
$$CH$$
XIII.

In order to obtain some evidence on this point, the substance was oxidised with potassium permanganate, when a mixture of dimethylmalonic and as-dimethylsuccinic acids was obtained. These acids could only be formed from a substance of formula XII, for from a

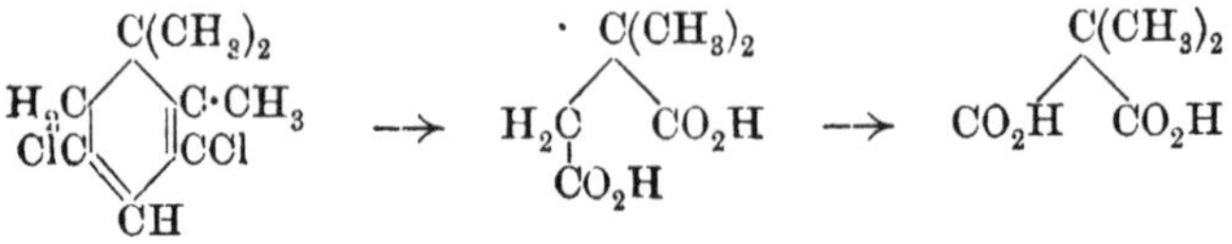

$$C(CH_3)_2$$
$$H_2C \diagdown C\cdot CH_3$$
$$ClC \diagup CCl$$
$$CH$$

$$\longrightarrow$$

$$C(CH_3)_2$$
$$H_2C \diagdown CO_2H$$
$$CO_2H$$

$$\longrightarrow$$

$$C(CH_3)_2$$
$$CO_2H \diagup \diagdown CO_2H$$

substance of formula XIII trimethylsuccinic acid should have re-sulted on oxidation, but no evidence of its formation could be obtained.

The solid melting at 76·5°, produced by the action of phosphorus pentachloride on trimethyldihydroresorcin would appear to be 3 : 5-di-chloro-1 : 2 : 6-trimethylbenzene, for though there does not seem to be any ready method for directly orientating this substance, yet, arguing from analogy and taking into account its chemical behaviour, there can be little doubt that the above expression accurately describes its constitution.

In the first place, as pointed out on page 875, phosphorus penta-chloride acts on dimethyldihydroresorcin to give 3 : 5-dichloro-1 : 1-di-methyl-Δ^2 ^{4}-dihydrobenzene (IV), which substance can be partially converted into 3 : 5-dichloro-o-xylene (V) by prolonged heating with phosphorus pentachloride. For this transformation to take place, a methyl group must wander into either an $ortho$- or a $para$-position :

$$
\begin{array}{ccc}
\text{C·CH}_3 & \text{C(CH}_3)_2 & \text{C·CH}_3 \\
\text{HC} \diagup \diagdown \text{CH} & \text{H}_2\text{C} \diagup \diagdown \text{CH} & \text{HC} \diagup \diagdown \text{C·CH}_3 \\
\text{ClC} \diagdown \diagup \text{CCl} \quad \longleftarrow & \text{ClC} \diagdown \diagup \text{CCl} \quad \longrightarrow & \text{ClC} \diagdown \diagup \text{CCl} \\
\text{C·CH}_3 & \text{CH} & \text{CH}
\end{array}
$$

but both in this case and in those of other derivatives of dimethyl-dihydroresorcin which have been examined (Trans., 1904, **85**, 264), no instance has been observed in which a methyl group has wandered into any but an $ortho$-position. It would seem therefore legitimate to con-clude that in the case of 3 : 5-dichloro-1 : 1 : 2-trimethyl-$\Delta^{2:4}$-dihydro-benzene (VI), where a methyl group might also wander into an $ortho$- or a $para$-position, the former would be selected and the resulting substance would be 3 : 5-dichloro-1 : 2 : 6-trimethylbenzene (VII).

One striking difference is observed in the behaviour of dimethyl- and of trimethyl-dihydroresorcin towards phosphorus pentachloride, and that is the comparative ease with which the corresponding hydro-aromatic dichlorides are transformed into the aromatic derivatives. Thus, dimethyldihydroresorcin, when heated for two to three hours with phosphorus pentachloride, gives about 75 per cent. of the theoretical yield of the hydroaromatic and a comparatively small yield of the aromatic dichloride. The formation of the latter sub-stance was indeed only observed after very careful fractionation of the accumulated high boiling residues from a large number of experiments.

Trimethyldihydroresorcin, on the other hand, under exactly the same conditions gives a product consisting of about one part of the liquid hydroaromatic and two parts (nearly 50 per cent. of the theoretical) of the solid aromatic dichloride.

3 M 2

The reaction was carried out under various conditions with the object of obtaining a better yield of the hydroaromatic dichloride, which was required for another research. The desired result was, however, not obtained, for if the amount of phosphorus pentachloride be decreased, or if the mixture be not heated for so long, the reaction does not complete itself, and the production of large quantities of dichlorotrimethyldibydrobenzene would appear to be impossible by this method, as the conditions necessary for its formation are also those under which it is transformed into dichlorotrimethylbenzene.

The last-mentioned substance is readily acted on by chlorine or bromine to give respectively trichloro- or dichloro-bromotrimethyl-benzene ; but it is extremely difficult to nitrate, probably owing to the fact that, no matter what conditions are adopted, the nitric acid acts principally as an oxidising agent.

3 : 5-Dichloro-4-nitro-1 : 2 : 6-trimethylbenzene,

$$CH_3 \cdot C \underset{C(CH_3):CCl}{\overset{C(CH_3) \cdot CCl}{<}} \!\!\!\!\! > C \cdot NO_2,$$

can be obtained in small amount by pouring a solution of dichloro-trimethylbenzene in glacial acetic acid into fuming nitric acid. If, however, the conditions are slightly altered and fuming nitric acid is gradually added to an acetic acid solution of the aromatic derivative, a curious reaction takes place, resulting in the formation of a very small quantity of dichloronitrotrimethylbenzene and a much larger amount of trichlorotrimethylbenzene, $CH_3 \cdot C \underset{C(CH_3):CCl}{\overset{C(CH_3) \cdot CCl}{<}} \!\!\!\!\! > CCl$ (0·5 gram from 2 grams). The identity of this substance was established by analysis and by a mixed melting point with pure trichlorotrimethyl-benzene, prepared by the direct chlorination of dichlorotrimethyl-benzene. Presumably some of the dichlorotrimethylbenzene is first oxidised by the nitric acid, hydrochloric acid resulting, which in contact with nitric acid liberates chlorine, and this then causes substitution to take place.

When dichlorotrimethylbenzene is oxidised by nitric acid (sp. gr. 1·15) in sealed tubes, it is converted into a dichlorobenzenetricarb-oxylic acid, apparently 3 : 5-dichlorohemimellitic acid,

$$CO_2H \cdot C \underset{C(CO_2H):CCl}{\overset{C(CO_2H) \cdot CCl}{<}} \!\!\!\!\! > CH,$$

which resembles hemimellitic acid very closely in chemical properties (compare Graebe and Leonhardt, *Annalen*, 1896, **290**, 217). Thus, on simply melting, it is transformed into its anhydride, which gives a particularly brilliant fluorescein· reaction and yields an imide when ammonia is passed into it in the molten condition.

Experiments were carried out on the direct esterification of dichloro-hemimellitic acid, as there seemed to be a possibility of thereby bringing some direct evidence to bear on the constitution of the dichloro-trimethylbenzene from which it is derived. The latter substance might have one of the following formulæ :

$$\begin{array}{ccc}
& \text{C·CH}_3 & \\
\text{CH}_3\text{·C} & & \text{C·CH}_3 \\
\text{ClC} & & \text{CCl} \\
& \text{CH} &
\end{array} \quad\text{or}\quad \begin{array}{ccc}
& \text{C·CH}_3 & \\
\text{HC} & & \text{C·CH}_3 \\
\text{ClC} & & \text{CCl} \\
& \text{C·CH}_3 &
\end{array}$$

according as to whether a methyl group had wandered into an *ortho*- or a *para*-position during its formation from dichlorotrimethyldihydro-benzene ; hence the acid obtained from dichlorotrimethylbenzene on oxidation might be represented by :

$$\begin{array}{ccc}
& \text{C·CO}_2\text{H} & \\
\text{HO}_2\text{C·C} & & \text{C·CO}_2\text{H} \\
\text{ClC} & & \text{CCl} \\
& \text{CH} &
\end{array} \quad\text{or}\quad \begin{array}{ccc}
& \text{C·CO}_2\text{H} & \\
\text{HC} & & \text{C·CO}_2\text{H} \\
\text{ClC} & & \text{C·Cl} \\
& \text{C·CO}_2\text{H} &
\end{array} \cdot$$

$$\text{XIV.} \qquad\qquad \text{XV.}$$

But an acid of formula **XIV** should not be capable of direct esterifica-tion, for all the ortho-positions relative to the carboxyl group are occupied, whereas an acid of formula **XV** should give a mono-ester on direct esterification.

When the above acid was dissolved in methyl alcohol, the solution saturated with hydrogen chloride and allowed to stand for twelve hours, nothing but unchanged material was recovered ; but on repeating the experiment and allowing to stand 120 hours, a few crystals (0·05 gram from 1 gram of acid) of a monomethyl ester were obtained. This ester was also produced in about the same amount by heating the acid for six hours with a four per cent. solution of sulphuric acid in methyl alcohol. It cannot therefore be said that dichlorohemimellitic acid is capable of direct esterification at all readily, and it would appear most probable that the formation of this small amount of ester is really due to the fact that the anhydride of the acid is first formed, and this is then converted into the ester by the boiling methyl alcohol. Graebe and Leonhardt (*ibid.*, p. 226) have shown that the best method for preparing the monomethyl ester of hemimellitic acid is by boiling the correspond-ing anhydride with methyl alcohol, and this also applies to the acid at present under consideration. When its anhydride is boiled with methyl alcohol, it is slowly converted into the monomethyl ester, one gram of anhydride giving 0·2 gram of the ester after boiling for thirty-two hours. All available evidence is therefore in favour of the substance being a dichlorohemimellitic acid having the constitution represented by formula **XIV**.

EXPERIMENTAL.

The action of phosphorus pentachloride on trimethyldihydroresorcin, as mentioned in the introduction, has been tried under very varied conditions, but it does not seem necessary to quote the details of more than one experiment.

Thirty grams (1 molecule) of trimethyldihydroresorcin and 80 grams of dry chloroform were placed in a flask attached to a reflux condenser, and 84 grams (2 molecules) of phosphorus pentachloride added, at first in small quantities, as a somewhat vigorous reaction sets in, and afterwards much more freely. The mixture was then heated on the water-bath for five hours, during which time a remarkable series of colour changes took place: the solution, at first yellowish-brown, became deep green, deep indigo-blue, bluish-black, and finally greenish-black. The reaction mixture was then slowly poured into 500 c.c. of water, being well shaken and cooled throughout, and the whole extracted three times with ether, the ethereal solution washed with dilute sodium hydroxide solution until no longer acid, then with water, dried over calcium chloride, and the ether evaporated.

The residue set to a mass of crystals (A = 8 grams), which were collected at the pump and the yellow oily filtrate distilled, when at 36 mm., 10·4 grams passed over between 126° and 131°. The higher boiling fractions solidified almost completely on standing and yielded 8 grams of solid (B). The fraction 126—131° was then twice distilled, when a further quantity of solid (C) was obtained from the higher fractions, and 6·5 grams of a liquid boiling constantly at 118—119° (33 mm.), which on analysis gave the following numbers :

0·1255 gave 0·2611 CO_2 and 0·0716 H_2O.   C = 56·74 ; H = 6·34.

0·1284 ,, 0·1948 AgCl.   Cl = 37·53

$C_9H_{12}Cl_2$ requires C = 56·54 ; H = 6·28 ; Cl = 37·17 per cent.

3 : 5-*Dichloro*-1 : 1 : 2-*trimethyl*-$\Delta^{2:4}$-*dihydrobenzene*,

$$(CH_3)_2 \cdot C \underset{CH_2 - CCl}{\overset{C(CH_3):CCl}{<}} \hspace{-0.3em} > CH,$$

is an oily liquid boiling at 118—119° (33 mm). and possessing a camphoraceous odour. It is colourless when freshly distilled, but quickly becomes yellow; the colour deepens on standing and the whole slowly resinifies. On adding bromine to a solution of the substance in an equal volume of dry chloroform, the colour of the former is rapidly discharged and quantities of hydrogen bromide are evolved. If the addition of bromine be stopped as soon as the first signs of excess appear and the chloroform be evaporated, the residue partially solidifies. After spreading on a porous plate, the solid crystallised from methyl alcohol in long, glistening needles, which melted at 76·5° and proved to be identical with the dichlorotrimethylbenzene described below. If, however, a large excess of bromine be added to the chloroform solution, or if bromine be added directly to dichlorotrimethyldihydrobenzene, it is slowly absorbed and hydrogen bromide evolved. The solid residue separated from absolute ethyl alcohol in silky needles melting at 222—223°, identical in every respect with the dichlorobromotrimethylbenzene described on page 882.

Oxidation of Dichlorotrimethyldihydrobenzene.—Nine grams of the dichloride were suspended in 150 c.c. of water, the whole heated on a water-bath in a flask attached to a reflux condenser, and a cold saturated solution of potassium permanganate gradually added. Oxidation took place very slowly and required about sixty hours for completion. The product was separated from manganese dioxide, evaporated to a small bulk, acidified with dilute sulphuric acid, and extracted six times with ether, when, after evaporation of the ether, a syrupy residue smelling strongly of acetic acid was obtained. After standing some time in a vacuum desiccator over sulphuric acid, it partially solidified and was spread on a porous plate. The white solid obtained in this way softened at 105°, melted at 115—120°, and evolved gas at 160—165°. This is the behaviour of a mixture of dimethylmalonic and *as*-dimethylsuccinic acids (compare Trans., 1905, **87**, 1499), of which the substance was proved to consist in the following manner. It was heated until no more carbon dioxide was evolved, during which time a liquid with the unmistakable odour of *iso*butyric acid was evolved; the residue was dissolved in boiling water and the solution saturated with hydrogen chloride, when needle-shaped crystals separated, which melted at 140°, nor was this melting point altered on mixing with pure *as*-dimethylsuccinic acid.

A portion of the acid was then boiled with acetyl chloride, and the anhydride so obtained converted into the anilic acid in the usual manner; this was found to crystallise from methyl alcohol in nacreous, scaly needles melting at 186—187° with evolution of gas, and after mixing the substance with pure dimethylsuccinanilic acid no change in the melting point was observed.

The crystalline residues A, B, and C (see page 880) were purified by crystallisation from absolute methyl alcohol and analysed:

0·1048 gave 0·2188 CO_2 and 0·0506 H_2O. $C = 56·94$; $H = 5·36$.
0·1376 „ 0·2872 CO_2 „ 0·0680 H_2O. $C = 56·92$; $H = 5·49$.
0·1410 „ 0·2147 AgCl. $Cl = 37·66$.
0·1610 „ 0·2439 AgCl. $Cl = 37·48$.

$C_9H_{10}Cl_2$ requires $C = 57·14$; $H = 5·30$; $Cl = 37·56$ per cent.

3 : 5-*Dichloro*-1 : 2 : 6-*trimethylbenzene*, $CH_3 \cdot C \underset{C(CH_3) \cdot CCl}{\overset{C(CH_3):CCl}{<}} \hspace{-1em} > CH$, is

readily soluble in the cold in chloroform, benzene, acetone, ethyl acetate, or light petroleum (b. p. 80—100°) and crystallises from ethyl or, better, methyl alcohol in glistening needles, frequently, when using the latter solvent, 3 to 4 inches in length, and melting at 76·5°. A chloroform solution of bromine was not decolorised by the addition of dichlorotrimethylbenzene, but on dissolving the latter substance in the smallest possible quantity of chloroform, adding a trace of iron filings and then bromine, substitution took place readily after slight warming, and the whole set to a solid mass, which was purified by crystallisation from ethyl alcohol and analysed :

0·1044 gave 0·1860 Ag haloids and 0·1266 Ag. Cl = 26·25 ; Br = 30·60.

$C_9H_9Cl_2Br$ requires Cl = 26·49 ; Br = 29·84 per cent.

3 : 5-*Dichloro*-4-*bromo*-1 : 2 : 6-*trimethylbenzene*,

$$CH_3 \cdot C \underset{C(CH_3) \cdot CCl}{\overset{C(CH_3):CCl}{<}} C \cdot Br,$$

is readily soluble in the cold in chloroform, somewhat less so in acetone, benzene, or ethyl acetate, and crystallises from absolute ethyl alcohol in long, silky needles melting at 222—223°. When acted on with fuming nitric acid, it yields a small amount of a nitrogenous body, crystallising from alcohol in yellow, flattened needles melting at 170—171°, which was thought to be 3 : 5-dichloro-4-nitro-1 : 2 : 6-trimethylbenzene (see page 883), but a mixture of these two substances melted at 140°, and as the body melting at 171° was produced in such small amount, it was impossible to investigate its nature further.

3 : 4 : 5-*Trichloro*-1 : 2 : 6-*trimethylbenzene* was obtained by dissolving dichlorotrimethylbenzene in the smallest possible quantity of chloroform, adding a trace of iron filings, and then passing dry chlorine into the solution. It is readily soluble in the cold in benzene or chloroform, crystallises from alcohol, acetone, or ethyl acetate in silky needles melting at 217—218°, and sublimes in long, glistening needles.

0·1004 gave 0·1931 AgCl. Cl = 47·58.

$C_9H_9Cl_3$ requires Cl = 47·65 per cent.

Action of Nitric Acid on Dichlorotrimethylbenzene.—Two grams of dichlorotrimethylbenzene were powdered and added to 15 c.c. of glacial acetic acid, an insufficient amount for complete solution in the cold, and 16 c.c. of fuming nitric acid gradually dropped into the solution, when a slight amount of heat was evolved, and nearly the whole of the solid dissolved. When all the nitric acid had been added, there was a sudden separation of solid matter (0·5 gram), which was filtered (filtrate = A) after allowing the whole to stand for several hours. This solid crystallised from ethyl acetate in silky needles melting at 217—218°, nor was this melting point altered on mixing with pure 3 : 4 : 5-trichloro-1 : 2 : 6-trimethylbenzene (see page 882). The identity of the two substances was further proved by a chlorine determination.

0·1016 gave 0·1952 AgCl. Cl = 47·52.

$C_9H_9Cl_3$ requires Cl = 47·65 per cent.

The filtrate A (see p. 882) was poured into water, when a very viscid semi-solid separated. This, together with material obtained in similar experiments, representing in all 8 grams of dichlorotrimethylbenzene, was dissolved in absolute alcohol, when, on cooling, 0·3 gram of silky needles separated, melting at 175—176°, and a mixed melting point proved that they were identical with 3 : 5-dichloro-4-nitro-1 : 2 : 6-trimethylbenzene described below.

In another experiment, 1 gram of dichlorotrimethylbenzene was dissolved in glacial acetic acid by slightly heating and the solution slowly poured into 10 c.c. of fuming nitric acid. After the addition, the whole was gently warmed on the water-bath until all the solid had gone into solution and then allowed to stand for twelve hours, when long, needle-shaped crystals separated (0·2 gram), which were filtered (filtrate = A), and after purification the nitrogen was determined :

0·2105 gave 10 c.c. moist nitrogen at 11° and 760 mm. N = 5·66.

$C_9H_9O_2NCl_2$ requires N = 5·98 per cent.

3 : 5-*Dichloro*-4-*nitro*-1 : 2 : 6-*trimethylbenzene*,

$$CH_3 \cdot C {\displaystyle \mathop{\lessgtr}^{C(CH_3) \cdot CCl}_{C(CH_3) : CCl}} C \cdot NO_2,$$

is readily soluble in the cold in acetone, benzene, or chloroform and crystallises from ethyl acetate or alcohol in long, faintly yellow, silky needles melting at 175—176°.

On pouring the above-mentioned filtrate A into water, a very viscous solid (0·2 gram) separated, from which, on treatment with alcohol, a further small quantity of the nitro-product was obtained. This derivative is also produced in minute amounts by the action of a nitrating mixture on dichlorotrimethylbenzene.

Oxidation of Dichlorotrimethylbenzene.

Dichlorotrimethylbenzene, in quantities of 2 grams at a time, was heated with 16 c.c. of nitric acid (sp. gr. 1·15) in a sealed tube for five to six hours at 170—180°. The liquid contents of the tube were then evaporated to dryness on the water-bath, purified by crystallisation from water saturated with hydrogen chloride, and analysed after drying in a vacuum over potassium hydroxide :

0·1548 gave 0·2189 CO_2 and 0·0244 H_2O.   C = 38·56 ; H = 1·75.
0·1332 ,,   0·1356 AgCl.   Cl = 25·18.
$C_9H_4O_6Cl_2$ requires C = 38·71 ; H = 1·43 ; Cl = 25 43 per cent.

Dichlorohemimellitic   acid   (3 : 5-dichlorobenzene-1 : 2 : 6-*tricarboxylic acid*), $CO_2H \cdot C {<}{<}^{C(CO_2H) \cdot CCl}_{C(CO_2H):CCl}{>}CH$, of which the yield is practically theoretical, is insoluble in benzene or chloroform, readily soluble in the cold in water, ethyl alcohol, acetone, ether, or ethyl acetate, and crystallises from water saturated with hydrogen chloride in clusters of short, colourless needles, melting at 226—227°, with previous softening and contraction in bulk. It can also be crystallised by the slow evaporation of its aqueous solution, when it separates in transparent, glistening, hexagonal plates, which finally melt at 226—227°, but with more softening and contraction in bulk than when the acid is crystallised from hydrochloric acid. These crystals contain two molecules of water, which they readily lose on heating to 100°.

0·5792 lost 0·0660. H_2O = 11·39.
$C_9H_4O_6Cl_2,2H_2O$ requires H_2O = 11·42 per cent.

The molecular weight of the acid was determined by titration with potassium hydroxide, using phenolphthalein as indicator, and also by analysis of its silver salt (see page 885) :

0·2817 required 30·6 c.c. N/10 KOH = 0·1713 KOH.
Molecular weight, found 276·2 ; calculated 279.

When the acid is heated with resorcinol and a drop of sulphuric acid, the mixture dissolved in sodium hydroxide and poured into water, a red solution with a particularly brilliant green fluorescence is obtained. If a solution of potassium chloride be added to a concentrated aqueous solution of the acid and the whole allowed to stand, well-formed, transparent, flattened needles separate, and it was thought that this would be the acid potassium salt, and that dichlorohemimellitic acid behaved like hemimellitic acid (compare Graebe and Leonhardt, *Annalen*, 1896, **290**, 217) itself in this respect. But although the crystals contained potassium, the amount was insufficient for even a monopotassium acid salt, and apparently a solution sufficiently concentrated for this salt to separate is also concentrated enough for the acid to crystallise out to some extent.

The *silver* salt, prepared in the usual manner, is a white, caseous precipitate.

0·1446 gave 0·0781 Ag.   Ag = 54·01.

$C_9HO_6Cl_2Ag_3$ requires Ag = 54·0 per cent.

Molecular weight of acid, found 278·9 ; calculated 279.

The *anhydride* is most easily obtained by heating the acid to 235—240° for a few minutes, and a specimen produced in this way was analysed without any further purification :

0·1154 gave 0·1740 CO_2 and 0·0124 H_2O.   C = 41·12 ; H = 1·19.

$C_9H_2O_5Cl_2$ requires C = 41·37 ; H = 0·76 per cent.

This substance can also be prepared by dissolving the acid in acetic anhydride, from which it crystallises on standing. The anhydride, which sublimes in fern-like aggregates of flattened needles, is readily soluble in the cold in water to give an acid solution, it is also soluble in ethyl alcohol, acetone, or ethyl acetate, and is insoluble in benzene, chloroform, or light petroleum. It crystallises from acetic anhydride in nodules consisting of stellar aggregates of very fine needles, which melt sharply at 227—228°· Evidently the final melting point of dichlorohemimellitic acid is really the melting point of its anhydride, the previous softening and contraction in bulk being due to anhydride formation.

The *imide* prepared by passing ammonia into the molten anhydride crystallised from water in nodules of microscopic crystals, which melted at 253—254°.

0·2407 gave 11·4 c.c. moist nitrogen at 12° and 770 mm.   N = 5 69.

$C_9H_3O_4NCl_2$ requires N = 5·39 per cent.

Esterification of Dichlorohemimellitic Acid.

The *trimethyl* ester may be prepared in quantitative amount by heating the silver salt in benzene suspension with the calculated amount of methyl iodide. It is insoluble in cold water or potassium hydroxide solution, readily soluble in the cold in chloroform, ethyl acetate or acetone, and crystallises from methyl alcohol or, better, from light petroleum (b. p. 40—60°) in glistening, transparent rhombohedra melting at 62—63°.

0·1154 gave 0·1902 CO_2 and 0 0371 H_2O.   C = 44·94 ; H = 3 57.

$C_{12}H_{10}O_6Cl_2$ requires C = 44·86 ; H = 3·11 per cent.

Action of Methyl Alcohol on the Anhydride.—A solution of 1 gram of the anhydride in methyl alcohol was heated to the boiling point for thirty-two hours, when the residue obtained on evaporation slowly solidified. It dissolved for the most part readily in water and probably consisted largely of unchanged anhydride, but a small portion would not dissolve without boiling the solution, which, when allowed to evaporate slowly in a vacuum, deposited radiating clusters of needles, readily soluble in cold potassium hydroxide solution.   These were purified by crystallisation from water, when nacreous, scaly needles separated, melting at 141—142°, with a previous distinct contraction in bulk at 88—90°.   On analysis, the following figures were obtained :

0·1008 gave 0·1500 CO_2 and 0·0262 H_2O.   C = 40·58 ; H = 2·88.

$C_{10}H_6O_6Cl_2$ requires C = 40·95 ; H = 2·05 per cent.

Although these numbers do not agree so closely as could be desired, they show conclusively that the substance is the monomethyl ester of dichlorohemimellitic acid ; but the amount obtained was so small as not to permit a further purification.

The mother liquors from the analysed material deposited a few small transparent cubes, which melted sharply at 141—142° without any previous contraction, and were probably the pure monomethyl ester.

This same ester was formed by submitting dichlorohemimellitic acid to Fischer and Speyer's esterification method, and also by saturating a methyl-alcoholic solution of the acid with hydrogen chloride and allowing it to stand for 120 hours.   The former method gave rather more of the ester than the latter, but in both cases the amount was very small and much less than is produced by boiling the anhydride with methyl alcohol.

RESEARCH LABORATORY, PHARMACEUTICAL SOCIETY,
17, BLOOMSBURY SQUARE, W.C.

XCIII.—The Densities of Liquid Nitrogen and Liquid Oxygen and their Mixtures.

By JOHN KENNETH HAROLD INGLIS and JOSEPH EDWARD COATES.

DETERMINATIONS of the densities of liquid nitrogen and liquid oxygen have been made by a number of investigators (Baly and Donnan, Trans., 1902, 81, 907), but in all cases the measurement of the temperature was not very accurate, so that some uncertainty attaches to the values given.   Now in certain cases it is convenient to measure the volume of a large quantity of gas by determining the volume of liquid formed from it at some known temperature, and this is particularly the case when dealing with the analysis of a mixture of gases of which the greater portion consists of nitrogen and oxygen (Inglis, J. Soc. Chem. Ind., 1906, 25, 149).   The following experiments were therefore undertaken to determine the densities of liquid nitrogen, liquid oxygen, and a few of their mixtures at two fixed temperatures.

Temperatures in the neighbourhood of $-190°$ C. are most conveniently and accurately measured by determining the vapour pressure of pure oxygen at the temperature in question and making use of the vapour pressure curve of oxygen as determined by Travers (*Phil. Trans.*, 1902, *A*, 152). This method was used in these experiments and measurements of the densities were made at temperatures at which the vapour pressure of pure oxygen was (*a*) 100 mm., (*b*) 200 mm. On the hydrogen scale, these vapour pressures correspond to temperatures $74·70°$ and $79·07°$ Abs. respectively or to $-198·33°$ and $-193·96°$.

The oxygen used in the experiments was prepared by heating potassium permanganate and was passed through soda-lime and phosphorus pentoxide tubes. The nitrogen was prepared by heating a mixture of solutions of ammonium sulphate and potassium nitrite. The crude nitrogen obtained in this way was fractionated at a low temperature and was also passed through a tube containing red-hot copper.

In order to determine the density of the liquefied gas, a known volume of liquid was allowed to evaporate into a space of known volume (273·3 c.c.) and the pressure at a known temperature was then measured. This space was almost completely (270·9 c.c.) surrounded by a water-jacket so that the temperature of the gas could be measured rapidly and accurately. The portion of the space not jacketed (2·4 c.c.) was so small that no sensible error was introduced in considering its temperature the same as that of the jacketed portion, for the difference in temperature never exceeded 5° C. After the volume of the gas had been determined, the whole of it was transferred to a gas holder, and from this a sample was taken for analysis. The analysis was carried out by removing the oxygen (by means of phosphorus) from a known volume of the gas and measuring the residual nitrogen. The error of an analysis carried out in this way did not exceed 0·1 per cent.

The density bulb itself was blown on one end of a piece of capillary tubing which was graduated for 30 mm. above the bulb. The bulb and capillary were calibrated in the usual way with mercury. Taking the density of mercury as 13·603 at 0° (this being the apparent density using brass weights), the volume of the bulb to zero graduation was found to be 0·27880 c.c. at 17·5°, and the volume of each 10 mm. of the capillary was 0·00266 c.c. at the same temperature. These volumes can be corrected for the contraction that takes place on cooling to $-197°$ by making use of Baly's value '(*Phil. Mag.*, 1900, [v], **49**, 518) for the mean coefficient of contraction of glass between

15° and −190°. This correction leads to the values 0·27747 and 0·00265 at −197°. The correction necessary for a change of temperature of less than 10° is too small to be considered, so that the volume of the bulb may be considered constant between the two temperatures used in the experiments.

In carrying out an experiment, the density bulb, together with a bulb containing liquid oxygen and connected to a manometer, was surrounded with liquid air boiling under diminished pressure, and by adjustment of this pressure the temperature was reduced to that required for the experiment. The gas was then admitted to the density bulb, and when the meniscus appeared on the graduated part of the capillary tube the supply of gas was cut off and the temperature kept constant until the meniscus remained at a constant level. This volume of liquid was evaporated off and measured as already described. The results are given in Tables I and II.

TABLE I.

Temp. 74·70° abs.   Oxygen, vap. press. 100 mm.

Nitrogen per cent. by vol. found.	Volume of liquid in c.c. found.	Volume of gas in c.c. found.	Volume of liquid from 200 c.c. of gas.		
			Calculated.	Smoothed.	Difference.
0·0	0·28332	242·1	0·2340		—
0·0	0·28580	243·3	0·2338	0·2338	—
0·0	0·27919	238·9	0·2337		—
1·8	0·28429	241·8	0·2352	0·2351	+0·0001
4·5	0·28205	238·3	0·2367	0·2368	−0·0001
9·8	0·28354	236·4	0·2399	0·2399	—
13·9	0·27959	230·6	0·2425	0·2426	−0·0001
26·4	0·28369	226·4	0·2506	0·2508	−0·0002
37·8	0·28301	218·9	0·2585	0·2586	−0·0001
49·1	0·28197	211·8	0·2663	0·2663	—
55·8	0·27983	206·6	0·2709	0·2710	−0·0001
59·3	0·28200	206·1	0·2736	0·2736	—
65·6	0·28343	204·0	0·2779	0·2779	—
69·2	0·28162	200·7	0·2807	0·2804	+0·0003
72·4	0·28144	199·2	0·2826	0·2826	—
86·8	0·28324	193·8	0·2923	0·2930	−0·0007
88·0	0·28436	193·3	0·2942	0·2940	+0·0002
98·8	0·28065	186·1	0·3016	0·3020	−0·0004
100·0	0·28182	185·9	0·3032	0·3030	±0·0002
100·0	0·28214	186·3	0·3028		

TABLE II.

Temp. 79·07° abs.   Oxygen, vap. press. 200 mm.

Nitrogen per cent. by vol. found.	Volume of liquid in c.c. found.	Volume of gas in c.c. found.	Volume of liquid from 200 c.c. of gas.		
			Calculated.	Smoothed.	Difference.
0·0	0·28206	237·3	0·2377	0·2376	+ 0·0001
0·0	0·28070	236·3	0 2376	0·2376	—
23·6	0·28232	223·8	0·2523	0·2523	—
47·5	0·28264	209·6	0·2697	0·2696	+ 0·0001
73·2	0·28383	196·2	0·2892	0·2891	+ 0·0001
100·0	0·28343	182·3	0·3109	0·3110	− 0·0001
100 0	0·28293	181·9	0·3111	0·3110	+ 0·0001

If two liquefied gases mix without change of volume, the volume of liquid formed from a fixed volume of gas has a linear relation to the composition of the gas. In the tables, therefore, the volume of liquid formed from 200 c.c. gas (N.T.P.) is given, and if these values are plotted against the molecular percentage of nitrogen (see figure) it is found that the deviation from a straight line is only slight. Thus, in Table I, if there were no contraction, the volume of liquid corresponding to 49·1 per cent. of nitrogen would be 0·2678 instead of 0·2663, the contraction being therefore only 0·6 per cent. For the percentage 47·5 in Table II, the value would be 0·2726 if there were no contraction; the observed value being 0·2697, the contraction is slightly over 1 per cent. As one would naturally expect, the contraction is greater at the higher temperature, but it is somewhat surprising that there is any contraction at all at so low a temperature.

The densities of the pure liquids may be calculated from the figures given in the tables, and the values found are given in Table III.

TABLE III.

Temperature.	Density of nitrogen.		Density of oxygen.	
	I. and C.	B. and D.	I. and C.	B. and D.
74·76° abs.	0·8297	0·8218	1·223	1·217
79·07 abs.	0 8084	0·8010	1·203	1·196

These values may be compared with those found by Baly and Donnan (loc. cit.), which are also given in the tables, and it is to be noticed that there are considerable differences which are probably due to uncertainty in the temperature measurements of these investigators.

In a paper published by one of us (Inglis, Phil. Mag., 1906, [vi], 11, 640) on the isothermal distillation of nitrogen and oxygen, it was shown that the partial pressure of nitrogen above a mixture of nitrogen and oxygen was approximately proportional to the concentration of nitrogen in the liquid, but that a similar relation in the case of oxygen did not hold. Owing to the densities of mixtures of nitrogen and oxygen not having been determined, the concentrations had to be calculated from Baly

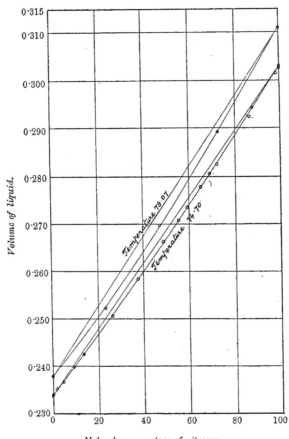

Molecular percentage of nitrogen.

and Donnan's results on the assumption that no contraction took place on mixing the two liquids. The results which have now been obtained enable us to calculate the concentrations more accurately, and the figures so obtained are given in Tables IV and V, the concentrations being given in grams per 100 c.c.

TABLE IV

Temp. 74·70° abs.   Oxygen, vap. press. 100 mm.

Nitrogen.			_Oxygen._		
Concentration. C.	Partial pressure, mm. P.	C/P.	Concentration. C.	Partial pressure, mm. P.	C/P.
0·0	—		122·3	100·0	1·223
5·6	34·5	0·1624	114·0	95·5	1·194
7·7	47·5	0·1620	111·1	93·5	1·188
11·9	72·7	0·1636	105·1	90·0	1·168
17·1	104·5	0·1634	97·6	85·5	1·141
21·3	129·5	0·1649	91·3	81·0	1·127
25·7	155·7	0·1651	85·0	77·0	1·104
30·2	182·5	0·1656	78·3	72·5	1·080
32·8	197·9	0·1659	74·5	69·9	1·065
36·5	218·6	0·1668	69·1	66·4	1·041
40·2	242·0	0·1662	63·7	61·7	1·032
42·5	255·4	0·1665	60·3	59·6	1·012
45·8	274·6	0·1666	55·5	55·9	0·992
48·9	293·3	0·1667	50·9	53·4	0·953
52·2	314·3	0·1662	45·9	48·7	0·943
53·0	318·8	0·1661	44·9	47·7	0·941
57·4	347·9	0·1650	38·2	42·1	0·907
61·6	374·6	0·1644	32·1	36·6	0·876
66·1	405·0	·0·1633	25·3	30·5	0·831
69·2	427·2	0·1621	20·7	25·8	0·801
72·6	450·7	0·1610	15·7	20·3	0·774
75·9	475·8	0·1595	10·7	14·2	0·757
79·2	500·8	0·1580	5·8	8·2	0·709
82·6	528·5	0·1564	(0·5)	(0·5)	(1·0)
83·0	(531·0)	(0·1563)	0·0	—	—

TABLE V.

Temp. 79·07° abs.   Oxygen, vap. press. 200 mm.

Nitrogen.			_Oxygen._		
Concentration. C.	Partial pressure, mm. P.	C/P.	Concentration. C.	Partial pressure, mm. P.	C/P.
0·0	0·0	—	120·3	200·0	0·601
3·6	37·0	0·0964	115·3	194·5	0·593
11·0	114·3	0·0965	104·6	181·2	0·577
19·8	205·8	0·0964	91·9	165·7	0·555
27·8	284·8	0·0975	82·0	150·7	0·544
38·5	402·2	0·0957	64·3	127·8	0·503
47·1	495·5	0·0951	51·5	108·5	0·474
54·3	577·6	0·0941	40·5	89·4	0·453
63·4	684·7	0·0926	26·9	65·3	0·412
70·2	773·6	0·0907	·16·6	42·4	0·391
75·3	845·6	0·0890	8·7	24·4	0·355
80·8	931·0	0·0868	0·0	0·0	—

These figures show only slight differences when compared with those obtained on the assumption that there is no contraction on mixing the liquefied gases, and they indicate conclusively that the law governing the relation of the partial pressure to the concentration is simpler in the case of nitrogen than in the case of oxygen. For moderate concentrations of nitrogen, the relation is that known as Henry's law, thus showing that liquid nitrogen has the normal molecular weight. For moderate concentrations of oxygen there is a deviation from this law, and this deviation indicates that liquid oxygen has a higher molecular weight than oxygen vapour. The association factor so indicated is about 1·09.

In conclusion, we wish to thank Sir William Ramsay for the kind interest he took in this work.

CHEMICAL DEPARTMENT,
    UNIVERSITY COLLEGE, LONDON.

XCIV.—*The Chemistry of Organic Acid "Thiocyanates" and their Derivatives.*

By AUGUSTUS E. DIXON, M.D.

So far as our knowledge extends at present, all acid chlorides containing the group –CO·Cl, on treatment with certain metallic thiocyanates, yield derivatives of the form R·CO(CNS).

Amongst the organic products obtained in this way, remarkable differences are encountered, particularly with regard to their behaviour with water or nitrogenous bases. For, whilst all are hydrolysed with more or less facility by contact with water, some members are said to yield carbon oxysulphide, but no thiocyanic acid :

$$\text{(i)} \quad R\text{·}CO(CNS) + H_2O = COS + R\text{·}CO\text{·}NH_2 \; ;$$

whereas others yield thiocyanic acid, but little or no carbon oxysulphide :

$$\text{(ii)} \quad R\text{·}CO(CNS) + H_2O = R\text{·}CO_2H + HSCN$$

(see, for example, Miquel, *Ann. chim. phys.*, 1877, [v], **11**, 299 and 302). Nitrogenous bases unite at the ordinary temperature with some of these by direct addition, forming disubstituted thiocarbamides :

$$\text{(iii)} \quad R\text{·}CO(CNS) + NH_2R' = R\text{·}CO\text{·}NH\text{·}CS\text{·}NHR' \; ;$$

with others, under like conditions, the chief result is a double decomposition, whereby substituted amide is formed, together with basic thiocyanate :

$$\text{(iv)} \quad R \cdot CO(CNS) + NH_2R' = R \cdot CO \cdot NHR' + HSCN.$$

Although these properties are by no means an exact parallel to what occurs amongst purely hydrocarbon thiocarbimides and thiocyanates respectively, they seemed at one time fairly consistent with the view that compounds yielding thiocyanic acid were really thiocyanates whilst compounds behaving according to equations (i) and (iii) above were thiocarbimides.

But a purely hydrocarbon thiocyanate (if not transformed by heat into the isomeric form) behaves in all its relations as such, and, similarly, a hydrocarbon thiocarbimide, once formed, acts consistently as thiocarbimide ; moreover, each form can be isolated and kept. In the class of acidic derivatives, however, it is different, no case yet being known where two distinct isomerides can be produced : one $R \cdot CO \cdot SCN$, and the other $R \cdot CO \cdot NCS$; besides, more recently, it has often been observed that a given specimen, when brought into contact with some one particular base, may lose practically all its sulphur as thiocyanic acid, whilst a different base may unite with it almost quantitatively to form a disubstituted thiocarbamide. If, therefore, the nature of the CNS derivative were to be defined according to its behaviour with bases, the definition must depend on what base is presented for interaction (compare Dixon, Trans., 1896, 69, 1599—1602) ; in other words, it could now no longer be maintained—at all events in some cases—that a particular substance was exclusively thiocarbimide or thiocyanate. And to such phenomena the term "tautomeric" was applied, with the reservation, however, that such tautomerism could not be explained by van Laar's hypothesis of the mobility of hydrogen within the molecule.

Next, a chance observation, made in preparing acetylphenylthiocarbamide, showed that, even with a given "thiocyanate" and a given base, the result might depend largely on the temperature at which interaction is brought about (Doran, Trans., 1905, 87, 331) ; this observation was followed up by a quantitative examination of the aniline-acetyl-" thiocyanate" system (Dixon and Hawthorne, ibid., 468), and it was now found that, with temperature increasing from $-12°$ to about $120°$, the yield of aniline thiocyanate together with acetanilide diminished from 94 to $2\frac{1}{2}$ per cent. of the total products obtained, the formation of acetylphenylthiocarbamide increasing at a corresponding rate, save that at the highest temperatures a slight irregularity occurred through (i) the production of a little hydrogen sulphide and (ii) the transforming effect of heat on the aniline thiocyanate.

In further pursuing these observations, it was next shown by Hawthorne (Trans., 1906, 89, 556) that the chemical behaviour of

3 N 2

propionyl " thiocyanate " with aniline at different temperatures is very similar to that of acetyl " thiocyanate" so far as the distribution of the sulphur with the base is concerned (a single measurement with stearyl " thiocyanate " gave a like result), but that, if o-toluidine be presented to acetyl " thiocyanate," the proportion of sulphur found in the products as acetyl-o-tolylthiocarbamide is largely increased for a given temperature. This fact, equally with certain others previously observed, could scarcely be reconciled with the view that acetyl " thiocyanate " consisted of an equilibrium-mixture of this substance with acetylthiocarbimide, in preportions variable with the temperature and determined by it alone, since in that case the distribution of the sulphur should be independent of whether aniline or toluidine was used ; and on examining the optical properties of the highly purified compound, he found the molecular refractive power to remain practically constant through an interval of temperature which must have influenced this property to a very considerable extent if any change corresponding to the variations of its chemical behaviour had occurred in the composition of the liquid. Furthermore, the value obtained for the molecular refractivity was not only constant, but agreed closely with that calculated (from various measurements with known thiocyanates and thiocarbimides) for the constitution $CH_3 \cdot CO \cdot NCS$, whilst diverging widely from that required for $CH_3 \cdot CO \cdot SCN$.

- The subject matter indicated above is still under investigation, but, so far as can be judged from these experiments, it would seem that true acetyl thiocyanate remains yet to be discovered ; that the compound so called is really acetylthiocarbimide ; and that its behaviour as thiocyanate is conditioned at the moment of interaction by at least two factors, namely, the character of the base presented and the temperature at which combination is effected. From some preliminary quantitative experiments recently conducted in this laboratory there is reason to believe that the order in which the constituents are mixed at a given temperature may affect the relative proportions of the pro- ducts : it is intended to make further inquiry regarding this.

In the case of other " thiocyanates " of the acidic class, there is yet another factor which sometimes plays a very important part, namely, the character of the radicle associated with the CNS group. Benzoyl-thiocarbimide, for instance—and the same is true as regards many other derivatives in which the radicle is more or less aromatic in structure— shows in the main the properties of a thiocarbimide, pure and simple. But even benzoylthiocarbimide, when treated with certain compounds containing sodium, can yield thiocyanic acid, and, in view both of this and of the fact that a distinctly thiocyanic isomeride is still unknown, it is plain that this substance will have to be examined at very low temperatures before the statement can confidently be made that it is a thiocarbimide in the same sense that "oil of mustard" is a thiocarbimide.

The mere presence of an aromatic nucleus in the electro-positive group is not, however, alone sufficient to determine essentially thiocarbimidic behaviour in the combination, although it certainly disposes in this direction : for the phenacetyl derivative,

$$PhCH_2 \cdot CO(CNS),$$

in which the CO is separated by CH_2 from the cyclic nucleus, when treated with ammonia at the ordinary temperature, undergoes change principally by double decomposition, yielding the amide of phenylacetic acid (Dixon, Trans., 1896, **69**, 868).

Our knowledge of these interactions with bases, so far as it can be generalised at present, may be stated approximately as follows :

(i) Purely aromatic "thiocyanates" (that is, those having the group CO attached directly to the nucleus) exhibit, within considerable ranges of temperature a strongly marked thiocarbimidic behaviour, which is but little dependent on the nature of the base presented for combination.

(ii) The behaviour of fatty acid "thiocyanates" with respect to bases is influenced by the temperature of interaction, and sometimes is determined practically by this alone : high temperature enhances thiocarbimidic power ; low temperature favours the development of thiocyanic character.

(iii) For any given moderate temperature, the function of the CNS group of a fatty acid "thiocyanate" varies principally according to the nature of the base submitted for interaction : ammonia tends strongly to develop its thiocyanic behaviour, whilst aromatic bases are most effective in deciding thiocarbimidic power; in this respect, o-toluidine is markedly superior to aniline.

(iv) The thiocyanic power of a fatty acid "thiocyanate" under given conditions is diminished by the substitution of an aromatic group for one of the hydrogen atoms of the fatty radicle ; concurrently, the thiocarbimidic power is increased.

Scarcely anything is yet known as regards combination with fatty bases, and very little in respect of bases such as methylaniline.

It has been pointed out above that the thiocarbimidic power of phenacetylthiocarbimide, in respect to its behaviour with ammonia at the ordinary temperature, is inconsiderable, the compound acting substantially as thiocyanate, and since that of benzoylthiocarbimide in like circumstances is very high, the question arose whether this difference could be conditioned through the separation of the –CO(CNS) group from the aromatic nucleus by any bivalent-linking group. For if the acquisition of thiocyanic properties is due solely to the absence of direct attachment of –CO(CNS) to the benzene ring, a similar effect might be anticipated if, instead of the group CH_2, an oxygen atom were interposed—for example, such a compound as $C_6H_5 \cdot O \cdot CO(CNS)$ might be expected to behave towards ammonia as a thiocyanate. Our knowledge of this class of oxy-aromatic derivatives is very scanty, being confined to the results of one or two preliminary experiments made by the late R. E. Doran (Trans., 1905, **87**, 342), without reference to the point in question ; I have therefore prepared a few of these "thiocyanates," $R \cdot O \cdot CO(CNS)$, in order to learn something of their behaviour with bases, and especially with ammonia. For the most essential materials required in conducting this work, I am greatly indebted to the courtesy of the Bayer Company, who prepared and presented to me liberal specimens of various aromatic chlorocarbonates, for which I desire to return my thanks. Incidentally, some other experiments with acidic "thiocyanates" have been made, which may find a place in this communication.

PART I.

Cyclic Oxythiocarbimides, $RO \cdot CO \cdot NCS$.

(i) *Thiocarbiminophenylcarbonate,* $C_6H_5 \cdot O \cdot CO \cdot NCS$.

The phenyl chlorocarbonate, $C_6H_5O \cdot CO \cdot Cl$, when purified by rectification, is a colourless liquid, boiling at 87° under a pressure of 25 mm. By allowing this substance, dissolved in benzene, to remain in contact with excess of dry finely-divided sodium thiocyanate (or by boiling the solution with dry lead thiocyanate), interaction took place, with formation of metallic chloride, the organic product remaining dissolved in the benzene. The process is slow, a week or more being required for the preparation in the cold ; if heat be employed, a certain amount of decomposition is apt to occur, the solution becoming brown after some hours, and before the interaction is yet complete ; except the coloration, however, there seemed to be little difference in the products. After filtering by the aid of the pump, the solution was usually distinctly yellow, had a slightly pungent thiocarbimidic odour, and responded freely to the desulphurisation tests, with ammoniacal silver nitrate or hot alkaline lead hydroxide, for a thiocarbimide. If shaken up with cold water, the aqueous portion was not affected by treatment with ferric chloride, but if well boiled with water, a faint red coloration was produced : in these circumstances, therefore, hydrolytic change into thiocyanic acid occurs to a very limited extent.

Action of Ammonia.—On treating the thiocarbimide in benzene solution with a considerable excess of alcoholic ammonia, the mixture became warm, and in a short time solid matter began to separate ; by concentration to a small bulk, the liquor gave another crop of the same material, and the filtrate contained a little ammonium thiocyanate. The main product, when recrystallised from alcohol, was obtained in snow-white, soft prisms, melting with effervescence at 175° (corr.). It was nearly insoluble in cold water, moderately easily soluble in boiling alcohol, rather sparingly so in the cold, and gave on analysis the figures required for an additive compound.

Found, $N = 14\cdot4$; $S = 16\cdot3$.

$C_8H_8O_2N_2S$ requires $N = 14\cdot29$; $S = 16\cdot32$ per cent.

Carbo-phenoxythiourea in alcoholic solution gives with alcoholic silver nitrate a curdy, white precipitate, which blackens quickly on the addition of ammonia ; it is desulphurised readily by heating with an alkaline solution of lead. The substance dissolves very easily in cold dilute caustic alkali ; if this solution be gently warmed and then acidified with hydrochloric acid, no precipitate is formed, but carbon dioxide and hydrogen sulphide escape with effervescence ; the residual liquid has an odour of phenol, and gives the reactions for this substance ; it gives a deep red coloration with ferric chloride, and on warming the alkaline solution, ammonia is evolved. These reactions may be explained in part by the equation

$$CSN_2H_3\cdot O\cdot COPh + H_2O = PhOH + CO_2 + NH_4\cdot SCN.$$

When the substance was kept at a temperature just above its melting point, gas escaped, containing hydrogen sulphide and carbon dioxide, and soon a white solid began to accumulate, which did not melt at 200° ; the cooled residue formed a paste, from which cold water dissolved phenol and some thiocyanic acid ; the undissolved portion, when recrystallised from boiling water, formed microscopic white needles, easily desulphurised by ammoniacal silver nitrate, but scarcely affected by boiling with alkaline lead solution. This product was not further examined.

It has been previously pointed out that, although thiocarbiminophenylcarbonate (or carbo-phenoxythiocarbimide) yields no measurable amount of thiocyanic acid when brought into contact with either hot or cold water, yet by the action of ammonia this acid is produced. A benzene solution of the thiocarbimide was treated with excess of alcoholic ammonia, the temperature, during the process, being kept in the neighbourhood of 0°. The total solid product amounted to 4·57 grams (from 3·9 grams of chloride), and this, when extracted by cold water, yielded 0·171 gram of ammonium thiocyanate, as determined by Barnes and Liddle's method, using $N/10$ copper sulphate. Of the total weight of sulphur found, therefore, rather less than 10 per cent. appears as thiocyanic acid, the remainder existing in the form of carbo-phenoxythiourea, and hence the capacity of the parent thiocarbimide, even in circumstances highly favourable to the development of thiocyanic character, to behave as thiocyanate is but feebly marked. From the foregoing result, it was to be anticipated that, with primary aromatic bases, little or no thiocyanate would be formed, and in effect this proved to be the case.

Action of Aniline.—This base was used first in benzene solution, and then in alcohol, the latter method giving a much cleaner product : the mixture became hot, and, on cooling, 72 per cent. of the theoretical yield was obtained (reckoned from the amount of chloride employed), and on concentration the mother liquor gave a further crop. The substance crystallised from alcohol in white prisms, melting with effervescence at 148°, the temperature recorded by Doran (*loc. cit.*), and estimations of sulphur and nitrogen gave 11·65 and 10·4 per cent., against 11·76 and 10·29 respectively, calculated for

$$C_6H_5 \cdot O \cdot CO \cdot NH \cdot CS \cdot NH \cdot C_6H_5.$$

It is thus isomeric with phenylthiouramidobenzoic acid,

$$PhNH \cdot CS \cdot NH \cdot C_6H_4 \cdot CO_2H.$$

When brought into contact with dilute caustic alkali, the crystals became opaque, and dissolved on warming ; at the same time, drops of oil, recognised easily by its odour as phenylthiocarbimide, made their appearance. This was expelled by boiling (no ammonia being detected), and the residual solution was acidified with hydrochloric acid, whereupon effervescence occurred, with evolution of carbon dioxide and hydrogen sulphide, and the liquid now had a strong odour of phenol, but gave with ferric chloride no trace of a red coloration, thus showing that no thiocyanic acid had been formed. Some aniline was, however, present. The decomposition is obviously somewhat complex, but the ease with which it is started depends probably on the ready elimination of the $-O \cdot CO-$ group.

Action of o-*Toluidine.*—Heat was evolved on mixing alcoholic o-toluidine with the benzene solution of the thiocarbimide, and almost immediately the product began to crystallise out in beautiful tufts of vitreous prisms ; when purified by crystallisation from alcohol, in which it is somewhat sparingly soluble, it melted at 164—165° (corr.) with effervescence.

Found, $S = 11·25$ and $11·35$.

$C_{15}H_{14}O_2N_2S$ requires $S = 11·19$ per cent.

Dilute caustic alkali dissolved the solid on gently warming, and the presence of o-tolylthiocarbimide was manifested both by the odour and by the separation of drops of oil in the steam when the mixture was boiled : no ammonia was detected in the escaping vapours. After expelling the tolylthiocarbimide in this way and acidifying the residue with hydrochloric acid, carbon dioxide and hydrogen sulphide were evolved, and the odour of phenol now became strong. This decomposi·

tion, as in the preceding case, occurs very readily; in fact, the odour of tolylthiocarbimide can be detected on mere contact of the solid with cold alkali.

Action of p-*Toluidine.*—The product formed a mass of white, woolly needles, melting, after crystallisation from alcohol, at 144—145° (corr.); like its congeners, it was insoluble in water and sparingly soluble in alcohol.

Found, S = 11·3. $C_{15}H_{14}O_2N_2S$ requires S = 11·19 per cent.

Alcoholic silver nitrate gave with the alcoholic solution a white precipitate, rapidly becoming black, and the substance was easily desulphurised by heating with alkaline lead tartrate. When warmed with dilute caustic alkali, it behaved like the *ortho*-compound, *p*-tolyl-thiocarbimide being formed; on acidifying, carbon dioxide and hydrogen sulphide escaped, and phenol remained in solution : no thio-cyanic acid was present.

Action of Benzylamine.—Beautiful, white crystals were deposited, moderately soluble in alcohol, insoluble in water; when recrystallised, they formed brilliant, slender, vitreous prisms, melting at 153—154° (corr.) and slowly evolving gas.

Found, S = 11·3. $C_{15}H_{14}O_2N_2S$ requires S = 11·19 per cent.

The substance dissolves moderately easily in warm dilute caustic potash with formation of benzylthiocarbimide, benzylamine, sulphide and carbonate of potassium, and phenol, and hence is readily desulphurised by treatment with alkaline lead, or ammoniacal silver solution.

Action of Alcohol.—On warming the constituents together, a vigorous action soon commenced, the mixture boiling freely. When cool, the brown, syrupy liquid set to a radiating, crystalline mass, and the filtrate contained but a trace of thiocyanic acid. The residue, after being washed with dilute spirit, became nearly white, and amounted to about 70 per cent. of the theoretical; on evaporation, the mother liquor yielded another viscid crop of crystals. By crystallising from alcohol, in which it is freely soluble, the product was obtained in white prisms, melting at 82—83°. A sulphur determination gave S = 14·5, the figure calculated for the thiourethane,

$$PhO \cdot CO \cdot NH \cdot CS \cdot OEt^*$$

($C_{10}H_{11}O_3NS$), being 14·22 per cent.

The substance dissolved readily in hot caustic alkali without evolu-tion of ammonia ; when acidified, the solution effervesced, carbon dioxide escaping, together with some hydrogen sulphide ; the residual liquid

* Otherwise, ethyl phenyl iminothiondicarbonate ; probably it has the constitu-tion $PhO \cdot CO \cdot N{:}C(SH) \cdot OEt$, that is, carbophenoxyiminoethylthiolcarbonic acid.

had an odour of phenol, and gave with ferric chloride an intense red coloration :

$$PhCO_2 \cdot NH \cdot CS \cdot OEt + H_2O = PhOH + CO_2 + HSCN + EtOH.$$

When boiled with alkaline lead solution, the mixture darkened gradually,[but the desulphurisation was only partial, for the filtrate, when acidified and treated with ferric chloride, gave a strong indication of thiocyanic acid. Boiling water dissolved the substance to a slight extent, apparently without causing decomposition.

(ii) *Thiocarbimino-o-tolylcarbonate*, $CH_3 \cdot C_6H_4 \cdot O \cdot CO \cdot NCS$.

This substance was prepared by allowing *o*-tolylcarbonic chloride, $MeC_6H_4 \cdot CO_2Cl$, dissolved in benzene, to remain in contact with dry powdered sodium thiocyanate ; the reaction was conducted, however, at a gentle heat (about 30—40°), whereby the process was accelerated without any material loss through decomposition of the organic product : in these circumstances, the operation was completed in four days.

The filtered solution was yellowish-brown, had a slightly pungent odour, and was readily desulphurised by warming with alkaline salts of lead or silver. But when shaken up thoroughly with cold water, the aqueous extract gave no reaction for thiocyanic acid, and even after boiling with water, ferric chloride produced a barely perceptible red coloration.

Action of Aniline.—Heat was developed on mixing, and the solution as it cooled became solid, the yield of dry product amounting to 77 per cent. of the theoretical, reckoning from the weight of chloride taken for experiment. The mother liquor, when tested with ferric chloride, gave no red coloration, thus showing the absence of thiocyanate.

Found, $S = 11 \cdot 2$. $C_{15}H_{14}O_2N_2S$ requires $S = 11 \cdot 19$ per cent.

The compound, *a*-carbo-*o*-tolyloxy-*b*-phenylthiocarbamide, was sparingly soluble in hot alcohol, from which it separated as a felted mass of long, flexible, white needles, becoming highly electrical on friction, and melting at 155—156° (corr.). It was insoluble in water, but dissolved readily in boiling dilute caustic alkali with evolution of phenylthiocarbimide, but no ammonia could be detected. After acidifying the residual liquid, carbon dioxide and hydrogen sulphide were evolved, and an oil was left having the odour of *o*-cresol. No thiocyanic acid was present in the products of this decomposition.

Action of o-*Toluidine.*—Combination occurred at once, and the yield of solid product amounted to 76 per cent. of the theoretical, a further quantity separating from the mother liquor as it evaporated. When crystallised from boiling alcohol, in which it is sparingly soluble, the product was obtained in brilliant, lozenge-shaped prisms melting at 142—143° (corr.) with slow evolution of gas. The solid dissolved easily in hot dilute caustic alkali with formation of o-tolylthiocarbimide : the residual liquid, after the thiocarbimide had been expelled by heating, effervesced on treatment with hydrochloric acid, giving off hydrogen sulphide and carbon dioxide, and had a strong odour of o-cresol; ferric chloride produced no red coloration in the acidified mixture.

An experiment was now made in order to learn whether the compound produced by substituting the group $C_6H_4(CH_3)\cdot O\cdot CO-$ for 'the acetyl group in acetyl-o-tolylthiocarbamide is identical with the above.

Pure acetyl-o-tolylthiocarbamide was mixed with o-tolyl chlorocarbonate, the latter being in slightly greater proportion than that required according to the equation

$$AcNH\cdot CS\cdot NH\cdot C_6H_4(CH_3) + C_6H_4(CH_3)\cdot CO_2Cl =$$
$$AcCl + C_6H_4(CH_3)\cdot NH\cdot CS\cdot NH\cdot CO_2\cdot C_6H_4(CH_3),$$

and the mixture was cautiously heated. At 107°, interaction commenced with evolution of gas containing some hydrogen sulphide ; the vapour fumed copiously, had an odour of acetyl chloride, and when led into ethyl alcohol gave an acid liquid smelling strongly of ethyl acetate. The temperature was now reduced to 100°, and the process continued until the turbid mixture became clear and no more gas was evolved. The residual brown oil was dissolved in hot alcohol, from which, on cooling, a crystalline solid was deposited ; after a couple of recrystallisations, this was obtained as colourless lozenges melting at 142—143° (corr.), having the same appearance and showing precisely the same reactions as the compound previously described. The formula was checked by a sulphur estimation, which gave $S = 10.7$ per cent., the calculated figure for $C_{16}H_{16}O_2N_2S$ being 10.66 per cent.

Action of p-*Toluidine.*—Operating as before, the yield of solid product amounted to 85½ per cent. of the theoretical, calculated from the weight of chloride employed. By crystallisation from boiling alcohol, in which it is sparingly soluble, the compound was obtained in beautiful, fern-like aggregates of white prisms melting at 150—151° (corr.). When boiled with dilute alkali, it behaved in exactly the same way as the *ortho*-compound, save that p-tolylthiocarbimide was expelled. As usual amongst compounds of this class, where two substituting groups are present, no ammonia could be detected in the vapour during the boiling with potash.

Found, $S = 10.7$. $C_{16}H_{16}O_2N_2S$ requires $S = 10.66$ per cent.

Action of Ammonia.—Excess of alcoholic ammonia was used, the mixture becoming hot at once, and presently long, silky needles were deposited, the weight of which amounted to about 75 per cent. of the theoretical. The mother liquor, after being acidified by hydrochloric acid, gave a decided red coloration with ferric chloride ; in this case, therefore, just as in the corresponding one already described, where ammonia was caused to interact with the phenylic homologue, a certain amount of thiocyanic acid was formed.

After crystallisation from alcohol, in which it is moderately soluble at the boiling point, the compound melted at 156—157° (corr.), and a sulphur determination gave 15·5 per cent. against 15·24 calculated for $C_9H_{10}O_2N_2S$.

Carbo-*o*-tolyloxythiourea is practically insoluble in water, but dissolves, apparently without decomposition, in cold dilute alkali, the solution giving off ammonia when heated ; on acidifying, carbon dioxide and hydrogen sulphide escape with effervescence ; the residual liquor has an odour of *o*-cresol, and gives a strong thiocyanic reaction with ferric chloride. The process in the main appears to be a simple hydrolysis :

$$C_6H_4(CH_3) \cdot CO_2 \cdot NH \cdot C(SH) \vdots NH + H_2O =$$
$$C_6H_4(CH_3)OH + CO_2 + NH_4 \cdot SCN,$$

but probably, in addition to the ammonium thiocyanate, a little thiocarbamide is formed, the decomposition of which by the caustic alkali would lead to the formation of alkali hydrosulphide. The desulphurisation by heating with alkaline solution of lead does not extract the whole of the contained sulphur in the form of lead sulphide, for if the latter be removed and the clear liquor acidified, ferric chloride now gives a strong reaction for thiocyanic acid.

(iii) *Thiocarbimino-p-tolylcarbonate*, $CH_3 \cdot C_6H_4 \cdot O \cdot CO \cdot NCS$.

The benzene solution, prepared as before, was a pale yellow liquid, having little characteristic odour in the cold. When shaken up with hot or cold water, or with dilute alcohol, the mixture gave no red coloration with ferric chloride, and hence underwent no perceptible hydrolysis into thiocyanic acid. With silver nitrate alone no precipitate was formed, but on adding ammonia and warming, the mixture was blackened ; desulphurisation also took place readily on heating with a lead salt and alkali.

Action of Aniline.—To 75 c.c. of benzene solution, representing 9·65 grams of thiocarbimide (assuming the chloride to have undergone change quantitatively into the latter), a trifle more than the calculated amount of aniline was added, dissolved in 20 c.c. of strong alcohol ; the temperature rose at once by 18°, and on concentration nearly white crystals were deposited, weighing when dry 13·05 grams, that is, more than 91 per cent. of the possible yield for an additive product. When purified by recrystallisation from boiling alcohol, in which it is only moderately soluble, the compound formed long, brilliant, vitreous prisms, insoluble in water and melting at 157—158° (corr.) with slight effervescence.

Found, $S = 11·3$. $C_{15}H_{14}O_2N_2S$ requires $S = 11·19$ per cent.

Towards dilute caustic alkali, its behaviour was precisely similar to that of the isomeric carbophenoxy-o-tolylthiocarbamide (m. p. 165°), save that phenylthiocarbimide separated in place of o-tolylthiocarbimide, and that subsequently, on acidifying, p-cresol instead of phenol was liberated.

Action of o-*Toluidine.*—Operating as just described, the temperature rose by 18°, and ultimately pearly leaves were obtained, the yield in this case amounting to about 95 per cent. of the theoretical.

By solution in warm alkali, the substance yielded o-tolylthiocarbimide, and thence by acidification p-cresol, together with hydrogen sulphide and carbon dioxide; the cold alcoholic solution gave with alcoholic nitrate of silver a white precipitate, changing rapidly to black. The melting point was found to be 160—161° (corr.), and a sulphur estimation gave 10·7 per cent. against 10·66 calculated for $C_{16}H_{16}O_2N_2S$.

(iv) *Thiocarbiminobenzylcarbonate*, $C_6H_5 \cdot CH_2 \cdot O \cdot CO \cdot NCS$.

The benzene solution, prepared in the cold, is a pale straw-yellow liquid, having little distinctive odour, but giving off a vapour which attacks the eyes. Whilst readily desulphurised by alkaline silver or lead salts, it yielded to cold water scarcely a trace of thiocyanic acid; if boiled with water, the aqueous extract when treated with ferric chloride gave a slight red coloration. When heated with dilute caustic alkali and then treated with hydrochloric acid, the solution effervesced with evolution of carbon dioxide and hydrogen sulphide, and the residual liquid gave a distinct but not intense reaction for thiocyanic acid.

Action of Aniline.—Two parallel experiments were made, one at the laboratory temperature, the other at 65—67°; the yields were identical, and no particular difference of any kind was noticed, save that the product at the higher temperature was somewhat cleaner looking; in neither case was aniline thiocyanate found. After a thorough washing with light petroleum, the solid was recrystallised from boiling alcohol—in which it is moderately freely soluble, although

only sparingly so in the cold—and was thus obtained in microscopic needles melting at 144—145° (corr.). The alcoholic solution gave with alcoholic silver nitrate a white precipitate, becoming black in a few seconds, and was moderately easily desulphurised by boiling with alkaline solution of lead.

Found, $S = 11.2$. $C_{15}H_{14}O_2N_2S$ requires $S = 11.19$ per cent.

This substance, a-*carbo-benzoxy*-b-*phenylthiocarbamide*, is isomeric with the a-carbo-phenoxy-b-benzylthiocarbamide (m. p. 154°) previously described.

Concerning the function of the CNS group in benzenoid derivatives of the form R·X·CO(CNS), it may safely be concluded from the above experiments that the power to develop thiocyanic character is not conferred upon this group through the mere separation of the atomic complex ·CO(CNS) from the benzenoid nucleus R by means of any bivalent nucleus X, since if X be represented by $-CH_2-$, the product may in some circumstances be highly thiocyanic in character, whilst, on the other hand, if X be represented by $-O-$, the power of the resultant molecule to behave as thiocyanate in the same circumstances as before may become almost negligible. In short, these compounds, although not incapable of manifesting the properties displayed by the " thiocyanates " of fatty acids, are nevertheless for all practical purposes true thiocarbimides in the same sense that benzoylthiocarbimide is admitted to be one. This, however, does not imply that either they or the benzoyl compound must necessarily be regarded as thiocarbimides, for in the present state of our knowledge the writer is disposed to the view that all " thiocyanates " belonging to the acidic class are potentially tautomeric.

As to the properties of the disubstituted thiocarbamides originating from these oxythiocarbimides, it was observed that in no case did the product, after treatment with caustic alkali, yield thiocyanic acid : nevertheless, the decomposition is profound, including, as it does, the formation of (1) the phenol corresponding to the thiocarbimide employed, (2) carbon dioxide, (3) hydrogen sulphide, (4) the thiocarbimide corresponding to the base employed, and (5) the base itself. With monosubstituted thioureas, if the decomposition were to run a similar course, product (4) must be represented by thiocyanic acid, of which, in effect, more or less was always found. In all these cases, the primary change no doubt consists in the withdrawal of the acidic group, RO·CO, with subsequent breaking down both of the resultant carbonate and of the ·nitrogenised residue, CSN_2H_3R' or CSN_2H_4, the latter of which represents thiourea in the act of being formed, with the potentiality of yielding the isomeric ammonium thiocyanate.

On the other hand, it was found that each of the foregoing thioureas and thiocarbamides, when heated at the melting point, or slightly beyond it, yields a considerable amount of thiocyanic acid. To ascertain whether this behaviour on heating is common to thioureas and thiocarbamides in general, apart from the nature of the groups included, a series of experiments was made. In each case, a quantity—a decigram or so—was cautiously melted in a dry tube over a small flame, and warmed until signs of boiling commenced ; the melt was cooled somewhat, treated with a drop of hydrochloric acid and a few c.c. of hot water, and then with ferric chloride. Broadly speaking, the results grouped themselves into two classes : either no indication whatever was obtained of the presence of thiocyanic acid (or a reaction so faint as to be barely noticeable), or else a very intense red coloration appeared ; in some few cases, however, rather indeterminate effects were produced, namely, a reddening of the mixture, varying from a slight, but not negligible, to a distinct, bright colour ; these ill-defined cases, it will be noticed, were met with solely amongst derivatives containing the benzyl group. The results are classified below.

(i) Monosubstituted Thioureas.

Phenyl, o-tolyl, p-tolyl, sec.-butyl, α-chloroallyl, acetyl, benzoyl, anisoyl, cinnamoyl : all these gave an intense red coloration with ferric chloride.

(ii) ab-Disubstituted Thiocarbamides.

(a) The following yielded either no thiocyanic acid, or else merely the faintest trace : diphenyl, di-p-hydroxyphenyl, di-o-tolyl, sec.-butylphenyl, α-chloroallyl-p-tolyl, phenyl-o-tolyl, phenyl-p-tolyl, methyl-o-tolyl, methyl-p-tolyl, benzyl-o-tolyl, benzyl-m-tolyl, benzyl-p-tolyl, (o- slight, p- distinct), phenyl-α-naphthyl, benzyl-β-naphthyl, dibenzyl (faint), benzyl-m-xylyl (green coloration).

(b) A distinctly marked coloration was obtained with ethylbenzylthiocarbamide, and a decided bright red with allylbenzylthiocarbamide.

(c) An intense thiocyanic reaction was produced by the following thiocarbamides : acetylphenyl, acetyl-o-tolyl, acetylbenzyl, propionylphenyl, propionyl-o-tolyl, propionyl-p-tolyl, butyryl-o-tolyl, butyryl-α-naphthyl, i-butyrylphenyl, valeryl-p-tolyl, valeryl-α-naphthyl, caproyl-o-tolyl, palmityl-o-tolyl, palmityl-p-tolyl, stearylphenyl, stearyl-p-tolyl, benzoyl-o-tolyl, anisoylethyl, anisoylphenyl, anisoyl o- and p-tolyl, cinnamoyl-o-tolyl, cinnamoyl-p-tolyl, cinnamoyl-α-naphthyl, phenacetyl-

o-tolyl, phenoxyacetylphenyl, carboxymethyl-p-tolyl.   Also by methyl-phenylcarbamyl-o-tolylthiocarbamide,

$$\text{MePhN·CO·NH·CS·NHC}_6\text{H}_4\text{(Me)},$$

and by succinyl-di-o-tolyldithiocarbamide,

$$\text{C}_2\text{H}_4\text{(CO·NH·CS·NHC}_6\text{H}_4\text{Me)}_2.$$

(iii) aa-*Disubstituted Thioureas.*

Two unsymmetrical disubstituted thioureas only were examined, aa-methylphenyl and piperidyl respectively; each gave a very intense thiocyanic reaction.

(iv) *Trisubstituted Thioureas.*

(a) Thiocyanate reaction negative or very slight : trimethyl, ethyl-phenylbenzyl, ethylpiperidyl, a-chloroallylphenylbenzyl (green colora-tion), phenyldimethyl, phenylmethylbenzyl, phenylethylbenzyl, benzylethylphenyl, phenylphenylbenzyl, benzylphenylbenzyl, ' tri-benzyl.

(b) Benzyldimethyl and benzylmethylbenzyl thioureas gave a distinct, although not strong, red coloration.

(c) Thiocyanate reaction intense : acetylphenylbenzyl, propionyl-phenylbenzyl, caproylphenylbenzyl,· palmitylmethylphenyl, benzoyl-dimethyl, benzoyldiethyl, benzoylpiperidyl, anisoylphenylbenzyl, phenacetylphenylbenzyl.

These results seem to justify the following conclusions :

(i) Little or no thiocyanic acid is produced by ab-disubstituted thio-carbamides or by trisubstituted thioureas, provided that in both cases purely hydrocarbon radicles alone are present. Apparent exceptions were found, however, amongst certain benzylated forms, which gave more or less distinct reactions for it : these cases were met with solely amongst products derivable from benzylthiocarbimide. But the results were so irregular, both as regards occurrence and the amount of acid formed, as to suggest the probability that these exceptions might disappear if the compounds were purified to a sufficiently high degree.

(ii) Much thiocyanic acid is produced from :

(a) All monosubstituted thioureas, whatever the nature of the substituent.

(b) The two unsymmetrical disubstituted thioureas examined.

(c) All thioureas or thiocarbamides (up to trisubstituted) con-taining an acid group.

The mechanism of these changes, primarily at least, appears to consist of a dissociation of each compound into its constituents ; thus, for example, when thiocarbanilide is heated, it decomposes mainly into phenylthiocarbimide and aniline. In what follows, the type $NH_2 \cdot CS \cdot NH_2$, for the sake of uniformity, is assumed to represent all derivatives.

(i) ab-*Disubstitution Derivatives*, $NHX \cdot CS \cdot NHY$.—Since these substances are produced both from $X \cdot NCS$ and $Y \cdot NH_2$, and from $Y \cdot NCS$ and $X \cdot NH_2$, it was to be anticipated that all four constituents would appear on decomposition, and on testing the phenyl-*p*-tolyl melt both phenylthiocarbimide and a little aniline were found. In like manner, acetyl-*o*-tolylthiocarbamide gave both acetyl- and *o*-tolyl-thiocarbimides; nevertheless, as might be expected, one thiocarbimide tends to be formed preferentially; thus, when phenyl-*p*-tolylthiocarbamide was heated to the minimum temperature required to effect decomposition, phenylthiocarbimide was evolved, but the odour of *p*-tolylthiocarbimide could not be detected, although, as just stated, aniline was found to be present. In such cases, if the sulphuretted products, that is, $X \cdot NCS$ or $Y \cdot NCS$, or both, are incapable of functioning as $X \cdot SCN$ or $Y \cdot SCN$ (or of reverting to such), no indication of the presence of thiocyanate is to be expected (it should perhaps be mentioned that several of these melts were treated with alcoholic sodium sulphide to ascertain if any cyclic thiocyanate had been formed, but with a negative result). But if either of them does possess such power, the decomposition product must show it if placed under suitable conditions. Thus, if acetylphenylthiocarbamide yields acetyl " thiocyanate," it should give with water and ferric chloride a red coloration, which was found to be the case.

(ii) *a*. All *monosubstitution derivatives*, by the above decomposition, must, through loss of ammonia, yield $R \cdot NCS$ (which may or may not itself behave as thiocyanate), and by loss of $R \cdot NH_2$ the residue $H \cdot NCS$, otherwise $H \cdot SCN$. This serves to explain why the thiocyanic reaction is always obtained, no matter what radicle be included.

(ii) *b*. An *unsymmetrical disubstitution derivative*, by loss of secondary amine, must necessarily yield thiocyanic acid.

(ii) *c*. *Trisubstitution Derivatives*.—By simple dissociation,

$$NHX \cdot CS \cdot NYZ$$

can yield only $X \cdot NCS$ and $NHYZ$, and hence the products will react for thiocyanate if $X \cdot NCS$ can behave as such (namely, if it be an acid " thiocyanate "), and not otherwise. Although a hydrocarbon group X does not attach itself to the radicle $\cdot NYZ$, it is possible, nevertheless, that an acid group might transfer itself, in which case thiocyanic acid would be formed, together with a trisubstituted amide. But it seems improbable that a whole acid group should migrate directly instead of the mobile hydrogen atom, and, since the characteristic odour of an acid " thiocyanate " is always noticeable when an acidic trisubstituted thiourea is heated, it is quite possible that the formation of thiocyanic acid may be due to the CNS group undergoing a certain amount of thiocyanic change with the secondary base, thus : X·SCN + NHYZ = H·SCN + X·NYZ.

Some interest attached to the decomposition of an acetylated derivative of this class, because owing to the mode of production the resultant acid "thiocyanate" would in this case presumably be true acetylthiocarbimide, CH_3·CO·NCS, whilst the usual method of preparing the compound (namely, from acetyl chloride and lead thiocyanate) does not necessarily of itself lead to the formation of thiocarbimide. In testing this point, 8 grams of pure acetylphenylbenzylthiourea were carefully heated in a small flask under a pressure of 22 mm., the flask being immersed in an oil-bath. A reddish-yellow liquid distilled over at 48—49°, which had a pungent odour of acetyl "thiocyanate"; a portion, treated in boiling benzene with excess of o-toluidine, gave acetyl-o-tolylthiocarbamide, melting at 183—184°, and the remainder gave the usual reactions of the known acetyl compound, including its ready hydrolysis by water, with abundant formation of thiocyanic acid. The results of this experiment seem to accord satisfactorily with the view advanced recently by Hawthorne (loc. cit., p. 566), that acetyl "thiocyanate" is but a tautomeric form of acetylthiocarbimide.

PART II.

Phenoxyacetylthiocarbimide, C_6H_5·O·CH_2·CO·NCS.

The few derivatives briefly mentioned below were obtained some years ago, not in connection with the present research, but merely to learn whether a thiocarbimide corresponding to phenoxyacetic acid could be produced; no attempt, therefore, was made to trace the course of the CNS group, save as regards its capacity to behave as thiocarbimide.

The solution, prepared by boiling phenoxyacetyl chloride in benzene with sodium thiocyanate, was reddish-brown; it behaved as thiocarbimide towards lead and silver salts, and when shaken up with water and treated with ferric chloride it gave the reaction for a thiocyanate. When aniline was added to the hot benzene solution, an oil was precipitated, consisting mostly of aniline thiocyanate, and the clear decanted liquor deposited after some time a solid, which, when repeatedly crystallised from alcohol, gave brilliant leaves melting at 112—113° (corr.).

Found, S = 11·1. $C_{15}H_{14}O_2N_2S$ requires S = 11·19 per cent.

This substance, a-phenoxyacetyl-b-phenylthiocarbamide, is isomeric with a-carbobenzoxy-b-phenylthiocarbamide, &c., described above.

The compound was insoluble in cold water, easily soluble in boiling alcohol, but rather sparingly so in cold. Alcoholic silver nitrate, when added to the alcoholic solution, yielded a white precipitate, changing rapidly to black ; desulphurisation was readily effected by hot alkaline lead tartrate. When heated in a dry tube, phenylthiocarbimide was evolved, and the residue, treated with hydrochloric acid, water, and ferric chloride, gave a strongly marked reaction for thiocyanic acid.

o-*Toluidine* gave an oil which solidified in a few days ; after two recrystallisations from hot alcohol, in which it was very freely soluble, although but sparingly so in the cold, the product formed long, white needles melting at 100—101°.

Found, $S = 10\cdot9$. $C_{16}H_{16}O_2N_2S$ requires $S = 10\cdot66$ per cent.

Using p-*toluidine*, a solid was obtained melting at 129—130°, and giving $S = 11\cdot0$ per cent., against $10\cdot66$ calculated. Both the latter compounds had properties similar to those of the phenyl analogue : they are isomeric with the tolyl derivatives of the thiocarbiminotolyl carbonates described above.

An attempt was made to isolate the thiocarbimide by distilling the product obtained from 33 grams of phenoxyacetyl chloride under diminished pressure ; decomposition occurred, however, with evolution of much gas, and the residue consisted of a black tar.

PART III.

Phenyl Chlorocarbonate and Thiourea.

A few years ago it was shown (Dixon, Trans., 1903, 83, 550) that ethyl- and methyl-chlorocarbonates unite additively with thiourea, yielding the hydrochlorides of basic compounds, such as

$$NH\dot{:}C(NH_2)\cdot S\cdot CO_2Me,$$

in which the carboalkyloxy-group is united directly with sulphur ; when these substances are warmed, carbon dioxide is evolved, the alkyl group attaching itself to the sulphur atom, with formation of salts of iminothiocarbamic acid, for example, $NH\dot{:}C(NH_2)\cdot SMe$. It was hoped, therefore, that by the action of phenyl chlorocarbonate a similar compound would result, from which, by elimination of carbon dioxide, an isomeride of phenylthiourea might be obtained, namely, phenyl-pseudo-thiourea, $NH\dot{:}C(NH_2)\cdot SPh$, the synthesis of which the writer has already attempted (Trans., 1903, 83, 551), but unsuccessfully, by a different method. As will be seen below, the additive compound may easily be produced, but, strange to say, not the corresponding base nor the pseudo-thiourea.

Pure finely-divided thiourea, suspended in dry benzene, was mixed with about 10 per cent. mœ phenyl chlorocarbonate than that required from the equation

$$NH{:}C(NH_2){\cdot}SH + C_6H_5OOO{\cdot}Cl = NH{:}C(NH_2){\cdot}S{\cdot}O{\cdot}COC_6H_5,HCl \; ;$$

union occurred spontaneous, without any marked evolution of heat, and after twenty-four hous the product was collected, washed with benzene, followed by light ptroleum, and dried. It formed a bulky, snow-white, apparently cstalline powder, melting, with copious effervescence, at about 13(freely soluble in water, and giving the reactions of a hydrochloriß. The combined hydrochloric acid was measured (i) by neutralisabn with $N/10$ alkali, and (ii) by precipitation with $N/10$ silver nitrie in presence of nitric acid : the former estimation gave $HCl = 15{\cdot}9$ the latter $= 15{\cdot}5$ per cent., against $15{\cdot}63$ calculated for $C_8H_8O_2N_2S,Cl$.

When heated to near the boiling point, the aqueous solution effervesced freely, giving off cibon dioxide (but no hydrogen sulphide), and the odour of phenol beame strong ; the residual liquid contained hydrochloric acid, gave a red coloration with ferric until treated with ethyl nitrite, was readily desulpharised aline lead solution, with formatn of a mirror of galena, contained thiourea :

$$NH{:}C(NH_2){\cdot}S{\cdot}CO \ldots$$

Next, 11·6 grams of the hydro dissolved i water, were neutralised wth alkali ; phenol was noticed, and mast rated, hard ; the solution reacte ve for th when slowly evaporated to dry acted crystals of thiourea. The oph only twice recrystallised from alig alcoh pearly, flattened prisms ahtin Co tion, this product containd n itro decomposed by heating wit al to and phenolate, and benccetm phenoxy-ψ-thiourea, but f di interaction probably take plac

$$2CSN_2H_3{\cdot}CO_2Pb,HCl + 2KOH = 2CSN_2 \ldots$$

When heated in a dry the, the the liberation of hydrog chlorid which had an odour of phnol, whe potash, evolved ammonia but not mixture, on treatment wth a lead hence, this new attempt unite the thiourea was no more sucessful than tho one referred to above (p. 909). It is curious th. this should be so, for in the case of the analogous derivatives, $CSNI_3 \cdot CO_2Et, HCl$ and $CSN_2H_3 \cdot CO_2Me, HCl$, it is only by taking precautions against tho loss of carbon dioxide that the direct union of sulphur with the alkyl group can be prevented.

Conclusion.

The phenoxyacetyl compund, $PhO \cdot CH_2 \cdot CO(CNS)$, like its phonacetyl analogue, $PhCH_2 \cdot C(CNS)$, behaves in ordinary circumstances not only as thiocarbimide, bt also to a marked extent as thiocyanate ; under like conditions, its imeride, $Ph \cdot CH_2 \cdot O \cdot CO(CNS)$, shows very little thiocyanic character. Similar combinations to this, in which the benzyl group is representedby phenyl, o-tolyl, and p-tolyl, resemble it in being almost purely thicarbimidic in function.. Moreover, whilst the acetyl and propionyl erivatives, $MeCO(CNS)$ and $EtCO(CNS)$, tend generally towards tiocyanic behaviour, the methoxy- and ethoxy-analogues, $MeO \cdot CO(CNS)$ and $EtO \cdot CO(CNS)$, of Doran (Trans., 1896, **69**, 324 ; 1901, **79**,)6) are but little disposed to act, save as thiocarbimides. So far, terefore, as can be judged at present, it would seem that the powr to behave as thiocyanate, which is conferred upon the group $-C(CNS)$ by direct association with a fatty radicle, is greatly reduce (and in some circumstances practically inhibited) by the interpositon of an oxygen atom, provided that the 'atter is directly united to ie CO group.

When substituted thiocaoamides are heated, they tend to decompose all their possible constuents, trisubstitution derivatives yielding le thiocarbimide, ad disubstitution derivatives two ; mono-ed thioureas giv the thiocarbimide of the substituting ether with thicyanic acid. In all cases where an acidic up is includl, the products of heating react for thio-ourea itself, vhen carefully heated, undergoes a similar y ammoniaind thiocyanic acid are produced ; it has ever, by eynolds and Werner (Trans., 1903, **83**, 1) is revrsible, equilibrium becoming established thiocynate on the one hand and thiourea on the hiocarimide has not yet been obtained in a free ard) the facts mentioned above, it seems not erinteresting case of " dynamic isomerism " ring that, at a higher temperature, thiourea) ammonia and thiocyanic acid, and (2) carbimide, so that equilibrium eventually uretted products and the base with which ilst, at low temperatures, ammonium

with about 10 per cent. more phenyl chlorocarbonate than that required from the equation

$$NH{:}C(NH_2){\cdot}SH + C_6H_5O{\cdot}CO{\cdot}Cl = NH{:}C(NH_2){\cdot}S{\cdot}O{\cdot}COC_6H_5{,}HCl\ ;$$

union occurred spontaneously, without any marked evolution of heat, and after twenty-four hours the product was collected, washed with benzene, followed by light petroleum, and dried. It formed a bulky, snow-white, apparently crystalline powder, melting, with copious effervescence, at about 130°, freely soluble in water, and giving the reactions of a hydrochloride. The combined hydrochloric acid was measured (i) by neutralisation with $N/10$ alkali, and (ii) by precipitation with $N/10$ silver nitrate in presence of nitric acid : the former estimation gave HCl $= 15{\cdot}9$, the latter $= 15{\cdot}5$ per cent., against $15{\cdot}63$ calculated for $C_8H_8O_2N_2S{,}HCl$.

When heated to near the boiling point, the aqueous solution effervesced freely, giving off carbon dioxide (but no hydrogen sulphide), and the odour of phenol became strong ; the residual liquid contained hydrochloric acid, gave no red coloration with ferric chloride until treated with ethyl nitrite, was readily desulphurised by hot alkaline lead solution, with formation of a mirror of galena, and hence contained thiourea :

$$NH{:}C(NH_2){\cdot}S{\cdot}CO_2C_6H_5{,}HCl + H_2O = HCl + CO_2 + C_6H_5OH + CSN_2H_4.$$

Next, 11·6 grams of the hydrochloride, dissolved in 200 c.c. of cold water, were neutralised with $N/5$ caustic alkali ; a slight odour of phenol was noticed, and a pasty solid separated, which soon became hard ; the solution reacted very distinctly for thiocyanic acid, and when slowly evaporated to dryness and extracted with alcohol gave crystals of thiourea. The solid, amounting only to 3·2 grams, when twice recrystallised from slightly diluted alcohol, gave beautiful, pearly, flattened prisms melting at 78—79°. Contrary to anticipation, this product contained no sulphur or nitrogen ; it was readily decomposed by heating with alcoholic potash into potassium carbonate and phenolate, and hence consisted, not of the expected base, carbophenoxy-ψ-thiourea, but of diphenyl carbonate. In the main, this interaction probably takes place as follows:

$$2CSN_2H_3{\cdot}CO_2Ph{,}HCl + 2NaOH =$$
$$2CSN_2H_4 + 2NaCl + H_2O + CO_2 + CO(OPh)_2.$$

When heated in a dry tube, the hydrochloride effervesced, owing to the liberation of hydrogen chloride and carbon dioxide ; the residue, which had an odour of phenol, when heated with aqueous or alcoholic potash, evolved ammonia, but not phenyl mercaptan, and the warm mixture, on treatment with a lead salt, was readily desulphurised ; hence, this new attempt to unite the phenyl group with the sulphur of thiourea was no more successful than the one referred to above (p. 909). It is curious that this should be so, for in the case of the analogous derivatives, $CSN_2H_3 \cdot CO_2Et,HCl$ and $CSN_2H_3 \cdot CO_2Me,HCl$, it is only by taking precautions against the loss of carbon dioxide that the direct union of sulphur with the alkyl group can be prevented.

Conclusion.

The phenoxyacetyl compound, $PhO \cdot CH_2 \cdot CO(CNS)$, like its phenacetyl analogue, $PhCH_2 \cdot CO(CNS)$, behaves in ordinary circumstances not only as thiocarbimide, but also to a marked extent as thiocyanate; under like conditions, its isomeride, $Ph \cdot CH_2 \cdot O \cdot CO(CNS)$, shows very little thiocyanic character. Similar combinations to this, in which the benzyl group is represented by phenyl, o-tolyl, and p-tolyl, resemble it in being almost purely thiocarbimidic in function.. Moreover, whilst the acetyl and propionyl derivatives, $MeCO(CNS)$ and $EtCO(CNS)$, tend generally towards thiocyanic behaviour, the methoxy- and ethoxy-analogues, $MeO \cdot CO(CNS)$ and $EtO \cdot CO(CNS)$, of Doran (Trans., 1896, 69, 324; 1901, 79, 906) are but little disposed to act, save as thiocarbimides. So far, therefore, as can be judged at present, it would seem that the power to behave as thiocyanate, which is conferred upon the group $-CO(CNS)$ by direct association with a fatty radicle, is greatly reduced (and in some circumstances practically inhibited) by the interposition of an oxygen atom, provided that the latter is directly united to the CO group.

When substituted thiocarbamides are heated, they tend to decompose into all their possible constituents, trisubstitution derivatives yielding a single thiocarbimide, and disubstitution derivatives two; monosubstituted thioureas give the thiocarbimide of the substituting radicle, together with thiocyanic acid. In all cases where an acidic (oxidised) group is included, the products of heating react for thiocyanate. Thiourea itself, when carefully heated, undergoes a similar change, whereby ammonia and thiocyanic acid are produced; it has been shown, however, by Reynolds and Werner (Trans., 1903, 83, 1) that this change is reversible, equilibrium becoming established between ammonium thiocyanate on the one hand and thiourea on the other. Hydrogen thiocarbimide has not yet been obtained in a free state, but, having regard to the facts mentioned above, it seems not improbable that this very interesting case of "dynamic isomerism" may be explained by supposing that, at a higher temperature, thiourea tends to dissociate into (1) ammonia and thiocyanic acid, and (2) ammonia and hydrogen thiocarbimide, so that equilibrium eventually exists between the two sulphuretted products and the base with which either can combine. And whilst, at low temperatures, ammonium thiocyanate behaves purely as such, it is quite conceivable that by the aid of heat, which would tend to dissociate it into ammonia and thiocyanic acid, the effect of increased temperature on this may be, as in the case of fatty acid thiocyanates, to dispose towards thio-carbimidic power ; so that ultimately, and in presence of ammonia, the incapacity of hydrogen thiocarbimide to maintain wholly its con-figuration, even in the presence of base, might be balanced by the tendency of thiocyanic acid to acquire either the form of thiocarbimide or—however it be conditioned—the power to behave as such.

CHEMICAL DEPARTMENT,
QUEEN'S COLLEGE, CORK.

XCV.—The Action of Light on Potassium Ferrocyanide.

By GLYN WILLIAM ARNOLD FOSTER, B.Sc. (1851 Exhibition Scholar, Manchester University).

ALTHOUGH the decomposition of potassium ferrocyanide by light has long been known, no close investigation of the reactions involved appears to have been made. The experiments described in this paper are a continuation of those mentioned by Prof. F. Haber (Zeit. Elektro-chem., 1905, 10, 847) in his note to the Bunsen Gesellschaft in 1905.

Berthelot (Ann. Chim. Phys., 1900, [vii], 21, 204) noticed that ferrous sulphide is precipitated from a solution containing potassium ferrocyanide and an alkali sulphide. He makes no mention of the action of light, but explains the reaction by supposing that potassium ferrocyanide tends to decompose slowly into potassium cyanide and ferrous cyanide ; the latter, reacting with the alkali sulphide present, gives alkali cyanide and ferrous sulphide. He says, further, that the presence of the alkali cyanide diminishes the dissociation of the ferro-cyanide and a state of equilibrium is soon reached, so that the amount of sulphide precipitated is very small.

This cannot be the correct explanation, since ferrous sulphide dis-solves quantitatively in potassium cyanide, forming potassium ferro-cyanide.

Matuschek (Chem. Zeit., 1901, 25, 565) remarked that potassium ferrocyanide precipitates ferric hydroxide on exposure to sunlight and that potassium cyanide is afterwards present in the solution. He did not study the reaction further. ·

Haber has proposed the view that potassium ferrocyanide not only dissociates in the usual way, thus :

$$K_4Fe(CN)_6 \rightleftarrows 4K^{\cdot} + Fe(CN)_6''''',$$

but that on exposing the solution to light the complex ion, $Fe(CN)_6''''$, dissociates to some extent into iron-ions and cyan-ions :

$$Fe(CN)_6'''' \underset{\text{dark}}{\overset{\text{light}}{\rightleftarrows}} Fe^{\cdot\cdot} + 6CN'.$$

In the absence of light, the latter dissociation proceeds from right to left. The fundamental action thus appears to be reversible and photochemical.

The iron-ions set free in this dissociation are precipitated by an alkali sulphide or by oxygen and water as ferrous sulphide or ferric hydroxide.

The experiments were made with alkaline solutions of potassium ferrocyanide. As will be seen later, in neutral or acid solutions interference due to the formation of ferricyanide may be expected.

The energy of the reaction given by the equation

$$4K_4Fe(CN)_6 + O_2 + 2H_2O \rightleftarrows 4K_3Fe(CN)_6 + 4KOH$$

is obtained by the use of van't Hoff's expression :

$$A = RT\log_e K - RT\log_e \frac{C^4_{Fe(CN)_6'''}.C^4_{HO}}{C^4_{Fe(CN)_6''''}.C_{O_2}.C^2_{H_2O}}.$$

Here A may be regarded as a measure of the "tendency" of the reaction to take place in the direction left to right, that is, towards the formation of ferricyanide. If now $C_{OH'}$ is diminished, A is increased, and hence the tendency towards the formation of ferricyanide increases. If $C_{OH'}$ increases, A decreases, and ferricyanide is less liable to be formed.

By taking the numerical values (Haber, *Thermodynamik Technischer Gasreactionen*, 1905, 160) for the oxygen electrode (Nernst and Wartenberg, *Göttinger Nachrichten*, 1905, No. 1) and for the ferri-, ferrocyanide electrode, the latter from determinations by Schaum and von der Linde (Haber, *Zeit. Elektrochem.*, 1901, 7, 1043), it may be shown that A is large in acid solutions and has a small value only in alkaline solutions. Hence, in the latter, ferrocyanide is oxidised to a very small extent.

EXPERIMENTAL.

In the first experiments, the source of light was a "Uviol" lamp, made by Schott and Genossen, of Jena ; details of this lamp and of the experiments are given in Prof. Haber's note (*loc. cit.*). Some experiments were also made in sunlight. In the later experiments, a much more powerful lamp, kindly lent by W. C. Heraeus, of Hanau, was used. This is a mercury vapour lamp of quartz, and is shown in the accompanying sketch.

The essential part is a small bore tube, A, of quartz, bent into a U-shape with the arms close together. The upper ends are bent over and end in bulbs, which are filled with pure mercury up to the level shown. Platinum contact wires pass down two other tubes which are sealed into the bulbs at the back.

The bulbs, inlet tubes for the wires, and part of the U-tube are all enclosed in a brass casing, C, which has a glass front not shown in the sketch. The projecting part of the U-tube is enclosed in a thin-walled quartz tube, B, which is cemented into the brass casing. Close behind the U-tube a nickel tube enters, through which water flows. The water fills the whole of B and C and is carried away by another tube

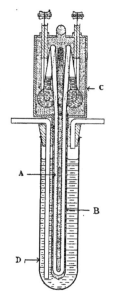

at the top, so that a water-jacket surrounds the lamp and keeps it quite cool. It is interesting to note that the ultra-violet light or radiation passes easily through this water-jacket.

The lamp is lighted by inverting it slowly, so that the two threads of mercury in the U-tube meet at the bend. On turning the lamp in the reverse direction, the arc is struck, and fills the whole of the U-tube up to the mercury in the bulbs. A starting resistance is necessary owing to the momentary short circuit which occurs when the mercury threads meet. The lamp requires 230—240 volts, and uses 5—6 amperes.

The solutions were contained in a large tube, D, similar to a large test-tube, which was fitted to B by means of an india-rubber stopper. Two glass tubes passing through this allowed gases to be bubbled through the solutions.

The lamp gives a very large amount of ultra-violet radiation; when burning in air, a very powerful odour of ozone is perceived, and the radiation can be distinctly felt on the hand. If a $N/1000$ solution of potassium permanganate containing 9 per cent. of sulphuric acid is exposed to this light in the tube D, it is completely decolorised in ten minutes with evolution of oxygen. A solution of potassium iodide $(0·09N)$ and starch becomes absolutely black in two minutes.

Action of Light on Potassium Ferrocyanide Solution in Presence of Oxygen.

The experiments mentioned in Haber's note to the Bunsen Gesellschaft were all carried out with the "Uviol" lamp. In each case, 300 c.c. of $N/10$ potassium ferrocyanide solution were used. The results are shown in the following table:

No. of experiment.	Strength of potash solution. Normal=1.	Time of exposure in hours.	Iron, gram, precipitated as Fe(OH)₃.	(CN) as KCN, gram.	Fe : CN
1	0·1	4·5 ⎫ 4·5 ⎭ 9·0	0·0196 ⎫ 0·0194 ⎭ 0·0390	0·1039	1 : 5·72
2	0·5	3·0	0·01756	0·04855	1 : 5·95
3	1·0	3·0	0·0169	0·04529	1 : 5·75

The equation requires the proportion 1 : 6 for Fe : CN.

In the first experiment, after exposure for four and a half hours, the solution was filtered and exposed for a further period of four and a half hours. The iron hydroxide was collected, washed, and dissolved in dilute sulphuric acid and estimated by the method of Storch (*Zeit. Elektrochem.*, 1901, **7**, 715).

The iron precipitate always dissolved without any odour of prussic acid, blue coloration, or any other sign of the presence of cyanide. The cyanide in the filtrate was estimated by titration with $N/10$ silver nitrate after addition of a small quantity of potassium iodide as indicator; the exactness of this method was tested by the titration of a mixture of potash and potassium ferrocyanide containing a known quantity of potassium cyanide.

From the foregoing table, it will be seen that roughly 5 milligrams of iron per hour are precipitated as hydrate. On repeating these experiments, using the quartz lamp, a much stronger action takes place, as the following numbers show:

Fe precipitated
as Fe(OH)₃,

	gram.	Fe : CN.
	⎧ 0·0132	1 : 4·9
Exposure, one hour .,........⎨ 0·0097	1 : 5·1
	⎩ 0·0083	1 : 4·1

The solutions of ferrocyanide and caustic potash were decinormal, and in each experiment about 100 c.c. of solution were exposed for one hour in a stream of oxygen. Some of the ferric hydroxide is deposited as a film on the wall of the lamp; this reduces the intensity of the light to some extent, and thus causes variations in the amount of hydroxide precipitated.

In these experiments, the relation of iron to cyanide is considerably below the theoretical 1 : 6. The experiments were made in strong alkaline solution, and at the ordinary temperature, so that the loss of cyanogen cannot be accounted for by supposing it to be expelled by the stream of oxygen. It appeared probable that hydrolysis of the cyanide took place under the influence of light. The following experiments were then carried out:

(*a*) One hundred c.c. of a solution containing 30 c.c. of $N/10$ potassium cyanide and 10 c.c. of $N/10$ potassium hydrate were exposed for one hour in a stream of oxygen. The quantity of cyanide in the solution diminished by 15 per cent.

(*b*) Another experiment was made, using a solution of the same composition. After exposure, the solution was divided into two equal parts. The first part was titrated with silver nitrate, and it was found that 23·5 per cent. of the cyanide had disappeared. The other half of the solution was acidified and distilled into a solution of caustic potash. The same amount of cyanide was found in this potash solution as in the first. half, showing that all cyanide had distilled over. The residue was made strongly alkaline, and again distilled into water. A considerable quantity of ammonia passed over, and was titrated with $N/10$ hydrochloric acid. The amount of cyanide corresponding to this ammonia was 20·9 per cent. of the original quantity used ; hence nearly all the missing cyanide was accounted for.

(*c*) Control experiments were made by passing a stream of oxygen through the solutions for an hour in absence of light ; the correct quantity of cyanide was found in each case and no ammonia was detected. The conclusion is that the cyanide is oxidised by oxygen under the influence of light with formation of cyanate, which on boiling with acid is decomposed, yielding the ammonium salt of the acid used.

Further experiments were then made in order to ascertain whether compounds more complex than cyanates were also produced. The solutions were exposed to light and then distilled as before, being first acidified and then rendered alkaline. The residue after these distillations was twice evaporated with concentrated hydrochloric acid. By this means, tricyanate and other polymerides would be decomposed with formation of ammonium chloride. On making alkaline and again distilling, ammonia should pass over. Repeated experiments with solutions exposed for three hours always gave ammonia at this third distillation, though the amount was small.

On comparing the experiments made with the " Uviol " lamp with those with the quartz lamp it is seen that in the former case the rela-tion Fe : CN is very near the theoretical value, although the exposure was longer. Thus the oxidation of cyanide to cyanate and other compounds takes place appreciably only in very strong light.

The foregoing experiments do not, however, explain how the ferric hydroxide is precipitated. Either the ferrocyanide may be decomposed, and the ferro-ion oxidised and precipitated, or the ferrocyanide may first be oxidised to ferricyanide, and the iron precipitated from this by hydrolysis, according to the equation :

$$K_3Fe(CN)_6 + 3H_2O = 3KCN + Fe(OH)_3 + 3HCN.$$

That this hydrolysis takes place to a certain extent seems to be proved by the fact that a neutral solution of ferricyanide precipitates a small amount of ferric hydroxide on exposure in a stream of oxygen ; no ferrocyanide is formed. This hydrolytic effect may be expected to be the same in an atmosphere of hydrogen, but in this case, owing to the reduction of the ferricyanide to ferrocyanide and the formation of acid, blue and green compounds are formed which prevent the detection of ferric hydroxide.

The hydrolysis of ferricyanide must proceed very slowly, otherwise a much larger amount of ferric hydroxide would be precipitated on exposure to light in presence of oxygen. If the formation of ferricyanide is an intermediate step in the precipitation of ferric hydroxide from ferrocyanide in alkaline solution, such a solution would contain ferricyanide after exposure : this is, however, not the case. Solutions of alkaline ferrocyanide, after exposure, were filtered to remove the precipitated ferric hydroxide [$0\cdot0015$ gram (calculated as Fe) per 130 c.c.], acidified, and then tested for ferricyanide with pure ferrous sulphate or silver nitrate, but no trace of ferricyanide was found. We may thus conclude that ferric hydroxide is precipitated from the ferrocyanide without intermediate formation and subsequent hydrolysis of ferricyanide.

In order to ascertain if oxygen is necessary for the precipitation of ferric hydroxide, the alkaline solution was made up as before, and, after having been boiled and cooled in an atmosphere of hydrogen, was exposed. No trace of hydroxide was found, and the solution remained clear. On admitting oxygen in place of hydrogen, the solution at once became turbid, and in a few minutes deposited a considerable quantity of ferric hydrate.

This experiment also shows that no ferrous hydrate is precipitated in absence of oxygen. The dissociation represented by the equation $Fe(CN)_6'''' \rightleftharpoons Fe^{\cdot\cdot} + 6CN'$ would lead to the precipitation of ferrous hydrate if the concentration of the Fe-ions became large enough to make the product ($C_{Fe^{\cdot\cdot}} \times C^2_{OH'}$) greater than the solubility product of ferrous hydrate. From the experiment, it is evident that the dissociation does not go far enough, and even in very strong light is too small. Haber (*loc. cit.*) has explained the theory which connects this view with the high velocity of ionic reactions.

It was also found that exposure to light does not increase the conductivity of the alkaline ferrocyanide solution to any appreciable extent. Different methods of determining the conductivity were tried, but all gave the same values, both when the solution was exposed to and protected from light.

Experiments with Alkali Sulphide and Ferrocyanide.

If a solution containing ferrocyanide, potassium sulphide, and an alkali is exposed to sunlight or the light from the quartz lamp, it turns brown or green, depositing after some time a black precipitate of ferrous sulphide. Here we have undoubtedly a reaction of ferrocyanide, since any ferricyanide would be immediately reduced by the potassium sulphide to ferrocyanide. It is well known that the green solution contains colloidal ferrous sulphide.

The experiments were all made in an atmosphere of hydrogen.

It often happened that, on exposure to light, a brown solution was obtained, which, only after being further exposed for a considerable time, turned green and gave the precipitate of ferrous sulphide. It was found that a higher temperature always accelerated the change from brown to green, and the precipitate was formed in much less time. A brown solution may be obtained if a very dilute solution of ferrous sulphate is mixed with an alkali sulphide and alkali. The brown solution is quite stable at $0°$, but on warming the colour changes to dark green, and finally a precipitate of ferrous sulphide is formed. If freshly precipitated ferric hydroxide is dissolved in very strong alkali and an alkali sulphide added, a clear red solution is obtained (*Zeit. Elektrochem.*, 1901, **7**, 724), which slowly changes to brown, and ferrous sulphide is precipitated. These brown solutions, in whichever way prepared, all turn green and deposit ferrous sulphide.

The red solution obtained by dissolving ferric hydroxide in strong alkali and adding an alkali sulphide most probably contains a double sulphide of iron and potassium. The brown solutions may contain either a double salt or a colloidal form of ferrous sulphide; most probably, however, they consist of a mixture of the two.

In solutions exposed to the lamp, the brown colour is only formed at temperatures below $30°$. For this reason, and in order to obtain an appreciable precipitate of ferrous sulphide, the tube containing the solution was, in the later experiments, immersed in a water-bath and kept warm. A temperature of $30—40°$ was found to be sufficient, but about $80°$ is most advantageous. At this temperature, very little colloidal green solution is formed, and a considerable precipitate of ferrous sulphide is obtained.

Even at this increased temperature, the reaction is photochemical. One hundred and thirty c.c. of a solution containing $N/10$ ferrocyanide, $N/2$ alkali, and $N/2$ alkali sulphide gave, after three-quarters of an hour's exposure at $42°$, $0·012$ gram Fe precipitated as ferrous sulphide, whilst a similar solution heated for one and a half hours to $75°$ without exposure gave no trace of ferrous sulphide, but remained absolutely clear and colourless.

Ferrous sulphide dissolves easily in a solution of potassium cyanide forming ferrocyanide : hence it was to be expected that the green solutions on standing would become clear, owing to the solution of the sulphide by the potassium cyanide formed in the reaction. This did not occur, and on standing twenty-four hours the solutions were still green, although a considerable quantity of sulphide had precipitated. If ferrous sulphide, prepared by precipitation of ferrous sulphate, is shaken with excess of potassium cyanide in alkaline solution, it is completely dissolved. If the sulphide is in excess, the cyanide disappears. Hence, if cyanide is added to the solution of ferrocyanide, alkali, and alkali sulphide, no precipitation should take place on exposure. After prolonged exposure at 80°, no trace of ferrous sulphide was formed. This showed that cyanide, formed from the ferrocyanide during precipitation of sulphide, must be removed in some secondary reaction. On distilling the exposed solutions after the separation of the sulphide, in most cases no cyanide was found, and in others only very little. The distillation was carried out by Feld's method (Feld, *Journ. für Gasbeleucht.*, 1903, 564) with lead nitrate. Control experiments with a mixture of ferrocyanide, a known amount of cyanide, and alkali sulphide gave the correct amount of cyanide.

In the gas issuing from the solution, ammonia in sufficient quantity to react with litmus paper was found. Part of the cyanide is thus hydrolysed with formation of this gas. The exposed solutions were also tested for sulphocyanide in the following manner. The ferrocyanide, alkali, and alkali sulphide were precipitated as Prussian blue, iron hydroxide, and iron sulphide by excess of ferric chloride ; after separation, the filtrate showed the deep red colour of iron sulphocyanide.

In order to investigate this formation of sulphocyanide further, an experiment was made in which a solution composed of $N/10$ potassium cyanide, $N/2$ potassium sulphide, and $N/10$ caustic potash was exposed in a current of hydrogen for three hours. It was found that 32 per cent. of the cyanide had disappeared and the solution had become pale yellow. On extraction with ether and subsequent evaporation, a red liquid was obtained having a strong and unpleasant odour of onions. This suggests the formation of complex sulphocyanides by interaction of the cyanide and alkali sulphide under the influence of light. No cyanide is found in the green solutions, since it is all used in the formation of sulphocyanides or is hydrolysed.

In conclusion, it may be stated that the precipitation of ferric hydroxide or ferrous sulphide from potassium ferrocyanide by alkali or alkali sulphide under the influence of light takes place in the following manner :

Potassium ferrocyanide is dissociated in solution in absence of light into potassium-ions and the ferrocyan-ions, thus :

$$K_4Fe(CN)_6 \rightleftharpoons 4K^{\cdot} + Fe(CN)_6''''.$$

Under the influence of light, the complex ferrocyan-ion dissociates into iron-ions and cyan-ions,

$$Fe(CN)'''' \rightleftharpoons Fe^{\cdot\cdot} + 6CN'.$$

The existence of these iron-ions can be demonstrated in the ordinary manner by precipitation with alkali and oxygen or alkali sulphide. The concentration of $Fe^{\cdot\cdot}$ ions is too small to exceed the solubility product of $Fe(OH)_2$ or to be found directly by conductivity measurements. It is ferrocyanide which is decomposed in this manner, not ferricyanide. The small quantity of ferric hydrate obtained from neutral solutions of ferricyanide is produced by hydrolysis, which may take place according to the equation :

$$K_3Fe(CN)_6 + 3KOH = Fe(OH)_3 + 6HCN.$$

Potassium cyanide is easily oxidised in presence of light by oxygen, forming cyanate and polymerides. The cyanide which is formed when iron is precipitated as sulphide from ferrocyanide under the influence of light, reacts with alkali sulphide present, producing sulphocyanide and complex sulphocyano-compounds.

I wish to express my heartiest thanks to Prof. F. Haber for his kind interest and most valuable help.

TECHNISCHE HOCHSCHULE,
KARLSRUHE,
GERMANY.

XCVI.—*The Action of Magnesium Methyl Iodide on dextro-Limonene Nitrosochlorides.*

By WILLIAM AUGUSTUS TILDEN and FREDERICK GEORGE SHEPHEARD, B.Sc.

THE products of the action of, magnesium methyl iodide on pinene nitrosochloride were an oxime and a base resulting from the exchange of the oxygen of the nitroso-group for methyl (Tilden and Stokes, Trans., 1905, **87**, 836), but the same reagent in contact with the limonene nitrosochlorides gives rise to changes of quite a different character. Limonene α- and β-nitrosochlorides were prepared and separated by the application of Wallach's method (*Annalen*, 1889, 252, 108) and were separately treated with the reagent.

Fifteen grams of the nitrosochloride suspended in dry ether were mixed gradually with a quantity of the reagent prepared by adding 3·5 grams of magnesium to a solution of 25 grams of methyl iodide in 75 c.c. of dehydrated ether. After twelve hours, the product was decomposed by the cautious addition of water and sufficient hydrochloric acid to dissolve the precipitated magnesium hydroxide. The ethereal solution was separated and the greater part of the ether distilled off. On adding an equal volume of alcohol and cooling, a colourless, crystalline substance was deposited, the weight of which amounted altogether to about 25 per cent. of the weight of nitrosochloride taken. The aqueous solution separated from the ether, rendered alkaline by the addition of soda and extracted with ether, yielded a yellow, viscid substance which soon darkened and became tarry.

The crystalline product from the β-nitrosochloride after recrystallisation from alcohol melts at 150°. It is insoluble in alkalis and mineral acids, very soluble in ether, chloroform, or benzene, but less readily so in alcohol and petroleum. Bromine water and potassium permanganate have no action on this substance in the cold. Boiling concentrated aqueous solution of caustic potash produces no change, but alcoholic potash removes hydrogen chloride, and a yellow oil is formed which rapidly becomes brown and tarry. This compound is not reduced by zinc and acetic acid, but sodium and alcohol convert it into a yellow oil which darkens on exposure to air and from which no definite product could be obtained.

Limonene α-nitrosochloride yields by the action of magnesium methyl iodide a product which is in all respects similar to the product just described. The melting point is, however, much lower, namely, 42°, and it is more soluble in petroleum. The following are the results of analysis and estimation of molecular weight by observations of the freezing point of solutions in benzene.

	Compound from β-nitrosochloride.		Compound from α-nitrosochloride.
	I.	II.	
C per cent.	61·62	61·75	—
H ,, ,,	8·35	8·26	—
N ,, ,,	7·25	—	7·13
Cl ,, ,,	18·66	—	18·31
M.W. mean	317	—	327
[α]$_D$	+130·5°		+213°
M. p.	150°	—	42°

The formula $C_{20}H_{32}ON_2Cl_2$ requires $C = 62.02$; $H = 8.27$; $N = 7.23$; $Cl = 18.34$ per cent., with a molecular weight 387.

These compounds are, therefore, derived from the corresponding bimolecular nitrosochloride, $(C_{10}H_{16}ONCl)_2$, by removal of one atom of oxygen, and they are distinguished from it by a remarkable stability. No reagent has been found to produce from them definite derivatives except pentachloride of phosphorus, which exchanges chlorine for the remaining atom of oxygen.

Five grams of the α-compound were intimately mixed with 5 grams of the pentachloride and left at the common temperature for a day. Water and ice were then added and the semi-solid mass was crystallised from acetic acid and from petroleum. The crystals form small, feathery clusters and melt at 139°.

Analysis gave 6.42 per cent. of nitrogen, 31.65 per cent. of chlorine, and a molecular weight 388 as the mean of three experiments. The formula $C_{20}H_{32}N_2Cl_4$ requires $N = 6.35$, $Cl = 32.08$ per cent., and $M.W. = 442$. $[a]_D = +220°$ in chloroform.

The crystals are insoluble in acids and in alkalis, but are soluble in the usual organic solvents. This compound does not react as a nitrogen chloride, as it seems to possess no oxidising power towards potassium iodide or alcohol.

The corresponding compound derived from the β-nitrosochloride was obtained in the same way, but it could not be readily purified owing to the tendency to give up the elements of hydrogen chloride when heated with any solvent. It was obtained most nearly pure when crystallised from petroleum. It then melted at 164° and contained 30.22 per cent. of chlorine, theory requiring 32.08.

The product resulting from the loss of hydrogen chloride from this substance was prepared by heating it with alcohol for some hours. It forms needles which melt at 148°, and contains 7.59 per cent. of nitrogen and 19.51 per cent. of chlorine. The molecular weight estimated from the freezing point of the benzene solution was found to be 356 as the mean of three concordant experiments. The formula $C_{20}H_{30}N_2Cl_2$ requires $N = 7.60$; $Cl = 19.21$, and $M.W. = 369$. $[a]_D = +134.8°$ in chloroform.

The corresponding α-compound was obtained in the same way and crystallised from alcohol in small prisms melting at 113°. $[a]_D = +206.5°$.

The constitution of these compounds is not easily accounted for in view of the fact that the product obtained by the action of phosphorus pentachloride is saturated and that it does not give the reactions of a nitrogen chloride. Assuming that in the parent substance the two nitrogen atoms are linked together somewhat as follows:

$$C_{10}H_{16} \left< \begin{matrix} Cl & Cl \\ N = N \end{matrix} \right> C_{10}H_{16},$$
$$\overset{..}{O} \quad \overset{..}{O}$$

the loss of one atom of oxygen would leave one of these nitrogen atoms quinquevalent, while the other would be trivalent or, what is perhaps more probable, the valency of the remaining oxygen atom would be divided between them thus :

$$C_{10}H_{16} \left< \begin{matrix} Cl & Cl \\ N - N \end{matrix} \right> C_{10}H_{16}.$$
$$O$$

The action of pentachloride of phosphorus on such a compound might be expected to give a nitrogen chloride, but apparently the propenyl group on the other side of the limonene molecule is involved in the change and a closed ring probably results.

ROYAL COLLEGE OF SCIENCE,
    LONDON.

XCVII.—*Dinitroanisidines and their Products of Diazotisation (Second Communication).*

By RAPHAEL MELDOLA, F.R.S., and FRANK GEORGE C. STEPHENS.

IN the last communication dealing with this subject (Trans., 1905, **87**, 1199), two dinitroanisidines were described, and the products of diazotisation of one of these, namely, *iso*picramic methyl ether, studied in detail. The investigation of the other dinitroanisidine (m. p. 163°) is not yet complete, as the preparation of this compound has presented greater difficulties than were anticipated. In the meantime we have continued our experiments with the object of obtaining a dinitro-anisidine corresponding to dinitro-*m*-aminophenol. As these experiments have been successful, we make known the results in the present paper.

Nitration of Diacetyl-m-aminophenol.

Neither *m*-aminophenol nor its monoacetyl derivative can be nitrated, as nitric acid gives rise to oxidation products and colouring matters with these compounds. The diacetyl derivative, first prepared by Ikuta (*Amer. Chem. J.*, 1893, **15**, 41), can, however, be nitrated by the following method.

Ordinary concentrated nitric acid (sp. gr. 1·4) cooled in ice is saturated with the crystalline diacetyl derivative by adding the latter in small portions to the extent of about one gram per two c.c. of acid. To the solution thus obtained there is then added one-half the volume of fuming nitric acid (sp. gr. 1·5), and, after standing for three hours, the solution is poured on to ice. A crystalline deposit soon separates, and after filtration the mother liquor on standing for a few days deposits a second crop. The two compounds thus obtained proved to be isomeric mononitroacetaminophenols corresponding respectively to 4-nitro-3-aminophenol (first crop) and 6-nitro-3-aminophenol (second crop). The acetyl derivatives were purified by crystallisation from alcohol with animal charcoal and then from glacial acetic acid until the melting points were constant.

4-*Nitro-3-acetaminophenol*, obtained as above, crystallises in silvery scales or clusters of ochreous, prismatic needles, melting at 266°. It dissolves in aqueous alkali with a yellow colour.

0·0963 gave 11·7 c.c. moist nitrogen at 13° and 759·4 mm. N = 14·33.

$C_8H_8O_4N_2$ requires N = 14·28 per cent.

4-*Nitro-3-aminophenol*, obtained from its acetyl derivative by hydrolysis with sulphuric acid, crystallises from hot water or very dilute alcohol in slender, orange needles melting at 185—186°.

0·0865 gave 13·6 c.c. moist nitrogen at 17° and 759·7 mm. N = 18·21.

$C_6H_6O_3N_2$ requires N = 18·18 per cent.

The above nitroaminophenol does not appear to have been described before. Its constitution is proved by its giving *p*-nitrophenol when diazotised and boiled with absolute alcohol. The diazonium sulphate also on heating with hydriodic acid gives 4-nitro-3-iodophenol, m. p. 123°, which crystallises from hot water in the form of pale ochreous needles.

The methyl ether of this nitroaminophenol has been described by Bantlin (*Ber.*, 1878, **11**, 2106), who obtained it from 3:4-dinitroanisole by the action of alcoholic ammonia. We have prepared the same compound by methylating 4-nitro-3-acetaminophenol with dimethyl sulphate in presence of alkali. The methylation is very incomplete by this process, and the nitroacetanisidide undergoes partial hydrolysis by the alkali, so the product was completely hydrolysed by sulphuric acid after the removal of the phenolic (unmethylated) portion. The nitroanisidine thus obtained had the properties ascribed to it by Bantlin, only the melting point was found to be 131° (Bantlin gives

129°). The compound was acetylated by boiling for a short time with acetic anhydride, and the product, after crystallisation from alcohol, consisted of silky-white needles melting at 124°.

0·0765 gave 8·6 c.c. moist nitrogen at 14° and 769·9 mm. N = 13·41.

$C_9H_{10}O_4N_2$ requires N = 13·33 per cent.

6-*Nitro-3-acetaminophenol* (second crop) is much more soluble in all solvents than the isomeride. It crystallises in pale ochreous needles melting at 221° and dissolving in aqueous alkali with a yellow colour.

0·0792 gave 9·7 c.c. moist nitrogen at 13° and 764·1 mm. N = 14·52.

$C_8H_8O_4N_2$ requires N = 14·28 per cent.

6-*Nitro-3-aminophenol* was prepared from its acetyl derivative by hydrolysis with sulphuric acid and neutralisation of the solution of the sulphate with ammonia. It crystallises from hot water in dull orange needles melting at 158°.

0·0964 gave 15 c.c. moist nitrogen at 19° and 762 mm. N = 17·94.

$C_6H_6O_3N_2$ requires N = 18·18 per cent.

This nitroaminophenol appears to be new. On diazotisation and decomposition of the diazonium salt with hydriodic acid, an iodonitrophenol was obtained, crystallising from boiling water in straw-coloured scales melting at 92—93°, and probably identical with the compound described by Busch (*Ber.*, 1874, 7, 462), who gives the melting point 90—91°. An attempt to replace the amino-group by hydrogen by boiling the diazonium sulphate with absolute alcohol gave a product which distilled in a current of steam and which crystallised in yellow needles melting at 75°. This compound is phenolic, but is not, as we anticipated, *o*-nitrophenol. It is probably nitroresorcinol ethyl ether $(NO_2 : HO : OC_2H_5 = 6 : 1 : 3)$, but this part of the subject requires further investigation.

4 : 6-*Dinitro-3-aminophenol and Derivatives.*

The isomeric nitroacetaminophenols above described both yield the same dinitroacetaminophenol on further nitration. The nitration is effected by dissolving the acetyl derivative in fuming nitric acid cooled to – 10°; 15 c.c. of acid will nitrate about three grams of the acetyl derivative. The solution is poured on to ice and the dinitro-compound collected and washed and crystallised from alcohol or glacial

acetic acid, from which solvents it separates in white needles melting
at 168°.

0·1259 gave 19 c.c. moist nitrogen at 15° and 761 mm.   N = 17·67.

$C_8H_7O_6N_3$ requires N = 17·43 per cent.

The above acetyl derivative is readily hydrolysed by sulphuric acid
and yields 4 : 6-dinitro-3-amin'ophenol. The latter compound crystallises from alcohol in dull orange needles melting at 231° and dissolving
in aqueous alkali with an orange colour.

0·1139 gave 21·1 c.c. moist nitrogen at 18° and 757·8 mm.   N = 21·3.

$C_6H_5O_5N_3$ requires N = 21·1 per cent.

This dinitroaminophenol may be identical with that obtained by
Lippmann and Fleissner by the action of an alcoholic solution of
potassium cyanide on 2 : 4-dinitroaniline (Monatsh., 1886, 7, 95). These
authors, however, give a lower melting point, namely, 225°, although
in other respects the description of their compound agrees with the
properties of ours. On diazotisation and boiling the diazonium salt
with water, 4-: 6-dinitroresorcinol of m. p. 212—213° (Typke, Ber., 1883,
16, 552) is formed. The identity of the latter compound was
established by comparison with a specimen of dinitroresorcinol prepared
by the nitration of resorcinol diacetate. In order further to
characterise the resorcinol derivative, it was methylated by dimethyl
sulphate and alkali. Both preparations gave the same dimethyl
derivative, crystallising from dilute alcohol in white needles melting at
157° (Meldola and Eyre, Proc., 1901, 17, 132).* From these data,
combined with the fact that the two isomeric nitroacetaminophenols
both give the same product on further nitration, the constitution of
the dinitroaminophenol is established :

M. p. 185—186°.   M. p. 231°.   M. p. 158°.

* A dinitroresorcinol dimethyl ether to which the same constitution is assigned
(Beilstein, II, 925) is said to have been obtained by König (Ber., 1878, 11, 1042) by
the nitration of resorcinol dimethyl ether. The description of this compound by
König, however, as crystallising in reddish-brown crystals melting at 67° shows that
his preparation was very impure. We have obtained a compound (Meldola and
Eyre, loc. cit.) identical with the above by the direct nitration of resorcinol dimethyl
ether in glacial acetic acid, but the product thus obtained is always mixed with
colouring matters and requires much purification. The melting point given in the
paper referred to is 154°; by repeated crystallisation we have brought it up to 157°.
The compound is doubtless identical with that obtained by Jackson and Warren
(Amer. Chem. J., 1891, 13, 164), although these authors give a melting point of
167°, and also by Blanksma and Terwogt (Rec. Trav. Chim., 1902, 21, 286).

4 : 6-*Dinitro-3-aminophenol methyl ether* = 4 : 6-*Dinitro-m-anisidine.*
—This compound was prepared by the action of methyl iodide on the
silver salt of 4 : 6-dinitro-3-acetaminophenol and hydrolysis of the
product. The yield, however, is not very good by this method and the
same compound was more conveniently prepared by the direct nitration
of *m*-acetaminophenol methyl ether. In order to obtain the latter substance, *m*-aminophenol was converted into its monoacetyl derivative and
the latter methylated by dimethyl sulphate in presence of alkali. The
m-acetanisidide separates out as an oil which solidifies to a crystalline
cake on standing. The crude product can, if necessary, be purified by
crystallisation from very dilute alcohol, and, after purification, has the
melting point 80—81° assigned to it by its discoverers (Körner and
Wender, *Gazzetta*, 1887, **17**, 493).

The nitration of the *m*-acetanisidide was effected by dissolving the
compound in small quantities in fuming nitric acid (sp. gr. 1·52)
cooled in ice. One gram of the substance requires about 5 c.c. of acid
of the strength mentioned in order to convert it into a dinitro-derivative. The product was isolated by pouring the solution on to ice,
collecting and washing the precipitate, and crystallising from glacial
acetic acid. The dinitro-*m*-acetanisidide crystallises in small, white
needles melting at 146°.

0·0762 gave 10·8 c.c. moist nitrogen at 15° and 756·8 mm. N = 16·5.
$C_9H_9O_6N_3$ requires N = 16·5 per cent.

The dinitro-*m*-anisidine obtained from the acetyl derivative by
hydrolysis with sulphuric acid or dilute sodium hydroxide crystallises
from alcohol in canary-yellow needles melting at 208°.*

0·0937 gave 15·4 c.c. moist nitrogen at 11° and 768·1 mm. N = 19·84.
$C_7H_7O_5N_3$ requires N = 19·71 per cent.

This dinitroanisidine is not diazotisable in acetic acid solution. In
the presence of mineral acids, however, a diazonium salt is formed
which undergoes the usual decompositions, giving rise to products
which show that the reaction is in this case quite regular, neither the

* A dinitroanisidine is described in a paper by Blanksma (*Rec. Trav. Chim.*,
1904, **23**, 121) to which that author assigns the same constitution. The melting
point assigned to this compound is, however, quite different, namely, 156°, a discrepancy which requires explanation. The evidence adduced by Blanksma in
support of his formula appears to be quite sound, but we may point out that the
melting point of the corresponding dinitromonomethylanisidine prepared by him
(198°) is hardly in harmony with the view that it is a methyl derivative of a dinitroanisidine melting at 156°. The effect of the introduction of the methyl group is
generally to lower the melting point. We learn from a private communication with
which we have been favoured by M. Blanksma that the melting point 156° is
probably erroneous.

nitro-group nor the alkyl radicle being displaced. Thus on boiling the diazonium sulphate with hydriodic acid there is formed 4 : 6-*dinitro-3-iodoanisole*, which crystallises from alcohol in small, ochreous scales melting at 119°.

0·0914 gave 6·6 c.c. moist nitrogen at 11° and 767·9 mm. $N = 8.71$. $C_7H_5O_5N_2I$ requires $N = 8.67$ per cent.

The diazonium salt (sulphate or chloride) combines at once with β-naphthol in alkaline solution to form an azo-compound which is non-phenolic, and in which the methoxy-group is therefore present. After crystallisation from boiling aniline, the compound was obtained in brick-red needles melting at 257°.

0·1073 gave 13·6 c.c. moist nitrogen at 11° and 765·3 mm. $N = 15.24$. $C_{17}H_{13}O_6N_4$ requires $N = 15.2$ per cent.

The azo-compound dissolves in strong sulphuric acid with a violet colour, becoming redder in shade on dilution with water. The constitution of this azo-derivative and of the related compounds is shown by the formulæ :

$$
\begin{array}{cccc}
\text{OCH}_3 & \text{OCH}_3 & \text{OCH}_3 & \text{OCH}_3 \\
\text{NO}_2\underset{\text{NO}_2}{\bigcirc}\text{NHAc} & \text{NO}_2\underset{\text{NO}_2}{\bigcirc}\text{NH}_2 & \text{NO}_2\underset{\text{NO}_2}{\bigcirc}\text{I} & \text{NO}_2\underset{\text{NO}_2}{\bigcirc}\text{N}_2{\cdot}\text{C}_{10}\text{H}_6(\text{OH})\beta \\
\text{M. p. } 146°. & \text{M. p. } 208°. & \text{M. p. } 119°. & \text{M. p. } 257°.
\end{array}
$$

The foregoing result taken in conjunction with the observations recorded in previous papers leads to the conclusion that the displacement of the nitro-group on diazotisation is not determined solely by this group being ortho or para with respect to the diazonium group, since both these conditions are fulfilled in the dinitroanisidine described in the present paper and yet no displacement occurs. The condition of displacement appears to be the proximity of another nitro-group in the ortho-position to the nitro-group, which is ortho or para with respect to the diazonium group. Other negative substituents, such as halogens, may also be capable of loosening the attachment of the displaceable nitro-group. This point will be made the subject of further investigation.

FINSBURY TECHNICAL COLLEGE.

XCVIII.—*Electrolysis of Potassium Ethyl Dipropylmalonate.*

By DAVID COWAN CRICHTON, M.A., B.Sc. (Carnegie Research
Scholar).

CRUM-BROWN and Walker (*Annalen*, 1893, 274, 5) have described the
products of the electrolysis of alkyl-substituted malonic acids. In the
case of diethylmalonic acid, saponification of the electro-synthetic ester
yielded a neutral substance instead of the tetraethylsuccinic acid
expected. This substance was investigated by J. Walker and A. P.
Walker (Trans., 1905, 87, 961), who found it to be the anhydride of
tetraethylsuccinic acid, distinguished by its extraordinary stability
towards alkalis.

At the suggestion of Professor Walker, the present research was
undertaken in order to determine if the corresponding tetrapropylsuccinic anhydride possessed the great stability and resistance to
ordinary reagents of the tetraethyl compound.

An investigation of the product of the electrolysis of potassium
ethyl dipropylmalonate showed that the principal reactions during
the electrolysis take place in accordance with the following equations :

(1) $2CO_2Et \cdot CPr_2 \cdot CO_2' = CO_2Et \cdot CPr_2 \cdot CPr_2 \cdot CO_2Et + 2CO_2$.

Tetrapropylsuccinic ester.

(2) $CO_2Et \cdot CPr_2 \cdot CO_2' + OH' = CO_2Et \cdot CPr_2 \cdot OH + CO_2$.

Dipropylglycollic ester.

(3) $2CO_2Et \cdot CPr_2 \cdot CO_2' = CO_2Et \cdot CPr:CH \cdot Et + CO_2Et \cdot CPr_2H + 2CO_2$.

α-Propyl-β-ethylacrylic ester. Dipropylacetic ester.

The two latter substances were not isolated in the pure state, but
sufficient indication of their formation was obtained.

The half-saponification of the diethyl ester of dipropylmalonic acid
used as the original material offered little difficulty, thus differing
from the dimethyl and diethyl compounds. It was found that by
warming the alcoholic solution of the ester with potash, the saponification could be carried nearly half-way. The method employed was as
follows : a solution of potash in methyl · alcohol was accurately
titrated and a volume of this solution containing a quarter of the
amount of potash required for the complete saponification was added
to the alcoholic solution of the ester. The mixture was then heated
for a few hours on the water-bath. After removal of alcohol by
distillation, water was added and the unsaponified ester removed by
extraction with ether. The ester, thus recovered, was treated with a

quantity of the potash solution sufficient to complete the half-saponification. In this way a further quantity of impure potassium ethyl salt was obtained and added to the first portion. The combined aqueous solutions of the impure potassium ethyl salt were then evaporated to dryness. The dry substance, which was found to consist of potassium ethyl salt contaminated with about 12 per cent. of dipotassium salt, was several times extracted with hot absolute alcohol. The dipotassium salt, being insoluble in the alcohol, was left behind, whilst the soluble potassium ethyl salt was obtained in a pure state by evaporating off the alcohol from the filtrate.

In this way, about 121 grams of the pure potassium ethyl salt were obtained from 187 grams of the original diethyl ester.

A 50 per cent. solution of potassium ethyl dipropylmalonate in water was electrolysed in a nickel crucible, a spiral of platinum wire being used as the anode. The current was regulated to about 2·5 amperes, and the temperature was not allowed to exceed 30°.

As the electrolysis proceeded, the esters which were formed separated as an oily layer floating on the aqueous layer. On completion of the electrolysis, the oily product was subjected to steam distillation, and was in this way separated into two portions, namely, a portion of low boiling point, volatile with steam (about 50 c.c.) and a portion of high boiling point remaining in the distilling flask (about 22 c.c.).

α-Propyl-β-ethylacrylic Acid.

The ester of low boiling point was fractionally distilled at the ordinary pressure. The main fraction, which passed over between 190° and 194°, decolorised alkaline permanganate and absorbed bromine freely, thus behaving as an unsaturated compound. A combustion yielded results agreeing roughly with the formula $C_{10}H_{18}O_2$. A small portion was brominated in a solution of carbon disulphide and the pale yellow liquid obtained in this way was found to approximate to the composition expressed by the formula $C_{10}H_{18}O_2Br_2$. The results seemed to indicate that the substance was the ethyl ester of α-propyl-β-ethylacrylic acid, $CO_2Et·CPr:CH·Et$, a normal product of the electrolysis (see equation, p. 929).

The main portion of this fraction was then saponified with alcoholic potash and the acid, on being liberated from the aqueous solution of the potassium salt by means of hydrochloric acid, was at once extracted with ether.

The acid obtained from the washed and dried ethereal solution by evaporation was a viscous, yellow liquid, which showed no signs of crystallisation even on cooling in a freezing mixture or after long standing. A combustion of this acid indicated that an impurity, presumably dipropylacetic acid, was mixed with the unsaturated acid. The *calcium* salt was prepared by boiling the acid with freshly-precipitated calcium carbonate in water. The filtrate, after concentrating on the water-bath until signs of crystallisation appeared, was allowed to evaporate at the ordinary temperature. The salt was less soluble in hot than in cold water and crystallised on the surface of the solution.

The crystals obtained consisted of a monohydrate, as shown by the following analysis :

0·3283 lost, at 105°, 0·0176. Water of crystallisation = 5·36.

$C_{16}H_{26}O_4Ca, H_2O$ requires 5·29 per cent.

The calcium was determined in another portion dried at 120° :

0·2044 gave 0·0862 $CaSO_4$. Ca = 12·40.

$C_{16}H_{26}O_4Ca$ requires Ca = 12·42 per cent.

The *magnesium* and *cadmium* salts were prepared, but these sparingly soluble salts were useless for the purification of the acid owing to their viscid nature, probably caused by the separation of free acid. The *ammonium* salt, on warming in aqueous solution, gave off ammonia very readily, the acid appearing as an oil on the surface.

The *silver* salt formed by heating the acid with precipitated silver carbonate in water crystallised in feathery, white clusters on cooling the filtrate. This salt was also prepared by precipitating a solution of the ammonium salt with silver nitrate solution. The thick, curdy precipitate was dissolved in boiling water, the undissolved precipitate rapidly turning black in the boiling solution even in the dark. The filtrate yielded the white, crystalline clusters already mentioned, which turned dark when exposed to light. By repeated crystallisation, this salt was obtained moderately free from the impurity, and gave the following results on combustion :

0·3985 gave 0·5618 CO_2, 0·1957 H_2O, and 0·1725 Ag. C = 38·45 ; H = 5·45 ; Ag = 43·29.

$C_8H_{13}O_2Ag$ requires C = 38·56 ; H = 5·22 ; Ag = 43·37 per cent.

A combustion of the acid obtained from this salt resulted as follows :

I. 0·1389 gave 0·3440 CO_2 and 0·1309 H_2O. C = 67·53 ; H = 10·47.
$C_8H_{14}O_2$ requires C = 67·60 ; H = 9·86 per cent.

A portion of the acid obtained from the silver salt was treated with bromine in carbon disulphide solution. The bromo-acid, after being freed from carbon disulphide and excess of bromine by a current of dry air, was analysed by Carius' method.

0·1037 gave 0·1152 AgBr.   Br = 47·28.

$C_8H_{14}O_2Br_2$ requires Br = 52·96 per cent.

Another specimen of the acid was prepared from a second portion of the ester of low boiling point and was subjected to careful fractionation in a current of steam.   The acid boiled at 232—233° and gave the following results on analysis :

II. 0·1927 gave 0·4773 CO_2 and 0·1785 H_2O.   C = 67·55 ;   H = 10·29.

$C_8H_{14}O_2$ requires C = 67·60 ;   H = 9·86 per cent.

When an aqueous solution of the sodium salt of the bromo-acid obtained as above mentioned was gently warmed, it became turbid from separation of an oil insoluble in acids and in alkalis.   This reaction is characteristic of the dibromides of $a\beta$-unsaturated acids, and thus determines the position of the double bond in the unsaturated acid, which is doubtless a-propyl-β-ethylacrylic acid, the alkali salt of the dibromide of which would decompose according to the equation :

$$CHEtBr \cdot CPrBr \cdot CO_2Na = CHEt \vdots CPrBr + NaBr + CO_2.$$

Dipropylglycollic Acid.

The residual oil in the distillation flask, after the ester of low boiling point had been removed by steam, was, after separation and drying, subjected to distillation under reduced pressure.   The main fraction, which passed over between 160° and 215° under 40 mm. pressure, was heated with alcoholic potash for two days on the water-bath.   After this treatment, only a portion of the oil was saponified. The aqueous solution of the salt was evaporated and an acid was obtained from it, which formed a viscous, yellow liquid.   This crystallised on standing for a few days in a desiccator.   The crystals, after being purified by draining on a porous tile, were recrystallised from light petroleum.   The melting point of the recrystallised substance was 77—78°, and a combustion yielded the following results :

0·0703 gave 0·1533 CO_2 and 0·0639 H_2O.   C = 59·70 ; H = 10·10.

$C_8H_{16}O_3$ requires C = 60·00 ;   H = 10·00 per cent.

The acid thus has the composition of dipropylglycollic acid.   The ester is no doubt formed according to the equation :

$$CO_2Et \cdot CPr_2 \cdot CO_2' + OH' = CO_2Et \cdot CPr_2 \cdot OH + CO_2.$$

Dipropylglycollic acid, $Pr_2C(OH) \cdot CO_2H$, has been prepared by Rafalsky, who named it dipropyloxalic acid (J. Russ. Phys. Chem. Soc., 1881, 13, 237), and by Basse and Klinger (Ber., 1898, 31, 1219).   Rafalsky prepared the acid by synthesis from oxalic ester, and zinc propyl, obtaining a substance which melted at 80—81°.

Klinger and Schmitz, and later Basse and Klinger, prepared it from butyroin and found a melting point of 78°. Basse and Klinger, on repeating Rafalsky's synthesis, failed to obtain a crystalline acid. Since it was extremely probable that the acid obtained from the electrolysis was dipropylglycollic acid, it was resolved to prepare the acid by Rafalsky's method and to compare the properties of the two. No difficulty was experienced in obtaining a crystalline acid by following his instructions, but the yield was poor. The synthetic dipropylglycollic acid thus obtained behaved in all respects in the same way as the acid obtained electrolytically. Dipropylglycollic acid is very soluble in ether or light petroleum; the crystals obtained from the latter solvent melted at 78°, which corresponds with the value obtained by Basse and Klinger. It is moderately soluble in water, yielding crystals which melt at 79—80°, a melting point approaching that given by Rafalsky. The aqueous solution behaves characteristically as regards change of temperature; it was found that on allowing a boiling saturated solution of the acid to cool, the liquid became turbid at 72—74°, due to separation of the liquid acid, but became clear again at 35°. On further cooling, long needles were obtained. The same phenomena in the reverse order were apparent on reheating.

Tetrapropylsuccinic Acid.

A portion of the oil of high boiling point which resisted the action of potash was analysed :

0·1284 gave 0·3279 CO_2 and 0·1261 H_2O.　$C = 69·65$; $H = 10·90$.

$C_{20}H_{38}O_4$ requires $C = 70·18$; $H = 11·11$ per cent.

This analysis, together with a determination of the molecular weight by Beckmann's method, indicated that the oil was the somewhat impure diethyl ester of tetrapropylsuccinic acid, $CO_2Et \cdot CPr_2 \cdot CPr_2 \cdot CO_2Et$. The remainder of the oil was sealed up in a glass tube with a solution of hydrobromic acid (saturated at 0°), and heated for fifteen hours with a gradual increase of temperature from 50° to 120°. On opening the tube, the two layers seemed apparently unaltered, except that the colour was rather darker, but the oil, after extraction and drying, set in a desiccator to a hard, crystalline mass of melting point 35°. After recrystallisation from a mixture of alcohol and water, fine needles melting at 37·5° separated. The solution was neutral and the following analytical results showed that the substance was *tetrapropylsuccinic anhydride*, corresponding to the very stable anhydride obtained by Crum-Brown and Walker.

　I. 0·1185 gave 0·3130 CO_2 and 0·1138 H_2O.　$C = 72·04$; $H = 10·67$.

　II. 0·1023 　,,　 0·2666 CO_2 　,,　 0·0960 H_2O.　$C = 71·10$; $H = 10·43$.

　$C_{16}H_{28}O_3$ requires $C = 71·64$; $H = 10·45$ per cent.

This anhydride is almost entirely insoluble in water, very soluble in benzene, ether, or alcohol, and moderately so in formic acid, from which it may be readily crystallised. It shows the same resistance to alkalis as the corresponding tetraethyl compound, and generally exhibits extreme stability. By the action of sodium methoxide, the sodium methyl salt of tetrapropylsuccinic acid was formed from the anhydride by addition according to the equation :

$$\begin{array}{c} Pr_2C\cdot CO \\ | \\ Pr_2C\cdot CO \end{array}\!\!>\!\!O + NaOMe = \begin{array}{c} Pr_2C\cdot CO_2Na \\ | \\ Pr_2C\cdot CO_2Me \end{array}.$$

Methyl hydrogen tetrapropylsuccinate, obtained from this salt by acidification, is easily crystallised from a mixture of alcohol and water, from which it separates in small tables melting at 77—78°. On heating to its melting point and cooling quickly, it shows the same melting point as before when reheated. It was found, however, that the ester, when kept at a temperature of 80° for an hour, decomposed into methyl alcohol and the anhydride according to the equation :

$$\begin{array}{c} Pr_2C\cdot CO_2H \\ | \\ Pr_2C\cdot CO_2Me \end{array} = \begin{array}{c} Pr_2C\cdot CO \\ | \\ Pr_2C\cdot CO \end{array}\!\!>\!\!O + MeOH,$$

the melting point of the cooled solid being now 37·5°. A titration yielded the following results :

0·1240 required 8·25 ˙c.c. of $N/20$ NaOH. Equivalent = 300.
$C_{17}H_{32}O_4$ requires equivalent = 300.

It was found that the hydrogen methyl compound was gradually transformed into the anhydride in alcoholic solution, and even in the dry solid state it reverted in time to the anhydride.

Tetrapropylsuccinic acid was obtained by warming the hydrogen methyl compound with a 5 per cent. aqueous solution of potash, and acidifying and extracting as usual. The solid acid, when recrystallised from a mixture of alcohol and water, melted at 137°. It was found to be very soluble in ether and methyl alcohol, sparingly so in light petroleum and benzene, and practically insoluble in water. A small quantity was titrated, giving the following results :

0·0790 required 11·20 c.c. of $N/20$ NaOH. Equivalent = 141.
$CO_2H\cdot C_{14}H_{28}\cdot CO_2H$ requires equivalent = 143.

Analysis gave the following results :

0·1298 gave 0·3191 CO_2 and 0·1217 H_2O. C = 67·05 ; H = 10·42.
$C_{16}H_{30}O_4$ requires C = 67·13 ; H = 10·49 per cent.

The acid is quite stable in the dry state and showed no signs of passing into the anhydride even after a month ; in alcoholic or benzene solution, it reverted rapidly to the anhydride, but in ether under similar conditions no change was noticed. A small quantity of the acid left in contact with water gradually disappeared, and crystals of the anhydride began to form on other parts of the containing vessel. This change was not apparent until a month had elapsed.

The same phenomenon was noticed on warming the solution of the sodium salt of this acid with phenolphthalein as was observed with tetraethylsuccinic acid (Walker, *loc. cit.*, p. 965). The cold solution, which was quite neutral to this indicator, became deep pink on warming, and the colour disappeared after cooling.

In almost all respects, then, tetrapropylsuccinic acid and its derivatives resemble very closely the corresponding compounds of tetraethylsuccinic acid.

UNIVERSITY COLLEGE,
DUNDEE.

XCIX.—*Resolution of Lactic Acid by Morphine.*

By JAMES COLQUHOUN IRVINE, Ph.D., D.Sc. (Carnegie Fellow).

THE optically active lactic acids have been frequently obtained in quantity in this laboratory by resolving the inactive acid by crystallisation of the zinc ammonium salts, but so far the commoner method of resolution by means of alkaloidal salts has not proved applicable as a working process for preparing the active acids in question. The method of resolution by strychnine (Trans., 1895, **67,** 616) is tedious and imperfect, and the use of quinine, as described by E. Jungfleisch in a recent paper (*Compt. rend.*, 1904, **139,** 56), judging from the absence of exact data as to the yields and activities of the products obtained, does not appear to be more advantageous.

Experiments on the solubilities of the alkaloidal salts of *d*- and *l*-lactic acids showed that whereas morphine *l*-lactate crystallised readily from dilute aqueous solutions, the salt of the *d*-acid was exceedingly soluble, and only crystallised after standing for several weeks in a vacuum desiccator. Advantage was taken of this marked difference in solubility to resolve the inactive acid, and the process was found to give an almost quantitative yield of pure *l*-lactic acid and about 50 per cent. of the theoretical amount of the *d*-acid. The method, in addition, was found to possess the advantage of being very expeditious, as the morphine *l*-lactate separates quickly and can be readily purified, so that, in spite of the initial expense of the alkaloid, the method of resolution can be recommended when small quantities of active lactic acid displaying the maximum optical activity are required.

On neutralising an aqueous solution of fermentation lactic acid with morphine, the filtered liquid deposited the salt of the *l*-acid on cooling, whilst morphine *d*-lactate remained in solution. The re-crystallised salt was converted into zinc lactate, the purity of which was determined by observation of the specific rotation and esti-mation of the water of crystallisation. The analytical method in this case cannot, however, be regarded as a crucial test of the purity of the product, and as the metallic lactates display comparatively small rotations in solution, it is evident that the presence of traces of inactive material might readily escape detection. Thus, a saturated solution of crystallised zinc lactate gives a polarimetric reading of only 1·02° in a 2-dcm. tube (Trans., 1893, **63**, 1154), and the presence of 4 per cent. of the inactive salt would scarcely affect the rotation.

In the present instance, a more accurate index of the purity of the active lactic acid was obtained by converting it into methyl lactate and afterwards into methylic methoxypropionate (Trans., 1899, **75**, 485). The latter compound is highly active, and gives in a 1-dcm. tube a polarimetric reading of 95·21°. In determining a rotation of this magnitude, the experimental error is almost eliminated, and even traces of inactive material are readily detected. In this way, it was shown that the *l*-lactic acid obtained in the resolution displayed more than 99 per cent. of the maximum activity recorded for the substance.

EXPERIMENTAL.

Preparation of Morphine d- *and* l-*Lactates.*

The *d*- and *l*-lactic acids were obtained by the zinc ammonium salt method of resolution and the morphine salts prepared by neutralising hot aqueous solutions of the acids with the alkaloid. The salt of the *l*-acid crystallised readily in long prisms, which were recrystallised from dilute alcohol. The compound underwent no loss in weight when heated at 120° and was apparently anhydrous. The specific rotation in 5 per cent. aqueous solution was $[a]_D^{20°} - 91·8°$.

The salt of the *d*-acid readily formed supersaturated solutions, and only crystallised very slowly in clusters of radiating prisms. Analysis showed the compound to contain one molecule of water of crystallisa-tion (found, $H_2O = 4·40$; calculated $= 4·58$ per cent.). In 5 per cent. aqueous solution, the salt showed a specific rotation of $[a]_D^{20°} - 92·7°$.

Resolution of Inactive Lactic Acid by Morphine.

The lactic acid used (Kahlbaum's syrup) was, as already noted by McKenzie (Trans., 1905, **87**, 1375), slightly dextrorotatory, and gave in a 2-dcm. tube a rotation of $+1·85°$. As the activity of a 20 per cent. aqueous solution altered from dextro- to lævo-rotatory when maintained at 100°, this rotation was evidently due to the presence of a slight excess of the l-acid in the mixture. Titration of the syrup before and after hydrolysis showed it to contain 65·5 per cent. of free lactic acid and 23·2 per cent. of the monobasic anhydride.

Forty grams of the syrup were diluted to 400 c.c. with water and boiled under a condenser for six hours in order to hydrolyse the anhydro-acid. The hot solution was then neutralised with morphine and filtered. On cooling, an abundant crop of morphine l-lactate separated, which was filtered and washed successively with water, alcohol, and ether. A second crop was obtained after evaporating the mother liquor to 250 c.c., but after further concentration no more crystals could be obtained even when the liquid was cooled in a mixture of calcium chloride and ice and nucleated with morphine d-lactate. The preparation thus yielded the morphine salts of both the l- and d-acids, the former being obtained in the crystalline state and the latter as a syrupy solution. The two crops of crystals were united and recrystallised twice from 50 per cent. alcohol. The product (80 grams) was now quite white, and amounted to 95 per cent. of the theoretical yield.

The salt was dissolved in water and decomposed by the addition of ammonium hydrate, and the filtrate, after treatment with animal charcoal, was converted into the calcium salt. The free acid was then liberated by the exact addition of oxalic acid to an aqueous solution of the calcium salt and the filtrate evaporated to a syrup on a water-bath. The residue was extracted with a mixture of alcohol and ether, and, after separation from traces of calcium oxalate and removal of the solvents, pure l-lactic acid was obtained as a colourless syrup. A specimen of the crystallised zinc salt was prepared and the rotation determined.

$$c = 5\cdot6266,\ l = 2,\ a_D^{15^\circ} + 0\cdot78°,\ [a]_D^{15^\circ} + 6\cdot84°.\ Zn = 23\cdot39\ ;\ H_2O = 12\cdot98.$$
$$C_6H_{10}O_6Zn,2H_2O\ \text{requires}\ Zn = 23\cdot30\ ;\ H_2O = 12\cdot90\ \text{per cent.}$$

The remainder of the acid syrup was converted into the silver salt, and, by decomposing the latter with methyl iodide, a specimen of methylic l-lactate containing a small proportion of methylic l-methoxypropionate (Trans., 1898, 73, 296) was prepared. This was then methylated, as described in a previous paper, by means of the silver oxide reaction, with the production of pure methylic l-methoxypropionate. The product was entirely distilled without fractionation under diminished pressure [b. p. 61—62° (78 mm.)], and the rotation of the distillate determined.

$$l = 1,\ a_D^{20^\circ} + 94\cdot34°,\ d20°/4° = 0\cdot9967,\ [a]_D^{20^\circ} + 94\cdot7°.$$

The value already published for the pure compound is $[a]_D^{20°} + 95\cdot5°$, so that the specimen described above displays $99\cdot2$ per cent. of the maximum specific rotation recorded for this substance. As an obvious precaution in determining a rotation of this magnitude, the length of the tube was checked and found to be correct, whilst the temperature of the liquid was taken by a thermometer reading to $\frac{1}{50}°$. An accurate estimation of the activity of the lactic acid produced in the resolution was thus obtained, and, moreover, all possibility of the optical rotations being affected by the presence of traces of morphine is precluded by the preparation of a volatile derivative.

Reduction of Methylic l-Methoxypropionate by Hydriodic Acid.

The specimen of active methylic methoxypropionate described above was converted into lactic acid in order to ascertain if the configuration of methyl lactate remains unaltered during methylation by the silver oxide reaction. A similar experiment in which methylic dimethoxy-succinate was converted into tartaric acid is quoted by Purdie and Barbour (Trans., 1901, **79**, 972), but no such evidence has hitherto been obtained in the case of the lactates.

Ten grams of the methoxypropionate were introduced into the decomposition flask of a Zeisel apparatus, and heated with 100 c.c. of hydriodic acid (sp. gr. $1\cdot70$). The apparatus was connected in the usual way with flasks containing alcoholic silver nitrate solution, and the reduction was carried out at the minimum temperature ($85°$) at which silver iodide was formed. After four hours' treatment, the formation of methyl iodide ceased and the action was therefore complete. The bulk of the hydriodic acid was removed from the product by heating at $80°$ under diminished pressure, and the syrupy residue was diluted with water. The dark brown solution was then shaken with excess of silver oxide in order to remove iodine and hydriodic acid. The colourless filtrate, which contained silver lactate in solution, was saturated with sulphuretted hydrogen and the lactic acid thus liberated was converted into the zinc salt. The recrystallised salt (6 grams) was washed with alcohol to remove traces of zinc methoxypropionate and analysed :

Found $Zn = 23\cdot18$; $H_2O = 13\cdot14$.

$C_6H_{10}O_6Zn,2H_2O$ requires $Zn = 23\cdot30$; $H_2O = 12\cdot90$ per cent.

$c = 5\cdot0030$, $l = 2$, $a_D^{20°} + 0\cdot67°$, $[a]_D^{20°} + 6\cdot69°$.

The results show the product to be nearly pure zinc l-lactate, and the silver oxide method of alkylation is therefore without any disturbing effect on the configuration of an active lactate. The above method of carrying out such reactions obviates excessive heating with the hydriodic acid, and possible racemisation, or the reduction to propionic acid, is thus largely excluded.

Preparation of d-Lactic Acid from Morphine d-Lactate.

The syrupy mother liquor which had deposited the crops of morphine l-lactate, and in which the salt of the d-acid remained in solution, was utilised for the preparation of zinc d-lactate. The removal of the alkaloid and the liberation of the free acid was carried out as already described, and the whole of the product was converted into the zinc salt. After a series of recrystallisations, in which the purity of each crop was determined by means of the polarimeter, over 50 per cent. of the theoretical amount of pure zinc d-lactate was obtained.

$$c = 7\cdot1032, \; l = 2, \; a_D^{20°} - 0\cdot97°, \; [a]_D^{20°} - 6\cdot83°.$$
$$\text{Found } Zn = 23\cdot32 \; ; \; H_2O = 12\cdot96.$$
$$C_6H_{10}O_6Zn,2H_2O \text{ requires } Zn = 23\cdot30 \; ; \; H_2O = 12\cdot90 \text{ per cent.}$$

Reference may be made here to experiments which are in progress in this laboratory on the solubilities of the zinc lactates. It has been found that although active zinc lactate when crystallised in the usual manner readily loses its water of crystallisation at 110°, the residue left on evaporating an aqueous solution of the salt on a water-bath does not become completely anhydrous even at 150°. It is evident, therefore, that in estimating the specific rotation of active zinc lactate the concentration of the solution cannot be determined by the evaporation of a measured volume of the solution.

I desire to express my thanks to Professor Purdie, in whose laboratory the work was carried out, for providing the materials necessary for the research.

CHEMICAL RESEARCH LABORATORY,
UNITED COLLEGE OF ST. SALVATOR AND ST. LEONARD,
ST. ANDREWS UNIVERSITY.

C.—The Oxidation of Hydrocarbons by Ozone at Low Temperatures.

By JULIEN DRUGMAN.

THE present research has been carried out under the direction of Dr. W. A. Bone, with the object of defining the initial stages in the oxidation of a hydrocarbon at low temperatures by means of ozone.

Since the preliminary note on this subject (Proc., 1904, **20**, 127), the author has studied the process in greater detail.

Up to the time these experiments were undertaken, very little was definitely known concerning the action of ozone on gaseous hydrocarbons beyond the fact that in the case of ethylene rapid and even explosive action occurs, with the early formation of formaldehyde and formic acid.

Schönbein (*J. pr. Chem.*, 1868, **105**, 230), however, although without direct proof, expressed the belief that an ozonide is formed when ozone acts on an organic compound, and that such ozonides are unstable in the presence of water.

In 1898, Otto (*Ann. Chim. Phys.*, [vii], **13**, 116) published the results of some experiments with an improved form of apparatus, in which methane, ethane, and ethylene were used. In the case of methane, he detected formaldehyde and formic acid among the products of the reaction, and, from the smell of the liquid, also inferred the presence of methyl alcohol. In the case of ethane, he proved the formation of acetaldehyde and acetic acid, but not of ethyl alcohol. He altogether misinterpreted the results of the experiments with ethylene, stating that the chief products were acetaldehyde and acetic acid, a conclusion which is quite disproved by the author's experiments.

While this research was in progress, Harries and his co-workers (*Ber.*, 1903, **36**, 1933, 2998, 3658; 1904, **37**, 612, 839, 2708) published the results of an extensive investigation on the action of ozone on unsaturated organic compounds : chiefly hydrocarbons and alcohols. Their experiments did not, however, include gaseous hydrocarbons, which were only considered from a theoretical standpoint in a recent *résumé* of the results of their work (*Annalen*, 1905, **343**, 311). Harries' conclusions are, however, so largely borne out by the author's results, that it will be useful to outline them here. He finds that, as a general rule, where there is a double bond between adjacent carbon atoms ozone is added on directly, the resulting ozonide being a very unstable body which readily decomposes, giving aldehydes or ketones and acids, or, in the presence of water, aldehydes, ketones, and hydrogen peroxide. The author finds that this probably occurs in the case of ethylene, but the decomposition seems to be more complicated. Water is formed at an early stage in the process, with the result that the final products are a mixture of those which the direct decomposition of the ozonide would give, together with the products of its decomposition by water.

The formation of water is probably accompanied by the evolution of carbon monoxide, for this gas is also a constant product in the reaction. The following equations may represent the process :

According to Harries : $C_2H_4 + O_3 = C_2H_4O_3 = CH_2O + HCO_2H$.
The ozonide would also decompose directly to aldehyde, water, and
carbon monoxide : $C_2H_4O_3 = CH_2O + CO + H_2O$.
The water then reacts with some of the] ozonide, giving aldehyde
and hydrogen peroxide : $C_2H_4O_3 + H_2O = 2CH_2O + H_2O_2$.
The method employed consisted in bringing together, at as low a
temperature as was consistent with rapid interaction, dry oxygen rich
in ozone with the dry hydrocarbon and isolating the products,
either by solution in water, or, in the case of ethylene, by direct
condensation.

The diagram will explain the form of apparatus used. The
gases were stored in 18-litre, graduated holders, A and B. The
oxygen, before being ozonised, was thoroughly dried by passing
through sulphuric acid, followed by phosphoric anhydride. It then
entered [the ozonisers, C. These were thin concentric glass tubes,
the thickness and separation of which were adjusted according to
Otto's recommendations (*loc. cit.*). They were joined at their lower
ends in couples in U-form, and placed in a large vessel of cold copper
sulphate solution. The silent discharge was produced by a 10-inch
coil with mercury contact-break ; the current was obtained from the
200 volt main, and regulated by a resistance board. With a low-
contact frequency and a current of 3 amperes at 15 volts, a steady
supply of ozonised oxygen containing 10—12 per cent. of ozone
was obtained when passing through four ozonisers at the rate of
1 to 1·5 litres an hour. There was no appreciable heating of the
ozonisers.

The ozonised oxygen then mixed with the hydrocarbon, also well
dried, in the tube D, and the products were passed through worms, or,
as shown in the diagram, condensed directly in the tube E. The
gaseous products were drawn into the graduated receiver, R, a worm
containing a solution of potassium iodide, to destroy unchanged ozone,
and a phosphoric anhydride tube being inserted between R and E.

EXPERIMENTAL.

Methane and Ozone.—Two experiments, one at 15° and the other
at 100°, were carried out. The reaction was very slight at 15°, but at
100° formaldehyde and formic acid were present in appreciable
quantities. Otto's supposition that methyl alcohol was formed was,
however, not confirmed. The reaction in the case of methane is so
slow that any alcohol formed would be very rapidly oxidised to
formaldehyde.

Ethane and Ozone.—In the experiments described in the preliminary

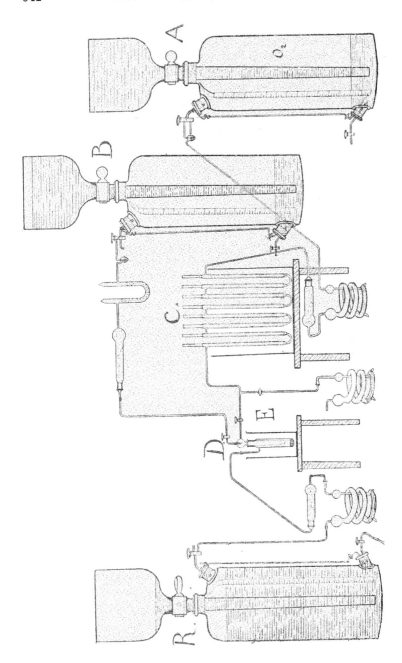

note, the percentage of ozone was only 2—3 per cent., and, besides acetaldehyde, acetic acid, and alcohol, traces of formaldehyde were found. The latter was, however, most likely due to oxidation of the acetaldehyde and acid, for in later experiments no formaldehyde was found.

At 15°, the reaction with 10 per cent. of ozone was slow, although much faster than in the case of methane. The products were ethyl alcohol, acetaldehyde, and acetic acid, besides traces of hydrogen peroxide. No oxy-acids, glycols, &c., could be detected.

At 100°, using a large excess of ethane, much less acid was formed, and the reaction was more rapid. Acetaldehyde was still the chief product, but ethyl alcohol was present in larger quantities than before. Hydrogen peroxide was also found, but here too only in traces : not enough to justify one in assigning it any place in the chief reaction.

It is clear, then, that ethyl alcohol is the first stage in the oxidation process, and that acetaldehyde and acetic acid are secondary products of the reaction.

Ethylene and Ozone.—In the first series of experiments, the gases were only partially dried and the products were absorbed by water. Under these conditions, formaldehyde, formic acid, and hydrogen peroxide were formed in large quantities, the ozone disappearing very rapidly. No explosions occurred under these conditions. Addition of caustic potash solution caused a sudden evolution of hydrogen. This reaction, as is well known, takes place when formaldehyde and hydrogen peroxide are together in solutions (*Ber.*, 1904, **37**, 515 ; also *J. Amer. Chem. Soc.*, 1905, **27**, 714). The volume of hydrogen liberated could be used as a measure of the hydrogen peroxide present, for formaldehyde was always present in excess. The hydrogen peroxide, it was thought, might be due to the presence of water ; in order to eliminate this cause, further experiments were carried out with the gases and apparatus thoroughly dried, and the products were condensed directly into a dry cooled tube.

At 15—18°, the gases combined at once, and a colourless liquid condensed. It was slightly viscous, reddened litmus paper, had a strong sharp aldehydic odour, and slowly evolved gas bubbles. When gentle heat was applied, the gas-evolution became more rapid, the gas proving to be pure hydrogen. Hydrogen peroxide, which was also found in the liquid prepared in this way, disappeared during the hydrogen evolution. At a higher temperature, rapid evolution of hydrogen was observed.

The liquid appeared too homogeneous to be a mixture, and several analyses and estimations of the amount of hydrogen evolved were made, but, as the following examples show, the composition was a variable one, and we came to the conclusion that the liquid was a mixture of formaldehyde, formic acid, hydrogen peroxide, and water in varying proportions.

	I.	II.	II (after 8 days).
C	23·55	21·85	24·4
H	7·39	8·11	6·08
O	69·06	70·03	69·52

Further examination of I showed that it was composed of 46 per cent. of formaldehyde, .23 per cent. of formic acid, 2·5 per cent. of hydrogen peroxide, and 28·5 per cent. of water. A large quantity of water is thus formed during the reaction, and analyses of the gases show that there is a correspondingly large quantity of carbon monoxide in the products. The reaction seems therefore to have passed through more than one stage. To obtain, if possible, the primary product, or at any rate some idea of its nature, experiments at lower temperatures were undertaken.

Using ice and salt as refrigerents, the reaction proceeded apparently as before, but on several occasions on stopping the experiment, and once or twice during its course, without apparent cause, violent explosions occurred. Reaction at lower temperatures seemed then to favour the production of an explosive compound.

Solid carbon dioxide was next used to cool both the gases before mixing and the products. Shortly after the gases had been allowed to enter, a sharp click was heard, accompanied by a flash and a slight separation of carbon. This was repeated at regular intervals as long as the experiment continued, but the latter had to be stopped after about an hour owing to some of the liquid product freezing in the ozone tube, and on attempting to clear this an explosion shattered the apparatus. A new experiment was also unsuccessful, an explosion occurring at the very beginning.

In a final experiment, ethylene cooled to nearly its liquefying point was very slowly led over a solution of ozone in liquid oxygen, cooled in a bath of liquid air. Immediate explosive interaction took place, accompanied by flame.

The initial product of the action of ozone on ethylene appears, therefore, to be an extremely unstable compound which can only accumulate in any quantity at low temperatures. In a concentrated condition, it is highly explosive. At the ordinary temperature, its rate of decomposition is comparable with its rate of formation, so that it does not accumulate and the reaction is a quiet one.

This initial product is probably a direct addition product, for analyses of the gases show that at least 65—70 per cent. of the ozone formed enters into combination, that is, a large proportion of it must take part as a whole in the reaction, whilst a smaller part only gives up the one atom of oxygen usually available for oxidation.

For the purpose of this paper, the isolation of this compound was not necessary, as a sufficient insight into the process had already been gained. In the oxidation of ethylene, it is clear that the first stage is immediately followed by a separation of the two carbon atoms, as oxidation products containing ono carbon atom are alone found.

Summary.

In the oxidation of hydrocarbons by ozone, there is a radical difference in the reaction in the case of a saturated and in that of an unsaturated hydrocarbon.

In the case of a saturated hydrocarbon, gradual hydroxylation of one carbon in the molecule takes place, alcohol being first formed. This is quickly oxidised further to tho relatively stable aldehyde, and more slowly to acid.

Unsaturated hydrocarbons, whether the initial product is an ozonide or a peroxide, decompose at the double bond, giving products containing a less number of carbon atoms. There is a weakness at this point, and oxygen enters between the carbon atoms.

Whether the analogy can be extended to oxidation of these two classes of hydrocarbons by oxygen is perhaps doubtful; formaldehyde is formed in large quantities in the combustion of ethylene, but the process is more probably ono of hydroxylation than of direct addition of oxygen.

Ozone is perhaps more comparable in its activity at the ordinary temperature with elements like chlorine than with molecular oxygen.

The author desires to express his best thanks to Dr. W. A. Bone for placing at his disposal a specially designed electrical installation, which had been purchased for such experiments out of a grant from the Government Grant Committee of the Royal Society.

FUEL AND METALLURGICAL LABORATORY,
VICTORIA UNIVERSITY OF MANCHESTER.

CI.—*Reactions Involving the Addition of Hydrogen Cyanide to Carbon Compounds. Part V. Cyanodihydrocarvone.*

By ARTHUR LAPWORTH.

CARVONE, the main constituent of oil of carraways, is perhaps the most prominent member of the class of unsaturated ketones which are related to the terpenes, and the study of its derivatives in tho hands of Baeyer, Wallach, and others has afforde i results of more than ordinary importance. The chemistry of this substance is so closely connected with that of the protean and much discussed compounds, pinene and camphor, that a special interest attaches to it on this account, but many reactions of carvone and its derivatives have a considerable interest of their own, and such are, for example, the curious structure-changes leading to the formation of carone, eucarvone, carvacrol, and *para*aminocarvacrol.

Carvone exhibits in a decided manner the general characters of $\alpha\beta$-unsaturated ketones, and forms with great readiness addition products in which the entrant groups occupy positions on the α- and β-carbon atoms, whilst, as is usual in such cases, the. carbonyl group itself manifests a comparatively feeble additive power. These characters are well-marked in the reactions of the ketone with metallic cyanides, for no cyanohydrin appears to be formed in the first instance, addition taking place exclusively at the $\alpha\beta$-carbon double-linking, and a cyanodihydrocarvone is formed. The production and properties of the latter compound, first referred to in a note by Hann and Lapworth (Proc., 1904, **20**, 54), form the subject of the present communication.

There can no longer be any reasonable doubt as to the mode in which such addition products are formed (compare Trans., 1904, **85**, 1218), and in the present instance the reaction is to be represented as follows :

$$CH_3 \cdot C \lessgtr^{CO \cdot CH_2}_{CH \cdot CH_2} > CH \cdot C \lessgtr^{CH_3}_{CH_2} + HCN =$$

$$CH_3 \cdot CH \lessgtr^{CO——CH_2}_{CH(CN) \cdot CH_2} > CH \cdot C \lessgtr^{CH_3}_{CH_2}.$$

Cyanodihydrocarvone is obtained at once under the experimental conditions adopted in a well-defined crystalline form and is quite homogeneous, although it was anticipated that a mixture of isomeric substances might be formed. The carvone used was the dextro-rotatory form of the ketone, which contains one asymmetric carbon atom ; the production of the hydrogen cyanide addition product involves the generation of two new asymmetric carbon atoms, so that four isomerides may be anticipated. Adopting Aschan's mode of representing the section of the ring-plane by a line (*Ber.*, 1902, **35**, 3 89), these four isomerides have the following configurations $\left(P = \cdot C \lessgtr^{CH_3}_{CH_2} \right)$:

P	P	P	P
CN	CN	CN	CN
Me	Me	Me	Me
(I.)	(II.)	(III.)	(IV.)

The four enantiomorphs of these, the configuration of which may be represented by placing P to the left of the line, are theoretically obtainable from l-carvone. That only one of the four was isolated is doubtless to be ascribed to the fact that the hydrogen atoms attached to the new asymmetric carbon atoms are labile, being in a-positions to a $-CN$ and a $-CO$ group respectively; in consequence, a condition of equilibrium will rapidly be attained in alkaline solution similar to that noticed by Kipping in the case of the a-halogen derivatives of camphor (Trans., 1905, 87, 628), and only the isomeride present in the amount which at equilibrium bears the highest ratio to its solubility will have opportunity to separate. Direct evidence indicating the occurrence of such an isodynamic change was obtained by observing the rotatory power of an alcoholic solution of the cyanoketone; this was constant until a trace of alkali was added, when a rapid and very considerable alteration took place, the reading quickly attaining a new constant value. However, repeated attempts to isolate from the resulting mixture any new isomeride were fruitless.

Satisfactory evidence that cyanodihydrocarvone is a cyanoketone with one ethylenic linking in the molecule is adduced in the experimental part of the paper. A brief reference, however, to the oxidation of the nitrile seems desirable. The only crystalline products of known constitution which could be isolated from the oxidation product were oxalic acid and tricarballylic acid, and comparison of the structure of the latter substance with that of the original compound appears to show that the skeleton common to both is that suggested by the following mode of comparing their formulæ :

$$
\begin{array}{cc}
CH_3\ CH_2 & \\
\diagdown\!\diagup & \\
\overset{|}{C} & \\
\overset{|}{CH} & CO_2H \\
\diagup\diagdown & \overset{|}{CH} \\
\overset{|}{CH_2}\ \overset{|}{CH_2} & \diagup\diagdown \\
CN\!\cdot\!CH\ CO & CH_2\ CH_2 \\
\diagdown\!\diagup & \overset{|}{CO_2H}\ \overset{|}{CO_2H} \\
\overset{|}{CH} & \\
\overset{|}{CH_3} &
\end{array}
$$

Cyanodihydrocarvone yields an addition product with one molecular proportion of bromine if the interaction is carried out in presence of sodium acetate so as to obviate the possibility of the formation of a substitution derivative (see Trans., 1904, 85, 38). The product,

$$
CH_3\!\cdot\!CH\!\!<\!\!{}^{CO\text{------}CH_2}_{CH(CN)\cdot CH_2}\!\!>\!\!CH\!\cdot\!CBr\!\!<\!\!{}^{CH_3}_{CH_2Br},
$$

is theoretically capable of existing in sixteen different optically active modifications, one-half of these being derived from d-carvone. The material examined by the author was prepared from one of the four possible cyanodihydrocarvones derived from d-carvone, and was consequently a mixture of two isomerides only, as the conditions were not conducive to isodynamic change. The two isomerides were separated by fractional crystallisation and their properties are described later.

Like other β-cyanoketones which have come under the author's observation, cyanodihydrocarvone may be resolved into hydrogen cyanide and the corresponding $\alpha\beta$-unsaturated ketone (carvone) by the action of alcoholic potassium or sodium hydroxide, especially if ferrous hydroxide be present at the same time so as to convert the cyanide formed into ferrocyanide. This reaction, which is the reverse of that to which the ketonitrile owes its origin, must be represented by inverting the sense of the equations which represent its synthesis; that is to say, it must be assumed that the cyanoketone (or its *enolic* form?) acts towards the alkali as an acid, and from the resulting negative ion the cyanion is eliminated, carrying away the charge and leaving the unsaturated ketone as the neutral component (compare Trans., 1904, **87**, 1215), or, as it may be represented,

$$R{<}^{H}_{CN}\overset{+}{H} \rightarrow \overset{-}{R}{<}_{CN} \rightarrow R + \overset{-}{C}N.$$

It will be seen that there is no essential difference between this change and the breaking down of a ketonecyanohydrin (Trans., 1902, **83**, 1002), the essential conditions being (1) the presence of a labile (ionisable) hydrogen atom and (2) the presence of a group capable of existence as a negative ion in such a position in the molecule that on its withdrawal from the complex ion the remaining aggregate can be represented by a legitimate structural formula by a mere rearrangement of the " bindings " in the molecule, whilst the component atoms retain their original relative positions.

The hydrolysis of cyanodihydrocarvone by acids in the cold leads to the formation of an amide, whilst if the action be carried out on the water-bath, or if the amide is hydrolysed by alkalis, a mixture of isomeric acids is obtained. These may be converted one into the other by the action either of acids or of alkalis, and doubtless differ only in the configuration of the molecule at one point, namely, at the carbon atom next the ketonic group, where the occurrence of isodynamic change is by far the .most likely. The configurations of the isomerides are probably thus to be represented,

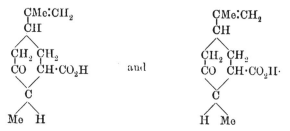

The possibility that the acids differed in the position of the double linking was at one time thought likely, as dihydrocarvone itself is readily converted by acids into carvenone by the migration of the double linking from the side-chain to the ring. The fact that the change of one acid into the other is reversible is in itself enough to dispose of this assumption; moreover, both acids are oxidised by permanganate and yield products which are rapidly attacked by lead peroxide in presence of acetic acid, yielding compounds which contain the group $-CO \cdot CH_3$, as they at once give bromoform with sodium hypo-bromite. These observations are consistent only with the view that both acids contain the grouping $-C{\mkern-2mu<\mkern-5mu\substack{CH_3 \\ CH_2}}$, converted by permanganate into $-C(OH){\mkern-2mu<\mkern-5mu\substack{CH_3 \\ CO_2H}}$, and the latter by lead peroxide and acetic acid into $-CO \cdot CH_3$.

EXPERIMENTAL.

Addition of Hydrogen Cyanide to Carvone.

Formation of Cyanodihydrocarvone [1-*Methyl*-4-iso*propenyl*-6-*cyano-cyclohexanone* (2)],

$$CH_3 \cdot CH{\mkern-2mu<\mkern-5mu\substack{CO \underline{\qquad} C \\ CH(CN) \cdot CH_2}}{\mkern-2mu>\mkern-2mu}CH \cdot C{\mkern-2mu<\mkern-5mu\substack{CH_3 \\ CH_2}} .$$

When carvone is left in contact with aqueous or alcoholic solutions of hydrogen cyanide or with the anhydrous substance, no appreciable quantity of any new compound is formed unless considerable quantities of bases or of potassium cyanide are present. In the presence of a small proportion of potassium cyanide, reaction occurs fairly rapidly at temperatures above 70°, and the product is mainly a cyanohydrin of cyanodihydrocarvone, and this is always produced if the free hydrogen cyanide originally present was in excess, or if the reaction is allowed to continue only for such a period that the carvone is not converted as fully as possible into the ketonitrile. As the conversion of the ketonitrile into its cyanohydrin is a rapid process compared with the first reaction, it is desirable to use a large excess of potassium cyanide which is able quickly to effect a transference of a molecule of hydrogen cyanide from the cyanohydrin to unaltered carvone, thus producing two molecules of the ketonitrile. If a sufficient concentration of potassium cyanide is used, the operation may be conducted in the cold, and the following method has been found to give very good results.

Potassium cyanide (15 grams) dissolved in water (35 c.c.) is mixed with 96 per cent. alcohol (80 c.c.) containing carvone (25 grams). To the cold green liquid contained in a narrow cylinder is then added gradually glacial acetic acid (10 grams) by means of a thistle funnel having a capillary stem, the end of which reaches to the bottom of the liquid. The whole is then allowed to remain for some hours (3—12, depending on the temperature) until a few drops of the liquid, mixed with two or three times its bulk of water, deposits a solid without any appreciable quantity of admixed oil. The crystalline material may be removed by filtration and washed with water. The mother liquor again yields a considerable quantity of solid or semi-solid matter on dilution with several times its bulk of water; this material should be shaken with a little dilute alcoholic potash, the dissolved portion being afterwards reprecipitated by means of water. The whole of the solid, the yield of which often exceeds 80 per cent. of that theoretically possible, may be purified by crystallisation from dilute alcohol.

0·1574 gave 0·4322 CO_2 and 0·1175 H_2O.   C = 74·8 ; H = 8·3.
$C_{11}H_{15}ON$ requires C = 74·6 ; H = 8·0 per cent.

Cyanodihydrocarvone is nearly insoluble in light petroleum or carbon disulphide, but dissolves somewhat readily in nearly all the other organic solvents. It crystallises from hot saturated solutions in benzene, methyl alcohol, or dilute ethyl alcohol in glistening, colourless crystals which melt sharply at 93·5—94·5°.

The crystals are long, flattened needles, in which the directions of greatest length and elasticity are at right angles ; the plane of the crystals is apparently parallel to the axial plane. When melted on a glass slide beneath a cover slip, it sets rapidly on cooling to a clear mass of large, flat needles ; through some of these forms, in convergent polarised light, the bisectrix of an axial interference figure of moderate angle can sometimes be seen. The double refraction here appears to be negative.

For the determination of its optical activity, a solution containing 0·4200 gram dissolved in alcohol (25 c.c.) was examined in a 2-dcm. tube at 12°. The observed rotation was + 0·445°, whence $[a]_D + 13·5°$.

As it was anticipated that two stereoisomeric nitriles would be formed in the synthesis of the substance, the rotatory powers of all the ractions obtained in one experiment were determined. The numbers found for $[a]_D$ were 13·25°, 13·50°, 13·30°, 12·9°, indicating that only one substance in varying degrees of purity was being dealt with. It was suspected that the explanation of this depended on the presence of alkali in the solution used, resulting in the rapid interconversion of the two forms, so that only one is deposited, and to test this the effect of alkali on the rotatory power of an alcoholic solution was investigated. It was found that a solution in alcohol having a rotatory power of +0·83° did not appreciably alter in properties in the course of forty-eight hours, but, on addition of a trace of sodium ethoxide, rose in rotatory power within a few seconds to a constant value of 1·00, a change doubtless due to partial conversion into the stereo-isomeric form (compare Kipping, Trans., 1905, 87, 628). With the object of obtaining the stereoisomeride, an alcoholic solution of the nitrile was rendered alkaline, allowed to stand some hours, then very carefully neutralised, and allowed to evaporate spontaneously. Crystals of the old form separated, and the residue remained oily after nearly all the alcohol had evaporated, but no new crystalline substance was detected. On shaking the oily material with dilute alkali, it rapidly solidified, reconversion into the pure original substance having taken place.

When strongly heated, the nitrile boils, and small quantities may be distilled without much decomposition at the ordinary pressure. It behaves as an unsaturated compound, instantly decolorising a solution of potassium permanganate in acetone or of bromine in chloroform. It exhibits faintly basic properties, and dissolves in sulphuric acid or in concentrated aqueous solutions of hydrochloric or hydrobromic acid, being reprecipitated unchanged if water is added immediately; if allowed to remain dissolved in the liquids for some time, it undergoes changes which are detailed elsewhere.

When boiled for some minutes with 3 per cent. alcoholic potash, it is not appreciably affected, being recovered almost quantitatively on dilution with water. If, however, the alkali is much more highly concentrated, decomposition occurs, and potassium cyanide may be detected in a very short time, whilst the odour of carvone becomes perceptible. By using boiling alcoholic potash containing suspended ferrous hydroxide, it was found easy almost entirely to decompose the substance into hydrogen cyanide and carvone, which was isolated by steam distillation and identified by the usual tests.

Many attempts were made to reduce the substance, but in no case was any definite reduction compound obtained. Zinc dust and cold glacial acetic acid was found not to affect the substance appreciably, and the use of tin and hydrochloric acid as well as of hydriodic acid only resulted in hydrolysis. On adding zinc dust to a solution of the substance in sulphuric, hydrobromic, or hydrochloric acid, a deep eosin-like colour was developed, but only oily products and unaltered nitrile were isolated. When sodium was added to a solution of the nitrile in boiling amyl or ethyl alcohol, hydrocyanic acid was removed by the resulting sodium alcoholate, and only carvone or its reduction products could be found.

When sodium amalgam was shaken with an alcoholic solution of the nitrile, reduction occurred if the solution was kept faintly acid; the product was oily, and, on analysis, was found to contain only about one atomic proportion of hydrogen more than the original nitrile. On heating this reduction product with acids, it was slowly converted into a substance having the consistency of gelatine, which decomposed on distillation, yielding a mixture of at least three sub-stances, and its investigation was therefore abandoned.

Cyanodihydrocarvonephenylhydrazone, $CN \cdot C_{10}H_{15} : N_2H \cdot C_6H_5$, was prepared by warming a mixture of equimolecular proportions of the nitrile with phenylhydrazone at 100° during half an hour and was purified by repeated crystallisation from dilute alcohol.

0·2448 gave 0·6740 CO_2 and 0·1700 H_2O.   C = 75·1 ; H = 7·7.

$C_{17}H_{21}N_3$ requires C = 75·4 ; H = 7·9 per cent.

The hydrazone dissolves very readily in most of the ordinary organic solvents with the exception of light petroleum, and separates from dilute alcohol in flat needles, having oblique extinctions in polarised light. It melts indefinitely with slight decomposition between 113° and 124° according to the rate of heating.

On continued boiling with dilute mineral acids, the substance is hydrolysed, phenylhydrazine and cyanodihydrocarvone or its hydrolytic products being formed.

Cyanodihydrocarvonesemicarbazone, $CN \cdot C_{10}H_{15} : N_2H \cdot CO \cdot NH_2$, is prepared by the ordinary methods; it was washed repeatedly with benzene and crystallised from methyl alcohol.

0·2828 gave 0·6371 CO_2 and 0·1990 H_2O.   C = 61·4 ; H = 7·8.

$C_{12}H_{18}ON_4$ requires C = 61·5 ; H = 7·7 per cent.

The substance is sparingly soluble in benzene or chloroform, and moderately so in hot ethyl formate, ethyl acetate, or acetone. It dissolves fairly readily, however, in ethyl or methyl alcohol and very readily in glacial acetic acid. It separates from hot methyl alcohol in needles ; these, when heated, show signs of incipient fusion at 195°, darken rapidly at 217°, and fuse completely at 224°. The crystals under the microscope in polarised light present the appearance of elongated rectangles having straight extinction, and through these the obtuse bisectrix of the crystal emerges perpendicularly, the axial plane being at right angles to the direction of greatest length. The double refraction is positive in sign and moderately strong.

The semicarbazone is rapidly hydrolysed when boiled with dilute mineral acids, the original nitrile being regenerated in nearly quantitative amount.

Cyanodihydrocarvoxime, $CN \cdot C_{10}H_{15} \colon NOH$.

The nitrile (5·9 grams) dissolved in alcohol (30 c.c.) was warmed on the water-bath during one and a half hours with a solution of hydroxylamine hydrochloride (4·6 grams) and potassium carbonate (2·3 grams) in water (15 c.c.). The product, obtained by pouring the resulting solution into 100 c.c. of water, was dried, and purified by crystallisation from absolute alcohol. On analysis:

0·2117 gave 0·5321 CO_2 and 0·1629 H_2O. $C = 68·5$; $H = 8·6$.

$C_{11}H_{16}ON_2$ requires $C = 68·7$; $H = 8·3$ per cent.

The substance is very sparingly soluble in light petroleum or cold carbon tetrachloride, somewhat more readily so in benzene, and dissolves freely in ethyl or methyl alcohol, ethyl acetate, acetone, and glacial acetic acid. It separates from hot carbon tetrachloride in platelike aggregates of flat needles and from absolute alcohol in rectangular plates. It melts sharply at 169—170°. In convergent polarised light the optic axis of a biaxial interference figure may be seen emerging obliquely through the large faces of most of the crystals. The double refraction is strong.

0·3974 gram, dissolved and made up to 25 c.c. with absolute alcohol, gave a mean rotation of $+0·450°$ at 17° in a 2-dcm. tube, whence $[a]_D + 14·2°$. The oxime is somewhat readily soluble in solutions of alkaline hydroxides, but dissolves only very sparingly in dilute acids. When it is boiled with dilute mineral acids, the clear liquid suddenly becomes turbid and the crystals finally change to a limpid liquid as the oxime becomes converted into the original nitrile and hydroxylamine salt.

Experiments were made with the object of inducing the substance to undergo isomeric change, to which many oximes of cyclic ketones are prone, but with no satisfactory result. Chlorides of phosphorus attack the solid substance very violently and only amorphous or charred products result; if the oxime is first dissolved in a solvent, the action can be moderated, but on pouring the resulting liquid into water, only the original oxime was obtained. Acetic chloride or anhydride attacked the compound, but the product always gave the original oxime when heated with dilute alkalis, and no better success attended experiments in which toluene sulphonic chloride was shaken or warmed with solutions of the oxime in alkaline hydroxides or in pyridine.

The substance is completely altered if it is dissolved in 90 per cent. sulphuric acid and the solution warmed on the water-bath for some minutes; on pouring the product into water and neutralising, an amorphous, yellow material separates, and a considerable quantity of this was prepared and examined. It was partially soluble in. alkali, whilst the insoluble portion was partially soluble in dilute acids. All the fractions were amorphous or microcrystalline and, on analysis, were found to have approximately the composition of the original oxime, and may have been mixtures of stereoisomeric "*isooximes*" or anhydrides of amido-acids. Attempts to convert them into the corresponding acids by the usual methods failed, and when they were heated in closed tubes with dilute mineral acids they were either not affected or at higher temperatures decomposed into ammonia, carbon dioxide, and hydrocarbons. The oxime was not reduced when sodium amalgam was left in contact with its alcoholic solution at temperatures up to 85°, nor could its reduction be effected by the use of zinc and acetic acid, aluminium amalgam, or sodium and boiling amyl alcohol. Attempts to effect the desired change by the use of agents in the presence of mineral acids invariably resulted in hydrolysis.

The *benzoyl* derivative, $CN \cdot C_{10}H_{15} \vdots N \cdot O \cdot CO \cdot C_6H_5$, was formed when benzoyl chloride was shaken with an ice-cold solution of the oxime in sodium hydroxide and was crystallised from glacial acetic acid.

0·2223 gave 0·5930 CO_2 and 0·1323 H_2O. C = 72·7; H = 6·6.

$C_{18}H_{20}O_2N_2$ requires C = 73·0; H = 6·7 per cent.

It dissolves readily in glacial acetic acid, moderately in hot benzene or ethyl acetate, is sparingly soluble in boiling ethyl or methyl alcohol, and nearly insoluble in light petroleum. It crystallises from glacial acetic acid in glistening plates, but separates from hot benzene in slender, matted needles. Both types of crystals melt and darken at 177—178°. The crystals, in polarised light, extinguish parallel to their longest edges, and are biaxial in type. The double refraction is strong and positive in sign. When boiled with dilute acids this substance yields benzoic acid, hydroxylamine, and cyanodihydrocarvone. Heated rapidly to a high temperature with pyridine, it yields a small quantity of a volatile oil which smells of raspberries.

Action of Bromine on Cyanodihydrocarvone. Formation of Stereo-isomeric Cyanodihydrocarvonedibromides,

$$CH_3 \cdot CH \begin{matrix} CO\text{---------}CH_2 \\ CH(CN) \cdot CH_2 \end{matrix} CH \cdot CBr \begin{matrix} CH_3 \\ CH_2Br \end{matrix}.$$

When bromine is added to a solution of cyanodihydrocarvone in carbon tetrachloride or carbon disulphide, instantaneous decolorisation occurs, but at the same time large quantities of hydrogen bromide are evolved. As the presence of free hydrogen bromide is conducive to substitution in ketones, the addition of bromine was carried out in glacial acetic acid containing excess of dissolved sodium acetate. Precisely one molecular proportion of bromine was absorbed before the colour of the halogen became permanent. On pouring the product into water, an oil separated which, after washing with dilute sodium carbonate solution and trituration with alcohol, became crystalline. The solid proved to be a mixture, and was therefore subjected to a process of fractional crystallisation, first from alcohol and then from ethyl acetate, and was finally resolved into two parts.

a-Cyanodihydrocarvonedibromide is the one formed in largest amount and is easily obtained quite pure.

0·2909 gave 0·3228 AgBr. Br = 47·2.

$C_{11}H_{15}ONBr_2$ requires Br = 47·5 per cent.

It is sparingly soluble in light petroleum and carbon disulphide, but dissolves somewhat freely in alcohol, chloroform, or benzene. It separates from warm alcohol in slender, colourless needles and melts at 146—147°.

The saturated character of the substance is shown by its behaviour towards bromine and potassium permanganate, by neither of which it is attacked in the cold.

0·3970 gram dissolved and made up to 25 c.c. with alcohol was examined in a 2-dcm. tube at 18°. A mean rotation of + 0·920° was observed, whence $[a]_D$ + 29·0°.

The compound is somewhat soluble in cold sulphuric acid, and, if the liquid is at once diluted with water, is recovered unchanged; if warmed with sulphuric acid, it loses hydrogen bromide, the solution turns red, and ammonium sulphate is produced.

Its behaviour towards alkalis is interesting; when it is covered with a strong solution of sodium hydroxide, it rapidly dissolves with production of heat. On diluting the liquid it remains clear, but if it is acidified, a colourless oil, readily soluble in alkalis, is precipitated; the formation of this acid, which has not yet been closely investigated, is attended by the elimination of hydrogen bromide. If the original solution of the bromo-compound in alkalis is heated, an odour resembling that of carvone becomes noticeable and large quantities of sodium cyanide as well as potassium bromide are afterwards found in the liquid.

β-Cyanodihydrocarvonedibromide is found in the mother liquors from the preceding substance, and when entirely freed from the isomeride forms crystals of considerable size.

0·2179 gave 0·2447 AgBr. Br = 47·8.

$C_{11}H_{15}ONBr_2$ requires Br = 47·5 per cent.

This substance is more readily soluble in most of the usual organic solvents than is the α-compound. It separates from alcohol on spontaneous evaporation in large, colourless, transparent prisms, which belong apparently to the monoclinic system. It melts at 91—92°.

0·3965 gram dissolved and made up to 25 c.c. with absolute alcohol was examined in a 2-dcm. tube at 18°. The mean observed rotation was +0·190°, whence $[\alpha]_D$ +6·0°.

In chemical properties, the substance is indistinguishable from the preceding substance

Action of Phosphorus Pentachloride on Cyanodihydrocarvone. Formation of Chlorocyanodipentene, $CH_3 \cdot C \underset{CH(CN) \cdot CH_2}{\overset{C(Cl)\text{----}CH_2}{<}} > CH \cdot C \underset{CH_3}{\overset{CH_2}{<}}$.

The ketonitrile is slowly attacked by phosphorus pentachloride in the cold, and the resulting liquid slowly loses hydrogen chloride, yielding a mixed product. No attempt, therefore, was made to isolate the initial compound, and equimolecular proportions of the substances were allowed to interact in dry ether on the water-bath for some hours, after which the whole was cooled, shaken repeatedly first with water, and then with dilute sodium hydroxide. The ether was then removed and the residue heated at 200° with excess of quinoline for fifteen minutes, extracted with dilute hydrochloric acid, dried, and fractionally distilled in a vacuum. A large fraction was obtained which distilled under the ordinary atmospheric pressure at 268—270° with slight decomposition.

0·2358 gave 0·1709 AgCl. Cl = 18·0.
0·2026 ,, 0·5039 CO_2 and 0·1336 H_2O. C = 67·8 ; H = 7·3.
$C_{11}H_{14}NCl$ requires Cl = 18·2 ; C = 67·5 ; H = 7·2 per cent.

The chloronitrile was a colourless, nearly odourless, slightly oily liquid, which had all the characters of an unsaturated nitrile. Attempts to obtain from it the corresponding carboxylic acid have not yet been successful, as chlorine appears to be removed very readily.

To obtain trustworthy evidence as to its constitution, a few grams of the compound were warmed in alcoholic solution with an equimolecular proportion of silver nitrate for an hour. Silver chloride was rapidly precipitated and an odour of carvone was noticeable ; on filtering the liquid and diluting with water, an oil was obtained which, after trituration with a little alkali, deposited crystals which were easily identified as cyanodihydrocarvone, evidently obtained by ketonisation of the hydroxycyanodipentene obtained as initial product in the hydrolysis.

Oxidation of Cyanodihydrocarvone.

Cyanodihydrocarvone is rapidly oxidised when shaken with an ice-cold solution of potassium permanganate, and in the attempt to obtain the products of this reaction 100 grams of the nitrile were shaken with exactly so much of this oxidising agent that on passing sulphur dioxide into a small portion of the liquid a clear solution free from undissolved nitrile was obtained. When the bulk of the filtered solution was reduced by evaporation to about half a litre, it was found that an increasing amount of potassium cyanide was being formed, and it was therefore cooled and divided into two portions. One part was saturated with hydrogen chloride in order to convert any nitriles present into carboxylic acids, but the solution became black and finally deposited a large amount of carbonaceous material and did not after-wards yield any appreciable quantity of substances soluble in organic solvents. The other part of the oxidation product was acidified with sulphuric acid and extracted repeatedly with ether, from which, on evaporation, about 45 grams of a clear oil were obtained; this did not deposit any crystalline material after some months, and could not be distilled in a vacuum. As portions of it did not yield any better results after acetylation or benzoylation, the remainder was oxidised with potassium dichromate and sulphuric acid at about 80°. The oxidation took place with considerable speed and large quantities of carbon dioxide were evolved. When the action finally ceased, the excess of potassium dichromate was reduced by means of sulphur dioxide and the whole extracted repeatedly, first with ether and then with ethyl acetate. The amount of oxidation product thus obtained weighed only about 12 grams, and doubtless the reaction had resulted largely in the destruction of the organic material, for a large amount of oxalic acid was afterwards found in the residue.

The extracted material was warmed for some days with strong hydrochloric acid and again recovered by evaporating the liquid and separating the acids from ammonium chloride by means of ethyl acetate. The material thus obtained yielded several grams of crystalline material at the end of some months in a bell-jar over water; this was separated, drained on porous earthenware, and recrystallised from a mixture of ethyl acetate and light petroleum.

0·1886 gave 0·2816 CO_2 and 0·0785 H_2O. C = 40·7; H = 4·6.
$C_6H_8O_6$ requires C = 40·9; H = 4·6 per cent.

The equivalent was determined by titration with alkali, using phenol-phthalein as indicator. 0·2012 required 348 c.c. $N/10$ for neutralisation, hence the equivalent of the acid was 58·1, the number calculated for a tribasic acid of the formula $C_6H_8O_6$ being 58·3.

As the acid did not lose carbon dioxide when strongly heated, it was not a malonic acid derivative and could only be tricarballylic acid, a conclusion which further examination of its properties confirmed.   It melted at 166°.

Hydrolysis of Cyanodihydrocarvone.   Dihydrocarvonecarboxylic

$$\textit{Amide, } CH_3 \cdot CH \overset{CO\text{---------}CH_2}{\underset{CH(CO \cdot NH_2) \cdot CH_2}{\diagdown}} CH \cdot C \overset{CH_2}{\underset{CH_3}{\diagdown}} .$$

Cyanodihydrocarvone (10 grams) is shaken in a stoppered bottle with 15 c.c. of an aqueous solution of hydrogen bromide (saturated at 0°) until dissolved and the solution is allowed to remain at the ordinary temperature until a drop, added to the water, gives no immediate precipitate, the time required for this varying from 12—48 hours, according to the temperature.   The whole is then poured into 100 c.c. of water and the liquid neutralised by addition of solid sodium carbonate, the solid which separates being collected on a filter, washed, and purified by crystallisation from glacial acetic acid.

0·1871 gave 0·4635 CO_2 and 0·1463 H_2O. $C = 67·5$; $H = 8·7$.

$C_{11}H_{17}O_2N$ requires $C = 67·7$; $H = 8·7$ per cent.

The compound thus has the composition of an amide corresponding with the original nitrile.   It is nearly insoluble in light petroleum, carbon tetrachloride, chloroform, and benzene, is rather sparingly soluble in ethyl acetate and hot water, more readily so in acetone and alcohol, and dissolves somewhat freely in acetic acid.   It separates from ethyl acetate, in compact prisms and from alcohol in large plates melting with slight decomposition at 228—230°.

The crystals are well-formed rectangular plates, having straight extinction in polarised light.   When fused between glass slips, the compound solidifies to a semi-transparent mass consisting of groups of parallel flat needles.   No optical figures were observed in convergent polarised light.

When the amide is heated with alcoholic potassium hydroxide it evolves large quantities of ammonia.   If at the end of a few minutes the liquid is poured into water and the resulting liquid acidified, a crystalline precipitate consisting partially of an acid and partially of unchanged amide is obtained; the acid may be separated from the neutral material by dissolving it in cold dilute sodium carbonate, and it is at once obtained in a solid form and thus, when once crystallised from ethyl acetate, melts at about 141° and has the properties of the α-dihydrocarvone carboxylic acid described later.   If, however, the action of the alcoholic potassium hydroxide be allowed to continue until ammonia ceases to be evolved, the product is a mixture of two isomeric acids. These observations indicate with some degree of certainty that the amide is that corresponding with the α-carboxylic acid.

It is interesting that this compound does not exhibit any appreciable tendency to unite with hydrogen cyanide under the conditions which have been successful with other ketonic compounds.

Isomeric Dihydrocarvonecarboxylic Acids,

$$CH_3 \cdot CH \begin{matrix} CO \\ CH(CO_2H) \cdot CH_2 \end{matrix} \begin{matrix} CH_2 \\ CH_2 \end{matrix} C_H \cdot C \begin{matrix} CH_2 \\ CH_3 \end{matrix}.$$

In order to obtain the carboxylic acids directly it is convenient to operate in the following manner. The nitrile is allowed to remain for several days at the ordinary temperature with about eight times its weight of hydrochloric acid (saturated at 5°), after which the whole is warmed over the water-bath for two or three days until a small quantity of the liquid, poured into excess of dilute alkali, gives only a very light permanent precipitate. The solution is now evaporated nearly to dryness, cooled, diluted, and shaken repeatedly with chloroform. The united chloroform extracts are washed with a little water, filtered, and the acid extracted from it with sodium carbonate solution. The latter is freed from traces of neutral material, soluble in chloroform, by extracting it once or twice with that solvent, and the acids may then be isolated by adding excess of hydrochloric acid and extracting with ether. The dried ethereal extract on evaporation yields an oily residue which becomes nearly solid in a few days; the mass should be ground with a little carbon disulphide and spread on porous porcelain to drain. The product thus obtained is a mixture of two stereoisomeric acids which must be separated by a long process of fractional crystallisation : first from carbon tetrachloride and afterwards from dry ethyl acetate. No salts were discovered which could be used for separating the two acids. The fractionation is rendered more tedious owing to the tardiness of crystallisation exhibited by the mixed material. The nature of the different fractions which separate during the process may roughly be gauged by means of a polarising microscope, as one acid is uniaxial and the other biaxial and the optical figures are seen without difficulty : fractions which consist largely of one form may then be mixed and recrystallised. It is useless to rely on melting-point determinations, and only when nearly pure do the two acids exhibit well-defined crystalline forms to the naked eye.

α-*Dihydrocarvonecarboxylic acid*, $C_{10}H_{15}O \cdot CO_2H$, is the acid which largely predominates when the hydrolysis of the nitrile has been incomplete. It is best purified finally by crystallisation from hot dry ethyl formate.

0·1618 gave 0·3987 CO_2 and 0·1174 H_2O. $C = 67·2$; $H = 8·1$.
$C_{11}H_{16}O_3$ requires $C = 67·4$; $H = 8·2$ per cent.

0·2335 gram required 11·9 c.c. of $N/10$ NaOH for neutralisation, with phenolphthalein as indicator, giving the equivalent for the acid 196, whilst the number calculated for a monobasic acid, $C_{11}H_{16}O_3$, is 196.

The acid dissolves to an appreciable extent in hot water and crystallises from it on cooling. It is also distinctly soluble in hot carbon disulphide and light petroleum, readily forming supersaturated solution in these liquids, and dissolves somewhat freely in most of the other common organic solvents. It separates from solution in hot ethyl acetate in flat needles and from a mixture of carbon tetra-chloride and light petroleum in plates ; both types of crystals melt at 97—98°.

The plates are thin, six-sided, apparently tetragonal forms, through which, in convergent polarised light, the figure characteristic of a uniaxial crystal with the axis emergent perpendicularly to the field may always be seen. The double refraction is strong and positive in sign.

The substance solidifies somewhat rapidly after fusion between glass slips to small patches of plates crystallographically identical with the foregoing form.

0·2435 gram dissolved and made up to 14·95 c.c. with ethyl acetate was examined in a 2-dcm. tube at 16·5°. The mean observed rotation was $+2·46°$, whence $[\alpha]_D$ $+49·9°$.

The acid has the properties of an unsaturated ketonic acid, reacts readily with phenyl hydrazine, yielding an oil, and gives a crystal-line semicarbazide and a crystalline oxime. It instantaneously decolorises a solution of biomine in chloroform, and its solution in ice-cold aqueous sodium carbonate at once reduces potassium permanganate. It gives a crystalline addition-compound when hydro-gen bromide is led into its solution in chloroform : this substance is unstable, however, and is rapidly decomposed by water, breaking up into its constituents. It dissolves in strong sulphuric acid, yielding a colourless solution which slowly develops a greenish-yellow fluorescence, especially when warmed.

The salts of the acid are mostly difficult to obtain in crystalline form. The potassium, sodium, ammonium, barium, calcium, zinc, and magnesium salts are very readily soluble in water. The *mercurous* salt and the *ferric* salts are sparingly soluble, the latter being buff-coloured. The *lead* salt is also sparingly soluble, and forms small, brilliant needles.

The *semicarbazone*, $CO_2H \cdot C_{10}H_{15} : N_2H \cdot CO \cdot NH_2$, of a-dihydrocarvonecarboxylic acid is purified by crystallisation from a large bulk of absolute alcohol.

0·2054 gave 0·4279 CO_2 and 0·1428 H_2O. $C = 56·7$; $H = 7·7$.
$C_{12}H_{19}O_3N_3$ requires $C = 56·9$; $H = 7·5$ per cent.

It is nearly insoluble in light petroleum, benzene, or chloroform, and is very sparingly soluble by hot water ; it is somewhat sparingly soluble also in ethyl acetate or acetone, but dissolves rather more readily in hot alcohol. It separates from hot alcohol in very small, slender needles which, when heated slowly, turn brown at about 210° and melt and decompose at 218—221°.

When boiled with dilute mineral acids, the compound dissolves rapidly, and is simultaneously decomposed, yielding pure α-dihydrocarvonecarboxylic acid melting at 97—98°.

The *oxime*, $CO_2H·C_{10}H_{15}:NOH$, is usually obtained as an oil which becomes crystalline when triturated with a little dilute acetic acid. It is best recrystallised from hot dilute alcohol.

0·1711 gave 0·3913 CO_2 and 0·1266 H_2O. $C = 62·4$; $H = 8·2$.
$C_{11}H_{17}O_3N$ requires $C = 62·5$; $H = 8·1$ per cent.

The oxime is very sparingly soluble in water, light petroleum, benzene, chloroform, or ether, but dissolves freely in ethyl formate cr acetate, acetone, or alcohol. It crystallises slowly from dilute acetic acid in small octahedra and from hot dilute alcohol in slender needles. When heated slowly it shows signs of incipient fusion at 180°, turning slightly brown, and finally melts and evolves gas at 193—194°.

In the form of its potassium salt, and in aqueous solution which is faintly alkaline to phenolphthalein, α-dihydrocarvonecarboxylic acid combines with hydrogen cyanide at the ordinary temperature. (The method adopted was similar to that described later in the case of the isomeric acid.) The addition product, however, was not, in this instance, obtained in a solid form, but as a colourless oil, which readily lost hydrogen cyanide when heated, and was slowly hydrolysed by fuming hydrochloric acid.

β-*Dihydrocarvonecarboxylic acid*, $C_{10}H_{15}O·CO_2H$, appears to form the main bulk of the mixture obtained when the hydrolysis of the original nitrile has been pushed to completion. When freed as far as possible from the isomeride by the method above suggested, it may finally be purified by crystallisation from hot ethyl acetate or benzene.

0·1530 gave 0·3785 CO_2 and 0·1109 H_2O. $C = 67·5$; $H = 8·1$.
$C_{11}H_{16}O_3$ requires $C = 67·4$; $H = 8·2$ per cent.

0·1662 gram required 8·5 c.c. of $N/10$ NaOH for complete neutralisation, phenolphthalein being used as indicator, whence the equivalent of the acid was 195, the number calculated for a monobasic acid of the above formula being 196.

The acid is sparingly soluble in light petroleum, carbon disulphide, or cold water, but is very appreciably soluble in hot water. It dissolves somewhat readily in ethyl bromide and freely in benzene, chloroform, ethyl acetate, acetone, or alcohol. It separates from cold ethyl acetate or formate on spontaneous evaporation in compact prisms, but appears in plates when crystallised from hot solvents. It melts at 142—143°.

The crystals from hot solvents are elongated, six-sided, orthorhombic plates, the axial angles being about 132°. Occasionally plates in the form of triangles broadly truncated at one apex are formed. The axial plane in the former is perpendicular at right angles to the direction of greatest length in these crystals, and the acute bisectrix always emerges perpendicularly through the large faces. The axial angle is moderately large and the dispersion small. The double refraction is positive in sign.

When fused between glass slips, the compound solidifies slowly in irregular transparent plates identical in character with those deposited from solvents, and disposed so that the acute bisectrices are at right angles to the field.

0·2421 gram dissolved and made up to 14·95 c.c. with ethyl acetate was examined in a 2-dcm. tube at 35°. The mean rotation observed was +0·936, whence $[a]_D + 28·8°$.

Like its isomeride, this compound exhibits all the characters of an unsaturated ketonic acid, yielding a crystalline semicarbazone ; the phenylhydrazone and the oxime, however, have not been obtained except as oils. Its behaviour towards bromine, potassium permanganate, and hydrogen bromide closely resembles that of the a-acid.

The salts of this acid, like those of the isomeric one, are usually difficult to obtain in crystalline form. The *mercurous* and *ferric* salts are sparingly soluble in water, and the *lead* salt, which forms fine needles, is somewhat more sparingly soluble in water than that from the a-acid.

The *semicarbazone*, $CO_2H \cdot C_{10}H_{15} : N_2H \cdot CO \cdot NH_2$, is best purified by crystallisation from alcohol

0·1892 gave 0·3929 CO_2 and 0·1287 H_2O. $C = 56·7$; $H = 7·5$.

$C_{12}H_{19}O_3N_3$ requires $C = 56·9$; $H = 7·5$ per cent.

In general characters and its solubility in organic solvents, the compound is scarcely to be distinguished qualitatively from the corresponding derivative of the isomeric acid. It separates from alcohol in small needles which melt and decompose at 235—236°. When boiled with dilute mineral acids, it is rapidly hydrolysed, yielding the β-dihydrocarvonecarboxylic acid melting at 142—143°.

Oxidation of α- and β-Dihydrocarvonecarboxylic Acids.

Both acids are rapidly attacked in hot alkaline solution by potassium ferricyanide and by mercuric oxide. In the latter instance, a pungent volatile oil is formed in small quantity; this has much the odour of formaldehyde, the presence of which, however, could not be detected. When dissolved in ice-cold sodium carbonate and mixed with a one per cent. solution of potassium permanganate at 0°, the colour of the latter disappears instantaneously. The product obtained in both cases was a liquid acid, but on prolonged heating on the water-bath deposited a small quantity of a solid substance, which was collected and purified by crystallisation from alcohol.

0·1972 gave 0·4513 CO_2 and 0·1304 H_2O.   C = 62·4 ; H = 7·3.

$C_{11}H_{16}O_4$ requires C = 62·3 ; H = 7·5 per cent.

The compound dissolved readily in alcohol, chloroform, or benzene, but was only sparingly soluble in water or in light petroleum. It separated from dilute alcohol in rectangular tables of considerable size and melted at 149—151°.

The substance dissolved slowly in a warm aqueous solution of potassium hydroxide and was reprecipitated unchanged on addition of mineral acids. The solution in alkalis did not at once discharge the colour of potassium permanganate. Unfortunately the small quantity of the compound obtained precluded a fuller examination of its nature.

The oxidation product from which the foregoing compound had been removed was very readily soluble in water and gave no crystalline material after fourteen months. When warmed with lead peroxide and acetic acid it was rapidly oxidised, carbon dioxide being copiously evolved, indicating that it consisted mainly of an α-hydroxy-carboxylic acid, and it is noteworthy that precisely the same results were obtained with both the isomeric dihydrocarvonecarboxylic acids. In both cases, also, the material, after being treated in this way and subsequently freed from lead, was much less readily soluble in water, and, moreover, behaved towards hypobromite as a substance containing the group $CO·CH_3$, for in alkaline solution it was rapidly attacked by this agent, yielding carbon tetrabromide and bromoform. Further oxidation of the resulting acids by means of nitric acid gave nothing definite except oxalic acid.

The *cyanohydrin*, $CO_2H·C_{10}H_5(OH)CN$, of β-dihydrocarvone-carboxylic acid is prepared without difficulty by the following method. The ketonic acid (1 mol.) is exactly neutralised by means of a 10 per cent. solution of sodium hydroxide, and to the resulting liquid is added pure potassium cyanide (1¼ mols.) dissolved in three times its weight of water. The resulting liquid is transferred to a flask which is capable of containing about twice the bulk of liquid used, and is provided with a tightly-fitting india-rubber stopper through which passes the tube of a dropping-funnel with a fine orifice. Into the dropping-funnel is put five per cent. dilute sulphuric acid (1 mol.) and the stop-cock is turned until it delivers the acid slowly, drop by drop. A few drops of the acid are allowed to run into the liquid in the flask, the stopper being withdrawn from the neck, and when a considerable quantity of air has been driven out by the hydrogen cyanide disengaged at the surface of liquid, the stopper is reintroduced so as to render the whole air-tight. On now allowing the apparatus to remain without shaking, the hydrogen cyanide slowly dissolves in the liquid and the acid consequently flows from the funnel at a corresponding rate, and requires no further attention. At the end of twenty-four hours, the action is complete and the cyanohydrin may be rendered stable by the further addition of enough mineral acid to make the solution distinctly acid to methyl-orange.* To isolate the product in this instance, the whole was poured into excess of dilute hydrochloric acid, the precipitated oil being extracted with ether, the residue from which rapidly solidified on trituration with a little strong hydrochloric acid and was crystallised from a mixture of ether and light petroleum.

0.1718 gave 0.4057 CO_2 and 0.1128 H_2O. $C = 64.4$; $H = 7.3$.

$C_{12}H_{17}O_3N$ requires $C = 64.6$; $H = 7.6$ per cent.

0.5872 gram of the compound was dissolved in excess of dilute sodium hydroxide and titrated with $N/10$ $AgNO_3$. The solution remained clear until 13.0 c.c. had been added, whence $CN = 11.5$ per cent. [calculated $= 11.7$].

The compound is sparingly soluble in water, benzene, or light petroleum, but dissolves readily in ether, ethyl acetate, and alcohol. It separates from a mixture of ether and light petroleum in white needles with low double refraction. The crystals are soft and aggregate on pressure to a waxy mass. It melts at $188—190°$.

Like other cyanohydrins it very easily loses hydrogen cyanide in the absence of free mineral acids, yielding the original ketonic acid. When heated with strong hydrochloric acid for some days on the water-bath, it is converted mainly into a syrupy acid, which has not yet been obtained in a crystalline form. A small quantity of a

* The above method is the only one which the author is able to recommend for preparing the cyanohydrins of ketonic acids, and is applicable to other compounds which react somewhat slowly with hydrogen cyanide. It allows of the addition of the acid, mechanically, at nearly the same speed as the addition process takes place, and obviates all chance of losing hydrogen cyanide, whilst the solution remains alkaline until the end, which is essential.

sparingly soluble acid melting at 280°, and apparently identical with one which will be described later, is sometimes produced as a by-product during the hydrolysis.

A large number of unsuccessful attempts were made to reduce the two isomeric dihydrocarvonecarboxylic acids to the corresponding hydroxy-acids, but in all cases without success. The ketonic acids were not affected by zinc and hot or cold glacial acetic acid, by sodium and boiling ethyl or amyl alcohol, by sodium and moist ether, or by zinc and hot sodium hydroxide solution. When left in aqueous solution with excess of sodium amalgam, they were both converted into compounds which were amorphous or indistinctly microcrystalline, very sparingly soluble in water, and decomposed by heat; these products were probably of high molecular weight and not closely investigated.

Intraconversion of the Isomeric Dihydrocarvonecarboxylic Acids.

It was at one time thought possible that the substance referred to as β-dihydrocarvonecarboxylic acid had been produced from the α-acid by the migration of the double linking, and, to test this, experiments on the intraconversion of the two-acids were undertaken.

Pure α-dihydrocarvonecarboxylic acid, melting at 97—98°, was dissolved in two or three times its weight of strong sulphuric acid and the colourless solution heated at 80°, whence in a short time a marked green fluorescence became perceptible. At the end of two hours, the liquid was diluted with water, shaken repeatedly with ether, and the extract examined. It remained oily for some days, but when triturated with a little carbon tetrachloride deposited crystals, among which biaxial ones were easily detected. The mixed solid was therefore drained and fractionally crystallised, and a considerable quantity of pure β-dihydrocarvonecarboxylic acid, melting at 140—141°, was finally isolated.

An attempt to effect a similar partial conversion of the β-acid into the α-compound by means of sulphuric acid was not altogether successful, although a few crystals of a uniaxial type were detected in the product. Fifty grams of purified β-acid were therefore heated on the water-bath with 30 per cent. aqueous solution of sodium hydroxide for twelve hours, the resulting liquid cooled, acidified, and extracted with ether. The residue obtained on the evaporation of the latter was very oily, and was therefore spread on filter paper and extracted in a large Soxhlet apparatus with carbon disulphide for two hours. The liquid was then evaporated until syrupy, cooled in a freezing mixture, and nucleated with a trace of the purified α-acid. A deposition of crystals took place, which were separated and recrystallised twice from ethyl formate, when nearly three grams of uniaxial crystals, melting at 95—96° and having all the properties of α-dihydrocarvone-carboxylic acid, were obtained. It may be safely concluded, therefore, that the two isomeric acids are mutually incontravertible.

The examination of cyanodihydrocarvonecyanohydrin and other derivatives of cyanodihydrocarvone has yielded results of some interest, and an account of the work will be communicated to the Society in due course.

The author's thanks are due to the Research Fund Committee of the Chemical Society for a grant which defrayed much of the cost of the investigation.

CHEMICAL DEPARTMENT,
GOLDSMITHS' COLLEGE,
NEW CROSS, S.E.

CII.—*The Relation between Absorption Spectra and Chemical Constitution. Part V. The isoNitroso-compounds.*

By EDWARD CHARLES CYRIL BALY, EFFIE GWENDOLINE MARSDEN, and ALFRED WALTER STEWART (Carnegie Research Fellow).

IT has been shown in previous papers that the yellow colour of the α-diketones and the quinones is due to the presence of a band in their absorption spectra, this band being situated in the visible blue region. The origin of this absorption band was shown to be due to the process of isorropesis, that is, an oscillation between the residual affinities of the carbonyl oxygen atoms. Furthermore, it was found to be necessary, that some disturbing influence should be present to start the isorrope-sis; in the α-diketones this disturbing influence is supplied by the potential tautomerism of the hydrogen atoms of the alkyl radicles adjacent to the carbonyl groups, whilst in the quinones it is sup-plied both by the hydrogen atoms and by the motions of the benzene ring.

In the present paper we deal with the constitution of the *iso*nitroso-compounds, which are peculiarly interesting in view of the theory that colour is due to isorropesis, as they are colourless in neutral solution and yellow in presence of alkali. It would appear that in the grouping usually attributed to these bodies,

$$R_1 \cdot \underset{O}{\overset{|}{C}} \cdot \underset{NOH}{\overset{|}{C}} \cdot R_2,$$

there is every condition for isorropesis between the residual affinities of the nitrogen and the ketonic oxygen atoms, for the necessary disturbing influence would be provided by the hydrogen atoms of the alkyl radicle R_1, exactly as in the case of ethyl pyruvate, which was dealt with in Part I of this series of papers. That isorropesis is possible between ketonic oxygen and unsaturated nitrogen atoms is

FIG. 1.

Oscillation frequencies.

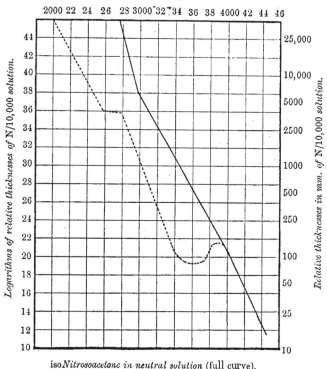

isoNitrosoacetone *in neutral solution* (full curve).
isoNitrosoacetone *in alkaline solution* (dotted curve).

proved by the yellow colour of quinone monoxime. We have examined the absorption spectra of the following compounds: *iso*nitrosoacetone, *iso*nitrosomethylacetone, ethyl *iso*nitrosoacetoacetate, *iso*nitrosoacetylacetone, *iso*nitrosoacetic acid, and ethyl *iso*nitrosomalonate, and their absorption in neutral and alkaline solutions is shown by the full and dotted curves respectively in Figs. 1—6. From these diagrams it will be seen that in every case there is no trace of any absorption band when the compounds are in neutral alcoholic solution, but that in presence of sodium ethoxide or hydroxide two bands are shown: one of these is the usual band in the ultra-violet due to enol-keto-dynamic isomerism, while the other in the visible region is due to isorropesis. From analogy with ethyl pyruvate and quinone monoxime it must be concluded that the *iso*nitroso-compounds do not possess, in

Fig. 2.

Oscillation frequencies.

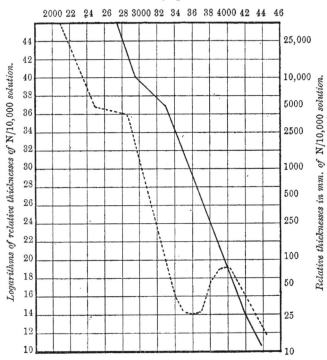

isoNitrosomethylacetone in neutral solution (full curve).
isoNitrosomethylacetone in alkaline solution (dotted curve).

neutral solution, the foregoing formula, since with this formula it is very difficult to conceive how isorropesis does not take place.

When these compounds are dissolved in sodium hydroxide, therefore, we must conclude that the hydrogen atom of the oximino-group is replaced by sodium and becomes labile. Since all these compounds in the presence of alkali exhibit the same absorption band in the ultra-violet, it is evident that the same type of isodynamic isomerism is common to them all, and, further, that the most probable type is that shown in the following example.

In the case of ethyl *iso*nitrosomalonate, it is evident that the only possibility of isodynamic isomerism is represented by the reversible equation :

$$\underset{\substack{|\\ \text{NONa}\\ a.}}{\text{EtO}_2\text{C} \cdot \text{C} \cdot \text{CO}_2\text{Et}} \quad \underset{\longleftarrow}{\longrightarrow} \quad \underset{\substack{|\\ \text{NO}\\ b.}}{\text{EtO}_2\text{C} \cdot \text{C Na} \cdot \text{CO}_2\text{Et}} \; .$$

Fig. 3.

Oscillation frequencies.

2000 22 24 26 28 3000 32 34 36 38 4000 42 44 46

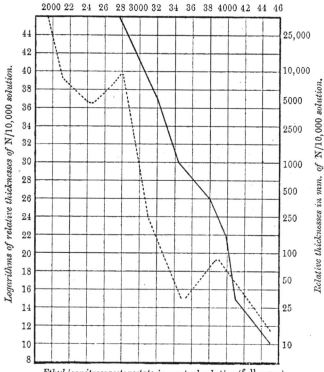

Ethyl isonitrosoacetoacetate *in neutral solution* (full curve).
Ethyl isonitrosoacetoacetate *in alkaline solution* (dotted curve).

The absorption band exhibited by these compounds in the presence of alkali hydroxide is the same as that given by ethyl acetoacetate and analogous compounds under the same conditions, and this has been shown to be due to the isodynamic isomerism expressed by the reversible equation :

$$\underset{\substack{|\\ \text{ONa}}}{\text{CH}_3 \cdot \text{C} \colon \text{CH}_2 \cdot \text{CO}_2\text{Et}} \quad \underset{\longleftarrow}{\longrightarrow} \quad \underset{\substack{||\\ \text{O}}}{\text{CH}_3 \cdot \text{C} \cdot \text{CHNa} \cdot \text{CO}_2\text{Et}} \; .$$

This reversible process is the origin of the absorption band in the ultra-violet. It has already been shown by Hartley (Trans., 1905, **87**, 1796) that such a process, involving change of linking round a nitrogen atom, gives rise to the ultra-violet absorption band, and indeed this direction is the only one possible in the case of ethyl *iso*nitroso-malonate. During the existence of this isodynamic isomerism, a

FIG. 4.

Oscillation frequencies.

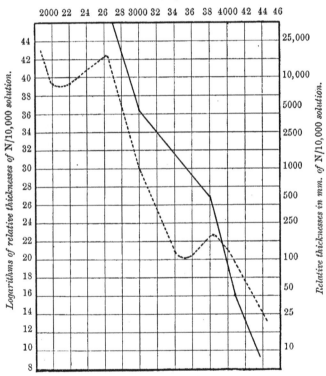

iso*Nitrosoacetylacetone in neutral solution* (full curve).
iso*Nitrosoacetylacetone in alkaline solution* (dotted curve).

certain portion of the compound exists in the form *a*, and when in this form isorropesis is started between the nitrogen and the ketonic oxygen atoms and the substance becomes yellow. Exactly the same arguments are applicable to the other compounds mentioned above, for the absorption curves are strikingly similar.

It will be noticed that in the cases of the *iso*nitroso-compounds of acetone and methylacetone the absorption band characteristic of the isorropesis is small. The persistence of this band is undoubtedly a measure of the extent to which this process occurs; as the extent shown by these two compounds is very small we may conclude that form a is statically existent in relatively small quantities.

It is very interesting to compare the position and character of the

FIG. 5.

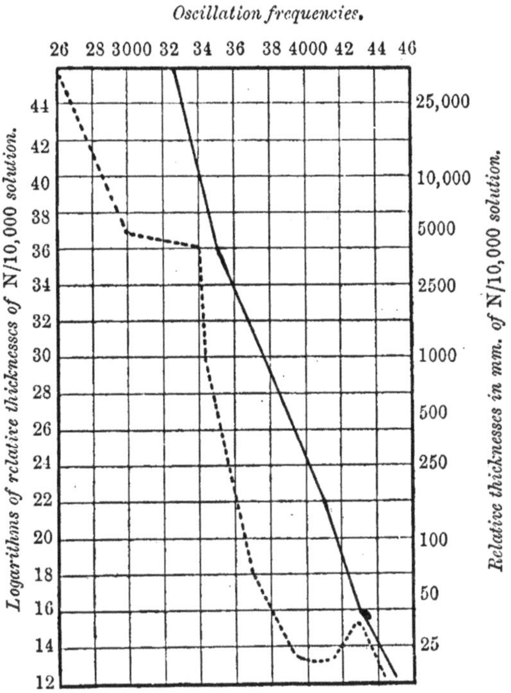

Oscillation frequencies.

26 28 3000 32 34 36 38 4000 42 44 46

iso*Nitrosoacetic acid in neutral solution* (full curve).
iso*Nitrosoacetic acid in alkaline solution* (dotted curve).

absorption bands due to isorropesis in the six compounds described. In the acetone and methyl acetone derivatives, the bands are small and the mean oscillation frequency is about 2650; in the case of *iso*nitroso-acetic acid, the mean frequency of the band is 3200. These compounds are very analogous, therefore, to diacetyl and ethyl pyruvate, for in the former case we have the isorropesis band at 2400 and in the latter at 3100. In *iso*nitrosoacetone and diacetyl, the isorropesis takes place with a true ketonic oxygen, but in ethyl pyruvate and *iso*nitrosoacetic

3 s

acid the isorropesis takes place with the ketonic oxygen of a carbonyl group.

· Again, in *iso*nitrosoacetylacetone the isorropesis band is much more strongly developed and is nearer to the red ; this, and also the increased breadth of the band, is doubtless due to this compound being a derivative of a triketone, there being in this case greater possibilities

FIG. 6.

Oscillation frequencies.

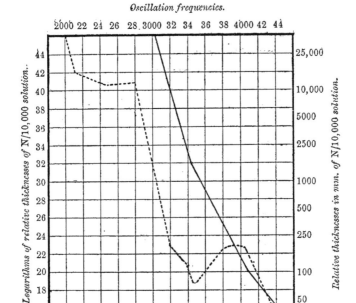

Ethyl isonitrosomalonate *in neutral solution* (full curve).
Ethyl isonitrosomalonate *in alkaline solution* (dotted curve).

for an increase of isorropesis both in character and quantity. Further, the shift of the band towards the red is to be noticed as the carboxyl carbonyl oxygen of the ethyl acetoacetate compound ·is replaced by the true ketonic oxygen of the acetylacetone derivative.

· We have also examined the spectrum of ordinary *iso*nitrosocamphor, and the absorption of this substance in neutral and alkaline solution is shown by the full and dotted curves respectively in Fig. 7. The former shows that in neutral solution *iso*nitrosocamphor differs entirely from the purely fatty compounds previously described, for considerable isorropesis is taking place. Now, on the addition of alkali, a shallow absorption band is developed; this band is very much broader than those exhibited by the derivatives of mono-ketones, and, moreover, the absorption band in the ultra-violet is absent. It is evident from these curves that isorropesis of somewhat different character must be taking place in ordinary *iso*nitrosocamphor in neutral and alkaline solution; it may be assumed at once that the process taking place in ordinary *iso*nitrosocamphor is not exactly the same as in the fatty compounds dealt with above. We may say, therefore, that the isodynamic isomerism of ordinary *iso*nitrosocamphor cannot entirely be represented by the reversible equation :

$$C_8H_{14}\!\!<\!\!{}^{CO}_{C:NOH} \quad \rightleftarrows \quad C_8H_{14}\!\!<\!\!{}^{CO}_{CH\cdot NO}$$
$$a. \qquad\qquad b.$$

It is, however, quite possible that a second type of isodynamic isomerism exists in this substance due to ring formation :

$$C_8H_{14}\!\!<\!\!{}^{CO}_{CH\cdot NO} \quad \rightleftarrows \quad C_8H_{14}\!\!<\!\!{}^{C\cdot OH}_{C\cdot NO}$$
$$b. \qquad\qquad c.$$

There is, indeed, every justification for our considering this second type of isodynamic isomerism, since when the compound exists in the enolic form c then we have a possibility of isorropesis, as follows :

$$C_8H_{14}\!\!<\!\!{}^{C\cdot OH}_{C\cdot NO} \quad \rightleftarrows \quad C_8H_{14}\!\!<\!\!{}^{C(OH)\cdot O}_{C=\!\!=N}$$
$$c. \qquad\qquad d.$$

The form represented in d may, or may not, be capable of static existence; clearly, however, when the labile hydroxyl hydrogen is replaced by an alkyl group, this form is stable, for we then have the oxygen alkyl ether of *iso*nitrosocamphor prepared and described by Forster. It is very probable, therefore, that the formation of Forster's oxygen ethers is due to the second type of isodynamic isomerism given above. The absorption spectrum of the oxygen methyl ether is shown by the dot and dash curve, Fig. 7 (p. 974).

Now Forster has described two series of derivatives from *iso*nitrosocamphor, a yellow and a colourless series. For example, there are two benzoyl derivatives, to which he attributes the two formulæ :

$$C_8H_{14}\!\!<\!\!{}^{CO}_{C:NOBz} \qquad\qquad C_8H_{14}\!\!<\!\!{}^{C(OBz)\cdot O}_{C=\!\!=N},$$
$$\text{Yellow.} \qquad\qquad\qquad \text{Colourless.}$$

3 s 2

and the two anhydrides with the formulæ:

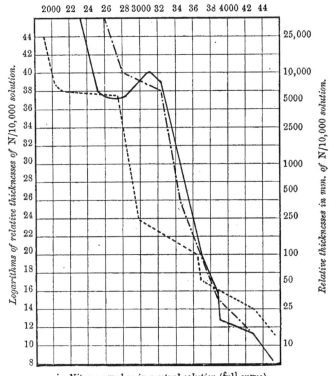

FIG. 7.

isoNitrosocamphor in neutral solution (full curve).
isoNitrosocamphor in alkaline solution (dotted curve).
isoNitrosocamphor o-methyl ether (dot and dash curve).

Whilst the formation of these two colourless compounds is to be readily explained by the second type of isodynamic isomerism suggested above, the formation of the two yellow compounds can only be accounted for by the assumption that in isonitrosocamphor the first type of isodynamic isomerism also is shown, namely,

$$C_8H_{14}{<}{\overset{CO}{\underset{C:NOH}{}}} \quad \rightleftarrows \quad C_8H_{14}{<}{\overset{CO}{\underset{CH \cdot NO}{}}} \ .$$

We can conclude, therefore, that in ordinary *iso*nitrosocamphor both these processes are taking place, as coloured and colourless derivatives are simultaneously obtained, and, further, that it is quite possible that there should be two modifications of *iso*nitrosocamphor, one in which the first type of isodynamic isomerism is mainly or entirely present, and a second in which the second type is the chief factor. Now Forster (Trans., 1905, **87**, 232) has described two modifications of this compound, a stable and an unstable variety. He has explained these as stereoisomerides :

$$\begin{array}{cc}
\underset{\substack{\diagup\diagdown \\ \text{C}\!-\!-\!\text{C} \\ \text{N·OH} \quad \text{O}}}{\text{C}_8\text{H}_{14}} &
\underset{\substack{\diagup\diagdown \\ \text{C}\!-\!-\!\text{C} \\ \text{HO·N} \quad \text{O}}}{\text{C}_8\text{H}_{14}} \\
\text{Stable.} & \text{Unstable.}
\end{array}$$

We venture to think, however, that this conception is inadequate, and prefer to consider that in the case of the unstable variety we are dealing in the main with the desmotropic forms,

$$\text{C}_8\text{H}_{14}{<}^{\text{CO}}_{\text{C:NOH}} \quad \rightleftarrows \quad \text{C}_8\text{H}_{14}{<}^{\text{CO}}_{\text{CH·NO'}}$$

and that in the stable form we are dealing with the isodynamic isomerism expressed by

$$\text{C}_8\text{H}_{14}{<}^{\text{CO}}_{\text{CH·NO}} \quad \rightleftarrows \quad \text{C}_8\text{H}_{14}{<}^{\text{C·O·H}}_{\text{C·NO}},$$

that is to say, we consider that Forster's unstable *iso*nitrosocamphor in alkaline solution consists of an equilibrium mixture of the two forms :

$$\text{C}_8\text{H}_{14}{<}^{\text{CO}}_{\text{C:NONa}} \quad \rightleftarrows \quad \text{C}_8\text{H}_{14}{<}^{\text{CO}}_{\text{CNa·NO'}}$$

and that his stable form in alkaline solution consists of an equilibrium mixture of the two forms :

$$\text{C}_8\text{H}_{14}{<}^{\text{CO}}_{\text{CNa·NO}} \quad \rightleftarrows \quad \text{C}_8\text{H}_{14}{<}^{\text{C·ONa}}_{\text{C·NO}}.$$

We have based our conclusions on the following evidence :

1. A pure specimen of the stable variety was prepared according to Forster's directions and its absorption spectrum observed. The curves given by this substance in neutral and alkaline solution are shown in Fig. 8, from which it will be noticed that the absorption is very different from that given by the ordinary *iso*nitrosocamphor (Fig. 7). Hartley (Trans., 1900, **77**, 509) has shown in the case of benzaldoxime

that the absorption of the *syn-* and *anti-*forms is identical, and, apart from this, it is in the highest degree improbable that the mere rotation of the nitrogen double-bond should produce such a profound change in the absorption as we have observed.

2. The unstable modification is prepared by the hydrolysis of the yellow form of the benzoyl derivative, and clearly this would tend to give the *iso*nitrosocamphor undergoing the first process of isodynamic isomerism :

$$C_8H_{14} <^{CO}_{C:NONa} \quad \rightleftarrows \quad C_8H_{14} <^{CO}_{CNa \cdot NO}$$

3. The unstable modification is oxidised much more easily than the stable modification; this is accounted for at once by the presence of isorropesis :

$$C_8H_{14} <^{CO}_{C:NONa} \quad \rightleftarrows \quad C_8H_{14} <^{C \cdot O}_{C \cdot NONa} ,$$

which would tend in the presence of an oxidising agent to give

$$C_8H_{14} <^{C \cdot O}_{C \cdot NONa} {>} O ,$$

owing to the extreme reactivity of the nitrogen and carbonyl oxygen atoms.

4. The absorption curve of the stable form of *iso*nitrosocamphor in alkaline solution (dotted curve, Fig. 8) is very similar to those of the *iso*nitroso-derivatives of acetone and methyl acetone, showing that doubtless a similar simple process is present. We may attribute the ultra-violet absorption band in this case to the second process of isodynamic isomerism given above and the isorropesis band to the following :

$$C_8H_{14} <^{C \cdot ONa}_{C \cdot NO} \quad \rightleftarrows \quad C_8H_{14} <^{C(ONa) \cdot O}_{C \underline{\quad\quad} N}.$$

In the absorption curve of ordinary *iso*nitrosocamphor in alkaline solution (dotted curve, Fig. 7) there is a very broad isorropesis band. Inasmuch as ordinary *iso*nitrosocamphor is a mixture of the stable and unstable forms, we may expect to find evidences of both forms in the absorption curve. The broad isorropesis band covers the region of the band present in the stable modification (Fig. 8), but also extends considerably nearer to the red. This extension nearer to the red is doubtless due to the isorropesis occurring in the unstable modification, namely,

$$C_8H_{14} <^{CO}_{C:NONa} \quad \rightleftarrows \quad C_8H_{14} <^{C \cdot O}_{C \cdot NONa}$$

On account of the readiness with which the unstable modification passes into the stable in the presence of light, we have not thought it worth while to examine its absorption spectrum. It is probable from the above that in alkaline solution it would show an isorropesis band with an oscillation frequency of 2200. For this reason the unstable

FIG. 8.

Oscillation frequencies.

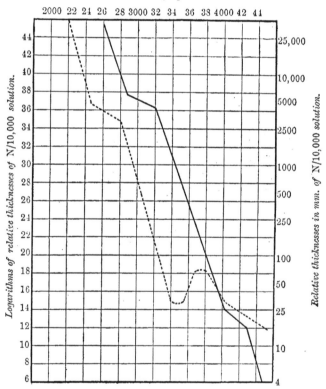

isoNitrosocamphor, *stable form in neutral solution* (full curve).
isoNitrosocamphor, *stable form in alkaline solution* (dotted curve).

modification should exhibit in alkaline solution a much more orange colour than the stable form, a fact which Forster has already observed.

Lastly, we have Forster's evidence (Trans., 1904, **85**, 892) that the yellow anhydride and benzoyl derivative are obtained from the unstable variety, whilst the stable form gives only the colourless isomerides.

This evidence appears to us to be conclusive in favour of the existence of two forms of *iso*nitrosocamphor such as we have described and their identity with Forster's two isomerides.

It has already been stated that it is possible that the formula for stable *iso*nitrosocamphor,

$$C_8H_{14} < \begin{matrix} C(OH)-O \\ | \quad | \\ C \!=\!=\!=\! N \end{matrix} ,$$

represents an unstable condition. A comparison, however, of the absorption curves of the stable compound in neutral solution (full curve, Fig. 8) and the oxygen methyl ether (dot and dash curve, Fig. 7) shows a striking similarity between them. It would appear from this that stable *iso*nitrosocamphor exists entirely in this form and is exactly analogous to the oxygen alkyl ethers; this would also account for the fact that the substance is quite white.

It would appear from these results that we have the clue to the form in which all the *iso*nitroso-bodies exist when in the free and colourless state. We have shown that in all probability stable *iso*nitrosocamphor in alkaline solution is an equilibrium mixture of the two forms

$$C_8H_{14} < \begin{matrix} CO \\ | \\ CNa \cdot NO \end{matrix} \quad \rightleftarrows \quad C_8H_{14} < \begin{matrix} C \cdot ONa \\ || \\ C \cdot NO \end{matrix} ,$$
$$\quad a. \qquad\qquad\qquad b.$$

and that the yellow colour is due to the isorropesis which is expressed by the reversible change

$$C_8H_{14} < \begin{matrix} C \cdot ONa \\ || \\ C \cdot NO \end{matrix} \quad \rightleftarrows \quad C_8H_{14} < \begin{matrix} C(ONa) \cdot O \\ | \quad | \\ C \!=\!=\!=\! N \end{matrix} ,$$
$$\quad b. \qquad\qquad\qquad c.$$

When the pure stable form is set free, it adopts the form *c*—that is to say, it is to be represented by

$$C_8H_{14} < \begin{matrix} C(OH)-O \\ | \quad | \\ C \!=\!=\!=\! N \end{matrix} .$$

We may argue from this at once that the unstable *iso*nitroso-camphor, which in alkaline solution consists of the equilibrium mixture

$$C_8H_{14} < \begin{matrix} CO \\ | \\ C:NONa \end{matrix} \quad \rightleftarrows \quad C_8H_{14} < \begin{matrix} CO \\ | \\ CNa \cdot NO \end{matrix} ,$$
$$\quad a. \qquad\qquad\qquad b.$$

and gives rise to the isorropesis expressed by

$$C_8H_{14} < \begin{matrix} CO \\ | \\ C:NONa \end{matrix} \quad \rightleftarrows \quad C_8H_{14} < \begin{matrix} C \cdot O \\ || \\ C \cdot NONa \end{matrix} ,$$

possesses in the free state the form

$$C_8H_4{<}^{\overset{\displaystyle C\cdot O}{|}}_{\underset{\displaystyle C\cdot N\cdot OH}{|}}.$$

Similarly, we may conclude that all the *iso*nitroso-compounds in the free state possess an analogous ring structure, namely,

$$\begin{array}{c} R_1\cdot \overset{\displaystyle C}{|}\overset{\displaystyle :}{}\overset{\displaystyle C}{|}\cdot R_2 \\ O\cdot N\cdot OH \end{array},$$

and that on the addition of alkali the hydrogen of the nitroso-group is replaced by the alkali metal, which becomes labile so that the ring is broken.

Two other formulæ for the colourless fatty *iso*nitroso-compounds suggest themselves : first, it is conceivable that they have the true nitroso-form :

$$R\cdot CO\cdot CH(NO)\cdot R,$$

but against this it may be argued that the bodies are quite colourless when molten, and do not show the reactions which would be expected in the case of a compound containing a static –NO group. The second formula is the *iso*oxime configuration :

$$\begin{array}{c} R\cdot CO\cdot C\cdot R \\ \overset{\displaystyle \triangle}{HN\!-\!O} \end{array}.$$

This, however, would not satisfy the facts in any way, as there is no possibility of the isodynamic isomerism we have found to exist.

We have been led from these results to consider the case of camphor itself, for Hartley (Trans., 1881, **39**, 153) has stated that camphor shows no absorption, being, indeed, quite diactinic. Now the reactivity of the carbonyl and of the neighbouring CH_2 group of this substance would lead us to expect that its absorption spectrum should show a strong band analogous to those shown by acetone and simple ketones. This band, however, should be evidenced in tenth-normal solutions, which is ten times the strength used by Hartley.

We have therefore examined the absorption spectrum of normal and tenth-normal solutions of camphor and found that, as we expected, a very well-marked band is developed. The absorption is shown by the full curve in Fig. 9, while the dotted curve is that of camphor monoxime, which, it will be seen, is a remarkably diactinic substance and shows no evidence of any tautomerism.

It is possible from these results to explain the colour of other analogous compounds, as, for example, certain oximino-ketones. One of the most interesting is violuric acid and its sodium salt, the absorption spectra of which were described recently by Hartley

(Trans., 1905, **87**, 1796). Violuric acid is colourless, but on the addition of sodium hydroxide the hydrogen of the nitroso-group is replaced by sodium and becomes labile, so that the two forms are in equilibrium :

$$CO<^{NH\cdot CO}_{NH\cdot CO}>CNa\cdot NO \quad \rightleftarrows \quad CO<^{NH\cdot CO}_{NH\cdot CO}>C\!:\!NONa.$$
$$a. \hspace{4cm} b.$$

This process will account for the band in the ultra-violet. The form b presents two possibilities of isorropesis, one between the

Fig. 9.

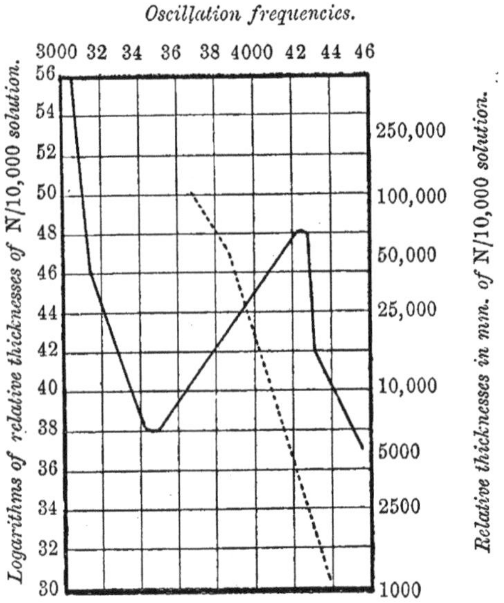

Oscillation frequencies.

Camphor (full curve).
Camphoroxime (dotted curve).

unsaturated nitrogen atom and the ketonic oxygen atoms on each side, and the other an isorropesis of the quinonoid type between the unsaturated nitrogen and the ketonic oxygen in the para-position, for if the two hydrogen atoms wander simultaneously, the compound would assume a pseudo-quinonoid form :

$$CO<^{N\!:\!C(OH)}_{N\!:\!C(OH)}>C\!:\!NONa.$$

The colour is due to these processes. Ostwald held the view that the colour was due to the ion of violuric acid, assuming that violuric acid is entirely undissociated in solution and that the sodium salt is dissociated. This conception is quite unnecessary, for all the phenomena observed with violuric acid are readily accounted for by the above explanation (compare also Baly and Desch, Trans., 1905, 87, 770).

Conclusions.

(1) The *iso*nitroso-derivatives of the aliphatic ketones in alkaline solution exist as an equilibrium mixture of the two forms

$$R_1 \cdot CO \cdot C(NOM) \cdot R_2 \text{ and } R_1 \cdot CO \cdot CM(NO) \cdot R_2,$$

where M denotes a monovalent metal.

(2) The colour of these compounds in alkaline solution is due to the isorropesis occurring with the form $R_1 \cdot CO \cdot C(NOM) \cdot R_2$ between the nitrogen and oxygen atoms.

(3) *iso*Nitrosocamphor exists in two modifications, the unstable, which in alkaline solution consists of an equilibrium mixture of the two forms $C_8H_{14} \diagdown \genfrac{}{}{0pt}{}{CO}{CNa \cdot NO}$ and $C_8H_{14} \diagdown \genfrac{}{}{0pt}{}{C \cdot ONa}{C \cdot NO}$, and the stable, which in alkaline solution consists of an equilibrium mixture of the two forms $C_8H_{14} \diagdown \genfrac{}{}{0pt}{}{CO}{C:NONa}$ and $C_8H_{14} \diagdown \genfrac{}{}{0pt}{}{CO}{CNa \cdot NO}$.

(4) Isorropesis occurs in both the forms $C_8H_{14} \diagdown \genfrac{}{}{0pt}{}{CO}{C:NONa}$ and $C_8H_{14} \diagdown \genfrac{}{}{0pt}{}{C \cdot ONa}{C \cdot NO}$, giving in the one case an orange, and in the other a yellow colour.

(5) In the free state, stable *iso*nitrosocamphor has the formula $C_8H_{14} \diagdown \genfrac{}{}{0pt}{}{C(OH) \cdot O}{C = N}$, and unstable *iso*nitrosocamphor the formula $C_8H_{14} \diagdown \genfrac{}{}{0pt}{}{C \cdot O}{C \cdot NOH}$.

(6) Similarly, the *iso*nitroso-derivatives of the aliphatic ketones, when in the free state, in all probability have the form

$$\underset{O \cdot NOH}{R_1 \cdot C : C \cdot R_2}.$$

(7) There are two types of isodynamic isomerism, one in which the absorption band appears with $N/10$ solutions; to this class belong camphor and the simple ketones. The other class, to which belong the β-diketones, exhibits the absorption band in $N/1000$ solutions. It is probable that in the second case we are dealing with a wandering

hydrogen atom, whilst in the first class a potential tautomerism only exists.

We have to record our indebtedness to Prof. Collie for the interest he has taken in these experiments, and our thanks are due also to the Chemical Society for a grant in aid of the work.

SPECTROSCOPIC LABORATORY,
　　UNIVERSITY COLLEGE, LONDON.

CIII.—*The Relation between Absorption Spectra and Chemical Constitution. Part VI. The Phenyl Hydrazones of Simple Aldehydes and Ketones.*

By EDWARD CHARLES CYRIL BALY and WILLIAM BRADSHAW TUCK.

IN a recent paper, Chattaway (Trans., 1906, 89, 462) suggested that the change of colour of benzaldehydephenylhydrazone from pale yellow or white to scarlet on exposure to light was due to the change into the azo-configuration, that is to say, to the formation of benzylphenyl-diazene.

E. Fischer has shown that benzeneazoethane, $C_6H_5 \cdot N \colon N \cdot CH_2 \cdot CH_3$, on being allowed to stand with 60 per cent. sulphuric acid, readily passes over into the phenylhydrazone of acetaldehyde,

$$C_6H_5 \cdot NH \cdot N \colon CH \cdot CH_3,$$

thus proving that the reverse change of azo-compound into hydrazone is easily produced by the action of acid. Several observers have also recorded that the phenylhydrazones of simple aldehydes and ketones become yellow on exposure to light, the explanation generally accepted being that oxidation has occurred. It has been noticed by E. Fischer that the melting point of acetaldehydephenylhydrazone when freshly prepared is 63—65°, but that after washing with light petroleum the melting point is raised to 80°. It occurred to us that the explanation of all these changes might be the assumption of the azo-configuration under the influence of light, and, having subjected the phenylhydr-azones of certain simple aldehydes and ketones to spectroscopic exami-nation, we have been able to find undoubted evidence that this ex-planation is the correct one.

The absorption of azobenzene has been described by Hartley (Trans., 1887, 51, 152), and we have reproduced the absorption curve of this substance in Fig. 1 (dotted curve); it will be noticed that a very large absorption band is exhibited at from 2 mm. to 0·5 mm. of a tenth-normal solution in the visible region. This band is the origin of the colour of this substance and is undoubtedly due to the linking:

it may be said that the band is of the same type as the bands already shown to be due to isorropesis in the case of the quinones and other similar benzenoid derivations. In a previous paper (Trans., 1906, 89, 514) the suggestion was put forward that the benzene ring is perpetually in a state of vibration or pulsation, the motions being analogous to those of a bell when struck. If we apply this motion to the two benzene-nuclei of azobenzene and assume that the vibrations take place along the dotted axes, as would undoubtedly be expected from the fact that the two rings are heavily weighted at one end,

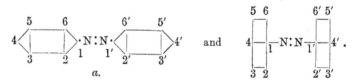

we shall have the vibration taking place between the extreme forms

In the form *a*, the four carbon atoms 1, 4, 1′, and 4′ are unsaturated, and we therefore have an unsaturated carbon atom on each side of the unsaturated nitrogen atoms, thus, $>\!C\!\cdot\!N\!:\!N\!\cdot\!C\!<$, if the unsaturated affinities of the carbon atoms be expressed by the dotted lines. It appears that in these circumstances oscillation between the unsaturated affinities is set up, a process which should be included under the term isorropesis. There is some analogy here with the condition obtaining in phorone, where unsaturated carbon atoms are situated on each side of the unsaturated ketonic group,

$$(Me)_2C\!:\!CH\!\cdot\!CO\!\cdot\!CH\!:\!C(Me)_2.$$

This compound is yellow, doubtless owing to a somewhat analogous process to that taking place in azobenzene. We propose to apply the term isorropesis to such types of oscillation, for it is undoubtedly a case in which an oscillation occurs between the residual affinities of adjacent atoms.

Apart from this suggested explanation of the colour of azobenzene, there is no doubt but that we may establish the absorption spectrum of this substance as a type of the absorption of azo-compounds. All true azo-compounds, therefore, should show an absorption spectrum of this type, that is to say, a band in the visible region exhibited by solutions of tenth-normal strength; it should be noticed here that the replacement of the phenyl group of azobenzene by an alkyl group would naturally tend to increase the concentration at which the band appears.

We have examined the hydrazones of formaldehyde, acetaldehyde, propylaldehyde, benzaldehyde, acetophenone, and diethylketone. Owing to the polymerisation which formaldehydephenylhydrazone undergoes (Walker, Trans., 1896, 69, 1282), there is no possibility of the change to the azo-configuration. The absorption spectrum of this substance or its polymeride, which is bright yellow, is shown in Fig. 1 (full curve), and, as can be seen, it is quite different from the absorption spectra of acetaldehyde and propylaldehydephenylhydrazones (Figs. 2 and 3).

The phenylhydrazone of acetaldehyde was next examined; when first prepared it is pale yellow, but after recrystallisation from light petroleum and several washings with the same solvent it can be obtained in snow-white crystals. A weighed quantity was dissolved in a mixture of alcohol and glacial acetic acid. The absorption spectrum of this solution is shown in Fig. 2 (full curve), whilst that of the acetaldehydephenylmethylhydrazone, a colourless liquid, is shown by the dotted curve. The resemblance between the two curves is sufficiently close to justify the assumption that both compounds have an analogous structure, which is doubtless expressed by the formulæ $C_6H_5NH\cdot N\colon CH\cdot CH_3$ and $C_6H_5N(CH_3)\cdot N\colon CH\cdot CH_3$. On exposing solutions of both these substances to sunlight, that of the phenylmethyl-hydrazone did not change in any way, but that of the phenylhydrazone slowly turned deep yellow. The absorption spectrum of the yellow solution now showed a band in the visible region at tenth-normal strength exactly typical of an azo-compound, Fig. 2 (dot and dash curve).

When the original solution of the phenylhydrazone was made in alcohol without the addition of the glacial acetic acid, this change took place so rapidly under the influence of the ultra-violet rays of the electric arc used for photographing the spectra that it was impossible to obtain the absorption of the true hydrazone; the curve obtained was identical with that of the acetic acid solution after exposure to light. It is evident, therefore, that the presence of the acid retards the change from the hydrazone- to the azo-configuration, which is to be expected from Fischer's observation of the action of sulphuric acid on benzeneazoethane.

We have also examined the absorption spectrum of acetaldehyde-p-bromophenylhydrazone, and the curve obtained shows that the constitution is the same as that of the hydrazone in acetic acid solution and the phenylmethylhydrazone ; it has therefore the hydrazone-configuration, $BrC_6H_4NH\cdot N\dot{:}CH\cdot CH_3$. On exposure to sunlight, the solution,

FIG. 1.

Oscillation frequencies.

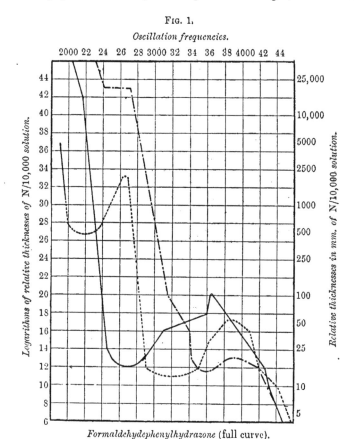

Formaldehydephenylhydrazone (full curve).
Azobenzene (dotted curve).
Benzeneazomethane (dot and dash curve).

like that of the hydrazone, turns yellow and shows the same absorption band as the hydrazone after exposure (Fig. 2, dash and two dots curve). The phenylhydrazone and the p-bromophenylhydrazone thus undergo the same change in solution on exposure to light; the former, however, changes much more readily than the latter, and in both cases the change is retarded by acetic acid.

It occurred to us that, since the tendency of acetic acid is to favour the existence of the hydrazone-configuration of these compounds, it would be of interest to investigate the interaction of formaldehyde and phenylhydrazine in the absence of acetic acid. Equivalent quantities

FIG. 2.

Oscillation frequencies.

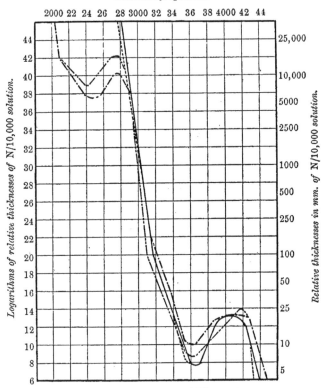

Acetaldehydephenylhydrazone in acetic acid (full curve).
Acetaldehydephenylhydrazone after exposure to light (dot and dash curve).
Acetaldehydephenylmethylhydrazone (dotted curve).
Acetaldehyde-p-bromophenylhydrazone (dash and two dots curve).

of these two compounds were therefore mixed together in alcoholic solution. In a few minutes an orange-yellow oil was precipitated, which very slowly set to a viscous, semi-crystalline mass.

It appeared to be possible, therefore, in this way to obtain a mixture of the azo- and the hydrazone-forms which polymerised on standing.

As the azo-compound, namely, $C_6H_5N:N\cdot CH_3$, is an oil volatile with steam (Tafel, *Ber.*, 1885, 18, 1740), a fresh quantity of the oil was prepared, separated, and at once distilled in a current of steam. A considerable quantity of a yellow oil passed over which boiled with slight decomposition at 150°; this agrees with Tafel's observation.

FIG. 3.

Oscillation frequencies.

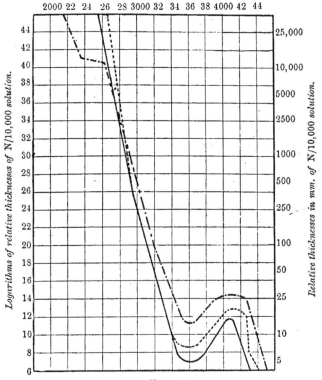

Propylaldehydephenylhydrazone (full curve).
Propylaldehydephenylhydrazone after exposure to light (dot and dash curve).
Propylaldehydephenylmethylhydrazone (dotted curve).

The oil was dissolved in a mixture of chloroform and alcohol and its absorption spectrum photographed. The curve obtained is shown in Fig. 1 (dot and dash curve), and is quite similar to the curves shown for the azo-configurations of the hydrazones already described.

On standing for some days, however, the oil began to crystallise, but

band is exhibited exactly in the same way as in the case of all the above hydrazones.

We have no hesitation in saying, therefore, that except in the case of formaldehydephenylhydrazone, which undergoes polymerisation, the phenylhydrazones of simple aliphatic aldehydes and ketones change in

Fig. 4.

Oscillation Frequencies

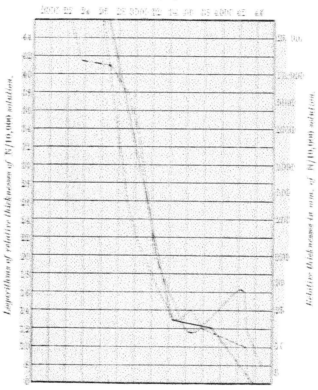

Acetophenylhydrazone in water and (full curve).
Acetophenylenehydrazone (dotted curve).
Acetophenylhydrazone after exposure to sunlight (dot and dash curve).

solution to the corresponding azo-compounds. This change is retarded by the presence of acetic acid and by the substitution of bromine in the phenylhydrazine nucleus.

In the aromatic series, we have examined the hydrazones of benzaldehyde and acetophenone, and find that the acetophenone compounds are

a small quantity obtained by the distillation of the substance was found to be quite stable. There seems little doubt, therefore, that it is possible to obtain a mixture of the azo- and hydrazone-compounds by the direct interaction of phenylhydrazine and formaldehyde, and to separate the azo-compound by distillation in a current of steam. The substance obtained in this way contains in all [probability a small quantity of the formaldehydehydrazone, which would account for its tendency to crystallise on standing owing to polymerisation taking place. The azo-compound obtained agrees in every respect with the benzeneazomethane prepared by Tafel, and we find also that on treatment with acid it assumes the hydrazone structure in the same way as observed by Fischer in the case of benzeneazoethane.

Exactly the same changes were observed in the case of the propylaldehyde compounds; the freshly-prepared phenylhydrazone is a colourless liquid which turned red on exposure to light. The absorption spectrum of the hydrazone in acetic acid is shown in Fig. 3 (full curve), while the dotted curve is that of the phenylmethylhydrazone. After exposure of the hydrazone solution to light, the typical azo-absorption band makes its appearance, and here again the change from the hydrazone- to the azo-configuration is produced by light.

Figs. 4 and 6 contain the curves we have obtained with the acetone and diethylketone compounds, and, as can be seen, are perfectly analogous to those of the aldehydephenylhydrazones. The only difference to be recorded is that the change from the hydrazone to the azo-configuration takes place even more readily than in the case of the aldehyde compounds. It is extremely difficult to prepare the hydrazones entirely free from colour in the case of the ketones, especially diethylketone, a few minutes' exposure to light producing a decided yellow tinge. If, however, the hydrazone be distilled under diminished pressure and the first runnings rejected, as they contain most of the azo-compound, a very faintly straw-yellow liquid is obtained, which, on being immediately dissolved in alcohol and glacial acetic acid, gives a colourless solution. The absorption spectra of these solutions were photographed as quickly as possible and the curves are shown in Figs. 4 and 6 (full curves). In less than a quarter of an hour the solutions became yellow, and in three hours the change was complete; the absorption spectra of the exposed solutions are shown by the dot and dash curves (Figs. 4 and 6), whilst the dotted curves represent the absorption of the phenylmethylhydrazones, which are perfectly stable substances, as in the case of the aldehyde compounds. The p-bromophenylhydrazone of acetone was also examined and was found to be quite analogous to the acetaldehyde-p-bromophenylhydrazone (see Fig. 5). In acetic acid solution, its spectrum is the same as that of the phenylmethylhydrazone, and after exposure to light the azo-absorption band is exhibited exactly in the same way as in the case of all the above hydrazones.

We have no hesitation in saying, therefore, that except in the case of formaldehydephenylhydrazone, which undergoes polymerisation, the phenylhydrazones of simple aliphatic aldehydes and ketones change in

FIG. 4.

Oscillation frequencies.

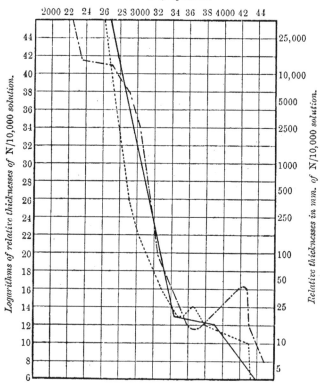

Acetonephenylhydrazone in acetic acid (full curve).
Acetonephenylmethylhydrazone (dotted curve).
Acetonephenylhydrazone after exposure to sunlight (dot and dash curve).

solution to the corresponding azo-compounds. This change is retarded by the presence of acetic acid and by the substitution of bromine in the phenylhydrazine nucleus.

In the aromatic series, we have examined the hydrazones of benzaldehyde and acetophenone, and find that the acetophenono compounds are exactly analogous to those of the aliphatic derivatives. The absorpon curves of acetophenonephenylhydrazone in acid solution before and ner expo-ure to light are shown in Fig. 8 (p. 993); the azo-band is well st wn in the last-mentioned curve. It is to be remarked that the presence of he two phenyl groups has decreased the concentration at which the ind

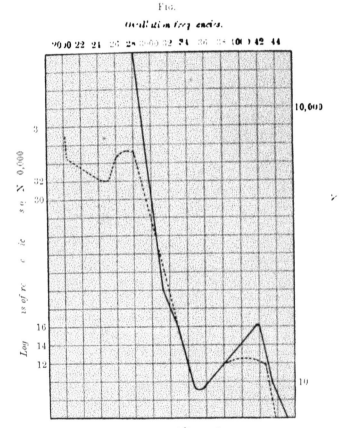

Acetone p-*bromophenylhydrazone* (full curve).
Acetone p-*bromophenylhydrazone after exposure to light* (dotted curve).

appears (see p. 984). Benzaldehydephenylhydrazone, on the oler hand, does not change on exposure to light when in solutior as already stated by Chattaway (*loc. cit.*). A solution was prepared nd its absorption spectrum photographed ; the solution was then exped to sunlight for a fortnight and its absorption spectrum again phlographed. No difference whatever was found, and the absorptio is very similar to the absorption of benzaldehydophenylmethylhydrazone (Fig. 7, p. 992). We must conclude, therefore, that benzaldehydophenyl-hydrazone in solution does not assume the azo-configuration. Now Chattaway states that pure benzaldehyde-benzylhydrazone, which has bee converted into the red variety by exposure to

FIG. 6.

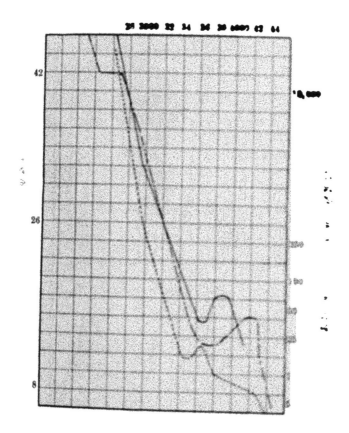

methylketonophenylhydrazone after exposure to light (dash dot and curve).

colorless on being dissolved. This fact, coupled with our observation that a spectroscopic change takes place on exposing the solution to light or a long time, leads us to the opinion that the change of the solid into the scarlet modification dealt with by Chattaway is not to be accunted for by a change into the azo-forms. Unfortunately we

exactly analogous to those of the aliphatic derivatives. The absorption curves of acetophenonephenylhydrazone in acid solution before and after exposure to light are shown in Fig. 8 (p. 993); the azo-band is well shown in the last-mentioned curve. It is to be remarked that the presence of the two phenyl groups has decreased the concentration at which the band

FIG. 5.

Oscillation frequencies.

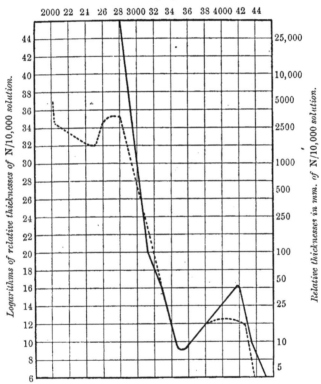

Acetone p-*bromophenylhydrazone* (full curve).
Acetone p-*bromophenylhydrazone after exposure to light* (dotted curve).

appears (see p. 984). Benzaldehydephenylhydrazone, on the other hand, does not change on exposure to light when in solution, as already stated by Chattaway (*loc. cit.*). A solution was prepared and its absorption spectrum photographed; the solution was then exposed to sunlight for a fortnight and its absorption spectrum again photographed. No difference whatever was found, and the absorption is very similar to the absorption of benzaldehydephenylmethylhydrazone (Fig. 7, p. 992). We must conclude, therefore, that benzaldehydephenyl-hydrazone in solution does not assume the azo-configuration. Now Chattaway states that pure benzaldehydephenylhydrazone, which has been converted into the red variety by exposure to light, becomes

Fig. 6.

Oscillation frequencies.

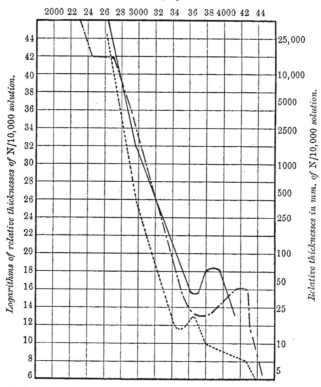

Diethylketonephenylhydrazone in acetic acid (full curve).
Diethylketonephenylmethylhydrazone (dotted curve).
Diethylketonephenylhydrazone after exposure to light (dash dot and curve).

colourless on being dissolved. This fact, coupled with our observation that no spectroscopic change takes place on exposing the solution to light for a long time, leads us to the opinion that the change of the solid into the scarlet modification dealt with by Chattaway is not to be accounted for by a change into the azo-forms. Unfortunately, we are unable to prove the point one way or the other, because no solution of the coloured form is obtainable.

On Figs. 7, 9, and 12 are shown the absorption curves of phenyl_hydrazine, phenylmethylhydrazine, and *p*-bromophenylhydrazine in

FIG. 7.

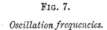

Oscillation frequencies.

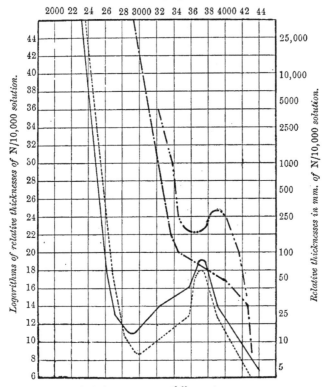

Benzaldehydephenylhydrazone (full curve).
Benzaldehydephenylmethylhydrazone (dotted curve).
Phenylhydrazine in neutral solution (dot and dash curve).
Phenylhydrazine with 5 eq. HCl (dash and two dots curve).

neutral and in acid solution. These curves show absorption bands which are of the same type as the bands of the hydrazones, so that we may attribute the latter to the phenylhydrazine residue in each case.

We have also examined the phenylhydrazones and the phenyl_

methylhydrazones of the three isomeric nitrobenzaldehydes. These
compounds, which are magnificently coloured, show absorption spectra
which are entirely different from those of the hydrazones of the fatty
aldehydes and ketones.   The curves are shown in Figs. 9, 10, and 11,
and, it will be seen, show large absorption bands at thousandth-normal

FIG. 8.

Oscillation frequencies.

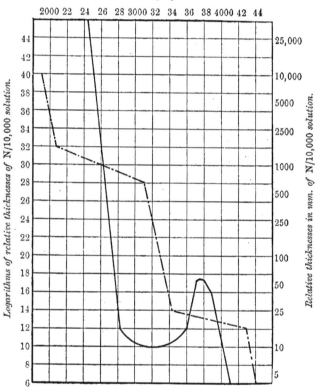

Acetophenonephenylhydrazone (full curve).
Acetophenonephenylhydrazone after exposure to light (dot and dash curve).

concentrations. This fact alone establishes a radical difference between
these bodies and the azo-form of the fatty hydrazones, for the bands
of the latter and of azobenzene always appear with tenth-normal
solutions.   This shows that the colour is not due to an azo-configuration, a conclusion which is confirmed by the fact that the phenylhydrazones and the phenylmethylhydrazones show the same absorption

band. It is clearly impossible that the phenylmethylhydrazone of nitrobenzaldehyde can exist in the azo-form, thus :

$$NO_2 \cdot C_6H_4 \cdot CH \colon N \cdot N(CH_3) \cdot C_6H_5.$$

The explanation of the colour must be sought elsewhere. A possible

Fig. 9.

Oscillation frequencies.

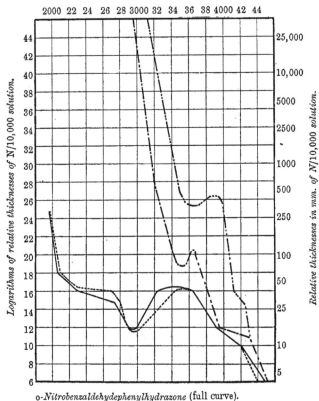

o-*Nitrobenzaldehydephenylhydrazone* (full curve).
o-*Nitrobenzaldehydcphenylmethylhydrazone* (dotted curve).
Phenylmethylhydrazine neutral solution (dot and dash curve).
Phenylmethylhydrazine with 2 eq. HCl (dash and two dots curve).

explanation is that these compounds exist in the quinonoid form and that they are analogous to the nitrophenols, that is, that the formula may be written $^{HO}_{O}{>}N \colon C_6H_4 \colon C \colon N \cdot NHC_6H_5.$ In order to test this, we have examined the absorption spectrum of the phenylmethylhydr.

azone of *m*-nitroacetophenone. This substance no doubt has the formula $NO_2 \cdot C_6H_4 \cdot C(CH_3) \colon N \cdot N(CH_3)C_6H_5$. It is snow-white, and exhibits an absorption spectrum shown by the dot and dash curve in Fig. 10. The replacement of the hydrogen atom of the benz-aldehyde carbon atom by the CH_3 group entirely removes the colour,

Fig. 10.

Oscillation frequencies.

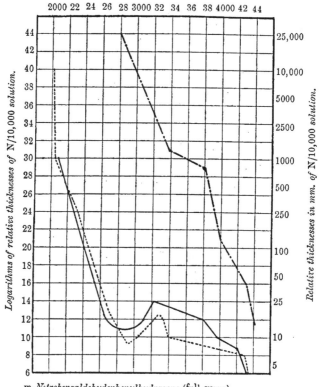

m-*Nitrobenzaldehydephenylhydrazone* (full curve).
m-*Nitrobenzaldehydephenylmethylhydrazone* (dotted curve).
m-*Nitroacetophenonephenylmethylhydrazone* (dot and dash curve).

so that the colour of the nitrobenzaldehydephenylhydrazone depends on this hydrogen atom. This is strongly in favour of the quinonoid hypothesis advanced above.

It would be expected if the phenylhydrazones of the nitrobenz-aldehydes possess partly or entirely the quinonoid structure suggested above that their absorption spectra would present some analogies to those of the nitrophenols in alkaline solution, since there is some considerable resemblance between the two configurations :

$$\underset{HO}{\overset{O}{>}}N:\left\langle\underset{}{\bigcirc}\right\rangle:C:N\cdot NHC_6H_5 \text{ and } \underset{HO}{\overset{O}{>}}N:\left\langle\underset{}{\bigcirc}\right\rangle:O.$$

A comparison between the absorption spectra of the phenylhydrazone of p-nitrobenzaldehyde and of p-nitrophenol in alkaline solution shows that the absorption bands are almost identical in position and in persistence. This affords a striking support of the quinonoid structure of the former substance. There is some difference in the

FIG. 11.

Oscillation frequencies.

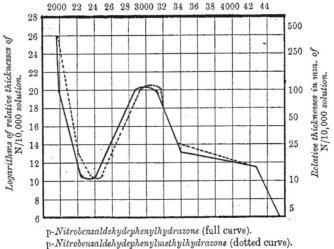

p-*Nitrobenzaldehydephenylhydrazone* (full curve).
p-*Nitrobenzaldehydephenylmethylhydrazone* (dotted curve).

persistence and position of the absorption bands, although they are very similar in character in the case of the o- and m-nitrobenzaldehyde-phenylhydrazones and of the o- and m-nitrophenols in alkaline solution; this difference is no doubt due to the fact that neither of the former substances is entirely in the quinonoid form. It is interesting to note that the addition of sodium ethoxide to the alcoholic solution of the o-nitrobenzaldehydephenylhydrazone deepens the colour very considerably. The evidence is therefore distinctly in favour of the quinonoid formula already suggested for the phenylhydrazones of the nitrobenzaldehydes, namely,

$$\underset{O}{\overset{O}{>}}N:C_6H_4:C:N\cdot NHC_6H_5.$$

Finally, we have examined the absorption spectra of p-nitrophenyl-hydrazine and acetone-p-nitrophenylhydrazone; the curves of these two compounds are shown on Fig. 12. We suggest that their colour is due to their existence in the quinonoid form:

$$\underset{HO}{\overset{O}{>}}N{:}C_6H_4{:}N{\cdot}N{\cdot}NH_2 \quad \text{and} \quad \underset{HO}{\overset{O}{>}}N{:}C_6H_4{:}N{\cdot}N{:}C(Me)_2.$$

Fig. 12.

Oscillation frequencies.

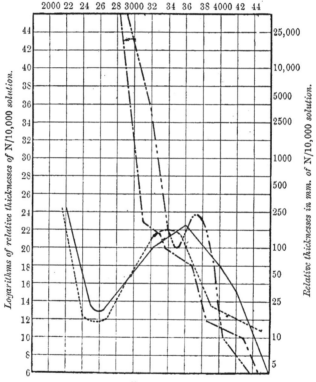

p-*Nitrophenylhydrazine* (full curve).
Acetone p-*nitrophenylhydrazone* (dotted curve).
p-*Bromophenylhydrazine in neutral solution* (dot and dash curve).
p-*Bromophenylhydrazine* + 2 eq. HCl (dash and two dots curve).

The absorption spectrum shows an absorption band of almost exactly the same persistence and in exactly the same position as that of p-nitroaniline; evidently, therefore, the two compounds must have an analogous configuration, and this is satisfied by the quinonoid structure given above, which is perfectly analogous to that of p-nitro-aniline.

Secondly, the absorption spectrum of the acetone-p-nitrophenylhydrazone is the same as that of the p-nitrophenylhydrazine, so that no change is introduced by the acetone residue.

Thirdly, Hyde (*Ber.*, 1889, **32**, 1813) has found that the p-nitrophenylhydrazones of acetaldehyde, benzaldehyde, acetophenone, and some others are soluble in an aqueous solution of sodium hydroxide, which is accounted for by the formation of the sodium salt. We have also found that acetone-p-nitrophenylhydrazone is soluble in sodium hydroxide solution with a bright red colour.

For these reasons, we are convinced that the quinonoid form represents the configuration of the p-nitrophenylhydrazine and its acetone derivative.

Conclusions.

(1) The phenylhydrazones of acetaldehyde, propylaldehyde, acetone, diethylketone, and acetophenone on exposure to light change into the azo-compounds, thus :

$$C_6H_5 \cdot NH \cdot N \colon C(Me)_2 \longrightarrow C_6H_5 \cdot N \colon N \cdot CH(Me)_2.$$

(2) This change is retarded by the presence of acetic acid and by the substitution of bromine in the phenylhydrazine nucleus.

(3) The phenylhydrazones of the three nitrobenzaldehydes exist partly or entirely in the quinonoid form, thus :

$$\frac{HO}{O}{>}N \colon C_6H_4 \colon C \colon N \cdot NHC_6H_5.$$

(4) Paranitrophenylhydrazine and its acetone derivative exist also in the quinonoid form, thus :

$$\frac{HO}{O}{>}N \colon C_6H_4 \colon N \cdot NH_2 \text{ and } \frac{HO}{O}{>}N \colon C_6H_4 \colon N \cdot N \colon C(Me)_2.$$

We wish to express our cordial thanks to Professor Collie and to Dr. Smiles for the interest they have taken in these experiments, and express our indebtedness to the Chemical Society for a grant in aid of the work.

SPECTROSCOPIC LABORATORY,
UNIVERSITY COLLEGE, LONDON.